From NEUROSCIENCE To NEUROLOGY

Neuroscience, Molecular Medicine, and the Therapeutic Transformation
of Neurology

Edited by
STEPHEN WAXMAN, MD, PhD
Yale University

ELSEVIER
ACADEMIC
PRESS

AMSTERDAM • BOSTON • HEIDELBERG • LONDON
NEW YORK • OXFORD • PARIS • SAN DIEGO
SAN FRANCISCO • SINGAPORE • SYDNEY • TOKYO

Elsevier Academic Press
200 Wheeler Road, 6th Floor, Burlington, MA 01803, USA
525 B Street, Suite 1900, San Diego, California 92101-4495, USA
84 Theobald's Road, London WC1X 8RR, UK

This book is printed on acid-free paper. ∞

Library of Congress Cataloging-in-Publication Data
Application submitted

British Library Cataloguing in Publication Data
A catalogue record for this book is available from the British Library

ISBN: 0-12-738903-2

For all information on all Academic Press publications
visit our Web site at www.academicpress.com

Printed in the United States of America
04 05 06 07 08 09 9 8 7 6 5 4 3 2 1

Contents

Preface

When I was a boy about five years old, I asked a grownup "will men ever be able to travel to the moon?" and the answer was "it's impossible."

Less than two decades later, as a medical student working in London, I watched on television as a man walked on the moon. The "impossible" had become possible.

This book focuses on the remarkable transformation that has occurred within neurology, a transformation from the "impossible" to the possible.

Twenty-five years ago computed tomography had just become available and was rapidly evolving, but magnetic resonance imaging was not yet feasible in the clinical realm. Medical students at some institutions were taught to try *not* to make the diagnosis of multiple sclerosis early in its course, since there was little that could be done. Some senior neurologists—good clinicians—made the diagnosis of MS and then walked away from the bedside, since there was little to offer in the way of effective therapy. Peripheral neuropathies were regarded as untreatable. Strokes were treated with bed rest, and not much else.

Now, only a few decades later, we have access to imaging techniques that portray the brain and spinal cord—noninvasively—in elegant detail. There are five approved drugs for MS, which favorably modify its course. Plasma exchange is routinely and effectively used to treat Guillain-Barré syndrome. Strokes are prevented by aggressive diagnosis and treatment of hypertension, and are sometimes reversed with thrombolysis. These changes in our therapeutic armamentarium are only the tip of an iceberg. New and more effective treatments for Parkinson's disease, migraine, spasticity, and dystonia at all at hand, and there is more to come.

The goal of this book is to capture the trajectory, and the creative processes, underlying the revolution that is going on in neurology. This book, rather than being a compendium of neurological therapeutics or of recent advances in neuroscience, tells the stories of *how neurological therapeutics were developed—or are being developed.* The first part of this book reviews the development of new therapies in neurology, from their inception in terms of basic science to their introduction into the clinical world. Within the first section, authors address the questions: *How has the revolution in neurology occurred? What lessons can we learn from past successes? Can these lessons help us to ensure progress in the future?* The second part of this volume explores evolving themes and new technologies. Changes in the practice of neurology have been driven by advances in the laboratory. But, in parallel with the transformation of clinical neurology, neuroscience and molecular medicine are moving forward rapidly and are generating new information and perspectives at an ever-increasing rate. The chapters in the second part of the book explore rapidly-advancing areas of research which may hold the key to still other therapeutic advances. Hopefully these chapters will provide a roadmap for the future.

Who knows what neurology will be like twenty-five years from now? There will undoubtedly be additional, and very powerful, diagnostic methods. And there will almost certainly be new, and more effective, therapies for patients with disorders of the nervous system. Some of these advances will be outgrowths of research that is ongoing at this time, while others may come from advances that have not yet been imagined. We cannot know for sure, at this time, which diseases of the nervous system will be cured, and which symptoms will be totally alleviated in two decades. We can, however, predict with a high degree of certainty that, as each year passes, the therapeutic transformation of neurology will continue so that, in each new decade, an increasing number of neurological disorders will yield to therapy. That is almost a certainty. And it is to that goal that this book is dedicated.

ACKNOWLEDGEMENTS

This book could not have been written without the efforts of its chapter authors and to them I am most grateful. On their behalf, I express deep thanks to the many individuals and organizations that have supported research on disorders of the nervous system. The Paralyzed Veterans of America, the United Spinal Association, and their leadership deserve special thanks for unwavering commitment and support. I am especially grateful to Dan Flaherty and Sandra Kulli who, behind the scenes, have been partners in the battle against disabling neurological disorders. I also extend my thanks to Sheila MacMillan who guided most of the chapters through early stages of the editing process, to my colleagues who encouraged me at every juncture, and to my family who endured the closed door of my study while I worked on this book. Finally, I am indebted to Academic Press and, particularly, to Karen Dempsey who very ably handled the editorial process at Academic Press, and Jasna Markovac whose encouragement helped make this book a reality.

Contributors

Amy Arnsten, Department of Neurobiology, Yale University School of Medicine, New Haven, Connecticut

Aviva Abosch, Department of Neurology and Surgery, Center for Neurodegenerative Disease, Emory University School of Medicine, Atlanta, Georgia

Arthur A. Asbury, Department of Neurology, Hospital of the University of Pennsylvania, Philadelphia, Pennsylvania

Hayrunnisa Bolay, Department of Neurology, Gazi University Hospitals, Ankara, Turkey

Sophia A. Colamarino, Laboratory of Genetics, The Salk Institute for Biological Studies, La Jolla, California

Cynthia L. Comella, Department of Neurology, Rush Presbyterian St. Lukes Medical Center, Chicago, Illinois

Theodore R. Cummins, Stark Neuroscience Research Institute, Indiana University School of Medicine, Indianapolis, Indiana

Gary H. Danton, Department of Neurological Surgery, University of Miami School of Medicine, Miami, Florida

W. Dalton Dietrich, The Miami Project and Department of Neurological Surgery, University of Miami School of Medicine, Miami, Florida

Jerome Engel, Reed Neurological Research Center, Neurology Department #1250, Los Angeles, California

Kenneth H. Fischbeck, National Institute of Neurological Disorders and Stroke, National Institutes of Health, Bethesda, Maryland

R.S.J. Frackowiak, Wellcome Department of Cognitive Neurology, Institute of Neurology, London, United Kingdom

Fred H. Gage, Laboratory of Genetics, The Salk Institute for Biological Studies, La Jolla, California

Daniel H. Geschwind, Department of Neurology, University of California Los Angeles School of Medicine, Los Angeles, California

Steve Goldman, Department of Neurology, University of Rochester Medical Center, Rochester, New York

Edward D. Hall, Spinal Cord and Brain Injury Research Center, University of Kentucky Chandler Medical Center, Lexington, Kentucky

Mark Hallett, Human Motor Control Section, National Institute of Neurological Disorders and Stroke, National Institutes of Health, Bethesda, Maryland

Jill Heemskerk, Technology Development, National Institute of Neurological Disorders and Stroke, National Institutes of Health, Bethesda, Maryland

George R. Heninger, Department of Psychiatry, Connecticut Mental Health Center, New Haven, Connecticut

Amie W. Hsia, Department of Neurology, Stanford University School of Medicine, Palo Alto, California

Ole Isacson, Neuroregeneration Laboratories, Harvard Medical School/McLean Hospital, Belmont, Massachusetts

Joseph Jankovic, Department of Neurology, Baylor College of Medicine, Houston, Texas

Dmitri M. Kullman, Experimental Epilepsy Group, Institute of Neurology, London, United Kingdom

D. Chichung Lie, Laboratory of Genetics, The Salk Institute for Biological Studies, La Jolla, California

Kenneth L. Marek, The Institute for Neurodegenerative Disease and Yale University School of Medicine, New Haven, Connecticut

Christiane Massicotte, Department of Clinical Studies, School of Veterinary Medicine, University of Pennsylvania, Philadelphia, Pennsylvania

John C. Mazziotta, Brain Mapping Center, University of California Los Angeles School of Medicine, Los Angeles, California 90095

Stephen B. McMahon, Center for Neuroscience, King's College—London, London, United Kingdom

Michael A. Moskowitz, Stroke and Neurovascular Regulation Laboratory, Department of Neurosurgery, Massachusetts General Hospital, Charlestown, Massachusetts

Beth Murinson, Department of Neurology, Johns Hopkins School of Medicine, Baltimore, Maryland

James H. Park, Department of Neurology, Yale University School of Medicine, New Haven, Connecticut

Donald L. Price, Department of Pathology, The Johns Hopkins University, Baltimore, Maryland

Richard A. Rudick, The Mellen Center, Cleveland Clinic Foundation, Cleveland, Ohio

Robert L. Ruff, Neurology Service, Cleveland Veteran's Affairs Medical Center, Cleveland, Ohio

Peter D. Schellinger, Stroke Diagnostics and Therapeutics, National Institute of Neurological Disorders and Stroke, National Institutes of Health, Bethesda, Maryland

Steven Scherer, Department of Neurology, The University of Pennsylvania Medical Center, Philadelphia, Pennsylvania

Lorelei D. Shoemaker, Neuroscience Interdepartmental PhD Program, University of California Los Angeles, Los Angeles, California

Lance Simpson, Department of Biochemistry and Molecular Pharmacology, Jefferson Medical College, Philadelphia, Pennsylvania

Hongjun Song, Institute for Cell Engineering, Department of Neurology, Johns Hopkins University School of Medicine, Baltimore, Maryland

Stephen Strittmatter, Department of Neurology, Yale University School of Medicine, New Haven, Connecticut

Patrick G. Sullivan, Spinal Cord and Brain Injury Research Center, University of Kentucky Chandler Medical Center, Lexington, Kentucky

Mark H. Tuszynski, Department of Neurology, University of California, San Diego, La Jolla, California

Jerrold L. Vitek, Center for Neurological Restoration, Cleveland Clinic Foundation, Cleveland, Ohio

Bernhard Voller, Human Motor Control Section, National Institute of Neurological Disorders and Stroke, National Institutes of Health, Bethesda, Maryland

Benjamin L. Walter, Center for Neurological Restoration, Cleveland Clinic Foundation, Cleveland, Ohio

Steven Warach, Stroke Diagnostics and Therapeutics, National Institute of Neurological Disorders and Stroke, National Institutes of Health, Bethesda, Maryland

Steven G. Waxman, Department of Neurology and the Center for Neuroscience and Regeneration Research, Yale University School of Medicine, New Haven, Connecticut

John N. Wood, Department of Anatomy and Developmental Biology, University College, London, United Kingdom

Clinical Rewards From Neuroscience, Molecular Medicine, and Translational Research

1

Seeing the Brain So We Can Save It: The Evolution of Magnetic Resonance Imaging as a Clinical Tool

Peter D. Schellinger, MD, PhD

Steven Warach, MD, PhD

"MRI researchers win Nobel Prize for medicine" – CBC News 10/2003

STOCKHOLM, SWEDEN -Two scientists whose discoveries led to the development of medical imaging of the body's inner organs have won the 2003 Nobel Prize for medicine. The Swedish Academy said the discoveries of American Paul C. Lauterbur and Briton Sir Peter Mansfield represent "a breakthrough in medical diagnostics and research." The last time the medicine prize was awarded in the field of diagnostics was for the development of computer assisted tomography in 1979. (See Figure 1.1.)

The last, but most important, step for a new technology is its establishment in clinical practice. The evolution from tentative reports and early research activity to the definition of possible clinical benefits and acceptance into practice (Jackson, 2001) is often protracted in complex technologies such as magnetic resonance imaging (MRI). Lauterbur (1973) published his landmark article about the derivation of position-dependent information by nuclear magnetic resonance and magnetic gradients in the early 1970s. Despite the work that followed by Sir Peter Mansfield (Mansfield and Maudsley, 1976, 1977) and other groups, the first clinical MRI system was not available until 1984 followed in 1987 by the first paramagnetic MRI contrast agent. The last 20 years have realized several advances in computer technology and software development, and these changes have continuously improved the accessibility and quality of MRI. The earliest scans—T2-weighted, T1-weighted, and (proton density) PD-weighted sequence without contrast—took hours to complete. Today, multiparametric protocols can assess within 15 minutes the most complex pathophysiological processes, which has allowed a dramatic shift in the evaluation and treatment of neurological illness. As the clinical discipline of neurology has evolved from a diagnostic to a therapeutic specialty, MRI is being transformed into a clinical tool that has an impact on neurology at the bedside. This chapter illustrates this transformation through three classical neurological diseases that have gained growing interest with regard to therapeutic options only within the last 30 years: acute ischemic stroke, multiple sclerosis, and brain tumors. It is exciting to observe that, with an increasing number of clinical therapeutic trials being designed, MRI may not only function as a diagnostic tool but may also have prognostic strength and thus serve as a surrogate endpoint for the development of new therapies (Warach, 2001b).

ADVANCES IN MRI

Once a technology is accepted into clinical practice, it is inevitable that the demands of clinical specialties stimulate the evolution of the technology and the extension of its functionality to new applications (Jackson, 2001). For instance, interest in functional imaging led to developments such as higher gradients, allowing rapid and repetitive image acquisition with high spatial and temporal resolution. Echo planar imaging (EPI) was found useful for expanding the capabilities of MRI in the routine acquisition of T2-weighted images (Heiland *et al.*, 1998; Pierpaoli *et al.*, 1996; Warach *et al.*, 1995) and allowing formerly time-consuming sequences such as fat- (STIR, short tau inversion recovery) and fluid-suppressed sequences (FLAIR, fluid attenuated inversion recovery) to become practical in the clinical setting (Bydder and

Paul C. Lauterbur   Sir Peter Mansfield

FIGURE 1.1   Paul C. Lauterbur and Sir Peter Mansfield.

Young, 1985; De Coene *et al.*, 1992). FLAIR images have substantially improved the diagnostic yield in diseases such as stroke and multiple sclerosis, where chronic vascular lesions and foci of demyelination are in close proximity to cerebrospinal fluid (CSF) spaces (Hajnal *et al.*, 1992). Other examples of technical progress stimulated by clinical needs include magnetic resonance angiography (Brant-Zawadzki and Heiserman, 1997; Fujita *et al.*, 1994; Johnson *et al.*, 1994) and magnetization transfer imaging (Haehnel *et al.*, 2000; Knauth *et al.*, 1996; Kurki *et al.*, 1994).

Two techniques that have received much attention in the last 15 years are diffusion- and perfusion-weighted imaging (DWI and PWI) (Heiland *et al.*, 1997; Moseley *et al.*, 1991; Ostergaard *et al.*, 1996; Warach *et al.*, 1992). The phenomenon of indirectly measuring brownian molecular motion of water with DWI was first described in 1965 (Stejskal and Tanner, 1965). The measurement of water diffusion renders pathophysiological tissue information, especially in ischemic, inflammatory, and neoplastic disease, that cannot be obtained with standard MRI sequences (Le Bihan *et al.*, 1986; Le Bihan *et al.*, 1992). In addition to the qualitative measurements of diffusion, the apparent diffusion coefficient (ADC) can be quantified from measurements at different b-values (Tanner and Stejskal, 1968) on a per pixel basis, within a region of interest, or to generate ADC parameter maps. For structures, such as nerve fibers or tracts, that have a predominant direction in space, the ADC depends on the direction of the diffusion gradient. Whereas water diffu-

sion is only mildly impaired in the longitudinal direction, water diffusion in a perpendicular direction to the main axis is decreased (Basser and Pierpaoli, 1996); this is referred to as anisotropy. In the brain, anisotropy is most pronounced in the white matter, especially in densely packed fiber structures such as the corpus callosum and the corona radiata. The effects of anisotropy can be used to obtain information of the regional fascicular or fiber tract anatomy, which may be helpful for neurosurgical procedures. The expansion of these principles has led to the development of diffusion tensor imaging, which has been used to characterize myelination and demyelination (Le Bihan *et al.*, 2001; Neil *et al.*, 1998; Ono *et al.*, 1995; Pierpaoli *et al.*, 1996).

PWI uses a dynamic bolus (paramagnetic contrast) tracking technique to derive concentration time curves, allowing the qualitative assessment of various hemodynamic parameters relative to areas of normal parenchyma (Rosen *et al.*, 1989, 1990; Villringer *et al.*, 1988). Typical parameters are mean transit time (MTT), cerebral blood volume (CBV), cerebral blood flow, and time to peak (TTP). The actual quantification of hemodynamic information has been limited by postprocessing capabilities and methods of calculating the arterial input function. Nevertheless, areas of relative hypoperfusion can be reliably demonstrated by current methods, which is of special relevance in ischemic stroke (Schlaug *et al.*, 1999). Other applications, such as functional imaging with the BOLD (blood oxygen level dependence) technique and spectroscopic imaging, exceed the scope of this chapter. It suffices to say that substantial technical and programming refinements have supplied neuroscience with a comprehensive diagnostic armament to guide our daily clinical practice. The growing interest in neurology for physiological data, the development of new treatment paradigms, and the definition of parameters that predict disease course will continue to promote the development and application of new imaging technology.

MRI IN ACUTE ISCHEMIC STROKE

Stroke is the third leading cause of death after myocardial infarction and cancer, the leading cause of permanent disability in western countries (WHO Guidelines Subcommittee, 1999), and the leading cause of disability-adjusted loss of independent life years. In addition to the tragic consequences for patients and their families, the socioeconomic impact of disabled stroke survivors is estimated to be between $35,000 and $50,000 per survivor, per year. In the face of our aging population, the incidence and prevalence of stroke are expected to rise. Therefore, an effective and widely available treatment for this devastating disease is desperately needed.

Before the development of MRI and continuing until the early 1990s, the stroke field was dominated by therapeutic nihilism. Patients were treated conservatively and sent to rehabilitation units or nursing homes. Major stroke trials performed during that time focused on secondary prevention. These included two trials of carotid endarterectomy for stroke attributed to internal carotid artery stenosis (European Carotid Surgery Trialists (ECST) Collaborative Group, 1991; North American Symptomatic Carotid Endarterectomy Trial (NASCET) Collaborators, 1991), one small negative trial for extracranial/intracranial bypass (EC/IC Bypass Study Group, 1985), and trials testing the efficacy of anticoagulation (warfarin) and antiplatelet agents (acetylsalicylic acid, ticlopidine) for stroke prevention (Antiplatelet Trialists Collaboration, 1994; Antithrombotic Trialists Collaboration, 2002). Heparin, low-molecular-weight heparins, and heparinoids were tested and found ineffective for early or late secondary prevention of stroke (Albers *et al.*, 2001). Still, trials for acute stroke therapy were lacking. Anecdotal reports of acute treatment with thrombolytics were first described in the early 1960s (Meyer *et al.*, 1963). Yet small trials focused on their use for subacute stroke (Schellinger *et al.*, 2001b); pilot trials for acute stroke did not occur until the early 1990s (Mori *et al.*, 1992; Yamaguchi *et al.*, 1993). No diagnostic imaging modality, including angiography, was used to establish the diagnosis of stroke in any of these studies.

The pilot study of Haley *et al.* (1993) was the first study to use a cross-sectional imaging method, noncontrast computed tomography (CT), before therapy. After several negative acute stroke trials with streptokinase, The National Institute of Neurological Disorders and Stroke (NINDS) demonstrated that intravenous thrombolysis with recombinant tissue plasminogen activator (rt-PA) is safe and effective in patients with acute ischemic stroke after exclusion of intracerebral hemorrhage (ICH) by noncontrast CT and when given within 3 hours of symptom onset. The NINDS trial was published in 1995 (The National Institutes of Neurological Disorders and Stroke rt-PA Stroke Study Group, 1995), and the U.S. Food and Drug Administration (FDA) approved the drug in 1996. Three more trials assessed the efficacy of rt-PA beyond the 3-hour window but yielded negative results (Clark *et al.*, 1999; Hacke *et al.*, 1995; Hacke *et al.*, 1998). While the NINDS trial applied CT only for the purpose of ICH exclusion, the later trials established criteria for the diagnosis of early *ischemic* stroke in a widely available cross-sectional imaging method used in the routine clinical management of neurological diseases (von Kummer *et al.*, 1995). Despite the poor sensitivity of noncontrast CT within the therapeutic time window (at best 45% to 65% among specialists) (von Kummer *et al.*, 1996) and the fact that meta-analyses suggested a therapeutic effect of rt-PA beyond the 3-hour window in some patients (Brott, February 7-9, 2002; Hacke *et al.*, 1999; Wardlaw *et al.*, 2002), it is surprising that modern techniques such as CT-angiography (Knauth *et al.*, 1997) or MRI have not had a role in acute stroke imaging until recently (Mohr *et al.*, 1995).

In parallel to the design and completion of the trials of thrombolytic therapy, the advent of new MRI techniques such as PWI and DWI has added another dimension to diagnostic imaging in stroke (Hacke and Warach, 2000; Schellinger *et al.*, 2003; Warach *et al.*, 1995). A growing body of evidence has documented the usefulness of these methods in the clinical setting (Davis *et al.*, 2003; Fiebach and Schellinger, 2003). Several investigators have described significant correlations of DWI and PWI changes with follow-up imaging and neurological outcome (Barber *et al.*, 1998; Lovblad *et al.*, 1997; Warach *et al.*, 1997; Warach *et al.*, 1996). Others have proposed a more rational selection of therapeutic strategies based on the presence or absence of a tissue at risk, as identified by means of DWI and PWI patterns. The supposed lack of feasibility and practicality of MRI as a diagnostic tool in hyperacutely ill patients (Powers, 2000; Powers and Zivin, 1998; Zivin and Holloway, 2000) has been consistently disproved. Many centers have shown that logistical obstacles can be overcome (Buckley *et al.*, 2003; Schellinger *et al.*, 2000b). DWI identifies ischemic tissue changes within minutes after vessel occlusion, with a sensitivity of approximately 90%, and is far superior to noncontrast CT (Fiebach *et al.*, 2002) and conventional MRI (Mohr *et al.*, 1995) in this regard. In the clinical setting of a hyperacute stroke, a lesion on DWI with a reduction of the ADC (Le Bihan *et al.*, 1986) and a normal T2-weighted image or FLAIR favor an acute and, in most instances, irreversible lesion. As more patients have been evaluated with stroke MRI, and at early time points, partial reversal of the initial DWI lesion has been reported. These findings suggest that the DWI lesion should be included in the therapeutic target, not as the definite infarct core (Fiehler *et al.*, 2002, 2003; Kidwell *et al.*, 2000, 2003).

PWI allows the relative measurement of capillary perfusion (Rosen *et al.*, 1990). Although it is not yet clinically feasible to obtain quantitative PWI parameters that differentiate oligemia from relevant hypoperfusion and infarct core (Yamada *et al.*, 2002), most authors agree that parameters such as MTT or TTP give the best prognostic information (Baird *et al.*, 2000; Schellinger *et al.*, 2001a) for patients with acute ischemic stroke. Further PWI research is aimed at differentiating oligemia from critical ischemia (Kidwell *et al.*, 2003), with the goal of better guiding clinical management. Also, the source images from PWI (susceptibility-weighted images or $T2^*$-weighted images) have been shown to reliably depict ICH (Fiebach *et al.*, 2004; Linfante *et al.*, 1999; Schellinger *et al.*, 1999a), achieving a sensitivity comparable to that of noncontrast CT.

The volume difference of DWI and PWI, also termed PWI/DWI-mismatch, gives an approximate measure of the tissue at risk of infarction (Figure 1.2) (Warach, 2001a). This model does not take into account that the PWI lesion includes areas of benign oligemia and that DWI abnormalities do not necessarily turn into infarction (Kidwell *et al.*, 2003). Also, the assessment of mismatch as a percentage may be less reliable than previously thought (Coutts *et al.*, 2003). However, this simple model of PWI/DWI-mismatch is sufficiently accurate in most acute stroke patients, and the stroke MRI findings are consistent with our understanding of the pathophysiology of acute ischemia (Schellinger *et al.*, 2003; Schellinger *et al.*, 2001a). According to unpublished data (Chalela *et al.*, 2004; Todd *et al.*, 2004), early reversal of PWI is a much better predictor of clinical and imaging outcome than partial or complete DWI lesion reversal. Applying the mismatch concept may identify the individual time window for the patient and thus allow therapeutic decision making based on an individual vascular and hemodynamic situation, rather than elapsed time. Additional MRI findings not captured by CT, such as early blood-brain barrier disruption (Latour *et al.*, 2004, submitted for publication) and old microbleedings (Kidwell *et al.*, 2002; Nighoghossian *et al.*, 2002) may predict a poor outcome after thrombolysis and therefore can be used to improve patient selection.

A prospective, open-label, nonrandomized multicenter trial examined the MRI baseline characteristics of 139 patients who presented with acute ischemic stroke within 6 hours of symptom onset and studied the influence of intravenous rt-PA on MRI parameters and functional outcome (Röther *et al.*, 2002). There was a significantly higher occurrence of independent outcome after treatment with rt-PA (76 patients vs 63 control subjects) despite the significantly worse baseline NIHSS score and the larger volume of tissue at risk in these patients. Another study treated patients according to the NINDS protocol within the first 3 hours (N = 115) and according to stroke MRI findings in the 3- to 6-hour window (N = 48). Baseline National Institutes of Health Stroke Scale (NIHSS) scores and the rate of symptomatic ICH did not differ between groups, but mortality did trend in favor of the patients receiving late rt-PA (3- to 6-hour window) according to stroke MRI criteria (16.5% vs 6.3%, P = .08). Interestingly, the outcome (independent vs dependent or dead [mRS 0-2 vs mRS 3-6]) of the stroke MRI group was better than the outcome of the group who received early intervention (47% vs 62.5%, OR 0.54, CI 0.27-1.06) based on CT and NINDS criteria. These numbers suggest that with a selection tool such as stroke MRI, the time window for treating stroke with thrombolytic therapy can be substantially expanded, with improved results as compared to historical studies. However, this has yet to be firmly established in a randomized trial.

Despite its potential, only 1% to 2% of all stroke patients receive rt-PA (Kaste, 2003). Among the major problems are the fact that relatively few candidates present within the time window and meet the clinical criteria. Educating the general public to regard stroke as a treatable emergency and training emergency caregivers in the use of thrombolysis can decrease these problems but demands a continuous effort. Health care institutions should be made aware of the potential in long-term cost savings once stroke management is optimized and thrombolytic therapy is more widely available. Stroke physicians are frequently confronted with stroke patients who awaken with a deficit (Fink *et al.*, 2002) or are unable to provide the required information as a result of aphasia or disorientation. At present, these patients are excluded from thrombolytic therapy, even if a CT scan is normal or has only minor ischemic changes. Thrombolysis is often withheld in these patients even though they might profit from rapid recanalization. In contrast, MRI presently is used in many centers worldwide in the management of acute stroke, beyond the NINDS time criteria, with thrombolysis rates ranging from 10% to 25% of the stroke patients in these centers. In due time, MRI may become not only the most powerful but also the most widely and potentially uniformly used tool to guide therapy based on individual pathophysiology rather than

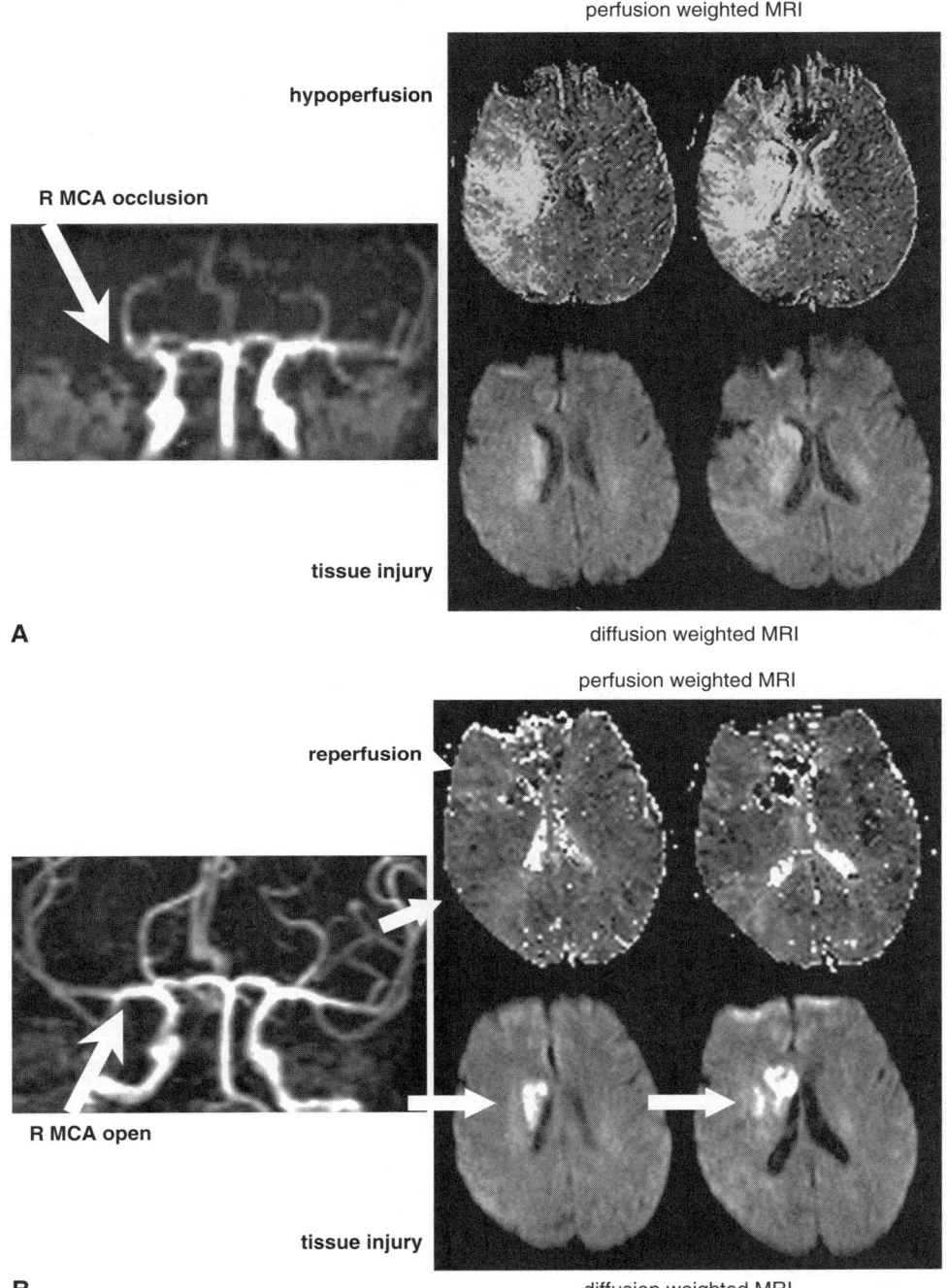

FIGURE 1.2 PWI-DWI mismatch in a patient with an acute right MCA mainstem occlusion 2 hours after symptom onset. **A:** The area of hypoperfusion accounts for most of the right MCA territory, whereas on DWI there is only a small periventricular area of reduced diffusion. **B:** Twenty-four hours after thrombolytic therapy with intravenous rt-PA, the MCA is recanalized, the brain is reperfused, and the damage is substantially smaller than the initial perfusion deficit.

a surrogate parameter such as elapsed time. New trials for thrombolysis in acute stroke have started to apply multiparametric MRI as a screening tool for patient inclusion, including DEFUSE (USA) and EPITHET (Australia) (Parsons *et al.*, 2002), which perform therapy with rt-PA in the 3- to 6-hour window. A Phase II trial for another thrombolytic drug (desmoteplase) has been completed, although results are not yet published. The combination of rt-PA with glycoprotein IIb/IIIa antagonists may be a promising approach and results in increased vessel patency rates in accordance with cardiological studies. These trials currently are only in Phases I and II, but will include patients in late time windows up to 24 hours after stroke onset, using stroke MRI findings as inclusion criteria (Tirofiban plus rt-PA – SATIS, Mario Siebler personal communication (2003); Abciximab plus retavase – ROSIE; Eptifibatide plus ASA, LMWH, rt-PA – ROSIE 2, Steven Warach, personal communication (2004)). The ongoing trials section in *Stroke* lists most of the new and ongoing trials (2003). Once Phase III MRI-based trials are available and yield positive results, MRI will become the required imaging modality as CT has been for nearly a decade.

MRI IN MULTIPLE SCLEROSIS

Multiple sclerosis (MS) is the most common demyelinating disorder and cause of neurological disability in young adults between 20 and 40 years old. MS has a higher prevalence in Caucasians from northern temperate climates (Noseworthy, 2003) and females, and affects up to 350,000 individuals in the United States (Anderson *et al.*, 1992). This chronic inflammatory disease of the central nervous system (CNS) is pathologically characterized by perivenous immune cell infiltration and myelin destruction, most likely due to autoimmune reactions against basic myelin protein, focal demyelination, and subsequent loss of oligodendrocytes and axons (Bruck *et al.*, 1997). Different clinical courses of MS have been defined, including relapsing remitting (RR) MS, secondary progressive (SP) MS, primary progressive (PP) MS, and progressive relapsing (PR) MS (Lublin and Reingold, 1996). RRMS is characterized by discrete episodes of neurological symptoms followed by a variable degree of recovery and accounts for up to 85% of the initial presentations. This group of patients has demonstrated the best response to treatment; however, more than 50% of RRMS patients display continuous, progressive symptoms characteristic of SPMS within 10 years (up to 90% within 25 years). By contrast, PPMS is characterized by a steady decline in neurological function from onset, without superimposed attacks, and is present in 10% of MS patients. MS often presents at a time when the clinical extent of disease is apparently limited, although, even at this early stage of disease, substantial damage may have already occurred. Thus, approximately 50% to 80% of individuals who present with a clinically isolated syndrome (CIS) already have lesions on MRI, consistent with prior (occult) disease activity (Jacobs *et al.*, 1986).

Diagnosis during a single acute episode is difficult to establish because of the variability of focal and diffuse symptoms, as well as exclusion of MS variants such as acute disseminated encephalomyelitis (Schwarz *et al.*,

2001). All existing diagnostic criteria for RRMS, including the early ones reported by Schumacher *et al.* (1965) and revised by Poser *et al.*, (1983) and a recent consensus statement (McDonald *et al.*, 2001) require two or more distinct events separated in time (generally by more than 1 month) in addition to involvement of at least two distinct areas of the CNS (Tables 1.1 and 1.2). This is referred to as the "criteria of dissemination in time and space" (Frohman *et al.*, 2003). Also of importance, all of the existing diagnostic schemes require the exclusion of alternative diagnoses (among others human immunodeficiency virus, stroke, neoplastic disease, infections) by appropriate laboratory and radiographic means.

Although published in 1983, the Poser criteria already acknowledged MRI as a "paraclinical" tool for the diagnosis of MS. In those early days, MRI was a much less sensitive tool, especially with the lack of contrast agents. With advances in technology, MRI reveals ten times as many lesions as CT (Bergers and Barkhof, 2001; Mushlin *et al.*, 1993) and provides a means to characterize demyelination in the spine. As MRI technology has progressed, criteria for the MRI diagnosis of MS have been published (Barkhof *et al.*, 1997; Fazekas *et al.*, 1988; Paty *et al.*, 1988). The topic has been reviewed elsewhere in detail (Bergers and Barkhof, 2001). In brief, although T2-weighted image demonstrates foci of chronic demyelination and new lesions with edema, the diagnostic yield of FLAIR is substantially better because of the predominantly periventricular localization of MS lesions and CSF artifact (Hashemi *et al.*, 1995). FLAIR studies have been espe-

TABLE 1.1 Poser Criteria for the Diagnosis of Multiple Sclerosis (MS)

Clinically definite MS:
2 Attacks[a] + 2 lesions on examination[b]
2 Attacks + 1 lesion on examination + 1 paraclinical lesion[c]

Laboratory-supported definite MS:
2 Attacks + 1 lesion on examination or 1 paraclinical lesion + abnormal cerebrospinal fluid (CSF)[d]
1 Attack + 2 lesions on examination + abnormal CSF[d]
1 Attack + 1 lesion on examination + 1 paraclinical lesion[c] + abnormal CSF[d]

Clinically probable MS:
2 Attacks + 1 lesion on examination
1 Attack + 2 lesions on examination
1 Attack + 1 lesion on examination + 1 paraclinical lesion[c]

Laboratory-supported probable MS:
2 Attacks + abnormal CSF[d]

From Frohman *et al.*, 2003; Poser *et al.*, 1983.
[a]Symptoms > 24h including subjective and anamnestic lesions.
[b]Separated by 1 month.
[c]MRI and evoked potentials.
[d]Oligoclonal bands or increased CNS IgG synthesis.

TABLE 1.2 McDonald Criteria for the Diagnosis of Clinically Definite[a] Multiple Sclerosis (CDMS)

2 Attacks[b] + 2 lesions on examination[c]

2 Attacks + 1 lesion on examination + [MRI (a)[d] or 2 MRI lesions + abnormal CSF]

1 Attack + 2 lesions on examination + MRI (b)[e]

1 Attack + 1 lesion on examination + [MRI (a)[d] or 2 MRI lesions + abnormal CSF] + MRI (b)[e]

Progressive disease + abnormal CSF + MRI (b)[e] +

   a) [≥9 T2 MRI lesions or ≥2 spinal cord lesions or 4–8 brain MRI lesions + 1 spinal cord lesion]

OR

   b) Abnormal (Visually Evolved Potentials) VEP[c] + [≥4 brain MRI lesions or <4 brain MRI lesions +1 spinal cord lesion]

From Frohman *et al.*, 2003; McDonald *et al.*, 2001.

[a]If the criteria are fulfilled and met, the diagnosis is MS. If the criteria are not completely met, the diagnosis is possible MS. If the criteria are fully explored and not met, the diagnosis is not MS.

[b]Symptoms lasting >24 hours excluding pseudoattacks.

[c]Objective evidence of 2 lesions on neurological examination disseminated in space. A subclinical lesion by VEP can substitute for one clinical lesion.

[d]MRI (a): Dissemination in space: 3 of 4 criteria need to be met:
   1. ≥1 enhancing lesion or ≥9 T2-weighted image lesions
   2. ≥1 infratentorial MRI lesion
   3. ≥1 juxtacortical lesion
   4. ≥3 periventricular lesions
1 spinal cord lesion can substitute for 1 brain lesion, presence of oligoclonal bands: or increased IgG index.

[e]MRI (b): Dissemination in time:
   1. If original MRI performed < 3 months after the clinical attack, follow-up MRI ≥ 3 months with new enhancing lesion or second follow-up ≥ 3 months with new enhancing or new T2-weighted image lesion necessary.
   2. If original MRI performed ≥ 3 months after the clinical attack, enhancing lesion at other site than clinical event or second follow-up ≥ 3 months with new enhancing or new T2-weighted image lesion necessary.

cially good in depicting callosal lesions in sagittal images; however, they are not very sensitive for infratentorial lesions (Bastianello *et al.*, 1997). Precontrast and postcontrast T1-weighted images can differentiate chronic from active lesions. By combining the findings on T2- and T1-weighted images, the sensitivity for the diagnosis of clinically definite MS can reach 80% by imaging criteria alone (Barkhof *et al.*, 1997).

The Poser scheme was developed at a time when MRI was in its infancy, and it may permit a diagnosis of MS based on unspecific white matter lesions in circumstances where it is unjustified (Frohman *et al.*, 2003) (Figure 1.3). It also does not allow for sequential paraclinical testing to establish dissemination in time and cannot diagnose PPMS. A new set of diagnostic criteria relying even more on MRI parameters has been proposed by McDonald *et al.* (2001). These criteria are more stringent than those of Poser *et al.*, as they allow for MRI criteria of spatial and temporal dissemination and the diagnosis of PPMS.

According to the published report of the Therapeutics and Technology Assessment subcommittee of the American Academy of Neurology (Frohman *et al.*, 2003), the evidence with respect to MRI in suspected MS is as follows:

1. The presence of three or more white matter lesions on T2-weighted image MRI predicts a diagnosis of clinically definite multiple sclerosis (CDMS) within the next 7 to 10 years with an 80% sensitivity (Class I-III evidence, Type A recommendation).
2. New T2-weighted image lesions or new enhancing lesions documented 3 or more months after a clinically isolated demyelinating episode and a baseline MRI are highly predictive for near term development of CDMS (Type A recommendation).
3. The presence of two or more enhancing lesions at baseline is highly predictive of future development of CDMS (Type B recommendation).
4. Diagnosis other than MS with a CIS with any MRI abnormality of 1–3, after exclusion of alternative diagnoses is unlikely (Type A recommendation).
5. MRI features for PPMS cannot be determined from the existing evidence (Type U recommendation).

The authors recommend that studies be undertaken to assess the use of surveillance MRI as a means of monitoring disease activity to assess the response to different therapies and therefore guide differential immunomodulatory therapy (Frohman *et al.*, 2003).

Although the use of conventional MRI has had a great impact on the diagnosis of MS, the heterogeneity of tissue damage and repair occurs at least partially beyond the detection sensitivity of conventional MRI (Miller *et al.*, 1998). As the primary pathological changes in this disease are inflammation and neurodegeneration, investigators have turned to quantitative MRI techniques including volumetric atrophy assessment, DWI, diffusion tensor imaging with tractography, magnetization transfer imaging, and MRI spectroscopy to help characterize the evolution and extent of tissue damage and repair. These newer MRI applications, however, currently play more of a scientific role than a diagnostic and therapeutic one (Bergers and Barkhof, 2001; Filippi, 2003; Reidel *et al.*, 2003).

In summary, MRI was suggested as an additional diagnostic tool for MS shortly after its introduction as a clinical imaging modality (Poser *et al.*, 1983). After modification of the diagnostic criteria (McDonald *et al.*, 2001), it has been formally and firmly established as a tool in MS, which

clinicians can rely on for differential diagnosis. The application of immunosuppressive drugs (corticosteroids, cyclophosphamide, azathioprine, mitoxantrone) and initiation of modern immunomodulating medication such as β1-a or β1-b interferons and glatiramer acetate can be justified according to the preceding MRI criteria. Also, the most recent drug trials (interferons and glatiramer acetate) used MRI criteria for inclusion as well as a surrogate parameter in secondary endpoint analyses. Therefore, present and future management of possible or definite MS without MRI as a clinical tool to guide therapeutic decisions for acute or prophylactic treatment is almost unimaginable.

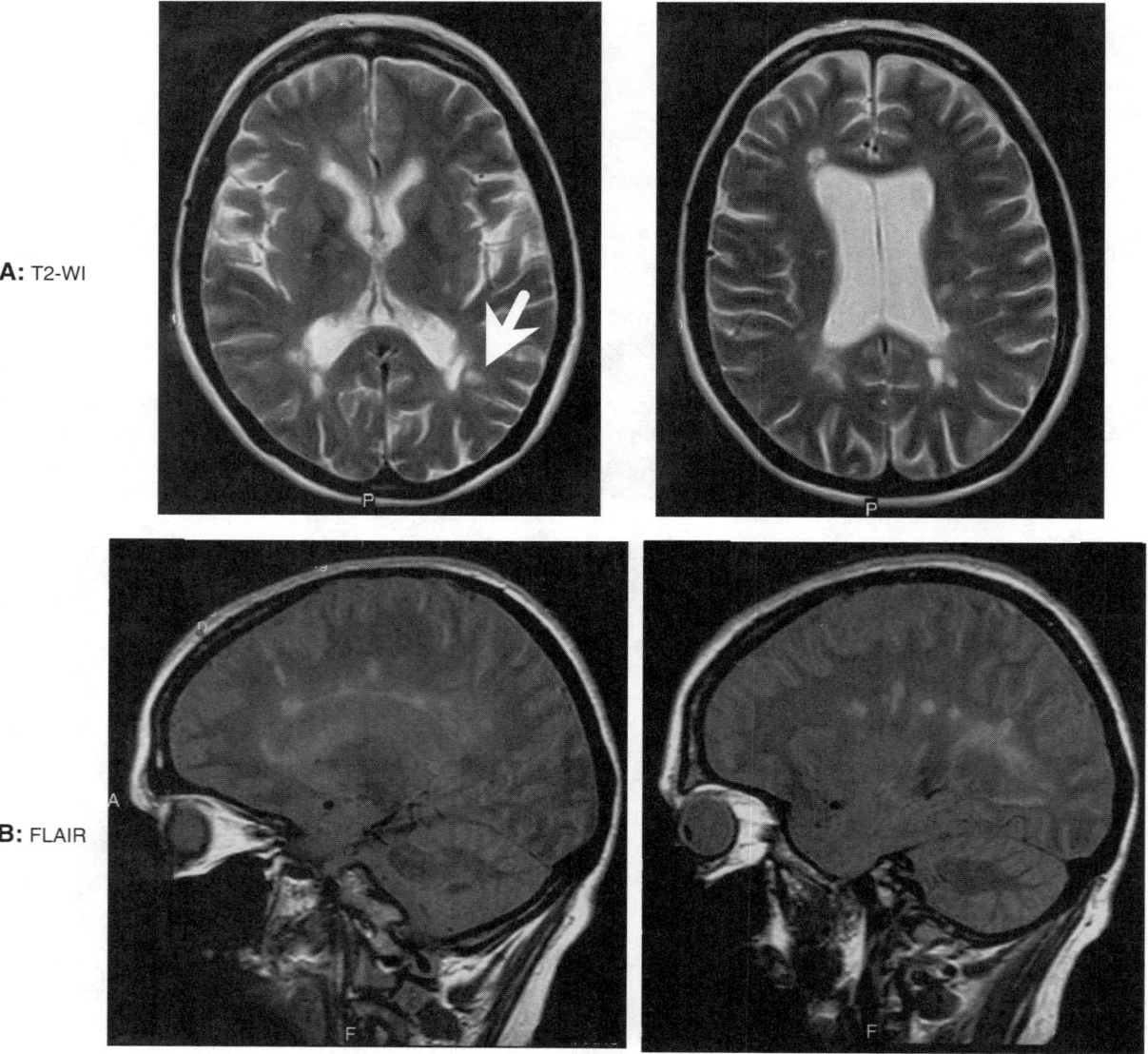

A: T2-WI

B: FLAIR

FIGURE 1.3   A 50-year-old patient with multiple sclerosis. **A:** Axial slices on T2-weighted image with periventricular hyperintensities. **B:** Sagittal FLAIR images show association with the corpus callosum.

MRI IN BRAIN TUMORS

The World Health Organization's GLOBOCAN 2000 study of cancer prevalence and mortality reported more than 17,000 new cases and nearly 13,000 deaths annually from primary CNS tumors in the United States alone (Parkin *et al.*, 2001). These data do not include the prevalence of benign lesions or the estimated 100,000 cases of brain metastases from a non-CNS primary cancer. The classification and prognosis of intracranial neoplasms are based on histological features including cell type, degree of anaplasia, rate of growth and invasion, and clinical factors such as patient age, onset, and progression of symptoms. In the early 1980s, the imaging gold standard for brain tumors was contrast-enhanced CT. Even in these early precontrast days of MRI, however, the ability to conduct multiplanar imaging and generate reconstructable three-dimensional datasets, reduced posterior fossa and bone artifacts, good gray/white matter contrast, sensitivity to edema (T2-weighted images), and good depiction of anatomical detail (T1-weighted images) established MRI's considerable diagnostic strength. Once

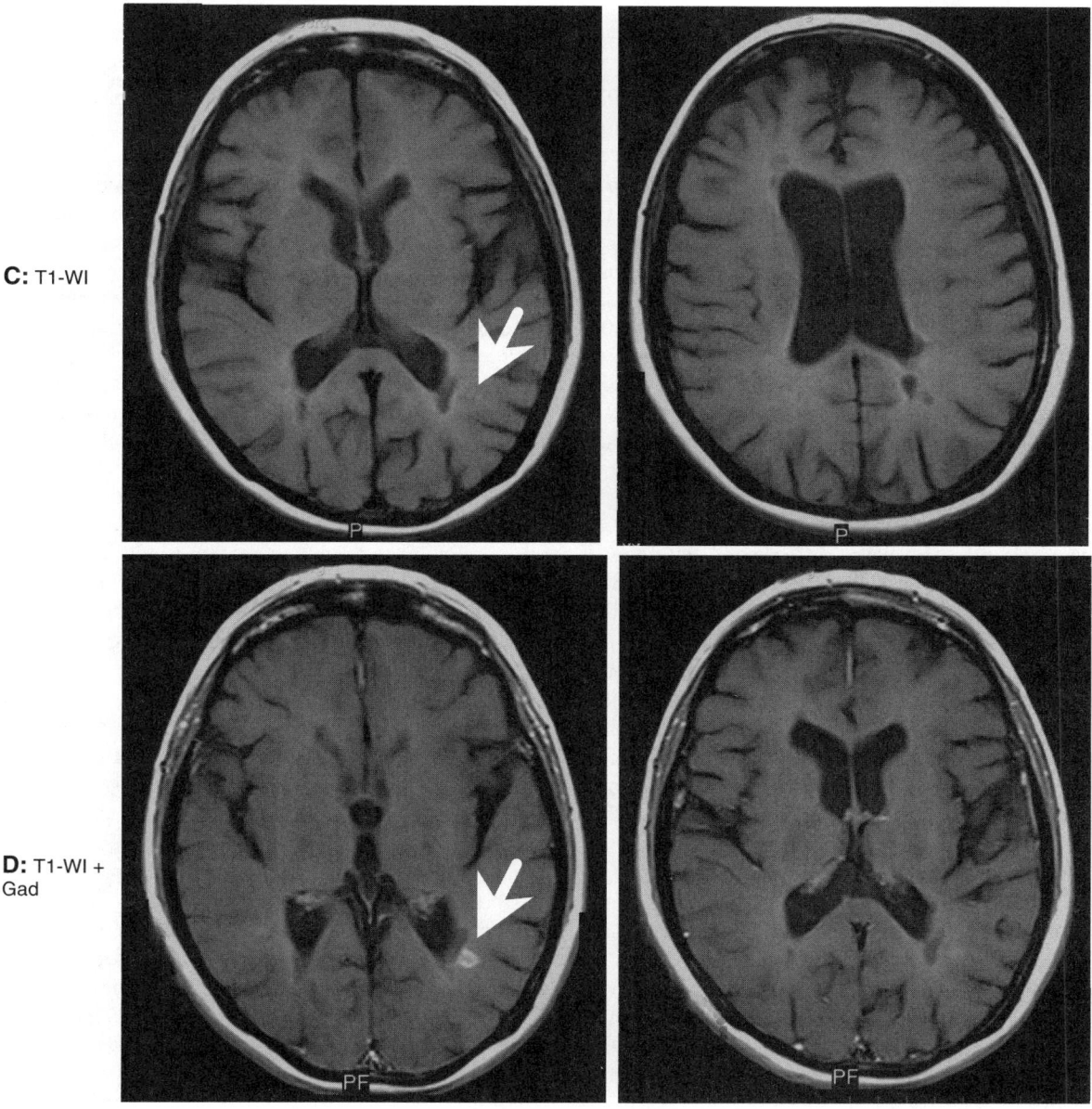

C: T1-WI

D: T1-WI + Gad

FIGURE 1.3, cont'd   **C:** One hyperintensity at the left dorsal horn of the ventricle can be identified as hypointensity on T1-weighted intensity and **D:** enhancing active lesion after gadolinium administration (*arrows*).

intravenous paramagnetic contrast agents were developed, enhancing lesions were clearly defined against normal tissue (Wilms *et al.*, 2001). In 1993, the WHO revised the histological typing of CNS tumors to include new entities described in the era of MRI such as neuroectodermal and dysembryoblastic neuroepidermal tumors, ganglioglioma, and neurocytoma (Kleihues *et al.*, 1993).

As brain metastases are the first sign of disease in 10% of all non-CNS primary neoplasms (up to 25% of lung cancers) and are multiple in up to 50% of patients (Posner and Chernik, 1978), the application of MRI to evaluating brain tumors substantially changed clinical procedures. Approximately half the patients with presumed solitary metastases are candidates for surgical excision of brain metastases. With adjuvant whole-brain irradiation, median survival times of 14 months can be achieved (Patchell *et al.*, 1986, 1990; Smalley *et al.*, 1992). As the optimum therapy depends on the number and location of brain metastases, the diagnostic accuracy of a neuroradiological technique is of utmost importance. Contrast-enhanced MRI is far superior to contrast-enhanced CT in detecting multiple lesions. In 32% of patients with a contrast-enhanced CT-defined solitary metastasis, there will be multiple lesions on contrast-enhanced MRI (CI = 16.2% to 49.0%, P = .01) (Schellinger *et al.*, 1999b). As in MS, magnetization transfer imaging substantially improves the diagnostic yield for primary and secondary brain tumors and is at least as sensitive as triple dose Gd-DTPA (Knauth *et al.*, 1996), which, because of potential side effects and cost, should no longer be applied.

Newer imaging techniques such as DWI and PWI have been applied to aid in the differential diagnosis of brain tumors from non-neoplastic lesions such as tumefactive demyelinating lesions (TDL), abscesses, benign cystic lesions, and radiation necrosis, which require different treatment or for which surgical intervention is not indicated. As DWI and PWI techniques reflect changes such as ischemia, necrosis, and angiogenesis, the information from these studies is interesting for the application to the diagnosis of intracranial mass lesions, optimization of biopsy selection, and as a means of noninvasive surveillance during treatment. The distinction of an abscess from a cystic or necrotic tumor, one that is difficult using only conventional imaging and clinical symptoms, is more reliably made using DWI and ADC values (Guzman *et al.*, 2002). Both types of lesions can appear as solitary or multiple lesions with low T1 and high T2 signal intensities with ring enhancement after contrast administration. However, as DWI reflects the restriction of free water movement, this method is well suited to detect differences in lesion viscosity. Purulent abscesses contain a highly viscous combination of inflammatory cells, debris, and bacteria,

whereas in cystic and necrotic areas of tumors, there is reduced restriction as a result of the disruption of the intrinsic physical and chemical barriers to water diffusion (Wilms *et al.*, 2001).

As the recruitment of vascular networks is crucial for tumor extension and proliferation, vascular morphology is a critical parameter in determining malignant potential. Dynamic blood volume imaging of patients with brain tumors can provide unique information about tumor grade and vascularity that is not appreciated from CT, T2-weighted images, or contrast-enhanced T1-weighted images. TDL can mimic high-grade neoplasms, as both enhance with contrast on T1-weighted imaging and have perilesion edema and variable mass effect and central necrosis (Zagzag *et al.*, 1993). There have been numerous reports of unnecessary surgery and radiation therapy in these patients as a result of the difficulty of diagnosing this process from primary glioma or lymphoma. PWI may play a helpful role in the characterization of these processes, as the vasculature of TDL remains normal in contrast to the neovascular proliferation of tumors (Cha *et al.*, 2001). MRI measurements of CBV have been shown to correlate with conventional angiographic assessments of tumor vascularity and histological specimens of tumor neovascularization (Knopp *et al.*, 1999). Although elevated CBV is highly suggestive of neoplasm, increased vascularity alone does not confirm malignancy. Benign extra-axial neoplasms such as meningioma and choroid plexus papilloma are highly vascular. However, these tumors often can be identified by their anatomical location and features on conventional MRI.

MRI has changed not only clinical management because of a higher diagnostic yield but also daily neurological and neurosurgical practice. Not all tumors are candidates for surgical resection; however, confirmation of the pathology is critical for treatment decisions or enrollment in a clinical trial. Imaging and the development of intraoperative MRI suites (Tronnier *et al.*, 1997) have played an increasing role in assisting with biopsy guidance as part of frameless navigation systems. Intraoperative shifts and resulting inaccuracies have been a concern in frame-based and frameless stereotactically guided interventions, particularly in open microsurgical procedures. MRI data sets, acquired during surgery and used to update neuronavigation, have shown promise to compensate for brain shift and may increase the number of cases where radiologically complete resection can be achieved (Wirtz *et al.*, 1997). In high-grade gliomas, two consecutive series showed that intraoperative MRI can be performed without complications, with a good image quality and significant improvement in surgical resection (Knauth *et al.*, 1999; Wirtz *et al.*, 2000b). For high-grade gliomas, intraoperative MRI reduced the percentage of cases in which residual tumor was identified from 62% to 33%, paralleled by a significant increase in survival times for patients without residual tumor (Wirtz *et al.*, 2000a). Operating times were identical in a matched group analysis of patients with glioblastoma operated with and without intraoperative MRI (Wirtz *et al.*, 2000b). Radiological radicality was achieved in 31% of navigation cases vs 19% in conventional operations and again radical tumor resection was associated with a highly significant prolongation in survival (median 18.3 vs 10.3 months, $P < 0.0001$). Survival was also longer in patients operated on using neuronavigation (median 13.4 vs 11.1 months) (Wirtz *et al.*, 2000a). (See Figure 1.4.)

Functional MRI is another technique that aids in neurosurgical therapy, as it can correlate the MR images with the observed neurological deficit (Wilms *et al.*, 2001). In the same way it can predict functional deficit if an aggressive surgical therapy involving sensitive brain areas is planned (Stippich *et al.*, 2002, 2003). In the presence of potential misinterpretation of anatomical landmarks because of tissue shift from mass effect and edema, intraoperative functional MRI (fMRI) allows the exact localization of tumor with respect to eloquent functional brain areas (Wilms *et al.*, 2001). Therefore, fMRI can contribute to more efficient surgical removal of both benign and malignant brain tumors, with an increase in patient survival and a decrease in surgical morbidity (Wilms *et al.*, 2001).

In summary, MRI has added to diagnostic sensitivity and classification of primary, metastatic, and recurrent brain tumors. As a result of this improved sensitivity and with the use of modern MRI sequences, therapeutic decision making and procedures for the treatment of patients with CNS tumors have been influenced. As in the fields of stroke and MS, larger trials of chemotherapy may in the future also use surrogate MRI parameters for inclusion criteria and as secondary endpoints.

MRI IN THE DESIGN OF THERAPEUTIC TRIALS IN NEUROLOGY

The disappointingly slow progress in developing effective therapies in neurological disease has led to a reevaluation of the strategies for clinical trials in stroke (thrombolytics, neuroprotectants), MS (immunosuppressors, immunomodulators), and brain tumors (chemotherapy, radiation therapy, neurosurgery). As described previously, MS trials have used MRI as a tool for inclusion into studies as well as a surrogate parameter for secondary endpoint analyses. This section illustrates the relevance of MRI for therapeutic trials using stroke trials as the example. To demonstrate efficacy in a clinical trial, the features of trial design need to be optimized. Proof of pharmacological activity in Phase II is needed before lengthy, expensive, labor-intensive, and potentially risky Phase III clinical trials are undertaken. The requirement of proof of concept Phase II studies will prevent the wastefulness of Phase III trials that are doomed to futility before they begin. Image-guided Phase II studies may answer the question of tar-

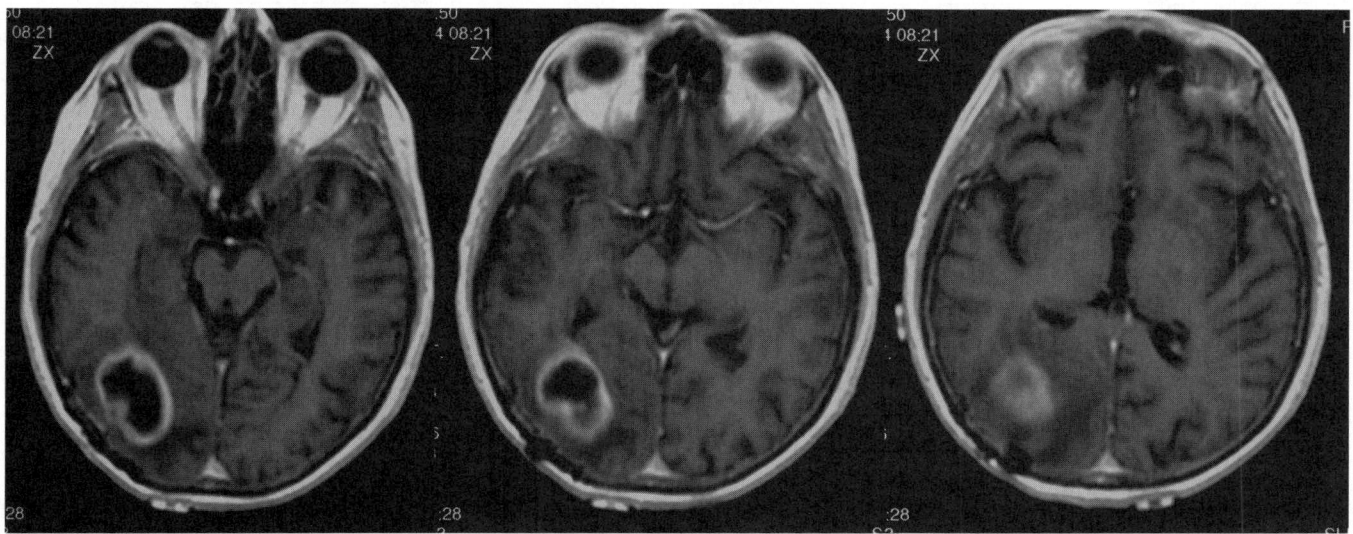

A

FIGURE 1.4 A 54-year-old patient with biopsy confirmed diagnosis of glioblastoma in the right temporooccipital region. **A:** Baseline T1-weighted image postcontrast images at 1.5 T show an enhancing lesion with central necrosis and perifocal edema.

Continued

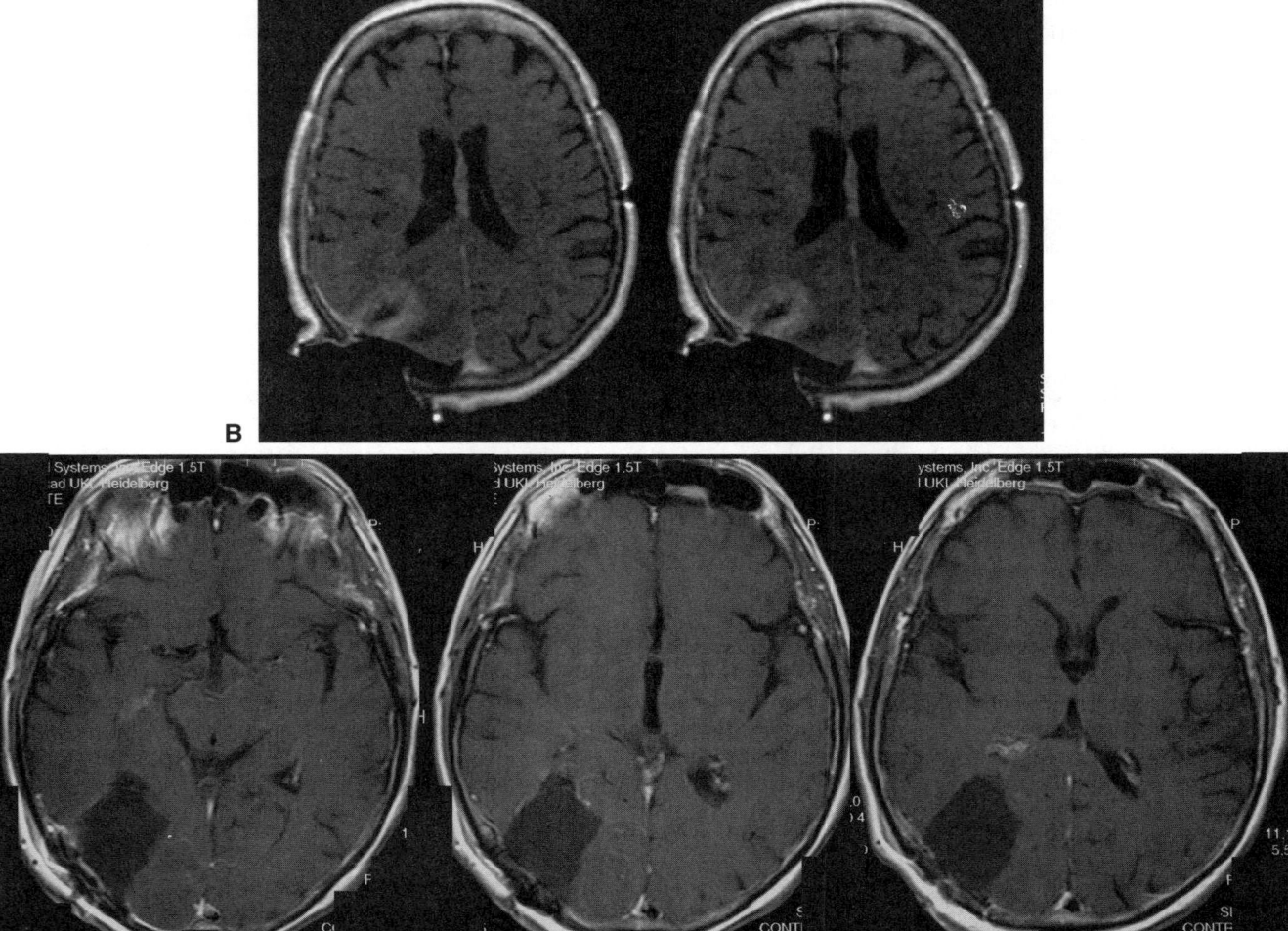

FIGURE 1.4, cont'd **B:** Intraoperative MRI after resection (T1-weighted image postcontrast, 0.2 T) shows residual enhancing tumor below the craniotomy defect. **C:** The final postoperative T1-weighted image postcontrast images show no residual enhancing tumor. The mild periventricular and intraventricular hyperintensity not seen on the baseline scan is probably due to contrast agent in the dorsal horn and choroid plexus of the right ventricle.

get biological activity in fewer than 200 patients, which is the sample size typical for Phase II trials. This is important because trends toward benefit using clinical scales at Phase II have in the past been notoriously poor predictors of clinical outcomes in Phase III trials on much larger samples, especially in stroke but also in MS trials.

It would be unthinkable for clinical trials in cardiology or oncology to enroll patients by bedside clinical impression alone, without objective evidence from diagnostic testing confirming the pathology. Yet this has been the traditional standard by which clinical trials for ischemic stroke have been conducted because, until recently, there was no practical alternative. Stroke MRI provides that alternative. In using MRI as a selection criterion, the goal would be a sample selection based on a positive imaging diagnosis of a pathology rationally linked to the drug's mechanisms of action. Requiring a positive diagnosis of acute ischemic injury by DWI and PWI would ideally ensure that no patients with diagnoses mimicking stroke are included in the sample, a desirable objective unachievable in trials using bedside impression and nonhemorrhagic CT as the basis of inclusion. The goal of image-based patient selection is to narrow the range of patient characteristics, leading to a more homogeneous sample, reducing within group variance, and increasing the statistical power (lowering sample size requirement) of the experimental design to demonstrate efficacy. For reperfusion therapies, the optimal target of therapy would be patients with evidence of an arterial occlusion or hypoperfusion, with the greatest territory at risk for infarction (i.e., the diffusion-perfusion mismatch) (Marks *et al.*, 1999; Schellinger *et al.*, 2000a; Warach *et al.*, 1996).

For neuroprotective drugs acting only in cortical areas of ischemic penumbra, optimal selection of patients could be based on demonstration of acute lesions involving the cerebral cortex or deep gray matter. For some neuroprotective drugs, the diffusion-perfusion mismatch may be the optimal target. For other agents that would protect against reperfusion injury or would not achieve sufficient concentration in oligemic tissue, the optimal target would be a patient in whom reperfusion has occurred. Patients may be excluded from the trial at screening if subacute or chronic lesions are found, which could confound measurements of lesion volumes or clinical severity as outcome variables.

Furthermore, clear definition of lesion volume at enrollment would provide an opportunity to evaluate lesion growth and may better differentiate an effective treatment from placebo. Before an experimental stroke therapy is brought from the laboratory to clinical trial, it is necessary to demonstrate that the treatment causes reduction in lesion volume in experimental models. The fundamental premise of drug discovery and development in acute stroke is that treatments that reduce lesion size are those most likely to lead to clinical benefit. In clinical trial programs that depend solely on clinical endpoints as indices of benefit, drugs may be brought to Phase III testing without evidence that the drug will have the therapeutic effect observed in the experimental model. In practice, this traditional approach to stroke trials has been unsuccessful and often misleading. Four major factors are hypothesized to predict tissue response and clinical efficacy in stroke trials: time to treatment, the salvageable tissue-at-risk, the relevance of the patient sample to the treatment, and the intrinsic effectiveness of the therapeutic strategy (Warach, 2001b). Time is an important factor (Marler *et al.*, 2000), but it is not the only important factor hypothesized to affect response to therapy. The clinical trial optimized on *all* features would be expected to have the most robust response.

Stroke is a special case among brain diseases because: (1) it is a single event that is not progressive beyond the initial hours and days, (2) there is a high rate of spontaneous clinical recovery (implying that clinical improvement is less reflective of drug effect), (3) it requires rapid diagnosis under emergency conditions (the diagnostic certainty is less), and (4) a single discrete lesion fully captures the pathology (the clinical manifestations result from the size and location of the ischemic damage). For stroke the true clinical endpoint, disability, is difficult to measure and is only approximated by clinical scales. Furthermore, experts do not agree on

how to measure outcome using clinical scales, and the criterion of "complete recovery" used in many trials may include patients with significant disability. For these reasons, lesion volume as a biomarker is likely to be more helpful for stroke than in cancer and cardiac disorders.

Ordinarily a drug must have a beneficial effect on a clinical endpoint or on a validated surrogate endpoint to demonstrate effectiveness and lead to registration. Current FDA regulations stipulate that for unmet medical needs for serious and life-threatening conditions such as stroke, a drug may be approved based on a nonvalidated surrogate endpoint if it is reasonably likely to have clinical benefit. The internationally accepted regulatory standards of the International Conference on Harmonization also state that surrogate endpoints may be used as primary endpoints when the surrogate is reasonably likely to predict clinical outcome. Much of the thought on the use of biomarkers as measures of drug activity and potential surrogates have come from fields such as oncology or cardiology, where death or a comparably objective and reliable assessment is the relevant clinical endpoint. For these disease categories, the biomarker does not fully capture the pathology of the disease as well as the clinical endpoint. However, the use of biomarkers and potential surrogates are different for brain disorders in which disability (defined by imperfect clinical rating scales) rather than death is the relevant clinical variable, and in which the biomarker (macroscopic brain lesion) more fully captures the pathology than the clinical scales.

If the results on clinical and imaging endpoints were discordant, what are the possible explanations? If lesion volume decreases, without clinical improvement, the most likely explanations are the trial design or the choice of clinical endpoints that is insensitive to the drug effect, or that a toxicity affecting the clinical outcome offsets any neuroprotective effect of the therapy. If clinical improvement occurs without lesion volume changes, the imaging methods may be insensitive to the drug effect, or the clinical benefit is not mediated by a direct effect on the evolving infarct. The latter possibility may apply to classes of drugs that, for example, treat post-stroke mood disorders or would lead to enhanced recovery through functional reorganization. These comparisons are only meaningful if studies are optimally designed and equally powered to show effect on their respective outcome measures, i.e., the optimal sample size for imaging studies may be too small to show clinical effects. An imaging benefit may never stand alone as a surrogate, without evidence of clinical benefit. While full validation must eventually be proven, MR surrogate measures may help in achieving that goal. One could imagine a small but statistically significant volume reduction that would have a trivial or undetectable clinical effect. Nonetheless, a benefit on the surrogate may be acceptable as an independent source of confirmatory data in support of benefit seen in clinical endpoint trials. However, an application for registration of a stroke drug using MRI outcomes has not yet been submitted to regulatory agencies.

The pharmaceutical industry has taken the initiative in investigating this final step in validation. The results of several industry-sponsored drug trials using MRI as a surrogate will be known over the next several years, and those studies should provide the most decisive information regarding the utility of MRI as a measure in stroke trials. MRI-based recruitment trials with a window of 6 hours have proven feasible, as have selection based on lesion size, location, and the PWI/DWI-mismatch. The field of stroke clinical trials continues to examine opportunities for improving trial design, using imaging to confirm, rather than exclude, pathology, and in treatment assessment; these are obvious, useful roles. That MRI is increasingly used as a selection tool and an outcome measure in stroke trials reflects the growing recognition that direct pathophysiological imaging may provide a more rational approach to stroke therapeutics. Patient selection and outcomes based exclusively on clinical assessment and nonhemorrhagic CT scans may no longer be appropriate for all stroke trials.

CONCLUSION

Most neurologists and neurosurgeons are convinced that MRI is at least as effective as CT in most diseases of the CNS, even if there are only few directly comparative studies. A new and better imaging technique should result in both improvement in quality of care and reduction of health care costs (Demaerel, 2001). Although sufficiently large and methodologically sound studies with regard to cost effectiveness of MRI are still lacking, the overall impression is that MRI has revolutionized not only the diagnosis but also the management of neurologically ill patients. Contemporary management of MS patients and patients with neoplastic disease of the CNS now firmly incorporates the use of MRI. MRI has also become a major player in the field of stroke, which is of extraordinary relevance given its high disease burden. MRI does not yet have such an impact on management for hereditary, degenerative, and metabolic neurological diseases with no or poor therapeutic options. However, given the rapid evolution of the diagnostic strength of MRI and the advantages of a lack of harmful radiation, the use of a well-tolerated contrast agent, the overall availability of the modality as a result of an ever-increasing number of scanners, and the continuous reduction of imaging times as a result of technological progress, this too is likely to change.

References

Albers GW et al. (2001). Antithrombotic and thrombolytic therapy for ischemic stroke, Chest 119:300S-320S.

Anderson DW et al. (1992). Revised estimate of the prevalence of multiple sclerosis in the United States, Ann Neurol 31:333-336.

Antiplatelet Trialists' Collaboration. (1994). Warfarin versus aspirin for prevention of thromboembolism in atrial fibrillation: Collaborative overview of randomised trials of antiplatelet therapy. I. Prevention of death, myocardial infarction, and stroke by prolonged antiplatelet therapy in various categories of patients. Stroke Prevention in Atrial Fibrillation II Study, Nouv Rev Fr Hematol 36:213-228.

Antithrombotic Trialist's Collaboration. (2002). Collaborative meta-analysis of randomised trials of antiplatelet therapy for prevention of death, myocardial infarction, and stroke in high risk patients, Br Med J 324:71-86.

Baird AE et al. (2000). Clinical correlations of diffusion and perfusion lesion volumes in acute ischemic stroke, Cerebrovasc Dis 10:441-448.

Barber PA et al. (1998). Prediction of stroke outcome with echoplanar perfusion- and diffusion-weighted MRI, Neurology 51:418-426.

Barkhof F et al. (1997). Comparison of MRI criteria at first presentation to predict conversion to clinically definite multiple sclerosis, Brain 120(Pt 11):2059-2069.

Basser PJ, Pierpaoli C. (1996). Microstructural and physiological features of tissues elucidated by quantitative-diffusion-tensor MRI, J Magn Reson B 111:209-219.

Bastianello S et al. (1997). Fast spin-echo and fast fluid-attenuated inversion-recovery versus conventional spin-echo sequences for MR quantification of multiple sclerosis lesions, AJNR Am J Neuroradiol 18:699-704.

Bergers E, Barkhof F. (2001). Multiple sclerosis. In Demaerel P, editor: Recent advances in diagnostic neuroradiology, Berlin, Heidelberg: Springer-Verlag.

Brant-Zawadzki M, Heiserman JE. (1997). The roles of MR angiography, CT angiography, and sonography in vascular imaging of the head and neck, AJNR Am J Neuroradiol 18:1820-1825.

Brott TG. (2002). A combined meta-analysis of NINDS, ECASS I and II, ATLANTIS. International Stroke Conference, San Antonio, TX, Feb 7-9.

Bruck W et al. (1997). Inflammatory central nervous system demyelination: correlation of magnetic resonance imaging findings with lesion pathology, Ann Neurol 42:783-793.

Buckley BT, Wainwright A, Meagher T, Briley D. (2003). Audit of a policy of magnetic resonance imaging with diffusion-weighted imaging as first-line neuroimaging for in-patients with clinically suspected acute stroke, Clin Radiol 58:234-237.

Bydder GM, Young IR. (1985). MR imaging: clinical use of the inversion recovery sequence, J Comput Assist Tomogr 9, 659-675.

Cha S et al. (2001). Dynamic contrast-enhanced T2*-weighted MR imaging of tumefactive demyelinating lesions, AJNR Am J Neuroradiol 22:1109-1116.

Chalela JA et al. (2004). Early magnetic resonance imaging findings in patients receiving tissue plasminogen activator predict outcome: insights into the pathophysiology of acute stroke in the thrombolysis era, Ann Neurol 55:105-112.

Clark WM et al. (1999). Recombinant tissue-type plasminogen activator (Alteplase) for ischemic stroke 3 to 5 hours after symptom onset. The ATLANTIS Study: a randomized controlled trial. Alteplase Thrombolysis for Acute Noninterventional Therapy in Ischemic Stroke, JAMA 282:2019-2026.

Coutts SB et al. (2003). Reliability of assessing percentage of diffusion-perfusion mismatch, Stroke 34:1681-1683.

Davis S, Fisher M, Warach S. (2003). Magnetic resonance imaging in stroke, Cambridge: University Press.

De Coene B *et al.* (1992). MR of the brain using fluid-attenuated inversion recovery (FLAIR) pulse sequences, *AJNR Am J Neuroradiol* 13:1555-1564.

Demaerel P. (2001). Guidelines for brain MR imaging. In Demaerel P, editor: *Recent advances in diagnostic neuroradiology*, Berlin, Heidelberg: Springer-Verlag.

EC/IC Bypass Study Group. (1985). Failure of extracranial-intracranial artery bypass to reduce the risk of ischemic stroke. Results of an international randomized trial, *N Engl J Med* 313:1191-1200.

European Carotid Surgery Trialist's (ECST) Collaborative Group, (1991). MRC European Surgery Trial. Interim results for symptomatic patients with severe (70-99%) or with mild (0-29%) carotid stenosis, *Lancet* 337:235-1243.

Fazekas F *et al.* (1988). Criteria for an increased specificity of MRI interpretation in elderly subjects with suspected multiple sclerosis, *Neurology* 38:1822-1825.

Fiebach JB, Schellinger PD. (2003). *Stroke MRI*, Darmstadt: Steinkopff Verlag.

Fiebach JB *et al.* (2004). Stroke magnetic resonance imaging is accurate in hyperacute intracerebral hemorrhage. A multicenter study on the validity of stroke imaging, *Stroke* 35(2):502-507.

Fiebach JB *et al.* (2002). CT and diffusion-weighted MR-Imaging in randomized order: DWI results in higher accuracy and lower interrater variability in the diagnosis of hyperacute ischemic stroke, *Stroke* 33:2206-2210.

Fiehler J *et al.* (2002). Severe ADC decreases do not predict irreversible tissue damage in humans, *Stroke* 33:79-86.

Fiehler F *et al.* (2004). Predictions of apparent diffusion, coefficient neuralization in stroke patients, *Stroke* 35(2):516-519.

Filippi M. (2003). Magnetization transfer MRI in multiple sclerosis and other central nervous system disorders, *Eur J Neurol* 10:3-10.

Fink JN *et al.* (2002). The stroke patient who woke up: clinical and radiological features, including diffusion and perfusion MRI, *Stroke* 33:988-993.

Frohman EM *et al.* (2003). The utility of MRI in suspected MS: report of the Therapeutics and Technology Assessment Subcommittee of the American Academy of Neurology, *Neurology* 61:602-611.

Fujita N *et al.* (1994). MR imaging of middle cerebral artery stenosis and occlusion: value of MR angiography, *AJNR Am J Neuroradiol* 15:335-341.

Guzman R *et al.* (2002). Use of diffusion-weighted magnetic resonance imaging in differentiating purulent brain processes from cystic brain tumors, *J Neurosurg* 97:1101-1107.

Hacke W *et al.* (1999). Thrombolysis in acute ischemic stroke: controlled trials and clinical experience, *Neurology* 53:S3-14.

Hacke W *et al.* (1995). Intravenous thrombolysis with recombinant tissue plasminogen activator for acute hemispheric stroke. The European Cooperative Acute Stroke Study, *JAMA* 274:1017-1025.

Hacke W *et al.* (1998). Randomised double-blind placebo-controlled trial of thrombolytic therapy with intravenous alteplase in acute ischaemic stroke (ECASS II), *Lancet* 352:1245-1251.

Hacke W, Warach S. (2000). Diffusion-weighted MRI as an evolving standard of care in acute stroke, *Neurology* 54:1548-1549.

Haehnel S *et al.* (2000). Magnetisation transfer ratio is low in normal-appearing cerebral white matter in patients with normal pressure hydrocephalus, *Neuroradiology* 42:174-179.

Hajnal JV *et al.* (1992). Use of fluid attenuated inversion recovery (FLAIR) pulse sequences in MRI of the brain, *J Comput Assist Tomogr* 16:841-844.

Haley EC Jr *et al.* (1993). Pilot randomized trial of tissue plasminogen activator in acute ischemic stroke. The TPA Bridging Study Group, *Stroke* 24:1000-1004.

Hashemi RH *et al.* (1995). Suspected multiple sclerosis: MR imaging with a thin-section fast FLAIR pulse sequence, *Radiology* 196:505-510.

Heiland S *et al.* (1998). Comparison of different EPI-sequence types in perfusion-weighted MR imaging. Which one is the best?, *Neuroradiology* 40:216-222.

Heiland S, Reith W, Forsting M, Sartor K. (1997). Perfusion-weighted magnetic resonance imaging using a new gadolinium complex as contrast agent in a rat model of focal cerebral ischemia, *J Magn Reson Imaging* 7:1109-1115.

Jackson A. (2001). Technical progress in neuroradiology and its application. In Demaerel P, editor: *Recent advances in diagnostic neuroradiology*, Berlin, Heidelberg: Springer-Verlag.

Jacobs L, Kinkel PR, Kinkel WR. (1986). Silent brain lesions in patients with isolated idiopathic optic neuritis. A clinical and nuclear magnetic resonance imaging study, *Arch Neurol* 43:452-455.

Johnson BA, Heisermann JE, Drayer BP, Keller PJ. (1994). Intracranial MR angiography: its role in the integrated approach to brain infarction, *AJNR Am J Neuroradiol* 15:901-908.

Kaste M. (2003). Approval of alteplase in Europe: will it change stroke management?, *Lancet Neurol* 2:207-208.

Kidwell CS, Alger JR, Saver JL. (2003). Beyond mismatch: evolving paradigms in imaging the ischemic penumbra with multimodal magnetic resonance imaging, *Stroke* 34:2729-2735.

Kidwell CS *et al.* (2000). Thrombolytic reversal of acute human cerebral ischemic injury shown by diffusion/perfusion magnetic resonance imaging, *Ann Neurol* 47:462-469.

Kidwell CS *et al.* (2002). Magnetic resonance imaging detection of microbleeds before thrombolysis: an emerging application, *Stroke* 33:95-98.

Kleihues P, Burger PC, Scheithauer BW. (1993). The new WHO classification of brain tumours, *Brain Pathol* 3:255-268.

Knauth M *et al.* (1996). MR enhancement of brain lesions: increased contrast dose compared with magnetization transfer, *AJNR Am J Neuroradiol* 17:1853-1859.

Knauth M *et al.* (1997). Potential of CT angiography in acute ischemic stroke, *AJNR Am J Neuroradiol* 18:1001-1010.

Knauth M *et al.* (1999). Intraoperative MR imaging increases the extent of tumor resection in patients with high-grade gliomas, *AJNR Am J Neuroradiol* 20:1642-1646.

Knopp EA *et al.* (1999). Glial neoplasms: dynamic contrast-enhanced T2*-weighted MR imaging, *Radiology* 211:791-798.

Kurki T, Niemi P, Valtonen S. (1994). MR of intracranial tumors: combined use of gadolinium and magnetization transfer, *Am J Neuroradiol* 15:1727-1736.

Lauterbur P. (1973). Image formation by induced local interactions: examples employing nuclear magnetic resonance, *Nature* 242:190-191.

Le Bihan D *et al.* (1986). MR imaging of intravoxel incoherent motions: application to diffusion and perfusion in neurologic disorders, *Radiology* 161:401-407.

Le Bihan D *et al.* (2001). Diffusion tensor imaging: concepts and applications, *J Magn Reson Imaging* 13:534-546.

Le Bihan D, Turner R, Douek P, Patronas N. (1992). Diffusion MR imaging: clinical applications, *AJNR Am J Roentgenol* 159:591-599.

Linfante I, Llinas RH, Caplan LR, Warach S. (1999). MRI features of intracerebral hemorrhage within 2 hours from symptom onset, *Stroke* 30:2263-2267.

Lovblad KO *et al.* (1977). Ischemic lesion volumes in acute stroke by diffusion-weighted magnetic resonance imaging correlate with clinical outcome, *Ann Neurol* 42:164-170.

Lublin FD, Reingold SC. (1996). Defining the clinical course of multiple sclerosis: results of an international survey. National Multiple Sclerosis Society (USA) Advisory Committee on Clinical Trials of New Agents in Multiple Sclerosis, *Neurology* 46:907-911.

Mansfield P, Maudsley AA. (1976). Line scan proton spin imaging in biological structures by NMR, *Phys Med Biol* 21:847-852.

Mansfield P, Maudsley AA. (1977). Medical imaging by NMR, *Br J Radiol* 50:188-194.

Marks MP *et al.* (1999). Evaluation of early reperfusion and IV rt-PA therapy using diffusion- and perfusion-weighted MRI, *Neurology* 52:1792-1798.

Marler JR *et al.* (2000). Early stroke treatment associated with better outcome: the NINDS rt-PA stroke study, *Neurology* 55:1649-1655.

(2003). Major ongoing stroke trials, *Stroke* 34:61-72.

McDonald WI *et al.* (2001). Recommended diagnostic criteria for multiple sclerosis: guidelines from the International Panel on the Diagnosis of Multiple Sclerosis, *Ann Neurol* 50:121-127.

Meyer JS, Gilroy J, Barnhart MI, Johnson JF. (1963). Therapeutic thrombolysis in cerebral thromboembolism, *Neurology* 13:927-937.

Miller DH, Grossman RI, Reingold SC, McFarland HF. (1998). The role of magnetic resonance techniques in understanding and managing multiple sclerosis, *Brain* 121(Pt 1):3-24.

Mohr JP *et al.* (1995). Magnetic resonance versus computed tomographic imaging in acute stroke, *Stroke* 26:807-812.

Mori E *et al.* (1992). Intravenous recombinant tissue plasminogen activator in acute carotid artery territory stroke, *Neurology* 42:976-982.

Moseley ME, Wendland MF, Kucharczyk J. (1991). Magnetic resonance imaging of diffusion and perfusion, *Top Magn Reson Imaging* 3:50-67.

Mushlin AI *et al.* (1993). The accuracy of magnetic resonance imaging in patients with suspected multiple sclerosis. The Rochester-Toronto Magnetic Resonance Imaging Study Group, *JAMA* 269:3146-3151.

Neil JJ *et al.* (1998). Normal brain in human newborns: apparent diffusion coefficient and diffusion anisotropy measured by using diffusion tensor MR imaging, *Radiology* 209:57-66.

Nighoghossian N *et al.* (2002). Old microbleeds are a potential risk factor for cerebral bleeding after ischemic stroke: a gradient-echo t2*-weighted brain MRI study, *Stroke* 33:735-742.

North American Symptomatic Carotid Endarterectomy Trial (NASCET) Collaborators. (1991). Beneficial effect of carotid endarterectomy in symptomatic patients with high-grade carotid stenosis, *N Engl J Med* 325:445-453.

Noseworthy JH. (2003). Management of multiple sclerosis: current trials and future options, *Curr Opin Neurol* 16:289-297.

Ono J *et al.* (1995). Differentiation between dysmyelination and demyelination using magnetic resonance diffusional anisotropy, *Brain Res* 671:141-148.

Ostergaard L *et al.* (1996). High resolution measurement of cerebral blood flow using intravascular tracer bolus passages, Part I. Mathematical approach and statistical analysis, *Magn Reson Med* 36:715-725.

Parkin DM, Bray F, Ferlay J, Pisani P. (2001). Estimating the world cancer burden: Globocan 2000, *Int J Cancer* 94:153-156.

Parsons MW *et al.* (2002). Diffusion- and perfusion-weighted MRI response to thrombolysis in stroke, *Ann Neurol* 51:28-37.

Patchell RA *et al.* (1986). Single brain metastases: surgery plus radiation or radiation alone, *Neurology* 36:447-453.

Patchell RA *et al.* (1990). A randomized trial of surgery in the treatment of single metastases to the brain, *N Engl J Med* 322:494-500.

Paty DW *et al.* (1988). MRI in the diagnosis of MS: a prospective study with comparison of clinical evaluation, evoked potentials, oligoclonal banding, and CT, *Neurology* 38:180-185.

Pierpaoli C *et al.* (1996a). High temporal resolution diffusion MRI of global cerebral ischemia and reperfusion, *J Cereb Blood Flow Metab* 16:892-905.

Pierpaoli C *et al.* (1996b). Diffusion tensor MR imaging of the human brain, *Radiology* 201:637-648.

Poser CM *et al.* (1983). New diagnostic criteria for multiple sclerosis: guidelines for research protocols, *Ann Neurol* 13:227-231.

Posner JB, Chernik NL. (1978). Intracranial metastases from systemic cancer, *Adv Neurol* 19:579-592.

Powers WJ. (2000). Testing a test: a report card for DWI in acute stroke, *Neurology* 54:1549-1551.

Powers WJ, Zivin J. (1998). Magnetic resonance imaging in acute stroke: not ready for prime time, *Neurology* 50:842-843.

Reidel MA *et al.* (2003). Differentiation of multiple sclerosis plaques, subacute cerebral ischaemic infarcts, focal vasogenic oedema and lesions of subcortical arteriosclerotic encephalopathy using magnetisation transfer measurements, *Neuroradiology* 45:289-294.

Rosen BR, Belliveau JW, Chien D. (1989). Perfusion imaging by nuclear magnetic resonance, *Magn Reson Q* 5:263-281.

Rosen BR, Belliveau JW, Vevea JM, Brady TJ. (1990). Perfusion imaging with NMR contrast agents, *Magn Reson Med* 14:249-265.

Röther J *et al.* (2002). Effect of intravenous thrombolysis on MRI parameters and functional outcome in acute stroke <6h, *Stroke* 33:2438-2445.

Schellinger PD, Fiebach JB, Hacke W. (2003). Imaging-based decision making in thrombolytic therapy for ischemic stroke: present status, *Stroke* 34:575-583.

Schellinger PD *et al.* (2001a). Stroke magnetic resonance imaging within 6 hours after onset of hyperacute cerebral ischemia, *Ann Neurol* 49:460-469.

Schellinger PD *et al.* (2001b). Thrombolytic therapy for ischemic stroke: a review. Part I. Intravenous thrombolysis, *Crit Care Med* 29:1812-1818.

Schellinger PD *et al.* (1999). A standardized MRI stroke protocol: comparison with ct in hyperacute intracerebral hemorrhage, *Stroke* 30:765-768.

Schellinger PD *et al.* (2000a). Monitoring intravenous recombinant tissue plasminogen activator thrombolysis for acute ischemic stroke with diffusion and perfusion MRI, *Stroke* 31:1318-1328.

Schellinger PD *et al.* (2000b). Feasibility and practicality of MR imaging of stroke in the management of hyperacute cerebral ischemia, *AJNR Am J Neuroradiol* 21:1184-1189.

Schellinger PD, Meinck HM, Thron A. (1999b). Diagnostic reach of MRI compared to CCT in patients with brain metastases, *J Neurooncol* 44:275-281.

Schlaug G *et al.* (1999). The ischemic penumbra: operationally defined by diffusion and perfusion MRI, *Neurology* 53:1528-1537.

Schumacher GA *et al.* (1965). Problems of experimental trials of therapy in multiple sclerosis: report by the panel on the evaluation of experimental trials of therapy in multiple sclerosis, *Ann N Y Acad Sci* 122:552-568.

Schwarz S *et al.* (2001). [Acute disseminated encephalomyelitis (ADEM)], *Nervenarzt* 72:241-254.

Smalley SR *et al.* (1992). Resection for solitary brain metastasis. Role of adjuvant radiation and prognostic variables in 229 patients, *J Neurosurg* 77:531-540.

Stejskal EO, Tanner JE. (1965). Spin diffusion measurements: spin echoes in the presence of a time-dependent field gradient, *J Chem Phys* 42:288-292.

Stippich C *et al.* (2003). Robust localization and lateralization of human language function: an optimized clinical functional magnetic resonance imaging protocol, *Neurosci Lett* 346:109-113.

Stippich C, Ochmann H, Sartor K. (2002). Somatotopic mapping of the human primary sensorimotor cortex during motor imagery and motor execution by functional magnetic resonance imaging, *Neurosci Lett* 331:50-54.

Tanner JE, Stejskal EO. (1968). Restricted self-diffusion of protons in colloidal systems by the pulsed-gradient, spin-echo method, *J Chem Phys* 49:1768-1777.

The National Institute of Neurological Disorders and Stroke rt-PA Stroke Study Group. (1995). Tissue plasminogen activator for acute ischemic stroke, *N Engl J Med* 333:1581-1587.

Todd JW *et al.* Reversal of ischemic injury after standard intravenous alteplase therapy predicts favorable clinical recovery in stroke, *Neurology* (submitted for press).

Tronnier VM *et al.* (1997). Intraoperative diagnostic and interventional magnetic resonance imaging in neurosurgery, *Neurosurgery* 40:891-900; discussion 900-892.

Villringer A *et al.* (1988). Dynamic imaging with lanthanide chelates in normal rat brain: contrast due to magnetic susceptibility effects, *Magn Reson Med* 6:164-174.

von Kummer R, Bozzao L, Manelfe C. (1995). Early CT diagnosis of hemispheric brain infarction, 1-95.

von Kummer R *et al.* (1996). Detectability of cerebral hemisphere ischaemic infarcts by CT within 6 h of stroke, *Neuroradiology* 38:31-33.

Warach S. (2001a). Tissue viability thresholds in acute stroke: the 4 factor model, *Stroke* 32:2460-2461.

Warach S. (2001b). Use of diffusion and perfusion magnetic resonance imaging as a tool in acute stroke clinical trials, *Curr Control Trials Cardiovasc Med* 2:38-44.

Warach S, Boska M, Welch KM. (1997). Pitfalls and potential of clinical diffusion-weighted MR imaging in acute stroke, *Stroke* 28:481-482.

Warach S *et al.* (1992). Fast magnetic resonance diffusion-weighted imaging of acute human stroke, *Neurology* 42:1717-1723.

Warach S, Dashe JF, Edelman RR. (1996). Clinical outcome in ischemic stroke predicted by early diffusion-weighted and perfusion magnetic resonance imaging: a preliminary analysis, *J Cereb Blood Flow Metab* 16:53-59.

Warach S *et al.* (1995). Acute human stroke studied by whole brain echo planar diffusion-weighted magnetic resonance imaging, *Ann Neurol* 37:231-241.

Wardlaw JM, del Zoppo G, Yamaguchi T. (2002). Thrombolysis for acute ischaemic stroke (Cochrane Review). In *The Cochrane Library*.

WHO Guidelines Subcommittee. (1999). 1999 World Health Organization-International Society of Hypertension Guidelines for the Management of Hypertension, *J Hypertens* 17:151-183.

Wilms G, Sunaert S, Flamen P. (2001). Recent developments in brain tumour diagnostics. In Demaerel P, *Recent advances in diagnostic neuroradiology*, Berlin, Heidelberg: Springer-Verlag.

Wirtz CR *et al.* (2000a). The benefit of neuronavigation for neurosurgery analyzed by its impact on glioblastoma surgery, *Neurol Res* 22:354-360.

Wirtz CR *et al.* (2000b). Clinical evaluation and follow-up results for intraoperative magnetic resonance imaging in neurosurgery, *Neurosurgery* 46:1112-1120; discussion 1120-1112.

Wirtz CR *et al.* (1997). Image-guided neurosurgery with intraoperative MRI: update of frameless stereotaxy and radicality control, *Stereotact Funct Neurosurg* 68:39-43.

Yamada K *et al.* (2002). Magnetic resonance perfusion-weighted imaging of acute cerebral infarction: effect of the calculation methods and underlying vasculopathy, *Stroke* 33:87-94.

Yamaguchi T, Hayakawa T, Kiuchi H, Japanese Thrombolysis Study Group. (1993). Intravenous tissue plasminogen activator ameliorates the outcome of hyperacute embolic stroke, *Cerebrovasc Dis* 3:269-272.

Zagzag D *et al.* (1993). Demyelinating disease versus tumor in surgical neuropathology. Clues to a correct pathological diagnosis, *Am J Surg Pathol* 17:537-545.

Zivin JA, Holloway RG. (2000). Weighing the evidence on DWI: caveat emptor, *Neurology* 54:1552.

2

Advances in the Acute Treatment and Secondary Prevention of Stroke

Amie W. Hsia, MD
Gregory W. Albers, MD

Stroke continues to be the leading cause of disability in industrialized nations and is one of the most difficult therapeutic challenges for physicians. Developments in neuroimaging and other diagnostic tests have improved our ability to identify and localize ischemic brain lesions and clarify their underlying etiologies. Until recently, however, physicians had no effective means of treating acute stroke. The success of thrombolytic therapy for acute stroke has invigorated clinical research aimed at identifying additional treatment options to expand and improve on intravenous thrombolysis. In the first part of this chapter, we focus on the large clinical trials of thrombolytic therapy for stroke and the clinical trial methodology that led to the success of the National Institutes of Neurologic Disorders and Stroke (NINDS) IV tPA trial.

Since most stroke patients are not eligible for acute treatment, improving our ability to prevent stroke remains of critical importance. The second part of this chapter highlights the developments of specific antihypertensive, lipid-lowering, and antiplatelet agents that have expanded our armamentarium to diminish stroke risk.

Looking to the future, we discuss the development of a new oral anticoagulant and ongoing challenges in the quest for a clinically effective neuroprotectant.

ACUTE TREATMENT: THROMBOLYTICS

The use of thrombolytic therapy for the treatment of acute ischemic stroke is the consequence of recent advances in clinical stroke trial design. The pivotal study from the NINDS recombinant tissue plasminogen activator (rt-PA) Stroke Study Group (1995) documented a significant benefit of IV rt-PA treatment of acute ischemic stroke within 3 hours of symptom onset. Based on this report, the U.S. Food and Drug Administration (FDA) approved rt-PA for the treatment of acute ischemic stroke within 3 hours of symptom onset.

Background

The investigation of thrombolytic agents for stroke treatment developed because the majority of ischemic strokes are caused by thrombotic or thromboembolic occlusions of cerebral vasculature. Without thrombolytic therapy, early spontaneous recanalization occurs in only a minority of patients (Del Zoppo et al., 1998; Kassem-Moussa and Graffagnino, 2002). Animal stroke models have demonstrated clot lysis with thrombolytic agents and improved neurological outcome following thrombolysis (Zivin et al., 1985; Del Zoppo et al., 1986). Initiation of human stroke trials was also encouraged by the success of streptokinase (SK) for the treatment of myocardial infarction first reported in 1988 in the Second International Study of Infarct Survival (ISIS-2) (1988).

Clinical Trials (Table 2.1)

Streptokinase

SK was studied for use in acute ischemic stroke in three large randomized trials that were conducted in Europe and Australia (Donnan et al., 1995, 1996). All three of the trials were terminated early because of safety concerns. The Multicenter Acute Stroke Trial-Italy

(MAST-I) (1995) and the MAST-Europe trial (1996) randomized patients to a fixed dose of SK within 6 hours of stroke onset, whereas the Australian Streptokinase Trial (ASK) (Donnan *et al.*, 1996) randomized patients to a fixed dose of SK within 4 hours of onset. The fixed dose used in all three stroke trials was the same as that used for the treatment of myocardial infarction (MI). No studies designed to clarify the optimal dose of SK for stroke patients were undertaken before embarking on the large treatment trials. An indirect analysis using data from the MAST-E trial showed a trend for increased risk of early death in patients with lower body weight and thus a higher relative dose of SK (Cornu *et al.*, 2000). This may be one explanation for the failed trials of SK treatment for stroke: The dose used may have been too high and was not adjusted for patient weight. SK is also frequently associated with hypotension when used for MI (10% of treated patients in ISIS-2 compared to 2% treated with placebo) and variably reported in stroke (20% of treated patients in the ASK trial, 1.9% in MAST-I, and 0.6% in MAST-E). In stroke patients acute hypotension could theoretically cause impaired blood flow to the ischemic penumbra, leading to poorer outcomes (Cornu *et al.*, 2000). Finally, because SK has more prolonged fibrinogen depletion and anticoagulant effects compared to rt-PA, these differences may have contributed to the higher rates of hemorrhagic transformation (Cornu *et al.*, 2000).

Recombinant Tissue Plasminogen Activator (rt-PA)

Multiple safety and feasibility studies investigating the potential use of rt-PA in acute stroke were conducted before the large-scale stroke trials (Brott *et al.*, 1992; Haley *et al.*, 1992). Based on dose-finding studies, a dose substantially less than that used for treatment of MI (0.9 mg/kg) was selected as the optimal dose for the large clinical trials.

The NINDS t-PA Stroke Study Group performed a randomized, double-blind, placebo-controlled trial of intravenous tPA for ischemic stroke within 3 hours of symptom onset (1995). A total of 624 patients were enrolled to receive either intravenous tPA 0.9 mg/kg (maximum 90 mg) or placebo. Therapy with tPA was initiated with a bolus (10% of the total dose) infused over 1 minute, and the remainder of the total dose infused over 60 minutes. A pretreatment computed tomography (CT) scan was required to exclude the presence of intracerebral hemorrhage (ICH). To minimize the risk of treatment-related ICH, strict inclusion and exclusion criteria were adhered to. After treatment, blood pressure was strictly maintained within prespecified values, and the use of antiplatelets and anticoagulants was not permitted for 24 hours.

The trial was conducted in two parts. Part 1 included 291 patients and was designed to assess the early clinical efficacy of tPA measured by an improvement in the National Institutes of Health Stroke Scale (NIHSS) score by four or more points or complete neurological recovery 24 hours after stroke onset. Part 2 included 333 patients and was designed to evaluate clinical outcomes at 3 months, focusing on the percentages of patients with minimal or no deficits.

In Part 1, no significant difference was detected in the percentages of patients with neurological improvement at 24 hours as previously defined. However, a secondary analysis did find significant improvement in median NIHSS scores among the tPA group.

In Part 2, the results of all outcome measures favored the tPA group. Benefits were consistent regardless of patient age, stroke subtype, and stroke severity. Treated patients were 30% more likely to have minimal or no disability at 3 months compared to placebo-treated patients,

TABLE 2.1　Data From the Four Major Trials of IV tPA for Stroke, Comparing Dose, Therapeutic Window, Mortality, and OR for Benefit of tPA in the Incidence of Death and Dependency

Study	Patients No.	Dose, mg (Maximum)	Window h	Symptomatic ICH		Mortality		Benefit
				tPA %	Placebo %	tPA %	Placebo %	Death or Dependency OR (95% CI)
NINDS	624	0.9 (90)	≤ 3	6.4	0.6	17.4	20.6	0.49 (0.35–0.69)
ECASS-I	620	1.1 (100)	≤ 6	19.8[*]	6.5[*]	22	15.6	0.68 (0.55–0.95)
ECASS-II	800	0.9 (90)	≤ 6	8.8	3.4	10.5	10.7	0.72 (0.55–0.95)
ATLANTIS-B	547	0.9 (90)	3–5	7.0	1.1	11.0	6.9	1.04 (—)

[*]Parenchymal hematoma (symptomatic ICH not reported in ECASS-I).
Reproduced from Albers *et al.* (2001). Antithrombotic and thrombolytic therapy for ischemic stroke, *Chest* 119:300-320.

with an 11% to 13% absolute increase in the number of patients with excellent outcomes. Symptomatic ICH occurred in 6.4% of treated patients vs 0.6% for placebo ($P < .001$); however, there was no difference in mortality. The benefit of IV tPA seen at 3 months in the NINDS study was sustained over the long term. During a 12-month follow-up evaluation of these patients, the benefit of tPA over placebo remained virtually identical, with an 11% to 13% absolute increase in the number of patients achieving excellent outcomes (Kwiatkowski et al., 1999).

The European Cooperative Acute Stroke Study (ECASS) trial (Hacke et al., 1995) was a multicenter, double-blind, placebo-controlled trial that randomized 620 patients within 6 hours of stroke onset to treatment with IV tPA at a dose of 1.1 mg/kg (total dose limit 100 mg) or placebo. The protocol excluded patients with the most severe hemispheric stroke symptoms and patients with major early infarct signs on CT scan exceeding one third of the MCA territory. Primary outcome measures were the Barthel Index (BI) and modified Rankin Scale (mRS) at 3 months after treatment. Both an intention-to-treat (ITT) analysis and a target population (TP) (per-protocol) analysis were performed. A total of 109 patients were excluded from the TP analysis primarily because of the presence of extensive early ischemic changes on CT scan. There was no difference in BI scores at 3 months for either the ITT or TP groups. In the TP analysis, there was a significant difference in the mRS favoring tPA. There was no difference in mortality at 30 days; however, the incidence of parenchymal hemorrhages was significantly more frequent in the tP-treated patients (19.8% vs 6.5% in the placebo group). In an explanatory analysis of the ECASS data, advanced age was associated with an increased risk of parenchymal hemorrhage, while time-to-treatment was not related (Larrue et al., 1997). The initial clinical stroke severity and the presence of early ischemic changes on CT scan were associated with increased risk of hemorrhagic infarction. The investigators concluded that tPA may have a net benefit if patient selection could be improved to exclude patients at higher risk for complications, particularly those with major early infarct signs on CT scan.

The ECASS investigators also suggested that the higher dose of tPA used may have contributed to the increased hemorrhagic complications, a relationship supported by data from the myocardial infarction trials. Furthermore, strict blood pressure parameters were not included in this protocol. Therefore, ECASS II (Hacke et al., 1998) was designed with a lower dose of tPA (0.9 mg/kg to match the NINDS protocol) given within 6 hours of symptom onset, strict guidelines for blood pressure control, and strict adherence to CT criteria, including investigator participation in CT training courses before and during the course of the study. ECASS II also found no significant difference in the primary outcome (the percentage of patients with a favorable outcome [score 0 to 1] on mRS at 3 months). However, for the secondary post-hoc outcome of functional independence (score 0 to 2) on mRS, there was a significant difference in favor of tPA. Although there was a 2.5-fold increase in the symptomatic intracerebral hemorrhage rate for tPA-treated patients vs placebo, there was no difference in mortality. Patients in ECASS II had less severe strokes, with baseline median NIHSS scores of 11 in both groups, compared to 14 and 15 in the NINDS trial. This finding may account for the better outcomes in the ECASS II placebo group compared to the other tPA studies and may contribute to the lack of a substantial treatment effect on the primary outcome. Also, most patients (642 of 800) were treated in the 3- to 6-hour window, whereas in the NINDS trial, all patients were treated within 3 hours, and half within 90 minutes of stroke onset.

The Alteplase Thrombolysis for Acute Noninterventional Therapy in Ischemic Stroke (ATLANTIS) trial (Albers et al., 2002) had several similarities to ECASS II. The patients enrolled had milder strokes (median NIHSS score 11) and were treated at a later time (median time to treatment 4 hours 35 minutes) than the NINDS study. The ATLANTIS study design was similar to the NINDS study except for the time windows. ATLANTIS Part A began in 1991 with a 0- to 6-hour time window, which was changed in 1993 to 0 to 5 hours (Part B) because of safety concerns in the 5- to 6-hour window. There was no benefit in the treatment group for outcome measures at 3 months either for the target population treated in the 3- to 5-hour window or in the ITT analysis. However, a prespecified analysis of the 61 patients in ATLANTIS who were enrolled within 3 hours of stroke onset did find that the tPA-treated patients were more likely to have a very favorable outcome (NIHSS = 1) at 3 months ($P = .01$), supporting the conclusions of the NINDS study.

Primarily based on the success of the NINDS studies, in 1996 the FDA approved tPA for use in early acute ischemic stroke. After the publication of the NINDS trials, reports of clinical practice experience with IV tPA both in academic and community settings have demonstrated similar safety and clinical outcomes when the NINDS criteria were strictly used (Albers et al., 2000). The advent of an effective treatment for acute ischemic stroke has energized clinicians and spurred local and national stroke education campaigns. It has also stimulated ongoing research, including efforts to expand the therapeutic window of tPA using advanced neuroimaging techniques for patient selection, endovascular delivery of thrombolytics, and coupling of tPA with neuroprotective therapies.

STROKE PREVENTION

Treatment of Hypertension

The latest report of the Joint National Committee on Prevention, Detection, Evaluation, and Treatment of High Blood Pressure emphasizes the need for more aggressive blood pressure (BP) control to prevent cardiovascular disease, including stroke (Chobanian et al., 2003). Normal BP is now defined as <120/80. For those 40 to 70 years old, each increment of 20 mm Hg in systolic BP or 10 mm Hg in diastolic BP doubles the risk of cardiovascular disease across the entire BP range from 115/75 to 185/115 (Lewington et al., 2002). To achieve adequate BP control, clinical trials have demonstrated that most patients will require two or more antihypertensive medications.

For primary and secondary stroke prevention, the effectiveness of angiotensin-converting enzyme inhibitors and angiotensin receptor blockers has recently gained attention. Why have these particular classes of antihypertensives been of interest? There is both theoretical and clinical evidence to support the vascular and cardiac benefits of blocking the renin-angiotensin-aldosterone system (RAAS). This system controls systemic blood pressure through multiple mechanisms including modulation of the sympathetic nervous system, as well as direct effects on the heart, kidneys, and blood vessels (Weir and Henrich, 2000). The RAAS maintains salt and water homeostasis but may also promote chronic hypertension (Ruland and Gorelick, 2003). In this system, angiotensin I is converted to angiotensin II by angiotensin-converting enzyme (ACE). Angiotensin II, a peptide hormone, has many vascular effects including vasoconstriction, inflammation, vascular remodeling, thrombosis, and plaque rupture primarily via activation of angiotensin II type 1 receptors (Schiffrin, 2002; McFarlane et al., 2003). Many of these effects are mediated via direct effects on endothelial cells and vascular smooth muscle cells. Angiotensin II and aldosterone increase the production of plasminogen activator inhibitor type I (PAI-1), the most important physiological inhibitor of tissue-type plasminogen activator (TPA) in plasma, and promote platelet aggregation (Lonn et al., 1994; Ruland and Gorelick, 2003). Angiotensin II is a major contributor to the generation of reactive oxygen species that oppose the vascular effects of nitric oxide, including the inhibition of the growth, remodeling, and migration of vascular smooth muscle cells, as well as the expression of proinflammatory molecules (McFarlane et al., 2003). Therefore, an imbalance between reactive oxygen species and nitric oxide leads to activation of endothelins (potent vasoconstrictors) and up regulation of proinflammatory mediators that contribute to vascular disease (Schiffrin, 2002).

The exact mechanisms by which ACE inhibitors may prevent vascular disease are not completely understood; however, multiple potential mechanisms have been hypothesized based on the actions of angiotensin II described previously (Figure 2.1). By blocking the conversion of angiotensin I to angiotensin II, ACE inhibitors may improve vascular compliance, reduce vascular smooth muscle proliferation, and have plaque-stabilizing as well as antithrombotic and antiinflammatory effects. They improve renal blood flow and reduce aldosterone secretion, thus reducing reabsorption of sodium (Ruland and Gorelick, 2003).

ACE inhibitors also prevent the breakdown of bradykinin, and it has been suggested that many of the known clinical benefits of ACE inhibitors may be to a large extent related to the increased concentrations of bradykinin in serum and possibly tissue (Weir and Henrich, 2000). Bradykinin is a powerful vasodilator that also secondarily augments the production of other vasodilators such as nitric oxide and cyclic GMP (Weir and Henrich, 2000). It causes a cascade of vasodilatory and antithrombotic effects. Human clinical studies have shown that plasma angiotensin II levels remain at or above pretreatment levels with chronic dosing of an ACE inhibitor, localizing the effects of ACE inhibitors perhaps to the vascular tissue (Weir and Henrich, 2000). Furthermore, it has been shown that a bradykinin receptor antagonist can inhibit the blood pressure-lowering effects of an ACE inhibitor, thus pointing to bradykinin as a potent mediator in the setting of ACE inhibitors (Weir and Henrich, 2000).

Angiotensin II acts via type 1 and type 2 receptors. The type 1 receptors mediate the known effects of angiotensin II. The current angiotensin receptor (AT_1) blockers specifically block these type I receptors (Figure 2.1). There is evidence that the clinically observed benefits of AT_1 blockade may, to a large extent, be due to stimulation of the type 2 receptors, which are expressed more commonly in diseased and damaged tissues, rather than direct antagonism of the type 1 receptors (Weir and Henrich, 2000; Schiffrin, 2002). While the function of the type 2 receptors is less well defined, it appears to be antagonistic to type 1, thereby stimulating vasodilation via the bradykinin-nitric oxide-cyclic guanine monophosphate cascade as well as inhibiting smooth muscle cell proliferation and inflammatory responses (Weir and Henrich 2000; Schiffrin 2002).

There have been several recent large randomized trials that have demonstrated the effectiveness of ACE inhibitors or angiotensin receptor blockers (ARBs) in primary or secondary stroke prevention. Debate persists as to whether this benefit is due purely to the BP-lowering effect of these agents or to additional vascular protective effects described previously, which may be specific to these

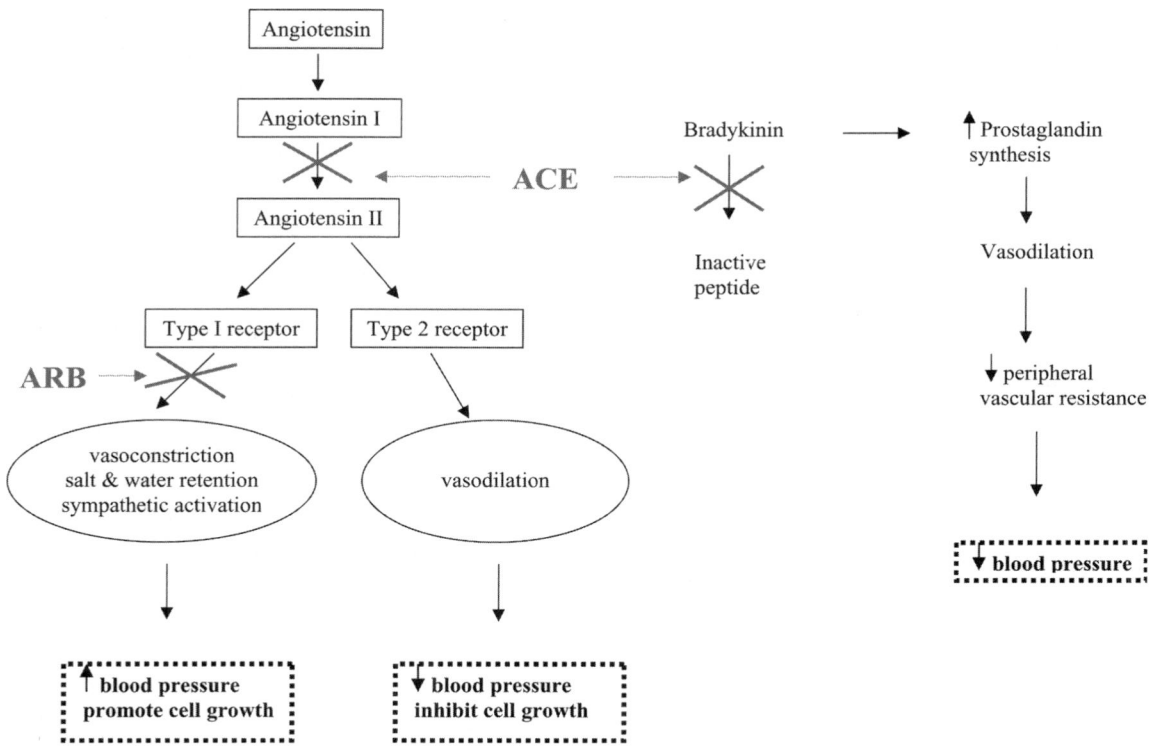

FIGURE 2.1 Mechanisms of action of ACE inhibitors and angiotensin II receptor subtype 1 blockers. ACE, angiotensin-converting enzyme; ARB, angiotensin receptor blocker.

particular classes of antihypertensive agents (Anderson, 2003; Bath, 2003; Davis and Donnan, 2003).

The Heart Outcomes Prevention Evaluation (HOPE) (The Heart Outcomes Prevention Evaluation Study Investigators, 2000) was a double-blind, randomized, placebo-controlled trial of the use of an ACE inhibitor, ramipril, for the prevention of the composite outcome of MI, stroke, and cardiovascular death in high-risk cardiovascular disease patients. Of the 9541 patients randomized, 1013 had a previous stroke or transient ischemic attack. The mean BP at enrollment was 139/79. There was a 22% relative risk reduction (RRR) in the primary composite outcome. In fact, the study was terminated early because of the clear benefit of ramipril. There was also a 32% RRR of any stroke (3.4% vs 4.9% stroke incidence for ramipril and placebo, respectively) and a 61% RRR of fatal stroke (0.4% vs 1%) in the ramipril group (Bosch *et al.*, 2002).

Despite these large cardiovascular outcome benefits, there was only a small reduction in BP (3/2 mm Hg) with treatment. Based on data derived from the World Health Organization and the International Society of Hypertension, the expected relative risk of MI and stroke from this small BP reduction would have been 5% and 13%, respectively, rather than the observed reductions of 20% and 32%, respectively (Sleight *et al.*, 2001). Therefore, it has been hypothesized that much of the beneficial effect of ramipril on vascular events may be independent of its BP-lowering effect (Sleight *et al.*, 2001). Alternatively, it has been suggested that the study's daytime BP measurements may have underestimated the 24-hour BP reduction of ramipril, which was administered as a nighttime dose. A small substudy of HOPE monitored 38 patients with peripheral vascular disease who underwent 24-hour ambulatory BP (ABP) monitoring before randomization and after 1 year (Svensson *et al.*, 2001). This substudy found that the 24-hour ABP was significantly reduced (10/4 mm Hg) primarily because of a more pronounced BP-lowering effect during nighttime (17/8 mm Hg). Therefore, the benefits of ramipril may be more attributable to the reduction in BP than previously concluded, although these findings are drawn from a very small sample size.

The benefit of BP lowering for reducing the risk of recurrent stroke was definitively shown in the Perindopril Protection Against Recurrent Stroke Study (PROGRESS) (2001). This investigator-initiated trial was a double-blind, randomized, placebo-controlled study of the effects of a flexible BP-lowering regimen based on an ACE inhibitor (perindopril) with or without the addition of a diuretic (indapamide) on the primary outcome of total stroke (fatal or nonfatal). Of the 6105 patients enrolled, there was a 28% RRR for recurrent stroke for those assigned active treatment over placebo (10% vs 14% stroke incidence). The overall reduction in recurrent stroke risk was comparable to the reduction in primarily initial stroke risk seen in the HOPE study (32%) and included reduction in the outcome endpoints of nonfatal or disabling stroke (24%), ischemic stroke (24%), and cerebral hemorrhage (50%). Combination therapy with perindopril plus indapamide reduced BP by 12/5 mm Hg and stroke risk by 43%, whereas perindopril alone reduced BP by 5/3 mm Hg and had no significant effect on stroke risk. There were similar benefits for both the hypertensive and nonhypertensive subgroups, although hypertension was defined as a systolic BP >160 or a diastolic BP >90 at baseline, a cutoff level quite high compared with current guidelines of 120/80 (Chobanian et al., 2003). The mean baseline BP of those classified as hypertensive was 159/94 mm Hg and for the nonhypertensive was 137/79 mm Hg. In addition to the stroke risk reduction, there was a 26% reduction in major vascular events including a 38% reduction in nonfatal MI.

Combination therapy of perindopril plus indapamide clearly resulted in larger BP reductions and larger risk reductions than perindopril alone. PROGRESS demonstrated that aggressive BP lowering can significantly reduce the risk of recurrent stroke and major vascular events, even in patients who may already be treated with other stroke prevention therapy such as antiplatelets or anticoagulants. It was not designed, however, to test for potential benefits of the ACE inhibitor independent of its BP-lowering effect, as the study was powered to detect a 30% reduction in total stroke risk among those receiving active treatment over placebo.

While PROGRESS did not seek to establish a benefit of the ACE inhibitor perindopril beyond its BP-lowering effect, the Losartan Intervention for Endpoint reduction (LIFE) trial (Dahlof et al., 2002) did aim to determine whether the ARB, losartan, had cardiovascular benefits independent of BP-lowering. LIFE was an investigator-initiated, double-blind, randomized, controlled trial of losartan (the first angiotensin II type 1 receptor blocker) vs atenolol in patients with hypertension and left ventricular hypertrophy for the prevention of cardiovascular morbidity and death. Atenolol was chosen as the comparator because it was recognized worldwide as a first-line treatment for hypertension with similar antihypertensive efficacy to losartan. Hydrochlorothiazide and additional antihypertensive agents (except ACE inhibitors, ARBs, and β-blockers) could be added to the blinded study medications if needed to achieve a goal BP <140/90. Blood pressures were reduced substantially in both groups (by 30.2/16.6 and 29.1/16.8 mm Hg in the

losartan and atenolol groups, respectively). There was equal use of additional antihypertensive agents between the two groups. During a mean follow-up period of 4.8 years, there was a 13% RRR for the primary composite endpoint of cardiovascular death, MI, and stroke for losartan over atenolol. Furthermore, there was a 25% RRR for stroke for losartan-treated patients; of the 9193 subjects who participated, 232 losartan patients and 309 atenolol patients experienced a stroke. There were fewer MIs than strokes in both groups and no difference in the incidence of MI between the two groups. There was a very small difference between groups in systolic and diastolic blood pressures but not in mean arterial pressure. Adjustment of the primary outcome for changes in systolic, diastolic, or mean arterial pressure did not appreciably affect the outcome results. Therefore, the cardiovascular benefit, and particularly the stroke benefit, seen with losartan over atenolol supports a BP-independent effect of this angiotensin II type 1 receptor blocker in high-risk hypertensive patients.

Treatment of Hypercholesterolemia

Although epidemiological studies have not established as clear a link between serum cholesterol and stroke as they have for cholesterol and MI (Thomas et al., 1966), this failure may be due to several study limitations. A meta-analysis of 45 prospective observational cohorts, including 450,000 subjects and 13,000 strokes, found no association between total cholesterol and stroke (1995). The Multiple Risk Factor Intervention Trial (MRFIT) (Iso et al., 1989) did find a positive association between serum cholesterol levels and death from nonhemorrhagic stroke in men 35 to 57 years old, but also found a negative association between intracranial hemorrhage and serum cholesterol levels <160 mg/dl (4.1 mmol/L) in hypertensive men. Several possible explanations have been proposed for why the observational studies to date have failed to show a clear association between cholesterol and stroke. The primary limitation of these studies has been that the cohorts of patients have not been representative of the population at risk for ischemic stroke (Amarenco, 2001). The studies focused on middle-aged subjects at risk for MI, when the incidence of stroke is known to rise one to two decades later than coronary heart disease (CHD). When strokes do occur in this younger population, the cause is less likely due to atherothrombosis and instead is more commonly due to causes unrelated to cholesterol such as heart valve disease or carotid/vertebral artery dissection. In many studies the strokes were not differentiated by subtype to possibly detect an association between cholesterol and ischemic atherthrombotic strokes separated from hemorrhagic or cardioembolic strokes (Amarenco, 2001).

Despite the lack of epidemiological evidence to confirm cholesterol as a marker for stroke risk, several landmark trials have demonstrated a beneficial effect of 3-hydroxy-3-methylglutaryl-coenzyme A (HMG-CoA) reductase inhibitors ("statins") in reducing stroke risk in patients with CHD. Statins inhibit the rate-limiting enzyme in cholesterol biosynthesis. By inhibiting this enzyme, statins lead to an up regulation in the expression of low-density lipoprotein (LDL) receptors on hepatocytes and therefore enhanced clearance of circulating LDL (Hess *et al.*, 2000). However, beyond lowering serum cholesterol levels, statins possess additional mechanisms of action that may contribute to their beneficial effects in stroke and may be independent of cholesterol-lowering effects (Table 2.2). Statins have antithrombotic effects by blocking platelet activation and increasing endothelial cell fibrinolytic activity. They have anti-inflammatory effects by blocking macrophages, reducing matrix metalloproteinase secretion, lowering levels of C-reactive protein, and inhibiting the activation of inflammatory cytokines. Furthermore, they decrease smooth muscle cell migration and proliferation and, in vitro, induce vascular smooth muscle cell apoptosis, which may be one of the mechanisms by which statins reduce intima-media wall thickness in patients with carotid atherosclerosis (Hess *et al.*, 2000).

A meta-analysis of the 13 randomized, placebo-controlled, double-blind trials of statins reporting on stroke from 1980 to 1996 found an overall stroke risk reduction of 31% (Blauw *et al.*, 1997). The mean age in these trials ranged from 55 to 68 years. This analysis included the Scandinavian Simvastatin Survival Study (4S) (1994) in which 4444 patients with CHD and a mean total cholesterol of 263 mg/dl were randomized to simvastatin or placebo. In a post-hoc analysis, there was found to be a 28% RRR for fatal and nonfatal strokes or transient

ischemic attacks (Pedersen *et al.*, 1998). The meta-analysis also included the Cholesterol and Recurrent Events (CARE) trial (Sacks *et al.*, 1996), which was a randomized, controlled trial of pravastatin in 4159 patients with recent MI and average cholesterol levels (mean total cholesterol 209 mg/dl, mean LDL 139 mg/dl). The treatment group had a 31% RRR (incidence 2.6% vs 3.8%) of stroke, a prespecified endpoint of the trial. There was a treatment-associated reduction across all stroke subtypes, although there was a small number of outcome events in each subtype (Plehn *et al.*, 1999). There was also no increase in intracerebral hemorrhages with pravastatin, although there were only eight hemorrhagic events (Plehn *et al.*, 1999).

The Long-Term Intervention with Pravastatin in Ischemic Disease (LIPID) study (The Long-Term Intervention with Pravastatin in Ischaemic Disease (LIPID) Study Group, 1998) was published after the previously mentioned meta-analysis but supported the findings of both the meta-analysis and the 4S and CARE trials. LIPID was a randomized, controlled trial of pravastatin in 9014 patients with CHD, with a somewhat broader range of total cholesterol levels (range 155 to 271 mg/dl). There was a 19% RRR for total stroke, with a 23% RRR for ischemic stroke (incidence 3.4% vs 4.4%) and a consistent effect across all ischemic stroke subtypes. Again, there was no increase in intracerebral hemorrhage, although the total number of hemorrhagic events (28) was too small to draw any conclusions (White *et al.*, 2000).

The most recent clinical evidence establishing the benefit of lipid modification for stroke prevention in high-risk patients is the Heart Protection Study (2002), which reported its results of a randomized, controlled trial of simvastatin in 20,536 patients with vascular risk factors. This study further established that there are benefits irrespective of initial cholesterol concentrations. Subjects included men and women 40 to 80 years old (28% older than age 70) with total cholesterol ≥135 mg/dl (3.5 mmol/L), and a history of CHD, other occlusive arterial disease, or diabetes. There was a 25% RRR (4.3% vs 5.7%) for first stroke, primarily resulting from a 30% RRR in ischemic stroke (2.8% vs 4%). There was no difference in the incidence of hemorrhagic strokes, with a total of 104 hemorrhagic stroke events. Because approximately one sixth of the treatment group stopped taking statin therapy and one sixth of the placebo group initiated a statin during the study, the ITT analysis actually reflects the effects of about two thirds of the treated group actually taking simvastatin. Therefore, the actual reduction in stroke rate may be greater than the 25% RRR seen in the study. Furthermore, this benefit was additive to the other preventive treatments continued during the study including antiplatelet and antihypertensive agents.

TABLE 2.2 Beneficial Effects of Statin Agents

Lipid-modifying
  ↓LDL
  ↓Triglycerides
  ↑HDL

Antithrombotic
  Block platelet activation
  ↑Endothelial cell fibrinolytic activity

Anti-inflammatory
  Block macrophages
  ↓Matrix metalloproteinase secretion
  ↓Decrease C-reactive protein
  Inhibit inflammatory cytokine activation

Vasomotor
  ↓Smooth muscle cell migration and proliferation
  Induces vascular smooth muscle cell apoptosis
  →decrease intima-media wall thickness

The proportional reduction in LDL with statin treatment in the Heart Protection Study was independent of initial cholesterol level. There was also no lower threshold seen below which lowering cholesterol did not reduce risk. Even among subjects with "normal" LDL levels of <100 mg/dl (2.6 mmol/L), reducing the average LDL to 65 mg/dl (1.7 mmol/L) in the treated group was safe and resulted in a risk reduction similar to those with higher LDL levels.

These large clinical trials definitively establish that middle-aged and older high-risk patients achieve substantial benefit in ischemic stroke risk reduction with statin therapy, specifically pravastatin and simvastatin, even with normal or moderate baseline cholesterol levels. Based on these studies, statins do not appear to increase hemorrhagic stroke incidence, although the total number of observed hemorrhagic events has been small. Most of these studies have focused on patients with CHD rather than those with primarily cerebrovascular disease. The Stroke Prevention by Aggressive Reduction of Cholesterol Levels (SPARCL) trial (Amarenco, 1999) has been designed to determine whether aggressive cholesterol-lowering therapy with atorvastatin, 80 mg, can reduce the incidence of stroke in patients without CHD but with a history of stroke or transient ischemic attack. This first prospective study of statins in secondary stroke prevention is currently ongoing.

A neuroprotective mechanism of statins has been described (Endres et al., 1998). Mediated by up regulation of endothelial nitric oxide synthase rather than by the decrease in cholesterol levels, infarct size in mice with normal cholesterol levels who were pretreated with simvastatin was significantly reduced. This protection may be related to enhanced blood flow or inhibition of platelet aggregation or leukocyte adhesion, all known nitric oxide-mediated effects. The prophylactic use of statins to decrease the severity of ischemic injury deserves further investigation.

Antiplatelet Therapy: Dipyridamole

In the quest to improve the effectiveness of antiplatelet therapy beyond aspirin in secondary stroke prevention, dipyridamole has been investigated alone and in combination with aspirin. It was initially thought that the antithrombotic effects of dipyridamole were purely platelet-mediated via phosphodiesterase inhibition. However, early platelet aggregometry studies, which separated platelets from other blood cells, showed only weak antiplatelet effects of dipyridamole (Eisert, 2001a, 2001b). The development of whole-blood impedance aggregometry, which better approximates *in vivo* thrombus formation, demonstrated that dipyridamole inhibits platelet aggregation more effectively in whole blood than in platelet-rich plasma. This suggested that dipyridamole had additional antithrombotic mechanisms beyond its direct antiplatelet effects. Other newer laboratory techniques have corroborated this hypothesis. A process was developed to create a subendothelial matrix covered with endothelial cells to simulate the vascular environment. With this model, dipyridamole has been found to enhance the indirect (near-field) antithrombotic action of the endothelium through multiple possible mechanisms (Eisert, 2001a, 2001b). These mechanisms include the inhibition of the uptake of adenosine, a potent endogenous inhibitor of platelet aggregation; potentiation of endogenous prostacyclin, a potent antithrombotic substance released from the vessel wall; and enhancement of endothelium-derived relaxing factor or nitric oxide, an inhibitor of platelet aggregation and adhesion as well as a vasodilator (Eisert, 2001a and b). Dipyridamole also has antioxidant properties that may contribute to its benefit in atherosclerosis (Eisert, 2001b) (Figure 2.2).

Early clinical studies of dipyridamole as an antiplatelet agent were discouraging. The Antiplatelet Trialists' Collaboration (1994) performed a meta-analysis of 14 trials that compared the combination of dipyridamole and aspirin vs aspirin alone for prevention of nonfatal stroke, nonfatal MI, or vascular death. They found that dipyridamole provided no additional reduction in vascular events (316/2661 with aspirin plus dipyridamole vs 312/2656 with aspirin alone), but they concluded that a moderate difference had not been excluded by the studies. The European Stroke Prevention Study (ESPS-1) compared aspirin (325 mg three times a day) plus dipyridamole standard-release formulation (75 mg three times a day) with placebo in patients with previous transient ischemic attack (TIA) or stroke and demonstrated a 38% relative risk reduction of recurrent stroke (33% RRR in stroke or death) (ESPS-1, 1987). However, ESPS-1 did not compare the combination of dipyridamole plus aspirin to aspirin alone. The results of ESPS-2, reported in 1996, were consistent with ESPS-1 and it was the first study to demonstrate a significant benefit of dipyridamole plus aspirin over aspirin alone (Diener et al., 1996). It was a randomized, placebo-controlled, double-blind trial involving 6602 patients with prior TIA or stroke within the preceding three months. It compared aspirin (25 mg twice a day) plus modified-release dipyridamole (200 mg twice a day) to each medication alone and to placebo. The combination of dipyridamole plus aspirin achieved a 37% RRR in recurrent stroke over placebo and a 23% RRR in recurrent stroke (3% absolute risk reduction) over aspirin alone. Why was the ESPS-2 trial able to demonstrate a significant benefit of combination therapy

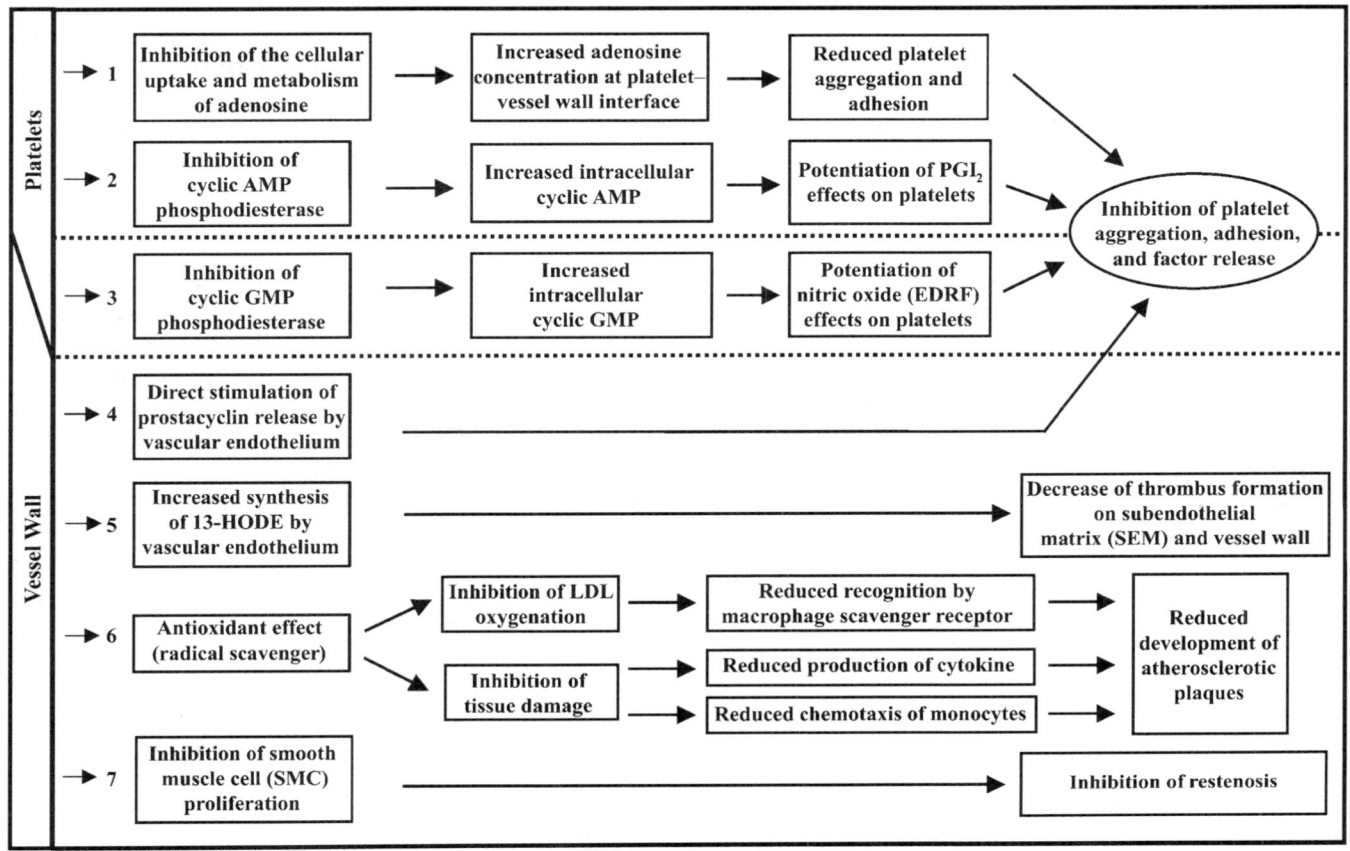

FIGURE 2.2 The antithrombotic mechanisms of action of dipyridamole. AMP, adenosine monophosphate; GMP, guanine monophosphate; PGI$_2$, prostaglandin I$_2$; EDRF, endothelium-derived relaxing factor; 13-HODE, 13-hydroxyoctadecadienoic acid. Reproduced from Eisert W. (2002). Dipyridamole. In Michelson A, editor, *Platelets,* New York, 2002, Academic Press.

dipyridamole plus aspirin while the previous 14 trials analyzed by the Antiplatelet Trialists' Collaboration had not? One important factor may be the limited number of events captured in those earlier trials: 142 nonfatal strokes compared to the 323 nonfatal strokes in ESPS-2 alone (Wilterdink and Easton, 1999). Furthermore, the Antiplatelet Trialists' Collaboration compared the reduction in combined vascular events (stroke, MI, and vascular death). When one excludes the trials in which no strokes occurred or in which stroke outcomes were not studied, data from the remaining nine trials combined favor dipyridamole plus aspirin over aspirin alone for prevention of nonfatal stroke, although results were not statistically significant (Wilterdink and Easton, 1999). The total number of nonfatal strokes in the other nine trials was 142; therefore, the 323 strokes in ESPS-2 provides more than twice as much outcome data as those previous nine trials combined. ESPS-2 was the only one of the trials discussed to use the sustained-release formulation, which provides better absorption of dipyridamole and allows twice-a-day (rather than three to four times per day) dosing.

FUTURE DIRECTIONS

Direct Thrombin Inhibitors

With respect to oral anticoagulation therapy, the armamentarium to date has been limited to warfarin and related vitamin K antagonists, which limit the synthesis of γ-glutamyl carboxylated forms of coagulation factors, factors II, VII, IX, X, protein C, and protein S, thereby impairing their function (Elg *et al.*, 1999). Clinicians are intimately familiar with the limitations of warfarin including its narrow therapeutic window—a doubling of the dose of warfarin in an arterial thrombosis model in rats increased the antithrombotic effect of warfarin from 23% to 81%—and its large individual patient variability in effect, necessitating close monitoring and dose adjustments.

Warfarin also has a delayed onset of action dependent on the turnover rate of the coagulation factors and numerous food and drug interactions.

Low-molecular-weight direct thrombin inhibitors (e.g., melagatran and inogatran) are under investigation as alternative oral anticoagulants. These agents are direct inhibitors of thrombin; therefore, they inhibit only one factor in the coagulation cascade, the final common pathway for fibrin formation. A rat arterial thrombosis model demonstrated shallow dose- and plasma concentration-response curves for the direct thrombin inhibitors (for the purposes of the experiment they were administered as continuous infusions) vs the steeper curve for warfarin. It has been suggested that the dose-response curves for each individual factor will superimpose on each other, resulting in a steeper dose-response curve and thus the narrow therapeutic window exhibited by warfarin (Elg et al., 1999).

Bleeding complications are a constant concern in the clinical use of anticoagulants. A tail transection bleeding time experiment in a rat model showed significant prolongation of the bleeding time for heparin and warfarin at the doses necessary to achieve an 80% antithrombotic effect, while melagatran did not prolong the bleeding time (Elg et al., 1999). Bleeding time for melagatran did not increase until twice the dose necessary to achieve an 80% antithrombotic effect, suggesting a wider therapeutic window. With regard to laboratory assays, there was little prolongation of activated prothrombin time and prothrombin time with melagatran treatment in the rat model despite therapeutic antithrombotic effect. Instead, thrombin time was the assay prolonged in a dose-dependent manner with melagatran treatment.

With parenteral or oral administration of melagatran, therapeutic effect is achieved within minutes to hours (Bredberg, 1999). Whereas parenteral administration provides complete bioavailability and low variability, oral administration does not. To overcome this obstacle and optimize the gastrointestinal absorption of melagatran, the first prodrug direct thrombin inhibitor was developed (Gustafsson et al., 2001). The prodrug is uncharged at intestinal pH and 170 times more lipophilic than melagatran and, as a result, has much greater bioavailability and lower variability; therefore, the pharmacodynamic benefits of melagatran are preserved. In addition, the minimal antithrombin effects of the prodrug against thrombin should decrease the risk of major bleeding in patients with undiagnosed gastrointestinal bleeding (Gustafsson et al., 2001).

The melagatran prodrug, called ximelagatran, is currently being investigated in two long-term Phase III human clinical trials for stroke prevention (SPORTIF III and V: Stroke Prevention Using Oral Thrombin Inhibitor in Atrial Fibrillation) (Halperin, 2003b). Enrolled patients have atrial fibrillation and at least one additional risk factor for stroke. SPORTIF III is a randomized, open-label, parallel-group study with blinded event assessment involving 3407 patients in 23 countries; SPORTIF V is similar but with double-blind treatment allocation involving 3922 patients in North America. Given the established impressive efficacy of warfarin in stroke prevention in patients with nonvalvular atrial fibrillation (68% relative risk reduction in one meta-analysis) (Atrial Fibrillation Investigator, 1994), the objective of the phase III SPORTIF trials is to establish the noninferiority of ximelagatran relative to warfarin in this patient population. The safety and efficacy of fixed-dose ximelagatran (36 mg twice a day) are being compared with dose-adjusted warfarin (INR 2.0-3.0) with the primary endpoint being the incidence of all strokes and systemic embolic events.

Results of SPORTIF III were recently presented (Halperin 2003a). In the ITT analysis, 56 primary events occurred in the warfarin group (2.3% per year) and 40 occurred in the ximelagatran group (1.6% per year). This relative risk reduction of 29% and absolute risk reduction of 0.7% per year showed ximelagatran's noninferiority relative to warfarin. There was also no significant difference between the two agents in the event rate for intracranial hemorrhage or major bleeding. Results of SPORTIF V have not yet been reported. Should they confirm the findings of SPORTIF III, one would anticipate that ximelagatran will become first-line treatment for stroke prevention in patients with atrial fibrillation.

Neuroprotectives

Despite the available antiplatelet, antihypertensive, and lipid-lowering agents for stroke prevention, as well as thrombolytic therapy for acute stroke treatment, minimizing ischemic injury during a stroke remains an important challenge. Although several neuroprotective drugs have been successfully developed in stroke animal models (Sydserff et al., 1995; Lesage et al., 1996), human clinical trials have been disappointing. What lessons can be learned from the shortcomings of the negative clinical trials when designing future neuroprotective trials?

Time Window

Although animal studies have involved drug administration before or soon after the onset of ischemia, none of the clinical trials published between 1995 and 1999 enrolled patients within 3 hours of symptom onset (Kidwell et al., 2001). In fact, the median time to entry was 12 hours, with a median time to treatment of 14 hours. To maximize the chances for detecting a treatment benefit, the time window should be as soon as possible after stroke onset and should reflect data from preclinical models.

Dose and Duration

As with the selection of time window, the drug dose chosen should reflect data from the preclinical and preliminary human studies. In some instances, the dose of a neuroprotective drug found to be effective in animals may be associated with adverse effects in clinical application. This can lead to suboptimal dosing to avoid even potentially acceptable side effects and result in a negative clinical trial (De Keyser et al., 1999). In general, Phase II trials have been inadequately designed to determine the most likely effective and safe doses to focus on in Phase III trials. Optimal duration of treatment for neuroprotective drugs is also not well established. Side effects can limit treatment duration. Based on evidence such as the prolonged elevation of excitatory amino acids after stroke in some patients (Bullock et al., 1995) and magnetic resonance spectroscopy suggesting ongoing neuronal loss over many days after stroke (Saunders et al., 1995), an extended treatment duration may be required (Gladstone et al., 2002). Certain drugs such as NMDA antagonists, benzodiazepines, or barbiturates may be beneficial if given early after ischemia but may impair recovery if given late (Gladstone et al., 2002). This issue will need to be addressed on an individual drug or drug class basis.

Stroke Standardization

Animal stroke models usually involve young, healthy animals with a middle cerebral artery occlusion, while neuroprotective clinical trials have typically enrolled a variety of stroke subtypes of varying severity (Gladstone et al., 2002). Depending on the mechanism of action, some drugs may have more of an effect on cortical than on white matter infarcts, or vice versa. Furthermore, in strokes in certain territories and of an intermediate severity, one may be more likely to be able to detect and measure benefit (Gladstone et al., 2002; Grotta, 2002). Perhaps the use of advanced neuroimaging techniques such as magnetic resonance diffusion and perfusion-weighted imaging can be used to identify patients who are most likely to benefit from acute stroke therapies.

Combination Therapies

Most likely the future use of neuroprotective drugs in acute ischemic stroke will be in the setting of combination therapy, such as two neuroprotective agents with synergistic mechanisms of action, the use of thrombolytic therapy preceded and/or followed by neuroprotection, the use of two or more agents during different therapeutic windows, or the use of a neuroprotective with nonpharmacological interventions such as hypothermia (De Keyser et al., 1999; Gladstone et al., 2002; Grotta, 2002). Although these trials will require increased complexity in design and conduct, they offer the potential to demonstrate efficacy with lower drug doses and therefore decreased toxicity and adverse effects.

CONCLUSIONS

We have illustrated the developments of several recent advances in acute stroke treatment and secondary stroke prevention. Clearly, the ability to effectively treat a patient having an acute stroke and reverse the neurological deficit can be dramatic and rewarding. The use of IV tPA has empowered physicians when they encounter eligible stroke patients. To expand our treatment abilities, many avenues remain open for exploration including neuroimaging for patient selection, neurointerventional techniques, and neuroprotectants. To diminish stroke risk, aggressive control of traditional risk factors continues to be the key element, and it is in the treatment of these risk factors that we are elucidating some of the mechanisms underlying cerebrovascular disease such as inflammation. Elucidating mechanisms that allow traditional therapies, such as antihypertensives and cholesterol-lowering agents to provide additional vascular protective effects will lead to the development of more effective pharmacological agents. Continued clinical investigations will enable us to build on our current successes treating and preventing stroke and its long-term effects.

References

Albers GW et al. (2000). Intravenous tissue-type plasminogen activator for treatment of acute stroke: the Standard Treatment with Alteplase to Reverse Stroke (STARS) study, *JAMA* 283: 1145-1150.

Albers GW et al. (2002). ATLANTIS Trial: Results for Patients Treated Within 3 Hours of Stroke Onset *Editorial Comment: Results for patients treated within 3 hours of stroke onset, *Stroke* 33:493-496.

Amarenco P. (2001). Hypercholesterolemia, lipid-lowering agents, and the risk for brain infarction, *Neurology* 57:35S-44.

Amarenco PBJ et al. (1999). Effect of atovastatin compared with placebo on cerebrovascular endpoints in patients with previous stroke or transient ischemic attack: the SPARCL study. 8th European Stroke Conference, Venice, Italy.

Anderson C. (2003). Blood pressure-lowering for secondary prevention of stroke: ACE inhibition is the key, *Stroke* 34:1333-1334.

Antiplatelet Trialists' Collaboration. (1994). Collaborative overview of randomised trials of antiplatelet therapy: Prevention of death, myocardial infarction, and stroke by prolonged antiplatelet therapy in various categories of patients, *BMJ* 308:81-106.

The Atrial Fibrillation Investigators. (1994). Risk factors for stroke and efficacy of antithrombotic therapy in atrial fibrillation: analysis of pooled data from five randomized controlled trials. *Arch Intern Med* 154: 1449-1457.

Bath P. (2003). Blood pressure-lowering for secondary prevention of stroke: ACE inhibition is not the key, *Stroke* 34:1334-1335.

Blauw GJ *et al.* (1997). Stroke, statins, and cholesterol : a meta-analysis of randomized, placebo-controlled, double-blind trials with HMG-CoA reductase inhibitors, *Stroke* 28:946-950.

Bosch J *et al.* (2002). Use of ramipril in preventing stroke: double blind randomised trial, *BMJ* 324:699.

Bredberg U *et al.* (1999). Pharmacokinetics of melagatran, a novel thrombin inhibitor, in healthy volunteers following intravenous, subcutaneous and oral administration. *Blood* 94 (Suppl 1):110.

Brott TG *et al.* (1992). Urgent therapy for stroke. Part I. Pilot study of tissue plasminogen activator administered within 90 minutes, *Stroke* 23:632-640.

Bullock R *et al.* (1995). Massive persistent release of excitatory amino acids following human occlusive stroke, *Stroke* 26:2187-2189.

Chobanian AV *et al.* (2003). The Seventh Report of the Joint National Committee on Prevention, Detection, Evaluation, and Treatment of High Blood Pressure: the JNC 7 report, *JAMA* 289:2560-2572.

Cornu C *et al.* (2000). Streptokinase in acute ischemic stroke: an individual patient data meta-analysis : The Thrombolysis in Acute Stroke Pooling Project, *Stroke* 31:1555-1560.

Dahlof B *et al.* (2002). Cardiovascular morbidity and mortality in the Losartan Intervention For Endpoint reduction in hypertension study (LIFE): a randomised trial against atenolol, *Lancet* 359:995-1003.

Davis SM, Donnan GA. (2003). Blood pressure reduction and ace inhibition in secondary stroke prevention: mechanism uncertain, *Stroke* 34:1335-1336.

De Keyser J *et al.* (1999). Clinical trials with neuroprotective drugs in acute ischaemic stroke: are we doing the right thing?, *Trends Neurosci* 22:535-540.

Del Zoppo GJ *et al.* (1986). The beneficial effect of intracarotid urokinase on acute stroke in a baboon model, *Stroke* 17:638-643.

Del Zoppo GJ *et al.* (1998). PROACT: a phase II randomized trial of recombinant pro-urokinase by direct arterial delivery in acute middle cerebral artery stroke. PROACT Investigators. Prolyse in Acute Cerebral Thromboembolism, *Stroke* 29:4-11.

Diener HC *et al.* (1996). European Stroke Prevention Study. 2. Dipyridamole and acetylsalicylic acid in the secondary prevention of stroke, *J Neurol Sci* 143:1-13.

Donnan GA *et al.* (1995). Trials of streptokinase in severe acute ischaemic stroke, *Lancet* 345:578-579.

Donnan GA *et al.* (1996). Streptokinase for acute ischemic stroke with relationship to time of administration: Australian Streptokinase (ASK) Trial Study Group, *JAMA* 276: 961-966.

Eisert WG. (2001a). How to get from antiplatelet to antithrombotic treatment, *Am J Ther* 8:443-449.

Eisert WG. (2001b). Near-field amplification of antithrombotic effects of dipyridamole through vessel wall cells, *Neurology* 57:S20-23.

Elg M *et al.* (1999). Antithrombotic effects and bleeding time of thrombin inhibitors and warfarin in the rat, *Thrombos Res* 94:187-197.

Endres M *et al.* (1998). Stroke protection by 3-hydroxy-3-methylglutaryl (HMG)-CoA reductase inhibitors mediated by endothelial nitric oxide synthase, *PNAS* 95:8880-8885.

ESPS Group. (1987). The European Stroke Prevention Study (ESPS): principal end-points. *Lancet* 2: 1351-1354.

Gladstone DJ *et al.* (2002). Toward wisdom from failure: lessons from neuroprotective stroke trials and new therapeutic directions, *Stroke* 33:2123-2136.

Grotta J. (2002). Neuroprotection is unlikely to be effective in humans using current trial designs, *Stroke* 33:306-307.

Gustafsson D *et al.* (2001). The direct thrombin inhibitor melagatran and its oral prodrug H 376/95: intestinal absorption properties, biochemical and pharmacodynamic effects, *Thromb Res* 101: 171-181.

Hacke W *et al.* (1995). Intravenous thrombolysis with recombinant tissue plasminogen activator for acute hemispheric stroke. The European Cooperative Acute Stroke Study (ECASS), *JAMA* 274:1017-1025.

Hacke W *et al.* (1998). Randomised double-blind placebo-controlled trial of thrombolytic therapy with intravenous alteplase in acute ischaemic stroke (ECASS II). Second European-Australasian Acute Stroke Study Investigators, *Lancet* 352:1245-1251.

Haley EC Jr *et al.* (1992). Urgent therapy for stroke. Part II. Pilot study of tissue plasminogen activator administered 91-180 minutes from onset, *Stroke* 23:641-645.

Halperin JL. (2003a). SPORTIF III: a long-term randomized trial comparing ximelagatran with warfarin for prevention of stroke and systemic embolism in patients with nonvalvular atrial fibrillation, American College of Cardiology 52nd Annual Scientific Session, Chicago, IL.

Halperin JL. (2003b). Ximelagatran compared with warfarin for prevention of thromboembolism in patients with nonvalvular atrial fibrillation: rationale, objectives, and design of a pair of clinical studies and baseline patient characteristics (SPORTIF III and V), *Am Heart J* 146:431-438.

Hess DC *et al.* (2000). HMG-CoA reductase inhibitors (statins): a promising approach to stroke prevention, *Neurology* 54:790-796.

Iso H *et al.* (1989). Serum cholesterol levels and six-year mortality from stroke in 350,977 men screened for the multiple risk factor intervention trial, *N Engl J Med* 320:904-910.

Kassem-Moussa H, Graffagnino C. (2002). Nonocclusion and spontaneous recanalization rates in acute ischemic stroke: a review of cerebral angiography studies, *Arch Neurol* 59: 1870-1873.

Kidwell CS *et al.* (2001). Trends in acute ischemic stroke trials through the 20th century, *Stroke* 32:1349-1359.

Kwiatkowski TG *et al.* (1999). Effects of tissue plasminogen activator for acute ischemic stroke at one year. National Institute of Neurological Disorders and Stroke Recombinant Tissue Plasminogen Activator Stroke Study Group, *N Engl J Med* 340: 1781-1787.

Larrue V *et al.* (1997). Hemorrhagic transformation in acute ischemic stroke: potential contributing factors in the European Cooperative Acute Stroke Study, *Stroke* 28:957-960.

Lesage AS *et al.* (1996). Lubeluzole, a novel long-term neuroprotectant, inhibits the glutamate-activated nitric oxide synthase pathway, *J Pharmacol Exp Ther* 279:759-766.

Lewington S *et al.* (2002). Age-specific relevance of usual blood pressure to vascular mortality: a meta-analysis of individual data for one million adults in 61 prospective studies, *Lancet* 360: 1903-1913.

Lonn EM *et al.* (1994). Emerging role of angiotensin-converting enzyme inhibitors in cardiac and vascular protection, *Circulation* 90:2056-2069.

McFarlane SI *et al.* (2003). Mechanisms by which angiotensin-converting enzyme inhibitors prevent diabetes and cardiovascular disease, *Am J Cardiol* 91:30-37.

MRC/BHF Heart Protection Study. (2002). Heart protection study of cholesterol lowering with simvastatin in 20, 536 high-risk individuals: a randomised placebo-controlled trial, *The Lancet* 360:7-22.

The National Institutes of Neurological Disorders and Stroke rt-PA Stroke Study Group. (1995). Tissue plasminogen activator for acute ischemic stroke, *N Engl J Med* 333:1581-1587.

Pedersen MD *et al.* (1998). Effect of simvastatin on ischemic signs and symptoms in the Scandinavian Simvastatin Survival Study (4S), *Am J Cardiol* 81:333-335.

Plehn JF *et al.* (1999). Reduction of stroke incidence after myocardial infarction with pravastatin: The Cholesterol and Recurrent Events (CARE) Study, *Circulation* 99:216-223.

PROGRESS -Perindopril Protection Against Recurrent Stroke Study. (1999). Characteristics of the study population at baseline. Progress Management Committee, *J Hypertens* 17:1647-1655.

Ruland S, Gorelick PB. (2003). Are cholesterol-lowering medications and antihypertensive agents preventing stroke in ways other than by controlling the risk factor?, *Curr Atheroscler Rep* 5:38-43.

Sacks FM *et al.* (1996). The effect of pravastatin on coronary events after myocardial infarction in patients with average cholesterol levels, *N Engl J Med* 335:1001-1009.

Saunders DE *et al.* (1995). Continuing ischemic damage after acute middle cerebral artery infarction in humans demonstrated by short-echo proton spectroscopy, *Stroke* 26:1007-1013.

The Scandinavian Simvastatin Survival Study (4S). (1994). Randomised trial of cholesterol lowering in 4444 patients with coronary heart disease, *Lancet* 344:1383-1389.

Second International Study of Infarct Survival Collaborative Group. (1988). Randomised trial of intravenous streptokinase, oral aspirin, both, or neither among 17,187 cases of suspected acute myocardial infarction: ISIS-2, *Lancet* 2:349-360.

Schiffrin EL (2002). Vascular and cardiac benefits of angiotensin receptor blockers, *Am J Med* 113:409-418.

Sleight P *et al.* (2001). Blood-pressure reduction and cardiovascular risk in HOPE study, *Lancet* 358:2130-2131.

Svensson P *et al.* (2001). Comparative effects of ramipril on ambulatory and office blood pressures: a HOPE substudy, *Hypertension* 38:28e-32.

Sydserff SG *et al.* (1995). The effect of chlormethiazole on neuronal damage in a model of transient focal ischaemia, *Br J Pharmacol* 114:1631-1635.

The Heart Outcomes Prevention Evaluation Study Investigators. (2000). Effects of an angiotensin-converting-enzyme inhibitor, ramipril, on cardiovascular events in high-risk patients, *N Engl J Med* 342:145-153.

The Long-Term Intervention with Pravastatin in Ischaemic Disease (LIPID) Study Group. (1998). Prevention of cardiovascular events and death with pravastatin in patients with coronary heart disease and a broad range of initial cholesterol levels, *N Engl J Med* 339:1349-1357.

Thomas HE Jr *et al.* (1966). Cholesterol-phospholipid ratio in the prediction of coronary heart disease. The Framingham Study, *N Engl J Med* 274:701-705.

Weir MR, Henrich WL. (2000). Theoretical basis and clinical evidence for differential effects of angiotensin-converting enzyme inhibitors and angiotensin II receptor subtype 1 blockers, *Curr Opin Nephrol Hypertens* 9:403-411.

White HD *et al.* (2000). Pravastatin therapy and the risk of stroke, *N Engl J Med* 343:317-326.

Wilterdink JL, Easton JD. (1999). Dipyridamole plus aspirin in cerebrovascular disease, *Arch Neurol* 56:1087-1092.

Zivin JA *et al.* (1985). Tissue plasminogen activator reduces neurological damage after cerebral embolism, *Science* 230:1289-1292.

Preserving Function in Acute Nervous System Injury

Edward D. Hall, PhD
Patrick G. Sullivan, PhD

Traumatic brain injury (TBI) and spinal cord injury (SCI) are two of the most catastrophic medical events that human beings can experience. In most cases of moderate or severe TBI, and in many mild cases, the neurological, economic, and social consequences are devastating to the patient, his or her family, and to society as a whole. In the United States, there are approximately 1.2 million TBIs that occur yearly, of which about 58,000 are severe (Glasgow Coma Score = 3 to 8) and 64,000 moderate (Glasgow Coma Score = 9 to 12). In the case of SCI, there are about 11,000 each year in the United States. Although TBI and SCI can victimize active individuals at any age, most occur in young adults in the second and third decades of life. Those who survive their initial injuries can now expect to live long lives as a result of improvements in medical and surgical care. Nevertheless, the need for intensive rehabilitation and prolonged disability exacts a significant toll on the individual, his or her family, and society. Effective ways of maintaining or recovering function could markedly improve the outlook for those with TBI or SCI by enabling higher levels of independence and productivity.

The potential for pharmacological intervention to preserve neurological function after TBI and SCI exists because most of the neurodegeneration that follows these injuries is not due to the primary mechanical insult (i.e., shearing of blood vessels and nerve cells) but rather to secondary injury events. For instance, most SCIs do not involve actual physical transection of the cord, but rather the spinal cord is damaged as a result of a contusive, compressive, or stretch injury. Some residual white matter, containing portions of the ascending sensory and descending motor tracts, remains intact, which allows for

the possibility of neurological recovery. During the first minutes and hours after injury, however, a secondary degenerative process is initiated by the primary mechanical injury that is proportional to the magnitude of the initial insult. Nevertheless, the anatomical continuity of the injured spinal cord in the majority of cases, together with our present knowledge of many of the factors involved in the secondary injury process, has led to the notion that pharmacological treatments that interrupt the secondary cascade, if applied early, could improve spinal cord tissue survival and thus preserve the necessary anatomical substrates for functional recovery to take place. Similarly, the outcome after TBI is mainly determined by the extent of the potentially treatable secondary pathophysiology and neurodegeneration.

PATHOGENESIS OF SECONDARY POSTTRAUMATIC CNS INJURY

Overview

Several reviews of post-TBI or SCI secondary injury have been published to which the reader is referred (Anderson and Hall, 1993; Faden, 1997; Faden and Salzman, 1992; Hall, 1995; Hall and Braughler, 1989; McIntosh et al., 1997; McIntosh and Raghupathi, 1997; Tator and Fehlings, 1991). Figure 3.1 displays the key players and the complex interrelationships involved in the secondary cascade of events occurring during the first minutes, hours, and days after traumatic central nervous system (CNS) injury. The most immediate event is mechanically induced depolarization and the consequent opening of voltage-dependent ion channels (i.e., Na⁺, K⁺, Ca⁺⁺). This leads to massive release of a variety of neurotransmitters including glutamate, which can cause the opening of glutamate receptor-operated ion channels (e.g., NMDA, AMPA). The most important consequence of these rapidly evolving ionic disturbances is the accumulation of intracellular Ca^{++} (i.e., Ca^{++} overload), which initiates several damaging effects. The first of these is mitochondrial dysfunction, which leads to a failure of aerobic energy metabolism, shift to glycolytic (i.e., anaerobic) metabolism, and the accumulation of lactate. The second is activation of mitochondrial and cytoplasmic nitric oxide synthase (NOS) and nitric oxide production. The third is activation of phospholipase A_2, which liberates arachidonic acid (AA), which is then converted by cyclooxygenases (COX 1, 2) to a number of deleterious prostanoids. These include the potent vasoconstrictor prostaglandin $F_{2\alpha}$ ($PGF_{2\alpha}$) and the vasoconstrictor/platelet aggregation promoter thromboxane A_2 (TXA_2). In addition, activated lipoxygenases lead to an increase in tissue leukotrienes (LTs), some of which are chemoattractants for polymorphonuclear leukocyte (PMN) and macrophage influx. The fourth consequence of intracellular Ca^{++} overload is the activation of the calcium-activated protease calpain, which degrades a variety of cellular substrates including cytoskeletal proteins.

One of the by-products of mitochondrial dysfunction, COX and lipoxygenase activity, and NOS activation is the formation of reactive oxygen species (ROS) including peroxynitrite (PON; ONOO⁻). Peroxynitrite is a product of the reaction of superoxide and nitric oxide radicals. Although PON can trigger cellular damage by a variety of mechanisms, cell membrane (plasma and organellar) lipid peroxidation (LP) has been conclusively demonstrated to be a key mechanism (Braughler and Hall, 1989; Hall, 1995; Hall and Braughler, 1989, 1993). However, iron is a powerful catalyst that accelerates the propagation of LP reactions. Glycolytically derived lactate promotes LP by stimulating the release of iron from storage sites (e.g., ferritin). In addition, primary and secondary petechial hemorrhages supply hemoglobin-bound iron. Lipid peroxidation occurs in neurons and blood vessels, directly impairing neuronal and axonal membrane function and integrity and causing microvascular damage and secondary ischemia that indirectly contribute to the secondary neuronal injury. Trauma-induced release of endogenous opiates, especially dynorphin A, exacerbates the secondary injury process by stimulating NMDA receptors and by activating opiate receptors, the latter contributing to vascular dysfunction and ionic and metabolic disturbances (Faden, 1997; Faden and Holaday, 1981; Faden and Salzman, 1992).

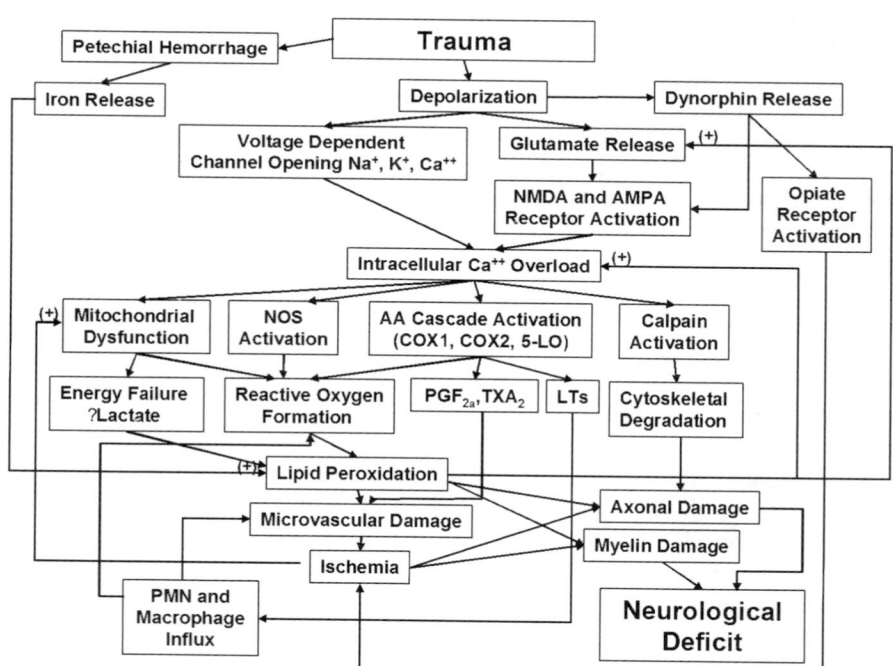

FIGURE 3.1   Pathophysiology of secondary injury in the injured central nervous system. NOS = nitric oxide synthase; COX = cyclooxygenase; 5-LO = 5-lipoxygenase; ONOO⁻ = peroxynitrite anion; $PGF_{2\alpha}$ = prostaglandin $F_{2\alpha}$; TXA_2 = thromboxane A_2; LTs = leukotrienes; PMN = polymorphonuclear leukocyte.

In the case of SCI, the secondary events occur initially in the central gray matter and then spread to the surrounding white matter. As implied previously, the key issue in predicting recovery of function is the degree of preservation of the ascending and descending white matter tracts. Many of the axons that do survive, however, do not conduct impulses as a result of posttraumatic demyelination. Therefore, the goal of neuroprotective pharmacotherapy is to preserve as many of the white matter axons and as much of their investing myelin as possible. In TBI, a key determinant in neurological recovery is also the loss of axons. Based on the often widespread loss of axons in the injured brain, this phenomenon is referred to as diffuse axonal injury. It should be realized, however, that a significant factor in influencing the extent of neural injury both in TBI and SCI is a decrease in brain or spinal cord microvascular perfusion (i.e., secondary ischemia). When this occurs, the result is an exacerbation of the injury process resulting from superimposed tissue ischemic hypoxia. Moreover, deficiencies in CNS hypoperfusion can be aggravated by systemic hypotension and/or hypoxia. Thus, it is important to note that secondary injury involves both neuronal and microvascular events.

Role of Reactive Oxygen-Induced Oxidative Damage

There is compelling experimental support for an important role of ROS in the pathophysiology of acute TBI and SCI. Kontos and colleagues showed an almost immediate postinjury increase in brain microvascular superoxide radical production, together with a compromise of autoregulatory function in fluid percussion TBI models (Kontos and Povlishock, 1986; Kontos and Wei, 1986). Scavengers of superoxide radical (O_2^-) reduce the posttraumatic superoxide levels and protect against the loss of autoregulatory competency. Using the salicylate trapping method, a rise in brain hydroxyl radical ($\cdot OH$) levels has also been documented in mouse diffuse and rat focal TBI models (Globus et al., 1995; Hall and Braughler, 1993; Hall et al., 1994; Smith et al., 1994). As with the work of Kontos, the cerebral microvasculature appears to be the initial source of posttraumatic radical production. More recent work shows that a major source of posttraumatic brain radical production is increased ROS leakage from injury brain mitochondria (Azbill et al., 1997; Matsushita and Xiong, 1997; Sullivan et al., 1999a; Sullivan et al., 1999b). Elegant microdialysis studies using salicylate trapping have also demonstrated an increase in hydroxyl radical levels in the injured rat spinal cord during the first minutes after contusion (Liu et al., 2001).

The most studied mechanism of oxidative damage in models of TBI concerns ROS-induced LP. Work using the rat focal contusion injury model has shown an increase in brain LP product (lipid hydroperoxide) levels that is measurable within 30 minutes after injury, following closely behind the increase in $\cdot OH$ (Smith et al., 1994). Moreover, on the heels of the increase in LP markers, there is an opening of the blood-brain barrier (BBB), suggesting that the initial site of ROS-induced LP is the microvascular endothelium. Others have confirmed the posttraumatic increase in LP products in rats after focal contusion injury and its association with brain edema mechanisms (Nishio et al., 1997). A role of LP in neuronal dysfunction has also been demonstrated in synaptosomes from the injured hemisphere where an increase in LP products occurs coincidently with an impairment of glutamate and glucose uptake (Sullivan et al., 1998). Biochemical indices of early oxygen radical reactions in the bluntly injured spinal cord (contusion or compression injuries) have also been demonstrated using a variety of analytical techniques (Hall and Braughler, 1993).

Several years ago, Beckman and co-workers introduced the theory that the principal ROS involved in producing tissue injury in a variety of neurological disorders is PON ($ONOO^-$), which is formed by the combination of NOS-generated $\cdot NO$ radical and superoxide radical (Beckman, 1991). Since that time, the biochemistry of PON, which is often referred to as a reactive nitrogen species, has been clarified. PON-mediated oxidative damage is actually caused by PON decomposition products that possess potent free radical characteristics. These are formed in one of two ways. The first involves the protonation of PON to form peroxynitrous acid (ONOOH), which can undergo hemolytic decomposition to form the highly reactive nitrogen dioxide radical ($\cdot NO_2$) and hydroxyl radical ($\cdot OH$). Probably more important physiologically, PON will react with carbon dioxide (CO_2) to form nitrosoperoxocarbonate ($ONOOCO_2$), which can decompose into $\cdot NO_2$ and carbonate radical ($\cdot CO_3$). Figure 3.2 displays the biochemistry of PON formation as well as other iron-dependent mechanisms of oxygen radical formation.

Each of the PON-derived radicals ($\cdot OH$, $\cdot NO_2$, and $\cdot CO_3$) can initiate LP cellular damage by abstraction of an electron from a hydrogen atom bound to an allylic carbon in polyunsaturated fatty acids or cause protein carbonylation by reaction with susceptible amino acids (e.g., lysine, cysteine, arginine). Moreover, the aldehydic LP products malondialdehyde (MDA) and 4-hydroxynonenal (4-HNE) can bind to cellular proteins also compromising their structural and functional integrity. 4-HNE is the more interesting of the two aldehydes in that it is itself neurotoxic (Kruman et al., 1997). Also, $\cdot NO_2$ can nitrate the 3 position of tyrosine

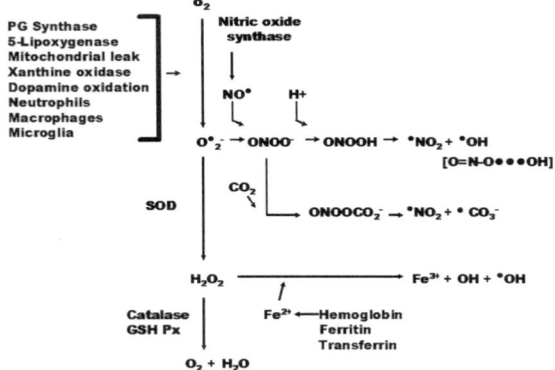

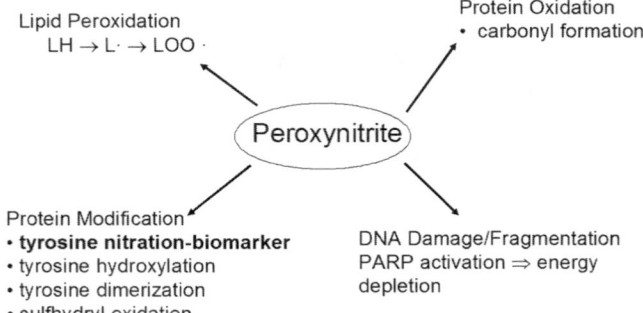

FIGURE 3.2 Biochemistry of reactive oxygen species generation and free radical formation in the injured CNS. PG synthase = prostaglandin synthase; O_2^- = superoxide radical; NO· = nitric oxide; ONOO⁻ = peroxynitrite anion; ONOOH = peroxynitrous acid; ·NO_2 = nitrogen dioxide radical; ·OH = hydroxyl radical; $ONOOCO_2$ = nitrosoperoxocarbonate; ·CO_3 = carbonate radical; SOD= superoxide dismutase; Fe⁺⁺ = ferrous iron; Fe⁺⁺⁺ = ferric iron; GSH PX = glutathione peroxidase.

FIGURE 3.3 Mechanisms of peroxynitrite-induced cellular damage. Tyrosine nitration (i.e., 3-nitrotyrosine) is a specific biomarker for peroxynitrite-mediated injury.

residues in proteins; 3-NT is a specific footprint of PON-induced cellular damage. Collectively, these oxidative mechanisms, which are summarized in Figure 3.3, underlie the demonstrated neurotoxic effects of PON reported in neuronal cell culture models (Kruman *et al.,*. 1997; Neely *et al.*, 1999). Increased 3-NT has been found in injured brain (Mesenge *et al.*, 1998) and spinal cord (Bao and Liu, 2003) indicative of a role of PON in posttraumatic neurodegeneration.

Role of Mitochondrial Dysfunction

Mitochondrial dysfunction plays a key role in the posttraumatic death cascade (Finkel, 2001; Hunot and Flavell, 2001). It is clear that this is directly related to Ca⁺⁺ ions that alter mitochondrial function and increase ROS production (Kristal and Dubinsky, 1997; Mattson *et al.*, 1995; Nicholls and Budd, 2000; Rego *et al.*, 2001; Stout *et al.*, 1998; Verweij *et al.*, 1997; Wang and Thayer, 1996; Ward *et al.*, 2000; White and Reynolds, 1997). After TBI, loss of mitochondrial homeostasis, increased mitochondrial ROS production, and disruption of synaptic homeostasis have been shown to occur (Azbill *et al.*, 1997; Matsushita and Xiong, 1997; Sullivan *et al.*, 1999a, 1999b), implicating a pivotal role for mitochondrial dysfunction in the neuropathological sequelae of TBI. Of importance, several reports have solidified this theory by demonstrating that therapeutic intervention with cyclosporin A after experimental TBI significantly reduces mitochondrial dysfunction (Sullivan *et al.*, 1999b) and cortical damage (Scheff and Sullivan, 1999; Sullivan *et al.*, 2000b, 2000c), as well as cytoskeletal changes and axonal dysfunction (Okonkwo *et al.*, 1999; Okonkwo and Povlishock, 1999). Furthermore, maintaining mitochondrial bioenergetics by dietary

supplementation with creatine has also proved effective in ameliorating neuronal cell death by reducing mitochondrial ROS production and maintaining adenosine triphosphate (ATP) levels after TBI (Sullivan *et al.*, 2000a).

Evidence has begun to accumulate that the particular ROS being formed by mitochondria is PON. Nitric oxide has been shown to be present in mitochondria (Lopez-Figueroa *et al.*, 2000; Zanella *et al.*, 2002), and a mitochondrial NOS isoform (mtNOS) has been isolated. Although probably playing a physiological role in mitochondria, dysregulation of mitochondrial ·NO generation, and the aberrant production of its toxic metabolite PON, appear to play a role in many, if not all, of the major acute and chronic neurodegenerative conditions (Heales *et al.*, 1999). Exposure of mitochondria to high Ca⁺⁺, leads to PON generation, which in turn triggers mitochondrial Ca⁺⁺ release (i.e., limits their Ca⁺⁺ uptake or buffering capacity) (Bringold *et al.*, 2000). Both PON forms, ONOO⁻ and $ONOOCO_2$, have been shown to deplete mitochondrial antioxidant stores and to cause protein nitration (Valdez *et al.*, 2000). The relatively long half-life of PON in comparison to most other short-lived ROS also allows for mtNOS derived PON to diffuse from one cell to another. Accordingly, co-culture studies have shown that astrocyte-derived ·NO (probably due to PON formation) can bring about damage to neuronal mitochondrial respiratory complexes II, III, and IV (Stewart *et al.*, 2000, 2002).

Figure 3.4 illustrates the source of PON formation in dysfunctional mitochondria in either the injured brain or spinal cord. As shown, O_2^- radical production is a by-product of the mitochondrial electron transport chain during ATP generation. Electrons escape from the chain and reduce O_2 to O_2^-. Normally, cells convert

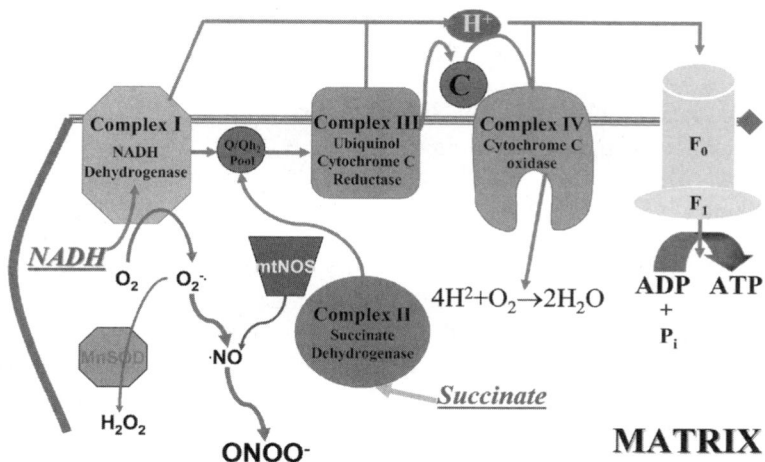

FIGURE 3.4 Schematic diagram showing the mitochondrial source of superoxide (O_2^-) from Complex I and ·NO from mitochondrial nitric oxide synthase (mtNOS), which together give rise to peroxynitrite anion ($O_2^- + \cdot NO \rightarrow ONOO^-$). This reaction has a very fast rate constant of $> 10^7$ moles/sec, which exceeds the rate constant for dismutation of O_2^- by MnSOD.

O_2^- to H_2O_2 utilizing both manganese superoxide dismutase (MnSOD), which is also localized to the mitochondria, and copper/zinc superoxide dismutase (Ca/ZnSOD). However, if pathophysiological insults such as mechanical trauma trigger an increase in intracellular Ca^{++}, causing an increase in mtNOS activity and ·NO liberation, PON formation is a certainty since the rate constant for reaction of ·NO with O_2^- greatly exceeds the rate constant for dismutation of O_2^- by MnSOD (Beckman, 1991). PON can then damage mitochondria by tyrosine nitration and by causing LP and the production of 4-HNE that conjugates to mitochondrial membrane proteins, impairing their function (Keller *et al.*, 1997a, 1997b; Mark *et al.*, 1997; Sullivan *et al.*, 1998). Such oxidative injury results in significant alterations in neuronal function. In particular, ROS induction of LP and protein oxidation products may be particularly important in SCI and TBI (Braughler *et al.*, 1985; Braughler and Hall, 1989, 1992; Sullivan *et al.*, 1998).

Role of Calcium Homeostatic Dysregulation and Calpain-Mediated Cytoskeletal Degradation as a Final Common Neurodegenerative Pathway

ROS-initiated oxidative damage to membrane proteins and lipids aggravates glutamate release and compromises reuptake mechanisms and worsens the intracellular calcium overload by inhibition of the key mechanisms for controlling intracellular Ca^{++} levels (Hall, 1998). Disruption of intracellular Ca^{++} homeostasis is a critical issue in the secondary neurodegenerative pathophysiology of TBI. After an injurious event, there is a massive posttraumatic Ca^{++} influx initially caused by depolarization-induced glutamate release and the opening of glutamate receptor-operated and voltage-dependent Ca^{++} channels (Hall, 1995; McIntosh and Raghupathi, 1997). Evidence of increased intracellular Ca^{++} concentrations in TBI models has been demonstrated by several laboratories using each of the mainstream rodent TBI models (Fineman *et al.*, 1993; McIntosh and Raghupathi, 1997; Siesjo and Bengtsson, 1989; Verity, 1992). Excessive intracellular Ca^{++} accumulation then leads to neuronal degeneration by activation of various enzymes including proteases, kinases, phosphatases, and phospholipases (Fineman *et al.*, 1993; McIntosh and Raghupathi, 1997; Siesjo and Bengtsson, 1989; Verity, 1992); induction of additional ROS release; and detrimental changes in gene expression (Bading *et al.*, 1993; Rink *et al.*, 1995; Raghupathi, 1997).

The principal mechanism of posttraumatic Ca^{++}-mediated neuronal injury involves the activation of the neutral proteases known as calpains (Bartus, 1997; Kampfl *et al.*, 1997; McCracken *et al.*, 1999; McIntosh and Raghupathia, 1997). When activated, calpains are known to degrade cytoskeletal proteins, receptor proteins, signal transduction enzymes, and transcription factors (Kampfl *et al.*, 1997; Yuen, 1996). In the case of cytoskeleton proteins, α-spectrin (a 280 kDa protein that provides structural support to membranes) can be cleaved by calpain at tyrosine 1176 to yield a 150 kDa fragment (SBDP150) or at glycine 1230 to yield a 145 kDa fragment (SBDP145) (Harris and Morrow, 1988). Consistent with the concept that calpain-mediated cytoskeletal degeneration is an important pathway of posttraumatic neurodegeneration, prototypical inhibitors of calpain reduce cytoskeletal degradation and/or improve neurological recovery after experimental TBI including leupeptin, antipain, calpain inhibitors I and II, calpeptin, E64, AK295, MDL28170, SJA6017 and PD150606 (Fukiage *et al.*, 1997; Kupina *et al.*, 2001; Li *et al.*, 1998; Markgraf *et al.*, 1998; Posmantur *et al.*, 1997; Saatman *et al.*, 1996, 2000; Yuen, 1996, 1998). This is consistent with the notion that cytoskeletal damage is the final common mechanism of posttraumatic neurodegeneration. Therefore, its inhibition is a meaningful biochemical measure of the efficacy of upstream neuroprotective strategies. However, it should also be noted that TBI-induced proteolysis of the cytoskeleton can be mediated by caspase-3 activation (Pike *et al.*, 1998; Rink *et al.*, 1995; Yakovlev *et al.*, 1997).

Multiple laboratories and studies have demonstrated an important role of calpain activation in mediating posttraumatic axonal damage in acute SCI models (Banik and Shields, 2000; Ray *et al.*, 2001a, 2001b, 2002, 2003; Shields *et al.*, 2000; Wingrave *et al.*, 2003; Zhang *et al.*, 2003). Several of these studies have reported neuroprotective efficacy of prototype calpain inhibitors. However, the translation of calpain inhibition into neuroprotective clinical trials has been precluded by the lack of small molecule inhibitors with sufficient CNS penetration and appropriate pharmaceutical and pharmacokinetic properties. Figure 3.5 shows a schematic integration of posttraumatic calcium accumulation triggered mitochondrial dysfunction, reactive oxygen generation, calcium overload, calpain activation, cytoskeletal degradation, and neurodegeneration.

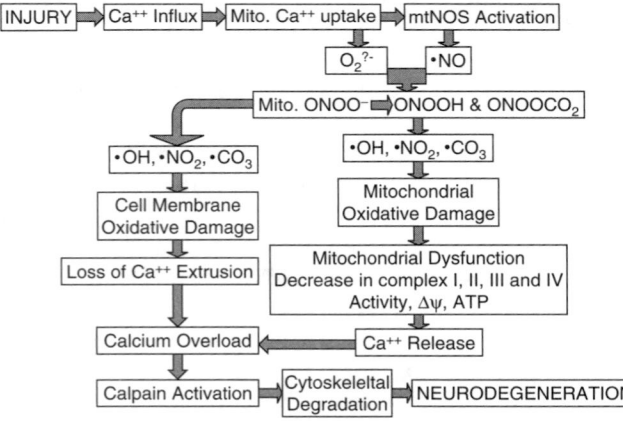

FIGURE 3.5 Hypothesis of Ca^{++}-induced mitochondrial peroxynitrite production, mitochondrial and cell membrane lipid and protein oxidative damage, Ca^{++} overload, calpain activation, cytoskeletal degradation and ultimately neurodegeneration.

DEVELOPMENT OF NEUROPROTECTIVE PHARMACOTHERAPY FOR SPINAL CORD INJURY

Early Use of Glucocorticoid Steroids

The glucocorticoid steroids, mainly dexamethasone and methylprednisolone, were extensively used in the clinical treatment of spinal cord trauma beginning in the mid 1960s and throughout the 1970s. The mechanistic rationale for their use initially centered on the expectation that they would reduce posttraumatic spinal cord edema. This notion was based on the rather remarkable reduction of peritumoral brain edema that glucocorticoids can induce in brain tumor patients (Reulen and Schurmann, 1972). Furthermore, steroid pretreatment became a standard of care before neurosurgical procedures to prevent intraoperative and postoperative brain swelling. A limited amount of experimental evidence supported the possibility that glucocorticoid dosing in animal SCI models might be neuroprotective (Reulen and Schurmann, 1972).

NASCIS I Clinical Trial

In the mid-1970s, a randomized, multicenter clinical trial was organized to try to determine if steroid dosing was beneficial in improving neurological recovery in humans after SCI. This trial was named the National Acute Spinal Cord Injury Study (NASCIS I). It compared the efficacy of "low-dose" methylprednisolone (100 mg intravenous bolus/day for 10 days) and "high-dose" methylprednisolone (1000 mg intravenous bolus/day for 10 days) in affecting outcome after SCI (Bracken *et al.*, 1984, 1985). The trial, which began in 1979, did not involve a placebo group because of the prevailing belief that glucocorticoid dosing probably was beneficial and could not be ethically withheld; however, results failed to show any difference between the low- and high-dose groups at either 6 months (Bracken *et al.*, 1984) or 1 year (Bracken *et al.*, 1985). Based on this result, the investigators concluded that steroid dosing was of little benefit. Also, there was a suggestion that the 10-day high-dose regimen increased the risk of infections, a predictable side effect of sustained glucocorticoid dosing. Based on the negative results of NASCIS I, as well as waning neurosurgical enthusiasm for steroid treatment of CNS injury in general, the majority of neurosurgeons concluded after NASCIS I that the conventional use of steroids in the acute management of spinal trauma was not beneficial while at the same time being fraught with the potential for serious side effects.

Inhibition of Lipid Peroxidation by High-Dose Methylprednisolone

Increasing knowledge of the posttraumatic LP mechanism in the 1970s and early 1980s prompted the search for a neuroprotective pharmacologic strategy aimed at antagonizing oxygen radical-induced LP in a safe and effective manner. Attention was focused on the hypothetical possibility that glucocorticoid steroids might be effective inhibitors of posttraumatic LP based on their high lipid solubility and known ability to intercalate into artificial membranes between the hydrophobic polyunsaturated fatty acids of the membrane phospholipids and to thereby limit the propagation of LP chain reactions throughout the phospholipid bilayer (Demopoulos *et al.*, 1980; Hall, 1992; Hall and Braughler, 1981, 1982).

The lead author and his colleagues became interested in the LP hypothesis of secondary SCI during parallel investigations of the effects of high-dose methylprednisolone (MP) (15 to 90 mg/kg IV) on spinal cord electrophysiology, as those might serve to improve impulse conduction and recovery of function in the injured spinal cord (Hall, 1982). Consequently, it was decided to test the possibility that a similar high dose of MP, which enhanced spinal neuronal excitability and impulse transmission, might also be required to inhibit posttraumatic spinal cord LP. In an initial set of experiments in cats, it was observed that the administration of an IV bolus of MP could indeed inhibit posttraumatic LP in spinal cord tissue (Hall and Braughler, 1981), but the doses required for this effect were much higher (30 mg/kg) than previously hypothesized or than those empirically used in the clinical treatment of acute CNS injury or tested in the NASCIS trial. Further experimental studies, also conducted in cat SCI models, showed that the 30-mg/kg dose of MP not only prevented LP, but in parallel inhibited posttraumatic spinal cord ischemia (Hall *et al.*, 1984; Young and Flamm, 1982), supported aerobic energy metabolism (i.e., reduced lactate and improved ATP and energy charge) (Anderson *et al.*, 1982; Braughler and Hall, 1983a, 1984), improved recovery of extracellular calcium (i.e., reduced intracellular overload) (Young and Flamm, 1982), and attenuated calpain-mediated neurofilament loss (Braughler and Hall, 1984). However, the central effect in this protective scenario is the inhibition of posttraumatic LP (Figure 3.6). With many of these therapeutic parameters (LP, secondary ischemia, aerobic energy metabolism), the dose-response for MP follows a sharp U-shaped pattern (Figure 3.7). The neuroprotective and vasoprotective effect is partial with a dose of 15 mg/kg, optimal at 30 mg/kg, and diminishes at higher doses (60 mg/kg) (Hall, 1992).

The antioxidant neuroprotective action of MP is closely linked to the drug's tissue pharmacokinetics (Braughler and Hall, 1982, 1983a, 1983b; Hall, 1992). For instance, when MP tissue levels are at their peak after administration of a 30 mg/kg IV dose, lactate levels in the injured cord are suppressed. When tissue MP levels decline, spinal tissue lactate rises (Figure 3.8). However,

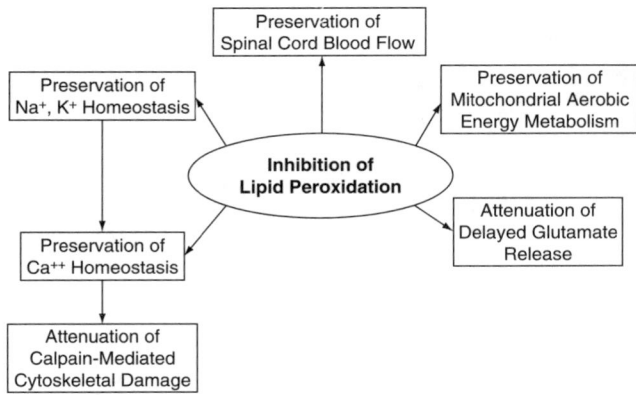

FIGURE 3.6 Hypothesized central role of inhibition of lipid peroxidation in the neuroprotective effects of high-dose methylprednisolone in acute SCI.

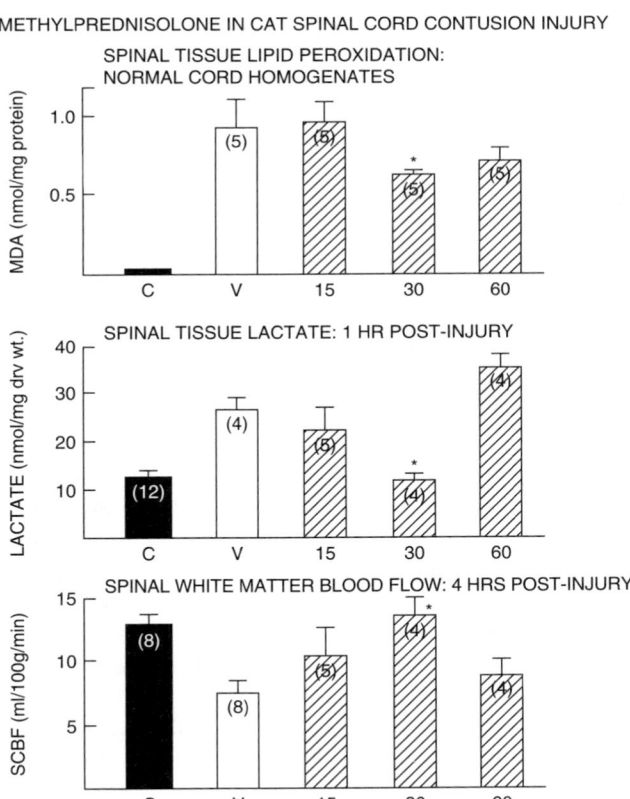

FIGURE 3.7 Dose-response correlation in cats for the effects of MP to inhibit *ex vivo* LP (Hall and Braughler, 1981) to antagonize posttraumatic lactic acid accumulation (Braughler and Hall, 1983a) and to prevent posttraumatic white matter ischemia (Hall *et al.*, 1984). All values = mean ± standard error. The number of animals is parenthetically given in each bar. $^*P < .05$ vs vehicle-treated group.

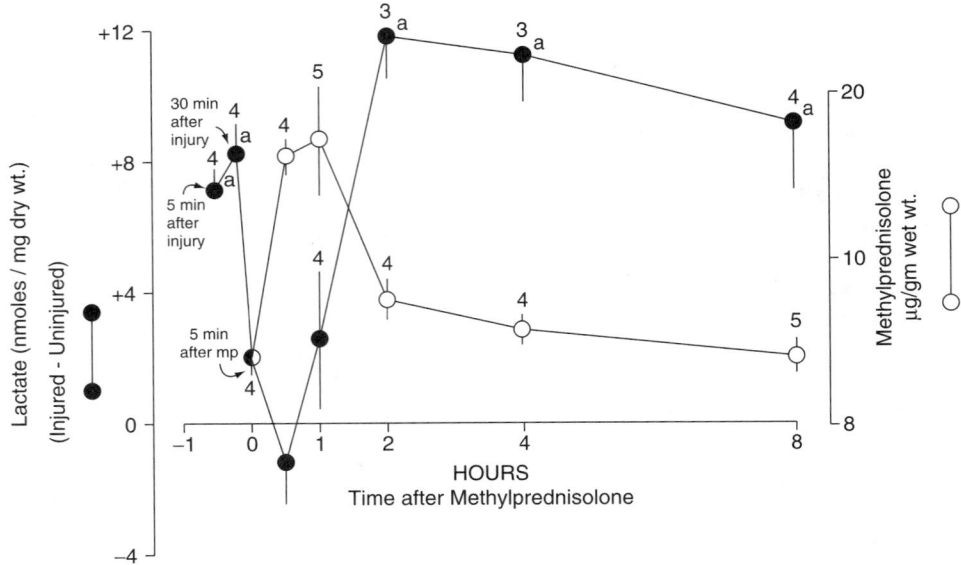

FIGURE 3.8 Comparison of the uptake and elimination of MP from the injured cat spinal cord with the suppression of posttraumatic lactate accumulation. During the steep phase of MP elimination from spinal cord tissue between 1 and 2 hours after injection, the lactate levels show a secondary rise.

the administration of a second dose (15 mg/kg IV), at the point at which the levels after the first dose have declined by 50%, maintains the suppression of lactate seen at the peak of the first dose and more effectively maintains ATP generation and energy charge and protects spinal cord neurofilaments from degradation (Figure 3.9) (Braughler and Hall, 1983a, 1984). This finding prompted the hypothesis that prolonged MP therapy might better suppress the secondary injury process and lead to better outcomes compared to the effects of a single large IV dose. Indeed, subsequent experiments in a cat spinal injury model demonstrated that animals treated with MP, using a 48-hour antioxidant dosing regimen, had improved recovery of motor function over 4 weeks (Anderson et al., 1985; Braughler et al., 1987). Table 3.1 summarizes the neuroprotective pharmacology of high-dose MP derived from acute SCI models.

The preclinical studies defining the antioxidant neuroprotective pharmacology showing that high-dose MP could exert antioxidant and related neuroprotective effects were conducted in cat models of blunt (nontransecting) SCI, which were the standard in the experimental SCI field before the 1990s. Since that time, rat contusion and compression models have become the standard, and several investigators have tested the ability of high-dose MP (usually 30 mg/kg IV as a starting dose) to lessen posttraumatic pathophysiology and neurodegeneration and/or to improve neurological recovery. Several of these studies have replicated in some fashion the neuroprotective properties of MP in the injured rat spinal cord. Specifically, high-dose MP has been reported in rat SCI

models to attenuate posttraumatic LP (Taoka et al., 2001), decrease lactate accumulation (Farooque et al., 1996), prevent hypoperfusion (Holtz et al., 1990), attenuate vascular permeability (Xu et al., 1992), decrease inflammatory markers (Xu et al., 2001), and improve neurological recovery (Behrmann et al., 1994; Holtz et al., 1990), reminiscent of similar effects shown earlier in cat models (Hall, 1992; Hall and Braughler, 1982). In contrast, failures of high-dose MP to improve neurological recovery in rat SCI models have also appeared in the literature (Koyanagi and Tator, 1997; Rabchevsky et al., 2002). Of considerable concern in the extrapolation of the cat MP dosing parameters to the rat, however, is the lack of any definition thus far of the relative pharmacokinetics of MP in rats. This is in striking contrast to the documentation of the uptake and elimination of MP from the cat spinal cord and correlation of plasma and spinal tissue levels with the neuroprotective actions (Braughler and Hall, 1982, 1983a, 1983b; Hall 1992). The likelihood that the precise dose-response relationship and requirements for repeated dosing defined in cats is also optimal for the rat is exceedingly small. Thus the interpretation of rat SCI studies with MP, whether positive or negative, is difficult without the necessary pharmacokinetic correlation.

Comparison of Different Glucocorticoids

The early empirical treatment of peritumoral edema and acute SCI with glucocorticoid steroids was heavily weighted toward the use of dexamethasone based on the fact that it was, and is, the most potent synthetic

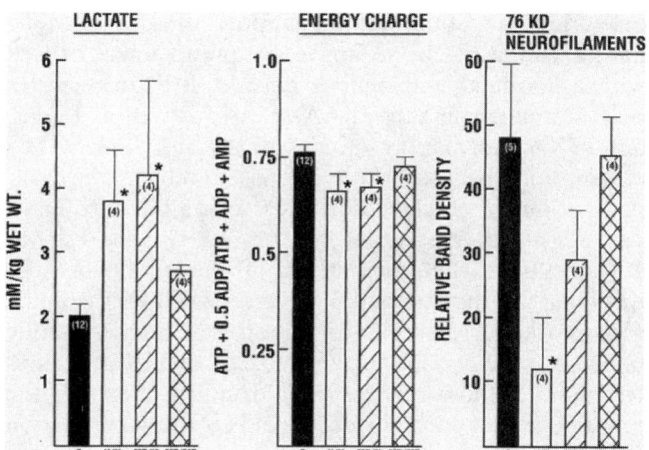

FIGURE 3.9 Comparative effects in spinal-cord-injured cats of a single 30 mg/kg IV dose of MP given at 30 minutes postinjury vs a double-dose regimen of 30 mg/kg IV at 30 minutes postinjury plus a second dose of 15 mg/kg IV at 2.5-hours postinjury on the 4-hour postinjury levels of lactate, energy charge (ATP+ADP+AMP/ADP+AMP), and neurofilament preservation (Braughler and Hall, 1984). All values = mean ± standard error. Numbers of animals are given in parentheses in each bar.

TABLE 3.1 Pharmacological Characteristics of High-Dose Methylprednisolone (MP) Therapy for Acute SCI

Inhibition of posttraumatic lipid peroxidation appears to be the principal neuroprotective mechanism and this is unrelated to glucocorticoid receptor-mediated actions.

Microvascular and neuroprotective effects are both involved.

Large intravenous doses are required (30 mg/kg).

Antioxidant protective effects of MP follow a biphasic (U-shaped) dose-response curve; doubling the dose from 30 to 60 mg/kg results in a loss of the protective efficacy.

Early treatment is required because lipid peroxidation develops rapidly and is irreversible.

Time course of antioxidant protection parallels the spinal cord tissue pharmacokinetics; there is consequently a need for constant IV infusion to maintain effective tissue concentrations.

Optimal treatment duration is uncertain but needs to be continued as long as conditions within the injured spinal cord favor lipid peroxidative reactions (i.e., at least 24-48 hours).

Glucocorticoid receptor-mediated anti-inflammatory effects play only a minor role compared to lipid antioxidant effects.

glucocorticoid steroid available for parenteral use. Dexamethasone is about five times more potent than MP in regard to glucocorticoid receptor affinity and anti-inflammatory potency (Schimmer and Parker, 2001). However, it has been found that the antioxidant efficacy of MP is unrelated to its glucocorticoid steroid receptor activity (Hall et al., 1987). Indeed, a careful concentration-response study has compared the ability of different glucocorticoid steroids to inhibit oxygen radical-induced LP damage in rat brain synaptosomal preparations and confirmed that LP-inhibiting potencies and anti-inflammatory potencies do not correlate. Although dexamethasone is five times more potent than MP as a glucocorticoid, it is only slightly more potent than MP as an inhibitor of LP (Braughler, 1985). Furthermore, the maximal antioxidant activity of MP appears to be superior to that for dexamethasone. The prototype glucocorticoid hydrocortisone is completely lacking in ability to inhibit oxygen radical damage in CNS tissue. Thus, the choice of a steroid for its potential antioxidant neuroprotective activity should not be predicated on glucocorticoid receptor-mediated anti-inflammatory actions. In addition, selection of the most potent glucocorticoid would logically carry the greatest potential for concomitant steroid-related side effects.

Neuroprotective Effects of the Opiate Receptor Antagonist Naloxone

At the same time as the discovery of the neuroprotective efficacy of high-dose MP in animal models of SCI, others were demonstrating the ability of the opiate receptor antagonist naloxone to improve recovery after experimental SCI. The rationale for the study of naloxone in acute SCI models was based on its positive effects in experimental shock models (Holaday and Faden, 1981a, 1981b). Studies in cat SCI models, carried out in two separate laboratories, documented the ability of naloxone to maintain spinal cord blood flow (Flamm et al., 1982) and to improve neurological recovery after contusion SCI (Faden and Holaday, 1981; Faden et al., 1981, 1982; Flamm et al., 1982). Furthermore, a phase I study in acute SCI patients showed that the drug was safe in that population (Flamm et al., 1985).

NASCIS II Clinical Trial

The preceding experimental studies with high-dose MP and naloxone inspired the second National Acute Spinal Cord Injury Study (NASCIS II) (Bracken et al., 1990) even though the earlier NASCIS trial, which came to be known as NASCIS I, had failed to show any efficacy of lower MP doses, even when administered over 10 days (Bracken et al., 1984, 1985). The NASCIS II trial compared 24 hours of dosing with MP or naloxone with placebo for the treatment of acute SCI. A priori trial hypotheses included the prediction that SCI patients treated within the first 8 hours after injury would respond better to pharmacotherapy than patients treated after 8 hours. Results showed the effectiveness of 24 hours of intensive MP dosing (30 mg/kg IV bolus plus a 23-hour infusion at 5.4 mg/kg per hour) when treatment was initiated within 8 hours. Significant benefit was observed in individuals with both neurologically complete (i.e., "plegic") and incomplete (i.e., "paretic") injuries. Moreover, the functional benefits were sustained at 6-week, 6-month, and 1-year follow-up evaluations (Bracken, 1993; Bracken and Holford, 1993; Bracken et al., 1990, 1992). The high-dose regimen actually improved function below the level of the injury and lowered the level of the functional injury (Bracken and Holford, 1993). Although predictable side effects of steroid therapy were noted, including GI bleeding, wound infections, and delayed healing, these were not significantly more frequent than those recorded in placebo-treated patients (Bracken et al., 1990). Another finding was that delay in the initiation of MP treatment until after 8 hours is actually associated with decreased neurological recovery (Bracken and Holford, 1993). Thus, treatment within the 8 hour window is beneficial, whereas dosing after 8 hours may be detrimental. Possible explanations for this latter effect are discussed later.

The original NASCIS II publications (Bracken et al., 1990, 1992) stated that in contrast to the beneficial actions of high dose MP, the opiate receptor naloxone did not significantly improve the return of sensory or motor function. In a subsequent analysis, however, naloxone improved neurological function below the lesion in patients with incomplete injuries (Bracken and Holford, 1993). Thus, in the case of both high-dose MP and naloxone, at least a partial validation of the positive effects of these two therapeutic approaches and compounds in animal models of SCI was achieved in the placebo-controlled NASCIS II trial.

Anti-Inflammatory Effects Play a Minor Role in Methylprednisolone's Neuroprotective Efficacy

The rationale for the high-dose MP arm of the NASCIS II trial was derived from the animal studies showing that the steroid can inhibit posttraumatic LP and associated pathophysiological events (Hall, 1992). Positive effects of high-dose MP obtained in NASCIS II were at least tentatively viewed as a validation of the LP hypothesis (Bracken and Holford, 1993; Bracken et al., 1990, 1992). However, after the publication of the NASCIS II results, it was suggested that a complete assignment of the mechanism of the MP neuroprotective effect to the inhibition of LP was premature. After all, the glucocorticoid steroid MP possesses various glucocorticoid receptor-mediated anti-inflammatory actions that could reasonably be expected to play a neuroprotective role in addition to the nonglucocorticoid LP inhibition. In view of the known importance of certain prostanoids ($PGF_{2\alpha}$ and TXA_2) in posttraumatic pathophysiology (Figure 3.1), it is conceivable that MP, by virtue of its phospholipase A_2 inhibitory action, might protect the injured spinal cord by inhibiting AA release, and consequently attenuating the formation of these deleterious prostanoids. In support of this possible mechanism, it has been shown that pretreatment of cats with a 30-mg/kg intravenous dose of MP does reduce posttraumatic spinal cord AA release, as well as $PGF_{2\alpha}$ and TXA_2 formation (Anderson et al., 1986). The assumption that this effect is due entirely to the well-known ability of glucocorticoids to inhibit phospholipase A_2, however, is not tenable for three reasons. First of all, the glucocorticoid inhibition of phospholipase A_2 is mediated via the interaction of the steroid with specific glucocorticoid receptors, an action that most certainly does not require a 30-mg/kg IV dose to be fully manifested; much lower doses should suffice. Second, the reported spinal cord glucocorticoid receptor population (Clark et al., 1981) involved in this activity would be saturated at much lower glucocorticoid doses. Third, the reduction of posttraumatic spinal cord AA release (and reduced prostanoid formation downstream) may simply be secondary to an attenuation of LP since peroxidative reactions can liberate arachidonate independent of phospholipase A_2 activity. In other words, the MP effect on prostanoid buildup in the injured spinal cord may be an indirect result of lessened membrane LP. Fairly convincing support for this view is that pretreatment of cats with the lipid antioxidant vitamin E also reduced posttraumatic AA liberation and accumulation of $PGF_{2\alpha}$ and TXA_2 (Anderson et al., 1986). Furthermore, subsequent studies showed that the administration of antioxidant neuroprotective doses of MP (i.e., 30 mg/kg IV) at 30 minutes after moderately severe compression SCI in cats did not attenuate the posttraumatic elevation in $PGF_{2\alpha}$ or TXA_2 (Hall et al., 1995). Therefore, it is unlikely that the protective effects of MP on the injured cord are more a reflection of the phospholipase A_2-inhibiting activity than its antioxidant property. Similarly, administration of neuroprotective doses of MP in a rat model of contusion SCI also failed to significantly reduce the infiltration of PMN into the injured cord (Xu et al., 1992). On the other hand, another group has reported that a single high dose of MP can significantly suppress PMN and macrophage influx into the rat cord after a partial transaction injury, although this effect was not associated with reduction of secondary neurodegeneration (Bartholdi and Schwab, 1995).

Despite this confusing body of data concerning the possible involvement of anti-inflammatory mechanisms for MP neuroprotection in acute SCI, inflammatory processes play an increasingly appreciated role in post-SCI secondary injury (Popovich and Jones, 2003; Popovich et al., 1997a). Moreover, glucocorticoid steroids, including MP, possess a wide array of anti-inflammatory mechanisms and are probably the most potent and reliable anti-inflammatory agents in the current pharmacological armamentarium (Schimmer and

Parker, 2001). These facts support the possibility that particular anti-inflammatory mechanisms may play some role in the MP neuroprotective activity. In that regard, high doses of MP have been more recently reported to attenuate the posttraumatic expression of the pro-inflammatory mediators activator protein-1 (AP-1) and nuclear factor κB (NFκB) and matrix metalloproteinase 1 and 9 in the injured spinal cord, although this study used a relatively mild level of injury (Xu *et al.*, 2001). Recently, a role of delayed influx of T cells has been suggested to be involved in secondary SCI (Popovich *et al.*, 1996, 1997b). Considering that glucocorticoids also suppress T-cell-mediated hypersensitivity in a variety of conditions (Schimmer and Parker, 2001), there is a strong possibility that high-dose MP may have an impact on this immune response in the injured spinal cord, although this has not been investigated in SCI models.

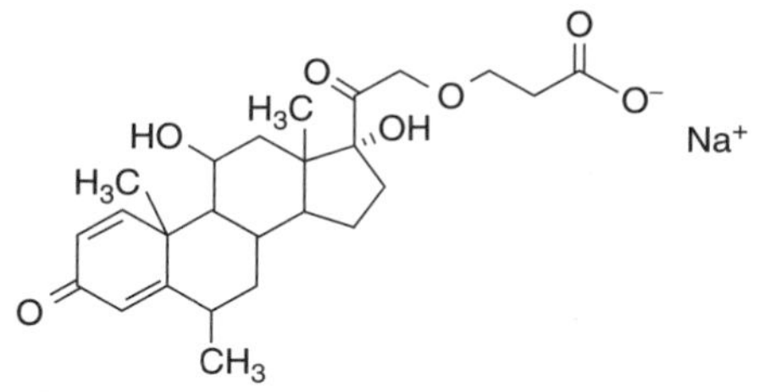

Discovery of the Nonglucocorticoid Steroid Tirilazad

Methylprednisolone is a potent glucocorticoid that possesses a number of glucocorticoid receptor-mediated anti-inflammatory actions. Despite the role of anti-inflammatory effects of MP in the injured spinal cord, the principal neuroprotective mechanism appears to be the inhibition of posttraumatic LP that is not mediated via glucocorticoid receptor-mediated activity (Braughler *et al.*, 1988; Hall, 1997; Hall *et al.*, 1994). This prompted speculation that modifying the steroid molecule to enhance the anti-LP effect, while eliminating the steroid's glucocorticoid effects, would result in more targeted antioxidant therapy devoid of the typical side effects of steroid therapy. This rationale led to the development of more potent LP inhibitors, the 21-aminosteroids or "lazaroids," which lack the glucocorticoid receptor-mediated side effects that limit the clinical utility of high-dose MP. One of these, tirilazad, was selected for development. Figure 3.10 compares the structures of the glucocorticoid, MP, and the nonglucocorticoid 21-aminosteroid, tirilazad. Tirilazad was extensively evaluated in animal models of spinal cord injury (SCI), traumatic brain injury (TBI), ischemic stroke, and subarachnoid hemorrhage (SAH) and was shown to exert a variety of neuroprotective and vasoprotective effects (Hall, 1997; Hall *et al.*, 1994). Based on these preclinical studies, clinical trials of tirilazad were conducted in SCI, TBI, SAH; and ischemic stroke.

NASCIS III Clinical Trial

The demonstrated efficacy of a 24-hour dosing regimen of MP in human SCI in NASCIS II (Bracken *et al.*, 1990) and the discovery of tirilazad (Braughler *et al.*, 1988; Hall, 1997; Hall *et al.*, 1994) led to the organization

Methylprednisolone sodium succinate

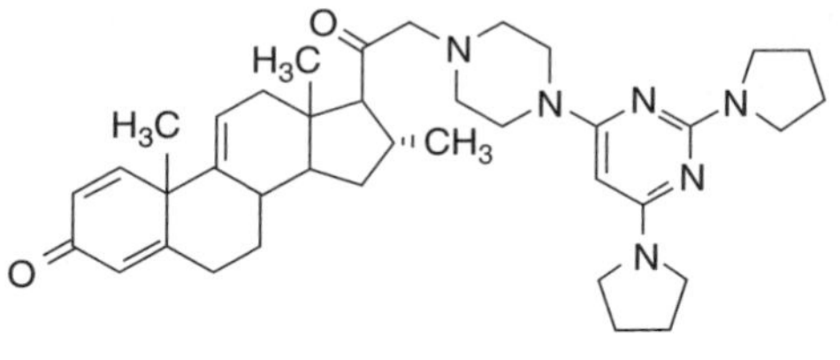

Tirilazad Mesylate U-74006F

FIGURE 3.10 Chemical structures of the glucocorticoid steroid methylprednisolone shown as the sodium salt of the 21-hemisuccinate ester and the nonglucocorticoid 21-aminosteroid tirilazad mesylate.

and conduct of NASCIS III (Bracken *et al.* 1997, 1998). In the NASCIS III trial, three groups of patients were evaluated. The first (active control) group was treated with the 24-hour MP dosing regimen that had previously been shown to be effective in NASCIS II. The second group was also treated with MP, except that the duration of MP infusion was prolonged to 48 hours. The purpose was to determine whether extension of the MP infusion from 24 to 48 hours resulted in greater improvement in neurological recovery in acute SCI patients. The third group of patients was treated with a single 30 mg/kg IV bolus of MP followed by the 48-hour administration of tirilazad. No placebo group was included because it was deemed ethically inappropriate to withhold at least the initial large bolus of MP. Another objective of the study was to ascertain whether treatment initiation within 3 hours after injury was more effective than when therapy was delayed until 3 to 8 hours after SCI.

On completion of the NASCIS III trial, it was found that all three treatment arms produced comparable degrees of recovery when treatment was begun within the shorter 3-hour window. When the 24-hour dosing of MP was begun more than 3 hours after SCI, recovery was poorer in comparison to the cohort treated within 3 hours after SCI. However, in the 3-to-8-hr post-SCI cohort, when MP dosing was extended to 48 hours, significantly better recovery was observed than with the 24-hour dosing. In the comparable tirilazad cohort (3 to 8 hours post-SCI), recovery was slightly, but not significantly, better than in the 24-hour MP group, and poorer than in the 48-hour MP group.

These results showed that: (1) initiation of high-dose MP treatment within the first 3 hours is optimal; (2) the nonglucocorticoid tirilazad is as effective as 24-hour MP therapy; and (3) if treatment is initiated more than 3 hours post-SCI, extension of the MP dosing regimen is indicated, from 24 hours to 48 hours. Compared with the 24-hour dosing regimen, however, significantly more glucocorticoid-related immunosuppression-based side effects were seen with more prolonged dosing (i.e., the incidence of severe sepsis and pneumonia significantly increased). In contrast, tirilazad showed no evidence of steroid-related side effects, suggesting that this nonglucocorticoid 21-aminosteroid would be safer for extension of dosing beyond the 48-hour limit used in NASCIS III (Bracken et al., 1997, 1998). Tirilazad's ability to improve neurological recovery in NASCIS III at least as well as MP (i.e., treatment within 3 hours after injury) while producing fewer side effects (Bracken et al., 1997, 1998) strongly suggests that it is worthy of additional trials in acute SCI that could ultimately show greater efficacy and safety compared with high-dose MP.

Limitations to the Use of High-Dose Methylprednisolone

The use of large doses of glucocorticoid steroids in SCI patients has appropriately been referred to as a "two-edged sword" (Galandiuk et al., 1993). This is derived from the fact that the neuroprotective properties of MP can be offset by the potential of the necessary high doses of MP to elicit glucocorticoid, receptor-mediated side effects that could compromise the neurological outcome and even the survival of acute SCI patients. This is most apparent when treatment is extended beyond the apparently safe limit of 24 hours demonstrated in NASCIS II. Complications of high-dose MP, including increased incidence of pneumonias, pressure sores, GI bleeding, and deep vein thrombosis, have also been documented when the NASCIS II dosing is used as a prophylactic neuroprotectant for spinal surgery (Molano Del Rey et al., 2002). However, there are other glucocorticoid-related actions, that although not necessarily life-threatening, can potentially complicate the steroid's proper usage in SCI patients and/or counteract its neuroprotective actions.

1. **Biphasic (U-Shaped) Neuroprotective Dose Response:** The sharp U-shaped dose-response curve for the neuroprotective properties of MP and the need for repeated dosing (i.e., continuous infusion) to maintain these effects (Hall, 1992) make the administration of MP in SCI patients difficult, and even tricky. For instance,

optimal doses of MP can lessen posttraumatic anaerobic metabolism in the injured spinal cord and lessen lactate accumulation (Braughler and Hall, 1983a, 1983b, 1984). Inadequate or excessive MP dosing has been shown to aggravate posttraumatic lactate accumulation (Braughler and Hall, 1983a, 1983b, 1984). This is a consequence of glucocorticoid receptor-mediated gluconeogenesis stimulatory actions of MP and similar steroids. As a result, there is an opposition between the beneficial effects of optimal MP doses to support aerobic metabolism and to lessen lactate accumulation while nonoptimal doses can result in lactate production being enhanced via an increase in blood glucose levels. These can potentially drive up anaerobic glycolysis in the injured cord, which will aggravate, rather than inhibit, LP reactions. Thus considerable care is required in dose calculation whether it is based on body weight or mass because too little will not serve to effectively protect and too much will result in a loss of the antioxidant neuroprotection. During the last 13 post-NASCIS II years of widespread use of high-dose MP (NASCIS II 24-hour or NASCIS III 48-hour protocols), there have undoubtedly been many dose miscalculations that have resulted in less than or more than optimal MP dosing in many SCI patients. As Bracken (2001) notes, there are only ad hoc reports of compliance to the 24-hour NASCIS II MP dosing protocol in routine practice; however, these include reports of therapy commencing later than the recommended 8-hour cutoff, discontinuation of therapy before 24 hours, accidental administration of the full 24-hour dose within the first hour, the maintenance infusion being at rates faster or slower than the recommended 5.4 mg/kg per hour and inaccurate estimates of patient body weight. Similarly, the first author has listened to several anecdotes over the past several years concerning MP dose miscalculations either in regard to the magnitude of the initial bolus loading dose or the rate and duration of the maintenance infusion. The rather narrow U-shaped dose-response curve dictates that great care be taken in calculating and administering the NASCIS II or III MP protocols. Based on the experimentally determined U-shaped dose-response curve, the need for initiation of treatment as soon as possible, and the requirement for maintenance dosing to maintain the neuroprotective effects during the first 24 to 48 hours, it is likely that MP will have no benefit if the protocol is erroneously administered.

2. **Duration of Treatment Dependent on Treatment** Initiation Time: The results of the NASCIS III study show that if treatment cannot be initiated until after the first 3 posttraumatic hours, then it would be better to extend the infusion phase of the treatment from 24 to 48 hours. On the other hand, if begun within the first 3 hours, the 24-hour NASCIS II protocol is suffi-

cient (Bracken *et al.*, 1997, 1998). Thus, attention to the time of injury vs treatment initiation time is an important consideration that adds a small, but significant complication to the therapy.

3. **Late Treatment Initiation May Exacerbate Damage:** Initiation of MP therapy beyond the 8-hour therapeutic window determined in NASCIS II may actually exacerbate the posttraumatic secondary injury and lessen the expected neurological outcome compared to no therapy (Bracken and Holford, 1993). Therefore, the need to take into account the exact time of injury in making the decision to treat or not to treat the SCI patient with high-dose MP is yet another complicating factor. Treatment later than 8 hours after injury should be avoided.

4. **Possible Negative Effects on Neuronal Survival and Plasticity:** Glucocorticoids have been repeatedly shown to inhibit axonal sprouting and synaptogenesis in various CNS areas (Scheff *et al.*, 1980; Scheff and Cotman, 1982; Scheff and DeKosky, 1983, 1989). Although it is not known whether these actions will occur concomitantly with the acute antioxidant neuroprotective effects, the potential of the steroid to attenuate posttraumatic plasticity mechanisms is perhaps the most serious concern regarding the administration of high doses of MP for any longer than necessary. Although the beneficial neurological effects of 24, and even 48, hours of dosing seen in NASCIS II and III, respectively, might suggest that the antiplasticity effects are not a problem with short-term, high dose MP therapy; this possibility has not been investigated.

Other neurotoxic actions documented in certain neuronal populations (e.g., hippocampus) with either physiological or pharmacological levels of glucocorticoids (Dinkel *et al.*, 2003; Elliott and Sapolsky, 1992; Landfield and Eldridge, 1994; McIntosh and Sapolsky, 1996; Morse and Davis, 1990; Sapolsky, 1985; Sapolsky *et al.*, 1986; Shin *et al.*, 2001; Stein-Behrens *et al.*, 1992) give one pause when administering high doses of a glucocorticoid steroid in the context of the injured CNS. Although the sensitivity of spinal cord neurons to these detrimental actions of glucocorticoids has not been specifically investigated, the possibility that high-dose MP therapy may aggravate posttraumatic neuronal damage begs for further investigation. If in fact the plasticity-inhibiting and neurotoxic actions of glucocorticoids are shown to occur in the injured cord coincident with the neuroprotective antioxidant actions, this will be yet another example of how MP can constitute a "two-edged sword" in the treatment of acute SCI. In any event, the potential for steroid side effects, inhibition of plasticity mechanisms, and even neurotoxic actions underscores the fact that glucocorticoids such as MP are a far from ideal approach to dealing with the posttraumatic oxidative stress and LP-related damage and consequent need for antioxidant dosing that continues beyond the first 24 to 48 hours.

Despite the concerns about the potential inhibition of axonal sprouting and other neurotoxic actions of glucocorticoids discussed previously, however, the safety and neuroprotective value of the ongoing clinical use of high doses of MP in SCI therapy is sufficiently encouraging that high doses of MP have actually been reported to lessen axonal dieback of vestibulospinal fibers and to promote their terminal sprouting in transected rat spinal cords (Oudega *et al.*, 1999). The same group has shown that identical high doses of MP can improve axonal regeneration into Schwann cell grafts (Chen *et al.*, 1996) and lessen caspase 3 activation (Li *et al.*, 2000) in transected rat spinal cords. In regard to the latter antiapoptotic action, other laboratories have also shown the high-dose MP therapy can lessen apoptotic neurodegeneration after traumatic (Ray *et al.*, 1999) or ischemic (Kanellopoulos *et al.*, 1997) SCI in rats. These effects were achieved with acute MP administration limited to the first 4 hours after injury. In contrast, a recent report has shown that MP can exacerbate retinal ganglion cell apoptosis in the context of autoimmune CNS inflammation (Diem *et al.*, 2003). Thus, the question of whether high-dose MP is neuroprotective or neurotoxic may depend on dose selection, timing and duration of administration, and the particular neuronal population in question.

The Methylprednisolone SCI Controversy and Criticism of the NASCIS Trials

Before completion of NASCIS II, no treatment for acute SCI had ever been demonstrated in a placebo-controlled clinical trial to have a beneficial effect on neurological outcome. Therefore, the revelation of the beneficial effects of high-dose MP was a surprise to many in the medical community, including some, if not all, of the NASCIS II clinical investigators. So surprising and gratifying were the results that the NIH sanctioned a prepublication press release of the results to inform the neurosurgical community so that the NASCIS II MP dosing protocol would be immediately available to SCI patients rather than having to wait for the publication of the initial trial manuscript that had been accepted by the *New England Journal of Medicine*. The editorial office of the journal uncharacteristically approved the prepublication release of the findings based on their view that this indeed represented a breakthrough treatment for acute SCI and it should be immediately available to SCI victims. For the NIH and the NASCIS II group to have done otherwise would undoubtedly have been viewed by many physicians and patients as unethical. However, the negative response of a few neurosurgeons to the idea of publishing clinical trial results in the popular press before publication and before presentation to, and careful scrutiny by, the neurosurgical community represented the beginning of a heated controversy about the design, analysis, and reporting of the NASCIS II (Bracken and Holford, 1993; Bracken et al., 1990, 1985) and later the NASCIS III (Bracken and Holford, 2002; Bracken et al., 1997, 1998) trials. This controversy has raged off and on for the past 13 years.

A few years ago, three highly critical reviews appeared in the literature concerning the NASCIS trials and the resulting view that high-dose MP therapy should be considered the standard of care for acute SCI (Coleman et al., 2000; Hurlbert, 2000; Short et al., 2000). The first critical review (Hurlbert, 2000) was written by a single author who is a member of the "American Association of Neurological Surgeons/Congress of Neurological Surgeons practice guidelines subcommittee for the pharmacological treatment of SCI." However, he was the sole author and a Disclosure was included at the end of the article stating that *The views expressed in this paper do not represent the final consensus of the committee.* Therefore, it could be perceived that the other AANS/CNS committee members did not agree with the negative assessment of high-dose MP therapy for SCI patients. In any case, Hurlbert (2000) leveled six criticisms of the NASCIS II study. The first of these was that the overall analysis of the effects of either MP or naloxone vs placebo was negative when all patients were included in the analysis (i.e., no split of patients treated within 8 hours vs after 8 hours). The further claim was made that the stratified analysis on the basis of time to treatment was a post hoc decision. However, this was not the case in the design of NASCIS II. The rule for subgroup analysis is that if the subgroup is prespecified in the study protocol, then that becomes the primary endpoint. Indeed, the NINDS proposal that funded NASCIS II specified that two dichotomies would be studied. First of all, the effects of early vs late treatment with MP or naloxone would be analyzed with the split being made at the median time to treatment. Secondly, an analysis of the effects of either drug would be assessed separately in incomplete and complete SCI patients (Bracken, 2001). The precise time of the data cut was not prespecified because going into the trial, there was no way to know what the median treatment initiation time would be. After the final patient was enrolled, it turned out that the 8 hours was the nearest whole number to the median treatment initiation time (8.5 hours), and thus 8 hours was selected for the data split (Bracken, 2001). This approach has been used and widely accepted in establishing the 3-hour cutoff for the safe and effective use of tissue plasminogen activator in acute stroke therapy (Kwiatkowski et al., 1999; Marler et al., 2000). Therefore, the claim that the NASCIS II trial did not reach its primary endpoint in

regard to the efficacy of MP is not consistent with the documented trial design in which the plan to look at early vs late treatment and the influence of injury severity was, in fact, preplanned.

Additional criticisms of NASCIS II put forth by Hurlbert (2000) were that the MP effect sizes were small, their functional importance was uncertain due to the reliance on neurological sensory and motor assessments rather than functional improvement scales, and that effects on long tract function were not established. Each of these has been addressed by additional analyses of the NASCIS II data set carried out by Bracken and colleagues. Concerning the size of the effect, it has indeed been acknowledged that the benefits of MP in the context of NASCIS II were "modest, but [MP] does appear to have the potential to result in important clinical recovery in some patients" and "Even small changes in motor recovery, typically assessed in the MPSS trials on one side of the body, have the potential to be amplified into meaningful improvements in the quality of life" (Bracken, 2001). Regarding the issue of long tract vs segmental recovery effects, further analysis of the NASCIS II data revealed that the greatest degree of motor function recovery within patients receiving MP within the 8-hour window occurred in long spinal tracts as recorded below the level of the injury, although significant segmental recovery was also found at the level of the lesion (Bracken, 2001; Bracken and Holford, 1993). The criticism that the clinical significance of the functional improvement was not assessed by inclusion of functional or quality of life measures such as the Functional Independence Measure (FIM) in NASCIS II is valid. Recently, the NASCIS II principal investigator has published an estimate of functional recovery in NASCIS II from results modeled in the NASCIS III study that did include FIM assessments (Bracken and Holford, 2002). In that report, it is claimed: "The extent of MP therapy-related motor function recovery observed in NASCIS II predicted clinically important recovery in the FIM."

The final NASCIS II critique by Hurlbert (2000) concerning the questionable safety of high-dose MP therapy in SCI patients is difficult to reconcile with the NASCIS II data set, which found no significant differences in adverse reactions between placebo- and MP-treated patients (Bracken et al., 1990, 1992). Since NASCIS II represents the only placebo-controlled trial of a 24-hour high-dose regimen in SCI patients, it is difficult to understand the claim that this particular regimen is unsafe. In contrast, the NASCIS III trial showed that extension of the duration of dosing from 24 to 48 hours resulted in an increased incidence of pneumonias and a nearly significant ($P < .07$) increase in the occurrence of severe sepsis, both of which are undoubtedly manifestations of the immunosuppressive properties of intensive glucocorticoid dosing (Bracken et al., 1997). Therefore, the available data that directly address the safety of high-dose MP in SCI patients in a controlled context shows that the 24-hour regimen is not associated with an increase in adverse reactions, whereas the 48-hour regimen does carry some risk of serious glucocorticoid-related problems. This risk must be balanced against the results of NASCIS III showing that in patients not treated until after the first 3 post-SCI hours, the extension of dosing to 48 hours produces a significantly better neurological outcome compared to the 24-hour regimen. Considering the devastating nature of most SCIs, it would seem that the risk of MP treatment would be outweighed by the potential benefits to the patient's neurological recovery.

Hurlbert (2000) criticizes the NASCIS III trial on five counts: (1) lack of a placebo group, (2) the claim that the overall analysis was negative, (3) the stratification of patients treated within 3 hours vs between 3 and 8 hours was a post hoc analysis, (4) improvement in the FIM score in 48-hour MP-treated patients compared to the 24-hour MP cohort was modest, and (5) MP treatment is unsafe based on the increased incidence of sepsis and pneumonias. The first of these concerning the lack of a placebo group has already been addressed. The next two regarding the statistical analysis are identical to those leveled against NASCIS II. Here again, the NASCIS III investigators have gone on record as stating that the dichotomized analysis of the NASCIS III data on the basis of time to treatment (early vs late) was a preplanned and not a post hoc analysis (Bracken, 2001; Bracken et al., 1997). The issue of the significance of the FIM score improvement shown in 48-hour MP patients in NASCIS III requires a careful viewing of the data. Hurlbert (2000) focuses on the fact that in the intent to treat analysis, the improvement in the total FIM score only reached a level of $P < .08$ (Bracken et al., 1997). Looking at the subscores, however, reveals a significantly better self-care score in the 48-hour treated patients ($P < .03$) and sphincter control score ($P < .01$). If, however, the analysis is redone including only those patients in which the dosing protocols were completely followed, even the total FIM score in the 48-hour treated patients reaches the .05 level vs the 24-hour MP cohort. In regard to the safety criticism, the 48-hour MP group displayed a significant increase in pneumonias and a nearly significant increase in severe sepsis compared to the 24-hour group. As noted previously, these are predictable risks of high-dose glucocorticoid therapy in critically ill patients. The risk/benefit ratio is a matter of personal opinion.

The other two critiques (Coleman et al., 2000; Short et al., 2000) have multiple authors, and similar to Hurlbert (2000), intensely criticize not only the NASCIS trial design, analysis, and reporting, but also that of the preclinical studies that inspired the conduct of NASCIS II and III. As previously discussed, the principal investigator

of the NASCIS trials has addressed the various criticisms and misunderstandings in the form of a recent meta-analysis of the NASCIS and non-NASCIS trials of MP in acute SCI, as well as their overall statistically significant support of the efficacy and safety of high-dose MP therapy for acute SCI (Bracken, 2001). He concludes from the meta-analysis that "high dose MP given within 8 hours of acute spinal cord injury is a safe and modestly effective therapy that may result in important clinical recovery for some patients. Further trials are needed to identify superior pharmacological therapies and to test drugs that may sequentially influence the postinjury cascade."

This view is generally endorsed in a recent summary statement in the journal *Spine* from the Spine Focus Panel (Fehlings, 2001b) in which Dr. Michael Fehlings stated (Toronto, Ontario):

Members of the Spine Focus Panel extensively discussed the role of methylprednisolone in acute spinal cord injury but could not reach agreement on key points. While it acknowledged that the NASCIS trials represent landmark clinical studies, no clear consensus could be reached on the appropriate use of steroids in acute spinal cord injury. Many members of the Spine Focus Panel acknowledged that although methylprednisolone is only modestly neuroprotective, this drug is clearly indicated in acute spinal cord injury because of its favorable risk/benefit profile and the lack of alternative therapies. However, a significant minority was of the opinion that the evidence supporting the use of steroids in spinal cord injury was weak and did not justify the use of this medication. The Spine Focus Panel did agree that given the devastating impact of spinal cord injury and the modest efficacy of methylprednisolone, clinical trials of other therapeutic interventions are urgently needed. Many panel members also felt that a reanalysis of the NASCIS data might help to resolve the controversies surrounding the use of steroids in acute spinal cord injury. Of note, Dr. Bracken has agreed to release the data for "protocol-driven research by qualified investigators."

Following the Summary Statement, Dr. Fehlings presented an editorial: Recommendations Regarding the Use of Methylprednisolone in Acute Spinal Cord Injury—Making Sense Out of the Controversy (Fehlings, 2001a) in which he first summarizes the criticisms of NASCIS II and III, and then presents the "Suggested Indications for the Use of Methylprednisolone in Acute SCI:"

- For acute nonpenetrating SCI (< 3 hours after injury), MP should be given as per NASCIS II protocol (i.e., 24 hours of treatment)
- For acute nonpenetrating SCI (> 8 hours after injury), MP should not be used
- For acute nonpenetrating SCI (after 3 hours, within 8 hours), MP should be given as per NASCIS III protocol (i.e., 48 hours of treatment)
- For acute penetrating SCI, MP is not recommended

He closes the editorial by stating:

Given the devastating impact of SCI and the evidence of a modest, beneficial effect of MP, clinicians should consciously consider using this drug despite the well-founded criticisms that have been directed against the NASCIS II and III trials. With great understanding of the biomolecular events contributing to the pathogenesis of SCI, it is hoped that other neuroprotective agents will enter into clinical practice in the next 5-10 years.

Clinical Testing of Monosialoganglioside GM1

Besides MP and naloxone, the only other drug tested in Phase III trials in SCI patients is the monosialoganglioside compound GM1. This compound was shown as far back as the 1970s to exert neuroprotective and neuroregenerative effects in a variety of animal models of neural injury (Geisler et al., 2001c; Gorio, 1986, 1988; Gorio and Vitadello, 1987), although it was never tested in an acute SCI model per se. Nevertheless, it captured the interest of non-NASCIS investigators to conduct clinical trials of its potential benefits in acute SCI. An initial small single-center Phase II trial of 37 patients suggested that the administration of GM1 to SCI patients within 24 hours after injury might lead to an improvement in neurological recovery (Geisler et al., 1990, 1991). Subsequently, the same investigator led a multicenter Phase III trial of GM1 in acute SCI patients. Since the organization of this trial occurred after the widespread acceptance of high-dose MP as the standard of care for acute SCI, the effects of GM1 were examined on top of the 24-hour NASCIS II high-dose MP protocol (Geisler et al., 2001a, 2001b, 2001c). Administration of GM1 was begun after completion of the NASCIS II 24-hour MP dosing protocol (Geisler et al., 2001a, 2001b, 2001c). The use of GM1 after MP therapy resulted in faster achievement of peak neurological recovery, although the extent of functional improvement was not greater than that observed in patients who received only MP. This suggests that although GM1 may have some trophic effects that stimulate recovery, the compound is not neuroprotective in the context of acute SCI.

DEVELOPMENT OF NEUROPROTECTIVE PHARMACOTHERAPY FOR BRAIN INJURY

Drug Discovery and Development for Traumatic Brain Injury

During the past 15 to 20 years, there has been intense activity in the pharmaceutical industry and academia to discover and develop pharmacological agents for acute treatment of TBI. Several glutamate receptor antagonists and release inhibitors, calcium channel blockers, and free radical scavengers/antioxidants have been shown to be effective neuroprotectants in a variety of preclinical TBI models. As a result, at least five separate Phase II and III clinical development programs have been carried out over the last 15 years to evaluate the potential benefits of the calcium channel blocker nimodipine, the glutamate antagonists selfotel (CGS 19755) and aptiganel (CNS 1102), the free radical scavenger polyethylene glycol-conjugated superoxide dismutase (PEG-SOD), and the LP inhibitor tirilazad (U-74006F) (Langham et al., 2000; Marshall et al., 1998; Narayan et al., 2002). However, these trials, as a whole, have been therapeutic failures in that no overall benefit has been documented in moderate and severe TBI patient populations. These failures can be attributed to several factors. Perhaps most important, the preclinical assessment of compounds destined for acute TBI trials has been woefully inadequate in regard to the definition of neuroprotective dose-response relationships, pharmacokinetic-pharmacodynamic correlations, therapeutic window, and optimum dosing regimen, and treatment duration (Narayan et al., 2002).

Effects of Tirilazad in TBI

The protective efficacy of tirilazad has also been evaluated in multiple animal models of acute head injury. For instance, the compound significantly improves the early (1 to 4 hours) neurological recovery and survival of mice subjected to severe concussive head injury (Hall et al., 1988). Tirilazad has similarly been shown to exert beneficial effects on motor recovery and survival in a rat model of moderately severe fluid percussion head injury (McIntosh et al., 1992). In a cat model of severe concussive head injury, tirilazad significantly reduced posttraumatic lactic acid accumulation in both the cerebral cortex and subcortical white matter, indicative of a positive action on aerobic energy metabolism (Dimlich et al., 1990). The locus of this particular effect is unclear. It most likely involves a protection of neuronal mitochondrial function. While the compound is largely localized in the microvascular endothelium, the posttraumatic disruption of the BBB is known to allow the successful penetration of tirilazad into the brain parenchyma, as noted earlier (Hall et al., 1992). Other mechanistic data derived from the rat-controlled cortical impact and the mouse concussive head injury models have definitively shown that a major effect of tirilazad is to lessen posttraumatic BBB opening (Hall et al., 1992; Smith et al., 1994). In the mouse model, the BBB protection occurred together with an attenuation of the early posttraumatic rise in brain hydroxyl radical levels measured via the salicylate trapping method (Hall et al., 1992; Smith et al., 1994). Thus the effect of tirilazad to protect the BBB is paralleled by a reduced formation of hydroxyl radicals and/or a protection of the microvascular endothelium from hydroxyl radical-induced LP.

Based on the strength of these data, tirilazad was evaluated in two Phase III multicenter clinical trials for its ability to improve neurological recovery in moderate and closed head injury patients, one in North America and the other in Europe. In both trials, TBI patients were treated within 4 hours after injury with vehicle or tirilazad (2.5 mg/kg IV q6h for 5 days). The North American trial was never published because of a major confounding imbalance in the randomization of the patients to placebo or tirilazad in regard to injury severity and pretreatment neurological status. In contrast, the European trial had much better randomization balance and has been published (Marshall *et al.*, 1998). Results failed to show a significant beneficial effect of tirilazad in either moderate or severe patients. However, male TBI patients with traumatic SAH showed significantly less mortality after treatment with tirilazad (34%) compared to placebo (43%, *P* < .026) (Lyons *et al.*, 1994; Marshall *et al.*, 1998). This result is consistent with the fact that this compound is highly effective in animal models of SAH (Hall *et al.*, 1994).

Effects of High-Dose Methylprednisolone in TBI

As discussed previously, MP has been shown to be neuroprotective in large part via inhibition of posttraumatic LP in the injured cord. Despite a long and confused history of steroid usage in acute TBI patients, success with high-dose MP in SCI has caused some to reconsider the possible application of identical treatment in human TBI. This view is supported by a recent meta-analysis of trials of steroids in TBI patients, which shows a small but statistically significant benefit (Alderson and Roberts, 1997). Although the extension of high-dose MP treatment from SCI to TBI has a strong rationale, the potential benefits vs detriments of MP have been inadequately studied in TBI models in terms of a careful mechanistically driven pharmacological approach. Thus, the efficacy vs safety of this treatment requires further study in animal models of diffuse and focal TBI. Nevertheless, as high-dose MP constitutes the only clinically demonstrated neuroprotective agent for acute CNS injury, full exploration of its neurotherapeutic potential, the validity of LP inhibition as a mechanistic hypothesis for its protective effects, and its potential glucocorticoid-related drawbacks is warranted.

There is experimental evidence that the neuroprotective properties of MP, extensively studied in SCI, may also be applicable to TBI. The steroid has been shown to enhance the early recovery of mice subjected to a moderately severe concussive head injury when administered at 5 minutes postinjury. The dose-response curve for this effect is remarkably similar to that discussed above for

SCI (Hall, 1985). A 30-mg/kg IV dose was observed to be optimum, whereas lower (15-mg/kg) and higher (60- and 120 mg/kg) doses were ineffective. In the same study, two other glucocorticoids were compared with MP. Prednisolone, given in steroid-equivalent doses with MP, was equally efficacious but half as potent, in that a 60-mg/kg IV dose was optimal. Hydrocortisone was ineffective in improving neurological recovery in injured mice at any dose up to 120 mg/kg. This structure activity relationship parallels the relative abilities and potencies of these three steroids as inhibitors of lipid peroxidation (Braughler, 1985).

Methylprednisolone and prednisolone are equally efficacious as lipid antioxidants *in vitro*, but the latter is half as potent. Hydrocortisone is ineffective as an inhibitor of lipid peroxidation, even at exceedingly high concentration. Dexamethasone, a steroid widely used in neurosurgery, also possesses lipid antioxidant activity, but it is slightly less effective than MP or prednisolone.

Consistent with the beneficial effect of high-dose MP in experimental head injury, one controlled clinical trial has shown that a high-dose regimen begun within 6 hours of severe injury (Glasgow Coma Scale scores 3 to 8) with a 30-mg/kg IV bolus can significantly increase survival and the recovery of speech (Giannotta *et al.*, 1984). However, while these data are suggestive, flaws in the study prevent it from being definitive. For instance, there were three groups of patients: a placebo group, a low-dose MP group (100-mg bolus plus 8 days of tapered maintenance dosing), and the high-dose group (30-mg/kg bolus plus tapered dosing). There was clearly no difference between the placebo and low-dose groups, but only when these two groups were combined after the study and statistically compared to the high-dose group could the significant effect on survival and speech recovery be demonstrated. Furthermore, the effect was only demonstrable if patients over 40 years old were excluded, also after data collection. Thus, an additional trial would be required to determine the efficacy of MP in brain injury. Nevertheless, preliminary results with early administration of antioxidant doses of MP are more encouraging than various studies of treatment of head injury with conventional steroid dosing regimens, which show no apparent effect.

CRASH Trial

A few years ago, a group of investigators based in the United Kingdom, supported by the Medical Research Council, carried out a meta-analysis of all steroid TBI trials that had been carried out over the last 30 years. That analysis, which lumped together studies that used a variety of glucocorticoids, doses, and dosing regimens, showed that the risk of death was 2% less in patients treated with steroids than with placebo (Alderson and

Roberts, 1997). The impression was that this small, but significant overall effect was due to those few studies that involved high doses of steroids. This, together with the benefits of high-dose MP in experimental and clinical SCI, inspired a 20,000 patient, multicenter, multinational clinical trial of the 48 hour NASCIS III MP dosing protocol within the first 8 hours after "TBI associated with impaired loss of consciousness" (i.e., moderate and severe TBI). This trial, the Corticosteroid Randomization After Significant Head Injury (CRASH) Trial has currently enrolled 9920 TBI patients (www.crash.lshtm.ac.uk).

Although an extrapolation of the apparent efficacy of antioxidant dosing with MP to TBI is supported by its efficacy in acute SCI, the only studies that have examined the efficacy of high-dose TBI are those carried out by the lead author several years ago and discussed previously (Hall, 1985; Hall *et al.*, 1987). Although these studies demonstrated that high-dose MP could enhance neurological recovery, the only time point examined was 1 hour postinjury. An investigation of its ability to reduce posttraumatic oxidative damage or neurodegeneration or to improve chronic neurological recovery has not been carried out. Moreover, the possible detrimental effects of high-dose MP have not been evaluated in TBI models. Consequently, the ongoing CRASH Trial lacks any mechanism-based preclinical experimentation with which to interpret whatever efficacy, or lack thereof, may be observed. Furthermore, even if 48-hour high-dose MP treatment is shown to be significantly beneficial, there is presently no preclinical analysis of the pharmacology of MP in TBI models with which to optimize steroid treatment in regard to mechanism of action, dose-response, pharmacodynamic-pharmacokinetic correlation, therapeutic window (i.e., 8-hour window seen in SCI may not be the same in TBI), and optimal treatment duration (48 hours may not be ideal).

Effects of Cyclosporin A in TBI

Mitochondrial dysfunction has been documented to occur in several experimental models of TBI, including controlled cortical impact (Sullivan *et al.*, 1998, 1999a, 1999b; Xiong *et al.*, 1997), fluid percussion (Lifshitz *et al.*, 2003; Sullivan *et al.*, 2000b), and human head injury patients (Verweij *et al.*, 2000). Increased mitochondrial ROS production (Sullivan *et al.*, 1998, 1999b), reduced mitochondrial respiration (Xiong *et al.*, 1997), disruption of Ca^{++} buffering, and reduced thresholds for mitochondrial permeability transition (mPT) (Sullivan *et al.*, 1999b, 2000a; Xiong *et al.*, 1997) have all been documented both at acute (15 to 30 minutes) and chronic time (7 to 14 days postinjury) points postinjury. The prominent role that mitochondria and possibly the mPT play in neuronal cell survival and death after TBI have recently

come to light as a result of several studies that have targeted mitochondrial dysfunction after TBI.

The inner mitochondrial membrane is normally impermeable to solutes that are not specifically transported. The mPT is defined as the sudden increase of inner membrane permeability to solutes of molecular mass less than 1500 daltons (Bernardi, 1996; Bernardi *et al.*, 1994; Nicolli *et al.*, 1996). This loss of impermeability is due to the opening of a proteinaceous pore and is a catastrophic event that results in swelling of the mitochondrial matrix, rupture of the outer mitochondrial membrane, and a complete uncoupling of the electron transport system from ATP production. The first definitive reports of modulation of the mitochondrial Permeability Transition Pore (mPTP) came in the late 1980s and demonstrated that even sub-μM concentrations of cyclosporin A (CsA) transiently blocked calcium-induced mitochondrial surveilling. CsA is an undecapeptide of fungal origin that is used clinically as an immunosuppressive agent. CsA inhibits the activation of T cells by inhibiting the protein phosphatase 2B, also know as calcineurin, and the subsequent transcription of the interleukin (IL)-2 gene. CsA inhibits the opening of the mPTP by inhibiting the interaction of matrix cyclophilin with the adenine nucleotide translocator, thus preventing mPT-induced loss of mitochondrial homeostasis (Broekemeier *et al.*, 1989; Broekemeier and Pfeiffer, 1995; Szabo and Zoratti, 1991).

Okonkwo and colleagues (1999) showed that an intrathecal bolus of CsA (20 mg/kg), preinjury, reduced diffuse axonal injury and Ca^{++}-induced cytoskeletal damage following TBI in rats. The clinically relevant question of CsA efficacy postinjury was then tested in two different strains of mice and in rats by Scheff and Sullivan (1999). CsA significantly reduced cortical damage (50%) after TBI when administered intraperitoneally 15 minutes and 24 hours after injury. There appeared to be a biphasic dose response for the neuroprotective benefits of CsA, with the greatest protection afforded by the lowest dose (20 mg/kg intraperitoneally) (Scheff and Sullivan, 1999). This study also showed that the immunosuppressive properties of CsA were most likely not responsible for the neuroprotection, as the more potent immunosuppressor FK506 afforded no neuroprotection in this paradigm. In a follow-up study, mitochondrial function was directly assessed after TBI (Sullivan *et al.*, 1999b). These results suggested that the neuroprotective properties of CsA were mediated through modulation of the mPTP and maintenance of mitochondrial homeostasis after TBI because it was demonstrated that administration of CsA (20 mg/kg intraperitoneally) 15 minutes after injury significantly attenuated mitochondrial dysfunction measured using several biochemical assays of mitochondria integrity and bioenergetics.

Although CsA is FDA-approved and routinely used in the clinic as an immunosuppressor, it has been shown to have significant dose-dependent neurological effects, including seizures and cell death (Berden *et al.*, 1985; de Groen *et al.*, 1987; Famiglio *et al.*, 1989; Kahan, 1989; Walker and Brochstein, 1988). Therefore it was critical to determine the complete dose curve for CsA neuroprotection after TBI. To address this question, injured animals were administered various dosages (40 mg/kg, 20 mg/kg, 10 mg/kg, 5 mg/kg, 1 mg/kg) of CsA or vehicle 15 minutes after injury with a subsequent injection 24 hours later. The most advantageous CsA therapy was a 20 mg/kg bolus intraperitoneal injection initiated 15 minutes after injury. Although other doses of CsA ranging from 40 mg/kg to 1 mg/kg significantly reduced cortical damage compared to vehicle-treated animals, subjects treated with 20 mg/kg were afforded the greatest neuroprotection (Scheff and Sullivan, 1999; Sulhran *et al.*, 2000b). The extent of the postinjury therapeutic window was also shown in this study by administering the optimal dose of CsA (20 mg/kg) at various times after injury. Results demonstrate that CsA therapy was effective when initiated at 15 minutes, 1 hour, or 24 hours after injury. Paradoxically, the efficacy was lost if therapy with CsA was initiated at 6 hours after injury. That onset of treatment 24 hours after injury was as effective as initiation at 1 hour illustrating that the mechanisms responsible for the tissue destruction were active and amenable to pharmacological manipulation for at least 24 hours after injury. Lack of neuroprotection by CsA if administration is delayed for 6 hours after injury may highlight the role of the BBB in designing TBI therapeutics. CsA does not freely cross the intact BBB (Shiga *et al.*, 1992; Uchino *et al.*, 1995, 1998) so after TBI, this allows targeting of CsA to the injured tissue, however dynamic fluctuations in the permeability of the BBB post-injury would also block CsA uptake.

Since the time frame of secondary injury resulting from TBI ranges from minutes to days, continuous dosing of CsA could be advantageous. To address this question, animals were administered a 20 mg/kg intraperitoneal bolus of CsA or vehicle 15 minutes postinjury followed by continuous delivery of CsA (4.5 or 10 mg/kg per day) for 7 days postinjury. All animals receiving CsA showed a significant reduction in lesion volume, with the highest dose offering the most neuroprotection (74% reduction in lesion volume) (Sullivan *et al.*, 2000c). These data highlight the possible need for multiple dosing regimens after TBI. While all animals receiving CsA were significantly better than vehicle-treated controls, injured animals receiving only a single bolus injection were afforded the least protection and administering a second bolus injection 24 hours after injury afforded greater neuroprotection. Continuous infusion of CsA for 7 days resulted in the greatest sparing of tissue. The therapeutic window appeared to be within the first 7 days, as continuous infu-

sion at 10mg/kg per day almost completely eliminated tissue loss in some animals. That animals administered a second bolus after 24 hours were noticeably better than the single injection group suggests that the first 24 hours is perhaps a critical window for CsA administration.

These studies show that CsA has emerged as a promising therapeutic agent for TBI, although the drug displays complex pharmacokinetics, even in the tightly controlled laboratory setting. Currently, two Phase IIa single center clinical trials are being conducted at the University of Kentucky (B. Young, P.I.) and the Medical College of Virginia (R. Bullock, P.I.) to evaluate the safety and pharmacokinetics of CsA, including CSF penetration of CsA, as a prelude to an evaluation of its neuroprotective efficacy.

HOW DO WE BUILD ON PAST SUCCESS?

This chapter has reviewed the ability to translate neuroprotective effects of pharmacological agents in TBI and SCI models into successful clinical trials. Thus far, the discovery and development of high dose MP therapy in acute SCI are arguably the most successful demonstration of clinical neuroprotection to date (NASCIS II and III), although its use is not without controversy. Also, there are perhaps enough data in SCI trials (NASCIS III) to show that the neuroprotective effects of the glucocorticoid steroid MP, which are believed to be largely due to antioxidant actions, can be duplicated by the nonglucocorticoid steroid tirilazad. Additional trials with tirilazad would seem warranted, as this steroid is devoid of steroid side effect potential and is therefore safe for more prolonged administration compared to MP. It would seem that naloxone dosing in NASCIS II may have produced enough of an effect to validate its neuroprotective efficacy shown in experimental SCI models. Although the effects of GM-1 are similarly modest, this compound does seem to improve the rate, although not the extent, of post-SCI neurological recovery in SCI patients.

Clinical trials are in progress to evaluate whether the neuroprotective effects of high-dose MP shown in SCI can be extrapolated to TBI (CRASH Trial) and whether the immunosuppressive agent CsA, which is neuroprotective via its ability to block mitochondrial damage, can improve neurological recovery in TBI patients.

Not discussed in this chapter are the many promising pharmacotherapies and mechanisms that are being evaluated at the preclinical level. Based on past experience with neuroprotective drug discovery and development, however, several considerations are apparent in regard to how best to evaluate neuroprotective agents before attempting translation into clinical trials:

- Demonstration of the time course of the target patho-physiological mechanism in relevant animal models. This is needed to determine when treatment needs to begin and how long it must be maintained.
- Rigorous dose-response analysis in regard to effects on the target mechanism, ability to reduce posttraumatic neurodegeneration and improve behavioral recovery.
- Correlation of neuroprotective action with plasma and CNS tissue pharmacokinetics.
- Correlation of plasma and CNS pharmacokinetics with plasma or CNS biomarker (e.g., lipid peroxidation products).
- Comparison of single vs multiple-dose regimens to establish optimum treatment regimen (NASCIS experience suggests IV bolus plus infusion makes the most sense).
- Determination of therapeutic window in order to know how early treatment must begin.
- Comparison of neuroprotective pharmacology in multiple injury models to determine whether the agent in question works only in certain types of injuries.
- Determination of pharmacodynamic and pharmacokinetic interactions with other commonly used ancillary treatments (e.g., anticonvulsants).

The detailed preclinical evaluation of neuroprotective agents in TBI or SCI models should then be followed by a clinical trial design that is consistent with the results of preclinical trials. One of the most glaring examples of how this has not been done concerns the issue of therapeutic window. Even with agents for which the therapeutic efficacy window had been determined, the subsequent clinical trials typically involved an enrollment window that far exceeded the time at which a particular agent could have been expected to retain neuroprotective potential. It has been argued, in these instances, that if a particular agent only has a 1-hour window in a rat TBI or SCI model, the window in humans with the corresponding condition is likely to be much longer; however, there is absolutely no evidence to support this assumption. On the contrary, the therapeutic window shown for high-dose (antioxidant) MP therapy in humans (8 hours) (Bracken *et al.*, 1990) is reminiscent of the 8-hour window shown for the antioxidant tirilazad in a cat SCI model (Anderson *et al.*, 1991). Thus, the time course of posttraumatic LP may not be all that different between cats and humans. Whether this is consistently the case across pathophysiological mechanisms requires further study. At this point, it would be prudent to take the magnitude of the therapeutic window determined in animal CNS injury models literally when planning clinical trials. Consequently, from a practical standpoint, it is no longer appropriate to advance compounds into clinical trials in SCI or TBI that have not been demonstrated to have a therapeutic window of at least 2 or more hours. Ideally agents with at least an 8 hour window would carry the greatest clinical practicality.

Key to improving the successful translation of preclinical neuroprotective effects into successful mechanism-based clinical trials is the identification of measurable biomarkers that allow the pathophysiology and drug effects thereon to be determined. Three examples of promising CNS injury biomarkers are the LP-related isoprostanes (Pratico *et al.*, 2002) and neuroprostanes (Morrow and Roberts, 2002), degradation products of the cytoskeletal protein α-spectrin (Pike *et al.*, 2001), and the astrocytic injury marker S100β (Pelinka *et al.*, 2003; Wunderlich, 2003). The validation of these for monitoring of posttraumatic injury to the spinal cord or brain will lead to a quantifiable means for the more efficient clinical evaluation of neuroprotective pharmacotherapies in regard to dose-response, therapeutic window, and optimal dosing regimen.

References

Alderson P, Roberts I. (1997). Corticosteroids in acute traumatic brain injury: systematic review of randomised controlled trials, *BMJ* 314:1855-1859.

Anderson DK, Hall ED. (1993). Pathophysiology of spinal cord trauma, *Ann Emerg Med* 22:987-992.

Anderson DK *et al.* (1991). Effect of delayed administration of U74006F (tirilazad mesylate) on recovery of locomotor function after experimental spinal cord injury, *J Neurotrauma* 8:187-192.

Anderson DK, Means ED, Waters TR, Green ES. (1982). Microvascular perfusion and metabolism in injured spinal cord after methylprednisolone treatment, *J Neurosurg* 56:106-113.

Anderson DK *et al.* (1985). Lipid hydrolysis and peroxidation in injured spinal cord: partial protection with methylprednisolone or vitamin E and selenium, *Cent Nerv Syst Trauma* 2:257-267.

Azbill RD *et al.* (1997). Impaired mitochondrial function, oxidative stress and altered antioxidant enzyme activities following traumatic spinal cord injury, *Brain Res* 765:283-290.

Bading H, Ginty DD, Greenberg ME. (1993). Regulation of gene expression in hippocampal neurons by distinct calcium signaling pathways, *Science* 260:181-186.

Banik NL, Shields DC. (2000). The role of calpain in neurofilament protein degradation associated with spinal cord injury, *Methods Mol Biol* 144:195-201.

Bao F, Liu D. (2003). Peroxynitrite generated in the rat spinal cord induces apoptotic cell death and activates caspase-3, *Neuroscience* 116:59-70.

Bartholdi D, Schwab ME. (1995). Methylprednisolone inhibits early inflammatory processes but not ischemic cell death after experimental spinal cord lesion in the rat, *Brain Res* 672:177-186.

Bartus R. (1997). The calpain hypothesis of neurodegeneration: evidence for a common cytotoxic pathway, *The Neuroscientist* 3:314-327.

Beckman JS. (1991). The double-edged role of nitric oxide in brain function and superoxide-mediated injury, *J Dev Physiol* 15:53-59.

Behrmann DL, Bresnahan JC, Beattie MS. (1994). Modeling of acute spinal cord injury in the rat: neuroprotection and enhanced recovery with methylprednisolone, U-74006F and YM-14673, *Exp Neurol* 126:61-75.

Berden JH, Hoitsma AJ, Merx JL, Keyser A. (1985). Severe central-nervous-system toxicity associated with cyclosporin, *Lancet* 1:219-220.

Bernardi P. (1996). The permeability transition pore. Control points of a cyclosporin A-sensitive mitochondrial channel involved in cell death, *Biochim Biophys Acta* 1275:5-9.

Bernardi P, Broekemeier KM, Pfeiffer DR. (1994). Recent progress on regulation of the mitochondrial permeability transition pore; a cyclosporin-sensitive pore in the inner mitochondrial membrane, *J Bioenerg Biomembr* 26:509-517.

Bracken MB. (1993). Pharmacological treatment of acute spinal cord injury: current status and future projects, *J Emerg Med* 11(Suppl 1):43-48.

Bracken MB. (2001). Methylprednisolone and acute spinal cord injury: an update of the randomized evidence, *Spine* 26:S47-54.

Bracken MB *et al.* (1984). Efficacy of methylprednisolone in acute spinal cord injury, *JAMA* 251:45-52.

Bracken MB, Holford TR. (1993). Effects of timing of methylprednisolone or naloxone administration on recovery of segmental and long-tract neurological function in NASCIS 2, *J Neurosurg* 79:500-507.

Bracken MB, Holford TR. (2002). Neurological and functional status 1 year after acute spinal cord injury: estimates of functional recovery in National Acute Spinal Cord Injury Study II from results modeled in National Acute Spinal Cord Injury Study III, *J Neurosurg* 96:259-266.

Bracken MB *et al.* (1990). A randomized, controlled trial of methylprednisolone or naloxone in the treatment of acute spinal-cord injury. Results of the Second National Acute Spinal Cord Injury Study, *N Engl J Med* 322:1405-1411.

Bracken MB *et al.* (1992). Methylprednisolone or naloxone treatment after acute spinal cord injury: 1-year follow-up data. Results of the Second National Acute Spinal Cord Injury Study, *J Neurosurg* 76:23-31.

Bracken MB *et al.* (1985). Methylprednisolone and neurological function 1 year after spinal cord injury. Results of the National Acute Spinal Cord Injury Study, *J Neurosurg* 63:704-713.

Bracken MB *et al.* (1997). Administration of methylprednisolone for 24 or 48 hours or tirilazad mesylate for 48 hours in the treatment of acute spinal cord injury. Results of the Third National Acute Spinal Cord Injury Randomized Controlled Trial. National Acute Spinal Cord Injury Study, *JAMA* 277:1597-1604.

Bracken MB *et al.* (1998). Methylprednisolone or tirilazad mesylate administration after acute spinal cord injury: 1-year follow up. Results of the Third National Acute Spinal Cord Injury Randomized Controlled Trial, *J Neurosurg* 89:699-706.

Braughler JM. (1985). Lipid peroxidation-induced inhibition of gamma-aminobutyric acid uptake in rat brain synaptosomes: protection by glucocorticoids, *J Neurochem* 44:1282-1288.

Braughler JM *et al.* (1988). A new 21-aminosteroid antioxidant lacking glucocorticoid activity stimulates adrenocorticotropin secretion and blocks arachidonic acid release from mouse pituitary tumor (AtT-20) cells, *J Pharmacol Exp Ther* 244:423-427.

Braughler JM, Duncan LA, Chase RL. (1985). Interaction of lipid peroxidation and calcium in the pathogenesis of neuronal injury, *Cent Nerv Syst Trauma* 2:269-283.

Braughler JM, Hall ED. (1982). Correlation of methylprednisolone levels in cat spinal cord with its effects on (Na+ + K+)-ATPase, lipid peroxidation, and alpha motor neuron function, *J Neurosurg* 56:838-844.

Braughler JM, Hall ED. (1983a). Lactate and pyruvate metabolism in injured cat spinal cord before and after a single large intravenous dose of methylprednisolone, *J Neurosurg* 59:256-261.

Braughler JM, Hall ED. (1983b). Uptake and elimination of methylprednisolone from contused cat spinal cord following intravenous injection of the sodium succinate ester, *J Neurosurg* 58:538-542.

Braughler JM, Hall ED. (1984). Effects of multi-dose methylprednisolone sodium succinate administration on injured cat spinal cord neurofilament degradation and energy metabolism, *J Neurosurg* 61:290-295.

Braughler JM, Hall ED. (1989). Central nervous system trauma and stroke. I. Biochemical considerations for oxygen radical formation and lipid peroxidation, *Free Radic Biol Med* 6:289-301.

Braughler JM, Hall ED. (1992). Involvement of lipid peroxidation in CNS injury, *J Neurotrauma* 9 Suppl 1:S1-7.

Braughler JM *et al.* (1987). Evaluation of an intensive methylprednisolone sodium succinate dosing regimen in experimental spinal cord injury, *J Neurosurg* 67:102-105.

Bringold U, Ghafourifar P, Richter C. (2000). Peroxynitrite formed by mitochondrial NO synthase promotes mitochondrial Ca2+ release, *Free Radic Biol Med* 29:343-348.

Broekemeier KM, Dempsey ME, Pfeiffer DR. (1989). Cyclosporin A is a potent inhibitor of the inner membrane permeability transition in liver mitochondria, *J Biol Chem* 264:7826-7830.

Broekemeier KM, Pfeiffer DR. (1995). Inhibition of the mitochondrial permeability transition by cyclosporin A during long time frame experiments: relationship between pore opening and the activity of miotchondrial phospholipases, *Biochemistry* 34:16440-16449.

Chen A, Xu XM, Kleitman N, Bunge MB. (1996). Methylprednisolone administration improves axonal regeneration into Schwann cell grafts in transected adult rat thoracic spinal cord, *Exp Neurol* 138:261-276.

Clark CR, Maclusky NJ, Naftolin F. (1981). Glucocorticoid receptors in the spinal cord, *Brain Res* 217:412-415.

Coleman WP *et al.* (2000). A critical appraisal of the reporting of the National Acute Spinal Cord Injury Studies (II and III) of methylprednisolone in acute spinal cord injury, *J Spinal Disord* 13:185-199.

de Groen PC *et al.* (1987). Central nervous system toxicity after liver transplantation. The role of cyclosporine and cholesterol, *N Engl J Med* 317:861-866.

Demopoulos HB, Flamm ES, Pietronigro DD, Seligman ML. (1980). The free radical pathology and the microcirculation in the major central nervous system disorders, *Acta Physiol Scand Suppl* 492:91-119.

Diem R *et al.* (2003). Methylprednisolone increases neuronal apoptosis during autoimmune CNS inflammation by inhibition of an endogenous neuroprotective pathway, *J Neurosci* 23:6993-7000.

Dimlich RV *et al.* (1990). Effects of a 21-aminosteroid (U-74006F) on cerebral metabolites and edema after severe experimental head trauma, *Adv Neurol* 52:365-375.

Dinkel K, MacPherson A, Sapolsky RM. (2003). Novel glucocorticoid effects on acute inflammation in the CNS, *J Neurochem* 84:705-716.

Elliott EM, Sapolsky RM. (1992). Corticosterone enhances kainic acid-induced calcium elevation in cultured hippocampal neurons, *J Neurochem* 59:1033-1040.

Faden AI. (1997). Therapeutic approaches to spinal cord injury, *Adv Neurol* 72:377-386.

Faden AI, Holaday JW. (1981). A role for endorphins in the pathophysiology of spinal cord injury, *Adv Biochem Psychopharmacol* 28:435-446.

Faden AI, Jacobs TP, Holaday JW. (1981). Opiate antagonist improves neurologic recovery after spinal injury, *Science* 211:493-494.

Faden AI, Jacobs TP, Holaday JW. (1982). Comparison of early and late naloxone treatment in experimental spinal injury, *Neurology* 32:677-681.

Faden AI, Salzman S. (1992). Pharmacological strategies in CNS trauma, *Trends Pharmacol Sci* 13:29-35.

Famiglio L *et al.* (1989). Central nervous system toxicity of cyclosporine in a rat model, *Transplantation* 48:316-321.

Farooque M, Hillered L, Holtz A, Olsson Y. (1996). Effects of methylprednisolone on extracellular lactic acidosis and amino acids after severe compression injury of rat spinal cord, *J Neurochem* 66:1125-1130.

Fehlings MG. (2001a). Editorial: recommendations regarding the use of methylprednisolone in acute spinal cord injury: making sense out of the controversy, *Spine* 26:S56-57.

Fehlings MG. (2001b). Summary statement: the use of methylprednisolone in acute spinal cord injury, *Spine* 26:S55.

Fineman I *et al.* (1993). Concussive brain injury is associated with a prolonged accumulation of calcium: a 45Ca autoradiographic study, *Brain Res* 624:94-102.

Finkel E. (2001). The mitochondrion: Is it central to apoptosis?, *Science* 292:624-626.

Flamm ES *et al.* (1985). A phase I trial of naloxone treatment in acute spinal cord injury, *J Neurosurg* 63:390-397.

Flamm ES *et al.* (1982). Experimental spinal cord injury: treatment with naloxone, *Neurosurgery* 10:227-231.

Fukiage C *et al.* (1997). SJA6017, a newly synthesized peptide aldehyde inhibitor of calpain: amelioration of cataract in cultured rat lenses, *Biochim Biophys Acta* 1361:304-312.

Galandiuk S, Raque G, Appel S, Polk HC, Jr. (1993). The two-edged sword of large-dose steroids for spinal cord trauma, *Ann Surg* 218:419-425; discussion 425-417.

Geisler FH, Coleman WP, Grieco G, Poonian D. (2001a). Measurements and recovery patterns in a multicenter study of acute spinal cord injury, *Spine* 26:S68-86.

Geisler FH, Coleman WP, Grieco G, Poonian D. (2001b). Recruitment and early treatment in a multicenter study of acute spinal cord injury, *Spine* 26:S58-67.

Geisler FH, Coleman WP, Grieco G, Poonian D. (2001c). The Sygen multicenter acute spinal cord injury study, *Spine* 26:S87-98.

Geisler FH, Dorsey FC, Coleman WP. (1990). GM1 gangliosides in the treatment of spinal cord injury: report of preliminary data analysis, *Acta Neurobiol Exp (Warsz)* 50:515-521.

Geisler FH, Dorsey FC, Coleman WP. (1991). Recovery of motor function after spinal-cord injury: a randomized, placebo-controlled trial with GM-1 ganglioside, *N Engl J Med* 324:1829-1838.

Giannotta SL, Weiss MH, Apuzzo ML, Martin E. (1984). High dose glucocorticoids in the management of severe head injury, *Neurosurgery* 15:497-501.

Globus MY *et al.* (1995). Glutamate release and free radical production following brain injury: effects of posttraumatic hypothermia, *J Neurochem* 65: 1704-1711.

Gorio A. (1986). Ganglioside enhancement of neuronal differentiation, plasticity, and repair, *CRC Crit Rev Clin Neurobiol* 2:241-296.

Gorio A. (1988). Gangliosides as a possible treatment affecting neuronal repair processes, *Adv Neurol* 47:523-530.

Gorio A, Vitadello M. (1987). Ganglioside prevention of neuronal functional decay, *Prog Brain Res* 71:203-208.

Hall ED. (1982). Glucocorticoid effects on central nervous excitability and synaptic transmission, *Int Rev Neurobiol* 23:165-195.

Hall ED. (1985). High-dose glucocorticoid treatment improves neurological recovery in head-injured mice, *J Neurosurg* 62:882-887.

Hall ED. (1992). The neuroprotective pharmacology of methylprednisolone, *J Neurosurg* 76:13-22.

Hall ED. (1995). Mechanisms of secondary CNS injury. Edited by J. D. Palmer. *Neurosurgery 96: Manual of neurosurgery,* New York: Churchill-Livingstone.

Hall ED. (1997). Lazaroid: mechanisms of action and implications for disorders of the CNS, *The Neuroscientist* 3:42-51.

Hall ED. (1998). Antioxidant pharmacotherapy. In Bogusslavsky MDGJ, editor: *Pathophysiology, diagnosis, and management,* Malden, MA: Blackwell Science.

Hall ED, Braughler JM. (1981). Acute effects of intravenous glucocorticoid pretreatment on the in vitro peroxidation of cat spinal cord tissue, *Exp Neurol* 73:321-324.

Hall ED, Braughler JM. (1982). Glucocorticoid mechanisms in acute spinal cord injury: a review and therapeutic rationale, *Surg Neurol* 18:320-327.

Hall ED, Braughler JM. (1989). Central nervous system trauma and stroke. II. Physiological and pharmacological evidence for involvement of oxygen radicals and lipid peroxidation, *Free Radic Biol Med* 6:303-313.

Hall ED, Braughler JM. (1993). Free radicals in CNS injury, *Res Publ Assoc Res Nerv Ment Dis* 71:81-105.

Hall ED *et al.* (1987). A nonglucocorticoid steroid analog of methylprednisolone duplicates its high-dose pharmacology in models of central nervous system trauma and neuronal membrane damage, *J Pharmacol Exp Ther* 242:137-142.

Hall ED, McCall JM, Means ED. (1994). Therapeutic potential of the lazaroids (21-aminosteroids) in acute central nervous system trauma, ischemia and subarachnoid hemorrhage, *Adv Pharmacol* 28:221-268.

Hall ED, Wolf DL, Braughler JM. (1984). Effects of a single large dose of methylprednisolone sodium succinate on experimental posttraumatic spinal cord ischemia. Dose-response and time-action analysis, *J Neurosurg* 61:124-130.

Hall ED *et al.* (1992). Biochemistry and pharmacology of lipid antioxidants in acute brain and spinal cord injury, *J Neurotrauma* 9 (Suppl 2):S425-442.

Hall ED, Yonkers PA, McCall JM, Braughler JM. (1988). Effects of the 21-aminosteroid U74006F on experimental head injury in mice, *J Neurosurg* 68:456-461.

Hall ED, Yonkers PA, Taylor BM, Sun FF. (1995). Lack of effect of postinjury treatment with methylprednisolone or tirilazad mesylate on the increase in eicosanoid levels in the acutely injured cat spinal cord, *J Neurotrauma* 12:245-256.

Harris AS, Morrow JS. (1988). Proteolytic processing of human brain alpha spectrin (fodrin): identification of a hypersensitive site, *J Neurosci* 8:2640-2651.

Heales SJ *et al.* (1999). Nitric oxide, mitochondria and neurological disease, *Biochim Biophys Acta* 1410:215-228.

Holaday JW, Faden AI. (1981a). Naloxone reverses the pathophysiology of shock through an antagonism of endorphin systems, *Adv Biochem Psychopharmacol* 28:421-434.

Holaday JW, Faden AI. (1981b). Naloxone treatment in shock, *Lancet* 2:201.

Holtz A, Nystrom B, Gerdin B. (1990). Effect of methylprednisolone on motor function and spinal cord blood flow after spinal cord compression in rats, *Acta Neurol Scand* 82:68-73.

Hunot S, Flavell RA. (2001). APOPTOSIS: death of a monopoly?, *Science* 292:865-866.

Hurlbert RJ. (2000). Methylprednisolone for acute spinal cord injury: an inappropriate standard of care, *J Neurosurg* 93:1-7.

Kahan BD. (1989). Cyclosporine, *N Engl J Med* 321:1725-1738.

Kampfl A *et al.* (1997). Mechanisms of calpain proteolysis following traumatic brain injury: implications for pathology and therapy: a review and update, *J Neurotrauma* 14:121-134.

Kanellopoulos GK *et al.* (1997). Neuronal cell death in the ischemic spinal cord: the effect of methylprednisolone, *Ann Thorac Surg* 64:1279-1285; discussion 1286.

Keller JN *et al.* (1997a). 4-Hydroxynonenal, an aldehydic product of membrane lipid peroxidation, impairs glutamate transport and mitochondrial function in synaptosomes, *Neuroscience* 80:685-696.

Keller JN *et al.* (1997b). Impairment of glucose and glutamate transport and induction of mitochondrial oxidative stress and dysfunction in synaptosomes by amyloid beta-peptide: role of the lipid peroxidation product 4-hydroxynonenal, *J Neurochem* 69:273-284.

Kontos HA, Povlishock JT. (1986). Oxygen radicals in brain injury, *Cent Nerv Syst Trauma* 3:257-263.

Kontos HA, Wei EP. (1986). Superoxide production in experimental brain injury, *J Neurosurg* 64:803-807.

Koyanagi I, Tator CH. (1997). Effect of a single huge dose of methylprednisolone on blood flow, evoked potentials, and histology after acute spinal cord injury in the rat, *Neurol Res* 19:289-299.

Kristal BS, Dubinsky JM. (1997). Mitochondrial permeability transition in the central nervous system: induction by calcium cycling-dependent and -independent pathways, *J Neurochem* 69:524-538.

Kruman I et al. (1997). Evidence that 4-hydroxynonenal mediates oxidative stress-induced neuronal apoptosis, J Neurosci 17:5089-5100.

Kupina NC et al. (2001). The novel calpain inhibitor SJA6017 improves functional outcome after delayed administration in a mouse model of diffuse brain injury, J Neurotrauma 18:1229-1240.

Kwiatkowski TG et al. (1999). Effects of tissue plasminogen activator for acute ischemic stroke at one year. National Institute of Neurological Disorders and Stroke Recombinant Tissue Plasminogen Activator Stroke Study Group, N Engl J Med 340:1781-1787.

Landfield PW, Eldridge JC. (1994). The glucocorticoid hypothesis of age-related hippocampal neurodegeneration: role of dysregulated intraneuronal calcium, Ann N Y Acad Sci 746:308-321; discussion 321-306.

Langham J et al. (2000). Calcium channel blockers for acute traumatic brain injury, Cochrane Database Syst Rev:CD000565.

Li PA et al. (1998). Postischemic treatment with calpain inhibitor MDL 28170 ameliorates brain damage in a gerbil model of global ischemia, Neurosci Lett 247:17-20.

Li X, Oudega M, Dancausse HA, Levi AD. (2000). The effect of methylprednisolone on caspase-3 activation after rat spinal cord transection, Restor Neurol Neurosci 17:203-209.

Lifshitz J et al. (2003). Structural and functional damage sustained by mitochondria after traumatic brain injury in the rat: evidence for differentially sensitive populations in the cortex and hippocampus, J Cereb Blood Flow Metab 23:219-231.

Liu D, Li L, Augustus L. (2001). Prostaglandin release by spinal cord injury mediates production of hydroxyl radical, malondialdehyde and cell death: a site of the neuroprotective action of methylprednisolone, J Neurochem 77:1036-1047.

Lopez-Figueroa MO et al. (2000). Direct evidence of nitric oxide presence within mitochondria, Biochem Biophys Res Commun 272:129-133.

Lyons WE et al. (1994). Immunosuppressant FK506 promotes neurite outgrowth in cultures of PC12 cells and sensory ganglia, Proc Natl Acad Sci U S A 91:3191-3195.

Mark RJ et al. (1997). A role for 4-hydroxynonenal, an aldehydic product of lipid peroxidation, in disruption of ion homeostasis and neuronal death induced by amyloid beta-peptide, J Neurochem 68:255-264.

Markgraf CG et al. (1998). Six-hour window of opportunity for calpain inhibition in focal cerebral ischemia in rats, Stroke 29:152-158.

Marler JR et al. (2000). Early stroke treatment associated with better outcome: the NINDS rt-PA stroke study, Neurology 55:1649-1655.

Marshall LF et al. (1998). A multicenter trial on the efficacy of using tirilazad mesylate in cases of head injury, J Neurosurg 89:519-525.

Matsushita M, Xiong G. (1997). Projections from the cervical enlargement to the cerebellar nuclei in the rat, studied by anterograde axonal tracing, J Comp Neurol 377:251-261.

Mattson MP, Barger SW, Begley JG, Mark RJ. (1995). Calcium, free radicals, and excitotoxic neuronal death in primary cell culture, Methods Cell Biol 46:187-216.

McCracken E et al. (1999). Calpain activation and cytoskeletal protein breakdown in the corpus callosum of head-injured patients, J Neurotrauma 16:749-761.

McIntosh LJ, Sapolsky RM. (1996). Glucocorticoids may enhance oxygen radical-mediated neurotoxicity, Neurotoxicology 17:873-882.

McIntosh T, Saatman KE, Raghupathi R. (1997). Calcium and the pathogenesis of traumatic CNS injury: cellular and molecular mechanisms, The Neuroscientist 3:169-175.

McIntosh TK, Raghupathi R. (1997). Calcium and the pathogenesis of traumatic CNS injury: cellular and molecular mechanisms, The Neuroscientist 3:169-175.

McIntosh TK, Thomas M, Smith D, Banbury M. (1992). The novel 21 aminosteroid U74006F attenuates cerebral edema and improves survival after brain injury in the rat, J Neurotrauma 9:33-46.

Mesenge C et al. (1998). Reduction of tyrosine nitration after N(omega)-nitro-L-arginine-methylester treatment of mice with traumatic brain injury, Eur J Pharmacol 353:53-57.

Molano Mdel R, Broton JG, Bean JA, Calancie B. (2002). Complications associated with the prophylactic use of methylprednisolone during surgical stabilization after spinal cord injury, J Neurosurg 96:267-272.

Morrow JD, Roberts LJ. (2002). The isoprostanes: their role as an index of oxidant stress status in human pulmonary disease, Am J Respir Crit Care Med 166:S25-30.

Morse JK, Davis JN. (1990). Regulation of ischemic hippocampal damage in the gerbil: adrenalectomy alters the rate of CA1 cell disappearance, Exp Neurol 110:86-92.

Narayan RK et al. (2002). Clinical trials in head injury, J Neurotrauma 19:503-557.

Neely MD, Sidell KR, Graham DG, Montine TJ. (1999). The lipid peroxidation product 4-hydroxynonenal inhibits neurite outgrowth, disrupts neuronal microtubules, and modifies cellular tubulin, J Neurochem 72:2323-2333.

Nicholls DG, Budd SL. (2000). Mitochondria and neuronal survival, Physiol Rev 80:315-360.

Nicolli A et al. (1996). Interactions of cyclophilin with the mitochondrial inner membrane and regulation of the permeability transition pore, a cyclosporin A-sensitive channel, J Biol Chem 271:2185-2192.

Nishio S et al. (1997). Detection of lipid peroxidation and hydroxyl radicals in brain contusion of rats, Acta Neurochir Suppl (Wien) 70:84-86.

Okonkwo DO, Buki A, Siman R, Povlishock JT. (1999). Cyclosporin A limits calcium-induced axonal damage following traumatic brain injury, Neuroreport 10:353-358.

Okonkwo DO, Povlishock JT. (1999). An intrathecal bolus of cyclosporin A before injury preserves mitochondrial integrity and attenuates axonal disruption in traumatic brain injury, J Cereb Blood Flow Metab 19:443-451.

Oudega M et al. (1999). Long-term effects of methylprednisolone following transection of adult rat spinal cord, Eur J Neurosci 11:2453-2464.

Pelinka LE, Toegel E, Mauritz W, Redl H. (2003). Serum S 100 B: a marker of brain damage in traumatic brain injury with and without multiple trauma, Shock 19:195-200.

Pike BR et al. (2001). Accumulation of non-erythroid alpha II-spectrin and calpain-cleaved alpha II-spectrin breakdown products in cerebrospinal fluid after traumatic brain injury in rats, J Neurochem 78:1297-1306.

Pike BR et al. (1998). Regional calpain and caspase-3 proteolysis of alpha-spectrin after traumatic brain injury, Neuroreport 9:2437-2442.

Popovich PG, Jones TB. (2003). Manipulating neuroinflammatory reactions in the injured spinal cord: back to basics, Trends Pharmacol Sci 24:13-17.

Popovich PG, Stokes BT, Whitacre CC. (1996). Concept of autoimmunity following spinal cord injury: possible roles for T lymphocytes in the traumatized central nervous system, J Neurosci Res 45:349-363.

Popovich PG, Wei P, Stokes BT. (1997a). Cellular inflammatory response after spinal cord injury in Sprague-Dawley and Lewis rats, J Comp Neurol 377:443-464.

Popovich PG, Yu JY, Whitacre CC. (1997b). Spinal cord neuropathology in rat experimental autoimmune encephalomyelitis: modulation by oral administration of myelin basic protein, J Neuropathol Exp Neurol 56:1323-1338.

Posmantur R et al. (1997). A calpain inhibitor attenuates cortical cytoskeletal protein loss after experimental traumatic brain injury in the rat, Neuroscience 77:875-888.

Pratico D et al. (2002). Local and systemic increase in lipid peroxidation after moderate experimental traumatic brain injury, J Neurochem 80:894-898.

Rabchevsky AG *et al.* (2002). Efficacy of methylprednisolone therapy for the injured rat spinal cord, *J Neurosci Res* 68:7-18.

Ray SK, Hogan EL, Banik NL. (2003). Calpain in the pathophysiology of spinal cord injury: neuroprotection with calpain inhibitors, *Brain Res Brain Res Rev* 42:169-185.

Ray SK *et al.* (2000). Increased calpain expression is associated with apoptosis in rat spinal cord injury: calpain inhibitor provides neuroprotection, *Neurochem Res* 25:1191-1198.

Ray SK *et al.* (2001a). Cell death in spinal cord injury (SCI) requires de novo protein synthesis. Calpain inhibitor E-64-d provides neuroprotection in SCI lesion and penumbra, *Ann N Y Acad Sci* 939:436-449.

Ray SK *et al.* (2001b). Inhibition of calpain-mediated apoptosis by E-64 d-reduced immediate early gene (IEG) expression and reactive astrogliosis in the lesion and penumbra following spinal cord injury in rats, *Brain Res* 916:115-126.

Ray SK *et al.* (1999). Calpeptin and methylprednisolone inhibit apoptosis in rat spinal cord injury, *Ann N Y Acad Sci* 890:261-269.

Rego AC, Ward MW, Nicholls DG. (2001). Mitochondria control ampa/kainate receptor-induced cytoplasmic calcium deregulation in rat cerebellar granule cells, *J Neurosci* 21:1893-1901.

Reulen H and Schurmann K. (1972). *Steroids and brain edema*, Berlin: Springer-Verlag.

Rink A *et al.* (1995). Evidence of apoptotic cell death after experimental traumatic brain injury in the rat, *Am J Pathol* 147:1575-1583.

Saatman KE *et al.* (1996). Calpain inhibitor AK295 attenuates motor and cognitive deficits following experimental brain injury in the rat, *Proc Natl Acad Sci U S A* 93:3428-3433.

Saatman KE, Zhang C, Bartus RT, McIntosh TK. (2000). Behavioral efficacy of posttraumatic calpain inhibition is not accompanied by reduced spectrin proteolysis, cortical lesion, or apoptosis, *J Cereb Blood Flow Metab* 20:66-73.

Sapolsky RM. (1985). Glucocorticoid toxicity in the hippocampus: temporal aspects of neuronal vulnerability, *Brain Res* 359:300-305.

Sapolsky RM, Krey LC, McEwen BS. (1986). The neuroendocrinology of stress and aging: the glucocorticoid cascade hypothesis, *Endocr Rev* 7:284-301.

Scheff SW, Benardo LS, Cotman CW. (1980). Hydrocortisone administration retards axon sprouting in the rat dentate gyrus, *Exp Neurol* 68:195-201.

Scheff SW, Cotman CW. (1982). Chronic glucocorticoid therapy alters axon sprouting in the hippocampal dentate gyrus, *Exp Neurol* 76:644-654.

Scheff SW, DeKosky ST. (1983). Steroid suppression of axon sprouting in the hippocampal dentate gyrus of the adult rat: dose-response relationship, *Exp Neurol* 82:183-191.

Scheff SW, Dekosky ST. (1989). Glucocorticoid suppression of lesion-induced synaptogenesis: effect of temporal manipulation of steroid treatment, *Exp Neurol* 105:260-264.

Scheff SW, Sullivan PG. (1999). Cyclosporin A significantly ameliorates cortical damage following experimental traumatic brain injury in rodents, *J Neurotrauma* 16:783-792.

Schimmer BP and Parker KL. (2001). Adrenocorticotropic hormone; adrenocortical steroids and their synthetic analogs: inhibitors of the synthesis and actions of adrenocortical hormones. In Hardman JG, Limberd LE, Gilman AG, editors: *Goodman and Gilman's the pharmacological basis of therapeutics* , New York: McGraw-Hill.

Shields DC, Schaecher KE, Hogan EL, Banik NL. (2000). Calpain activity and expression increased in activated glial and inflammatory cells in penumbra of spinal cord injury lesion, *J Neurosci Res* 61:146-150.

Shiga Y, Onodera H, Matsuo Y, Kogure K. (1992). Cyclosporin A protects against ischemia-reperfusion injury in the brain, *Brain Res* 595:145-148.

Shin CY *et al.* (2001). Glucocorticoids exacerbate peroxynitrite mediated potentiation of glucose deprivation-induced death of rat primary astrocytes, *Brain Res* 923:163-171.

Short DJ, El Masry WS, Jones PW. (2000). High dose methylprednisolone in the management of acute spinal cord injury: a systematic review from a clinical perspective, *Spinal Cord* 38:273-286.

Siesjo BK, Bengtsson F. (1989). Calcium fluxes, calcium antagonists, and calcium-related pathology in brain ischemia, hypoglycemia, and spreading depression: a unifying hypothesis, *J Cereb Blood Flow Metab* 9:127-140.

Smith SL, Andrus PK, Zhang JR, Hall ED. (1994). Direct measurement of hydroxyl radicals, lipid peroxidation, and blood-brain barrier disruption following unilateral cortical impact head injury in the rat, *J Neurotrauma* 11:393-404.

Stein-Behrens BA *et al.* (1992). Glucocorticoids exacerbate kainic acid-induced extracellular accumulation of excitatory amino acids in the rat hippocampus, *J Neurochem* 58:1730-1735.

Stewart VC *et al.* (2002). Nitric oxide-dependent damage to neuronal mitochondria involves the NMDA receptor, *Eur J Neurosci* 15:458-464.

Stewart VC, Sharpe MA, Clark JB, Heales SJ. (2000). Astrocyte-derived nitric oxide causes both reversible and irreversible damage to the neuronal mitochondrial respiratory chain, *J Neurochem* 75:694-700.

Stout AK *et al.* (1998). Glutamate-induced neuron death requires mitochondrial calcium uptake, *Nat Neurosci* 1:366-373.

Sullivan PG *et al.* (1999a). Exacerbation of damage and altered NF-kappaB activation in mice lacking tumor necrosis factor receptors after traumatic brain injury, *J Neurosci* 19:6248-6256.

Sullivan PG, Geiger JD, Mattson MP, Scheff SW. (2000a). Dietary supplement creatine protects against traumatic brain injury, *Ann Neurol* 48:723-729.

Sullivan PG, Keller JN, Mattson MP, Scheff SW. (1998). Traumatic brain injury alters synaptic homeostasis: implications for impaired mitochondrial and transport function, *J Neurotrauma* 15:789-798.

Sullivan PG *et al.* (2000b). Dose-response curve and optimal dosing regimen of cyclosporin A after traumatic brain injury in rats, *Neuroscience* 101:289-295.

Sullivan PG, Thompson M, Scheff SW. (2000c). Continuous infusion of cyclosporin A postinjury significantly ameliorates cortical damage following traumatic brain injury, *Exp Neurol* 161:631-637.

Sullivan PG, Thompson MB, Scheff SW. (1999b). Cyclosporin A attenuates acute mitochondrial dysfunction following traumatic brain injury, *Exp Neurol* 160:226-234.

Szabo I, Zoratti M. (1991). The giant channel of the inner mitochondrial membrane is inhibited by cyclosporin A, *J Biol Chem* 266:3376-3379.

Taoka Y, Okajima K, Uchiba M, Johno M. (2001). Methylprednisolone reduces spinal cord injury in rats without affecting tumor necrosis factor-alpha production, *J Neurotrauma* 18:533-543.

Tator CH, Fehlings MG. (1991). Review of the secondary injury theory of acute spinal cord trauma with emphasis on vascular mechanisms, *J Neurosurg* 75:15-26.

Uchino H *et al.* (1998). Amelioration by cyclosporin A of brain damage in transient forebrain ischemia in the rat, *Brain Res* 812:216-226.

Uchino H *et al.* (1995). Cyclosporin A dramatically ameliorates CA1 hippocampal damage following transient forebrain ischaemia in the rat, *Acta Physiol Scand* 155:469-471.

Valdez LB *et al.* (2000). Reactions of peroxynitrite in the mitochondrial matrix, *Free Radic Biol Med* 29:349-356.

Verity MA. (1992). Ca(2+)-dependent processes as mediators of neurotoxicity, *Neurotoxicology* 13:139-147.

Verweij BH *et al.* (1997). Mitochondrial dysfunction after experimental and human brain injury and its possible reversal with a selective N-type calcium channel antagonist (SNX-111), *Neurol Res* 19:334-339.

Verweij BH *et al.* (2000). Impaired cerebral mitochondrial function after traumatic brain injury in humans, *J Neurosurg* 93:815-820.

Walker RW, Brochstein JA. (1988). Neurologic complications of immunosuppressive agents, *Neurol Clin* 6:261-278.

Wang GJ, Thayer SA. (1996). Sequestration of glutamate-induced Ca2+ loads by mitochondria in cultured rat hippocampal neurons, *J Neurophysiol* 76:1611-1621.

Ward MW, Rego AC, Frenguelli BG, Nicholls DG. (2000). Mitochondrial membrane potential and glutamate excitotoxicity in cultured cerebellar granule cells, *J Neurosci* 20:7208-7219.

White RJ, Reynolds IJ. (1997). Mitochondria accumulate Ca2+ following intense glutamate stimulation of cultured rat forebrain neurones, *J Physiol (Lond)* 498:31-47.

Wingrave JM *et al.* (2003). Early induction of secondary injury factors causing activation of calpain and mitochondria-mediated neuronal apoptosis following spinal cord injury in rats, *J Neurosci Res* 3:95-104.

Wunderlich MT. (2003). Head injury outcome prediction in the emergency department: a role for protein S-100B?, *J Neurol Neurosurg Psychiatry* 74:827-828; author reply 828.

Xiong Y *et al.* (1997). Mitochondrial dysfunction and calcium perturbation induced by traumatic brain injury, *J Neurotrauma* 14:23-34.

Xu J, *et al.* (2001). Glucocorticoid receptor-mediated suppression of activator protein-1 activation and matrix metalloproteinase expression after spinal cord injury, *J Neurosci* 21:92-97.

Xu J, Qu ZX, Hogan EL, Perot PL, Jr. (1992). Protective effect of methylprednisolone on vascular injury in rat spinal cord injury, *J Neurotrauma* 9:245-253.

Yakovlev AG *et al.* (1997). Activation of CPP32-like caspases contributes to neuronal apoptosis and neurological dysfunction after traumatic brain injury, *J Neurosci* 17:7415-7424.

Young W, Flamm ES. (1982). Effect of high-dose corticosteroid therapy on blood flow, evoked potentials, and extracellular calcium in experimental spinal injury, *J Neurosurg* 57:667-673.

Yuen PW WK. (1996). Therapeutic potential of calpain inhibitors in neurodegenerative disorder, *Exp Opin Invet Drugs* 5:1291-1304.

Yuen PW WK. (1998). Calpain inhibitors: novel neuroprotectants and potential anticataract agents, *Drugs of the Future* 23:741-749.

Zanella B *et al.* (2002). Mitochondrial nitric oxide localization in H9c2 cells revealed by confocal microscopy, *Biochem Biophys Res Commun* 290:1010-1014.

Zhang SX, Bondada V, Geddes JW. (2003). Evaluation of conditions for calpain inhibition in the rat spinal cord: effective postinjury inhibition with intraspinal MDL28170 microinjection, *J Neurotrauma* 20:59-67.

Slowing the Progression of Multiple Sclerosis

Richard A. Rudick, MD

This book tracks the therapeutic transformation of neurology and the rapidly evolving understanding of disease mechanisms on which progress is based. Multiple sclerosis (MS) is a chronic disease of the central nervous system (CNS) characterized by inflammation, myelin and axonal destruction, intermittent disability, and eventual chronic progressive neurological deterioration. The rapid transformation of MS from untreatable to partially controllable and concurrent advances in understanding its underlying mechanisms over a 20-year period have been remarkable. Events over 20 years also exemplify the interdependence between therapeutic advances and basic neurobiological research. These two seemingly independent approaches to the study of human disease have become intertwined in the MS field. Clinical trials are indispensable for confirming mechanistic hypotheses, and mechanistic research has become indispensable for designing and testing clinical hypotheses. In this chapter, I review major historical developments in MS therapeutics and changing concepts of MS, and will emphasize the interdependence between these two lines of work. During the past 20 years, a short interval relative to the history of MS knowledge, we have developed extraordinary new insights into MS pathogenesis, as well as therapies that meaningfully ameliorate the disease. While current therapies are measurably effective, they are by no means adequate. It is hoped that advances in understanding pathology and pathogenesis will lead to even more rapid progress over the next 20 years.

BRIEF HISTORY OF MS EXPERIMENTAL THERAPEUTICS

Before the 1980s, MS therapeutics was relatively undeveloped as a discipline. There were no accepted methods for testing putative therapies empirically, and therapeutic claims without scientific evidence dominated (van den Noort, 1983). A notable exception was the study by Rose and colleagues (1968, 1970). That study set a standard for testing therapeutic interventions in MS and for doing outcomes research in MS generally. During the clinical trial, the Standard Neurologic Examination, the Kurtzke Disability Status Scale, and a 7-day symptom score were evaluated and each was found to be a reliable indicator of neurological status (Kuzma *et al.*, 1969). Rose and colleagues documented short-term benefits of ACTH in resolving disability related to relapse. It also documented the feasibility of multicenter, placebo-controlled clinical trials, and foreshadowed an era of much more scientific approaches to MS therapeutics.

In April 1982, the MS Societies of the United States and Canada convened the first international conference on therapeutic trials in MS on Grand Isle, NY (Herndon and Murray, 1983). Conference participants were skeptical about the feasibility of MS clinical trials, voicing concerns that MS was too variable and too unpredictable, that there were no reliable outcome measures, and little understanding of the disease mechanisms. The participants advocated basic research in preference to clinical trials. The important topics discussed at the conference included the rationale for therapeutic approaches, scoring techniques and problems in evaluating patients, design problems, interpretation of contemporary trial results, financing for clinical trials, patient selection, simple protocol development, and ethical or legal aspects of clinical trials. The Grand Isle conference was a watershed event that ended an inactive, nihilistic period, and ushered in a 20-year period of unprecedented advances in MS therapeutics.

The most significant development in the 1980s in MS experimental therapeutics was not a clinical trial; it was the application of magnetic resonance imaging (MRI) to MS patients. Initially a diagnostic tool (Johnson *et al.*, 1984),

MRI quickly became entrenched as a tool to understand evolution and pathogenesis of MS and to monitor therapy. Also in the 1980s, a large multicenter trial of cyclosporine was conducted for progressive MS (The Multiple Sclerosis Study Group, 1990). The study found an unfavorable ratio of efficacy to toxicity, so cyclosporine did not achieve widespread use. Analogous to the ACTH trial, pivotal outcomes research was done as part of the cyclosporine trial. Syndulko and Tourtellotte applied quantitative evaluation of neurological function to a subset in the cyclosporine study. This approach provided sensitive and reliable markers of disease progression and foreshadowed development of the multiple sclerosis functional composite (MSFC), which was recommended as a clinical trial outcome measure 10 years later (Rudick et al., 1996, 1997).

In the 1990s, pivotal trials of disease-modifying drugs were conducted, described later, leading to approval and marketing of partially effective therapies for relapsing remitting MS (RRMS). This formally ended the era of MS as an untreatable disease, and stimulated intense interest in MS by the pharmaceutical industry. Also, in the mid-1990s, the underlying basis for MS clinical trials was advanced by another pivotal workshop, this one on outcome measures (Whitaker et al., 1995). The scientific methods for testing experimental therapies were examined and critiqued, and recommendations were provided for future progress. As a direct outgrowth of this workshop, task forces were organized to develop recommendations on the use of MRI as a clinical trial outcome measure (Miller et al., 1996) and to recommend optimal strategies for clinical outcome assessment in future trials (Rudick et al., 1997).

Into the late twentieth century, MS was viewed as a disease characterized by episodic relapses, occurring about once per year, interspersed with disease remission. Relapses were assumed to be caused by immune attack on CNS tissue, and MS became the prototypic organ-specific autoimmune disease, a distinction that remains. As implied by the name relapsing remitting MS, the disease was assumed to be quiescent between relapses. This explains the near-exclusive focus on relapses (also known as exacerbations) through the 1980s. Relapse count or rate was considered a direct and straightforward disease measure and was the most common outcome measure for clinical trials in RRMS. To some degree, this view still holds. It was recognized that many patients eventually developed gradually progressive neurological disability, a disease stage termed secondary progressive MS (SPMS). This occurred most commonly 15 to 20 years after symptom onset. This phase of the disease was considered to be different from RRMS, but the mechanisms leading to transition from RRMS to SPMS were not known. During the progressive stage of MS, neurological disability was viewed as a more appropriate measure of the disease

process, and the Kurtzke Disability Status Scale (Kurtzke, 1955) and subsequently the Kurtzke Expanded Disability Status Scale (Kurtzke, 1983) were accepted as direct measures of the underlying disease. This describes the prevailing view of MS and its measurement at the time of the Grand Isle international conference on MS Clinical Trials.

MRI studies in the 1980s were changing the traditional view of MS, however. New brain lesions were observed in clinically stable patients, dispelling the myth of remission between relapses. The ratio of new MRI lesions to clinical relapses was reported as 4:1 (Paty, 1993) and later increased to 10:1 (McFarland et al., 1996). An early pioneer in the application of MRI to MS, Don Paty, commonly stated that MRI visualized the pathology and was a much more direct disease measure than clinical outcomes. As a result of the MRI studies, MS was more accurately perceived as active during the early stages, and relapses were increasingly seen as only a loose reflection of underlying pathology. Disease activity and progression during the relapsing stage of MS were viewed increasingly as subclinical, and MRI was incorporated into MS clinical trials.

At the present time, MS is viewed as a largely subclinical inflammatory disease in its early stage. Periodic relapses punctuate the early course in most cases, but do not accurately reflect the severity of the underlying pathology. As viewed by imaging studies, MS is a continuously active, albeit fluctuating disease process, which results in irreversible brain tissue injury and brain atrophy alarmingly early in the disease course. After years of fluctuating inflammation with related tissue injury, progressive neurological disability ensues relatively late in the disease. The pathological targets in MS are now known to include neurons and axons in addition to myelin, and progressive axonal loss is thought responsible for progressive neurological disability commonly observed in later stages. In the tradition of Rose and colleagues in the 1970s, and Syndulko, Tourtellotte, and colleagues in the 1980s, attention has turned toward outcomes assessment, both through rapidly evolving imaging techniques and improved clinical assessement methodology, and to therapeutic targets.

INTERFERON THERAPY FOR MS

Interferon therapy is the most widely used disease therapy for MS. The history of its development is described elsewhere (Jacobs and Johnson, 1994). The reader may wish to consult various detailed reviews of interferon clinical trials (Calabresi, 2002; Grigoriadis, 2002; Horowski, 2002; Hughes, 2003; Noseworthy, 2003; Noseworthy et al., 2000; Rolak, 2001; Rudick, 2001; Vartanian, 2003; Wiendl and Kieseier, 2003). In this section, key trials will be highlighted to point out results of historical significance and to relate interferon therapy to disease pathogenesis.

Rationale and Early Trials

The rationale for testing interferon was the view that MS might be caused by a viral infection and the observation that lymphocytes from MS patients produced subnormal amounts of interferon in response to viral or mitogen challenge (Neighbour *et al.*, 1981). Early studies used natural interferon preparations produced by fibroblast cultures. Natural interferon-β was administered by intrathecal injection (Jacobs *et al.*, 1981), and natural interferon-α by subcutaneous injection (Knobler *et al.*, 1984) in independent studies that were started around the same time. Results suggested that interferon α or β (both type I interferons) reduced the frequency of clinical relapses. Definitive interpretation was difficult because the studies were small (10 patients in the Jacobs study and 24 in the Knobler study) and because of methodological limitations.

Recombinant DNA technology developed during the 1980s eventuated in recombinant interferon preparations (rIFN) in sufficient purity and quantity to allow large scale testing. A pilot study of intravenous rIFNγ (type II interferon) was initiated because IFNγ shared antiviral, antiproliferative, and certain immunomodulatory activities with type I interferons. Unexpectedly, patients treated with rIFNγ exhibited an alarmingly high relapse rate during the initial month of treatment, leading to early termination of the study (Panitch *et al.*, 1987). The explanation for worsening disease with IFN-γ but amelioration with IFN-β or IFN-α has never been fully clarified, but all subsequent interferon studies have used type I interferon. Some investigators suggested that differential immune effects of type I and type II interferon might underlie different clinical effects, but this remains conjectural. Efficacy with IFN-α was confirmed in a larger trial (Durelli *et al.*, 1994), but significant toxicity discouraged further development. Definitive trials of interferon for MS have therefore used recombinant preparations of IFN-β.

Definitive Trials

In the most remarkable 10-year period in the history of MS therapeutics, pivotal trials with three preparations of IFN-β were planned, conducted, and successfully completed between 1987 and 1997. The studies led to approval for RRMS by regulatory agencies for IFN-β1b (Betaseron) (The IFN-β Multiple Sclerosis Study Group, 1993), IFN-β1a (Avonex) (Jacobs *et al.*, 1996), and IFN-β1a (Rebif) (PRISMS Study Group, 1998).

The IFN-β1b (Betaseron) study was historical for two main reasons. First, Betaseron was the initial drug approved to modify the natural history of MS. Its approval officially ended the era of MS as an untreatable disease. Second, approval was dependent on the observation that Betaseron reduced new MRI lesions. Betaseron was the first drug that achieved U.S. Food and Drug Administration (FDA) approval based on expedited review authority allowing use of a nonclinical surrogate marker. Since there were no statistically significant benefits of Betaseron on the disability endpoints, the drug may not have been approved were it not for prominent reduction in new T_2 lesions. After the FDA approved Betaseron in spring 1993, all MS clinical trials incorporated cranial MRI scans. Interestingly, the pivotal study of glatiramer acetate (GA, Copaxone), designed before the end of the Betaseron study, did not include MRI scans. Glatiramer acetate was approved for use on the basis of a beneficial relapse effect. A study was conducted subsequently to evaluate the effects of GA on MRI lesions (Comi *et al.*, 2001b).

The second pivotal interferon trial tested IFN-β1a (Avonex) (Jacobs *et al.*, 1996). This trial was of historical significance for several reasons. It was the first major RRMS trial targeting disability progression as the primary endpoint. Relapses were demoted to a secondary outcome status. This resulted from the NIH peer review. In the context of new MRI studies showing subclinical disease activity, the study section viewed relapses as a suboptimal outcome measure. Although disability progression had not been a focus of prior trials in RRMS, members of the study section recommended that the investigation focus on disability progression. Disability progression was defined as ≥1.0 point worsening from baseline EDSS, persisting for at least 6 months. The Avonex trial also incorporated a time-to-failure (survival curve) analysis for the primary outcome, a method not commonly used in MS trials. The study demonstrated a 37% reduction in the probability of sustained disability progression with IFN-β1a, which satisfied the prospectively selected primary outcome measure. While the drug was approved by regulatory agencies, the study results were controversial, as there was no consensus in the MS field on the proper definition of disability progression in RRMS. The Avonex study stimulated much lively debate about the significance of the endpoint used in the trial and about optimal clinical outcome measures for RRMS trials generally.

The PRISMS trial of IFN-β1a (Rebif) (PRISMS Study Group, 1998) compared placebo with two doses of IFN-β1a—22 μg and 44 μg—administered by subcutaneous injection three times per week. Relapse count per patient was the primary outcome in this trial, but disability progression, defined as a 1-point worsening from baseline EDSS lasting at least 3 months, was included as a secondary outcome measure. As with the Avonex trial, the PRISMS study used time-to-failure analysis for the disability outcome. Results on this outcome were similar to the Avonex trial. At the end of the originally planned 2-year PRISMS study, patients on IFN-β1a were continued on drug, and placebo patients were randomized to either of the two doses. Patients were then monitored an additional 2 years, and about 80% of the original study population was available for analysis at 4 years. This interesting design permitted evaluation delayed compared with early treatment. This study has provided some of the best evidence suggesting that early therapy is beneficial.

The three pivotal RRMS interferon studies recruited patients with similar characteristics: disease duration between 4 and 8 years, relatively high prestudy relapse rates, and mild-moderate disability at study entry (Table 4.1). Disease duration was similar but by no means the same in the different trials. For example, disease duration was about 50% longer in the Avonex compared with the Betaseron study. Since disease duration may be an important determinant of therapeutic outcome (see SPMS discussion later), this difference in the study populations may be important. Second, there is variation in prestudy relapse rate between the trials. Patients in the Betaseron study had higher relapse rates than Avonex patients. While this may relate to methodological differences in defining and recording relapses, it highlights a major problem in MS clinical trials: lack of standard definitions or standard procedures for defining or measuring relapses. Consequently, comparing trial results from the separate studies is hazardous and probably not meaningful.

Table 4.2 lists the effects of IFN-β on relapses or new MRI activity. There is a dose effect for the Betaseron study—the rate ratio for relapses was 0.87 with the lower dose and 0.69 with the higher dose. However, there was no dose effect in the Rebif study—the rate ratio was 0.71 for the lower dose and 0.68 for the higher dose. This suggests that increasing the dose beyond a ceiling does not translate into additional therapeutic benefits. The rate

ratio in the Avonex trial was 0.70 for patients studied for 2 years, but 0.82 when all patients were included. The difference has not been fully explained. One possibility is that patients monitored for short intervals may artificially impact relapse rate calculations. Another possibility is that patients monitored for less than 2 years (who were entered toward the end of the Avonex trial) had less disease activity than patients entered earlier in the study, thereby reducing the relapse rate in the placebo patients. Irrespective of the correct explanation, the relapse rate reduction in the whole Avonex trial population raised the possibility that a suboptimal dose of IFN-β1a was used. Because of this concern, a study was conducted to determine if 60 μg IFN-β1a (Avonex) intramuscularly weekly was more effective than 30 μg as used in the original trial. That study enrolled over 800 patients, so it was adequately powered to determine differences. There was no benefit of the higher dose, however (Clanet et al., 2002).

Second, Table 4.2 shows that rate ratios were lower for MRI activity measures compared with clinical relapses, indicating a higher therapeutic effect size for MRI activity compared with relapses. This interesting finding has stimulated debate about the significance of new MRI lesions compared with relapses in RRMS studies. Which parameter is more meaningful? To date, there are no data regarding how these two outcomes compare as predictors for clinically meaningful disability in future years.

Direct comparison across studies is also difficult for disability measures. The designs of the studies, duration of follow-up period, and application of the EDSS scale differed in each study. Also, the distribution of patients at each EDSS level differed. This alone would result in different progression rates from study to study, as the amount of time patients remain at each EDSS step differs (Weinshenker et al., 1991). Despite these caveats, EDSS was applied in all three pivotal trials. For all patients and all time on study in the Betaseron trial, there was no reduction in 3-month sustained EDSS worsening for the lower dose group,

TABLE 4.1   Characteristics of Patients in Pivotal Trials Leading to Approval of IFN for Relapsing Remitting Multiple Sclerosis

	Disease Duration (Yrs)	RR* in Prior 2 Years	EDSS[†] at Study Entry	EDSS Range
IFN-β1b (Betaseron)				
Placebo	3.9	1.8	2.8	0-5.5
8 MIU SC QOD	4.7	1.7	3.0	
IFN-β1a (Avonex)				
Placebo	6.4	1.2	2.3	1-3.5
30 μg IM QWk	6.6	1.2	2.4	
IFN-β1a (Rebif)				
Placebo	6.1	1.5	2.4	
22 μg SC 3 times per wk	7.7	1.5	2.5	0-5.5
44 μg SC 3 times per wk	7.8	1.5	2.5	

* RR=relapse rate
[†] EDSS=Expanded Disability Status Score

TABLE 4.2 Characteristics of Patients in Pivotal Trials Leading to Approval of IFN for Relapsing Remitting Multiple Sclerosis (Modified from Kappos, 2003)

	Relapse Outcome*			MRI Activity Measure[†]		
	IFN1[‡] vs Plc	IFN2[§] vs Plc	IFN2 vs IFN1	IFN1 vs Plc	IFN1 vs Plc	IFN2 vs IFN1
IFN-β1b (Betaseron) 1.6 MIU SC QOD 8 MIU SC QOD	0.87	0.69	0.79	0.37	0.41	1.1
IFN-β1a (Avonex) 30 µg IM QWk	0.82[‖]	NA	NA	0.67	NA	NA
IFN-β1a (Rebif) 22 µg SC TIW 44 µg SC TIW	0.71	0.68	0.95	0.19	0.13	0.65

*Expressed as a rate ratio of IFN-treated patients compared to placebo-treated patients.
[†]Based on different methods for detecting and reporting new lesions.
[‡]IFN1 = lower dose IFN.
[§]IFN2 = higher dose IFN.
[‖]Based on the entire cohort. Rate ratio in 2-year cohort was 0.70.

and a 29% statistically nonsignificant reduction in the higher dose group. For the Avonex study there was a 37%, statistically significant, reduction in 6-month sustained EDSS worsening in the active arm. For the Rebif study, there was a 23% statistically significant reduction in 3-month sustained EDSS worsening with the lower dose, and a 31% statistically significant reduction with the higher dose. EDSS change from baseline was not published for the Betaseron study. For the 2-year cohort in the Avonex study, mean EDSS worsening was 0.74 in placebo patients, and 0.25 in IFN-β1a patients. In the Rebif study, mean EDSS worsening was 0.48 in placebo patients, reduced to 0.23 in the lower dose patients and 0.24 in the higher dose patients. The explanation for beneficial effects on EDSS in the IFN-β1a studies, but not in the IFN-β1b study is not clear. This may relate to methodological differences related to EDSS or may reflect lower efficacy with the IFN-β1b drug.

Subsequent Studies in Other MS Disease Categories

The pivotal RRMS interferon trials changed the perception of MS from an untreatable to a treatable condition, which has had profound scientific, psychological, and practical implications. The "market" for MS disease therapy quickly developed, with worldwide sales of interferon products now exceeding $2.5 billion. This has driven pharmaceutical industry investment in MS research. Studies were initiated to evaluate the effects of interferon in SPMS, and at the time of a clinically isolated syndrome, as well as numerous head-to-head comparison studies, most designed to show that one product was "superior" to another, as reviewed by Kappos (2003).

Two studies addressed the effects of IFN-β1a at the time of a clinically isolated syndrome. In the CHAMPS study (Jacobs et al., 2000), 383 patients with an acute demyelinating event, together with brain MRI lesions, were randomized to receive weekly IFN-β1a (Avonex) or placebo, after initial treatment with high-dose corticosteroids. During a follow-up period of up to 2 years, the cumulative probability of developing clinically definite MS was lower in the IFN-β1a than the placebo group (rate ratio, 0.56). The IFN-β1a group also had significantly reduced volumes of T_2-weighted MRI brain lesions, fewer new or enlarging T_2 lesions, and fewer gadolinium-enhancing lesions at 18 months. Comi and colleagues (2001a) reported results from the ETOMS study, in which patients with clinically isolated syndromes were randomized to IFN-β1a (Rebif) 22 µg once weekly, or placebo. Fewer patients on Rebif developed clinically definite MS (34% vs 45%; $P < .05$). Both the CHAMPS and ETOMS studies demonstrated that IFN-β1a treatment at the first MS clinical symptom had significant positive effects on clinical and MRI outcomes. These studies were important for the following reasons:

1. Both studies demonstrated that patients with clinically isolated syndromes and ≥ 2 cranial MRI lesions had a high risk of developing relapses or new MRI lesions within 2 years.
2. The studies provide large cohorts in which to determine the relationship between baseline characteristics and future disability. Such studies will be useful in identifying patients at first presentation who are at high risk for disability progression who can be selected for more aggressive medical management.

3. Both studies document beneficial effects of IFN-β at first presentation. They provide a potential opportunity to determine long-term benefits of treatment started at first presentation, and may be useful in determining individual therapeutic response predictors.

The second major initiative following the RRMS trials was application of IFN-β to SPMS. Four major trials were conducted, as listed in Tables 4.3 and 4.4. In the first study, The European SP MS Study (European Study Group on interferon beta-1b in secondary progressive MS, 1998; Miller et al., 1999), IFN-β1b (Betaferon) 8 MIU (Million International Units) SC QOD (every other day) was compared with placebo in a randomized, double-blind clinical trial. There was a significant benefit of treatment on time to EDSS progression (≥ 1.0 worsening from baseline EDSS, confirmed after 3 months). The benefit was apparent across EDSS levels, and in patients with or without superimposed relapses. As in the RRMS trials, the modest clinical benefits were accompanied by more pronounced MRI benefits. A subset of 95 patients was studied over 36 months to determine the effect of IFN-β1b on MRI markers of tissue destruction (Molyneux et al., 2000). Cerebral volumes declined by 3.9% in the placebo group and 2.9% in the treated group by month 36. The differences were not statistically significant. There was, however, a therapeutic effect of IFN-β1b on T_1 black holes (Barkhof et al., 2001) in the same patients. Baseline T_1 black hole lesion load was matched at baseline in the two groups; it increased by a median of 14% per year in the placebo group and 7.7% per year in IFN-β1b group ($P < 0.01$). Because of the beneficial effects of IFN-β1b on EDSS progression and T_1 black hole progression, these encouraging results suggested that IFN-β1b may have reduced axonal injury in SPMS.

This was a very hopeful, but transient, moment in the history of IFN-β therapy for MS; unfortunately, it was not replicated in other interferon SPMS studies. A second trial of the same drug, IFN-β1b (Betaseron) (Goodkin & The North American Study Group on Interferon beta-1b in Secondary Progressive MS, 2000) in SPMS used a similar study design. Disappointingly, the North American Study found positive results only on measures of disease activity—relapses and MRI lesions—and failed to replicate the disability results from the European Study. This perplexing result underscored the importance of replicating research findings generally, but also led to discussions about differences between the European and North American patient populations, discussed later.

A third study in SPMS, the SPECTRIMS Study, tested IFN-β1a (Rebif), 22 μgm and 44 μg SC three times per week compared with placebo (Li et al., 2001; Secondary Progressive Efficacy Clinical Trial of Recombinant Interferon-beta-1a in MS (SPECTRIMS) Study Group, 2001). In a fourth study, the IMPACT Study, IFN-β1a (Avonex) 60 μg IM QW was tested against placebo (Cohen et al., 2002). Both the SPECTRIMS Study and the IMPACT study failed to demonstrate therapeutic benefits

TABLE 4.3 Randomized, Double-Blind, Placebo-Controlled Trials of Interferon-β in Secondary Progressive MS (Modified from Marrie and Cohen, 2003)

	Eu IFN-β1b	NA IFN-β1b	SPECTRIMS	IMPACT
Centers	32	35	22	42
Subjects	718	939	618	436
Entry criteria	Age 18-55	Age 18-65	Age 18-55	Age 18-60
	EDSS 3.0-6.5	EDSS 3.0-6.5	EDSS 3.0-6.5 with Pyramidal Functional System Score ≥ 2	EDSS 3.5-6.5
Treatment	IFN-β1b (Betaferon)	IFN-β1b (Betaferon)	IFN-β1a (Rebif)	IFN-β1a (Avonex)
	8MIU	8 MIU	44 μg	60 μg
	Placebo	IFN-β1b 5 MIU/m²	22 μg	Placebo
		Placebo	Placebo	
	SC alternate days	SC alternate days	SC 3/week	IM 1/week
Primary outcome	Time to EDSS progression:	Time to EDSS progression:	Time to EDSS progression:	2-year change in MSFC
	1.0 step for EDSS 3.0-5.5 or	1.0 step for EDSS 3.0-5.5 or	1.0 step for EDSS 3.0-5.0 or	
	0.5 step for EDSS 6.0-6.5	0.5 step for EDSS 6.0-6.5	0.5 step for EDSS 5.5-6.5	
	sustained 3 months	sustained 6 months	sustained 3 months	

TABLE 4.4 Baseline Characteristics and Results from Phase 3 Studies of Interferon-β in SPMS. (Modified from Marrie and Cohen, 2003)

	Eu IFN-β1b	NA IFN-β1b	SPECTRIMS	IMPACT
Mean age (years)	41	47	42.8	47.6
Mean MS duration (years)	13.1	14.7	13.3	16.5
Mean prestudy relapse rate (year^{-1})	0.87 (2 years)	0.41 (2 years)	0.45 (2 years)	0.47 (3 years)
Mean baseline EDSS	5.1	5.1	5.4	5.2
Mean placebo on-study relapse rate (year^{-1})	0.64	0.28	0.71	0.30
Primary outcome result	EDSS (Positive)	EDSS (Negative)	EDSS (Negative)	MSFC (Positive)
Secondary outcomes with positive results	Relapses T2 lesion burden New/enlarging T2 lesions GdE lesions	Relapses T2 lesion burden New/enlarging T2 lesions GdE lesions	Relapses T2 lesion burden New/enlarging T2 lesions GdE lesions	Relapses T2 lesion burden New/enlarging T2 lesions GdE lesions Quality of life
Secondary outcomes with negative results				EDSS

of IFN-β1a on disability progression measured with EDSS. In the IMPACT Study, the primary endpoint for MS-related disability was the MSFC, although EDSS was included as a secondary endpoint. Active treatment resulted in significantly lower deterioration in MSFC scores in IFN-β1a patients compared with placebo patients. No significant benefit was found for the timed 25-foot walk and only a tendency in favor of active treatment for the PASAT (Paced Auditory Serial Addition Test), a measure of neuropsychological function. The treatment difference in the MSFC was mainly due to the 9-hole-peg-test, a measure of arm function. Because the MSFC is new, neurologists have less familiarity with the measure, and the clinical implications of the observed benefits are currently a matter of debate.

Across all four SPMS studies, significant beneficial effects were observed for relapses and MRI markers of inflammation, but the effects on disability progression were mixed. Although the studies had similar inclusion criteria (Table 4.3), demographic and disease characteristics at baseline were different (Table 4.4). The European SPMS Study recruited younger patients and more patients with relapses superimposed on a progressive course. Only about 30% of the patients in this study were relapse-free during the 2 years preceding randomization, compared with around 50% for the other studies. Despite the differences in the proportions of relapse-free patients in the European SPMS Study and the SPECTRIMS Study (30% vs 55%), the proportion of relapse-free in the placebo groups during these two studies was nearly identical (36% vs 37%). On the other hand, the proportion of relapse-free placebo patients during the North American

Study and the IMPACT Study was much higher (62% vs 63%) (Table 4.4). It is impossible to determine how much of the difference is due to the patient population and how much is due to differences in methods for counting relapses. While both are likely to contribute, these findings again point out the limitations of relapses as an outcome measure, and further suggest that comparing relapse results across studies is not meaningful.

One likely reason for greater efficacy observed in the European SPMS Study relative to other SPMS studies is the relatively short disease duration in the European SPMS Study. Mean disease duration ranged from 13.1 years in the European Study to 16.5 years in the IMPACT Study. Increasingly, there is evidence that inflammation may decline as a pathogenic mechanism in later stages of the disease, and that destructive and degenerative processes may become dominant. Therefore, it seems plausible that efficacy of an anti-inflammatory drug may decline as disease duration increases.

The SPMS studies raise important questions about the impact of IFN-β on disability progression. The therapeutic effect size of IFN-β in SPMS on markers of inflammation (relapses and new MRI lesions) is similar to that seen in the RRMS studies, but there is relatively little impact on EDSS progression at this stage of the disease. Since the trials are of short duration, it is possible that anti-inflammatory effects might translate into meaningful benefits on disability progression eventually, and that prolonged follow-up time would be required to measure this. Another possibility is that inflammation is no longer the predominant pathogenic factor at this stage of the disease, and inflammation-independent mechanisms are driving disability progression. Still another possibility is that more robust benefits of IFN-β on disability progression in RRMS are artifactual, and based on differences in the scale performance in mild vs more severely affected patients. Distinguishing among these possibilities would provide significant insights into MS pathogenesis.

Mechanism of Action

The mechanisms of action of interferon in MS are unknown. A number of laboratories have shown that interferon has beneficial effects in experimental autoimmune encephalomyelitis, focusing attention on the immunomodulatory or anti-inflammatory properties of interferon that may underlie therapeutic benefits in MS. The biological response to interferon is complex, however, and it is not a simple matter to determine the effects most relevant to therapeutic benefits.

Interferons generate biological responses by transcriptional regulation of a large family of genes. Hundreds of genes are known to be regulated by type 1 interferon. There is cell-type specificity to the interferon response, and the state of cell activation also determines responsiveness. It is not currently known which genes underlie therapeutic effects in MS, or what cell populations constitute the therapeutic target. Ransohoff (2003) reviewed biological responses to interferon as they relate to therapeutic efficacy, and Karp (2001) listed the most relevant immune effects (Table 4.5).

Interferon is presumed to have anti-inflammatory effects in MS, as demonstrated by reduced gadolinium enhancement, and by CSF testing in the IFN-β1a (Avonex) trial. Patients in that study had CSF analysis at study entry and after 24 months. Cell counts were significantly reduced in IFN-β1a but not placebo recipients (Rudick et al., 1996). Effects that may explain the anti-inflammatory outcome include: (1) interferon therapy was shown to increase soluble VCAM (Calabresi et al., 1997), and this correlated with decreased gadolinium-enhancing lesions. Soluble levels of VCAM in the blood may compete with endothelial VCAM for VLA-4 on the surface of activated T cells, decreasing entry of activated T cells into the CNS; (2) interferon therapy was shown to increase intracellular levels of interleukin (IL)-10 (Liu et al., 2001), and IL-10 levels in CSF (Rudick et al., 1998b). Increased CSF IL-10 levels correlated with favorable clinical outcomes in the IFN-β1a study; (3) type 1 interferon was shown to lower levels of IL-12, and to

TABLE 4.5   Selected Effects of Interferon β on the Immune System, Adapted from Karp et al. (2001)

Parameter	Reported Effects
MHC class I expression	Increased
MHC class II expression	Increased/decreased/no change
B7-1 expression by B cells	Decreased (Liu et al., 2001)
B7-2 expression by macrophages	Increased (Liu et al., 2001)
CD40 expression by T cells, B cells; monocytes	Decreased; increased (Liu et al., 2001)
CD40 ligand expression by T cells	Decreased (Teleshova et al., 2000)
FcR expression, antibody-dependent cellular cytotoxicity	Increased
Apoptosis of T cells	Decreased (Marrack et al., 1999)
	Increased (Kaser et al., 1999)
Fas expression by T cells	Increased (Kaser et al., 1999; Rep et al., 1999)
Adhesion molecule expression	Decreased (Muraro et al., 2000)
Matrix metalloproteinase-9 activity	Decreased (Leppert et al., 1997)
Tissue inhibitor of matrix metalloproteinase expression	Increased (Ozenci et al., 2000)
Interferon-γ production	Increased (Becher et al., 1999) Decreased (Furlan et al., 2000)
Interleukin-1 receptor antagonist production	Increased (Sciacca et al., 2000)
RANTES, MIP-1α expression	Decreased (Iarlori et al., 2000; Zang et al., 2001)
CCR5 expression	Decreased (Zang et al., 2001)
Nitric oxide secretion	Decreased (Hua et al., 1998)
Interleukin-10 production	Increased (Byrnes et al., 2002; Rudick et al., 1998b)
Interleukin-12 production	Decreased (Byrnes et al., 2002)

decrease ratios of IL-12 to IL-10 (Byrnes *et al.*, 2002; Karp *et al.*, 2001). This might promote an anti-inflammatory environment within the CNS; and (4) type 1 interferon was shown to inhibit inducible Class II expression, possibly reducing antigen presentation within the CNS (Lu *et al.*, 1995).

Newer molecular approaches may shed light on therapeutic mechanisms and individual response variability. Techniques such as microarrays (Sturzebecher *et al.*, 2003) and macroarrays consisting of gene transcripts known to be regulated by interferon (Schlaak *et al.*, 2002) are just beginning to be used for this purpose.

Neutralizing Antibodies

As is the case with many protein therapies, individual interferon recipients may develop neutralizing antibodies (NABs) to the drug. This was reported in the original Betaseron study, where development of NABs was associated with relapse rates and MRI lesion rates similar to placebo patients (The IFN-b Multiple Sclerosis Study Group, 1996). In the late 1990s, several groups demonstrated that interferon NABs were associated with blunted or absent *in vivo* responsive to interferon injections (Rudick *et al.*, 1998c; Deisenhammer *et al.*, 1999; Bertolotto *et al.*, 2003). These studies indicate that interferon NABs inhibit the biological response to therapy, and this explains why interferon patients who develop NABs have relapse and MRI disease activity similar to placebo-treated patients. The risk of developing NAB is higher with IFN-β1b compared with IFN-β1a, and lower with IFN-β1a (Avonex) compared with IFN-β1a (Rebif). The differences are presumed due to different doses (lower with Avonex), different physical chemical properties (Betaseron is nonglycosylated), and different routes of administration (Avonex is given by IM injection; the other products are given subcutaneously). The risk of NAB is one factor to be considered in selecting an interferon product for use in MS patients. Identifying individual risk for NAB, optimal monitoring strategies, and managing patients with NABs are important research topics.

Is One IFN Drug "Better"?

Several companies market IFN-β products for MS. Based on public reports on second quarter sales in 2003, annual worldwide sales of IFN-β products are currently estimated to be $2.5 billion. Since the market is split between three drugs, there has been intense marketing for market share by pharmaceutical companies. Much of the marketing has focused on whether one of the products is superior, and this has taken many forms: physician and lay advocates, advertising, educational programs, and postmarketing studies. There has been a bewildering array of claims and counterclaims.

The EVIDENCE Study (Panitch *et al.*, 2002) is a significant postmarketing study. Two preparations of IFN-β1a, Rebif (44 µg TIW) and Avonex (30 µg QW) were compared in an open-label 6-month study. Patients with active RRMS who had not previously used an interferon product were randomized to one of the arms and monitored to determine relapse frequency and new MRI lesions. There were fewer relapse-free patients in the IFN-β1a (Avonex) treated patients, and more new MRI lesions. However, the unblinded study design raised some concerns about the accuracy of the relapse findings, and the short study design raised questions about the significance of the findings (Kieburtz and McDermott, 2002). The observed benefits were considerably smaller in a subsequent 6-month observation period, suggesting that more frequent injections had a faster onset of action. Despite these issues, the EVIDENCE Study was important, because the FDA approved Rebif based on this study despite marketing exclusivity for Avonex under the Orphan Drug Act. The decision was based on FDA opinion that the EVIDENCE Study confirmed a significant advantage of Rebif compared with Avonex.

The question of the optimal interferon product remains highly controversial in the MS field, despite the EVIDENCE Study. Comparisons across the three pivotal trials suggest similar therapeutic effects, although precise comparisons are difficult. It is unclear whether short-term advantages of Rebif demonstrated in the EVIDENCE Study translate into long-term advantages, or whether they justify additional injections and toxicity.

What Is the Impact of IFN-β on Inflammation/Destructive Pathology, and Does Interferon Therapy Translate into Long-term Clinically Relevant Benefits?

The demonstration that acute MS lesions contain large numbers of transected axons (Trapp *et al.*, 1998) and the findings that IFN-β inhibits new MRI lesions and reduces clinical relapses and CSF cell counts (Rudick *et al.*, 1998a) raise the possibility that IFN-β may be neuroprotective. IFN-β may induce biological responses consistent with neuroprotective properties. For example, type 1 IFN reduces astrocytic production of nitric oxide, and hence may inhibit neuronal damage as has been shown in cell-culture (Schlaak *et al.*, 2002; Stewart *et al.*, 1997). IFN-β also induces expression of nerve growth factor by astrocytes (Boutros *et al.*, 1997).

Indirect evidence based on MRI findings suggests that IFN-β may lessen destructive pathology in RRMS patients. Simon and colleagues reported that IFN-β1a reduced accumulation of T_1 black holes in the Avonex

RRMS study (Simon *et al.*, 2000). IFN-β1a also had a stabilizing effect on T_1 black holes in the RRMS patients in the PRISMS study (Gasperini *et al.*, 1999). In patients with SPMS, IFN-β1b was found to reduce the number of T_1 black holes. This was related to reduced new lesion formation, as the proportion of new lesions that evolved into T_1 black holes was similar in IFN-β1b and placebo patients (Brex *et al.*, 2001).

Rudick *et al.* (1999) reported that IFN-β1a slowed whole brain atrophy progression in the second year of the original Avonex trial, as measured by the brain parenchymal fraction (Fisher *et al.*, 1997). Hardemeier *et al.* (2003) used the same BPF software in the dose comparison study of IFN-β1a (Avonex) for RRMS (Clanet *et al.*, 2002). This group found IFN-β1a treatment was associated with significantly slower rates of whole brain atrophy progression. In that study, brain atrophy progressed at a rate of approximately 1% per year from 4 months before treatment through 4 months after treatment onset. Between months 4 and 36 of therapy, atrophy progressed at a significantly lower rate. BPF values and the rates of atrophy on IFN-β1a therapy were quite similar in the original Avonex RRMS study and in the dose comparison study. Jones and colleagues (2001), on the other hand, applied a similar method to patients in the PRISMS trial, but failed to demonstrate a therapeutic benefit of IFN-β1a (Rebif) on atrophy progression. Also, none of the studies of INF-β therapy in progressive MS has shown benefits on brain atrophy. Thus, the anti-inflammatory effects of IFN-β provide a basis for a neuroprotective effect, and MRI data are suggestive. However, not all RRMS studies have demonstrated a beneficial effect on brain atrophy, and studies in SPMS have thus far failed to demonstrate an effect of IFN-β on brain atrophy progression.

Despite promising results from MRI studies and the logical conclusion that interferon treatment is neuroprotective, the effects of IFN on clinically significant long-term disability progression remains controversial. An approach to this problem would be longer randomized, placebo-controlled clinical trials, but this is not practical for two reasons. First, prolonged use of a placebo group in RRMS is ethically questionable, because multiple randomized controlled clinical trials have demonstrated therapeutic benefits. Second, patients in a clinical trial are most likely to withdraw from the trial if they are doing poorly. If the more severely affected patients are lost to follow-up evaluation, and this occurs more in one arm of the study than the other, the results are increasingly unreliable. Consequently, it is unlikely that long-term, randomized, placebo-controlled studies will be available to definitively assess the long-term impact of DMDT (disease modifying drug therapy). The PRISMS-4 study demonstrated that those patients who were initially

treated with placebo for 2 years failed to "catch up" to those initially treated with INF-β-1a (Rebif) after switching to Rebif for the subsequent 2 years (PRISMS Study Group, 2001). The publication was accompanied by an editorial advocating early rather than delayed treatment (Schwid and Bever Jr, 2001). Therapeutic benefits beyond 4 years are not known, but follow-up data from the original IFN-β1a (Avonex) study (Rudick *et al.*, 2001a) suggested that benefits are still evident 8.1 years after randomization in those patients initially randomized to active treatment.

GLATIRAMER ACETATE (GA) FOR MULTIPLE SCLEROSIS

Rationale and Early Trials

Animal models of autoimmune demyelination have formed the basis for decades of productive research into mechanisms of autoimmunity relevant to MS pathogenesis. The most important animal model, experimental autoimmune encephalomyelitis (EAE), is regarded as a relevant animal model for MS. EAE is a T-cell-mediated disease induced in susceptible animals by inoculation with components of the central nervous system. Purified myelin proteins, such as myelin basis protein (MBP), proteolipid protein, or even peptides derived from them, are encephalitogenic in susceptible animals.

Glatiramer acetate (GA) was developed based on this model of the human disease. Dr. Ruth Arnon and colleagues (1996) synthesized 11 copolymers with amino acid compositions similar to MBP, in order to obtain synthetic encephalitogens for EAE studies. The copolymers did not induce EAE, but unexpectedly several prevented EAE or reduced its severity. Copolymer 1 (Cop1), composed of L-glutamate, L-lysine, L-alanine, and L-tyrosine, was the most potent EAE antagonist, reducing the incidence of EAE in MBP-challenged guinea pigs by as much as 75% (Arnon, 1996). Cop1 also suppressed EAE in other species and biochemical studies demonstrated that the degree of cross-reactivity between Cop1 and MBP correlated with the ability of Cop1 to suppress EAE (Teitelbaum *et al.*, 1991).

In the first clinical study, Bornstein and colleagues randomized 48 patients to Cop1 or placebo (Bornstein *et al.*, 1987) and determined the proportion relapse free. This study may have been the first to use an evaluating neurologist masked to treatment arm assignment to improve the objectivity of the outcome assessment. Sixteen relapses occurred in Cop1 patients, with 62 in placebo patients. More Cop1 patients were relapse-free; placebo patients were more likely to have three or more relapses. Disability progression was defined as worsening by ≥ 1.0 point on the disability status scale, sustained for 3 months. Cop1 treatment slowed disability progression using this definition. The drug was well tolerated, although 2 Cop1-treated patients reported flushing and chest tightness, sometimes accompanied by anxiety and dyspnea after their injections. These symptoms resolved within 30 minutes without sequelae. This exciting result set the stage for definitive studies of Cop1 in RRMS, but it would take more than 5 years to generate adequate drug of sufficient purity to initiate definitive studies.

Definitive Trial in RRMS

Teva Pharmaceuticals entered into an agreement to further develop Cop1 with the Weitzman Institute, where prior Cop1 work had been conducted. Considerable effort was required to standardize the manufacturing methods to provide the quantities of drug needed to conduct a large-scale trial. The final product, thereafter called glatiramer acetate (GA), consisted of random, synthetic polypeptide chains ranging in molecular weight from 4000 to 13,000 daltons consisting of L-alanine, L-glutamate, L-lysine, and L-tyrosine in molar ratios of 4.2, 1.4, 3,4, to 1.0. A dose of 20 mg administered daily by subcutaneous injection was selected for the double-blind, placebo-controlled trial, which began in 1992.

The trial was conducted at 11 MS centers in the United States (Johnson et al., 1995). The primary outcome measure was the mean number of relapses in the study arms. Other prespecified outcome measures included other relapse-based endpoints, mean change in EDSS and ambulation index from baseline, and proportion of patients with sustained EDSS progression, defined as worsening by ≥ 1.0 from baseline EDSS persisting for at least three months. MRI scans were not included in the protocol. A total of 251 patients were randomized to GA or placebo, and the two groups were well matched for demographic and disease factors at baseline (Table 4.6). The patients in this study were similar to patients in the pivotal interferon studies.

There were 161 relapses in GA patients, and 210 in placebo patients. The mean annualized relapse rate was 0.59 for GA patients and 0.84 for placebo patients, a 29% reduction ($P < .01$). Mean EDSS change was lower for the GA group compared to placebo ($P = .02$), but there were no differences in ambulation index change, or in the number of patients with sustained EDSS change. Injection site reactions were the most reported common adverse event. The transient postinjection reaction first observed in the pilot trial occurred in 15% of GA patients, usually within seconds or minutes of an injection. This 2-year pivotal trial confirmed that daily, subcutaneous injections of GA reduced relapses in patients with RRMS. On this basis, the U.S. Food and Drug Administration (FDA) approved GA (Copaxone) for RRMS in December 1997.

Because patients in the blinded study were evaluated until the last patient completed 4 months, there was an average of 5.5 months additional data from the double-blind portion of the study. The 24-month data and the extension phase data were combined in a second report (Johnson et al., 1998). In all, 215 (86%) of 251 randomized patients completed 24 months, and 203 of the 215 entered the extension phase. In the extended study analysis, the mean annual relapse rate was 0.67 for GA patients and 0.99 for placebo patients, a reduction of 32% ($P < .01$). At the end of the extension phase, 24.6% of the placebo group and 33.6% of the GA group were relapse-free from study onset ($P < .05$). The extension study also failed to show significant benefits on the predetermined EDSS-based outcome, but a post hoc analysis showed that more patients developed a 1.5 point worsening from baseline EDSS at some point during the study.

Studies in Progressive MS

One trial for chronic progressive MS was conducted during the 1980s, using Cop1 obtained from the Weitzman Institute (Bornstein et al., 1991). A total of 106 patients with documented disability progression were studied. The primary study endpoint was the time to confirmed worsening ≥ 1.0 point from baseline EDSS for patients with baseline EDSS of 5.0 or greater and ≥ 1.5 points for patients with entry EDSS of less than 5.0. There was a trend for less progression in Cop1 patients compared to placebo (17.6% vs 25.5%), but this was not statistically significant. Further studies in SPMS have not been conducted.

The PROMiSe trial (Wolinsky et al., 2002) was a double-blind, placebo-controlled trial designed to evaluate the effect of GA in patients with primary progressive MS (PPMS). Accrual of 943 patients with PPMS for this trial represents a monumental accomplishment, given the relative scarcity of patients with PPMS. Patients were 49% male, average age 50.4, and symptomatic an average of 10.9 years at study entry. They were moderately disabled with average EDSS of 4.9. Insights into PPMS will likely

TABLE 4.6 Patient Characteristics in the Pivotal Glatiramer Acetate Study in RRMS

	Glatiramer Acetate	Placebo
	(n = 125)	(n = 126)
Age in years (SD)	34.6 (6.0)	34.3 (6.5)
Disease duration in years (SD)	7.3 (4.9)	6.6 (5.1)
Relapse rate prior 2 years (SD)	2.9 (1.3)	2.9 (1.1)
EDSS (SD)	2.8 (1.2)	2.4 (1.3)

emerge from the study, but the clinical trial was discontinued because interim analysis demonstrated no hint of efficacy and a futility analysis demonstrated a low likelihood of a positive outcome. Thus, at present, there are no data suggesting a therapeutic effect of GA on disease progression in SPMS or PPMS.

Mechanism of Action

Possible mechanisms of action of GA have been reviewed (Neuhaus et al., 2001) (Table 4.7). Commonly cited possibilities include: 1) competitive binding to the major histocompatibility complex (MHC) in preference to myelin protein antigens, 2) preferred binding of GA-MHC complexes over MBP-MHC complexes to T-cell receptors, 3) induction of tolerance in MBP-specific T-cells; and 4) induction of GA-specific T cells expressing anti-inflammatory Th2 cytokines. GA treatment increases GA-specific T cells in peripheral blood, which then decrease with continued GA administration (Duda et al., 2000). GA-specific T cells may be polarized toward a Th2 bias, and secrete anti-inflammatory cytokines. Conceivably, GA-specific Th2-polarized T cells may recognize myelin antigens within the CNS and mediate bystander suppression. This mechanism was suggested in EAE studies (Aharoni et al., 1998), and a study of eight MS patients initiating therapy with GA supports this possible mode of action (Qin et al., 2000). GA and MBP reactive T-cell lines from GA-treated patients were characterized, using the ratio of IFN-γ to IL-5 as a measure of Th2 bias. GA-reactive lymphocytes had a significant Th2 bias compared to MBP-reactive cells.

Thus, the prevailing view is that GA functions as an antigen and induces proliferation of GA-reactive T cells. Thus, GA may be the first treatment of autoimmune disease thought to work via modulating specific T cell responses to autoantigen.

TABLE 4.7   Potential Mechanisms of Action of Glatiramer Acetate in Multiple Sclerosis

Parameter	Reported Effects
Binding to MHC class II	Compete with MBP or other myelin antigens for T-cell binding (Fridkis-Hareli et al., 1994; Fridkis-Hareli et al., 1995)
Binding to T-cell receptors	Compete with myelin peptides for MHC binding (Aharoni et al., 1999)
Induction of tolerance in MBP-specific T-cells	Reduced T-cell-mediated brain inflammation
Induction of GA-specific Th2 T cells	Secretion of anti-inflammatory cytokines in the brain

Antibodies to GA form during treatment, but the impact is not known. A total of 130 patients from the RRMS GA clinical trial were studied for antibodies (Brenner et al., 2001). Antibodies developed in all patients, peaked at 3 months of therapy, and declined to near baseline levels subsequently. The relevance of antibodies to GA with respect to the in vivo biological response to treatment or to treatment efficacy is currently not known.

Impact on Inflammation / Destructive Pathology

Compared with the interferons, relatively little is known about the impact of GA on destructive pathology in MS. The European/Canadian MRI Study was initiated in 1987 as a randomized, double-blind, placebo-controlled trial designed to determine the MRI effect of GA (Comi et al., 2001b). A total of 239 patients with active RRMS and at least one gadolinium-enhancing lesion at study entry (selected from 485 potential subjects) were randomized to GA or placebo and studied under double-blind conditions for 9 months. After this phase, all patients were treated with GA in an open-label study for an additional 9 months. MRI scans were done monthly for 9 months, then every 3 months for the remaining 9 months. During the 9-month, double-blind phase, there were 29% fewer gadolinium-enhancing lesions in the GA treated group ($P < .01$), and more gadolinium volume decline ($P = .01$). The groups began to separate at month 5 and were significantly different in months 6 through 9. T_2 lesion accrual between entry and 9 months was 33% less in the GA group ($P < .01$). Relapse rate, a secondary outcome in the study, was reduced by 33% ($P < .05$).

Compared to INF-β, GA effects on gadolinium-enhancing MRI lesions are modest and delayed, suggesting that GA has a different mechanism of action than interferon. Because the significance of MRI effects in terms of long-term clinical benefits is unknown, it is not valid to conclude that GA is less effective than IFN-β, although it suggests that the anti-inflammatory effect has a slower onset or is more modest. Also, the difference between the GA and IFN-β provides a rationale for combining the two treatments in combination protocols.

The European/Canadian MRI Study examined the effect of GA on T_1 black holes (Filippi et al., 2001). Each of 1722 gadolinium-enhancing MRI lesions from 239 patients was followed prospectively to determine the persistence of a T_1 black hole and the frequency of reenhancement. Over the 9-month double-blind phase of the study, new gadolinium lesions could be tracked for up to 8 months. Fewer lesions evolved into T_1 black holes in GA-treated patients compared with placebo-treated patients at 7 months (18.9 vs 26.3%; $P = .04$) and 8 months (15.6 vs 31.4%; $P = .002$). This finding was interpreted as showing a neuroprotective effect of GA. However, another analysis failed to show benefits of GA on brain volume change (Rovaris *et al.*, 2001). Using a semiautomated segmentation technique, brain volume was measured from seven contiguous periventricular slices from the scans obtained at baseline, the end of the double-blind phase, and the end of the study. Brain volume correlated with disability at each time-point, but no significant differences between placebo- and GA-treated patients were found for baseline brain volumes or for rate of brain volume change during the study. A trend toward slower atrophy progression during the second 9 months of the study (Rovaris *et al.*, 2001) is consistent with a delayed therapeutic benefit of GA on brain atrophy, but the differences in the study were not statistically significant, and the effects of GA on destructive pathology remain uncertain.

Controversies and Future Directions with Antigen-Specific Therapy

Double blind, placebo-controlled GA studies have not demonstrated significant benefit on sustained EDSS change (Bornstein *et al.*, 1987; Comi *et al.*, 2001b; Johnson *et al.*, 1995), but a meta-analysis, conducted by pooling data on 540 subjects within these clinical trials, demonstrated a significantly reduced odds ratio (OR = 0.57, $P < .02$) for ≥ 1.0 EDSS point worsening sustained for 90 days (Wolinsky *et al.*, 2003). In the meta-analysis time to sustained EDSS worsening was longer in GA-treated patients (ratio estimate 1.88, $P < .02$). Although not definitive, the meta-analysis suggests that GA may have therapeutic benefits on sustained EDSS worsening in RRMS.

GA effects on disease progression have important implications for "antigen-specific" therapy in MS generally. Two Phase II trials were conducted of an altered peptide ligand, synthesized based on encephalitogenic determinants of MBP (Bielekova *et al.*, 2000; Kappos *et al.*, 2000). These studies produced equivocal or negative results. The altered peptide ligand was accompanied by expansion of T cells reactive with the APL and with MBP (Kim *et al.*, 2002), and one of the trials (Bielekova *et al.*, 2000) suggested immune activation and increasing brain inflammation in some of the patients.

HOW HAVE THESE STUDIES CHANGED MS, AND WHAT ARE KEY QUESTIONS RELATED TO FUTURE PROGRESS?

The clinical trials described in this chapter have changed the face of the MS field. No longer is MS considered "untreatable," and no longer are potential therapies considered "untestable." Clinical trial methodology has advanced with new disability measures, new MRI approaches for identifying destructive pathology, and new study designs. We have entered an era of active arm comparison studies in MS and trials of drugs used in combination. Most of the effort still focuses on anti-inflammatory therapies. However, anti-inflammatory drugs are being tested earlier in the disease process, based on the concept that inflammation is central to pathogenesis early in MS, but that pathogenic factors not dependent on inflammation may play a larger role in later stages. Second, anti-inflammatory drugs are being used in combination. This approach is based on experience in other fields, (e.g., cancer therapeutics) that suggests a single drug may not be adequate to interrupt a complex disease. Attention is being directed at mechanisms of axonal and neuronal pathology in MS and at the use of neuroprotection strategies.

What Is the Best Way to Measure the Effects of MS Therapy?

There is still no consensus about the best method for measuring therapeutic effects in MS, but it is abundantly clear that the traditional outcome measures—relapses and EDSS—are imprecise, used differently by different investigators, and do not allow interpretable comparisons of separate studies. Relapses are dependent on patient reporting and neurologist interpretation, and are particularly questionable in studies that are not effectively blinded. This is a major problem for open-label studies comparing the effects of different interferon preparations on relapses, which can be influenced by expectations about the outcome by patients and investigators. As the field of experimental therapeutics in MS has matured, studies have gradually incorporated more meaningful clinical measures of disability progression, and more meaningful global MRI parameters such as atrophy, but acceptance of more meaningful measures has not been universal.

A clear consensus was reported that traditional outcome measures are inadequate (Whitaker *et al.*, 1995). This led to an MRI task force (Miller *et al.*, 1996) that recommended protocols to standardize analysis of gadolinium-enhancing lesions in monitoring therapy in Phase I and Phase II studies. This recommendation was generally accepted, and many trials have incorporated the recommendations. However, the relationship between gadolinium-enhancing lesion frequency and subsequent disability status was found to be fairly weak (Kappos *et al.*, 1999), as was the relationship between gadolinium enhancement and subsequent brain atrophy (Fisher *et al.*, 2000). Consequently, gadolinium enhancement has not been accepted as a validated surrogate for clinically meaningful outcomes, although it is widely used to screen anti-inflammatory therapies, and as a secondary outcome measure in pivotal trials.

The MS outcomes assessment workshop (Whitaker et al., 1995) also led to a clinical outcomes assessment task force (Rudick et al., 1997). The MS functional composite (MSFC) (Rudick et al., 1996, 1997) was recommended as an improved clinical outcome measure for future clinical trials. The MSFC contains a walking test, a test of arm dexterity, and a cognitive test of sustained attention. Each of the three measures is standardized with scores from a reference population to create a z-score, and the z-scores for the three tests are averaged to create a single patient score. MSFC change over time along a continuous scale is easy to calculate. The MSFC was found to predict future disability status in patients with RRMS (Rudick et al., 2001b) and in patients with SPMS (Cohen et al., 2001). MSFC scores were found to correlate with MRI lesions and brain atrophy more strongly than does EDSS (Fisher et al., 2000; Rudick et al., 2002). The MSFC was reproducible in the context of a multinational trial (Cohen et al., 2001), and was used as the primary outcome measure in the IMPACT Trial of IFN-β1a in SPMS (Cohen et al., 2002b). In that study, IFN-β1a slowed MSFC worsening, but no benefits were observed using EDSS, presumably because EDSS is a less sensitive measure. Numerous ongoing studies have incorporated MSFC into the study designs, and much data are likely to emerge regarding the usefulness of MSFC in the near future. One potential advantage of MSFC is that it may be feasible to compare results across studies more meaningfully than with current measures. Because the MSFC component scores are standardized to a reference population and expressed as z-scores, and because consistent MSFC use is easy to achieve (Fischer et al., 1999), it may be possible to directly compare therapeutic benefits from one study to another using MSFC. Another advantage of the MSFC is its sensitivity or responsiveness. This advantage may be particularly useful in detecting small incremental benefits, which will be useful in active arm comparison studies.

Increasingly, there is a tendency to include newer MRI measures of destructive pathology in MS trials, and brain atrophy has received a lot of attention (reviewed in Miller 1998; Miller et al., 2002). Many studies have been completed using atrophy as one of the MRI outcomes (Rovaris and Filippi, 2003). As with MRI, there are many methods for demonstrating atrophy, and it is not prudent to compare different atrophy results derived from clinical trials using different measurement methods. Also, atrophy progresses slowly over the course of years, so atrophy may not be a practical measure for short-term studies. Recently, Hardemeir et al. (2002) reported significant atrophy progression during a 4-month interval before the IFN-β1a (Avonex) dose comparison study. That result suggests that a reasonably short MS study targeting brain atrophy may be feasible.

Other MRI methods, such as spectroscopy, magnetization transfer imaging, diffusion tensor imaging, and functional imaging have not been widely applied to clinical trials, but hold great promise.

What Is the Significance of Relapses with Respect to Eventual Disability Progression?

The interferon studies demonstrated that anti-inflammatory therapy in RRMS has significant beneficial effects on measures of brain inflammation: reduced frequency of relapses, reduced gadolinium-enhancing lesion frequency, and reduced T_2 lesions. These benefits are accompanied by less EDSS worsening in short-term clinical trials, raising the possibility that these short-term anti-inflammatory effects will significantly improve the long-term outcome in MS patients. However, the relationship between relapse reduction and long-term benefits has not been demonstrated empirically.

Confavreux reported the relationship between relapses and EDSS change observed in a large database of MS patients studied longitudinally (Confavreux et al., 2000, 2003). Worsening from EDSS scores of 4 to higher levels did not differ in groups with or without prior relapses, or in patients with or without continuing relapses. These studies suggest that patients who enter the progressive disability stage worsen whether or not they continue to have relapses, and irrespective of the prior occurrence of relapses. This study has stimulated debate about the relevance of relapses to disability progression, or even the relevance of inflammation to ultimate disease pathogenesis. It seems likely that brain inflammation is pathogenic and contributes to eventual disability progression, but that brain inflammation during the RRMS stage eventually leads to a self-perpetuating neurodegenerative process that becomes independent from continuing inflammation. In this model, patients in the progressive stage of disease would deteriorate irrespective of ongoing anti-inflammatory therapy, which is consistent with the observations from the SPMS clinical trials. Another possibility is that brain inflammation might be secondary to an undefined underlying pathological process. If that were true, studies using more aggressive anti-inflammatory therapy early in the disease may not prevent progressive disability progression. Currently, ongoing or planned trials of aggressive early anti-inflammatory drug therapy will help distinguish between these possibilities.

Do Short-term Effects of Current Therapy on Measures of Inflammation Translate into Meaningful Long-term Benefits in MS?

There are no long-term follow-up studies to answer this question. However, in a group of patients reevaluated approximately 6 years after the end of a 2-year clinical trial, clinically meaningful benefits were evident in the original IFN-β1a group compared with the original placebo group (Rudick *et al.*, 2001a). Also, systematic follow-up evaluation over 4 years in another study of IFN-β1a (PRISMS Study Group, 2001; Schwid and Bever, 2001) demonstrated that earlier treatment resulted in significant benefits at the final follow-up evaluation compared with treatment delayed 2 years. The clinical significance of sustained EDSS progression during the relapsing stage of MS was assessed in the original IFN-β1a (Avonex) study (Rudick *et al.*, 2001b). Of 172 patients randomized to the trial early enough to complete 2-year follow-up evaluation, data were available for 160 patients 8.1 years after randomization. Among the 160 patients, there were 45 patients with disability progression during the 2-year clinical trial, and 115 without disability progression, defined as 6-month sustained EDSS worsening from baseline by ≥ 1.0. Table 4.8 shows the status at the 8-year follow-up evaluation in these patients. Patients with disability progression during the 2-year clinical trial had worse EDSS at the follow-up evaluation (6.1 vs 3.7; $P < .001$); worse MSFC scores (-2.66 vs -0.44; $P < .001$); and more brain atrophy (BPF 0.79 vs 0.81; $P = .06$). This study supports the clinical meaningfulness of disability progression as defined in this study, but such validation studies have not been reported for other definitions of disability progression or for relapse rate reduction. It seems likely, however, that clinical benefits of IFN-β1a on sustained EDSS progression during the short-term trials would translate into clinically significant benefits over the longer term.

While there is evidence that IFN-β can slow whole brain atrophy, and that IFN-β reduces EDSS worsening in short-term studies in RRMS, it is also clear that many patients on IFN-β therapy continue to deteriorate. Also, IFN-β does not have a robust disability benefit in patients with SPMS. Why is this? One possibility is that IFN-β has only partial anti-inflammatory effects. If this were the case, combinations of anti-inflammatory drugs would be expected to more significantly affect disease progression, and such studies are in progress. Another possibility is that there is a window of opportunity for anti-inflammatory drug therapy, beyond which a neurodegenerative process becomes somewhat autonomous. This possibility would predict that adequate anti-inflammatory therapy administered early enough would abort the progressive stage of the disease. Follow-up evaluation of the patients entering trials at the first clinical episode (CHAMPS and ETOMS Studies) may provide insight into this hypothesis, as will aggressive immunosuppression with mitoxantrone, cyclophosphamide, CAMPATH, or stem cell transplantation at very early stages of MS. Another less optimistic hypothesis is that inflammation is a response to underlying neuronal or axonal pathology. If this were the case, fully effective anti-inflammatory therapy applied at disease onset may not prevent eventual disability progression.

Axonal or Neuronal Pathology Has Emerged as a Key Pathological Process. Is This a Result of Brain Inflammation, or Are There Other Mechanisms?

Considerable attention has focused on axonal transection and chronic axonal degeneration since the observations by Trapp *et al.* (1998) and commentary by Waxman (1998). Axonal transection and degeneration cannot be measured directly *in vivo*. However, proton MR spectroscopy allows *in vivo* measurement of n-acetylaspartate (NAA), providing a biochemical index of axonal integrity, reviewed in Stefano *et al.* (2000). Several MS studies demonstrated sustained decreases in NAA in white matter lesions and in brain regions appearing normal on conventional MRI, suggesting diffuse axonal pathology in brain that appears otherwise normal. Studies demonstrated clinical relevance of decreased NAA (Pan *et al.*, 2001) and suggested that diffuse axonal pathology may exist at an early MS stage (Cifelli *et al.*, 2002; De Stefano *et al.*, 2001; Filippi *et al.*, 2003). Some (Narayanan *et al.*, 2001) but not other (Parry *et al.*, 2003) studies suggest that IFN-β treatment improves NAA levels. Studies focusing on the mechanisms of neuronal injury in MS, methods for measuring and monitoring this in patients, and methods for determining the effects of neuroprotective therapy are urgently needed.

The Concepts of Subclinical Disease and Disability Threshold Affect Future Designs and Clinical Care

During the past 20 years, the concept of subclinical disease and the "disability threshold" have been generally accepted. Subclinical disease has been amply demonstrated by MRI studies showing new lesion formation and brain atrophy progression during early stages of MS in patients without relapses or other

TABLE 4.8 Status of Patients from the IFN-β1a (Avonex) Study at 8-Year Follow-Up Evaluation Depending on Progressor Status During 2-Year Clinical Trial

	Non-Progressors	Progressors*	P Value
EDSS	3.69 +/− 2.0 (n = 115)	6.13 +/− 2.2 (n = 45)	< .001
MSFC	− 0.44 +/− 1.5 (n = 98)	− 2.66 +/− 2.8 (n = 37)	< .001
BPF	0.81 +/− 0.03 (n = 96)	0.79 +/− 0.03 (n = 38)	= .06

*Patients who worsened by ≥ 1.0 from baseline EDSS, sustained for at least 6 months during original clinical trial.

obvious clinical manifestations. This clearly implies that an MS patient may be clinically stable despite considerable disease activity and progression. The implications for clinical trials and patient management are clear: One can be lulled into a false sense of security by monitoring only clinical manifestations. Clinical trials in this stage of MS must include imaging studies, and imaging is increasingly used for managing individual patients.

The disability threshold is illustrated in Figure 4.1. In this conception, MS is a continuously progressive disease, but progressive neurological disability ensues only after the disease burden has surpassed a threshold. Beyond that threshold, further pathology progression directly translates into progressive clinical deterioration.

Where Are We Going with Clinical Trials after the ABCR Drugs?

The ethical basis for placebo-controlled trials in RRMS must be given careful consideration (Lublin and Reingold, 2001). Important issues include informed consent, availability of effective drugs, duration of the trial, and adequate monitoring for disease worsening. In lieu of placebo-controlled trials, combinations of drugs will be compared with active therapy. Some such trials are underway (Rudick *et al.*, 2003). They require very large sample sizes, more sensitive outcome measures, or some combination of the two. Trials of intensive immune suppression or bone marrow transplantation may be initiated at very early stages of MS.

It is hoped that outcome assessment in MS will advance to the point that the specific effects of neuroprotective agents, alone or in combination with anti-inflammatory drugs, can be determined. Transplantation of progenitor or stem cells will be possible, but it will be necessary to have more sensitive and specific methods to determine the result.

Studies in PPMS and SPMS will continue to be significant, but therapeutic interventions other than anti-inflammatories will be needed. These alternatives should be based on emerging mechanistic insights into neurodegeneration in MS.

Finally, there is a strong need for skilled clinical investigators in the MS field. Improved outcome measures and their validation will not be provided by the pharmaceutical industry. Ideally, collaborations among clinical investigators, government, and the pharmaceutical industry will promote refinement of MS trial methodology, so that as our understanding of disease mechanisms improves, we will have more useful methods to study the effects of attractive therapies.

References

Aharoni R, Teitelbaum D, Arnon R, Sela M. (1999). Copolymer 1 acts against the immunodominant epitope 82-100 of myelin basic protein by T cell receptor antagonism in addition to major histocompatibility complex blocking, *Proc Natl Acad Sci USA* 96: 634-639.

Aharoni R, Teitelbaum D, Sela M, Arnon R. (1998). Bystander suppression of experimental autoimmune encephalomyelitis by T cell lines and clones of the Th2 type induced by copolymer 1, *J Neuroimmunol* 91: 135-146.

Arnon R. (1996). The development of Cop 1 (Copaxone), an innovative drug for the treatment of multiple sclerosis: personal reflections, *Immunol Lett* 50: 1-15.

Barkhof F, *et al.* (2001). T(1) hypointense lesions in secondary progressive multiple sclerosis: effect of interferon beta-1b treatment, *Brain* 124: 1396-1402.

Bertolotto A, *et al.* (2003). Persistent neutralizing antibodies abolish the interferon beta bioavailability in MS patients, *Neurology* 60: 634-639.

Bielekova B, *et al.* (2000). Encephalitogenic potential of the myelin basic protein peptide (amino acids 83-99) in multiple sclerosis: results of a phase II clinical trial with an altered peptide ligand, *Nat Med* 6: 1167-1175.

Bornstein MB, *et al.* (1987). A pilot trial of Cop 1 in exacerbating-remitting multiple sclerosis, *N Engl J Med* 317: 408-414.

Bornstein MB, *et al.* (1991). A placebo-controlled, double-blind, randomized, two-center, pilot trial of Cop 1 in chronic progressive multiple sclerosis, *Neurology* 41: 533-539.

Boutros T, Croze E, Yong VW. (1997). Interferon-beta is a potent promoter of nerve growth factor production by astrocytes, *J Neurochem* 69: 939-946.

Brenner T, *et al.* (2001). Humoral and cellular immune responses to Copolymer 1 in multiple sclerosis patients treated with Copaxone, *J Neuroimmunol* 115: 152-160.

Brex PA, *et al.* (2001). The effect of IFNbeta-1b on the evolution of enhancing lesions in secondary progressive MS, *Neurology* 57: 2185-2190.

Byrnes AA, McArthur JC, Karp CL. (2002). Interferon-beta therapy for multiple sclerosis induces reciprocal changes in interleukin-12 and interleukin-10 production, *Ann Neurol* 51: 165-174.

Calabresi PA, *et al.* (1997). Increases in soluble VCAM-1 correlate with a decrease in MRI lesions in multiple sclerosis treated with interferon β-1b, *Ann Neurol* 41: 669-674.

Calabresi PA. (2002). Considerations in the treatment of relapsing-remitting multiple sclerosis, *Neurology* 58: S10-S22.

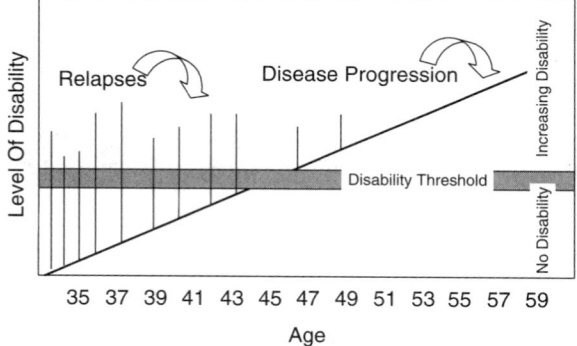

FIGURE 4.1   The disability threshold.

Cifelli A, *et al.* (2002). Thalamic neurodegeneration in multiple sclerosis, *Ann Neurol* 52: 650-653.

Clanet M, *et al.* (2002). A randomized, double-blind, dose-comparison study of weekly interferon beta-1a in relapsing MS, *Neurology* 59: 1507-1517.

Cohen JA, *et al.* (2001a). Predictive validity of the ms functional composite in progressive MS, *Multiple Sclerosis* 7 (Suppl 1): S15.

Cohen JA, *et al.* (2001b). Use of the multiple sclerosis functional composite as an outcome measure in a phase 3 clinical trial, *Arch Neurol* 58: 961-967.

Cohen JA, *et al.* (2002). Benefit of interferon beta-1a on MSFC progression in secondary progressive MS, *Neurology* 59: 679-687.

Comi G, *et al.* (2001a). Effect of early interferon treatment on conversion to definite multiple sclerosis: a randomised study, *Lancet* 357: 1576-1582.

Comi G, Filippi M, Wolinsky JS. (2001b). European/Canadian multicenter, double-blind, randomized, placebo-controlled study of the effects of glatiramer acetate on magnetic resonance imaging–measured disease activity and burden in patients with relapsing multiple sclerosis. European/Canadian Glatiramer Acetate Study Group, *Ann Neurol* 49: 290-297.

Confavreux C, Vukusic S, Moreau T, Adeleine P. (2000). Relapses and progression of disability in multiple sclerosis, *N Engl J Med* 343: 1430-1438.

Confavreux C, Vukusic S, Adeleine P. (2003). Early clinical predictors and progression of irreversible disability in multiple sclerosis: an amnesic process, *Brain* 126: 770-782.

De Stefano ND, *et al.* (2000). Proton MR spectroscopy to assess axonal damage in multiple sclerosis and other white matter disorders. [Review] [57 refs], *J Neurovirol* 6 (Suppl 2):S121-S129.

De Stefano ND, *et al.* (2001). Evidence of axonal damage in the early stages of multiple sclerosis and its relevance to disability, *Arch Neurol* 58: 65-70.

Deisenhammer F, *et al.* (1999). Bioavailability of interferon beta 1b in MS patients with and without neutralizing antibodies, *Neurology* 52: 1239-1243.

Duda PW, *et al.* (2000). Glatiramer acetate (Copaxone) induces degenerate, Th2-polarized immune responses in patients with multiple sclerosis, *J Clin Invest* 105: 967-976.

Durelli L, *et al.* (1994). Chronic systemic high-dose recombinant interferon alfa-2a reduces exacerbation rate, MRI signs of disease activity, and lymphocyte interferon gamma production in relapsing-remitting multiple sclerosis, *Neurology* 44: 406-413.

European Study Group on interferon beta-1b in secondary progressive MS [see comments]. (1998). Placebo-controlled multicentre randomised trial of interferon beta-1b in treatment of secondary progressive multiple sclerosis, *Lancet* 352: 1491-1497.

Filippi M, *et al.* (2003). Evidence for widespread axonal damage at the earliest clinical stage of multiple sclerosis, *Brain* 126: 433-437.

Filippi M, *et al.* (2001). Glatiramer acetate reduces the proportion of new MS lesions evolving into "black holes," *Neurology* 57: 731-733.

Fischer JS, *et al.* (1999). Administration and Scoring Manual for the Multiple Sclerosis Functional Composite Measure (MSFC). New York: Demos Medical Publishing.

Fisher E, *et al.* (1997). Knowledge-based 3D segmentation of MR images for quantitative MS lesion tracking, *SPIE Medical Imaging* 3034: 599-610.

Fisher E, *et al.* (2000). Relationship between brain atrophy and disability: an 8-year follow-up study of multiple sclerosis patients, *Mult Scler* 6: 373-377.

Fridkis-Hareli M, Teitelbaum D, Arnon R, Sela M. (1995). Synthetic copolymer 1 and myelin basic protein do not require processing prior to binding to class II major histocompatibility complex molecules on living antigen-presenting cells, *Cell Immunol* 163: 229-236.

Fridkis-Hareli M, *et al.* (1994). Direct binding of myelin basic protein and synthetic copolymer 1 to class II major histocompatibility complex molecules on living antigen-presenting cells–specificity and promiscuity, *Proc Natl Acad Sci USA* 91: 4872-4876.

Furlan R, *et al.* (2000). Interferon-beta treatment in multiple sclerosis patients decreases the number of circulating T cells producing interferon-gamma and interleukin-4, *J Neuroimmunol* 111: 86-92.

Gasperini C, *et al.* (1999). Interferon-beta-1a in relapsing-remitting multiple sclerosis: effect on hypointense lesion volume on T1 weighted images, *J Neurol Neurosurg Psychiatry* 67: 579-584.

Goodkin DE, The North American Study Group on Interferon beta-1b in Secondary Progressive MS. (2000). Interferon beta-1b in secondary progressive MS: clinical and MRI results of a 3-year randomized controlled trial, *Neurology* 54: 2352.

Grigoriadis N. (2002). Interferon beta treatment in relapsing-remitting multiple sclerosis. A review, *Clin Neurol Neurosurg* 104: 251-258.

Hardmeier M, *et al.* (2002). Detectable atrophy occurs within three months in relapsing-remitting MS (RRMS), *Neurology* 58: A503.

Hardmeier M, *et al.* (2003). Predictors of atrophy during treatment with rIFN-beta 1a IM in relapsing multiple sclerosis: a three year follow-up, *Neurology* 60: A301.

Herndon RM, Murray TJ. (1983). Proceedings of the international conference on therapeutic trials in multiple sclerosis, *Arch Neurol* 40: 663-710.

Horowski R. (2002). Multiple sclerosis and interferon beta-1b, past, present and future, *Clin Neurol Neurosurg* 104: 259-264.

Hua LL, Liu JS, Brosnan CF, Lee SC. (1998). Selective inhibition of human glial inducible nitric oxide synthase by interferon-beta: implications for multiple sclerosis, *Ann Neurol* 43: 384-387.

Hughes RA. (2003). Interferon beta 1a for secondary progressive multiple sclerosis, *J Neurol Sci* 206: 199-202.

Iarlori C, *et al.* (2000). RANTES production and expression is reduced in relapsing-remitting multiple sclerosis patients treated with interferon-beta-1b, *J Neuroimmunol* 107: 100-107.

Jacobs L, Johnson KP. (1994). A brief history of the use of interferons as treatment of multiple sclerosis, *Arch Neurol* 51: 1245-1252.

Jacobs L, *et al.* (1981). Intrathecal interferon reduces exacerbations of multiple sclerosis, *Science* 214: 1026-1028.

Jacobs LD, *et al.* (2000). Intramuscular interferon beta-1a therapy initiated during a first demyelinating event in multiple sclerosis. CHAMPS Study Group, *N Engl J Med* 343: 898-904.

Jacobs LD, *et al.* (1996). Intramuscular interferon beta-1a for disease progression in relapsing multiple sclerosis. The Multiple Sclerosis Collaborative Research Group (MSCRG) [see comments] [published erratum appears in Ann Neurol 1996 Sep;40(3):480], *Ann Neurol* 39: 285-294.

Johnson KP, *et al.* (1995). Copolymer 1 reduces relapse rate and improves disability in relapsing-remitting multiple sclerosis: results of a phase III multicenter, double-blind placebo-controlled trial, *Neurology* 45: 1268-1276.

Johnson KP, *et al.* (1998). Extended use of glatiramer acetate (Copaxone) is well tolerated and maintains its clinical effect on multiple sclerosis relapse rate and degree of disability, *Neurology* 50: 701-708.

Johnson MA, Li DK, Bryant DJ, Payne JA. (1984). Magnetic resonance imaging: serial observations in multiple sclerosis, *AJNR Am J Neuroradiol* 5: 495-499.

Jones CK, *et al.* (2001). MRI cerebral atrophy in relapsing-remitting MS: Results from the PRISMS trial, *Neurology* 56: A379.

Kappos L. (2003). Interferons in relapsing remitting multiple sclerosis. In Cohen JA, Rudick RA, editors: *Multiple sclerosis therapeutics*, ed 2, London: Martin Dunitz.

Kappos L, *et al.* (2000). Induction of a non-encephalitogenic type 2 T helper-cell autoimmune response in multiple sclerosis after administration of an altered peptide ligand in a placebo-controlled, randomized phase II trial [In Process Citation], *Nat Med* 6: 1176-1182.

Kappos L, *et al.* (1999). Predictive value of gadolinium-enhanced magnetic resonance imaging for relapse rate and changes in disability or impairment in multiple sclerosis: a meta-analysis. Gadolinium MRI Meta-analysis Group, *Lancet* 353: 964-969.

Karp CL, Boxel-Dezaire AH, Byrnes AA, Nagelkerken L. (2001). Interferon-beta in multiple sclerosis: altering the balance of interleukin-12 and interleukin-10?, *Curr Opin Neurol* 14: 361-368.

Kaser A, Nagata S, Tilg H. (1999). Interferon alpha augments activation-induced T cell death by upregulation of Fas (CD95/APO-1) and Fas ligand expression, *Cytokine* 11: 736-743.

Kieburtz K, McDermott M. (2002). Needed in MS: evidence, not EVI-DENCE, *Neurology* 59: 1482-1483.

Kim HJ, *et al.* (2002). Persistence of immune responses to altered and native myelin antigens in patients with multiple sclerosis treated with altered peptide ligand, *Clin Immunol* 104: 105-114.

Knobler RL, *et al.* (1984). Clinical trial of natural alpha interferon in multiple sclerosis, *Ann N Y Acad Sci* 436: 382-388.

Kurtzke JF. (1983). Rating neurologic impairment in multiple sclerosis: an expanded disability status scale (EDSS), *Neurology* 33: 1444-1452.

Kurtzke JF. (1955). A new scale for evaluating disability in multiple sclerosis, *Neurology* 5: 580-583.

Kuzma JW, *et al.* (1969). An assessment of the reliability of three methods used in evaluating the status of multiple sclerosis patients, *J Chronic Dis* 21: 803-814.

Leppert D. *et al.* (1997). Interferon beta-1b inhibits gelatinase secretion and in vitro migration of human T cells: a possible mechanism for treatment efficacy in multiple sclerosis, *Ann Neurol* 40: 846-852.

Li DK, Zhao GJ, Paty DW. (2001). Randomized controlled trial of interferon-beta-1a in secondary progressive MS: MRI results, *Neurology* 56: 1505-1513.

Liu Z, *et al.* (2001). Immunomodulatory effects of interferon beta-1a in multiple sclerosis [In Process Citation], *J Neuroimmunol* 112: 153-162.

Lu HT, *et al.* (1995). Interferon (IFN) beta acts downstream of IFN-gamma-induced class II transactivator messenger RNA accumulation to block major histocompatibility complex class II gene expression and requires the 48-kD DNA-binding protein, ISGF3-gamma, *J Exp Med* 182: 1517-1525.

Lublin FD, Reingold SC. (2001). Placebo-controlled clinical trials in multiple sclerosis: ethical considerations. National Multiple Sclerosis Society (USA) Task Force on Placebo-Controlled Clinical Trials in MS, *Ann Neurol* 49: 677-681.

Marrack P, Kappler J, Mitchell T. (1999). Type I interferons keep activated T cells alive, *J Exp Med* 189: 521-530.

Marrie RA, Cohen JA. (2003). Interferons in secondary progressive multiple sclerosis. In Cohen JA, Rudick RA. editors: *Multiple sclerosis therapeutics*, ed 2, London: Martin Dunitz.

McFarland HF, *et al.* (1996). MRI studies of multiple sclerosis: implications for the natural history of the disease and for monitoring effectiveness of experimental therapies, *Mult Scler* 2: 198-205.

Miller DH. (1998). Multiple sclerosis: use of MRI in evaluating new therapies, *Semin Neurol* 18: 317-325.

Miller DH *et al.* (1996). Guidelines for the use of magnetic resonance techniques in monitoring the treatment of multiple sclerosis. US National MS Society Task Force, *Annals of Neurol* 39: 6-16.

Miller DH *et al.* (2002). Measurement of atrophy in multiple sclerosis: pathological basis, methodological aspects and clinical relevance, *Brain* 125:1676-1695.

Miller DH *et al.* (1999). Effect of interferon-beta1b on magnetic resonance imaging outcomes in secondary progressive multiple sclerosis: results of a European multicenter, randomized, double-blind, placebo-controlled trial. European Study Group on Interferon-beta1b in secondary progressive multiple sclerosis, *Ann Neurol* 46:850-859.

Molyneux PD *et al.* (2000). The effect of interferon beta-1b treatment on MRI measures of cerebral atrophy in secondary progressive multiple sclerosis. European Study Group on Interferon beta-1b in secondary progressive multiple sclerosis, *Brain* 123 (Pt 11):2256-2263.

Muraro PA, Leist T, Bielekova B, McFarland HF. (2000). VLA-4/CD49d downregulated on primed T lymphocytes during interferon-beta therapy in multiple sclerosis, *J Neuroimmunol* 111:186-194.

Narayanan S *et al.* (2001). Axonal metabolic recovery in multiple sclerosis patients treated with interferon beta-1b, *J Neurol* 248:979-986.

Neighbour PA, Miller AE, Bloom BR. (1981). Interferon responses of leukocytes in multiple sclerosis, *Neurology* 31:561-566.

Neuhaus O, Farina C, Wekerle H, Hohlfeld R. (2001). Mechanisms of action of glatiramer acetate in multiple sclerosis, *Neurology* 56:702-708.

Noseworthy JH. (2003). Treatment of multiple sclerosis and related disorders: what's new in the past 2 years?, *Clin Neuropharmacol* 26:28-37.

Noseworthy JH, Lucchinetti C, Rodriguez M, Weinshenker BG. (2000). Multiple sclerosis, *N Engl J Med* 343:938-952.

Ozenci V *et al.* (2000). Multiple sclerosis: pro- and anti-inflammatory cytokines and metalloproteinases are affected differentially by treatment with IFN-beta, *J Neuroimmunol* 108:236-243.

Pan JW, Krupp LB, Elkins LE, Coyle PK. (2001). Cognitive dysfunction lateralizes with NAA in multiple sclerosis, *Appl Neuropsychol* 8:155-160.

Panitch H *et al.* (2002). Randomized, comparative study of interferon beta-1a treatment regimens in MS: The EVIDENCE Trial, *Neurology* 59:1496-1506.

Panitch HS, Hirsch RL, Haley AS, Johnson KP. (1987). Exacerbations of multiple sclerosis in patients treated with gamma interferon, *Lancet* 1:893-895.

Parry A *et al.* (2003). Beta-Interferon treatment does not always slow the progression of axonal injury in multiple sclerosis, *J Neurol* 250:171-178.

Paty DW. (1993). Magnetic resonance in multiple sclerosis, *Curr Opin Neurol Neurosurg* 6:202-208.

PRISMS (Prevention of Relapses and Disability by Interferon beta-1a Subcutaneously in Multiple Sclerosis) Study Group [see comments]. (1998). Randomised double-blind placebo-controlled study of interferon beta-1a in relapsing/remitting multiple sclerosis, *Lancet* 352:1498-1504.

PRISMS (Prevention of Relapses and Disability by Interferon beta-1a Subcutaneously in Multiple Sclerosis) Study Group. (2001). PRISMS-4: Long-term efficacy of interferon-beta-1a in relapsing MS (2001), *Neurology* 56:1628-1636.

Qin Y *et al.* (2000). Characterization of T cell lines derived from glatiramer-acetate-treated multiple sclerosis patients, *J Neuroimmunol* 108:201-206.

Ransohoff RM. (2003). Biologic response to type I interferons: relationship to therapeutic effects in multiple sclerosis. In Cohen JA, Rudick RA, editors: *Multiple sclerosis therapeutics*, ed 2, London: Martin Dunitz.

Rolak LA. (2001). Multiple sclerosis treatment 2001, *Neurol Clin* 19:107-118.

Rose AS *et al.* (1970). Cooperative study in the evaluation of therapy in multiple sclerosis: ACTH vs placebo. Final report, *Neurology* 20:1-59.

Rose AS *et al.* (1968). Cooperative study in the evaluation of therapy in multiple sclerosis: ACTH vs placebo in acute exacerbations. Preliminary report, *Arch Neurol* 18:1-20.

Rovaris M *et al.* (2001). Short-term brain volume change in relapsing-remitting multiple sclerosis: effect of glatiramer acetate and implications, *Brain* 124:1803-1812.

Rovaris M, Filippi M. (2003). Interventions for the prevention of brain atrophy in multiple sclerosis : current status, *CNS Drugs* 17:563-575.

Rudick R et al. (1996). Clinical outcomes assessment in multiple sclerosis, *Ann Neurol* 40:469-479.

Rudick R et al. (1997). Recommendations from the National Multiple Sclerosis Society Clinical Outcomes Assessment Task Force, *Ann Neurol* 42:379-382.

Rudick RA. (2001). Contemporary immunomodulatory therapy for multiple sclerosis, *J Neuroophthalmol* 21:284-291.

Rudick RA et al. (1996). CSF abnormalities in a phase III trial of AvonexTM (IFNβ-1a) for relapsing multiple sclerosis, *Ann Neurol* 40:516.

Rudick RA et al. (1999a). Cerebrospinal fluid abnormalities in a phase III trial of Avonex (IFNβ-1a) for relapsing multiple sclerosis, *J Neuroimmunol.* 93:8-14.

Rudick RA et al. (2001a). Estimating effects of disease modifying therapy in patients with multiple sclerosis followed longitudinally after a controlled clinical trial, *Neurology* 56:A353.

Rudick RA et al. (2001b). Use of the Multiple Sclerosis Functional Composite to predict disability in relapsing MS, *Neurology* 56:1324-1330.

Rudick RA, Cutter G, Reingold S. (2002). The multiple sclerosis functional composite: a new clinical outcome measure for multiple sclerosis trials, *Mult Scler* 8:359-365.

Rudick RA et al. (1999). Use of the brain parenchymal fraction to measure whole brain atrophy in relapsing remitting MS, *Neurology* 53:1698-1704.

Rudick RA et al. (1998b). In vivo effects of interferon beta-1a on immunosuppressive cytokines in multiple sclerosis, *Neurology* 50:1294-1300.

Rudick RA, Sandrock A, Panzara M, Polman C. (2003). Study designs of two phase III trials to determine the safety and efficacy of natalizumab (Antegren) alone and when added to interferon beta 1a (Avonex) in patients with relapsing-remitting multiple sclerosis, *Neurology* 60:A479.

Rudick RA et al. (1998c). Incidence and significance of neutralizing antibodies to interferon beta-1a in multiple sclerosis. Multiple Sclerosis Collaborative Research Group (MSCRG), *Neurology* 50:1266-1272.

Schlaak JF et al. (2002). Cell-type and donor-specific transcriptional responses to interferon-alpha. Use of customized gene arrays, *J Biol Chem* 277:49428-49437.

Schwid SR, Bever CT Jr. (2001). The cost of delaying treatment in multiple sclerosis: what is lost is not regained, *Neurology* 56:1620.

Sciacca FL, Canal N, Grimaldi LM. (2000). Induction of IL-1 receptor antagonist by interferon beta: implication for the treatment of multiple sclerosis, *J Neurovirol* 6 (Suppl 2):S33-S37.

Secondary Progressive Efficacy Clinical Trial of Recombinant Interferon-beta-1a in MS (SPECTRIMS) Study Group. (2001). Randomized controlled trial of interferon-beta-1a in secondary progressive MS. Clinical results, *Neurology* 56:1496-1504.

Simon JH et al. (2000). A longitudinal study of T1 hypointense lesions in relapsing MS: MSCRG trial of interferon beta-1a. Multiple Sclerosis Collaborative Research Group, *Neurology* 55:185-192.

Stewart VC et al. (1997). Pretreatment of astrocytes with interferon-alpha/beta impairs interferon-gamma induction of nitric oxide synthase, *J Neurochem* 68:2547-2551.

Sturzebecher S et al. (2003). Expression profiling identifies responder and non-responder phenotypes to interferon-beta in multiple sclerosis, *Brain* 126:1419-1429.

Teitelbaum D, Aharoni R, Sela M, Arnon R. (1991). Cross-reactions and specificities of monoclonal antibodies against myelin basic protein and against the synthetic copolymer 1, *Proc Natl Acad Sci USA* 88:9528-9532.

Teleshova N et al. (2000). Elevated CD40 ligand expressing blood T-cell levels in multiple sclerosis are reversed by interferon-beta treatment, *Scand J Immunol* 51:312-320.

The IFN-b Multiple Sclerosis Study Group, T. (1993). Interferon beta-1b is effective in relapsing-remitting multiple sclerosis. I. Clinical results of a multicenter, randomized, double-blind, placebo-controlled trial, *Neurology* 43:656-661.

The IFNB Multiple Sclerosis Study Group and the University of British Columbia MS/MRI Analysis Group (1996). Neutralizing antibodies during treatment of multiple sclerosis with interferon beta-1b: experience during the first three years, *Neurology* 47:889-894.

The Multiple Sclerosis Study Group. (1990). Efficacy and toxicity of cyclosporine in chronic progressive multiple sclerosis: a randomized, double-blinded, placebo-controlled clinical trial, *Ann Neurol* 27:591-605.

Trapp BD et al. (1998). Axonal transection in the lesions of multiple sclerosis, *N Engl J Med* 338:278-285.

van den Noort S. (1983). Therapeutic fads and quack care, *Arch Neurol* 40:673-674.

Vartanian T. (2003). An examination of the results of the EVIDENCE, INCOMIN, and phase III studies of interferon beta products in the treatment of multiple sclerosis, *Clin Ther* 25:105-118.

Waxman SG. (1998). Demyelinating diseases—new pathological insights, new therapeutic targets [editorial; comment], *N Engl J Med* 338:323-325.

Weinshenker BG et al. (1991). The natural history of multiple sclerosis: a geographically based study. 4. Applications to planning and interpretation of clinical therapeutic trials, *Brain* 114:1057-1067.

Whitaker JN, McFarland HF, Rudge P, Reingold SC. (1995). Outcomes assessment in multiple sclerosis clinical trials: a critical analysis, *Mul Scer Clin Issues* 1:37-47.

Wiendl H, Kieseier BC. (2003). Disease-modifying therapies in multiple sclerosis: an update on recent and ongoing trials and future strategies, *Exp Opin Invest Drugs* 12:689-712.

Wolinsky JS et al. (2003). Glatiramer acetate slows sustained accumulated disability in relapsing multiple sclerosis: meta-analysis results of three double-blind, placebo-controlled clinical trials, *Neurology* 60:A480.

Wolinsky JS, Narayana PA, He R, The PROMiSe Study Group. (2002). Overview of treatment trials in primary progressive multiple sclerosis: early baseline clinical and MRI data of the PROMiSe trial. In Filippi M, Comi G, editors: *Primary progressive multiple sclerosis*, Milan: Springer-Verlag Italia.

Zang YC et al. (2001). Regulation of chemokine receptor CCR5 and production of RANTES and MIP-1alpha by interferon-beta, *J Neuroimmunol* 112:174-180.

5

The Emergence of Neurosurgical Approaches to the Treatment of Epilepsy

Jerome Engel, Jr, MD, PhD

The modern era of epilepsy surgery—that is, surgical intervention designed to correct natural rather than supernatural causes of epilepsy—began in the latter half of the nineteenth century. The history of the development of surgical treatment for epilepsy, based on a scientific understanding of epileptogenic etiologies, is in essence the history of localization of function within the brain (Engel, 1993a). Early scientific investigations into the organization of brain activity informed neurologists and neurosurgeons about the areas of cortex to be resected, while the study of focal epilepsy and of the brain during surgical treatment for epilepsy provided much of the information that enabled early neuroscientists to determine the topography of essential human cerebral functions. This sociohistorical perspective provides several lessons that are generally applicable to the emergence of multidisciplinary collaborative neuroscientific research today. To understand the enormous contributions of the early pioneers of epilepsy surgery, not only to the treatment of epileptic seizures, but also to the development of the field of neuroscience, it is important to appreciate the primitive context in which these conceptual advances were made.

In the mid-nineteenth century, people with epilepsy were still viewed as possessed and suffered as much from the occult treatments of well-meaning physicians as they did from their seizures (Temkin, 1945). Trephining, cauterization, and castration were commonly used and were still mentioned as credible interventions in William Gowers' (1881) textbook *Epilepsy and Other Chronic Convulsive Disorders*. Most physicians only recognized generalized tonic-clonic ictal events as epileptic seizures, and these were believed to originate in the medulla oblon-gata. Other types of epileptic seizures were referred to as "epileptiform" or considered manifestations of mental illness. The first antiseizure drug, bromide, was introduced as a treatment for women with epilepsy by a society physician, Charles Locock (1857), because it was believed to suppress libidinous behavior and hysteria. Phenobarbital did not appear as a treatment until 1912 (Hauptmann, 1912). Indeed, until the 1980s, all antiepileptic drugs were discovered serendipitously, or by trial and error. Only recently have antiepileptic compounds been designed to treat specific cellular and subcellular epileptogenic mechanisms, revealed to a large extent by basic research on the human epileptic brain carried out in an epilepsy surgery setting (Meldrum, 1997).

Prevalent beliefs about localization of brain function at the end of the eighteenth century were influenced by the introduction of phrenology in Vienna by Franz Josef Gall (1800) (Figure 5.1A). Gall maintained that compartmentalization of cerebral activity produced bumps on the skull and that personality types could be diagnosed by palpation of the head (Figure 5.1B). Phrenology was subsequently so vehemently debunked that the scientific community rejected any attempt to resurrect concepts of cerebral localization for almost half a century. It was a British philosopher, Herbert Spencer (Figure 5.1C), who powerfully restored respectability to the view that function might be localized within the human brain: "Localization of function is the law of all organizations whatever: separation of duty is universally accompanied with separateness of structure; and it would be marvelous were an exception to exist in the cerebral hemispheres" (Spencer, 1855). Spencer's writings on this subject had a major influence on John Hughlings

Jackson (Fig. 5.2A), generally recognized to be the father of modern epileptology (Young, 1990).

HUGHLINGS JACKSON AND THE CONCEPT OF LOCALIZATION OF FUNCTION

The English literature credits Jackson, who began his appointment at London's National Hospital for the Paralyzed and Epileptic at Queen Square in 1862, with the recognition that focal seizures are also epilepsy, that these ictal events result from localized discharging cortex, and that the best way to understand the fundamental mechanisms of epilepsy is to study their focal manifestations. Others before Jackson, including the Englishmen Pritchard (1822) and Bright (1831), and the German Griesinger (1867), had studied focal "epileptiform" events, but did not make the clear connection with epilepsy. The Frenchman Bravais (1827) was the first to describe localized postictal paresis, although the phenomenon was subsequently named after the Irishman Todd (1856). Jackson's contributions were seminal because he was a keen observer of clinical behavior, and because he correlated his detailed descriptions of focal ictal signs and symptoms with the anatomical location of specific lesions of the brain identified postmortem (Jackson, 1880; Jackson and Colman, 1898; Taylor, 1958). These studies were the foundation for early understanding of the location of normal functions in the human brain, but they also had important clinical applications. Eventually, by direct observation and detailed descriptions of the initial ictal clinical features of patients with focal epilepsy, Jackson was often able to predict where in the brain the seizures began, and therefore where the offending pathological lesion could be found if a neurosurgeon were brave enough to open the skull.

The first report of successful surgical removal of epileptogenic lesions, at a time when antiseptic and anesthetic techniques were all but nonexistent, was that of Benjamin Winslow Dudley (1828) of the Transylvania University Medical School in Lexington, Kentucky. Dudley reported on five surgical procedures performed between 1818 and 1828 to correct palpable traumatic skull defects that were causing epileptic seizures. All five patients survived, three became seizure free, and two had marked improvement, which Dudley attributed to the clean air of Kentucky. He commented that these results would not have been possible in an unhealthy urban environment.

The concept of utilizing neurological signs and symptoms to localize an otherwise "invisible" cerebral lesion was first used for surgical intervention by the French surgeon Pierre-Paul Broca (1861) (Fig. 5.2B). Based on his postmortem examination of brains from patients with expressive aphasia, Broca localized motor language to the area of the left frontal operculum now named after him, and later correctly located and removed an "invisible" extradural abscess in an aphasic patient; however, the patient died.

Faradic stimulation in animals also contributed importantly to nineteenth century concepts of localization of cerebral function and to confirmation of observations made by clinical neuroscientists. The Germans Gustav Theodor Fritsch and Edward Hitzig (1870) used faradic stimulation to map the dog motor cortex, and, in London, David Ferrier (1873, 1874a, 1874b, 1875, 1883) (Figure 5.2C) carried out faradic stimulation of cortex in several animal species, including monkey, to reproduce the ictal behavior described by Jackson in patients. In a preview of things to come, the ethics of Ferrier's research on monkeys was viciously attacked by British antivivisectionists; however, he urged surgeons to use his data to localize resectable epileptic lesions (Ferrier, 1883).

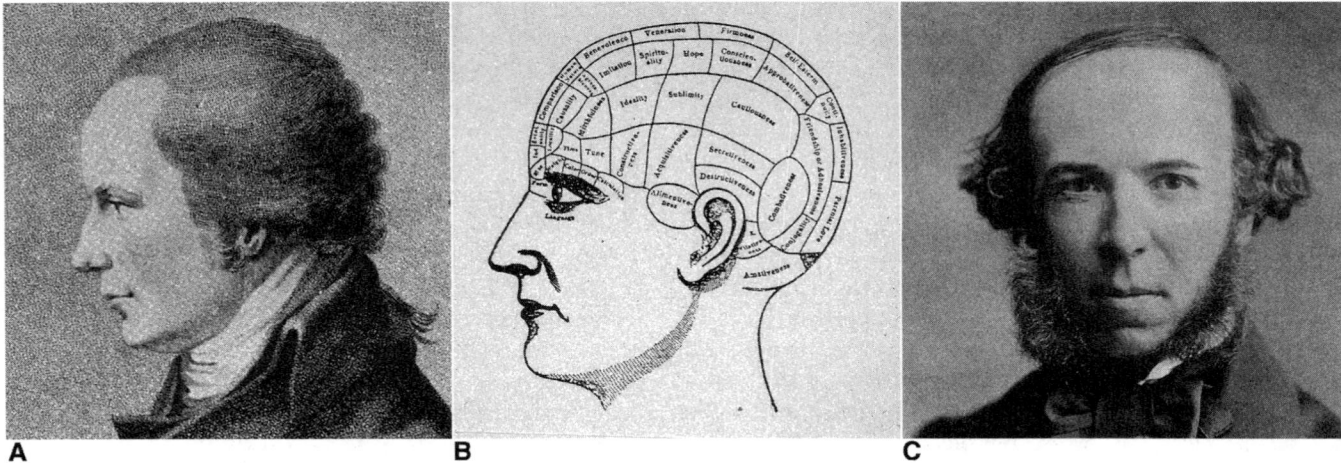

A   **B**   **C**

FIGURE 5.1 **A:** Franz Josef Gall (1758-1828). **B:** Phrenology head showing location of different functions within the brain circa 1791. **C:** Herbert Spencer (1820-1903).

FIGURE 5.2 **A:** John Hughlings Jackson (1820-1903). **B:** Pierre-Paul Broca (1824-1880). **C:** David Ferrier (1843-1928). **D:** William Macewen (1848-1924). **E:** Victor Horsley (1857-1916).

Perhaps the most bizarre report of early stimulation was that of Robert Bartholow (1874) of Cincinnati, who stimulated the exposed motor cortex of his house servant through a skull defect caused by cancer of the scalp. Bartholow was also severely criticized for his ethics.

By the late 1800s, when advances in antiseptic and anesthetic techniques made surgery on the brain more commonplace, William Macewen (1879) (Figure 5.2D) of Glasgow reported the first resection of an "invisible" lesion to treat epilepsy, based on Jackson's clinical approach to localization. The patient was relieved of his focal motor seizures after successful removal of a frontal meningioma, and Macewen (1881, 1888) went on to publish a series of neurosurgical procedures guided entirely by clinical findings. Although Bennett and Godlee (1884) in London and Durante (Horsley, 1884) in Italy had also performed surgery for epilepsy earlier, Victor Horsley (1886) (Figure 5.2E) is credited with initiating the modern era of epilepsy surgery, after operations on three patients with partial seizures at London's National Hospital in 1886, which he published that same year. Horsley had also carried out stimulation experiments on monkeys and worked closely with Jackson and Ferrier, who were both present in the operating theater for these landmark procedures. These three men heralded the essential multidisciplinary epilepsy surgery team of a neurosurgeon, neurologist, and electrophysiologist, which still exists today. Why Horsley, and not Macewen, is generally recognized as the initiator of modern epilepsy surgery may be due to his important teamwork with Jackson and Ferrier, who also deserve credit for bringing this new therapeutic approach to fruition, or it may merely reflect the British disdain for Scots at the time. The latter hypothesis is perhaps supported by Jackson's quip to Ferrier after closure of the cranial vault on the first patient, who

came from Scotland, that Horsley had missed the opportunity to put a joke in a Scotsman's head (Taylor, 1987).

EARLY APPROACHES AND LESION-DIRECTED SURGERY

By the turn of the twentieth century, a few European surgeons were operating on carefully selected patients with intractable focal epilepsy, following the course set by Macewen and Horsley. The most active epilepsy surgery programs were in Germany, where Feodor Krause (1909; Krause and Heyman, 1914) in Berlin and Otfrid Foerster (1925) (Figure 5.3A) in Breslau used Jacksonian localization concepts to determine where to perform the craniotomy, and intraoperative electrical stimulation to help determine the extent of neocortical removal. Surgical resection, however, was always primarily aimed at removing a visible cortical scar or other obvious lesion.

Foerster is particularly recognized in the history of epilepsy surgery for his most famous pupil, Wilder Penfield (Fig. 5.3B), who brought these concepts of lesion-directed surgery for intractable epilepsy to Montreal in 1928, and founded the Montreal Neurological Institute (MNI) in 1934. Penfield not only introduced epilepsy surgery to the Western Hemisphere, but also realized the importance of creating a multidisciplinary team of neuroscientists who were among the first to use the potential of the epilepsy surgery setting to perform invasive research on human brain function. In addition to carrying out intraoperative electrocortical stimulation for localizing the epileptogenic region, Penfield and his colleagues (Penfield and Erickson, 1941; Penfield and Jasper, 1954) also used stimulation as a research tool to explore normal cortical function. They

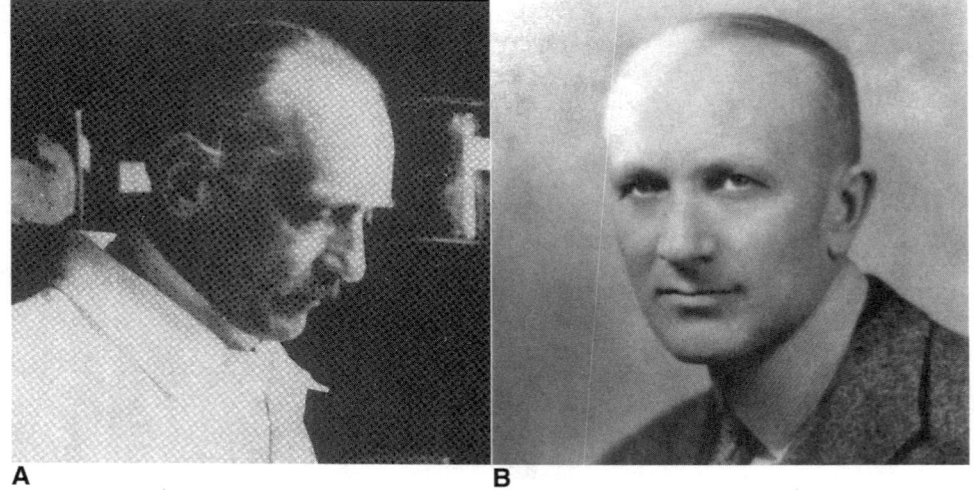

A   **B**

FIGURE 5.3   **A:** Otfrid Foerster (1873-1941). **B:** Wilder Penfield (1891-1976).

created detailed maps of the motor and sensory homunculi (Figure 5.4) and contributed importantly to early investigations into the localization of function in the human brain. The MNI was also to become the most important center for clinical advances in epilepsy surgery for many decades.

Lesion-directed epilepsy surgery was greatly enhanced by the development of roentgenographic techniques, particularly pneumoencephalography (Dandy, 1919) and cerebral angiography (Moniz, 1934). With the advent of these diagnostic tests, some "invisible" lesions of the brain could actually be seen before surgery, and their identification served not only to confirm localization obtained from ictal semiology but also to provide crucial localizing information, even when ictal and interictal signs and symptoms were equivocal.

During the first half of the twentieth century, localized resective surgery for epilepsy was confined to small numbers of well-selected patients with clear evidence of a visible focal lesion; however, a few neurologists and surgeons pursued hypotheses that other surgical interventions might be beneficial for patients with intractable epilepsy. Among the various interventions attempted, corpus callosotomy, first performed by William van Wagenen and Yorke Herren (1940) in the United States, and hemispherectomy, first performed by Kenneth McKenzie (1938) of Canada (who did not fully publish his work), followed by Rowland Krynauw (1950) in South Africa, were proven to be effective and have endured until the present. Krynauw was also the first to pursue surgical therapy for childhood epilepsies.

THE IMPACT OF ELECTROENCEPHALOGRAPHY AND THE EMERGENCE OF TEMPORAL LOBE SURGERY

Arguably, the most important advance in the development of modern surgical treatment for epilepsy was the advent of electroencephalography (EEG) as a laboratory

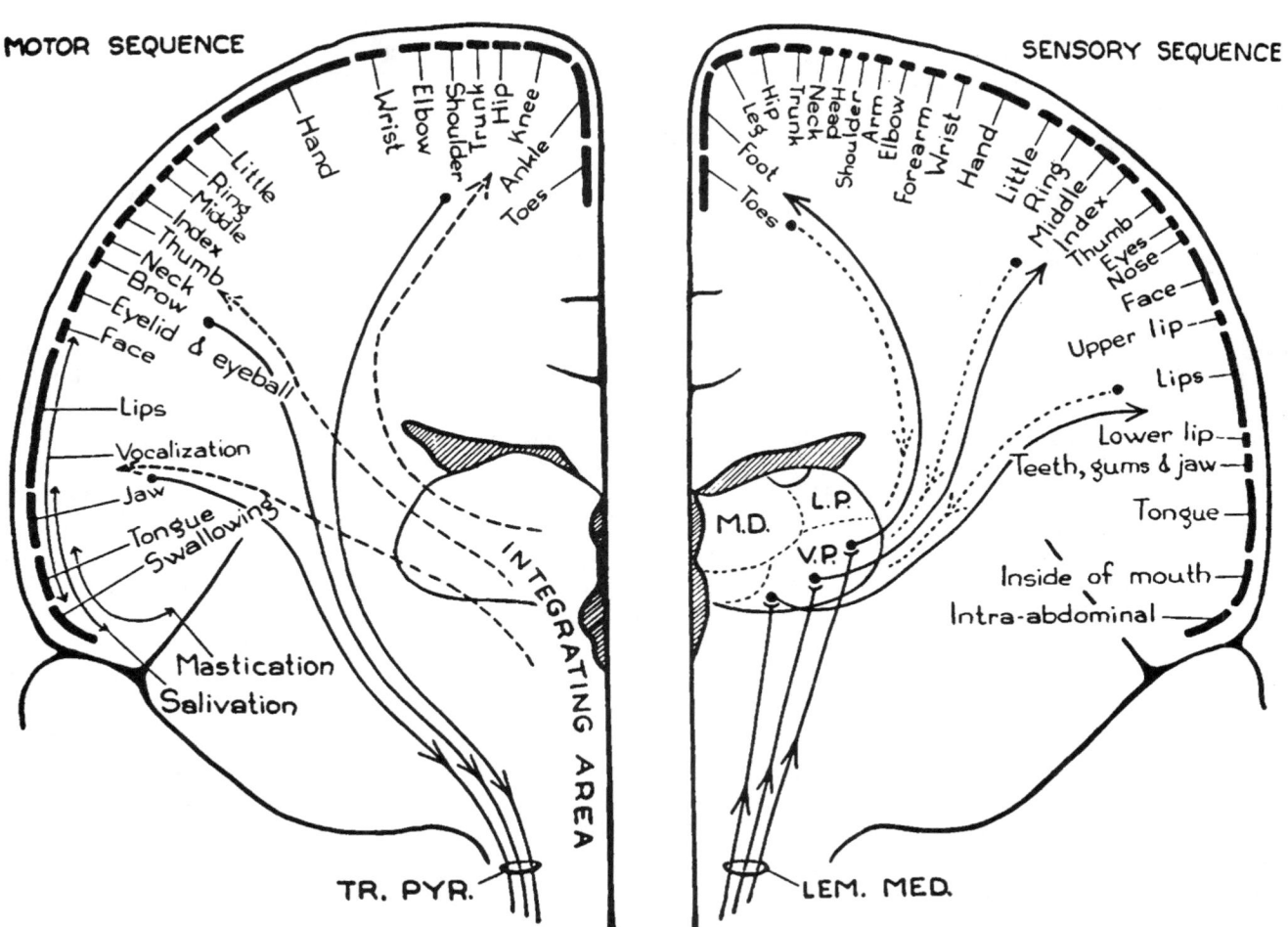

FIGURE 5.4 Location of the motor (*left*) and sensory (*right*) homunculus on a cross-section of the human brain. From Penfield and Jasper (1954).

tool for diagnosing epileptic disorders and localizing focal epileptogenic abnormalities. Richard Caton (1875), a physiologist in Liverpool, first showed that electrical activity could be recorded directly from the brains of animals, but no one published attempts to obtain similar recordings from the intact scalp until Hans Berger (1929) (Figure 5.5A) of Jena published his first article on the human EEG. Perhaps because Berger worked alone in an isolated environment (Gloor, 1969; Millett, 2001), few believed the data reported by the German psychiatrist, and even fewer realized the potential diagnostic value of the first objective laboratory measures of cerebral function. Not until Lord Adrian (Adrian and Matthews, 1934), an internationally renowned neurophysiologist, confirmed Berger's work and demonstrated the alpha rhythm live on stage during a scientific meeting, did the medical community eventually accept this ground-breaking discovery. Berger never received the credit he deserved for this landmark achievement during his lifetime and eventually committed suicide.

It is difficult to appreciate today the impact EEG must have had on the fledgling field of neuroscience at a time when objective measures of brain activity in intact awake and performing animals and humans was thought to be an unobtainable dream. Electrophysiology rapidly became the preeminent discipline for basic research into mechanisms of brain function, and EEG became the symbol for this entire field of investigation. Indeed, before the founding of the Society for Neuroscience in 1970, the American EEG Society was the premier society for neuroscience research in the world, and similarly *Electroencephalography and Clinical Neurophysiology* was the premier neuroscience journal. In both cases, the terms *EEG* and *electroencephalography* were used not to specify a particular diagnostic test, but to symbolize an entire field of modern functional neuroscience (Herbert Jasper, personal communication).

The pioneer electroencephalographers Frederick and Erna Gibbs (Figure 5.5B), along with one of the most prominent early American epileptologists, William Lennox (Figure 5.5C), first described EEG patterns that clearly distinguished among three ictal events: grand mal, petit mal, and psychomotor seizures (Gibbs *et al.*, 1938). The Gibbses, however, used only linked-ear common reference recording for their EEG investigations and therefore believed that all three patterns were generalized. In Montreal, Herbert Jasper (Figure 5.5D) and John Kershman recognized the temporal location of epileptiform EEG events recorded from patients with psychomotor epilepsy because they used bipolar chain montages. They correctly concluded that for psychomotor seizures, and for other focal seizures, it was the initial location and not the pattern of the EEG discharges that provided the most important diagnostic information.

Although temporal lobe epilepsy is the most common form of human epilepsy, among the most refractory to pharmacotherapy (Semah *et al.*, 1998; Engel, 1998a), and the most amenable to surgical therapy (Engel *et al.*, 1993), until the mid-twentieth century, epilepsy surgery was limited to resections of lesions of neocortex. Ammon's horn sclerosis, however, had been recognized more than a century earlier by C. Bouchet and Jean-Baptiste Cazauvieilh (1825) of Paris, who reported that it was a common postmortem finding in patients with "mental alienation" seizures. Jackson (Jackson, 1880; Jackson and Colman, 1898) knew by the end of the nineteenth century that epilepsy with an "intellectual aura" or "dreamy state" and "tasting movements" was associated with lesions in mesial temporal structures. This prescient observation was in part the result of Jackson's study of his most famous patient and friend, Dr. Z (Taylor and Marsh, 1980). At about the same time, the Germans Wilhelm Sommer (1880) and E. Bratz (1899) independently described the cellular pathology of hippocampal sclerosis (Figure 5.6), but proposed that this was the result, not the cause, of epileptic seizures. Gowers (1881), who exercised considerable influence over the neurological community, at least in England, maintained that the hippocampus had nothing to do with the cause of epilepsy. Thus, no early neurosurgeons had ever attempted to remove mesial temporal structures, which they considered *terra incognita*. It was the EEG that eventually led the field of epilepsy surgery to the mesial temporal lobe.

Based on scalp and intraoperative direct-brain EEG recordings, Jasper (1941) concluded that the ictal manifestations of psychomotor seizures originate in the temporal lobe and wrote this in a chapter of a textbook by Penfield and Erickson (1941), in which Penfield, in a separate chapter of the same book, stated that the origin of psychomotor seizures is unknown and that: "When an encephalograph" (meaning pneumoencephalogram) "shows no evidence of a lesion, either atrophic or expanding . . . , any operative procedure is probably doomed to turn out to be a negative exploration." This clear evidence that Penfield did not seriously apply the observations of his colleague Jasper in his first published series of surgery for temporal lobe epilepsy (Penfield and Flanigin, 1950) explains why the article by Frederic Gibbs, who had moved to Chicago, and the neurosurgeon Percival Bailey (Figure 5.5E), published a year later, is considered to be the first series of temporal lobe surgeries for epilepsy based on scalp EEG evidence alone (Bailey and Gibbs, 1951).

Penfield and Flanigin (1950) reported on 68 patients who underwent temporal lobe resections for epilepsy and, a year later, Jasper *et al.* (1951) published the complete electrophysiological findings in these patients,

FIGURE 5.5 **A:** Hans Berger (1873-1941). **B:** Erna L. (1904-1988) and Frederick A. (1903-1992) Gibbs. **C:** William G. Lennox (1884-1960). **D:** Herbert H. Jasper (1906-1999). **E:** Percival Bailey (1892-1973).

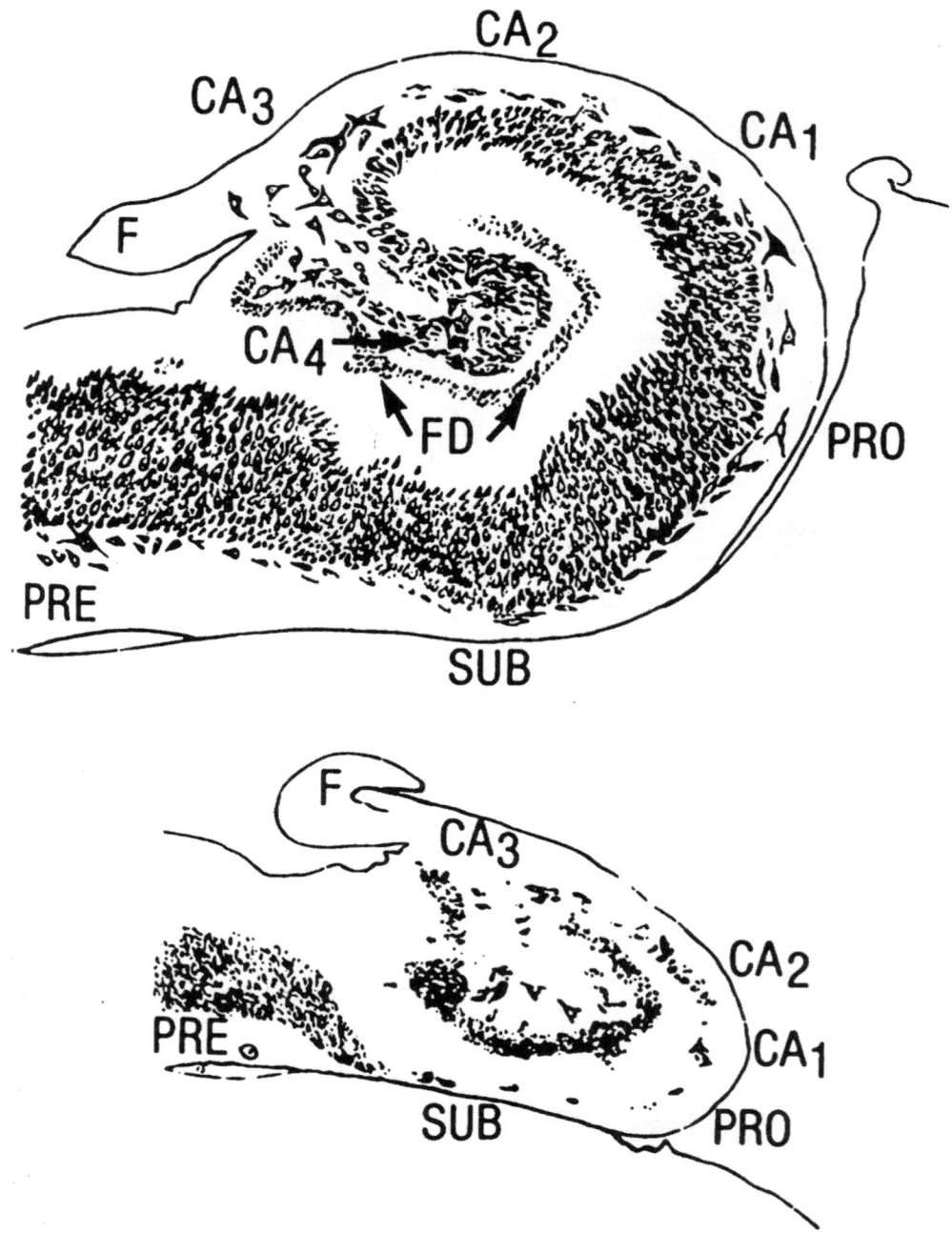

FIGURE 5.6   Wood-carved plate from Bratz (1899) showing cells in a normal (*top*) and sclerotic epileptic (*bottom*) hippocampus.

including intraoperative recording of ictal onset from the uncus (Figure 5.7). In the same year, Bailey and Gibbs published results from a series of 25 patients who received temporal lobe resections based entirely on scalp EEG evidence of localization (Bailey and Gibbs, 1951). By this time the Gibbses (Gibbs *et al.*, 1948) had been using unilateral referential electrodes and had replaced the active ear, so they could lateralize and localize parox-

ysmal EEG events to one temporal lobe. They were then able to convince Bailey to operate on the basis of EEG evidence alone. Although his colleague Jasper had made the same EEG observations during the period of Penfield's 68 temporal lobe resections reported in 1950, Penfield continued to rely on evidence of structural abnormalities for localization (Herman Flanigin, personal communication). Again, credit appears to go to the group that exhibited teamwork between a surgeon and an electrophysiologist. After the Bailey and Gibbs demonstration that EEG activity could be used to localize the epileptogenic region in this most common form of epilepsy, there was an explosion in surgeries for temporal lobe epilepsy in Europe (Falconer, 1953; Guillaume *et al.*, 1953; Paillas *et al.*, 1953; Petit-Dutallis *et al.*, 1953; Maspes and Marossero, 1953; Obrador, 1953; Krayenbuhl *et al.*, 1953), the United States (Green *et al.*, 1951; Morris, 1956), and Cuba (Picaza and Gumá, 1956). Anterior temporal resections became, and remain, the most common surgical procedure used as a treatment for epilepsy.

In 1950, the neurosurgeon Arthur A. Morris (1950) presented a series of five patients at a meeting in Washington, DC, who underwent temporal resections based on localization from EEG. This preliminary presentation may have been an incentive for Bailey and Gibbs to publish their surgical series quickly. More important,

however, is that Morris (1956) removed the anterior hippocampus and amygdala, whereas previous surgeons had been afraid to injure the hippocampus. All of the patients in the Bailey and Gibbs (1951) series, and all but two in the Penfield and Flanigin (1950) series, underwent bigyrectomies or trigyrectomies, leaving the mesial structures intact, with relatively poor results by present-day standards; only about one third of patients became seizure free. It might be argued, in retrospect, that these results were largely a placebo effect, given that surgery, even outside the brain, made seizures better (Penfield, 1936). Perhaps if it had not been for the placebo effect, temporal lobe surgery would not have been pursued. In any event, once it was demonstrated that it was safe to remove the hippocampus and perihippocampal structures, which Jasper had shown were most often the generators of interictal and ictal epileptiform discharges in patients with psychomotor seizures (Jasper and

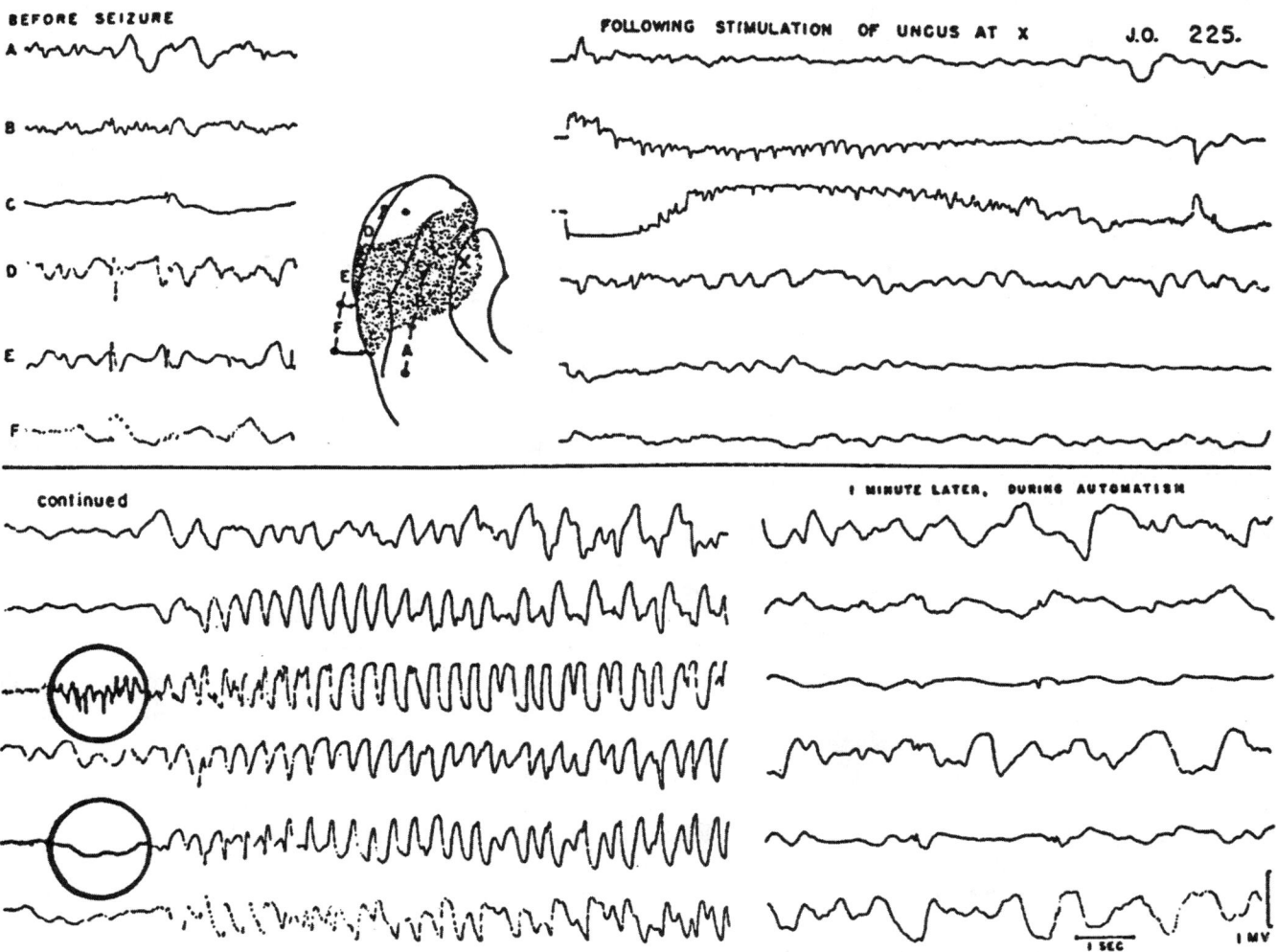

FIGURE 5.7 Direct intraoperative recording from the mesial temporal lobe during a clinical complex partial seizure produced by electrical stimulation of the uncus. From Jasper, Pertuisset, and Flanigin (1951).

Kershman, 1941; Jasper, 1941; Jasper *et al.*, 1951), Penfield reoperated to resect mesial temporal structures in his surgical failures and included mesial temporal structures in subsequent resections, with much better results (Penfield and Baldwin, 1952; Earle *et al.*, 1953). An unfortunate historical footnote is the logical extension of this surgical direction, when Scoville treated his famous patient, H.M., with bilateral hippocampal resections, rendering him globally amnesic (Scoville and Milner, 1957).

Resected mesial temporal specimens, although removed piecemeal, appeared to reveal hippocampal sclerosis (Earle *et al.*, 1953). Subsequently, most neurosurgeons included mesial temporal structures in their anterior temporal resections and Murray Falconer (1953) (Figure 5.8A), in London, introduced a standardized *en bloc* anterior temporal lobectomy procedure that provided large intact tissue specimens for pathologists to examine. The resulting clinicopathological correlations not only demonstrated that a high percentage of patients with temporal lobe epilepsy have hippocampal sclerosis but also that this pathological finding predicted an excellent postoperative outcome (Falconer and Serafetinides, 1963; Falconer and Taylor, 1968). Although evidence continued to mount that hippocampal sclerosis could result from epileptic seizures, such as prolonged febrile convulsions in early childhood, the fact that its removal could cure epilepsy eventually convinced most epileptologists that hippocampal sclerosis was also a cause of temporal lobe epilepsy (Falconer, 1974).

Various noninvasive EEG techniques using nasopharyngeal (Gastaut, 1948; Grinker, 1938; MacLean, 1949; Roubicek and Hill, 1948), tympanic (Arellano, 1949), and sphenoidal (Jones, 1951; Pampiglione and Kerridge, 1956; Pertuiset and Capdevielle-Arfel, 1951; Rovit and Gloor, 1960) electrodes were used to distinguish mesial temporal from lateral temporal interictal and ictal events more accurately; however, direct intraoperative recording from the brain was considered to be more reliable, and EEGers began to pursue approaches to extraoperative invasive recording. Reginald Bickford and Hugh Cairns first carried out such direct recordings from the human brain at Oxford in 1944 by inserting multistranded insulated wires into a cerebral bullet track. Bickford *et al.* (1953) later was the first to publish the results of depth electrode recording from the Mayo Clinic, when he placed electrodes freehand through the skull for both recording and stimulation. Jean Talairach (Figure 5.8B) and Jean Bancaud (Figure 5.8C) introduced stereotactic depth electrode placement in Paris in the late 1950s, but French law permitted them to record for only several hours (Talairach *et al.*, 1958, 1974). Consequently, localization with this stereo EEG technique was based on interictal spike activity, as well as ictal activity induced by electrical stimulation and convulsant drugs.

Paul Crandall and colleagues at the University of California, Los Angeles (UCLA) (Crandall *et al.*, 1963) (Figure 5.8*D*), were the first to report on the use of truly chronic stereotactically implanted depth electrode recordings to localize the onset of spontaneous ictal EEG events. Crandall combined analyses of the electrophysiological features of spontaneous temporal lobe seizures with the *en bloc* anterior temporal resection of Falconer, permitting detailed electropathological correlations. This investigative approach became the basis for the innovative multidisciplinary collaborative research on fundamental mechanisms of human epilepsy that is productively pursued at many epilepsy surgery centers today (Crandall and Babb, 1993; Engel, 1998b; Engel *et al.*, 1987; Schwartzkroin, 1993a). The investigative team involves not only neurosurgeons, neurologists, electrophysiologists, and neuropathologists, but also basic scientists committed to finding rational therapies for epilepsy that will make surgery no longer necessary. Furthermore, Crandall deserves credit for recognizing the potential of EEG telemetry technology, which was being developed at the time at UCLA by Ross Adey and colleagues (1956) to record from chimpanzees orbiting the earth as part of the U.S. space program. With Richard Walter, Crandall demonstrated the ability of telemetry to yield artifact-free continuous EEG recordings from depth electrodes over long periods and established the first epilepsy EEG telemetry unit (Dymond *et al.*, 1971). Crandall's electronic engineer, Tony Dymond, eventually founded BioMedical Sciences, the first manufacturer of commercial EEG-telemetry systems.

Subsequent advances in EEG presurgical evaluation included the use of photography, cinematography, and then video, for simultaneously monitoring behavior and obtaining detailed second-by-second electroclinical correlations of ictal onset and propagation (Ajmone-Marsan and Ralston, 1957; Hunter and Jasper, 1949; Penry *et al.*, 1975; Schwab *et al.*, 1953; Goldensohn, 1966). This was followed by the application of long-term video and EEG monitoring techniques to obtain noninvasive scalp and sphenoidal recordings (Engel *et al.*, 1981; Ives and Gloor, 1978), semi-invasive foramen ovale (Wieser, 1986) and peg (Barnett *et al.*, 1990) electrode recordings, and a variety of other chronic invasive recordings using epidural (Goldring and Gregorie, 1984) and subdural (Lüders *et al.*, 1987) electrodes.

Although EEG has excellent temporal resolution, spatial resolution is severely limited. Localization within centimeters can be inferred from scalp and sphenoidal EEG recordings, but epileptiform activity in the depths of the brain, particularly in mesial temporal structures, is not recorded from the scalp unless it propagates to superficial cortex. Intraoperative and chronic extraoperative intracranial recordings using intracerebral depth

FIGURE 5.8   **A:** Murray A. Falconer (1910-1977). **B:** Jean Talairach (1911-). **C:** Jean Bancaud (1921-1994); **D:** Paul Crandall (1923-).

electrodes or electrode arrays over the cortical surface monitor only limited areas of the brain and are subject to sampling error. A variety of computerized techniques were developed to automatically detect epileptiform events (Gotman, 1981) and to localize the sources of their dipoles noninvasively (Ebersole, 1997). The accuracy of dipole source localization remains limited by the facts that (1) multiple dipole sources can give rise to identical surface-recorded signals, (2) dipole models do not account for distributed sources, and (3) epileptiform events that cannot be recorded from the scalp remain undetected. Magnetoencephalography (MEG) is gaining increasing application for presurgical evaluation as a result of certain advantages over routine EEG for calculating the location of dipole sources for epileptiform events in three dimensions, because magnetic signals are not altered by passage through structures of different density, such as brain, cerebral spinal fluid, bone, and skin (Barth *et al.*, 1984). Dipole source localization of interictal and ictal epileptiform abnormalities obtained

by MEG and superimposed on the patient's own magnetic resonance imaging (MRI) scan now reveal, in some patients, that apparently diffuse or multifocal scalp EEG events have a single site of origin (Eliashiv *et al.*, 2002). MEG, therefore, is becoming increasingly useful for planning chronic intracranial electrode explorations when multiple source localization hypotheses are possible, and perhaps, in a few cases, MEG data can be used to avoid the need for invasive recording.

THE IMPACT OF ASSESSING NONEPILEPTIC ASPECTS OF CEREBRAL FUNCTION

Neurocognitive testing is an important part of modern presurgical evaluation. Neuropsychological batteries are based largely on Brenda Milner's (Figure 5.9AB) description of material-specific functions of the left and right temporal lobes obtained from studies of large numbers of patients who underwent anterior temporal lobe resections at the MNI (Milner, 1958, 1971; Milner and Penfield, 1955). The discipline of neuropsychology has benefited enormously from studies of patients with focal epilepsy in an epilepsy surgery setting, both before and after surgical resection. Neurocognitive test batteries, therefore, are used to provide evidence of focal functional deficit, confirming mesial temporal or neocortical localization obtained from clinical, EEG, and neuroimaging evaluations.

Extraoperative lateralization of hemispheric dominance for language was first demonstrated in Japan by Juhn Wada (1949) (Figure 5.9B), who injected sodium amobarbital into one carotid artery, anesthetizing half of the forebrain so that the other half could be examined independently. The intracarotid amobarbital procedure, or Wada test as it is commonly called, was not known or accepted in the West for more than a decade, until Wada repeated his studies at the MNI (Wada and Rasmussen, 1960). A modified version of the intracarotid amobarbital procedure that permits independent evaluation of memory functions of the dominant and nondominant mesial temporal structures has become important in the presurgical evaluation for intractable temporal lobe epilepsy. This test is used to predict the risk of global amnesia resulting from the inability of the contralateral hemisphere to support memory (Klove *et al.*, 1969; Milner *et al.*, 1962) and to confirm dysfunction in the mesial temporal structures intended for resection (Engel *et al.*, 1981).

Psychiatrists have long been involved in the care of patients with epilepsy, particularly in the United Kingdom, where epileptologists tended to be psychiatrists, rather than neurologists, until very recently. David Taylor (Figure 5.9C), a British psychiatrist who also was the first to describe focal cortical dysplasia as a cause of neocortical epilepsy (Taylor *et al.*, 1971), deserves credit for observing that people with epilepsy seek treatment not merely to eliminate their epileptic seizures but to relieve the predicament caused by these recurrent ictal events (Taylor, 1993). To assess the outcome of surgery in 100 patients operated on by Falconer, Taylor actually visited their homes and assessed the impact of seizure freedom or reduction on activities of daily living (Taylor, 1972), a tour-de-force that has been repeated only once since (Bladin, 1992). Taylor emphasized the need to listen to patients' reasons for wanting surgical treatment and to require realistic expectations as part of the decision-making process.

Taylor's work and subsequent studies by many others have led to the conclusion that effects of surgical intervention, for better or worse, should not be measured by seizure outcome alone, but rather by the impact of seizure freedom or reduction on quality of life. Although most studies of surgical treatment for epilepsy now incorporate quantitative measure of health-related quality of life (Vickrey, 1992; Devinsky *et al.*, 1995), neuropsychological and psychiatric evaluations, as well as a number of other assessments of activities of daily living such as employment and schooling, domiciliary status, and ability to drive, are necessary to determine why scores on quality of life tests change after surgery. Consequently, neuropsychologists, psychiatrists, and health outcome researchers are an essential part of today's epilepsy surgery teams; their contributions help to predict which patients are likely to benefit from surgery, to asses the full impact of surgical outcome on patients' lives, and ultimately to optimize opportunities for improved quality of life in patients who become seizure free. Furthermore, research by many psychiatrists in the epilepsy surgery setting has greatly advanced our understanding of the biological basis of behavioral disorders (Taylor, 2003).

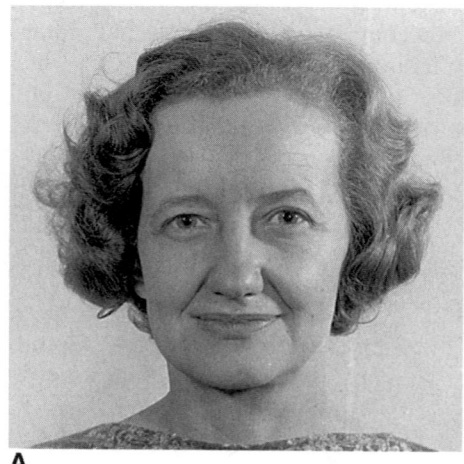

A   B   C

FIGURE 5.9   **A:** Brenda Milner (1918-). **B:** Juhn Wada (1924-). **C:** David C. Taylor (1933-).

Other interictal functional disturbances can be important in the decision-making process before surgical treatment for epilepsy. Persistent neurological signs and symptoms elicited during history and a detailed neurological examination can suggest the presence of a structural lesion that could be epileptogenic. Transient postictal focal neurological dysfunction lasting hours to days provides insight into the location of ictal onset. Intermittent and continuous focal EEG slowing and focal attenuation of faster frequencies, seen on scalp and sphenoidal EEG and more commonly with intracerebral EEG recording, can be important for defining the location and extent of the epileptogenic region. The intravenous thiopental test, developed by W.A. Kennedy and Dennis Hill in London, was used in association with sphenoidal EEG recording to identify attenuation of barbiturate-induced beta activity in the epileptogenic mesial temporal lobe (Kennedy and Hill, 1958). This technique is still used occasionally during intraoperative electrocorticography to define the extent of a neocortical functional disturbance.

Intraoperative functional cortical mapping is necessary to avoid introducing new neurological deficits when suspected neocortical epileptogenic regions involve, or are adjacent to, areas of essential function such as language or motor cortex (Ojemann et al., 1993). In the early days of epilepsy surgery, the operation was performed under local anesthesia because cortical mapping was necessary before cortex could be removed. Today, intraoperative cortical mapping using direct cortical stimulation and recording of evoked potentials can be carried out with the patient under general anesthesia or, depending on the modalities under investigation, with the patient briefly awakened in the operating room. Cortical mapping is also commonly an extraoperative procedure using chronic subdural grid electrodes (Lüders et al., 1987). Occasionally, these studies also provide evidence of aberrant localization of function, which can help define the abnormal epileptogenic neocortex. Since the early work of Penfield and Jasper (1954), intraoperative functional mapping of neocortex in the course of surgical treatment for epilepsy has continued to provide research opportunities for important investigations into the anatomical substrates of uniquely human behaviors such as language (Ojemann, 1991).

CONCEPTS OF THE EPILEPTOGENIC ZONE: SETTING THE STAGE FOR COMPUTERIZED NEUROIMAGING

By 1980, strategies for epilepsy surgery were guided by a number of independent diagnostic approaches for determining the area to be resected (Engel et al., 1981).

Surgical treatment for focal epilepsy requires not only accurate localization of the epileptogenic region but also demarcation of the extent of tissue necessary for generation of spontaneous seizures to determine the boundaries of the resection. Electropathological correlations made possible by the combination of depth electrode studies and en bloc anterior temporal resections led to the description of a common pathophysiological substrate for mesial temporal lobe epilepsy, which resulted in the widespread acceptance of standardizing the surgical procedure for this condition (Crandall, 1987). Although the size and shape of the standardized anterior temporal resection has varied over time and from one surgical program to another ranging from selective amygdalohippocampectomy (Niemyer, 1958; Wieser, 1986) to a slightly larger anteromesial temporal resection (Spencer et al., 1984), to the older Falconer resection (Falconer, 1953; Crandall, 1987), which involves more lateral temporal cortex, the presurgical evaluation for this procedure merely needs to confirm that the epileptogenic region is within the area of standard resection (Engel, 1987). Localized neocortical resections, on the other hand, are always tailored, and the presurgical evaluation must also reveal the extent of the epileptogenic region (Engel, 1987b). The epileptogenic zone, defined as the area of brain tissue necessary and sufficient for generating spontaneous seizures and therefore the minimal resection required to achieve freedom from seizures, is a hypothetical concept determined from a variety of presurgical studies (Lüders et al., 1993).

Extensive electroclinical correlations, made possible by video-EEG telemetry, greatly improved on Jackson's original observations, and clues to the epileptogenic zone can be obtained from analysis of initial ictal semiology. The cortical areas giving rise to initial ictal signs and symptoms, the symptomatic zone, however, is not necessarily the site of ictal onset (Lüders et al., 1993). Seizures often begin in silent areas of the brain, and initial signs and symptoms reflect propagation. For instance, patients and observers usually are unaware of hippocampal ictal discharges until they propagate to neocortical or diencephalic areas.

Not only is the spatial resolution of EEG relatively poor, but areas generating interictal epileptiform events, the irritative zone, are usually much larger than the epileptogenic zone (Lüders et al., 1993). The exact site of ictal onset, the ictal onset zone, can be difficult to determine even with intracranial recording and usually provides insufficient information concerning the actual volume of tissue necessary for spontaneous seizure generation (Lüders et al., 1993).

Before computerized functional neuroimaging, the focal functional deficit zone, determined by interictal neurological examination and neurocognitive testing, could provide useful information only when the epileptogenic region involved cortical areas responsible for behaviors that can be measured such as motor, language, and memory function (Lüders *et al.*, 1993). Large association areas were not testable and spatial resolution was also poor.

Before computerized structural neuroimaging, routine radiological studies, including plain skull films, cerebral angiography, and pneumoencephalography, provided important information regarding the location of a structural lesion in some patients, but the evidence from these tests was often circumstantial and not all structural lesions are necessarily epileptogenic.

Although the presurgical evaluation effectively used all of these diagnostic approaches to approximate the location and boundaries of the epileptogenic zone, there was a great need for new diagnostic techniques that could display structural and functional features of the entire brain, in three dimensions, with high spatial resolution. This was achieved with the advent of computerized neuroimaging.

THE IMPACT OF COMPUTERIZED NEUROIMAGING AND THE RETURN TO LESION-DIRECTED SURGERY

X-ray computed tomography (CT) was the first computerized tomographic approach to visualizing the entire brain in three dimensions. Application to patients with focal epilepsy revealed localized structural lesions in many (Gastaut and Gastaut, 1976), and greatly improved selection of potential surgical candidates for localized resection, particularly of alien tissue lesions. The development of magnetic resonance imaging (MRI) greatly improved resolution of discrete structural lesions that might be epileptogenic, particularly disorders of cortical malformation, such as focal cortical dysplasia, making lesion-directed surgery increasingly possible in patients who, in the past, would have been diagnosed as having cryptogenic focal epilepsy and considered to be poor surgical candidates. The resolution of MRI, however, was considered to be insufficient to image hippocampal sclerosis, the most common pathological substrate of epilepsy, until a small, hyperintense hippocampus was convincingly observed at the MNI in 1985 (Andermann, 1987) (Figure 5.10A) but not published until much later (Jackson *et al.*, 1990). Subsequently, tremendous improvements in identifying potentially epileptogenic structural abnormalities in patients with intractable focal epilepsy, using modern structural neuroimaging, greatly increased the number of patients who could be considered surgical candidates (Cascino and Jack, 1996; Kuzniecky and Jackson, 1995). EEG continues to be necessary, however, to show that an identified structural lesion is, in fact, epileptogenic. Patients with medically intractable focal epilepsy occasionally have structural lesions that are unrelated to the epileptogenic region, or multiple or diffuse structural abnormalities of which only a small part is epileptogenic and requires excision. With the ready availability and increasingly high resolution of MRI, the focus of epilepsy surgery is again becoming lesion-directed, as it was in the early part of the twentieth century, with EEG assuming a confirmatory, although still essential, role.

The first computerized neuroimaging technique to play a major role in presurgical evaluation for temporal lobe epilepsy was not a structural imaging technique, but a functional imaging technique. At UCLA beginning in 1978, positron emission tomography (PET) with 18F-fluorodeoxyglucose (FDG), which measures local cerebral glucose metabolism, revealed that most patients with mesial temporal lobe epilepsy had unilateral temporal hypometabolism (Engel *et al.*, 1981, 1982a, 1982b, 1982c; Kuhl *et al.*, 1980) (Figure 5.10B). Unilateral temporal lobe hypometabolism corresponded to the site of ictal EEG onset, as well as a mesial temporal structural lesion identified postoperatively, usually hippocampal sclerosis. This finding not only helped to identify surgical candidates but obviated the need for invasive studies in many patients (Engel *et al.*, 1981). Unilateral temporal hypometabolism, in association with an ipsilateral sphenoidal ictal EEG onset, and confirmatory localization from tests of focal functional deficit, in the absence of conflicting findings from seizure semiology and structural and functional investigations, permitted effective temporal lobe surgery without depth electrode recording. FDG-PET, therefore, increased the numbers of patients who could be treated surgically and greatly reduced the cost and risk of presurgical evaluation.

FDG-PET was also useful in identifying large unilateral areas of cortical dysplasia in infants and small children with apparent catastrophic secondary generalized epilepsy and a negative MRI (Chugani *et al.*, 1988). Multilobar resections in these children, who are otherwise destined for institutionalization, can eliminate life-threatening seizures and reverse developmental delay (Mathern *et al.*, 1999). FDG-PET, however, did not often reveal areas of focal hypometabolism in adult patients with neocortical epilepsy who did not already have obvious structural lesions on MRI (Henry *et al.*, 1991). Improved PET localization in cryptogenic focal neocortical epilepsy has been achieved using flumazenil, a benzodiazepine receptor ligand, as a tracer; however, this technique is not widely available (Savic *et al.*, 1988). Another tracer, α-methyl-tryptophan (AMT), may also be useful in localizing certain epileptogenic areas in patients with widespread or multiple structural lesions (Chugani *et al.*, 1998). For instance, in tuberous sclerosis

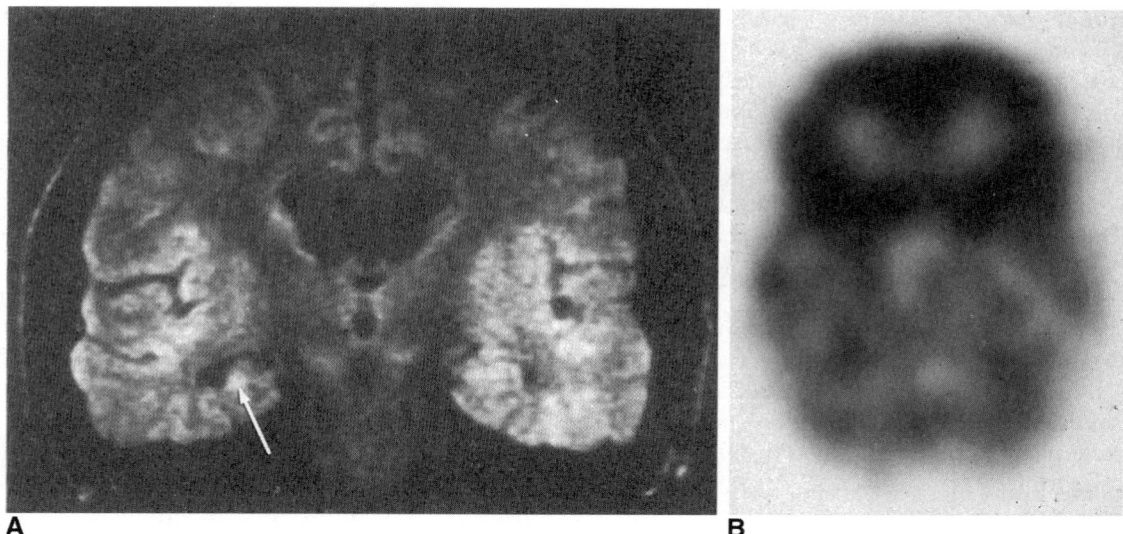

FIGURE 5.10 **A:** One of the first MRI scans to reveal mesial temporal sclerosis as a hyperintense atrophic hippocampus. This patient underwent temporal lobe resection at the Montreal Neurological Institute in 1985 and has been seizure free since (unpublished figure courtesy of Samuel Berkovic). **B:** One of the first PET scans to show temporal lobe hypometabolism in association with mesial temporal sclerosis. This patient underwent temporal lobe resection at UCLA in 1979 and has been seizure free since.

with multiple tubers, only the tuber giving rise to spontaneous seizures appears to take up AMT.

Single photon emission computed tomography (SPECT) revealed decreased cerebral perfusion in epileptogenic temporal lobes with somewhat lower spatial resolution than PET, but this finding may be falsely localizing (Bonte *et al.*, 1983; Sanabria *et al.*, 1983). The demonstration of unilateral temporal lobe hyperperfusion on ictal SPECT, however, more reliably localized the epileptogenic region in mesial temporal lobe epilepsy; and with injection later during the postictal period, there is profound lateral temporal hypoperfusion with residual mesial temporal hyperperfusion (Lee *et al.*, 1986; Rowe *et al.*, 1991). Localized areas of hyperperfusion on ictal SPECT scans during focal neocortical seizures are more difficult to interpret but have become more reliable with the use of statistical parametric mapping to subtract ictal from interictal scans, and to superimpose these data onto the patient's own MRI scan (SISCOM) (O'Brien *et al.*, 1998).

Functional MRI (fMRI) makes use of the natural differences in magnetic properties between oxygenated and deoxygenated hemoglobin to localize activity-related changes in cerebral blood oxygenation with high spatial resolution and second to millisecond temporal resolution. Consequently, it is possible to image functional activation in discretely localized areas of the brain during seizures (Jackson *et al.*, 1994) and the occurrence of interictal EEG spikes (Krakow *et al.*, 1999; Warach *et al.*, 1996). The ability to record EEG during MRI acquisition permits MRI data acquired during interictal spikes to be subtracted from MRI data acquired

between interictal spikes, and the difference to be superimposed on the patient's own structural MRI scan. Interictal EEG spikes are of limited localizing value because some interictal spikes represent propagated activity, and others represent epileptiform events originating in brain areas that are not sufficiently epileptogenic to generate spontaneous seizures; however, there is reason to believe that fMRI may be able to distinguish between spikes originating in the epileptogenic zone (red spikes) and interictal spikes that are falsely localizing (green spikes). *In vivo* microelectrode recordings from epileptic hippocampi of patients with temporal lobe epilepsy and animal models of this disorder have revealed that interictal spikes originating from the area of seizure generation are associated with very high frequency (250-600 Hz) oscillations, termed Fast Ripples. These do not appear with interictal spikes that are propagated or arise from areas where spontaneous seizures have not been recorded (Bragin *et al.*, 1999; Staba *et al.*, 2002). This indicates that the fundamental neuronal mechanisms underlying red spikes may be sufficiently different from the mechanisms causing green spikes to permit their metabolic correlates to be distinguished with fMRI. EEG-fMRI remains a research technique because of the limited number of centers with the ability to record EEG during MRI at the present time; however, specialized EEG equipment for this purpose may soon become commercially available, and much more extensive clinical application will then be possible.

Magnetic resonance spectroscopy (MRS) is yet another application of MRI to localize areas of structural and functional disturbances in the brain (Duncan, 1997). MRS can be used to measure a variety of metabolic processes with a temporal resolution of a few minutes, much better than PET, but not as good as fMRI. It is most often used in epilepsy to identify areas of reduced n-acetyl-aspartate (NAA), which commonly accompanies the cell loss of hippocampal sclerosis. There also is evidence to suggest that reduced NAA is reversible in some circumstances and therefore a reflection of functional disturbances associated with the epileptogenic region as well (Guye *et al.*, 2002). The use of a number of MRS measurements, including NAA, glutamine, glutamate, γ-aminobutyric acid transaminase, phosphorous, lactate, and pH to localize areas of epileptogenic dysfunction is currently under investigation.

Optical imaging is a dynamic invasive technique that makes use of the fact that light passing through brain tissue is altered by changes in neuronal activity. Optical imaging of the human neocortex during epilepsy surgery has revealed areas involved in electrical stimulation-induced afterdischarges as they propagate across the brain surface, with remarkable structural and temporal resolution (Haglund *et al.*, 1992a). Optical imaging can be accomplished in animals through a thin skull, and it is conceivable that techniques eventually will be developed that overcome the thickness of the human skull, permitting noninvasive imaging of ictal and interictal epileptiform abnormalities. Clinical application of optical imaging in epilepsy, however, remains hypothetical.

IMPACT OF THE CREATION OF AN INTERNATIONAL EPILEPSY SURGERY COMMUNITY

The first Palm Desert conference on Surgical Treatment of the Epilepsies, held in 1986, brought together representatives from 53 epilepsy surgery programs in 17 countries (Engel, 1987a). They represented essentially all active epilepsy centers offering surgery as a treatment for epilepsy. At the time, only one textbook on epilepsy surgery was available (Purpura *et al.*, 1975), and there were several different schools of thought concerning approaches to surgical therapy (Engel, 1987b), and very little collaboration or even exchange of information among centers comparing these various approaches. Furthermore, there was no standardized approach to the evaluation of surgical outcome, and results reported in the few published articles of epilepsy surgery series could not be easily correlated. The charge of admission to the conference was data from each epilepsy surgery program in a relatively uniform format which, despite the diverse

approaches, revealed consistent results with respect to surgery for temporal lobe epilepsy and focal neocortical epilepsy (Engel, 1987c) (Table 5.1).

Most epilepsy surgery teams pursued only one particular approach, depending on their lineage (Appendix II, 1987). Those who trained at the MNI under Penfield and Jasper and their successors (Rasmussen, 1983), and later at the University of Washington (Dodrill *et al.*, 1986), based surgical excisions largely on the results of intraoperative electrocorticography, and all resections were tailored. Those who trained in Paris under Bancaud and Talairach and their colleagues (Munari *et al.*, 1985) carried out extensive, predominantly unilateral, stereotactic depth electrode evaluations in almost every patient; and all resections were tailored based on results of invasive studies, except in Zürich (Wieser, 1986), where only amygdalohippocampectomies were performed after invasive presurgical evaluation. Those who trained at the Maudsley in London under Falconer and his colleagues (Powell *et al.*, 1985) performed standardized anterior temporal resections based largely on the results of interictal noninvasive EEG recordings, and although they routinely performed intraoperative electrocorticography, they rarely used these data to alter the temporal resection. Those who trained at UCLA with Crandall and his colleagues (Cahan *et al.*, 1984) also performed standardized anterior temporal resections, but these were based on results of ictal EEG onset patterns recorded with bilateral chronic depth electrodes, and intraoperative recordings were not used for temporal lobe surgery. Later, however, before the Palm Desert meeting, the UCLA protocol changed considerably to rely more heavily on scalp and sphenoidal video-EEG telemetry, FDG-PET, and other tests of focal functional deficit to perform standardized anterior temporal resections without requiring invasive recordings (Engel *et al.*, 1981). At Yale almost all patients continued to be studied with depth electrodes before standardized anterior temporal resection (Spencer *et al.*, 1982), and at the Cleveland Clinic, tailored resections were based on interictal and ictal data obtained from chronic subdural grid electrode recordings (Lesser *et al.*, 1987).

Clinical pathological correlations were difficult at centers where tailored resections were performed and tissue from the epileptogenic region often was removed by suction. Centers that performed *en bloc* surgical resections had better opportunities for pathological evaluation, but their findings could not be universally confirmed.

From the discussions at this conference, it became apparent that no particular approach was necessarily better than another; rather, each had advantages and disadvantages that made it superior for certain types

TABLE 5.1 Outcomes before 1985 and from 1986 to 1990

	Seizure-free	Improved	Not Improved	Total
Limbic resections				
Before 1985	1296 (55.5)	648 (27.7)	392 (16.8)	2336 (100)
1986-1990 ATL	2429 (67.9)	860 (24.0)	290 (8.1)	3579 (100)
AH	284 (68.8)	92 (22.3)	37 (9.0)	413 (100)
Neocortical resections				
Before 1985	356 (43.2)	229 (27.8)	240 (29.1)	825 (100)
1986-1990 ETR	363 (45.1)	283 (35.2)	159 (19.8)	805 (100)
L	195 (66.6)	63 (21.5)	35 (11.9)	293 (100)
Hemispherectomies				
Before 1985	68 (77.3)	16 (18.2)	4 (4.5)	88 (100)
1986-1990 H	128 (67.4)	40 (21.1)	22 (11.6)	190 (100)
MR	75 (45.2)	59 (35.5)	32 (19.3)	166 (100)
Corpus callosotomies				
Before 1985	10 (5.0)	140 (71.0)	47 (23.9)	197 (100)
1986-1990	43 (7.6)	343 (60.9)	177 (31.4)	563 (100)

ATL, anterior temporal lobectomy; AH, amygdalohippocampectomy; ETR, extratemporal resection; L, lesionectomy; H, hemispherectomy; MR, large multilobar resection.
Number of patients (percent).
Reprinted from Engel *et al.*, 1993c, with permission.

of epilepsy and inferior for others. As a result, between 1986 and 1992, most epilepsy surgery programs began to adopt approaches used at other centers for particular purposes (Appendix II, 1993). For instance, centers that always performed chronic invasive recording began to limit these studies to patients with specific diagnostic problems, while centers that never performed chronic invasive recording began to do so when indicated. Most centers began to use stereotactically implanted depth electrodes for mesial temporal epilepsy but would often use subdural grid electrodes for focal neocortical epilepsy when it was necessary to determine not only the location, but the extent of the epileptogenic region.

Although computerized structural and functional neuroimaging was just beginning, the exceptional potential of these diagnostic techniques for localizing epileptogenic abnormalities in patients with medically intractable focal epilepsy was readily appreciated. The creation of an epilepsy surgery community in 1986 resulted in a much more universal collaborative approach to the application of these diagnostic technologies among epilepsy surgery centers than had occurred with the application of other new diagnostic techniques in the past. Neuropsychologists also began to compare notes on the use and interpretation of specific neurocognitive batteries and the intracarotid amobarbital procedure, and shared information led to increasingly similar strategies from one epilepsy surgery center to another. Finally, a standardized classification for surgical outcome with respect to epileptic seizures was

agreed on and subsequently used for most published reports (Engel, 1987c).

Interest in epilepsy surgery burgeoned in the late 1980s, spurred in part by the enthusiasm for more proactive and cooperative efforts engendered at the first Palm Desert conference, and in part by the advent of higher resolution structural and functional neuroimaging that greatly improved the identification of potential surgical candidates. At the same time, there was a resurgence of interest in some older surgical techniques such as corpus callosotomy (Spencer *et al.*, 1987) and hemispherectomy (Rasmussen, 1987), which were discussed at the first Palm Desert conference, and the introduction of two new surgical procedures, multiple subpial transection (Morrell *et al.*, 1989) and lesionectomy (Kelly, 1988), which offered a reasonable probability of seizure remission without additional neurological deficit when the epileptogenic region involved essential cortical areas.

The second Palm Desert conference on Surgical Treatment of the Epilepsies, held in 1992, documented the tremendous progress made in the 5 years between 1985 and 1990 with respect to improved diagnostic technology, as well as microsurgical techniques (Engel, 1993b). The increased application of epilepsy surgery was reflected by representation from 118 epilepsy surgery centers in 25 countries, essentially a doubling of active epilepsy surgery programs in 6 years. Each center again brought its data, and the number of patients undergoing surgery between 1985 and 1990 was 8234, compared to 3446 patients for the entire period before

1985 (Engel and Shewmon, 1993). Furthermore, the percentage of patients who became free of disabling seizures after temporal lobe resections improved from 56% to 68% between 1985 and 1990, and many more patients than in the past were undergoing localized neocortical resections, hemispherectomies, multilobar resections, and corpus callosotomies, with excellent results (Engel et al., 1993b, c) (Table 5.1).

The designation of surgically remediable epilepsy syndromes (Engel, 1996) was a major conceptual advance, introduced at the second Palm Desert conference at the suggestion of the psychiatrist David Taylor. Given that the number of antiepileptic drugs was increasing to the point where it might take years, and possibly a lifetime, to prove medical intractability, it was no longer reasonable to consider surgery for epilepsy as a treatment of last resort. Rather, early surgery was recommended for surgically remediable epilepsy syndromes, which are conditions with a relatively well-known pathological substrate, a natural history characterized by medical refractoriness after failure of the first few appropriate drug trials, and a progressive course including development of behavioral disturbances if seizures are not controlled (Engel, 1996). The final criterion for a surgically remediable epilepsy syndrome is a high probability of complete control of disabling seizures, on the order of 70% to 90%, with surgical intervention, which can also prevent development of irreversible adverse psychological and social consequences, including developmental delay in infants and small children, if surgery is performed early.

Early surgical intervention for surgically remediable epilepsy syndromes is not only desirable to achieve the best outcome but is also cost-effective because the presurgical evaluation can usually be performed noninvasively, and few patients fail to benefit from treatment. The prototype of a surgically remediable epilepsy syndrome is mesial temporal lobe epilepsy, more often due to hippocampal sclerosis than to other mesial temporal lesions (Engel et al., 1997). Patients with this condition are among the most refractory to medical treatment and at risk for disabling interictal behavioral disturbances, either as a result of the social and psychological consequences of frequent seizures, or because direct biological effects of repeated ictal events cause progressive memory deterioration, as well as depression and other mental disorders (Engel et al., 1991). Patients with medically intractable focal seizures resulting from discrete resectable neocortical lesions also have a surgically remediable epilepsy syndrome.

A number of surgically remediable epilepsy syndromes in infants and small children are due to diffuse structural abnormalities limited to one hemisphere, such as Sturge-Weber syndrome, hemimegencephaly, large porencephalic cysts, and Rasmussen's encephalitis (Andermann et al., 1993). In these conditions, antiepileptic medications are ineffective in controlling frequent seizures, which cause progressive behavioral disturbances and can be life-threatening. If these children survive, they commonly develop mental retardation and require permanent institutionalization; however, early hemispherectomy or hemispherotomy can completely eliminate disabling seizures and reverse developmental delay, without introducing additional unacceptable neurological deficits in most patients who already have hemiparesis and a useless hand (Schramm, 2002). In some patients with extensive structural abnormalities that do not involve the motor cortex, and who do not have hemiparesis, large multilobar resections can achieve the same results without introducing a new motor deficit (Leiphart et al., 2001).

EPILEPSY SURGERY TODAY AND IN THE FUTURE

Since 1986 there have been numerous other conferences on epilepsy surgery, and many books and monographs have been published on this topic (e.g., Apuzzo, 1991; Bingaman, 2002a, 2002b; Dam et al., 1988, 1994; Duchowny et al., 1990; Elisevich and Smith, 2002; Engel, 1987a, 1993b; Lüders, 1992; Lüders and Comair, 2001; Mathern, 1999; Pickard et al., 1990; Silbergeld and Ojemann, 1993; Spencer and Spencer, 1991; Theodore, 1992; Tuxhorn et al., 1997; Wieser and Elger, 1987; Wyler and Hermann, 1994; Zentner and Seeger, 2003). A randomized controlled trial of surgery for temporal lobe epilepsy was finally completed in 2001 (Wiebe et al., 2001), and in 2002 the American Academy of Neurology issued a practice parameter recommending surgery as treatment for this disorder (Engel et al., 2003). The presurgical evaluation for surgically remediable epilepsy syndromes, particularly mesial temporal lobe epilepsy, has been simplified to a point where it can be used in regions with relatively limited resources. Surgical therapy is now being offered in a number of developing countries (Wieser and Silfvenius, 2000), where it may be more cost-effective than continued antiepileptic drug treatment for well-selected patients in these areas.

Emphasis placed on surgical treatment for the recognized surgically remediable epilepsy syndromes in recent years should not detract from the value of surgical intervention for other intractable epileptic conditions that may not fall into this category. This includes patients with cryptogenic epilepsy who have electroclinical features suggestive of a single epileptogenic zone, although there is no history or imaging evidence of a structural lesion, and patients with diffuse or multifocal structural abnormalities in whom a smaller resectable area might be responsible for disabling spontaneous seizures. These patients currently require invasive evaluation with either depth or subdural electrodes, or both, and the chances for complete freedom from disabling seizures is not as good as it is for those with surgically remediable syndromes (Engel, 1996). Despite increased cost and risk, the reasonable expectation of better seizure control, if not complete seizure control, and an improved quality of life dictates the need to consider surgical intervention in these patients as well. However, further trials of antiepileptic drugs may be more appropriate before considering surgery than in patients with surgically remediable epilepsy syndromes, while hasty, inappropriate surgery can preclude a future effective intervention if new experimental diagnostic techniques, such as fMRI and MEG, eventually can localize the epileptogenic region more accurately.

Not only has multiple subpial transection and lesionectomy increased opportunities for surgical treatment in patients whose epileptogenic zone involves essential cortical areas, but also there is some evidence that multiple subpial transection, performed unilaterally or bilaterally, may prevent or reverse the disabling progressive verbal agnosia and other language disturbances associated with the Landau-Kleffner syndrome (Morrell et al., 1995). Another rare, highly refractory epileptic condition is gelastic epilepsy, which is often associated with a hypothalamic hamartoma. Recent studies now suggest that these seizures actually begin within the hamartoma (Munari et al., 1995) and that they may be controlled by removal of this tumor (Rosenfeld et al., 2001). Gamma knife surgery is now being used for some types of focal epilepsy, and although results may not be as good as those of resective surgery and it can take up to a year for seizures to resolve, this intervention offers hope for patients with difficult-to-approach epileptic lesions and medical conditions that contraindicate craniotomy (Regis et al., 1995). Research on new surgical techniques involving deep brain stimulation, delivered either repetitively (Velasco et al., 1987) or triggered by computerized EEG seizure detection (Milton and Jung, 2002), could provide seizure relief in the future for patients who do not have identifiable epileptogenic regions that can be surgically resected.

Despite tremendous advances in the accuracy and cost-effectiveness of presurgical evaluation, and improvements in the safety and efficacy of microsurgical approaches for the treatment of epilepsy, epilepsy surgery remains, arguably, the most underused acceptable treatment in the field of medicine. It has been estimated that there are 100,000 to 200,000 patients with intractable epilepsy in the United States who are potential candidates for epilepsy surgery; yet only 500 procedures were performed in this country in 1985, and 1500 in 1990

(Engel and Shewmon, 1993). More surgery is done today, but not much more. Similar statistics can be found in all countries of the industrialized world, and surgical treatment is still offered in only a few developing countries. When surgical treatment is performed, patients typically experience a delay of 20 to 22 years between the onset of epilepsy and referral to an epilepsy surgery program (Berg et al., 2003; Benbadis et al., 2003). In a recent review of patients undergoing surgical treatment for focal epilepsy, it took an average of 9 years to determine that seizures did not respond to two antiepileptic drugs (Berg et al., 2003).

Why surgery continues to be considered a last resort, or is not considered at all, even in patients with surgically remediable epilepsy syndromes is difficult to explain (Swarztrauber et al., 2003). Patients and their physicians fear the possibility of surgical complications that could result in new neurological deficits or death; however, studies have documented that the morbidity and mortality associated with recurrent seizures is considerably greater than that associated with surgical treatment (Sperling et al., 1999). Cost may be a factor, but most insurance companies now reimburse for surgery, and the total cost for surgical treatment for epilepsy is much less than the cost to the patient and to society of a lifetime of disability. For years, absence of a randomized controlled trial of epilepsy surgery was a factor that led many physicians to continue to question the efficacy and lack of adverse consequences associated with removing parts of the brain as a treatment for epilepsy. Results of the recent randomized controlled trial of surgery for temporal lobe epilepsy should overcome this concern (Wiebe et al., 2001). Finally, it is possible that the community of epileptologists involved in surgical treatment for epilepsy have not adequately educated their colleagues and the general public about the place of surgical intervention in the overall management of epileptic disorders, despite the extensive literature in recent years.

Surgical intervention should play a greater role in the treatment of epilepsy than it does at present, particularly for surgically remediable epilepsy syndromes such as mesial temporal lobe epilepsy, as recommended in the AAN practice parameters (Engel et al., 2003). Furthermore, if surgical treatment is to be effective, it should be performed early enough to avoid irreversible disabling consequences of recurrent seizures. Infants and small children with intractable epilepsy that causes developmental delay end up in institutions. Recurrent disabling epileptic seizures during adolescence and early adulthood compromise acquisition of interpersonal and vocational skills. Too often, patients with uncontrolled seizures during adolescence and early adulthood who have surgery performed in their mid-twenties or later and become free of seizures are too socially and psychologically disabled to leave home, get a job, get married, raise a family, and pay taxes; they remain dependent on their families and on society. For many such patients, successful early surgical intervention might have rescued them from a lifetime of disability. Consequently, it is important to establish guidelines for deciding when to abandon additional trials of antiepileptic drugs and refer patients to an epilepsy surgery facility. One study has provided evidence that medical intractability can be predicted with a high degree of confidence after two appropriate trials of antiepileptic drugs have failed due to inefficacy, not intolerance (Kwan and Brodie, 2000).

The National Institute of Neurological Disorders and Stroke (NINDS) supported multicenter Early Randomized Surgical Epilepsy Trial (ERSET) is currently attempting to compare surgical intervention with continued optimal medical treatment for patients with mesial temporal lobe epilepsy who have not had disabling seizures for more than 2 years and who have failed adequate trials of at least two antiepileptic drugs (Engel, 2003). Outcome measures include not only seizure control but also quality of life, cognitive function, and activities of daily living. ERSET, which involves 19 epilepsy centers around the United States, is not about surgery per se, but about eliminating seizures without side effects as soon as possible, which should be the treatment goal of choice for epilepsy. Whether surgery or optimal trials of new antiepileptic drugs is more successful in achieving this goal is unknown, but whatever the outcome, ERSET could greatly influence physicians' attitudes toward early aggressive treatment for epilepsy in the future.

LESSONS FOR OTHER DISORDERS OF THE NERVOUS SYSTEM

Epilepsy is an ancient affliction. By the beginning of modern neurology, it had already been recognized and described for almost 3000 years (Kinnier et al., 1990) and attributed to disorders of the brain for almost 2500 years (Temkin, 1945). As the fields of neuroscience and neurology began to share concepts and inform each other, epilepsy had already been long established as a disorder worthy of investigations. Research on epilepsy further benefited from the relatively early realization that symptoms might be alleviated by resection of focal pathology, which provided surgeons with opportunities for examining the living human brain directly. Neuroscientists interested in normal brain function benefited as much from research involving patients with epilepsy, particularly in the operating room, as clinical neurologists and surgeons interested in understanding epilepsy benefited from the advances of neuroscience.

Epileptologists were among the first to take advantage of each new clinical neuroinvestigative tool, whether it was EEG, neurocognitive evaluation, or neuroimaging, because surgical treatment of epilepsy required better understanding of the functional and anatomical substrates of localized brain activity. Perhaps for this reason, epilepsy surgeons have permitted neurophysiologists to carry out recording and stimulation experiments intraoperatively and through chronically implanted intracranial electrodes, as well as to perform microdialysis studies, have given resected surgical specimens to neuropathologists, neuroanatomists, neurochemists, molecular biologists, and electrophysiologists for *in vitro* investigations, and allowed neuropsychologists to carry out detailed evaluations of their patients prior to and following localized brain excisions. Even disconnective surgery, such as corpus callosotomy, provided material of great interest to basic neuroscientists, contributing, for instance, to landmark research on hemispheric lateralization that helped earn Roger Sperry a Nobel Prize (Sperry, 1982). Research-oriented epilepsy surgery programs, such as those at the MNI, UCLA, the University of Bonn, and Yale, continue to exemplify the best tradition of multidisciplinary biomedical science first manifest by the team of Jackson, Ferrier, and Horsley in 1886. Today, multidisciplinary epilepsy surgery teams involve not only neurosurgeons, neurologists, and clinical neurophysiologists, but also neuropsychologists, neuroradiologists, nuclear medicine specialists, psychiatrists, and neuropathologists, along with a large number of diverse basic scientists who recognize the unique opportunity epilepsy surgery offers to investigate the fundamental neuronal mechanisms of both normal and abnormal functions of the human brain. Their objective is to eventually make it possible to cure epilepsy without surgery and to understand aspects of brain function well enough to realize currently unimaginable treatments for other neurological and mental illnesses.

Advances in the field of neurosurgery for epilepsy, and its contributions to our knowledge of the brain, would not have occurred had it not been for the willingness of clinical and basic neuroscientists to collaborate, share their expertise and their data, and create a whole that in every way is greater than the sum of its parts. Throughout the history related here, multidisciplinary collaborative teams, beginning with Jackson, Ferrier, and Horsley, through the extensive epilepsy surgery programs of today, have been responsible for rapid progress in the field, whereas the applications of contributions by isolated investigators, like Berger, were inappropriately delayed. The building of multidisciplinary collaborative teams is not easy, and the continuing success of productive epilepsy surgery programs is particularly remarkable at a time when academic institutions and funding agencies do not always give appropriate credit to individuals who selflessly commit themselves to group efforts (Editorial, 2003).

A few other neurological disorders may be amenable to similar surgical interventions, but the lessons of the successes experienced in the field of epilepsy surgery should not be applied superficially to concepts of identifying and excising an offending piece of brain. Rather, the lessons are more sociopolitical, emphasizing the importance of teamwork in identifying and effectively applying the seminal insights that move a field forward. The complexity of the tools available to us to investigate the nervous system for clinical and research purposes is increasing exponentially, and no single person could conceivably learn enough to take advantage of more than a small percentage of them. The future of biomedical science in general, and neurology specifically, lies with the ability of collaborative multidisciplinary teams to comprehensively exploit the highest technology for focused investigations of clinical relevance, which ultimately can be identified only by clinical neurologists. The clinical neurologist whose primary concern is the welfare of patients, therefore, remains an essential component of such a team, and traditional clinical neurology should never be overshadowed but, rather, continually enhanced by future technological advances in neuroscience investigation.

ACKNOWLEDGMENTS

Original research by the author and the Palm Desert conferences were supported by grants NS02808, NS33310, NS42372, NS15654, NS21444, and NS29615 from the National Institutes of Health. The author is grateful to Dr. David Millett for critical review of the manuscript and historical advice, to Dr. Jason Soss for helping prepare the photographic montages, and to Dale Booth for literature searches and manuscript preparation.

References

Adey WR, Hanley J, Kado RT, Zweizig JR. (1956). A multichannel telemetry system for EEG Recording, *Proc Symp Biomed Engineering Marquette University* 1:36-39.

Adrian ED, Matthews BHC. (1934). The Berger rhythm: potential changes from the occipital lobes in man, *Brain* 57:355-385.

Ajmone-Marsan C, Ralston BL. (1957). *The epileptic seizure, its functional morphology and diagnostic significance: a clinical-electrographic analysis of metrazol-induced attacks*, Springfield, IL: Charles C. Thomas.

Andermann F. (1987). Identification of candidates for surgical treatment of epilepsy. In Engel J Jr, editor: *Surgical treatment of the epilepsies*, New York: Raven Press.

Andermann F, Freeman JM, Vigevano F, Hwang ALS. (1993). Surgically remediable diffuse hemispheric syndromes. In Engel J Jr, editor: *Surgical treatment of the epilepsies*, ed 2, New York: Raven Press.

Appendix II: Presurgical evaluation protocols. (1987). In Engel J Jr, editor: *Surgical treatment of the epilepsies*, New York: Raven Press.

Appendix II: Presurgical evaluation protocols. (1993). In Engel J Jr, editor: *Surgical treatment of the epilepsies*, ed 2, New York: Raven Press.

Apuzzo MLJ. (1991). *Neurosurgical aspects of epilepsy*, Park Ridge, IL: American Association of Neurological Surgeons.

Arellano ZAP. (1949). A tympanic lead, *EEG Clin Neurophysiol* 1:112-113.

Bailey P, Gibbs FA. (1951). The surgical treatment of psychomotor epilepsy, *JAMA* 145:365-370.

Barnett GH et al. (1990). Epidural peg electrodes for the presurgical evaluation of intractable epilepsy, *Neurosurgery* 27:113-115.

Barth DS, Sutherling W, Engel J Jr, Beatty J. (1984). Neuromagnetic evidence of spatially distributed sources underlying epileptiform spikes in the human brain, *Science* 223:293-296.

Bartholow R. (1874). Experimental investigations into the functions of the human brain, *Am J Med Sci* 67:305-313.

Benbadis SR, Heriaud I, Tatum WO, Vale FL. (2003). Epilepsy surgery delays, and referral patterns, *Seizure* 12:167-170.

Bennett AH, Godlee RJ. (1884). Excision of a tumour from the brain, *Lancet* 2:1090-1091.

Berg AT et al. (2003). How long does it take for partial epilepsy to become intractable?, *Neurology* 60:186-190.

Berger H. (1929). Über das elektrenkephalogram des menschen, *Arch Psychiatry Nervenkr* 87:527-570.

Bickford RG, Dodge HW Jr, Sem-Jacobsen CW, Petersen MC. (1953). Studies on the electrical structure and activation of an epileptogenic focus, *Proc Staff Meetings Mayo Clinic* 28:175-181.

Bingaman WE, Guest Ed. (2002a). *Neurosurgery clinics of North America, cortical dysplasias*, Philadelphia: WB Saunders.

Bingaman WE, Guest Ed. (2002b). *Neurosurgery clinics of North America, hemispherectomy techniques*, Philadelphia: WB Saunders.

Bladin PF. (1992). Psychosocial difficulties and outcome after temporal lobectomy, *Epilepsia* 33:898-907.

Bonte FJ, Stokely EM, Devous MD Sr, Homan RW. (1983). Single-photon tomographic study of regional cerebral blood flow in epilepsy: a preliminary report, *Arch Neurol* 40:267-270.

Bouchet C, Cazauvieilh JB. (1825). De l'épilepsie considérée dans ses rapports avec l'aliénation mentale, *Arch Gen Med* 9:510-542.

Bragin A et al. (1999). Hippocampal and entorhinal cortex high frequency oscillations (100-500 Hz) in kainic acid-treated rats with chronic seizures and human epileptic brain, *Epilepsia* 40:127-137.

Bratz E. (1899). Ammonshornbefunde der epileptischen, *Arch Psychiatr Nervenkr* 31:820-836.

Bravais L-F. (1827). Recherches sur les symptômes et le traitement de l'épilepsie hémiplégique, Paris (Thèse de Paris No. 118).

Bright R. (1831). *Reports of medical cases*, Vol. II, Part II. London: Longmans.

Broca PP. (1861). Nouvelle observation d'aphémie produite par une lésion de la moitié postérieure des deuxième et troisième convolutions frontales gauches, *Bull Soc Anat* 36:398-407.

Cahan L et al. (1984). Review of the 20-year UCLA experience with surgery for epilepsy, *Cleve Clin Q* 51:313-318.

Cascino GD, Jack CR Jr, editors. (1996). *Neuroimaging in epilepsy: principles and practice*, Boston: Butterworth-Heinemann.

Caton R. (1875). The electrical currents of the brain, *Br Med J* 2:278.

Chugani DC et al. (1998). Imaging epileptogenic tubers in children with tuberous sclerosis complex using alpha-[11C]methyl-L-tryptophan positron emission tomography, *Ann Neurol* 44: 858-866.

Chugani HT et al. (1988). Surgical treatment of intractable neonatal-onset seizures: the role of positron emission tomography, *Neurology* 38:1178-1188.

Crandall PH. (1987). Cortical resections. In Engel J Jr, editor: *Surgical treatment of the epilepsies*, New York: Raven Press.

Crandall PH, Babb TL. (1993). The UCLA epilepsy program: historical review 1960-1992, *J Clin Neurophysiol* 10:226-238.

Crandall PH, Walter RD, Rand RW. (1963). Clinical applications of studies on stereotactically implanted electrodes in temporal lobe epilepsy, *J Neurosurg* 20:827-840.

Dam M, Andersen AR, Rogvi-Hansen B, Jennum P, editors. (1994). Epilepsy surgery: non-invasive versus invasive focus localization, *Acta Neurol Scand* 152(Suppl):218 entire vol.

Dam M, Gram L, Schmidt K, editors. (1988). Surgical treatment of epilepsy, *Acta Neurol Scand* 117(Suppl):154 entire vol.

Dandy, W. E. (1919). Roentgenography of the brain after injection of air into the spinal canal. *Ann. Surg* 70, 397-403.

Devinsky O *et al.* (1995). Development of the quality of life in epilepsy inventory, *Epilepsia* 36:1089-1104.

Dodrill CB *et al.* (1986). Multidisciplinary prediction of seizure relief from cortical resection surgery, *Ann Neurol* 20:2-12.

Duchowny M, Resnick T, Alvarez L, editors. (1990). Pediatric epilepsy surgery, *J. Epilepsy* 3(Suppl. 1):141-155.

Dudley BW. (1828). Observations on injuries of the head, *Transylvania J Med* 1:9-40. As cited in Patchell RA, Young AB, Tibbs PA. (1987). Benjamin Winslow Dudley and the surgical treatment of epilepsy, *Neurology* 37:290-291.

Duncan JS (1997). Magnetic resonance spectroscopy. In Engel J Jr, Pedley TA: *Epilepsy: a comprehensive textbook*, Philadelphia: Lippincott-Raven.

Dymond AM, Sweizig JR, Crandall PH, Hanley J. (1971). Clinical application of an EEG radiotelemetry system, *Proc Rocky Mt Bioengineer Symp* 16-20.

Earle KM, Baldwin M, Penfield W. (1953). Incisural sclerosis and temporal lobe seizures produced by hippocampal herniation at birth, *Arch Neurol Psychiatry* 69:27-42.

Ebersole JS. (1997). EEG and MEG dipole source modeling. In Engel J Jr, Pedley TA, editors: Epilepsy: a comprehensive textbook, Philadelphia: Lippincott-Raven.

Editorial. (2003). Who'd want to work in a team?, *Nature* 424:1.

Eliashiv DS *et al.* (2002). Ictal magnetic source imaging as a localizing tool in partial epilepsy, *Neurology* 59:1600-1610.

Elisevich K, Smith BJ, editors. (2002). *Epilepsy surgery: case studies and commentaries*, Philadelphia: Lippincott Williams & Wilkins.

Engel J Jr, editor. (1987a). *Surgical treatment of the epilepsies*, New York: Raven Press.

Engel J Jr. (1987b). Approaches to localization of the epileptogenic lesion. In Engel J Jr, editor: *Surgical treatment of the epilepsies*, New York: Raven Press.

Engel J Jr. (1987c). Outcome with respect to epileptic seizures. In Engel J Jr, editor: *Surgical treatment of the epilepsies*, New York: Raven Press.

Engel J Jr. (1993a). Historical perspectives. In Engel J Jr, editor: *Surgical treatment of the epilepsies*, ed 2, New York: Raven Press.

Engel J Jr, editor. (1993b). *Surgical treatment of the epilepsies*, ed 2, New York: Raven Press.

Engel J Jr. (1993c). Update on surgical treatment of the epilepsies, *Neurology* 43:1612-1617.

Engel J Jr (1996). Current concepts: surgery for seizures, *N Engl J Med* 334:647-652.

Engel J Jr. (1998a). Etiology as a risk factor for medically refractory epilepsy: a case for early surgical intervention, *Neurology* 51:1243-1244.

Engel J Jr. (1998b). Research on the human brain in an epilepsy surgery setting, *Epilepsy Res* 32:1-11.

Engel J Jr. (2003). A greater role for surgical treatment of epilepsy: why and when?, *Epilepsy Currents* 3:37-40.

Engel J Jr, Bandler R, Griffith NC, Caldecott-Hazard S. (1991). Neurobiological evidence for epilepsy-induced interictal disturbances. In Smith D, Treiman D, Trimble M, editors: *Advances in neurology*, vol. 55, New York: Raven Press.

Engel J Jr *et al.* (1982a). Pathological findings underlying focal temporal lobe hypometabolism in partial epilepsy, *Ann Neurol* 12:518-528.

Engel J Jr, Kuhl DE, Phelps ME, Crandall PH. (1982b). Comparative localization of epileptic foci in partial epilepsy by PCT and EEG, *Ann Neurol* 12:529-537.

Engel J Jr, Kuhl DE, Phelps ME, Mazziotta JC. (1982c). Interictal cerebral glucose metabolism in partial epilepsy and its relation to EEG changes, *Ann Neurol* 12:510-517.

Engel J Jr, Ojemann G, Lüders H, Williamson PD, editors. (1987). *Fundamental mechanisms of human brain function*, New York: Raven Press.

Engel J Jr *et al.* (1981). Correlation of criteria used for localizing epileptic foci in patients considered for surgical therapy of epilepsy, *Ann Neurol* 9:215-224.

Engel J Jr, Shewmon DA. (1993). Overview: who should be considered a surgical candidate?, In Engel J Jr, editor: *Surgical treatment of the epilepsies*, ed 2, New York: Raven Press.

Engel J Jr, Van Ness P, Rasmussen TB, Ojemann LM. (1993). Outcome with respect to epileptic seizures. In Engel J Jr, editor: *Surgical treatment of the epilepsies*, ed 2, New York: Raven Press.

Engel J Jr *et al.* (2003). Practice parameter: temporal lobe and localized neocortical resections for epilepsy, *Neurology* 60:538-547.

Engel J Jr, Williamson PD, Wieser H-G. (1997). Mesial temporal lobe epilepsy. In Engel J Jr, Pedley TA, editors: *Epilepsy: a comprehensive textbook*, Philadelphia: Lippincott-Raven.

Falconer MA. (1974). Mesial temporal (Ammon's horn) sclerosis as a common cause of epilepsy: aetiology, treatment and prevention, *Lancet* 2:767-770.

Falconer MA. (1953). Discussion on the surgery of temporal lobe epilepsy: surgical and pathological aspects, *Proc R Soc Med* 46:971-974.

Falconer MA, Serafetinides EA. (1963). A follow-up study of surgery in temporal lobe epilepsy, *J Neurol Neurosurg Psychiatry* 26:154-165.

Falconer MA, Taylor DC. (1968). Surgical treatment of drug-resistant epilepsy due to mesial temporal sclerosis: etiology and significance, *Arch Neurol* 19:353-361.

Ferrier D. (1873). Experimental researches in cerebral physiology and pathology. *The West Riding Lunatic Asylum Medical Reports* 3:30-96.

Ferrier D. (1874b). The localisation of function in the brain. *Proc R Soc. Lond* 22:229-232.

Ferrier D. (1874a). On the localisation of the functions of the brain, *Br Med J* 2:766-767.

Ferrier D. (1875). The Croonian Lecture: experiments on the brains of monkeys. *Philos Trans R Soc London* 165, 433-488.

Ferrier D. (1883). An address on the progress of knowledge in the physiology and pathology of the nervous system, *Br Med J* ii:805-808.

Foerster O. (1925). Pathogenese und Chirurgischer Behandlung der Epilepsie, *ZBL Chir* 52:531-549.

Fritsch G, Hitzig E. (1870). Ueber die elektrische Erregbarkeit des Grosshirns. Berlin, n.d. Reprinted from *Reichert's und de Bois-Reymond's Archiv*, Heft 3.

Gall F. (1800). *Philosophisch-medizinische Untersuchungen über Natur und Kunst im kranken und gesunden Zustände des Menschen*, Leipzig: Baumgärtner.

Gastaut H. (1948). Présentation d'une électrode pharyngée bipolaire, *Rev Neurol* 80:623-624.

Gastaut H, Gastaut JL. (1976). Computerized transverse axial tomography in epilepsy, *Epilepsia* 17:325-336.

Gibbs EL, Gibbs FA, Fuster B. (1948). Psychomotor epilepsy, *Arch Neurol Psychiatry* 60:331-339.

Gibbs FA, Gibbs EL, Lennox WG. (1938). Cerebral dysrhythmias of epilepsy, *Arch Neurol Psychiatry* 39:298-314.

Gloor P. (1969). Hans Berger on the electroencephalogram of man, *EEG Clin Neurophysiol* 28:1-36.

Goldensohn, ES. (1966). Simultaneous recording of EEG and clinical seizures using kinescope, *Electroencephalogr Clin Neurophysiol* 21:623.

Goldring S, Gregorie EM. (1984). Surgical management of epilepsy using epidural recordings to localize the seizure focus. Review of 100 cases, *J Neurosurg* 60:457-466.

Gotman J. (1981). Interhemispheric relations during bilateral spike and wave activity, *Epilepsia* 22:453-466.

Gowers WR. (1881). *Epilepsy and other chronic convulsive diseases*, London: J. A. Churchill.

Green JR, Duisberg REH., McGrath WB. (1951). Focal epilepsy of psychomotor type: a preliminary report of observations on effects of surgical therapy, *J Neurosurg* 8:157-172.

Griesinger W. (1867). Mental pathology and therapeutics. Translated by C L Robertson and J Rutherford, London: New Sydenham Society.

Grinker RR. (1938). A method for studying and influencing cortico-hypothalamic relations, *Science* 87:73-74.

Guillaume J, Mazars G, Mazars Y. (1953). Indications chirurgicales dans les épilepsies dites "temporales," *Rev Neurol* 88:461-501.

Guye M *et al.* (2002). Metabolic and electrophysical alterations in sub-types of temporal lobe epilepsy: a combined proton magnetic resonance spectroscopic imaging and depth electrodes study, *Epilepsia* 43:1197-1209.

Haglund MM *et al.* (1992a). Changes in gamma-aminobutyric acid and somatostatin in epileptic cortex associated with low grade gliomas, *J Neurosurg* 77:209-216.

Hauptmann A. (1912). Luminal bei epilepsie, *Munch Med Wochenschr* 59:1907-1909.

Henry TR *et al.* (1991). Interictal cerebral metabolism in partial epilepsies of neocortical origin, *Epilepsy Res* 10:174-182.

Horsley V. (1884). Case of occipital encephalocoele in which a correct diagnosis was obtained by means of the induced current, *Brain* 7:228-243.

Horsley V. (1886). Brain surgery, *Br Med J* 2:670-675.

Hunter J, Jasper HH. (1949). A method of analysis of seizure pattern and electroencephalogram, *Electroencephalogr Clin Neurophysiol* 1:113-114.

Ives JR, Gloor P. (1978). A long-term time-lapse video system to document the patient's spontaneous clinical seizure synchronized with the EEG, *Electroehcephalogr Clin Neurophysiol* 45:412-416.

Jackson GD *et al.* (1990). Hippocampal sclerosis can be reliably detected by magnetic resonance imaging, *Neurology* 40:1869-1875.

Jackson GD *et al.* (1994). Functional magnetic resonance imaging of focal seizures, *Neurology* 44: 850-856.

Jackson JH. (1880). On a particular variety of epilepsy ("intellectual aura"): one case with symptoms of organic brain disease, *Brain* 11:179-207.

Jackson JH, Colman WS. (1898). Case of epilepsy with tasting movements and "dreaming state": very small patch of softening in the left uncinate gyrus, *Brain* 21:580-590.

Jasper H, Kershman J. (1941). Electroencephalographic classification of the epilepsies, *Arch Neurol Psychiatry* 45:903-943.

Jasper H, Pertuisset B, Flanigin H. (1951). EEG and cortical electro-grams in patients with temporal lobe seizures, *Arch Neurol Psychiatry* 65:272-290.

Jasper HH. (1941). Electroencephalography. In Penfield W, Erickson TC, editors: *Epilepsy and cerebral localization*, Springfield, IL: Charles C. Thomas.

Jones DP. (1951). Recording of the basal electroencephalogram with sphenoidal needle electrodes, *Electroenceph Clin Neurophysiol* 3:100.

Kelly PJ. (1988). Volumetric stereotactic surgical resection of intra-axial brain mass lesions, *Mayo Clin Proc* 63:1186-1198.

Kennedy WA, Hill D. (1958). The surgical prognostic significance of the electroencephalographic prediction of Ammon's horn sclerosis in epileptics, *J Neurol Neurosurg Psychiatry* 21:24-30.

Kinnier Wilson JV, Reynolds EH. (1990). Translation and analysis of a cuneiform text forming part of a Babylonian treatise on epilepsy, *Med Hist* 34:185-198.

Klove H, Trites RL, Grabow JD. (1969). Evaluation of memory functions with intracarotid sodium amytal, *Trans Am Neurol Assoc* 94:76-80.

Krakow K *et al.* (1999). EEG-triggered functional MRI of interictal epileptiform activity in patients with partial seizures, *Brain* 122:1679-1688.

Krause F. (1909). Die operative Behandlung der Epilepsie, *Med Klin Berl* 5:1418-1422.

Krause F, Heyman F. (1914). Lehrbuch der Chirurgischen Operationen, Berlin: Urban und Schwarzenberg.

Krayenbuhl H, Hess R, Weber G. (1953). Enseignements de l'élec-troencéphalographie et corticographie et thérapeutique chirurgicale d'apres 21 cas d'épilepsie temporale traités par excision corticale ou par lobectomie, *Rev Neurol* 88:564-567.

Krynauw RA. (1950). Infantile hemiplegia treated by removing one cerebral hemisphere, *J Neurol Neurosurg Psychiatry* 13:243-267.

Kuhl DE, Engel J Jr, Phelps ME, Selin C. (1980). Epileptic patterns of local cerebral metabolism and perfusion in humans determined by emission computed tomography of ^{18}FDG and ^{13}NH$_3$, *Ann Neurol* 8:348-360.

Kuzniecky RI, Jackson GD. (1995). *Magnetic resonance in epilepsy*, New York: Raven Press.

Kwan P, Brodie MJ. (2000). Early identification of refractory epilepsy, *N Engl J Med* 342:314-319.

Lee BI *et al.* (1986). Single photon emission computed tomography (SPECT) brain imaging using N,N,N'-(2-hydroxy-3-methyl-5-^{123}I-iodobenzyl)-1,3-propanediamine 2 HCl (HIPDM): intractable com-plex partial seizures, *Neurology* 36:1471-1477.

Leiphart JW, Peacock WJ, Mathern GW. (2001). Lobar and multilobar resections for medically intractable pediatric epilepsy, *Pediatr Neurosurg* 34:311-318.

Lesser RP *et al.* (1987). Extraoperative cortical functional localization in patients with epilepsy, *J Clin Neurophysiol* 4:27-53.

Locock C. (1857). Discussion of a paper by EH Sieveking. Analysis of 52 cases of epilepsy observed by the author, *Lancet* 1:527.

Lüders HO, editor. (1992). *Epilepsy surgery*, New York: Raven Press.

Lüders HO, Comair YG, editors. (2001). *Epilepsy surgery,* ed 2, Philadelphia: Lippincott, Williams & Wilkins.

Lüders HO, Engel J Jr, Munari C. (1993). General principles. In Engel J Jr, editor: *Surgical treatment of the epilepsies*, ed 2, New York: Raven Press.

Lüders H *et al.* (1987). Commentary: chronic intracranial recording and stimulation with subdural electrodes. In Engel J Jr, editor: *Surgical treatment of the epilepsies*, New York: Raven Press.

Macewen W. (1879). Tumour of the dura mater removed during life in a person affected with epilepsy, *Glasgow Med J* xii:210.

Macewen W. (1881). Intra-cranial lesions: illustrating some points in connexion with the localisation of cerebral affections and the advantages of antiseptic trephining, *Lancet* ii:581-583.

Macewen W. (1888). On the surgery of the brain and spinal cord, *Br Med J* ii:302.

MacLean PD. (1949). A new nasopharyngeal lead, *EEG Clin Neurophysiol* 1:110-111.

Maspes PE, Marossero F. (1953). Considérations sur 28 cas d'épilepsie temporale operés, *Rev Neurol* 88:578-580.

Mathern GW, editor. (1999). Pediatric epilepsy and epilepsy surgery, *Dev Neurosci* 21:159-408.

Mathern GW *et al.* (1999). Post-operative seizure control and anti-epileptic drug usage in pediatric epilepsy surgery patients: the UCLA experience, 1986-1997, *Epilepsia* 40:1740-1749.

McKenzie KG. (1938). The present status of a patient who had the right cerebral hemisphere removed, *JAMA* 111:168.

Meldrum BS. (1997). Current strategies for designing and identifying new antiepileptic drugs. In Engel J Jr, Pedley TA, editors: *Epilepsy: a comprehensive textbook*, Philadelphia: Lippincott-Raven.

Millett D. (2001). Hans Berger: from psychic energy to the EEG, *Perspect Biol Med* 44:522-542.

Milner B. (1958). Psychological defects produced by temporal-lobe excision, *Res Publ Assoc Nerv Ment Dis* 36:244-257.

Milner B. (1971). Interhemispheric differences in the localization of psychological processes in man, *Br Med Bull* 27:272-277.

Milner B, Branch C, Rasmussen T. (1962). Study of short-term memory after intracarotid injection of sodium amytal. *Trans Am Neurol Assoc* 87:224-226.

Milner B, Penfield W. (1955). The effect of hippocampal lesions on recent memory, *Trans Am Neurol Assoc* 80:42-48.

Milton J, Jung P, editors. (2002). *Epilepsy as a dynamic disease*, New York: Springer Verlag.

Moniz E. (1934). *L'angiographie cérébrale*, Paris: Masson et Cie.

Morrell F, Whisler WW, Bleck TP. (1989). Multiple subpial transection, a new approach to the surgical treatment of focal epilepsy, *J Neurosurg* 70:231-239.

Morrell F *et al.* (1995). Landau-Kleffner syndrome: treatment with subpial intracortical transection, *Brain* 118:1529-1546.

Morris AA. (1950). The surgical treatment of psychomotor epilepsy, *Med Ann Dist Col* XIX:121-131.

Morris AA. (1956). Temporal lobectomy with removal of uncus, hippocampus and amygdala, *Arch Neurol Psychiatry* 76:479-496.

Munari C *et al.* (1995). Role of the hypothalamic hamartoma in the genesis of gelastic fits (a video-stereo-EEG study), *Electroencephalogr Clin Neurophysiol* 95:154-160.

Munari C *et al.* (1985). Sémiologie électroclinique des crises temporales subintrantes, *Rev EEG Neurophysiol* 15:289-298.

Niemyer P. (1958). The transventricular amygdalohippocampectomy in temporal lobe epilepsy. In Baldwin M, Bailey P, editors: *Temporal lobe epilepsy*, Springfield, IL: Charles C. Thomas.

O'Brien TJ *et al.* (1998). Subtraction ictal SPECT co-registered to MRI improves clinical usefulness of SPECT in localizing the surgical seizure focus, *Neurology* 50:445-454.

Obrador S. (1953). Personal experience in the surgical treatment of epilepsy, *J Neurosurg* 10:52-63.

Ojemann G. (1991). Cortical organization of language, *J Neurosci* 11:2281-2287.

Ojemann GA *et al.* (1993). Cortical stimulation. In Engel J Jr, editor: *Surgical treatment of the epilepsies*, ed 2, New York: Raven Press.

Paillas J-E, Gastaut H, Bonnal J, Vigouroux R. (1953). Corrélations anatomo-électrocliniques dans l'epilepsie temporale; à propos des résultats obtenus chez 38 opérés, *Rev Neurol* 88:568-574.

Pampiglione G, Kerridge J. (1956). E.E.G. abnormalities from the temporal lobe studied with sphenoidal electrodes, *J Neurol Neurosurg Psychiatry* 19:117-129.

Penfield W. (1936). Epilepsy and surgical therapy, *Arch Neurol Psychiatry* 36:449-484.

Penfield W, Baldwin M. (1952). Temporal lobe seizures and the technic of subtotal temporal lobectomy, *Ann Surg* 136:625-634.

Penfield W, Erickson TC. (1941). *Epilepsy and cerebral localization*, Springfield, IL: Charles C. Thomas.

Penfield W, Flanigin H. (1950). Surgical therapy of temporal lobe seizures, *Arch Neurol Psychiatry* 64:491-500.

Penfield W, Jasper H. (1954). *Epilepsy and the functional anatomy of the human brain*, Boston: Little Brown & Co.

Penry JK, Porter RJ, Dreifuss FE. (1975). Simultaneous recording of absence seizures with video tape and electroencephalography: a study of 374 seizures in 48 patients, *Brain* 98:427-440.

Pertuiset B, Capdevielle-Arfel G. (1951). Deux techniques particulières d'exploration basale. Les électrodes sphéno-ptérygoïdiennes et orbitaires, *Rev Neurol* 84:606-612.

Petit-Dutallis D, Christophe J, Pertuiset B, Dreyfus-Brisac C. (1953), *Semin Hôp Paris* 29:3858.

Picaza JA, Gumá J. (1956). Experience with the surgical treatment of psychomotor epilepsy, *Arch Neurol Psychiatry* 75:57-61.

Pickard JD, Trojanowski T, Maira G, Polkey CE, editors. (1990). *Neurosurgical aspects of epilepsy*, New York: Springer-Verlag.

Powell GE, Polkey CE, McMillan T. (1985). The new Maudsley series of temporal lobectomy. 1: short-term cognitive effects, *Br J Clin Psychol* 24:109-124.

Pritchard JC. (1822). *A treatise on diseases of the nervous system*, part the first, London.

Purpura DP, Penry JK, Walter RD, editors. (1975). Neurosurgical management of the epilepsies, *Adv Neurol* 8:356. New York: Raven Press.

Rasmussen T. (1987). Commentary: extratemporal cortical excisions and hemispherectomy. In Engel J Jr, editor: *Surgical treatment of the epilepsies*, New York: Raven Press.

Rasmussen TB. (1983). Surgical treatment of complex partial seizures: results, lessons and problems, *Epilepsia* 24 (Suppl 1):65-76.

Regis J *et al.* (1995). First selective amygdalohippocampal radiosurgery for "mesial temporal lobe epilepsy," *Stereotact Funct Neurosurg* 64(Suppl 1):193-201.

Rosenfeld JV *et al.* (2001). Transcallosal resection of hypothalamic hamartomas, with control of seizures, in children with gelastic epilepsy, *Neurosurgery* 48:108-118.

Roubicek J, Hill D. (1948). Electroencephalography with pharyngeal electrodes, *Brain* 71:77-87.

Rovit RL, Gloor P. (1960). Temporal lobe epilepsy: a study using multiple basal electrodes. II. Clinical EEG findings, *Neurochirurgia* 3:19-34.

Rowe CC *et al.* (1991). Patterns of postictal cerebral blood flow in temporal lobe epilepsy: qualitative and quantitative analysis, *Neurology* 41:1096-1103.

Sanabria E *et al.* (1983). Single photon emission computed tomography (SPECT) using ^{123}I-isopropyl-iodo-amphetamine (IAMP) in partial epilepsy. In Baldy-Moulinier M, Ingvar D-H, Meldrum BS, editors: *Current problems in epilepsy, I. Cerebral blood flow, metabolism and epilepsy*, London: John Libbey.

Savic I *et al.* (1988). In-vivo demonstration of reduced benzodiazepine receptor binding in human epileptic foci, *Lancet* 8616:863-866.

Schramm J. (2002). Hemispherectomy techniques. In Bingaman WE, editor: *Neurosurg Clin North Am* 3:113-134.

Schwab RS, Schwab MW, Withee D, Chock YC. (1953). Synchronized moving picture of patient with his EEG, Proceedings of the Third International Congress, on EEG Clin Neurophysiol 47.

Schwartzkroin PA. (1993a). Basic research in the setting of an epilepsy surgery center. In Engel J Jr, editor: *Surgical treatment of the epilepsies*, ed 2, New York: Raven Press.

Scoville WB, Milner B. (1957). Loss of recent memory after bilateral hippocampal lesions, *J Neurol Neurosurg Psychiatry* 20:11-21.

Semah F *et al.* (1998). Is the underlying cause of epilepsy a major prognostic factor for recurrence?, *Neurology* 51:1256-1262.

Silbergeld DL, Ojemann GA, editors. (1993). Epilepsy surgery, *Neurosurg Clin North Am* 4:356 entire vol.

Sommer W. (1880). Erkrankung des ammonshorns als aetiologisches moment der epilepsie, *Arch Psychiar Nervenkr* 10:631-675.

Spencer DD, Spencer SS, editors. (1991). *Surgery for epilepsy*, Cambridge: Blackwell.

Spencer DD *et al.* (1984). Access to the posterior medial temporal lobe structure in surgical treatment of temporal lobe epilepsy, *Neurosurgery* 15:667-671.

Spencer H. (1855). *The principles of psychology*, London: Langman, Brown, Green, and Langman.

Spencer SS *et al.* (1987). Corpus callosum section. In Engel J Jr, editor. *Surgical treatment of the epilepsies*, New York: Raven Press.

Spencer SS, Spencer DD, Williamson PD, Mattson RH. (1982). The localizing value of depth electroencephalography in 32 patients with refractory epilepsy, *Ann Neurol* 12:248-253.

Sperling MR *et al.* (1999). Seizure control and mortality in epilepsy, *Ann Neurol* 46:45-50.

Sperry RW. (1982). Some effects of disconnecting the cerebral hemispheres, *Science* 217:1223-1226.

Staba RJ *et al.* (2002). Quantitative analysis of high frequency oscillations (80-500 Hz) recorded in human epileptic hippocampus and entorhinal cortex, *J Neurophysiol* 88:1743-1752.

Swarztrauber K, Dewar S, Engel J Jr. (2003). Patient attitudes about treatments for intractable epilepsy, *Epilepsy Behav* 4:19-25.

Talairach J *et al.* (1974). Approche nouvelle de la neurochirurgie de l'épilepsie. Méthodologie stéréotaxique et résultats thérapeutiques, *Neurochirurgie* 20(Suppl 1):240 entire vol.

Talairach J, David M, Tournoux P. (1958). *L'exploration chirugicale stéréotaxique du lobe temporale dans l'épilepsie temporale*, Paris: Masson et Cie.

Taylor DC. (1972). Mental state and temporal lobe epilepsy. A correlative account of 100 patients treated surgically, *Epilepsia* 13:727-765.

Taylor DC. (1993). Epilepsy as a chronic sickness. In Engel J Jr, editor: *Surgical treatment of the epilepsies*, ed 2, New York: Raven Press.

Taylor DC. (1987). One hundred years of epilepsy surgery: Sir Victor Horsley's contribution. In Engel J Jr, editor. *Surgical treatment of the epilepsies*, New York: Raven Press.

Taylor DC. (2003). Schizophrenia and epilepsies: why? when? how? Originally prepared for the Bethel/Cleveland Meeting, Biefeld, May 2003.

Taylor DC, Falconer MA, Bruton CJ, Corsellis JAN. (1971). Focal dysplasia of the cerebral cortex in epilepsy, *J Neurol Neurosurg Psychiatry* 34:369-387.

Taylor DC, Marsh SM. (1980). Hughlings Jackson's Dr. Z: the paradigm of temporal lobe epilepsy revealed, *J Neurol Neurosurg Psychiatry* 43:758-767.

Taylor J, editor. (1958). *Selected writings of John Hughlings Jackson*, vol 1, New York: Basic Books Inc.

Temkin O. (1945). *The falling sickness: a history of epilepsy from the Greeks to the beginnings of modern neurology*, Baltimore: Johns Hopkins University Press.

Theodore WH, editor. (1992). *Surgical treatment of epilepsy*, Amsterdam: Elsevier.

Todd RB. (1856). *Clinical lectures on paralysis, certain diseases of the brain, and other affections of the nervous system*, ed 2, London: Churchill

Tuxhorn I, Holthausen H, Boenigk H, editors. (1997). *Paediatric epilepsy syndromes and their surgical treatment*, London: John Libbey.

van Wagenen WP, Herren RY. (1940). Surgical division of commissural pathways in the corpus callosum: relation to spread of an epileptic attack, *Arch Neurol Psychiatry* 44:740-759.

Velasco F, Velasco M, Ogarrio C, Fanghanel G. (1987). Electrical stimulation of the centromedian thalamic nucleus in the treatment of convulsive seizures: a preliminary report, *Epilepsia* 28:421-430.

Vickrey BG *et al.* (1992). A health-related quality of life instrument for patients evaluated for epilepsy surgery, *Med Care* 30:299-319.

Wada J. (1949). A new method for determination of the side of cerebral speech dominance: a preliminary report on the intra-carotid injection of sodium amytal in man. Igaku to Seibutsugaki, *Med Biol* 14:221-222.

Wada J, Rasmussen T. (1960). Intracarotid injection of sodium amytal for the lateralization of cerebral speech dominance: experimental and clinical observations, *J Neurosurg* 17:266-282.

Warach S *et al.* (1996). G-triggered echo-planar functional MRI in epilepsy, *Neurology* 47:89-93.

Wiebe S, Blume WT, Girvin JP, Eliasziw M. (2001). A randomized, controlled trial of surgery for temporal lobe epilepsy, *N Engl J Med* 345:311-318.

Wieser HG. (1986). Selective amygdalohippocampectomy: indications, investigative technique and results. In Symon L *et al.*, editors. *Advances and technical standards in neurosurgery*, vol. 13, Vienna: Springer-Verlag.

Wieser HG, Elger CE, editors. (1987). *Presurgical evaluation of epileptics: basics, techniques, implications*, Berlin: Springer-Verlag.

Wieser H-G, Silfvenius H, editors. (2000). Epilepsy surgery in developing countries, *Epilepsia* 41(Suppl 4):550.

Wyler AR, Hermann BP, editors. (1994). *The surgical management of epilepsy*, Boston: Butterworth-Heinemann.

Young RM. (1990). *Mind, brain and adaptation in the nineteenth century: cerebral localization and its biological context from Gall to Ferrier*, Oxford: Oxford University Press.

Zentner J, Seeger W, editors. (2003). *Surgical treatment of epilepsy*, New York: Springer-Verlag.

6

The Neurobiology of Migraine and Transformation of Headache Therapy

Hayrunnisa Bolay, MD, PhD

Michael A. Moskowitz, MD

The molecular and cellular origins of migraine headache are among the most complex problems in contemporary neurology. To meet these challenges, researchers have successfully applied the tools of neuroimaging, neurogenetics, neuropharmacology, and neurophysiology. With recent advances, we now have a clearer description of cellular events that characterize the migraine visual aura and emerging knowledge about how these events promote the development of headache. We have achieved greater insights into drug-receptor interactions at the neuronal level, as well as insights into central mechanisms regulating pain and its response to treatment. Together with the discovery that mutated P/Q calcium channels underlie a rare migraine headache subtype, these important developments raise expectations that identifying genetic mutations and their neurobiological consequences may help to decipher molecular and cellular mechanisms underlying more typical migraine types. Hence, this chapter highlights recent advances in the neurobiology of migraine headache with the hope that a more coherent scientific basis will be achieved.

CLINICAL CHARACTERISTICS AND NEUROIMAGING OF VISUAL AURA

Migraine is an episodic often disabling disorder accompanied by headache and complex heterogeneous clinical features. The prodromal phase precedes headache by up to 24 hours and is experienced by approximately one third of migraineurs. Symptoms include disturbances in mood (e.g., euphoria, depression, irritation), alertness (e.g., hyperactivity, fatigue, drowsiness), and heightened perception of sound, light, or smell. Also, cravings for food or drink or yawning are more common (Headache Classification Committee, 1988). These symptoms may reflect activation of hypothalamus, limbic structures, frontal lobes, and/or locus ceruleus, but this assumption has not been validated by functional imaging as yet.

The aura is typified by transient focal neurological dysfunction affecting half the visual field or body in about one fifth of migraineurs. It anticipates headache usually by 15 to 40 minutes, but may accompany headache onset and lasts 5 to 60 minutes. Aura symptoms can vary widely from one attack to the next in the same patient and between migraineurs. Some auras are not accompanied by headache. Visual disturbances are the most common symptom, typically characterized by a serrated arc of shimmering, crenellated shapes. Scintillations usually appear as formless flashing lights, or sometimes as zig-zags (fortification spectra) that typically propagate and expand from the paracentral region to the periphery in one visual field often followed by a scotoma. In other auras, sensory disturbances (typically tingling affecting distal parts of the extremities and perioral region that may slowly migrate to adjacent areas), motor symptoms, or dysphasia is less commonly experienced (Headache Classification Committee, 2004).

It was assumed for some time that visual auras originate from propagating events within somatotopic cortical regions (i.e., primary visual cortex). Recent magnetic resonance imaging (MRI) studies using high field strength near-continuous recordings have helped to validate this assumption and suggest a strong relationship between the visual aura and cortical spreading depression within the occipital lobe. Cortical spreading depression (CSD) is a transient phenomenon affecting one hemisphere (lissencephalic brain) or cerebral gyrus (gyrencephalic brain) characterized by slowly (mm per min) propagating excitation (depolarization) followed by a long-lasting depression (hyperpolarization) of neuronal and glial neurophysiological activity (Leao, 1944). Calcium waves may also contribute to the propagation. In lissencephalic brain, hemodynamic events are quite prominent during early and late phases and include an initial hyperemia followed by long lasting oligemia in cortical gray matter coincident with spreading depolarization and prolonged hyperpolarization, respectively. The association of CSD and migraine visual aura was posited initially by Leao and also by Milner based on the slow propagation rate and reinforced by the calculated velocity of spreading oligemia in migraineurs using low resolution SPECT studies (Lauritzen and Olesen, 1984).

More recently, migraine visual aura was explored using fMRI neuroimaging tools with high spatial and temporal resolution (Hadkikhani et al., 2001; Cao et al., 1999). This approach revealed multiple neurovascular events in the occipital cortex that resemble cortical spreading depression (Hadjikhani et al., 2001). In subjects with exercise or light-induced visual auras, an initial increase in blood oxygen level-dependent (BOLD) signal (possibly reflecting vasodilation) was followed by a decrease in the baseline signal (possibly reflecting vasoconstriction) propagating within primary visual cortex from posterior to anterior (i.e., from cortical representation of paracentral to peripheral eccentricities) (Figure 6.1; see also the Color Plate section). The most likely source of the initial BOLD response was an increase in blood flow and volume caused by heightened neural activity reflected perceptually by the shining, scintillating migrating visual aura. The increase in BOLD signal was not confined to a single vascular territory. Hence, these studies suggested that visual auras were accompanied by focal occipital cortical blood flow changes propagating along primary visual cortex temporally and somatotopically congruent with the visual percept (Figure 6.1; see also the Color Plate section). In human studies, the duration of hyperemia, velocity of propagation, prolonged period of hypoperfusion, and transient suppression of responsiveness to visual activation (as reported in rabbit after CSD) are consistent with the known properties of CSD (Figure 6.2; see also the Color Plate section) (Hadjikhani et al., 2001). Furthermore, the propagating BOLD signal change was restricted to gray matter and did not cross major sulci; it was also consistent with CSD in lower animals. To summarize, functional imaging studies implicate CSD or a closely related neurophysiological event in the generation of the migraine visual aura.

The relationship between CSD and migraine visual aura raises several interesting possibilities. For example, do CSDs underlie other types of migraine auras (e.g., somatosensory auras)? Might CSDs emanate from "silent" brain areas and cause migraine headache without an aura (Kunkler and Kraig, 2003), or could spreading events restricted to a few cortical laminae (instead of transcortically) cause so-called silent or undetectable auras (by the patient) or underlie other migraine subtypes in which aura is not perceived to anticipate headache (e.g., migraine without aura)? If cortical spreading depression provides the neurophysiological substrate for headache, how does it relate to the genesis of pain (see later)? Finally, does the link between CSD and aura present new therapeutic targets and opportunities? The fact that CSD was found recently in traumatized human cortex (Strong et al., 2002) and normal feline cortex by electrophysiological and MRI techniques reinforces the concept that the gyrencephalic mammalian cortex can generate a CSD-like event.

HEADACHE

Migraineurs usually experience unilateral, severe throbbing pain that is worsened by routine physical activity. Headache is accompanied by increased sensitivity to light or sound that improves in a dark quiet room. Anorexia, nausea, rarely vomiting, and sensitivity to smell and sound are other symptoms associated with headache. Headache lasts 4 to 72 hours without treatment or insufficient therapy, and sleep resolves or diminishes the severity of headache in many cases (Headache Classification Committee, 2004). Perception of headache requires activation of caudal trigeminal nuclei and subsequent transmission of pain impulses rostrally to thalamus and associated cortical areas (e.g., somatosensory cortex). Insula, cingulate cortex, frontal cortex, and thalamus are among the structures that have been implicated in pain processing during headache (Bolay and Moskowitz, 2002). Moreover, the red nucleus and substantia nigra are reportedly activated during the early phases of headache (Welch et al., 1998; Cao et al., 2002). Functional imaging studies have implicated other brain regions (dorsal raphe nucleus and locus ceruleus) in migraine pathophysiology hours after headache onset (Weiller et al., 1995). However, once the headache phase begins, it becomes difficult to decipher temporal sequences or the ways in which brainstem nuclei participate in the genesis or processing of pain or in migraine-associated symptoms. Hence, the importance of these specific nuclei and their activation remain for further study.

Trigeminal Innervation of Pain-Sensitive Intracranial Structures

Symptoms of migraine headache just summarized are triggered after activation of the trigeminovascular system (Moskowitz and Macfarlane, 1993; Moskowitz, 1984, 1991).

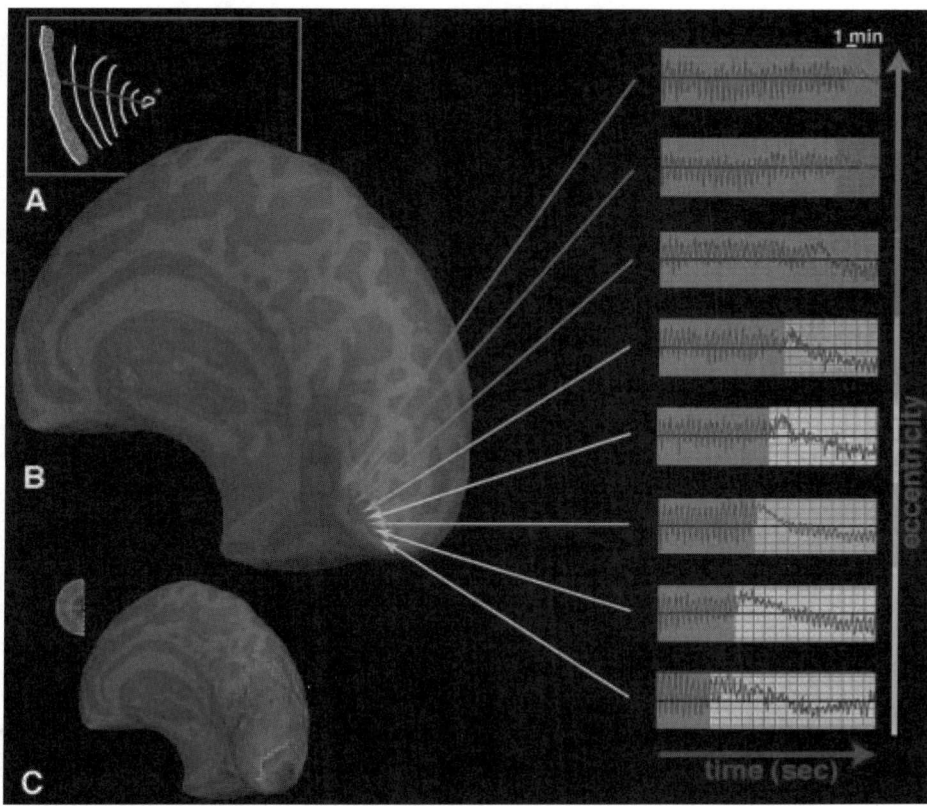

FIGURE 6.1 Spreading suppression of cortical activation during migraine visual aura. **A:** Drawing (*top*) showing the propagation (*red arrow*) of scintillations and scotoma in the left hemifield (*red arrow*) over 20 minutes during visual aura; cross indicates the fixation point. **B:** An inflated reconstruction of the migraineur's brain showing the cortical sulci (*dark*) and gyri (*lighter gray*). fMRI signal changes are shown on the right sampled from primary visual cortex from occipital pole anteriorly during the visual aura. Initial focal increase in BOLD signal followed by diminished BOLD response to light stimulation was observed in all extrastriate areas starting from occipital pole (central visual fields represented) toward more anterior regions representing central and peripheral eccentricities, respectively, and differing only in time of onset. The retinotopic progression of the BOLD signal was congruent with spread of visual aura (**A** and **C**). **D:** The maps of retinotopic eccentricities from the same subject acquired interictally. Activation within fovea are coded in red; parafoveal eccentricities are shown in blue and more peripheral eccentricities are shown in green (Hadjikhani *et al.*, 2001). (See also the Color Plate section.)

The trigeminal nerve transmits nociceptive information from cephalic structures via small caliber C fibers containing neuropeptides. Peripheral afferent fibers innervate the adventitia and perivascular space and are found densely in major vessels of dura mater and leptomeninges (pia mater and arachnoid) that cover the brain parenchyma, particularly over the basal surface (Mayberg *et al.*, 1981; Liu-Chen *et al.*, 1983; Suzuki *et al.*, 1989).

Trigeminal nerve fibers are strategically situated at the interface between blood and cerebrospinal fluid or brain parenchyma, possibly to detect the presence of any noxious substances originating from these sources. Sensitization or activation of trigeminovascular afferents releases vasoactive neuropeptides (calcitonin gene-related peptide [CGRP] substance P [SP] and neurokinin A [NKA]) that mediate vasodilation and edema. The

trigeminal innervation appears remarkably similar in species such as the rat, cat, and human, despite differences in brain size and organizational complexity (Mayberg *et al.*, 1981; Suzuki *et al.*, 1989; Feindel *et al.*, 1960). Axons project to pain sensitive (meningeal blood vessels) but not insensate (brain parenchyma) structures, although trigeminal fibers are often found within the cranium in areas considered insensate (e.g., unassociated with vessels or surrounding smaller blood vessels). It seems reasonable to assume that the meninges contain silent nociceptors that discharge upon sensitization during inflammatory conditions.

The dura mater and its large vessels (middle meningeal artery, MMA) are one of the most pain-sensitive intracranial structures (Penfield and McNaughton, 1940; Ray and Wolff, 1940). Neurogenic inflammation within dura

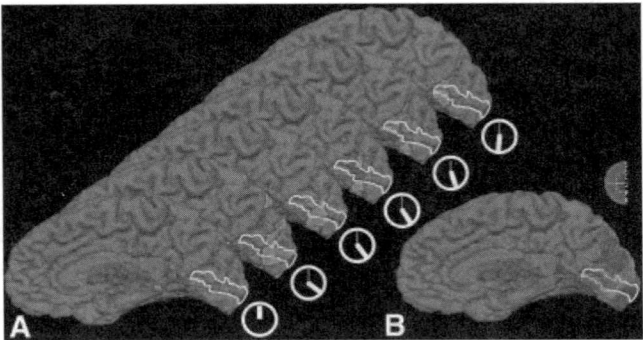

FIGURE 6.2 Progression of scintillations in the dark without visual stimuli during visual aura. **A:** A series of MR images over time (clock) showing posterior to anterior propagation of enhanced BOLD signal. The primary visual cortex (V1) is outlined with a white line. Initially no activation is seen in V1. However, with the onset of scintillations (after 20 minutes, see clock), activation appears that progresses from the parafoveal representation to more peripheral representations, paralleling the progression of the scintillations described by the subject. **B:** A medial view of the subject's brain with the MR maps of retinotopic eccentricity acquired during interictal scans (see Figure 6.1) (Hadjikhani *et al.*, 2001). (See also the Color Plate section.)

mater has been proposed as fundamental to the pathogenesis of migraine headaches (Moskowitz, 1984, 1992). Wolff and colleagues performed a series of experiments on humans and demonstrated that electrical or mechanical stimulation of large vessels in the dura and pia mater generates throbbing unilateral migraine-like pain (Ray and Wolff, 1940). Some trigeminal neurons in rodents project bifurcating axons to innervate both the middle cerebral (pial) and middle meningeal (dural) arteries (Mayberg *et al.*, 1981). The existence of axonal bifurcations predicts that discharge of perivascular afferents, surrounding pial-arachnoid arteries, may modulate the diameter and permeability of overlying vessels within dura mater (Bolay *et al.*, 2002; Iedecola, 2002) (Figure 6.3).

Activation of the trigeminovascular system is posited to coincide with development of headache. Among others, CGRP is the prominent neuropeptide that is released from activated trigeminovascular afferents as evidenced by the finding that plasma CGRP levels become increased during migraine headache and decreased after aborting the attack (Goadsby *et al.*, 1990). The release of neuropeptides promoting neurogenic vasodilation and plasma protein extravasation within ipsilateral dura mater might further induce perivascular nociceptive afferents to facilitate headache. Supporting the latter notion, CGRP infusion can induce a migraine headache in humans (Lassen *et al.*, 2002). Because of the release of neuropeptides exerting their effect on postjunctional vascular wall and development of primary afferent sensitization (see later), the vascular pulsations and venous pressure changes are perceived as noxious (Longmore

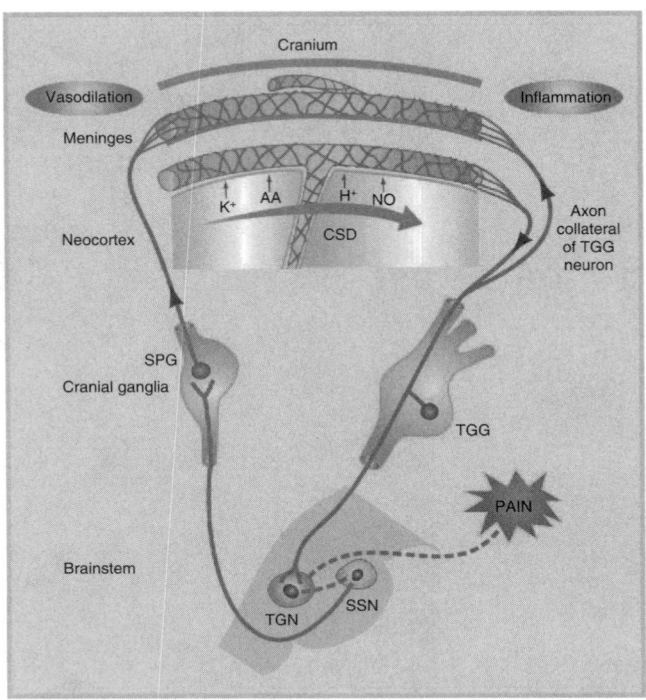

FIGURE 6.3 Proposed mechanism suggesting a link between CSD and trigeminovascular activation. Intense neurometabolic cortical activity such as CSD (illustrated by DC shift in cortex) releases potassium (K^+) and hydrogen (H^+) ions, neurotransmitters, and metabolites such as nitric oxide (NO) adenosine, and arachidonate (AA) into the extracellular and perivascular space. Within the perivascular space, these molecules activate/sensitize perivascular trigeminal afferents and transmit impulses centrally to trigeminal ganglia and second-order neurons in nucleus caudalis (TGN). Activation of trigeminal axonal collaterals innervating dura mater release vasoactive neuropeptides (substance P, calcitonin gene-related peptide, neurokinin A) mediating perivascular neurogenic plasma protein extravasation. CSD-induced TGN activation causes long-lasting vasodilation and blood flow increase in dura mater through a pathway originating from superior salivatory nucleus and reaching meningeal blood vessels via sphenopalatine ganglia. Postganglionic parasympathetic agents promote vasodilation and augmented flow by releasing vasoactive intestinal polypeptide, nitric oxide, and acetylcholine into the dura mater. As evidence for a central trigeminal-parasympathetic "reflex," lesioning trigeminal nerve significantly suppressed prolonged blood flow elevation in the middle meningeal artery (MMA) (P2), edema, and c-fos expression after CSD. The MMA response was dependent upon trigeminal brainstem inputs based on results after trigeminal rhizotomy. The perception of pain is mediated by rostral projections from TGN. The data are consistent with the formulation that events within the brain stimulate trigeminovascular afferents to generate headache and promote MMA vasodilation in an animal model of aura-induced migraine (Iedecola, 2002).

et al., 1997) and cause axonal discharge during pulsations in ways that remain obscure. Nevertheless, these features can enlighten the characteristic pulsating nature of migraine headache that is worsened by daily activities such as coughing, bending, or climbing stairs, which normally lead to innocuous changes in intracranial pressure.

Molecular Dissection of the Trigeminovascular System

Trigeminal afferents surround the wall of meningeal vessels and express several unique receptors such as 5-HT$_{1B/D/F}$ subtypes (Longmore *et al.*, 1997). In other respects they are presumed to show characteristics of other small unmyelinated nociceptive C fibers. They would therefore possess numerous prejunctional endings containing various receptors such as prostaglandin receptors (EP1), vanilloid receptors (VR1), nerve growth factor receptors (trκA), adenosine receptors (A1), proteinase activated receptor (PAR2), purinergic receptors (P2X3), or specific ion channels differentially expressed in nociceptive pathways such as sensory neuron specific Na$^+$ channels (SNS or NaN) or the K$^+$ channel (Kv1.4) (Waxman *et al.*, 1999; Gold *et al.*, 1996; Woolf and Salter, 2000; Stock *et al.*, 2000) (Figure 6.4). This assumption is not unreasonable because topically applied potassium, protons, and prostaglandins sensitize and discharge unmyelinated C fibers in dura mater as they do other primary afferent axons (Strassman *et al.*, 1996; Burstein *et al.*, 1998). On binding of molecules such as acetylcholine, prostaglandin, bradykinin, protons, glutamate, adenosine, or nerve growth factor, those receptors either activate nociceptive terminals and initiate an action potential, or reduce the amount of depolarization required for activation (sensitization) (Woolf and Salter, 2000). Sensitization of nociceptors could also be mediated by repeated exposure to weaker stimuli. Based on their sensitivity to capsaicin, trigeminovascular axons probably express vanilloid receptors. Vanilloid receptors are nonselective cation channels residing on small C fibers that can be sensitized by repeated heat, capsaicin, bradykinin, or proton exposure (Figure 6.4) (Bolay and Moskowitz, 2002; Woolf and Salter, 2000).

Activation/sensitization is mediated through either increases in intracellular calcium levels or activation of intracellular kinases (e.g., protein kinase C, protein kinase A, or tyrosine kinases), some of which may phosphorylate sensory neuron-specific channels and VR1 receptors. Post receptor events appear critical because protons and prostaglandin (PG) E$_2$ sensitize peripheral afferents by phosphorylation-dependent mechanisms. Phosphorylation of sensory neuron-specific channels could lower the activation threshold and rate of inactivation and increase the magnitude of sodium current. The agonistic activity of ligands such as bradykinin can be potentiated by PKC-dependent vanilloid receptor phosphorylation (Bolay and Moskowitz, 2002).

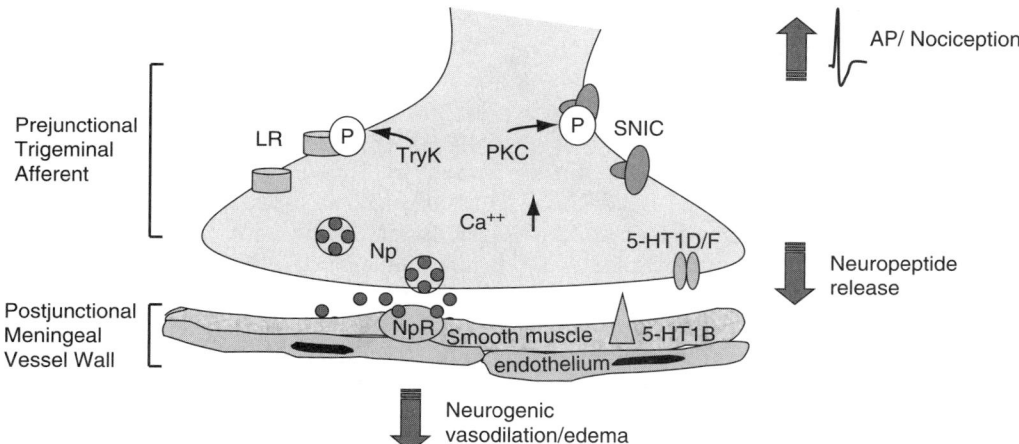

FIGURE 6.4 Trigeminal afferent fibers surround the wall of meningeal vessels and possess numerous prejunctional endings that express various receptors such as serotonin 5-HT$_{1D/F}$ subtypes, ligand-activated receptors (LR) (e.g., prostaglandin, nerve growth factor, adenosine, proteinase-activated, purinergic, or vanilloid receptors) or sensory neuron-specific ion channels (SNIC) differentially expressed in nociceptive pathways (e.g., sensory neuron-specific Na channel, K$^+$ channel. On binding of molecules such as prostaglandin, bradykinin, heat, protons, glutamate, adenosine, nerve growth factor, purines, or plasminogen those receptors either initiate an action potential (AP) or sensitize the nociceptive prejunctional terminal. Intracellular Ca^{++} increase, activation of intracellular kinases (protein kinase C [PKC], tyrosine kinases [TryK]), and phosphorylation play a major role in the sensitization/activation process. Activated prejunctional trigeminal afferents mediate vesicular release of vasoactive neuropeptides (Np) such as calcitonin gene-related peptide, substance P, and neurokinin A acting on their corresponding receptors (NpR) on the vessel wall and leading to neurogenic vasodilation and blood flow increase and edema. Antimigraine efficacy of triptans and ergots may be mediated via serotonin 5-HT$_{1D/F}$ receptors on prejunctional side and central endings within trigeminal nucleus caudalis, as well as 5-HT$_{1B}$ subtype on cerebral vasculature.

Serotonin $5\text{-}HT_1$ receptors have a major impact on migraine pathophysiology and are expressed on both primary afferent terminals and vascular smooth muscle promoting vasoconstriction (Johnson et al., 1998). $5\text{-}HT_{1B,D}$ and $5\text{-}HT_{1F}$ receptors are expressed on trigeminal neurons, whereas $5\text{-}HT_{1B}$ is predominant within the cranial vasculature. Agonist activation of $5\text{-}HT_{1D}$ receptors inhibits trigeminal discharges and may be responsible for antimigraine efficacy of triptans and dihydroergotamine (Moskowitz, 1992; Buzzi and Moskowitz, 1991). Of course, $5\text{-}HT_{1B/D/F}$ receptors are expressed on prejunctional nerve endings within trigeminal nucleus caudalis (TNC), where they may inhibit transmitter release. Their presence on postsynaptic neurons within TNC seems doubtful.

SENSITIZATION AND MIGRAINE

Sensitization has been documented in experimental headache models and in migraineurs. In headache models, sensitization develops in meningeal nociceptors and central trigeminal neurons after chemical irritation or electrical stimulation of dura mater. The chemical substances such as K^+, H^+, and inflammatory substances (e.g., prostaglandins) applied to dura mater activate trigeminovascular afferents and sensitize them to mechanical stimuli (Strassman et al., 1996; Burstein et al., 1998). Dural inputs and cutaneous trigeminal axons converge on single central trigeminal neurons within brainstem. After stimulation of dura mater with proinflammatory substances, those TNC neurons respond to subsequent "subthreshold" mechanical or thermal stimulation of periorbital skin. Cutaneous allodynia has also been described in patients during migraine headache. Allodynia, a sensation of pain evoked by a non-noxious stimulus, has been detected in 79% of migraneurs in the ipsilateral head or extending beyond the referred pain area (Burstein et al., 2000). Allodynia is transient, occurring during the migraine attack. The phenomenon is limited to the painful area in the ipsilateral head, reflecting a transient increase in the sensitivity of peripheral (first order) and central (second order) trigeminal neurons that receive convergent input from intracranial structures, such as blood vessels, meninges, and extracranial structures, such as skin and hair follicles. Allodynia can affect areas outside the trigeminal system. The latter may indicate the development of sensitization of at least third-order neurons (i.e., thalamic neurons) that receive convergent input from distant body sites, as well as from the dura and periorbital skin (Figure 6.5) (Burstein et al., 2000). Augmentation of the nociceptive blink reflex response may also reflect sensitization of brainstem trigeminal neurons during migraine headache (Kaube et al., 2002). Development of central sensitization depends on the

impulses coming from trigeminal nociceptors and indicates the importance of early treatment of migraine headaches.

As a potential trigger for headache, cortical spreading depression causes release of ions and neurotransmitters from neurons and glia (e.g., potassium ions, protons, arachidonic acid, nitric oxide, and adenosine) and is hypothesized to depolarize trigeminal axons and dilate meningeal blood vessels (Bolay et al., 2002; Iedecola, 2002; Lauritzen et al., 1990; Kraig and Nicholson, 1978; Read et al., 1997). Although the route taken by these ions and molecules to access the pial surface is not known exactly, gap junctional communications between astrocyte-pia-arachnoidal cells and the meningial glial network (Grafstein et al., 2000) play a role in transmitting intraparenchymal signals to meningeal perivascular trigeminal axons (Figure 6.3). Venous channels from the cerebral cortex (draining into the dural sinuses) may contribute. Recently, matrix metallaproteases (MMPs) have been implicated. Within 15 minutes after CSD, MMP activity increases and changes in vascular permeability documented using Evans blue extravasation. Changes in vascular permeability are abrogated after MMP-9 inhibition and do not develop in mice lacking MMP-9 gene expression (Gursoy-Ozdemir et al., 2004).

The potential role for nitric oxide (NO) as a mediator of first-and/or second-order trigeminal neuronal sensitization has been proposed (Jones et al., 2001). Neurogenic inflammation mediated by vasoactive peptides such as CGRP and SP may also promote sensitization and a lowered threshold for depolarization (Moskowitz and Macfarlane, 1993; Moskowitz, 1992; Markowitz et al., 1987). Proinflammatory molecules such as PGE_2, serotonin, bradykinin, epinephrine, adenosine, or nerve growth factor are also potential mediators, as is mast cell degranulation, and possibly macrophage activation.

TRIGGERING/ANTICIPATING EVENTS (INTRINSIC BRAIN DERIVED AND EXTRINSIC)

Triggering mechanisms and cascades of events leading to the development of headache are poorly understood. Genetic susceptibility is important, but genetic factors alone cannot explain the significant variability from one exposure to the next in the same patient. On exposure to a trigger, acute migraine headache usually develops within 24 hours. Common triggers include altered brain activity such as stress, emotional upset or sleep changes, aura, circulating chemicals (nitroglycerin, monosodium glutamate, aspartame), food (alcoholic or caffeinated beverages, cheese, chocolate), environmental factors (whether changes, bright lights, odors), allergens, and hormones.

A. Intracranial hypersensitivity

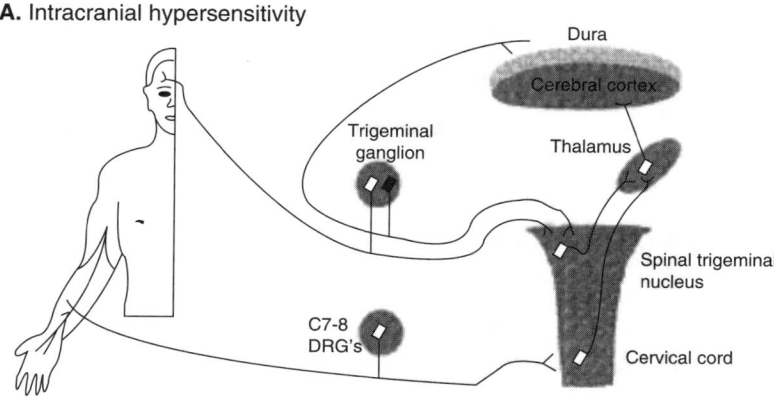

B. Intracranial hypersensitivity and ipsilateral extracranial allodynia

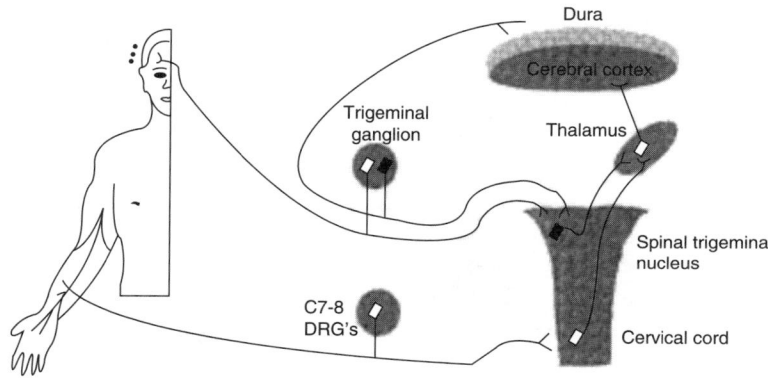

C. Intracranial hypersensitivity and extended extracranial and extracephalic allodynia

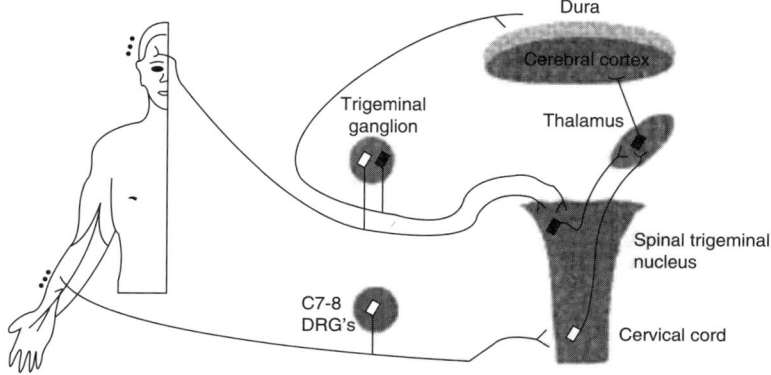

FIGURE 6.5 Cutaneous allodynia during migraine attacks. Patients with acute migraine headache showed cutaneous allodynia in the ipsilateral head or extending beyond the referred pain area. **A:** Intracranial hypersensitivity is explained by sensitization of dural nociceptors (first-order neurons) in the trigeminal ganglion. **B:** Cutaneous allodynia, limited to the pain area on the ipsilateral head, reflects a transient increase in the sensitivity of central (second-order) trigeminal neurons in brainstem that receive convergent inputs from intracranial structures, such as blood vessels, meninges, as well as extracranial structures, such as skin and hair follicles. **C:** Intracranial hypersensitivity and extracephalic allodynia are explained by sensitization of third-order neurons that receive convergent inputs from the head and forearms. Black diamonds represent sensitized nociceptive neurons. DRG = dorsal root ganglion (Burstein *et al.*, 2000).

Intrinsic Triggers and the Link Between Aura and Headache

Recent animal experimental studies shed light on the mechanism of headache development following two different triggers: an endogenous trigger, visual aura (Bolay *et al.*, 2002), and exogenous trigger, nitroglycerin (Reuter *et al.*, 2001, 2002). It was shown that CSD is a sufficient stimulus to trigger trigeminovascular activation (Bolay *et al.*, 2002). Using a novel technique called laser speckle imaging (Dunn *et al.*, 2001), it was demonstrated that CSD triggered long-lasting neurogenic blood flow responses and plasma protein leakage in dura mater. CSD caused a slow and sustained (45 minutes) MMA blood flow increase and vasodilation (Figure 6.6; see also Color Plate section) that was mediated by trigeminal-parasympathetic reflex activation (Bolay *et al.*, 2002). Transection of the trigeminal nerve (nasociliary nerve), trigeminal root, or postganglionic parasympathetic afferent fibers abolished the sustained blood flow increase in the MMA, indicating that this vascular response was neurogenically mediated and required central brainstem transmission. CSD also caused protein leakage and edema in the meninges, and this response was blocked by selective trigeminal nerve transection. Besides meningeal changes, CSD also activated ipsilateral trigeminal nucleus caudalis (TNC) neurons as reflected by expression of early immediate response gene, c-fos within TNC (Bolay *et al.*, 2002; Moskowitz *et al.*, 1993). Trigeminal nerve transection or sumatriptan pretreatment abrogated the c-fos response (Bolay *et al.*, 2002; Moskowitz *et al.*, 1993). These findings collectively demonstrate that intense neurometabolic activity within the cerebral cortex can evoke meningeal and brainstem trigeminal activation to promote vasodilation and events consistent with the development of head pain.

Chemical Triggers

The role of nitroglycerin as a trigger depends on several observations that implicate a role for NO in migraine pathogenesis (Olesen *et al.*, 1993; Lassen *et al.*, 1997). Nitroglycerin, an NO donor, has been known to induce severe headaches for many years, and early reports came from patients working in factories to make explosives. Administration of nitroglycerin induces immediate transient headache in individuals; however, typical migraine headaches develop only in migraineurs and only hours later (Olesen *et al.*, 1993). Migraine headache starting after a long delay (4 to 6 hours) was observed after nitroglycerin administration, far exceeding the half-life of the infused drug.

Recent studies investigating the molecular mechanisms of nitroglycerin-induced headache shed light on our understanding of this relationship (Reuter *et al.*, 2002; Lambert *et al.*, 2000). Nitroglycerin administration promoted a cascade of inflammatory signaling events exclusively within the dura mater (and not in the cerebral cortex); interleukin (IL)-1β IL-6 production and inducible nitric oxide synthase (iNOS) induction were measured in dura mater, with a time course consistent with nitroglycerin-induced migraine in humans. The nitroglycerin effects were dose-dependent, and the threshold dose in the experimental animal was similar to that in the human (Reuter *et al.*, 2001). The inflammatory responses were delayed and distributed most prominently within resident macrophages around large dural vessels. The time for gene up regulation could explain the delayed headache after nitroglycerin administration in susceptible individuals (Figure 6.7).

NFκB, a transcriptional factor, mediates nitroglycerin-induced inflammation. Its activation and nuclear translocation was shown one hour after nitroglycerin infusion, and one hour before iNOS gene and protein expression (Reuter *et al.*, 2002). Parthenolide, the active ingredient in Feverfew (an old herbal migraine remedy), blocks iNOS expression plus NFκB translocation, indicating the potential therapeutic importance of this transcription factor in migraine headache (Figure 6.7).

The nitroglycerin experiments reinforce the importance of inflammation and NO to trigeminal activation and migraine. It is well known that NO exerts its main physiological effects via formation of a second messenger, cyclic guanosine monophosphate (cGMP) as a consequence of activating soluble guanylate cyclase. In fact, sildenafil (Viagra), an inhibitor of cyclic nucleotide phosphodiesterase-5 (breakdown cGMP), caused typical migraine attack (with nausea, photophobia, phonophobia) in 11 of 12 patients (Kruuse *et al.*, 2003).

GENETIC MODULATORS

Migraine has a complex and heterogeneous phenotype, as well as multifactorial genetics, and exogenous triggers are known to play an important role as well. Recent identification of mutations in specific genes linked to rare types of migraine contributes significantly to our understanding of the pathogenesis of the disease, although the emerging picture appears quite complex and is far from complete.

Twin studies show a high concordance rate and there is a sixfold increased risk in first-degree relatives of probands in most common migraine types. So far, a simple mendelian inheritance pattern for common types has not been identified.

As noted previously, less common variants have provided more fruitful pathophysiological clues. One such

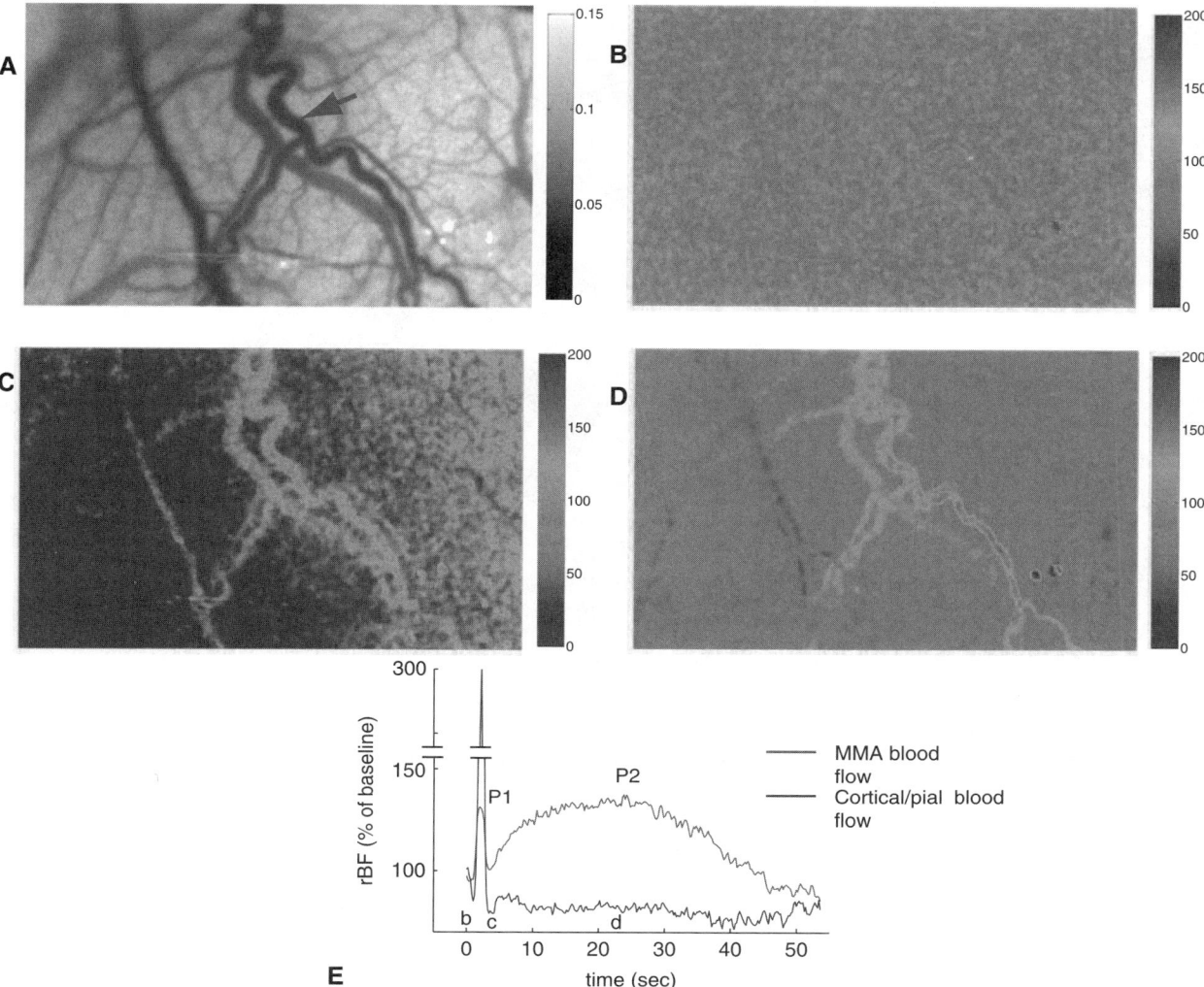

FIGURE 6.6 Blood flow images of MMA following CSD. Laser speckle images of blood flow in middle meningeal artery (MMA) and underlying parietal cortex and pial vessels following a single CSD. **A,** Speckle contrast image demonstrating the spatial flow heterogeneities across the imaged area at a single time illustrating MMA (*arrow*; proximal vessel is at top) and associated dural vein plus pial vessels and cerebral cortex. The darker values correspond to higher blood flow. **B-D:** Relative blood flow maps expressed as percentage of baseline, during the temporal evolution of CSD and represented by pseudo-colored images at the time points shown in **E**. **B:** Imaged blood flow after CSD induction, but before propagation of cortical hyperemia into the imaged area showing no immediate changes. **C:** Cortical hyperemia (*dark red*) spreading from left to right during CSD. Note that the marked flow increase in cortex and overlying pial vessels was comparable during CSD. A smaller transient rise in MMA blood flow (125% of baseline) was detected at this time. **D:** Blood flow map taken 20 minutes after CSD induction showing selective elevation of flow in MMA (140% of baseline), as well as in a dural vein. The relative blood flow in pial vessels and cortex is below baseline at this time (80% of baseline). **E:** The time course of the change in relative blood flow in MMA (*red line*) and cerebral cortex (*blue line*) demonstrates that CSD induces biphasic blood flow elevations within MMA with an early transient peak (P1) coincident with cortical hyperemia. A slower and more sustained increase in MMA blood flow (P2) is observed for approximately 45 minutes. These images clearly demonstrate that CSD-induced trigeminal activation causes long-lasting vasodilation selectively in dura mater; oligemia was prominent in cerebral cortex (Bolay *et al.*, 2002). (See also the Color Plate section.)

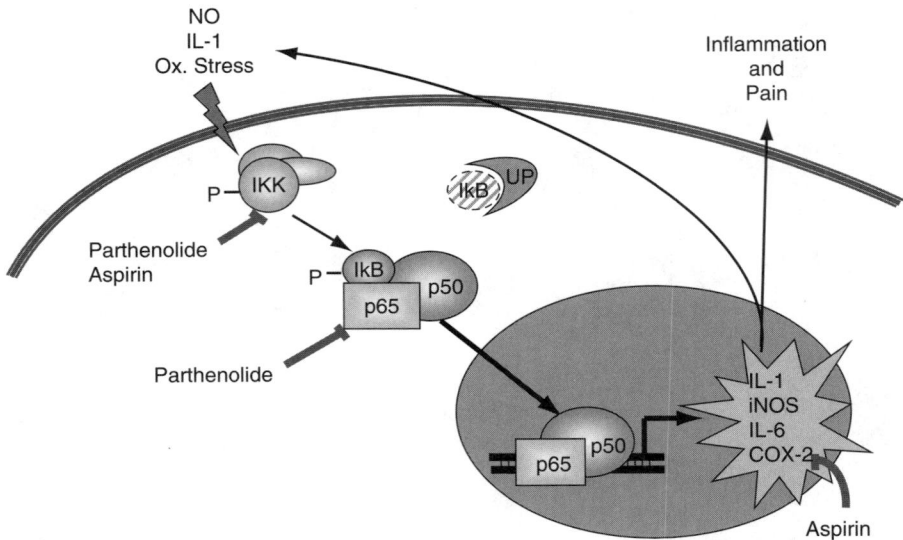

FIGURE 6.7 A proposed molecular model for nitroglycerin-induced migraine headache. Under resting conditions, the transcription factor nuclear factor-κ B (NF-κb) resides in the cytoplasm of quiescent dural macrophages as a p50-p65 dimer inactivated by inhibitory proteins-kb (Iκb). Noxious stimuli, such as proinflammatory cytokines, oxygen radicals, or NO (e.g., after GTN infusion an NO donor) indirectly activate Iκb kinase complex (IKK) to phoshorylate Iκb proteins. Once phoshorylated Iκb family members dissociate from NF-κb (p65-p50) and translocate to the nucleus, binding specific recognition elements in the promoter regions of inflammatory and stress-related genes, thereby activating transcription. Inducible nitric oxide synthase (iNOS), interleukin (IL)-1β, and IL-6 contribute to proinflammatory signaling that ultimately leads to increased blood vessel permeability, tissue edema, and pain sensitization, providing in part the molecular and functional mechanisms related to migraine pain in dura mater. Molecules such as aspirin or parthenolide inhibit the NF-κb-dependent transcriptional activation at multiple sites to prevent meningeal inflammation and migraine headache (Reuter *et al.*, 2002).

variant, familial hemiplegic migraine (FHM), a rare autosomal dominant form of migraine with aura, was linked to chromosome 19 and missense mutations in genes CACAN1A, encoding the P/Q Ca^{++} channel protein, Cav2.1 (type 1) (Ophoff *et al.*, 1996). Recently, a polymorphism in the Na$^+$, K$^+$-ATPase gene was also reported in other pedigrees (type 2) (De Fusco *et al.*, 2003). Because the P/Q type Ca^{++} channel is expressed primarily in neurons (predominantly glutamatergic neurons), whereas the affected Na$^+$, K$^+$-ATPase 2 is mainly expressed on astrocytes, factors contributing to the clinical phenotype are rather complex and implicate both neurons and glia, and perhaps even reinforce the importance of an event such as cortical spreading depression.

Glutamate cycling has recently been implicated. Efficient Ca^{++}-dependent vesicular release of glutamate depends on the preferential location of Cav2.1 channels adjacent to SNARE proteins at presynaptic terminals (Qian and Noebels, 2001). Postsynaptic Cav2.1 channels on dendrites and cell soma appear critical for neuronal excitability and Ca^{++}-dependent intracellular events such as gene expression. Mutations at one or both sites may be important. In FHMI, mutations in CACNA1A are heterogeneous and at least 14 different missense mutations have been identified. The functional importance of

Cav2.1 mutations and consequent loss or gain of function have been established only for some point mutations. Recent data overexpressing the human mutant Cav2.1α1 subunit in cells null for the CACNA1A gene revealed an increased channel opening probability over a wide range of voltages, and activation of the mutant channel that was insufficient to open wild-type channels in response to small depolarizations (Tottene *et al.*, 2002). FHM mutations in cerebellar granular cells caused an increase of single channel Ca^{++} influx during more hyperpolarized voltages, and decreased density of functional channels along with maximal Cav2.1 current. Such gain of function mutations at presynaptic terminals might lead to increased Ca^{++} influx through channels strategically located at the vesicular release site. Enhanced glutamate release could account for increased cortical excitability of migraineurs and possibly susceptibility to spreading depression (Figure 6.8) (Moskowitz *et al.*, 2004).

Leaner mice expressing naturally occurring missense mutation in the Cav2.1 channel gene show an altered CSD phenotype. The mutated channel in the leaner mice showed a reduced open probability, and activation was shifted to more depolarized voltages (Dove *et al.*, 1998). From the biophysical characteristics of the expressed mutant channel (i.e., loss of function of Cav2.1), tottering

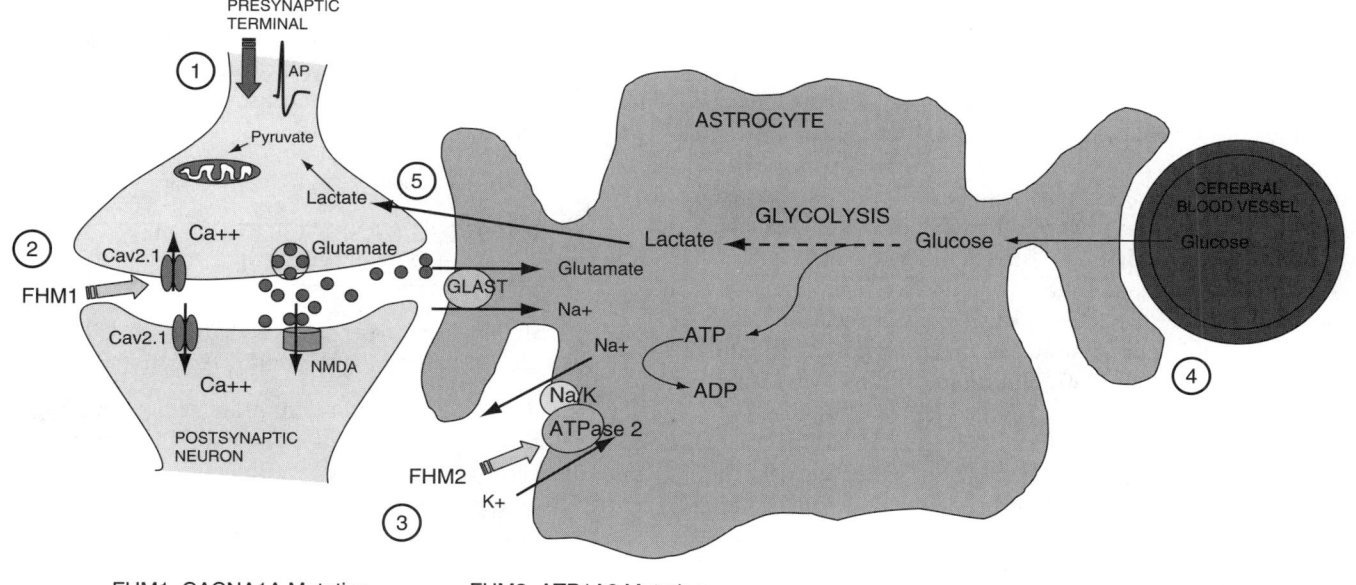

FIGURE 6.8 Proposed effects of FHM mutations on neuronal metabolic coupling (Magistretti, 1999) and aura symptoms. (1) An action potential (AP) at presynaptic terminals promotes Ca^{++} channel opening and influx through P/Q type Ca^{++} channels (Cav2.1) followed by vesicular release of glutamate into the synaptic cleft. (2) Glutamate's synaptic activity is terminated predominantly by astrocytic uptake by glutamate transporters in astrocytes, (GLAST/GLT1). (3) Na^+ ions co-transported with glutamate increase intra-astrocytic Na^+ and this gradient is maintained by activation of Na^+, K^+-ATPase pump. (4) Activation of Na^+, K^+-ATPase requires ATP produced by anaerobic glycolysis; this drives uptake of glucose from blood vessels. (5) Synaptic glutamate-induced anaerobic glycolysis results in lactate production in astrocytes, lactate is preferred by neurons as metabolic substrate after intense activity, though direct glucose uptake occurs under basal conditions. Lactate is aerobically metabolized via pyruvate within neurons. Results of missense mutations in CACNA1A gene coding Cav2.1 (gain of function) and ATP1A2 gene coding astrocytic Na^+, K^+-ATPase (loss of function) underlying FHM1 and FHM2 phenotype, respectively, are shown.

mice are refractory to both potassium and shock-induced CSD; they also show suppressed glutamate release when challenged by high potassium (Ayata *et al.*, 2000). This study also showed an impairment of glutamate release. In this mutation, it remains unclear as to whether the behavior of the channel under physiological conditions is different from that measured in response to prolonged cortical depolarization evoked either by K^+ application or faradic stimuli. Nevertheless, the paroxysmal nature of migraine attacks followed by normal neurological function interictally is consistent with neurological disturbances in which one or more ion channels or ion pumps have been implicated.

The FHM2 mutation occurs in the Na^+, K^+-ATPase$_{\alpha 2}$ subunit and suggests the importance of maintaining and preserving ionic gradients by glia. Astrocytes restore extracellular glutamate and K^+ to normal levels during neuronal depolarization. The mutated protein in FHM2 may cause pump failure and enhance excitability and also enhances levels of intracellular calcium as a result of the close association of Na^+, K^+-ATPase with the Na^+/Ca^{++} exchanger.

Astrocytic Na^+, K^+-ATPase also plays a critical role in driving glycolysis and coupling neuronal activity to metabolism (Figure 6.8) (Magistretti and Pellerin, 1999). Reuptake of glutamate from the synaptic cleft occurs primarily by astrocytes via glutamate transporters (GLAST and GLT-1) depending on electrochemical Na^+ gradients. Co-transport of 2-3 Na^+ ions along with one glutamate into astrocytes drives Na^+, K^+-ATPase by increasing intracellular Na^+ concentration. Activity of Na^+, K^+-ATPase is linked to glucose uptake, glycolysis, and production of lactate in astrocytes. Therefore glutamate uptake is indirectly coupled to glycolysis. Hydrolysis of 2 ATP molecules produced by glycolysis drives Na^+, K^+-ATPase pump that removes 6 intra-astrocytic Na^+ ions in exchange for 4 extracellular K^+ ions. Lactate released from astrocytes has been convincingly shown to provide the main energy substrate for synaptic activity (in the absence of glucose), but even preferred by the neuron, particularly during intense activity or stressful conditions such as ischemia or hypoxia (Schurr et al., 1997). We propose that any disturbance of such a critically positioned Na^+, K^+-ATPase would lead to slow clearance of glutamate and K^+ from the synaptic cleft, glutamate spillover, and inadequate glycolysis and lactate production required for recovery of neuronal function. The latter aspect may explain long-lasting disturbances of synaptic activity (as evidenced by evoked potential studies) and uncoupling after CSD. Those notions may also account for prolonged neurological dysfunction during aura in patients with FHM and generation of migraine pain, although the exact mechanism is still unclear.

CADASIL (cerebral autosomal dominant arteriopathy with subcortical infarcts and leukoencephalopathy) is an inherited migraine variant wherein 30% of patients experience migraine with aura. Recurrent subcortical strokes, dementia, and depression are other symptoms and MRIs show hyperintense symmetrical lesions in subcortical white matter affecting basal ganglia. The arteriopathy within small cerebral arteries is not related to atherosclerosis or amyloidosis. Notch 3, a member of a notch family involved in cell fate with epidermal growth factor-like domain, is encoded by a defective gene. The link between cell survival genes, arteriopathy, and migraine aura in these families is unknown but could relate to a damage-induced response. Considering that migraineurs often exhibit white matter abnormalities on brain imaging studies, however, it is intriguing to search for a specific mechanism by which this vascular disorder can initiate aura symptoms. It is not unreasonable to speculate that neurons and glia as well as blood vessels (and blood components) can trigger CSD and migraine under the appropriate conditions in susceptible individuals.

The first attempts to tie FHM or CADASIL mutations with more common migraine types failed to reveal a striking relationship; however, other genes have been implicated. Dopamine receptors, serotonin receptors, serotonin transporters, chloride channels, angiotensin-converting enzyme, nitric oxide synthase, and phospholipase A2 were investigated. Linkage studies excluded the possible involvement of serotonin receptors such as 5-HT$_{1B}$, 5-HT$_{1D}$, 5-HT$_{2\alpha}$, and 5-HT transporter in migraineurs with or without aura. Nevertheless, results of association studies are difficult to interpret, and sampling high numbers of patients is needed to draw

definite conclusions. The insulin receptor gene that is located within the same region with CACNA1A is associated with migraine aura, although no functional abnormality was detected in this mutated protein in preliminary studies.

SCIENTIFIC BASIS FOR TREATMENT WITH TRIPTANS

Serotonin and 5-HT$_{1B/D}$ Receptors and the Development of Specific Agonists

The concept behind today's migraine treatment began to develop three decades ago with early, albeit imperfect and often contradictory, observations on the importance of serotonin in migraine pathogenesis. For example, it was shown that platelet serotonin decreased rapidly during migraine headaches, whereas urinary excretion of 5-hydroxyindoleacetic acid, a main serotonin metabolite, was increased (Curran et al., 1965). Furthermore, migraineurs developed headache after reserpine or fenfluramine treatment, drugs that release serotonin from neurons and platelets. Supporting a role of serotonin receptors in migraine, parenteral serotonin administration aborted headache in some patients. Because 5-HT reportedly drops during migraine attacks and 5-HT infusion alleviates migraine headaches, Humphrey and colleagues suggested that targeting selective serotonin receptor subtypes might abort migraine headache. Despite the unequivocal and dramatic success of targeting serotonin receptors in migraine, we still have only limited information about the role of serotonin in migraine pathogenesis.

Classification of serotonin receptors has received considerable attention. On the basis of ligand binding studies and molecular cloning techniques, 5-HT receptors are divided into seven major subgroups (Hoyer et al., 1994). Most are G protein-coupled receptors. Members of the 5-HT$_1$ family of inhibitory G protein-linked receptors appearing most relevant to migraine are composed of five subgroups, A through F. 5-HT$_{1D}$ and 5-HT$_{1B}$ receptors and their mRNAs are expressed within human trigeminal ganglia, as are 5-HT$_{1F}$ receptors, specifying their prejunctional location. However, immunoreactive 5-HT$_{1D}$ is present on trigeminal axonal endings, whereas 5-HT$_{1B}$ immunoreactivity is restricted to cranial blood vessels (Longmore et al., 1997). Ergot alkaloids and triptans have high affinities for both 5-HT$_{1D}$ and 5-HT$_{1B}$ receptors. Drug binding to 5-HT$_{1B/D}$ receptors block peripheral (and probably central) trigeminal neurotransmitter release and also enhance closure of putative cranial arteriovenous anastomoses (Buzzi and Moskowitz, 1991; Buzzi et al., 1992). Hence, triptans possess serotonin $5\text{-HT}_{1B/D}$ agonist activity that mediates their primary therapeutic effect; they also activate 5-HT_{1A} or 5-HT_{1F} receptors to a lesser extent. However these drugs are not specific to 5-HT_{1D} and also activate 5-HT_{1B} receptors that mediate vasoconstriction (Johnson et al., 1998). The extent to which triptans require brain penetration to demonstrate therapeutic efficacy is currently debated. Also debated is whether the blood-brain barrier remains intact during a migraine attack.

5-HT_{1F} is another member of the serotonin receptor-1 subtype family that is implicated in migraine. 5-HT_{1F} receptors are distributed within trigeminal nucleus caudalis and substantia gelatinosa with a greater density than 5-HT_{1D} receptors. Selective 5-HT_{1F} receptor agonists inhibit dural plasma protein extravasation and block expression of the early immediate response gene, c-fos, within TNC after intracisternal capsaicin; 5-HT_{1F} agonists do not constrict blood vessels, supporting the notion that migraine efficacy does not necessarily require vasoconstriction (Shepheard et al., 1999). 5-HT_{1F} can abort headaches in migraineurs, as demonstrated in a clinical study. Therefore the 5-HT_{1F} receptor may be another promising therapeutic target. However, 5-HT_{1F} may not be essential for triptan activity, as almitidan, a new triptan with negligible 5-HT_{1F} ligand binding activity, is more potent than sumatriptan in blocking neurogenic protein extravasation and is effective clinically.

Effectiveness of Triptans in Experimental Migraine Models

The most effective mechanism in pain relief in migraine appears to be the ability to inhibit neuronal firing in TNC and consequently neurogenic edema and dilation of meningeal vessels. Several animal models such as trigeminal ganglionic stimulation, sagittal sinus stimulation, intracisternal capsaicin administration, and cortical spreading depression have been used to investigate migraine pathophysiology and the efficacy of treatment (Markowitz et al., 1987; Buzzi and Moskowitz, 1991; Bolay et al., 2002). Electrical stimulation of the trigeminal ganglion results in neurogenic plasma protein extravasation in dura mater that can be effectively blocked by $5\text{-HT}_{1B/1D}$ receptor agonist's triptans, as well as ergot alkaloids but not by simple vasoconstrictor agents (Ray and Wolff, 1940; Buzzi and Moskowitz, 1991; Buzzi et al., 1992). Plasma extravasation was reported to be accompanied by mast cell degranulation and platelet aggregation in dura mater (Buzzi et al., 1992). Neuropeptide release (such as CGRP and substance P) following trigeminal afferent stimulation also leads to vasodilation of the middle meningeal artery and its branches in dura mater that

can be used as a model for drug screening. Effective antimigraine compounds abort the vasodilation in response to electrical or chemical stimulation of dural afferents. CGRP infusion itself is enough to induce vasodilation and increase firing of second-order neurons in TNC, indicating the importance of this neuropeptide for both vasodilation in meninges and activation of the trigeminovascular system.

Activation of TNC neurons in response to nociceptive stimuli was also investigated by examining expression of the immediate early gene c-fos. Stimulation of sagittal sinus, dural vessels, or MMA activates trigeminal complex induced c-fos expression that was also suppressed by $5\text{-HT}_{1B/1D}$ receptor agonists. Cortical spreading depression has been shown to induce neurogenic plasma extravasation, vasodilation in dura mater, and activation of brainstem nociceptive neurons (Bolay et al., 2002; Moskowitz et al., 1993). CSD-induced neurogenic edema in rat was partially suppressed by an NK1 receptor antagonist implicating the role of substance P release. Sumatriptan in a dosage comparable to that used in humans was able to block c-fos induction.

Neurogenic inflammation has been postulated as a mechanism relevant to vascular headache pathogenesis and treatment in addition to central processing within the trigeminal nucleus caudalis (Moskowitz, 1992). Its inhibition alone, if not accompanied by an inhibitory effect on second-order trigeminal neurons, does not seem to be predictive for clinical efficacy against central sensitization and cutaneous allodynia. Antagonists of endothelin receptor A/B and NK-1 receptors failed in clinical trials despite their potent inhibitory effect on neurogenic edema.

In summary, $5\text{-HT}_{1B/1D/1F}$ receptors effectively inhibit neurogenic plasma extravasation and vasodilation in dura mater and inhibit neuropeptide release from activated trigeminovascular nerve endings in animal models. These drugs block transmitter release from central endings of primary afferents and therefore inhibit transmission of nociceptive impulses to second-order neurons within trigeminal nucleus caudalis (Levi et al., 2004).

Triptans are the first antimigraine drugs developed specifically to activate 5-HT_1 receptors and are a milestone in migraine treatment. However, failure of those agents to abort migraine in one fifth of patients, headache recurrences, and common side effects (such as chest tightness, dizziness) have led researchers to look for more potent and better tolerated new therapeutic strategies. Considering that the action of other abortive and prophylactic medications for migraine are nonspecific and unsatisfactory in many patients, the necessity for developing new treatment strategies seems imperative.

OTHER THERAPEUTIC APPROACHES

Lessons from Treatment Strategies Other Than Triptans

Neuropeptide Receptor Antagonists

Vasoactive neuropeptides, released from activated trigeminal nerve fibers, have been targeted in migraine treatment based on data demonstrating that (1) neuropeptides, CGRP, substance P, or NKA mediate sterile inflammation in dura mater and many prejunctional $5\text{-}HT_{1B/1D}$ agonist-acting drugs (triptans and ergots) block neuropeptide release, vasodilation, and protein extravasation in dura mater (neurogenic inflammation); and (2) neuropeptides and probably glutamate are released from central axons of primary afferents on TNC neurons to potentiate nociceptive signaling. Although synaptic release by prejunctional receptor occupation is the preferred method for new migraine drugs, drugs that block postjunctional receptors have a role as well.

CGRP is the most potent vasodilator neuropeptide in the cerebral circulation. Vasodilator effects of CGRP are not dependent on endothelium and can be inhibited by a peptide antagonist, CGRP (8-37). CGRP antagonists also inhibit the increase in dural blood flow evoked by electrical stimulation of meninges. In addition to vasodilation and blood flow increase, some recent data suggest a role for CGRP in neurogenic edema as well. Those preclinical data promoted the development of CGRP antagonists for clinical trials with expectation of blocking cerebrovascular postjunctional CGRP receptors. The initial pharmacological properties of such compounds showed their high selectivity for human CGRP receptors. The first clinical investigation of an injected CGRP antagonist BIBN4096BS for acute migraine showed a favorable 66% response rate without cardiovascular side effects (Olesen *et al.*, 2004).

Adenosine Receptors

There is evolving evidence suggesting that adenosine A1 receptors may be a target. The selective adenosine A1 receptor agonist, GR79236, inhibits neuronal activity within the trigeminocervical complex and reduces the increased CGRP level in response to superior sagittal sinus stimulation (Goadsby *et al.*, 2002). The inhibitory effect of GR79236 on neurogenic dural vasodilation, but not by intravenous CGRP administration, suggests that the effect of GR79236 is mediated via the activation of prejunctional adenosine A1 receptors.

Antiepileptics (Valproate, Gabapentin, Lamotrigine)

In recent years, antiepileptic agents have been among the most promising drugs for the prevention of migraine because of their action on enhancing γ-aminobutyric acid (GABA)-mediated inhibition and interaction with voltage gated Na^+ channels, although the exact mechanism is not understood. There is evidence that valproate aborts neurogenic dural extravasation and inhibits the central transmission within the trigeminal nucleus caudalis through GABA-A receptors. Sodium valproate is the only antiepileptic drug approved by the U.S. Food and Drug Administration for migraine prevention. Newer antiepileptics, including gabapentin, topiramate, and lamotrigine, are being evaluated. Lamotrigine might be useful for migraine aura by decreasing glutamate release.

Botulinum Toxin

Botulinum toxin blocks vesicular release of neuropeptides by interfering with vesicular- and membrane-bound synaptic proteins. Botulinum toxin reduces substance P release from dorsal horn neurons. Antinociceptive effects of botulinum toxin A have been demonstrated in a formalin model; Botox A inhibited peripheral sensitization by blocking neurotransmitter release. It was postulated that Botox A would be useful in migraine preventive therapy. Although preliminary reports are encouraging, randomized controlled studies with greater number of patients are awaited.

Opioid Receptors

Opioid analgesics are also potent antimigraine agents and have been used to treat severe cases of migraine attack, particularly when other agents are ineffective. Opioid receptors are predominantly concentrated at brainstem trigeminal nucleus caudalis and are present presynaptically and postsynaptically on the trigeminal sensory Aδ and C fiber primary afferent terminals. Opioids inhibit neurogenic plasma extravasation and neurogenic vasodilation and exert their inhibitory effect through prejunctional OP3 (μ) receptor activation. Besides OP3 (OP2(μ) receptor activation, OP2(κ) receptors also play a role, as evidenced by the effectiveness of butorphanol. Spinal opioid receptor activation decreases release of neurotransmitters from presynaptic sites while hyperpolarizing the postsynaptic neuron, thereby blocking at two distinct sites. Opioid receptors are also found in periaqueductal gray (PAG), brainstem, amygdala, and hypothalamus, and descending pathways from PAG and medullary reticular formation and contribute to central effects of opioids. Blockade of spinal 5-HT3 receptors reduce thermal and mechanical responses after nerve injury, and some preliminary reports indicate its effectiveness in patients with fibromyalgia (Hrycaj *et al.*, 1996). Further studies are needed to develop selective OP3 agonists interfering with nociception without physical dependence or tolerance.

Vanilloid Receptors

Vanilloid receptors residing on trigeminal peripheral axons are another potential target. Civamide is a vanilloid receptor agonist chemically related to capsaicin that depletes trigeminal neurons of their neurotransmitter content via desensitizing receptors. In a preliminary study, intranasal civamide was effective in aborting acute attack; however, unwanted effects such as nasal burning were unpleasant (Diamond *et al.*, 2000).

Future Prospects

Future directions for prevention and treatment of migraine should target inhibition of both peripheral and central terminals and transmission of nociceptive impulses.

Glutamatergic receptors NMDA and AMPA do augment nociception at the spinal level and also mediate central sensitization. New NMDA and AMPA receptor antagonists with fewer psychomotor side effects will be tested for their efficacy in migraine pain. The NMDA receptor NR2B subunit is an attractive target for pain management based on (1) its co-localization within CGRP positive sensory neurons, and (2) the distinctive pattern of receptor distribution within superficial lamina and within pain associated forebrain structures such as anterior cingulate and insular cortex. Transgenic mice overexpressing NR2B selectively within these forebrain structures exhibit enhanced pain behavior, indicating the significance of the cortical NMDA receptor NR2B subunit in augmenting the nociception (Wei *et al.*, 2001).

Prostaglandins are products of the cyclo-oxygenase pathway of arachidonic acid metabolism. Inhibition of the cyclooxygenase pathway mediates the main effect of nonsteroidal anti-inflammatory agents that are widely used in migraine treatment. The PGE2 receptors EP1 and EP3 mediate nociception and represent a selective target for interfering with a migraine attack (Stock *et al.*, 2000).

Aborting the migraine attack during the prodrome is a challenging direction that could be targeted at several distinct stages such as by (1) inhibiting susceptibility to spreading depression or blocking spreading depression itself (e.g., possibly by enhancing the activity of the Na$^+$, K$^+$-ATPase pump, inhibition of Ca^{++} entry from Cav2.1 channels, NMDA receptor antagonists, NR2B subunit antagonists [Menniti *et al.*, 2000], or antiepileptic agents [lamotrigine, tonabersat]); (2) blocking factors that link or enhance intrinsic triggers to trigeminovascular activation (e.g., TNFα, IL-1β, NFkB, atrial natriuretic factor or other proinflammatory molecules); and (3) interfering with mechanisms that lead to the passage of nociceptive molecules from vascular or parenchymal compartments to the perivascular space (e.g., matrix metalloproteinase inhibitors, gap junctional blockers). Once an attack is underway, strategies such as targeting purinergic P2X3 and P2X2 receptors, SNS, or proteinase-activated receptors, in addition to those approaches discussed previously, could prove promising.

ACKNOWLEDGMENTS

Some of the studies described in this chapter were supported by an NIH-sponsored Interdepartmental Migraine Program Project grant NS-35611 (MAM). We also wish to thank Professor Turgay Dalkara, Hacettepe University, for helpful discussion.

References

Ayata C *et al.* (2000). Impaired neurotransmitter release and elevated threshold for cortical spreading depression in mice with mutations in the 1A subunit of P/Q type calcium channels, *Neuroscience* 95:639-645.

Bolay H, Moskowitz MA. (2002). Mechanisms of pain modulation in chronic syndromes, *Neurology* 59(5 Suppl 2):S2-7.

Bolay H *et al.* (2002). Intrinsic brain activity triggers trigeminal meningeal afferents in a migraine model, *Nature Med* 8(2):136-142.

Burstein R, Yamamura H, Malick A, Strassman AM. (1998). Chemical stimulation of the intracranial dura induces enhanced responses to facial stimulation in brain stem trigeminal neurons, *J Neurophysiol* 79:964-982.

Burstein R *et al.* (2000). An association between migraine and cutaneous allodynia, *Ann Neurol* 47:614-624.

Buzzi MG, Moskowitz MA. (1991). Evidence for 5-HT1B/1D receptors mediating the antimigraine effect of sumatriptan and dihydroergotamine, *Cephalalgia* 11:165-168.

Buzzi MG, Dimitriadou V, Theoharides TC, Moskowitz MA. (1992). 5-Hydroxytryptamine receptor agonists for the abortive treatment of vascular headaches block mast cell, endothelial and platelet activation within the rat dura mater after trigeminal stimulation, *Brain Res* 583:137-149.

Cao Y, Welch KM, Aurora S, Vikingstad EM. (1999). Functional MRI-BOLD of visually triggered headache in patients with migraine, *Arch Neurol* 56:548-554.

Cao Y *et al.* (2002). Functional MRI-BOLD of brainstem structures during visually triggered migraine, *Neurology* 59:72-78.

Curran DA, Hinterberger H, Lance JW. (1965). Total plasma serotonin, 5-hydroxyindoleacetic acid and p-hydroxy-m-methoxymandelic acid excretion in normal and migrainous subjects, *Brain* 88:997-1010.

De Fusco M *et al.* (2003). Haploinsufficiency of *ATP1A2* encoding the Na+/K+ pump α2 subunit gene is responsible for familial hemiplegic migraine type 2, *Nature Genet* 33:192-196.

Diamond S *et al.* (2000). Intranasal civamide for the acute treatment of migraine headache, *Cephalalgia* 20:597-602.

Dove LS, Abbott LC, Griffith WH. (1998). Whole-cell and single-channel analysis of P-type calcium currents in cerebellar Purkinje cells of leaner mutant mice, *J Neurosci* 18:7687-7699.

Dunn AK, Bolay H, Moskowitz MA, Boas DA. (2001). Dynamic imaging of cerebral blood flow using laser speckle, *J Cereb Blood Flow Metab* 21:195-201.

Feindel W, Penfield W, McNaughton F. (1960). The tentorial nerves and localization of intracranial pain in man, *Neurology* 10:555-563.

Goadsby PJ, Edvinsson L, Ekman R. (1990). Vasoactive peptide release in the extracerebral circulation of humans during migraine headache, *Ann Neurol* 28:183-187.

Goadsby PJ *et al.* (2002). Adenosine A1 receptor agonists inhibit trigeminovascular nociceptive transmission, *Brain* 125:1392-1401.

Gold M, Reichling DB, Shuster MJ, Levine JD. (1996). Hyperalgesic agents increase a tetrodotoxin-resistant Na+ current in nociceptors, *Proc Natl Acad Sci U S A* 93:1108-1112.

Grafstein B *et al.* (2000). Meningeal cells can communicate with astrocytes by calcium signaling, *Ann Neurol* 47:18-25.

Gursoy-Ozdemir Y *et al.* (2004). Cortical spreading depression activates and upregulates MMP-9. *J Clin Invest.* 113:1447-1455.

Hadjikhani N *et al.* (2001). Mechanisms of migraine aura revealed by functional MRI in human visual cortex, *Proc Natl Acad Sci U S A* 98:4687-4692.

Headache Classification Committee of the International Headache Society. (2004). The International Classification of Headache Disorders 2nd edition. *Cephalagia* 24 (suppl 1):24-36.

Hoyer D *et al.* (1994). VII. International Union of Pharmacology classification of receptors for 5-hydroxytryptamine (serotonin), *Pharmacol Rev* 46:157-203.

Hrycaj P *et al.* (1996). Pathogenetic aspects of responsiveness to ondansetron (5-HT3 antagonist) in patients with primary fibromyalgia syndrome, *J Rheumatol* 23:1418-1423.

Iedecola C. (2002). From CSD to headache: a long and winding road, *Nature Med* 8(2):110-112.

Johnson KW, Phebus LA, Cohen ML. (1998). Serotonin in migraine: theories, animal models and emerging therapies, *Prog Drug Res* 51:221-244.

Jones MG *et al.* (2001). Nitric oxide potentiates response of trigeminal neurones to dural or facial stimulation in the rat, *Cephalalgia* 21:643-655.

Kaube H *et al.* (2002). Acute migraine headache: possible sensitization of neurons in the spinal trigeminal nucleus? *Neurology* 58:1234-1238.

Kraig RP, Nicholson C. (1978). Extracellular ionic variations during spreading depression, *Neuroscience* 3:1045-1059.

Kruuse C, Thomsen LL, Birk S, Olesen J. (2003). Migraine can be induced by sildenafil without changes in middle cerebral artery diameter, *Brain* 126:241-247.

Kunkler PE, Kraig RP. (2003). Hippocampal spreading depression bilaterally activates the caudal trigeminal nucleus in rodents, *Hippocampus* 2003

Lambert GA, Donaldson C, Boers PM, Zagami AS. (2000). Activation of trigeminovascular neurons by glyceryl trinitrate, *Brain Res* 887(1):203-210.

Lassen LH *et al.* (1997). Nitric oxide synthase inhibition in migraine, *Lancet* 349:401-402.

Lassen LH *et al.* (2002). CGRP may play a causative role in migraine, *Cephalalgia* 22:54-61.

Lauritzen M, Hansen AJ, Kronborg D, Wieloch T. (1990). Cortical spreading depression is associated with arachidonic acid accumulation and preservation of energy charge, *J Cereb Blood Flow Metab* 10:115-122.

Lauritzen M, Olesen J. (1984). Regional cerebral blood flow during migraine attacks by Xenon-133 inhalation and emission tomography, *Brain* 107:447-461.

Leao A. (1944). Spreading depression of activity in the cerebral cortex, *J Neurophysiol* 7:359-390.

Levy D, Jakubowski M, Burstein R. (2004). Disruption of communication between peripheral and central trigeminovascular neurons mediates the antimigraine action of 5HT 1B/1D receptor agonists. *Proc Natl Acad Sci U S A* 101(12):4274-9.

Liu-Chen LY, Han DH, Moskowitz MA. (1983). Pia arachnoid contains substance P originating from trigeminal neurons, *Neuroscience* 9:803-808.

Longmore J *et al.* (1997). Differential distribution of 5-HT1D and 1B immunoreactivity within human trigeminocerebrovascular system: implications for the discovery of new antimigraine drugs, *Cephalagia* 17:833-842.

Magistretti PJ, Pellerin L. (1999). Astrocytes couple synaptic activity to glucose utilisation in the brain, *News Physiol Sci* 14: 177-182.

Markowitz S, Saito K, Moskowitz MA. (1987). Neurogenically mediated leakage of plasma protein occurs from blood vessels in dura mater but not brain, *J Neurosci* 7:4129-4136.

Mayberg M, Langer RS, Zervas NT, Moskowitz MA. (1981). Perivascular meningeal projections from cat trigeminal ganglia: possible pathway for vascular headaches in man, *Science* 213:228-230.

Menniti FS *et al.* (2000). CP-101,606, an NR2B subunit selective NMDA receptor antagonist, inhibits NMDA and injury induced c-fos expression and cortical spreading depression in rodents, *Neuropharmacology* 39:1147-1155.

Moreno MJ. (2002). Efficacy of the non-peptide CGRP receptor antagonist BIBN4096BS in blocking CGRP-induced dilations in human and bovine cerebral arteries: potential implications in acute migraine treatment, *Neuropharmacology* 42(4):568-576.

Moskowitz MA, Bolay H, Dalkara T. (2004). Deciphering migraine mechanisms: clues from familial hemiplegic migraine genotypes. *Ann Neurol* 55:276-280.

Moskowitz MA. (1984). The neurobiology of vascular head pain, *Ann Neurol* 16:157-168.

Moskowitz MA. (1991). The visceral organ brain: implications for the pathophysiology of vascular head pain, *Neurology* 41:182-186.

Moskowitz MA. (1992). Neurogenic versus vascular mechanisms of sumatriptan and ergot alkaloids in migraine, *Trends Pharmacol Sci* 13:307-311.

Moskowitz MA, Macfarlane R. (1993). Neurovascular and molecular mechanisms in migraine headaches, *Cerebrovasc Brain Metab Rev* 5:159-177.

Moskowitz MA, Nozaki K, Kraig RP. (1993). Neocortical spreading depression provokes the expression of c-fos protein-like immunoreactivity within trigeminal nucleus caudalis via trigeminovascular mechanisms, *J Neurosci* 13:1167-1177.

Olesen, J. *et al.* (2004). Calcitonin gene-related peptide receptor antagonist BIBN409BS for the acute treatment of migraine. *N Engl J Med* 350(11):1104-1110.

Olesen J, Iversen HK, Thomsen LL. (1993). Nitric oxide supersensitivity: a possible molecular mechanism of migraine pain, *Neuroreport* 4:1027-1030.

Ophoff RA *et al.* (1996). Familial hemiplegic migraine and episodic ataxia type-2 are caused by mutations in the calcium channel gene *CACNL1A4, Cell* 87:543-552.

Penfield W, McNaughton F. (1940). Dural headache and innervation of the dura mater, *Arch Neurol Psychiatry* 44:43-75.

Qian J, Noebels JL. (2001). Presynaptic Ca2+ channels and neurotransmitter release at the terminal of a mouse cortical neuron, *J Neurosci* 21:3721-3728.

Ray BS, Wolff HG. (1940). Experimental studies on headache: pain sensitive structures of the head and their significance in headache, *Arch Surg* 41:813-856.

Read SJ, Smith MI, Hunter AJ, Parsons AA. (1997). Enhanced nitric oxide release during cortical spreading depression following infusion of glyceryl trinitrate in the anaesthetized cat, *Cephalalgia* 17:159-165.

Reuter U *et al.* (2001). Delayed inflammation in rat meninges: implications for migraine pathophysiology, *Brain* 124(Pt 12):2490-2502.

Reuter U, Chiarugi A, Bolay H, Moskowitz MA. (2002). NF-κB as a molecular target for migraine therapy, *Ann Neurol* 51(4):507-516.

Schurr A, Payne RS, Miller JJ, Rigor BM. (1997). Brain lactate is an obligatory aerobic energy substrate for functional recovery and after hypoxia: further in vitro validation, *J Neurochem* 69: 423-426.

Shepheard S *et al.* (1999). Possible antimigraine mechanisms of action of the 5HT1F receptor agonist LY334370, *Cephalalgia* 19:851-858.

Stock JL *et al.* (2000). The prostaglandin E2EP1 receptor mediates pain perception and regulates blood pressure, *J Clin Invest* 107:325-331.

Strassman AM, Raymond SA, Burstein R. (1996). Sensitization of meningeal sensory neurons and the origin of headaches, *Nature* 384:560-564.

Strong AJ *et al.* (2002). Spreading and synchronous depressions of cortical activity in acutely injured human brain, *Stroke* 33:2738-2743.

Suzuki N, Hardebo JE, Owman C. (1989). Origins and pathways of cerebrovascular nerves storing substance P and calcitonin gene-related peptide in rat, *Neuroscience* 31:427-438.

Tottene A *et al.* (2002). Familial hemiplegic migraine mutations increase Ca2+ influx through single human Cav2.1 channels and decrease maximal Cav2.1 current density in neurons, *Proc Natl Acad Sci U S A* 99:13284-13289.

Vergnolle N *et al.* (2001). Proteinase-activated receptor-2 and hyperalgesia: a novel pathway, *Nature Med* 7(7):821-826.

Waxman SG, Dib-Hajj S, Cummins TR, Black JA. (1999). Sodium channels and pain, *Proc Natl Acad Sci U S A* 96:7635-7639.

Wei F *et al.* (2001). Genetic enhancement of inflammatory pain by forebrain NR2B overexpression, *Natl Neurosci* 4:164-169.

Weiller C *et al.* (1995). Brain stem activation in spontaneous human migraine attacks, *Nature Med* 1:658–660.

Welch KM *et al.* (1998). MRI of the occipital cortex, red nucleus, and substantia nigra during visual aura of migraine, *Neurology* 51:1465-1469.

Woolf CJ, Salter MW. (2000). Neuronal plasticity: increasing the gain in pain, *Science* 288:1765-1768.

7

Botulinum Toxins: Transformation of a Toxin into a Treatment

Cynthia L. Comella, MD

Lance L. Simpson, PhD

Joseph Jankovic, MD

The history of botulinum toxin (BoNT) is marked on the one hand by its power to cause sudden and deadly disease and on the other by it role as a potent therapeutic agent. The unique capacity of this neurotoxin to survive the harsh environment of the gut, bind to specific acceptor sites, translocate into a cell, and enzymatically cleave specific proteins involved in membrane fusion sets it apart from any therapy used in medicine today. BoNT has opened a window to the molecular mechanics of cellular secretion. Although BoNT remains a potent poison that has potential as a weapon of bioterrorism, its simultaneous development as a therapy has grown to include more than 50 conditions, covering an array of disorders, including dystonia, spasticity, ophthalmological, gastrointestinal, urological, orthopaedic, and secretory problems, as well as pain and cosmetics. These therapeutic applications have emerged over the short span of 25 years since BoNT first became available for clinical use. In the future, the possibility of developing hybrid molecules of BoNT that could selectively inhibit secretion or deliver desired molecules through the gastrointestinal tract and into selected cells will expand the use of BoNT at an exponential rate. The application of BoNT as a therapy has only just begun.

BOTULINUM TOXINS

BoNT is the consummate example of a deadly biological agent that has been transformed into a major therapeutic drug. The toxin has long been recognized as the cause of a potentially severe neurological disease, botu-

lism, but in the recent past it has also come to be accepted as a highly beneficial agent for the treatment of neurological disorders. The toxin was originally introduced for the treatment of certain neuroophthalmological problems, but it is now being used for the treatment of a wide range of disorders. Regardless of whether the toxin is acting as an etiological agent to cause disease or as a therapeutic agent to relieve disease, its site and mechanism of action remain the same. BoNT acts selectively on peripheral cholinergic nerve endings to block acetylcholine release.

Botulinum is derived from the Latin word *botulus*, meaning sausage, and botulism was originally called "sausage poisoning" because it occurred after ingestion of poorly prepared blood sausage. Although BoNT poisoning has been occurring for centuries, the first major breakthrough in understanding the disease came with the publication of a series of cases that detailed the clinical manifestations of the disease. Justinus Kerner (1786-1862) was the first to describe the features of botulism, and he went on to collect additional cases and to conduct experiments on animals and on himself (Ergbuth *et al.*, 1999; Pearce, 1999). Kerner was also the first to suggest that the toxin interrupted peripheral nerve transmission, while sparing the brain, and to speculate as to the future therapeutic applications of this agent (Ergbuth *et al.*, 1999).

The second major historical event was the isolation of the causative agent of "sausage poisoning." In 1895, Van Ermengem isolated an anaerobic bacillus from the remnants of a meal that sickened and killed some of the members of a party of musicians in Ellezelle, Belgium

(Sakaguchi, 1983). He named the organism *Bacillus botulinus* (Caya, 2001).

The characteristic signs and symptoms of botulism include a descending flaccid paralysis with prominent cranial features (i.e., blurred vision, mydriasis, diplopia, ptosis, dysphagia, and dysarthria). Initially, the disease was viewed solely as a form of food poisoning (e.g., "sausage poisoning"). As knowledge about the causes of botulism has grown, the etiology of poisoning has been divided into five categories: food-borne, infant, wound, hidden, and inadvertent. More recently, and particularly in the wake of concern about bioterrorism, there has been growing awareness of the phenomenon of inhalation poisoning.

Although it is true that the level of understanding of the disease has grown, it is also true that the nomenclature for identifying etiologies has become somewhat confused. For example, "food-borne" refers to the origin of the toxin, "infant" identifies the susceptible population, "wound" refers to an antecedent event, "inhalation" conveys information about how the toxin was dispersed, etc. There is no reason to believe that physicians would have any greater difficulty diagnosing or treating the disease if it were described in more conventional terms such as primary infection and primary intoxication. Nevertheless, a historical survey is somewhat compelled to use the terms that were used in the past (Centers for Disease Control, 1998; Cherrington, 1998; Arnon, 2002).

Food-borne botulism was the first form of the disease to be recognized. BoNT exists in seven different serotypes, designated A, B, C, D, E, F, and G. Almost all food-borne disease can be attributed to serotypes A, B, and E. Serotype F only rarely causes human disease (Centers for Disease Control, 1998) and serotypes C, D, and G are not typically associated with human illness. Food-borne disease arises from ingestion of preformed toxin from poorly preserved food. Thus, food-borne botulism is an illustration of a primary intoxication. Absorption of BoNT through the gut after ingestion of tainted food causes symptoms in approximately 2 to 48 hours, with the major determinants of onset being the concentration of toxin in the suspect food and the amount of this food consumed. The initial symptom after ingestion is usually gastrointestinal, with nausea, vomiting, and abdominal upset that arise from the contamination of the food with other byproducts of bacterial growth. Certain other gastrointestinal problems, such as paralytic ileus, are due to toxin-induced blockade of cholinergic transmission.

Initial neurological symptoms involve the bulbar muscles with blurred vision, diplopia, ptosis, ophthalmoplegia, dysarthria, and dysphagia. This can progress to weakness in the arms and torso, weakness in the legs, and in extreme cases, loss of respiratory function. All of these symptoms are directly attributable to blockade of cholinergic transmission by the toxin (Cherrington, 1998).

Outbreaks of food-borne botulism were fairly common in the United States in the early twentieth century because of the consumption of home products that were poorly prepared or preserved, as well as to suspect commercial products (Meyer, 1956). Many patients who contracted the disease eventually died. The mortality resulting from the disease early in the century sometimes exceeded 50%. Although outbreaks continue to be reported (Wainwright *et al.*, 1988), with 724 cases of food-borne botulism in the United States from 1973 through 1996 (Shapiro *et al.*, 1998), survival has improved. Three reasons can explain this increased survival: (1) an increased awareness of the disease, (2) the development and use of antitoxins, and (3) advances in medical technology, allowing for improved life support.

Infant botulism, described in 1976 (Pickett *et al.*, 1976), occurs when spores of *Clostridium botulinum* colonize the gut, particularly the intestine. Clostridial spores do not ordinarily colonize the human gut because they are unable to compete against normal gut flora. There are, however, two notable exceptions: (1) infants of less than 1 year in whom the normal flora has not yet developed, and (2) persons of any age in whom antibiotic use has cleared the gut flora. *C. botulinum* spores have been found in food, such as honey, soil, and dust. The Centers for Disease Control and Prevention (CDC) (CDC, 2003) currently recommends avoiding honey in infants younger than 1 year. Most cases of infant botulism, however, are not traceable to a single food source. Currently, infant botulism is the most common form of botulism in the United States with 1444 cases of infant botulism reported between 1973 and 1996. (Shapiro *et al.*, 1998; CDC, 2003). From 2001 to 2002, the CDC estimated the incidence of infant botulism to be 2 cases per 100,000 live births, with the highest occurrence of 68 per 100,000 found in Staten Island (CDC, 2003).

Wound botulism, initially recognized in 1946, occurs as a result of bacterial colonization of deep flesh wounds (Merson and Dowell, 1973). Wound botulism was rare, but increased dramatically in the early 1990s when users of illicit drugs, including injectable drugs and cocaine inhalation, became the most common group affected (Cherrington, 1998; Arnon, 2002). "Hidden botulism" includes those patients with adult botulism in whom no source of toxin can be determined. This type of botulism is likely to be derived primarily from intestinal colonization with *C. botulinum* or *C. baratti* and subsequent production and absorption of the toxin (CDC, 1998; Caya, 2001). Most at risk for hidden botulism are those with preexisting gastrointestinal pathology (Chia *et al.*, 1986; Cherrington, 1998).

The most recently recognized form of botulism is "inadvertent botulism" arising from systemic spread after therapeutic injections of BoNT. Fortunately, these reports are rare, but they suggest that even with submaximal therapeutic doses of BoNT, clinically detectable systemic spread can occur. Thus far, all patients with inadvertent botulism have made complete recovery (Bhatia *et al.*, 1999; Bakheit *et al.*, 1997). BoNT treatment has also infrequently been associated with the exacerbation of previous neuromuscular junction diseases such as myasthenia gravis and Eaton-Lambert syndrome (Tarsy *et al.*, 2000).

Inhalational botulism, although rarely reported (Holzer, 1962), may be potentially the most devastating route if the toxin were to be used as a weapon of bioterrorism (Arnon *et al.*, 2001; Park and Simpson, 2003; Simpson, 2004). BoNT received serious consideration as a biological weapon during World War II, and it has subsequently been developed as a weapon of mass destruction in many countries throughout the world, including the United States. The U.S. program was discontinued in 1969 to 1970. The possibility of using BoNT to kill large populations has been limited by technical constraints in maintaining potency and stabilizing the toxin. However, an aerosolized form of BoNT has the estimated potential to kill a significant number of people within an estimated 0.5 kilometer distance of release if these technical limitations are overcome (Arnon *et al.*, 2001).

In stark contrast to the grim picture of botulism and bioterrorism, BoNT has emerged over the past two decades as an important medical treatment, approved by the U.S. Food and Drug Administration (FDA) for use in several dystonic and nondystonic neurological disorders, including blepharospasm, hemifacial spasm, strabismus, and cervical dystonia (Botox® Package Insert, 2000; Myobloc™ Package insert, 2000). The first speculation that BoNT could be used as a therapeutic agent was made in the late 1800s by Justinius Kerner (Ergbuth and Naumann, 1999). The crystallization and purification of BoNT A in 1946 opened the door for realization of its clinical potential (Lamanna *et al.*, 1946; Lamanna, 1959; Schantz and Johnson, 1997). The initial use of BoNT as a therapeutic tool was pioneered by Alan Scott, MD, through a collaboration with Edward Schantz, PhD, beginning in the late 1960s (Schantz and Johnson, 1997). In 1973, the first use of BoNT to provide a graded reduction of muscle activity in nonhuman primates was reported (Scott *et al.*, 1973). The first use in humans was reported by Scott in 1980, when 56 injections of botulinum A toxin were given to humans for correction of strabismus (Scott, 1980). The first double-blind, placebo-controlled study of BoNT, published in 1987, established this treatment modality as safe and effective in cranial-cervical dystonia (Jankovic and Orman, 1987). Since then, the uses of BoNT have expanded to other forms of dystonia and to other neurological and non-neurological disorders (TTA, 1990; Jankovic and Brin, 1991). There are now more than 50 "off-label" uses of BoNT (Jost and

TABLE 7.1 Time Line of Major Events in the Transformation of Botulinum Toxin into a Therapeutic Agent

1820	Justinius Kerner describes clinical features of "sausage poisoning" and postulates the peripheral action of BoNT
1895	Van Ermengem isolates the causative organism, calling it *Bacillus botulinus*
1930s	Investigation of BoNT as a weapon of warfare
1946	Crystallization of BoNT serotype A
1946	Wound botulism described
1970s	Description of pharmacology and structure of botulinum toxin serotypes
1976	Infant botulism described
1973	First report of BoNT A application as a potential therapeutic agent in nonhuman primates
1979	Production of 200 g of crystallized BoNT A for use in humans
1980	First report of BoNT A as a treatment for strabismus and blepharospasm
1984-1989	Investigational use of BoNT A (Oculinum, Smith Kettlewell) for treatment of focal dystonias, hemifacial spasm, and spasticity
1987	First double-blind, placebo-controlled study on BoNT
1989	FDA orphan drug approval in the United States of BoNT A as Botox (Allergan) for blepharospasm, strabismus, and other disorders of the seventh cranial nerve
1991	Approval of BoNT A as Dysport (Ipsen) in the United Kingdom
2000	FDA approval in U.S. of BoNT A as Botox (Allergan) and BoNT B as Myobloc (Elan) for use in cervical dystonia
2000	Recognition that antibody-mediated resistance to botulinum toxin therapy was a major cause of nonresponse to treatment
2000-2003	Continued investigation into other uses Approval for cosmetic application

Kohl, 2000; Jankovic, 2004). Current applications of BoNT span a broad spectrum of disorders, from neurological to cosmetic (Table 7.1). This review focuses on the important events in understanding the pharmacology of BoNT and its evolution from a potent poison to an effective treatment for various neurological conditions (Table 7.2).

PHARMACOLOGICAL ACTIONS OF BOTULINUM TOXIN

In the mid-part of the last century, two significant advances were made in our understanding of botulinum toxin. First, the peripheral cholinergic nerve ending was

TABLE 7.2 Clinical Uses of Botulinum Toxins

Ophthalmological	Pain
Strabismus	Headache
Nystagmus	Migraine
Apraxia of eye lid opening	Tension headache
Protective ptosis	Fibromyalgia
Dystonia	Low back pain
Blepharospasm	Painful muscle spasm
Cervical dystonia	Radiculopathy
Spasmodic dysphonia	Gastrointestinal disorders
Oromandibular dystonia	Achalasia
Limb dystonia	Anal sphincter spasm
Occupational dystonia	Constipation
Spasticity	Urological
Poststroke	Vaginismus
Multiple sclerosis	Urinary sphincter spasm
Cerebral palsy	Spastic bladder
Spinal cord injury	Cosmetic
Other neurological disorders	Brow furrows
Hemifacial spasm	Wrinkles
Palatal myoclonus	Miscellaneous
Tremor	Essential hyperhidrosis
Tics	Sialorrhea
Parkinson's disease	Stuttering
Tremor	
Freezing gait	
Clenched fist	
Bruxism	
Myokymia	

identified as the target organ for toxin action; furthermore, the transmitter release mechanism was identified as the specific site of toxin action. Second, the botulinum toxin molecule was isolated and characterized. This work revealed that the toxin is a protein with a molecular mass of approximately 150,000. This early work on botulinum toxin was reviewed by a number of authors (Sugiyama, 1980; Gunderson, 1980; Simpson, 1981; DasGupta, 1994).

During the latter half of the last century, there were again two major advances. In one line of investigation, botulinum toxin action was fractionated into several steps, including binding, internalization, and intracellular poisoning (Simpson, 1980, 1981). The toxin was shown to block transmitter release by acting as a zinc-dependent endoprotease to cleave specific polypeptides that are essential for exocytosis (Montecucco and Shiavo, 1993). In another line of investigation, considerable progress was made in defining the structure of the toxin molecule. This included molecular biology studies in which the genes encoding the toxin were sequenced, and from which amino acids could be deduced, as well as crystallization studies in which three-dimensional structures of the toxin were determined. This work, too, has been reviewed by various authors (Minton, 1995; Popoff and Marvaud, 1999; Lacy and Stevens, 1999).

BoNT is synthesized by *C. botulinum*, *C. butyricum*, and *C. baratii*, all of which are anaerobic spore-forming bacilli. *C. botulinum* is a common bacterium that is widely dispersed in the environment. The spores are heat resistant, and they can germinate to produce toxin in the appropriate environment of anaerobic conditions, low acidity, and liquid medium, as found in some foods (Cherington, 1998). The toxin is ingested and absorbed through the gastrointestinal tract into the systemic circulation (Maksymowych et al., 1999).

BoNT is synthesized in seven immunologically distinct serotypes, designated A to G (Hathaway, 1990). The serotypes most often associated with human illness are A, B, and E (Shapiro *et al.*, 1998). Serotype F only rarely causes human disease, and serotypes C, D, and G are not typically associated with human disease. It should be noted, however, that serotype C does paralyze human neuromuscular transmission (Coffield *et al.*, 1997). The clinical effects caused by poisoning by the various serotypes are similar, although there may be individual distinctions. For example, BoNT/B appears to have more profound autonomic effects than other serotypes, causing intense dry mouth, loss of visual accommodation, and papillary abnormalities, even when muscle weakness is minimal (Jenzer *et al.*, 1975; Racette *et al.*, 2002; Dressler and Benecke, 2003).

BoNT is synthesized and released from bacteria in a complex multimeric form. BoNT/A, for example, is released as part of a noncovalent complex that includes other proteins, such as hemagglutinin (HA), and nontoxin, nonhemagglutinin (NTNH) (Simpson, 2000). These auxiliary proteins play a key role in the phenomenon of oral poisoning. The complex of BoNT, HA, and NTNH is surprisingly resistant to the harsh conditions of low pH and proteolytic enzymes found in the gut. Thus, in the absence of HA and NTNH, BoNT is much less active as an oral poison (Maksymowych *et al.*, 1999). On the other hand, these auxiliary proteins play no role in the ability of the toxin to block cholinergic transmission.

The BoNT molecule itself is synthesized as a relatively inactive single chain polypeptide with a molecular mass of approximately 150,000. Enzymatic nicking of the polypeptide by bacterial or gut proteases results in an activated di-chain molecule, composed of a heavy chain (approximately 100,000) and a light chain (approximately 50,000) linked by a disulfide bond. The carboxyterminal half of the heavy chain (H_c) plays a critical role in toxin binding to the cholinergic membrane, the aminoterminal half of the heavy chain (H_n) is essential for translocation in the cytosol, and the light chain is an enzyme with endoprotease activity (Simpson *et al.*, 1993, 2001).

Each serotype of BoNT binds to serotype-specific acceptor sites on the presynaptic nerve terminal (Black and Dolly, 1986; Schiavo et al., 1993a). The fact that BoNT acts preferentially on cholinergic nerve terminals is thought to be due to the enrichment of these acceptor sites on cholinergic membranes. The identity of the binding sites for botulinum toxin has not been conclusively established, although there is mounting evidence for the involvement of sialic acid-containing molecules.

BoNT enters the nerve terminal through a progression of two events. The toxin crosses the plasma membrane by receptor-mediated endocytosis, and it subsequently penetrates the endosome membrane by pH-induced translocation (Coffield et al., 1994). It is a certainty that the light chain reaches the cytosol, because it is this portion of the holotoxin that possesses enzymatic activity. However, it is unclear whether the heavy chain escapes the endosome to reach the cytosol.

The light chain of BoNT is a highly specific endoprotease (Rosetto et al., 2002) that cleaves peptides involved in vesicle fusion. The proteins involved in vesicle docking and fusion for neurotransmitter release, the SNARE (soluble N-ethlymaleimide sensitive factor attachment protein receptor) proteins, were identified in the early 1990s (Barinaga, 1993; Sollner and Rothman, 1994). The preferred substrates for BoNT action are synaptosomal-associated protein of 25kDa (SNAP-25), vesicle-associated membrane protein (VAMP, but also known as synaptobrevin), and syntaxin (Catsicas et al., 1994). Serotypes A and E cleave SNAP-25, serotypes B, D, F, and G cleave VAMP, and serotype C cleaves both syntaxin and SNAP-25 (Blasi et al.,1993; Schiavo et al., 1993a and b). These three polypeptides, in association with several other intraneuronal polypeptides, are essential for vesicle-mediated transmitter release. Therefore, toxin action on these polypeptides produces blockade of exocytosis, which in turn accounts for neuromuscular and autonomic blockade (Simpson et al., 1993, 2001; Humeau et al., 2000).

SNAP-25, VAMP, and syntaxin are not unique to peripheral cholinergic nerve endings (MacKenzie et al., 1982; Linial, 1997). To the contrary, these polypeptides appear to be involved in release of most neurotransmitters. They are also integral to exocytosis in non-neuronal cells, such as gland cells. The presence of the high affinity H_c acceptor sites on cholinergic nerve terminals accounts for the selective intoxication of these neurons (Schiavo et al., 2000). The introduction of BoNT into noncholinergic cells, which can be accomplished by a variety of experimental techniques, blocks release of many chemical mediators.

Botulinum toxin does not appear to cross the blood-brain barrier. The effects of BoNT are predominantly at cholinergic synapses outside the central nervous system (Grandas et al., 1998); however, central nervous system effects cannot be excluded. Labeled BoNT/A injected into muscle has been shown to penetrate into the spinal cord via retrograde neuronal axonal transport into the ventral roots and then alter the discharge patterns of the abducens nucleus (Wiegand et al., 1976; Moreno-Lopez et al., 1997). This is similar to the pathway used by tetanus toxin, but the effect of BoNT on spinal neurons is not marked and does not reproduce the effects of tetanus toxin. BoNT has been shown to transiently reorganize inhibitory and excitatory intracortical circuits. This cortical effect likely arises indirectly from peripheral mechanisms (Berardelli et al., 2002), although the clinical effects at this time remain to be investigated.

The action of BoNT at the neuromuscular junction is to interrupt transmission and in effect to denervate muscle (Montecucco et al., 1996). This chemodenervation effect persists for weeks to months. The duration of effect may depend on serotype (Sloop et al., 1997; Dolly et al., 2002; Dolly, 2003). The mechanism for this extended duration has been hypothesized to arise from either continued protease activity within the cell or from persistent interference by cleaved substrate with normal membrane fusion (Dolly et al., 2002). Currently, there is no known way to reverse the paralytic effects of BoNT after it has been internalized. Both active and passive immunization can inactivate toxin in the circulation, but antibody cannot enter nerves to neutralize internalized toxin.

Recovery from BoNT occurs spontaneously and may take months to be complete (Hambleton, 1992). At the cellular level, 3 to 4 weeks after a single injection of BoNT/A in mice, there is sprouting of new processes along the nerve axon, with formation of multiple synapses with the muscle and up regulation of the muscle nicotinic receptors. Subsequently, the neuronal sprouts undergo regression and the original synaptic connection is restored, with restoration of the original neuromuscular junction (Aoki, 2001; dePaiva et al., 1999).

THERAPEUTIC APPLICATIONS OF BoNT

One of the interesting aspects of BoNT as a therapeutic modality has been its development and initial approval as an orphan drug by the U.S. FDA in 1989. The Orphan Drug Act provides an avenue for approval of a product that treats a rare disease affecting fewer than 200,000 Americans. Despite the fact that BoNT was the most potent neurotoxin known, the initial approval of BoNT as BOTOX® was based on one placebo-controlled study (Jankovic and Orman, 1987) and a number of open-label uncontrolled studies and published clinical observations in 1684 blepharospasm and 2322 strabismus patients (Walton, 1999; BOTOX® Package Insert, 2000).

Once approved by the FDA, the "off label" uses of BoNT rapidly increased, using doses far exceeding those initially approved (TTA, 1990; NIH Consensus Statement, 1991). This gave rise to concerns regarding safety and immunogenicity and led to the more stringent requirements by the FDA for controlled studies before approval of BOTOX® and MyoBloc™ for the indication of cervical dystonia in 2000. Yet most disorders now treated with BoNT do not have adequate assessments of dosing or appropriate techniques of administration, and the efficacy and safety data remain largely derived from uncontrolled clinical experience (Jost and Kohl, 2000). However, clinical experience with BoNT for a variety of disorders has largely shown long-term efficacy and safety (Jankovic and Schwartz, 1993; Kessler et al., 1999; Hsiung et al. 2002; Jankovic, 2004; Naumann and Jankovic, 2004).

BOTULINUM TOXIN TREATMENT OF BLEPHAROSPASM AND HEMIFACIAL SPASM

The first clinical applications of BoNT as a therapeutic intervention were in strabismus, hemifacial spasm, and blepharospasm, a form of focal dystonia. Dystonia is a syndrome characterized by involuntary sustained muscle contractions that usually cause twisting and repetitive movements or abnormal postures. (Fahn et al., 1998). Focal dystonia involves a single body region, with limited muscle involvement making it most amenable to chemodenervation with BoNT. Focal dystonia affects the neck (cervical dystonia), eyes (blepharospasm), jaw and oral muscles (oromandibular dystonia), larynx and vocal cords (spasmodic dysphonia), and limbs (arm/leg dystonia, writer's cramp). Botulinum toxin has been used extensively as a symptomatic treatment in each of these conditions.

Before the introduction of BoNT injections, the treatment of dystonia was rarely successful. Although approximately 23% to 55% of dystonia patients benefit from oral pharmacological agents (Greene et al., 1988), the benefits from oral medications are usually transient, and frequently complicated by systemic side effects. (Jankovic and Orman, 1984; Grandas et al., 1988). Botulinum toxin injections were first introduced as a treatment for blepharospasm in the early 1980s (Scott et al., 1985). Subsequent studies showed successful control of facial spasms in 75% to 100% of patients (Mauriello, 1985; Jankovic and Orman, 1987; Mauriello and Coniaria, 1987; Dutton and Buckley, 1988; Mauriello et al., 1996a; Jitpimolmard et al., 1998; Mauriello, 2002). Although many of these were observational studies, a limited number of controlled studies in a small cohort of patients confirmed the results of open trials (Jankovic

and Orman 1987; Mezaki et al., 1999). Long-term follow-up evaluation of patients receiving multiple treatments showed sustained improvement, with reduction in spasm intensity (Dutton, 1996). The mechanism of the clinical improvement was largely due to peripheral muscle denervation: this notion is supported by the absence of normalization of blink reflexes after injection (Valls-Sole et al., 1991; Girlanda et al., 1996).

Subsequent studies were directed toward the refinement of injection techniques and reduction of the adverse effects after BoNT injection. The adverse effects of BoNT for blepharospasm largely arose from the localized diffusion from the injection site. These local side effects included diplopia, ptosis, lid entropion, and epiphora (Dutton and Buckley, 1988; Patrinely et al., 1988; Mauriello and Aljian, 1991). Ptosis was one of the most frequent complications of BoNT (Dutton, 1996). The anatomy of the eyelid and face, with the proximity of the levator superioris muscle to the orbicularis oculi muscle, was thought to underlie this frequent effect. The orbicularis oculi muscle is approximately 1 mm thick, with a loose connective tissue fascia plane below it that allows for easy diffusion of the toxin. The orbital septum lies deep to this facial plane. This septum resists diffusion of injected liquid, but may be attenuated in older patients or punctured, allowing free access to the levator muscle. Medial and lateral pretarsal injections (avoiding the midline) replaced those into the orbital part of the orbicularis oculi muscle, reducing the occurrence of this adverse event (Aramideh et al., 1995; Albanese et al., 1996; Jankovic et al., 1996). Diplopia, a less frequent complication, was usually due to BoNT injections into the medial aspect of the lower eyelid, and diffusion of toxin through the orbital septum into the inferior oblique muscle (Wutthiphan et al., 1997). By omitting injections into this area of the lower lid, diplopia was minimized (Frueh et al., 1988). BoNT was also found to diffuse into the lower facial muscles, with a reduction of compound muscle action potentials and motor evoked potentials occurring in lower facial muscles (Eleopra et al., 1996; Girlanda et al., 1996). If pronounced, this local diffusion also caused facial weakness and asymmetry.

The use of BoNT for blepharospasm provided an opportunity to assess both the electrophysiological and pathological time course of BoNT effect in human muscle. After treatment with BoNT into the orbicularis oculi muscle, serial electromyographical examinations showed marked denervation changes with increased jitter and blocking appearing at 1 week. These changes were associated with clinical improvement. After a mean of 116 days, denervation of the muscle persisted but at a reduced level, despite a return to baseline clinical status (Bogucki et al., 1999). Despite prolonged denervation after repeated injections, changes in the orbicularis oculi muscle were minimal, with few showing any evidence of fibrosis, indicating that the effects of BoNT injections, while beneficial for months after each injection, do not result in irreversible muscle atrophy or degenerative changes if administered chronically (Borodic and Ferrante, 1992). The FDA approved BoNT serotype A in 1989 for the indications of blepharospasm, strabismus, and other disorders of the facial nerve (including hemifacial spasm).

BOTULINUM TOXIN TREATMENT OF CERVICAL DYSTONIA

Approximately 5 years after BoNT treatment was used in low doses for the treatment of blepharospasm, investigations into the usefulness of BoNT in higher doses for cervical dystonia began. Over the past 2 decades, open-label and controlled studies have shown the effectiveness of BoNT A as a treatment for cervical dystonia (Ceballos-Baumann, 2001).

As in blepharospasm, the early studies were mostly open-label, uncontrolled investigations (Tsui et al., 1985; Stell et al., 1988; Lorentz et al., 1990; Blackie and Lees, 1990), but taken together demonstrated the promise of a new therapy for this disorder, with benefit that could be sustained over repeated injections. Furthermore, the nonmotor features of cervical dystonia, including disability, depression, and quality of life, also showed significant improvement with BoNT treatment (Jahanshahi and Marsden, 1991; Brans et al., 1998; Brefel-Courbon et al., 2000; Jankovic et al. 2004).

The first controlled study of BoNT A was conducted in 21 patients with cervical dystonia using a crossover design of active BoNT A at 100 Units and placebo. In the 19 patients who completed the trial, there was significant but modest improvement (Tsui et al., 1986). However, another controlled study in 20 patients found only subjective benefit without any objective changes based on the Tsui scale (Gelb et al., 1989), and an additional study observed no improvement in any measure when administering a fixed dose of BoNT into the same muscles of all patients (Koller et al.,1990). The need to develop appropriate methodology to assess this new therapy was clear. Using a different outcome measure, the Columbia Torticollis Rating Scale, Greene and associates (1990) showed both objective and subjective improvement following BoNT, with an open-label extension demonstrating that additional benefit accrued if higher dosages of BoNT were used. This positive outcome was found by a number of additional investigators (Jankovic and Orman, 1987; Moore and Blumhardt, 1991; Lorentz et al., 1991).

The insights gained during this early period of investigation were used in developing the large, pivotal clinical trials of BoNT A and B to obtain approval by the FDA for these drugs. BoNT A (BOTOX) was assessed using a run-in, two-treatment period design to determine the optimal pattern of injection, with flexibility for dosing and muscle selection. The outcome measure, the Cervical Dystonia Severity Scale, was tested for validity before its use in the study. A large number of patients (N = 170) were enrolled in the randomized, double-blind, placebo-controlled treatment period and monitored for 10 weeks after treatment. This study showed that there was significant improvement of BoNT A group over placebo for all outcome measures. It further demonstrated that the occurrence of adverse effects across multiple centers was small, with the most common being muscle weakness (7.9%) and dysphagia (6.8%) (BOTOX® package insert). Although only reported in an abstract, this study served as the pivotal clinical trial for approval by the FDA of BoNT for cervical dystonia in the United States.

BoNT serotype B was introduced as a treatment for cervical dystonia in 1997 (Lew et al., 1997). The pivotal controlled clinical trials were published in 1999 (Brashear et al., 1999; Brin et al., 1999). The methodology used in these two pivotal studies took into consideration many of the limitations in prior studies, allowing a flexible injection pattern and using a previously established rating scale for cervical dystonia, the Toronto Western Spasmodic Torticollis Rating Scale (TWSTRS). In addition, these studies included a method to prevent possible unblinding during the course of investigation by appointing a separate rating investigator who did not administer the BoNT nor record the adverse events. The positive outcome of both of these studies definitively confirmed the efficacy of BoNT for the treatment of cervical dystonia, and provided a methodology that could be useful in future studies of other agents for this disorder. In one of the studies (Brin et al., 1999), only patients with cervical dystonia who had lost sensitivity to BoNT A were enrolled. The benefit seen in the BoNT A-resistant patients using a different serotype provided strong evidence that the BoNT serotypes were distinct. The availability of the two serotypes provided clinicians with valuable treatment options, especially for those patients who had ceased to benefit from treatment with one serotype.

Optimal treatment using BoNT is largely dependent on appropriate clinical assessment, knowledge of functional anatomy, selection of the most involved muscles as targets for injection, and the use of appropriate dosage and injection techniques (Koller et al., 1990; Gelb et al., 1991; Odergren et al., 1998; Poewe et al., 1998). The long clinical experience using BoNT for cervical dystonia has provided extraordinary insights into the use of neurotoxins as a treatment. In one clinical series, Brashear and colleagues (2000) retrospectively monitored 151 cervical dystonia patients who had received at least one treatment. 20% to 33% of their patients discontinued treatment. Several reasons were cited: (1) a lack of benefit from treatment was the primary reason, (2) the occurrence of adverse effects after injection was second, and (3) the expense of continued treatment. Treatment failure can arise for several reasons. Apart from immunological resistance to BoNT, which will be discussed below, there have been changes in the technique of administration that have improved the outcome. Cervial dystonia is a dynamic disorder that can change in the quality of the movement and in the occurrence of pain. This has been shown by several investigators (Gelb et al., 1991; Munchau et al., 2001). Munchau et al., (2001) showed that using synchronized video and electromyographical recordings, approximately half of cervical dystonia patients will have spontaneously changing patterns of muscle activation. These patients did not show an equivalent benefit after injection as those patients with unchanging muscle activation. Injection of uninvolved muscles may reduce or negate the effect of treatment. Although objective techniques have been successfully applied to assess head position in cervical dystonia, the additional equipment, training, and time needed to use these methods preclude routine application in a clinical setting (Albani et al., 2001). The clinical examination remains the gold standard for assessment in CD and other disorders when choosing appropriate muscles for treatment.

The use of EMG in conjunction with the clinical examination in both muscle selection and targeting injections has been evaluated in CD and can be extrapolated to other disorders with complex muscle activation patterns. Investigators using EMG guidance have reported increased benefits in their patients and the potential to use smaller doses of toxin for treatment (Dubinsky et al., 1991; Brans et al., 1995; Comella et al.,1992; Ostergaard et al., 1994). A single-blinded study of consecutive cervical dystonia patients randomized to injection with or without EMG assistance demonstrated that there was a greater magnitude of improvement and a greater number of patients with a marked improvement when EMG was used (Comella et al., 1992). EMG is thought to increase accuracy of injection into muscle, although alternatively, it may be in enhancing the clinical examination in the process of muscle selection (Gelb et al., 1991; Brans et al., 1995; Speelman and Brans, 1995; Ostergaard et al., 1996; Finsterer et al., 1997; Ajax et al., 1998; Dressler, 2000; Van Gerpen et al., 2000). Although the deep cervical muscles have been shown to be spontaneously overactive in 68% of consecutive cervical dystonia patients, needle placement in these muscles using clinical landmarks alone is accurate in only 50% of attempts (Speelman and Brans, 1995).

The success of the cervical dystonia studies provided the impetus for a broader application of BoNT not only as a treatment for cervical dystonia, but also in other disorders in which muscle overactivity was the predominant disorder.

BOTULINUM TOXIN TREATMENT OF SPASTICITY

With the observation that BoNT could improve muscle spasms related to dystonia, studies were undertaken to assess whether similar efficacy could be demonstrated in spasticity (Simpson, 2002). Spasticity is a velocity-dependent increase in muscle tone and deep tendon reflexes as a part of the upper motor neuron syndrome that arises from numerous etiologies (Lance, 1990). Spasticity is more common than dystonia and currently represents the largest population with a neurological disorder that may be amenable to treatment with BoNT. In contrast to dystonia, however, spasticity is a complex syndrome, combining elements of weakness, impaired dexterity, pain, stiffness, and muscle spasms. With chronic spasticity, there can be shortening and fibrosis in the muscle, causing contractures. The varied manifestations of spasticity and the diverse populations in which it is found have presented a challenge to clinicians to identify the particular patient group that may benefit most from BoNT.

The initial goal of BoNT treatment for spasticity was a reduction in muscle tone, as measured by the Ashworth scale or goniometry. The concept that these quantitative goals may not be sufficient to assess the effects of BoNT on spasticity led to the development of additional goals that focused on improving motor function by balancing muscle forces across joints: health-related quality of life, decreasing caregiver burden, decreasing pain from spasticity, and improving self-esteem (Koman et al., 2003).

Spasticity disorders that have been treated with BoNT include multiple sclerosis, spinal cord injury, cerebral palsy, and stroke. Published results were variable, and most studies were either open-label or controlled studies in small numbers of patients. Patient selection, methodology, and outcome measures differed among studies. A consistent finding among these studies was a reduction in muscle tone after BoNT injection, but improvement in functional outcomes was not consistent and proved difficult to demonstrate convincingly (van Kuijk et al., 2002; Burbaud et al., 1996; Pittock et al., 2003; Graham, 2000; Corry et al., 1997).

In selected patients, there has been evidence that BoNT improves function. A large placebo-controlled study demonstrated improvement in a simple functional scale after BoNT treatment for wrist and finger spasticity following stroke (Brashear et al., 2002). Controlled studies of cerebral palsy have shown improved gait (Sutherland et al., 1996; Baker et al., 2002), increased arm and hand function, (Fehlings *et al.*, 2000), reduction in pain, and increased comfort and tolerability of bracing (Pierson *et al.*, 1996). Other studies have shown improved ability to perform simple tasks (Hurvitz *et al.*, 2003) and improved ability to perform hygienic measures (Snow *et al.*, 1990). In some, but not all studies, improvement persisted for as long as 6 months after injection (Slawek and Klimont, 2003), although long-term improvement may be attributed to the simultaneous and judicious use of rehabilitation therapies. Currently, BoNT has been approved in some European countries for the treatment of spasticity.

BOTULINUM TOXIN TREATMENT OF TREMOR DISORDERS

Several studies have assessed the usefulness of BoNT for the treatment of essential hand tremor. Two controlled trials demonstrated improvement in tremor severity rating scales and tremor amplitude as measured using accelerometery. Both studies, however, failed to show consistent improvement in functional outcomes, possibly because of occurrence of adverse effects. A prominent adverse effect was the occurrence of a dose-dependent mild to moderate finger weakness or reduced grip strength in 30% to 70% of patients (Jankovic *et al.*, 1996; Brin *et al.*, 2001). However, each of these studies used fixed doses of BoNT and required injection into four specified forearm muscles: the flexor carpi radialis and ulnaris and extensor carpi radialis and ulnaris, regardless of specific tremor movement, and allowed injections without EMG guidance. In contrast, three open-label studies allowed injections into variable muscles depending on the tremor manifestations showed significant improvement in functional ability and mild to moderate improvement in tremor amplitude (Pachetti *et al.*, 2000). In these studies, finger weakness was much less common and less severe, partly as a result of exclusion of the extensor carpi muscles and adjustment of BoNT dosage.

BOTULINUM TOXIN TREATMENT OF HEADACHE AND PAIN

BoNT has been effective in reducing pain in cervical dystonia and spasticity. The current status of BoNT and its applications to headache and other pain disorders is promising but remains to be established (Brin *et al.*, 2002). Current evidence is largely anecdotal. One controlled study in migraine showed improvement (Silberstein *et al.*, 2000); another assessing tension headache did not (Rollnik *et al.*, 2000). The mechanism of analgesic action of BoNT is unknown but has been proposed to be related to yet poorly understood effects on sensory nociceptive systems, particularly on substance P (Van den Bergh *et al.*, 1996). Further carefully designed investigations are needed.

The collective experience gained through the application of BoNT treatment for cervical dystonia blepharospasm, spasticity, tremor, and other neurological disorders have clearly established BoNT as the treatment of choice in some disorders and a less reliable means of controlling others. The essential requirement for well-designed, controlled assessments of BoNT that use its unique mechanism of action and recognize its limitations has also become apparent.

IMMUNOLOGY AND RESISTANCE TO BoNT

The extensive use of BoNT in CD showed that a major limitation to repeated BoNT treatment was the development of resistance. Botulinum toxin is a protein that serves as an antigen when injected repeatedly into humans. The development of an antibody response to an antigen is dependent on several factors: the presence of an adjuvant (a substance that increases the immune response), the persistence of the antigen in the tissues, and the frequency of exposure to and quantity of the antigen. In addition, other factors, such as the genetic make-up of an individual, may increase the susceptibility for that individual to develop antibodies to a particular antigen (Critchfield, 2000; Jankovic, 2002). Antibody formation in response to BoNT is a desirable event in people with frequent exposure to the toxin and risk of contracting botulism. Hence, in endemic areas and in laboratory personnel, vaccinations using toxoid are administered. However, in patients whose clinical condition responds to BoNT, the formation of neutralizing antibodies results in a clinical resistance to the beneficial effects.

In 1979, Shantz and colleagues at the University of Wisconsin prepared 200 mg of crystallized BoNT A. This small amount was the only commercially available source of toxin in the United States for almost a decade (Schantz and Johnson, 1997). BoNT then became available in two serotypes: BoNT type A, distributed as BOTOX® (Allergan) and Dysport (Ipsen), and BoNT B as Neurobloc and MyoBloc™ (Elan). Production practices have been refined and BOTOX® has been reformulated to provide a lower protein load (25 ng protein/100 units to 5 ng protein/100 units) (Borodic *et al.*, 1996; Jankovic *et al.*, 2003). In available serotypes, there is no established dosing equivalency (Sloop *et al.*, 1997).

The factors that predispose to development of antibodies have not been identified. Large doses of botulinum toxin ($\geq$ 250 units botulinum toxin A BOTOX®), larger cumulative doses, and injections administered at less than 3-month intervals ("booster" injections) have been identified as possible risk factors for the development of resistance (Greene et al., 1994). Botulinum toxin resistance primarily occurs in patients receiving higher doses of toxin for the treatment of disorders such as cervical dystonia, and is rare in disorders treated with lower doses, such as blepharospasm or hemifacial spasm.

Initially, BoNT resistance was not considered to be common. Early reports suggested a frequency of approximately 5% resistance in cervical dystonia patients receiving repeated injections (Jankovic and Schwartz, 1993; Kessler and Benecke, 1997). This observation was based on retrospective assessments of patients at single locations. However, the clinical trials of BoNT serotypes A and B published as the package inserts for the drugs when FDA approval was obtained revealed antibody formation to be much more frequent than previously thought. Prospective studies of both Botox and Myobloc are underway to assess the frequency and factors associated with resistance.

Antibodies to BoNT are detected using various methods (Goschel et al., 1997). The mouse protection assay (MPA) is the gold standard method. This assay evaluates the ability of increasing dilutions of a patient's serum to protect experimental mice from lethal test doses of BoNT. Recently, the mouse hemidiaphragm model with phrenic nerve intact has also been described. The immunoprecipitation assay (IPA) is a simple, rapid technique for detecting antibodies against BoNT. IPA has been found to be both sensitive and specific (Palace et al., 1998; Dressler and Dirnberger, 2001). Compared to the MPA, IPA was more sensitive than the MPA in detecting BoNT antibodies and shows a positive result earlier than the MPA, suggesting it may predict future unresponsiveness (Hanna and Jankovic, 1998). The in vitro assays listed previously are expensive and have not yet been shown to predict whether clinical resistance will occur. Using the MPA, Jankovic and colleagues (2003) showed that patients treated with the current BOTOX® (5 ng protein/100 units) had no evidence of blocking antibodies compared to 9.4% frequency of blocking antibodies in patients treated with the original BOTOX® (25 ng/100 units) for the same period.

In addition to the laboratory tests currently being assessed, a variety of clinical tests have been developed to ascertain whether resistance to the effects of toxin exists. The FTA (frontalis type A test) (Borodic et al., 1995), the unilateral brow injection (UBI) (Hanna et al., 1999), the SCM (sternocleidomastoid) test, and the EDB (extensor digitorum brevis) test have all been explored as sensitive methods to detect clinical resistance (but not antibody titers) in patients reporting the secondary failure of BoNT to improve symptoms (Kessler and Benecke, 1997; Sloop et al., 1997; Birklein and Engbutl, 2000; Dressler et al., 2000). The basic principle of these clinical tests was to determine whether an injection of BoNT into a specific muscle would cause denervation and atrophy in that muscle. The absence of changes in the muscle after injection suggested loss of sensitivity to BoNT. Although studies are in progress, the tests have not been validated.

Treatment of BoNT resistance is problematic. Depletion of the neutralizing antibodies through plasma exchange or immunosuppression using drugs such as mycophenolate has been proposed (Naumann et al., 1998; Duane et al., 2000). Because each serotype of BoNT is immunologically distinct, replacing one serotype with another was found to be effective, although the long-term sensitivity of the patient to an alternate serotype of BoNT has not been prospectively evaluated (Greene and Fahu, 1993, 1996; Sheehan and Lees, 1995; Truong et al., 1997; Chen et al., 1998; Houser et al., 1998; Brin et al., 1999; Aoki, 2001). Preliminary results, however, suggest that the development of blocking antibodies to one serotype increases the risk of blocking antibodies to another serotype (Atassi, 2002; Dressler et al., 2003).

FUTURE APPLICATIONS

Several distinctive features of BoNT account for its toxicity, but also provide potential avenues for novel uses as a therapeutic agent (Simpson, 2000). First, the ability of the toxin to escape degradation in the gut and be absorbed through binding to the gut epithelial cells allows an avenue for entry of other proteins through an oral route. Second, the absence of a significant immune-mediated response on the part of the host against circulating toxin opens the possibility of repeated exposure without resistance. Third, the presence of the high-affinity acceptor sites on preganglionic cholinergic neurons allows for the targeted binding of the heavy chain of BoNT on specific cells. Fourth, the enzymatic properties of the light chain that result in "multiplicative" rather than a "stochastic" mechanism of toxicity permits small doses to have extensive effects. Fifth, the selective nature of each serotype to cleave proteins involved in neurotransmitter release lays open the possibility of selectively inhibiting secretion from other cell types. Sixth, the absence of neuronal degeneration and the reversibility of toxin effects present the opportunity to affect cells, without permanent destruction of cell mechanisms. Some of these properties have already led to the use of BoNT as a potent treatment for disorders characterized by muscle overactivity, including dystonia and spasticity as discussed previously. These properties have also been used in investigations exploring mechanisms of neurotransmitter release (Simpson, 2004). Current research is investigating the possibilities of developing hybrid molecules of BoNT that exploit its unique features in order to design molecules that resist degradation within the gut, are not toxic, and deliver agents artificially attached to the BoNT molecule into the cytosol of specific cells (Simpson, et al., 1999; Simpson, 2000). These could be oral vaccines or other therapeutic drugs. For example, by producing a hybrid consisting of the light chain and H_n fraction of the heavy chain and conjugating with a second protein that binds to noncholinergic cell receptors, it may be possible to harness the enzymatic activity of the light chain in order to inhibit secretion from targeted cells by interrupting vesicle fusion (Foster, 2002). The potential of this powerful neurotoxin to serve as a beneficial treatment for a variety of neurological and non-neurological diseases is still being realized.

References

Ajax T, Ross MA, Rodnitzky RL. (1998). The role of electromyography in guiding botulinum toxin injections for focal dystonia and spasticity, *J Neuro Rehab* 12:1-4.

Albanese A *et al.* (1996). Pretarsal injections of botulinum toxin improve blepharospasm in previously unresponsive patients [letter], *J Neurol Neurosurg Psychiatry* 60:693-694.

Albani G *et al.* (2001). The position of the head in space: a kinematic analysis in patients with cervical dystonia treated with botulinum toxin, *Funct Neurol* 16:135-141.

Aoki KR. (2001). Pharmacology and immunology of botulinum toxin serotypes, *J Neurol* 248(suppl 1):I3-I10.

Aramideh M *et al.* (1995). Pretarsal application of botulinum toxin for treatment of blepharospasm, *J Neurol Neurosurg Psychiatry* 59:309-311.

Arnon SS *et al.* (2001). Botulinum toxin as a biological weapon: medical and public health management, *JAMA* 285:1059-1070.

Arnon SS. (2002). Clinical botulism. In Brin M, Hallett M, Jankovic J, editors: *Scientific and therapeutic aspects of botulinum toxin.* Philadelphia: Lippincott Williams & Wilkins.

Atassi MZ. (2002). Immune recognition and cross-reactivity of botulinum neurotoxins. In Brin MF, Hallett M, Jankovic J, editors: *Scientific and therapeutic aspects of botulinum toxin.* Philadelphia: Lippincott Williams & Wilkins.

Baker R *et al.* (2002). Botulinum toxin treatment of spasticity in diplegic cerebral palsy: a randomized, double-blind, placebo-controlled, dose-ranging study, *Dev Med Child Neurol* 44:666-675.

Bakheit AM, Ward CD, McLellan DL. (1997). Generalized botulism-like syndrome after intramuscular injections of botulinum toxin type A: a report of two cases [letter], *J Neurol Neurosurg Psychiatry* 62:198.

Barinaga M. (1993). Secrets of secretion revealed, *Science* 260:487-489.

Berardelli A, Gilio F, Curra A. (2002). Effects of botulinum toxin type A on central nervous system function. In Brin MF, Jankovic J, Hallett M, editors: *Scientific and therapeutic aspects of botulinum toxin.* Philadelphia: Lippincott Williams & Wilkins.

Bhatia K.P *et al.* (1999). Generalized muscular weakness after botulinum toxin injections for dystonia: a report of three cases, *J Neurol Neurosurg Psychiatry* 67:90-93.

Birklein F, Erbguth F. (2000). Sudomotor testing discriminates between subjects with and without antibodies against botulinum toxin A— a preliminary observation, *Mov Disord* 15:146-149.

Black JD, Dolly O. (1986). Interaction of 125I-labeled botulinum neurotoxins with nerve terminals. I. Ultrastructural autoradiographic localization and quantitation of distinct membrane acceptors for types A and B on motor nerves, *J Cell Biol* 103:521-534.

Blackie JD, Lees AJ. (1990). Botulinum toxin treatment in spasmodic torticollis, *J Neurol Neurosurg Psychiatry* 53:640-643.

Blasi J *et al.* (1993). Botulinum neurotoxin A selectively cleaves the synaptic protein SNAP-25, *Nature* 365:160-163.

Bogucki A. (1999). Serial SFEMG studies of orbicularis muscle after the first administration of botulinum toxin, *Eur J Neurol* 6:461-467.

Borodic G, Johnson E, Goodnough M, Schantz E. (1996). Botulinum toxin therapy, immunologic resistance, and problems with available materials, *Neurology* 46:26-29.

Borodic GE, Duane D, Pearce B, Johnson E. (1995). Antibodies to botulinum toxin, *Neurology* 45:204.

Borodic GE, Ferrante R. (1992). Effects of repeated botulinum toxin injections on orbicularis oculi muscle, *J Clin Neuroophthalmol* 12:121-127.

Brans JW *et al.* (1995). Botulinum toxin in cervical dystonia: low dosage with electromyographic guidance, *J Neurol* 242:529-534.

Brans JW, Lindeboom R, Aramideh M, Speelman JD. (1998). Long-term effect of botulinum toxin on impairment and functional health in cervical dystonia, *Neurology* 50:1461-1463.

Brashear A *et al.* (2000). Patients' perception of stopping or continuing treatment of cervical dystonia with botulinum toxin type A, *Mov Disord* 15:150-153.

Brashear A *et al.* (2002). Intramuscular injection of botulinum toxin for the treatment of wrist and finger spasticity after a stroke, *N Engl J Med* 347:395-400.

Brashear A *et al.* (1999). Safety and efficacy of NeuroBloc (botulinum toxin type B) in type A-responsive cervical dystonia, *Neurology* 53:1439-1446.

Brefel-Courbon C *et al.* (2000). A pharmacoeconomic evaluation of botulinum toxin in the treatment of spasmodic torticollis, *Clin Neuropharmacol* 23:203-207.

Brin MF, *et al.* (2002). Botulinum toxin type A BOTOX® for pain and headache. In Brin MF, Jankovic J, Hallett M, editors: *Scientific and therapeutic aspects of botulinum toxin,* Philadelphia: Lippincott Williams & Wilkins.

Brin MF *et al.* (1999). Safety and efficacy of NeuroBloc (botulinum toxin type B) in type A-resistant cervical dystonia, *Neurology* 53:1431-1438.

Brin MF *et al.* (2001). A randomized, double masked, controlled trial of botulinum toxin type A in essential hand tremor, *Neurology* 56:1523-1528.

Burbaud P *et al.* (1996). A randomized, double blind, placebo controlled trial of botulinum toxin in the treatment of spastic foot in hemiparetic patients, *J Neurol Neurosurg Psychiatry* 61:265-269.

Catsicas S, Grenningloh G, Pcij EM. (1994). Nerve-terminal proteins: to fuse to learn, *Trends Neurosci* 17:368-373.

Caya JG. (2001). *Clostridium botulinum* and the ophthalmologist: a review of botulism, including biological warfare ramifications of botulinum toxin [survey] *Ophthalmol* 46:25-34.

Ceballos-Baumann AO. (2001). Evidence-based medicine in botulinum toxin therapy for cervical dystonia, *J Neurol* 248(suppl 1): I/14-I/20.

Centers for Disease Control and Prevention (1998). *Botulism in the United States, 1899-1996. Handbook for epidemiologists, clinicians, and laboratory workers,* Atlanta: Centers for Disease Control and Prevention.

Centers for Disease Control. (2003). Infant botulism in New York City 2001-2001, *MMWR* 52:21-24.

Chen R, Karp BI, Hallett M. (1998). Botulinum toxin type F for treatment of dystonia: long-term experience, *Neurology* 51:1494-1496.

Cherrington M. (1998). Clinical spectrum of botulism, *Muscle Nerve* 21:701-710.

Chia JK, Clark JB, Ryan CA, Pollack M. (1986). Botulism in an adult associated with food-borne intestinal infection with *Clostridium botulinum*, *N Engl J Med* 315(4):239-241.

Coffield J, Considine RV, Simpson LL. (1994). The site and mechanism of action of botulinum neurotoxin. In Jankovic J, Hallett M, editors: *Therapy with botulinum toxin*, New York: Marcel Dekker.

Coffield JA *et al.* (1997). In vitro characterization of botulinum toxin types A, C, and D action on human tissues: combined electrophysiologic, pharmacologic and molecular biologic approaches, *J Pharmacol Exp Ther* 280:1489-1498.

Comella CL *et al.* (1992). Botulinum toxin injection for spasmodic torticollis: increased magnitude of benefit with electromyographic assistance, *Neurology* 42:878-882.

Corry IS *et al.* (1997). Botulinum toxin A in the hemiplegic upper limb: a double blind trial, *Dev Med Child Neurol* 39:185-193.

Critchfield J. (2000). Immunogenicity of botulinum toxin therapy. In Ginsberg DL, editor: *CNS Spectrums* 7 (suppl 6):51-58.

DasGupta BR. (1994). Structures of botulinum neurotoxin, its functional domains and perspectives on the crystalline type A toxin. In Jankovic J, Hallett M, editors: *Therapy with botulinum toxin*, New York: Marcel Dekker.

dePaiva A *et al.* (1999). Functional repair of motor endplates after botulinum neurotoxin type A poisoning: biphasic switch of synaptic activity between nerve sprouts and their parent terminals, *Proc Natl Acad Sci* 96:3200-3205.

Dolly JO *et al.* (2002). Insights into the extended duration of neuroparalysis by botulinum neurotoxin A relative to the other shorter-acting serotypes: differences between motor nerve terminals and cultured neurons. In Brin MF, Jankovic J, Hallett M, editors: *Scientific and therapeutic aspects of botulinum toxin*, Philadelphia: Lippincott Williams & Wilkins.

Dolly O. (2003). Synaptic transmission: inhibition of neurotransmitter release by botulinum toxins, *Headache* 43 (Suppl 1):S16-24.

Dressler D, Benecke R. (2003). Autonomic side effects of botulinum toxin type B treatment of cervical dystonia and hyperhidrosis, *Eur Neurol* 49:34-38.

Dressler D, Bigalke H, Rothwell JC. (2000). The sternocleidomastoid test: an in vivo assay to investigate botulinum toxin antibody formation in humans, *J Neurol* 247:630-632.

Dressler D, Dirnberger G. (2001). Botulinum toxin antibody testing: comparison between the immunoprecipitation assay and the mouse diaphragm assay, *Eur Neurol* 45:257-260.

Dressler D. (2000). Electromyographic evaluation of cervical dystonia for planning of botulinum toxin therapy, *Eur J Neurol* 7:713-718.

Dressler D, Bigalke H, Benecke R. (2003). Botulinum toxin type B in antibody-induced botulinum toxin type a therapy failure, *J Neurol* 250:967-969.

Duane DD, Monroe J, Morris RE. (2000). Mycophenolate in the prevention of recurrent neutralizing botulinum toxin A antibodies in cervical dystonia [letter], *Mov Disord* 15:365-366.

Dubinsky RM, Gray CS, Vetere-Overfield B, Koller WC. (1991). Electromyographic guidance of botulinum toxin treatment in cervical dystonia, *Clin Neuropharmacol* 14:262-267.

Dutton JJ, Buckley EG. (1988). Long-term results and complications of botulinum A toxin in the treatment of blepharospasm, *Ophthalmology* 95:1529-1534.

Dutton JJ. (1996). Botulinum-A toxin in the treatment of craniocervical muscle spasms: short- and long-term, local and systemic effects, *Surv Ophthalmol* 41:51-65.

Eleopra R, Tugnoli V, Caniatti L, De Grandis D. (1996). Botulinum toxin treatment in the facial muscles of humans: evidence of an action in untreated near muscles by peripheral local diffusion, *Neurology* 46:1158-1160.

Erbguth FJ, Naumann M (1999). Historical aspects of botulinum toxin: Justinus Kerner (1786-1862) and the "sausage poison." *Neurology* 53:1850-1853.

Fahn S, Bressman SB, Marsden CD. (1998). Classification of dystonia, *Adv Neurol* 78:1-10.

Fehlings D, Rang M, Glazier J, Steele C. (2000). An evaluation of botulinum-A toxin injections to improve upper extremity function in children with hemiplegic cerebral palsy, *J Pediatr* 137:331-337.

Finsterer J, Fuchs I, Mamoli B. (1997). Quantitative electromyography-guided botulinum toxin treatment of cervical dystonia, *Clin Neuropharmacol* 20:42-48.

Foster KA. (2002). Novel toxin developments: delivery of endopeptidase activity of botulinum neurotoxin to new target cells. In Brin M, Hallett M, Jankovic J, editors: *Scientific and therapeutic aspects of botulinum toxin*, Philadelphia: Lippincott Williams & Wilkins.

Frueh BR *et al.* (1988). The effect of omitting botulinum toxin from the lower eyelid in blepharospasm treatment, *Am J Ophthamol* 106:45-47.

Gelb DJ, Lowenstein DH, Aminoff MJ. (1989). Controlled trial of botulinum toxin injections in the treatment of spasmodic torticollis, *Neurology* 39:80-84.

Gelb DJ *et al.* (1991). Change in pattern of muscle activity following botulinum toxin injections for torticollis, *Ann Neurol* 29:370-376.

Girlanda P *et al.* (1996). Unilateral injection of botulinum toxin in blepharospasm: single fiber electromyography and blink reflex study, *Mov Disord* 11:27-31.

Goschel H *et al.* (1997). Botulinum A toxin therapy: neutralizing and nonneutralizing antibodies—therapeutic consequences, *Exp Neurol* 147:96-102.

Graham HK. (2000). Botulinum toxin A in cerebral palsy: functional outcomes, *J Pediatr* 137:300-303.

Grandas F, Elston J, Quinn N, Marsden CD. (1988). Blepharospasm, a review of 264 patients, *J Neurol Neurosurg Psychiatry* 51:767-772.

Grandas F, Traba A, Alonso F. (1998). Blink reflex recovery cycle in patients with blepharospasm unilaterally treated with botulinum toxin, *Clin Neuropharmacol* 21:307-311.

Greene P, Fahn S, Diamond B. (1994). Development of resistance to botulinum toxin type A in patients with torticollis, *Mov Disord* 9:213-217.

Greene P, *et al.* (1990). Double-blind, placebo-controlled trial of botulinum toxin injections for the treatment of spasmodic torticollis, *Neurology* 40:1213-1218.

Greene P, Shale H, Fahn S. (1988). Experience with high dosages of anticholinergic and other drugs in the treatment of torsion dystonia, *Adv Neurol* 50:547-556.

Greene PE, Fahn S. (1996). Response to botulinum toxin F in seronegative botulinum toxin A-resistant patients, *Mov Disord* 11:181-184.

Greene PE, Fahn S. (1993). Use of botulinum toxin type F injections to treat torticollis in patients with immunity to botulinum toxin type A, *Mov Disord* 8:479-483.

Gundersen CB. (1980). The effects of botulinum toxin on synthesis, storage and release of acetylcholine, *Prog Neurobiol* 14:99-119.

Hambleton P. (1992). *Clostridium botulinum* toxins: a general review of involvement in disease, structure, mode of action and preparation for clinical use, *J Neurol* 239:16-20.

Hanna PA, Jankovic J, Vincent A. (1999). Comparison of mouse bioassay and immunoprecipitation assay for botulinum toxin antibodies, *J Neurol Neurosurg Psychiatry* 66:612-616.

Hanna PA, Jankovic J. (1998). Mouse bioassay versus Western Blot assay for botulinum toxin antibodies, *Neurology* 50:1624-1629.

Hathaway CL. (1990). Toxigenic clostridia, *Clin Microbiol Rev* 3:66-98.

Holzer VE. (1962). Botulism from inhalation, *Med Klin* 57:1735-1738.

Houser MK, Sheean GL, Lees AJ. (1998). Further studies using higher doses of botulinum toxin type F for torticollis resistant to botulinum toxin type A, *J Neurol Neurosurg Psychiatry* 64:577-580.

Hsiung GY *et al.* (2002). Long-term efficacy of botulinum toxin A in treatment of various movement disorders over a 10-year period, *Mov Disord* 17:1288-1293.

Humeau Y, Doussau F, Grant NJ, Poulain B. (2000). How botulinum and tetanus neurotoxins block neurotransmitter release, *Biochimie* 82:427-446.

Hurvitz EA, Conti GE, Brown SH. (2003). Changes in movement characteristics of the spastic upper extremity after botulinum toxin injection, *Arch Phys Med Rehabil* 84:444-454.

Jahanshahi M, Marsden. (1992). Psychological functioning before and after treatment of torticollis with botulinum toxin, *J Neurol Neurosurg Psychiatry* 229-231.

Jahan Shahi M, Marsden CD. (1992). Psychological functioning before and after treatment of torticollis with botulinum toxin. *J Neurol Neurosurg Psychiatry*. 55: 229-31.

Jankovic J, Brin MF. (1991). Therapeutic uses of botulinum toxin, *N Engl J Med* 324:1186-1194.

Jankovic J, Orman J. (1987). Botulinum A toxin for cranial-cervical dystonia: a double-blind, placebo-controlled study, *Neurology* 37:616-623.

Jankovic J, Orman J. (1984). Blepharospasm: demographic and clinical survey of 250 patients, *Anna Ophthalmol.* 16(4):371-376.

Jankovic J *et al.* (1996). A randomized, double-blind, placebo-controlled study to evaluate botulinum toxin type A in essential hand tremor, *Mov Disord* 11:250-256.

Jankovic J, Schwartz KS. (1993). Longitudinal experience with botulinum toxin injections for treatment of blepharospasm and cervical dystonia, *Neurology* 43:834-836.

Jankovic J. (2002). Botulinum toxin: clinical implications of antigenicity and immunoresistance. In Brin MF, Hallett M, Jankovic J, editors: *Scientific and therapeutic aspects of botulinum toxin*, Philadelphia: Lippincott Williams & Wilkins.

Jankovic J, Vuong KD, Ahsan J. (2003). Comparison of efficacy and immunogenicity of original versus current botulinum toxin in cervical dystonia, *Neurology* 60:1186-1188.

Jankovic J. (2004). Botulinum toxin in clinical practice. *J Neurol Neurosurg Psychiatry* 75: 951-957.

Jankovic J et al. (2004). Evidence-based review of patient reported outcomes with botulinum toxin type A. *Clin Neuropharmacol* (in press).

Jenzer G, Mumenthaler M, Ludin HP, Robert F. (1975). Autonomic dysfunction in botulism B: a clinical report, *Neurology* 25:150-153.

Jitpimolmard S, Tiamkao S, Laopaiboon M. (1998). Long-term results of botulinum toxin type A (Dysport) in the treatment of hemifacial spasm: a report of 175 cases, *J Neurol Neurosurg Psychiatry* 64:751-757.

Jost WH, Kohl A. (2000). Botulinum toxin: evidence-based medicine criteria in rare conditions, *J Neurol* 248(Suppl 1):I39-I44.

Kessler KR, Benecke R. (1997). The EBD test—a clinical test for the detection of antibodies to botulinum toxin type A, *Mov Disord* 12:95-99.

Kessler KR, Skutta M, Benecke R. (1999). Long-term treatment of cervical dystonia with botulinum toxin A: efficacy, safety, and antibody frequency. German Dystonia Study Group, *J Neurol* 246:265-274.

Koller W, Vetere-Overfield B, Gray C, Dubinsky R. (1990). Failure of fixed-dose, fixed muscle injection of botulinum toxin in torticollis, *Clin Neuropharmacol* 13:355-358.

Koman LA, Paterson Smith B, Balkrishnan R. (2003). Spasticity associated with cerebral palsy in children: guidelines for the use of botulinum A toxin, *Paediatr Drugs* 5:11-23.

Lacy DB, Stevens RC. (1999). Sequence homology and structural analysis of the clostridial neurotoxins, *J Mol Biol* 291:1091-1104.

Lamanna C, Eklund HW, McElroy OE. (1946). Botulinum toxin (type A), *H Bacteriol* 73:42-47.

Lamanna C. (1959). The most poisonous poison, *Science* 130:763-772.

Lance JW. (1990). What is spasticity? *Lancet* 335:606.

Lew MF *et al.* (1997). Botulinum toxin type B: a double-blind, placebo-controlled, safety and efficacy study in cervical dystonia, *Neurology* 49:701-707.

Linial M. (1997). SNARE proteins-why so many, why so few? *J Neurochem* 69:1781-1792.

Lorentz IT, Subramaniam SS, Yiannikas C. (1990). Treatment of idiopathic spasmodic torticollis with botulinum-A toxin: a pilot study of 19 patients, *Med J Aust* 152:528-530.

Lorentz IT, Subramaniam SS, Yiannikas C. (1991). Treatment of idiopathic spasmodic torticollis with botulinum toxin A: a double-blind study on twenty three patients, *Mov Disord* 6:145-150.

MacKenzie I, Burnstock G, Dolly JO. (1982). The effects of purified botulinum neurotoxin type A on cholinergic, adrenergic and non-adrenergic, atropine-resistant autonomic neuromuscular transmission, *Neuroscience* 7:997-1006.

Maksymowych AB *et al.* (1999). Pure botulinum neurotoxin is absorbed from the stomach and small intestine and produces peripheral neuromuscular blockade, *Infect Immunol* 67:4708-4712.

Munchau A *et al.* (2001). Spontaneously changing muscular activation pattern in patients with cervical dystonia, *Mov Disord* 16:1091-1097.

Mauriello JA, Aljian J. (1991). Natural history of treatment of facial dyskinesias with botulinum toxin: a study of 50 consecutive patients over seven years, *Br J Ophthalmol* 75:737-739.

Mauriello JA, Coniaris H. (1987). Use of botulinum in the treatment of 100 patients with blepharospasm, *N J Med* 84:43-44.

Mauriello JA *et al.* (1996a). Treatment profile of 239 patients with blepharospasm and Meige syndrome over 11 years, *Br J Ophthalmol* 80:1073-1075.

Mauriello JA, *et al.* (1996b). Treatment choices of 119 patients with hemifacial spasm over 11 years, *Clin Neurol Neurosurg* 98:213-216.

Mauriello JA. (1985). Blepharospasm, Meige syndrome, and hemifacial spasm: treatment with botulinum toxin, *Neurology* 35:1499-500.

Mauriello JA. (2002). The role of botulinum toxin type A (BOTOX®) in the management of blepharospasm and hemifacial spasm. In Brin MF, Jankovic J, Hallett M, editors: *Scientific and therapeutic aspects of botulinum toxin*, Philadelphia: Lippincott Williams & Wilkins.

Merson MH, Dowell VR. (1973). Epidemiologic, clinical and laboratory aspects of wound botulism, *N Engl J Med* 289:1105-1110.

Meyer KF. (1956). The status of botulism as a world health problem, *Bull World Health Organization* 15:281-298.

Mezaki T *et al.* (1999). Combined use of type A and F botulinum toxins for blepharospasm: a double-blind controlled trial, *Mov Disord* 14:1017-1020.

Minton NP. (1995). Molecular genetics of clostridial neurotoxins, *Curr Top Microbiol Immunol* 195:161-194.

Montecucco C, Schiavo G, Rossetto O. (1996). The mechanism of action of tetanus and botulinum neurotoxins, *Arch Toxicol Suppl* 18:342-354.

Montecucco C, Shiavo G. (1993). Tetanus and botulism neurotoxins: a new group of zinc endoproteases, *Trends Biochem Sci* 18:324-327.

Moore AP, Blumhardt LD. (1991). A double blind trial of botulinum toxin "A" in torticollis, with one year follow up, *J Neurol Neurosurg Psychiatry* 54:813-816.

Moreno-Lopez B, Pastor AM, de la Cruz RR, Delgado-Garcia JM. (1997). Dose-dependent, central effects of botulinum neurotoxin type A: a pilot study in the alert behaving cat, *Neurology* 48:456-464.

National Institutes of Health Consensus Development Conference, (1991). Clinical use of botulinum toxin, *Arch Neurol* 48:1294-1298.

Naumann M *et al.* (1998). Depletion of neutralizing antibodies resensitizes a secondary non-responder to botulinum A neurotoxin, *J Neurol Neurosurg Psychiatry* 65:924-927.

Nauman M, Jankovic J. (2004). Safety of botulinum toxin type A: A systematic review and meta-analysis. *Current Medical Research and Opinion* (in press).

Odergren T et al. (1998). A double blind randomized parallel group study to investigate the dose equivalence of Dysport and Botox in the treatment of cervical dystonia, *J Neurol Neurosurg Psychiatry* 64:6-12.

Ostergaard L *et al.* (1996). Quantitative EMG in cervical dystonia, *Electromyogr Clin Neurophysiol* 36:179-185.

Ostergaard L *et al.* (1994). Quantitative EMG in botulinum toxin treatment of cervical dystonia. A double-blind, placebo-controlled study, *Electroencephalogr Clin Neurophysiol* 93:434-439.

Pacchetti C *et al.* (2000). Botulinum toxin treatment for functional disability induced by essential tremor, *Neurol Sci* 20:349-353.

Package insert, Botox, Allergan Pharmaceuticals, Dec 2000.

Package insert, Myobloc™, Elan Pharmaceutical, Dec 2000.

Palace J *et al.* (1998). A radioimmuno-precipitation assay for antibodies to botulinum A, *Neurology* 50:1463-1466.

Park J, Simpson LL. (2003). Inhalational poisoning by botulinum toxin and inhalation vaccination with its heavy-chain component, *Infect Immun* 71(3):1147-1154.

Patrinely JR, Whiting AS, Anderson RL. (1998). Local side effects of botulinum toxin injections, *Adv Neurol* 49:493-500.

Pearce JMS. (1999). A note on the use of botulinum toxin, *J Neurol Neurosurg Psychiatry* 67:230.

Pickett J, Berg B, Chaplin E, Brunstretter-Shafer MA. (1976). Syndrome of botulism in infancy: clinical and electrophysiologic study, *N Engl J Med* 295:770-772.

Pierson SH, Katz DI, Tarsy D. (1996). Botulinum toxin A in the treatment of spasticity: functional implications and patient selection, *Arch Phys Med Rehabil* 77:717-721.

Pittock SJ *et al.* (2003). A double-blind randomized placebo-controlled evaluation of three doses of botulinum toxin type A (Dysport) in the treatment of spastic equinovarus deformity after stroke, *Cerebrovasc Dis* 15:289-300.

Poewe W *et al.* (1998). What is the optimal dose of botulinum toxin A in the treatment of cervical dystonia? Results of a double blind, placebo controlled, dose ranging study using Dysport, *J Neurol Neurosurg Psychiatry* 64:13-17.

Popoff MR, Marvaud JC. (1999). Structural and genomic features of clostridial neurotoxins. In Alouf JE, Freer JH, editors: *The comprehensive sourcebook of bacterial protein toxins*, London: Academic Press.

Racette BA *et al.* (2002). Ptosis as a remote effect of therapeutic botulinum type B injections, *Neurology* 59:1445-1447.

Rollnik JD *et al.* (2000). Treatment of tension-type headache with botulinum toxin type A: a double-blind, placebo-controlled study, *Headache* 40:300-305.

Rossetto O *et al.* (2002). The metalloprotease activity of tetanus and botulinum neurotoxins. In Brin M, Hallett M, Jankovic J, editors: *Scientific and therapeutic aspects of botulinum toxin*, Philadelphia: Lippincott Williams & Wilkins.

Sakaguchi G. (1983). *Clostridium botulinum* toxins, *Pharm Ther* 19:165-194.

Schantz EJ, Johnson EA. (1997). Botulinum toxin: the story of its development for the treatment of human disease, *Perspect Biol Med* 40:317-327.

Schiavo G, Matteoli M, Montecucco C. (2000). Neurotoxins affecting neuroexocytosis, *Physiol Rev* 80:717-766.

Schiavo G *et al.* (1993a). Identification of the nerve terminal targets of botulinum neurotoxin serotypes A,D, and E, *J Biol Chem* 268:23784-23787.

Schiavo G *et al.* (1993b). Botulinum neurotoxin serotype F is a zinc endopeptidase specific for VAMP/synaptobrevin, *J Biol Chem* 268:11516-11519.

Scott AB, Kennedy RA, Stubbs HA. (1985). Botulinum toxin injection as a treatment for blepharospasm, *Arch Ophthamol* 103:347-350.

Scott AB, Rosenbaum A, Collins CC. (1973). Pharmacologic weakening of extraocular muscles, *Invest Ophthalmol Vis Sci* 12:924-927.

Scott AB. (1980). Botulinum toxin injection into extraocular muscles as an alternative to strabismus surgery, *J Pediatr Ophthalmol Strabismus* 17:21-25.

Shapiro RL, Hathaway C, Swerdlow DL. (1998). Botulism in the United States: a clinical and epidemiologic review, *Ann Intern Med* 129:221-228.

Sheean GL, Lees AJ. (1995). Botulinum toxin F in the treatment of torticollis clinically resistant to Botulinum toxin A, *J Neurol Neurosurg Psychiatry* 59:601-607.

Silberstein S, Mathew N, Saper J, Jenkins S. (2000). For the BOTOX Migraine Clinical Research Group. Botulinum toxin type A as a migraine preventive treatment, *Headache* 40(6):445-450.

Simpson LL. (1980). Kinetic studies on the interaction between botulinum toxin type A and the cholinergic neuromuscular junction, *J Pharmacol Exp Ther* 212:16-21.

Simpson LL. (1981). The origin, structure and pharmacologic activity of botulinum toxin, *Pharmacol Rev* 33:155-188.

Simpson LL, Coffield JA, Bakry N. (1993). Chelation of zinc antagonizes the neuromuscular blocking properties of the seven serotypes of botulinum neurotoxin as well as tetanus toxin, *J Pharmacol Exp Ther* 267:720-727.

Simpson LL, Maksymowych AN, Kiyatkin N. (1999). Botulinum toxin as a carrier for oral vaccines, *Cell Mol Life Sci* 56:47-61.

Simpson LL. (2000). Identification of the characteristics that underlie botulinum toxin potency: implications for designing novel drugs, *Biochimie* 82:943-953.

Simpson LL, Maksymowych AB, Hao S. (2001). The role of zinc binding in the biological activity of botulinum toxin, *J Biol Chem* 276:27034-27041.

Simpson DM. (2002). Clinical trials of botulinum toxin in the treatment of spasticity. In Mayer NH, Simpson DM, editors: *Spasticity: etiology, evaluation, management and the role of botulinum toxin.* New York: WE MOVE.

Simpson L. (2004). Identification of the major steps in botulinum toxin action, *Annu Rev Pharmacol Toxicol* 44:167-193.

Slawek J, Klimont L. (2003). Functional improvement in cerebral palsy patients treated with botulinum toxin A injections—preliminary results, *Eur J Neurol* 10:313-317.

Sloop RR, Cole BA, Escutin RO. (1997). Human response to botulinum toxin injection: Type B compared with type A, *Neurology* 49:189-194.

Snow BJ *et al.* (1990). Treatment of spasticity with botulinum toxin: a double-blind study, *Ann Neurol* 28:512-515.

Sollner T, Rothman JE. (1994). Neurotransmission: harnessing fusion machinery at the synapse, *Trends Neurosci* 17:344-348.

Speelman JD, Brans JW. (1995). Cervical dystonia and botulinum treatment: is electromyographic guidance necessary? *Mov Disord* 10:802.

Stell R, Thompson PD, Marsden CD. (1988). Botulinum toxin in spasmodic torticollis, *J Neurol Neurosurg Psychiatry* 51:920-923.

Sugiyama H. (1980). *Clostridium botulinum* neurotoxin, *Microbiol Rev* 44:419-448.

Sutherland DH, Kaufman KR, Wyatt MP, Chambers HG. (1996). Injection of botulinum A toxin into the gastrocnemius muscle of patients with cerebral palsy: a 3-dimensional motion analysis study, *Gait Posture* 4:269-279.

Tarsy D, Bhattacharyya N, Borodic G. (2000). Myasthenia gravis after botulinum toxin A for Meige syndrome, *Mov Disord* 15:736-738.

Therapeutics and Technology Assessment Subcommittee of the American Academy of Neurology. (1990). Assessment: the clinical usefulness of botulinum toxin-A in treating neurologic disorders, *Neurology* 40:1332-1336.

Truong DD *et al.* (1997). BotB (botulinum toxin type B): evaluation of safety and tolerability in botulinum toxin type A-resistant cervical dystonia patients (preliminary study), *Mov Disord* 12:772-775.

Tsui JK *et al.* (1985). A pilot study on the use of botulinum toxin in spasmodic torticollis, *Can J Neurol Sci* 12:314-316.

Tsui JK *et al.* (1986). Double-blind study of botulinum toxin in spasmodic torticollis, *Lancet* 245-246.

Valls-Sole J, Tolosa ES, Ribera G. (1991). Neurophysiological observations on the effects of botulinum toxin treatment in patients with dystonic blepharospasm, *J Neurol Neurosurg Psychiatry* 54:310-313.

Van den Bergh P, De Beukelaer M, Deconinck N. (1996). Effect of muscle denervation on the expression of substance P in the ventral raphe-spinal pathway of the rat, *Brain Res* 707:206-212.

Van Gerpen JA *et al.* (2000). Utility of an EMG mapping study in treating cervical dystonia, *Muscle Nerve* 23:1752-1756.

van Kuijk AA, Geurts AC, Bevaart BJ, van Limbeek J. (2002). Treatment of upper extremity spasticity in stroke patients by focal neuronal or neuromuscular blockade: a systematic review of the literature, *J Rehabil Med* 34:51-61.

Wainwright RB *et al.* (1988). Food-borne botulism in Alaska, 1947-1985: epidemiology and clinical findings, *J Infect Dis* 157(6):1158-1162.

Walton M. (1999). PLA 91-0184. Botox (Botulinum toxin type A) for the treatment of cervical dystonia—clinical review, http:www.fda.gov

Warrick P. (2000). Botulinum toxin for essential tremor of the voice with multiple anatomical sites of tremor: a crossover design study of unilateral versus unilateral injection, *Laryngoscope* 110:1266-1374.

Wiegand J, Erdmann G, Wellhoner HH. (1976). 125I-labelled botulinum A neurotoxin: pharmacokinetics in cats after intramuscular injection, *Arch Pharmacol* 292:161-165.

Wutthiphan S *et al.* (1997). Diplopia following subcutaneous injections of botulinum A toxin for facial spasms, *J Pediatr Ophthalmol Strabismus* 34:229-234.

8

Prospects for Slowing the Progression of Parkinson's Disease

Kenneth Marek, MD

Parkinson's disease (PD) is a slow, but inexorably progressive neurodegenerative disorder resulting in severe disability for people with the disease, unrelenting hardship for family members and caregivers, and tremendous cost to society at large (Dodel *et al.*, 2001; Guttman *et al.*, 2003). PD affects approximately 1 million people in North America, with an increasing incidence with age most marked after 60 (Bower *et al.*, 1999; Van Den Eeden *et al.*, 2003). The defining features of PD, resting tremor, rigidity, and bradykinesia or motor slowing, are generally well controlled with medications for several years (Lang and Lozano, 1998). However, most often within a decade of disease onset, progressive symptoms of disease such as freezing, falling, and dementia, and drug-induced side effects such as motor complications substantially reduce quality of life for PD patients (Olanow, 2003). Therefore the most important goal for PD therapeutics is the development of effective and enduring neuroprotective therapy to slow or stop disease early in its course or ideally even before the onset of symptoms during the preclinical disease phase.

During the past several decades molecular neuroscience and genetics have markedly increased our understanding of PD pathobiology and have dramatically accelerated development of PD therapeutics. PD has served as a model for neurodegenerative disorders in that advances in neuroscience have informed clinical therapeutics and in turn have been informed by clinical research. For example, PD is characterized by degeneration of the dopaminergic neurons in the substantia nigra pars compacta (Bernheimer *et al.*, 1973), and the mainstay of PD therapy remains dopamine replacement (Agid *et al.*, 1999; Ahlskog, 2001; Olanow *et al.*,

2001). Clinical studies in PD therapeutics continue to elucidate dopamine neurobiology and pharmacology. Conversely, neuropathological and neurochemical studies demonstrating the selective vulnerability of dopaminergic and other subcortical neurons in PD have led to potential therapies to replace those cells such as neurotransplantation for PD (Freed *et al.*, 2001; Olanow *et al.*, 2003a). Most recently basic and clinical scientists have each contributed to the challenge of neuroprotective therapy for PD, on the one hand identifying numerous targets for potential therapies and on the other hand working to define the clinical studies to demonstrate neuroprotection.

Neuroprotection for PD may be defined as a meaningful and enduring slowing in the progressive disability of PD associated with a slowing in the neuronal degeneration known to occur in PD (Figure 8.1). Neurorescue or neurorestoration refers to improvement in symptoms of PD associated with partial or complete restoration of neuronal function (Olanow *et al.*, 2003b; Shoulson, 1998c). Neuroprotection may have different meanings or at least the meaning may have a different emphasis for the various constituencies with a stake in PD research. For patients and families of patients, neuroprotection means a longer period of better quality of life. For clinicians, neuroprotection may mean slowed worsening of clinical symptoms, delay of disease milestones such as falling or dementia, and longer duration of effective treatment with symptomatic medications. For neuroscientists, neuroprotection is defined by slowing in the neuronal degeneration characteristic of PD. For regulatory agencies, neuroprotection may mean demonstrating slowed worsening of symptoms not explained by a symptomatic effect, associated with slowed loss of neurons. Studies assessing neuroprotection must balance the requirements of these different constituencies with what is practical in clinical research. This review focuses on recent studies evaluating disease-modifying drugs for PD and in particular the challenge of assessing and demonstrating neuroprotection.

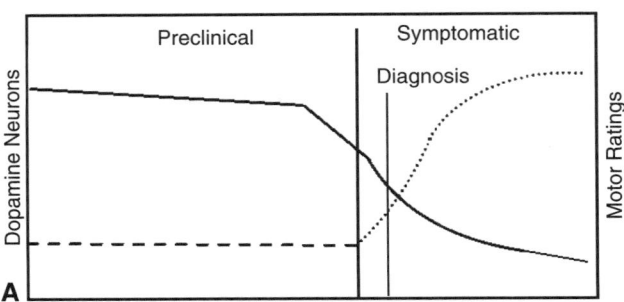

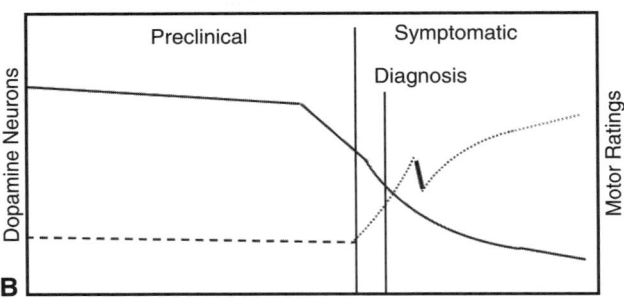

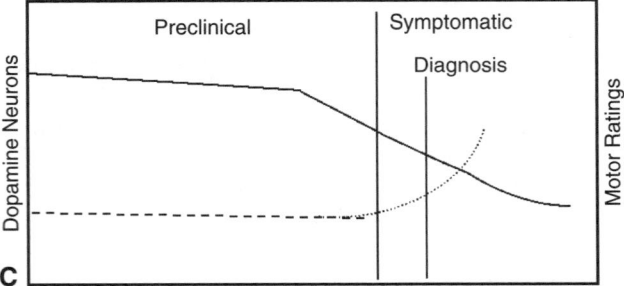

FIGURE 8.1 Model for progression loss of neuronal function in PD and development of clinical symptoms. Note that there is a prolonged period during which loss of neuronal function has occurred but symptoms had not yet appeared followed by diagnosis and subsequent disease progression. **A,** PD is modeled in untreated subjects, assuming symptomatic therapy with improvement in symptoms, **B,** and assuming neuroprotective therapy with slowed loss of neuronal degeneration and slowed development of clinical symptoms, **C.**

NEUROSCIENCE TARGETS AND ANIMAL MODELS

Molecular neuroscience has transformed the development of potential neuroprotective or disease-modifying drugs. Research has identified and has begun to elucidate numerous molecular targets and biochemical pathways potentially relevant to neurodegeneration and disease progression (Dawson and Dawson, 2002; Mandel *et al.*, 2003). These strategies, summarized briefly next, are directed at both dopaminergic degeneration, in particular, and neuronal degeneration, in general (Table 8.1).

Molecular genetic studies have been especially useful both in identifying genes that may be causative or associated with PD and in uncovering functional mechanisms subserved by products of those genes that may provide targets for neuroprotective drugs. Mutations in a-synuclein and parkin have both resulted in defects in the ubiqutin-proteasome system (UPS), now thought critical to neuronal degeneration (McNaught *et al.*, 2003; Miller and Wilson, 2003; Skipper and Farrer, 2002) and providing new targets for therapeutic intervention. Potential targets include drugs to reduce abnormal protein folding or vaccine strategies to reduce synuclein deposition. Further elucidation of some of the specific pathways leading to cell death in PD and related disorders has also provided several sites for pharmacological intervention. Specifically, drugs that inhibit the caspase cascade, thereby potentially slowing cell death mechanisms including CEP1347 and related compounds, are being evaluated in ongoing clinical trials (Harris *et al.*, 2002; Maroney *et al.*, 2001). The propargylamines, compounds including selegiline, rasagiline, and TCH346, which may reduce apoptotic cell death, are also being studied (Andringa *et al.*, 2003; Maruyama *et al.*, 2002; Parkinson Study Group, 2002a; Tatton *et al.*, 2002; Youdim *et al.*, 2003). Reducing potential glutamatergic excititoxicity is yet another approach to slowing cell death (Greenamyre *et al.*, 1992; Spooren *et al.*, 2001). Relatively weak glutamatergic drugs, including ramacemide and rulizole, have

TABLE 8.1 Potential Disease Modifying Targets and Drugs

Targets Pathways	Drugs
Antioxidants	Co-Q10, dopamine agonists
Growth factors	GDNF, immunophilin ligands
Apoptosis	Propargylamines, dopamine agonists
Caspase-inhibitors	MLK inhibitors
Glutamatergic agents	Receptor modulators
Mitochondrial drugs	Co-Q10
Inflammation	Nonsteroidal anti-inflammatory

been tested in PD but have not shown substantial clinical benefit for PD patients or any evidence of neuroprotection (Shoulson *et al.*, 2001). More recently modulators of glutamate receptors have been proposed as drugs that may modify the progression of PD cell death (Marino *et al.*, 2003). There is substantial evidence that mitochondrial function is abnormal in PD, specifically a defect in complex I in the electron transport chain, and that this bioenergetic deficit contributes to degeneration (Beal, 2003; Greenamyre *et al.*, 2001; Schapira *et al.*, 1998). Both laboratory and clinical studies have suggested that drugs designed to bypass complex I to restore mitochondrial function may slow the progression of PD (Shults and Schapira, 2001). Growing evidence has also indicated that PD may be associated with an inflammatory response in the brain marked by activated microglial cells (Hirsch *et al.*, 2003; McGeer *et al.*, 2001). This provides yet another potential therapeutic approach to slowing the degeneration of PD—using drugs aimed at reducing this inflammatory response as has been reported in Alzheimer's disease (Etminan *et al.*, 2003).

Growth factors and trophic factors may also slow loss of dopaminergic neurons and/or restore dopaminergic neuron function. In particular glial-derived neurotrophic factor (GDNF) appears to have specific effects on dopaminergic neurons (Brundin, 2002; Ugarte *et al.*, 2003), and studies of GDNF administered by surgical infusion are under way in PD patients (Gill *et al.*, 2003). In recent studies other factors such as stress or exercise appear to activate pathways leading to increased GDNF, providing another approach for slowing of PD progression (Cohen *et al.*, 2003). Other trophic factors including neurturin and sonic hedgehog may also have protective effects on dopamine neurons (Kim *et al.*, 2003; Oiwa *et al.*, 2002). Neuroimmunophilin ligands, small molecule growth factors, also have demonstrated both *in vitro* and *in vivo* changes on dopamine neurons, possibly consistent with neuronal sprouting (Gold and Villafranca, 2003; Zhang *et al.*, 2001). These orally active drugs have been well tolerated in PD patients, and clinical studies to assess the effects of neuroimmunophilin ligands in PD progression are in progress (Seibyl *et al.*, 2002).

Several recent studies have examined the potential disease-modifying effects of dopaminergic drugs, commonly used to treat PD for their symptomatic benefit. Preclinical and animal data have provided a clear rationale to consider whether levodopa, dopamine agonists, and monoamine oxidase inhibitors might influence PD progression. Levodopa, the mainstay of PD symptomatic treatment, has been postulated to have both neuroprotective and neurotoxic effects (Agid, 1998; Fahn, 1997). Recent studies have shown that levodopa-induced oxidative stress may hasten neurodegeneration and specifically may increase synuclein deposition *in vitro*

(Asanuma *et al.*, 2003; Conway *et al.*, 2001), but no evidence for levodopa toxicity has been found *in vivo* (Datla *et al.*, 2001; Murer *et al.*, 1998; Mytilineou *et al.*, 2003). Extensive *in vitro* and animal data suggest that dopamine agonists may slow neuronal degeneration via both dopamine receptor mediated and non-dopamine receptor-mediated mechanisms (Olanow *et al.*, 1998; Schapira, 2002). Clinical studies based on this rationale are discussed in detail later in this chapter. Monoamine oxidase B (MAO-B) inhibitors, including selegiline, were initially considered potentially neuroprotective because of their antioxidant effects, but as already mentioned, recent preclinical data suggest that these drugs with a propargyline structure may have a more direct effect on slowing apoptosis. The dopamine transporter (DAT), a recent target for symptomatic drugs for PD, is yet another target of potential neuroprotective medication. Studies show that DAT binding at the dopamine cell body may worsen neuronal degeneration (Blakely, 2001).

Translation of neuroscience targets into potential drugs requires informative *in vivo* models of disease. Although there is no adequate animal model for progressive PD, two toxic models of dopaminergic neuronal loss, MPTP and 6OH dopamine, have been crucial in assessing potential PD drugs. The most important animal model for PD has been the MPTP model, providing a primate model of selective dopaminergic cell loss. MPTP administration, both in primates and in rodents, has been widely used to test potential neuroprotective drugs (Jenner, 2003; Wichmann and DeLong, 2003). Although recovery from the behavioral effects of MPTP is a reliable predictor of symptomatic improvement in humans, it remains unclear whether recovery from an MPTP lesion will predict changes in PD progression. The loss of nigrostriatal neurons after treatment with 6OH dopamine in rats and the characteristic turning response after challenge with dopaminergic medications have also been widely used to understand mechanisms of dopaminergic cell death and potential interventions. The dopaminergic cell loss associated with rotenone, a common pesticide, has provided an alternative toxic model and appears to mediate its effect via a mitochondrial mechanism (Perier *et al.*, 2003; Sherer *et al.*, 2003). Invertebrate models of PD have also been developed. For example, *C. elegans* has provided a model to assess dopaminergic toxicity (Nass *et al.*, 2002). More recently, characterization of the genetic abnormalities associated with familial PD has focused attention on the need for models of synuclein and parkin overexpression. Drosophila models of synuclein and parkin mutations have been recently established and can be used to examine the pathologic consequences of these proteins (Feany and Bender, 2000; Greene *et al.*, 2003). Transgenic mice with a synuclein mutation have been developed and appear to demonstrate pathological features of PD and transgenics with impairment of the ubiquitin/proteasome system are also under study (Giasson et al., 2002; Lindsten et al., 2003).

CLINICAL STUDIES OF PD PROGRESSION

Many clinical studies have been designed to test drugs that may slow PD progression and/or restore neuronal function. In response to a recent request from the National Institutes of Health, 59 compounds were suggested by clinicians and researchers as possibly neuroprotective (Ravina et al., 2003). Many more drugs are in early or preclinical development by the pharmaceutical industry. Although numerous candidate drugs exist, the most effective study design and clinical endpoints to evaluate the effect of these drugs on disease progression remain uncertain. Several clinical endpoints have been used to assess progressive functional decline in PD. The most appropriate endpoint depends on the study question and in particular on the study population, the duration of the study follow-up period, the potential symptomatic effect of the study drug, and the effect of concomitant medications. The most commonly used clinical endpoint has been the change during the study period in the widely used clinical scale, the Unified Parkinson disease rating scale (UPDRS) (Fahn et al., 1987). Alternative clinical endpoints depend on the time to a disease milestone such as need for therapy, development of motor fluctuations, or a disabling feature of disease such as falls, hallucinations or dementia, or change in UPDRS after a drug washout (Fahn, 1999; Parkinson Study Group, 1993, 2000a; Rascol et al., 2000). Mixed endpoints combining disability scales and disease milestones have also been used (Shults et al., 2002). More recently, quality of life (QOL) measures have been added to the assessment of disease progression, but there is no uniformly accepted PD QOL scale (Welsh et al., 2003). Pharmacoeconomic measures have also been included to assess the impact of potential neuroprotection (Siderowf et al., 2000).

Caveats in Clinical Study Design

Identifying a clinical endpoint that reliably reflects PD progression and that can be reasonably and practically used in clinical studies remains challenging. Clinical studies of potential disease-modifying drugs are limited by the slow progression and marked variability in progression of the clinical features of PD. Variability in disease progression is likely explained by both the typical individual variability among a population in any disease phenotype and, in particular, the variability in individuals'

ability to compensate, using as yet unknown mechanisms, for the loss of dopaminergic neurons. The variability among PD patients also may be explained if, as is likely, currently defined PD is really several related disease entities with similar symptoms. The best current evidence that PD is really Parkinson syndromes are the several genetic parkinsonisms caused by recently identified different mutations (Gasser, 2003). For example, recent studies of families with the parkin mutation and with Park 7 mutation suggest a slow progression compared to idiopathic PD (Dekker et al., 2003; Khan et al., 2002).

The slow progression of PD presents the most difficult dilemma for clinical study design of neuroprotective PD drugs. The magnitude of the effect of a neuroprotective drug may be modest, but still clinically valuable. It is widely agreed among clinicians that a reduction in disease progression of greater than 25% would clearly be clinically important. However, given the slow and varied progression of PD, it would require a large sample size and long duration (typically at least 18 months) of follow-up time to show a 25% neuroprotective effect (Marek et al., 2003a). Typically during the study period, patients require initiation or modification of symptomatic medications to treat slowly worsening disability. The availability of effective symptomatic therapy for PD is a blessing for patients, but the remarkable symptomatic benefit of these drugs creates a serious confound when using clinical symptoms and scales to assess disease progression. For example the progressive disability of PD, as measured by the UPDRS, differs greatly at different study stages. In early untreated PD, the UPDRS scale increases at approximately 10 to 12 units/year in a linear manner (Parkinson Study Group, 1989, 1993, 2002a). It is generally possible to maintain a newly diagnosed PD patient without symptomatic medications for about 9 months (Parkinson Study Group, 1989, 2003). When dopaminergic treatment is initiated, the UPDRS is decreased by 8 to 13 units, and then once the patient has been treated, it is slowly increased by 1 to 2 units/year (Jankovic and Kapadia, 2001; Parkinson Study Group, 2000b). The symptomatic effect of commonly used drugs may overwhelm a more modest but ultimately more enduring protective effect.

The confound from symptomatic medications is further complicated when the potential neuroprotective drug also has symptomatic effects (Parkinson Study Group, 1993). Several approaches have been used to overcome the potential confound of symptomatic treatment in assessing disease-modifying drugs. The simplest approaches are either to use endpoints that occur before symptomatic therapy, such as time to require initial treatment or to wash out the symptomatic drug at the end of the study, but this has proved difficult because of both practical clinical needs of patients and unexpectedly long duration effects of medication (Hauser and Holford, 2002; Parkinson Study Group, 2003; Shults et al., 2002; Whone et al., 2002). Alternatively, endpoints independent of symptomatic benefit might be used, including those unaffected by the dopaminergic treatment (e.g., falls, dementia) (Kieburtz, 2003). Randomized start designs, in which both the placebo and treatment groups are eventually treated with the study drug but the duration of treatment differs between groups, have also been suggested to mitigate any symptomatic effect of the study drug (Leber, 1997).

THE ROLE OF BIOMARKERS IN STUDIES OF PD PROGRESSION

The slow and variable clinical progression of PD, coupled with the confounding effects of ultimately necessary and effective symptomatic treatment, has fostered the development of more objective clinical tools to assess PD progression and therefore potential PD neuroprotective drugs. These tools are biomarkers for neurodegeneration and/or for clinical manifestations of PD. Biomarkers are broadly defined as characteristics that are objectively measured and evaluated as indicators of normal biological processes, pathogenic processes, or pharmacological responses to a therapeutic intervention (Biomarkers Definitions Working Group, 2001). Given the likely multiple etiologies and pathological processes (e.g., protein aggregation, inflammation, oxidative stress) underlying neuronal degeneration in PD and the clear heterogeneity in the expression and progression of the clinical manifestations of PD, several biomarkers will likely be necessary to fully assess disease progression. Although biomarkers may be extraordinarily useful and new technologies very seductive, a number of caveats must be considered. First the performance characteristics of the marker must be established in the subject population under study. Second, the marker should be meaningful or relevant to the disease process. Third, it should be clear to what extent the marker is generalizable beyond a specific study population. Finally, the effect of factors such as age, medications, or environment on the biomarker must be clarified (DeKosky and Marek, 2003; Frank and Hargreaves, 2003). Current biomarkers for PD do not fully meet these criteria, and studies to further validate these markers are necessary (Brooks et al., 2003a). Yet, even as identification and validation of these tools is ongoing, biomarkers have already played an important role in clinical studies of PD progression.

Major motivations to use biomarkers in studies of PD progression are to shorten the study duration, reduce the study sample size and/or to provide information otherwise unavailable from clinical endpoints. Biomarkers for PD may be divided into four main categories: genetic, imaging, clinical, and biochemical. For example, genetic PD markers ultimately may be critical in clinical studies in selecting subjects and stratifying outcomes, providing information that cannot be obtained from clinical evaluation. Simple biochemical markers for PD degeneration are not yet available, but newer array-based technologies, including proteomics, transcriptomics, and metabolomics, offer the promise to target biochemical deficits in PD more effectively to identify serum and CSF markers of disease. Currently, imaging biomarkers have emerged as the most mature and widely used tools to assess PD progression and the effects of potential disease-modifying drugs in clinical studies. Imaging biomarkers provide a window to the degeneration of dopaminergic neurons and the potential to test the effects of drugs on the progressive loss of dopamine neuron function longitudinally in a living PD patient (Brooks et al., 2003a; Marek and Seibyl, 2000).

Imaging Biomarkers

During the past decade in vivo neuroreceptor imaging targeting nigral dopaminergic degeneration has been used to monitor onset, severity, and progression of PD. Recent advances in radiopharmaceutical development, imaging detector technologies, and image analysis software have expanded and accelerated the role of imaging in clinical research in PD in general, and neurotherapeutics for disease modifying drugs in particular. Several markers, most focused on the dopamine system, using single photon emission computed tomography (SPECT) and positron emission tomography (PET) technology have been used to assess PD. The most widely studied imaging biomarkers to assess PD progression have targeted presynaptic nigrostriatal function. Specifically, these ligands include PET/[18F]Dopa, which measures conversion of F-dopa to F-dopamine and therefore dopamine function; PET/[11C]VMAT2 (dihydrotetrabenazine), which tags the vesicular dopamine transporter; and several ligands including SPECT/[123I]B-CIT (2β-carboxymethoxy-3β(4-iodophenyl)tropane), [123I] FP-CIT, [123I] IPT, [99Tc]TRODAT-1, and 18F-CFT, which bind to the DAT (Figure 8.2; see also the Color Plate section) (Asenbaum et al., 1997; Booij et al., 1998; Brooks et al., 1990; Eidelberg et al., 1995; Frey et al., 1996; Innis et al., 1993; Mozley et al., 2000; Vingerhoets et al., 1994).

The properties of the radiopharmaceutical agent used to target the dopamine system is the most critical issue in developing a useful imaging tool for PD. Some of the key steps in development of a potential radioligand include assessment of the brain penetration of the radioligand, selectivity of the radioligand for the target site, binding properties of the radioligand to the site, and the metabolic

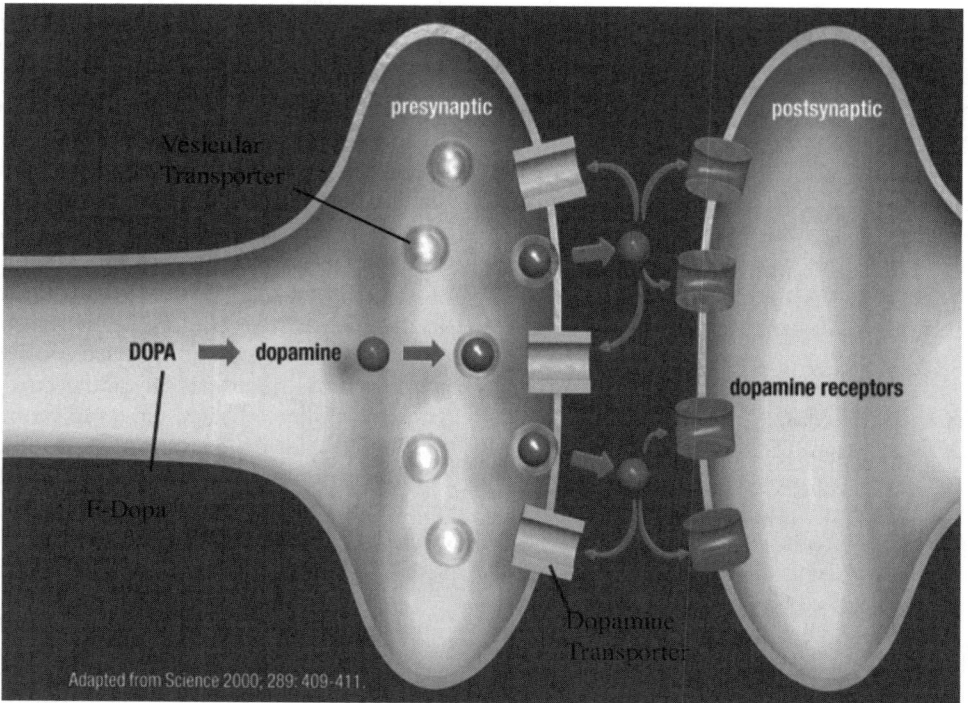

FIGURE 8.2 Dopamine synapse showing the presynaptic dopamine imaging targets (Marek and Seibyl, 2000). (See also the Color Plate section.)

fate of the radioligand. These properties help to determine the signal/noise ratio of the ligand and the ease of quantitation of the imaging signal. The quantitation of the signal is essential to use these imaging markers as clinical research tools.

Both PET (also called dual photon emission tomography) and SPECT are sensitive methods of measuring *in vivo* neurochemistry (Lassen and Holm, 1992; Phelps, 1992). The choice of imaging modality is ultimately determined by the specific study questions and study design. Although in general PET cameras have better resolution than SPECT cameras, SPECT studies may be technologically and clinically more feasible, particularly for the large clinical studies requiring rapid patient accrual and necessary to assess PD progression. PET studies may benefit from greater flexibility in the range of radiopharmaceuticals that can be tested, but SPECT studies have the advantage of longer half-life radiopharmaceuticals necessary for some studies.

Dopamine ligands are useful to assess PD insofar as they reflect the ongoing dopaminergic degeneration in PD. In the study most directly correlating changes in dopamine neuronal numbers and imaging outcomes, there is good correlation between dopamine neuron loss and [18F]DOPA uptake, although conclusions are limited by a very small sample size of only five subjects (Snow *et al.*, 1993). Numerous other studies have shown that the DAT is reduced in striatum in postmortem brain

from PD patients (Kaufman and Madras, 1991; Niznik *et al.*, 1991; Wilson *et al.*, 1996). In turn, numerous clinical imaging studies have shown reductions in [18F]DOPA and [C11]VMAT2 and DAT ligands uptake in PD patients and aging healthy subjects consistent with the expected pathology of PD and of normal aging. Specifically, these imaging studies show asymmetrical, putamen>caudate loss of dopaminergic uptake, and the imaging loss correlates with worsening clinical symptoms in cross-sectional evaluation (Booij *et al.*, 1997; Fischman *et al.*, 1998; Frey *et al.*, 1996; Huang *et al.*, 2001; Sawle *et al.*, 1994; Seibyl *et al.*, 1995). In addition DAT ligands show reductions in activity with normal aging (Mozley *et al.*, 1999; van Dyck *et al.*, 2002; Volkow *et al.*, 1994).

Monitoring PD Progression

In several studies, neuroreceptor imaging of the nigrostriatal dopaminergic system has been further developed as a research tool to monitor progressive dopaminergic neuron loss in PD. In longitudinal studies of PD progression [18F]DOPA, [11C]VMAT2, and dopamine transporter imaging (β-CIT and CFT) using both PET and SPECT have shown an annualized rate of reduction in striatal [18F]DOPA, [11C]VMAT2, [18F]CFT, or [123I]β-CIT uptake of about 4% to 13% in PD patients compared with 0% to 2.5% change in healthy control subjects (Brooks *et al.*, 2003b; Marek *et al.*, 2001;

Morrish *et al.*, 1998; Nurmi *et al.*, 2000b, 2001; Pirker *et al.*, 2002). Evidence from studies of hemi-PD subjects provides further insight into the rate of progression of disease. In early hemi-PD there is a reduction in [18F]DOPA, [C11]VMAT2, and DAT uptake of about 50% in the affected putamen and of 25% to 30% in the unaffected putamen. As most patients will progress clinically from unilateral to bilateral in 3 to 6 years, it is likely that the loss of these *in vivo* imaging markers of dopaminergic degeneration in the previously unaffected putamen will progress at about 5% to 10% per year (Booij *et al.*, 1998; Guttman *et al.*, 1997; Marek *et al.*, 1996; Sawle *et al.*, 1994).

Imaging studies assessing progression of disease have provided data to estimate sample sizes required to detect slowing of disease progression resulting from study drug treatment. The sample size required depends on the effect size of the disease-modifying drug and the duration of exposure to the drug. The effect of the drug is generally expressed as the percent reduction in rate of loss of the imaging marker in the group treated with the study drug vs a comparison or control group. More specifically, imaging studies have sought a reduction of between 25% and 50% in the rate of loss of [18F]DOPA, [C11]VMAT2, or [123I]β-CIT uptake (i.e., a reduction from 10% per year to 5% to 7.5% per year). The sample size needed to detect a 25% to 50% reduction in the rate of loss of F-dopa or β-CIT uptake during a 24-month interval ranges from approximately 30 to 150 research subjects in each study arm (Brooks *et al.*, 2003b). These estimates may be influenced substantially by the stage of PD and other symptomatic medications allowed in the study.

These data and the sample size estimates support the use of dopamine neuroreceptor imaging to assess the effects of potential neuroprotective drugs in PD, but there are several limitations in the study design and interpretation of these studies.

1. Imaging outcomes in studies of PD patients are biomarkers for brain activity, but are not true surrogates for drug effects or clinical outcomes in PD patients (De Gruttola *et al.*, 2001). Although numerous cross-sectional studies have shown a clear correlation between imaging measures such as [123I]β-CIT and [18F]Dopa and clinical outcomes such as the UPDRS, longitudinal studies have shown only minimal correlation at best between the change in imaging outcomes and the change in clinical outcomes (Brooks *et al.*, 2003b; Marek *et al.*, 2001; Morrish *et al.*, 1996). There are several explanations for the lack of correlation between [123I]β-CIT or [18F]DOPA uptake and UPDRS in longitudinal studies. First, the UPDRS is confounded by the effects of the patient's anti-Parkinson medica-

tions, both acutely after initiating therapy and with ongoing treatment. Even evaluation of the UPDRS in the "defined off" state (12 hours after last medication) or after prolonged washout does not eliminate the long duration symptomatic effects of these treatments (Hauser and Holford, 2002; Nutt and Holford, 1996). Second, the temporal patterns for rate of loss of DAT and the change in UPDRS may not be congruent. This is best illustrated in early PD by data demonstrating a loss of approximately 40% to 60% of striatal [123I]β-CIT or [18F]DOPA uptake at the time of diagnosis when clinical symptoms measured by the UPDRS may be minimal. This preclinical period is estimated to last 3 to 12 years based on back-extrapolating imaging data to calculate the time interval between start of dopamine degeneration and start of symptoms (Marek, 1999; Morrish *et al.*, 1996). Later in the disease course, after treatment with dopaminergic drugs, the UPDRS changes very slowly, 1.34/year in the "on" and 1.58/year in the "off" state in a recent report from a large clinic, differing from the 10 point/year change in untreated patients (Jankovic and Kapadia, 2001; Parkinson Study Group, 1989, 2002a). This slow change in UPDRS during the treatment phase of PD reflects treatment to mitigate disability, but imaging during this period shows the expected continued loss of [123I]β-CIT or [18F]DOPA uptake, again indicating that clinical and imaging outcomes provide complementary, but not necessarily correlative, data. Third, the UPDRS is a measure of dopaminergic and non-dopaminergic symptoms and is vulnerable to both patient and evaluator subjectivity. Fourth, both imaging outcomes and UPDRS may not progress linearly. Evidence from longitudinal imaging studies suggests a relatively linear progression for the initial 5 years after diagnosis, but that later in disease the rate of loss of [123I]β-CIT uptake may slow (Jennings *et al.*, 2000; Pirker *et al.*, 2000).

2. Like clinical outcomes, imaging outcomes of disease progression may be confounded by pharmacological effects of dopaminergic drugs used to treat PD symptoms and/or by other study drugs. Although most preclinical studies evaluating the regulation of imaging outcomes by dopamine agonists and antagonists and levodopa have not shown a consistent effect of these drugs, some studies have shown modulation of the DAT, the vesicular transporter, and dopamine turnover (Gnanalingham and Robertson, 1994; Moody *et al.*, 1996; Truong *et al.*, 2003; Vander Borght *et al.*, 1995; Zigmond *et al.*, 1990). These data raise concerns, although the relevance of these preclinical studies to human imaging studies is questionable because of the brief exposure time to drugs, suprapharmacological dosing, and species differences. In one of the few clinical

studies comparing imaging ligands within subjects, 35 PD patients and 16 age-matched control subjects imaged with [11C]methylphenidate (a dopamine transporter ligand), [18F]Dopa, and [11C]VMAT2 showed reduction in DAT > vesicular transporter > F-dopa uptake. These data suggest that differential regulation of these imaging targets might occur in a progressively denervated striatum (Lee *et al.*, 2000). Other studies have more directly assessed the potential short-term regulation of imaging ligands by common PD medications. In the CALM-PD CIT study (detailed later), there was no significant change in β-CIT uptake after 10 weeks of treatment with either pramipexole (dosage 1.5 to 4.5 mg) or levodopa (dose 300 to 600 mg), consistent with previous studies evaluating levodopa effects after 6 to 12 weeks (Innis *et al.*, 1999; Nurmi *et al.*, 2000a; Parkinson Study Group, 2002b). In a similar study, treatment with pergolide for 6 weeks also showed no significant changes in [123I]β-CIT striatal, putamen, or caudate uptake, but an insignificant trend toward increased [123I]β-CIT uptake (Ahlskog *et al.*, 1999). Data assessing RTI-32, another dopamine transporter ligand, demonstrated significant reductions from baseline in striatal DAT after 6 weeks of treatment with both levodopa and pramipexole, but also with placebo, and this pilot study could not detect differences between the treatment and placebo (Guttman *et al.*, 2001). There was no effect on [18F]Dopa uptake in a study of 5 patients with restless legs syndrome who had been treated with levodopa (Turjanski *et al.*, 1999). Although these clinical studies do not demonstrate significant regulation of the dopamine transporter or [18F]Dopa uptake, they do not exclude a significant short-term treatment-induced change in imaging outcomes, nor do they address the possibility that pharmacological effects may emerge in longer term studies.

3. The rate of change in imaging outcomes used to measure disease progression is very slow, reflecting the slow clinical progression in PD requiring the duration of these progression studies to be at least 18 to 24 months. Furthermore, it is crucial to assess whether changes in loss of imaging outcomes are enduring and whether they are ultimately associated with slowing of clinical outcomes. Therefore the study design should include the potential for continued follow-up evaluation to monitor long-term changes in disease progression in the study cohort.

4. Progressive loss in brain DAT imaging activity also occurs in aging healthy individuals, consistent with pathology, although at a rate approximately one tenth that of PD patients (Marek *et al.*, 2001). No change in F-Dopa uptake with aging has been shown, consistent with the presumed up regulation of dopamine turnover in normal aging (Morrish *et al.*, 1996).

CLINICAL STUDIES OF NEUROPROTECTION IN PD

Despite the difficulties in developing clinical outcome measures and the limits of imaging as a biomarker, several clinical studies have been completed or are underway to assess potentially neuroprotective drugs in PD. Although no drug has been established as neuroprotective, these studies have provided important data and incremental experience in using clinical and imaging outcomes. These studies, several of which are detailed next, have educated investigators and informed clinical study design so that clinical studies are increasingly more likely to convincingly demonstrate meaningful neuroprotection.

DATATOP

The DATATOP study pioneered clinical study design to assess neuroprotective drugs. In this study 800 subjects with early untreated PD were randomized to treatment with either or both selegiline and tocopherol. The study was prompted by basic science evidence suggesting that these drugs might slow neurodegeneration by reducing oxidative neuronal stress. The study used a novel clinical endpoint, the time to need for levodopa, as a practical and easily definable clinical measure of disease progression. Results from this study and other similarly designed smaller studies showed that selegline treatment significantly delayed the time to need for levodopa compared to placebo treatment; tocopherol had no effect (Parkinson Study Group, 1989; Tetrud and Langston, 1989). However, the data also showed that selegiline had a small but significant beneficial effect on PD symptoms as measured by the UPDRS and that this improvement lasted longer than the 1-month washout used in this study (Parkinson Study Group, 1993). Therefore, although the study clearly showed that selegline was effective in slowing disease progression as measured by the primary clinical outcome, selegiline's symptomatic effect on PD created a confound that could fully explain this effect.

The results of the DATATOP study remain controversial. Subsequent analyses and studies have suggested that there may be both a symptomatic and protective effect of selegiline (Langston and Tanner, 2000; Larsen *et al.*, 1999; Shoulson *et al.*, 2002). Further preclinical studies have suggested that selegiline may exert neuroprotective benefit through its effects on the apoptotic pathway and not as an antioxidant (Maruyama *et al.*, 2002; Tatton *et al.*, 2002). However, the symptomatic confound noted in the DATATOP study remains unchallenged, and the extent of any possible neuroprotective effect of selegiline is uncertain.

Although the DATATOP study did not demonstrate neuroprotection of either selegline or tocopherol, it was a critical step in the study design of neuroprotective clinical studies for PD (Shoulson, 1998a). It established need for levodopa as an important clinical endpoint, developed a consortium of clinical centers able to recruit and study 800 research subjects, and provided crucial data on the progression of disease of a large cohort of early PD subjects. Data from the DATATOP study have been crucial in establishing the progression of PD symptoms in an untreated placebo group and the duration that an early PD placebo group can remain untreated with symptomatic medication, both important in estimating sample size for further studies. The DATATOP study also highlighted the importance of developing clinical studies that could distinguish between a small symptomatic effect and a neuroprotective effect.

Dopamine Agonist/Levodopa Studies

Several similar clinical studies have compared the effect of treatment with the dopamine agonists, pramipexole (CALM-PD), ropinirole (056 Study), cabergoline, and pergolide (PELMOPET) to treatment with levodopa on PD progression (Parkinson Study Group, 2000b; Rascol *et al.*, 2000; Rinne *et al.*, 1998). These clinical studies have been designed to examine the policy of initial treatment with dopamine agonists vs levodopa in early PD patients during a 2- to 5-year period, with the primary clinical endpoint being the development in motor fluctuations. Motor fluctuations, when patients experience periods of relatively poor function resulting from slow movements often cycling with involuntary abnormal movements, generally occur in PD patients after treatment with dopaminergic therapies. Fluctuations consist of "wearing off" of medication, abnormal involuntary movements called dyskinesias, and on-off phenomena (very rapid cycling between off and functional periods). Although it is unclear whether motor fluctuations reflect progressive loss of dopaminergic neurons, it is certain that development of motor fluctuations is an important clinical milestone of disease progression. The primary clinical data in these studies demonstrated that pramipexole, ropinirole, cabergoline, or pergolide delayed the onset of dopaminergic motor complications, particularly dyskinesia, compared with levodopa therapy, but that initial levodopa therapy was more effective than the dopamine agonist in ameliorating signs and symptoms of PD as measured by the UPDRS.

In parallel with these clinical outcomes, *in vivo* imaging using either [123I]β-CIT/SPECT or [18F]Dopa/PET has been used to compare the progressive loss of dopaminergic neurons in early PD patients initially treated with either dopamine agonists or levodopa. Initial results from studies comparing pramipexole (CALM-PD CIT), ropinirole (056), or pergolide (PELMOPET) to levodopa

showed nonsignificant trends for a reduction in the loss of [123I]β-CIT uptake (pramipexole) and [18F]Dopa uptake (ropinirole and pergolide) compared to levodopa treatment (Parkinson Study Group, 2000b). Based on these studies, the blinded imaging assessments in the pramipexole (CALM-PD CIT) study were extended to 46 months, and a second ropinirole study focused on neuroimaging was initiated called the REAL-PET study (Parkinson Study Group, 2002b; Whone *et al.*, 2002).

In CALM-PD CIT, a parallel-group, double-blind randomized study conducted by the Parkinson Study Group and sponsored by The Pharmacia Corp. and Boehringer-Ingelheim, 82 PD patients enrolled in the CALM-PD study were imaged with β-CIT/SPECT at baseline and again 22 (n = 78), 34 (n = 71), and 46 (n = 65) months after initial treatment. Patients were randomly assigned to receive pramipexole 0.5 mg three times per day (n = 42) or levodopa 100 mg three times per day (n = 40). For patients with residual disability, the dosage was escalated up to 1.5 mg pramipaxole or 200 mg levodopa three times daily during the first 10 weeks and subsequently open label levodopa could be added. After 24 months of follow-up evaluation, the dosage of study drug could be further modified. The primary outcome variable was the percent change from baseline in striatal [123I]β-CIT uptake after 46 months. Comparison of the treatment groups in CALM-PD CIT showed that the percent loss in striatal [123I]β-CIT uptake from baseline was significantly reduced in the group initially treated with pramipexole compared to the group initially treated with levodopa at each time point: $-7.1 \pm 9.0\%$ vs $-13.5 \pm 9.6\%$ at 22 months, $P = .004$; $-10.9 \pm 11.8\%$ vs $-19.6 \pm 12.4\%$ at 34 months, $P = .009$; and $-16.0 \pm 13.3\%$ vs $-25.5 \pm 14.1\%$ at 46 months, $P = .01$. Approximately 75% of those subjects initially treated with pramipexole alone were also treated with supplemental levodopa by 46 months after baseline. Analysis of the putamen and caudate data separately was similar to the combined striatal region of interest (Parkinson Study Group, 2002b).

In the REAL-PET study, sponsored by GlaxoSmithKline, 186 denovo PD patients were treated in a randomized, double-blinded multicenter design with either ropinirole or levodopa and imaged with [18F]dopa on study drug and 24 months after initial treatment. Patients were randomized in equal numbers to each treatment group. The mean daily doses after 2 years were 12.2 ± 6.2 mg ropinirole and 558.7 ± 180.8 mg levodopa. Supplemental levodopa could be added to subjects with insufficient therapeutic benefit from study drug during the study. The primary outcome variable was the percent change from baseline in the putamen [18F]Dopa uptake. Note that baseline imaging was on study drug, but the dose was increased during the 24-month study period. Comparison of the treatment groups showed that the percent loss from baseline in putamen [18F]Dopa in the ropinirole group was significantly reduced compared with the levodopa group: −13% vs −20%, P = .022. An alternative analysis of these data using statistical parametric mapping, a technique designed to take advantage of the brain imaging data in all brain regions, similarly showed reduction in loss of activity in putamen in the ropinirole vs levodopa groups: −14% vs −20%, P = .034 (Whone *et al.*, 2002).

These two clinical imaging studies targeting dopamine function with different imaging ligands and technology both demonstrate slowing in the rate of loss of [123I]β-CIT or [18F]Dopa uptake, in early PD patients treated with dopamine agonists compared to levodopa. These studies evaluated two related, predominantly D2 dopamine receptor agonists, suggesting that the results may indicate a class effect. The relative reduction in the percent loss from baseline of [123I]β-CIT uptake in the pramipexole vs the levodopa group was 47% at 22 months, 44% at 34 months, and 37% at 46 months after initiating treatment. The relative reduction of [18F]Dopa uptake in the ropinirole group vs the levodopa group was 35% at 24 months. These data suggest that treatment with the dopamine agonists, pramipexole and ropinirole, and/or with levodopa may either slow or accelerate the dopaminergic degeneration of PD. Furthermore, these studies demonstrate that *in vivo* imaging can be used to assess potential disease-modifying drugs in well-controlled, blinded clinical studies.

Although both the CALM-PD CIT and the REAL-PET demonstrate a robust and remarkably consistent reduction in imaging uptake in the dopamine agonist vs levodopa groups, several issues in the study design limit interpretation of these data regarding neuroprotection.

Lack of Placebo

Both studies compared two active medications without a placebo group. Therefore, these data cannot directly distinguish whether the difference in the rate of loss of [123I]β-CIT or [18F]Dopa uptake in the treatment groups results from a reduction due to pramipexole or ropinirole, an increase due to levodopa, or both. Long-term data from the CALM-PD CIT study may be consistent with levodopa toxicity. In the CALM-PD CIT study, the percent reduction of striatal [123I]β-CIT uptake from baseline in the pramipexole vs levodopa groups differs significantly at 22, 34, and 46 months; however, the difference in the rate of change between the two groups is significant between baseline and 22 months, but not 22 to 46 months. This finding may be explained by the increasing percentage of subjects initially treated with pramipexole who are supplemented

with levodopa with disease progression (75% by month 46), possibly blunting the pramipexole effect evident at 22 months (Parkinson Study Group, 2002b).

Possible Regulation of Imaging Outcomes

Another possible explanation for the difference in the loss of [123I]β-CIT and [18F]Dopa uptake in the dopamine agonist vs levodopa groups in these studies is that results were due to a pharmacological interaction between pramipexole or ropinirole, levodopa, or both and the DAT or dopamine turnover, rather than slowed or accelerated neuron degeneration (Ahlskog, 2003; Albin and Frey, 2003; Wooten, 2003). These concerns highlight the need to further investigate the effect of short-term treatment with dopamine agonists and levodopa on dopamine imaging outcomes with a well-powered study. While the possibility of pharmacological regulation remains, this explanation must account for the data from both the CALM-PD CIT and REAL-PET studies.

Clinical Imaging Correlation

In the CALM-PD CIT and REAL-PET studies, there was no correlation between the percent change from baseline in the imaging outcome and the change from baseline in UPDRS at 22 to 24 months. However, the loss of striatal [123I]β-CIT uptake from baseline was significantly correlated (r = −0.40, P = .001) with the change in UPDRS from baseline at the 46-month evaluation, suggesting that the correlation between clinical and imaging outcomes begins to emerge with longer monitoring. These data underscore that particularly in early PD, clinical and imaging outcomes provide complementary data and that long-term follow-up will be required to correlate changes in clinical and imaging outcomes. Slowing the loss of imaging outcomes in PD is relevant only if these imaging changes ultimately result in meaningful, measurable, and persistent changes in clinical function in PD patients.

ELLDOPA Study

The ELLDOPA (Earlier vs Later L-DOPA Therapy in Parkinson Disease) study compared treatment of early untreated PD patients with carbidopa/levodopa (37.5 mg/150 mg, 75 mg/300 mg, or 150 mg/600 mg) or placebo for 40 weeks (Fahn, 1999). Several preclinical studies have indicated that levodopa may be toxic to dopaminergic neurons *in vitro,* whereas other studies have not shown any toxicity in animals, even in the presence of oxidative stress (Fahn, 1999; Mytilineou *et al.*, 2003). Moreover, other studies have suggested that levodopa may protect against antiapoptotic proteins or may be trophic and potentially promote functional recovery of damaged nigral neurons (Datla *et al.*, 2001; Murer *et al.*, 1998). The goal of the ELLDOPA study was to examine whether levodopa therapy influences the rate of progression of PD using both clinical and imaging outcomes. The study is especially clinically relevant, as carbidopa/ levodopa remains the mainstay of symptomatic therapy for PD.

In ELLDOPA, 361 subjects were recruited and randomized to the four treatment arms. After 40 weeks of treatment, the study drug was discontinued and patients were evaluated 1 and 2 weeks later. The primary outcome was the change in UPDRS between baseline and the 2-week washout period (42 weeks). All doses of levodopa produced clinical benefit compared with placebo throughout the study and after discontinuing treatment for 2 weeks. Changes of UPDRS scores between baseline and 2 weeks after withdrawal of active treatment were less severe than in the placebo group (change of 7.8 ± 9.0, 1.9 ± 6.0, 1.9 ± 6.9, and −1.4 ± 7.7 for placebo, 150 mg/day, 300 mg/day, and 600 mg/day, respectively, $P >$ .0001) (Parkinson Study Group, 2003).

A subset of the ELLDOPA subjects (N = 135) underwent [123I]β-CIT imaging at baseline and after 40 weeks to assess the effect of levodopa on dopaminergic function as measured by this imaging biomarker. Comparison of the treatment groups in ELLDOPA showed that the percent loss in striatal [123I]β-CIT uptake from baseline was reduced in the groups treated with levodopa compared to placebo, but this reduction was not statistically significant. However, 19 of the 135 subjects (14%) enrolled in the imaging substudy had both baseline and 40-week [123I]β-CIT scans with β-CIT uptake > 75% age-expected putaminal uptake (Marek *et al.*, 2003b). These scans without dopaminergic deficit (SWEDD) suggest these subjects are unlikely to have PD. Similar scans in the "normal" range were reported in the REAL-PET study in 11% of enrolled subjects (Whone *et al.*, 2002). Analysis of the [123I]β-CIT SPECT results excluding the subjects with SWEDDs (N = 116) showed a statistically greater decrease in [123I]β-CIT uptake in the levodopa (−6 ± 10.3%, −4 ± 9.4%, −7.2 ± 7.6%, for 150 mg/days, 300 mg/days, and 600 mg/days, respectively) than in the placebo group (−1.4 ± 10.0%), $P = .036$ for a dose response (Parkinson Study Group, 2003).

The ELLDOPA study, like all clinical neuroprotection studies to date, has not provided definitive answers, has raised many critical questions, and has taught us many important lessons. The clinical outcome in the ELLDOPA study, the persistent and dose-related improvement in UPDRS after a 2-week washout of levodopa, suggests two potential explanations. First, levodopa may be neuropro-

tective, slowing the loss of disability in PD as measured by the UPDRS even after its symptomatic effect is eliminated. Second, the 2-week washout of levodopa was too short to fully eliminate a symptomatic effect of levodopa. This study cannot differentiate between these possibilities. The ELLDOPA study has shown that while a 2-week washout of a powerful symptomatic drug such as levodopa is possible, it is unlikely to distinguish between a lingering symptomatic and an ongoing neuroprotective effect. In the future, study designs such as randomized start designs, which reduce reliance on a washout period, will be used in neuroprotective studies.

The imaging outcome from the ELLDOPA study has also raised several questions. The increased loss of [123I]β-CIT uptake in the levodopa treated groups suggests that levodopa may accelerate the loss of dopamine neurons, despite its remarkable symptomatic effect in PD. These results are consistent with the imaging results in the CALM-PD and REAL-PET studies, all suggesting that levodopa treatment may accelerate the loss of imaging markers of dopamine neuronal function in comparison to either placebo or dopamine agonists (Figure 8.3). An alternative explanation for these imaging results, however, is that the change in [123I]β-CIT uptake in the levodopa, dopamine agonist, and placebo-treated groups may be due to a direct regulatory effect of these drugs on the DAT or dopamine turnover without any change in neuronal degeneration. As detailed previously, the studies that have addressed this question are inconclusive because of small sample size. The ELLDOPA study now coupled with the dopamine agonist studies has provided a powerful rationale to initiate well-powered clinical imaging studies with sufficient sample size to answer the crucial question as to whether the study drug-related difference in imaging outcomes in these studies is explained by a short-term regulatory effect in DAT and F-DOPA imaging or whether it reflects differential progressive loss of dopaminergic function.

Coenzyme Q10 Study

In the coenzyme Q10 study, known as QE-2, early untreated PD patients were treated with either coenzyme Q10 at doses of 300, 600, or 1200 mg/day or placebo for up to 16 months (Shults *et al.*, 2002). Numerous studies have demonstrated mitochondrial dysfunction in PD patients with a clear deficit in complex I (Beal, 2003; Greenamyre *et al.*, 2001; Schapira *et al.*, 1998). It remains uncertain whether this mitochondrial dysfunction results from and/or contributes to the neuronal cell loss of PD. Coenzyme Q10 (ubiquinone) serves as the electron acceptor for complexes I and II of the mitochondrial electron transport chain and also acts as an antioxidant. Several preclinical studies have suggested that coenzyme

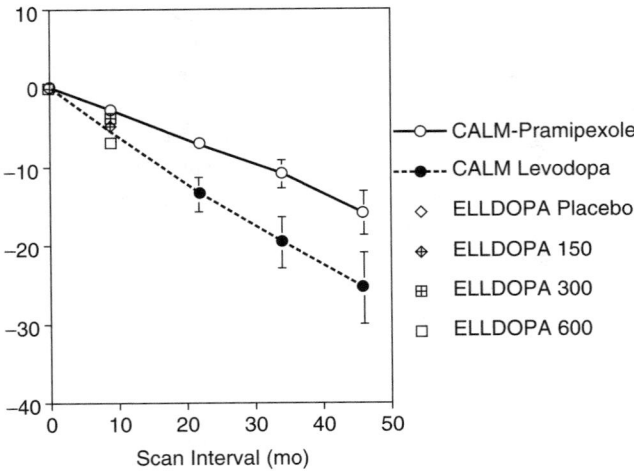

FIGURE 8.3   Percent loss of [123I]β-CIT uptake from baseline in the CALM-PD CIT and ELLDOPA CIT. Note the reduction in the loss of [123I]β-CIT uptake in subjects treated with either pramipexole or placebo compared to levodopa (Parkinson Study Group, 2002b, 2003).

NEUROPROTECTION IN PD: LESSONS LEARNED

Slowing the progression of PD remains a crucial unmet need for PD patients and their families (Ravina *et al.*, 2003; Shoulson, 1998b). During the past 15 years, beginning with the pioneering DATATOP study, several clinical studies (some detailed previously) have been conducted in an attempt to demonstrate neuroprotection in PD. Although none of these studies have as yet proven that any study drug modifies PD progression, they have provided a crucial evolving understanding of how to assess potentially neuroprotective medications. Fueled by an increasing number of biochemical targets and an ever-present human need to solve this problem, new studies of neuroprotective drugs are underway and continue to develop (Table 8.1). The lessons learned from prior clinical studies, some listed later, must be applied to these newer studies to optimize their potential success.

Q10 might improve mitochondrial function and therefore might reduce the vulnerability of dopamine neurons (Shults, 2003). In addition, a large study assessing the effect of coenzyme Q10 in Huntington disease has shown a trend for reduction in the rate of clinical deterioration (Huntington Study Group, 2001).

In QE-2, 80 subjects were recruited and randomized to the four treatment arms and treated for 16 months or until they required treatment with levodopa. The primary outcome, a novel clinical measure known as the Oakes design, was the change in UPDRS between baseline and the last visit, thereby combining the change in UPDRS, a continuous measure of progressive disability, with need for levodopa, a disease milestone. The adjusted mean total UPDRS changes were +11.99 for the placebo group, +8.81, +10.82, +6.69 for the 300 mg/day, 600 mg/day, and 1200 mg/day group. The change in UPDRS was significantly greater in the placebo group than in the 1200mg/day group, $P = .04$, and there was a trend for a dose response among the groups, $P = .09$ (Shults *et al.*, 2002).

The QE-2 study used a novel combination clinical endpoint to assess neuroprotection. The study results are provocative, but the sample size of this Phase II study does not allow definitive conclusions. Even in this study of a nondopaminergic drug, the data raise the question of a potential symptomatic rather than neuroprotective effect of coenzyme Q10. The possibility of an unexpected symptomatic effect of coenzyme Q10 is suggested by the short-term improvement (after 1 month) in treated patients in clinical scales reflecting activities of daily living. A larger more definitive study of the effect of high doses of coenzyme Q10 on the progressive disability of PD is planned to address this question.

1. *Reduce expectations:* Clinical studies and clinical outcomes are an imperfect compromise between the necessary scientific rigor of clinical trials and the clinical needs of the participating subjects. It is unlikely that a single study will prove that a drug is neuroprotective. Rather a cumulative body of evidence from several preclinical and clinical studies will be required. Therefore, clinical studies should focus on achievable outcomes such as whether there is slowing of clinical disability or reduction in the loss of a relevant biomarker.

2. *Expect surprises:* Clinical studies of PD neuroprotection have consistently demonstrated surprising results. The most frequent unexpected result has been a previously unrecognized or underestimated symptomatic effect of study drugs obscuring a potential protective effect. For example, the long-lasting symptomatic effect of selegiline in the DATATOP study was not expected based on data from smaller studies. Similarly the short-term improvement in activities of daily living scales in patients treated with coenzyme Q10 was unexpected and has confused the results. In the ELLDOPA study, the marked improvement in UPDRS after a 2-week washout has raised the possibility of a previously unrecognized long-lasting effect of levodopa vs a surprisingly prominent neuroprotective effect.

New study designs must identify and/or control for potential symptomatic effects. Studies should be adequately powered to identify small symptomatic effects of the study drug. Studies should also be powered to detect potential confounding effects of the study drug on biomarkers such as imaging outcomes used in the study. Washout periods should be long enough to eliminate lingering symptomatic effects or, if this is not practically possible, should be discarded in favor of other study designs. Designs such as randomized start studies in which all subjects are ultimately treated with study drug may minimize any symptomatic effect of the study drug in the treatment groups. For example, a recent report of the TEMPO study, evaluating the symptomatic effect of rasagiline in early PD patients, showed that the magnitude of UPDRS improvement in rasagiline-treated patients after 1 year was greater in those treated at study start than after 6 months on placebo, suggesting that the improvement in UPDRS is not explained as solely a symptomatic benefit, but possibly a disease-modifying effect. While this study was not designed as a randomized start, it provides clues and points out the challenges of using this type of study design to assess disease progression when a symptomatic effect is also present. Another study strategy to reduce the importance of any relatively short-term symptomatic effects is a large simple trial design to assess long-term effects of study drugs on disease progression in a real-life setting using long-term outcomes such as institutionalization or mortality.

3. *Understand the disease:* The study design depends on the disease stage, clinical needs of the patient population, medications allowed, and duration of follow-up period. One simple approach to disease stage is to divide PD patients into untreated patients, treated with stable response to treatment, treated with a fluctuating response to treatment, and end-stage patients without an effective treatment effect. The rate of disease progression may depend on disease stage, with relative slowing of disease progression as PD continues (Jennings *et al.*, 2000). Although studying early untreated subjects focuses on patients with the most neuronal reserve remaining and avoids confounding concomitant medications, it may limit the duration of the study, as most subjects will require symptomatic medication within 12 months (Parkinson Study Group, 1989, 2002a). The randomized start approach may enable early-stage patients to remain in the study even after treatment. Another approach has been to study patients who are treated with stable response to treatment. Addition of study drug to these patients already treated may be useful if these subjects remain on the same symptomatic medication throughout the study or if change in medication is a clinical outcome. Yet another critical practical concern is that treated patients are much easier to identify and recruit into clinical studies than early untreated patients.

Clinical studies are often a compromise between the most comprehensive study and the most practical study. Even in a very practical study, however, both the sample size and study duration must be sufficient to answer the study question. In PD, the slow and variable course of disease progression tends to increase the necessary sample size and prolong the study duration. In a recent example of an underpowered and short duration study, the effect of the neuroimmunophilin ligand (NILA) on PD progression was assessed after only 6 months and the results were equivocal (Seibyl *et al.*, 2002). This has resulted in a second study of 2-years duration to answer this question.

4. *Understand the use of biomarkers:* Although biomarkers, in particular imaging markers, have provided important results in neuroprotection studies, they must be used with caution. Biomarkers may be specific to study stage and/or confounded by study medication or symptomatic medications. In addition, recent studies using imaging biomarkers have also clearly demonstrated that there may be a poor correlation between imaging and clinical measures of disease progression, especially in short-term studies allowing symptomatic treatment. Biomarkers and clinical outcomes provide complementary information regarding disease progression, but not necessarily correlative information in disease progression studies.

5. *Increase expectations:* Neuroprotection must be clinically meaningful and clinical endpoints must clearly address the major causes of long-term disability of PD. In addition to slowing of motor symptoms, studies should focus on slowing cognitive loss, gait and balance disturbance, and autonomic symptoms. These studies will likely require patients with moderate to later stage disease and prolonged follow-up time. Delaying or preventing these disease milestones would provide unequivocal evidence of meaningful disease modification.

Therefore studies to show neuroprotection must include assessment of subjects at various clinical stages of disease, slowing of both motor and nonmotor clinical disability, and evidence of slowing of dopamine degeneration. It is likely that neuroprotection studies will require two phases, the first in early PD subjects to demonstrate slowed worsening in accepted clinical scales and slowed loss of a biomarker of neurodegeneration and the second in later PD subjects to show meaningful delay of milestones of PD disability.

NEUROPROTECTION IN PD: FUTURE PROSPECTS

The goal of neuroprotective therapy is to provide treatment for PD patients as early as possible after the start of neurodegeneration. Like many other neurodegenerative disorders, PD is characterized by a prolonged preclinical period during which dopamine neurons are degenerating,

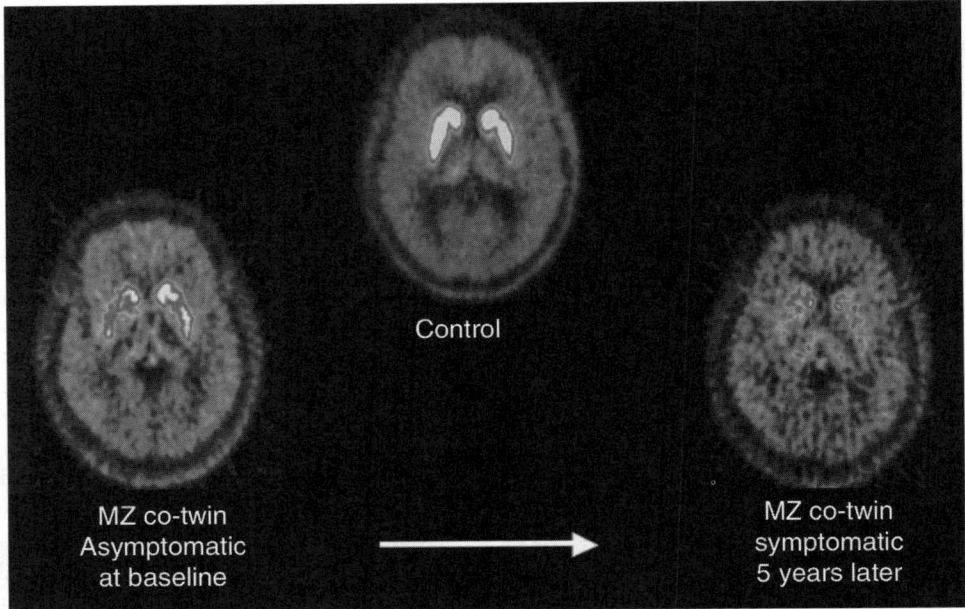

FIGURE 8.4 [18F]Dopa PET of an initially asymptomatic co-twin of a PD patient at baseline and after 5 years. Note that baseline scan shows loss of [18F]Dopa uptake compared to control and that scans show progressive loss of [18F]Dopa uptake activity consistent with onset of symptoms of PD at time of follow-up scan (Figure courtesy of Prof. David Brooks, Hammersmith Hospital). (See also the Color Plate section.)

but symptoms have not yet begun (Figure 8.1). This preclinical period, likely lasting years, provides an opportunity to initiate early therapy and to protect neurons. However, neuroprotective treatments can be initiated only if tools are available to accurately identify subjects either at risk for dopamine degeneration and/or who are already experiencing preclinical dopaminergic degeneration. Studies assessing family members of PD patients as an at-risk group have demonstrated neuronal loss in unaffected family members (Figure 8.4; see also the Color Plate section), but the relative contribution of genetics in sporadic PD has been questioned in recent twin studies and in epidemiological surveys (Elbaz *et al.*, 2003; Marder *et al.*, 2003; Piccini *et al.*, 1997, 1999; Tanner *et al.*, 1999). Studies to develop and validate screening tools for PD will be crucial to the early use of neuroprotective drugs.

Prospects for neuroprotective therapy in PD continue to expand. The studies to date have provided exciting results, but also notes of caution regarding the difficulties in clearly demonstrating neuroprotection. Ultimately it may be necessary to combine drugs to develop a neuroprotective cocktail for PD patients or to tailor drugs for subsets of PD patients, Nonetheless, the explosion of information in basic and clinical neuroscience providing new biochemical targets, additional *in vivo* models, improved clinical designs, and better validated biomarkers for PD continues to energize the field and accelerate future research.

References

Agid Y. (1998). Levodopa: Is toxicity a myth? *Neurology* 50:858-863.

Agid Y *et al.* (1999). Levodopa in the treatment of Parkinson's disease: a consensus meeting, *Mov Disord* 14:911-913.

Ahlskog JE. (2001). Parkinson's disease: medical and surgical treatment, *Neurol Clin* 19:579-605, vi.

Ahlskog JE. (2003). Slowing Parkinson's disease progression: recent dopamine agonist trials, *Neurology* 60:381-389.

Ahlskog JE *et al.* (1999). The effect of dopamine agonist therapy on dopamine transporter imaging in Parkinson's disease, *Mov Disord* 14:940-946.

Albin RL, Frey KA. (2003). Initial agonist treatment of Parkinson disease: a critique, *Neurology* 60:390-394.

Andringa G *et al.* (2003). TCH346 prevents motor symptoms and loss of striatal FDOPA uptake in bilaterally MPTP-treated primates, *Neurobiol Dis* 14:205-217.

Asanuma M, Miyazaki I, Ogawa N. (2003). Dopamine- or L-DOPA-induced neurotoxicity: the role of dopamine quinone formation and tyrosinase in a model of Parkinson's disease, *Neurotox Res* 5:165-176.

Asenbaum S *et al.* (1997). Imaging of dopamine transporters with iodine-123-B-CIT and SPECT in Parkinson's disease, *J Nucl Med* 38:1-6.

Beal MF. (2003). Mitochondria, oxidative damage, and inflammation in Parkinson's disease, *Ann N Y Acad Sci* 991:120-31.

Bernheimer H *et al.* (1973). Brain dopamine and the syndromes of Parkinson and Huntington, clinical, morphological, and neurochemical correlates, *J Neurol Sci* 20:415-455.

Biomarkers Defintions Working Group. (2001). Biomarkers and surrogate endpoints: preferred definitions and conceptual framework, *Clin Pharmacol Ther* 69:89-95.

Blakely RD. (2001). Neurobiology. Dopamine's reversal of fortune, *Science* 293:2407-2409.

Booij J et al. (1998). Imaging of dopamine transporters with iodine-123-FP-CIT SPECT in healthy controls and patients with Parkinson's disease, J Nucl Med 39:1879-1884.

Booij T, Tissingh G, Boer G. (1997). [123I]FP-SPECT shows a pronounced decline of striatal dopamine transporter labelling in early and advanced Parkinson's disease, J Neurol Neurosurg Psychiatry 62:133-140.

Bower JH et al. (1999). Incidence and distribution of parkinsonism in Olmsted County, Minnesota, 1976-1990, Neurology 52:1214-1220.

Brooks D et al. (2003a). Assessment of neuroimaging techniques as biomarkers of the progression of Parkinson's disease, Exp Neurol 184:S68-79.

Brooks DJ et al. (2003b). Assessment of neuroimaging techniques as biomarkers of the progression of Parkinson's disease, Exp Neurol 184(Suppl 1):68-79.

Brooks et al. (1990). Differing patterns of striatal 18F-dopa uptake in Parkinson's disease, multiple system atrophy, and progressive supranuclear palsy, Ann Neurol 28:547-555.

Brundin P. (2002). GDNF treatment in Parkinson's disease: time for controlled clinical trials?, Brain 125:2149-2151.

Cohen AD et al. (2003). Neuroprotective effects of prior limb use in 6-hydroxydopamine-treated rats: possible role of GDNF, J Neurochem 85:299-305.

Conway KA et al. (2001). Kinetic stabilization of the alpha-synuclein protofibril by a dopamine-alpha-synuclein adduct, Science 294:1346-1349.

Datla K, Blunt S, Dexter D. (2001). Chronic L-DOPA administration is not toxic to the remaining dopaminergic nigrostriatal neurons, but instead may promote their functional recovery, in rats with partial 6-OHDA or FeCl3 nigrostriatal lesions, Mov Disord 16:424-434.

Dawson TM, Dawson VL. (2002). Neuroprotective and neurorestorative strategies for Parkinson's disease, Nat Neurosci 5(Suppl):1058-1061.

De Gruttola VG et al. (2001). Considerations in the evaluation of surrogate endpoints in clinical trials. Summary of a National Institutes of Health workshop, Control Clin Trials 22(5):485-502.

Dekker M et al. (2003). Clinical features and neuroimaging of PARK7-linked parkinsonism, Mov Disord 18:751-757.

DeKosky S, Marek K. (2003). Looking backward to move forward: early detection of neurodegenerative disorders, Science 302:830-834.

Dodel RC, Berger K, Oertel WH. (2001). Health-related quality of life and healthcare utilisation in patients with Parkinson's disease: impact of motor fluctuations and dyskinesias, Pharmacoeconomics 19:1013-1038.

Eidelberg D et al. (1995). Early differential diagnosis of Parkinson's disease with 18F-fluorodeoxyglucose and positron emission tomography, Neurology 45:1995-2005.

Elbaz A et al. (2003). Validity of family history data on PD: evidence for a family information bias, Neurology 61:11-17.

Etminan M, Gill S, Samii A. (2003). Effect of non-steroidal anti-inflammatory drugs on risk of Alzheimer's disease: systematic review and meta-analysis of observational studies, Br Med J 327:128.

Fahn S. (1997). Levodopa-induced neurotoxicity: does it represent a problem for the treatment of Parkinson's disease?, CNS Drugs 8:376-393.

Fahn S. (1999). Parkinson disease, the effect of levodopa, and the ELLDOPA trial, Arch Neurol 56:529-535.

Fahn S, Elton RL, Members of the UPDRS Development Committee. (1987). Unified Parkinson's disease rating scale. In Fahn CD et al, editors: Recent developments in Parkinson's disease, vol. 2, Florham Park, NJ: Macmillan Healthcare Information.

Feany MB, Bender WW. (2000). A Drosophila model of Parkinson's disease, Nature 404:394-398.

Fischman AJ et al. (1998). Rapid detection of Parkinson's disease by SPECT with altropane: a selective ligand for dopamine transporters, Synapse 29:128-141.

Frank R, Hargreaves R. (2003). Clinical biomarkers in drug discovery and development, Nature Reviews Drug Discovery 2:566-580.

Freed CR et al. (2001). Transplantation of embryonic dopamine neurons for severe Parkinson's disease, N Engl J Med 344:710-719.

Frey KA et al. (1996). Presynaptic monoaminergic vesicles in Parkinson's disease and normal aging, Ann Neurol 40:873-884.

Gasser T. (2003). Overview of the genetics of parkinsonism, Adv Neurol 91:143-152.

Giasson BI et al. (2002). Neuronal alpha-synucleinopathy with severe movement disorder in mice expressing A53T human alpha-synuclein, Neuron 34:521-533.

Gill S et al. (2003). Direct brain infusion of glial cell line-derived neurotrophic factor in Parkinson disease, Nat Med 9:589-595.

Gnanalingham KK, Robertson RG. (1994). The effects of chronic continuous versus intermittent levodopa treatments on striatal and extrastriatal D1 and D2 dopamine receptors and dopamine uptake sites in the 6-hydroxydopamine lesioned rat: an autoradiographic study, Brain Res 640:185-194.

Gold BG, Villafranca JE. (2003). Neuroimmunophilin ligands: the development of novel neuroregenerative/ neuroprotective compounds, Curr Top Med Chem 3:1368-1375.

Greenamyre J et al. (1992). Glutamate receptor antagonism as a novel therapeutic approach in Parkinson's disease. In Simon R, editor: Excitatory amino acids, vol. 9, New York: Thieme Medical Publishers.

Greenamyre JT et al. (2001). Complex I and Parkinson's disease, IUBMB Life 52:135-141.

Greene JC et al. (2003). Mitochondrial pathology and apoptotic muscle degeneration in Drosophila parkin mutants, Proc Natl Acad Sci U S A 100:4078-4083.

Guttman M et al. (1997). [11C]RTI-32 PET studies of the dopamine transporter in early dopa-naive Parkinson's disease, Neurology 48:1578-1583.

Guttman M et al. (2003). Burden of parkinsonism: a population-based study, Mov Disord 18:313-319.

Guttman M et al. (2001). Influence of L-dopa and pramipexole on striatal dopamine transporter in early PD, Neurology 56:1559-1564.

Harris CA et al. (2002). Inhibition of the c-Jun N-terminal kinase signaling pathway by the mixed lineage kinase inhibitor CEP-1347 (KT7515) preserves metabolism and growth of trophic factor-deprived neurons, J Neurosci 22:103-113.

Hauser R, Holford N. (2002). Quantitative description of loss of clinical benefit following withdrawal of levodopa-carbidopa and bromocriptine in early Parkinson's disease, Mov Disord 17:961-968.

Hirsch EC et al. (2003). The role of glial reaction and inflammation in Parkinson's disease, Ann N Y Acad Sci 991:214-228.

Huang WS et al. (2001). Evaluation of early-stage Parkinson's disease with 99mTc-TRODAT-1 imaging, J Nucl Med 42:1303-1308.

Huntington Study Group. (2001). A randomized, placebo-controlled trial of coenzyme Q10 and remacemide in Huntington's disease, Neurology 57:397-404.

Innis RB et al. (1999). Effect of treatment with L-dopa/carbidopa or L-selegiline on striatal dopamine transporter SPECT imaging with [123I]beta-CIT, Mov Disord 14:436-442.

Innis RB et al. (1993). Single photon emission computed tomographic imaging demonstrates loss of striatal dopamine transporters in Parkinson disease, Proc Natl Acad Sci U S A 90:11965-11969.

Jankovic J, Kapadia AS. (2001). Functional decline in Parkinson disease, Arch Neurol 58:1611-1615.

Jenner P. (2003). The contribution of the MPTP-treated primate model to the development of new treatment strategies for Parkinson's disease, Parkinsonism Relat Disord 9:131-137.

Jennings D et al. (2000). [123I]β-CIT and SPECT assessment of progression in early and late Parkinson's disease, Neurology 56 (Suppl 3):A74.

Kaufman M, Madras B. (1991). Severe depletion of cocaine recognition sites associated with the dopamine transporter in Parkinson's diseased striatum, *Synapse* 9:43-49.

Khan NL *et al.* (2002). Progression of nigrostriatal dysfunction in a parkin kindred: an [18F]dopa PET and clinical study, *Brain* 125:2248-2256.

Kieburtz K. (2003). Designing neuroprotection trials in Parkinson's disease, *Ann Neurol* 53(Suppl 3):S100-107; discussion S107-109.

Kim TE *et al.* (2003). Sonic hedgehog and FGF8 collaborate to induce dopaminergic phenotypes in the Nurr1-overexpressing neural stem cell, *Biochem Biophys Res Commun* 305:1040-1048.

Lang AE, Lozano AM. (1998). Parkinson's disease. First of two parts, *N Engl J Med* 339:1044-1053.

Langston JW, Tanner CM. (2000). Selegiline and Parkinson's disease: it's deja vu-again, *Neurology* 55:1770-1771.

Larsen JP, Boas J, Erdal JE. (1999). Does selegiline modify the progression of early Parkinson's disease? Results from a five-year study. The Norwegian-Danish Study Group, *Eur J Neurol* 6:539-547.

Lassen N, Holm S. (1992) Single photon emission computerized tomography (SPECT). In Mazziota J, Gilman S, editors: *Clinical brain imaging: principles and applications*, Philadelphia: FA Davis.

Leber P. (1997). Slowing the progression of Alzheimer disease: methodologic issues, *Alzheimer Dis Assoc Disord* 11(Suppl 5):S10-21; discussion S37-39.

Lee CS *et al.* (2000). In vivo positron emission tomographic evidence for compensatory changes in presynaptic dopaminergic nerve terminals in Parkinson's disease, *Ann Neurol* 47:493-503.

Lindsten K *et al.* (2003). A transgenic mouse model of the ubiquitin/proteasome system, *Nat Biotechnol* 21:897-902.

Mandel S *et al.* (2003). Neuroprotective strategies in Parkinson's disease : an update on progress, *CNS Drugs* 17:729-762.

Marder K *et al.* (2003). Accuracy of family history data on Parkinson's disease, *Neurology* 61:18-23.

Marek K. (1999). Dopaminergic dysfunction in Parkinsonism: new lessons from imaging, *Neuroscientist* 5:333-339.

Marek K *et al.* (2001). [123I]beta-CIT SPECT imaging assessment of the rate of Parkinson's disease progression, *Neurology* 57:2089-2094.

Marek K, Jennings D, Seibyl J. (2003a). Dopamine agonists and Parkinson's disease progression: what can we learn from neuroimaging studies, *Ann Neurol* 53(Suppl 3):160-166.

Marek K, Seibyl J. (2000). Imaging: a molecular map for neurodegeneration, *Science* 289:409-411.

Marek K, Seibyl J, Parkinson Study Group. (2003b). β-CIT Scans without evidence of dopaminergic deficit (SWEDD) in the ELLDOPA-CIT and CALM-CIT study: long-term imaging assessment, *Neurology* 60(Suppl 1):A298.

Marek K *et al.* (1996). [I-123]CIT Spect imaging demonstrates bilateral loss of dopamine transporters in hemi-Parkinson's disease, *Neurology* 46:231-237.

Marino MJ *et al.* (2003). Allosteric modulation of group III metabotropic glutamate receptor 4: a potential approach to Parkinson's disease treatment, *Proc Natl Acad Sci U S A* 100:13668-13673.

Maroney AC *et al.* (2001). Cep-1347 (KT7515), a semisynthetic inhibitor of the mixed lineage kinase family, *J Biol Chem* 276:25302-25308.

Maruyama W *et al.* (2002). Neuroprotection by propargylamines in Parkinson's disease: suppression of apoptosis and induction of prosurvival genes, *Neurotoxicol Teratol* 24:675-682.

McGeer PL, Yasojima K, McGeer EG. (2001). Inflammation in Parkinson's disease, *Adv Neurol* 86:83-89.

McNaught S *et al.* (2003). Altered proteasomal function in sporadic Parkinson's disease, *Exp Neurol* 179:38-46.

Miller RJ, Wilson SM. (2003). Neurological disease: UPS stops delivering! *Trends Pharmacol Sci* 24:18-23.

Moody CA, Granneman JG, Bannon MJ. (1996). Dopamine transporter binding in rat striatum and nucleus accumbens is unaltered following chronic changes in dopamine levels, *Neurosci Lett* 217:55-57.

Morrish P *et al.* (1998). Measuring the rate of progression and estimating the preclinical period of Parkinson's disease with [18F}dopa PET, *J Neurol Neurosurg Psychiatry* 64:314-319.

Morrish PK, Sawle GV, Brooks DJ. (1996). An [18F]dopa-PET and clinical study of the rate of progression in Parkinson's disease, *Brain* 119(Pt 2):585-591.

Mozley PD *et al.* (1999). Effects of age on dopamine transporters in healthy humans, *J Nucl Med* 40:1812-1817.

Mozley PD *et al.* (2000). Binding of [99mTc]TRODAT-1 to dopamine transporters in patients with Parkinson's disease and in healthy volunteers, *J Nucl Med* 41:584-589.

Murer M *et al.* (1998). Chronic levodopa is not toxic for remaining dopamine neurons, but instead promotes their recovery, in rats with moderate nigrostriatal lesions, *Ann Neurol* 43:392-398.

Mytilineou C *et al.* (2003). Levodopa is toxic to dopamine neurons in an in vitro but not an in vivo model of oxidative stress, *J Pharmacol Exp Ther* 304:792-800.

Nass R *et al.* (2002). Neurotoxin-induced degeneration of dopamine neurons in *Caenorhabditis elegans*, *Proc Natl Acad Sci U S A* 99:3264-3269.

Niznik HB *et al.* (1991). The dopamine transporter is absent in Parkinsonian putamen and reduced in the caudate nucleus, *J Neurochem* 56:192-198.

Nurmi E *et al.* (2000a). Reproducibility and effect of levodopa on dopamine transporter function measurements: a [18F]CFT PET study, *J Cereb Blood Flow Metab* 20:1604-1609.

Nurmi E *et al.* (2001). Rate of progression in Parkinson's disease: A 6-[18F]fluoro-L-dopa PET study, *Move Disord* 16:608-615.

Nurmi E *et al.* (2000b). Progression in Parkinson's disease: a positron emission tomography study with a dopamine transporter ligand [18F]CFT, *Ann Neurol* 47:804-808.

Nutt J, Holford N. (1996). The response of levodopa in Parkinson's disease: imposing pharmacological law and order, *Ann Neurol* 39:561-573.

Oiwa Y *et al.* (2002). Dopaminergic neuroprotection and regeneration by neurturin assessed by using behavioral, biochemical and histochemical measurements in a model of progressive Parkinson's disease, *Brain Res* 947:271-283.

Olanow C, Jenner P, Brooks D. (1998). Dopamine agonists and neuroprotection in Parkinson's disease, *Ann Neurol* 44(suppl 1):167-174.

Olanow CW. (2003). Present and future directions in the management of motor complications in patients with advanced PD, *Neurology* 61:S24-33.

Olanow CW *et al.* (2003a). A double-blind controlled trial of bilateral fetal nigral transplantation in Parkinson's disease, *Ann Neurol* 54:403-414.

Olanow CW, Schapira AH, Agid Y. (2003b). Neuroprotection for Parkinson's disease: prospects and promises, *Ann Neurol* 53(Suppl 3):S1-2.

Olanow CW, Watts RL, Koller WC. (2001). An algorithm (decision tree) for the management of Parkinson's disease (2001): treatment guidelines, *Neurology* 56:S1-S88.

Parkinson Study Group. (1989). Effect of deprenyl on the progression of disability in early Parkinson's disease, *N Engl J Med* 321:1364-1371.

Parkinson Study Group. (1993). Effects of tocopherol and deprenyl on the progression of disability in early Parkinson's disease, *N Engl J Med* 328:176-183.

Parkinson Study Group. (2000a). Design of a clinical trial comparing pramipexole to levodopa in early PD (CALM-PD). *Clin Neuropharmacol* 23:34-44.

Parkinson Study Group. (2000b). Pramipexole vs levodopa as initial therapy for Parkinson's disease, *JAMA* 284:1931-1938.

Parkinson Study Group. (2002a). A controlled trial of rasagiline in early Parkinson disease: the TEMPO Study, *Arch Neurol* 59: 1937-1943.

Parkinson Study Group. (2002b). Dopamine transporter brain imaging to assess the effects of Pramipexole vs levodopa in Parkinson disease progression, *JAMA* 287:1653-1661.

Parkinson Study Group. (2003). Does levodopa slow or hasten the rate of progression of Parkinson Disease? The results of the ELLDOPA trial, *Neurology* 60(Suppl 1):A80-81.

Perier C *et al.* (2003). The rotenone model of Parkinson's disease, *Trends Neurosci* 26:345-346.

Phelps M. (1992). Positron emission tomography (PET). In Mazziota J, Gilman S, editors: *Clinical brain imaging: principles and applications,* Philadelphia: FA Davis.

Piccini P *et al.* (1999). The role of inheritance in sporadic Parkinson's disease: evidence from a longitudinal study of dopaminergic function in twins, *Ann Neurol* 45:577-582.

Piccini P *et al.* (1997). Dopaminergic function in familial Parkinson's disease: a clinical and 18F-dopa positron emission tomography study, *Ann Neurol* 41:222-229.

Pirker W *et al.* (2000). [123I]beta-CIT SPECT in multiple system atrophy, progressive supranuclear palsy, and corticobasal degeneration, *Mov Disord* 15:1158-1167.

Pirker W *et al.* (2002). Progression of dopaminergic degeneration in Parkinson's disease and atypical parkinsonism: a longitudinal β-CIT SPECT study, *Mov Disord* 17:45-53.

Rascol O *et al.* (2000). A five-year study of the incidence of dyskinesia in patients with early Parkinson's disease who were treated with ropinirole or levodopa, *N Engl J Med* 342:1484-1491.

Ravina BM *et al.* (2003). Neuroprotective agents for clinical trials in Parkinson's disease: a systematic assessment, *Neurology* 60: 1234-1240.

Rinne UK *et al.* (1998). Early treatment of Parkinson's disease with cabergoline delays the onset of motor complications. Results of a double-blind levodopa controlled trial. The PKDS009 Study Group, *Drugs* 55(Suppl 1):23-30.

Sawle G *et al.* (1994). Separating Parkinson's disease from normality: discriminant function analysis of [18F] dopa PET data, *Arch Neurol* 51:237-243.

Schapira AH. (2002). Neuroprotection and dopamine agonists, *Neurology* 58:S9-S18.

Schapira AH *et al.* (1998). Mitochondria in the etiology and pathogenesis of Parkinson's disease, *Ann Neurol* 44:S89-98.

Seibyl JP *et al.* (2002). 123 β-CIT SPECT Imaging assessment of Parkinson disease patients treated for six months with a neuroimmunoplilin ligand, Nil A, *Neurology* 58(suppl3):A203.

Seibyl JP *et al.* (1995). Decreased single-photon emission computed tomographic [123I]beta-CIT striatal uptake correlates with symptom severity in Parkinson's disease, *Ann Neurol* 38:589-598.

Sherer TB *et al.* (2003). Mechanism of toxicity in rotenone models of Parkinson's disease, *J Neurosci* 23:10756-10764.

Shoulson I. (1998a). DATATOP: a decade of neuroprotective inquiry. Parkinson Study Group. Deprenyl and tocopherol antioxidative therapy of parkinsonism. *Ann Neurol* 44:S160-166.

Shoulson I. (1998b). Experimental therapeutic of neurodegenerative disorders: unmet needs, *Science* 282:1072-1074.

Shoulson I. (1998c). Where do we stand on neuroprotection? Where do we go from here?, *Mov Disord* 13(Suppl 1):46-48.

Shoulson I *et al.* (2002). Impact of sustained deprenyl (selegiline) in levodopa-treated Parkinson's disease: a randomized placebo-controlled extension of the deprenyl and tocopherol antioxidative therapy of parkinsonism trial, *Ann Neurol* 51:604-612.

Shoulson I *et al.* (2001). A randomized, controlled trial of remacemide for motor fluctuations in Parkinson's disease, *Neurology* 56:455-462.

Shults CW. (2003). Coenzyme Q10 in neurodegenerative diseases, *Curr Med Chem* 10:1917-1921.

Shults CW *et al.* (2002). Effects of coenzyme Q10 in early Parkinson disease: evidence of slowing of the functional decline, *Arch Neurol* 59:1541-1550.

Shults CW, Schapira AH. (2001). A cue to queue for CoQ?, *Neurology* 57:375-376.

Siderowf AD, Holloway RG, Stern MB. (2000). Cost-effectiveness analysis in Parkinson's disease: determining the value of interventions, *Mov Disord* 15:439-445.

Skipper L, Farrer M. (2002). Parkinson's genetics: molecular insights for the new millennium, *Neurotoxicology* 23:503-514.

Snow B *et al.* (1993). Human positron emission tomographic [18F]flouordopa studies correlate with dopamine cell counts and levels, *Ann Neurol* 34:324-330.

Spooren WP *et al.* (2001). Novel allosteric antagonists shed light on MGLU(5) receptors and CNS disorders, *Trends Pharmacol Sci* 22:331-337.

Tanner CM *et al.* (1999). Parkinson disease in twins: an etiologic study, *JAMA* 281:341-346.

Tatton WG *et al.* (2002). Propargylamines induce antiapoptotic new protein synthesis in serum- and nerve growth factor (NGF)-withdrawn, NGF-differentiated PC-12 cells, *J Pharmacol Exp Ther* 301:753-764.

Tetrud JW, Langston JW. (1989). The effect of deprenyl (selegiline) on the natural history of Parkinson's disease, *Science* 245:519-522.

Truong JG *et al.* (2003). Pramipexole increases vesicular dopamine uptake: implications for treatment of Parkinson's neurodegeneration, *Eur J Pharmacol* 474:223-226.

Turjanski N, Lees AJ, Brooks DJ. (1999). Striatal dopaminergic function in restless leg syndrome: 18F-dopa and 11C-raclopride PET studies, *Neurology* 52:932-937.

Ugarte SD *et al.* (2003). Effects of GDNF on 6-OHDA-induced death in a dopaminergic cell line: modulation by inhibitors of PI3 kinase and MEK, *J Neurosci Res* 73:105-112.

Van Den Eeden SK *et al.* (2003). Incidence of Parkinson's disease: variation by age, gender, and race/ethnicity, *Am J Epidemiol* 157: 1015-1022.

van Dyck CH *et al.* (2002). Age-related decline in dopamine transporters: analysis of striatal subregions, nonlinear effects, and hemispheric asymmetries, *Am J Geriatr Psychiatry* 10:36-43.

Vander Borght T *et al.* (1995). The vesicular monoamine transporter is not regulated by dopaminergic drug treatments, *Eur J Pharmacol* 294:577-583.

Vingerhoets FJ *et al.* (1994). Positron emission tomographic evidence for progression of human MPTP-induced dopaminergic lesions, *Ann Neurol* 36:765-770.

Volkow ND *et al.* (1994). Decreased dopamine transporters with age in healthy human subjects, *Ann Neurol* 36:237-239.

Welsh M *et al.* (2003). Development and testing of the Parkinson's disease quality of life scale, *Mov Disord* 18:637-645.

Whone A *et al.* (2002). The REAL-PET study: Slower progression in early Parkinson's disease treated with ropinirole compared with L-Dopa, *Neurology* 58(suppl 3):A82-A83.

Wichmann T, DeLong, M. R. (2003). Pathophysiology of Parkinson's disease: the MPTP primate model of the human disorder, *Ann N Y Acad Sci* 991:199-213.

Wilson JM *et al.* (1996). Differential changes in neurochemical markers of striatal dopamine nerve terminals in idiopathic Parkinson's disease, *Neurology* 47:718-726.

Wooten GF. (2003). Agonists vs levodopa in PD: the thrilla of whitha, *Neurology* 60:360-362.

Youdim MB *et al.* (2003). The essentiality of Bcl-2, PKC and proteasome-ubiquitin complex activations in the neuroprotective-antiapoptotic action of the anti-Parkinson drug, rasagiline, *Biochem Pharmacol* 66:1635-1641.

Zhang C *et al.* (2001). Regeneration of dopaminergic function in 6-hydroxydopamine-lesioned rats by neuroimmunophilin ligand treatment, *J Neurosci* 21:RC156.

Zigmond MJ *et al.* (1990). Compensations after lesions of central dopaminergic neurons: some clinical and basic implications, *Trends Neurosci* 13:290-296.

9

Current Neurosurgical Treatments for Parkinson's Disease: Where Did They Come From?

Benjamin L. Walter, MD
Aviva Abosch, MD, PhD
Jerrold L. Vitek, MD, PhD

The history of surgery for Parkinson's disease (PD) provides a window from which we can view the evolution of our understanding of the pathophysiology of the disease, advances in surgical technique, and the development of imaging technology. Improvement in surgical outcomes has occurred as a direct result of the union of these disciplines, which has led to the development of current surgical approaches for the treatment of PD. Surgical therapy for PD began as a desperate attempt to improve the quality of life of those afflicted with this disease. Early surgeries were based on a rudimentary understanding of the motor system and resulted from the earnest attempt by physicians and surgeons to aid those most severely afflicted with this disorder. Success was rare, but frequent enough to warrant further efforts. Surgery for PD eventually evolved on the shoulders of scientific studies, which both exposed the problems associated with these approaches and led to the development of new solutions. While we stand in the twenty-first century and marvel at the work of our predecessors, wondering what they were thinking when extirpating the motor cortex, yet marveling at the insights they established for those to follow, we can only wonder what our colleagues of the future will say when examining our work 100 years from now. This chapter reviews early empirical studies of surgical therapy and the evolution that has led to present neurosurgical approaches for the treatment of PD. We hope to provide a picture of the remarkable progression that has taken place in this field, the unique mixture of disciplines that have led to current therapy, and the subsequent application of these same techniques for other disorders of neurological function.

EMPIRICALLY BASED SURGICAL STRATEGIES

The earliest surgeries for Parkinson's disease were predominantly empirically based, derived from a limited understanding of the pathogenesis of parkinsonism and the neuroanatomical basis for motor control. The driving force behind attempting such therapies lay in the desperate need of patients, and the willingness on the part of patients and their physicians to explore experimental therapeutic approaches. At the turn of the twentieth century, the understanding of motor systems focused on peripheral mechanisms and was derived from Sherrington's work describing rigidity and the role of propriospinal reflexes in the decerebrate animal model (Sherrington, 1898). This led Foerster in 1911 to use bilateral dorsal rhizotomy to treat spasticity in a series of children with cerebral palsy, with beneficial results (Foerster, 1913). Following this work, Leriche, reported the first cases of parkinsonian rigidity treated with dorsal rhizotomy; however, results were mixed (Leriche, 1912). Nonetheless, the work of Foerster and Leriche fueled further investigations and continued the focus on the peripheral nervous system until 1930, when Pollock and Davis (1930) reported additional cases of dorsal rhizotomy in parkinsonian patients. Although dorsal rhizotomy reduced rigidity in some muscles, tremor remained, and some patients developed permanent flexion of the wrist, biceps, and neck, leading most to conclude it was not an effective therapy.

After the 1916-1927 epidemic of encephalitis lethargica of von Economo, other peripheral surgeries including neurectomy, sympathetic ramisectomies, and ganglionectomies were performed with the presumed notion that these structures may harbor residual infection contributing to the pathogenesis of PD (The Matheson Commission, 1929). Results were not encouraging and the subsequent use of peripheral nervous system surgeries became increasingly infrequent (The Matheson Commission, 1929, 1932).

Although there was early interest in the peripheral nervous system, surgical therapies for movement disorders that focused on the central nervous system (CNS) were slow to develop. Victor Horsley, in 1908, used an open approach to excise the precentral gyrus of a patient with severe hemiathetosis (Horsley, 1909). Horsley was aware of the role of the precentral gyrus in motor function but expected a certain degree of redundancy in the control of volitional movements. Indeed, the patient experienced complete postoperative resolution of the disabling spasmodic movements, albeit with transient monoplegia followed by a gradual return of some function. After this result, various parts of the pyramidal motor system, including motor and premotor cortex and the corticospinal tracts, were ablated by several groups, guided by the conviction that it was necessary to interrupt pathological signals between the cortex and the periphery. The observation that parkinsonian tremor and rigidity improved after stroke was thought to support this idea. Bucy and Case in 1937 first reported the resection of motor cortex for the treatment of PD tremor (Bucy and Case, 1939) (Figure 9.1). Symptoms typically improved after resection or lesioning of the pyramidal system, but the consequent motor deficit resulted in significant disability. Shortly thereafter, Klemme reported a series of cases with premotor cortex extirpation accompanied by relief of tremor in over half of patients, without paralysis. However, there was still an unacceptably high (17%) rate of postoperative mortality (Klemme, 1940). Surgeons were still reluctant to target the basal ganglia, as a prominent neurosurgeon of the time, Walter Dandy, was convinced that lesions in the extrapyramidal system would result in disorders of consciousness. Corticectomy was ultimately abandoned in the early 1950s because of inconsistent postoperative improvement, as well as the substantial rate of associated complications.

The extrapyramidal system became a focus of movement disorder surgery in the late 1930s. This change in focus was driven by refined anatomical knowledge, in particular by Hassler's demonstration of a role for the substantia nigra pars compacta (SNc) in PD (Hassler, 1938), and by Spatz's characterization of the extrapyramidal motor system and suggestion that it may play a role in the generation of movement disorders (Spatz, 1927). Influenced by these findings, Russell Meyers in 1939 targeted the basal ganglia deliberately and directly.

He hoped to disrupt motor circuits thought to mediate the motor symptoms of Parkinson's disease without causing significant paralysis. In his first case, he operated on a patient with postencephalitic parkinsonism using an open transventricular approach to the most accessible part of the basal ganglia—the caudate nucleus (Meyers, 1951). The patient had nearly complete resolution of her tremors and improved functional status, allowing her to return to work. Subsequent cases were not as promising, and, citing the lack of a parkinsonian animal model, he went on to systematically target various extrapyramidal structures, including the anterior limb of the internal capsule and the pallidofugal fibers (Meyers, 1951). In the end, he concluded that sectioning pallidofugal fibers was the most effective; however, all of these approaches carried an unacceptably high mortality rate of 16% to 41%, likely a consequence of the open approach. Nevertheless, this experience established the necessary groundwork for the less invasive approach to the basal ganglia afforded by "closed" stereotactic surgery and demonstrated that it was possible to target the extrapyramidal system in the treatment of movement disorders without impairing consciousness or trading involuntary movements for motor deficits. A major first step had been taken.

In 1952, Irving Cooper accidentally ligated the anterior choroidal artery during an attempted mesencephalic pedunculotomy on a patient with postencephalitic tremor and rigidity (Cooper, 1953). Cooper aborted the procedure without having performed the pedunculotomy, but the patient awoke from surgery with a significant improvement in contralateral tremor and rigidity, and no motor deficit. This observation led to Cooper's hypothesis that occlusion of the anterior choroidal artery had resulted in a medial pallidal infarct, and the resulting improvement in the patient's symptoms—a hypothesis that was consistent with anatomical data of the era. Cooper proceeded to ligate the anterior choroidal artery deliberately in a series of 55 patients with movement disorders before ultimately abandoning the procedure because the variable territory irrigated by the anterior choroidal artery led to unintended neurological deficits, including one patient with a postoperative visual field deficit, one with aphasia, and three with contralateral hemiplegia (Cooper and Bravo, 1958). Although results were inconsistent and associated with a high incidence of side effects, the globus pallidus and ansa lenticularis remained favorite targets for the treatment of Parkinson's disease during the 1950s.

Thus within 50 years, the focus of surgery had evolved from the peripheral nervous system to the CNS, and within the CNS the focus had shifted from the pyramidal motor system to the extrapyramidal system.

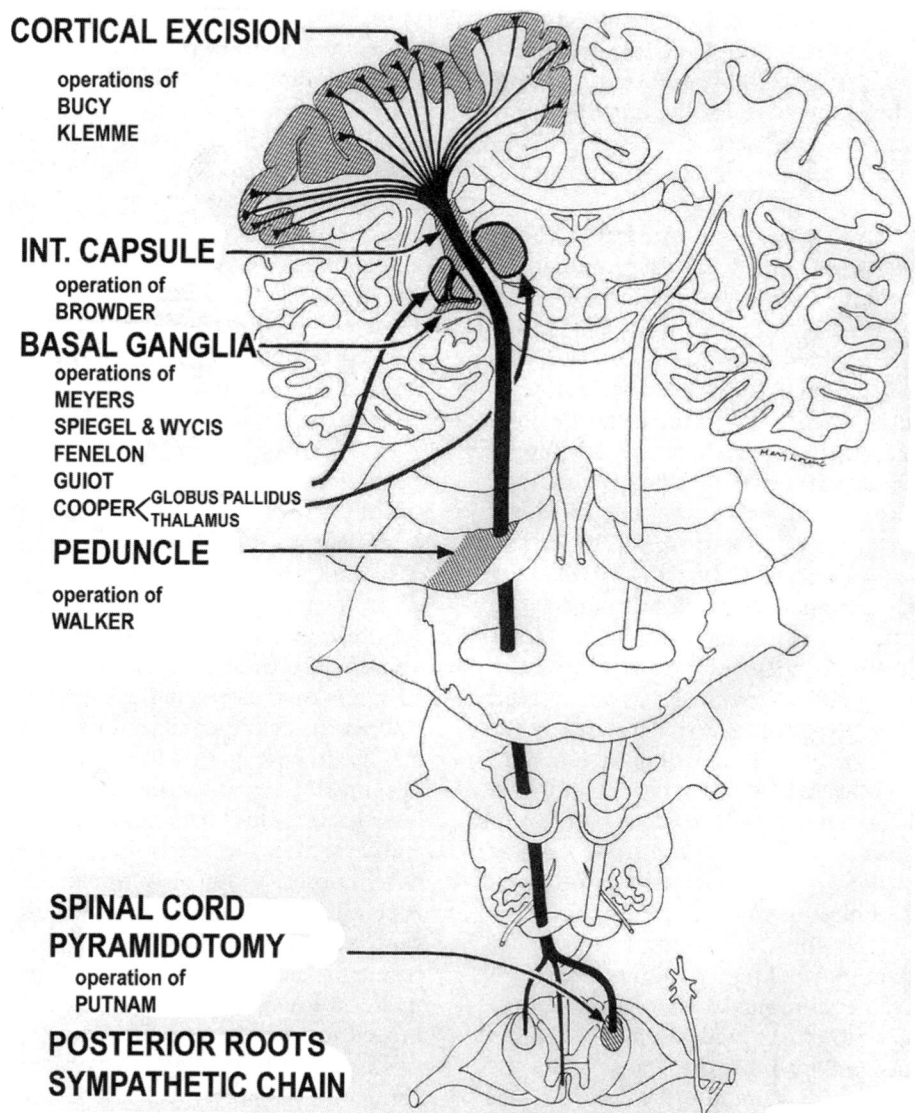

CORTICAL EXCISION

operations of
BUCY
KLEMME

INT. CAPSULE

operation of
BROWDER

BASAL GANGLIA

operations of
MEYERS
SPIEGEL & WYCIS
FENELON
GUIOT
COOPER ⟨GLOBUS PALLIDUS
       ⟨THALAMUS

PEDUNCLE

operation of
WALKER

**SPINAL CORD
PYRAMIDOTOMY**

operation of
PUTNAM

**POSTERIOR ROOTS
SYMPATHETIC CHAIN**

FIGURE 9.1   Sites of central and peripheral nervous system surgeries attempted for PD. As the surgical treatment of PD evolved, surgeons targeted sympathetic chain ganglia, dorsal roots, basal ganglia, and corticospinal fibers at the level of the pyramids, cerebral peduncle, internal capsule, primary motor cortex, or premotor cortex. Adapted from Cooper, 1960, with permission.

At the same time, understanding of the pathogenesis and pathophysiology of PD had made similar transformations and became focused on the extrapyramidal system and on the basal ganglia, in particular. Still, little was known about the actual function of the basal ganglia, their role in generating the symptoms of Parkinson's disease, or the functional arrangement of basal ganglia circuits. The surgical techniques were still crude, depended on direct visualization of the target structure, and resulted in a wide variety of outcomes. Technological advances were needed and would soon be realized with the application of animal stereotaxis to the human CNS.

EARLY STEREOTACTIC SURGERY

Horsley and Clarke (1908) described the first animal stereotaxic apparatus, by which Cartesian coordinates (x, y, and z) could be assigned to every point within the brain. Their system was based on three Cartesian planes (midsagittal; basal, running through the external auditory canals; and coronal, perpendicular to the other two planes and also running through the external auditory canals). Such stereotaxic systems allowed investigators to insert an electrode or cannula accurately into a desired subcortical target area, with minimal damage to the overlying cortex and white matter. However, variability between the human skull and underlying intracranial anatomy was far too great to attempt the use of such a technique in the human. Another 40 years elapsed before the combination of intraoperative radiography and the development of a human stereotactic atlas would allow for successful stereotactic surgery in the treatment of human diseases.

In 1947, Spiegel and Wycis used intraoperative radiography to delineate the pineal gland and the foramen of Munro by injecting contrast into the lumbar thecal sac and then obtaining skull x-ray films (Spiegel et al., 1947a). This technique enabled them to assign Cartesian coordinates based not on skull landmarks, but on deep intracerebral landmarks, and served as the basis for publishing a human stereotactic atlas (Spiegel and Wycis, 1952). Spiegel and Wycis performed their first stereotactic surgeries in human patients in 1947, using the dorsomedial thalamotomy as an alternative to frontal lobotomy for psychiatric patients (Spiegel et al., 1947b). A year later, they performed the first stereotactic pallidotomy to treat chorea in a patient with Huntington's disease (Spiegel, 1966). Interestingly, they did this on the premise that pallidotomy would cause hypokinesis and increased tone. Contrary to their expectations, rigidity was never found to be a complication of pallidal lesions and, in fact, when they began to use lesions of the pallidum and ansa lenticularis to treat "choreatic hyperkinesis" in parkinsonism, no such rigidity resulted. It was not until later that it became clear that pallidotomy was actually effective in improving rigidity. Notably, operative mortality using the stereotactic apparatus was only 2.0%, compared to 15.7% using Meyers' open approach (Meyers, 1951). The development of stereotaxis revolutionized functional surgery by providing a coordinate system with which the human brain could be mapped and by improving the safety of the procedures. As such, human stereotactic technique was the necessary second step toward modern functional neurosurgery.

Shortly thereafter, Talairach and colleagues in Paris (Talairach et al., 1949), Hassler and Reichert in Freiburg (Reichert and Wolf, 1951), Narabayashi in Japan (Narabayashi and Okuma, 1954), and Leksell in Sweden (Leksell, 1949) all developed stereotactic devices and began performing stereotactic pallidotomies. The Paris and Freiburg groups began using the intercommissural line (connecting the anterior and posterior commissures) as the reference line for assigning Cartesian coordinates. Lars Leksell's device was the first arc-centered (vs patient-centered) stereotactic apparatus (Leksell, 1949). Despite these advances, targeting accuracy remained inconsistent and the optimal target for ablative surgery for PD remained unclear.

Early stereotactic pallidotomies were generally placed in the anterior dorsal portion of the pallidum. The results of these anterodorsal pallidotomies were mixed, with some benefit to rigidity, but no long-lasting effect on tremor (Gillingham et al., 1960; Hassler and Reichert, 1954; Hassler et al., 1960; Spiegel and Wycis, 1954). Hassler was disappointed with the meager results achieved for tremor following pallidotomy and found that significant tremor suppression in his hands required a very large lesion, which also produced postoperative confusion, sedation, and weakness, likely from lesion extension into the posterior limb of the internal capsule (Hassler et al., 1960). He subsequently performed detailed retrograde tracing studies and characterized the pallidothalamic and thalamocortical projections. This led him to explore ablative therapy in thalamic areas V.o.a., V.o.p and L.p.o.—areas that receive afferents from the pallidum and cerebellum and project to area 6 and motor cortex (Hassler, 1949, 1953). Hassler's subsequent work demonstrated that lesions of V.o.a./V.o.p. had some effect on rigidity and lesions of V.o.p./V.i.m. were the best target for tremor (Hassler et al., 1965, 1970). He felt that thalamic lesions were superior to pallidal lesions for tremor suppression and, at the same time, he was able to avoid making lesions in proximity to the internal capsule. Cooper also concluded that thalamic lesions were better, but he arrived at this in a more serendipitous fashion. Specifically, Cooper attempted to lesion globus pallidus interna (Gpi) in a patient and achieved excellent postoperative tremor suppression. When the patient died from other causes, an autopsy revealed that the targeting had been off by centimeters and the lesion was actually in the ventrolateral thalamus (Guridi and Lozano, 1997). As a result of these findings, and the inconsistent results of pallidotomy using the techniques of the era, most groups moved to lesioning the thalamus.

Leksell was similarly frustrated by the inconsistent results of pallidotomy for tremor, but instead of abandoning the pallidum, he varied the location of the lesion within the pallidum. Svennilson, a neurologist working with Leksell, systematically analyzed lesion location and outcome in 81 patients who underwent pallidotomy. Leksell varied the lesion site in the first third of this series, finding that the optimal site for reduction of tremor and rigidity was in the ventral posterior pallidum, just anterior to the internal capsule and 20 mm lateral to midline—a target that was substantially posterior to that reported in previous studies (Figure 9.2). After optimizing lesion size in the middle third of patients, Leksell kept the site and size constant in the final 20 patients and achieved a remarkable reduction in tremor and rigidity in 19/20 (95%) of these patients (Svennilson et al., 1960). Unfortunately, these results did not draw much attention at the time, as many practitioners, swayed by Hassler's results, had moved to thalamic surgery. Shortly thereafter, advances in medical therapeutics would provide an effective pharmacological therapy for the treatment of PD and arrest advances in surgical therapy for PD for nearly three decades.

THE INFLUENCE OF MEDICAL THERAPEUTIC ADVANCES

In 1959, Hornykiewicz and Ehringer found that dopamine was decreased in striatal brain homogenates from patients with PD (Ehringer and Hornykiewicz, 1960). Hornykiewicz then convinced the neurologist Birkmayer in 1961 to inject his PD patients with L-dopa, and found striking short-term amelioration of akinesia (Birkmayer and Hornykiewicz, 1961). This was followed by another pivotal report by Cotzias in 1967, documenting the efficacy of oral doses of D, L-dopa (Cotzias et al., 1967). When the neurosurgeon, Irving Cooper, heard of the effectiveness of levodopa therapy, he actually diverted 700 of his patients scheduled for surgery to Dr. Cotzias' clinic for enrollment in this pivotal study. These initial results transformed the treatment of PD, leading to the multiple formulations of levodopa and dopamine agonists in use today. With the advent of effective medical therapies, surgery for PD all but vanished.

REVIVAL OF SURGICAL THERAPIES

Before long, however, the shortcomings of levodopa in the treatment of PD became obvious. By the early 1970s, it was clear that the response to levodopa became less predictable and increasingly short-lived with continued therapy. Despite advances in carboxylase inhibitors and dopamine agonists, medical therapy had not provided an effective sustainable response. Management of levodopa-induced dyskinesia proved even more problematic. Meanwhile, thalamotomies were continued by a handful of surgeons including Hassler's and Cooper's groups (Hassler et al., 1979; Scott et al., 1970). These groups predominately operated on patients with severe, medically refractory tremor and those who could not tolerate medical therapies. At the same time, some groups were beginning to discover that thalamotomy, when extended anteriorly, was effective for reducing rigidity and suppressing levodopa-induced dyskinesias (Cooper, 1977; Matsumoto et al., 1976; Narabayashi et al., 1984). Thalamic lesions generally have not been reported to improve bradykinesia or akinesia and have been associated with the postoperative worsening of speech and gait disorders in some patients (Selby, 1967; Speelman, 1991).

Although surgical therapy continued at a less than prolific pace through the 1960s, 1970s, and 1980s, significant advancements continued to occur, setting the stage for a revival of pallidotomy and later for the introduction of chronic stimulation therapies in the 1990s. Human microelectrode recording (Figure 9.3) was pioneered in France by Albe-Fessard and Guiot in 1960 (Albe-Fessard et al., 1963, 1966, 1967), allowing for physiological confirmation of the target and surrounding structures during surgery. Microelectrode recording has also allowed for the intraoperative testing of physiological hypotheses regarding the basal ganglia and the structures with which they are interconnected.

Technology for radiographic targeting also made critical advances during the later half of the twentieth century. In 1972, Hounsfield (Ambrose and Hounsfield, 1973) and Cormack independently developed computed tomography (CT), with installation of the first commercial scanners between 1974 and 1976. In 1973, Lauterbur

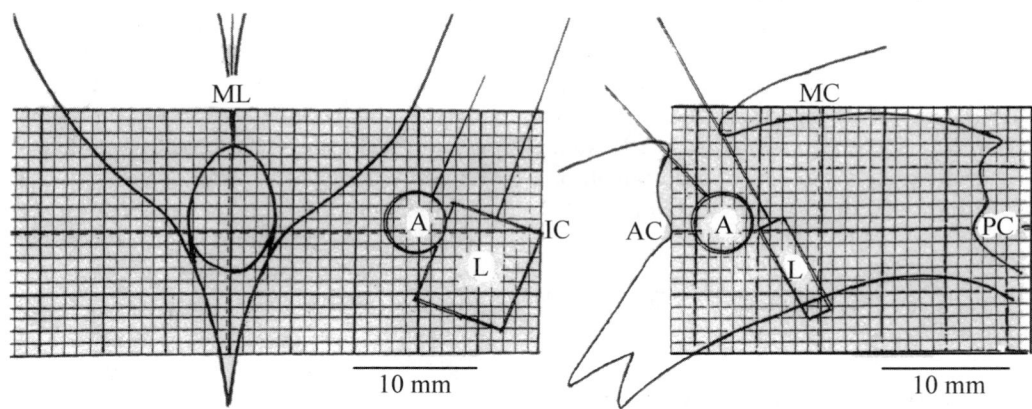

FIGURE 9.2 Leksell's target in GPi. Leksell found that lesions in the ventroposterolateral pallidum resulted in dramatic reduction of tremor and rigidity. Leksell's target (L) was centered at 20 mm lateral to the AC-PC line and is shown relative to the target that most others at that time had used (A). AC, anterior commissure; PC, posterior commissure; MC, mid-commissure; ML, midline. From Laitinen et al., 1992, with permission.

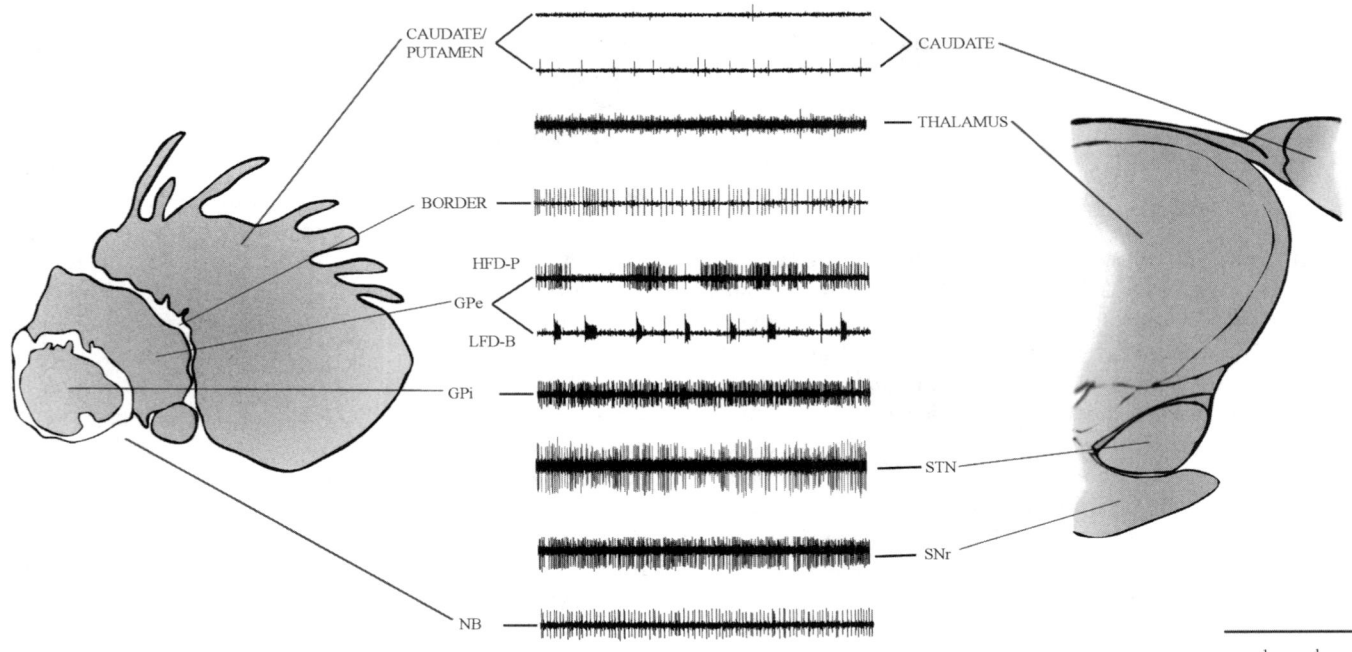

FIGURE 9.3 Microelectrode physiology encountered when targeting GPi and STN. Typical neural activity is shown for striatum (putamen and caudate), thalamus, GPe (globus pallidus externa), GPi (globus pallidus interna), subthalamic nucleus (STN), substantia nigra pars reticulata (SNr), and NB (nucleus basalis). Populations of high frequency discharge pausing cells (HFD-P) and low frequency discharge bursting cells (LFD-B) are found in GPe. A line drawing shows a parasagittal representation of these structures from the Shaltenbrand and Bailey Atlas (Schaltenbrand and Bailey, 1959).

(1980) developed magnetic resonance imaging (MRI) and in 1983, the first commercial MRI scanners were available in Europe. Although these technologies were crude at first, they allowed for the visualization of lesion size and location, provided critical feedback on targeting accuracy, and ushered in the era of CT- and MRI-based stereotactic targeting systems.

In 1982, another major advance occurred in PD research, with the discovery that 1-methyl-4-phenyl-1, 2,3,6-tetrahydropyridine (MPTP) was the etiological agent responsible for the development of parkinsonism in a group of California heroin addicts who used tainted recreational drugs (Langston *et al.*, 1983). This discovery led to the development of the research tool that Russell Myers longed for in the early 1950s—the development of a nonhuman primate model for PD (Burns *et al.*, 1983).

MODEL-BASED APPROACHES: REVIVAL OF PALLIDOTOMY

The MPTP primate model of PD closely resembles the cardinal motor symptoms, as well as the biochemical and pathological profile of the idiopathic human condition. It was not long before Fillion *et al.*, and Miller and DeLong used microelectrode recording techniques to

study the physiological changes in the experimental model. These studies demonstrated increased tonic neuronal firing rates in the subthalamic nucleus (STN) and in GPi, as well as decreased tonic firing rates in globus pallidus externa (GPe) (Filion *et al.*, 1985; Miller and DeLong, 1987b).

The understanding gained from microelectrode studies, together with a plethora of new neuroanatomical tract tracing studies, led to new hypotheses about the functional organization of the basal ganglia. Alexander, DeLong, and Strick summarized the early anatomical studies and concluded that instead of corticostriatal inputs converging or "funneling" into the basal ganglia, evidence suggested that cortico-striato-pallido-thalamo-cortico circuits remain segregated in parallel pathways (Alexander *et al.*, 1986). They proposed the existence of separate "motor," "prefrontal," "oculomotor," and "limbic" circuits with similar organization but with each circuit projecting from different cortical areas to separate areas of striatum, pallidum/SNr, and thalamus. Further studies led to the hypothesis that each of these circuits projected through a "direct" striatal-GPi/SNr-thalamus loop and an "indirect" striatal-GPe-STN-GPi/SNr-thalamus loop (Figure 9.4) (Alexander *et al.*, 1990). In the motor circuit, supplementary motor area (SMA), arcuate premotor, primary motor and somatosensory cortex have somatotopically organized projections to the putamen, which projects to the caudal and ventrolateral two thirds of GPe and GPi and caudolateral portion of the SNr. The sensorimotor portion of GPi then projects to the ventralis lateralis pars oralis (VLo;Voa, Vop human analog) and lateral ventral anterior pars caudalis (VApc;VA human analog), which then project back to the SMA and premotor cortex, respectively (Alexander *et al.*, 1990). Models of basal ganglia function were proposed first by Albin *et al.* (1989) and later by DeLong (1990b). These investigators proposed a model for hypokinetic disorders based on changes in mean discharge rates of basal ganglia structure, which became known as the rate model for PD.

The rate model of PD suggests that the loss of dopaminergic cells in the SNc causes a differential effect on neuronal activity in the direct and indirect pathways, and the consequent change in neuronal discharge rates in these pathways leads to the motor symptoms characteristic of this disease (Figure 9.4). Loss of dopaminergic input at excitatory striatal D1 receptors was proposed to lead to a decrease in inhibitory activity from the striatum to GPi in the direct pathway. In the indirect pathway, loss of dopamine at inhibitory D2 striatal synapses was proposed to produce a relative increase in activity of inhibitory putaminal neurons projecting to GPe, thus leading to decreased neuronal activity in GPe. Resultant decreased inhibitory output from GPe to STN leads to increased mean discharge rates in the STN and excessive excitation of GPi. Thus, by both decreased inhibition of GPi via the direct pathway and increased excitation of

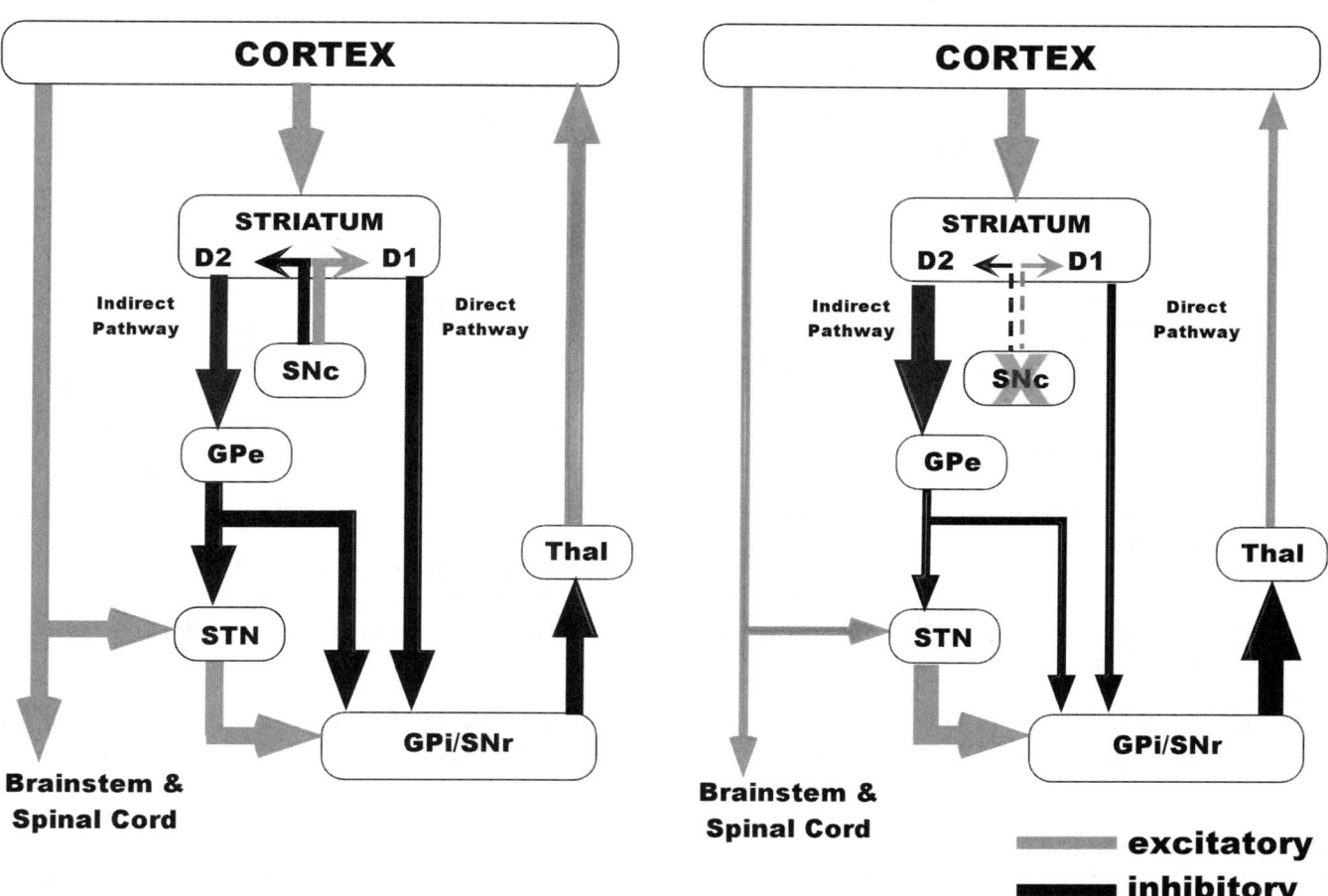

FIGURE 9.4 Schematic illustration of the rate model for PD. Indirect and direct pathways from the striatum to basal ganglia output nuclei are represented by the arrows. GPe, globus pallidus pars externa; GPi, globus pallidus pars interna; STN, subthalamic nucleus; SNr, substantia nigra pars reticulata; SNc, substantia nigra pars compacta; Thal, thalamus. Dopamine (DA) receptor subtypes 1 and 2 are represented as D1 and D2, respectively. Wider lines represent an increase while thinner lines represent a decrease in neuronal activity.

GPi by STN in the indirect pathway, GPi inhibitory output is excessively increased, leading to suppression of activity in thalamic and brainstem nuclei receiving afferent input from GPi. Therefore, in the motor pathway, there is inhibition of thalamocortical activity projecting to the SMA and other cortical motor areas, which was considered the underlying factor for the development of hypokinesia and other motor signs associated with PD. In support of this circuit theory, Bergman, Wichmann, and DeLong found that the cardinal parkinsonian symptoms were ameliorated by excitotoxic (ibotenic acid) STN lesions and could be temporarily improved with reversible inactivation of the STN with the γ-aminobutyric acid-A agonist, muscimol (Bergman et al., 1990; Wichmann et al., 1994).

During these investigations into the pathophysiology of PD, Laitinen reintroduced pallidotomy into the surgical arena by publishing a series of patients he operated on from 1985 to 1990 using Leksell's successful, but forgotten techniques (Laitinen et al., 1992). Unlike Leksell in his series, Laitinen and colleagues now also had the aid of CT-guided stereotactic targeting. They reported that 92% of patients had near complete loss of rigidity, 81% of those with tremor had near complete tremor reduction, and improvements in dystonia, gait, and levodopa-induced dyskinesia were also noted. From the MPTP-primate work, it was clear that improvement of parkinsonian motor symptoms required targeting of the sensorimotor circuit. This circuit passes ventroposterolaterally in GPi, accounting for the success of Leksell's empirically chosen target and the failure of earlier anterodorsal GPi lesions. Furthermore, electrophysiological mapping of the basal ganglia allowed for confirmation of the location of the sensorimotor region. Several subsequent studies used improved MRI-based stereotactic targeting and intraoperative electrophysiological mapping to place lesions in the sensorimotor portion of GPi with consistently beneficial results (Baron et al., 1996, 2000; Dogali et al., 1995; Lozano et al., 1995).

Contemporary surgical studies have also benefited from more rigorously blinded study design and standardized rating scales (Fahn and Elton, 1987). A prospective single-blind trial randomizing 36 patients to unilateral pallidotomy vs best medical therapy showed that unilateral pallidotomy improved the total Unified Parkinson's Disease Rating Scale (UPDRS) score by 32% over the presurgical baseline while patients with medical therapy alone worsened by 5%. There was significant improvement in tremor, rigidity, bradykinesia, dyskinesias, gait, and freezing (Vitek et al., 2003) in the pallidotomy group.

Although clearly more effective than medical therapy alone in appropriately selected patients, pallidotomy still had several limitations. While unilateral procedures provided relief from all of the cardinal symptoms on the contralateral side with modest ipsilateral effect, gait and balance did not show consistent long-term improvements. Bilateral lesions were required to substantially ameliorate gait and balance disturbances for the majority of patients (De Bie et al., 2002; Favre et al., 2000; Intemann et al., 2001; Merello et al., 2001). Unfortunately, bilateral lesions were associated with an increase in the rate of side effects. Most studies reported significant hypophonia, dysarthria, and worsening cognitive and neuropsychiatric function after bilateral pallidotomy (De Bie et al., 2002; Favre et al., 2000; Intemann et al., 2001; Merello et al., 2001). Although cognitive side effects may occur as a result of encroachment on the more anterior nonmotor areas, hypophonia has been a persistent side effect reported by most studies. Similar to bilateral pallidotomy, bilateral thalamotomy was also associated with a high incidence of side effects that precluded the use of such a procedure except under the most dire of circumstances (Matsumoto et al., 1976; Schuurman et al., 2000). Neurorestorative approaches were in development at this time, with tremendous enthusiasm for adrenal and fetal tissue transplantation. This enthusiasm leads physicians to take a more cautious approach toward ablative procedures, not wanting to preclude the patient from benefiting from the future use of these technologies. During this time, work with electrical stimulation was underway and was on course to provide a viable alternative to ablative therapies.

DEEP BRAIN STIMULATION

High frequency stimulation was found to relieve parkinsonian tremor as early as the 1960s, when it was used during intraoperative evaluations by Hassler to test for the optimal site for thalamotomy (Hassler et al., 1960). Cooper, noting Sherrington's early work showing cerebellar stimulation in the decerebrate cat reduced extensor tone (Sherrington, 1897), implanted chronic cerebellar stimulators in patients with spasticity and cerebral palsy beginning in 1973 (Cooper, 1973a, 1973b). He also implanted cerebellar stimulators for the control of epilepsy, based on the theory that cerebellar stimulation increased CNS inhibition (Cooper, 1974). Results of his studies were mixed, and drew criticism for poor experimental design and nonstandard evaluation scales (Rosenow et al., 2002). In 1979, he used deep brain stimulation (DBS) with a quadripolar Medtronic electrode in the pulvinar and internal capsule in a patient, with reported relief of her chronic pain and spasticity. He subsequently tried thalamic and internal capsule DBS for dystonia, dysarthria, and torticollis (Cooper et al., 1980).

In 1987, as Laitinen was reexploring pallidotomy as an alternative to thalamotomy, Benabid et al. and Blond *et al.*, looking for a way to relieve parkinsonian tremor bilaterally, implanted the Medtronic quadripolar electrode in the Vim thalamus in patients who already had a contralateral thalamotomy (Benabid *et al.*, 1987; Blond and Siegfried, 1991). Benabid reported that 88% of patients had improvement in their tremor (Benabid *et al.*, 1996). However, many of these patients eventually required further surgery because of progression of akinetic-rigid symptoms or dyskinesias for which Vim-DBS was ineffective (Benazzouz *et al.*, 2002). The lack of a reliable effect on the cardinal parkinsonian symptoms other than tremor has limited the use of Vim-DBS for PD and led to the exploration of other targets.

By 1992, several critical events had taken place that would transform the field: (1) there was a new understanding of the organization and function of the basal ganglia motor circuits with the proposed rate model, (2) there was a renewed interest in the pallidal target brought about by Laitinen's publication of his experience with pallidotomies for PD, and (3) there was evidence from work with thalamic Vim DBS, that deep brain stimulation may produce a similar effect to ablative therapy but one that is reversible and without destruction of the target. It was at this time that Sigfried and Lippitz placed bilateral DBS electrodes in Leksell's target, the ventroposterolateral GPi, in three patients (Siegfried and Lippitz, 1994). Their results were good, and paralleling the effects of pallidotomy, seemed to show improvement in bradykinesia, rigidity, tremor, gait, and speech.

About the same time, and armed with new evidence that STN lesions also relieved parkinsonian symptoms in the MPTP primate model (Bergman *et al.*, 1990), Benazzouz *et al.* (1993) showed that STN stimulation also improved parkinsonism in the MPTP primate model of PD. This finding led to a collaboration with Benabid's group and the first STN DBS implantation in human patients, with excellent results (Pollak *et al.*, 1993). In a large, multicenter, prospective, double-blind, but not randomized study, both the STN and GPi were targeted in patients with advanced PD. Both targets showed significant benefit with bilateral GPi DBS, producing 37% improvement in the "off-medicine" UPDRS motor subscore and STN DBS producing 49% (The Deep-Brain Stimulation for Parkinson's Disease Study Group, 2001). The only trial completed to date that randomized patients to either target involved only 10 patients and showed no significant difference between groups (Burchiel *et al.*, 1999). Different centers use widely different targeting techniques; most implantations in STN are bilateral, while many in GPi are unilateral and there is also great variability in lead placement within each target, making comparison between centers difficult. Currently, although both STN and GPi are used to treat PD, the majority of centers performing these procedures target the STN exclusively.

REEVALUATION OF THE MODELS

With some exceptions, the *rate model* successfully predicts many of the clinical and electrophysiological changes seen before and after functional neurosurgical interventions for PD. Evidence from patients and from the MPTP nonhuman primate model of parkinsonism demonstrates decreased discharge rates in GPe and VLo and increased mean discharge rates in STN and GPi, consistent with the model (DeLong, 1990a; Filion and Tremblay, 1991; Miller and DeLong, 1987a; Vitek *et al.*, 1994, 1998a). However, other observations from the "pre-DBS" era contradict the rate model. In particular, lesions in the motor thalamus do not exacerbate or induce parkinsonism but rather improve tremor and rigidity (Hassler and Dieckmann, 1970; Narabayashi, 1982; Narabayashi *et al.*, 1984). Second, the rate model predicts that a pallidotomy would produce excessive involuntary movements or dyskinesias by disinhibiting the thalamus; however, this is not the case. In fact, pallidotomy is very effective in alleviating drug-induced dyskinesia, hemiballismus, and chorea (Speigel, 1966; Suarez *et al.*, 1997; Vitek *et al.*, 2000). Albin *et al.*, readdressed their model in a 1995 publication and, while pleased with its overall applicability, cited some shortcomings, particularly the ability of the rate model to explain the pathophysiology of dyskinesia and dystonia (Albin *et al.*, 1995). Vitek *et al.* (1999) reported that in patients with dystonia and hemiballism, there were decreased mean discharge rates of neurons in GPe and GPi. More important, he reported that neurons in GPe and GPi had an irregular pattern of neural activity characterized by irregularly grouped discharges separated by intermittent pauses. The characterization of physiological changes in these disorders has led to the development of an alternative pattern model of basal ganglia function (Vitek *et al.*, 1994, 1998b, 1999; Vitek and Giroux, 2000).

The *pattern model* of PD (Figure 9.5) postulates that clinical features in movement disorders do not result exclusively from a change in rate of firing but rather from a change in a combination of physiological characteristics including pattern, somatosensory responsiveness, degree of synchronization, and rate. In the monkey model, 40% of pallidal neurons develop oscillatory behavior after treatment with MPTP, and these oscillations are often at the frequency of the monkey's tremor (Nini *et al.*, 1995; Raz *et al.*, 2000). There is also evidence of increased synchronization of pallidal neurons, which disappears with dopamine replacement (Havazelet Heimer *et al.*, 2001; Raz *et al.*, 1996, 2000). There is also evidence of increased and less specific somatosensory responsiveness of pallidal and thalamic neurons in the MPTP monkey as compared to normal control subjects (Filion *et al.*, 1988; Vitek *et al.*, 1990).

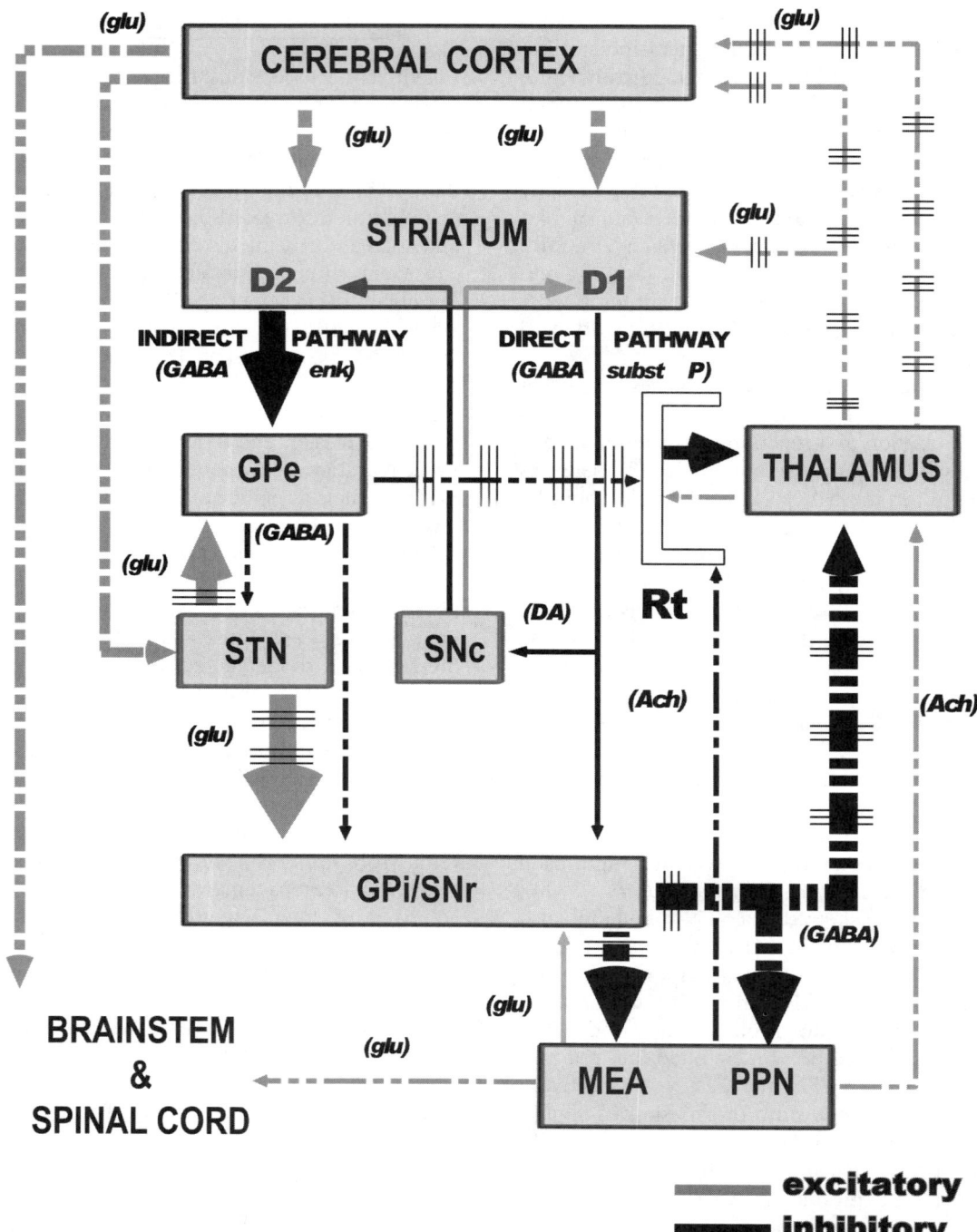

FIGURE 9.5   Revised pattern model. Schematic illustration of the basal ganglia thalamocortical motor circuit incorporating changes in pattern of neuronal activity and alternative pathways from the external segment of the globus pallidus (GPe) and pedunculopontine nucleus (PPN). Interrupted arrows represent an alternation in the pattern of neuronal activity. The grouped lines crossing the projections from the subthalamic nucleus (STN), the internal segment of the globus pallidus (GPi), and thalamus represent the presence of bursting activity. SNc, substantia nigra pars compacta; SNr, substantia nigra pars reticulata; D1, D2, dopamine 1 and 2 receptor subtypes; Rt. reticularis nucleus of the thalamus; GABA, γ-aminobutyric acid; MEA, midbrain extrapyramidal area; glu, glutamate; enk, enkephalin; subst P; substance P; DA, dopamine; Ach, acetylcholine.

MECHANISM OF DEEP BRAIN STIMULATION

With the advent of deep brain stimulation, it has become increasingly clear that in many cases, the rate model may be too simplistic and cannot explain the pathophysiological basis for the development of parkinsonian motor signs nor the mechanism underlying the improvement in these signs after surgical intervention, particularly with DBS. There is an evolving consensus that an alteration in the *pattern* of neural firing plays a key role in the mechanism by which DBS alleviates the motor symptoms of PD and other movement disorders. Electrophysiological studies during STN stimulation in parkinsonian monkeys have documented a net increase in the mean neural discharge rate in GPi cells, suggesting activation of excitatory efferent projections to GPi (Hashimoto *et al.*, 2003). This effect has been noted only at stimulation rates that are therapeutic; at subtherapeutic rates there is no significant difference in the number of neurons with increased, decreased, or unchanged firing rates (Vitek, 2002). Moreover, the pattern of neural activity in GPi was stimulus-locked to STN stimulation, with an increased regularity occurring in GPi neurons during stimulation. These findings are antithetical to the predictions of the rate model; as STN output is increased from clinically effective stimulation, the rate model would suggest a net excitation of GPi neurons, which in turn would inhibit thalamocortical activity causing a contradictory worsening of motor function. This has not been observed during STN stimulation. Some theories suggest that stimulation might "act like a lesion" by depolarization blockade, synaptic exhaustion, or activation of presynaptic inhibitory afferents to the target structure. While neuronal activity may be suppressed in the STN by one or more of these mechanisms, the net output from the STN is increased as evidenced by increased mean discharge rates in GPi during STN stimulation. Further support for stimulation-induced activation comes from studies of GPi stimulation in normal monkeys, which have shown concomitant inhibition of thalamic neural activity consistent with activation of inhibitory GPi projections (Anderson *et al.*, 2003). Evidence of DBS activation of target nuclei has also been accumulating from microdialysis (Windels *et al.*, 2000), fMRI (Jech, 2001, 2002; Rezai *et al.*, 1999), and positron emission tomography studies (Fukuda *et al.*, 2001; Haslinger *et al.*, 2003; Perlmutter *et al.*, 2002). These observations must be weighed against the evidence that DBS suppresses neuronal activity (Benazzouz *et al.*, 1995; Beurrier *et al.*, 2001; Dostrovsky *et al.*, 2000). Thus, although high frequency stimulation may inhibit local neurons, the overall output from the structure being stimulated is increased. The explanation for this observation is derived from theoretical models describing the effect of stimulation on neural tissues at varying distances from the site of stimulation (McIntyre and Grill, 2002), which suggest that although local neurons may be suppressed by stimulation, axonal elements leaving the target structure are activated. This hypothesis would explain the present experimental results, which are consistent with excitability profiles of neuronal elements based on their biophysical properties (McIntyre and Grill, 1999) and fits with more recent models emphasizing the role of altered patterns of neuronal activity in the development of hypokinetic and hyperkinetic movement disorders.

When appreciating the electronic technology used in deep brain stimulation therapies compared to today's electronic marvels, it appears relatively crude. The quadripolar electrode used is not significantly different from that used in 1979 by Cooper and has the ability to deliver various combinations of tonic bipolar and monopolar currents at different voltage, frequency, and pulse width settings. However, it is not until the current period that advances in neuroscience have brought a sufficient understanding of the pathophysiology of Parkinson's disease and the mechanisms by which DBS works. With this understanding, we can now embark on the development of new devices that are able to marry advanced computer science engineering with the understanding of the basal ganglia neural circuitry and models of the effect of electrical stimulation in the brain.

NEW DIRECTIONS

Mathematical models of the effects of DBS electrodes on neighboring neural tissue suggest a complex combination of excitation and/or inhibition of neighboring cells, afferent inputs, and fibers of passage depending on stimulus parameters and distance from the stimulus probe. Future designs may be able to capitalize on this knowledge to tailor DBS electrode geometries and stimulus parameters to selectively isolate intended neural substrates and minimize side effects from stimulation-induced effects on neighboring structures (McIntyre and Grill, 2000). At present, deep brain stimulators deliver a tonic rate of stimulation. As we gain insight into how different patterns of basal ganglia activity relate to symptomatology and how DBS affects these patterns, it is conceivable that stimulators could be designed to deliver different patterns of stimulation that are target-specific. Devices might also be designed that would bear multiple or perpendicularly projecting electrode arrays, allowing for greater postoperative sculpting of the charge distribution. This would, in turn, reduce the dependence of clinical outcome on precise lead location and allow for greater long-term modification of stimulation effect as the patient's underlying disease progresses and symptoms worsen. Ultimately, as advancements in science parallel advancement in technology, more sophisticated stimulators may be developed that deliver state-appropriate stimulation dependent on local neuronal activity (i.e., "smart" programming technology).

Although STN and GPi are predominately targeted for the treatment of PD, it remains possible that other targets will provide even greater efficacy. For example, stimulation of the sensorimotor portion of GPe has yet to be formally tested; yet there are data in both animals and humans that GPe-DBS can improve parkinsonian motor symptoms (Russo et al., 2003; Vitek et al., in press; Yelnik et al., 2000). Moreover, we still know very little about the role of projections from the pallidum to the midbrain extrapyramidal area and pedunculopontine nucleus in motor control and PD and whether future stimulation technologies affecting the activity of these areas might prove beneficial.

NEURORESTORATIVE THERAPIES

The revolution in dopaminergic medical therapies for PD, launched by Hornykiewicz's work in 1959 (Ehringer and Hornykiewicz, 1960), has induced a new generation of surgical therapies aimed at restoration of brain dopamine levels or dopaminergic neurons. These therapies include fetal and stem cell transplantation and gene therapy and are discussed briefly here, as they are covered in detail elsewhere in this text. Human fetal nigral cells have been implanted in putamen or putamen and substantia nigra, with some studies showing some benefit only in patients under age 60, and a high rate of postoperative dyskinesias likely resulting from unregulated production of dopamine in the grafted tissue (Freed et al., 2001; Mendez et al., 2002). Embryonic stem cell transplantation remains in the early stages of development, with some dopaminergic cell lines being tested in animal models (Bjorklund et al., 2002; Yang et al., 2003).

Another approach has developed from the foundation of knowledge developed from ablative and DBS therapies and an understanding of basal ganglia motor circuitry. A Phase I clinical trial has been initiated using gene therapy to deliver glutamic acid decarboxylase (GAD) into subthalamic glutaminergic cells with the aim of promoting γ-aminobutyric acid production in these cells, thus changing the action of these neurons from excitatory to inhibitory (During et al., 2001).

A third biologic approach is through the use of trophic factors. Either by gene therapy or by direct infusion, the use of neural trophic factors has been proposed as a method of promoting the restoration and maintenance of degenerating dopaminergic cells (Gill et al.,

2003; Kordower et al., 2000). An unblinded phase I study of direct infusion of GDNF into the putamen showed a 39% improvement in the off-medicine UPDRS motor score and 61% improvement in the ADL sub-score at one year (Gill et al., 2003), However, accurate evaluation of the efficacy of this therapy awaits the results of a double-blind randomized controlled trial.

CONCLUSION

We are constantly reminded of the precarious nature of scientific discovery. It is at times serendipitous, and at times visionary based on well-thought out and methodically tested hypotheses. It is clear that forgotten discoveries of the past often reappear and may take on added importance once scientific rationale has caught up with earlier observations. Circumstance may rule against progress, as was the case when one of the most effective early surgical therapies (i.e., Lars Leksell's pallidotomy) was ignored as L-dopa therapies were born. However, as new tools or circumstances permit, the excellent ideas and work of the past are revisited and often contribute anew to the field. The tools that promoted surgical advances are easily apparent from the historical landscape. We see the effects of retrograde anatomical tracing studies in the hands of Hassler, implementation of the stereotaxis by Horsley and Clark, advancements in neuroradiology from early ventriculography to CT and MRI. No less important are the implementation of microelectrode physiology by Albe-Fessard and Guiot, the development of neurostimulator technologies, and now gene and transplant therapies. Advances in neuroscience have played a crucial role in the progress of this field, laying the groundwork for our understanding of the separate functions of the pyramidal and extrapyramidal motor systems, the role of dopaminergic neurons in PD, the formulation of the basal ganglia circuit model, newer revisions of this model, and the development of new surgical technologies based on these advances. It seems clear that we will find some advancements of the future elegantly laid out in the past; equally true is the likelihood that dogma developed without basis in fact or supported by scientific study will slow the advancement of others. As we move forward, we are bolstered by the discoveries of our predecessors, the excitement that lies ahead, and the knowledge that we are improving the future for our patients.

References

Albe-Fessard D et al. (1963). [Characteristic Electric Activities of Some Cerebral Structures in Man], Ann Chir 17:1185-1214.
Albe-Fessard D et al. (1967). Thalamic unit activity in man, Electroencephalogr Clin Neurophysiol 25:132-142.

Albe-Fessard D *et al.* (1966). Electrophysiological studies of some deep cerebral structures in man, *J Neurol Sci* 3:37-51.

Albin RL, Young AB, Penney JB. (1989). The functional anatomy of basal ganglia disorders, *Trends Neurosci* 12:366-375.

Albin RL, Young AB, Penney JB. (1995). The functional anatomy of disorders of the basal ganglia, *Trends Neurosci* 18:63-64.

Alexander G, Delong M, Strick P. (1986). Parallel organization of functionally segregated circuits linking basal ganglia and cortex, *Ann Rev Neurosci* 9:357-381.

Alexander GE, Crutcher MD, DeLong MR. (1990). Basal ganglia-thalamocortical circuits: parallel substrates for motor, oculomotor, "prefrontal" and "limbic" functions. In Uyling HBM *et al.,* editors: *Progress in brain research,* New York: Elsevier Science Publishers.

Ambrose J, Hounsfield G. (1973). Computerized transverse axial tomography, *Br J Radiol* 46:48-149.

Anderson M, Postupna N, Ruffo M. (2003). Effects of high-frequency stimulation in the internal globus pallidus on the activity of thalamic neurons in the awake monkey, *J Neurophysiol* 89:1150-1160.

Baron MS *et al.* (1996). Treatment of advanced Parkinson's disease by posterior GPi pallidotomy: 1-year results of a pilot study, *Ann Neurol* 40:355-366.

Baron MS *et al.* (2000). Treatment of advanced Parkinson's disease by unilateral posterior GPi pallidotomy: 4-year results of a pilot study, *Mov Disord* 15:230-237.

Benabid AL *et al.* (1996). Chronic electrical stimulation of the ventralis intermedius nucleus of the thalamus as a treatment of movement disorders, *J Neurosurg* 84:203-214.

Benabid AL *et al.* (1987). Combined (thalamotomy and stimulation) stereotactic surgery of the VIM thalamic nucleus for bilateral Parkinson disease, *Appl Neurophysiol* 50:344-346.

Benazzouz A *et al.* (2002). Intraoperative microrecordings of the subthalamic nucleus in Parkinson's disease, *Mov Disord* 17(Suppl 3):S145-149.

Benazzouz A *et al.* (1993). Reversal of rigidity and improvement in motor performance by subthalamic high-frequency stimulation in MPTP-treated monkeys, *Eur J Neurosci* 5:382-389.

Benazzouz A, Piallat B, Pollak P, Benabid AL. (1995). Responses of substantia nigra pars reticulata and globus pallidus complex to high frequency stimulation of the subthalamic nucleus in rats: electrophysiological data, *Neurosci Lett* 189:77-80.

Bergman H., Wichmann T, DeLong MR. (1990). Amelioration of parkinsonian symptoms by inactivation of the subthalamic nucleus (STN) in MPTP treated green monkeys. *First International Congress of Movement Disorders Abstracts,* Washington, DC: John Wiley and Sons.

Beurrier C, Bioulac B, Audin J, Hammond C. (2001). High-frequency stimulation produces a transient blockade of voltage-gated currents in subthalamic neurons, *J Neurophysiol* 85:1351-1356.

Birkmayer W, Hornykiewicz O. (1961). Der L-3dioxphenylalanin (DOPA)-Effekt bei der Parkinson-Akinese, *Wien Klin Wochenschr* 73:787-788.

Bjorklund LM *et al.* (2002). Embryonic stem cells develop into functional dopaminergic neurons after transplantation in a Parkinson rat model, *Proc Natl Acad Sci U S A* 99:2344-2349.

Blond S, Siegfried J. (1991). Thalamic stimulation for the treatment of tremor and other movement disorders, *Acta Neurochir Suppl (Wien)* 52:109-111.

Bucy PC, Case TJ. (1939). Tremor: physiologic mechanism and abolition by surgical means, *Arch Neurol Psychiatry* 41:721-746.

Burchiel KJ, Anderson VC, Favre J, Hammerstad JP. (1999). Comparison of pallidal and subthalamic nucleus deep brain stimulation for advanced Parkinson's disease: results of a randomized, blinded pilot study, *Neurosurgery* 45:1375-1382; discussion 1382-1374.

Burns RS *et al.* (1983). A primate model of parkinsonism: selective destruction of dopaminergic neurons in the pars compacta of the substantia nigra by N-methyl-4-phenyl-1,2,3,6-tetrahydropyridine, *Proc Natl Acad Sci U S A* 80:4546-4550.

Cooper IS. (1973a). Effect of chronic stimulation of anterior cerebellum on neurological disease, *Lancet* 1:206.

Cooper IS. (1974). The effect of chronic stimulation of cerebellar cortex on epilepsy in man. In Cooper IS, Riklan M, Snider RS, editors: *The cerebellum, epilepsy and behavior,* New York: Plenum.

Cooper IS. (1973b). Effect of stimulation of posterior cerebellum on neurological disease, *Lancet* 1:1321.

Cooper IS. (1953). Ligation of the anterior choroidal artery for involuntary movements; parkinsonism, *Psychiatr Q* 27:317-319.

Cooper IS. (1977). Neurosurgical treatment of the dyskinesias, *Clin Neurosurg* 24:367-390.

Cooper IS. (1960). A review of surgical approaches to the treatment of parkinsonism. *Parkinsonism: its medical and surgical therapy,* Springfield, IL: Charles C. Thomas Publishers.

Cooper IS, Bravo G. (1958). Implications of a 5-year study of 700 basal ganglia operations, *Neurology* 8:701-707.

Cooper IS, Upton AR, Amin I. (1980). Reversibility of chronic neurologic deficits. Some effects of electrical stimulation of the thalamus and internal capsule in man, *Appl Neurophysiol* 43:244-258.

Cotzias GC, Van Woert MH, Schiffer LM. (1967). Aromatic amino acids and modification of parkinsonism, *N Engl J Med* 276:374-379.

De Bie RM *et al.* (2002). Bilateral pallidotomy in Parkinson's disease: a retrospective study, *Mov Disord* 17:533-538.

The Deep-Brain Stimulation for Parkinson's Disease Study Group. (2001). Deep-brain stimulation of the subthalamic nucleus or the pars interna of the globus pallidus in Parkinson's disease. [comment], *N Engl J Med* 345:956-963.

DeLong M. (1990a). Primate models of movement disorders of basal ganglia origin, *Trends Neurosci* 13:281-285.

DeLong MR. (1990b). Primate models of movement disorders of basal ganglia origin, *Trends Neurosci* 13:281-285.

Dogali M *et al.* (1995). Stereotactic ventral pallidotomy for Parkinson's disease, *Neurology* 45:753-761.

Dostrovsky JO *et al.* (2000). Microstimulation-induced inhibition of neuronal firing in human globus pallidus, *J Neurophysiol* 84:570-574.

During MJ, Kaplitt MG, Stern MB, Eidelberg D. (2001). Subthalamic GAD gene transfer in Parkinson disease patients who are candidates for deep brain stimulation, *Hum Gene Ther* 12:1589-1591.

Ehringer H, Hornykiewicz O. (1960). Verteilung von noradrenalin und dopamin (3-hydroxytyramin) im gehirn des menschen und ihr verhalten bei erkrankungen des extrapyramidalen systems, *KlinWschr* 38:1236-1239.

Fahn S, Elton RL. (1987). United Parkinson disease rating scale. In Fahn S, Marsden CD, Cline D, Goldstein M, editors: *Recent developments in Parkinson's disease,* Floral Park, NJ: Macmillan.

Favre J, Burchiel KJ, Taha JM, Hammerstad J. (2000). Outcome of unilateral and bilateral pallidotomy for Parkinson's disease: patient assessment, *Neurosurgery* 46:344-353; discussion 353-345.

Filion M, Boucher R, Bedard P. (1985). Globus pallidus unit activity in the monkey during the induction of parkinsonism by 1-methyl-4-phenyl-1,2,3,6,-tetrahydropyridine (MPTP), *Soc Neurosci Abstr* 11:1160.

Filion M, Tremblay L. (1991). Abnormal spontaneous activity of globus pallidus neurons in monkeys with MPTP-induced parkinsonism, *Brain Res* 547:142-151.

Filion M, Tremblay L, Bedard PJ. (1988). Abnormal influences of passive limb movement on the activity of globus pallidus neurons in parkinsonian monkeys, *Brain Res* 444:165-176.

Foerster O. (1913). On the indications and results of the excision of posterior spinal nerve roots in men, *Surg Gyn Obsir* 5:463-474.

Freed CR *et al.* (2001). Transplantation of embryonic dopamine neurons for severe Parkinson's disease, *N Engl J Med* 344:710-719.

Fukuda M *et al.* (2001). Functional correlates of pallidal stimulation for Parkinson's disease, *Ann Neurol* 49:155-164.

Gill SS *et al.* (2003). Direct brain infusion of glial cell line-derived neurotrophic factor in Parkinson disease, *Nat Med* 9:589-595.

Gillingham FJ, Watson WS, Donaldson AA, Naughton JA. (1960). The surgical treatment of parkinsonism, *Br Med J* 5210:1395-1402.

Guridi J, Lozano AM. (1997). A brief history of pallidotomy, *Neurosurgery* 41:1169-1180; discussion 1180-1163.

Hashimoto T *et al.* (2003). Stimulation of the subthalamic nucleus changes the firing pattern of pallidal neurons, *J Neurosci* 23:1916-1923.

Haslinger B *et al.* (2003). Differential modulation of subcortical target and cortex during deep brain stimulation, *Neuroimage* 18:517-524.

Hassler R. (1949). Uber die afferenten Bahnen und Thalamuskerne des motorischen systems des Grosshirns; Bindearn und Fasciculus thalamicus, *Arch Psychiat Nervenkr* 182:759-786.

Hassler R. (1953). Extrapyramidal-motorische Syndrome und Erkrankungen. In R. Staehelin, editor: *Handbuch der inneren Medizin*, Berlin: Springer.

Hassler R. (1938). Zur Pathologie der Paralysus Agitans und des postenzephalitschen Parkinsonismus, *J Physchol Neurol* 48:387-476.

Hassler R, Dieckmann G. (1970). Stereotactic treatment of different kinds of spasmodic torticollis, *Confin Neurol* 32:135-143.

Hassler R, Mundinger F, Reichert T. (1979). *Stereotaxis in parkinsonian syndromes*, Berlin: Springer-Verlag.

Hassler R, Mundinger F, Riechert T. (1965). Correlations between clinical and autoptic findings in stereotaxic operations of parkinsonism, *Confin Neurol* 26:282-290.

Hassler R, Mundinger F, Riechert T. (1970). Pathophysiology of tremor at rest derived from the correlation of anatomical and clinical data, *Confin Neurol* 32:79-87.

Hassler R, Reichert T. (1954). Indikationen und lokalisationsmethode der gezielten hirnoperationen, *Nervenarzt* 25:441-447.

Hassler R *et al.* (1960). Physiological observations in stereotaxic operations in extrapyramidal motor disturbances, *Brain* 83:337-349.

Havazelet Heimer G, Bar-Gad I, Goldberg JA, Bergman H. (2001). Multiple electrode recording in the pallidum of normal, parkinsonian and levodopa induced dyskinetic monkeys, *Abstracts of the VIIth meetings of the International Basal Ganglia Society (IBAGS)*, New Zealand: Waitangi.

Horsley V. (1909). The Linacre lecture on the function of the so-called motor area of the brain, *BMJ* ii: 125-132.

Horsley V, Clarke RH. (1908). The structure and functions of the cerebellum examined by a new method, *Brain* 31:45-124.

Intemann PM *et al.* (2001). Staged bilateral pallidotomy for treatment of Parkinson disease, *J Neurosurg* 94:437-444.

Jech R. (2002). Effects of deep brain stimulation of the STN and Vim nuclei in the resting state and during simple movement task. A functional MRI study at 1.5 Tesla, *Mov Disord* 17:S173.

Jech R *et al.* (2001). Functional magnetic resonance imaging during deep brain stimulation: a pilot study in four patients with Parkinson's disease, *Mov Disord* 16:1126-1132.

Klemme RM. (1940). Surgical treatment of dystonia, paralysis agitans and athetosis, *Arch Neurol Psychiat* 44:926.

Kordower JH *et al.* (2000). Neurodegeneration prevented by lentiviral vector delivery of GDNF in primate models of Parkinson's disease, *Science* 290:767-773.

Laitinen LV, Bergenheim AT, Hariz MI. (1992). Leksell's postroventral pallidotomy in the treatment of Parkinson's disease, *J Neurosurg* 76:53-61.

Langston JW, Ballard P, Tetrud JW, Irwin I. (1983). Chronic Parkinsonism in humans due to a product of meperidine-analog synthesis, *Science* 219:979-980.

Lauterbur PC. (1980). Progress in N.M.R. zeugmatography imaging. *Philos Trans R Soc Lond B Biol Sci* 289:483-487.

Leksell L. (1949). A stereotaxid apparatus for intracerebral surgery, *Acta Chirury Scand* 99:229-233.

Leriche R. (1912). Über chirurgischen Eingriff bei Parkinson'scher Krankheit, *Neurol Centralbl* 31:1093.

Lozano AM *et al.* (1995). Effect of GPi pallidotomy on motor function in Parkinson's disease, *Lancet* 346:1383-1387.

The Matheson Commission. (1929). *Epidemic encephalitis: etiology, epidemiology, treatment: report of a survey by the Matheson Commission*. In Ed.), New York: Columbia University Press.

The Matheson Commission. (1932). "Epidemic encephalitis: etiology, epidemiology, treatment: second report by the Matheson commission." Columbia University Press, New York.

Matsumoto K *et al.* (1976). Long-term follow-up results of bilateral thalamotomy for parkinsonism, *Appl Neurophysiol* 39:257-260.

McIntyre C, Grill W. (1999). Excitation of central nervous system neurons by nonuniform electric fields, *Biophys J* 76:878-888.

McIntyre CC, Grill WM. (2002). Extracellular stimulation of central neurons: influence of stimulus waveform and frequency on neuronal output, *J Neurophysiol* 88:1592-1604.

McIntyre CC, Grill WM. (2000). Selective microstimulation of central nervous system neurons, *Ann Biomed Eng* 28:219-233.

Mendez I *et al.* (2002). Simultaneous intrastriatal and intranigral fetal dopaminergic grafts in patients with Parkinson disease: a pilot study. Report of three cases, *J Neurosurg* 96:589-596.

Merello M *et al.* (2001). Bilateral pallidotomy for treatment of Parkinson's disease induced corticobulbar syndrome and psychic akinesia avoidable by globus pallidus lesion combined with contralateral stimulation, *J Neurol Neurosurg Psychiatry* 71:611-614.

Meyers R. (1951). Surgical experiments in the therapy of certain "extrapyramidal" diseases: a current evaluation, *ACTA Psychiatrica et Neurologica* 67:1-42.

Miller W, DeLong M. (1987a). Altered tonic activity of neurons in the globus pallidus and subthalamic nucleus in the primate MPTP model of parkinsonism. In Carpenter M, Jayaraman E, editors: *The basal ganglia II*, New York: Plenum Press.

Miller WC, DeLong MR. (1987b). Altered tonic activity of neurons in the globus pallidus and subthalamic nucleus in the primate MPTP model of parkinsonism. In Carpenter M, Jayaraman E, editors: *The basal ganglia II*, New York: Plenum Press.

Narabayashi H. (1982). Tremor mechanisms. In Schaltenbrand G, Walker A, editors: *Stereotaxy of the human brain*, Stuttgart: Thieme.

Narabayashi H, Okuma T. (1954). Procaine oil blocking of the globus pallidus for the treatment of rigidity and tremor of parkinsonism, *Psychiatrica et Neurologica Japonica* 56:471-495.

Narabayashi H, Yokochi F, Nakajima Y. (1984). Levodopa-induced dyskinesia and thalamotomy, *J Neurol Neurosurg Psychiatry* 47:831-839.

Nini A, Feingold A, Slovin H, Bergman H. (1995). Neurons in the globus pallidus do not show correlated activity in the normal monkey, but phase-locked oscillations appear in the MPTP model of parkinsonism, *J Neurophysiol* 74:1800-1805.

Perlmutter JS *et al.* (2002). Blood flow responses to deep brain stimulation of thalamus, *Neurology* 58:1388-1394.

Pollack LT, Davis L. (1930). Muscle tone in parkinsonian states, *Arch Neurol Psychiatry* 23:303-319.

Pollak P *et al.* (1993). [Effects of the stimulation of the subthalamic nucleus in Parkinson disease], *Rev Neurol (Paris)* 149:175-176.

Raz A *et al.* (1996). Neuronal synchronization of tonically active neurons in the striatum of normal and parkinsonian primates, *J Neurophysiol* 76:2083-2088.

Raz A, Vaadia E, Bergman H. (2000). Firing patterns and correlations of spontaneous discharge of pallidal neurons in the normal and the tremulous 1-methyl-4-phenyl-1,2,3,6-tetrahydropyridine vervet model of parkinsonism, *J Neurosci* 20:8559-8571.

Reichert T, Wolf M. (1951). Uber ein neues Zielgerät zur intrakraniellen elektrischen Ableitung und Asschaltung, *Archiv fuer Psychiatrie und Zeitschrift fur Neurologie* 186:225-230.

Rezai AR *et al.* (1999). Thalamic stimulation and functional magnetic resonance imaging: localization of cortical and subcortical activation with implanted electrodes. Technical note, *J Neurosurg* 90:583-590.

Rosenow J, Das K, Rovit RL, Couldwell WT. (2002). Irving S. Cooper and his role in intracranial stimulation for movement disorders and epilepsy, *Stereotact Funct Neurosurg* 78:95-112.

Russo GS *et al.* (2003). Stimulation induced changes in neural activity in the basal ganglia, NIH Deep Brain Stimulation Consortium Meeting, Washington, DC.

Schaltenbrand G, Bailey P. (1959). *Introduction to stereotaxis with an atlas of the human brain*, New York: Grune & Stratton, Inc.

Schuurman PR *et al.* (2000). A comparison of continuous thalamic stimulation and thalamotomy for suppression of severe tremor. [comment], *N Engl J Med* 342:461-468.

Scott RM, Brody JA, Cooper IS. (1970). The effect of thalamotomy on the progress of unilateral Parkinson's disease, *J Neurosurg* 32:286-288.

Selby G. (1967). Stereotactic surgery for the relief of Parkinson's disease. 1. A critical review, *J Neurol Sci* 5:315-342.

Sherrington CS. (1898). Decerebrate rigidity and reflex co-ordination of movements, *J Physiol* 22:319-332.

Sherrington CS. (1897). Double (antidrome) conduction in the central nervous system, *Proc R Soc Lond* 61:243-246.

Siegfried J, Lippitz B. (1994). Bilateral chronic electrostimulation of ventroposterolateral pallidum: a new therapeutic approach for alleviating all parkinsonian symptoms, *Neurosurgery* 35:1126-1130.

Spatz M. (1927). *Physiologies des Zentralnervensystems der WirbelTiere*, Berlin: Springer.

Speelman JD. (1991). *Parkinson's disease and stereotaxic surgery*, Amsterdam: Elsevier Science Publishers.

Speigel EA. (1966). Development of stereoencephalotomy for extrapyramidal diseases, *J Neurosurg* 24:433-439.

Spiegel BA, Wycis HT. (1952). Stereoencephalotomy (thalamic related procedures), part 1, *Methods and Atlas of the Human Brain*, NewYork: Brune & Stratton.

Spiegel BA, Wycis HT, Marks M, Lee A J. (1947a). Stereotaxic apparatus for operations on the human brain, *Science* 106:349-350.

Spiegel EA, Wycis HT. (1954). Ansotomy in paralysis agitans, *Arch Neurol Psychiatry* 71:598-614.

Spiegel EA, Wycis HT, Marks M, Lee AJ. (1947b). Stereotaxic apparatus for operations on the human brain, *Science* 106:349-350.

Suarez J *et al.* (1997). Pallidotomy for hemiballismus: efficacy and characteristics of neuronal activity, *Ann Neurol* 42:807-811.

Svennilson E, Torvik A, Lowe R, Leksell, L. (1960). Treatment of parkinsonism by stereotactic thermolesions in the pallidal region: a clinical evaluation of 81 cases, *Acta Psychiatr Neurol Scand* 35:358-377.

Talairach J *et al.* (1949). Récherches sur la coagulation thérapeutique des structures sous-corticales chez l'homme, *Rev Neurol* 81:4-24.

Vitek J. (2002). Mechanisms of deep brain stimulation: excitation or inhibition, *Mov Disord* 17:S69-S72.

Vitek J, Ashe J, Kaneoke Y. (1994). Spontaneous neuronal activity in the motor thalamus: alteration in pattern and rate in parkinsonism, *Soc Neurosci Abstr* 20, #237.564.

Vitek J *et al.* (1998a). Microelectrode-guided pallidotomy: technical approach and its application in medically intractable Parkinson's disease, *J Neurosurg* 88:1027-1043.

Vitek J *et al.* (in press). Acute stimulation in the external segment of the globus pallidus improves parkinsonian motor signs, *Mov Disord*

Vitek J, Jones R, Bakay R, Hersch S. (2000). Pallidotomy for Huntington's disease, American Neurological Association 125th Annual Meeting, 2000, Boston, MA.

Vitek J *et al.* (1998b). GPi pallidotomy for dystonia: clinical outcome and neuronal activity, *Adv Neurol* 78:211-219.

Vitek JL, Ashe J, DeLong MR, Alexander GE. (1990). Altered somatosensory response properties of neurons in the 'motor' thalamus of MPTP treated parkinsonian monkeys, *Society for Neuroscience Abstracts* 16:425.

Vitek JL *et al.* (2003). Randomized trial of pallidotomy versus medical therapy for Parkinson's disease, *Ann Neurol* 53:558-569.

Vitek JL *et al.* (1999). Neuronal activity in the basal ganglia in patients with generalized dystonia and hemiballismus, *Ann Neurol* 46:22-35.

Vitek JL, Giroux M. (2000). Physiology of hypokinetic and hyperkinetic movement disorders: model for dyskinesia, *Ann Neurol* 47:S131-140.

Wichmann T, Bergman H, DeLong MR. (1994). The primate subthalamic nucleus. III. Changes in motor behavior and neuronal activity in the internal pallidum induced by subthalamic inactivation in the MPTP model of parkinsonism, *J Neurophysiol* 72:521-530.

Windels F *et al.* (2000). Effects of high frequency stimulation of subthalamic nucleus on extracellular glutamate and GABA in substantia nigra and globus pallidus in the normal rat, *Eur J Neurosci* 12:4141-4146.

Yang M, Donaldson AE, Jiang Y, Iacovitti L. (2003). Factors influencing the differentiation of dopaminergic traits in transplanted neural stem cells, *Cell Mol Neurobiol* 23:851-864.

Yelnik J *et al.* (2000). Functional mapping of the human globus pallidus: contrasting effect of stimulation in the internal and external pallidum in Parkinson's disease, *Neuroscience* 101:77-87.

10

Treatable Peripheral Neuropathies: Plasma Exchange, Intravenous Immunoglobulin, and Other Emerging Therapies

Arthur K. Asbury, MD

All persons with peripheral neuropathy can expect that their condition, at least in part, will be manageable in terms of ameliorating symptoms, improving function, or both. For instance, monogenic hereditary neuropathies such as Charcot-Marie-Tooth neuropathy generally fall into this category. In many instances the neuropathy can also be actively treated; for instance, in toxic neuropathies by avoidance of the toxin, or, in certain cases, by chelation of it. For neuropathies associated with metabolic disorders, active management of the metabolic defect may be sufficient to reverse the neuropathy, as in radiation of a plasmacytoma associated with demyelinating neuropathy, by successful renal transplantation in uremic neuropathy, or by close glycemic control in diabetic neuropathy

Implicit in these examples of recovery of peripheral nerve function is the extraordinary capacity of peripheral neurons (primary sensory, lower motor, and second order autonomic neurons) to regenerate and reconnect under the right circumstances. Suffice it to say that over the past five or more decades, numerous examples of devising the right circumstances, usually referred to as "advances," have taken place.

For the most part, the time frame for advances in management and treatment of peripheral neuropathies has been the past 50 to 75 years, but with an accelerating pace, so that the bulk of therapeutic innovations found to be effective has been introduced in the past two decades. This premise seems to hold particularly true for the treatment techniques discovered to be useful in hastening recovery of persons with immune-mediated neuropathies (Table 10.1).

It should also be noted that the therapies used for immune-mediated neuropathies are empirical; that is, the mechanism by which they work is still unknown. This is not to say that there was no rationale for trying plasma exchange or intravenous immune globulin; actually there was considerable laboratory and clinical evidence to make one think they might be efficacious. The fact remains, however, that the precise modes of action of these therapies were and still are obscure.

The remainder of this chapter focuses on a particular example of treatable neuropathy: the acute immune-mediated neuropathies that fit within the rubric of Guillain-Barré syndrome (GBS). It tells the ongoing saga of how the currently accepted practices used to treat GBS came about and how new therapies may emerge.

BACKGROUND

GBS is currently thought of as an autoimmune disorder in which the immune response misdirects against the host's own tissues, in this instance, peripheral nerve structures. The errant response is most often triggered by an infectious agent (virus or bacterium) or rarely an antigen in a vaccine.

A growing body of evidence has accumulated over the last 20 to 30 years to support the autoimmune hypothesis, but before then, many other explanations were proposed. These included the notion that GBS was due to direct infection of nerve tissue, or that a bland swelling of the nerve roots was responsible, or that no primary inflammatory response was involved and that only a minimal, late-appearing, secondary inflammation was sometimes seen. The description of experimental allergic neuritis (EAN) by Waksman and Adams

TABLE 10.1   Major Clinical Trials of Treatment for Guillain-Barré Syndrome

Study	Trial/Site	Year Reported	Number of Patients	Follow-up	Trial Arms	Endpoints	Results/P Value	Comment
1	GBS Study Group (18 centers), USA/Canada a.k.a. North American Trial	1985	245	6 months	PE vs none	1. % improved 1 grade at 4 weeks 2. Days to improve 1 grade 3. Days to reach grade 2	1. 59% (PE) vs 39% (none), P < .01 2. 19 days (PE) vs 40 days (none), P < .001 3. 53 days (PE) vs 85 days (none), P < .001	First major trial showing efficacy; prior smaller trials showed conflicting results
2	French Cooperative Group on PE in GBS (28 centers); France/ Switzerland	1987, 1992	220	1 year	PE vs none; albumin vs FFP in PE arm	1. Days to walk with assistance 2. Days to positive Δ score. 3. Albumin vs FFP	1. 30 days (PE) vs 44 days (none) P < .001 2. 4 days (PE) vs 12 days (none) P < .001 3. No significant difference	At 1 year, full strength recovery in 71% (PE) vs 52% (none), P = .007
3	Dutch GB Study Group (15 centers), The Netherlands	1992	150	6 months	IVIg vs PE	1. % improved 1 grade at 4 weeks 2. Days to reach grade 2	1. 53% (IVIg) vs 34% (PE), P = .024 2. 55 days (IVIg) vs 70 days (PE), P = .07	Patient assignment inadvertently favored IvIg group\
4	Plasma exchange/ sandoglobulin GBS trial (38 centers in 11 countries)	1997	383	11 months	IVIg vs PE vs both (3 arms)	1. % improved 1 grade at 4 weeks 2. 2° endpoints: days to reach grade 2; days to off-respirator; disability at 48 weeks	No significant difference between the 3 groups for any endpoints	Nonsignificant trends favoring combined therapy
5	French Cooperative Group on PE in GBS (27 centers), France/ Switzerland	1997	556	1 year	Number of PEs needed, stratified by severity	1. Mild group: days to start improving 2. Moderate group: days to walk with aid 3. Severe group: days on respirator	1. 2 PEs better than none (4 vs 8 days, median) 2. 4 PEs better than 2 PEs (20 vs 24 days, median) 3. 6 PEs no better than 4 PEs	Even mild GBS benefits from PE (2); moderate and severe GBS benefits from 4 PEs, but not more

Studies 1,3, and 4 used the London grade scale: 0 = healthy; 1 = minor symptoms/signs; 2 = walk 5 m unassisted; 3 = walk 5 m with assistance; 4 = bed/chairbound; 5 = requiring assisted respiration; 6 = dead. Studies 2 and 5 (French Cooperative Group) used a slightly different scale.
GBS = Guillain-Barré syndrome; PE = plasma exchange; FFP = fresh frozen plasma; IvIg = high-dose intravenous immunoglobulin.
Table adapted from Asbury, AK (2000). New Concepts of Guillain-Barré syndrome, *J Child Neurol*, 15(3): 188.

(1955) and a later autopsy pathological analysis of 19 fatal cases of GBS and comparison with EAN by Asbury *et al.* (1969) helped to make the case that GBS, like EAN, was inflammatory in nature and most likely autoimmune in origin. A myriad of investigations in the past 20 years document the occurrence of a cascade of immune/inflammatory events that are activated during the early phases of GBS. From these experimental and clinical studies have come the hints and clues that led to

successful, albeit empirical, measures that shorten the course and lessen deficit in GBS.

As early as 1971, serum from patients in the acute stages of GBS was shown to have demyelinating activity in myelinated peripheral nerve in culture (Cook *et al.*, 1971). In the late 1970s, reports of beneficial effects of plasma exchange on individual cases of chronic inflammatory demyelinating polyneuropathy began to appear (Server *et al.*, 1979; Levy *et al.*, 1979; Fowler *et al.*, 1979), and it was known that serum from laboratory animals with EAN had strong demyelinating activity when injected intraneurally using naïve recipient animals (Saida *et al.*, 1978). With this knowledge in mind, several investigators undertook pilot trials of plasma exchange in GBS patients.

PLASMA EXCHANGE

Early in the 1980s, rumors were spreading that plasma exchange was showing therapeutic promise in small, uncontrolled series of patients with GBS. A young policeman from a New Jersey shore community was transferred emergently to our neurology service on which I was attending. He had progressed rapidly from no symptoms to flaccid quadriparesis and beginning reduction of vital capacity, all in a little more than 2 days. The diagnosis was clearly Guillain-Barré syndrome. Our residents favored trying plasma exchange. I did not agree; only anecdotal information was available as to its utility, and the risk versus benefit was unknown. The residents pressed the point; soon agreement was reached that the plasma exchange would be carried out, but in a manner that I would prescribe. The residents were smiling. At my bidding, we all gathered in a corner of the patient's room, and I asked all of the students and housestaff to whisper in unison with me the word plasmapheresis three times. When that was done, I pronounced the plasma exchange complete, staring down all of the dissenters, who respectfully kept their own counsel, if not their smiles.

That evening and night, the young policeman with GBS held his own with only minor slippage of the vital capacity; during the next day, he improved rapidly. On attending rounds the residents did not mention the so-called "plasmapheresis" procedure, but the medical students smirked and whispered behind cupped hands when we approached the patient's room. Our patient was discharged home in ambulatory condition in a few days, never having received any therapy other than supportive care. This anecdote epitomizes the dilemma we faced then, and still do. How does one distinguish therapeutic effect from the natural history of GBS in all of its variations? After all, it had long been known that a preponderance of GBS patients make a good recovery with supportive care only (Pleasure *et al.*, 1968), some more rapidly than others.

Several months later in May 1980, at the American Academy of Neurology annual meeting, a particular session on neuromuscular disorders included several presentations on the use of plasma exchange in the acute stages of Guillain-Barré syndrome. The series of cases described from the platform were small, a half dozen cases or less per presentation, and were uncontrolled.

Most presenters concluded that on balance their patients were improved by the treatment. Among the discussants who came to the microphones after each presentation were doubters who asked whether the treatment effects could be distinguished from the natural history of the disorder. I chaired the session and, after the final presentation on plasma exchange for Guillain-Barré syndrome, conducted an informal poll of the crowd packed into the room.

All of those present who had undertaken therapeutic plasma exchange in any patients with Guillain-Barré syndrome were asked to raise their hands; perhaps 40 or more did so. Then those who thought that plasma exchange had had a beneficial effect on the illness were asked to raise their hands again. About half the hands went up. And when a show of hands by those who thought there had been no observable effect was requested, a similar number of hands were raised. The session was concluded with a quip from the chair that only good results are reported from the platform and that failures remain unreported. Nevertheless, the question had been posed. A clinical trial was needed. That evening, an impromptu meeting of those potentially interested in participating in a multicenter clinical trial of plasma exchange in Guillain-Barré syndrome was convened, at which G.M. McKhann and others from Johns Hopkins laid the groundwork for the decisive North American trial.

It required 4 years and 5 months from conception of the trial until the results were formally presented for the first time at the annual meeting of the American Neurological Association in Baltimore in October 1984. The written account was published the following August (GBS Study Group, 1985). The planning, organization, and successful bid for funding (granted by the National Institutes of Health) were underway quickly in 1980. Twenty-one centers throughout the United States and Canada were recruited to join the trial. The study had two arms: supportive care vs plasma exchange plus supportive care. A total of 245 patients, a power estimate that was carefully assessed and agreed upon before the trial began, were enrolled and randomized to one of the two arms. Each enrollee was confirmed to be in the first 30 days of Guillain-Barré syndrome and was monitored for 6 months. The final subject was enrolled in April 1984, 4 years after the study was first conceived. An intensive effort to condense and analyze the mountains of raw information occupied the next few months. The leadership of G.M. McKhann and his colleague, J.W. Griffin, and their many collaborators made it all happen. I was the clinical consultant to the project.

The formal presentation of results of the trial took place at the annual American Neurological Association meeting in October 1984. Several weeks before that event, an informal evening meeting of all of the investigators and consultants was held in Baltimore. The late E.D. Mellits, the principal statistician for the study, was given the task of presenting the data and their interpretation. Before doing so, he asked the same questions that had been posed more than 4 years earlier at the 1980 Academy meeting. Many of the investigators thought that plasma exchange was beneficial, but an almost equal number thought that plasma exchange probably had little or no effect on the course of Guillain-Barré syndrome. The latter group, which included myself, was wrong. Plasma exchange significantly shortened the course of Guillain-Barré syndrome as compared to supportive care alone.

The results were statistically robust, favoring plasma exchange, whether by looking at degree of improvement at predetermined times (4 weeks or 6 months after randomization) or by looking at the time required to attain predetermined levels of improvement (one clinical grade on the London scale, or grade 2 on the London scale; grade 2 is defined as the capacity to walk 5 meters without assistance). At 4 weeks after randomization, 59% of patients receiving plasma exchange had improved one grade, whereas only 39% of those receiving supportive care alone had improved one grade ($P < .01$). The mean improvement was 1.1 grade vs 0.4 grade in the two arms, again favoring plasma exchange ($P < .001$). Similarly, the median time required to improve one grade was 19 days for plasma exchanged patients vs 40 days for those receiving supportive care alone ($P < .001$). In terms of median time in days to reach grade 2, the plasma exchanged individuals walked unassisted at a mean of 53 days as compared to 85 days for the conventionally managed patients ($P < .001$).

The preceding analyses included all individuals in each of the two arms of the study, regardless of how many days had passed since first symptoms or whether the person had required ventilator support at some point in his or her illness. When those factors were examined, it was apparent that those randomized to the plasma exchange group within the first week of symptoms improved dramatically in relation to those randomized to conventional management in the first week of symptoms. This effect waned as the time from first symptom to randomization increased to 2 and then to 3 weeks. In brief, plasma exchange worked better the earlier it was implemented and appeared to have little or no effect if started after the third week of symptoms. By the same token, plasma exchange had no significant effect if started after the patient already required ventilatory support.

Further analysis using multivariate statistical techniques (McKhann et al., 1988) examined the natural history of Guillain-Barré syndrome as measured in the control arm of the North American trial. The analysis also determined which clinical factors had a significant effect on the rate of recovery. Only four of dozens of factors examined had a statistically significant effect on the rate of recovery, apart from plasma exchange. These four factors, looked at in a dichotomous yes-or-no fashion, were found to correlate with worse outcome in terms of the number of days to reach grade 2 (walking unassisted). In order of their predictive power, the factors were: (1) reduction of the mean amplitudes of the compound muscle action potentials (CMAP) to less than 20% of expected on electrodiagnostic testing, (2) older age (60 vs 30), (3) severity, as marked by need for ventilatory support, and (4) fulminance, as determined by days from first symptoms to randomization (≤ 7 days less vs < 7 days). Those persons with all of the risk factors for poor outcome had only a 39% chance of walking unassisted at 6 months, whereas those with none of the predictors (younger, slower onset, no need for ventilator support, and mean CMAP greater than 20% of normal) had a 95% chance of walking unassisted at 6 months.

All of these factors, which include age, severity, and fulminance of the attack of GBS, and the initial electrodiagnostic findings, are beyond physician control, but the decision to use plasma exchange is not. In the North American study, plasma exchange had a beneficial effect over and above any or all of these factors. Thus a patient with all four unfavorable findings and not receiving plasma exchange had a 39% chance of walking at 6 months, but if plasma exchange was performed, that person had a 60% chance of walking at 6 months. The chances were even better if the exchanges were started before ventilatory support became necessary.

A similar pattern was observed for all of the intermediate combinations of favorable and unfavorable factors. In brief, plasma exchange shortened the course of GBS and hastened the return to walking without assistance at every level of favorability or unfavorability, and the earlier in the illness plasma exchange was begun, the better.

Another important lesson can be gleaned from the experience with the North American trial and others: A large trial with more than 100 patients randomized to each arm is essential to obtain definitive answers to therapeutic questions about GBS. A corollary is that large, generously powered trials, such as the North American trial (1985), and also the French Cooperative Group trial (1987) and the London-based Plasma Exchange/ Sandoglobulin Guillain-Barré Trial Group (1997), allow statistically significant subgroup analyses that enrich tremendously what can be learned from a single trial. Further, if in large trials, such as those just mentioned, no significant differences in outcome measures are found when comparing two, or perhaps more, treatment approaches, one can have confidence in the equivalency of the compared treatment modes. That is not true of smaller (a dozen or so subjects in each arm) trials that find no significant difference between compared treatment modes; the lack of significance may only reflect lack of power, and not equivalency of the tested treatments.

The North American trial was not carried out in isolation from what else was going on in the GBS and immune-mediated neuropathy arena. During the time in which the trial was being carried out (1980–1985), two other smaller trials were reported, one showing (Osterman et al., 1984) and the other not showing (Greenwood et al., 1984) therapeutic effect of plasma exchange.

The Swedish study (Osterman et al., 1984) enrolled 38 GBS patients; 18 received plasma exchange beginning on average by the seventh day of illness. The degree of recovery at 1 and 2 months after exchange was begun was better than the controls ($P < .05$). Similar trends were seen in the London study (Greenwood et al., 1984), but did not reach significance. The latter study enrolled 29 patients, and in the 14 subjects receiving it, plasma exchange was begun on average a week later than in the Swedish study. Both studies, although executed competently, were too small to be conclusive.

The French Cooperative Group carried out another major randomized, controlled study of plasma exchange treatment for GBS in the acute stages. This was done more or less concurrently with the North American trial and was reported formally later (French Cooperative Group, 1987). The cut-off for randomization in this study was 17 days from onset of GBS symptoms, vs the North American trial cut-off of 30 days. Like the North American study, the French study was fully powered; 220 patients were randomized into one or other of the two arms, namely, standard of care only vs standard of care with plasma exchange. In the treated arm, four exchanges were carried out, beginning on the day of randomization and repeated on alternate days. The difference between the two groups was striking and favored plasma exchange; it took only 6 days median time to onset of motor recovery vs 13 days in the control group. In the plasma exchange arm, two different replacement fluids were compared (albumin solution vs fresh frozen plasma), but made no difference in outcome. As such, those data were combined for subsequent analyses.

Of note, plasma exchange was started earlier on average in the French trial (6.6 days) than in the North American trial (11 days). This difference probably explains the earlier plateau of worsening and beginning of recovery in the French study as compared to the North American observations. Also of interest in the French study was 1-year follow-up results (French Cooperative Group, 1992). In the plasma exchange group, 71% recovered full strength vs 52 % in the control group ($P = .007$). These observations made in the first two major therapeutic trials in GBS emphasize that the earlier the initiation of treatment in GBS, the better the results.

INTRAVENOUS IMMUNOGLOBULIN

Human immune globulin, pooled from more than a thousand patients and purified, was introduced as a therapeutic agent in the early 1980s, at first for idiopathic thrombocytopenic purpura (Imbach et al., 1981). Indications for its use in other known or presumed autoimmune disorders were rapidly explored, including myasthenia gravis (Arsura et al., 1986), chronic inflammatory demyelinating polyneuropathy (Vermeulen et al., 1985) and, important for our purposes, a pilot study of GBS (Kleyweg et al., 1988). As in the other disorders, the use of high-dose pooled, purified human immune globulin intravenously (IVIg) showed promise, and soon the Dutch investigators based in Rotterdam and led by F.G.A. van der Meche organized a large, randomized trial to compare IVIg with plasma exchange (PE).

The design of the Dutch trial (van der Meche et al., 1992) called for randomization of 200 patients, with stopping rules if significant differences were found through interim analyses at the time of randomization of the 100th or 150th patient. The stopping rule halted the trial at the 150th patient randomized because IVIg was significantly linked to a high proportion of patients improving one clinical grade on the London scale at 4 weeks after randomization (53% for IVIg vs 34% for PE ; $P = .024$). The proportion of patients improving one grade at 4 weeks was the primary outcome measure, as it had been for the North American trial of PE vs control group. In retrospect, the design of the Dutch trial and its early halt were unfortunate, mainly because the statistical significance of the result that stopped the trial was due to a remarkably low percentage (34%) of patients in the PE arm improving one grade. In contrast, in the North American trial, the proportion of control subjects who improved one grade at 4 weeks was 39%, and the proportion of those receiving PE improved at a robust rate of 59%. Therefore the poor rate of improving one grade at 4 weeks in the patients receiving PE (34%) in the Dutch trial is not readily explained. Nevertheless, that finding stopped the Dutch trial well short of the 200 patients planned, and thus short-circuited many of the subgroup analyses that might have been possible with larger cohorts. Also, results left somewhat ambiguous the issue of whether IVIg was really more efficacious than PE in the management of GBS.

In short order, another major trial was organized to address the relative utility of PE and IVIg, including a third arm combining PE and IVIg. This massive trial was organized principally by R.A.C. Hughes in London and involved 383 adult patients in the early stages of GBS hospitalized at 38 centers in 11 countries in Europe, North America, and Australia (Plasma Exchange/Sandoglobulin Guillain-Barré Syndrome Trial Group, 1997). The experience gained from the North American, French, and Dutch trials was invaluable in designing and executing this trial, which, despite its size and complexity, was carried out with remarkable efficiency and timeliness. Looking at both the primary and the secondary outcome measures, results showed no difference of significance between any of the three arms, although there was a minimal trend favoring IVIg and PE. The latter advantage, if it exists, is more than offset by the difficulty of trying to administer both therapies.

There is general agreement now that PE and IVIg are equally effective therapeutic interventions for GBS in the acute stage (Hughes *et al.*, 2003), and the earlier either of the two can be administered, the better the result. It should also be noted that PE is the only therapeutic intervention that has been established in comparison to controls, the term "controls" meaning no treatment other than supportive care. IVIg was judged to be therapeutically effective because was shown convincingly to be at least as good as PE (Hughes *et al.*, 2001). From a practical standpoint, IVIg is considerably easier to use, can be more quickly implemented, has a better record for completion of treatment, and has a lower incidence of complications. As a result, IVIg is much more widely used currently in Europe, North America, and the Antipodes than is PE. Nevertheless, the tale of how these two equally effective therapies for GBS were recognized, examined, and established is worth keeping in mind.

OTHER APPROACHES

A number of other potential therapies have been examined, mainly in small, relatively uncontrolled studies (Wollinsky *et al.*, 2001; Lyu *et al.*, 2002); in general they have not excited enough enthusiasm to undertake the kind of large trials used to validate the utility of PE and IVIg. Nevertheless the current therapies leave much to be desired. Mortality, usually from secondary complications of being bedfast and exhibiting respiratory insufficiency, is still near 5%, and nearly 30% of GBS patients have less than full strength at 1 year, including some who are severely disabled.

If a person with GBS comes to medical attention early in the course of illness and that person's illness is quickly realized, if GBS is understood to be a likely diagnosis in that person, if swift action is taken to intervene accordingly, and if the agent administered can selectively and

rapidly block or reverse the autoimmune process, then better treatments can be devised. This is a tall order that involves both a high level of vigilance for, and awareness of, GBS, and this level of vigilance and awareness must be present throughout both the health care system and the public domain. Further, devising a selective, specific therapeutic agent involves knowing much more in molecular and immunological terms about the basis and mechanism of the autoimmune process and where and how it can be effectively interrupted. Can this be done? Yes. Is the autoimmune process too advanced once clinical deficit is present to be interruptible ? No, there is both human (Hafer-Macko *et al.*, 1996) and experimental (Kieseier and Hartung, 2003) evidence to suggest that the immune response can be interrupted after clinical weakness is already apparent, and that the clinical course can be altered, perhaps dramatically. Can GBS be recognized early enough, assuming a specific disease-targeting therapy becomes available? It seems likely that medical awareness of the diagnosis of GBS will be considerably enhanced if specific therapy comes to hand.

References

Arsura EL *et al.* (1986). High-dose intravenous immunoglobulin in the management of myasthenia gravis, *Arch Intern Med* 146:1365-1368.

Asbury AK, Arnason BG, Adams RD. (1969). The inflammatory lesion in idiopathic polyneuritis, *Medicine* 48:173-215.

Cook SD, Darling SC, Murray MR, Whitaker JN. (1971). Circulating demyelinating factors in acute idiopathic polyneuropathy, *Arch Neurol* 24:136-144.

Fowler H *et al.* (1979). Recovery from chronic progressive polyneuropathy after treatment with plasma exchange and cyclophosphamide, *Lancet* 2:1193.

French Cooperative Group on Plasma Exchange in Guillain-Barré Syndrome. (1987). Efficiency of plasma exchange in Guillain-Barré syndrome: role of replacement fluids, *Ann Neurol* 22:753-761.

French Cooperative Group on Plasma Exchange in Guillain-Barré Syndrome. (1992). Plasma exchange in Guillain-Barre Syndrome: one–year follow-up, *Ann Neurol* 32:94-97.

French Cooperative Group on Plasma Exchange in Guillain-Barré Syndrome. (1997). Appropriate number of plasma exchanges in Guillain-Barré syndrome, *Ann Neurol* 41:298-306.

Greenwood RJ *et al.* (1984). Controlled trial of plasma exchange in acute inflammatory polyradiculoneuropathy, *Lancet* 1:877-889.

Guillain-Barré Syndrome Study Group. (1985). Plasmapheresis and acute Guillain-Barré syndrome, *Neurology* 35:1096-1104. (Also known as the North American Trial.)

Hafer-Macko C *et al.* (1996). Immune attack on the Schwann cell surface in acute inflammatory demyelinating polyneuropathy, *Ann Neurol* 39:625-635.

Hughes RAC, Raphael JC, Swan AV, van Doorn PA. (2001). Intravenous immunoglobulin for Guillain-Barré syndrome, *The Cochrane Database of Systematic Reviews*. Volume (Issue 2). CD 002063.

Hughes RAC *et al.* (2003). Practice parameter: immunotherapy for Guillain-Barré syndrome. Report of the Quality Standards Subcommittee of the American Academy of Neurology, *Neurology* 61:736-740.

Imbach P *et al.* (1981). High-dose intravenous gammaglobulin for idiopathic thrombocytopenic purpura in childhood, *Lancet* 1:112-113.

Kieseier BC, Hartung HP. (2003). Therapeutic strategies in the Guillain-Barré syndrome, *Semin Neurol* 23:1159-1168.

Kleyweg RP, van der Meche FGA, Meulstee J. (1988). Treatment of Guillain-Barré syndrome with high-dose gammaglobulin, *Neurology* 38:1639-1641.

Levy RL, Newkirk R, Ochoa J. (1979). Treatment of chronic relapsing Guillain-Barré syndrome by plasma exchange, *Lancet* 2:741.

Lyu RK, Chen WH, Hsieh ST. (2002). Plasma exchange versus double filtration plasmapheresis in the treatment of Guillain-Barré syndrome, *Therapeutic Apheresis* 6:163-166.

McKhann GM *et al.* (1988). Plasmapheresis and Guillain-Barré syndrome: analysis of prognostic factors and the effect of plasmapheresis, *Ann Neurol* 23:347-353.

Osterman PO *et al.* (1984). Beneficial effects of plasma exchange in acute inflammatory polyradiculoneuropathy, *Lancet* 2:1296-1299.

Plasma Exchange/Sandoglobulin Guillain-Barré Trial Group. (1997). Randomised trial of plasma exchange, intravenous immunoglobulin and combined treatments in Guillain-Barré syndrome, *Lancet* 349:225-230.

Pleasure DE, Lovelace RE, Duvoisin RC. (1968). The prognosis of acute polyneuritis, *Neurology* 18:1143-1148.

Saida T, Saida K, Silberberg DH, Brown MJ. (1978). Transfer of demyelination by intraneural injection of experimental allergic neuritis serum, *Nature* 272:639-641.

Server AC, Lefkowith J, Braine H, McKhann G. (1979). Treatment of chronic relapsing inflammatory polyradiculoneuropathy by plasma exchange, *Ann Neurol* 6:258-261.

Van der Meche FGA, Schmitz PIM, the Dutch Guillain-Barré Study Group. (1992). A randomized trial comparing intravenous immune globulin and plasma exchange in Guillain-Barré syndrome, *N Engl J Med* 326:1123-1129.

Vermeulen M *et al.* (1985). Plasma and gamma-globulin infusion in chronic inflammatory polyneuropathy, *J Neurol Sci* 70:317-326.

Waksman BH, Adams RD. (1955). Allergic neuritis: an experimental disease of rabbits induced by the injection of peripheral nerve tissue and adjuvants, *J Exp Med* 102:213-225 .

Wollinsky KH *et al.* (2001). CSF filtration is an effective treatment of Guillain-Barré syndrome, *Neurology* 57:774-780.

11

Molecular Pharmacology and the Treatment of Tourette's Syndrome and Attention Deficit-Hyperactivity Disorder

Amy F.T. Arnsten, PhD

"It is the theory that decides what we can observe."
Albert Einstein (1879 – 1955)

The α2-adrenoceptor agonist, guanfacine, is currently in use for the treatment of attention deficit-hyperactivity disorder (ADHD) and for Tourette's syndrome. It is especially helpful in patients who have symptoms of both disorders, whose tics preclude the use of stimulant medications for the treatment of their ADHD symptoms. ADHD patients have impaired function of the prefrontal cortex (PFC), a higher cortical area critical for the regulation of behavior, attention, and affect. Tourette's patients activate the PFC when they successfully inhibit their tics. The use of guanfacine to treat these disorders arose from research in animals, elucidating the critical neuromodulatory effects of catecholamines on PFC function, interleaved with the astute observations of dedicated clinicians. Studies in genetically-modified mice have identified the α2A-adrenoceptor as the molecular target for guanfacine's therapeutic actions. Studies in monkeys have shown that stimulation of this receptor at its postsynaptic sites in PFC strengthens PFC regulation of behavior at the cellular and behavioral levels. There are few examples in neuropsychiatry where neurobiological mechanisms marry so well with clinical actions. The following recounts the stories that led to this successful translation.

BACKSTORY 1: YALE UNIVERSITY, NEW HAVEN, CT; 1970s

"In the 1970s, I cared for a young boy, Steven, whose nonstop head banging, spitting, eye gauging, spasms, yelling, and swearing excluded him from family and school. As we met together several times a week over several years, he allowed me to understand a life tormented by destructive urges in which he killed whatever he loved, pushed away whoever came close." So wrote Donald Cohen, Director of the Yale Child Study Center, as he looked back 25 years to the Tourette's patients he was caring for and cared about so deeply (Cohen, 2001). There were no effective treatments for Tourette's Syndrome at the time, and Cohen dedicated himself to finding one. He wondered about the neurobiology of the disease. What changes in the brain could lead to these torturous disruptions of a young life?

A few blocks away, at the Connecticut Mental Health Center, George Aghajanian and his graduate student, Jay Cedarbaum, were recording from the noradrenergic (NE) cells of the locus coeruleus (LC) in the brainstem of rats. They had read of work in Europe by Langer, showing that α-adrenergic agonists could reduce sympathetic nervous activity through stimulation of presynaptic receptors on NE neurons, and wondered if the same sort of negative feedback could be found in the regulation of central NE neurons. Aghajanian's group had recently found that the α-agonist, clonidine, could inhibit the firing of LC neurons (Svensson *et al.*, 1975), and thought this finding deserved further scrutiny. α-Adrenergic receptors were just beginning to be differentiated into α–1 vs α–2 receptors and so were still distinguished by their location. Cedarbaum and Aghajanian performed a classical pharmacological study, examining the effects of a series of compounds with differing affinities for α- vs β-adrenergic receptors on LC firing, including clonidine. Clonidine dramatically reduced LC cell firing. They wrote in their seminal 1977 paper: "The imidazoline compound clonidine . . . was such an extremely potent agonist that it was impossible to test it using the same current and time parameters as NE. LC cells are so sensitive to the inhibitory effects of clonidine that it had to be carried in a solution ionically diluted 10-fold with NaCl" (Cedarbaum and Aghajanian, 1977). These data had a powerful effect on the neuroscience and medical communities. They created the paradigm for how the field views α–2 adrenoceptor function: exclusively presynaptic, powerfully reducing LC firing and NE release—a view that pervades today. However, Cedarbaum and Aghajanian were not so simplistic in their thinking, and even in their landmark paper wrote "On the other hand, postsynaptic receptors similar to the LC α-receptor may exist elsewhere in the brain." (Cedarbaum and Aghajanian, 1977). Thus, these authors recognized from the beginning that α–2 receptors likely had differing functions as postsynaptic receptors. It took many years for the neuroscience community to similarly attain this balanced view.

Yale Medical School and the Connecticut Mental Health Center are situated in the inner city of New Haven, and in the 1970s heroin addiction was a major focus of drug abuse research. The Aghajanian lab discovered that opiates reduced LC firing, much like clonidine, and that opiate-addicted rats had a tremendous increase in LC activity when they were withdrawn from opiate treatment (Aghajanian, 1982). Similar results were replicated in monkeys (Roth et al., 1982). Aghajanian and colleagues wondered if excessive LC activity could contribute to the symptoms of opiate withdrawal in humans, and whether clonidine might allay some of these symptoms. Based on the animal research, Gold and colleagues tested clonidine in opiate addicts and found it to alleviate many of the symptoms of withdrawal (Gold et al., 1978).

Donald Cohen was intrigued by this research and asked Aghajanian to round with him to experience the patients with profound Tourette's symptoms. Might they too have excessive NE activity? Might clonidine have soothing effects in these patients as well? For the first trial, they selected Steven, the boy described previously.

We estimated the dose and then gathered around his bedside. Forty minutes after the first dose, he fell asleep. When he awoke in one hour, he was already calmer. Over the next days, his symptoms abated, and he became more easily engaged socially and in treatment. The clinical benefits of clonidine in the treatment of the disorder seemed apparent in other patients as well. (Cohen and Leckman, 1999).

Thus, clonidine became one of the first successful treatments for Tourette's Syndrome (Cohen et al., 1979).

A few years later, Robert Hunt, working with Cohen, tested clonidine in patients with Attention Deficit Hyperactiviy Disorder (ADHD), another syndrome characterized by disinhibited behavior. Again, clonidine ameliorated symptoms of hyperactivity (Hunt et al., 1985). Hunt and Cohen hypothesized that clonidine had its therapeutic effects by reducing excessive LC activity, sedating hyperaroused patients to a more optimal level of arousal. Although clonidine's profound sedative effects (and its severe hypotensive actions) were a problem clinically, it was thought that the sedative effects were inherent to its therapeutic actions, and thus a necessary side effect if the drug was to work (Hunt et al., 1990).

BACKSTORY 2: THE NATIONAL INSTITUTE OF MENTAL HEALTH, BETHESDA, MD; 1970S

Patricia Goldman (later Goldman-Rakic) began her career as a Fellow in the Section of Neuropsychology at the NIMH in 1965 in the lab of Haldor Enger Rosvold and later served as the Chief of the Section on Developmental Neurobiology. Rosvold and his colleague, Mort Mishkin, were engaged in lesion studies of the prefrontal cortex (PFC) in monkeys. The very first such studies had been done at Yale in the 1930s, examining the contributions of the dorsolateral PFC to memory. Jacobsen and Fulton (Jacobsen, 1936) created the spatial delayed response task, a test of spatial memory that required constant updating of memory for spatial position. Jacobsen found a striking loss of short-term spatial memory in monkeys with large dorsolateral frontal lesions. It was research far ahead of its time, and World War II brought it to a quick demise. Studies of frontal lobe began again at Yale in the 1950s, but this time with a very different motivation. Lobotomies were being performed at a clipping pace nearby at Connecticut Valley Hospital by surgeons such as Scoville. A few neuropsychologists, including Rosvold, were trying to understand how lobotomies worked and did not work, trying to document what had changed. The monkey studies were reinitiated to aid that process. Karl Pribram performed the neurosurgeries in monkeys. Pribram had been trained as a neurosurgeon by Paul Bucy (of the Kluver-Bucy Syndrome), and Pribram in turn trained Mishkin (and later Hebb and Lashley). Rosvold brought Mishkin with him to NIMH, who there trained Patricia Goldman when she arrived in the Rosvold lab. It was an extraordinary dynasty of neurosurgeons, spanning almost a century, and influencing a foundation of research on the neural basis of memory.

Goldman refined the earlier work of Jacobsen; she performed careful lesion studies and defined the critical region for delayed response deficits: the cortex surrounding the caudal two-thirds of the principal sulcus (Goldman and Rosvold, 1970). She showed that this cortex was needed only if the task required both visuospatial and memory processing. At the same time, Fuster (Fuster, 1973), in Los Angeles, and Kubota and Niki (Kubota and Niki, 1971) in Japan were recording from neurons in the PFC as monkeys performed delayed response tasks, and found cells that fired during the delay period when no cue was present. Thus, there was a cellular basis for the memory processes viewed at the behavioral level. This type of memory has come to be called working memory, based on the theories of the psychologist, Baddeley, who derived his concepts of working memory during this same time period (Baddeley, 1981). A few years later, Goldman-Rakic defined the anatomical inputs to the PFC, showing that the principal sulcal cortex receives its visuospatial information from parietal cortex, while other regions of PFC receive visuo-feature, auditory, or affective information (Goldman-Rakic, 1987). Thus, she revealed parallel circuits for guiding behavior based on representations of distinct types of information.

Goldman was also interested in other types of inputs to the PFC: the catecholamine-containing axons that arise from brainstem. The catecholamines had been discovered at NIMH in the1940s and 1950s by Axelrod and colleagues, and remained a focus of research in neuropsychiatry. Aware of this work, Goldman recruited Roger Brown from Carlson's lab to perform a new technique, high-pressure liquid chromatography (HPLC), so that they could measure catecholamines in the primate cortex for the first time. They found significant and heterogeneous levels of NE and dopamine (DA) throughout most of the cortex, and found that the PFC had surprisingly high levels of DA (Brown et al., 1979). With a psychopharmacologist named Tom Brozoski, , Goldman set out to learn how this DA innervation contributed to PFC working memory function.

In their landmark study, Brozoski et al. (1979) selectively depleted catecholamines from the dorsolateral prefrontal cortex by infusing the catecholamine neurotoxin, 6-OHDA, into the principal sulcal cortex. They attempted to make selective depletions of DA vs NE, by treating one group with DMI, a drug that blocks uptake of catecholamines and 6-OHDA into NE terminals (the so-called DA-depleted group). Another group received 6-OHDA without DMI (the so-called NE-depleted group). A control group received infusions of the ascorbic acid vehicle, the ascorbic acid being necessary to prevent oxidation of the 6-OHDA. A fourth group received 5.6-DHT+DMI to selectively deplete serotonin. Finally, there were two additional control groups: one that received an ablation of the principal sulcal cortex, and an untreated control group. Following cognitive assessment, detailed biochemical measurements were performed of the principal sulcus and surrounding tissues. Brozoski et al. reported that the DA-depleted group, but not the NE or serotonin-depleted animals, exhibited working memory deficits. The impairment was as marked as that observed with ablation of the tissue itself, and could be ameliorated with systemic treatment of the catecholamine precursor, L-dopa, or the DA agonist, apomorphine. A footnote stated that "Clonidine enhanced spatial delayed alternation performance of virtually all animals" and thus the effect was considered nonspecific. Brozoski et al. concluded that only DA had significant effects on PFC function. This study became the basis for all neuropharmacological studies of PFC. The finding that DA has profound neuromodulatory effects on PFC function continues to shape most theories on the neurobiological bases of neuropsychiatric disorders.

Thus, as I arrived at Yale in 1982, the prevailing views were that DA, but not NE, had important modulatory effects on PFC function, that α-2 receptors were exclusively presynaptic, and that clonidine had powerful effects at these receptors to reduce NE transmission.

NOREPINEPHRINE ENHANCES PFC FUNCTIONS VIA ACTIONS AT POSTSYNAPTIC α–2A-ADRENOCEPTORS

Patricia Goldman-Rakic moved to Yale in 1979. I joined her shortly thereafter as a postdoctoral fellow to examine the contribution of DA loss to age-related cognitive decline. Brown and Goldman-Rakic had shown that there is an extensive depletion of DA from the PFC with advancing age, while other DA terminal fields show more subtle changes (Goldman-Rakic and Brown, 1981). I wanted to determine whether the same compounds that had improved the young, 6-OHDA-depleted monkeys would also improve the working memory abilities of the aged monkeys. I found that apomorphine and L-dopa had very mixed effects, but that clonidine had surprisingly robust beneficial effects on working memory performance in the aged monkeys. I remembered the footnote about clonidine in the Brozoski paper, and gave Tom a call. He very generously sent me all the biochemical and cognitive data from the study.

The data were extraordinary. They showed that clonidine's enhancing effects were not nonspecific: there was a pattern. The vehicle-infused animals were improved by clonidine, but their levels of NE were not normal. The ascorbic acid in the vehicle solution had induced a depletion of NE from the principal sulcal cortex. What's more, the degree of improvement with clonidine directly related to the amount of NE depletion: those monkeys with the greatest NE loss (the 6-OHDA group) were the most improved, those with smaller NE depletion (6-OHDA+DMI; acsorbic acid) were modestly improved, and those with normal NE levels (unoperated or preoperated) showed little improvement (Figure 11.1; Arnsten and Goldman-Rakic, 1985). It was a classic example of postsynaptic supersensitivity. But this meant two things: (1) α–2 receptors must be localized postsynaptically as well as presynaptically, especially in the PFC, and (2) NE, as well as DA, must influence PFC cognitive function.

Postsynaptic Actions

Working with Jing Xia Cai, Jenna Steere Franowicz, and others, we built on this initial research in several directions. Cai showed that clonidine's beneficial effects were enhanced rather than diminished in monkeys treated with reserpine, a compound that depletes the presynaptic terminal of monoamines (Cai et al., 1993). This was a classic test of pre- vs postsynaptic function, confirming our original finding. Thus, animals with catecholamine depletion, either due to normal aging or a neurotoxin, were improved by clonidine, and those with the greatest depletion were improved the most by the drug. Similar findings were emerging in humans: Clonidine improved the performance of patients with Korsakoff's amnesia on tests of recall memory and the Stroop interference task (Mair and McEntee, 1986), a test that relies on PFC function, and these beneficial effects of clonidine directly correlated with indices of NE loss as measured by CSF MHPG (McEntee and Mair, 1990). Thus, as with the monkeys, those with the greatest NE loss were the most improved by clonidine. These data encouraged our views that clonidine acted postsynaptically, and that the findings we saw in animals were

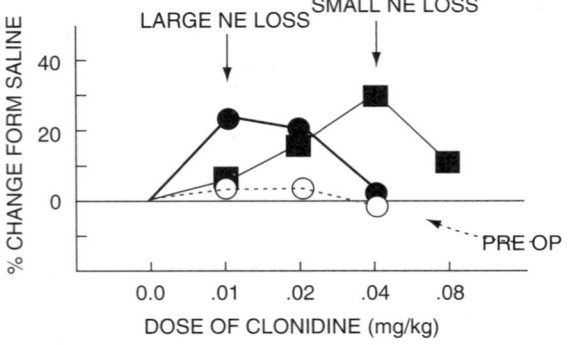

FIGURE 11.1   The potency of clonidine in enhancing spatial working memory performance in monkeys is directly related to the degree of noradrenergic (NE) depletion in the principal sulcal prefrontal cortex. These findings are consistent with clonidine acting at supersensitive, postsynaptic α-2 receptors in the prefrontal cortex. (Adapted from Arnsten and Goldman-Rakic, 1985).

relevant to humans. A few years later molecular biological tools became available, demonstrating that the vast majority of mRNA for α–2 receptors resides in noncatecholamine neurons (Scheinin et al., 1994), confirming the important role of postsynaptic α–2 receptors in the brain.

Dissociating Cognitive and Sedative Actions

Clonidine produced dramatic improvements in working memory, but it also produced pronounced sedative and hypotensive side effects. The aged monkeys would look wan and sleepy while performing near perfectly on the task. Although clonidine was tried in a variety of clinical disorders, it was clear that these side effects would severely limit its utility. Was there any way to dissociate the sedation and hypotension from the working memory enhancement? In the mid-1980s, the labs of Frances Leslie (Boyajian and Leslie, 1987) and David Bylund (Bylund, 1985) independently provided evidence for subtypes of α–2 receptors. I was close friends with the Leslie lab, and they showed me their autoradiographs while the study was still in progress. I wondered if we could see evidence of subtypes at the behavioral level. The Leslie lab described two sites in brain, the rauwolscine-sensitive site (Rs, now known to be the α–2C receptor), which was dense in striatum, and the rauwolscine-insensitive site (Ri, now known to be the α–2A receptor), which had a broader distribution. Agonists had varying affinities for these sites: guanfacine preferred the Ri site, while clonidine and BHT920 preferred the Rs site. Working with Jing Xia Cai, who was visiting from China, we assessed these compounds in our aged monkeys for their effects on working memory, blood pressure, and sedation. At first, the compounds all seemed rather similar; high doses of guanfacine improved working memory and produced some hypotension and sedation. To be rigorous, we decided to explore a wider dose range, and include very low doses as well as the high doses. We always test the monkeys "blind" to drug condition. One evening I was testing a particularly distractible aged monkey named "Chat;" she could only remember the correct spatial position over a few seconds before diverting herself with something in the test cage or grooming. This evening Chat was totally different. When the screen came up she was not picking at her foot or manipulating the caging, she was completely focused on the task. Her eyes were like laser beams, without a hint of sedation. She performed near perfectly and I knew something important had happened. After testing I broke the blind: it was a very low dose of guanfacine. A comparable profile was observed in the other aged monkeys. The low doses had no effect on blood pressure, and no signs of sedation, yet improved cognitive performance (Arnsten *et al.*, 1988). Shortly thereafter, we saw similar effects with UK14304, another agonist that preferred the Ri site as assessed by the Leslie lab (Arnsten and Leslie, 1991). We concluded that low doses of agonists with higher affinity for the Ri site such as guanfacine and UK14304 could improve working memory without hypotensive or sedative side effects, whereas drugs like clonidine and BHT920 with higher affinity for the Rs site only improved working memory at higher doses that induced a prominent side effect profile (Figure 11.2). Thus, it was possible in an aged animal with naturally occurring catecholamine depletion to completely dissociate the sedative and cognitive-enhancing effects of these compounds. The enhanced performance on the task was not simply due to sedating the animal to a more optimal level of arousal, but to strengthening PFC regulation of behavior.

Identifying the Molecular Target

Within a few years cloning became possible and three cloned subtypes of α–2 receptor were identified in humans: the α–2A, α–2B and α–2C subtypes. Using an antibody directed to the α–2A subtype, Chiye Aoki demonstrated that the α–2A receptor was localized over the postsynaptic density on dendritic spines of pyramidal cells in monkey prefrontal cortex (Aoki *et al.*, 1994; Aoki *et al.*, 1998a; Aoki *et al.*, 1998b; Arnsten *et al.*, 1996). She also saw the receptor at other cellular locations, including presynaptic sites, but there was no doubt that the α–2A receptor was in the ideal position to influence working memory function.

The Ri was now known to be the α–2A subtype, but we did not know for certain that this was the critical receptor for guanfacine's beneficial effects on working memory. Guanfacine prefers this receptor, but by rigorous definition it is not truly selective for this site. The advent of genetically altered mice allowed us to stringently test the role of the α–2A subtype for the first time. Both the α–2A and α–2C subtypes had been identified in the PFC. The Tanila lab in Finland examined wild-type vs α–2C knockout mice in a spatial working memory task. The α–2C knockout mice performed normally on the task, and showed enhancement with an α–2 agonist just like the wild-type animals (Tanila *et al.*, 1999). At the

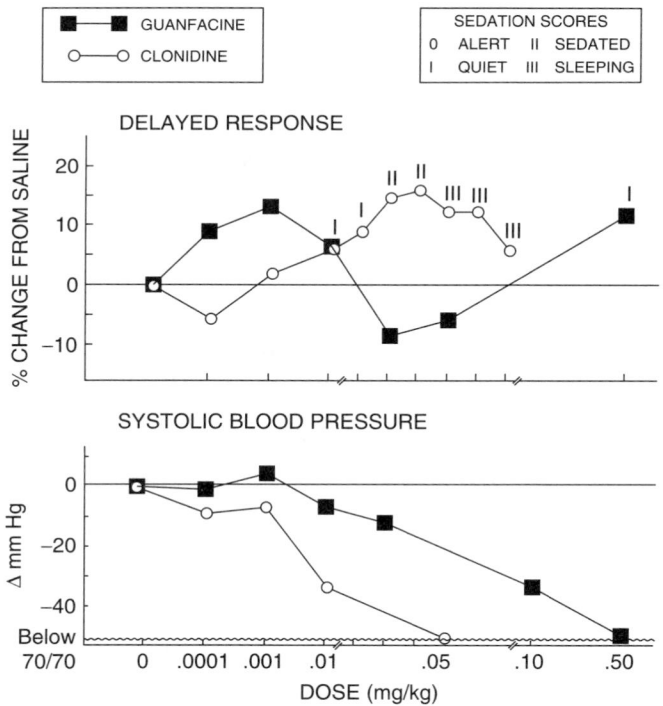

FIGURE 11.2   In aged monkeys with naturally occurring catecholamine depletion, guanfacine is able to improve working memory abilities without inducing sedation or hypotension. In contrast, clonidine improves working memory only at higher doses that produce prominent sedation and hypotension. (Adapted from Arnsten *et al.*, 1988).

same time, Jenna Franowicz and I were examining mice with a functional knockout of the α–2A receptor, mice created by the Limbird lab at Vanderbilt and donated to us by the Kobilka lab at Stanford. This work was part of Jenna's PhD thesis. She found that the wild-type mice were improved by guanfacine just like our monkeys, but the α–2A mutants showed no effect with the drug (Figure 11.3; (Franowicz et al., 2002)). Furthermore, although the α–2A mutant mice learned the spatial working memory task as fast as the wild-type animals, they had weaker working memory abilities and were not able to perform well with delays of more than a few seconds (Franowicz et al., 2002). Thus, this work definitively showed that the α–2A subtype was the molecular target for guanfacine's beneficial effects on working memory. Parallel studies in knockout mice have shown an important role for this subtype in the sedating and hypotensive effects of α–2 agonists (MacDonald et al., 1997). However, the sedative effects likely involve a wide interplay of α–2 receptors, including receptors in locus coeruleus (α–2A and C), cortex (α–2A and C), thalamus (α–2B), basal forebrain, and spinal cord (α–2A and C?).

Specificity of Actions in Prefrontal Cortex

The enhancing effects of α–2 agonists are specific to PFC cognitive functions. Systemic administration of clonidine, guanfacine and other α2 agonists to monkeys or rats improves the performance of a variety of PFC tasks: spatial as well as object working memory tasks dependent on lateral PFC, as well as reversal tasks dependent on ventral PFC that require behavioral inhibition (reviewed in Arnsten and Robbins, 2002). The α–2

agonists are especially helpful under distracting conditions when interference is high and PFC functions are challenged (Arnsten and Contant, 1992). Conversely, NE depletion, like PFC lesions, increases distractibility (Carli et al., 1983; Roberts et al., 1976). In contrast, α–2 agonists have no effect on non-PFC tasks such as recognition memory, visual pattern discrimination, or covert attentional orienting dependent on medial temporal, inferior temporal, and parietal cortices, respectively (Arnsten and Robbins, 2002). We performed a SPECT imaging study of monkeys administered guanfacine vs saline while performing a spatial working memory task and found that guanfacine improved performance and increased regional cerebral blood flow in the same region of PFC that Goldman-Rakic had earlier shown to be most critical to spatial working memory, the caudal two-thirds of the principal sulcal cortex (Figure 11.4A; See also the Color Plate section; (Avery et al., 2000)). Guanfacine had no effect on regional cerebral blood flow in auditory association cortex, a region unchallenged by the task. Thus, specificity was observed with systemic administration in both cognitive and imaging studies.

We have collaborated with Bao Ming Li's lab in China to show that α–2 agonists act directly within the PFC to enhance working memory. Thus, infusions of guanfacine directly into the prinicpal sulcal PFC produced a delay-related improvement in working memory (Mao et al., 1999), while blockade of α–2 receptors in this same region with yohimbine dramatically impaired working memory performance in a delay-related manner (Li and Mei, 1994). More recently the Li lab has shown that yohimbine infusion into the PFC induces impulsive errors on a go–no-go task similar to that used in ADHD children (Ma et al., 2003). They have also observed that intra-PFC yohimbine infusions induce locomotor hyperactivity (B.M. Li, personal communication). Thus, the cardinal symptoms of ADHD can be reproduced by blocking α–2 receptors in PFC. Bao Ming Li has shown similar effects at the level of the single cell. Iontophoretic application of yohimbine onto PFC neurons in monkeys performing working memory tasks reduced delay-related activity, the cellular measure of working memory, without effecting spontaneous cell activity (Li et al., 1999). Similar effects were observed in Japan by Sawaguchi (Sawaguchi, 1998). Conversely, Li et al. showed that iontophoretic application of clonidine strengthened delay-related activity (schematically represented in Figure 11.4B; see also the Color Plate section), and that this increase in delay-related firing could be revered by iontophoretic application of yohimbine (Li et al., 1999). Systemic administration of clonidine at a dose that improves delayed response performance also increased delay-related firing of PFC neurons (ibid). It is remarkable to have such consistency at the cellular and behavioral levels.

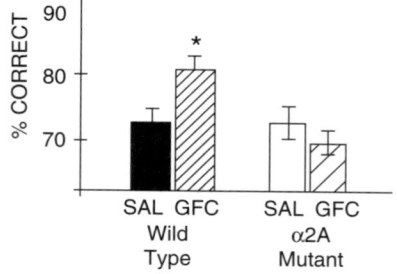

GUANFACINE IMPROVES WORKING MEMORY
IN WILD TYPE, BUT NOT α2A MUTANT MICE

FIGURE 11.3 The α-2 adrenoreceptor agonist, guanfacine, improves delayed alternation performance in wild-type mice, but has no effect in mice with a functional knockout of the α-2 A receptor. The mutant mice also showed evidence of weaker working memory abilities, highlighting the importance of endogenous α-2 A receptor stimulation in the maintenance of working memory abilities. (Adapted from Franowicz et al., 2002).

A. SPECT IMAGING

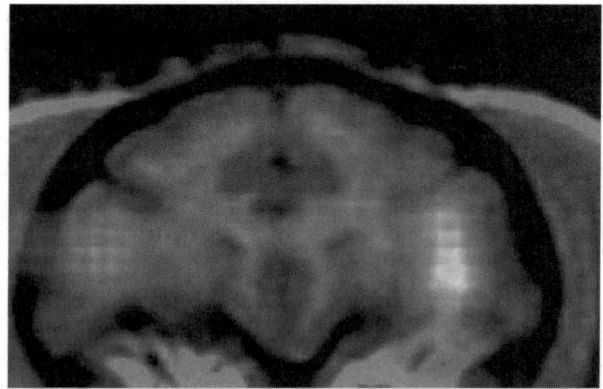

B. PHYSIOLOGY

SALINE

α2 AGONIST

FIGURE 11.4 **A:** A single photon emission computed tomography (SPECT) imaging study of rhesus monkeys performing a spatial working memory task found that guanfacine significantly increased regional cerebral blood flow in the principal sulcal prefrontal cortex. The difference between the guanfacine and saline SPECT scans is overlaid on a coronal section of the structural MRI for a representative rhesus monkey. (Adapted from Avery *et al.*, 2000). **B:** Many prefrontal cortical neurons exhibit increased firing during the delay period while monkeys perform a delayed response task. These cells are often tuned (e.g., firing only if the cue had been on the right but not on the left side). Bao Ming Li has shown that either systemic or iontophoretic application of clonidine significantly increases delay-related firing of prefrontal cortical neurons without altering spontaneous firing rate (Li *et al.*, 1999). This finding is schematically depicted in Figure 11.4B. (See also the Color Plate section.)

Second Messenger Mechanisms

Current research is examining the second messenger mechanisms underlying the enhancing effects of α2A receptor stimulation in PFC. These receptors are commonly coupled to Gi proteins which inhibit cAMP/protein kinase A intracellular signaling. Most research on second messenger mechanisms in mammals has focused on the hippocampus, where activation of cAMP/protein kinase A is needed to sustain long term potentiation (LTP) and consolidate memory (Frey *et al.*, 1993). However, our data indicate that activation of protein kinase A in PFC *impairs* working memory (Taylor *et al.*, 1999), perhaps consistent with the constant updating and fluidity required of working memory function.

Our recent data suggest that protein kinase A signaling is activated by stress exposure (Arnsten, 2000), and becomes disinhibited with advancing age (Ramos *et al.*, 2003). Therefore, some of the beneficial effects of α–2A agonists in stressed and aged animals may result from inhibition of detrimental protein kinase A signaling.

Beneficial Influences of Presynaptic α-2 Receptors on PFC function

Stress also impairs PFC function through activation of protein kinase C intracellular signaling (Arnsten, 2000). The doctoral thesis of Shari Birnbaum has shown that high levels of NE released during stress engage α–1 receptors, which in turn activate protein kinase C and impair working memory (Birnbaum *et al.*, 2004). Thus, α–2 agonists may have a protective effect on PFC function during stress by reducing high levels of NE release and decreasing high levels of tonic LC cell firing via presynaptic α–2A and α–2C receptors, thus reducing harmful α–1 receptor actions in the PFC. However, Birnbaum observed that guanfacine was more potent than clonidine in protecting working memory from stress, even though clonidine is 10 times more potent than guanfacine in reducing LC firing and decreasing NE release (Birnbaum *et al.*, 2000). Thus, postsynaptic α-2 receptor actions likely play an important role even under conditions of high NE release.

In summary, a highly consistent story has emerged from labs in China, Finland, England, Japan, and the United States, demonstrating that NE strengthens PFC function at the cellular and behavioral levels through activation of postsynaptic α–2A receptors.

TRANSLATION TO THE CLINIC

Guanfacine translated from the lab to the clinic due to discussions with physicians deeply interested in α-2 adrenoceptor mechanisms for childhood disorders. In the early 1990s, I was asked to give a seminar at Vanderbilt University. Robert Hunt was in the audience, and we spoke at length after the seminar, planning an open trial of guanfacine in ADHD. Donald Cohen also asked me to give a seminar at the Yale Child Study Center, presenting our findings that low doses of guanfacine could improve working memory and protect performance from distraction with few sedative or hypotensive effects. Donald later worked with Mark Ruddle and Phil Chappell to assess guanfacine in an open label study of children with both tics and ADHD symptoms who could not take stimulant medications. The Hunt (Hunt *et al.*, 1995) and Chappell (Chappell *et al.*, 1995) papers were published in 1995, along with a third open trial

(Horrigan and Barnhill, 1995). These positive results encouraged more rigorous analysis of guanfacine.

The first double-blind, placebo-controlled trial of guanfacine in children with ADHD and tics was carried out at the Yale Child Study Center by Scahill et al. (2001). This study showed that guanfacine significantly improved teacher ratings of both inattention and hyperactivity/impulsivity. It also improved performance of the Connors Continuous Performance Task, a test that requires sustained attention and behavioral inhibition, functions reliant on PFC. Guanfacine also reduced the number of tics. The Tourette's Center at Yale had shown in functional imaging studies that the dorsolateral PFC is activated when Tourette's patients successfully inhibit their tics (Peterson et al., 1998). Thus, it is possible that guanfacine helps reduce tics through a similar mechanism: strengthening PFC inhibition of an abnormal process in deeper basal ganglia structures.

Another controlled study has shown that guanfacine can ameliorate ADHD symptoms in adults with ADHD (Taylor and Russo, 2001). Interestingly, this study found that guanfacine was superior to stimulants in enhancing performance of the Stroop interference task, again pointing to a PFC mechanism. A similar pattern has emerged in normal adults given acute guanfacine. Reminiscent of the results in monkeys, guanfacine improved performance of tasks dependent on the PFC such as spatial working memory, paired associates and planning, but did not improve recognition memory (Jakala et al., 1999a,1999b). The same battery of tasks has recently been given to ADHD patients, and interestingly, they are impaired on the very same PFC tasks that are improved by guanfacine (McLean et al., 2003). An extensive literature now documents that ADHD patients are impaired on a host of PFC tasks (reviewed in Arnsten et al., 1996), and both functional and structural imaging studies have demonstrated that PFC-striatal-cerebellar circuits are disrupted in patients with this disorder (reviewed in Arnsten and Castellanos, 2002). Thus, guanfacine may provide a rational therapy, strengthening PFC regulation of behavior and attention. Guanfacine and similar agents are being tested in other patient populations (schizophrenia, posttraumatic stress disorder) where PFC cognitive deficits contribute to disability (Friedman et al., 2001; Horrigan, 1996).

The pioneering studies of Jacobsen, Kubota, Fuster, and Goldman-Rakic have given us the paradigms to analyze our highest cognitive abilities, subserved by prefrontal cortical circuits. We have learned that prefrontal cortical neurons can fire to stimuli that are no longer in the environment, allowing us to hold in mind representations of faces, events and places, and rewards and punishments from the distant or recent past. These representations can inhibit our thoughts, feelings and actions, and direct us toward worthier goals: a cohesive sentence, a math assignment completed, a horrible memory suppressed, so that we may go on living. These are the cognitive processes most assaulted in neuropsychiatric illness. Our ability to understand the neurochemical processes that strengthen or weaken prefrontal cortical abilities will give us the tools to treat these illnesses, and the wisdom to view these disorders with compassion rather than disdain.

It is my hope that future strategies for the treatment of neuropsychiatruc illness will be sufficiently informed by basic neuroscience to permit a wholly rational approach: (1) A sound understanding of the brain circuits underlying distinct neuropsychiatric symptoms, (2) a knowledge of the neurochemical needs of those circuits and if possible, the neurochemical changes in those circuits associated with the disorder, and (3) the ability to optimize the neurochemical environment for that circuit with the appropriate medication, ultimately with compounds that are delivered to that circuit selectively. This strategy would only be effective if the substrate for drug actions is present (i.e., drugs like guanfacine do not work if the PFC is ablated) (Arnsten and van Dyck, 1997). However, this strategy may be very effective in disorders such as ADHD, bipolar disorder and posttraumatic stress disorder where the cerebral architecture remains generally intact.

THE PROCESS OF DISCOVERY

Several themes recur as one examines the elements that facilitate the discovery process. Perhaps most evident is the importance of an intellectually rich and adventurous research community that promotes the cross-pollination of ideas. New discoveries are generally made by physicians and scientists who are both curious and open-minded about research outside their own area of expertise and who actively seek out novel perspectives. It is certainly important to disregard false boundaries such as neurological vs psychiatric disorders, molecular vs behavior research, and to recognize that insights can emerge from a variety of frameworks. These associations between seemingly disparate arenas are certainly among the most joyful of scientific experiences. They are the reward of disciplined study: "In the field of observation, chance favors only the prepared mind" (Louis Pasteur 1854).

Discovery also seems to be facilitated by immersion in the raw data (be it the symptoms of a disorder or the performance of a monkey on a task). Highly digested or preinterpreted data can remove the very patterns most needed to detect new mechanisms. And it is critical to be rigorously objective and thus adventurous in one's thinking, to try not to be contaminated by current fashions in thinking. Perhaps most difficult of all is to see the data for what they are, free of one's own wishes for them. One must have a guiding vision, and yet the ability to discard that vision when the data disagree (a mental flexability that requires PFC function!). Theories allow us to see patterns and realize that they are clues, but they can also prejudice our vision and create inertia. We are always fooling ourselves in some manner, and yet, there is no better way to proceed.

References

Aghajanian GK (1982). Central noradrenergic neurons: a locus for the functional interplay between alpha-2 adrenoceptors and opiate receptors, *J Clin Psychiatry* 43:20-24.

Aoki C, Go C-G, Venkatesan C, Kurose H. (1994). Perikaryal and synaptic localization of alpha-2A-adrenergic receptor-like immunoreactivity, *Brain Res* 650:181-204.

Aoki C *et al.* (1998a). Cellular and subcellular sites for noradrenergic action in the monkey dorsolateral prefrontal cortex as revealed by the immunocytochemical localization of noradrenergic receptors and axons, *Cerebral Cortex* 8:269-277.

Aoki C, Venkatesan C, Kurose H. (1998b). Noradrenergic modulation of the prefrontal cortex as revealed by electron microscopic immunocytochemistry, *Adv Pharmacol* 42:777-780.

Arnsten AFT. (2000). Stress impairs PFC function in rats and monkeys: role of dopamine D1 and norepinephrine alpha-1 receptor mechanisms, *Prog Brain Res* 126:183-192.

Arnsten AFT, Cai JX, Goldman-Rakic PS. (1988). The alpha-2 adrenergic agonist guanfacine improves memory in aged monkeys without sedative or hypotensive side effects, *J Neurosci* 8:4287-4298.

Arnsten AFT, Castellanos FX. (2002). Neurobiology of attention regulation and its disorders. In Martin A, Scahill L, Charney D, Leckman J, editors: *Textbook of child and adolescent psychopharmacology*, New York: Oxford University Press.

Arnsten AFT, Contant TA. (1992). Alpha-2 adrenergic agonists decrease distractability in aged monkeys performing a delayed response task, *Psychopharmacology* 108:159-169.

Arnsten AFT, Goldman-Rakic PS. (1985). Alpha-2 adrenergic mechanisms in prefrontal cortex associated with cognitive decline in aged nonhuman primates, *Science* 230:1273-1276.

Arnsten AFT, Leslie FM. (1991). Behavioral and receptor binding analysis of the alpha-2 adrenergic agonist, UK-14304 (5 bromo-6 [2-imidazoline-2-yl amino] quinoxaline): evidence for cognitive enhancement at an alpha-2 adrenoceptor subtype, *Neuropharmocology* 30:1279-1289.

Arnsten AFT, Robbins TW. (2002). Neurochemical modulation of prefrontal cortical function in humans and animals. In Stuss DT, Knight RT, editors: *Principles of frontal lobe function*, New York: Oxford University Press.

Arnsten AFT, Steere JC, Hunt RD. (1996). The contribution of alpha-2 noradrenergic mechanisms to prefrontal cortical cognitive function: potential significance to attention deficit hyperactivity disorder, *Arch Gen Psychiatry* 53:448-455.

Arnsten AFT, van Dyck CH. (1997). Monoamine and acetylcholine influences on higher cognitive functions in nonhuman primates: relevance to the treatment of Alzheimer's disease. In Brioni JD, Decker MW: *Pharmacological treatment of Alzheimer's disease: molecular and neurobiological foundations*, New York: John Wiley and Sons.

Avery RA *et al.* (2000). The alpha-2A-adenoceptor agonist, guanfacine, increases regional cerebral blood flow in dorsolateral prefrontal cortex of monkeys performing a spatial working memory task, *Neuropsychopharmacology* 23:240-249.

Baddeley A. (1981). The concept of working memory: a view of its current state and probable future development, *Cognition* 10:17-23.

Birnbaum SG, Podell DM, Arnsten AFT. (2000). Noradrenergic alpha-2 receptor agonists reverse working memory deficits induced by the anxiogenic drug, FG7142, in rats, *Pharmacol Biochem Behav* 67:397-403.

Birnbaum SG *et al.* (2004). Protein kinase C overactivity impairs prefrontal cortical regulation of behavior. In press, *Science*.

Boyajian CL, Leslie FM. (1987). Pharmacological evidence for alpha-2 adrenoceptor heterogeneity: differential binding properties of [3H]rauwolscine and [3H]idazoxan in rat brain, *J Pharmacol Exp Ther* 241:1092-1098.

Brown RM, Crane AM, Goldman PS. (1979). Regional distribution of monoamines in the cerebral cortex and subcortical structures of the rhesus monkey: concentrations and in vivo synthesis rates, *Brain Res* 168:133-150.

Brozoski T, Brown RM, Rosvold HE, Goldman PS. (1979). Cognitive deficit caused by regional depletion of dopamine in prefrontal cortex of rhesus monkey, *Science* 205:929-931.

Bylund DB. (1985). Heterogeneity of alpha-2 adrenergic receptors, *Pharmacol Biochem Behav* 22:835-843.

Cai JX, Ma Y, Xu L, Hu, X. (1993). Reserpine impairs spatial working memory performance in monkeys: reversal by the alpha-2 adrenergic agonist clonidine, *Brain Res* 614:191-196.

Carli M, Robbins TW, Evenden JL, Everitt BJ. (1983). Effects of lesions to ascending noradrenergic neurons on performance of a 5-choice serial reaction task in rats: implications for theories of dorsal noradrenergic bundle function based on selective attention and arousal, *Behav Brain Res* 9:361-380.

Cedarbaum JM, Aghajanian GK. (1977). Catecholamine receptors on locus coeruleus neurons: pharmacological characterization, *Eur J Pharmacol* 44:375-385.

Chappell PB *et al.* (1995). Guanfacine treatment of comorbid attention deficit hyperactivity disorder and Tourette's syndrome: preliminary clinical experience, *J Am Acad Child Adolesc Psychiatry* 34:1140-1146.

Cohen DJ. (2001). Sterling Lecture, February 27, 2001, Into life: autism, Tourette's syndrome and the community of clinical research, *Isr J Psychiatry Relat Sci* 38:226-234.

Cohen DJ, Leckman JF. (1999). Introduction: the self under siege. In Leckman JF, Cohen DJ, editors: *Tourette's syndrome: tics, obessions, compulsions—developmental psychopathology and clinical care*, New York: John Wiley and Sons.

Cohen DJ, Young JG, Nathanson JA, Shaywitz BA. (1979). Clonidine in Tourette's syndrome, *Lancet* 2:551-553.

Franowicz JS *et al.* (2002). Mutation of the alpha2A-adrenoceptor impairs working memory performance and annuls cognitive enhancement by guanfacine, *J Neurosci* 22:8771-8777.

Frey U, Huang Y-Y, Kandel ER. (1993). Effects of cAMP simulate a late stage of LTP in hippocampal CA1 neurons, *Science* 260:1661-1664.

Friedman JI *et al.* 2001. Guanfacine treatment of cognitive impairment in schizophrenia: a pilot study, *Neuropsychopharmacology* 25:402-409.

Fuster JM. (1973). Unit activity in prefrontal cortex during delayed response performance: Neuronal correlates of transient memory, *J Neurophysiol* 36:61-78.

Gold MS, Redmond DEJ, Kleber HD. (1978). Clonidine in opiate withdrawal, *Lancet* 29:929-930.

Goldman PS, Rosvold HE. (1970). Localization of function within the dorsolateral prefrontal cortex of the rhesus monkey, *Exp Neurol* 27:291-304.

Goldman-Rakic PS. (1987). Circuitry of the primate prefrontal cortex and the regulation of behavior by representational memory. In Plum F, editor: *Handbook of physiology: the nervous system, higher functions of the brain*, Bethesda: American Physiological Society.

Goldman-Rakic PS, Brown RM. (1981). Regional changes of monoamines in cerebral cortex and subcortical structures of aging rhesus monkeys, *Neuroscience* 6:177-187.

Horrigan JP. (1996). Guanfacine for PTSD nightmares, *J Am Acad Child Adolesc Psychiatry* 35:975-976.

Horrigan JP, Barnhill LJ. (1995). Guanfacine for treatment of attention-deficit-hyperactivity disorder in boys, *J Child Adolescent Psychopharmacol* 5:215-223.

Hunt RD, Arnsten AFT, Asbell, MD. (1995). An open trial of guanfacine in the treatment of attention deficit hyperactivity disorder, *J Am Acad Child Adolesc Psychiatry* 34:50-54.

Hunt RD, Capper L, O'Connell, P. (1990). Clonidine in child and adolescent psychiatry, *J Child Adolesc Psychiatry* 1:87-102.

Hunt RD, Mindera RB, Cohen DJ. (1985). Clonidine benefits children with attention deficit disorder and hyperactivity: reports of a double-blind placebo-crossover therapeutic trial, *J Am Acad Child Psychiatry* 24:617-629.

Jacobsen CF. (1936). Studies of cerebral function in primates, *Comp Psychol Monogr* 13:1-68.

Jakala P *et al.* (1999a). Guanfacine, but not clonidine, improves planning and working memory performance in humans, *Neuropsychopharmacology* 20:460-470.

Jakala P *et al.* (1999b). Guanfacine and clonidine, alpha-2 agonists, improve paired associates learning, but not delayed matching to sample, in humans, *Neuropsychopharmacol* 20:119-130.

Kubota K, Niki H. (1971). Prefrontal cortical unit activity and delayed alternation performance in monkeys, *J Neurophysiol* 34:337-347.

Li B-M, Mao Z-M, Wang M, Mei Z-T (1999). Alpha-2 adrenergic modulation of prefrontal cortical neuronal activity related to spatial working memory in monkeys, *Neuropsychopharmacol* 21: 601-610.

Li, B-M, Mei, Z-T. (1994). Delayed response deficit induced by local injection of the alpha-2 adrenergic antagonist yohimbine into the dorsolateral prefrontal cortex in young adult monkeys, *Behav Neural Biol* 62:134-139.

Ma CL, Qi XL, Peng JY, Li BM. (2003). Selective deficit in no-go performance induced by blockade of prefrontal cortical alpha2-adrenoceptors in monkeys, *Neuroreport* 14:1013-1016.

MacDonald E, Kobilka BK, Scheinin M. (1997). Gene targeting: homing in on alpha-2-adrenoceptor subtype function, *Trends Pharmacol Sci* 18:211-219.

Mair RG, McEntree WJ. (1986). Cognitive enhancement in Korsakoff's psychosis by clonidine: a comparison with l-dopa and ephedrine, *Psychopharmacology* 88:374-380.

Mao Z-M, Arnsten AFT, Li B-M. (1999). Local infusion of alpha-1 adrenergic agonist into the prefrontal cortex impairs spatial working memory performance in monkeys, *Biol Psychiatry* 46:1259-1265.

McEntee WJ, Mair RG. (1990). The Korsakoff syndrome: a neuro-chemical perspective, *Trends Neurosci* 13:340-344.

McLean A, Dowson J, Robbins TW, Sahakian BJ. (2004). Characteristic neurocognitive profile associated with adult attention-deficit disorder, *Psychol Med* 34:681-92.

Peterson BS *et al.* (1998). A functional magnetic resonance imaging study of tic suppression in Tourette syndrome, *Arch Gen Psychiatry* 54:326-333.

Ramos B *et al.* (2003). Dyregulation of protein kinase A signaling in the aged prefrontal cortex: new strategy for treating age-related cognitive decline, *Neuron* 40:835-45.

Roberts DC, Price MT, Fibiger HC. (1976). The dorsal tegmental noradrenergic projection: an analysis of its role in maze learning, *J Comp Physiol Psychol* 90:363-372.

Roth RH, Elsworth JD, Redmond DEJ. (1982). Clonidine suppression of noradrenergic hyperactivity during morphine withdrawal by clonidine: biochemical studies in rodents and primates, *J Clin Psychiatry* 43:42-46.

Sawaguchi T. (1998). Attenuation of delay-period activity of monkey prefrontal cortical neurons by an alpha-2 adrenergic antagonist during an oculomotor delayed-response task, *J Neurophysiol* 80:2200-2205.

Scahill L *et al.* (2001). Guanfacine in the treatment of children with tic disorders and ADHD: A placebo-controlled study, *Am J Psychiatry* 158:1067-1074.

Scheinin M *et al.* (1994). Distribution of alpha-2-adrenergic receptor subtype gene expression in rat brain, *Brain Res* 21:133-149.

Svensson TH, Bunney BS, Aghajanian GK. (1975). Inhibition of both noradrenergic and serotonergic neurons in brain by the alpha-adrenergic agonist clonidine, *Brain Res* 92:291-306.

Tanila H *et al.* (1999). Role of alpha-2C-adrenoceptor subtype in spatial working memory as revealed by mice with targeted disruption of the alpha-2C-adrenoceptor gene, *Eur J Neurosci* 11:599-603.

Taylor FB, Russo J. (2001). Comparing guanfacine and dextroamphetamine for the treatment of adult attention deficit-hyperactivity disorder, *J Clin Psychopharmacol* 21:223-228.

Taylor JR, Birnbaum SG, Ubriani R, Arnsten AFT. (1999). Activation of protein kinase A in prefrontal cortex impairs working memory performance, *J Neuroscience* (Online) 19:RC23.

12

Neuroscience, Molecular Medicine, and New Approaches to the Treatment of Depression and Anxiety

George R. Heninger, MD

Until the present, progress in the understanding of the causes of depression and anxiety has been slow, primarily because diagnosis has relied only on descriptive methods. This chapter discusses the ways in which neuroscience and neuropsychopharmacology are beginning to provide a new improved understanding of these disorders. The delineation of the underlying neurobiological abnormalities will provide the methods to define more homogenous diagnostic subgroups. This in turn will allow a greater opportunity for the development of direct, targeted, and more efficacious treatments. Functional abnormalities in the corticotropin-releasing hormone and hypothalamic pituitary adrenal axis have been observed in patients with depression and anxiety and molecular abnormalities, as possible causes of these abnormalities are now being evaluated. Imaging studies have disclosed atrophy in the hippocampus in depression that is reversed with antidepressant treatment. This finding is supported by powerful evidence from preclinical research showing that stress decreases neurogenesis and cell survival in the hippocampus, which is reversed with antidepressant treatment. In addition, the molecular and cellular pathways involved in neuroplasticity and cell survival have been found to be influenced by several other drugs effective in the treatment of mood disorders. The pathways and cellular constituents involved in conditioned fear are being elucidated, and the molecular and pharmacological factors underlining fear acquisition and extinction are being delineated. This information will allow the development of more physiologically relevant pharmacological and behavioral treatments for anxiety disorders. Candidate genes are being investigated and initial results are promising. Individuals homozygous for the short allele of the serotonin transporter have been shown to be particularly vulnerable to the stressful life events and childhood maltreatment that can cause depression. The new understanding of molecular and neurobiological pathogenesis will allow the definition of new clinical subgroups where rational treatments can be directed at the cause of illness. This in turn will improve efficacy and ultimately lead to prevention of these disabling disorders.

BACKGROUND

Epidemiology

There is a high prevalence of depression in the community and when other medical illness is present, prevalence rates are much higher. The most recent national survey reports a 16.2% lifetime and 6.6% 12-month prevalence of major depressive disorder in the United States (Kessler et al., 2003). The prevalence is increasing with each new birth cohort. As can be seen in Figure 12.1, 25% of individuals who are currently in their twenties can expect to have a major depressive episode during their lifetime. Because of its high prevalence and disabling symptoms, depression causes the fourth highest number of disability adjusted life-years of all medical conditions occurring worldwide. Thus, depression is listed as the fourth highest worldwide cause of burden of disease, with only lower respiratory infections, diarrheal diseases, and perinatal disorders ranked higher (Murray and Lopez, 1997).

There is a high comorbidity of depression and anxiety, with 59% of depressed individuals having a comorbid anxiety disorder and also a 25% rate of substance use and a 25% rate of impulse control disorder, with a 70% rate for any other disorder (Kessler et al., 2003). The use of formal diagnostic criteria that requires an individual to manifest a high level of symptoms to be labeled a case

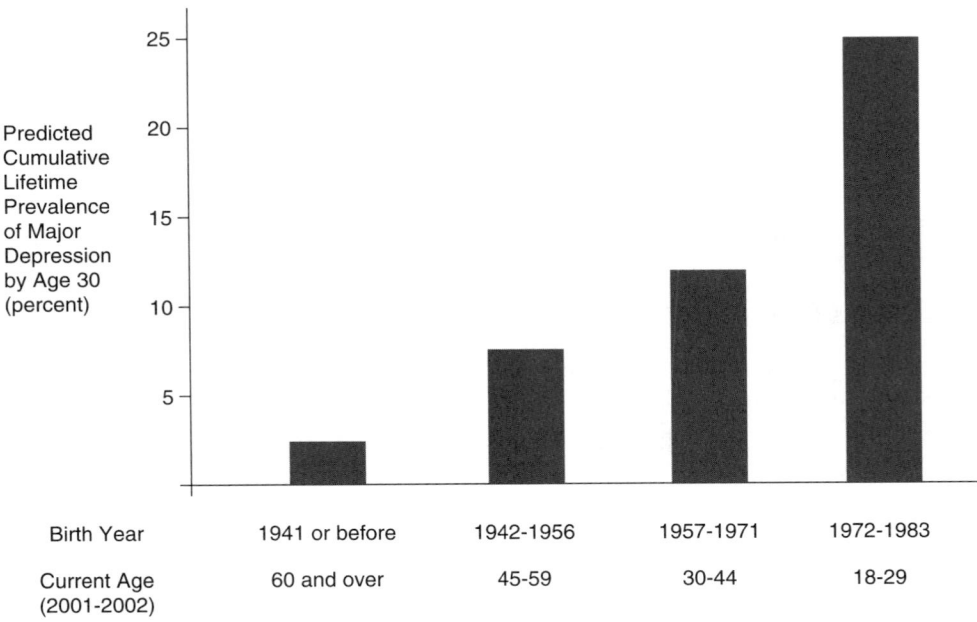

FIGURE 12.1 Depression is increasing in young people. From the National Comorbidity Survey Replication conducted in 2001-2002. (Constructed from data in Kessler RC *et al.* (2003). *JAMA* 289:3095.)

underestimates the magnitude of the problem. For instance, there is a more than sixfold increase in prevalence of panic attack symptoms, as the criteria for diagnosis are less stringent (Figure 12.2) (Eaton *et al.*, 1994). This is important because symptom levels too low to meet full criteria for a disorder according to the American Psychiatric Association Diagnostic and Statistical Manual, Fourth Edition (DSM IV), can still have major effects in terms of disability and mortality. For example, it has been known for some time that the presence of major depression, as defined in DSM IV, increases mortality in individuals with cardiac disease. However, as illustrated in Figure 12.3, even low levels of depression that are usually not thought to be clinically relevant are also associated with a fivefold to sixfold increase in percent mortality (Bush *et al.*, 2001). The importance of the comorbidity of depression on overall subjective well-being is illustrated in individuals with cardiac disease in Figure 12.4. Severity of impairment of left ventricular ejection fraction, which directly relates to mortality, did not relate to subjective symptom burden, physical limitation, quality of life, or overall health. Ischemia, as measured by exercise capacity on the treadmill test, did significantly relate to all four subjective categories, especially physical limitation; and, most important, depressive symptoms related the strongest to the four subjective categories, even more strongly to physical limitation than did exercise capacity. This illus-

trates the pervasive effect that depression can have on quality of life (Ruo *et al.*, 2003).

There is at least a doubling of the rate of depression in most medical illness, and there is an increased rate of depression the more severe the illness. For example, in a mail survey of individuals with multiple sclerosis, 42% reported clinically significant depressive symptoms. There was a 58% increase in severity of depressive symptoms reported by those individuals who had advanced disease vs those who had minimal disease (Chwastiak *et al.*, 2002). Individuals reporting a chronic painful condition seem to have an especially high rate of depression. In a large phone survey in Europe, compared to individuals reporting no pain and no illness, those reporting a medical illness had an odds ratio of 1.2 of having depression, those reporting a chronic painful condition had an odds ratio of 3.4, and those reporting both a medical condition and a chronic painful condition had an odds ratio of 5.1. Thus, the effect of a medical condition and chronic painful condition appears to be additive (Ohayon and Schatzberg, 2003).

Taken together the epidemiological findings indicate a high prevalence of depression and anxiety disorders in the community, and even higher rates of depression are associated with a wide range of medical illness. The importance of depression on mortality and quality of life is well illustrated in cardiovascular disease, where the

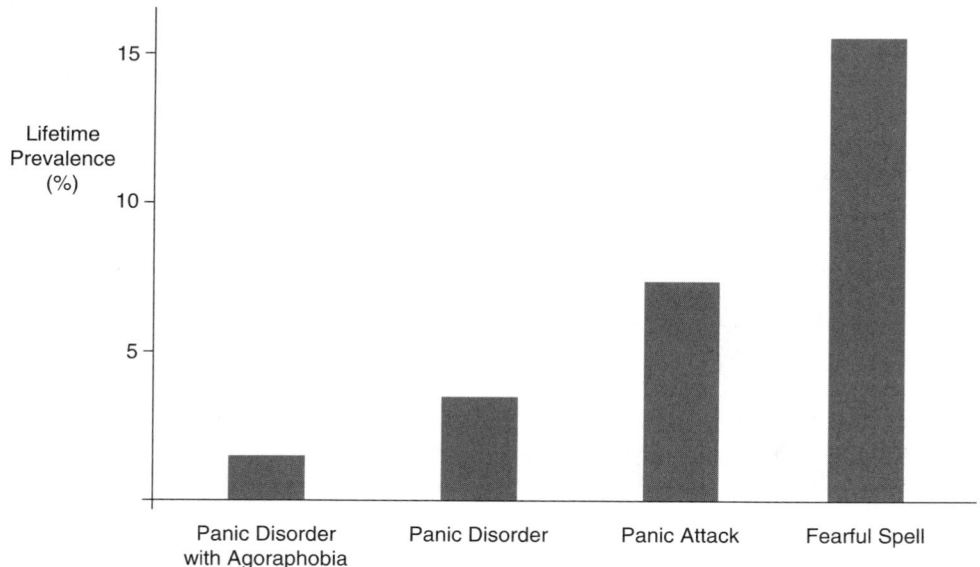

FIGURE 12.2 Panic symptoms are much more prevalent than a specific diagnosis. Estimated prevalence of DSM III-R diagnosis and related experience, U.S. population, National comorbidity study. (Constructed from data in Eaton WW *et al.* (1994). *Am J Psychiatry* 151:413.)

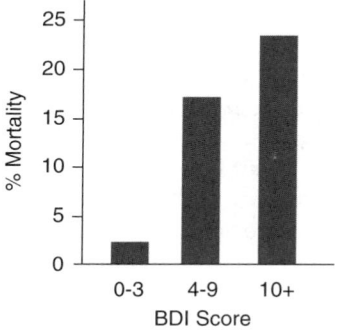

FIGURE 12.3 Low levels of depression increase mortality risk after acute myocardial infraction. Relationship of Beck Depression Inventory Scores (BDI) to mortality rate at 4 months in 144 patients over 65 years old who survived until discharge. BDI scores < 10 are often though not to be clinically meaningful, but there is a 6.5-fold increase in percent mortality scores 4 to 9 compared to scores 0 to 3. (Constructed from data in Bush DE *et al.* (2001). *Am J Cardiol* 88:337.)

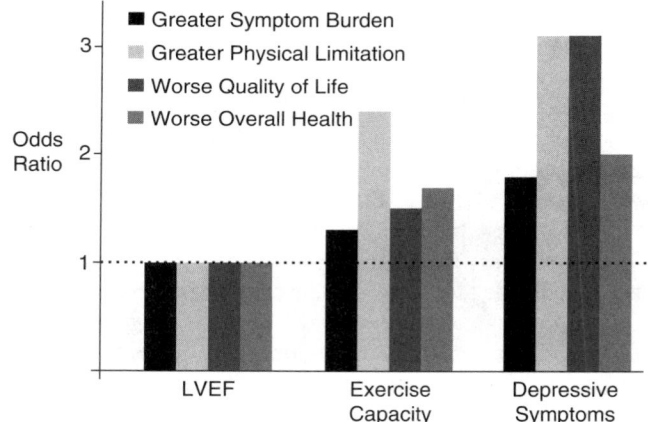

FIGURE 12.4 Subjective health status is associated with depression not cardiac function. Study conducted with 1024 participants who all had coronary artery disease. *LVEF* = left ventricular ejection fraction measured with echocardiogram, Exercise capacity measured with treadmill test, Depression measured by 9-item patient health questionnaire, score of 10 or greater. Odds ratio from ordinal logistic regression comparing high-risk vs low-risk outcome categories. (Constructed from data in Ruo B *et al.* (2003). *JAMA* 290:215.)

presence of depression is associated with increased mortality, as well as greater symptom burden, greater physical limitation, worse quality of life, and worse overall health.

Historical Emergence of Neuropsychopharmacology and Neuroscience

Descriptions of depression and anxiety date back to the earliest written records, and the interaction of innate factors within the individual with environmental

demands has always been recognized. As an example, in 800 BC, Saul, King of Israel, developed depression, guilt, and incapacity. Among the treatments was placing a young woman in his bed and playing soothing music. He later became "psychotic" and attempted to kill members of the royal household.

In both depression and anxiety, the subjective nature of the symptoms without reliable abnormalities in structural anatomy or laboratory measures has usually always resulted in the symptoms or the syndromes or the diseases being categorized as "functional." At the present time the diagnosis of depression and anxiety disorders outlined in the DSM IV is based solely on the description of magnitude and duration of symptoms. There are no objective measures from laboratory chemistries, imaging, electrophysiological studies, or functional tests included to help make a positive diagnosis of these conditions. During the 26 years from the publication of DSM II in 1968 to the publication of DSM IV in 1994, the number of descriptive diagnoses listed in DSM IV has doubled from 182 to 365. The number of anxiety diagnoses has increased fourfold, and the number of diagnoses in depression increased from 8 in 1968 to a possibility of more than 40 in 1994 depending on the specifiers used. As described later, this proliferation of diagnostic subtypes does not mean that we have a better understanding of pathophysiology with the consequent improvement of specificity and efficacy of treatment.

The neuropsychopharmacology revolution starting in the mid 1950s introduced an entirely new dimension of treatment for depression and anxiety. Previously, "moral" and psychosocial treatments (with some use of electroconvulsive treatment [ECT], and sedatives) were the only treatments available. The finding in the 1950s that lysergic acid diethylamide (LSD) at doses as low as 0.1 μg/kg could alter normal consciousness and produce reliable alterations in perception, cognition, and affect was a major catalyst to the neuropsychopharmacological revolution that has dominated progress during the last 50 years. During this time period the substantial advances in the methods and ideas of neuroscience have swept neuropsychopharmacology forward. Considering that serotonin (5-HT) was just being identified in brain in the 1960s, the recent report in 2003 that a specific 5-HT-4 agonist can overcome opiate-induced respiratory depression without loss of analgesic effect illustrates the high degree of specificity with which brain circuits and receptor systems can now be targeted with selective drugs (Manzke *et al.*, 2003).

Some of the major events in our understanding of treatments for depression and anxiety are listed in Table 12.1. The initial serendipitous discovery that iproniazid, used in the treatment of tuberculosis, was antidepressant led to the development of monoamine oxidase inhibitors (MAOI) as antidepressant treatments. This was followed 4 years later by the discovery that imipramine, which was later found to inhibit monoamine uptake, was also an effective antidepressant. These findings led to the monoamine theories of depression in 1963 and 1964, which hypothesized that depression was the result of deficiency of monoamines in brain, and that antidepressants act by increasing synaptic monoamine

levels. By 1980 additional drugs that were not effective MAOIs or monoamine uptake inhibitors were found to be effective antidepressants. In addition, MAOIs and monoamine uptake inhibiting drugs produced an immediate increase in brain monoamine levels, but clinical antidepressant effects were delayed for 10 to 14 days. This led to the monoamine receptor sensitivity theory of depression, which hypothesized that it was slower changes in brain receptor systems that were important in producing the antidepressant effect and that it was the net increase of 5-HT transmission that was central to antidepressant efficacy.

A major advance occurred when the molecular and cellular theory of depression was proposed in 1997. This hypothesis was derived from the observations that antidepressant treatments (ADTs) result in the sustained activation of the cyclic adenosine 3-5 monophosphate (cAMP) system, which in turn activates the cAMP response element-binding protein (CREB). This results in increased expression of brain-derived neurotrophic factor (BDNF) in hippocampus and cortex. Since stress decreases levels of BDNF expression and increases cell atrophy in the dentate gyrus and pyramidal cell layer of the hippocampus and this is prevented by chronic (but not acute) administration of all classes of antidepressant treatments, it was hypothesized that the reversal of cellular atrophy by increasing BDNF was a key component of effective antidepressant drug action (Duman *et al.*, 1997). The more recent findings that a substance P antagonist was antidepressant and that cytokine therapies such as interferon-α can regularly cause depression broadened the possibilities that depression was not just the result of monoamine abnormalities. It has focused attention on intracellular pathways as the area where both the cause of illness and mechanism of therapy can best be understood (Manji *et al.*, 2001; Nestler *et al.*, 2002; Coyle and Duman, 2003). Recently, during the past few years, research has taken advantage of progress in molecular genetics to identify gene variants that relate to various behaviors relevant to depression and anxiety, and there is currently a concerted effort to identify genes of major effect in the cause of these disorders.

Problems in Relating Clinical Descriptive Diagnosis to Neuroscience

In most areas of medicine, the current predictive and therapeutically relevant diagnostic structure has evolved out of a murky past, and this is the same with depression and anxiety; but in these disorders, the progress of understanding has lagged far behind other areas of medicine (by some estimates by as much as 50 to 60 years). In part, this slow progress has occurred because it is not clear whether the patient is reporting nonspecific

TABLE 12.1   Brief History of Major Events in the Understanding and Treatment of Depression and Anxiety

Year	Discovery or Theory
1798	Pinel's classification of mental illness reflecting a functional view based on phenomenology resulted in a prescription of "moral" therapy rather than "physical" therapy.
1921	Kraepelin characterized phenomenology of bipolar disorder.
1944	Electroconvulsive treatment effective in treating depression.
1945–1955	Psychosocial model—Psychiatric disorders were a failure of individual to adapt at their environment.
1954	Birth of psychopharmacology. Discovery that monoamine oxidase inhibitors (MAOI) have antidepressant effects.
1958	Discovery that the tricyclic drug imipramine was an effective antidepressant.
1962	Imipramine effective in treatment of agoraphobia.
1963	*Serotonin Theory of Depression*: MAOIs act by increasing serotonin in brain.
1964	*Catecholamine Theory of Depression*: Antidepressant treatments (ADTs) act by increasing catecholamines in brain. Introduction of lithium treatment in USA.
1969	Increased acceptance of the utility of descriptive diagnosis as guide to treatment. Increasing scientific rigor of evaluating behavioral treatments.
1977–1980	*DSM III created and published.*
1981–1987	*Monoamine Receptor Sensitivity Theory of Depression*: ADTs act by altering the sensitivity of several monoamine receptor subtypes in brain, and to increase over all efficacy in serotonergic transmission. Use of benzodiazepines to treat panic disorder. Increasing use of anticonvulsants in the treatment of mania.
1994	*DSM IV published.*
1996	*A Molecular and Cellular Theory of Depression*: ADTs act by producing a sustained activation of the cyclic adenosine 3′, 5′ monophosphate cAMP system which increases brain levels of neurotrophic factors that reverse the effects of stress in certain brain areas. Rise of molecular psychiatry.
1998	Discovery that a substance P antagonist, where the initial effect does not involve monoamine systems, is as effective an antidepressant, and that cytokine treatments can cause depression.
1999	Increased progress of molecular psychiatry where genetic polymorphisms are related to diagnosis and/or neurobiologic and behavioral function.
2000–2003	Emergence of the current multifaceted view that genetic and neurobiologic vulnerabilities interact with environmental precipitants to produce a diverse array of unstable syndromes. Both pharmacologic and behavioral treatments are effective, but often in different subtypes.

symptoms, has a syndrome, or is manifesting a disease. Clearly, the symptoms of depression and anxiety can be elicited in normally encountered situations (e.g., depressive symptoms during loss and grief or anxiety symptoms when the individual is under threat). Large questions remain as to when depression and anxiety symptom clusters remain stable enough to constitute a syndrome and when the findings are reliable enough to constitute a disease. In contrast to other areas of medicine and neurology where more reliable symptoms, signs, and objective laboratory measures are available, the psychiatric assessment of depression and anxiety is based only on the history of symptom presentation.

Another confounding issue is the large variability in the behavioral repertoire of humans. Throughout the history of psychopharmacology, it has been well known that there is a wide range of behavioral response to all types of psychoactive compounds, ranging from hallucinogens, through anxiogenic compounds, to drugs of abuse. These behavioral variations depend on the subject's history, expectations, and physiological response. Thus, even in experimental laboratories with healthy humans, it has been difficult to predict the wide range of behavioral responses from experimentally administered agents. In patients with depression and anxiety, it has been even more difficult to establish diagnostic categories that will reliably predict drug response.

The sole reliance on description for diagnosis in psychiatry creates a serious dilemma: Since the descriptively defined diagnostic groups are heterogeneous as to genetic etiology, structural pathology, and pathophysiology, attempts to discover a common neurobiological abnormality for any single diagnostic group are doomed to meet with failure. Because of this situation, it has been difficult to build a ladder of progress by sequentially defining sub groups and refining treatments that are mo e targeted, specific, and efficacious. Thus, at the presen. time, drugs such as the tricyclic drug imipramine and the selective 5-HT uptake inhibitors (selective serotonin reuptake inhibitor, SSRI) are equally effective treatments for descriptively different diagnostic groups such as depression, panic disorder, and obsessive compulsive disorder.

Neuroscience and Molecular Medicine

The role that modern neuroscience and molecular medicine can play in solving the problems of descriptive diagnosis is illustrated in Figure 12.5. Even though the diagnostic subgroups for anxiety and depression have multiplied in the past 30 years, their specificity in predicting treatment response has not increased proportionately. In the model presented in Figure 12.5, each new descriptively described subdivision of the initial heterogeneous diagnostic group includes a second heterogeneous subgroup of patients based on structural abnormalities, pathophysiology, and genetic contributions to the illness. The descriptive separation into subgroups simply perpetuates the heterogeneity, as no methods are used to stratify it. In contrast, there are many examples from modern molecular medicine where specific genetic, pathophysiological, or structural abnormalities have served to define important subgroups. This has allowed the development of rational, focused, and specifically directed treatment; and when this has occurred, the treatments have been much more efficacious. The model presented in Figure 12.5 outlines the role that neuroscience and modern molecular medicine will play in psychiatry. The remaining review attempts to assess how far along we are in applying this model.

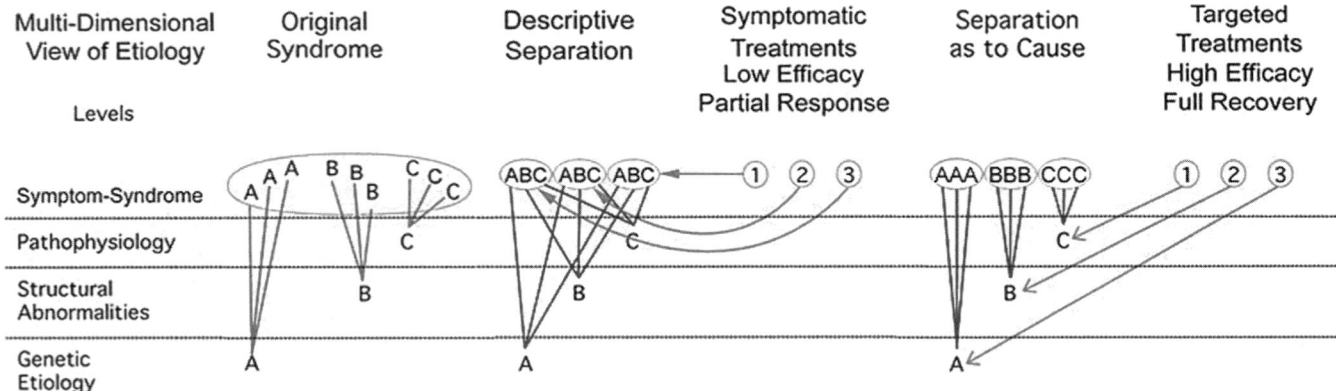

FIGURE 12.5 When the original syndrome was described it contained a heterogeneous group of patients where the cause of illness was either (A) genetic, (B) structural, or (C) pathophysiologic (e.g. and environmental stress, etc.). A descriptive separation of the original syndrome results in 3 new heterogeneous groups that each contain patients with (A) genetic, and (B) structural, and (C) pathophysiologic causes of illness. Thus, when the specific treatments directed at (1) genetic, (2) structural, and (3) pathophysiologic causes are used in each of these 3 new heterogeneous groups, there is low efficacy because the specific treatment is not matched to the cause of illness in a homogenous group of patients—thus only some patients show full recovery. In contrast, with modern molecular medicine where the original syndrome is separated by cause of illness into 3 separate homogenous groups based on (A) genetic, or (B) structural, or (C) pathophysiologic causes, each of the specific treatments can be directed at the correct and homogenous group of patients that results in high efficacy and full recovery in all patients.

MODELS OF PATHOGENESIS FOR DEPRESSION AND ANXIETY

Background

Progress in the understanding of pathogenesis of disease has been critically dependent on the development of appropriate animal models. Only in the animal models can there be the rigorous experimentation and experimental control required to dissect the biological factors involved in disease production. Obviously, clinical studies are also important, as they can validate the disease pathogenesis in humans through careful clinical observation and experimental studies involving symptom provocation and treatment. In understanding the pathogenesis of depression and anxiety, there has been more difficulty using animal models than in other medical specialties where more simple and valid animal models are available (e.g., stroke, coronary artery disease, renal failure, bone fractures). Many aspects of the presentation of depression and anxiety symptoms are uniquely human. Thus it has been difficult to develop the predictive power of animal models seen in other areas of medicine; however, there has been considerable progress. Animal models of fear as it relates to human anxiety disorder have provided an extremely effective avenue of research. In the area of stress, animal models have been extremely useful in delineating the pathways involved and for discovering preventative or remedial treatment. To a lesser extent animal models of depression that involve impaired escape behaviors such as the forced swim test or learned helplessness have also been useful.

The most important aspect of progress has been the robust rise in power and efficiency of neuroscience, neuropsychopharmacology, and now molecular medicine. The growth and improvement of neuroscience methods, understanding, and applications have been interactive throughout with neuropsychopharmacology, so at present the vast majority of even purely physiological studies use some form of neuroactive substances as part of their experimental design. The advent of molecular biology and molecular medicine has increased the opportunities and effectiveness of neurobiological research by orders of magnitude. For example, before 1999 in the field of neuropsychiatry, the pathogenesis of narcolepsy was quite vague and confusing. With the discovery through positional cloning of a narcolepsy gene in dogs (a mutation in the hypocretin II receptor), the field experienced an unprecedented leap forward. Today there is a rich matrix of information on the causes and subtypes of narcolepsy involving the hypocretin system and other clinical and genetic variables (Nishino, 2003). In recent years molecular neurobiology has been essential for many of the major discoveries that are leading the field forward. These methods are now being applied to the understanding of depression and anxiety where a brief survey is described next.

Stress and Corticotropin-Releasing Hormone and the Hypothalamic, Pituitary Adrenal Axis

Stress has always been thought to play a prominent role in the cause of depression and anxiety, which is related to the widespread view that depression and anxiety simply reflect acute and chronic failure of stress related control systems. At the clinical level the hypothesis that depression and anxiety are caused by failures in the regulation of the stress-response system has the most data to support it. There are high rates of depression (up to 85%) in Cushing's disease, which has high circulating cortisol levels, and severely depressed individuals have been shown to excrete abnormally high levels of corticosteroids and their metabolites. The resistance of plasma cortisol to suppression by the potent synthetic corticosteroid dexamethasone has been the basis for a widely used test that has generated a great deal of clinical data. Individuals whose cortisol levels are resistant to suppression by dexamethasone have a higher than average rate of relapse, even with apparently effective symptomatic treatment. This hypothesis has been extended to corticotropin-releasing hormone (CRH). Levels of CRH in the cerebral spinal fluid are elevated in depression. At autopsy, patients have been found to have an increased number of CRH secreting neurons in limbic brain regions and the number of CRH binding sites in frontal cortex is reduced, presumably secondary to increased CRH concentrations. When CRH is infused into depressives, they have a blunted adrenocorticotropic hormone (ACTH) response, which suggests a desensitized pituitary CRH receptor system. Recently the combined dexamethasone CRH test has been applied. The CRH elicited ACTH response is blunted in depressives, but dexamethasone pretreatment produces an opposite effect; compared to control subjects, there is actually an enhanced ACTH release following CRH in the depressed patients. This result is thought to be related to increased numbers of vasopressin-expressing neurons in the parvocellular part of the hypothalamic paraventricular nucleus, which has been confirmed on postmortem examination (Holsboer, 2000).

Like the dexamethasone test, the dexamethasone-CRH test has shown that normalization of the CRH–hypothalamic, pituitary adrenal (HPA) axis is necessary for full clinical remission. Patients showing abnormalities in the test at the time of hospital discharge have a high propensity to relapse. The combined dexamethasone-CRH test is also able to differentiate healthy individuals who are at high risk for developing depression from healthy control subjects. In Figure 12.6, it can be seen that the healthy individuals who have a family history of depression but who have not yet developed depression, have a dexamethasone-CRH test intermediate between the healthy individuals from families with no history of depression and depressed patients. This finding suggests that there is an abnormality in the corticosteroid receptor signaling pathways involving the regulation of the HPA axis in individuals vulnerable to depression (Holsboer, 2000).

Current research using molecular methods to manipulate the status of the various corticosteroid regulatory receptor systems has produced promising leads, but they have not been validated in clinical populations (Holsboer, 2000). The clinical relevance of abnormalities in the CRH-HPA axis is highlighted by the finding that

the dexamethasone-CRH test produces higher cortisol concentrations in patients with multiple sclerosis. Those patients with relapsing-remitting multiple sclerosis had the least abnormality, intermediate were patients with secondary progressive multiple sclerosis, and the most marked abnormality was seen in patients with primary progressive multiple sclerosis. The HPA axis abnormality correlated with neurological disability but not with evaluations of depressed mood (Then Bergh *et al.*, 1999). Abnormalities in the CRH-HPA system have also been found in anxiety disorders where there is evidence for CRH overdrive and negative glucocorticoid feedback in patients with posttraumatic stress disorder (PTSD) (Kellner, *et al.*, 2000). Even though abnormalities in HPA axis regulation are present in depression and anxiety, and it is clear that this is an important part of these syndromes, it remains to be shown that these abnormalities are the primary cause of the disorder rather than a secondary consequence of dysregulation in other systems.

The Molecular and Cellular Theory of Depression

Since this hypothesis was proposed (Duman *et al.*, 1997), there have been extensive additional data that support and extend the idea that factors influencing neuroplasticity, neurogenesis, and cell survival are critical in the pathogenesis and treatment of depression. Currently, this theory offers the most promise for understanding the cause and treatment of depression. In addition to the monoamine-cAMP-CREB pathway, additional pathways have been delineated that could be related to the cause of mood disorders and subserve the mechanism of action of drugs used to treat them.

The initial observations—that stress decreased neurogenesis in the hippocampus and that this was reversed by antidepressant treatment—have been extended by the finding that several different classes of antidepressants increase hippocampal neurogenesis after chronic but not after acute treatment (Malberg *et al.*, 2000). This finding is consistent with the clinical data, which show that chronic but not acute treatment is effective. In addition, studies have shown that antidepressant treatment increases the proliferation of hippocampal cells and that these new cells mature and become neurons. Studies have shown that newly generated cells in hippocampus display passive membrane properties, action potentials, and functional synaptic inputs similar to those found in mature dentate granular cells (van Praag *et al.*, 2002). In addition, there is evidence that a subgroup, approximately 14% of newly generated neurons in an adult rat, become γ-aminobutyric acidergic (GABAergic) basket cells that are capable of forming inhibitory synapses with principal excitatory granular cells (Liu *et al.*, 2003). This demonstrates that in hippocampus, neurogenesis can

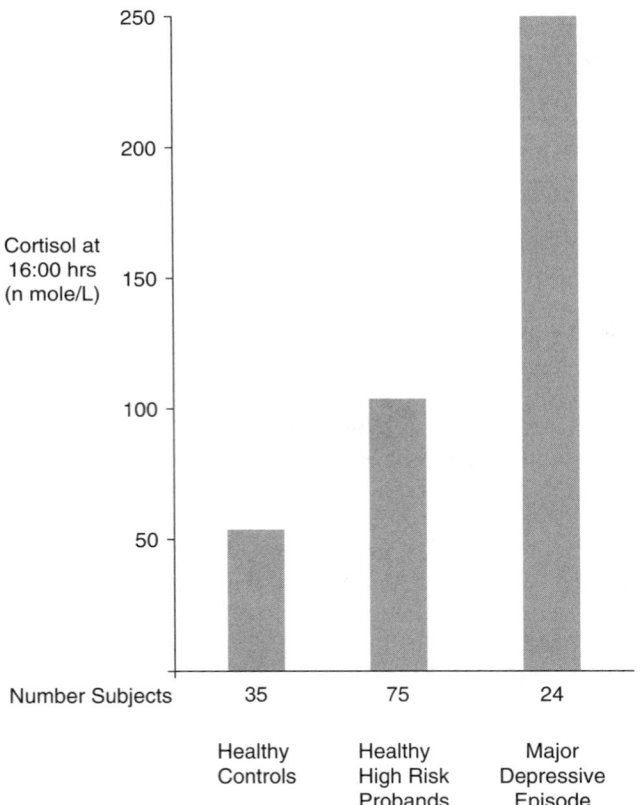

FIGURE 12.6 Healthy individuals with high risk for depression have an elevated cortisol response to the combined dexamethasone-CRH test. Healthy high-risk probands were selected from families with a high genetic load for depression. Matched controls from families without a history of psychiatric morbidity and matched patients with major depressive episode. dexamethasone at 1.5 mg, at 23:00 day before. Human corticotrophin-releasing hormone (CRH) 100 μg IV at 15:00. Cortisol peak in major depressive episode at 16:00. (Constructed from data in Holsboer F. (2000). *Neuropsychopharmacology* 23:477.)

produce both functional excitatory and inhibitory cells in adult animals. It has been shown that disrupting hippocampal neurogenesis blocks the behavioral responses to antidepressants (Santarelli *et al.*, 2003). In addition, this work provides important new data on differences between antidepressants. Mice with the 5-HT 1A receptor knocked out were insensitive to the neurogenic and behavioral effects of the SSRI fluoxetine, but the tricyclic antidepressant imipramine, which can act via the noradrenergic system, did produce neurogenesis and behavioral effects in the 5-HT-1A knockout mice. X-irradiation of the hippocampus prevented the neurogenic and behavioral effects of both types of antidepressants.

It is of major interest that early maternal deprivation reduces the expression of BDNF in the hippocampus of adult rats. After 24 hours of maternal separation on postnatal day 9, rats studied at postnatal day 72 had reduced BDNF expression and an altered BDNF response to stress (Roceri *et al.*, 2002). In rat pups undergoing maternal separation, treatment with fluoxetine prevents the separation-induced decrease in neurogenesis (Lee *et al.*, 2001). Because early life stress is a documented risk factor for the development of affective disorders, it is of further interest that prenatal restraint stress of the mother on day 18 after mating reduces the number of granule cells in hippocampus of female but not male offspring when they were 75 days old (Schmitz *et al.*, 2002). This has considerable relevance for a possible explanation for the 1.5 to 3 times more frequent occurrence of depression in women than in men.

The idea that BDNF is a target for antidepressant drug action has clinical support in that BDNF expression in dentate gyrus, hilus, and super granular regions was higher in humans treated with antidepressant medications at the time of their death compared to subjects not treated with antidepressants (Chen *et al.*, 2001). Extensive imaging studies of mood disorders have revealed structural brain changes within a neuroanatomical circuit termed the limbic-cortico-striatal-pallidal-thalamic tract (Sheline, 2003). The imaging studies are also supported by a series of postmortem studies where major depressive disorder and bipolar illness are characterized by alterations in the density and size of neuronal and glial cells in limbic brain regions (Rajkowska, 2003). In addition to the increased BDNF expression observed at postmortem in patients treated with antidepressants, another powerful line of evidence supporting the hypothesis is that in depressed patients, reduced hippocampal volume can be related to the number of days of untreated depression (Sheline *et al.*, 2003). This is strong but indirect evidence that antidepressant treatment increases hippocampal volume, presumably through neurogenesis. Since lithium treatment is reported to increase human brain gray matter (Moore *et al.*, 2000), other mechanisms

other than neurogenesis in hippocampus may also be involved.

Several pathways are involved in increasing neuroplasticity, neurogenesis, and survival. In Table 12.2, four of the pathways that are known to be effected by treatments for mood disorders are illustrated. Antidepressants, ECT, lithium, valproic acid, carbamazapine, and lamotrigine, all of which have proven to be effective in the treatment of mood disorders, have effects on the pathways that increase neuroplasticity, neurogenesis, and cell survival (Coyle and Duman, 2003). It is of interest that drugs that have such different pharmacologic profiles overlap in this manner. Extensive cross talk and interaction between the pathways may account for the lack of specificity seen in the disorder and the widely different but effective treatments.

There have been a wide number of family twin and other genetic studies of depression and bipolar illness, and it is of interest that mutations in the area surrounding the BDNF gene have been identified as associated with bipolar disorder (Sklar *et al.*, 2002). The role of the cAMP signaling pathway in mood disorders is strengthened by the finding that variations in CREB1 cosegregate with depression in women (Zubenko *et al.*, 2003). Another recent report is that 10% of bipolar disorder cases have a polymorphism in the G protein receptor kinase 3 (Barrett *et al.*, 2003). Studies such as these will need to be replicated, but they offer considerable promise for the identification of patient subgroups that could be targeted with specific treatments.

Neuroimmune Pathways in Depression

Another line of evidence for heterogeneity in the pathways leading to depression involves the immune system. For the past several years it has been known that treatment with interferon-α and interleukin 2-based immunotherapy is regularly accompanied by increased depression, reaching clinical levels in up to 50% of patients. Figure 12.7 illustrates the change in depression ratings after 4 weeks of interferon-α treatment in 10 patients. All but one patient had an increase in depression during treatment, and the increase was greater in those patients who had higher baseline ratings before treatment (Capuron and Ravaud, 1999). The initial injection of interferon-α produces an increase in cortisol and ACTH in all patients, but those patients with the highest increase of both cortisol and ACTH are the ones who develop clinical depression later in treatment (Figure 12.8) (Capuron *et al.*, 2003). Interferon-α has been found to lower serum dipeptidyl peptidase IV activity, which is proportional to the increase in depressive and anxiety symptoms seen with interferon-α treatment. It is also related to the immune activation (Maes *et al.*, 2001). Immune activation is also associated with a

TABLE 12.2   Intracellular Pathways for Mood Disorder Treatments

Type of Pathway	Monoamine	Neurotrophins	WnT	Myoinositol
Treatment Effect on:				
Important Intermediate Component	↑ cAMP	↑ ERK	↓ GSK-3B ↑ B, Catenin	↓ PKC
Transcription Factor	↑ CREB ↑ Neurogenesis ↑ Cell Survival	↑ CREB	↑ Tcf/Lef-1	↓ MARCKS
Neurobiologic Effects	↑ Growth Mature Neurons ↑ Neuronal Sprouting ↑ Synaptic Strength	↑ Cell Survival ↑ Synaptic Plasticity	↑ Antiapoptotic ↑ Axongenesis	↑ Growth Cones ↑ Neurite Extension
Treatments	Antidepressants ECT	Lithium VPA ECT	Lithium VPA Lamotrigine	Lithium VPA Carbamazapine

cAMP: cyclic Adenosine 3-5 MonoPhosphate
CREB: Cyclic AMP Response Element Binding protein
ERK: Extracellular signal Regulated Kinase
GSK-38: Glycogen Synthese Kinase—3B
MARCKS: Myristoyleted Alanine-Rich C Kinase Substrate
Tcf/Lef-1: Tcf/Lef Transcription factor
PKC: Phosphokinase C
VPA: ValProic Acid
WnT: Wingless signaling pathway

decrease in serum tryptophan, the concentration of which relates to the increased depressive symptoms (Capuron *et al.*, 2002a). Depression after interferon-α treatment can be prevented by treatment with the SSRI paroxetine. It is of interest that the paroxetine treatment prevented the development of depressive symptoms and cognitive symptoms, but it had much less of an effect on the neurovegetative symptoms such as abnormal

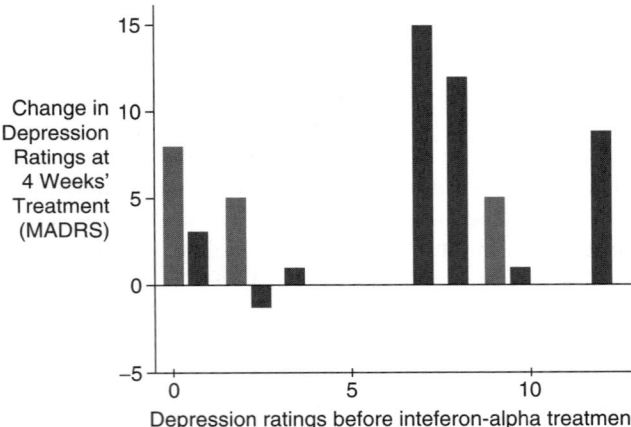

FIGURE 12.7 Inteferon-α increases depression. There is a larger effect in individuals who are depressed before treatment. 9 of 10 patients had an increase in depression that was greatest in those patients who had higher depression before treatment. *MADRS* = Montgomery-Asberg Depression Rating Scale. (Constructed from data in Capuron L, Revaud A. (1999). *N Engl J Med* 340:1370.)

appetite, abnormal sleep, fatigue, psychomotor retardation, pain, gastrointestinal symptoms, and bowel problems (Capuron *et al.*, 2002b).

Neuropeptides in Depression

Although initially antidepressants were thought to work solely through the monoamine systems, there is now evidence that neuropeptides are also involved and that neuropeptide antagonists can be antidepressant. In an initial study, a substance P antagonist was more effective than placebo and equally effective as paroxetine in reducing depressive symptoms (Kramer *et al.*, 1998). A second clinical study of a substance P antagonist as a treatment for depression has also shown similar efficacy to paroxetine. In a study of 23 depressed patients where substance P was measured before and after antidepressant treatment, mean substance P levels in serum were higher than in control subjects before treatment. Although in the patients there was no overall change in mean substance P levels between baseline and 4-weeks of treatment, there was a correlation. Approximately one third of the patients who had a decrease in substance P during treatment had a better drug response that those with an increase (Bondy *et al.*, 2003). The infusion of substance P into healthy volunteers is reported to produce a small but reliable worsening of mood.

Clinical findings have been given additional support from studies in mice where the Tac-1 gene that encodes

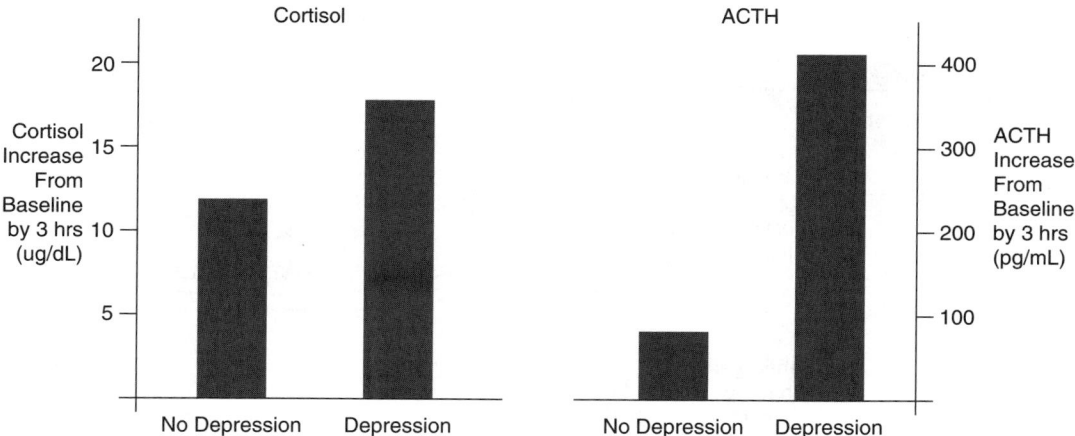

FIGURE 12.8 A larger increase in ACTH and cortisol after an initial injection of interferon-α is associated with the development of depression during interferon-α treatment. Interferon-α at 20 million units/meter squared administered intravenously. Seven patients developed depression and seven did not during 12 weeks of interferon-α therapy. Cortisol and ACTH at 3 hours significantly increased $P < .01$ in patients with depression vs no depression. (Constructed from data in Capuron L *et al.* (2003). *Am J Psychiatry* 160:1342.)

the neuropeptides substance P and neurokinin A has been deleted. The Tac-1–deficient mice were more active in the porsolt swim test and did not become hyperactive after bulbectomy, thus demonstrating a resistance to depression-like effects. In addition, they were less fearful in several models of anxiety (Bilkei-Gorzo *et al.*, 2002). These findings demonstrate that tachykinins are powerful mediators of depression-like or anxiety-like behavior in mice and support the continued development of substance P antagonist as treatments. In addition, neuropeptide Y has been implicated as having antidepressant actions in an animal model by acting through the NPY Y-1 receptor subtype (Redrobe *et al.*, 2002).

Neurobiology of Anxiety

Progress in understanding the neurobiological pathogenesis of anxiety has been greater than in any other diagnostic category in psychiatry. This is primarily due to the major advances in the use of animals models, which has allowed the discovery of the neuropathways subserving conditioned fear and the neurobiological factors subserving anxiety-like behavior. At the clinical level there is a great deal of ambiguity regarding the diagnosis of anxiety disorder in patients (i.e., terms and concepts put forward to describe anxiety disorders range from anxiety neurosis and anxiety hysteria to street fear and soldiers heart to conditions like autonomic epilepsy and vasomotor neurosis). The current DSM IV diagnostic system has improved this, but problems still remain. Anxiety symptoms accompany a large range of other medical illnesses including cardiac arrhythmias, hypothyroidism, anaphy-

laxis, asthma, seizure disorders, and after intoxication with or withdrawal from numerous drugs. The use of benzodiazepines and antidepressants and the development of more reliable behavioral therapies now make treatment of anxiety disorders the most effective in the field of psychiatry.

Stress has always been seen as a central factor in the production of fear and anxiety, and the discovery and use of specific CRH agonists and antagonists have validated this idea. CRH agonists have been found to produce a variety of fear and anxiety-like behaviors, and all of these effects can be reduced or abolished by centrally administered CRH antagonists. CRH receptor-deficient mice have impaired spatial recognition memory but also demonstrate decreased anxiety-like behaviors (Contarino *et al.*, 1999). This finding is consistent with the anxiogenic effect of CRH. It is of interest that conditional mutagenesis of the glucocorticoid receptor produced the expected impairment of HPA axis regulation but also produced impaired behavioral response to stress and reduced anxiety (Tronche *et al.*, 1999). Thus, the data support the idea that the CRH-HPA axis not only provides a response to fear and anxiety but also plays a role in initiating and maintaining these states.

Some of the most significant advances in the field have derived from the discovery of neuroautonomic circuits subserving conditioned fear. Figure 12.9 outlines auditory fear conditioned pathways. The basal lateral nucleus of the amygdala plays a central role in integrating auditory input and in sending information to the central nucleus of the amygdala, which orchestrates the behavioral, autonomic, and endocrine outputs. It has

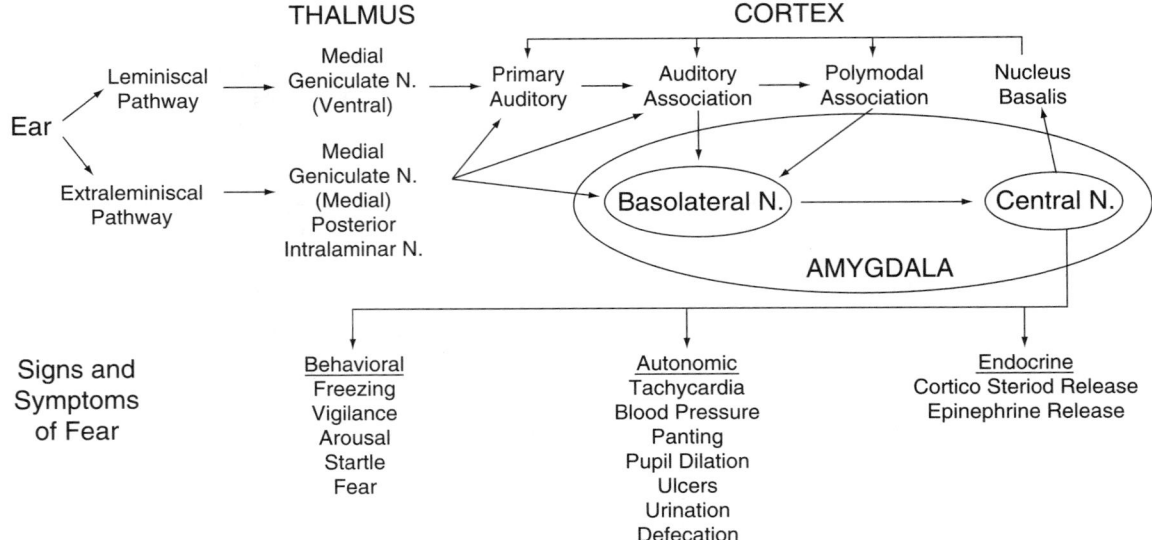

FIGURE 12.9 Auditory conditioned fear pathways. Sensory information is transmitted to the thalamus via the leminiscal pathway involving the ventral medial geniculate nucleus, which projects only to the primary auditory cortex. Pathways thought to be involved in emotional learning involve extraleminiscal pathways, which project to the medial division of the medial geniculate nucleus, and the posterior intralaminar nucleus, which project to primary auditory cortex, auditory association cortex, and the basolateral nucleus of the amygdala, which also receives inputs via the auditory association and polymodal association cortex. The basolateral nucleus projects to the central nucleus of the amygdala, a major output of the amygdala, which projects to brain areas that produce the behavioral, autonomic, and endocrine manifestations of fear. The central nucleus also feeds back to the cortex via the nucleus basalis.

been shown that the amygdala is required for both the acquisition and expression of the learned fear response. The cortex is also involved. It has been shown that damage to the dorsal part of the medial prefrontal cortex enhances fear to a conditioned tone. Prefrontal neurons reduce their spontaneous activity in the presence of a conditioned aversive tone as a function of the degree of fear, and this is related to amygdala activity, indicating that the amygdala controls both fear expression and prefrontal neuronal activity (Garcia *et al.*, 1999). This relationship is reciprocal; the medial prefrontal cortex is involved in the formation of new memories that extinguish prior conditioning. The new extinction memories stored in the medial prefrontal cortex inhibit fearful memories that most likely reside in subcortical structures (Milad and Quirk, 2002). This fits with the data that extinction does not erase conditioning but instead forms a new memory that supercedes it. The delineation of fear circuits has facilitated the discovery of new molecular mechanisms involved in fear-related memory. Receptors for gastrin-releasing peptide have been found on GABA interneurons of the lateral nucleus of the amygdala. When the gastrin-releasing peptide is deficient, mice demonstrate more long-term fear memory (Shumyatsky *et al.*, 2002). Similarly, it has been shown that endoge-

nous cannabinoids are also involved in the extinction of long-term fear memory; knock out of the cannabinoid 1 receptor and an antagonist of this receptor impairs short- and long-term extinction of fear conditioning (Marsicano *et al.*, 2002).

Benzodiazepines, which modulate the GABA receptor, are widely used in the treatment of panic and anxiety disorders. In patients suffering from panic attacks, a deficit of GABAa receptors has been identified in brain using imaging studies. These data indicate that variations in GABAa receptor subtype composition within patients could be a contributor to the lower threshold for anxiety seen in these individuals. These fine differences in the GABAa subtypes may provide an advantage in the design of new therapeutic agents. Along this line, it has been found that mice with mutation in the α-1 type GABAa receptor demonstrated anxiolytic, myorelaxant, motor impairing, and ethanol potentiating effects of drugs acting on the receptor while the sedative, amnestic, and some of the anticonvulsant effects were abolished (Rudolph *et al.*, 1999).

Another dimension of the GABAa system is its interaction with neurosteroids. Patients with panic disorder have greater concentrations of the neurosteroids that positively enhance GABAa function and a lower concentration of the neurosteroids that are antagonist to GABAa (Strohle *et al.*, 2002). When panic attacks were induced in patients with panic disorder with sodium lactate and choleystokinin, patients had pronounced decreases in the neurosteroids that were GABAa agonists and an increase in the concentrations of the functional antagonists. Similar changes were not seen in control subjects (Strohle *et al.*, 2003). These investigations illustrate the opportunities for neurochemical dissection of the factors controlling the low threshold for panic attacks in patients with panic disorder.

Improved animal models have allowed investigation of the developmental aspects of anxiety disorders. It is known that maternal separation increases hypothalamic CRH gene expression, HPA activity, and behavioral responses to stress in adult animals. Environmental enrichment during the peripubertal period reverses the effects of maternal separation on HPA and behavioral responses but does not effect CRH-mRNA expression in adult animals (Francis *et al.*, 2002). This finding has implications for treatment, as it demonstrates in this model that the functional behavioral effects can be modified while the fundamental neurobiological defect remains unchanged. Mice lacking the 5HT 1-A receptor have increased anxiety-like behavior, but it has been shown using a tissue-specific conditional rescue strategy that receptor expression during early postnatal period, but not in the adult, is necessary to rescue the increased

anxiety-like behavior (Gross *et al.*, 2002). Findings like this demonstrate that the postnatal processes are important in establishing anxiety-like behavior and that there may be different functions and consequences of these receptor systems throughout development.

Development is important as it relates to critical issues in genetic vulnerability of humans to stress and the manifestation of anxiety disorders. The initial findings demonstrating that humans with PTSD have reduced hippocampal volume fit nicely with the hypothesis that the reduced volume was secondary to stress-induced atrophy. However, in a pivotal study, monozygotic twins who were discordant for trauma exposure had smaller hippocampi than control subjects (Figure 12.10) (Gilbertson *et al.*, 2002). Figure 12.11 illustrates that the severity of the PTSD in the exposed twin correlated with the decrease in that twin's own hippocampal size; however, the same intensity of exposure correlated with the decreased hippocampal size in the brother who was never exposed. This points to critical developmental factors in the predisposition to develop PTSD.

The rich matrix of findings from preclinical and clinical studies now offers many opportunities for development of new and more specific treatments for anxiety disorders. Since conditioned fear in laboratory animals and human PTSD share many of the same mechanisms, application of the preclinical information should result in improved treatments for patients.

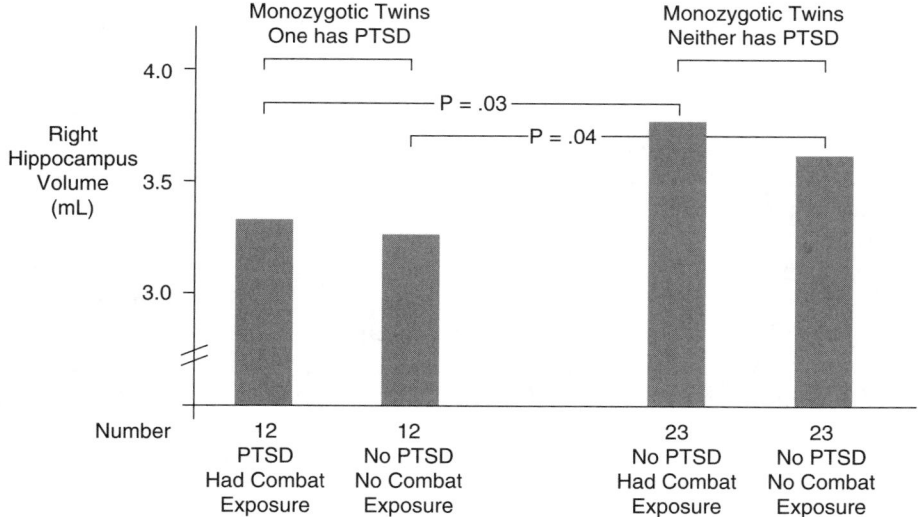

FIGURE 12.10 Hippocampal volume is equally reduced in monozygotic twins who had no combat exposure as the co-twin who did have combat exposure. Hippocampal volume determined by MRI. PTSD determined by clinical administered PTSD scale (CAPS) score greater than 65. Findings were similar for left hippocampus and total hippocampus, but not as robust. Amygdala size did not differ among the four groups. From 35 Vietnam veterans and their 35 identical twins who did not go to Vietnam. (Constructed from data in Gilbertson MW *et al.* (2002). *Nat Neurosci* 5:1242.)

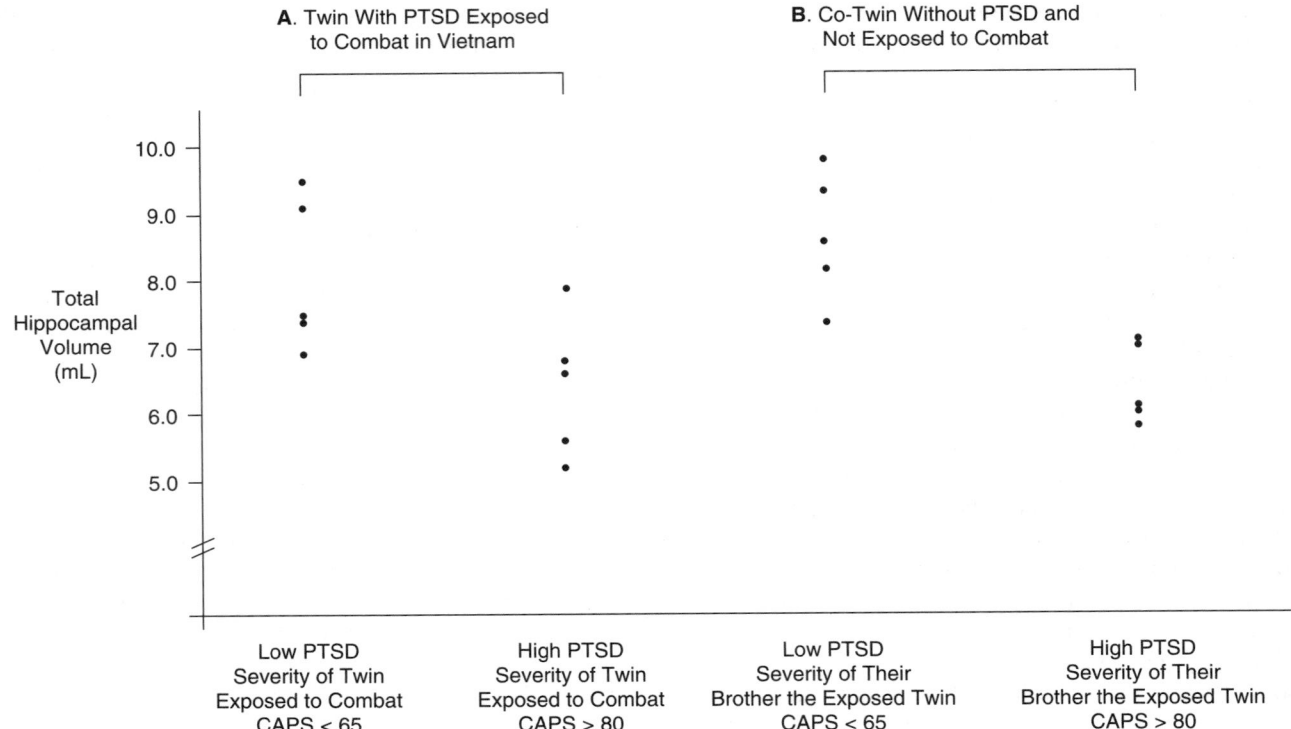

FIGURE 12.11   **A**: Reduced hippocampal volume correlates with increased PTSD severity in the monozygotic twins exposed to combat. However, the same strong correlation is seen in the unexposed twins without PTSD. **B**: A reduced hippocampal volume not seen in the unexposed twins without PTSD when the twin brother had high PTSD severity. (Constructed from data in Gilbertson MW *et al.* (2002). *Nat Neurosci* 5:1242.)

RECENT PROGRESS IN MOLECULAR MEDICINE: GENETICS AND BEHAVIOR

Allelic Variation of Serotonin Transporter Expression

The 5HT transporter (5-HTT) is involved in the reuptake of serotonin at brain synapses and is a target of a variety of antidepressant drug treatments. The promoter activity for the 5-HTT gene is located on 17q 11.2. and is modified by sequence elements within the proximal 5′ regulatory region. This is designated the 5-HTT gene linked polymorphic region (5-HTTLPR). The short allele has been associated with lower transcriptional efficiency and reduced 5HT transport compared to the long allele.

It has been shown through a series of studies that reducing plasma tryptophan reduces brain serotonin levels. When this test is applied to depressed patients recently recovered on treatment, there has been a brief increase in depressive symptoms. In Figure 12.12 the left panel illustrates that when serotonin is depleted when using the tryptophan depletion test, there was a larger increase in depressed ratings in patients treated with

SSRIs than in patients treated with catecholamine uptake inhibitors (Delgado *et al.*, 1999). In contrast, (right panel) when recently recovered depressed patients were given the catecholamine depleting drug, alpha methyl paratyrosine, patients on catecholamine uptake inhibitors had a higher depression score, and little effect was seen in patients treated with SSRIs (Miller *et al.*, 1996). Symptom relapse was specific to the type of antidepressant treatment and the catecholamine system impaired. Since antidepressant treatment with SSRIs or catecholamine uptake inhibitors appears equally efficacious in depressed patients, the specificity illustrated in Figure 12.12 was not easily explained. Now it appears that the 5HT pathway and the catecholamine pathway involve separate mechanisms that both increase neurogenesis in the hippocampus. The 5HT 1A knockout mice did not showed neurogenesis when treated with a catecholamine uptake inhibitor, but not an SSRI (Santarelli *et al.*, 2003). In addition, it has now been found that symptom increase after serotonin depletion has a genetic basis. Nondepressed healthy women were given a tryptophan depletion test and patients with the l/l allele showed little effect. Patients with the s/s allele showed the largest increase in depressive symptoms and patients with l/s

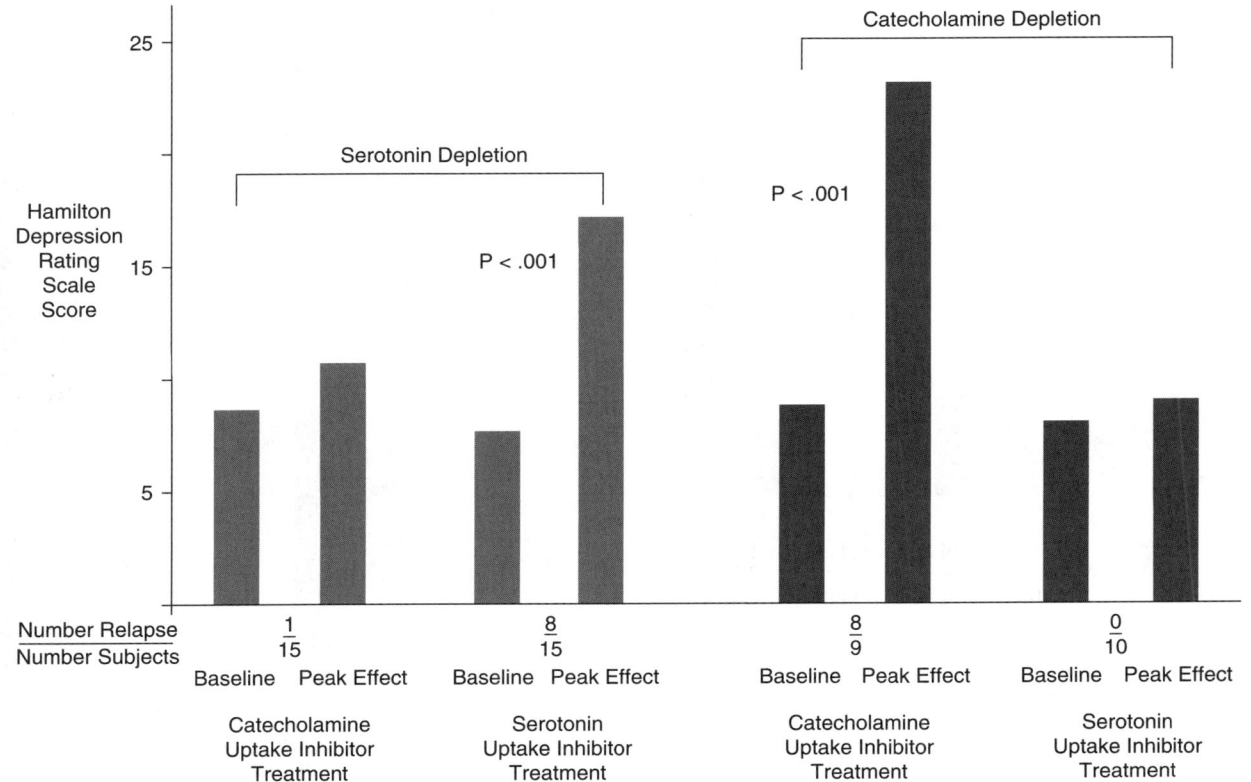

FIGURE 12.12 Relapse after serotonin or catecholamine depletion is specific to serotonin uptake inhibitor or catecholamine uptake inhibitor treatment, respectively. (Tryptophan depletion constructed from data in Delgado PL *et al.* (1999). *Biol Psychiatry* 46:212.) (Catecholamine depletion constructed from data in Miller *et al.* (1996). *Arch Gen Psychiatry* 53:117.)

allele experienced immediate symptoms, with those patients with a family history of depression showing a larger increase in depressive symptoms (Figure 12.13) (Neumeister *et al.*, 2002). Thus, in this instance, findings from the use of 5HT 1A knockout mice and the findings of the genetic contribution to a clinically useful provocative test have added new understanding as to the possible mechanisms of antidepressant action.

In contrast to studies where the 5HT polymorphism has been inconsistently associated with different diagnostic groups of patients, recent studies of genetic and environmental interactions have been much more informative. In a unique initial cohort of 1037 children who have been monitored until age 26 years, it has been possible to relate the number of stressful life events occurring after their twenty-first birthday to depression at age 26. Figure 12.14 shows that the s/s genotype has almost a doubling of probability of major depression over the l/l genotype. In addition, when childhood maltreatment and the occurrence of depression are examined in these individuals, it can be seen (Figure 12.15) that the s/s genotype had again an approximately doubling of depression compared to the l/l genotype (Caspi *et al.*, 2003). The s/s

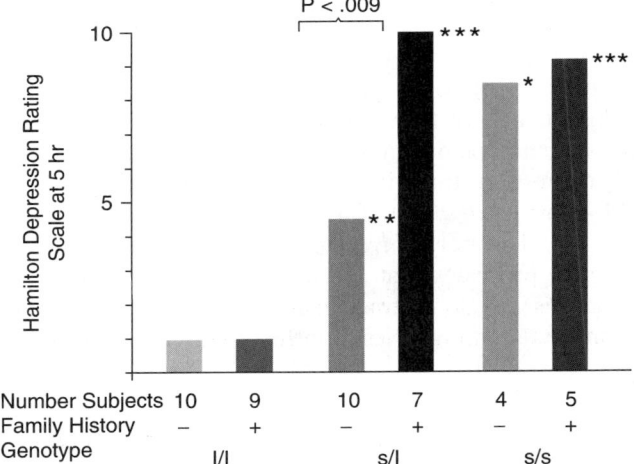

FIGURE 12.13 Serotonin transporter gene polymorphism and depressive response to tryptophan depletion in healthy women. Mean scores on 21-item version of Hamilton Depression Rating Scale were maximal at 5 hours after tryptophan depleting drink. Hamilton Depression scores on a separate day after a drink that did not deplete tryptophan were less than 2, and baseline scores before the drinks were 1 or less. Depletion-induced increases in depression $^*P < .007$, $^{**}P < .002$, $^{***}P < .001$. (Constructed from data in Neumeister A *et al.* (2002). *Arch Gen Psychiatry* 59:613.)

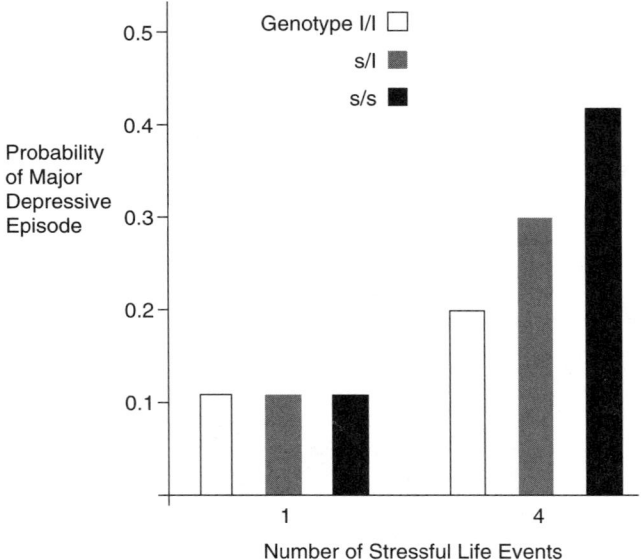

FIGURE 12.14 Stressful life events produce depression in genetically vulnerable individuals. Life events measured during 21, 22, 23, 24, and 25 years of age included employment, financial, housing, health, and relationship stressors. Depression evaluated for the last year of age 26. Life events predicted depression for s/l and s/s genotypes ($P < .001$) but not l/l genotype ($P = .24$). (Constructed from data in Caspi A et al. (2003). Science 301:386.)

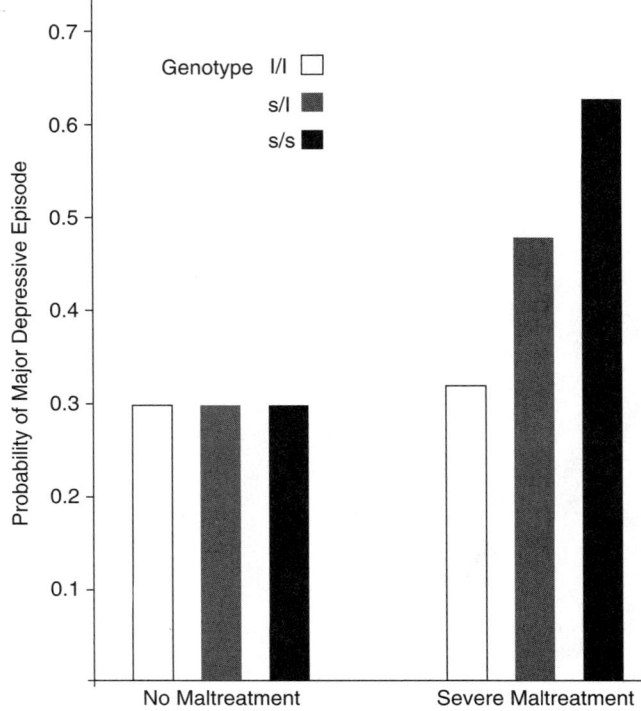

FIGURE 12.15 Childhood maltreatment relates to depression in genetically vulnerable individuals. Childhood maltreatment assessed for the first 10 years of life. Depression assessed between 18 and 26 years of life. Maltreatment predicted depression in s/s $P = .02$, and s/l $P = .01$ genotypes but not l/l genotype ($P = .99$). (Constructed from data in Caspi A et al. (2003). Science 301:386.)

genotype did not relate to depression in the absence of childhood maltreatment and the l/l genotype appears resistant to the severe maltreatment. This is one of the best studies to illustrate the fundamentally important concept, that the interaction of the genotype with the environment generates the clinical syndrome. The use of these genetic methods may have clinical applications (Figure 12.16) where patients with Parkinson's disease (which is known to be associated with depression) with the s/s genotype had more than four times the percentage of depression than individuals with the l/l genotype (Mossner et al. 2001). Taken together, the data suggest that the l/l-s/s polymorphisms reflect important neurobiological processes that are not directly associated with diagnosis but do interact in important ways with environmental contingencies to influence disease expression.

NEUROSCIENCE AND CURRENT TREATMENTS

Monoamines

Imipramine was discovered to be an effective antidepressant more than 40 years ago, and no new drugs have been developed to demonstrate unequivocal superior efficacy. The development of large numbers of drugs, with varying degrees of specificity in blocking monoamine uptake and antagonism of various monoamine receptor

subtypes, has resulted in considerable differences in side effect profiles. At the present time the physician's choice of the drug's side effect profile primarily determines which drug treatment is used. In this regard, the SSRIs are widely prescribed because of their relative lack of side effects compared to the older trycyclic drugs. There have been inconsistent data regarding whether different antidepressants are more efficacious in different patient subgroups; however, there are consistent data to show that psychotic depression requires the addition of a neuroleptic in addition to standard antidepressant treatment. Atypical depression may respond better to monoamine oxidase inhibitors. Reversible inhibitors of monoamine oxidase have been developed that do not carry some of the side effect liabilities of the older drugs. Overall, there has been no major shift in the paradigm of monoamine uptake inhibitors or monoamine oxidase inhibitors as treatments, although there has been a massive effort by the pharmaceutical industry to improve side effect profiles. A wide range of compounds has been used to augment the antidepressant response in treatment nonresponders. These include the use of lithium, thyroid, and neuroleptics, especially respiridone.

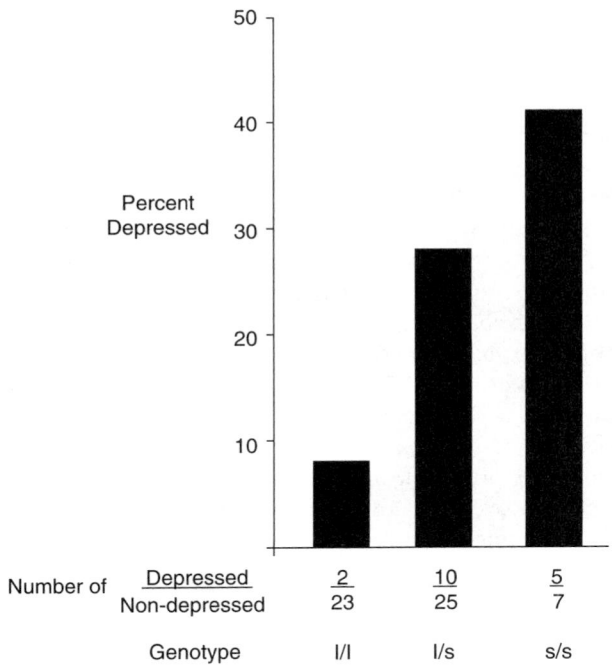

FIGURE 12.16 Reduced function of serotonin transporter gene is associated with increased depression in individuals with Parkinson's disease. Depression defined as 10 or higher on Hamilton Depression Rating Scale. A total of 72 patients with idiopathic Parkinson's disease, 59 males, 33 females, mean disease duration 9.9 years. Severity of Parkinson's symptoms did not vary by genotype. (Constructed from data in Mossner R *et al.* (2001). *Mol Psychiatry* 6:350.)

Advances with ECT have primarily reflected better anesthetic practice and the use of unilateral treatment to reduce memory impairment. The advent of transcranial magnetic stimulation has shown some antidepressant efficacy, but it has yet to find a main role in common clinical practice. For patients very resistant to treatment, vagal nerve stimulation has been used, but with only partial improvements.

One area where practice has improved is the use of antidepressants in patients with comorbid illness. With the use of SSRIs and their reduced number of cardiovascular side effects, it is now possible to engage in active treatment in patients with cardiac disease and depression. In a recent study of behavioral intervention that improved psychosocial function and depressive symptoms, behavioral treatment did not impact on cardiovascular mortality (Writing Committee, 2003). However, emerging data show that treatment with SSRIs can help reduce mortality in cardiovascular disease, especially with the use of paroxetine (Sauer *et al.*, 2003).

It is important to note that up to 50% of antidepressant trials are failures, as the rate of improvement of patients on the test drug is not significantly different than the rate of patients on placebo. This indicates that there is large opportunity for further progress.

CRH and the HPA Axis

A number of CRH antagonists have been under development, and in many animal models they show clear antidepressant and antianxiety efficacy. In one of the first clinical studies, a human CRH-1 receptor antagonist was studied in 24 patients with a dose ranging study. Significant reductions in depression and anxiety scores were observed, with some worsening of affective symptomatology after drug discontinuation (Zobel *et al.*, 2000). This result supports the principle that a CRH-1 antagonist could be a new type of antidepressant and anxiolytic treatment.

A study of mifepristone (a corticosteroid antagonist) in psychotic depression was conducted in 30 patients. A placebo control was not used, and when the high-dose group was compared to the low-dose group, 68% met the criterion of a 30% or greater decline in the depression rating scale, whereas only 36% of the low-dose group met this threshold. Although results were not statistically significant, the number of subjects was small, and if that number were doubled at the same rate of improvement, statistically significant results would have been observed (Belanoff *et al.*, 2002).

Excitatory and Inhibitory Neurotransmission

A large number of clinical trials have investigated the use of GABA agonists in the treatment of depression. These drugs have shown some partial efficacy, but as yet they have not demonstrated superior efficacy over the standard tricyclic or SSRI treatments. Studies have shown that GABA levels in the occipital lobe of depressed patients are lower in some patient groups and that this increases to normal levels with effective treatment. The most promising studies appear to be related to the NMDA receptor system. In preclinical studies, modulators of the glycine side on this receptor were highly correlated to antidepressant-like effects on a wide range of animal models. It is of considerable interest that the NMDA antagonist ketamine is an antidepressant. When a single dose was administered intravenously to depressed patients, beginning at 4 hours and extending to 72 hours, there was a clear antidepressant response, reaching a drop of 13 points on the Hamilton rating scale at 72 hours (Berman *et al.*, 2000).

Behavioral Treatments

As opposed to the pharmacological treatments, with behavioral treatments it has been extremely difficult or impossible to identify the active therapeutic factor and to quantify its dosage and ensure its administration; however, there have been major advances in this field over the past 50 years. At the present time, therapists follow manuals and receive reliability training so there is increasing assurance that the expected behavioral treatment is actually being delivered.

When cognitive behavioral therapy and antidepressant therapy are used in combination, some advantages have emerged. However, in the treatment of panic disorder, even though there is an advantage of the combined treatment over cognitive behavioral treatment (CBT) alone at the end of treatment, there was a greater relapse in the combined group at follow-up evaluation (Foa *et al.*, 2002). Thus, there are some data to indicate that the pharmaceutical therapy and CBT may address quite different aspects of cerebral function and that, depending on the specific circumstances, this may be synergistic or antagonistic.

Behavioral therapies have special indications where pharmacotherapy is questioned, such as in pregnancy. Up to 20% of pregnant women can have elevated depressive symptoms, and a trial in interpersonal psychotherapy in this patient group has shown that up to 60% of women can meet recovery criteria, which is significantly better than a parenting educational control program (Spinnelli and Endicott, 2003). Psychotherapy treatments

may also be useful in situations where pharmacotherapy is not available. A well-controlled trial of group interpersonal therapy in rural Uganda showed markedly greater antidepressant effect in the interpersonal therapy group than the control group (i.e., a reduction of 17 vs 3 points, respectively, in the measure of the severity of depression) (Bolton *et al.*, 2003).

The role of exercise in the treatment of depression has been of considerable interest. A number of preclinical studies demonstrate that increased motor activity in rodents is related to an increase in BDNF in hippocampus. In a study of depressed individuals over age 50 who were randomly assigned to a 4-month course of aerobic exercise, sertraline therapy, or a combination of exercise and sertraline, at the end of the treatment protocol, all three groups exhibited significant improvement and the proportion of individuals who met remission criteria was comparable across the treatment conditions. However, at follow-up evaluation after 10 months, subjects in the exercise group had significantly lower relapse rates than subjects in the medication group (Babyak *et al.*, 2000). In Figure 12.17 it can be seen that the percentage of patients at 10 months who were continuing to exercise was about the

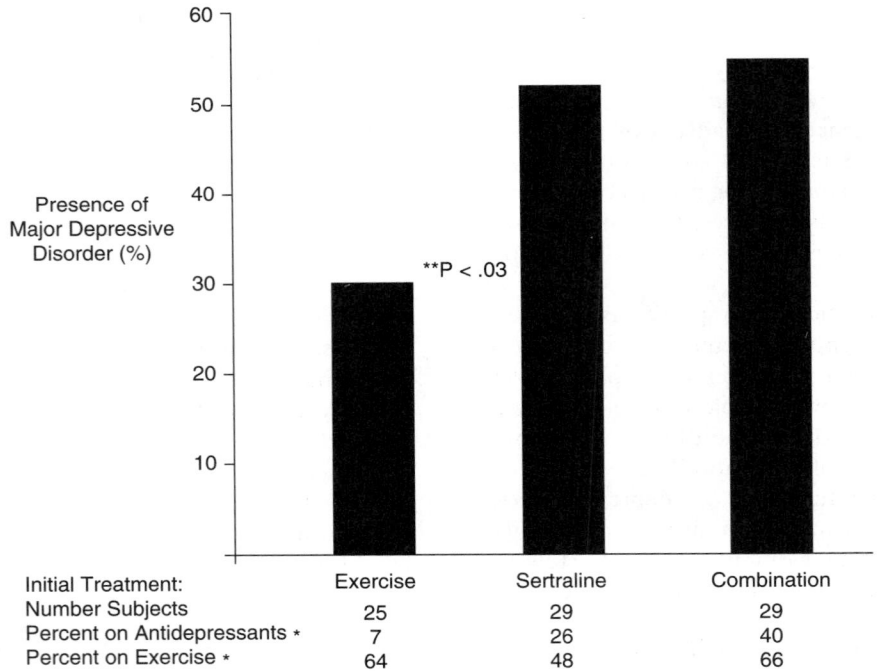

FIGURE 12.17 Exercise treatment alone results in lower rates of depression at 10 months after initial treatment with exercise, sertraline, or the combination. During 6-month follow-up period, depressed individuals over 50 years of age treated with exercise, sertraline, or a combination for 4 months and then evaluated at 10 months. *Percent on antidepressants and doing exercise at 10 months. Depression defined as Hamilton Depression Rating score > 7. (Constructed from data in Babyak M *et al.* (2000). *Psychosom Med* 62:63.)

same in the exercise and combination group, although the exercise group had a clearly reduced prevalence of depression.

As delineated previously, the animal models for PTSD are the most congruent to human anxiety disorders, as conditioned fear in animals and PTSD fear in humans demonstrate many symptomatic and other characteristic similarities. It is of interest in this regard that some of the behavioral treatments for PTSD have been found to be quite efficacious. In a study of sixth grade students who were exposed to violence and had clinical levels of PTSD, those who had a 10-session course of standardized cognitive behavioral therapy had lower symptom scores than the control group (8.9 vs 15.5, respectively) (Stein *et al.*, 2003). Other treatments of PTSD in adults have shown that exposure and cognitive restructuring are effective in reducing symptoms and that they are better than relaxation.

The behavioral treatment literature has little specific data from animal models to design treatment. However, our current understanding of the mechanisms that produce conditioned fear may now be applied to improve the treatment of PTSD. In particular, exposure therapy to

stimuli that are identical to the original trauma could be used to facilitate extinction of traumatic memories. Individuals in Turkey who had PTSD consequent to an earthquake, when given 1 hour of extinction treatment in an earthquake simulator (which was similar to the original trauma), showed significant improvement most evident during the subsequent 4 weeks (Figure 12.18) (Basoglu *et al.*, 2003).

The high order of cognitive abilities of humans compared to other laboratory animals is both an advantage and disadvantage when developing behavioral treatments for depression and anxiety. The advantages are that these capacities can be used to discriminate between threatening and nonthreatening situations, increase control, avoid symptoms, and generate adaptations and alternative coping strategies. The disadvantage can be that these cognitive elements, if uncontrolled, can negate or allow escape from some of the behavioral interventions. As understanding of the molecular, cellular, and physiological control of depressive and anxiety behaviors improves it would be expected that the animal models may be used to build a foundation and framework for human studies. Because of the high order of cognitive processing in

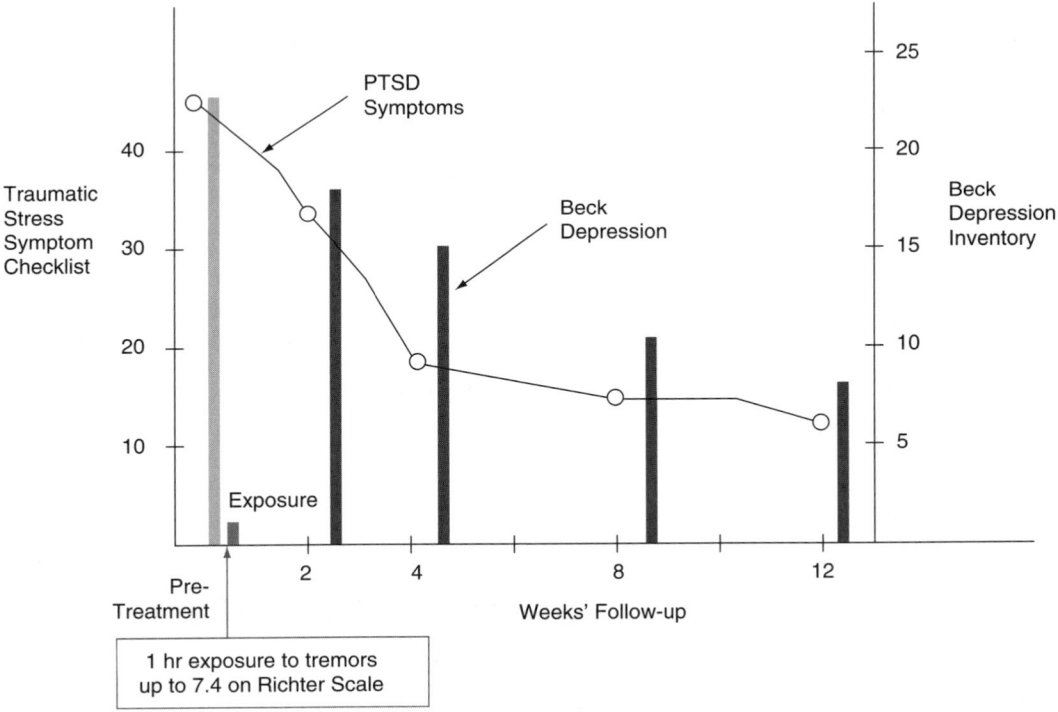

FIGURE 12.18 A single session of extinction training reduces PTSD symptoms in earthquake survivors. Ten survivors of earthquakes in Turkey controlled the intensity of shaking of a small furnished house they were in. They were encouraged to increase the intensity to a level they could tolerate, and the session ended when they had a reduction in and control over their distress. (Constructed from data in Basoglu M *et al.* (2003). *Am J Psychiatry* 160:788.)

humans, the new methods and techniques will always have to be evaluated in the clinical populations.

FUTURE DIRECTIONS

Diagnostic Specificity

The reliance on descriptive diagnosis is expected to remain in place until practical and trustworthy alternatives are demonstrated. As in other areas of medicine (e.g., diabetes, epilepsy), the clinical diagnostic categories lag far behind our understanding of the specific avenues of pathogenesis. It is hoped that there will be an increasing acceptance of the diffuseness of the disorders and acceptance of the graded and blurred boundaries between symptoms, syndromes, and diseases. It is clear no single lesion will be discovered for any diagnostic

group. As in other areas of medicine, multiple pathways of pathogenesis will be delineated and their relationship to clinical manifestations will become more clear.

Neuroscience and Molecular Medicine

Understanding of the mechanisms of brain function and specific influences of genetic variations on behavior will continue to improve. In particular, the opportunity to identify specific genetic abnormalities in families and then correlate the behavioral and clinical manifestations of illness to this abnormality will shed a new perspective on pathogenesis and diagnosis of anxiety and depression. A major influence will be the role of the genetic abnormalities as risk factors and a high emphasis on environmental interactions. Even today the short and long alleles of the 5-HT transporter can be used in dealing with the causes and treatment interventions for

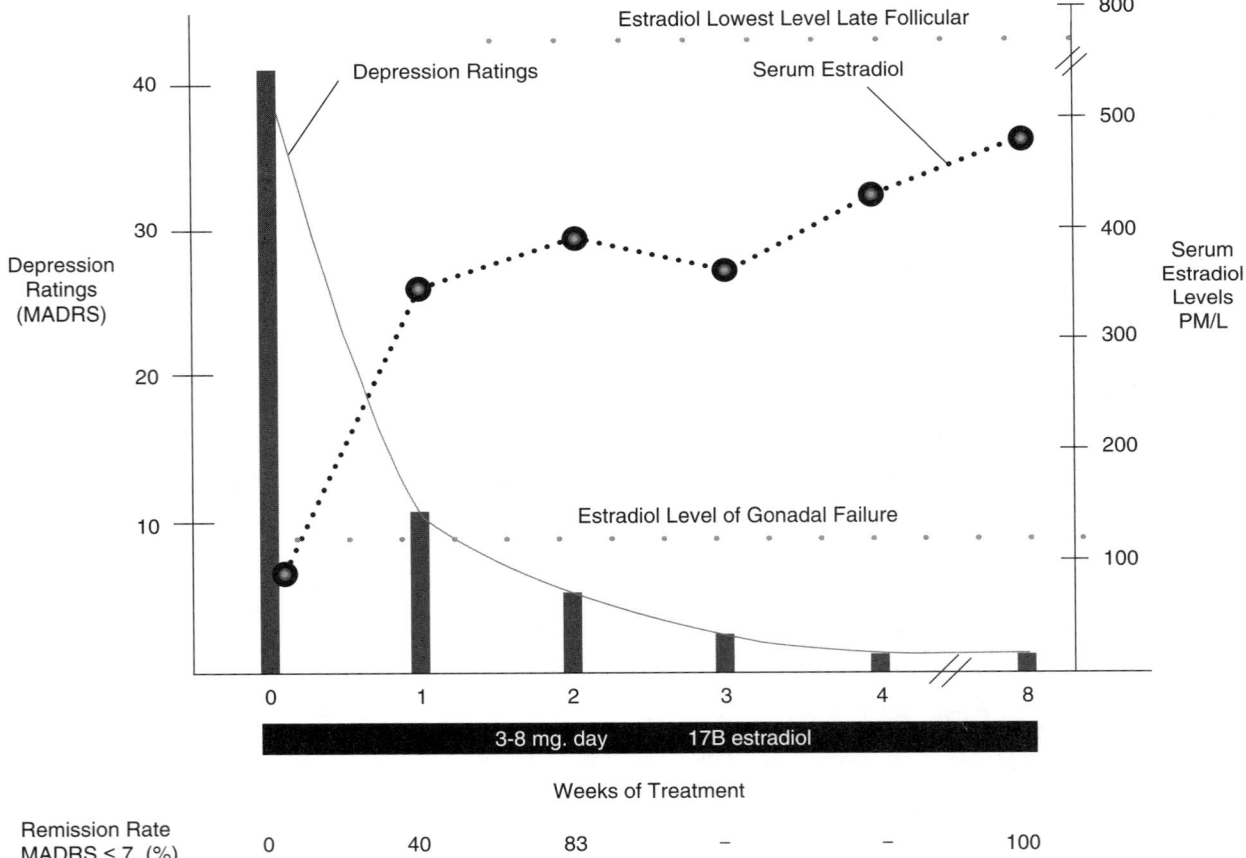

FIGURE 12.19 Estrogen treatment for postpartum depression in 23 women. A total of 23 women with postpartum depression were consecutively recruited from a psychiatric emergency unit. They had low estrogen levels and received sublingual estrogen for 8 weeks. (Constructed from data in Ahokas A *et al.* (2001). *J Clin Psychiatry* 62:332.)

childhood maltreatment. Clinicians will need a much more far-ranging concept of disease pathogenesis. They will need to come to grips with the power of the environmental interaction given the particular genetic vulnerabilities of individuals.

An Example from Obstetrics

At the descriptive level, a great deal has been written about postpartum depression, its existence, causes, and possible treatments. The data in Figure 12.19 present a striking result. Replacement of estrogen deficiency in women presenting with postpartum psychosis produces an 83% remission rate in 2 weeks, with 100% remission rate by 8 weeks (Ahokas *et al.*, 2001). This finding illustrates three points: (1) A behavioral diagnosis missed the etiological variable; (2) when the etiological variable was directly corrected, a complete recovery was obtained; (3) the pathogenesis within the diagnostic subgroup needs to be homogeneous and exactly corrected by the treatment. In this group of women homogeneous for estrogen deficiency, the result of estrogen replacement was total recovery in 100% of patients.

Long-Term Future

With the advances stemming from neurosciences and molecular medicine, it is expected that the specificity of diagnosis relative to pathogenic pathways will be increasingly congruent. Thus treatments will be directed at the cause of illness and will thereby be much more effective. There is no reason to expect that the high prevalence and destructive disability of depressive and anxiety disorders will remain at its current level if this formula can be effected. On the contrary, it is probable that these disorders will be more effectively and more frequently treated in the future.

References

Ahokas A, *et al.* (2000). Estrogen deficiency in severe postpartum depression: successful treatment with sublingual physiologic 17B estradiol: a preliminary study, *J Clin Psychiatry* 42:332-336.

Babyak M *et al.* (2000). Exercise treatment for major depression: maintenance of therapeutic benefit at 10 months, *Psychosom Med* 62:633-638.

Barrett TB *et al.* (2003). Evidence that a single nucleotide polymorphism in the promoter of the G protein receptor kinase 3 gene is associated with bipolar disorder, *Mol Psychiatry* 8:546-557.

Basoglu M, Livanou M, Salcioglu E. (2003). A single session with an earthquake simulator for traumatic stress in earthquake survivors, *Am J Psychiatry* 160:788-790.

Belanoff JK *et al.* (2002). An open label trial of C-1073 (mifepristone) for psychotic major depression, *Biol Psychiatry* 52:386-392.

Berman RM *et al.* (2000). Antidepressant effects of ketamine in depressed patients, *Biol Psychiatry* 47:351-354.

Bilkei-Gorzo A, Racz I, Michel K, Zimmer A. (2002). Diminished anxiety- and depression-related behaviors in mice with selective deletion of the Tac1 gene, *J Neurosci* 22:10046-10052.

Bolton P *et al.* (2003). Group interpersonal psychotherapy for depression in rural Uganda, *JAMA* 289:3117-3124.

Bondy B *et al.* (2003). Substance P serum levels are increased in major depression: preliminary results, *Biol Psychiatry* 53:538-542.

Bush DE *et al.* (2001). Even minimal symptoms of depression increase mortality risk after acute myocardial infarction, *Am J Cardiol* 88:337-341.

Capuron L, Ravaud A. (1999). Prediction of the depressive effects of interferon alfa therapy by the patient's initial affective state, *N Engl J Med* 340:1370.

Capuron L *et al.* (2002a). Association between decreased serum tryptophan concentrations and depressive symptoms in cancer patients undergoing cytokine therapy, *Mol Psychiatry* 7:468-473.

Capuron L *et al.* (2002b). Neurobehavioral effects of interferon-a in cancer patients: phenomenology and paroxetine responsiveness of symptom dimensions, *Neuropsychopharmacology* 26:643-652.

Capuron L *et al.* (2003). Association of exaggerated HPA axis response to the initial injection of interferon-alpha with development of depression during interferon-alpha therapy, *Am J Psychiatry* 160:1342-1345.

Caspi A *et al.* (2003). Influence of life stress on depression: moderation by a polymorphism in the 5-HTT gene, *Science* 301:386-389.

Chen B *et al.* (2001). Increased hippocampal BDNF immunoreactivity in subjects treated with antidepressant medication, *Biol Psychiatry* 50:260-265.

Chwastiak L *et al.* (2002). Depressive symptoms and severity of illness in multiple sclerosis: epidemiologic study of a large community sample, *Am J Psychiatry* 159:1862-1868.

Contarino A *et al.* (1999). Reduced anxiety-like and cognitive performance in mice lacking the corticotropin-releasing factor receptor 1, *Brain Res* 835:1-9.

Coyle JT, Duman RS. (2003). Finding the intracellular signaling pathways affected by mood disorder treatments, *Neuron* 38:157-160.

Delgado PL *et al.* (1999). Tryptophan-depletion challenge in depressed patients treated with desipramine or fluoxetine: implications for the role of serotonin in the mechanism of antidepressant action, *Biol Psychiatry* 46:212-220.

Duman RS, Heninger GR, Nestler EJ. (1997). A molecular and cellular theory of depression, *Arch Gen Psychiatry* 54:597-606.

Eaton WW, Kessler RC, Wittchen HU, Magee WJ. (1994). Panic and panic disorder in the United States, *Am J Psychiatry* 151:413-420.

Foa EB, Franklin ME, Moser J. (2002). Context in the clinic: how well do cognitive-behavioral therapies and medications work in combination?, *Biol Psychiatry* 52:987-997.

Francis DD, Diorio J, Plotsky PM, Meaney MJ. (2002). Environmental enrichment reverses the effects of maternal separation on stress reactivity, *J Neurosci* 22:7840-7843.

Garcia R, Vouimba R-M, Baudry M, Thompson RF. (1999). The amygdala modulates prefrontal cortex activity relative to conditioned fear, *Nature* 402:294-296.

Gilbertson MW *et al.* (2002). Smaller hippocampal volume predicts pathologic vulnerability to psychological trauma, *Nat Neurosci* 5:1242-1247.

Gross C *et al.* (2002). Serotonin$_{1A}$ receptor acts during development to establish normal anxiety-like behaviour in the adult, *Nature* 416:396-400.

Holsboer F. (2000). The corticosteroid receptor hypothesis of depression, *Neuropsychopharmacology* 23:477-501.

Kellner M *et al.* (2000). Behavioral and endocrine response to cholecystokinin tetrapeptide in patients with posttraumatic stress disorder, *Biol Psychiatry* 47:107-111.

Kessler RC *et al.* (2003). The epidemiology of major depressive disorder: results from the national comorbidity survey replication (NCS-R), *JAMA* 289:3095-3105.

Kramer MS *et al.* (1998). Distinct mechanism for antidepressant activity by blockade of central substance P receptors, *Science* 281:1640-1645.

Lee HJ *et al.* (2001). Fluoxetine enhances cell proliferation and prevents apoptosis in dentate gyrus of maternally separated rats, *Mol Psychiatry* 6:725-728.

Liu SH *et al.* (2003). Generation of functional inhibitory neurons in the adult rat hippocampus, *J Neurosci* 23:732-736.

Maes M *et al.* (2001). Treatment with interferon-alpha (IFNa) of hepatitis C patients induces lower serum dipeptidyl peptidase IV activity, which is related to IFNa-induced depressive and anxiety symptoms and immune activation, *Mol Psychiatry* 6:475-480.

Malberg JE, Eisch AJ, Nestler EJ, Duman RS. (2000). Chronic antidepressant treatment increases neurogenesis in adult rat hippocampus, *J Neurosci* 20:9104-9110.

Manji HK, Drevets WC, Charney DS. (2001). The cellular neurobiology of depression, *Nat Med* 7:541-547.

Manzke T *et al.* (2003). 5-HT$_{4(a)}$ Receptors avert opioid-induced breathing depression without loss of analgesia, *Science* 301:226-229.

Marsicano G *et al.* (2002). The endogenous cannabinoid system controls extinction of aversive memories, *Nature* 418:530-534.

Milad MR, Quirk GJ. (2002). Neurons in medial prefrontal cortex signal memory for fear extinction, *Nature* 420:70-74.

Miller HL *et al.* (1996). Effects of alpha-methyl-para-tyrosine (AMPT) in drug-free depressed patients, *Neuropsychopharmacology* 14:151-157.

Moore GJ *et al.* (2000). Lithium-induced increase in human brain grey matter, *Lancet* 356:1241-1242.

Mossner R *et al.* (2001). Allelic variation of serotonin transporter expression is associated with depression in Parkinson's disease, *Mol Psychiatry* 6:350-352.

Murray CJ, Lopez AD. (1997). Global mortality, disability, and the contribution of risk factors: global burden of disease study, *Lancet* 349:1436-1442.

Nestler EJ *et al.* (2002). Neurobiology of depression, *Neuron* 34:13-25.

Neumeister A *et al.* (2002). Association between serotonin transporter gene promoter polymorphism (5HTTLPR) and behavioral responses to tryptophan depletion in healthy women with and without family history of depression, *Arch Gen Psychiatry* 59:613-620.

Nishino S. (2003). The hypocretin/Orexin system in health and disease, *Biol Psychiatry* 54:87-95.

Ohayon MM, Schatzberg AF. (2003). Using chronic pain to predict depressive morbidity in the general population, *Arch Gen Psychiatry* 60:39-47.

Rajkowska G. (2003). Depression: what we can learn from postmortem studies, *Neuroscientist* 9:273-284.

Redrobe JP, Dumont Y, Fournier A, Quirion R. (2002). The neuropeptide Y (NPY) Y1 receptor subtype mediates NPY-induced antidepressant-like activity in the mouse forced swimming test, *Neuropsychopharmacology* 26:615-624.

Roceri M *et al.* (2002). Early maternal deprivation reduces the expression of BDNF and NMDA receptor subunits in rat hippocampus, *Mol Psychiatry* 7:609-616.

Romeo E *et al.* (1998). Effects of antidepressant treatment on neuroactive steroids in major depression, *Am J Psychiatry* 155:910-913.

Rudolph U *et al.* (1999). Benzodiazepine actions mediated by specific GABA$_a$-aminobutyric acid$_A$ receptor subtypes, *Nature* 401:796-800.

Ruo B *et al.* (2003). Depressive symptoms and health-related quality of life: the heart and soul study, *JAMA* 290:215-221.

Santarelli L *et al.* (2003). Requirement of hippocampal neurogenesis for the behavioral effects of antidepressants, *Science* 301:805-809.

Sauer WH, Berlin JA, Kimmel SE. (2003). Effect of antidepressants and their relative affinity for the serotonin transporter on the risk of myocardial infarction, *Circulation* 108:32-36.

Schmitz C *et al.* (2002). Depression: reduced number of granule cells in the hippocampus of female, but not male, rats due to prenatal restraint stress, *Mol Psychiatry* 7:810-813.

Sheline YI. (2003). Neuroimaging studies of mood disorder effects on the brain, *Biological Psychiatry* 54:338-352.

Sheline YI, Gado MH, Kraemer HC. (2003). Untreated depression and hippocampal volume loss, *Am J Psychiatry* 160:1516-1518.

Shumyatsky GP *et al.* (2002). Identification of a signaling network in lateral nucleus of amygdala important for inhibiting memory specifically related to learned fear, *Cell* 111:905-918.

Sklar P *et al.* (2002). Family-based association study of 76 candidate genes in bipolar disorder: BDNF is a potential risk locus, *Mol Psychiatry* 7:579-593.

Spinelli MG, Endicott J. (2003). Controlled clinical trial of interpersonal psychotherapy versus parenting education program for depressed pregnant women, *Am J Psychiatry* 160:555-562.

Stein BD *et al.* (2003). A mental health intervention for schoolchildren exposed to violence, *JAMA* 290:603-611.

Strohle A *et al.* (2002). GABA$_A$ receptor-modulating neuroactive steroid composition in patients with panic disorder before and during paroxetine treatment, *Am J Psychiatry* 159:145-147.

Strohle A *et al.* (2003). Induced panic attacks shift g-Aminobutyric acid type A receptor modulatory neuroactive steroid composition in patients with panic disorder, *Arch Gen Psychiatry* 60:161-168.

Then Bergh F *et al.* (1999). Dysregulation of the hypothalamo-pituitary-adrenal axis is related to the clinical course of MS, *Neurology* 53:772-777.

Tronche F *et al.* (1999). Disruption of the glucocorticoid receptor gene in the nervous system results in reduced anxiety, *Nat Genet* 23:99-103.

van Praag H *et al.* (2002). Functional neurogenesis in the adult hippocampus, *Nature* 415:1030-1034.

Writing Committee for the ENRICHD Investigators. (2003). Effects of treating depression and low perceived social support on clinical events after myocardial infarction. The enhancing recovery in coronary heart disease patients (ENRICHD) randomized trial, *JAMA* 289:3106-3116.

Zobel AW *et al.* (2000). Effects of the high-affinity corticotropin-releasing hormone receptor 1 antagonist R121919 in major depression: the first 20 patients treated, *J Psychiatr Res* 34:171-181.

Zubenko GS *et al.* (2003). Sequence variations in CREB1 co-segregate with depressive disorders in women, *Mol Psychiatry* 8:611-618.

Moving Toward the Clinic:
Evolving Themes and Technologies

Genomics, Proteomics, and Neurology

Lorelei D. Shoemaker
Daniel H. Geschwind, MD, PhD

THE SUCCESS OF THE GENETICS PARADIGM

The use of genetic methods to further our understanding of disease has provided major advances during the last decade and promises more in the near future. The last century of progress in genetics has been impressive. The word *gene* was first used in the early 1900s by William Bateson. The discovery that chromosomes contained the genetic material occurred in the next decade in Thomas Hunt Morgan's laboratory and firmly established the fruit fly as a model organism in genetics. The structure of DNA was solved by Watson and Crick in 1953 about three decades later. Now scientists have essentially sequenced the entire human and mouse genomes, in addition to other invertebrates, and the chimpanzee is close behind (Adams *et al.*, 2000; Lander *et al.*, 2001; Venter *et al.*, 2001; Waterston *et al.*, 2002; Wolfsberg *et al.*, 2002). All of this, less than 50 years after the structure of DNA was first published.

Although knowledge of the human and other genomes provides an important road map, we still have a long way to go to translate this windfall into new therapeutics. Thus far, the field of neurogenetics has had most of its success in finding genes for relatively rare conditions with mendelian inheritance. This is not to diminish these seminal contributions, as finding these genes has already led to major advances in our understanding of several common neurological diseases such as Alzheimer's disease (Tanzi and Bertram, 2001), Parkinson's disease, and frontotemporal dementia (Clark *et al.*, 1998; Hong *et al.*, 1998), as well as rarer conditions such as the triplet repeat diseases (Paulson and Fischbeck, 1996; Rosenberg, 1996; Paulson, 2000; Taylor *et al.*, 2002). Several of these genetic discoveries have already led to therapeutic clinical trials. We are now beginning to understand how genes contribute to normal human development and individual variability in common disease susceptibility, and cognitive and behavioral phenotypes.

The ability to study individual genes afforded by advances in molecular biology and genetics that have so propelled our knowledge of neuronal development and functioning over the last two decades has also contributed to an increasing gap between molecular and systems neuroscientists. How do we move from the individual gene to an understanding of brain function or dysfunction at a systems level? In addition, how do we reconcile the classic, hypothesis driven, neurobiological approach with the discovery-based methodology on which most modern genetic research is based? Knowledge of the genome sequence of humans and experimental animals invites us to use this genome knowledge and attempt to integrate it with our knowledge of the neural systems that underlie disease. To do so, we must be ready to accept research methods that rely primarily on large-scale discovery, rather than the more standard approach of testing one hypothesis at a time (Geschwind, 2000; Lockhart and Barlow, 2001; Mirnics *et al.*, 2001; Geschwind and Gregg, 2002). This chapter emphasizes two such methods, gene expression analysis with DNA microarrays and proteomics.

GENOMICS

Why Gene Expression?

About 1% to 2% of the human genome codes for expressed genes, or messenger RNA (mRNA), which is subsequently translated into protein within cells (Figure 13.1).

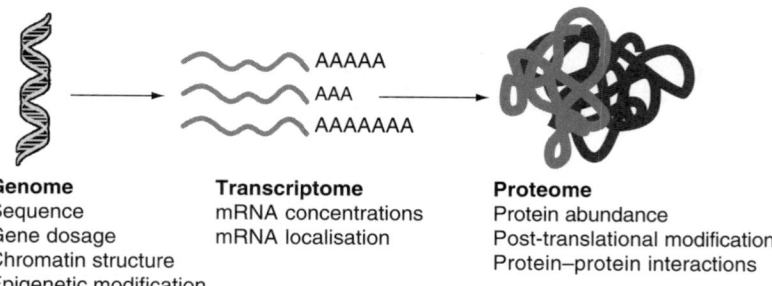

Genome   **Transcriptome**   **Proteome**
Sequence   mRNA concentrations   Protein abundance
Gene dosage   mRNA localisation   Post-translational modification
Chromatin structure   Protein–protein interactions
Epigenetic modification

FIGURE 13.1 Levels of biological system understanding from genome to the proteome. Neurological disease can be studied at each level, all of which should be considered to develop appropriately targeted therapeutics. This chapter focuses on how we can assess the functional consequences arising from the genome or the interactions of the genome and the environment in a high throughput manner at the transcriptome (microarray) and proteome (proteomics) level. Figure and legend reprinted with permission from *Lancet Neurology*, 2003 2:275-282, Elsevier Press.

Estimates based on bioinformatics and expressed sequence tags suggested that there are about 30,000 genes in the human genome, about twice that in the fruit fly (Adams *et al.*, 2000; Venter *et al.*, 2001). Given this modest increase in the number of expressed genes vs an almost exponential rise in organismal complexity from drosophila to human, it is clear that it is the regulation of genes through the timing and levels of expression and alternative splicing of these genes that underlie cellular homeostasis and signaling within the nervous system. Therefore, to understand the role of genes in development, functioning, and diseases of the nervous system, we need to be able to measure the dynamic patterns of gene expression within specific brain regions and cell types. To perform this task efficiently requires methods allowing study of the expression patterns of large numbers of genes (mRNA) and their regulation at the translational (protein) and posttranslational level (e.g., protein phosphorylation or glycosylation). Microarrays permit tour de force monitoring of the entire transcriptome (all mRNAs in a cell) of a given neuronal or glial cell or brain region. Rapidly evolving proteomics techniques, such as mass spectrometry (MS) coupled to two-dimensional gel electrophoresis (2DGE) permit similar assessment of the proteome (all proteins in a cell). These and other methods focusing on different levels, from gene to RNA to protein, and their relationships are depicted in Figure 13.1. Microarrays and proteomic techniques provide a striking improvement in power and throughput over conventional techniques such as reverse transcription-polymerase chain reaction (RT-PCR), and Northern or Western blotting, which allow the simultaneous assessment of only single or small groups of genes and proteins.

DNA Microarrays

DNA microarrays, also referred to as gene chips, are composed of DNA attached to a solid surface in an ordered fashion at high density (Schena *et al.*, 1995; Lipshutz *et al.*, 1999). About half of the genome can be arrayed on a surface the size of a glass microscope slide or other similar surface, such as a silicon wafer or mirror. Each spot on the array is a distinct DNA species that provides a specific target for a hybridization experiment. In a typical microarray experiment, one wants to measure the amount of a given mRNA species in a specific brain region or cell type. As there are thousands of gene spots on each array, all can be queried in a single hybridization step because of two basic principles of nucleic acid hybridization: (1) DNA and RNA bind specifically to their complementary sequence due to Watson-Crick base pairing, and (2) binding will occur in proportion to the nucleic acid species abundance in a mixture. The more abundant a given mRNA (or cDNA) is in a cell or tissue, the more binding to its complement on the array will occur. Thus, by hybridizing the array with mRNA or its more stable counterpart, cDNA, from a region or cell of interest, one gets a reading of the abundance of every gene represented on an array in the hybridized sample. The availability of microarray sets containing more than 80% of the genes in the genome of interest makes it conceivable to identify all of the important target disease-causing genes in a few hybridization steps (Lipshutz *et al.*, 1999; Lockhart and Winzeler, 2000; Geschwind and Gregg, 2002).

Array Platforms

Currently there are two major array formats, oligonucleotide arrays and cDNA arrays (Schena *et al.*, 1995; Lipshutz *et al.*, 1999; Lockhart and Winzeler, 2000; Luo and Geschwind, 2001; Geschwind and Gregg, 2002). Price, availability, and the presence of genes of interest are reasonable criteria to apply when choosing between these platforms. As more comparative performance data become available, this additional factor can be considered as well. Some of the choice may come down to factors related to individual study design. For example, to study alternative splicing, oligonucleotide arrays are preferable to cDNA-based arrays (Shoemaker *et al.*, 2001).

Most commercial platforms involve short (22- to 25-mer; Affymetrix) or longer (30- to 70-mer; Codelink/Amersham; Agilent; BD Biosciences) oligonucleotides representing unique sequences from individual genes. Affymetrix arrays are manufactured using a unique *in situ* synthesis of short oligonucleotides using a patented photolithography method (Lipshutz *et al.*, 1999). Agilent arrays are made using ink jet printing technology for *in situ* synthesis of 60-mer oligonucleotides on the array (Hughes *et al.*, 2001), whereas Amersham Codelink arrays are composed of covalently attached 30-mer oligonucleotides (Ramakrishnan *et al.*, 2002). All of the major platforms use some form of fluorescent labeling of the applied probe to detect gene expression (Karsten and Geschwind, 2002).

In general, oligonucleotides are easier to scale to mass production and afford easier quality control than cDNA. cDNA arrays involve more steps, including large-scale PCR amplification of clone inserts, insert purification, resuspension in printing buffer, and gridding onto a nonporous surface, most commonly a 30×15 mm glass microscope slide (Luo and Geschwind, 2001; Karsten and Geschwind, 2002). A typical cDNA array is printed using a computer-controlled robotic cantilever arm; each spot is about 50 to 150 μM in diameter. To print an oligonucleotide array, one can use the same gridding robot, arraying purchased sets of oligonucleotides, omitting the time consuming step of clone insert amplification and purification. Other flexible and porous surfaces such as a nylon membrane can also be used as array surfaces, and these can be probed with radioactive targets (Whitney *et al.*, 1999; Chiang *et al.*, 2001).

The major advantage to in-house manufactured cDNA or oligonucleotide arrays is their reduced cost. However, this advantage has to be balanced with the improved reliability and batch-to-batch consistency obtained with many more expensive, commercial arrays. Printing consistent, high quality cDNA arrays is difficult and has been accomplished by only a few centers. To address this problem, NINDS and NIMH have funded a microarray consortium consisting of three centers to provide consistent low-cost arrays and expertise to the neuroscience community (http://arrayconsortium.cnmc research.org/NINDS/jsp/overview.jsp). The reliability of these arrays remains to be demonstrated, but they should serve as a reasonable alternative to commercial arrays.

Few studies of array reliability have been published (Ramakrishnan *et al.*, 2002; Yuen *et al.*, 2002; Barczak *et al.*, 2003), and whether the commercial sets of partial

or whole genome oligonucleotide sets (e.g., Sigma, Illumina, Operon, MWG) for in-house printing perform as well as cDNA arrays or commercial oligonucleotide arrays from Agilent, Codelink, or Affymetrix is not firmly established; however, initial publications assessing printed oligonucleotide arrays are promising (Barczak *et al.*, 2003). Furthermore, in most published studies, when gene expression changes that are identified as significantly differentially expressed with either a single cDNA or oligonucleotide platform are checked using alternative methods such as RT-PCR, Northern blotting, or *in-situ* hybridization, the majority of such changes are usually confirmed (Mirnics *et al.*, 2000; Geschwind *et al.*, 2001; Ramakrishnan *et al.*, 2002; Iwashita *et al.*, 2003; Karsten *et al.*, 2003). This attests to the reliability of the standard microarray platforms in general.

Whatever platform is used, microarray technology offers the ability to rapidly survey all of the genes expressed in the genome (transcriptome) in parallel for a very low pergene cost. Several hundred articles have been published using microarrays to study a variety of problems in basic and clinical neuroscience (Geschwind, 2003). Some of the more notable early studies include a survey of gene expression across mouse strains and brain regions (Sandberg *et al.*, 2000; Mody *et al.*, 2001; Pavlidis and Noble, 2001; Brown *et al.*, 2002a, 2002b; Su *et al.*, 2002), neural stem cell biology (Geschwind *et al.*, 2001; Terskikh *et al.*, 2001; Karsten *et al.*, 2003), human postmortem schizophrenic brain (Mirnics *et al.*, 2000), multiple sclerosis (Whitney *et al.*, 1999, 2001; Lock *et al.*, 2002), as well as the downstream effects of single gene defects in mouse models of neurodevelopmental disorders such as Fragile X and Rett syndrome (Brown *et al.*, 2001; D'Agata *et al.*, 2002; Tudor *et al.*, 2002). We believe that the application of microarrays and proteomics techniques to study both human tissues and disease models will lead to the discovery of numerous novel pathways for drug development in neurological disease. From this perspective, this technology truly fulfills the promise of translating the genome from the bench to the bedside (Geschwind, 2003).

A major issue in microarray experimentation is the need for statistical and computational know how, which is often underestimated by the first-time user. The actual array hybridization experiment takes very little time; most of the time is consumed in analyzing the large volume of data produced. Data management and bioinformatic analysis is a key component of microarray experiments (Kohane *et al.*, 2002). Numerous tools are available to annotate gene expression lists and to connect these genes with biological pathways (Geschwind, 2002). A detailed discussion of this issue is beyond the scope of this chapter, but a number of helpful reviews, books, and websites are available in this area (Karsten and Geschwind, 2002;

Kohane *et al.*, 2002; Nadon and Shoemaker, 2002). The growth of microarray data repositories will allow for innovative methods of data analysis using existing data sets, and it is clear that data sharing greatly increases the utility of the genomic approach using microarrays (Geschwind, 2001b).

A Roadmap: Basic Issues of Experimental Design and Limitations

Whether or not one uses microarrays or other methods that measure gene expression in a particular experiment depends on many factors (Geschwind, 2001a; Luo and Geschwind, 2001; Karsten and Geschwind, 2002). If one is interested in studying only a handful of genes in a large number of samples, quantitative RT-PCR may be optimal. In-depth analysis of a single tissue may be best served by serial analysis of gene expression (SAGE) and massively parallel signature sequencing, which both involve large-scale sequencing of small sequence tags that allow relatively unambiguous and efficient identification of genes (Brenner *et al.*, 2000; Velculescu *et al.*, 2000). The frequency of each fragment is determined by counting, so as to assess the relative abundance of that particular gene in a particular library. For example, genes expressed in different cancer cells have been identified using SAGE libraries derived from those cell lines. The sensitivity of these techniques is directly related to the amount of sequencing done and, in many cases with deep sequencing efforts, will be more sensitive than microarrays (Velculescu *et al.*, 1995, 2000; Brenner *et al.*, 2000; Datson *et al.* 2001; Evans *et al.*, 2002). However, these techniques are expensive and time consuming and are best suited for an in-depth analysis of all genes expressed in only a few (one to three) libraries from tissues or cells. So, if one wants to identify every gene expressed in a given tissue, these clone and count methods are ideal. Databases of SAGE-derived gene abundances derived from the efforts of many investigators, each studying a few cell types or tissues, are available and will be increasingly useful as their number expands (Velculescu *et al.*, 2000). Microarrays may not be quite as sensitive, but are more rapid, and allow nearly two orders of magnitude more comparisons for the same price as these sequencing-based methods. Thus, microarrays are optimal for a high throughput design, when more than a few samples are compared, and biological replicates will be studied (Karsten and Geschwind, 2002).

Which Array?

The most important issue in this regard (after asking, Is it affordable?) is whether some of the genes of potential interest are on the array. If one has a few candidate genes, it is helpful to ensure that these genes are present. Until now, one has needed two arrays from major vendors to cover the entire genome. Single slide whole genome arrays from most major vendors cover about 80% to 90% of the transcriptome. In addition, no assessment of alternative splicing can be made using these commercial arrays. We have approached the issue of gene coverage in two ways. The first way is to use subtraction to create arrays of subtracted libraries that are designed to study specific biological processes of interest (Dougherty and Geschwind, 2002). For example, we have used representational difference analysis subtraction to create microarrays to study neural and other stem cell populations (Geschwind *et al.*, 2001; Terskikh *et al.*, 2001), as well as cerebral patterning during embryogenesis in mice and humans (Geschwind and Miller, 2001). The value of this combined approach is evidenced by the success of these studies in identifying a number of novel genes likely to be critical regulators of neural stem cell self-renewal and pluripotentiality, as well as novel patterning genes.

The second method to enhance for genes of interest on microarrays is to use bioinformatics to identify the majority of the genes expressed in the brain, or system of interest, and make an array using these genes. Preliminary neuroarrays of this type have been made by BD Biosciences/Clontech and Kevin Becker at NIA and his colleagues (Barrett *et al.*, 2001). We have used a combination of bioinformatics and discrete expert knowledge to create a larger neuroarray containing about 11,000 genes expressed in the nervous system (www.brainmapping.com). We anticipate making a large scale oligonucleotide-based neuroarray and testing it during 2004, making it available through the NIH consortium in late 2004.

Replication and Statistical Analysis

Early microarray studies used few samples and few biologic replicates because of the cost of arrays. It is now clear that even with the most reliable array platforms, the use of biological replicates is important (Karsten and Geschwind, 2002). This necessitates careful attention to experimental design and choice of an array platform that will afford the ability to perform the number of experiments required to identify reliable changes in gene expression (Yang and Speed, 2002). A minimum of three replicates is usually required, although this depends on the purpose of the study and amount of downstream confirmation to be done. For example, in an elegant study, Zirlinger and colleagues were able to get by using few arrays in their study of amygdala genes, because the arrays were only being used as an initial screening device to identify genes for large scale *in situ* hybridization experiments (Zirlinger *et al.*,

2001). In this study, there was no attempt to identify statistically significant changes—a goal that requires careful replication. Although initial studies relied little on statistics, identifying genes meeting a certain threshold (Schena et al., 1995; Geschwind et al., 2001), or belonging to a particular cluster (Iyer et al., 1999; Zhang and Zhao, 2000), statistically defined cutoffs are now more or less required to define differentially expressed genes (Nadon and Shoemaker, 2002; Sabatti et al., 2002; Irizarry et al., 2003; Smyth et al., 2003). In most cases, the major statistical problem is not whether the genes identified are false positives, but how many genes have been missed (α error) because of the large number of multiple comparisons being done relative to the number of independent replicates. Many new methods that help surmount this problem have been described (Long et al., 2001; Nadon and Shoemaker, 2002; Sabatti et al., 2002), and are now in widespread use.

Even with careful replication, some degree of confirmation using an alternative method is usually required. In most cases, this method is RT-PCR, Northern blotting, or in situ hybridization of a few of the key changes identified. In our initial study of the genetic program of neural stem cells and progenitors, we performed a large number of Northern blots and in situ hybridizations because of the few replicates performed. This technique demonstrated that in spite of the lack of statistics, the threshold methods used were robust, and the majority of genes identified as differentially expressed by arrays were confirmed (Geschwind et al., 2001).

Limitations and Technical and Methodological Issues

RNA Quality

The most important consideration is RNA quality. Only the highest quality RNA should be used. In most cases, using total RNA is preferable to mRNA, given the fewer processing steps and excellent results obtained with total RNA (Karsten et al., 2002). Small variability in conditions can also induce large changes in gene expression (Karsten and Geschwind, 2002). In tissue culture experiments, factors such as media batch, or CO_2 or temperature variation in the incubator can confound results. Other common sources of variability include any cDNA amplification methods used, probe labeling, hybridization conditions, and washing (Karsten and Geschwind, 2002). The same methods should always be used to compare two conditions. When these issues of experimental design are considered and proper technique is applied, microarray experiments provide reliable and consistent data on gene expression at a systems level.

Human Disease

When considering the use of human tissue, these variables include tissue preservation methods, postmortem interval, dissection methods, and RNA quality (Ginsberg and Che, 2002; Karsten et al., 2002; VanDeerlin et al., 2002). Frozen tissue generally provides the best material, although ethanol fixation can be used. Differences in gene expression between samples may be due to differences in genetic background (ethnicity) and sex, and also must be considered, necessitating a larger number of samples than with inbred mouse strains or cell culture (Geschwind, 2000).

One critical issue that is often not considered when studying human disease is the stage of disease in the individual from whom the tissue has been obtained. In most cases, these are postmortem, or chronic cases, where the pathological changes are in their end stages. For example, in neurodegenerative disease, such as amyotrophic lateral sclerosis, Parkinson disease, or Alzheimer's disease, most of the critically affected neurons or glia are already lost. In this case, a microarray study is only able to elucidate the changes in gene expression accompanying neuronal loss, rather than the initial metabolic insults. So at a biological level, what one observes are changes in gene expression that signify loss of particular cells. Unfortunately, this does little to enhance our understanding of the mechanism of disease. In the same vein, study of neurodevelopmental diseases such as autism, during embryonic or fetal development, is precluded, so we are left with assessing the outcome rather than the cause. However, in a disease like autism, or schizophrenia, where relatively little is known about the regions or cell types affected, this kind of information is far more valuable than in a disease such as Parkinson's disease, where the pathological alterations are relatively well documented.

Cellular Heterogeneity

A major concern of many neuroscientists is the detection limit of microarrays. This is an issue in neuroscience where most tissues being studied are relatively heterogeneous, in contrast to cell lines or cancer cells, which are simpler systems. Most array platforms allow reliable detection of species at 1/100,000 the level of low abundance mRNAs within a given cell. However, central nervous system tissues can be composed of dozens or hundreds of cell types, so that some low abundance genes will not be detected in heterogeneous tissues. This makes it a reasonable consideration of experimental design to focus on small brain regions or individual cells. Methods such as single cell PCR and microdissection (Eberwine et al., 1992; Crino et al., 1996; Ginsberg et al.,

2000) and laser capture microdissection (LCM) (Luo et al., 1999; Kamme et al., 2003) allow for capture of cDNA from single cells, permitting the study of gene expression in a single cell type or group of cells (Tietjen et al., 2003). These methods facilitate the study of epileptic foci, tuberous sclerosis lesions (Crino et al., 1996), neurofibrillary tangles, and a number of other focal or cell type-specific lesions observed in disease (Ginsberg et al., 2000; VanDeerlin et al., 2002). For practical reasons, however, it is often preferable to start the gene expression profiling study using a particular brain region or tissue, which can be monitored using methods that provide more microscopic cellular resolution. Again, this goal will depend on the precise line of study. To understand the vulnerability of particular cell classes to disease-causing mutations, such as dopaminergic neurons in Parkinson's disease, specific motor neurons in amyotrophic lateral sclerosis, hippocampal neurons in Alzheimer's disease, study at the level of the cell is likely to be required.

Preliminary Success in Studies of Disease: Biomarkers and Disease Pathophysiology

We view the use of cDNA microarray technology to study disease as falling into two major classes: (1) those that attempt to identify causal molecular alterations occurring in disease, and (2) those trying to identify biomarkers that will aid in diagnosis and patient classification. The main goal of the first group of studies is to advance knowledge as to disease pathophysiology. The main goal(s) of the second group of studies is to improve the accuracy of diagnosis or to identify subsets of patients most likely to respond to specific therapies. Several studies have taken the first approach, attacking diseases such as schizophrenia (Mirnics et al., 2000; Hakak et al., 2001; Hemby et al., 2002; Middleton et al., 2002; Vawter et al., 2002), multiple sclerosis (Whitney et al., 1999, 2001; Lock et al., 2002; Mycko et al., 2003), alcoholism (Lewohl et al., 2000; Mayfield et al., 2002), and brain tumors (Rickman et al., 2001; Pollack et al., 2002; Pomeroy et al., 2002). Complex genetic diseases, or those that are sporadic such as Multi-System Atrophy, are excellent choices for microarray analysis because other genetic approaches, although direct, require enormous coordinated effort from many investigators and will not help with sporadic diseases (Risch and Merikangas, 1996; Nadeau and Frankel, 2000). The use of arrays offers an alternative to positional cloning by simultaneously assaying both the effects of genes and the environment that lead to alterations in gene expression that eventually cause disease. In most common diseases, which are genetically complex, these methods offer a manner for identifying new pathways involved in disease pathophysiology, with the hope that these pathways will be good therapeutic targets.

Microarrays have also been successfully used in mouse models of developmental defects or disease, including Freidreich's ataxia (Tan et al., 2003), fragile X (Brown et al., 2001), and Rett syndrome (Colantuoni et al., 2001; Tudor et al., 2002). Using microarrays to identify the critical pathways altered early in disease in transgenic or knockout models where a single gene or combination of genes has been altered is an especially compelling approach (Livesey et al., 2000). In this case, the issues of postmortem artifacts, stage of disease, and genetic background, which can plague human postmortem studies, can be avoided (Geschwind, 2000). Tissues can be studied early, berfore neuronal loss or the downstream effects of disease. Pathways altered in animal models can be identified and then tested to see whether they are similarly altered in humans, providing confirmation of their relevance to the human disease.

Biomarkers of Disease

Most published studies using microarrays have been directed at identifying mechanisms causing disease. We are aware of only one major study that has looked for biomarkers that could aid in diagnosis (Tang et al., 2001). The lack of published articles in this area is probably due to the intense criticism that these studies will not work, because they do not look at brain tissue, but peripheral tissues such as blood. We believe that this is a specious and overly conservative criticism that underestimates the power of both gene expression profiling and proteomics techniques. Evidence supporting this position is provided by the important study of Tang et al. (2001). Tang and colleagues compared gene expression profiles in blood using Affymetrix arrays in rats subjected to acute models of stroke, hypoxia, hypoglycemia, and seizures. Specific patterns of gene expression were identified that allowed distinguishing between these neurologic insults, 24 hours after the insult occurred. These data provide a proof of principle that complex patterns of peripheral blood gene expression can accurately predict the cause of a previous acute brain injury. The obvious next step is to see whether these patterns are useful in diagnosis of human patients in clinical settings. While this approach is risky in that it may fail, it poses little risk to patients, and its benefits are clear. We anticipate a time in the future when gene expression profiling in peripheral blood will be used as an additional piece of data to diagnose patients in an acute setting (Geschwind, 2003).

Perhaps more valuable will be the use of these techniques in the diagnosis of chronic or subacute illnesses. Already, microarrays have been used to accurately classify brain tumors, adding precision beyond classic pathological grading (Rickman *et al.*, 2001; Pomeroy *et al.*, 2002; Mischel *et al.*, 2003; Shai *et al.*, 2003). It is unknown whether gene expression in peripheral blood allows differentiation between patients with different neurodegenerative diseases such as Alzheimer's disease, Lewy body disease, Creutzfeldt-Jakob disease, or frontotemporal dementia. Such approaches could allow for early diagnosis and hence presymptomatic treatment (Figure 13.2; see also the Color Plate section). This is a huge issue, as once symptoms appear, it may be too late to regain previous functional status. In addition, gene expression profiles could predict drug response or toxicity, helping the clinician to optimize or individualize therapy (Figure 13.2). This is especially important in cancer therapy, where chemotherapeutic response is often hard to predict, and toxicities are serious. In addition, could a subset of the gene expression alterations observed in the studies of brain tumors themselves be identified in peripheral blood or other peripheral cell lines derived from patients, so as to allow screening? The issue of using peripheral blood cells or other peripheral cells for screening or disease biomarkers is an important clinical issue that begs for pilot studies looking at gene expression in patients with chronic neurological disease. And while microarrays are optimal for studying mRNA expressed in cells, certain obtainable biological specimens such as cerebrospinal fluid or serum, may be hard to study using microarrays, as they are protein rich and mRNA poor. These tissues are perhaps better studied using proteomics techniques, which offer a similar promise of identifying biomarkers for chronic neurological diseases.

PROTEOMICS

Why Proteins?

With an estimated 30,000 genes in the human genome, which could translate into an order of magnitude greater number of proteins when alternative splicing and post-translational modifications are considered, a natural, if not necessary, complement to gene expression patterns is an understanding of the protein products of these genes. Although studying gene expression at the mRNA level has the advantage of being more scalable and is currently easier than working with protein, mRNA does not

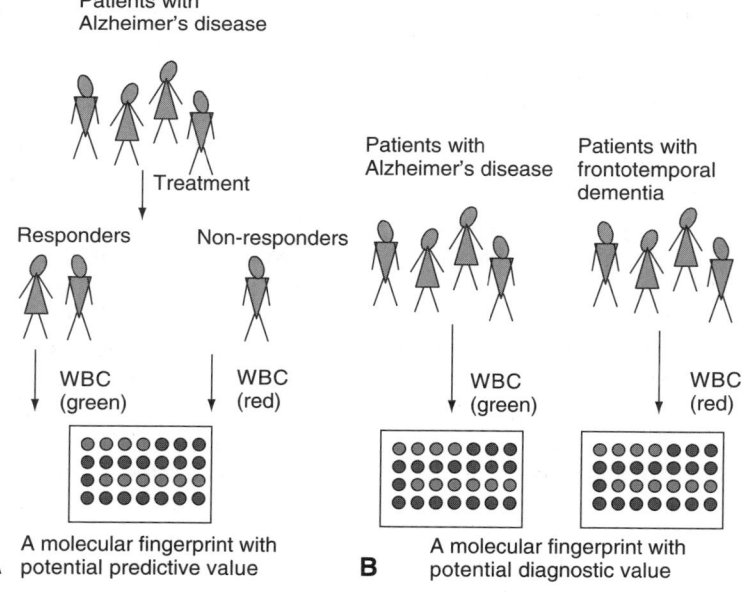

FIGURE 13.2 The use of microarrays for disease classification. **A:** Microarrays can be used to detect single nucleotide differences (polymorphisms) or identify gene expression patterns that predict treatment outcome or adverse effects in patients. **B:** Oligonucleotide arrays that allow polymorphism detection in a number of candidate genes simultaneously, such as cytochrome p450, that may affect drug metabolism, hence predisposing to treatment failure or adverse effects, will likely soon be widely adopted. Microarrays may also be used to augment diagnostic protocols. In the speculative example shown, mRNA from peripheral blood leukocytes is hybridized onto a cDNA array and the pattern indicates whether the patient is more likely to have Alzheimer's disease or frontotemporal dementia. Perhaps, more readily adopted into clinical practice will be the use of arrays to classify tumor samples from patients to both aid in diagnosis and predict optimal treatment. Figure and legend reprinted with permission from *Lancet Neurology*, 2003, 2:275-282, Elsevier Press. (See also the Color Plate section)

accurately predict the level, function, subcellular location, and possible posttranslational modifications (PTMs) of protein products, as demonstrated in previous work in yeast (Gygi *et al.*, 1999b). There are estimated to be more than 300 different posttranslational modifications including glycosylation, nitration, oxidation, and phosphorylation, highlighting the diversity of protein products possible from a single gene (Aebersold and Goodlett, 2001). Altered protein expression or modification is likely to provide valuable clues to understanding the disease state, aiding diagnosis, and targeting drug therapy. Clinically relevant biological samples such as serum, cerebrospinal fluid, saliva, and urine are protein rich (Bergquist *et al.*, 2002) and RNA poor, rendering them inferior candidates for DNA array studies but excellent for proteomics-based studies. By combining proteomic and genomic techniques, investigators can overcome the drawbacks of genomic approaches described previously, and ultimately provide a foundation for correlating protein function or protein modification with disease states.

What Is Proteomics?

Proteomics is broadly defined as the identification and analysis of sets of proteins expressed within anything from a cell through to an organism, and encompasses the determination of protein-protein interactions, protein localization (such as within an organelle or aggregate), identification of posttranslational modifications, temporal expression patterns, and function. Characterization of the proteome is an attempt to describe *all* proteins within a cell or organism in a defined state and several proteomes relevant to neuroscience are being analyzed, including the mouse and rat cerebellum (Taoka *et al.*, 2000; Beranova-Giorgianni *et al.*, 2002) and human cortex (Langen *et al.*, 1999). In the context of neurological disease, however, the focus is rather on differential protein profiling to identify changes in protein expression or modification between normal and abnormal states, to understand the basic biology, or to develop diagnostic tools and treatments (Petricoin *et al.*, 2002; Hanash, 2003; Petricoin and Liotta, 2003). Proteomics strategies have been successfully used in the area of Alzheimer's disease to identify oxidized proteins in postmortem diseased brain (Castegna *et al.*, 2003), in Down syndrome (Yoo *et al.*, 2001; Weitzdoerfer *et al.*, 2002), and brain tumors (Zhang *et al.*, 2003) to identify disease-related proteins, and in Huntington's disease (HD) to identify both proteins associated with intracellular aggregates (Suhr et al, 2001) and differentially expressed proteins in an HD mouse model and postmortem human HD tissue (Zabel *et al.*, 2002).

Traditional approaches to studying proteins include Western blot, immunocytochemistry, immunoprecipitation,

and yeast two-hybrid analyses. These techniques generally require probing single proteins at a time, demand a prior knowledge of the protein (such as the molecular weight or amino acid sequence), or are dependent on the use of existing antibodies. Similar to their RNA counterparts of Northern blotting, RNAse protection, or *in situ* hybridization, these traditional protein approaches also preclude discovery of novel but potentially clinically relevant proteins.

Comparable to microarrays, protein profiling is used as a discovery tool because of the ability to analyze many proteins at one time without previous knowledge or bias. Profiling of complex mixtures of proteins progressed significantly in the 1970s with the application of two-dimensional gel electrophoresis (2DGE) (O'Farrell, 1975), whereby proteins are separated according to charge by isoelectric focusing in the first dimension, and according to size using standard sodium dodecye sulfate-polyacrylamide gel electrophesis (SDS-PAGE) in the second dimension. Identification of proteins was limited, however, and was based on isoelectric point and molecular weight estimations (Skene and Willard, 1981; Bellatin *et al.*, 1982; Schubert *et al.*, 1986; Castellucci *et al.*, 1988; Kennedy *et al.*, 1988; Geschwind and Hockfield, 1989). Sequencing of the protein through Edman degradation provided additional information but was an option only for higher abundance proteins, as it requires substantial amounts of pure protein. This was a considerable limitation to the downstream analysis of differentially expressed proteins (Geschwind *et al.*, 1996).

A Roadmap for Mass Spectrometry–Based Proteomics: Basic Issues of Experimental Design and Limitations

We can now separate proteins using gel- and non-gel-based systems and identify them by MS, greatly improving the power and efficiency of proteomics research (Aebersold and Goodlett, 2001; Mann *et al.*, 2001). Contemporary proteomics is composed of three components (Figure 13.3): (1) analytical separation to reduce the complexity of the protein matrix; (2) MS analysis to measure the mass of either the peptides from the digested protein or the intact protein; and (3) bioinformatic analysis for protein and/or PTM identification and ultimately for generation of protein networks and signaling pathways. There is a multitude of technical options within each component, discussed in detail later, enabling the researcher to custom design experiments best suited for the biological sample and the scientific question. Automation is possible at essentially every point in the process and generally allows for medium to high throughput analysis, although the technology can be expensive, is not currently widely available, and, in some cases, is still being developed. Although the basic

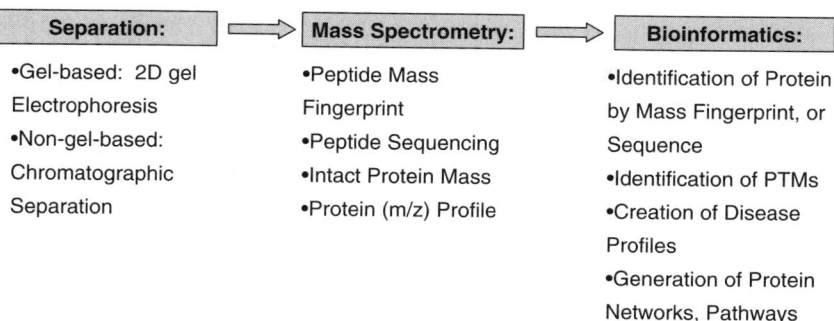

FIGURE 13.3 The three principal components of a contemporary proteomics experiment. Listed below each component are the basic experimental options available. Separation of proteins is accomplished by either gel or chromatography based methods. Mass spectrometry (MS) is used to measure the masses of whole proteins and peptides, to provide peptide sequence, and to generate m/z profiles. Bioinformatics is concerned with interpreting MS data and to ultimately understanding the information in the context of the biological system.

proteomics experiment is illustrated in Figure 13.4, it is important to note that no one experimental design will provide complete coverage of the proteome, necessitating a combination of techniques both at the level of separation and of MS analysis.

Proteomics Platforms: Separating Proteins

Two-Dimensional Gel Electrophoresis

The most common approach to the separation of complex protein mixtures is 2DGE, and within the last decade, there has been marked improvement in resolution, reliability, and MS compatibility (Gorg *et al.*, 2000; Ong and Pandey, 2001). Gel-based protein separations are best suited for protein identification by peptide mass and sequencing experiments, as intact proteins are difficult to extract from gels (Nesatyy and Ross, 2002). Proteins are separated by 2DGE, detected with a protein stain such as Coomassie Blue or the more sensitive fluorescent stains (for example, SyproRuby) and imaged. The most sensitive dyes are capable of detecting nanograms of protein per spot, so analysis of lower abundance proteins (1/10,000 to 1/100,000) require the application of hundreds of micrograms of sample protein.

Although it is possible to resolve several thousands of proteins by 2DGE, the technique is time-consuming and difficult to automate, and gel-to-gel comparisons can be challenging. Gel comparisons are more rapid and reliable with difference gel electrophoresis (DIGE) reagents that label each protein sample with a unique cyanine dye, allowing multiple samples to be analyzed within the same gel (Unlu *et al.*, 1997). Zhou *et al.* (2002) used DIGE reagents to compare total cell lysates of esophageal carcinoma cells and normal cells prepared from the same frozen tissue sample and were able to identify 165 cancer-specific proteins with a

greater than threefold difference. After image analysis, differentially expressed proteins are excised as individual gel pieces and digested in the gel with a protease such as trypsin. The peptides are extracted from the gel and the protein is identified by MS, the most common instruments used being matrix-assisted laser desorption ionization–time-of-flight (MALDI-TOF) and electrospray ionization-ion trap mass spectrometers, discussed in detail later. To identify oxidized proteins, Castegna *et al.* (2003) analyzed total cell lysates of both control and Alzheimer's disease postmortem human brains using 2DGE, matrix-assisted laser desorption ionization–time-of-flight (MALDI-TOF), and liquid chromatography-MS/MS to identify six differentially nitrated proteins, the roles of which remain to be determined (Castegna *et al.*, 2003).

Drawbacks to gel-based separations include less than maximal extractions of tryptic peptides from in-gel digestion and incomplete peptide map coverage of the protein, restricting protein identification (Gygi *et al.*, 2000). A major limitation to gel-based separations is an existing bias against lower abundance proteins and proteins in certain classes (hydrophobic, highly acidic, or basic). Membrane proteins, which are important for cell-to-cell communication, cellular transport, and drug targeting, pose a unique challenge, as they are likely present at a low copy number within the cell, in addition to being hydrophobic.

Chromatography

To analyze those classes of proteins not amenable to 2DGE, to increase throughput, and to improve reproducibility, chromatography involving a series of separation chemistries (such as ion exchange, reverse phase, size exclusion) is used (Link *et al.* 1999; Whitelegge *et al.*, 2002; Wu *et al.*, 2003). Fraction collection of the column eluant at any point in the process before MS

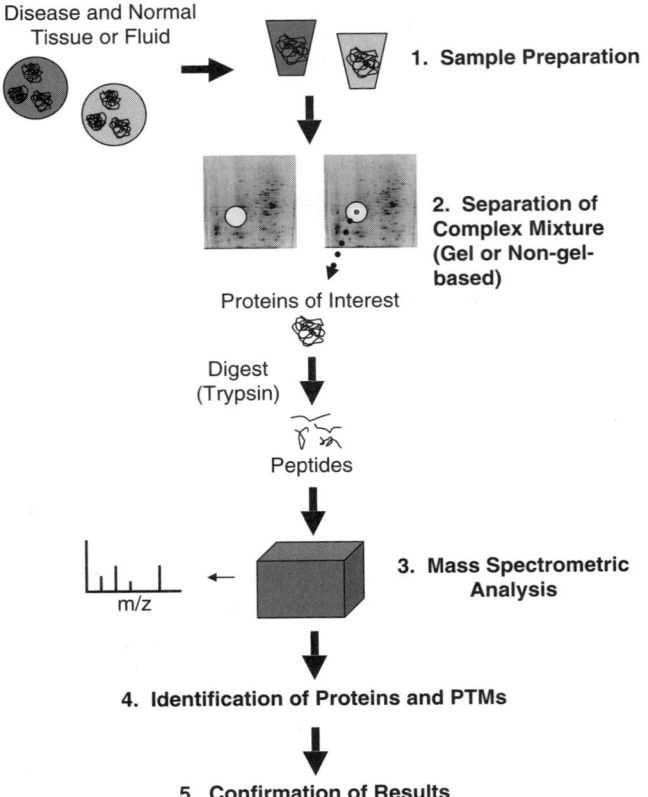

Disease and Normal Tissue or Fluid

1. **Sample Preparation**

2. **Separation of Complex Mixture (Gel or Non-gel-based)**

Proteins of Interest

Digest (Trypsin)

Peptides

3. **Mass Spectrometric Analysis**

m/z

4. **Identification of Proteins and PTMs**

5. **Confirmation of Results**

FIGURE 13.4 Flow diagram of a basic proteomics experiment. **1:** Diseased and normal tissue or fluid samples are prepared and simplified to concentrate or isolate the proteins of interest. **2:** The proteins in the samples are separated by gel electrophoresis. Differentially expressed proteins are identified and subjected to further analysis. Peptides resulting from treatment of the protein with a protease are used to identify the protein. **3:** Mass spectrometry (MS) is used to measure the peptide masses, and/or to determine peptide sequence. **4:** Using experimental peptide mass and/or sequence information, MS software programs predict protein identification using existing protein databases and identify possible posttranslational modifications (PTMs). **5:** Studies to confirm protein results include Western blot analysis, immunochemistry, and RT-PCR.

allows for enrichment of low abundance proteins or, in the case of intact proteins, an opportunity for intact mass measurement and off-line protein digestion for confident protein identification by peptide mass fingerprinting and sequence information. As with gel-based separations, however, certain proteins may not be represented in the analysis, although this is becoming less of an issue as these techniques are further refined. To exploit the improved coverage of membrane proteins with chromatographic separations, Schirmer *et al.* (2003) used multidimensional chromatography and ion trap MS of peptide digests to identify 67 previously unknown potential integral membrane proteins of the nuclear envelope, 23 of which mapped to chromosomal regions linked to various dystrophies. In another study

to investigate the effects of treatment with a novel anticancer drug, peninsularinone, Yan *et al.* (2003) studied total cell lysates from a treated and control human adenocarcinoma cell line. Cytosolic proteins were separated by ion exchange chromatography (isoelectric point) in the first dimension and by nonporous reverse phase chromatography (hydrophobicity) in the second dimension. Intact proteins were mass measured using electrospray-time-of-flight MS, and digested proteins were identified by MALDI-TOF MS, and by MALDI-QTOF if there was ambiguity in protein identification. Differential protein expression was evaluated by chromatographic subtraction software. This strategy, while also yielding an overwhelming amount of information on each protein (isoelectric point, hydrophobicity, intact mass for possible PTMs, protein identification, and relative abundance), offers an improvement on reproducibility over 2DGE. The features, strengths, and weaknesses of both gel and chromatography-based approaches to protein separation are summarized in Table 13.1.

Enhancing Quantitation with Isotope-Coded Affinity Tag Technology

To improve both the qualitative and quantitative analysis of protein expression, isotope-coded affinity tag (ICAT) technology is currently the best strategy and is used with both gel and chromatography separation techniques (Gygi *et al.*, 1999a). ICAT labels two protein samples separately with "heavy" and "light" isotope reagents and the samples are mixed. MS analysis of the modified peptide mix reveals a parallel heavy and light profile, the intensities of which vary according to protein (or peptide) abundance. In a study to identify and quantify the effects of an antitumor drug, camptothecin, on primary cultured cortical neurons, Yu and colleagues (2002) used ICAT reagents to label soluble proteins from treated and untreated cells. Using liquid chromatography-MS/MS for sequence information (and subsequent protein identification) and Fourier transform ion cyclotron (FTICR) MS for precise measurements of peptide mass and abundance, they identified 129 differentially expressed proteins. Although this particular drug may not be effective in humans, this work demonstrates the higher throughput, accuracy, and power gained by combining techniques to examine the effects of drug treatment at the level of the proteome.

The main disadvantage of ICAT is that the reagents only label cysteine-containing peptides, so if a particular protein or peptide does not happen to contain a cysteine residue (more likely for a peptide), it will not be represented. Development of additional reagents for labeling other functional groups is proceeding rapidly.

TABLE 13.1 Summary of the Two Broad Experimental Approaches to Separating Complex Mixtures of Proteins before Analysis by Mass Spectrometry

	Gel-Based Separations	Non-gel-based Separations
Example	2DGE	Multi-dimensional chromatography
Form of protein	Proteins applied as intact but are extracted from gel as peptides	Intact or digested proteins
Type of information	Protein and PTM identification; intact protein mass difficult to determine	Possibility for both intact protein mass and protein identification
Throughput	Low to medium, difficult to automate	High, amenable to automation
Qualitative protein expression	Based on gel image comparisons, improved with DIGE reagents	Based on chromatographic subtraction
Quantitative protein expression	Semi-quantitative, improved with DIGE and ICAT reagents	Yes, improved with ICAT reagents
Advantages	Well-established technique, ease of use, visual results	May offer better representation of low abundance, hydrophobic, highly basic and acidic proteins
Disadvantages	Low abundance, hydrophobic, highly basic and acidic proteins may be underrepresented, reproducibility issues	Certain proteins may be underrepresented, incomplete resolution of proteins/peptides
Types of samples	Tissue and cell lysates, biological fluids	Tissue and cell lysates, biological fluids

2DGE, two-dimensional gel electrophoresis; *PTM*, posttranslational modification; *DIGE*, difference gel electrophoresis; *ICAT*, isotope-coded affinity tag.

Proteomics Platforms: Protein Identification

Although the first instrument was built in the early 1900s by J. J. Thomson, MS was not effective for protein analysis until the development of soft ionization techniques in the 1980s. MS has now emerged as the method of choice for identifying and sequencing proteins. The fundamental advantage to coupling MS to protein analysis is the ability to obtain accurate mass measurements of biological molecules. This characteristic enables researchers to identify proteins from their peptide masses and sequence information to analyze intact protein mass (Gomez *et al.*, 2002), as well as to determine the nature and location of posttranslational modifications (Whitelegge *et al.*, 1999, 2000; McLachlin and Chait, 2001; Mann and Jensen, 2003). In addition, MS generally requires low picomoles (10^{-12} moles) of material and does not require 100% protein purity. Acquisition of MS data is highly automated, and the structural as well as the functional assessment of these data is facilitated by available software and databases, such as Sonar and Sequest, and the National Center for Biotechnology Information (NCBI).

Identification of Proteins

Protein identification based on intact mass alone is difficult, as several proteins may possess a similar mass, and theoretical mass calculated from gene sequence rarely matches the measured mass of the expressed protein. Identification of proteins is best approached by digesting the protein before MS analysis, using a site-specific cleavage reagent such as the protease, trypsin, and measuring the mass of the resulting peptides. The protein is identified by comparing the experimental "mass fingerprint," containing no fewer than four to five peptide masses, to theoretical "mass fingerprints" predicted from *in silico* digests of proteins within a database. The most confident approach to protein identification, however, is to obtain partial amino acid sequence of these peptides by fragmenting the peptide within the mass spectrometer (termed collision-induced dissociation) and mass measuring the products (a tandem MS or MS/MS experiment). The peptides break predictably at the peptide bond from both the N and C termini, resulting in the creation of a ladder of N-terminus (or b type) and C-terminus (or y type) product ion (m/z) peaks. Mass analysis of the resulting fragments provides sequence information, as the difference between each m/z peak is a reflection of the mass of the cleaved amino acid, each amino acid (with the exception of the isomers, leucine and isoleucine) having a unique mass. Proteins are confidently identified with as few as one or two peptides, again, by comparison with *in silico* digests and sequence of known proteins. Previously uncharacterized proteins for which there is no match in existing databases are analyzed in a similar manner to obtain peptide sequence but require increased accuracy and fragmentation (Shevchenko *et al.*, 1996). Identification of PTMs present within the peptide is made by incorporating the possibility into the MS analysis software. A phosphorylation event, for instance, is represented by an 80 dalton shift in the mass. Obviously, if a PTM is present in a peptide that was not analyzed (as a result of poor recovery from the digested gel, for example), it will not be identified.

Determining Intact Protein Mass

Intact mass measurements of proteins provide insight into protein heterogeneity resulting from PTMs (including those not identified in digested proteins), alternative splicing events and even DNA sequencing errors. The mass difference between the experimental and the theoretical mass is, again, a reflection of the modification. The identification and degree of PTMs, however, still present a challenge to the field of MS-based proteomics because of their diversity and stability, although PTM-specific methods are being developed (Mann and Jensen, 2003).

Which Mass Spectrometer?

The choice of mass spectrometer is frequently determined by availability, as researchers must normally rely on core proteomics facilities. As mentioned previously, the most confident protein identifications are based on peptide sequence information, but not all mass spectrometers are capable of providing this, as it requires the ability to mass measure and collide peptides, and mass measure the resulting fragments. Protein identification based on peptide mass alone, however, is a reasonable alternative.

Mass spectrometers share three fundamental components: (1) an ion source that generates gas phase ions; (2) a mass analyzer where ions are separated according to mass (m) and charge (z); and (3) an ion detector that records ion arrival and abundance as a mass-to-charge ratio (m/z) peak, from which mass information is obtained. Mass spectrometers differ in the type of ion source, as well as in the design of the mass analyzer, and each has its particular strengths and weaknesses, accuracy, and resolution. Thorough reviews describing mass spectrometers are available (McLuckey and Wells, 2001; Vestal, 2001; Aebersold and Mann, 2003). The amount of a specific protein required for MS analysis varies widely, depending on the specific molecule and the mass spectrometer, but is of the order of low picomoles (10^{-12} moles) of protein.

The two most common ion source/mass analyzer combinations are MALDI-TOF and electrospray ionization/ion trap mass spectrometers. Table 13.2 offers a brief comparison of these two instruments. MALDI-TOF MS requires spotting the proteins or peptides with an ionization-enhancing matrix onto a plate and ionization of the sample with a laser. Although MALDI-TOF is best suited for high throughput mass measurement of intact proteins or peptides, limited peptide sequence can also be obtained with a post-source decay experiment, although the fragmentation pattern is generally more difficult to interpret.

Electrospray ionization/ion trap MS requires introduction of the protein or peptide into the mass spectrometer as a fine liquid spray, often from a liquid chromatography system, and as such, requires more analysis time than MALDI-TOF. Ionization of the sample into multiply charged species occurs as a result of an applied voltage differential. Both mass and peptide sequence are obtained in a typical tandem MS experiment using this instrument, often described as liquid chromatography-MS/MS.

Other instruments exist, such as quadrapole and fourier transform ion cyclotron resonance (FT-ICR) mass spectrometers, as do hybrid instruments, where multiple mass analyzers are placed in series. FT-ICR MS currently offers the highest accuracy ($\pm$ 0.02 Da for a 20 kDa protein, for example) and resolution, but is costly and not widely available.

SELDI-TOF MS and the Quest for Disease Biomarkers

For the rapid generation of protein profiles that can be quickly compared for biomarker identification, surface-enhanced laser desorption ionization-time-of-flight (SELDI-TOF) MS combines separation of complex mixture of proteins and MS analysis all on one chip (Issaq *et al.*, 2003). As little as 1 µl of protein mixtures, such as saliva or serum, are applied directly to a plate that has a chemically (such as cation exchange) or biochemically (such as an antibody) modified surface. A subset of proteins interacts with these surfaces, allowing nonbinding proteins to be washed away. The plate is placed into the mass spectrometer and the bound proteins are mass measured. In a manner analogous to microarrays, profiles from disease and normal samples are compared to identify disease-specific species or a disease fingerprint that is diagnostically useful (see Figure 13.2).

A serious disadvantage to SELDI-TOF MS is the difficulty in identifying the altered proteins, caused mainly by the lower accuracy and resolution of the mass spectrometer and inability to obtain MS/MS peptide sequence data. Instruments are being developed to circumvent this issue. A comparison of SELDI-TOF to the two other MS instruments is outlined in Table 13.2.

SELDI-TOF MS was used to screen serum for detection of ovarian cancer by analyzing low molecular weight serum protein patterns in both diseased and healthy patients (Petricoin *et al.*, 2002). The identities of these differentially expressed proteins is unknown, however,

TABLE 13.2   Summary of Three Common Mass Spectrometry Instruments Used in the Field of Proteomics

	MALDI-TOF MS	Electrospray Ionization-Ion Trap MS	SELDI-TOF MS
Sample preparation required	Yes, although direct tissue analysis possible	Yes	Separation occurs directly on mass spectrometry (MS) plate
Measurement of intact protein or peptide mass	Both, especially valuable to examine heterogeneity of protein	Both, but better suited for peptides and small proteins	Generation of m/z profiles
Peptide sequencing	Yes, but more difficult to interpret	Yes	No
Advantages	Rapid sample analysis (sec/sample)	Can be coupled directly to chromatographic systems (LC)	Rapid sample analysis (sec/sample); m/z fingerprints offer visual results; separation and MS on one chip
Disadvantages	Cannot be coupled directly to chromatographic systems such as LC	Analysis time can be significant if using LC (tens of min per sample)	Difficult to identify differentially expressed proteins

leaving many unanswered questions regarding the disease biology, but it does demonstrate the potential power of protein profiling in disease detection. Zhang *et al.* (2002) used SELDI-TOF MS to analyze cellular secretions from CD8 T lymphocytes of immunologically stable HIV-1-infected patients and observed a unique cluster of peaks in the protein fingerprints. They were fortunate to be able to identify these peaks as α-Defensin 1, 2, and 3, based on intact mass alone, which was verified through the use of antibodies. More sophisticated strategies to the discovery of biomarkers are being developed and include the advanced chromatographic separation, as α-Defensin 1, 2, and 3, labeling technologies, and MS discussed previously.

Bioinformatics

Programs designed to interpret MS and MS/MS spectra and provide probable protein matches include Sequest (http://fields.scripps.edu/sequest), Sonar (www.genomicslutions.com), and Mascot (www.matrix-science.com), and use existing databases such as NCBI (www.ncbi.nlm.nih.gov) and SwissProt (www.ebi.ac.uk). Search strategies typically involve peptide mass fingerprint data, peptide sequence data, and/or raw MS/MS data (Yates, 1998; Perkins *et al.*, 1999). Peptide mass fingerprint relies strictly on the match of the mass of the fragments produced with those theoretically generated from an *in silico* digest of known proteins. This strategy relies heavily on mass accuracy, and protein identification is often confounded by PTMs, mutations, and sequencing errors, which may alter the experimental peptide masses. Modifying search conditions to allow for mass changes reflective of PTMs permits their identification in the peptide. Protein identification by amino acid sequence information, using the National Center for

Biotechnology Information's (NCBI's) Basic Local Alignment Search Tool (BLAST) search algorithm, for example, is straightforward, although its success in unambiguously identifying proteins is dependent on the amount of sequence queried. Raw MS/MS data offers searching based on peptide mass, as well as peptide sequence, and yields confident protein matches. An important consideration for any of these strategies is that protein identification is possible only if that protein, or one that is highly homologous, exists in the database. *De novo* sequencing for entirely uncharacterized and unpredicted proteins is generally more advanced and technically challenging because, unlike Edman degradation, the peptide will fragment simultaneously from both the C and N terminus, the directionality of which is initially unknown, complicating the interpretation (Shevchenko *et al.*, 1996).

Non-MS-Based Protein Profiling

Although the advantages of MS-based proteomics are clear, it is worth mentioning other important advances in protein profiling independent of MS. Protein array technology that is analogous to microarray is currently being developed and is either analytical or functional in nature (Phizicky *et al.*, 2003; Zhu and Snyder, 2003). Analytical chips use affinity compounds such as antibodies or antigens, whereas functional chips are composed of pure proteins covalently bound to the array surface. In a manner similar to microarrays, the protein arrays are exposed to a fluorescent-labeled protein sample, noninteracting proteins are removed, and interactions are detected. Anderson *et al.* (2003) used antibody arrays to compare proteins differentially expressed in primary muscle cell cultures from a patient with spinal muscular atrophy compared with control specimens and observed nine down regulated proteins involved in transcription, six of which were confirmed by oligonucleotide arrays.

Although these arrays offer a relatively quick analysis, they are generally dependent on the use of existing and known proteins and antibodies. Some of the challenges facing the protein chip field include optimizing the attachment chemistry of the protein such that neither function nor binding ability is compromised.

Importance of Posttranslational Modifications: Phosphorylation as an Example

Phosphorylation is an important regulator of protein function and activity and figures prominently in intracellular signaling cascades. Hyperphosphorylation of proteins is believed to play a role in the pathogenesis of many diseases, including cancer (Sebolt-Leopold, 2000) and neurodegenerative disorders such as Alzheimer's disease and frontotemporal dementia (Brion et al., 2001; Jackson et al., 2002; Agarwal-Mawal et al., 2003; Pei et al., 2003). MS-based approaches include 2DGE and ^{32}P or phosphospecific protein stains such as Pro-Q Diamond for global phosphoprotein analysis (McLachlin and Chait, 2001). For low abundance phosphorylation events, phosphorylated proteins or peptides are enriched by affinity purification, followed by identification by MS (Goshe et al., 2001; Oda et al., 2001; Zhou et al., 2001). Newly developed non-MS-based array technologies, although dependent on preexisting knowledge of kinases and substrates, are interesting and perhaps higher-throughput alternatives for more hypothesis-based testing. Zhu et al. (2000) arrayed 119 yeast protein kinases onto a slide and characterized their activities for 17 substrates by ^{33}Pγ-ATP and phosphoimaging. They discovered a number of unpredicted kinase activities, and the existence of many more tyrosine-phosphorylating kinases than predicted.

Downstream Confirmation of Results

Similar to microarray studies, independent confirmation of results is a crucial step in any proteomics experiment, and is often accomplished through conventional techniques, such as Western blot for semiquantitative and qualitative protein expression analysis, and immunocytochemistry for localization of the protein. This is augmented by in situ hybridization for localization of the mRNA and RT-PCR to correlate protein and transcript expression. These protein techniques represent the bottleneck of most studies, as they are generally time-consuming and rely on antibodies for further analysis.

Limitations, Technical and Methodological Issues

Human Disease

Biological fluids such as cerebrospinal fluid (Yuan et al., 2002; Ramstrom et al., 2003), urine (Pang et al., 2002), serum (Anderson and Anderson, 2002), and saliva (Yao et al., 2003), although not optimal for microarray studies owing to low mRNA levels, are excellent candidates for research and diagnosis in biomedical proteomics, as alterations in the physiology of cells are likely reflected in the circulating or excreted proteins. Biopsy and postmortem tissue, frozen or ethanol-fixed, is also a potential sample source (Ahram et al., 2003; Hynd et al., 2003), but is affected by variability of protein stability and preparation-related protein modifications, as well as those variables outlined in microarray studies, namely tissue preservation methods and prefixation times. Protein quality is as important in proteomics as mRNA quality is in array studies. Proteins can be easily altered (e.g., degraded or oxidized) through sample handling and analysis, often with a deleterious effect on reproducibility of results. Recently, cells and intact tissue have been analyzed directly by MALDI MS, circumventing reliance on any isolation and purification of the proteins (Schwartz et al., 2003). Stoeckli et al. (2001) have used this technique to examine, in situ, the spatial distribution of differentially expressed proteins in a human glioblastoma xenograft, demonstrating the ability to analyze protein differences without extensive extractions and separations, effectively decreasing analysis time but, more importantly, minimizing protein degradation or introduction of artifacts.

Protein Abundance and Issues of Heterogeneity

A successful proteomics strategy begins with simplification of the protein mixture, a necessary first step when one considers the thousands of proteins present, the abundances of which vary by several orders of magnitude (Corthals et al., 2000). Simplification is accomplished by subcellular fractionation, immunoaffinity purification, or isolation of protein-protein complexes. Tissue heterogeneity continues to be an issue, as it is with cDNA array studies, and small but significant protein expression patterns can be completely obscured by inappropriate choice of sample. There is also the issue of heterogeneity of subcellular localization or protein topography. A protein present at similar levels within two different cell states may be localized specifically to the membrane surface in one state compared to another, suggestive of altered protein function. Microarray experiments cannot yield this level of information, as this technique only surveys mRNA abundance.

The Transition to the Clinic

Despite its potential, MS-based proteomics has not yet fully made its way into the diagnostic clinic and is likely further from this application than arrays because clinic-based applications of proteomics for disease diagnosis requires quick, cost-effective, and user friendly methodology whose foundation is analysis of patient biomaterial samples for the generation of disease patterns. Most current approaches require specialized knowledge for use and interpretation, and are costly. Although biomedical proteomics has the potential to have a tremendous impact on disease diagnosis and treatment, it is an emerging clinical technology whose discoveries require further validation. Biomedical research has generally embraced MS-based proteomics, but state-of-the-art proteomics facilities are certainly not ubiquitous, even in research institutions. Basic research applications often demand greater accuracy and resolution, requiring more sophisticated instruments, and knowledgeable personnel to run and maintain them and to correctly analyze the data. In many respects, SELDI-TOF systems that offer "proteomics on a chip" (such as Ciphergen) are in the best position to enter the clinic by offering a low technology, but high throughput platform for disease biomarker profile discovery. The need continues, however, for the more advanced techniques of protein separation and MS analysis to identify these disease-related proteins, to understand their function, and ultimately to exploit them for uses beyond diagnosis to protein-based therapy. The field of proteomics has benefited greatly from the growing pains experienced by the field of genomics and shows great promise for revolutionizing the way in which we think about, diagnose, and treat disease.

MERGING GENOMICS AND PROTEOMICS APPROACHES: NEURAL STEM CELL BIOLOGY

Work in our laboratory, and that of our collaborators, provides one example of the complementary nature of genomics and proteomics approaches. DNA array studies of neural stem and progenitor cells in the Kornblum and Geschwind laboratories (Geschwind et al., 2001; Karsten et al., 2003) identified a number of novel genes likely to have a role in the regulation of neural stem cell proliferation and pluripotentiality. Surprisingly, few membrane proteins were identified, despite being likely regulators of extrinsic signals involved in central nervous system stem cell fate specification, proliferation, and differentiation. To address this, 2DGE is now being used to separate a membrane fraction obtained by differential centrifugation of neural stem and progenitor cells to identify differentially expressed membrane and membrane-associated proteins.

The differentially expressed proteins are cut from the gel, digested with trypsin and the peptides mass measured and sequenced by liquid chromatography-MS/MS. Using this standard proteomics strategy, more than 50 differentially expressed proteins were identified, a small number of which were also identified by microarray studies (Shoemaker et al., 2003). Most importantly, several membrane proteins were identified, including a novel 7 transmembrane protein. To expand the coverage of differentially expressed membrane proteins, alternative separation strategies such as multidimensional chromatography are being explored. Further integration of genomics and proteomics data will provide more insight into the intrinsic and extrinsic mechanisms underlying neural stem cell biology and lead to markers that will enable more specific studies in the area of neural repair (Kornblum and Geschwind, 2001).

CHALLENGES AND CONCLUSIONS

A daunting challenge facing the field of genomics and proteomics is the merging of microarray and proteomics data into meaningful datasets in order to enable the correlation of changes in mRNA with changes in protein profiles (Ideker et al., 2001). Microarray data are more amenable to sharing and several public databases exist (Geschwind, 2001b; Becker et al., 2002), but the inconsistency of cross-platform comparisons, or cross-laboratory comparisons even within one platform, pose a barrier (Mirnics et al., 2001). Data sharing presents certain difficulties, as protein and mRNA expression is dynamic, depending on, for example, time of day of sampling, disease stage, and access of the tissue or cells to nutrients and other growth variables such as temperature. Additional variability is introduced during sample preparation and analysis. Standardization of methodologies and result formats may go a long way to improving the technical issues of data sharing. Establishing a standard ontology for microarray experiments was a first critical step in achieving that goal (Brazma et al., 2001; Geschwind, 2001b). Although public 2DGE image repositories exist (for example www.expasy.ch), there is a need to establish comprehensive proteomics databases that contain not only the experimental data but also the information necessary to effectively evaluate and compare data sets, an important consideration because proteins, and the techniques used to study them, are so dynamic. To meet this need, Taylor et al. (2003) recently proposed a model of standardized proteomics data based on MIAME, again, borrowing from the field of genomics. Organizations such as the Human Proteome Organization (HUPO, www.hupo.org) may also play a role in advancing the field of proteomics by encouraging the sharing of data, technologies and bioformatics tools, as well as coordinating proteome initiatives.

Once these issues are addressed, one is left with the challenge of interpretation. As mentioned previously, there are likely to be biologically real variations between transcript and protein levels; understanding the physiological relevance of these differences is a slow process. For example, human glioblastomas have been studied using both cDNA array and proteomics techniques (Sallinen *et al.*, 2000; Zhang *et al.*, 2003). Zhang *et al.* (2003) used an epidermal growth factor receptor (EGFR)-mutant glioblastoma model cell line to identify 52 differentially expressed predicted and novel proteins by 2DGE, MALDI-TOF MS, and liquid chromatography-MS/MS. They confirmed five of these proteins by either Western blot, enzyme-linked immunosorbent assay, or RT-PCR, and found four that were also highly expressed in human glioblastoma tissue. The levels of three of the four proteins correlated with disease grade, suggesting these candidates might be useful to understand the biology, as well as to serve as disease markers.

In contrast, microarray studies of glioblastoma and other brain tumors have led to more of a global picture of the pathophysiological alterations, because they have identified large-scale changes in gene expression, leading to improved classification and prediction in many cases (Pomeroy *et al.*, 2002; Mischel *et al.*, 2003). One such study was able to predict the chromosomal origin of an amplification event because of the up regulation of contiguous genes as a separate event from EGFR amplification. In this study, investigation at the protein level in patient tumor specimens identified the EGFR expression changes, which could then be correlated with a particular gene expression profile in those tumors (Mischel *et al.*, 2003). This illustrates an optimal use of technology to integrate individual protein expression data with global gene expression data. Based on this and other work, it is likely that microarray profiling will soon be used to complement analysis of protein expression by immunohistochemistry and other methods in the clinic for diagnosis and prognostication in human brain cancers (Geschwind, 2003).

Systems biology will play an important role in developing and refining the tools to build, use, and interpret networks of transcript and protein data (Davidov *et al.*, 2003). Ultimately, we must rely on both genomics and proteomics approaches to understand the interplay of expression, translation, and posttranslational control within the context of disease biology. This level of understanding is only one, albeit, critical step in developing novel therapeutics.

ACKNOWLEDGMENTS

We are grateful to our colleagues in the Kornblum, Pasarow Mass Spectrometry and Geschwind laboratories, specifically Harley Kornblum, for his support and collaboration. We acknowledge funding from NIMH, NINDS, and NIA.

References

Adams MD *et al.* (2000). The genome sequence of *Drosophila melanogaster*, *Science* 287(5461):2185-2195.

Aebersold R, Goodlett DR. (2001). Mass spectrometry in proteomics, *Chem Rev* 101(2):269-295.

Aebersold R, Mann M. (2003). Mass spectrometry-based proteomics, *Nature* 422(6928):198-207.

Agarwal-Mawal A *et al.* (2003). 14-3-3 connects glycogen synthase kinase-3 beta to tau within a brain microtubule-associated tau phosphorylation complex, *J Biol Chem* 278(15):12722-12728.

Ahram M *et al.* (2003). Evaluation of ethanol-fixed, paraffin-embedded tissues for proteomic applications, *Proteomics* 3(4): 413-421.

Anderson K *et al.* (2003). Protein expression changes in spinal muscular atrophy revealed with a novel antibody array technology, *Brain* 126(Pt 9): 2052-2064.

Anderson NL, Anderson NG. (2002). The human plasma proteome: history, character, and diagnostic prospects, *Mol Cell Proteomics* 1(11):845-867.

Barczak A *et al.* (2003). Spotted long oligonucleotide arrays for human gene expression analysis, *Genome Res* 13(7):1775-1785.

Barrett T *et al.* (2001). Assembly and use of a broadly applicable neural cDNA microarray, *Restor Neurol Neurosci* 18(2-3):127-135.

Becker KG *et al.* (2002). cDNA array strategies and alternative approaches in neuroscience: focus on radioactive probes. In Geschwind DH, Gregg J, editors: *Microarrays for the neurosciences*, Boston: MIT Press.

Bellatin J *et al.* (1982). Changes in the relative proportion of transformation-sensitive polypeptides in giant HeLa cells produced by irradiation with lethal doses of x-rays, *Proc Natl Acad Sci U S A* 79(14): 4367-4370.

Beranova-Giorgianni S *et al.* (2002). Preliminary analysis of the mouse cerebellum proteome, *Brain Res Mol Brain Res* 98(1-2): 135-140.

Bergquist J *et al.* (2002). Peptide mapping of proteins in human body fluids using electrospray ionization Fourier transform ion cyclotron resonance mass spectrometry, *Mass Spectrom Rev* 21(1):2-15.

Brazma A *et al.* (2001). Minimum information about a microarray experiment (MIAME)-toward standards for microarray data, *Nat Genet* 29(4):365-371.

Brenner S *et al.* (2000). Gene expression analysis by massively parallel signature sequencing (MPSS) on microbead arrays, *Nat Biotechnol* 18(6):630-634.

Brion JP *et al.* (2001). Neurofibrillary tangles and tau phosphorylation, *Biochem Soc Symp* 67: 81-88.

Brown V *et al.* (2001). Microarray identification of FMRP-associated brain mRNAs and altered mRNA translational profiles in fragile X syndrome, *Cell* 107(4):477-487.

Brown VM *et al.* (2002a). High-throughput imaging of brain gene expression, *Genome Res* 12(2):244-254.

Brown VM *et al.* (2002b). Gene expression tomography, *Physiol Genom* 8(2):159-167.

Castegna A *et al.* (2003). Proteomic identification of nitrated proteins in Alzheimer's disease brain, *J Neurochem* 85(6):1394-1401.

Castellucci V F et al. (1988). A quantitative analysis of 2-D gels identifies proteins in which labeling is increased following long-term sensitization in Aplysia, Neuron 1(4):321-328.

Chiang LW et al. (2001). An orchestrated gene expression component of neuronal programmed cell death revealed by cDNA array analysis, Proc Natl Acad Sci U S A 98(5): 2814-2819.

Clark LN et al. (1998). Pathogenic implications of mutations in the tau gene in pallido-ponto-nigral degeneration and related neurodegenerative disorders linked to chromosome 17, Proc Natl Acad Sci U S A 95(22):13103-13107.

Colantuoni C et al. (2001). Gene expression profiling in postmortem Rett syndrome brain: differential gene expression and patient classification, Neurobiol Dis 8(5):847-865.

Corthals GL et al. (2000). The dynamic range of protein expression: a challenge for proteomic research, Electrophoresis 21(6):1104-1115.

Crino PB et al. (1996). Embryonic neuronal markers in tuberous sclerosis: single-cell molecular pathology, Proc Natl Acad Sci U S A 93(24): 14152-14157.

D'Agata V et al. (2002). Gene expression profiles in a transgenic animal model of fragile X syndrome, Neurobiol Dis 10(3):211-218.

Datson NA et al. (2001). Expression profile of 30,000 genes in rat hippocampus using SAGE, Hippocampus 11(4):430-444.

Davidov E et al. (2003). Advancing drug discovery through systems biology, Drug Discov Today 8(4):175-183.

Dougherty JD, Geschwind DH. (2002). Subtraction-coupled custom microarray analysis for gene discovery and gene expression studies in the CNS, Chem Senses 27(3):293-298.

Eberwine J et al. (1992). Analysis of gene expression in single live neurons, Proc Natl Acad Sci U S A 89(7):3010-3014.

Evans SJ et al. (2002). Evaluation of affymetrix gene chip sensitivity in rat hippocampal tissue using SAGE analysis. Serial Analysis of Gene Expression, Eur J Neurosci 16(3):409-413.

Geschwind DH. (2000). Mice, microarrays, and the genetic diversity of the brain, Proc Natl Acad Sci U S A 97(20):10676-10678.

Geschwind DH. (2001a). DNA microarrays: the new frontier in gene discovery and gene expression analysis, Washington, DC: Society for Neuroscience.

Geschwind DH. (2001b). Sharing gene expression data: an array of options, Natl Rev Neurosci 2(6):435-438.

Geschwind DH. (2002). Beyond the gene list: using bioinformatics to make sense out of array data. In Williams RW, Goldowitz D, editors: Bioinformatics 2002: a neuroscientist's guide to tools and techniques for mining and refining massive data sets, Washington, DC: Society for Neuroscience.

Geschwind DH. (2003). DNA microarrays: translation of the genome from laboratory to clinic, Lancet Neurol 2(5):275-282.

Geschwind DH, Gregg J, editors. (2002). Microarrays for the neurosciences. Cellular and molecular neuroscience, Cambridge: MIT Press.

Geschwind DH, Hockfield S. (1989). Identification of proteins that are developmentally regulated during early cerebral corticogenesis in the rat. J Neurosci 9(12):4303-4317.

Geschwind DH, Miller BL. (2001). Molecular approaches to cerebral laterality: development and neurodegeneration, Am J Med Genet 101(4):370-381.

Geschwind DH et al. (2001). A genetic analysis of neural progenitor differentiation, Neuron 29(2):325-339.

Geschwind DH et al. (1996). Changes in protein expression during neural development analyzed by two-dimensional gel electrophoresis, Electrophoresis 17(11):1677-1682.

Ginsberg SD, Che S. (2002). RNA amplification in brain tissues, Neurochem Res 27(10):981-992.

Ginsberg SD et al. (2000). Expression profile of transcripts in Alzheimer's disease tangle-bearing CA1 neurons, Ann Neurol 48(1):77-87.

Gomez SM et al. (2002). The chloroplast grana proteome defined by intact mass measurements from liquid chromatography mass spectrometry, Mol Cell Proteomics 1(1):46-59.

Gorg A et al. (2000). The current state of two-dimensional electrophoresis with immobilized pH gradients, Electrophoresis 21(6):1037-1053.

Goshe MB et al. (2001). Phosphoprotein isotope-coded affinity tag approach for isolating and quantitating phosphopeptides in proteome-wide analyses, Anal Chem 73(11):2578-2586.

Gygi SP et al. (2000). Evaluation of two-dimensional gel electrophoresis-based proteome analysis technology, Proc Natl Acad Sci U S A 97(17):9390-9395.

Gygi SP et al. (1999a). Quantitative analysis of complex protein mixtures using isotope-coded affinity tags, Nat Biotechnol 17(10): 994-999.

Gygi SP et al. (1999b). Correlation between protein and mRNA abundance in yeast, Mol Cell Biol 19(3):1720-1730.

Hakak Y et al. (2001). Genome-wide expression analysis reveals dysregulation of myelination-related genes in chronic schizophrenia, Proc Natl Acad Sci U S A 98(8):4746-4751.

Hanash S. (2003). Disease proteomics, Nature 422(6928):226-232.

Hemby SE et al. (2002). Gene expression profile for schizophrenia: discrete neuron transcription patterns in the entorhinal cortex, Arch Gen Psychiatry 59(7):631-640.

Hong M et al. (1998). Mutation-specific functional impairments in distinct tau isoforms of hereditary FTDP-17, Science 282(5395): 1914-1917.

Hughes TR et al. (2001). Expression profiling using microarrays fabricated by an ink-jet oligonucleotide synthesizer, Nat Biotechnol 19(4):342-347.

Hynd MR et al. (2003). Biochemical and molecular studies using human autopsy brain tissue, J Neurochem 85(3):543-562.

Ideker T et al. (2001). Integrated genomic and proteomic analyses of a systematically perturbed metabolic network, Science 292(5518):929-934.

Irizarry RA et al. (2003). Exploration, normalization, and summaries of high density oligonucleotide array probe level data, Biostatistics 4(2):249-264.

Issaq HJ et al. (2003). SELDI-TOF MS for diagnostic proteomics, Anal Chem 75(7):148A-155A.

Iwashita T et al. (2003). Hirschsprung disease is linked to defects in neural crest stem cell function, Science 301(5635):972-976.

Iyer VR et al. (1999). The transcriptional program in the response of human fibroblasts to serum, Science 283(5398):83-87.

Jackson GR et al. (2002). Human wild-type tau interacts with wingless pathway components and produces neurofibrillary pathology in Drosophila, Neuron 34(4):509-519.

Kamme F et al. (2003). Single-cell microarray analysis in hippocampus CA1: demonstration and validation of cellular heterogeneity, J Neurosci 23(9):3607-3615.

Karsten SL, Geschwind DH. (2002). Gene expression analysis using cDNA microarrays, in Crawley, JN et al., editors, Current Protocols in Neuroscience, Suppl. 20, Section 4:Unit 4.28.

Karsten SL et al. (2003). Global analysis of gene expression in neural progenitors reveals specific cell-cycle, signaling, and metabolic networks, Dev Biol 261(1):165-182.

Karsten SL et al. (2002). An evaluation of tyramide signal amplification and archived fixed and frozen tissue in microarray gene expression analysis, Nucleic Acids Res 30(2):E4.

Kennedy TE et al. (1988). Sequencing proteins from acrylamide gels, Nature 336(6198):499-500.

Kohane IS et al. (2002). Microarrays for an integrative genomics, Cambridge: MIT Press.

Kornblum HI, Geschwind DH. (2001). Molecular markers in CNS stem cell research: hitting a moving target, Nat Rev Neurosci 2(11):843-846.

Lander ES *et al.* (2001). Initial sequencing and analysis of the human genome, *Nature* 409(6822):860-921.

Langen H *et al.* (1999). Two-dimensional map of human brain proteins, *Electrophoresis* 20(4-5):907-916.

Lewohl JM *et al.* (2000). Gene expression in human alcoholism: microarray analysis of frontal cortex, *Alcohol Clin Exp Res* 24(12):1873-1882.

Link AJ *et al.* (1999). Direct analysis of protein complexes using mass spectrometry, *Nat Biotechnol* 17(7):676-682.

Lipshutz RJ *et al.* (1999). High density synthetic oligonucleotide arrays, *Nat Genet* 21(1 Suppl):20-24.

Livesey FJ *et al.* (2000). Microarray analysis of the transcriptional network controlled by the photoreceptor homeobox gene Crx, *Curr Biol* 10(6):301-310.

Lock C *et al.* (2002). Gene-microarray analysis of multiple sclerosis lesions yields new targets validated in autoimmune encephalomyelitis, *Nat Med* 8(5):500-508.

Lockhart DJ, Barlow C. (2001). Expressing what's on your mind: DNA arrays and the brain, *Nat Rev Neurosci* 2(1):63-68.

Lockhart DJ, Winzeler EA. (2000). Genomics, gene expression and DNA arrays, *Nature* 405(6788):827-836.

Long AD *et al.* (2001). Improved statistical inference from DNA microarray data using analysis of variance and a Bayesian statistical framework. Analysis of global gene expression in Escherichia coli K12, *J Biol Chem* 276(23):19937-19944.

Luo L *et al.* (1999). Gene expression profiles of laser-captured adjacent neuronal subtypes, *Nat Med* 5(1):117-122.

Luo Z, Geschwind DH. (2001). Microarray applications in neuroscience, *Neurobiol Dis* 8(2):183-193.

Mann M *et al.* (2001). Analysis of proteins and proteomes by mass spectrometry, *Annu Rev Biochem* 70:437-473.

Mann M, Jensen ON. (2003). Proteomic analysis of post-translational modifications, *Nat Biotechnol* 21(3):255-261.

Mayfield RD *et al.* (2002). Patterns of gene expression are altered in the frontal and motor cortices of human alcoholics, *J Neurochem* 81(4):802-813.

McLachlin DT, Chait BT. (2001). Analysis of phosphorylated proteins and peptides by mass spectrometry, *Curr Opin Chem Biol* 5(5):591-602.

McLuckey SA, Wells JM. (2001). Mass analysis at the advent of the 21st century, *Chem Rev* 101(2):571-606.

Middleton FA *et al.* (2002). Gene expression profiling reveals alterations of specific metabolic pathways in schizophrenia, *J Neurosci* 22(7):2718-2729.

Mirnics K *et al.* (2001). Analysis of complex brain disorders with gene expression microarrays: schizophrenia as a disease of the synapse, *Trends Neurosci* 24(8):479-486.

Mirnics K *et al.* (2000). Molecular characterization of schizophrenia viewed by microarray analysis of gene expression in prefrontal cortex, *Neuron* 28(1):53-67.

Mischel PS *et al.* (2003). Identification of molecular subtypes of glioblastoma by gene expression profiling, *Oncogene* 22(15):2361-2373.

Mody M *et al.* (2001). Genome-wide gene expression profiles of the developing mouse hippocampus, *Proc Natl Acad Sci U S A* 98(15):8862-8867.

Mycko MP *et al.* (2003). cDNA microarray analysis in multiple sclerosis lesions: detection of genes associated with disease activity, *Brain* 126(Pt 5):1048-1057.

Nadeau JH, Frankel WN. (2000). The roads from phenotypic variation to gene discovery: mutagenesis versus QTLs, *Nat Genet* 25(4):381-384.

Nadon R, Shoemaker J. (2002). Statistical issues with microarrays: processing and analysis, *Trends Genet* 18(5):265-271.

Nesatyy VJ, Ross NW. (2002). Recovery of intact proteins from silver stained gels, *Analyst* 127(9):1180-1187.

Oda Y *et al.* (2001). Enrichment analysis of phosphorylated proteins as a tool for probing the phosphoproteome, *Nat Biotechnol* 19(4):379-382.

O'Farrell PH. (1975). High resolution two-dimensional electrophoresis of proteins, *J Biol Chem* 250(10):4007-4021.

Ong SE, Pandey A. (2001). An evaluation of the use of two-dimensional gel electrophoresis in proteomics, *Biomol Eng* 18(5):195-205.

Pang JX *et al.* (2002). Biomarker discovery in urine by proteomics, *J Proteome Res* 1(2):161-169.

Paulson HL. (2000). Toward an understanding of polyglutamine neurodegeneration, *Brain Pathol* 10(2):293-299.

Paulson HL, Fischbeck KH. (1996). Trinucleotide repeats in neurogenetic disorders, *Annu Rev Neurosci* 19:79-107.

Pavlidis P, Noble WS. (2001). Analysis of strain and regional variation in gene expression in mouse brain, *Genome Biol* 2(10).

Pei JJ *et al.* (2003). Role of protein kinase B in Alzheimer's neurofibrillary pathology, *Acta Neuropathol (Berl)* 105(4):381-392.

Perkins DJ *et al.* (1999). Probability-based protein identification by searching sequence databases using mass spectrometry data, *Electrophoresis* 20(18):3551-3567.

Petricoin EF *et al.* (2002a). Use of proteomic patterns in serum to identify ovarian cancer, *Lancet* 359(9306):572-577.

Petricoin EF, Liotta LA. (2003). Clinical applications of proteomics, *J Nutr* 133(7 Suppl):2476S-2484S.

Petricoin EF *et al.* (2002b). Clinical proteomics: translating benchside promise into bedside reality, *Nat Rev Drug Discov* 1(9):683-695.

Phizicky E *et al.* (2003). Protein analysis on a proteomic scale, *Nature* 422(6928):208-215.

Pollack JR *et al.* (2002). Microarray analysis reveals a major direct role of DNA copy number alteration in the transcriptional program of human breast tumors, *Proc Natl Acad Sci U S A* 99(20):12963-12968.

Pomeroy SL *et al.* (2002). Prediction of central nervous system embryonal tumour outcome based on gene expression, *Nature* 415(6870):436-442.

Ramakrishnan R *et al.* (2002). An assessment of Motorola CodeLink microarray performance for gene expression profiling applications, *Nucleic Acids Res* 30(7):e30.

Ramstrom M *et al.* (2003). Protein identification in cerebrospinal fluid using packed capillary liquid chromatography Fourier transform ion cyclotron resonance mass spectrometry, *Proteomics* 3(2):184-190.

Rickman DS *et al.* (2001). Distinctive molecular profiles of high-grade and low-grade gliomas based on oligonucleotide microarray analysis, *Cancer Res* 61(18):6885-6891.

Risch N, Merikangas K. (1996). The future of genetic studies of complex human diseases, *Science* 273(5281):1516-1517.

Rosenberg RN. (1996). DNA-triplet repeats and neurologic disease, *N Engl J Med* 335(16):1222-1224.

Sabatti CS *et al.* (2002). Thresholding rules for recovering a sparse signal from microarray experiments, *Math Biosci* 176(1):17-34.

Sallinen SL *et al.* (2000). Identification of differentially expressed genes in human gliomas by DNA microarray and tissue chip techniques, *Cancer Res* 60(23):6617-6622.

Sandberg R *et al.* (2000). Regional and strain-specific gene expression mapping in the adult mouse brain, *Proc Natl Acad Sci U S A* 97(20):11038-11043.

Schena M *et al.* (1995). Quantitative monitoring of gene expression patterns with a complementary DNA microarray, *Science* 270(5235):467-470.

Schirmer EC *et al.* (2003). Nuclear membrane proteins with potential disease links found by subtractive proteomics, *Science* 301(5638):1380-1382.

Schubert D *et al.* (1986). Protein complexity of central nervous system cell lines, *J Neurosci* 6(10):2829-2836.

Schwartz SA *et al.* (2003). Direct tissue analysis using matrix-assisted laser desorption/ionization mass spectrometry: practical aspects of sample preparation, *J Mass Spectrom* 38(7):699-708.

Sebolt-Leopold JS. (2000). Development of anticancer drugs targeting the MAP kinase pathway, *Oncogene* 19(56):6594-6599.

Shai R, *et al.* (2003). Gene expression profiling identifies molecular subtypes of gliomas, *Oncogene* 22(31):4918-4923.

Shevchenko A *et al.* (1996). Linking genome and proteome by mass spectrometry: large-scale identification of yeast proteins from two dimensional gels, *Proc Natl Acad Sci U S A* 93(25):1440-1445.

Shoemaker DD *et al.* (2001). Experimental annotation of the human genome using microarray technology, *Nature* 409(6822):922-927.

Shoemaker LD *et al.* (2003). Differential expression of membrane proteins in neural stem and progenitor cells, Abstract, Society for Neuroscience Annual Meeting.

Skene JH, Willard M. (1981). Changes in axonally transported proteins during axon regeneration in toad retinal ganglion cells, *J Cell Biol* 89(1):86-95.

Smyth GK *et al.* (2003). Statistical issues in cDNA microarray data analysis, *Methods Mol Biol* 224:111-136.

Stoeckli M *et al.* (2001). Imaging mass spectrometry: a new technology for the analysis of protein expression in mammalian tissues, *Nat Med* 7(4):493-496.

Su AI *et al.* (2002). Large-scale analysis of the human and mouse transcriptomes, *Proc Natl Acad Sci U S A* 99(7):4465-4470.

Suhr ST *et al.* (2001). Identities of sequestered proteins in aggregates from cells with induced polyglutamine expression, *J Cell Biol* 153(2):283-294.

Tan G *et al.* (2003). Decreased expression of genes involved in sulfur amino acid metabolism in frataxin-deficient cells, *Hum Mol Genet* 12(14):1699-1711.

Tang Y *et al.* (2001). Blood genomic responses differ after stroke, seizures, hypoglycemia, and hypoxia: blood genomic fingerprints of disease, *Ann Neurol* 50(6):699-707.

Tanzi RE, Bertram L. (2001). New frontiers in Alzheimer's disease genetics, *Neuron* 32(2):181-184.

Taoka M *et al.* (2000). Protein profiling of rat cerebella during development, *Electrophoresis* 21(9):1872-1879.

Taylor CF *et al.* (2003). A systematic approach to modeling, capturing, and disseminating proteomics experimental data, *Nat Biotechnol* 21(3):247-254.

Taylor JP *et al.* (2002). Toxic proteins in neurodegenerative disease, *Science* 296(5575):1991-1995.

Terskikh AV *et al.* (2001). From hematopoiesis to neuropoiesis: evidence of overlapping genetic programs, *Proc Natl Acad Sci U S A* 98(14):7934-7939.

Tietjen I *et al.* (2003). Single-cell transcriptional analysis of neuronal progenitors, *Neuron* 38(2):161-175.

Tudor M *et al.* (2002). Transcriptional profiling of a mouse model for Rett syndrome reveals subtle transcriptional changes in the brain, *Proc Natl Acad Sci U S A* 99(24):15536-15541.

Unlu M *et al.* (1997). Difference gel electrophoresis: a single gel method for detecting changes in protein extracts, *Electrophoresis* 18(11):2071-2077.

VanDeerlin V *et al.* (2002). The use of fixed human postmortem brain tissue to study mRNA expression in neurodegenerative diseases: applications of microdissection and amplification. In Geschwind DH, Gregg J, editors: *Microarrays for the neurosciences: an essential guide,* Cambridge: MIT Press.

Vawter MP *et al.* (2002). Microarray analysis of gene expression in the prefrontal cortex in schizophrenia: a preliminary study, *Schizophr Res* 58(1):11-20.

Velculescu VE *et al.* (2000). Analysing uncharted transcriptomes with SAGE, *Trends Genet* 16(10):423-425.

Velculescu VE *et al.* (1995). Serial analysis of gene expression, *Science* 270(5235):484-487.

Venter JC *et al.* (2001). The sequence of the human genome, *Science* 291(5507):1304-1351.

Vestal ML. (2001). Methods of ion generation, *Chem Rev* 101(2): 361-375.

Waterston RH *et al.* (2002). Initial sequencing and comparative analysis of the mouse genome, *Nature* 420(6915):520-562.

Weitzdoerfer R *et al.* (2002). Reduction of actin-related protein complex 2/3 in fetal Down syndrome brain, *Biochem Biophys Res Commun* 293(2): 836-841.

Whitelegge JP *et al.* (1999). Toward the bilayer proteome, electrospray ionization-mass spectrometry of large, intact transmembrane proteins, *Proc Natl Acad Sci U S A* 96(19):10695-10698.

Whitelegge JP *et al.* (2000). Methionine oxidation within the cerebroside-sulfate activator protein (CSAct or Saposin B), *Protein Sci* 9(9):1618-1630.

Whitelegge JP *et al.* (2002). Full subunit coverage liquid chromatography electrospray ionization mass spectrometry (LCMS+) of an oligomeric membrane protein: cytochrome b(6)f complex from spinach and the cyanobacterium mastigocladus laminosus, *Mol Cell Proteomics* 1(10):816-827.

Whitney LW *et al.* (1999). Analysis of gene expression in mutiple sclerosis lesions using cDNA microarrays, *Ann Neurol* 46(3):425-428.

Whitney LW *et al.* (2001). Microarray analysis of gene expression in multiple sclerosis and EAE identifies 5-lipoxygenase as a component of inflammatory lesions, *J Neuroimmunol* 121(1-2):40-48.

Wolfsberg TG *et al.* (2002). A user's guide to the human genome, *Nat Genet* 32(Suppl):1-79.

Wu CC *et al.* (2003). A method for the comprehensive proteomic analysis of membrane proteins, *Nat Biotechnol* 21(5):532-538.

Yan F *et al.* (2003). A comparison of drug-treated and untreated HCT-116 human colon adenocarcinoma cells using a 2-D liquid separation mapping method based upon chromatofocusing PI fractionation, *Anal Chem* 75(10):2299-2308.

Yang YH, Speed T. (2002). Design issues for cDNA microarray experiments, *Nat Rev Genet* 3(8):579-588.

Yao Y *et al.* (2003). Identification of protein components in human acquired enamel pellicle and whole saliva using novel proteomics approaches, *J Biol Chem* 278(7):5300-5308.

Yates JR 3rd. (1998). Database searching using mass spectrometry data, *Electrophoresis* 19(6):893-900.

Yoo BC *et al.* (2001). Differential expression of molecular chaperones in brain of patients with Down syndrome, *Electrophoresis* 22(6):1233-1241.

Yu LR *et al.* (2002). Proteome analysis of camptothecin-treated cortical neurons using isotope-coded affinity tags, *Electrophoresis* 23(11):1591-1598.

Yuan X *et al.* (2002). Analysis of the human lumbar cerebrospinal fluid proteome, *Electrophoresis* 23(7-8):1185-1196.

Yuen T *et al.* (2002). Accuracy and calibration of commercial oligonucleotide and custom cDNA microarrays, *Nucleic Acids Res* 30(10):e48.

Zabel C *et al.* (2002). Alterations in the mouse and human proteome caused by Huntington's disease, *Mol Cell Proteomics* 1(5):366-375.

Zhang K, Zhao H. (2000). Assessing reliability of gene clusters from gene expression data, *Funct Integr Genomics* 1(3):156-173.

Zhang L *et al.* (2002). Contribution of human alpha-defensin 1, 2, and 3 to the anti-HIV-1 activity of CD8 antiviral factor, *Science* 298(5595):995-1000.

Zhang R *et al.* (2003). Identification of differentially expressed proteins in human glioblastoma cell lines and tumors, *Glia* 42(2):194-208.

Zhou G *et al.* (2002). 2D differential in-gel electrophoresis for the identification of esophageal scans cell cancer-specific protein markers, *Mol Cell Proteomics* 1(2):117-124.

Zhou H *et al.* (2001). A systematic approach to the analysis of protein phosphorylation, *Nat Biotechnol* 19(4):375-378.

Zhu H *et al.* (2000). Analysis of yeast protein kinases using protein chips, *Nat Genet* 26(3):283-289.

Zhu H, Snyder M. (2003). Protein chip technology, *Curr Opin Chem Biol* 7(1):55-63.

Zirlinger M *et al.* (2001). Amygdala-enriched genes identified by microarray technology are restricted to specific amygdaloid subnuclei, *Proc Natl Acad Sci U S A* 98(9):5270-5275.

Neuroprotection: Where Are We Going?

Gary H. Danton, MD, PhD

W. Dalton Dietrich, PhD

The study of brain and spinal cord injury pathogenesis identified many injury mechanisms that are potential targets for therapeutic interventions. Indeed, an explosion of information regarding the cellular and molecular events responsible for neuronal cell death and recovery has occurred over the last decade. These discoveries have clarified multiple injury cascades, including free radical generation, excitotoxicity, oxidative stress, apoptotic cell death, and inflammation, each of which is a prime target for the development of novel therapeutic interventions (McIntosh et al., 1989; Dirnagl et al., 1999).

An obvious approach to protecting the brain from irreversible injury as a consequence of cerebral ischemia or trauma is preventive measures. To this end, various stroke risk factors, including hypertension, diabetes, obesity, smoking, and surgical procedures, are just a few targeted factors discussed as preventive measures. In the area of traumatic brain injury (TBI), the use of automobile seat belts and airbags, as well as the use of motorcycle helmets, can help prevent or limit irreversible damage. However, after the primary insult has occurred, strategies, including intensive care procedures, to limit secondary insults and the use of protective strategies to inhibit secondary injury mechanisms may potentially limit damage and improve outcome. The ability of these measures to reduce structural damage and improve functional outcome depends on various factors that are a topic of the present discussion.

Stroke is the third most common cause of death in most industrialized countries, with an estimated global mortality of 4.7 million/year (American Heart Association, 2002). Stroke killed 283,000 people in 2000 and accounted for about 1 in every 14 deaths in the United States. Each year, about 700,00 people suffer new or recurrent stroke; this is a major cause of serious long-term disability, with more than 1 million reporting functional limitations after suffering from stroke. TBI is a leading cause of death and disability among children and young adults. Each year, an estimated 1.5 million Americans sustain head trauma and, as a consequence, 230,000 are hospitalized and survive. A total of 50,000 die, and 80,000 to 90,000 experience the onset of long-term disability. As a cumulative result of past TBI, an estimated 5.3 million men, women, and children in the United States are living with permanent TBI-related disability (Consensus Conference, 1999a). In the area of spinal cord injury (SCI), approximately 11,000 new cases occur each year in the United States. Current estimates are that approximately 250,000 patients are living with paralysis as a result of SCI. Thus continued investigations in the area of neuroprotection are extremely important with regard to limiting the detrimental consequences of each of these central nervous system (CNS) injuries.

A large amount of literature is available regarding neuroprotection strategies in various models of cerebral ischemia and trauma. Pharmacological treatments targeting specific injury cascades, as well as the use of therapeutic hypothermia, have provided compelling evidence for neuroprotection in preclinical studies. Nevertheless, few agents have been successfully moved to the clinic, and currently no pharmacological treatments have been proven to protect against the detrimental consequences of TBI (Doppenberg and Bullock, 1997; Maas, 2001; Narayan et al., 2002). With regard to stroke, there is one treatment that is somewhat successful in a subpopulation of victims. The thrombolytic tissue plasminogen activator has been proven effective in treating stroke when given within 3 hours of the onset of neurological symptoms (National Institute of Neurological Disorders and Stroke rt-PA Stroke Study Group,1995). Thus, successful thrombolysis after vascular thrombosis and embolization, if occurring early enough, can result in recanalization and reperfusion of previously ischemic areas. However, a current concern is the potential incidence of reperfusion injury, including hemorrhagic transformation. This potentially detrimental consequence of thrombolysis has severely limited the number of patients currently being treated with this therapy.

PATHOPHYSIOLOGICAL EVENTS

Stroke

A stroke occurs when blood is prevented from properly perfusing a region of the brain, resulting in cell death and neurological deficits. In all, 88% of strokes are ischemic (decreased blood flow) and 12% are hemorrhagic (owing to a ruptured vessel). Cerebral blood flow (CBF) may be reduced from the entire organ during cardiac arrest or severe hypotension. Flow reductions to part of the brain occur from acute vascular occlusion resulting from processes such as atherothrombosis of cerebral vessels (61% of all stroke excluding transitory ischemic attack) or emboli (24% of all stroke excluding transitory ischemic attack) from the heart, atherosclerosis of the carotids or aortic arch, and even fat from fractured bones (Wolf et al., 1977; American Heart Association, 2002). Under ischemic conditions, neurons die when delivery of oxygen and an energy source (e.g., glucose) do not meet the demands of neuronal metabolism. Therefore to preserve ischemic neurons, either blood flow needs to be restored or metabolic demand needs to decrease.

Approximately 75% of hemorrhagic strokes are due to intracerebral hemorrhage, with the remaining 25% attributed to subarachnoid hemorrhage (American Heart Association, 2002). There are similarities and differences between the physiological consequences of each particular etiology. Thus stroke is a heterogeneous group of disorders, and treatments that apply to one type of stroke may not apply to others.

Traumatic Brain Injury

Stroke and traumatic injuries arise from very different insults. Nevertheless, similarities exist between the molecular and cellular mechanisms underlying cell death. TBI results from mechanical impact acceleration-deceleration insults that cause skull fractures, compression of cerebral tissues, and tearing of white and gray matter with subsequent hemorrhage (Graham, 1996). TBI, like stroke, can lead to a spectrum of histopathological changes, including hemorrhagic contusion, intracerebral hemorrhage, subarachnoid hemorrhage, and widespread white matter damage (Adams, 1992; Graham and Gennarelli, 1997). As with cerebral ischemia, histopathological damage after TBI can be both focal and diffuse (Cortez et al., 1989; Dietrich et al., 1994). Also, countercoup contusions, most common after deceleration of the head, can occur remote from the point of injury.

After TBI, the primary insult initiates a wide range of secondary injury mechanisms that critically participate in the pathogenesis (McIntosh et al., 1998; Bramlett et al., 1999). Histopathological data from autopsy specimens after fatal head injury, for example, have shown evidence of ischemic processes contributing to the pathological process (Graham and Adams, 1971). The severity of the secondary injury mechanisms may depend on the injury severity (mild, moderate, severe) or location of the primary insult (Graham et al., 1995). In conditions of severe TBI, reductions in CBF have been reported to reach ischemic levels. Thus cerebral ischemia is discussed as one secondary injury mechanism that may participate in TBI (Miller, 1985).

Patterns of both diffuse and focal neuronal injury can be identified after human and experimental TBI (Christman et al., 1994; Graham et al., 1995). A common consequence of severe trauma is the generation of a hemorrhagic contusion that results from blunt head injury or a gliding insult. Both insults can directly damage blood vessels and neuronal membranes, including those of cell bodies and axonal processes (Peerless and Rewcastle, 1967). Regions exhibiting mild reductions in flow can surround focal areas of reduced CBF after brain trauma (DeWitt et al., 1986; Dietrich et al., 1996). This surrounding area, in some cases, is similar to the penumbral regions surrounding an ischemic core and may contain scattered damaged neurons with an intact neuropil. This area may be both sensitive to therapeutic interventions and at risk for secondary insults.

Damage to the hippocampus is commonly reported in autopsy studies of head-injured patients (Kotapka et al., 1992), and clinical and experimental studies describe cognitive abnormalities associated with hippocampal dysfunction. In an acceleration model of brain injury in human primates, CA1 hippocampal and histological damage was reported in the majority of animals (Gennarelli and Graham, 1998). Following parasagittal fluid percussion (F-P) brain injury, CA3 hippocampal neurons are reported to be selectively vulnerable, with bilateral damage to the dentate hilus (Lowenstein et al.,

1992). Thalamic damage after TBI is also described in clinical and experimental studies (Ross *et al.*, 1993; Bramlett *et al.*, 1997b). In human brain injury, the loss of inhibitory thalamic reticular neurons is proposed to underlie some forms of attention deficits. In experimental models, focal damage to the thalamic nuclei is seen with subacute and long-term F-P injury that may result from progressive circuit degeneration after axonal damage, neuronal cell death, or from the lack of neurotrophic delivery.

Delayed neuronal damage has been reported in the hippocampal CA1 sector after transient cerebral ischemia (Ito *et al.*, 1975; Pulsinelli *et al.*, 1982; Kirino *et al.*, 1984). Only recently has the progressive nature of the histopathological consequences of TBI been appreciated. Bramlett and colleagues (1997a) first reported that, two months after a moderate parasagittal F-P injury, significant atrophy of the cerebral cortex, hippocampus, and thalamus was apparent in histopathological sections. The progressive nature of tissue loss was subsequently documented by Smith and colleagues (1997a), where damage to cortex and hippocampus at various time periods up to 1 year after lateral F-P injury was reported. Dixon and colleagues (1999) reported similar findings in a model of controlled cortical impact injury. Most recently, Bramlett and Dietrich (2002) reported significant white matter loss at 1 year after TBI. Taken together, these studies emphasize the progressive nature of this pathology and emphasize the complexity of injury processes occurring long term after the primary insult has occurred.

DOMINANT MECHANISMS

Excitotoxicity

After ischemia and traumatic injury, neuronal depolarization permits sodium and calcium influx. Much of the initial depolarization is due to failure of membrane-bound ion pumps (Taylor and Meldrum, 1995; Dirnagl *et al.*, 1999; Endres and Dirnagl, 2002). Cell depolarization leads to glutamate release into the synaptic cleft through calcium-mediated exocytosis. Excessive extracellular glutamate binds to receptors on adjacent cells. Activated glutamate receptors are both metabotropic and inotropic, acting through G-protein coupled secondary messenger systems or ligand-gated ion channels, respectively. Subsequent neuronal activation leads to excess calcium entry, possibly the most important event in the excitotoxic cascade. Elevated intracellular calcium mediates second messenger systems that lead to activation of proteolytic enzymes, xanthine oxidases and phospholipases, and nitric oxide synthase (NOS); generates

free radicals; initiates apoptosis; and causes additional glutamate release. Inhibiting glutamate receptors, or sodium and calcium channels may also stop this excitotoxic cycle from continuing (Dirnagl *et al.*, 1999; Endres and Dirnagl, 2002).

Free Radicals

Free radicals are molecules with one or more unpaired electrons, making them capable of oxidation-reduction reactions. They are produced during ischemia-reperfusion (Kontos, 2001) as well as in TBI (Globus *et al.*, 1995). Superoxide (O_2^-) is a radical generated by oxidative enzymes during cellular reactions such as aerobic respiration and prostaglandin synthesis. Ordinarily, O_2^- production is minimal but during periods of metabolic stress, its production increases significantly. Superoxide is converted into the very reactive hydroxyl radical by the iron catalyzed Haber-Weiss reaction. Under normal conditions, superoxide dismutase (SOD) combines O_2^- and H^+ into molecular oxygen and hydrogen peroxide, thereby protecting cells from the harmful effects of O_2^-. The remaining hydrogen peroxide is broken down by catalase (Nelson *et al.*, 1992). Much of the damage caused by free radicals is thought to involve the highly reactive molecule peroxynitrite. The reaction between O_2^- and nitric oxide (NO), yielding peroxynitrite, is faster than the scavenging reaction between O_2^- and SOD (Beckman and Koppenol, 1996). Under normal physiological conditions, NO concentration is maintained by NO scavengers such as oxy-hemoglobin. During brain injury, however, up regulation of NOS produces excessive NO that is free to react with O_2^- (Stagliano *et al.*, 1997a; Wada *et al.*, 1998, 1999; Eliasson *et al.*, 1999). More detailed descriptions of free radical biology in brain injury have been published (Beckman and Koppenol, 1996; Kontos, 2001). These reactive molecules damage both elements of the CNS parenchyma and blood vessels resulting in cell death, blood-brain barrier (BBB) permeability, endothelial dysfunction, and inflammation.

Inflammation

Hematogenous inflammatory cells exacerbate injury by at least two mechanisms. They can form occlusive aggregates in blood vessels after initial reperfusion leading to secondary CBF reductions. They can also contribute directly to cell death through cytotoxin production (Danton and Dietrich, 2003). Both of these mechanisms require inflammatory cells to first bind endothelial cells in cerebral vessels, and both are more established in stroke models, but their involvement after TBI is still under investigation (Morganti-Kossmann *et al.*, 2001). Cell adhesion molecules such as integrins, selectins, and members of the immunoglobulin super-family (IgCAM) permit endothelial-inflammatory cell interactions. Integrins are transmembrane, heterodimeric molecules composed of an α and β subunit. They bind to extracellular matrix components and other adhesion molecules. Selectins are calcium dependent, transmembrane glycoproteins that bind carbohydrates. L-selectin (CD62-L) and E-selectin (CD62-E) are found on leukocytes and endothelial cells, respectively, whereas P-selectin (CD62-P) is found on both platelets and endothelial cells. IgCAMs are transmembrane proteins with an immunoglobulin-like domain that binds other adhesion molecules and extracellular matrix components (del Zoppo, 1997; del Zoppo and Hallenbeck, 2000; Danton and Dietrich, 2003). Evidence has mounted in both experimental and clinical literature that adhesion molecules increase and polymorphonuclear leukocytes (PMNs) accumulate in vessels. In some models and clinical situations, PMNs enter the parenchyma. Increased adhesion molecule expression was reported in rat models of thromboembolic and focal stroke in rats (Okada et al., 1994; Zhang et al., 1995b, 1998; Suzuki et al., 1998; McEver, 2001) and baboons (Okada et al., 1994). Reports of increased selectins, ICAMs, and vascular cell adhesion molecules (VCAM) are mixed in stroke patients, but these data are difficult to collect (reviewed in Danton and Dietrich, 2003). Inflammatory cell-mediated vascular occlusion was observed using in vivo microscopy through an open cranial window after transient middle cerebral artery occlusion (tMCAO). Fluorescent leukocytes adhered to vessels and were associated with decreased blood flow (Ritter et al., 2000). In a baboon model of MCAO, individual and PMN aggregates occluded capillaries and larger vessels (del Zoppo et al., 1991). Treatment with antibodies against integrins improved CBF in vessels with diameters between 7.5 and 50 μm. No difference was noted in neurological score 1 hour after reperfusion, but chronic outcomes were not studied (Mori et al., 1992).

Both clinical and experimental studies sought evidence for a cytotoxic role of PMNs. The hypothesis is that PMNs release proteases, phospholipases, NO, hydrogen peroxide, and O_2^- that damage viable tissue within the ischemic penumbra. Histological analysis of rat brains after pMCAO found that leukocytes do migrate into the infarct (Garcia et al., 1994). Epidemiological studies in patients found that the presence of generalized inflammation, whether from a recent infection or an elevated PMN count, worsens stroke outcome (Suzuki et al., 1995; Macko et al., 1996). Leukocyte infiltration was observed in stroke patients using technetium-99m hexamethylpropyleneamine oxime labeled leukocytes and single photon emission computed tomography. They observed increased signal representing inflammatory cells within brain images after acute stroke, but it is unclear whether these cells remained inside vessels or entered the parenchyma (Wang et al., 1993). Cerebrospinal fluid analyses from stroke patients demonstrate that leukocytes cross the BBB and that inflammatory response differs between stroke types. PMNs were observed in infarcts with collateral circulation, embolic etiology, and lobar hematomas, and not in infarcts without collaterals. Histopathological examination in deceased patients verified some of these findings (Kato et al., 1995).

Apoptosis

Apoptosis, or programmed cell death, is a natural process whereby cells die in an orderly fashion. The process allows organisms to get rid of unneeded cells without release of cytotoxins from dying tissue. Apoptotic cells can generally be observed undergoing chromatin condensation, DNA fragmentation in nucleosome linkage regions, cell shrinkage, and formation of apoptotic bodies (Majno and Joris, 1995; Wang, 2000). Necrosis occurs under toxic conditions in which damage is so severe that death is uncontrolled. Necrosis is characterized by ion influx, mitochondria and cell swelling, nonspecific DNA fragmentation, and cell rupture (Wang, 2000). Both necrosis and apoptosis occur in animal models of stroke, TBI, and SCI (Keane et al., 2001; Katz et al., 2001; Ozawa et al., 2002). In stroke models, the core infarct is thought to die via necrosis while cells in the penumbra may undergo apoptosis (Lipton, 1999). Since apoptosis occurs over a period of days through an orderly series of cellular cascades, inhibiting the process may prevent cell death and preserve function.

Initiation of apoptosis has been linked to oxidative stresses, DNA damage, and inflammation (Liu et al., 1999). Numerous members of apoptotic cascades have been identified (Liou et al., 2003), but caspases are particularly important and well characterized. Caspases are cysteine proteases activated by cleavage and separation of four domains followed by reassembly into a heterodimeric protein complex. At least 14 caspases have been identified. Some are considered initiators and others effectors of apoptosis. The intrinsic pathway is caspase-mediated and results from mitochondrial permeability and cytochrome c release. Cytochrome c associates with Apaf-1 to permit its association with procaspase-9. Once activated in the complex, caspase-9 activates caspase-3, which initiates the rest of the cascade. The extrinsic pathway is initiated by cell surface receptors known as death receptors. Once activated, they form a protein complex that involves activation of procaspase-8. Caspase-8 can activate both caspase-3 and a protein called Bid that induces mitochondrial permeability and the intrinsic cascade. Caspase-independent pathways have also been identified including those activated by proteins AIF, and proteases such as cathepsins, calpains, and granzymes (Phan *et al.*, 2002; Liou *et al.*, 2003).

ANIMAL MODELS

Cell Culture

Cells in culture can be made ischemic simply by decreasing oxygen and glucose in culture media. Individual cells, small groups of cultured cells, and organotypic slices can be studied to evaluate individual responses and interactions among neurons. Culture models are useful because they are well-controlled systems where experimental variables can be easily manipulated. However, structure, gene expression, and cellular responses may differ between cultured and *in vivo* cells. *In vitro* systems may be most useful for studying fundamental aspects of cell biology that are less likely to be altered by culture systems. Unfortunately, it is unclear how results in *in vitro* systems relate to human disease.

Animal models have advantages over *in vitro* systems because cells are in their natural state. Unfortunately, because manipulating individual variables is difficult, drawing conclusions from animal systems is challenging. To keep the number of variables to a minimum, animal models of human disease are designed to be easily reproducible among subjects. This often requires simplifying the insult, which results in the experimental disease being fundamentally different from the human condition. Models are designed so specifically that changing to another breed of the same species may yield different results (Markgraf *et al.*, 1993).

Global Ischemia

Global ischemia models reduce blood to the entire brain, mimicking cerebral ischemia from cardiac arrest or severe hypotension. In rats, global ischemia is produced by cardiac arrest or by two-vessel or four-vessel occlusion. Two-vessel occlusion is performed by tying ligatures around the carotid arteries and dropping mean arterial pressure to 50 mm Hg by controlled exsanguination. Blood is immediately reinfused after 10 minutes and the ligatures removed. Four-vessel occlusion is performed by tying off the vertebral arteries and temporarily occluding the carotid arteries. The cardiac arrest model induces ischemia in a more clinically relevant fashion by pharmacologically inducing cardiac arrest for 7-10 minutes followed by resuscitation with epinephrine and chest compressions. These injuries result in selective neuronal necrosis of the cortex, basal ganglia, and hippocampus about 3 to 7 days after injury. Because selective neuronal necrosis is delayed, therapies can be tested for their ability to prevent cell death (Dietrich *et al.*, 1999).

Focal Ischemia

Focal ischemia models reduce blood supply to specific brain regions by occluding cerebral vessels. A variety of techniques are performed to induce middle cerebral artery (MCA) occlusion. After surgical exposure by craniotomy, the MCA can be ligated with suture (Dietrich *et al.*, 1989). To perform MCA occlusion without a craniotomy, a stiffened suture is inserted into the external carotid, through the internal carotid and up to the MCA base (Belayev *et al.*, 1996). Photothrombotic MCA occlusion is produced by injecting rats with rose bengal or erythrosine B dye and focusing an argon laser onto the exposed MCA. The laser-dye interactions produce singlet oxygen, a reactive molecule that damages the wall of the vessel by initiating lipid peroxidation. Platelets activate in response to the damage and form an occlusive thrombus (Watson *et al.*, 1987). Suture models can be studied as permanent or transient occlusion by removing the suture after a specified amount of time. The photothrombotic model also has been studied in permanent and transient models by using an ultraviolet laser to dethrombose the occluded vessel (Watson *et al.*, 2002). The photothrombotic model was developed to more closely mimic human disease, as the occlusive lesion is a natural platelet clot.

Embolic Stroke

A variety of models have been developed to study embolic stroke in animals. Nonocclusive common carotid artery thrombosis (CCAT) is produced by concurrent injection of rose bengal or erythrosine B and focusing an argon laser onto the common carotid artery of rats and mice. Damage to the carotid wall activates platelets to produce a nonocclusive thrombus that showers the distal cerebral vasculature with emboli. Some of the emboli occlude distal cerebral vessels, restrict blood flow, and infarct the supplied tissue. Infarcts from CCAT are generally smaller than those in pure ischemic models and because the mechanism of injury is embolic, there is some variability between animals. CCAT has been used to study cerebrovascular consequences of embolic stroke (Danton *et al.*, 2002a) and the effects of multiple emboli over time (Danton *et al.*, 2002b) and to test neuroprotective agents (Stagliano *et al.*, 1997b). A second common model of thromboembolic stroke involves removing blood from animals and reinjecting the blood into the carotid circulation after permitting it to clot. This model produces primarily fibrin-based clots amenable to therapy with tPA (Zivin *et al.*, 1985).

Traumatic Brain Injury

Gennerelli (1994) classified models of head injury according to the method of producing injury. F-P and rigid indentation models are characterized by a percussion concussion, whereas inertial injury models and impact acceleration models are considered acceleration concussion. The central lateral and parasagittal F-P models are characterized by behavioral responsiveness, metabolic alterations, changes in local cerebral blood flow (lCBF), and blood brain barrier (BBB) permeability, resulting in behavioral deficits (Dixon *et al.*, 1987; McIntosh *et al.*, 1989; Dietrich *et al.*, 1994). The central F-P model tends to have variable and small contusions in the vicinity of the fluid pulse and scattered axonal damage, most limited to the brainstem. In the lateral and parasagittal F-P models (Cortez *et al.*, 1989; Dietrich *et al.*, 1994), cortical contusions remote from the impact site are commonly observed. Evidence of axonal damage is seen throughout the white matter tracks ipsilateral to the injury, with tissue tears seen in the gray-white interfaces (Bramlett *et al.*, 1997c; Graham *et al.*, 2000). Depending on the severity of the injury, the F-P model produces a range of disorders including contusion, widespread axonal damage, and selective neuronal necrosis. Controlled cortical impact (CCI) injury is another rodent TBI model that produces a well-demarcated cortical contusion with variable degrees of hippocampal involvement, depending on velocity and deformation (Dixon *et al.*, 1991). An advantage of the CCI injury is that it can be used in mice and allows for testing of genetically altered animals.

Inertial acceleration models produce pure acute subdural hematomas and diffuse axonal injury (Gennarelli, 1994). These models are characterized by a variable period of coma and axonal damage in the upper brainstem and cerebellum. An optic nerve model of traumatic axonal injury has been investigated (Saatman *et al.*, 2003). This model is advantageous in characterizing the pathophysiology of primary traumatic axonal injury. Although the total human injury condition cannot be addressed adequately with a single animal model, these models are advantageous for investigating a particular feature of human brain injury and uncovering the pathogenesis of the injury cascade. *In vitro* models have also been developed that can be used to critically investigate the consequences of trauma on a specific cell type in the absence of confounding and systemic cellular factors (Ellis *et al.*, 1995). Models range from scratching the cell culture with a pipette tip to inducing cellular deformation or stretching cultured cells. Also, combined mechan-

ical trauma and metabolic impairment *in vivo* have helped to clarify mechanisms underlying the cellular response to trauma and the role of specific cell types in these injury mechanisms (Allen *et al.*, 1999).

Spinal Cord Injury

Several established models of SCI have been used to clarify the pathophysiology and treatment of ischemia and traumatic injury. These include transection, compression, photochemical, and contusion injury models (Prado *et al.*, 1987; Gruner, 1992; Grill *et al.*, 1997; Murray *et al.*, 2002; Park *et al.*, 2003). The specific model that is chosen for a specific study is best determined by what scientific question is being addressed. For example, if evidence for axonal regeneration after SCI is being investigated, transection models (complete or incomplete) are advantageous to rigorously assess axonal growth. In contrast, in studies of neuronal protection, compression or contusion models may allow for specific vulnerable neuronal populations to be quantitatively assessed. Finally, SCI models have been developed in mice, taking advantage of genetically manipulated animals (Inman *et al.*, 2002).

NEUROPROTECTIVE DRUGS

As already emphasized, ischemic and traumatic insults share many of the same processes that are considered important to cell death mechanisms (Mendelow, 1993). Although the location and severity of the primary insult are important factors to determine the impact of the individual injury process, excitotoxicity, calcium-mediated events, free radicals, mitochondrial damage, inflammation, and apoptosis are commonly discussed as possible targets for therapeutic interventions (Miller, 1985).

Sodium Channel Blockers (Table 14.1)

Sodium influx occurs at a number of points along the ischemic cascade. Entry through voltage-sensitive channels occurs soon after Na/K pump failure-induced depolarization. Once sodium concentrations rise, they drive the Na/Ca exchange pump in reverse, allowing calcium to enter the cell as well. Furthermore, after glutamate-induced excitotoxicity, sodium channels open to propagate the action potential. These are all potential targets of these agents (Taylor and Meldrum, 1995). Experimental studies were conducted with phenytoin, carbamazepine, lamotrigine, sipatrigine (619C89), and riluzole. Members of this class were beneficial in gerbil models of global ischemia, rat models of MCAO, and improved both

TABLE 14.1 Neuroprotective Agents Targeting Ion Channels

Drug	References
Na Channel Blockers (Taylor and Meldrum, 1995)	
Phenytoin	(Taft *et al.*, 1989; Rataud *et al.*, 1994)
Carbamazepine	(Rataud *et al.*, 1994)
Sipatrigine	(Kawaguchi and Graham, 1997; Muir *et al.*, 2000; Dawson *et al.*, 2001)
Riluzole	(Pratt *et al.*, 1992)
K Channel Modulators	
BMS-204352	(Jensen, 2002) (Gribkoff *et al.*, 2001; Mackay, 2001)
Ca Channel Blockers (Horn and Limburg, 2001)	
(S)-emopamil	(Nakayama *et al.*, 1988; Lin *et al.*, 1990)
Nimodipine	(Vibulsresth *et al.*, 1987; Bailey *et al.*, 1991; Murray *et al.*, 1996)
Dantrolene	(Frandsen and Schousboe, 1991; Kross *et al.*, 1993; Zhang *et al.*, 1993b)
Flunarizine	(Limburg and Hijdra, 1990; Franke *et al.*, 1996)

histopathological and behavioral outcomes (Taft *et al.*, 1989; Pratt *et al.*, 1992; Rataud *et al.*, 1994; Smith *et al.*, 1997b). Sipatrigine demonstrated efficacy in pretreatment but not posttreatment studies (Kawaguchi and Graham, 1997; Smith *et al.*, 1997b). Clinical trials with sipatrigine and fosphenytoin were discontinued without success because of toxicity and lack of efficacy, respectively (Smith *et al.*, 1997b; Muir *et al.*, 2000).

Potassium Channel Modulators (Table 14.1)

Drugs such as BMS-204352 open postassium channels, hyperpolarizing neurons and protecting them from excitotoxicity. BMS-204352 opens the calcium-activated, BK channels and voltage-dependent, noninactivating KCNQ channels. Opening potassium channels hyperpolarizes cells, reversing the depolarized state induced by ischemia. BMS-204352 decreased infarct volume in a rat model of proximal MCAO (Gribkoff *et al.*, 2001) but showed no efficacy in Phase III trials (Mackay, 2001; Jensen, 2002).

Calcium Channel Blockers (Table 14.1)

Once depolarized, calcium enters damaged neurons through a variety of channels including voltage-gated L-type Ca channels (Pisani *et al.*, 1998), *N*-methyl-D-aspartate (NMDA)-sensitive ligand-gated channels, and release from intracellular stores. Calcium contributes to cell death by mediating a variety of toxic cascades. Activated proteolytic enzymes cleave cytoskeletal proteins that form intracellular aggregates. Calcium-medi-

ated lipolysis damages membranes and provides fatty acid substrates for free radical production. Increased intracellular calcium facilitates mitochondrial permeability, activating apoptotic cascades. Calcium-dependent exocytosis facilitates glutamate release, which forms the basis of the excitotoxicity hypothesis. As potential therapies toward all these processes, calcium channel blockers have been avidly pursued as neuroprotectives and tested in a variety of models.

(S)-emopamil is a phenylalkylamine with additional 5-HT2 activity. Pretreatment with (S)-emopamil improved neuronal survival by about twofold after global ischemia in rats, but was not effective when administered 30 minutes after treatment (Lin *et al.*, 1990). After pMCAO, (S)-emopamil decreased infarct volume when administered up to 1 hour after onset of ischemia (Nakayama *et al.*, 1988).

Nimodipine is a commonly used calcium channel blocker for treating hypertension and has been studied as a neuroprotectant in animal models (Bielenberg and Beck, 1991b; Yasui and Kawasaki, 1994) and clinical trials (Gelmers *et al.*, 1988). Nimodipine limits ischemic damage to neurons in culture (Pisani *et al.*, 1998) and in hippocampal slices (Yasui and Kawasaki, 1994; Wassmann *et al.*, 1996). It decreased infarct volume by 50% at 48 hours in a rat model when given in multiple doses starting before pMCAO (Bielenberg and Beck, 1991a). Experimental trials were mixed, however, as no benefit was observed after 20 minutes of four-vessel occlusion global ischemia (Vibulsresth *et al.*, 1987).

A variety of clinical trials were conducted with nimodipine after stroke and TBI. At least 14 trials of nimodipine in ischemic stroke and subarachnoid hemorrhage have been conducted since 1982. Nine ischemic stroke trials reported no effect, one reported short-term worse outcome with treatment, and four trials reported positive outcome. Positive outcomes included improved survival in men (Gelmers *et al.*, 1988), memory (Sze *et al.*, 1998), clinical course (Gelmers, 1984), and Mathew score among certain subgroups (Martinez-Vila *et al.*, 1990). Two clinical trials with flunarizine found no statistically significant improvement in outcome (Limburg and Hijdra, 1990; Franke *et al.*, 1996). A meta-analysis of 29 trials totaling 7665 patients treated with calcium channel blockers found no benefit for ischemic stroke (Horn and Limburg, 2001). At least three trials of nimodipine in TBI found no statistical benefit, except for the subgroup of patients with subarachnoid hemorrhage (Bailey *et al.*, 1991; Teasdale *et al.*, 1992; Murray *et al.*, 1996). Five trials designed to study subarachnoid hemorrhage found statistically significant benefits for nimodipine use in these patients. Improvements were noted in mortality and short- and long-term outcome (Allen *et al.*, 1983; Neil-Dwyer *et al.*, 1987; Jan *et al.*, 1988; Ohman *et al.*, 1991).

The calcium channel blocker dantrolene is being discussed as a possible neuroprotective agent (Ovbiagele et al., 2003). It prevents glutamate-induced excitotoxicity in neuron cultures by reducing intracellular calcium concentration (Frandsen and Schousboe, 1991). In animal models, it reduced cortical necrosis by 85% to 95% after MCAO in rats (Hong and Chiou, 1998) and protected against delayed neuronal death in gerbils (Zhang et al., 1993b), but had no effect after 11 minutes of complete ischemia in dogs (Kross et al., 1993).

NMDA Antagonists (Table 14.2)

The NMDA receptor is an inotropic, calcium-conducting receptor considered to be the main contributor to glutamate-mediated excitotoxicity (Dietrich et al., 1992a, 1992b; Nakamura et al., 1993). It rests on the postsynaptic membrane, and if the cell is already depolarized, glutamate binding opens the pore, allowing calcium entry. A number of antagonists have been studied in experimental and clinical trials (Lees, 1997; Clark and Lutsep, 1999; Jain, 2000; Ovbiagele et al., 2003). Although animal studies were very promising, most clin-

ical trials were stopped because of phencyclidine-like effects such as hallucinations and agitation.

The noncompetitive antagonist MK-801 (dizocilpine) improved outcome in models of focal ischemia, producing either no effect or between 13% and 78% reduction in infarct volume, depending on experimental conditions (Dirnagl et al., 1990; Buchan et al., 1992; McCulloch et al., 1993; Valtysson et al., 1994; Margaill et al., 1996). For example, age-related changes altered drug efficacy as MK-801 reduced infarct volume in young rats, but not in aged rats (Suzuki et al., 2003b). In models of global ischemia, significant improvements in CA1 neuron survival and behavioral improvement were observed, especially when combined with hypothermia (Lin et al., 1993; Green et al., 1995; Dietrich et al., 1995a). However, beneficial findings have not been observed consistently in all models. No benefit was observed in a rat model of photothrombotic MCA occlusion (Yao et al., 1993) or following transient focal ischemia in primates (Auer et al., 1996). Both MK-801 and dextromorphan, another noncompetitive NMDA inhibitor, exhibited protective effects in experimental studies, but clinical trials were terminated early because of phencyclidine-like side effects and lack of efficacy (Kermer et al., 1999; Jain, 2000).

Aptiganel (Cerestat, CNS1102) is a noncompetitive NMDA antagonist that decreases infarct volume and behavioral deficits when administered up to 120 minutes after MCAO (Pitsikas et al., 2001). Safety and tolerability studies in patients found neuroprotective plasma levels in healthy volunteers and patients with ischemic stroke (Dyker et al., 1999). Unfortunately, a Phase III clinical trial of aptiganel treatment within 6 hours of stroke onset was terminated early because of a lack of efficacy and a trend toward higher mortality in treated patients.

CP101,606 (ceresine) is a noncompetitive NMDA antagonist selective for the NR2B subunit. CP101,606 improved behavioral outcome and decreased infarction volume between 20% and 30% 2 hours after thromboembolic stroke in rats (Yang et al., 2003). A 63% reduction in infarct volume was observed in a feline model of focal ischemia with pretreatment and continuous infusion (Di et al., 1997). Initial trials found that CP101,606 was well tolerated in patients with TBI and intracerebral hemorrhage (Merchant et al., 1999; Bullock et al., 1999; Jain, 2000).

The NMDA antagonists selfotel, eliprodil, and NPS-1506 also demonstrated efficacy in animal models. Selfotel (CGS 19755) is a competitive NMDA antagonist that was discontinued after two Phase III trials found no beneficial effect (Davis et al., 1997). Eliprodil is a combined competitive NMDA antagonist and calcium channel blocker. It reduced neurological deficit and infarct volume by about 50% in rats with thromboembolic

TABLE 14.2 Neuroprotective Agents Targeting Excitotoxicity

Drug	Reference
NMDA Antagonists	
MK-801	(Dietrich et al., 1992a; Green et al., 1995; Dietrich et al., 1995a; De Vry et al., 2001)
Aptiganel	(Lees, 1997; Dyker et al., 1999)
CP101,606	(Di et al., 1997) (Bullock et al., 1999; Merchant et al., 1999)
Eliprodil	(Lees, 1997)
NPS-1506	(Mueller et al., 1999)
Magnesium	(Muir, 2002)
AMPA Antagonists (Akins and Atkinson, 2002)	
YM-872	(Kawasaki-Yatsugi et al., 2000; Suzuki et al., 2003b; Furukawa et al., 2003)
NAALADase Inhibitors	
GPI5232	(Williams et al., 2001)
2-PMPA	(Slusher et al., 1999; Vornov et al., 1999; Tortella et al., 2000)
GABA Agonists	
Clomethiazole	(Endres and Dirnagl, 2002) (Lyden et al., 2002)
Diazepam	(Endres and Dirnagl, 2002)
Glycine Antagonists	
Gavistinel	(Sacco et al., 2001)
Licostinel	(Albers et al., 1999)

stroke. The combination of eliprodil and tPA reduced deficit by 70% and infarct volume by 89%. A Phase III trial was terminated, however, because of a lack of efficacy (Jain, 2000). NPS-1506 is a noncompetitive NMDA receptor antagonist that inhibits active channels in a voltage-dependent fashion. It was neuroprotective in animal models for hemorrhagic stroke, tMCAO, and pMCAO, with between a 33% and 57% decrease in infarct volume at effective doses. Phase I trials were completed and found no PCP-like side effects at neuroprotective doses (Mueller et al., 1999).

Glycine Antagonists (Table 14.2)

Glycine is an obligatory coagonist at NMDA receptors. Instead of antagonizing the NMDA receptor at the glutamate site, the glycine-binding site can be blocked without causing phencyclidine-like side effects. Despite reducing infarct volume by up to 48% after MCAO in rats, gavistinel (GV150526) had no benefits in patient mortality in the GAIN Phase III clinical trial (Sacco et al., 2001). A second glycine antagonist, licostinel (ACEA 1021), was reported to be well tolerated in acute stroke patients (Albers et al., 1999), although there were questions of crystal formation in patients' urine.

AMPA Antagonists (Table 14.2)

AMPA receptors bind glutamate on the postsynaptic membrane and gate both sodium and potassium currents. Antagonists to this receptor might inhibit or reduce glutamate-mediated depolarization, thereby reducing NMDA receptor activation. YM-872 (zonampanel) administered up to 2 hours after pMCAO and tMCAO decreased infarct volume and neurological deficit by 30% to 40% at 1 week (Takahashi et al., 1998; Kawasaki-Yatsugi et al., 2000). YM-872 also improved outcome in a rat model of thromboembolic stroke and enhanced tPA efficacy (Suzuki et al., 2003b). Overall, animal models have reported between a 20% and 60% decrease in infarct volumes (Ovbiagele et al., 2003). After lateral fluid percussion injury in rats, YM-872 administered for 4 hours beginning 15 minutes after injury improved histological and behavioral outcome at 1 and 2 weeks (Furukawa et al., 2003).

YM-872 is being studied in two large clinical trials, the AMPA Receptor Antagonist Treatment in Ischemic Stroke Trial (ARTIST+) and AMPA Receptor Antagonist Treatment in Ischemic Stroke (ARTIST MRI). The ARTIST+ trial compares tPA with and without YM-872 treatment. The ARTIST MRI trial is measuring outcome at 28 days by T2-weighted MRI with FLAIR and at 90 days by clinical criteria.

N-acetylaspartylglutamate Inhibitors (Table 14.2)

N-acetylaspartylglutamate (NAAG) is a stored form of glutamate, released by the enzyme N-acetylated-α-linked acidic dipetidase (NAALADase). Inhibitors of NAALADase include 2-PMPA (2-(phosphonomethyl) pentanedioic acid and a more lipophilic analog GPI5232. By decreasing conversion of NAAG, they stoichiometrically decrease glutamate release and are neuroprotective in neuron culture as well after tMCAO (decreased infarct volume around 50%) (Slusher et al., 1999; Vornov et al., 1999; Tortella et al., 2000; Williams et al., 2001).

Metabotropic Glutamate Modulators (Table 14.2)

There are eight subtypes of metabotropic glutamate receptor (mGLUR), which are divided into three groups. The group I mGLURs (mGLUR 1 and 5) activate the phosphoinositide cascade, which leads to opening of Ca^{++} and K^+ channels. mGLUR5 potentiates NMDA receptor activation, which may be the mechanism of group I mGLUR antagonist neuroprotection. Groups II and III depress intracellular systems such as adenylate cyclase and Ca^{++} entry. Therefore, their agonists tend to be neuroprotective. Group II agonists require glia for their neuroprotective effects, whereas group III agonists work well in neuronal culture alone. Because group II receptor antagonists do not interfere with normal neurotransmission, they are not expected to have side effects similar to other glutamate antagonists. At least nine group I antagonists, six group II agonists, and three group III agonists demonstrated neuroprotective properties. BAY 36-7620 is a recently developed group I antagonist with oral bioavailability and good BBB penetration (Flor et al., 2002). It was neuroprotective after subdural hematoma in rats but statistical significance was not reached after MCAO (De Vry et al., 2001).

Magnesium (Table 14.2)

Magnesium essentially targets the rise in intracellular Ca^{++}. By blocking the gating mechanism of channels, it inhibits Ca^{++} influx through voltage-gated Ca channels and NMDA receptors. Magnesium also inhibits cortical spreading depression and neurotransmitter release, and it promotes vasodilation (Ovbiagele et al., 2003). It is particularly appealing because of its low cost, widespread availability, and relative safety. Magnesium was protective in models of thromboembolic, tMCAO, and pMCAO, with decreases in infarct volume between 25% and 61%. At least six clinical trials showed promise for magnesium therapy (Muir, 2002). Three large clinical trials in stroke are underway to further evaluate magnesium's efficacy. The Intravenous Magnesium Efficacy in

Stroke (IMAGES) trial is recruiting 2700 patients to assess death and disability when magnesium is administered within 12 hours after stroke onset. The Magnetic Resonance in Intravenous Magnesium Efficacy in Stroke Trial (MR Images) will determine the ability of Mg to reduce infarct volume determined by MRI. The Field Administration of Stroke Therapy-Magnesium Phase III Trial (FAST-MAG) will examine whether magnesium is effective when administered by ambulance personnel between 15 minutes and 2 hours after symptom onset. A randomized, double-blinded, placebo-controlled, parallel assignment trial is currently recruiting patients for a magnesium trial following TBI as well.

γ-Aminobutyric Acid Agonists (Table 14.2)

γ-Aminobutyric acid (GABA) is an inhibitory neurotransmitter thought to oppose glutamate-mediated excitotoxicity by opening a Cl-1 channel. GABA agonists were found to be neuroprotective in global and focal ischemia (reviewed in Schwartz-Bloom and Sah, 2001). Clomethiazole was the most recent candidate for GABA therapy. The Phase II Clomethiazole Acute Stroke Study (CLASS) found no benefit, but the drug was well tolerated. Phase III trials called the Clomethiazole Acute Stroke Study in Ischemic, Hemorrhagic, and tPA-treated Patients were discontinued after demonstrating no efficacy (Lyden et al., 2002).

Serotonin Antagonists (Table 14.3)

Serotonin (5-HT) is released from activated platelets and elevated levels were observed after CCAT and global ischemia (Wester et al., 1992). Atherosclerosis and carotid stenosis impair vascular dilation, making cerebral vessels susceptible to serotonin-mediated constriction (Ringelstein et al., 1992; Gur et al., 1996; White and Markus, 1997). Antagonists such as sarpogrelate, ketanserin, and ritanserin have been studied to prevent pathological vasoconstriction (Danton et al., 2002c). Ketanserin normalizes local CBF after photochemically induced cortical infarction (Dietrich et al., 1989a). Ritanserin improved histopathological outcome after transient global ischemia (Globus et al., 1992) and improved CBF after photothrombotic focal ischemia (Back et al., 1998), but it had no benefit during pMCAO (Takagi et al., 1994).

Clinical trials using serotonin antagonists suggest that they can be used safely. Ketanserin effectively reduced vascular morbidity by 23% in a trial of 3899 patients more than 40 years old with intermittent claudication. Unfortunately, there was an adverse reaction in patients using diuretics that resulted in a number of deaths

TABLE 14.3 Neuroprotective Agents Targeting Vascular Disease

Drug	References
Serotonin Antagonists	
Ketanserin	(Dietrich et al., 1989a)
Ritanserin	(Back et al., 1998)
Sarpogrelate	(Danton et al., 2002c) (Shouzu et al., 2000)
Adenosine Agonists	
GP683	(Tatlisumak et al., 1998b)
Acadesine	(Dietrich et al., 1995b)
Anti-Adhesion Molecules	
Enlimomab	(Use of anti-ICAM-1 therapy, 2001)
Matrix Mettaloproteinase Inhibitors	
Batimistat	(Asahi et al., 2000)
KB-R7785	(Jiang et al., 2001)
BB-1101	(Rosenberg et al., 1998)
HMG-CoA Reductase Inhibitors (Danton et al., 2002)	

(Prevention of Atherosclerotic Complications, 1989). Sarpogrelate decreased platelet-derived aggregates, soluble adhesion molecules, and thrombomodulin in patients with diabetes mellitus, suggesting that this treatment strategy may prevent thrombotic complications in at-risk patients (Shouzu et al., 2000). Patients with hemodynamic disturbances, risk factors for platelet aggregation and embolization, may be most likely to benefit from this treatment strategy.

Adenosine Agonists (Table 14.3)

Adenosine has antiplatelet properties and inhibits neuronal activity and excitotoxicity by binding to the A1 receptor. Adenosine-targeted drugs may act directly on adenosine receptors (A1, A2a, A2b, A3) or increase endogenous adenosine by inhibiting reuptake and breakdown. The adenosine-regulating agent acadesine reduced platelet deposition after CCAT and tended to decrease infarct areas (Dietrich et al., 1995b). The adenosine kinase inhibitor GP683 decreased infarct size in rats when administered 30 minutes after tMCAO, demonstrating its usefulness in focal ischemia (Tatlisumak et al., 1998a).

Adenosine agonists have been studied in clinical trials. A Phase III clinical trial with acadesine during coronary bypass surgery reported an 88% reduction in cardiovascular mortality (Mullane, 1993). The European Stroke Prevention Study-2 reported a reduction in stroke with either aspirin (18%), dipyridamole (16%), or combination therapy (37%). Both aspirin and dipyridamole are platelet inhibitors, but dipyridamole's effects may be due to an increased level of endogenous adenosine (Picano and Abbracchio, 1998).

Free Radical Scavengers (Table 14.4)

Free radicals are molecules with one or more unpaired electrons that participate in oxidation-reduction reactions. They are produced during ischemia-reperfusion, diffuse into neighboring tissue, and damage blood vessels, neurons, and glia. Vascular damage from these molecules may promote BBB permeability, endothelial dysfunction, and inflammation. Free radicals react with NO produced by NOS isoenzymes to form peroxynitrite, a highly reactive molecule. More detailed descriptions of free radical biology following ischemia and the role of NO in these reactions have been reviewed elsewhere (Beckman and Koppenol, 1996; Kontos, 2001).

Free radical scavengers protect both the cerebral vasculature and the brain parenchyma. Most agents have been successfully used as pretreatment or immediate posttreatment strategies. There are some experimental data on the use of these agents at longer intervals after injury. For example, U-101033E significantly reduced infarction size at 4 hours but not at 24 hours after pMCAO (Hall et al., 1996). NXY-059 reduced infarct size when administered 3 hours after tMCAO. Treatment at 6 hours, but NOT at 3 hours, improved behavioral deficits at 1 day but not at 2 days after injury. The scavenger PBN successfully reduced BBB permeability and infarct size 12 hours after focal ischemia (Cao and Phillis, 1994). Early treatment with the lipid peroxidation inhibitor LY341122 improved histopathological outcome after moderate fluid percussion brain injury in rats by up to 60% (Wada et al., 1999a). Although most agents were

tested in rodent models, NXY-059 has successfully improved the behavioral and histopathological consequences of pMCAO in marmosets (Marshall et al., 2001), raising hopes that these treatments will be clinically applicable.

Clinical trials have had limited success following acute ischemic stroke but have had more success treating subarachnoid hemorrhage. Patients administered nicaraven 5 days after subarachnoid hemorrhage had a reduced incidence of delayed ischemic neurological deficits. A number of trials examined tirilazad after stroke and a consensus group determined that patients were likely worse off in the treated groups. Numerous side effects may have complicated any benefit of the drug (Tirilazad International Steering Committee, 2000). The iron chelator deferoxamine improved endothelial-dependent dilation in patients with coronary artery disease (Duffy et al., 2001). Dysfunctional endothelial-dependent dilation of the MCA in rats was observed after CCAT and is a complication found in some patients with cerebrovascular disease. Selected groups of patients with these disease processes may benefit from free radical scavenger therapy.

Methylprednisolone

Clinically, the corticosteroid and anti-inflammatory agent, methylprednisolone (MP) is the recommended drug for administration following SCI in the United States (Bracken et al., 1990, 1992). Although the exact mechanism of action is unknown, MP has been reported to decrease oxygen radical-induced lipid peroxidation, inhibit eicosanoid formation, increase CBF, improve reversal of intracellular Ca^{++} accumulation, and enhance synaptic transmission (Hall, 1982, 1992; Young and Flamm, 1982; Anderson et al., 1985). In experimental and clinical studies, MP has been shown to improve outcome when given in the first hours after injury. However, there is a critical need to develop and test other novel therapeutic strategies that may be used in the clinical arena. Therapeutic strategies that are neuroprotective in the acute injury setting may also be used in conjunction with cellular transplantation and regeneration experiments to promote recovery in the chronic injury setting.

Anti-adhesion Molecule Therapy (Table 14.3)

Targeting the hematogenous inflammatory response has primarily focused on up regulated adhesion molecules. In models of tMCAO, pretreatment and posttreatment with antiselectin antibodies successfully decreased infarct volume up to 70% after 2 hours and improved CBF (Connolly et al., 1997; Goussev et al., 1998). The anti-ICAM-1 antibody 1A29 decreased infarct areas after

TABLE 14.4 Neuroprotective Agents Targeting Free Radicals

Drug	References
Free Radical Scavengers (Danton et al., 2002c)	
PBN	(Cao and Phillis, 1994)
Ebselen	(Imai et al., 2003)
MCI-186	(Kawai et al., 1997; Mizuno et al., 1998; Wu et al., 2000)
NXY-059	(Marshall et al., 2001)
PEG-SOD	(Muizelaar et al., 1993; Muizelaar, 1994)
Tirilazad	(Tirilazad International Steering Committee, 2000)
MDL 74,180	(Cowley et al., 1996)
LY341122	(Wada et al., 1999a)
Nicaraven	(Toyoda et al., 1997)

tMCA but not pMCAO. Cotreatment with the granulocyte-depleting antibody RB6-8C5 further improved outcome (Zhang et al., 1995a). Recently, a clinical trial of the murine anti-ICAM-1 antibody Enlimomab worsened neurological score and mortality in patients (Use of anti-ICAM-1 therapy, 2001). A follow-up study using the murine anti-rat ICAM-1 antibody in rats also found an increase in infarct volume and no efficacy. It is believed that immune activation in response to the foreign mouse protein probably accounted for the clinical and follow-up experimental results (Furuya et al., 2001). Modifications of the antiadhesion molecule therapy might make this strategy worthwhile.

Matrix Metalloproteinases (Table 14.3)

Matrix metalloproteinases (MMPs) are angiogenic enzymes that break down components of the extracellular matrix. Although angiogenesis seems like a potential benefit, MMPs may enhance BBB breakdown after stroke and increase inflammation. MMP inhibitors such as batimistat (BB-94) and KB-R7785 decreased infarct volume in mice following permanent focal ischemia (Asahi et al., 2000; Jiang et al., 2001). BB-1101 reduced edema 24 hours after focal ischemia in rats and reduced hemorrhage in both rats and rabbits (Rosenberg et al., 1998; Lapchak et al., 2000; Sumii et al., 2002). MMP inhibitors have been evaluated in patients for their antiangiogenesis properties in cancer treatment and are well tolerated in patients (Hidalgo and Eckhardt, 2001).

Antiapoptosis Agents (Table 14.5)

As previously described, caspases play a major role in both intrinsic and extrinsic apoptotic pathways and are a prime target for therapeutic development (Schulz et al., 1999). Caspase inhibitors have worked in a variety of brain injury models. In a model of neonatal hypoxia-ischemia, administration of the pan-caspase inhibitor boc-aspartyl(OMe)-fluoromethylketone was significantly neuroprotective when given up to 3 hours after the insult (Cheng et al., 1998). The caspase inhibitor N-benzyloxy-

TABLE 14.5   Other Neuroprotective Agents under
Investigation

Drug	References
Caffeine and Alcohol (Piriyawat et al., 2003) (Aronowski et al., 2003)	
Caspase inhibitors	(Schulz et al., 1999)
Cyclosporin	(Shiga et al., 1992; Uchino et al., 1995)
FK506	(Singleton et al., 2001; Furuichi et al., 2003)
cNOS inhibitors (Wada et al., 1998a, 1998b; Wada et al., 1999b)	

carbonyl-Asp(OMe)-Glu(OMe)-Val-Asp(OMe)-fluoromethylketone attenuated postischemic caspase-3-like enzymatic activity and blocked delayed neuron loss in the hippocampus (Gillardon et al., 1999). Treatment with Z-VAL-Ala-Asp (Ome)-FMK (zVAD) improved neurological outcome after cardiac arrest in rats (Katz et al., 2001). Improved outcomes have also been noted in models of tMCAO and pMCAO (Schulz et al., 1999).

Cyclosporin A and FK-506 are immunosuppressive agents that inhibit calcineurin and have mixed results suggesting antiapoptotic effects (Herr et al., 1999; Ovbiagele et al., 2003; Yardin et al., 1998). Using isolated brain mitochondria, cyclosporin A prevented mitochondrial permeability transition under partial but not strong depolarization. This suggests that cyclosporin A is an effective apoptosis inhibitor under limited conditions (Brustovetsky and Dubinsky, 2000). In neuron cultures, FK506 and not cyclosporin A blocked neuronal apoptosis induced by serum deprivation (Yardin et al., 1998). FK-506 decreased infarct volume in rodent models of pMCAO and tMCAO and improved cell density after global ischemia. The therapeutic window in these studies ranged from 60 to 240 minutes and benefit was noted between 1 and 2 weeks depending on the model (Furuichi et al., 2003). It also inhibited release of ceramide (a mediator of apoptosis) and apoptosis in both in vitro and in vivo stroke models (Herr et al., 1999). FK506 reduced axonal injury in a model of impact acceleration TBI in rats (Singleton et al., 2001). Cyclosporin A also reduced lesion volume in models of tMCAO, forebrain ischemia, and optic nerve crush injury (Shiga et al., 1992; Uchino et al., 1995; Rosenstiel et al., 2003). Although it is clear that these agents have neuroprotective properties, the mechanism is not conclusively understood.

Neutrophil activating protein (NAP) is a synthetic vasoactive intestinal peptide analog with sequences similar to neuroprotective proteins. NAP significantly reduced motor disability and infarct volumes 24 hours after stroke onset. NAP was effective when administered 4 but not 6 hours after pMCAO and was effective up to 30 hours. Histological analysis found that NAP significantly reduced the number of apoptotic cells (Leker et al., 2002).

Nitric Oxide Inhibitors (Table 14.5)

NOS consist of three isoenzymes, neuronal (nNOS or type I), inducible (iNOS or type II), and endothelial (eNOS or type III). nNOS is calcium activated, PDZ domain associated and found in all brain regions, as well as skeletal muscle, prostate, pancreas, and testis (Park et al., 2000). iNOS is calcium independent, soluble, and, with a few exceptions in peripheral tissues, is expressed when induced by cell stressors (Park et al., 2000). eNOS is calcium activated, membrane bound via palmitoylation and myristoylation sites, and associates with calveolin.

nNOS and iNOS are thought to be hazardous in the setting of stroke, TBI, and SCI, whereas eNOS has beneficial effects through vasodilation and maintaining perfusion. To evaluate the effect of eNOS inhibition on cerebral vessels, the nonspecific NOS inhibitor NG-nitro-L-arginine methyl ester hydrochloride (L-NAME) was infused intra-arterially and blood vessels visualized with *in situ* microscopy. Pial arteries constricted and cortical CBF decreased by 72.5%. Ultrastructurally, microvascular stasis and BBB disruption occurred as well (Prado *et al.*, 1992). L-NAME increased infarct volumes between 103% and 176%, depending on brain region after MCAO (Kuluz *et al.*, 1993). Following CCAT in rats, L-NAME worsened sensorimotor (forelimb placing) and cognitive (water maze) tasks, and the selective nNOS inhibitor 3-bromo-7-nitroindazole (7-NI) accelerated sensorimotor recovery and improved histopathological outcome (Stagliano *et al.*, 1997a). Using a functional assay, cNOS activity (eNOS and nNOS) increased within 5 minutes after TBI by 234% and returned to control values by 30 minutes. cNOS activity was reduced 1 day after TBI to about 50% and remained low for up to 7 days. Pretreatment but not posttreatment with 7-NI decreased contusion volume (Wada *et al.*, 1998a) and improved sensorimotor function after TBI in rats (Wada *et al.*, 1999b). These data strongly support beneficial roles for eNOS and detrimental roles for nNOS.

The iNOS inhibitor aminoguanidine was used to determine the effect of iNOS inhibition after TBI. Although aminoguanidine did not reduce contusion volume, it did reduce the number of necrotic neurons (Wada *et al.*, 1998b). In functional assays following TBI, iNOS activity increased more than 275% by 3 days and more than 600% by 7 days. Immunostaining for iNOS and glial fibrillary acidic protein to identify astrocytes at 3 and 7 days found that most iNOS expression was in both astrocytes and macrophages within the subarachnoid space. iNOS was first hypothesized to play a role in SCI when the transcription nuclear factor-(NF) κB activation was observed after SCI. NF-κB is required for iNOS expression (Bethea *et al.*, 1998). Recently cNOS activity was increased by 6 hours and then decreased. iNOS activity was elevated at all time points after SCI. Acute administration of aminoguanidine significantly improved functional and histopathological outcome (Chatzipanteli *et al.*, 2002).

Caffeine and Alcohol (Table 14.5)

Despite the ability of caffeine and alcohol to constrict cerebral blood vessels (Nehlig *et al.*, 1992; Zhang *et al.*, 1993a), the combination is neuroprotective after ischemic stroke. After tMCAO in rats, treatment with ethanol plus caffeine administered between 30 and 120 minutes postischemia decreases infarct volume by 71.7% and 47.1%, respectively. However, chronic daily oral ethanol plus caffeine before ischemia eliminated the neuroprotection seen with acute treatment (Strong *et al.*, 2000). Outcome was further improved when caffeine and alcohol at concentrations comparable to 2 to 3 cups of strong coffee and a cocktail were combined with hypothermia. In addition, there was no increase in hemorrhage when combined with tPA (Aronowski *et al.*, 2003). A pilot study in patients found that the combination can be safely given and neuroprotective concentrations were reached (Piriyawat *et al.*, 2003). An outcomes trial needs to be planned before instituting the public service announcement, "Rum and Coke when you're having a stroke."

PITFALLS OF CLINICAL TRIALS (TABLE 14.6)

Once evidence for a therapeutic strategy has mounted in animal experiments, it may be attempted in clinical trials. Kidwell *et al.* (2001) reviewed 178 clinical trials for ischemic stroke from 1990 to 1999, over half of which studied neuroprotective agents. Although 23% reported benefits, the investigators found that less than 2% met strict criteria defining them as having a positive outcome. According to Faden (2002), the past 50 years of clinical trials in TBI resulted in no convincing evidence of a beneficial neuroprotective agent. With these realizations, investigators began a great deal of introspection into fundamental methods of research to assess reasons for discrepancy between preclinical and clinical outcomes (Faden, 2001; Gladstone *et al.*, 2002). A number

TABLE 14.6 Why Neuroprotective Agents Work in Animal Models and Not in Patients

Experiments are not repeated by different investigators.

Many animal experiments do not evaluate the therapeutic window to determine if the treatment will be clinically feasible.

Animal experiments use histopathology while clinical investigations use functional outcomes.

Adverse side effects are not adequately studied.

Therapeutic doses are not achieved and may be further attenuated by blood-brain barrier impermeability.

Therapies target pathomechanisms that are not common to all stroke patients.

Patients with different types of stroke damage are combined in the same study.

Outcome measures are not standardized, sensitive, and well defined.

Therapies may be influenced by gender, age, and other individual differences.

of problems have been identified, some relating to pre-clinical experiments and others to clinical trial design.

Animal Models

Anesthesia

Almost all animal protocols require the use of anesthesia in any situation that causes pain to the experimental animal. Although the use of anesthesia is necessary for humanitarian reasons, volatile anesthetics such as isoflurane and halothane reduced infarct volumes early after focal ischemia and limited selective neuronal necrosis at 14 days (Kawaguchi et al., 2000; Xiong et al., 2003). Using anesthesia and a neuroprotective agent is a combination therapy that is unavoidable in most preclinical studies but must be kept in mind when translating results to patients.

Dosing

Drug doses used in clinical trials often are based on animal studies. Determining minimally effective doses entails generating dose-response curves in multiple injury models by different groups of investigators. Generating dose-response curves is frequently overlooked because doing so quickly increases experimental cost in terms of time, personnel, materials, and number of animals. However, this determination is critical as are detecting any hazards of using higher or lower doses. Small-animal studies are not ideal for detecting harmful side effects because small animals tolerate drugs very well, are not routinely examined for systemic side effects, and are frequently sacrificed early after treatment; thus long-term consequences are unknown. To address these issues, the Stroke Therapy Academic Industry Roundtable (STAIR) recommended generating dose-response curves in models of both small and large animals (NIH Consensus Development Panel, 1999b). The more species tested, the more likely investigators are to detect harmful side effects. In addition, larger animals might be more prone to developing side effects similar to those of humans.

Difficulties in dosing and side effects were particularly marked in studies with the glutamate receptor antagonists Selfotel (CGS 19755) and MK-801. Doses of Selfotel at approximately 40 mg/kg were neuroprotective in a variety of animals, including gerbil (Boast et al., 1988) and rabbit models of ischemia (Perez-Pinzon et al., 1995). However, Phase IIa trials found the limit of safe doses in patients with stroke to be 1.5 mg/kg (Grotta et al., 1995). Higher doses led to psychotomimetic effects characteristic of glutamate receptor antagonists. Trials with the agent continued and two Phase III trials were eventually suspended because of an early increase in mortality in the Selfotel-treated group and a trend toward increased 90-day mortality (Davis et al., 2000). Dawson et al. (2001) retrospectively studied side effects of glutamate receptor antagonists, including Selfotel in rats following pMCAO. They used a vestibulomotor test (rotorod) to evaluate adverse neurological effects. They determined that agents such as Selfotel have detectable side effects giving them therapeutic ratios of less than 1, while more promising neuroprotective therapies have ratios greater than 1. Careful detection of side effects and the generation of dose-response curves may prevent drugs that are bound for failure from going to clinical trials. Modifying neuroprotective strategies based on therapeutic ratios may improve clinical trial success rate.

It is hoped that improvements in clinical trial design will alleviate some of these problems. Lees (2002) summarized a new design for clinical trials. Adaptive randomization design is a technique based on Bayesian statistics (Malakoff, 1999) that allows trial designers to use fewer patients in determining minimal effective doses. Although these techniques are not appropriate for conclusions regarding efficacy they will reduce numbers of inadequately or over-treated patients (Lees, 2002).

Therapeutic Window

Therapeutic window is the time range between injury and treatment during which the treatment is still effective. Time windows for neuroprotective agents that target early injury mechanisms are frequently over before patients can reasonably get treatment. Many therapies work best when administered before and immediately after the insult. Experiments with animal models often begin with a pretreatment protocol. If the therapy works, it is tested at different intervals from injury onset. Most effective therapies work best within 15 to 30 minutes; rarely are they effective more than 3 hours after injury (Grotta, 1999). Although there is little reason to assume that the therapeutic window in small animals is the same as the window in humans, less than 3 hours seems to fit both paradigms. Unfortunately, it is difficult to get patients to the hospital, evaluated, and enrolled in a clinical trial within 3 hours of symptom onset. More than 17,324 patients were screened to enroll the 624 subjects in the NINDS rt-PA trial, and most were excluded because of the time window. In the European Cooperative Acute Stroke Study II (ECASS II) that looked at rt-PA therapy with a treatment window between 0 and 6 hours, more than 80% of patients were treated between 3 and 6 hours and no effect was detected (Hacke et al., 1998). Had the NINDS trial chosen a longer window, a difference might not have been detected, and rt-PA might not have made it to clinical use. To avoid erroneously discarding a beneficial therapy, future trials should either strictly enforce enrollment or stratify patients into groups based on time-to-treat. Conducting randomized, double-blinded, placebo-controlled studies that stratify patients based on time-to-treat will require larger numbers of subjects and take longer to complete. However, these studies may give investigators the best chance of detecting efficacious therapies.

Preclinical studies often begin with a pretreatment design to improve chances of detecting an effect. Clinical correlates to pretreatment studies include administering neuroprotective agents to chronically high-risk patients or during procedures that carry risk of stroke such as carotid endarterectomy. Most procedures have relatively small risks of complications, so these studies would likely take a long time to acquire sufficient numbers of subjects, but they may help establish the neuroprotective potential of new therapeutics.

Therapeutic Targets: The Ischemic Penumbra

The ischemic penumbra can be broadly defined as that region of the ischemic zone that is potentially salvageable (Astrup et al., 1981). Because saving the penumbra is a promising treatment strategy, it is the focus of current neuroprotective research in stroke. The penumbra resides around the core infarct and is characterized by hemodynamic, metabolic, and molecular alterations (Takagi et al., 1993; Sharp et al., 2000). Most of the biochemical and histological descriptions of the penumbra were studied in models of MCA occlusion where single large vessel occlusion permits collaterals to feed areas adjacent to the ischemic focus; this produces a large, well-characterized penumbra amenable to neuroprotective agents.

Reductions in CBF after an ischemic or traumatic injury can lead to the generation of propagated or nonpropagated depressions in electrical activity commonly called cortical spreading depression (Leao, 1944). These ionic events have been shown to contribute to neuronal injury in vulnerable brain regions such as the ischemic penumbra (Back et al., 1994; Busch et al., 1996; Hossmann, 1996). Therapeutic interventions, including glutamate receptor blockers and therapeutic hypothermia, may act by inhibiting these potentially deleterious ionic events. Strong et al. (2002, 2003) have provided clinical evidence for the presence of cortical spreading depressions in the human cortex after focal injury.

Embolic models such as CCAT produce thrombi in the common carotid artery that embolize and occlude both large and small distal vessels. It is unclear whether a significant penumbra exists in these models. Instead of occluding one large vessel and permitting collaterals to feed the "penumbral" region, collateral flow may be reduced by embolic occlusion of collateral vessels. In another rat model of embolic stroke, injected thrombotic clots were observed occluding large vessels at early time points and microvessels at later time points. The authors concluded that the initial clots in large vessels break up and embolize distally to smaller vessels (Wang et al., 2001).

Other subtle pathophysiological features of embolic models have been described. For instance, vasoactive factors produced from thrombotic processes occurring in damaged vessels may reduce flow to significant portions of the cerebral hemisphere, further limiting collateral flow and the penumbral territory (Dietrich et al., 1991). In addition, platelet embolization results in alterations of vascular reactivity through regulation of endothelial nitric oxide synthase, suggesting that vessels exposed to platelet emboli may have altered hemodynamic control (Danton et al., 2002a).

Preconditioning studies with embolic and ischemic stroke also have illustrated differences between models. After a brief episode of pure ischemia, molecular events occur that protect the brain from future insults in a phenomenon known as ischemic preconditioning. When CCAT was followed by global ischemia, the ischemic lesion was much more severe than global ischemia alone, even at sites distant from embolic infarcts (Dietrich et al., 1999b). In contrast to what is expected of pure ischemia models, these results suggest that platelet embolization makes the brain susceptible to future insults, even in locations that were not infarcted. Human stroke is more likely to have similar properties such as emboli, and it is reasonable to question how these events will affect the presence of a penumbra.

Translating experimental animal data to clinical use for humans is made difficult because of the following: (1) differences in anatomy and physiology, (2) differences in pathophysiological response to injury, and (3) differences between injury mechanisms in animal models vs human disease. The first two problems are being addressed by using multiple species in each model. Before going to clinical trials, primates are used to better predict response in humans. The third problem is addressed by using multiple models. Therapies that work in many models are more likely to target fundamental pathomechanisms and apply to human disease. Unfortunately, investigators often gain expertise in only a few similar models with questionable applicability to the broad range of human disease. Furthermore, having multiple, independent laboratories replicate studies provides a stronger rationale for moving a therapy to the clinic as the probability of having multiple false-positive studies is reduced.

Clinical Studies of the Penumbra

Many studies report evidence of penumbral regions in patients after stroke (Hakim *et al.*, 1989; Schlaug *et al.*, 1999; Baron, 1999, 2001; Heiss, 2000) and have begun to address these key questions:

1. How can the penumbra be identified?
2. What proportion of stroke patients has a viable penumbra?
3. How large is the ischemic penumbra?
4. How long does the penumbra last?
5. Will saving the penumbra improve outcome?

Answers to these basic questions will enable clinicians and investigators to identify patients most likely to bene-fit from neuroprotective therapies.

Radiological techniques visualize regions representing ischemic penumbra in some stroke patients. To define a region as the penumbra, hemodynamic and metabolic parameters associated with impending infarction were identified in both animal and human studies. Positron emission tomography (PET) identified tissue with reduced CBF and regional cerebral metabolic rate of oxygen (rCMRO2) that represented infarcted areas and surround-ing regions that may be the penumbra (Hakim *et al.*, 1989; Heiss, 2000). Baron and colleagues combined PET, repeated CT imaging, and clinical correlations to conclude that penumbral tissue is variably represented in different patients. Penumbra was detected up to 16 hours in some and was absent by 5 hours in other patients (Baron, 1999).

Heiss *et al.* (2001) used PET with 60mCi H215O and 20mCi [11C] used flumazenil to measure both CBF and benzodiazepine receptor (BDZR) density, respectively. BDZR density is used as a marker of neuronal integrity. They established 95% probability limits of infarction (4.8 ± 0.53 ml/100 g/min) and survival (14.1 ± 1.83 ml/100g/min) for CBF and for flumazenil binding. They concluded that measuring these parameters enables iden-tification of brain regions amenable to neuroprotective therapies. In their cohort of 10 patients, 13% of final infarct volume had sufficient CBF and neuronal integrity to be capable of rescue (Heiss *et al.*, 2001). In three patients, this area was more than 45%, suggesting that some patients are more likely to benefit from neuropro-tective therapies than others. A complication in this study was that patients were assessed an average of 6 hours after onset of symptoms, but one would expect the per-centage of salvageable tissue to decrease over time, par-ticularly after the first 3 hours (Baron, 2001).

White vs Gray Matter

In addition to damage of the gray matter structures, white matter vulnerability is observed in anoxic, ischemic, and traumatic conditions (Waxman and Ranson, 1997; Dietrich *et al.*, 1998; Dewar *et al.*, 1999). Because the normal function of the brain is dependent on the communication between brain regions, clarifica-tion of mechanisms contributing to acute and/or pro-gressive white matter damage is critical to our understanding of brain injury (Pantoni and Garcia, 1995; Pantoni *et al.*, 1999). Following focal cerebral ischemia, as well as TBI, damage to the white matter tracts can be observed using a variety of immunocyto-chemical markers (Gentleman *et al.*, 1995; Dietrich *et al.*, 1998; Gillard *et al.*, 2001).

Preclinical studies have largely focused on protecting gray matter, as rats have a higher ratio of gray to white matter than humans (Gladstone *et al.*, 2002). Neuropro-tective therapies that work well on gray matter do not necessarily protect white matter. In this regard, one explanation for the lack of successful human stroke trials is that neuroprotective strategies have primarily targeted gray but not white matter structures. For example, MK-801 was neuroprotective against selective neuronal necro-sis in photothrombotic stroke (Yao *et al.*, 1994) and in global ischemia (Lin *et al.*, 1993), but did not attenuate white matter damage in a model of focal ischemia in cats (Yam *et al.*, 2000). The AMPA antagonist SPD 502 reduced damage in both white and gray matter after focal ischemia (McCracken *et al.*, 2002). In the future, MRI techniques that distinguish gray and white matter injuries may be useful to stratify patients for particular therapies that are beneficial against the patient's particular lesion.

Although the majority of neuroprotection studies have concentrated on neuronal preservation, significant progress was also made in the pathophysiology and treat-ment of white matter injury (Fern *et al.*, 1996; Waxman and Ranson, 1997; Dewar *et al.*, 1999). A critical role of Ca^{++} influx involving persistent Na^+ channels and the Na^+-Ca^{++} exchanger has been demonstrated in anoxic CNS white matter injury (Stys *et al.*, 1992a). L-type and N-type Ca^{++} channels also appear to provide routes for the influx of Ca^{++} (Fern *et al.*, 1995). The neurotransmit-ter, GABA, and the neuromodulator, adenosine, have also been shown to play a role in white matter injury (Fern *et al.*, 1994, 1996).

In terms of white matter protection, Na^+ channel blocking drugs, including tetrodotoxin-TTX and saxi-toxin, provide protection from anoxic injury (Stys *et al.*, 1992a), as well as a number of antiepileptic drugs includ-ing phenytoin and carbamazepine (Fern *et al.*, 1993). Use-dependent blockage of Na^+ channels is a proposed mechanism for neuroprotective actions of antiarrhythmic drugs (Waxman and Ranson, 1997). L-type calcium channel antagonists, verapamil, diltiazem, and nifedipine have each been reported to protect the optic nerve from anoxic injury (Fern et al., 1995). Anoxic protection is also seen with drugs that block the Na^+-Ca^{++} exchanger (Kleyman and Cragoe, 1988; Kaczorowski et al., 1989).

In the area of TBI, recent studies have emphasized both acute and progressive injury mechanisms underlying diffuse axonal injury (Povlishock, 1992). These include increased permeability abnormalities (Pettus et al., 1994), calcium-induced calpain-mediated proteolysis (Buki et al., 1999; Wolf et al., 2001), as well as caspase activation (Ozawa et al., 2002). Mitochondrial abnormalities have also been implicated in axonal perturbations following trauma (Okonkwo and Povlishock, 1999). Based on current evidence, white matter changes as a result of anoxic/ischemic or traumatic insults play a critical role in the neuropathological and functional consequences of these insults and are important targets for therapeutic interventions (Waxman and Ranson, 1997; Dewar et al., 2003).

Gender-Specific Injury Response

Experimental studies frequently use only male animals to reduce variability between mixing genders and to avoid accounting for the estrus cycle in experimental design. Recent data have established gender differences following experimentally induced stroke and TBI (Hurn and Macrae, 2000; Roof and Hall, 2000; Stein, 2001). Models of embolic stroke and global and focal ischemia described reduced injury in intact female vs male rodents (Hall et al., 1991; Li et al., 1996; Alkayed et al., 1998). These data also suggest that ovariectomized females have injuries more consistent with males (Alkayed et al., 1998; Bramlett and Dietrich, 2001). Testosterone was studied by comparing castrated vs intact males and was reported to worsen outcome after focal ischemia (Hawk et al., 1998). The Women's Health Initiative (WHI) study reported that estrogen and progesterone hormone replacement therapy increase risk of stroke (Wassertheil-Smoller et al., 2003). The estrogen only arm of the study is awaiting completion. Because stroke patients are likely to be postmenopausal, it is unclear whether gender differences relating to estrogen introduce variability into clinical trials. Recent experimental data in a model of TBI reported that therapeutic hypothermia does not improve outcome in females as it does in males (Suzuki et al., 2003a). It is unclear whether these results are similar in stroke models. Thus gender differences might complicate neuroprotection trials with regard to both outcome and therapeutic response.

Patient Selection for Clinical Trials

In contrast to preclinical studies using young, healthy animals under controlled physiological conditions, patients present with various stroke etiologies and comorbidities. Therapies targeting penumbra will not work in patients without a penumbra. When patients without a penumbra are enrolled in a study, their lack of improvement may statistically hide any positive effects of treatment. Radiological techniques that can screen for penumbra are being validated for incorporation into clinical trial selection criteria. The combination of diffusion-weighted imaging lesion volume, NIHSS score, and time from symptom onset to scanning was recently validated as a predictor of patient outcome (Baird et al., 2001). The Alberta Stroke Program Early CT Score (ASPECTS) divides the MCA territory into 10 sections, and 1 point is subtracted for each section containing hypoattenuation or evidence of early ischemic changes. Thus a normal score is 10 and lesion of the entire MCA territory is 0. A score of less than 7 yielded a 14-fold increase in risk of symptomatic intracerebral hemorrhage in patients treated with the thrombolytic ateplase, compared with scores above 7 (Barber et al., 2000). Tools such as these may provide a rapid means of identifying candidate patients.

Schlaug et al. (1999) identified penumbra in 25 patients presenting within 24 hours of hemispheric stroke onset, and they operationally defined the penumbra by comparing diffusion and perfusion MRI. To determine parameters best able to identify penumbral regions at greatest risk of infarction, Schaefer et al. (2003) examined diffusion-weighted and perfusion-weighted images in 30 stroke patients between 1 and 12 hours after symptom onset. Regional CBF distinguished between penumbra that infarcts and hypoperfused tissue that recovers. The combined efforts of these and other investigators are establishing the utility of neuroradiological techniques to improve clinical trials and enhance stroke treatment.

Outcome Measures

Outcome measures are often different between experimental trials and clinical trials (Gladstone et al., 2002). Preclinical studies often evaluate acute injury response, whereas late outcomes are most clinically important. Infarct size may be decreased early after injury but just delayed in progression. Final lesion volume may be the same (Schaefer et al., 2003). Preclinical studies often use infarct volume to determine efficacy, whereas clinical studies rely on behavioral measures. Behavioral measures were developed to evaluate functional recovery in animals. Rats often regain much of their behavioral functioning over time, even without treatment, and a battery of tests is most effective at detecting persistent deficits. Following CCAT, for example, rats have sensorimotor (forelimb placing) deficits up to 3 days postinjury, no difference in vestibulomotor (beam balance) skills, and only subtle impairments in cognitive tasks (Morris water maze) at 48 hours, but not 21 days, post injury (Stagliano et al., 1997a). Following MCAO, sensorimotor deficits recovered by day 30, but cognitive deficits remained up to 5 weeks (Markgraf et al., 1992). Avoiding these pitfalls ensures that all neuroprotective agents have established acute and chronic histopathological and behavioral benefits (NIH Consensus Development Panel, 1999b).

Outcome measures are also problematic when comparing clinical trials. Duncan et al. (2000) reviewed outcome measures in 51 large neuroprotection trials. Only 29 had defined outcome measures and time frames for their endpoint. There was considerable variability between studies, including 15 different impairment measures, 11 activity measures, 1 quality of life/health status measure, and 8 miscellaneous measures designed by the investigators. Well-defined and standardized outcome measures would aid in meta-analysis and cross-comparisons between trials.

Clinical trials are beginning to use objective radiological data as outcome measures. Physicians agree that functional outcomes are most important, but reduced lesion volumes lead to functional improvement. Radiological outcomes could help steer research in appropriate directions by detecting subtle therapeutic benefits. Functional improvements may require substantial tissue salvage best accomplished by multiple therapies, each making small contributions to limit infarction. Therapies that yield a clear radiological benefit but no functional improvement should not be discarded. Reevaluating dosages and time of administration may improve efficacy. Flaws in trial design and patient selection may contribute to failures in detecting improved functional outcome. Warach (2001) argued that clinical trials of 100 to 200 patients using radiological outcomes might have sufficient power to evaluate efficacy, but 5 to 10 times as many patients are required in Phase III trials. Thus studies with radiological outcomes may be a cost-effective approach for deciding whether or not to plan a Phase III trial (Warach, 2001).

TEMPERATURE

Small variations in brain temperature have profound effects on histopathological and behavioral outcome in models of global and focal cerebral ischemia. Whereas a 2° C reduction in intraischemic brain temperature is neu-roprotective, a small increase in brain temperature worsens outcome (Dietrich et al., 1990a, 1990b; Dietrich, 2001; Minamisawa et al., 1990). In an early study by Busto and colleagues (1987), intraischemic hypothermia following transient global ischemia protected the CA1 hippocampus and dorsolateral striatum. In models of cardiac arrest and cardiopulmonary bypass, hypothermia protection has also been reported by various investigators (Leonov et al., 1990). These beneficial effects of cooling on histopathological outcome have been complemented by electrophysiological and functional outcome measures. For example, Green and colleagues (1992) reported that intraischemic hypothermia at 30° C attenuated the cognitive and sensory motor consequences of global ischemia. In dogs, mild hypothermia at 34° C resulted in significant improvement in neurological function at postarrest (Leonov et al., 1990). Postischemic hypothermia provides long-term benefits if cooling is initiated soon after injury and continued for 24 hrs (Dietrich et al., 1993; Colbourne et al., 1997).

Although an early study of mild hypothermia found no protection (Morikawa et al., 1992) others have shown that more profound temperature reductions (24° C) after permanent focal ischemia did provide protection (Onesti et al., 1991). In addition, extended periods of mild hypothermia have also been reported to reduce infarct volume (Zhang et al., 1993c). Thus, it appears that in conditions of permanent focal ischemia, profound degrees of hypothermia and/or extended hypothermic durations may be necessary to significantly protect the injured brain.

In contrast to permanent occlusion, significant improvements in outcome have been shown in various models of transient MCA occlusion (Huh et al., 2000; Kawai et al., 2000). Selective brain hypothermia during the 2-hour period of transient occlusion significantly reduced infarct volume (Chen et al., 1992). A brain temperature reduction of 3° C and 8° C, respectively, has been reported to provide dramatic and complete neuroprotection after 80 minutes of transient MCA occlusion (Barone et al., 1997). One of the limitations of postischemic hypothermia appears to be the therapeutic window. Although dramatic protection is observed if hypothermia is induced during or immediately after the ischemic insult, fewer degrees of protection are observed as a delay in the initiation of hypothermia is demonstrated. Based on current experimental data, a 2 to 3 hour hypothermic window has been consistenly reported in the literature (Coimbra and Wieloch, 1994), indicating that cooling needs to be initiated soon after the primary insult.

Brain temperature has also been shown to play an important role in TBI. Clifton first reported that immediate posttraumatic hypothermia for 1 hour improved behavioral outcome after F-P injury (1991). Subsequent histopathological studies showed that moderate hypothermia (30° C/3 hr) reduced the frequency of cortical neuronal damage and contusion volume (Dietrich et al., 1994a). Mild-to-moderate hypothermia provides neuroprotection of white matter structures after anoxic or traumatic injury (Stys et al., 1992b; Marion and White, 1996; Koizumi and Povlishock, 1998). Finally, cognitive function is improved in posttraumatic animals after hypothermic therapy (Bramlett et al., 1995). Various pathophysiological mechanisms, including excitotoxicity, free radical production, metabolic stress, and inflammation, have been reported to be reduced by therapeutic hypothermic interventions (Dietrich et al., 1996).

A number of single institute studies have reported a benefit of hypothermic therapy in head-injured patients (Marion et al., 1997; Jiang et al., 2000; Polderman et al., 2002). In these studies, hypothermia has commonly been shown to reduce intracranial pressure (ICP) and improve the rate of favorable outcome without increasing complication rates. However, a recent multicenter trial failed to demonstrate an overall improvement with hypothermia (Clifton et al., 2001), although post hoc analysis showed a benefit to patients who were less than 45 years old and were hypothermic on admission (Clifton et al., 2002). Thus more investigations are required to determine the limitations of hypothermic therapy and the identification of specific patient populations that can best benefit from this treatment.

Several clinical studies have been initiated to test the beneficial effects of therapeutic hypothermia in various patient populations (Marion and White, 1996; Schwab et al., 1998; Kammersgaard et al., 2000; Clifton et al., 2001). The results of two multicenter trials were published that reported benefits of therapeutic hypothermia (Hypothermia after Cardiac Arrest Study Group, 2002; Bernard et al., 2002). In the European study nine centers in five countries participated, with a total enrollment of 275 patients. Systemic hypothermia for 24 hours combined with passive rewarming resulted in a favorable neurological outcome in 55% of hypothermic patients compared to 39% of normothermic patients. That two independent studies resulted in basically the same positive findings and conclusions is extremely positive for the field of therapeutic hypothermia. Some have argued that these results may place hypothermia as part of expected standard of care.

Recently, feasibility studies have been initiated to determine whether systemic hypothermia can be produced in stroke patients (Schwab et al., 1998, 2001). Schwab and colleagues (1998) induced moderate hypothermia in 25 patients with severe ischemic stroke of the MCA territory. Patients were kept at 33° C for 48 to 72 hours by surface cooling. In that pilot study hypothermia was shown to reduce ICP; however, during the rewarming period, brain hemorrhage and secondary ICP were documented. Investigators are currently studying the rewarming phase to determine how to control ICP and cerebral perfusion pressure during the critical posthypothermic period (Steiner et al., 2001). Also, relevant to the preclinical data, studies have shown reduced levels of glutamate in the cerebrospinal fluid of stroke patients who were cooled compared to normothermic subjects (Berger et al., 2002).

An important area of investigation is the development of better methods of patient cooling. In the past, bags of ice and body blankets have been used to produce a surface cooling with some success. As the necessity for more critical control of a patient's temperature becomes apparent, other strategies must be developed to accomplish this goal. To this end, new strategies for surface cooling, as well as endovascular cooling catheters, are being developed and tested (Georgiadis et al., 2001). These cooling devices are demonstrating that periods of hyperthermia (fever) in patients after CNS injury can be reduced in frequency as well as being used to produce systemic hypothermia. The continued development of this important field will improve the way we administer hypothermia and provide better protection and treatment for patients with neurological disorders.

ENDOGENOUS REPARATIVE PROCESSES

One of the most exciting discoveries in recent years in the area of neuroscience has been the observation that neurogenesis continues in the adult CNS after injury (McKay, 1997; Doetsch et al., 1999; Gage, 2000; Alvarez-Buylla and Garcia-Verdugo, 2002; Garcia-Verdugo et al., 2002). Using a variety of labeling techniques and methods of tracing neurogenesis after CNS injection, several laboratories have reported neurogenesis in areas such as the subventricular zone and dentate gyrus of the hippocampus (Liu et al., 1998; Zhang et al., 2001; Dash et al., 2001; Parent et al., 2002). For example, after cerebral ischemia, seizures, and TBI, there is a massive proliferation of stem and progenitor cells (Mak et al., 2003; Chen et al., 2003). The identification of the fate of these cells is an area of intensive investigation. In a model of F-P brain injury, the total number of proliferating cells identified with 5-bromo-deoxyuridine (BrdU), a marker of metabolic activity, was shown to significantly increase in the area of subventricular zone and hippocampus (Chirumamilla et al., 2002; Chen et al., 2003). After transient cerebral ischemia, increased numbers of BrdU-labeled neurons were detected in the dentate gyrus (Liu et al., 1998). Targeting this endogenous proliferative response to injury, including the use of neuroprotective agents, may be one way to enhance recovery after brain injury.

WHERE ARE WE GOING?

Despite the lack of clinically successful neuroprotective strategies, there has never been more excitement in the field of drug development targeting death of gray and white matter. Indeed, the clarification of various injury mechanisms has now emphasized the shortcomings of previous clinical trials that it is hoped will improve future studies. Also, preclinical modeling has been significantly improved, whereby therapeutic windows, long-term protection, and functional outcome measures are now incorporated into neuroprotective studies. Pharmacokinetics and methods of drug delivery are now emphasized in both preclinical and clinical investigations.

The importance of temperature variations on injury mechanisms has led to the first successful neuroprotective strategy in patients following cardiac arrest. Future progress will involve the development of safe and efficient methods of inducing hypothermia so that multicenter trials can be more rigorously evaluated. These methods will enhance the ability to cool patients rapidly, critically maintain temperatures within predetermined ranges, and control rates of posthypothermia rewarming. Prevention of periods of CNS hyperthermia with these cooling devices should also reduce the incidence of secondary insults, possibly allowing pharmacological strategies to provide better protection. Indeed, the use of mild hypothermia in combination with drug therapy may lead to synergistic effects and better outcomes.

Endogenous ischemic tolerance or ischemic preconditioning is a phenomenon whereby an ischemic or hypoxic insult is so brief that it causes no pathological damage and results in neuroprotective molecular and cellular changes. This phenomenon is interesting for two reasons. First, by understanding the mechanism of endogenous neuroprotection, perhaps new therapeutic strategies can be developed to exploit these systems. Second, some investigators have suggested that brief, nonlethal episodes of hypoxia or ischemia may be induced in patients who are likely to experience an ischemic event, such as those undergoing high-risk surgical procedures. Although induction of hypoxia/ischemia in patients is probably premature at this point, research is ongoing into how this phenomenon can be exploited. Ischemic preconditioning works through antiexcitotoxic, antiapoptotic, and anti-inflammatory mechanisms, and clinical trials are being planned to test pharmacological ischemic preconditions in high-risk patients (Dirnagl et al., 2003).

Future directions in neuroprotection may also make use of multiple drugs targeting specific therapeutic windows and pathomechanisms. This "combination therapy" approach may find its way into clinical trials, although current support for such therapies by drug companies is limited. The ability of drugs to target multiple pathomechanisms should maximize the chances of CNS protection in the postinjury setting.

It has become increasingly apparent that therapeutic interventions must be, to some degree, customized for each patient. Every patient with CNS injury is unique with respect to the size and location of structural lesions, presence or absence of secondary injury mechanisms, and genetic or environmental factors that differ among patient populations. Basing treatment decisions on indicators of specific tissue injury mechanisms is an attractive and possibly beneficial practice. Development and use of biomarkers and imaging strategies that indicate injury cascade activity or the progression of secondary injury mechanisms will help make this possible.

In the future, genetic factors that may predispose or protect a patient to a certain type of injury may help guide therapeutic interventions. Genetic screening may be developed for use in new treatment protocols. The continued study of endogenous neuroprotection mechanisms such as hypoxic/ischemic preconditioning should reveal new therapeutic strategies.

Because of the restricted windows of opportunity for most therapeutic interventions, treatments need to be initiated at the earliest postinjury period. Thus the administration of drugs at the accident site or by emergency room staff should increase efficacy. This point is particularly important because an important reason why clinical trials fail is that the drug or therapy was not initiated soon enough to stop the injury cascade.

Finally, neuroprotection treatments should not be limited to the postinjury setting alone. Patients at risk during elective surgical procedures may benefit from pretreatment strategies. Drug therapies or mild hypothermia could be used during complications from invasive brain or spinal cord procedures. Also, as the field of transplantation strategies advances, it may be advantageous to use neuroprotective strategies in combination with transplantation and regeneration strategies. Many of these approaches will be invasive, and the use of a neuroprotective strategy may limit the damage produced by the procedure. Continued discussions between researchers in both the experimental and clinical arenas should continue to help advance the development and testing of novel therapeutic strategies in patients with a variety of CNS injuries.

References

Adams JH. (1992). Head injury. In Adams JH, Cuchenn LW, editors: *Greenfield's neuropathology*, New York: Oxford University Press.

Akins PT, Atkinson RP. (2002). Glutamate AMPA receptor antagonist treatment for ischaemic stroke, *Curr Med Res Opin* 18(Suppl 2):s9-13.

Albers GW *et al.* (1999). Dose escalation study of the NMDA glycine-site antagonist licostinel in acute ischemic stroke, *Stroke* 30:508-513.

Alkayed NJ *et al.* (1998). Gender-linked brain injury in experimental stroke, *Stroke* 29:159-165; discussion 166.

Allen GS *et al.* (1983). Cerebral arterial spasm: a controlled trial of nimodipine in patients with subarachnoid hemorrhage, *N Engl J Med* 308:619-624.

Allen JW, Knoblach SM, Faden AI. (1999). Combined mechanical trauma and metabolic impairment in vitro induces NMDA receptor-dependent neuronal cell death and caspase-3-dependent apoptosis, *FASEB J* 13:1875-1882.

Alvarez-Buylla A, Garcia-Verdugo JM. (2002). Neurogenesis in adult subventricular zone, *J Neurosci* 22:629-634.

American Heart Association. (2002). *Heart disease and stroke statistics — 2003 update,* Dallas, TX: American Heart Association.

Anderson DK. (1985). Lipid hydrolysis and peroxidation in injured spinal cord: partial protection with methylprednisolone or vitamin E and selenium, *Central Nervous System Trauma* 2:257-267.

Aronowski J, Strong R, Shirzadi A, Grotta JC. (2003). Ethanol plus caffeine (caffeinol) for treatment of ischemic stroke: preclinical experience, *Stroke* 34:1246-1251.

Asahi M *et al.* (2000). Role for matrix metalloproteinase 9 after focal cerebral ischemia: effects of gene knockout and enzyme inhibition with BB-94, *J Cereb Blood Flow Metab* 20:1681-1689.

Astrup J, Siesjo BK, Symon L. (1981). Thresholds in cerebral ischemia: the ischemic penumbra, *Stroke* 12:723-725.

Auer RN *et al.* (1996). Postischemic therapy with MK-801 (dizocilpine) in a primate model of transient focal brain ischemia, *Mol Chem Neuropathol* 29:193-210.

Back T, Kohno K, Hossmann KA. (1994). Cortical negative DC deflections following middle cerebral artery occlusion and KCl-induced spreading depression: effect on blood flow, tissue oxygenation, and electroencephalogram, *J Cereb Blood Flow Metab* 14:12-19.

Back T *et al.* (1998). Ritanserin, a 5-HT2 receptor antagonist, increases subcortical blood flow following photothrombotic middle cerebral artery occlusion in rats, *Neurol Res* 20:643-647.

Bailey I *et al.* (1991). A trial of the effect of nimodipine on outcome after head injury, *Acta Neurochir (Wien)* 110:97-105.

Baird AE *et al.* (2001). A three-item scale for the early prediction of stroke recovery, *Lancet* 357:2095-2099.

Barber PA, Demchuk AM, Zhang J, Buchan AM. (2000). Validity and reliability of a quantitative computed tomography score in predicting outcome of hyperacute stroke before thrombolytic therapy. ASPECTS Study Group. Alberta Stroke Programme Early CT Score, *Lancet* 355:1670-1674.

Baron JC. (1999). Mapping the ischaemic penumbra with PET: implications for acute stroke treatment, *Cerebrovasc Dis* 9:193-201.

Baron JC. (2001). Mapping the ischaemic penumbra with PET: a new approach, *Brain* 124:2-4.

Barone FC, Feuerstein GZ, White RF. (1997). Brain cooling during transient focal ischemia provides complete neuroprotection, *Neurosci Biobehav Rev* 21:31-44.

Beckman JS, Koppenol WH. (1996). Nitric oxide, superoxide, and peroxynitrite: the good, the bad, and ugly, *Am J Physiol* 271:C1424-1437.

Belayev L *et al.* (1996). Middle cerebral artery occlusion in the rat by intraluminal suture. Neurological and pathological evaluation of an improved model, *Stroke* 27:1616-1622; discussion 1623.

Berger C *et al.* (2002). Effects of hypothermia on excitatory amino acids and metabolism in stroke patients: a microdialysis study, *Stroke* 33:519-524.

Bernard SA *et al.* (2002). Treatment of comatose survivors of out-of-hospital cardiac arrest with induced hypothermia, *N Engl J Med* 346:557-563.

Bethea JR *et al.* (1998). Traumatic spinal cord injury induces nuclear factor-kappaB activation, *J Neurosci* 18:3251-3260.

Bielenberg GW, Beck T. (1991a). The effects of dizocilpine (MK-801), phencyclidine, and nimodipine on infarct size 48 h after middle cerebral artery occlusion in the rat, *Brain Res* 552:338-342.

Bielenberg GW, Beck T. (1991b). The effects of dizocilpine (MK-801), phencyclidine, and nimodipine on infarct size 48 h after middle cerebral artery occlusion in the rat, *Brain Res* 552:338-342.

Boast CA *et al.* (1988). The N-methyl-D-aspartate antagonists CGS 19755 and CPP reduce ischemic brain damage in gerbils, *Brain Res* 442:345-348.

Bracken MB *et al.* (1990). A randomized, controlled trial of methylprednisolone or naloxone in the treatment of acute spinal-cord injury. Results of the Second National Acute Spinal Cord Injury Study, *N Engl J Med* 322:1405-1411.

Bracken MB *et al.* (1992). Methylprednisolone or naloxone treatment after acute spinal cord injury: 1-year follow-up data. Results of the Second National Acute Spinal Cord Injury Study, *J Neurosurg* 76:23-31.

Bramlett HM, Dietrich WD. (2001). Neuropathological protection after traumatic brain injury in intact female rats versus males or ovariectomized females, *J Neurotrauma* 18:891-900.

Bramlett HM, Dietrich WD. (2002). Quantitative structural changes in white and gray matter 1 year following traumatic brain injury in rats, *Acta Neuropathol (Berl)* 103:607-614.

Bramlett HM, Dietrich WD, Green EJ, Busto R. (1997a). Chronic histopathological consequences of fluid-percussion brain injury in rats: effects of post-traumatic hypothermia, *Acta Neuropathol (Berl)* 93:190-199.

Bramlett HM, Green EJ, Dietrich WD. (1997b). Hippocampally dependent and independent chronic spatial navigational deficits following parasagittal fluid percussion brain injury in the rat, *Brain Res* 762:195-202.

Bramlett HM, Green EJ, Dietrich WD. (1999). Exacerbation of cortical and hippocampal CA1 damage due to posttraumatic hypoxia following moderate fluid-percussion brain injury in rats, *J Neurosurg* 91:653-659.

Bramlett HM *et al.* (1995). Posttraumatic brain hypothermia provides protection from sensorimotor and cognitive behavioral deficits, *J Neurotrauma* 12:289-298.

Bramlett HM, Kraydieh S, Green EJ, Dietrich WD. (1997c). Temporal and regional patterns of axonal damage following traumatic brain injury: a beta-amyloid precursor protein immunocytochemical study in rats, *J Neuropathol Exp Neurol* 56:1132-1141.

Brustovetsky N, Dubinsky JM. (2000). Limitations of cyclosporin A inhibition of the permeability transition in CNS mitochondria, *J Neurosci* 20:8229-8237.

Buchan AM, Slivka A, Xue D. (1992). The effect of the NMDA receptor antagonist MK-801 on cerebral blood flow and infarct volume in experimental focal stroke, *Brain Res* 574:171-177.

Buki A, Siman R, Trojanowski JQ, Povlishock JT. (1999). The role of calpain-mediated spectrin proteolysis in traumatically induced axonal injury, *J Neuropathol Exp Neurol* 58:365-375.

Bullock MR *et al.* (1999). An open-label study of CP-101,606 in subjects with a severe traumatic head injury or spontaneous intracerebral hemorrhage, *Ann N Y Acad Sci* 890:51-58.

Busch E *et al.* (1996). Potassium-induced cortical spreading depressions during focal cerebral ischemia in rats: contribution to lesion growth assessed by diffusion-weighted NMR and biochemical imaging, *J Cereb Blood Flow Metab* 16:1090-1099.

Busto R *et al.* (1987). Small differences in intraischemic brain temperature critically determine the extent of ischemic neuronal injury, *J Cereb Blood Flow Metab* 7:729-738.

Cao X, Phillis JW. (1994). Alpha-phenyl-tert-butyl-nitrone reduces cortical infarct and edema in rats subjected to focal ischemia, *Brain Res* 644:267-272.

Chatzipanteli K *et al.* (2002). Temporal and segmental distribution of constitutive and inducible nitric oxide synthases after traumatic spinal cord injury: effect of aminoguanidine treatment, *J Neurotrauma* 19:639-651.

Chen H, Chopp M, Zhang ZG, Garcia JH. (1992). The effect of hypothermia on transient middle cerebral artery occlusion in the rat, *J Cereb Blood Flow Metab* 12:621-628.

Chen XH *et al.* (2003). Neurogenesis and glial proliferation persist for at least one year in the subventricular zone following brain trauma in rats, *J Neurotrauma* 20:623-631.

Cheng Y *et al.* (1998). Caspase inhibitor affords neuroprotection with delayed administration in a rat model of neonatal hypoxic-ischemic brain injury, *J Clin Invest* 101:1992-1999.

Chirumamilla S, Sun D, Bullock MR, Colello RJ. (2002). Traumatic brain injury induced cell proliferation in the adult mammalian central nervous system, *J Neurotrauma* 19:693-703.

Christman CW *et al.* (1994). Ultrastructural studies of diffuse axonal injury in humans, *J Neurotrauma* 11:173-186.

Clark WM, Lutsep HL. (1999). Medical treatment strategies: intravenous thrombolysis, neuronal protection, and anti-reperfusion injury agents, *Neuroimag Clin North Am* 9:465-473.

Clifton GL *et al.* (2002). Hypothermia on admission in patients with severe brain injury, *J Neurotrauma* 19:293-301.

Clifton GL *et al.* (2001). Lack of effect of induction of hypothermia after acute brain injury, *N Engl J Med* 344:556-563.

Coimbra C, Wieloch T. (1994). Moderate hypothermia mitigates neuronal damage in the rat brain when initiated several hours following transient cerebral ischemia, *Acta Neuropathol (Berl)* 87:325-331.

Colbourne F, Sutherland G, Corbett D. (1997). Postischemic hypothermia. A critical appraisal with implications for clinical treatment, *Mol Neurobiol* 14:171-201.

Connolly ES Jr *et al.* (1997). Exacerbation of cerebral injury in mice that express the P-selectin gene: identification of P-selectin blockade as a new target for the treatment of stroke, *Circ Res* 81:304-310.

Cortez SC, McIntosh TK, Noble LJ. (1989). Experimental fluid percussion brain injury: vascular disruption and neuronal and glial alterations, *Brain Res* 482:271-282.

Cowley DJ, Lukovic L, Petty MA. (1996). MDL 74,180 reduces cerebral infarction and free radical concentrations in rats subjected to ischaemia and reperfusion, *Eur J Pharmacol* 298:227-233.

Danton G, Watson BD, Prado R, Dietrich WD. (2002c). Endothelium-targeted pharmacotherapeutics for the treatment of stroke, *Curr Opin Investig Drugs* 3:896-904.

Danton GH, Dietrich WD. (2003). Inflammatory mechanisms after ischemia and stroke, *J Neuropathol Exp Neurol* 62:127-136.

Danton GH *et al.* (2002a). Endothelial nitric oxide synthase pathophysiology after nonocclusive common carotid artery thrombosis in rats, *J Cereb Blood Flow Metab* 22:612-619.

Danton GH, Prado R, Watson BD, Dietrich WD. (2002b). Temporal profile of enhanced vulnerability of the postthrombotic brain to secondary embolic events, *Stroke* 33:1113-1119.

Dash PK, Mach SA, Moore AN. (2001). Enhanced neurogenesis in the rodent hippocampus following traumatic brain injury, *J Neurosci Res* 63:313-319.

Davis SM *et al.* (1997). Termination of acute stroke studies involving selfotel treatment. ASSIST Steering Committee, *Lancet* 349:32.

Davis SM *et al.* (2000). Selfotel in acute ischemic stroke : possible neurotoxic effects of an NMDA antagonist, *Stroke* 31:347-354.

Dawson DA, Wadsworth G, Palmer AM. (2001). A comparative assessment of the efficacy and side-effect liability of neuroprotective compounds in experimental stroke, *Brain Res* 892:344-350.

De Vry J, Horvath E, Schreiber R. (2001). Neuroprotective and behavioral effects of the selective metabotropic glutamate mGlu(1) receptor antagonist BAY 36-7620, *Eur J Pharmacol* 428:203-214.

del Zoppo GJ. (1997). Microvascular responses to cerebral ischemia/inflammation, *Ann N Y Acad Sci* 823:132-147.

del Zoppo GJ, Hallenbeck JM. (2000). Advances in the vascular pathophysiology of ischemic stroke, *Thromb Res* 98:73-81.

del Zoppo GJ *et al.* (1991). Polymorphonuclear leukocytes occlude capillaries following middle cerebral artery occlusion and reperfusion in baboons, *Stroke* 22:1276-1283.

Dewar D, Underhill SM, Goldberg MP. (2003). Oligodendrocytes and ischemic brain injury, *J Cereb Blood Flow Metab* 23:263-274.

Dewar D, Yam P, McCulloch J. (1999). Drug development for stroke: importance of protecting cerebral white matter, *Eur J Pharmacol* 375:41-50.

DeWitt DS *et al.* (1986). Effects of fluid-percussion brain injury on regional cerebral blood flow and pial arteriolar diameter, *J Neurosurg* 64:787-794.

Di X *et al.* (1997). Effect of CP101,606, a novel NR2B subunit antagonist of the N-methyl-D-aspartate receptor, on the volume of ischemic brain damage of cytotoxic brain edema after middle cerebral artery occlusion in the feline brain, *Stroke* 28:2244-2251.

Dietrich WD. (2001). Temperature changes and ischemic stroke. In Fisher M, Bogousslavsky J, editors: *Current review of cerebrovascular disease,*: Current Medicine, Inc.

Dietrich WD *et al.* (1994a). Post-traumatic brain hypothermia reduces histopathological damage following concussive brain injury in the rat, *Acta Neuropathol (Berl)* 87:250-258.

Dietrich WD *et al.* (1996). Widespread hemodynamic depression and focal platelet accumulation after fluid percussion brain injury: a double-label autoradiographic study in rats, *J Cereb Blood Flow Metab* 16:481-489.

Dietrich WD, Alonso O, Halley M. (1994b). Early microvascular and neuronal consequences of traumatic brain injury: a light and electron microscopic study in rats, *J Neurotrauma* 11:289-301.

Dietrich WD *et al.* (1992a). Intraventricular infusion of N-methyl-D-aspartate. 1. Acute blood-brain barrier consequences, *Acta Neuropathol (Berl)* 84:621-629.

Dietrich WD *et al.* (1993). Intraischemic but not postischemic brain hypothermia protects chronically following global forebrain ischemia in rats, *J Cereb Blood Flow Metab* 13:541-549.

Dietrich WD, Busto R, Bethea JR. (1999a). Postischemic hypothermia and IL-10 treatment provide long-lasting neuroprotection of CA1 hippocampus following transient global ischemia in rats, *Exp Neurol* 158:444-450.

Dietrich WD, Busto R, Ginsberg MD. (1989a). Effect of the serotonin antagonist ketanserin on the hemodynamic and morphological consequences of thrombotic infarction, *J Cereb Blood Flow Metab* 9:812-820.

Dietrich WD, Busto R, Globus MY, Ginsberg MD. (1996). Brain damage and temperature: cellular and molecular mechanisms, *Adv Neurol* 71:177-194; discussion 194-197.

Dietrich WD, Busto R, Halley M, Valdes I. (1990a). The importance of brain temperature in alterations of the blood-brain barrier following cerebral ischemia, *J Neuropathol Exp Neurol* 49:486-497.

Dietrich WD, Busto R, Valdes I, Loor Y. (1990b). Effects of normothermic versus mild hyperthermic forebrain ischemia in rats, *Stroke* 21:1318-1325.

Dietrich WD, Danton G, Hopkins AC, Prado R. (1999b). Thromboembolic events predispose the brain to widespread cerebral infarction after delayed transient global ischemia in rats, *Stroke* 30:855-861; discussion 862.

Dietrich WD *et al.* (1992b). Intraventricular infusion of N-methyl-D-aspartate. 2. Acute neuronal consequences, *Acta Neuropathol (Berl)* 84:630-637.

Dietrich WD, Kraydieh S, Prado R, Stagliano NE. (1998). White matter alterations following thromboembolic stroke: a beta-amyloid precursor protein immunocytochemical study in rats, *Acta Neuropathol (Berl)* 95:524-531.

Dietrich WD *et al.* (1995a). Effect of delayed MK-801 (dizocilpine) treatment with or without immediate postischemic hypothermia on chronic neuronal survival after global forebrain ischemia in rats, *J Cereb Blood Flow Metab* 15:960-968.

Dietrich WD *et al.* (1995b). Acadesine reduces indium-labeled platelet deposition after photothrombosis of the common carotid artery in rats, *Stroke* 26:111-116.

Dietrich WD, Nakayama H, Watson BD, Kanemitsu H. (1989b). Morphological consequences of early reperfusion following thrombotic or mechanical occlusion of the rat middle cerebral artery, *Acta Neuropathol (Berl)* 78:605-614.

Dietrich WD *et al.* (1991). Hemodynamic consequences of common carotid artery thrombosis and thrombogenically activated blood in rats, *J Cereb Blood Flow Metab* 11:957-965.

Dirnagl U, Iadecola C, Moskowitz MA. (1999). Pathobiology of ischaemic stroke: an integrated view, *Trends Neurosci* 22:391-397.

Dirnagl U, Simon RP, Hallenbeck JM. (2003). Ischemic tolerance and endogenous neuroprotection, *Trends Neurosci* 26:248-254.

Dirnagl U, Tanabe J, Pulsinelli W. (1990). Pre- and post-treatment with MK-801 but not pretreatment alone reduces neocortical damage after focal cerebral ischemia in the rat, *Brain Res* 527:62-68.

Dixon CE *et al.* (1991). A controlled cortical impact model of traumatic brain injury in the rat, *J Neurosci Methods* 39:253-262.

Dixon CE *et al.* (1999). One-year study of spatial memory performance, brain morphology, and cholinergic markers after moderate controlled cortical impact in rats, *J Neurotrauma* 16:109-122.

Dixon CE *et al.* (1987). A fluid percussion model of experimental brain injury in the rat, *J Neurosurg* 67:110-119.

Doetsch F, Garcia-Verdugo JM, Alvarez-Buylla A. (1999). Regeneration of a germinal layer in the adult mammalian brain, *Proc Natl Acad Sci U S A* 96:11619-11624.

Doppenberg EM, Bullock R. (1997). Clinical neuro-protection trials in severe traumatic brain injury: lessons from previous studies, *J Neurotrauma* 14:71-80.

Duffy SJ *et al.* (2001). Iron chelation improves endothelial function in patients with coronary artery disease, *Circulation* 103:2799-2804.

Duncan PW, Jorgensen HS, Wade DT. (2000). Outcome measures in acute stroke trials: a systematic review and some recommendations to improve practice, *Stroke* 31:1429-1438.

Dyker AG *et al.* (1999). Safety and tolerability study of aptiganel hydrochloride in patients with an acute ischemic stroke, *Stroke* 30:2038-2042.

Eliasson MJ *et al.* (1999). Neuronal nitric oxide synthase activation and peroxynitrite formation in ischemic stroke linked to neural damage, *J Neurosci* 19:5910-5918.

Ellis EF *et al.* (1995). A new model for rapid stretch-induced injury of cells in culture: characterization of the model using astrocytes, *J Neurotrauma* 12:325-339.

Endres M, Dirnagl U. (2002). Ischemia and stroke, *Adv Exp Med Biol* 513:455-473.

Enlimomab Acute Stroke Trial Investigators (2001). Use of anti-ICAM-1 therapy in ischenic stroke: results of the Enlimomab Acute Stroke Trial, *Neurology* 57:1428-1434.

The European Study Group on Nimodipine in Severe Head Injury. (1994). A multicenter trial of the efficacy of nimodipine on outcome after severe head injury, *J Neurosurg* 80:797-804.

Faden AI. (2001). Neuroprotection and traumatic brain injury: the search continues, *Arch Neurol* 58:1553-1555.

Faden AI. (2002). Neuroprotection and traumatic brain injury: theoretical option or realistic proposition. *Curr Opin Neurol* 15:707-712.

Fern R, Ransom BR, Stys PK, Waxman SG. (1993). Pharmacological protection of CNS white matter during anoxia: actions of phenytoin, carbamazepine and diazepam, *J Pharmacol Exp Ther* 266:1549-1555.

Fern R, Ransom BR, Waxman SG. (1995). Voltage-gated calcium channels in CNS white matter: role in anoxic injury, *J Neurophysiol* 74:369-377.

Fern R, Ransom BR, Waxman SG. (1996). Autoprotective mechanisms in the CNS: some new lessons from white matter, *Mol Chem Neuropathol* 27:107-129.

Fern R, Waxman SG, Ransom BR. (1994). Modulation of anoxic injury in CNS white matter by adenosine and interaction between adenosine and GABA, *J Neurophysiol* 72:2609-2616.

Flor PJ *et al.* (2002). Neuroprotective activity of metabotropic glutamate receptor ligands, *Adv Exp Med Biol* 513:197-223.

Frandsen A, Schousboe A. (1991). Dantrolene prevents glutamate cytotoxicity and Ca2+ release from intracellular stores in cultured cerebral cortical neurons, *J Neurochem* 56:1075-1078.

Franke CL *et al.* (1996). Flunarizine in stroke treatment (FIST): a double-blind, placebo-controlled trial in Scandinavia and the Netherlands, *Acta Neurol Scand* 93:56-60.

Furuichi Y *et al.* (2003). Neuroprotective action of tacrolimus (FK506) in focal and global cerebral ischemia in rodents: dose dependency, therapeutic time window and long-term efficacy, *Brain Res* 965:137-145.

Furukawa T *et al.* (2003). The glutamate AMPA receptor antagonist, YM872, attenuates cortical tissue loss, regional cerebral edema, and neurological motor deficits after experimental brain injury in rats, *J Neurotrauma* 20:269-278.

Furuya K *et al.* (2001). Examination of several potential mechanisms for the negative outcome in a clinical stroke trial of enlimomab, a murine anti-human intercellular adhesion molecule-1 antibody: a bedside-to-bench study, *Stroke* 32:2665-2674.

Gage FH. (2000). Mammalian neural stem cells, *Science* 287:1433-1438.

Garcia JH *et al.* (1994). Influx of leukocytes and platelets in an evolving brain infarct (Wistar rat), *Am J Pathol* 144:188-199.

Garcia-Verdugo JM *et al.* (2002). The proliferative ventricular zone in adult vertebrates: a comparative study using reptiles, birds, and mammals, *Brain Res Bull* 57:765-775.

Gelmers HJ. (1984). The effects of nimodipine on the clinical course of patients with acute ischemic stroke, *Acta Neurol Scand* 69:232-239.

Gelmers HJ, Gorter K, de Weerdt CJ, Wiezer HJ. (1988). A controlled trial of nimodipine in acute ischemic stroke, *N Engl J Med* 318:203-207.

Gennarelli TA. (1994). Animate models of human head injury, *J Neurotrauma* 11:357-368.

Gennarelli TA, Graham DI. (1998). Neuropathology of the head injuries, *Semin Clin Neuropsychiatry* 3:160-175.

Gentleman SM *et al.* (1995). Axonal injury: a universal consequence of fatal closed head injury?, *Acta Neuropathol (Berl)* 89:537-543.

Georgiadis D, Schwarz S, Kollmar R, Schwab S. (2001). Endovascular cooling for moderate hypothermia in patients with acute stroke: first results of a novel approach, *Stroke* 32:2550-2553.

Gillard JH *et al.* (2001). MR diffusion tensor imaging of white matter tract disruption in stroke at 3 T, *Br J Radiol* 74:642-647.

Gillardon F *et al.* (1999). Inhibition of caspases prevents cell death of hippocampal CA1 neurons, but not impairment of hippocampal long-term potentiation following global ischemia, *Neuroscience* 93:1219-1222.

Gladstone DJ, Black SE, Hakim AM. (2002). Toward wisdom from failure: lessons from neuroprotective stroke trials and new therapeutic directions, *Stroke* 33:2123-2136.

Globus MY et al. (1995). Glutamate release and free radical production following brain injury: effects of posttraumatic hypothermia, J Neurochem 65:1704-1711.

Globus MY, Wester P, Busto R, Dietrich WD. (1992). Ischemia-induced extracellular release of serotonin plays a role in CA1 neuronal cell death in rats, Stroke 23:1595-1601.

Goussev AV, Zhang Z, Anderson DC, Chopp M. (1998). P-selectin antibody reduces hemorrhage and infarct volume resulting from MCA occlusion in the rat, J Neurol Sci 161:16-22.

Graham DI. (1996). Neuropathology of head injury. In Narayan RK et al., editor: Neurotrauma, New York: McGraw Hill.

Graham DI, Adams JH. (1971). Ischaemic brain damage in fatal head injuries, Lancet 1:265-266.

Graham DI et al. (1995). The nature, distribution and causes of traumatic brain injury, Brain Pathol 5:397-406.

Graham DI, Gennarelli TA. (1997). Trauma. In Graham DH, Lantos PL, editors: Greenfield's neuropathology, New York: Oxford University Press.

Graham DI et al. (2000). Tissue tears in the white matter after lateral fluid percussion brain injury in the rat: relevance to human brain injury, Acta Neuropathol (Berl) 99:117-124.

Green EJ et al. (1992). Protective effects of brain hypothermia on behavior and histopathology following global cerebral ischemia in rats, Brain Res 580:197-204.

Green EJ et al. (1995). Combined postischemic hypothermia and delayed MK-801 treatment attenuates neurobehavioral deficits associated with transient global ischemia in rats, Brain Res 702:145-152.

Gribkoff VK et al. (2001). Targeting acute ischemic stroke with a calcium-sensitive opener of maxi-K potassium channels, Nat Med 7:471-477.

Grill R et al. (1997). Cellular delivery of neurotrophin-3 promotes corticospinal axonal growth and partial functional recovery after spinal cord injury, J Neurosci 17:5560-5572.

Grotta J et al. (1995). Safety and tolerability of the glutamate antagonist CGS 19755 (Selfotel) in patients with acute ischemic stroke. Results of a phase IIa randomized trial, Stroke 26:602-605.

Grotta JC. (1999). Acute stroke therapy at the millennium: consummating the marriage between the laboratory and bedside. The Feinberg lecture, Stroke 30:1722-172.

Gruner JA. (1992). A monitored contusion model of spinal cord injury in the rat, J Neurotrauma 9:123-126; discussion 126-128.

Gur AY, Bova I, Bornstein NM. (1996). Is impaired cerebral vasomotor reactivity a predictive factor of stroke in asymptomatic patients?, Stroke 27:2188-2190.

Hacke W et al. (1998). Randomised double-blind placebo-controlled trial of thrombolytic therapy with intravenous alteplase in acute ischaemic stroke (ECASS II). Second European-Australasian Acute Stroke Study Investigators, Lancet 352:1245-1251.

Hakim AM et al. (1989). The effect of nimodipine on the evolution of human cerebral infarction studied by PET, J Cereb Blood Flow Metab 9:523-534.

Hall ED. (1982). Glucocorticoid effects on central nervous excitability and synaptic transmission, Int Rev Neurobiol 23:165-195.

Hall ED. (1992). The neuroprotective pharmacology of methylprednisolone, J Neurosurg 76:13-22.

Hall ED et al. (1996). Neuroprotective efficacy of microvascularly-localized versus brain-penetrating antioxidants, Acta Neurochir Suppl (Wien) 66:107-113.

Hall ED, Pazara KE, Linseman KL. (1991). Sex differences in postischemic neuronal necrosis in gerbils, J Cereb Blood Flow Metab 11:292-298.

Hawk T et al. (1998). Testosterone increases and estradiol decreases middle cerebral artery occlusion lesion size in male rats, Brain Res 796:296-298.

Heiss WD. (2000). Ischemic penumbra: evidence from functional imaging in man, J Cereb Blood Flow Metab 20:1276-1293.

Heiss WD et al. (2001). Penumbral probability thresholds of cortical flumazenil binding and blood flow predicting tissue outcome in patients with cerebral ischaemia, Brain 124:20-29.

Herr I et al. (1999). FK506 prevents stroke-induced generation of ceramide and apoptosis signaling, Brain Res 826:210-219.

Hidalgo M, Eckhardt SG. (2001). Development of matrix metalloproteinase inhibitors in cancer therapy, J Natl Cancer Inst 93:178-193.

Hong SJ, Chiou GC. (1998). Effects of intracellular calcium reduction by dantrolene on prevention/treatment of ischemic stroke, J Cardiovasc Pharmacol Ther 3:299-304.

Horn J, Limburg M. (2001). Calcium antagonists for ischemic stroke: a systematic review, Stroke 32:570-576.

Hossmann KA. (1996). Periinfarct depolarizations, Cerebrovasc Brain Metab Rev 8:195-208.

Huh PW et al. (2000). Comparative neuroprotective efficacy of prolonged moderate intraischemic and postischemic hypothermia in focal cerebral ischemia, J Neurosurg 92:91-99.

Hurn PD, Macrae IM. (2000). Estrogen as a neuroprotectant in stroke, J Cereb Blood Flow Metab 20:631-652.

Hypothermia After Cardiac Arrest Study Group. (2002). Mild therapeutic hypothermia to improve the neurologic outcome after cardiac arrest, N Engl J Med 346:549-556.

Imai H, Graham DI, Masayasu H, Macrae IM. (2003). Antioxidant ebselen reduces oxidative damage in focal cerebral ischemia, Free Radic Biol Med 34:56-63.

Inman D, Guth L, Steward O. (2002). Genetic influences on secondary degeneration and wound healing following spinal cord injury in various strains of mice, J Comp Neurol 451:225-235.

Ito U, Spatz M, Walker JT Jr, Klatzo I. (1975). Experimental cerebral ischemia in mongolian gerbils. I. Light microscopic observations, Acta Neuropathol (Berl) 32:209-223.

Jain KK. (2000). Neuroprotection in cerebrovascular disease, Expert Opin Investig Drugs 9:695-711.

Jan M, Buchheit F, Tremoulet M. (1988). Therapeutic trial of intravenous nimodipine in patients with established cerebral vasospasm after rupture of intracranial aneurysms, Neurosurgery 23:154-157.

Jensen BS. (2002). BMS-204352: a potassium channel opener developed for the treatment of stroke, CNS Drug Rev 8:353-360.

Jiang J, Yu M, Zhu C. (2000). Effect of long-term mild hypothermia therapy in patients with severe traumatic brain injury: 1-year follow-up review of 87 cases, J Neurosurg 93:546-549.

Jiang X, Namura S, Nagata I. (2001). Matrix metalloproteinase inhibitor KB-R7785 attenuates brain damage resulting from permanent focal cerebral ischemia in mice, Neurosci Lett 305:41-44.

Kaczorowski GJ, Slaughter RS, King VF, Garcia ML. (1989). Inhibitors of sodium-calcium exchange: identification and development of probes of transport activity, Biochim Biophys Acta 988:287-302.

Kammersgaard LP et al. (2000). Feasibility and safety of inducing modest hypothermia in awake patients with acute stroke through surface cooling: a case-control study: the Copenhagen Stroke Study, Stroke 31:2251-2256.

Kato H, Kogure K, Araki T, Itoyama Y. (1995). Graded expression of immunomolecules on activated microglia in the hippocampus following ischemia in a rat model of ischemic tolerance, Brain Res 694:85-93.

Katz LM et al. (2001). Regulation of caspases and XIAP in the brain after asphyxial cardiac arrest in rats, Neuroreport 12:3751-3754.

Kawaguchi K, Graham SH. (1997). Neuroprotective effects of the glutamate release inhibitor 619C89 in temporary middle cerebral artery occlusion, Brain Res 749:131-134.

Kawaguchi M et al. (2000). Isoflurane delays but does not prevent cerebral infarction in rats subjected to focal ischemia, Anesthesiology 92:1335-1342.

Kawai H et al. (1997). Effects of a novel free radical scavenger, MCI-186, on ischemic brain damage in the rat distal middle cerebral artery occlusion model, J Pharmacol Exp Ther 281:921-927.

Kawai N, Okauchi M, Morisaki K, Nagao S. (2000). Effects of delayed intraischemic and postischemic hypothermia on a focal model of transient cerebral ischemia in rats, Stroke 31:1982-1989; discussion 1989.

Kawasaki-Yatsugi S et al. (2000). Neuroprotective effects of an AMPA receptor antagonist YM872 in a rat transient middle cerebral artery occlusion model, Neuropharmacology 39:211-217.

Keane RW et al. (2001). Apoptotic and anti-apoptotic mechanisms following spinal cord injury, J Neuropathol Exp Neurol 60:422-429.

Kermer P, Klocker N, Bahr M. (1999). Neuronal death after brain injury. Models, mechanisms, and therapeutic strategies in vivo, Cell Tissue Res 298:383-395.

Kidwell CS, Liebeskind DS, Starkman S, Saver JL. (2001). Trends in acute ischemic stroke trials through the 20th century, Stroke 32:1349-1359.

Kirino T, Tamura A, Sano K. (1984). Delayed neuronal death in the rat hippocampus following transient forebrain ischemia, Acta Neuropathol (Berl) 64:139-147.

Kleyman TR, Cragoe EJ Jr. (1988). Amiloride and its analogs as tools in the study of ion transport, J Membr Biol 105:1-21.

Koizumi H, Povlishock JT. (1998). Posttraumatic hypothermia in the treatment of axonal damage in an animal model of traumatic axonal injury, J Neurosurg 89:303-309.

Kontos HA. (2001). Oxygen radicals in cerebral ischemia: the 2001 Willis lecture, Stroke 32:2712-2716.

Kotapka MJ, Graham DI, Adams JH, Gennarelli TA. (1992). Hippocampal pathology in fatal non-missile human head injury, Acta Neuropathol (Berl) 83:530-534.

Kross J, Fleischer JE, Milde JH, Gronert GA. (1993). No dantrolene protection in a dog model of complete cerebral ischaemia, Neurol Res 15:37-40.

Kuluz JW et al. (1993). The effect of nitric oxide synthase inhibition on infarct volume after reversible focal cerebral ischemia in conscious rats, Stroke 24:2023-2029.

Lapchak PA, Chapman DF, Zivin JA. (2000). Metalloproteinase inhibition reduces thrombolytic (tissue plasminogen activator)-induced hemorrhage after thromboembolic stroke, Stroke 31:3034-3040.

Leao AAP. (1944). Spreading depression of activity in cerebral cortex, J Neurophysiol 7:359-390.

Lees KR. (1997). Cerestat and other NMDA antagonists in ischemic stroke, Neurology 49:S66-69.

Lees KR. (2002). Prospects for improved neuroprotection trials in stroke. In Cerebrovascular disease united kingdom, Cambridge: Cambridge University Press.

Leker RR et al. (2002). NAP, a femtomolar-acting peptide, protects the brain against ischemic injury by reducing apoptotic death, Stroke 33:1085-1092.

Leonov Y et al. (1990). Mild cerebral hypothermia during and after cardiac arrest improves neurologic outcome in dogs, J Cereb Blood Flow Metab 10:57-70.

Li K et al. (1996). Gender influences the magnitude of the inflammatory response within embolic cerebral infarcts in young rats, Stroke 27:498-503.

Limburg M, Hijdra A. (1990). Flunarizine in acute ischemic stroke: a pilot study, Eur Neurol 30:121-122.

Lin B et al. (1993). MK-801 (dizocilpine) protects the brain from repeated normothermic global ischemic insults in the rat, J Cereb Blood Flow Metab 13:925-932.

Lin BW, Dietrich WD, Busto R, Ginsberg MD. (1990). (S)-emopamil protects against global ischemic brain injury in rats, Stroke 21:1734-1739.

Liou AK et al. (2003). To die or not to die for neurons in ischemia, traumatic brain injury and epilepsy: a review on the stress-activated signaling pathways and apoptotic pathways, Prog Neurobiol 69:103-142.

Lipton P. (1999). Ischemic cell death in brain neurons, Physiol Rev 79:1431-1568.

Liu J, Solway K, Messing RO, Sharp FR. (1998). Increased neurogenesis in the dentate gyrus after transient global ischemia in gerbils, J Neurosci 18:7768-7778.

Liu PK, Hamilton WJ, Hsu CY. (1999). Apoptosis. In Miller LP, editor: DNA damage and repair in stroke therapy basic, preclinical and clinical directions,: Wiley-Liss.

Lowenstein DH, Thomas MJ, Smith DH, McIntosh TK. (1992). Selective vulnerability of dentate hilar neurons following traumatic brain injury: a potential mechanistic link between head trauma and disorders of the hippocampus, J Neurosci 12:4846-4853.

Lyden P et al. (2002). Clomethiazole Acute Stroke Study in ischemic stroke (CLASS-I): final results, Stroke 33:122-128.

Maas AI. (2001). Neuroprotective agents in traumatic brain injury, Expert Opin Investig Drugs 10:753-767.

Mackay KB. (2001). BMS-204352 (Bristol Myers Squibb), Curr Opin Investig Drugs 2:820-823.

Macko RF et al. (1996). Precipitants of brain infarction. Roles of preceding infection/inflammation and recent psychological stress, Stroke 27:1999-2004.

Majno G, Joris I. (1995). Apoptosis, oncosis, and necrosis. An overview of cell death, Am J Pathol 146:3-15.

Mak HK et al. (2003). Hypodensity of >1/3 middle cerebral artery territory versus Alberta Stroke Programme Early CT Score (ASPECTS): comparison of two methods of quantitative evaluation of early CT changes in hyperacute ischemic stroke in the community setting, Stroke 34:1194-1196.

Malakoff D. (1999). Bayes offers a 'new' way to make sense of numbers, Science 286:1460-1464.

Margaill I et al. (1996). Short therapeutic window for MK-801 in transient focal cerebral ischemia in normotensive rats, J Cereb Blood Flow Metab 16:107-113.

Marion DW et al. (1997). Treatment of traumatic brain injury with moderate hypothermia, N Engl J Med 336:540-546.

Marion DW, White MJ. (1996). Treatment of experimental brain injury with moderate hypothermia and 21-aminosteroids, J Neurotrauma 13:139-147.

Markgraf CG et al. (1992). Sensorimotor and cognitive consequences of middle cerebral artery occlusion in rats, Brain Res 575:238-246.

Markgraf CG et al. (1993). Comparative histopathologic consequences of photothrombotic occlusion of the distal middle cerebral artery in Sprague-Dawley and Wistar rats, Stroke 24:286-292; discussion 292-293.

Marshall JW, Duffin KJ, Green AR, Ridley RM. (2001). NXY-059, a free radical—trapping agent, substantially lessens the functional disability resulting from cerebral ischemia in a primate species, Stroke 32:190-198.

Martinez-Vila E et al. (1990). Placebo-controlled trial of nimodipine in the treatment of acute ischemic cerebral infarction, Stroke 21: 1023-1028.

McCracken E et al. (2002). Grey matter and white matter ischemic damage is reduced by the competitive AMPA receptor antagonist, SPD 502, J Cereb Blood Flow Metab 22:1090-1097.

McCulloch J et al. (1993). Glutamate receptor antagonists in experimental focal cerebral ischaemia, Acta Neurochir Suppl (Wien) 57:73-79.

McEver RP. (2001). Adhesive interactions of leukocytes, platelets, and the vessel wall during hemostasis and inflammation, Thromb Haemost 86:746-756.

McIntosh TK *et al.* (1998). The Dorothy Russell Memorial Lecture. The molecular and cellular sequelae of experimental traumatic brain injury: pathogenic mechanisms, *Neuropathol Appl Neurobiol* 24:251-267.

McIntosh TK *et al.* (1989). Traumatic brain injury in the rat: characterization of a lateral fluid-percussion model, *Neuroscience* 28:233-244.

McKay R. (1997). Stem cells in the central nervous system, *Science* 276: 66-71.

Mendelow AD. (1993). Mechanisms of ischemic brain damage with intracerebral hemorrhage, *Stroke* 24:I115-117; discussion I118-119.

Merchant RE *et al.* (1999). A double-blind, placebo-controlled study of the safety, tolerability and pharmacokinetics of CP-101,606 in patients with a mild or moderate traumatic brain injury, *Ann N Y Acad Sci* 890:42-50.

Miller JD. (1985). Head injury and brain ischaemia—implications for therapy, *Br J Anaesth* 57:120-130.

Minamisawa H, Smith ML, Siesjo BK. (1990). The effect of mild hyperthermia and hypothermia on brain damage following 5, 10, and 15 minutes of forebrain ischemia, *Ann Neurol* 28:26-33.

Mizuno A, Umemura K, Nakashima M. (1998). Inhibitory effect of MCI-186, a free radical scavenger, on cerebral ischemia following rat middle cerebral artery occlusion, *Gen Pharmacol* 30:575-578.

Morganti-Kossmann MC *et al.* (2001). Role of cerebral inflammation after traumatic brain injury: a revisited concept, *Shock* 16:165-177.

Mori E *et al.* (1992). Inhibition of polymorphonuclear leukocyte adherence suppresses no-reflow after focal cerebral ischemia in baboons, *Stroke* 23:712-718.

Morikawa E *et al.* (1992). The significance of brain temperature in focal cerebral ischemia: histopathological consequences of middle cerebral artery occlusion in the rat, *J Cereb Blood Flow Metab* 12:380-389.

Mueller AL *et al.* (1999). NPS 1506, a novel NMDA receptor antagonist and neuroprotectant. Review of preclinical and clinical studies, *Ann N Y Acad Sci* 890:450-457.

Muir KW. (2002). Magnesium in stroke treatment, *Postgrad Med J* 78:641-645.

Muir KW, Holzapfel L, Lees KR. (2000). Phase II clinical trial of sipatrigine (619C89) by continuous infusion in acute stroke, *Cerebrovasc Dis* 10:431-436.

Muizelaar JP. (1994). Clinical trials with Dismutec (pegorgotein; polyethylene glycol-conjugated superoxide dismutase; PEG-SOD) in the treatment of severe closed head injury. *Adv Exp Med Biol* 366:389-400.

Muizelaar JP *et al.* (1993). Improving the outcome of severe head injury with the oxygen radical scavenger polyethylene glycol-conjugated superoxide dismutase: a phase II trial, *J Neurosurg* 78:375-382.

Mullane K. (1993). Acadesine: the prototype adenosine regulating agent for reducing myocardial ischaemic injury, *Cardiovasc Res* 27:43-47.

Murray GD, Teasdale GM, Schmitz H. (1996). Nimodipine in traumatic subarachnoid haemorrhage: a re-analysis of the HIT I and HIT II trials, *Acta Neurochir (Wien)* 138:1163-1167.

Murray M *et al.* (2002). Transplantation of genetically modified cells contributes to repair and recovery from spinal injury, *Brain Res Brain Res Rev* 40:292-300.

Nakamura K, Hatakeyama T, Furuta S, Sakaki S. (1993). The role of early Ca2+ influx in the pathogenesis of delayed neuronal death after brief forebrain ischemia in gerbils. *Brain Res* 613, 181-92.

Nakayama H, Ginsberg MD, Dietrich WD. (1988). (S)-emopamil, a novel calcium channel blocker and serotonin S2 antagonist, markedly reduces infarct size following middle cerebral artery occlusion in the rat, *Neurology* 38:1667-1673.

Narayan RK *et al.* (2002). Clinical trials in head injury, *J Neurotrauma* 19:503-557.

National Center for Injury Prevention and Control. (2001). *Injury fact book 2001-2002*, Atlanta: Centers for Disease Control and Prevention.

The National Institute of Neurological Disorders and Stroke rt-PA Stroke Study Group. (1995). Tissue plasminogen activator for acute ischemic stroke, *N Engl J Med* 333:1581-1587.

Nehlig A, Daval JL, Debry G. (1992). Caffeine and the central nervous system: mechanisms of action, biochemical, metabolic and psychostimulant effects, *Brain Res Brain Res Rev* 17:139-170.

Neil-Dwyer G, Mee E, Dorrance D, Lowe D. (1987). Early intervention with nimodipine in subarachnoid haemorrhage, *Eur Heart J* 8(Suppl K):41-47.

Nelson CW *et al.* (1992). Oxygen radicals in cerebral ischemia, *Am J Physiol* 263:H1356-1362.

NIH Consensus Development Panel on Rehabilitation of Persons With Traumatic Brain Injury. (1999a). Consensus conference. Rehabilitation of persons with traumatic brain injury, *JAMA* 282:974-983.

NIH Consensus Development Panel on Rehabilitation of Persons With Traumatic Brain Injury. (1999b). Recommendations for standards regarding preclinical neuroprotective and restorative drug development, *Stroke* 30:2752-2758.

Ohman J, Servo A, Heiskanen O. (1991). Long-term effects of nimodipine on cerebral infarcts and outcome after aneurysmal subarachnoid hemorrhage and surgery, *J Neurosurg* 74:8-13.

Okada Y *et al.* (1994). P-selectin and intercellular adhesion molecule-1 expression after focal brain ischemia and reperfusion, *Stroke* 25:202-211.

Okonkwo DO, Povlishock JT. (1999). An intrathecal bolus of cyclosporin A before injury preserves mitochondrial integrity and attenuates axonal disruption in traumatic brain injury, *J Cereb Blood Flow Metab* 19:443-451.

Onesti ST, Baker CJ, Sun PP, Solomon RA. (1991). Transient hypothermia reduces focal ischemic brain injury in the rat, *Neurosurgery* 29:369-373.

Ovbiagele B, Kidwell CS, Starkman S, Saver JL. (2003). Neuroprotective agents for the treatment of acute ischemic stroke, *Curr Neurol Neurosci Rep* 3:9-20.

Ozawa H *et al.* (2002). Therapeutic strategies targeting caspase inhibition following spinal cord injury in rats, *Exp Neurol* 177:306-313.

Pantoni L, Garcia JH. (1995). The significance of cerebral white matter abnormalities 100 years after Binswanger's report. A review, *Stroke* 26:1293-1301.

Pantoni L *et al.* (1999). Role of white matter lesions in cognitive impairment of vascular origin, *Alzheimer Dis Assoc Disord* 13(Suppl 3):S49-54.

Parent JM, Valentin VV, Lowenstein DH. (2002). Prolonged seizures increase proliferating neuroblasts in the adult rat subventricular zone-olfactory bulb pathway, *J Neurosci* 22:3174-3188.

Park CS *et al.* (2000). Differential and constitutive expression of neuronal, inducible, and endothelial nitric oxide synthase mRNAs and proteins in pathologically normal human tissues, *Nitric Oxide* 4:459-471.

Park E, Liu Y, Fehlings MG. (2003). Changes in glial cell white matter AMPA receptor expression after spinal cord injury and relationship to apoptotic cell death, *Exp Neurol* 182:35-48.

Peerless SJ, Rewcastle NB. (1967). Shear injuries of the brain, *Can Med Assoc J* 96:577-582.

Perez-Pinzon MA *et al.* (1995). Correlation of CGS 19755 neuroprotection against in vitro excitotoxicity and focal cerebral ischemia, *J Cereb Blood Flow Metab* 15:865-876.

Pettus EH, Christman CW, Giebel ML, Povlishock JT. (1994). Traumatically induced altered membrane permeability: its relationship to traumatically induced reactive axonal change, *J Neurotrauma* 11:507-522.

Phan TG et al. (2002). Salvaging the ischaemic penumbra: more than just reperfusion?, Clin Exp Pharmacol Physiol 29:1-10.

Picano E, Abbracchio MP. (1998). European Stroke Prevention Study-2 results: serendipitous demonstration of neuroprotection induced by endogenous adenosine accumulation?, Trends Pharmacol Sci 19:14-16.

Piriyawat P et al. (2003). Pilot dose-escalation study of caffeine plus ethanol (caffeinol) in acute ischemic stroke, Stroke 34:1242-1245.

Pisani A et al. (1998). L-type Ca2+ channel blockers attenuate electrical changes and Ca2+ rise induced by oxygen/glucose deprivation in cortical neurons, Stroke 29:196-201; discussion 202.

Pitsikas N et al. (2001). Effects of cerestat and NBQX on functional and morphological outcomes in rat focal cerebral ischemia, Pharmacol Biochem Behav 68:443-447.

Polderman KH et al. (2002). Effects of therapeutic hypothermia on intracranial pressure and outcome in patients with severe head injury, Intensive Care Med 28:1563-1573.

Povlishock JT. (1992). Traumatically induced axonal injury: pathogenesis and pathobiological implications, Brain Pathol 2:1-12.

Prado R et al. (1987). Photochemically induced graded spinal cord infarction. Behavioral, electrophysiological, and morphological correlates, J Neurosurg 67:745-753.

Prado R, Watson BD, Kuluz J, Dietrich WD. (1992). Endothelium-derived nitric oxide synthase inhibition. Effects on cerebral blood flow, pial artery diameter, and vascular morphology in rats, Stroke 23:1118-1123; discussion 1124.

Pratt J et al. (1992). Neuroprotective actions of riluzole in rodent models of global and focal cerebral ischaemia, Neurosci Lett 140:225-230.

Prevention of Atherosclerotic Complications with Ketanserin Trial Group. (1989). Prevention of atherosclerotic complications: controlled trial of ketanserin, BMJ 298:424-430.

Pulsinelli WA, Brierley JB, Plum F. (1982). Temporal profile of neuronal damage in a model of transient forebrain ischemia, Ann Neurol 11:491-498.

Rataud J et al. (1994). Comparative study of voltage-sensitive sodium channel blockers in focal ischaemia and electric convulsions in rodents, Neurosci Lett 172:19-23.

Ringelstein EB, Van Eyck S, Mertens I. (1992). Evaluation of cerebral vasomotor reactivity by various vasodilating stimuli: comparison of CO2 to acetazolamide, J Cereb Blood Flow Metab 12:162-168.

Ritter LS et al. (2000). Leukocyte accumulation and hemodynamic changes in the cerebral microcirculation during early reperfusion after stroke, Stroke 31:1153-1161.

Roof RL, Hall ED. (2000). Gender differences in acute CNS trauma and stroke: neuroprotective effects of estrogen and progesterone, J Neurotrauma 17:367-388.

Rosenberg GA, Estrada EY, Dencoff JE. (1998). Matrix metalloproteinases and TIMPs are associated with blood-brain barrier opening after reperfusion in rat brain, Stroke 29:2189-2195.

Rosenstiel P et al. (2003). Differential effects of immunophilin-ligands (FK506 and V-10,367) on survival and regeneration of rat retinal ganglion cells in vitro and after optic nerve crush in vivo, J Neurotrauma 20:297-307.

Ross DT, Graham DI, Adams JH. (1993). Selective loss of neurons from the thalamic reticular nucleus following severe human head injury, J Neurotrauma 10:151-165.

Saatman KE et al. (2003). Traumatic axonal injury results in biphasic calpain activation and retrograde transport impairment in mice, J Cereb Blood Flow Metab 23:34-42.

Sacco RL et al. (2001). Glycine antagonist in neuroprotection for patients with acute stroke: GAIN Americas: a randomized controlled trial, JAMA 285:1719-1728.

Schaefer PW et al. (2003). Assessing tissue viability with mr diffusion and perfusion imaging, AJNR Am J Neuroradiol 24:436-443.

Schlaug G et al. (1999). The ischemic penumbra: operationally defined by diffusion and perfusion MRI, Neurology 53:1528-1537.

Schulz JB, Weller M, Moskowitz MA. (1999). Caspases as treatment targets in stroke and neurodegenerative diseases, Ann Neurol 45:421-429.

Schwab S et al. (2001). Feasibility and safety of moderate hypothermia after massive hemispheric infarction, Stroke 32:2033-2035.

Schwab S et al. (1998). Moderate hypothermia in the treatment of patients with severe middle cerebral artery infarction, Stroke 29:2461-2466.

Schwartz-Bloom RD, Sah R. (2001). Gamma-aminobutyric acid(A) neurotransmission and cerebral ischemia, J Neurochem 77:353-371.

Sharp FR, Lu A, Tang Y, Millhorn DE. (2000). Multiple molecular penumbras after focal cerebral ischemia, J Cereb Blood Flow Metab 20:1011-1032.

Shiga Y, Onodera H, Matsuo Y, Kogure K. (1992). Cyclosporin A protects against ischemia-reperfusion injury in the brain, Brain Res 595:145-148.

Shouzu A et al. (2000). Effect of sarpogrelate hydrochloride on platelet-derived microparticles and various soluble adhesion molecules in diabetes mellitus, Clin Appl Thromb Haemost 6:139-143.

Singleton RH et al. (2001). The immunophilin ligand FK506 attenuates axonal injury in an impact-acceleration model of traumatic brain injury, J Neurotrauma 18:607-614.

Slusher BS et al. (1999). Selective inhibition of NAALADase, which converts NAAG to glutamate, reduces ischemic brain injury, Nat Med 5:1396-1402.

Smith DH et al. (1997a). Progressive atrophy and neuron death for one year following brain trauma in the rat, J Neurotrauma 14:715-727.

Smith SE et al. (1997b). Long-term beneficial effects of BW619C89 on neurological deficit, cognitive deficit and brain damage after middle cerebral artery occlusion in the rat, Neuroscience 77:1123-1135.

Stagliano NE et al. (1997a). The role of nitric oxide in the pathophysiology of thromboembolic stroke in the rat, Brain Res 759:32-40.

Stagliano NE et al. (1997b). The effect of nitric oxide synthase inhibition on acute platelet accumulation and hemodynamic depression in a rat model of thromboembolic stroke, J Cereb Blood Flow Metab 17:1182-1190.

Stein DG. (2001). Brain damage, sex hormones and recovery: a new role for progesterone and estrogen?, Trends Neurosci 24:386-391.

Steiner T et al. (2001). Effect and feasibility of controlled rewarming after moderate hypothermia in stroke patients with malignant infarction of the middle cerebral artery, Stroke 32:2833-2835.

Strong AJ et al. (2002). Spreading and synchronous depressions of cortical activity in acutely injured human brain, Stroke 33:2738-2743.

Strong AJ, Hopwood SE, Boutelle MG. (2003). Measuring dynamic changes in perfusion in the penumbra with high spatial and temporal resolution using laser speckle imaging: comparison with indicator clearance, J Cereb Blood Flow Metab 12(Suppl 1):S300.

Strong R, Grotta JC, Aronowski J. (2000). Combination of low dose ethanol and caffeine protects brain from damage produced by focal ischemia in rats, Neuropharmacology 39:515-522.

Stys PK, Waxman SG, Ransom BR. (1992a). Ionic mechanisms of anoxic injury in mammalian CNS white matter: role of Na+ channels and Na(+)-Ca2+ exchanger, J Neurosci 12:430-439.

Stys PK, Waxman SG, Ransom BR. (1992b). Effects of temperature on evoked electrical activity and anoxic injury in CNS white matter, J Cereb Blood Flow Metab 12:977-986.

Sumii T, Lo EH. (2002). Involvement of matrix metalloproteinase in thrombolysis-associated hemorrhagic transformation after embolic focal ischemia in rats, Stroke 33:831-836.

Suzuki H et al. (1998). A change of P-selectin immunoreactivity in rat brain after transient and permanent middle cerebral artery occlusion, Neurol Res 20:463-469.

Suzuki M, Sasamata M, Miyata K (2003b). Neuroprotective effects of YM872 coadministered with t-PA in a rat embolic stroke model, *Brain Res* 959:169-172.

Suzuki S *et al.* (1995). Cerebrospinal fluid and peripheral white blood cell response to acute cerebral ischemia, *South Med J* 88:819-824.

Suzuki T, Bramlett H, Dietrich W. (2003a). The importance of gender on the beneficial effects of posttraumatic hypothermia. *Exp Neurol* 184:1017-1026

Suzuki Y *et al.* (2003c). Ability of NMDA and non-NMDA receptor antagonists to inhibit cerebral ischemic damage in aged rats, *Brain Res* 964:116-120.

Sze KH *et al.* (1998). Effect of nimodipine on memory after cerebral infarction, *Acta Neurol Scand* 97:386-392.

Taft WC, Clifton GL, Blair RE, DeLorenzo RJ. (1989). Phenytoin protects against ischemia-produced neuronal cell death, *Brain Res* 483:143-148.

Takagi K *et al.* (1994). The effect of ritanserin, a 5-HT2 receptor antagonist, on ischemic cerebral blood flow and infarct volume in rat middle cerebral artery occlusion, *Stroke* 25:481-485; discussion 485-486.

Takagi K *et al.* (1993). Changes in amino acid neurotransmitters and cerebral blood flow in the ischemic penumbral region following middle cerebral artery occlusion in the rat: correlation with histopathology, *J Cereb Blood Flow Metab* 13:575-585.

Takahashi M *et al.* (1998). Neuroprotective efficacy of YM872, an alpha-amino-3-hydroxy-5-methylisoxazole-4-propionic acid receptor antagonist, after permanent middle cerebral artery occlusion in rats, *J Pharmacol Exp Ther* 287:559-566.

Tatlisumak T *et al.* (1998a). Delayed treatment with an adenosine kinase inhibitor, GP683, attenuates infarct size in rats with temporary middle cerebral artery occlusion, *Stroke* 29:1952-1958.

Tatlisumak T, Takano K, Meiler MR, Fisher M. (1998b). A glycine site antagonist, ZD9379, reduces number of spreading depressions and infarct size in rats with permanent middle cerebral artery occlusion, *Stroke* 29:190-195.

Taylor CP, Meldrum BS. (1995). Na+ channels as targets for neuroprotective drugs, *Trends Pharmacol Sci* 16:309-316.

Teasdale G *et al.* (1992). A randomized trial of nimodipine in severe head injury: HIT I. British/Finnish Co-operative Head Injury Trial Group, *J Neurotrauma* 9(Suppl 2):S545-550.

Tortella FC *et al.* (2000). Neuroprotection produced by the NAALADase inhibitor 2-PMPA in rat cerebellar neurons, *Eur J Pharmacol* 402:31-37.

Toyoda T, Kassell NF, Lee KS. (1997). Attenuation of ischemia-reperfusion injury in the rat neocortex by the hydroxyl radical scavenger nicaraven, *Neurosurgery* 40:372-377; discussion 377-378.

Tirilazad International Steering Committee. (2000). Tirilazad mesylate in acute ischemic stroke: a systematic review, *Stroke* 31:2257-65.

Uchino H *et al.* (1995). Cyclosporin A dramatically ameliorates CA1 hippocampal damage following transient forebrain ischaemia in the rat, *Acta Physiol Scand* 155:469-471.

Valtysson J *et al.* (1994). Neuropathological endpoints in experimental stroke pharmacotherapy: the importance of both early and late evaluation, *Acta Neurochir (Wien)* 129:58-63.

Vibulsresth S, Dietrich WD, Busto R, Ginsberg MD. (1987). Failure of nimodipine to prevent ischemic neuronal damage in rats, *Stroke* 18:210-216.

Vornov JJ *et al.* (1999). Blockade of NAALADase: a novel neuroprotective strategy based on limiting glutamate and elevating NAAG, *Ann NY Acad Sci* 890:400-405.

Wada K *et al.* (1999a). Early treatment with a novel inhibitor of lipid peroxidation (LY341122) improves histopathological outcome after moderate fluid percussion brain injury in rats, *Neurosurgery* 45:601-608.

Wada K, Chatzipanteli K, Busto R, Dietrich WD. (1998a). Role of nitric oxide in traumatic brain injury in the rat, *J Neurosurg* 89:807-818.

Wada K, Chatzipanteli K, Busto R, Dietrich WD. (1999b). Effects of L-NAME and 7-NI on NOS catalytic activity and behavioral outcome after traumatic brain injury in the rat, *J Neurotrauma* 16:203-212.

Wada K *et al.* (1998b). Inducible nitric oxide synthase expression after traumatic brain injury and neuroprotection with aminoguanidine treatment in rats, *Neurosurgery* 43:1427-1436.

Wang CX *et al.* (2001). Patency of cerebral microvessels after focal embolic stroke in the rat, *J Cereb Blood Flow Metab* 21:413-421.

Wang KK. (2000). Calpain and caspase: can you tell the difference?, *Trends Neurosci* 23:20-26.

Wang PY, Kao CH, Mui MY, Wang SJ. (1993). Leukocyte infiltration in acute hemispheric ischemic stroke, *Stroke* 24:236-240.

Warach S. (2001). New imaging strategies for patient selection for thrombolytic and neuroprotective therapies, *Neurology* 57:S48-52.

Wassertheil-Smoller S *et al.* (2003). Effect of estrogen plus progestin on stroke in postmenopausal women: the Women's Health Initiative: a randomized trial, *JAMA* 289:2673-2684.

Wassmann H *et al.* (1996). Repetitive hypoxic exposure of brain slices and electrophysiological responses as an experimental model for investigation of cerebroprotective measurements, *Neurol Res* 18:367-369.

Watson BD, Dietrich WD, Prado R, Ginsberg MD. (1987). Argon laser-induced arterial photothrombosis. Characterization and possible application to therapy of arteriovenous malformations, *J Neurosurg* 66:748-754.

Watson BD *et al.* (2002). Cerebral blood flow restoration and reperfusion injury after ultraviolet laser-facilitated middle cerebral artery recanalization in rat thrombotic stroke, *Stroke* 33:428-434.

Waxman SG, Ranson BR. (1997). Neuroprotection of CNS white matter. In Bär PR, Beal MF, editors: *Neuroprotection in CNS diseases,*: Marcel Dekker, Inc.

Wester P *et al.* (1992). Serotonin release into plasma during common carotid artery thrombosis in rats, *Stroke* 23:870-875.

White RP, Markus HS. (1997). Impaired dynamic cerebral autoregulation in carotid artery stenosis, *Stroke* 28:1340-1344.

Williams AJ, Lu XM, Slusher B, Tortella FC. (2001). Electroencephalogram analysis and neuroprotective profile of the N-acetylated-alpha-linked acidic dipeptidase inhibitor, GPI5232, in normal and brain-injured rats, *J Pharmacol Exp Ther* 299:48-57.

Wolf JA *et al.* (2001). Traumatic axonal injury induces calcium influx modulated by tetrodotoxin-sensitive sodium channels, *J Neurosci* 21:1923-1930.

Wolf PA *et al.* (1977). Epidemiology of stroke. In Thompson RA, Green JR, editors:, New York: Raven Press.

Wu TW, Zeng LH, Wu J, Fung KP. (2000). MCI-186: further histochemical and biochemical evidence of neuroprotection, *Life Sci* 67:2387-2392.

Xiong L *et al.* (2003). Preconditioning with isoflurane produces dose-dependent neuroprotection via activation of adenosine triphosphate-regulated potassium channels after focal cerebral ischemia in rats, *Anesth Analg* 96:233-237, table of contents.

Yam PS *et al.* (2000). NMDA receptor blockade fails to alter axonal injury in focal cerebral ischemia, *J Cereb Blood Flow Metab* 20: 772-779.

Yang Y *et al.* (2003). Reduced brain infarct volume and improved neurological outcome by inhibition of the NR2B subunit of NMDA receptors by using CP101,606-27 alone and in combination with rt-PA in a thromboembolic stroke model in rats, *J Neurosurg* 98: 397-403.

Yao H *et al.* (1993). Failure of MK-801 to reduce infarct volume in thrombotic middle cerebral artery occlusion in rats, *Stroke* 24: 864-870; discussion 870-871.

Yao H *et al.* (1994). Glutamate antagonist MK-801 attenuates incomplete but not complete infarction in thrombotic distal middle cerebral artery occlusion in Wistar rats, *Brain Res* 642:117-122.

Yardin C *et al.* (1998). FK506 antagonizes apoptosis and c-jun protein expression in neuronal cultures, *Neuroreport* 9:2077-2080.

Yasui M, Kawasaki K. (1994). Vulnerability of CA1 neurons in SHRSP hippocampal slices to ischemia, and its protection by Ca2+ channel blockers, *Brain Res* 642:146-152.

Young W, Flamm ES. (1982). Effect of high-dose corticosteroid therapy on blood flow, evoked potentials, and extracellular calcium in experimental spinal injury, *J Neurosurg* 57:667-673.

Zhang A, Altura BT, Altura BM. (1993a). Ethanol-induced contraction of cerebral arteries in diverse mammals and its mechanism of action, *Eur J Pharmacol* 248:229-236.

Zhang L *et al.* (1993b).Dantrolene protects against ischemic, delayed neuronal death in gerbil brain, *Neurosci Lett* 158:105-108.

Zhang R *et al.* (1998). The expression of P- and E-selectins in three models of middle cerebral artery occlusion, *Brain Res* 785:207-214.

Zhang RL *et al.* (1995a). Anti-intercellular adhesion molecule-1 antibody reduces ischemic cell damage after transient but not permanent middle cerebral artery occlusion in the Wistar rat, *Stroke* 26:1438-1442; discussion 1443.

Zhang RL *et al.* (1995b). The temporal profiles of ICAM-1 protein and mRNA expression after transient MCA occlusion in the rat, *Brain Res* 682:182-188.

Zhang RL, Zhang ZG, Zhang L, Chopp M. (2001). Proliferation and differentiation of progenitor cells in the cortex and the subventricular zone in the adult rat after focal cerebral ischemia, *Neuroscience* 105:33-41.

Zhang ZG, Chopp M, Chen H. (1993c). Duration dependent postischemic hypothermia alleviates cortical damage after transient middle cerebral artery occlusion in the rat, *J Neurol Sci* 117:240-244.

Zivin JA *et al.* (1985). Tissue plasminogen activator reduces neurological damage after cerebral embolism, *Science* 230:1289-1292.

15

Prospects for Gene Therapy for Central Nervous System Disease

Mark H. Tuszynski, MD, PhD

WHY DELIVER GENES FOR NEUROLOGICAL DISEASE?

Substantial progress has been made over the last 20 years in understanding the basic biology and function of the *normal* nervous system, and in elucidating molecular and cellular mechanisms that underlie neurological disease. Together with these advances in understanding have come discoveries of novel genes and proteins, which collectively present an unprecedented opportunity to intervene in and treat a number of neurological disorders that heretofore have been untreatable.

These opportunities for treatment are matched by new practical challenges: How can we deliver novel therapeutics to the nervous system? For example, nervous system growth factors offer the potential to prevent cell death and stimulate cell function, but they are large protein molecules that do not penetrate the central nervous system (CNS). Further, growth factors cause adverse effects from stimulation of nontargeted systems if delivered into the CNS widely and without region-specific targeting. Gene delivery offers the potential to provide growth factors to specific regions of the CNS that contain degenerating neurons, thereby *bypassing* diffusion limitations of the blood-brain barrier while *restricting* delivery of growth factors to their intended targets. In a second example, effective treatment of developmental abnormalities of the nervous system caused by single gene mutations, such as the inherited mucopolysaccharidoses, would require insertion of a correct copy of a gene into many cells of the nervous system at a developmental or early postnatal time point. This, too, can theoretically be achieved by gene therapy by introducing gene delivery

"vectors" into multiple and broad sites of the CNS, thereby replacing defective gene copies with copies of natural, working genes.

Thus, the potential of gene therapy lies in practically targeting therapeutic substances to precise or broad CNS regions at effective concentrations for sustained time periods. This offers the possibility, depending on the nature of a given disease, to prevent cell death, augment cell function, or replace a developmentally defective gene. The field of gene therapy and its relevance to the treatment of neurological disease have come a long way in the past 20 years, and are likely to become a mainstay of neurological therapy in the next 20 years as we enter the era of "molecular medicine." This chapter describes the development of gene therapy for neurological disease and presents examples of its implementation to treat neurological disease together with some of the challenges that remain to be addressed.

DEVELOPMENT OF GENE THERAPY IN THE NERVOUS SYSTEM

The beginnings of gene therapy for diseases of the nervous system can be traced to any one of several potential time points: identification by early biologists of the critical importance of the nucleus of the cell to the survival and function of each cell; identification of the double helical structure of DNA in 1962; initial production and then routine use of "recombinant" DNA in the 1970s; first use of gene therapy to treat an animal model of neurological disease in 1988 (Rosenberg *et al.*, 1988); or the first human trial of gene therapy in 1990 (Culver *et al.*,

1991) (see also Osterman *et al.*, 1971; Friedmann and Roblin, 1972; Anderson and Fletcher, 1980). As noted previously, the real potential of gene therapy for nervous system disease lies in the ability to correct disorders resulting from genetic mutations at even early time points of development, or to deliver therapeutic genes to the nervous system that can in some way compensate, ameliorate or prevent cell loss or dysfunction resulting from disease. A third and more recently emerging use of gene therapy is to *block* the expression of deleterious genes that cause disease, such as Huntingtin or the amyloid precursor protein. Thus, in inborn errors of metabolism, the goal of gene therapy is to replace a mutated or deleted gene. In disorders such as Alzheimer's disease (AD), Parkinson's disease, or spinal cord injury, gene therapy can be used as a tool to deliver therapeutic substances such as growth factors to the nervous system to prevent cell death and stimulate cell function. Lastly, in diseases such as Huntington's disease or AD, expression of deleterious disease-causing genes could be blocked (by expressing small-interfering RNAs or antisense mRNAs via gene therapy).

To implement gene therapy, a gene of interest must be identified, sequenced, and, ideally, have a well-defined function. The replacement gene is then introduced into the nervous system and expressed for a sufficient time period (possibly indefinitely) to treat the disease.

Putting Genes into Cells: *Ex Vivo* vs *In Vivo* Gene Therapy

Gene therapy can be divided into two basic approaches: *ex vivo* and *in vivo* gene delivery (Figure 15.1). In *ex vivo* gene delivery, host cells are obtained from a biopsy and established as *in vitro* (or *ex vivo*) cultures, and are genetically modified using one of several potential methods described later. Production of the desired gene product by the modified cells can be quantified. Cells are then grafted into the nervous system where they act as localized biological pumps to deliver the gene product. Autologous cells that can act as vehicles of *ex vivo* gene delivery include fibroblasts (obtained from skin biopsies), bone marrow stromal cells, stem cells (e.g., from the bone marrow), clonal lymphocytes, or Schwann cells. Outside the nervous system, tumor cells have also frequently been targets of *ex vivo* genetic modification by enhancing their production of immune signaling molecules such as cytokines; once reintroduced into the body, the cells may more vigorously stimulate a response of the immune system to tumor antigens.

In vivo gene therapy, on the other hand, circumvents all *ex vivo* cell preparation by instead injecting genes, usually carried within viral vectors, directly into host cells *in vivo*. This results in the direct genetic modification of cells of the nervous system (Figure 15.1).

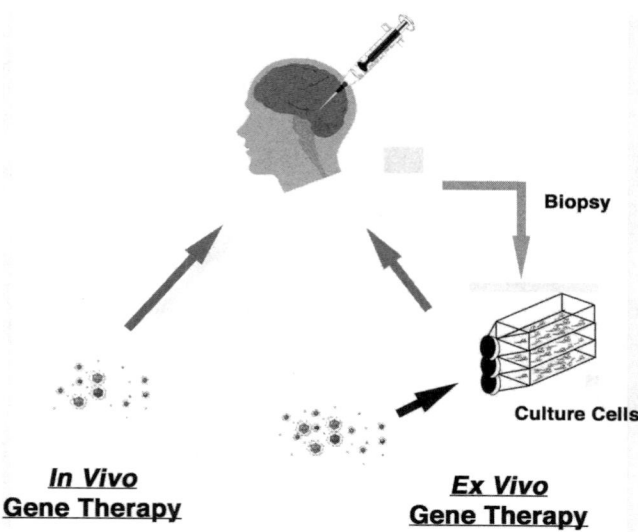

FIGURE 15.1   Schematic illustration of *in vivo* vs *ex vivo* gene therapy. *In vivo* gene therapy simply injects a therapeutic vector in the central nervous system (CNS) directly to modify native cells of the host brain. *Ex vivo* gene therapy obtains host cells from a biopsy, genetically modifies them *in vitro*, and then implants the host's own cells into the CNS to act as localized delivery agents for the therapeutic gene of interest.

Use of *ex vivo* vs *in vivo* Gene Therapy

In the early days of testing the potential of gene therapy for animal models of nervous system disease, *ex vivo* gene therapy vectors generally resulted in better levels of gene product delivery in the nervous system than *in vivo* methods. More recently, however, *in vivo* methods of gene delivery have advanced substantially and now appear to exceed the efficiency of *ex vivo* gene delivery. For this reason, *in vivo* gene delivery methods are likely to become the mainstay of gene therapy for the nervous system, except in cases where cells must also be introduced to treat disease. Examples of the latter could include spinal cord injury or multiple sclerosis, where not only a therapeutic gene would be beneficial (e.g., a growth factor to stimulate axon growth or myelin synthesis), but a replacement cell is also needed to provide a physical bridge in a lesion cavity (spinal cord injury) or to replace myelinating cells (e.g., a Schwann cell or neural stem cell, in multiple sclerosis). With few exceptions, however, most gene therapy applications will now use *in vivo* gene delivery.

Techniques of Gene Delivery into Cells

New genes can be introduced into cells using several methods. The process begins by constructing a "plasmid" that is composed of nucleic acids coding for the gene of interest. This gene of interest is expressed by a promoter that should be continuously active. The plasmid can also contain other nucleic acid sequences to regulate, stabilize, or augment gene expression or messenger RNAs produced by transcription. Next, the plasmid is introduced into a host cell using either mechanical techniques or modified viruses that are nonvirulent. Mechanical techniques of plasmid introduction include direct injection into cells ("gene gun"), mixture of the plasmid with calcium phosphate (opening pores or channels in the cell membrane for plasmid entry), liposomes (fat particles that surround the plasmid and directly fuse with the cell membrane due to their hydrophobicity), or electrical current ("electroporation," which opens pores in the membrane). Plasmid DNA can also be stabilized and placed directly in the extracellular environment (so-called naked DNA), allowing direct cell uptake of the DNA.

The most commonly used and efficient methods for plasmid introduction into cells use viruses, which, through natural selection, have adopted highly efficient means of delivering their genetic material into the nuclei of host cells. In viruses that have been modified for use in gene therapy applications, genes for viral replication and virulence are removed and replaced by therapeutic genes and constitutively active promoters (Figure 15.2). For *ex vivo* gene delivery, modified retroviral vectors have been used most commonly. For *in vivo* gene delivery, adenovirus, herpes virus, adenoassociated virus (AAV), and retroviruses (e.g., human immunodeficiency virus, HIV) have generated the most interest (Muzyczka, 1992; Naldini *et al.*, 1996b; Jacobs *et al.*, 1999; Rabinowitz and Samulski, 2000).

In vivo Gene Therapy Vectors

Retroviruses consist of single strands of RNA that, on entry into a cell, use a viral reverse transcriptase enzyme to synthesize viral DNA from RNA (hence the term *retro*virus). This viral DNA is then transcribed to synthesize proteins for viral replication, production of viral envelope or capsid proteins, packaging of viral particles into viral coats, and other virulence-mediating functions. Retroviruses can also code for integrase proteins that direct integration of viral nucleic acid into the host genome. Viral integration into the host genome can result in stable, long-term viral gene expression. Thus retroviruses have naturally evolved into potent biological mechanisms for expressing genetic material; the challenge of therapeutic gene delivery using viral vectors is to use viral mechanisms for expressing therapeutic genes while removing viral genes for virulence. This is generally accomplished by producing gene therapy vectors that are devoid of as many wild-type genes as possible, including all genes for viral replication, envelope production, and

FIGURE 15.2 General structure of *ex vivo* gene delivery vectors. A potent promoter drives the expression of a therapeutic gene of interest, such as a neurotrophic factor (NTF) gene. A second internal promoter drives the expression of a gene imparting neomycin resistance to cells. The 5′–3′ construct is terminated by the wild-type 3′-LTR region. *In vivo* gene delivery vectors express only a single novel gene because cells are not selected after *in vivo* transduction.

capsid production. When coupled with promoters that turn therapeutic gene expression "on" potently and persistently, a mechanism is created for expressing therapeutic mammalian genes at high levels over prolonged time periods in host cells.

One of the earliest retroviral vectors to be tested for use in therapeutic gene delivery was a modified form of the Moloney murine leukemia virus (MLV), in which wild-type genes for viral replication (*rep*), envelope (*env*) production, and capsid (*cap*) proteins were removed and replaced by a therapeutic gene of interest. To express the therapeutic gene, in many cases, the wild-type viral promoter contained in the 5'-long terminal repeat region of the virus was retained; in other instances, the wild-type MLV promoter was replaced by a promoter from cytomegalovirus or other viruses. This vector construct could only infect dividing cells because it entered cells exclusively during the "S" phase of cell division. For gene therapy application to the nervous system, this meant that MLV-based vectors could only be used for *ex vivo* therapeutic gene delivery, as there are few dividing cells in the adult nervous system. Thus host cells such as fibroblasts, Schwann cells, bone marrow stromal cells, or stem cells would be obtained from a host, cultured *in vitro*, exposed to the MLV-based vector, and then injected back into the host animal to achieve gene delivery. For non-nervous system applications, the retroviral vector itself could be injected *in vivo* and get taken up by actively dividing host cells (e.g., in the bone marrow) to express the gene of interest. In addition, the gene for neomycin resistance (*neo*) was often added to the vector construct to make cells that incorporated the transgene resistant to the antibiotic neomycin; in this manner, cells *in vitro* that successfully incorporated the therapeutic vector could be selected from nonmodified cells by adding antibiotics to the cell culture medium. Hence, a pure population of modified cells could be returned to the host. Using MLV-based retroviral vectors and *ex vivo* gene delivery to the nervous system, production of the protein product from the gene of interest, such as a growth factor, could be sustained for time periods of at least 1 year in the host brain.

Most recently, other RNA-based retroviral vectors have been used in gene therapy applications that do not require host cell division for viral entry into the cell. These include HIV-based vectors, feline immunodeficiency virus-based vectors, and lentiviral vectors (Naldini et al., 1996a). These viruses presumably bind to cell surface receptors to gain entry into cells. Because these vectors are capable of entering nondividing cells, they can be used for direct gene delivery into nondividing cells of the adult brain. As described previously for MLV-based gene delivery systems, these vectors are rendered safe by removing wild-type genes responsible for viral replication and particle production. The very few remaining wild-type genes in some of the vectors express no secreted proteins, and hence are not known to elicit an immune response from the host. When used for in vivo gene delivery to the nervous system, these vectors predominantly enter neurons (greater than 90% of modified cells in the brain are neuronal) and continue to express their gene product for at least one year in the primate brain, the longest time point examined to date. Like MLV vectors, HIV-based vectors integrate into the host genome.

A number of DNA-based viruses have also been adapted for use in gene therapy (Miller, 1990; Muzyczka, 1992; Glorioso et al., 1995; Naldini et al., 1996b, Rabinowitz and Samulski, 1998). One of the earliest viruses to be studied for therapeutic gene delivery was the common cold virus, adenovirus. Binding to cell surface-based receptors, this virus enters nondividing cells and can therefore be used in in vivo gene therapy paradigms inside and outside the nervous system. Indeed, this vector has been the most broadly tested of all in vivo gene therapy vectors in human clinical trials. However, the most commonly used version of the adenoviral vector expresses a number of wild-type viral proteins and, not surprisingly, has been plagued by resulting inflammatory and immune problems. These problems occur because of the existence of circulating neutralizing antibodies to adenoviral epitopes and as a result of the continued expression of wild-type adenoviral proteins from the vector after entry into host cells. Indeed, an immune response to adenovirus led to a patient death in a systemic (non-nervous system) gene therapy trial in 1999, representing a setback to the field of gene therapy. However, the availability of later-generation AAV and lentivirus vectors substantially reduced this risk of immune response because the vectors do not express wild-type viral genes, and, in the case of lentivirus, humans do not have preexisting circulating antibodies to the virus. The safety profile of the AAV and lentiviral vectors for use in the nervous system is further reinforced because the absolute quantity of vector delivered for nervous system applications will in general be far lower than quantities used for systemic (non-nervous system)

gene therapy. When adenovirus vectors are used for gene delivery in the nervous system, they do elicit immune responses (Cartmell et al., 1999). More recently, newer generation "gutless" adenovirus vectors have been developed that do not express wild-type viral genes, which may improve the safety profile of adenovirus-based vectors substantially. For the present, however, AAV- and HIV-based vectors appear to offer safety from the standpoint of an absence of significant host immune response to the vector and duration of in vivo gene expression. As mentioned previously, lentiviral vectors continue to express their therapeutic gene for at least 1 year after delivery to the primate brain, the longest time point examined to date (unpublished data). AAV vectors also appear to express their gene products for extended time periods in vivo, with persistent expression for at least 4.5 years in vivo (K. Bankiewicz, personal communication).

A potential drawback of some in vivo gene therapy vectors, however, is the following: Depending on the specific site of viral integration into a host genome, host gene expression may become disrupted. Hypothetically, if viral integration occurs in a host genomic sequence that codes for a tumor suppression gene, then tumor genes could become de-repressed, leading to malignant transformation. Alternatively, integration of viral promoters into the host genome could hypothetically activate the expression of certain host genes, which, if adjacent to coding sequences for oncogenes, could also lead to tumorigenesis. This is a greater hypothetical problem with in vivo gene delivery vectors that integrate into the host genome, such as retroviruses. AAV, which primarily remains intranuclear but extrachromosomal, is less likely to cause insertional mutagenesis in a host. Indeed, in a retroviral gene therapy trial for severe combined immunodeficiency in France, most patients showed a clear benefit from gene therapy to replace their defective immune system gene (McCormack and Rabbitts, 2004); however, 3 of 11 children in the trial subsequently developed a form of leukemia resulting from insertional mutagenesis. Work is underway to modify the retrovirus and alter its sites of integration into the host genome. Other retroviral vectors, such as HIV-based and lentiviral vectors may not have the same integration site and may therefore have less risk. And again, other efficient in vivo gene therapy vectors such as AAV show little integration into the host genome, and several clinical trials in humans using AAV to date have not reported subsequent development of cancers.

Comparison of in vivo and ex vivo Gene Therapy

For use in treating neurological disorders, ex vivo and in vivo gene therapy approaches each have distinct advantages and disadvantages, depending on the potential disease application (Table 15.1). Advantages of *in vivo* gene delivery include:

1. Simplicity. Gene delivery can be accomplished in a single step of direct vector injection into the desired brain target region, in contrast to the considerable cell processing that is required when using *ex vivo* gene delivery.
2. Minimal invasiveness. *In vivo* gene delivery is achieved by injection of several microliters of vector particles in a fluid solution to the brain and could be performed repeatedly if necessary; *ex vivo* gene therapy delivers cell suspensions that occupy some space in the brain and would be more difficult to repeat.

Potential disadvantages of *in vivo* gene therapy relative to *ex vivo* gene therapy include the following:

1. Nonspecificity of target cell infection. Neurons, glia, and vascular cells can become genetically modified when *in vivo* vectors are injected into the brain. Whether there might be deleterious effects of expressing certain genes in glia and vascular cells, such as growth factor or neurotransmitter genes, is unknown (although empirically, adverse effects have not been noted in prolonged primate expression studies to date). *Ex vivo* gene therapy allows specific selection of the cell type that expresses the gene(s) of interest before delivery to the brain.
2. Toxicity. As noted previously, some *in vivo* gene therapy vectors can be directly toxic to host cells, including herpes virus and rabies virus. Early adenovirus vectors elicited immune responses. HIV-based and AAV vector systems are generally safe and nontoxic and more recently have emerged as potential lead candidates for gene delivery in the nervous system. Newer generation herpes "amplicon" vectors (expressing no wild-type viral genes) and herpes virus/AAV hybrid "amplicon" vectors under development also offer potentially safe gene delivery to the nervous system (Maguir-Zeis *et al.*, 2001; Heister *et al.*, 2002; Wang *et al.*, 2002). "Gutless" adenoviruses that do not express wild-type viral genes could also allow the reemergence of adenovirus-based systems for nervous system application (Yant *et al.*, 2002).
3. Risk of malignant transformation. As mentioned previously, integrating vectors could cause cancer by interrupting a tumor suppressor gene or mutating an oncogene. This risk also exists for *ex vivo* gene delivery that uses vectors that integrate into the host genome. This risk is far less in nonintegrating *in vivo* vectors such as AAV.
4. Spontaneous recombination with wild-type viruses. It is hypothetically possible that an *in vivo* gene therapy

TABLE 15.1 Comparison of *in vivo* and *ex vivo* Gene Therapy

	in vivo Gene Therapy	*ex vivo* Gene Therapy
Advantages	Simple, effective minimally invasive	Targets specific cell types
		Safe (little risk of wild-type viral recombination)
Disadvantages	Modifies neurons, glia, vessels	Cumbersome
	Risk of cancer induction (retroviruses)	Risk of cancer induction
	Potential wild-type recombination	
	Absence of regulation	Absence of regulation

vector could recombine with a wild-type virus, resulting in replication-competent vectors or other mutated viruses with potential for virulence. This risk can be minimized to a range approaching near impossibility by removing most wild-type coding sequences in the therapeutic vector, by screening recipients of gene therapy for existing infection with wild-type virus such as HIV, and by the creation of hybrid vectors with little if any overlap in wild-type coding sequences.

Advantages of *ex vivo* gene delivery relative to *in vivo* gene delivery include the following:

1. Targeting. *Ex vivo* gene therapy can target selected cell types *in vitro* for subsequent grafting to the brain.
2. Safety. *Ex vivo* gene delivery does not introduce infectious viral particles into the brain; hence there is little risk of recombination with wild-type viruses. *Ex vivo* gene delivery uses cells from the host; thus there is no risk of immune rejection. Empirically, in a clinical trial of *ex vivo* nerve growth factor (NGF) gene delivery for AD, there have been no complications to date of the *ex vivo* gene delivery system.

Disadvantages of *ex vivo* gene therapy relative to *in vivo* gene therapy include the following:

1. Requirement for target cells to divide. As noted previously, for host cells to be maintained and genetically modified *in vitro*, they must be capable of dividing. Thus, most postmitotic cell populations such as neurons cannot act as vehicles of *ex vivo* gene therapy in the nervous system.
2. Tumor formation. It is possible that the dividing cells delivered to the brain could form tumors. However, this has not been observed when implanting primary (nonimmortalized) cells into the brain, although

grafts of immortalized cell lines have formed tumors. "Conditionally immortalized" cell lines that shut off immortalizing genes at 37°C have been successfully implanted into the body without forming tumors (McKay, 1992; Whittemore *et al.*, 1997). To date, tumor formation has not been observed in a human clinical trial of NGF gene delivery for AD. Insertional mutagenesis is also a risk when using viral vectors that integrate into the host genome in *ex vivo* gene therapy, as described previously under "disadvantages of *in vivo* gene delivery."

3. Complexity. The *ex vivo* gene therapy process of cell culture, gene modification, and preparation for implantation is cumbersome and time consuming, requiring 2 to 3 months. *In vivo* gene delivery can be performed immediately.

In vivo and *ex vivo* gene therapy both suffer from the drawback that they must be introduced into the brain by direct injection to be effectively administered to the CNS. The blood-brain barrier blocks passage of most vectors into the brain. Even if peripheral delivery routes for the CNS were available, they likely could not be used because most nervous system applications of gene delivery would require highly specific *and restricted* delivery of genes such as growth factors to specific regions of the brain. Production of the gene of interest in nontargeted regions of the brain could be deleterious. On the other hand, direct gene delivery into the CNS can be accomplished relatively simply and safely through a small burr hole.

Finally, whereas many gene therapy vectors now efficiently express their gene products for prolonged time periods (years in the case of AAV and lentivirus), the ability to turn "off" gene expression has not been perfected. Future development of a practical and effective gene delivery system that can be regulated is discussed later in more detail.

Growth Factor Gene Therapy for Neurological Disease: Alzheimer's Disease

Neurodegenerative diseases are attractive candidates for gene therapy. Several of these disorders, including AD, Huntington's disease, and amyotrophic lateral sclerosis (ALS), lack truly effective therapies. Other disorders, such as Parkinson's disease, have effective symptomatic therapies but no means to prevent disease progression. The discovery of a class of neuroprotective substances called nervous system growth factors offers the potential for the first time to reduce cell loss in neurological disease and to stimulate the function of remaining neurons. However, growth factors are large and polar molecules, and they do not cross the blood-brain barrier. Further, they exhibit toxicity if delivered to nonaffected

brain regions. Thus, effective harnessing of the potential of growth factors to treat neurological disease requires site-specific, intracranial delivery. Gene therapy has emerged as a leading method for achieving long-term, highly localized, and regionally restricted delivery of growth factors to the brain. Clinical trials of growth factor gene delivery are underway in the most common age-related neurodegenerative disorder, AD (Tuszynski, 2002), and will soon include Parkinson's disease, and possibly ALS and Huntington's disease.

Growth Factors

Several different families of growth factors exist, each of which affects the survival and function of various neurons throughout the nervous system. The first nervous system growth factor was identified more than 50 years ago by Rita Levi-Montalcini and Viktor Hamburger, and was aptly named *nerve growth factor*. Levi-Montalcini and Hamburger discovered that mouse sarcoma extracts contained a substance that promoted sensory and sympathetic neuron survival in the embryonic chick (Levi-Montalcini and Hamburger, 1951; Levi-Montalcini, 1987). Over subsequent decades, several important features surrounding the role of NGF during development were defined. Its structure was elucidated in the early 1970s (Angeletti *et al.*, 1973), and the gene was cloned in 1983 (Ullrich *et al.*, 1983). However, the startling discovery that NGF could also influence neuronal survival in the *adult* nervous system came nearly 40 years after its initial discovery.

In 1986 and 1987, three research groups independently reported that injections of NGF protein into the adult brain prevented the death of forebrain cholinergic neurons (Hefti, 1986; Williams *et al.*, 1986; Kromer, 1987). In 1987, it was also reported that NGF reversed *spontaneous* age-related morphological and behavioral decline in rats (Fischer *et al.*, 1987). These findings ushered in an era of intense interest in both basic mechanisms of growth factor action in the nervous system and an exploration of their therapeutic potential to prevent cell death in neurological disease. The ability of NGF to prevent the death of cholinergic neurons was extended to primate systems, where it promoted the survival of 80% to 100% of cholinergic neurons after lesions (Koliatsos *et al.*, 1990; Tuszynski *et al.*, 1990, 1991).

NGF for Alzheimer's Disease

The potential relevance of NGF actions to neuronal loss in AD was evident immediately (Hefti and Weiner, 1986; Phelps *et al.*, 1989). Cholinergic neurons of the basal forebrain undergo profound atrophy and death during the course of AD (Perry *et al.*, 1977; Whitehouse *et al.*, 1981), contributing to cognitive decline (Perry *et al.*, 1978; Bartus *et al.*, 1982; Whitehouse *et al.*, 1982; Candy *et al.*, 1983). NGF is normally produced throughout life in the brain and is believed to support the integrity of cholinergic systems (Sofroniew *et al.*, 1990). Notably, NGF transport and availability to support the function and survival of cholinergic neurons are defective in AD (Mufson *et al.*, 1995; Scott *et al.*, 1995). Thus, NGF could reduce or prevent the cholinergic component of cell loss in AD.

Delivery of NGF protein to the brain is difficult, however, because NGF is a medium-sized, charged protein that does not cross the blood-brain barrier. To reach the brain effectively in animal studies, NGF was pumped into the lateral ventricles where it diffused short distances to reach cholinergic cell bodies (Hefti, 1986; Williams *et al.*, 1986; Kromer, 1987; Emmett *et al.*, 1996). Intracerebroventricular infusion of NGF was the only available delivery route for a clinical trial in AD, and plans were formulated accordingly (Phelps *et al.*, 1989). However, preclinical distribution and toxicology studies of intracerebroventricular (ICV) infusions revealed several adverse effects of this route of NGF administration. All of these effects reflected the potency of NGF in stimulating the function of neurons of the adult brain, and all adverse effects were attributable to the broad, nontargeted distribution of NGF that resulted from intraventricular infusions. For example, Isaacson and Crutcher reported that NGF-sensitive sympathetic axons sprouted around the cerebral vasculature after ICV infusion (Isaacson *et al.*, 1990). Williams reported that ICV infusions of NGF caused weight loss in rats owing to a reduction in food intake (Williams, 1991). Perhaps most worrisome, Winkler and colleagues (Winkler *et al.*, 1997) as well as Emmett and colleagues (Emmett *et al.*, 1996) reported that Schwann cells, which bear receptors for NGF, migrated into an expanding cell layer in the subpial space surrounding the brainstem and spinal cord after ICV NGF infusions. This Schwann cell response reversed after discontinuation of NGF infusion. NGF infusions also led to sensory axon sprouting and a pain syndrome. These adverse effects all occurred in neuronal systems with known patterns of sensitivity to NGF, and all reflected stimulation of the relevant neuronal systems by NGF. Thus it was evident that central intraventricular infusions of NGF protein were an impractical delivery method for a clinical trial in AD. Indeed, three AD patients in Sweden received ICV infusions of NGF, and not surprisingly the trial was discontinued because of the development of weight loss and pain (Eriksdotter Jonhagen *et al.*, 1998).

NGF Gene Delivery

An alternative method was required that could deliver NGF across the blood-brain barrier, yet *restrict* its delivery to cholinergic cell targets to avoid adverse events. Gene therapy was already emerging as a potential means of achieving sustained and localized protein delivery to the brain. In 1988, the first successful use of *ex vivo* gene delivery to the brain was reported, using NGF gene delivery to rescue cholinergic neurons after injury (Rosenberg *et al.*, 1988). These early studies of gene therapy in the nervous system used *ex vivo* rather than *in vivo* gene delivery paradigms because the vectors for *ex vivo* gene delivery were superior at the time, both in terms of long-term gene expression in the brain and in safety (absence of immune responses). In the first study of gene delivery to the brain, rat fibroblasts were obtained from skin biopsies and genetically modified to produce and secrete human NGF using retroviral vectors. These cells were implanted into the septal nucleus of rats that had undergone fimbria-fornix lesions. Typically, fimbria-fornix lesions result in the degeneration and death of cholinergic neurons by depriving these cells of contact with their hippocampal targets. Rats that received *ex vivo* grafts of NGF-secreting fibroblasts exhibited highly significant rescue of cholinergic neurons compared to control subjects that received implants of cells expressing a control reporter gene, β-galactosidase. Subsequently, Chen and Gage (1995) reported that *ex vivo* gene delivery of NGF also reversed cholinergic neuronal atrophy and improved cognition in aged rats.

We then examined whether *ex vivo* NGF gene delivery could also prevent cholinergic neuronal degeneration in the brains of adult primates. Rhesus monkeys underwent fornix lesions followed by implants of primary autologous fibroblasts that were genetically modified to produce and secrete human NGF. Genetically modified cells were placed adjacent to degenerating cholinergic neuronal cell bodies in the basal forebrain. One month later, animals that received NGF-secreting cell implants exhibited a significant reduction in cholinergic degeneration compared to control monkeys that received implants of primary fibroblasts that did not secrete growth factors (Figure 15.3) (Tuszynski *et al.*, 1996). A total of 68% of neurons were rescued in monkeys that underwent *ex vivo* NGF gene delivery, whereas control monkeys showed survival of only 25% of cholinergic neurons ($P < .05$). Further, as we became more experienced in accurately placing *ex vivo* NGF-secreting cells adjacent to degenerating cholinergic neurons, the extent of neuronal protection increased to 92%. This neuronal protection was achieved using NGF doses roughly 500-fold lower than quantities of NGF protein used in previous ICV infusion studies (Koliatsos *et al.*, 1990; Tuszynski *et al.*, 1990).

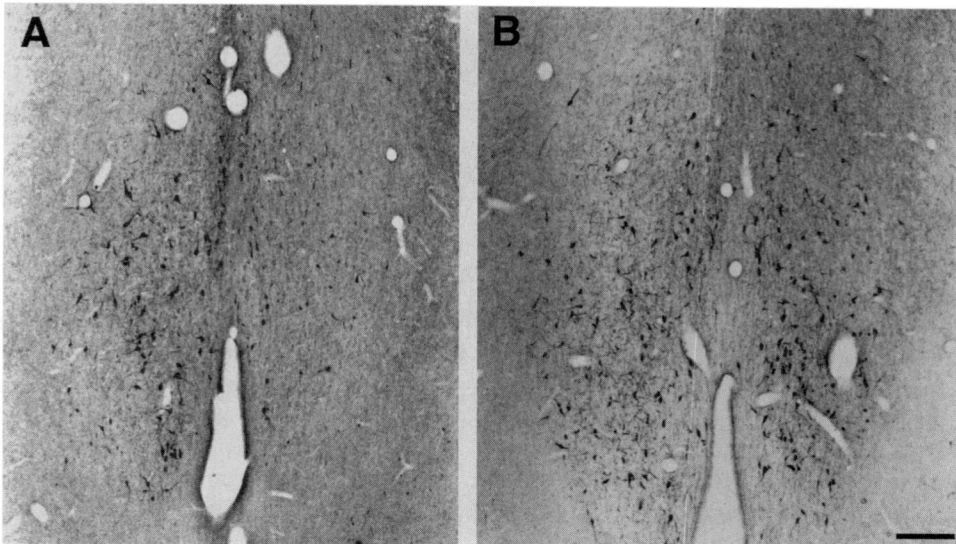

FIGURE 15.3 NGF prevents the death of adult primate cholinergic neurons. **A:** Following right-sided fornix lesions in adult primates, cholinergic neurons undergo degeneration, reflected by a reduction in the number of choline acetyltransferase-labeled neurons on the right side of the brain. Intact cholinergic neurons are seen on left half of the brain. **B:** NGF delivery prevents the lesion-induced loss of cholinergic neurons on the right side of the brain. Scale bar = 200 µm.

The requirement for accurate placement of genetically modified cells in the brain reflected potential improvements in the safety of the gene delivery approach for NGF administration to the brain: NGF diffused only 2 mm from sites of genetically modified cell implants. This limited diffusion substantially improved the potential safety profile of NGF, reducing the risk that adverse effects would result from protein spread beyond the targeted region. Emerich, Kordower, and colleagues also reported cholinergic neuronal rescue in monkeys after fornix lesions, using an alternative form of gene therapy in which encapsulated xenogenic cells were implanted into the brains of either young or aged monkeys (Emerich *et al.*, 1994; Kordower *et al.*, 1994).

At this point, it was clear that *ex vivo* NGF gene delivery could prevent lesion-induced degeneration of cholinergic neurons in the primate brain. Cholinergic neurons die in AD, although the precise mechanism whereby cholinergic neurons (or any neuron, for that matter) die in AD is unknown. Thus, we thought that it was important to demonstrate that cholinergic neurons degenerating as a result of another mechanism of neural damage could also respond to *ex vivo* NGF gene delivery. Aging is another form of damage to the nervous system, resulting in spontaneous atrophy of cholinergic neurons in rats. Thus we examined the brains of aged rhesus monkeys for sensitivity to *ex vivo* NGF gene therapy. Examination of the brains of unoperated, aged monkeys revealed that 40% of basal forebrain cholinergic neurons

are atrophic (but not dead) compared to young monkeys (Smith *et al.*, 1999). Notably, 3 months after undergoing *ex vivo* NGF gene delivery, aged rhesus monkeys exhibited a restoration of functional cholinergic neuronal markers to levels equivalent to young monkeys (Smith *et al.*, 1999). Cell size was also restored to levels indistinguishable from young monkeys. Control, aged monkeys that underwent *ex vivo* gene therapy containing a reporter gene, but not NGF, evidenced no neuroprotection. Further, aged monkeys undergoing *ex vivo* NGF gene delivery demonstrated a return to levels of young monkeys of cholinergic axon projections to the cortex (Conner *et al.*, 2001).

Dose-Escalation/Toxicity Studies of NGF *ex vivo* Gene Transfer in Primates

The preceding studies showed that *ex vivo* NGF gene delivery could prevent cholinergic neuronal degeneration and ameliorate age-related cholinergic neuronal atrophy in the brains of primates, suggesting that various mechanisms of cholinergic damage in primates were indeed sensitive to *ex vivo* NGF gene delivery. Rodent studies indicated that two additional mechanisms of cholinergic degeneration were also sensitive to NGF delivery: excitotoxic injury (Dekker *et al.*, 1991) and amyloid-induced degeneration of cholinergic neurons in trisomy-16 mutant mice (Holtzman *et al.*, 1993). Further, NGF diffused no more than 2 mm from brain sites of *ex vivo* gene delivery, in contrast to diffusion distances of several centimeters resulting from intraventricular or intraparenchymal NGF protein infusions (Conner and Tuszynski, unpublished observations). Thus the prospect that gene therapy could achieve localized and effective delivery of NGF to the brain was emerging.

To further test this possibility before embarking on a clinical trial, a dose escalation/toxicity study of *ex vivo* NGF gene delivery was performed in adult monkeys. Subjects received escalating volumes of autologous NGF-producing cells (autologous fibroblasts) into the nucleus basalis. Volumes ranged from 5 μl of cells per injection site to a maximum of 100 μl of cells per injection site. Extent and duration of gene expression were measured together with NGF levels in the cerebrospinal fluid, implanted graft size, weight, and general activity measures over a 1-year period. At the highest cell injection volumes, fibroblasts refluxed from injections sites, yet no resulting toxicity was observed. Weights of all grafted subjects remained stable, and there was no evidence of pain or discomfort. Implanted cells did not migrate and tumors did not form. NGF was not detectable by enzyme-linked immunosorbent assay in spinal fluid samples taken 6 months and 1 year after therapy, and Schwann cells did not migrate into the subpial space of the brainstem or spinal cord. Gene expression was present for at least 1 year in the primate brain, assessed from brain biopsies taken at graft sites; NGF levels exceeded physiological levels fivefold after 1 year *in vivo*. Thus, *ex vivo* NGF gene delivery in the primate brain could prevent cholinergic neuronal degeneration, reverse spontaneous age-related cholinergic atrophy, and safely sustain gene expression for 1 year with no evidence of toxicity.

A Phase I Clinical Trial of NGF Gene Therapy in Alzheimer's Disease

Based on the preceding safety and efficacy data, a rational basis was established for proceeding to a clinical trial to determine whether *ex vivo* NGF gene therapy could reduce cholinergic neuronal degeneration in AD. NGF could hypothetically benefit AD via two distinct mechanisms: by reducing cholinergic neuronal death, or by *augmenting* cholinergic neuronal function by stimulating NGF-mediated release of acetylcholine in the cortex. Two important questions remained to be tested in the clinical trial:

1. Could cholinergic cell loss in AD be prevented by NGF delivery? The underlying cause of neuronal death in AD remains unknown. Whereas extensive neurofibrillary degeneration occurs in cholinergic neurons in AD, there is little deposition of amyloid in the cholinergic nucleus basalis. As noted previously, NGF levels in the cholinergic basal forebrain in AD are significantly reduced compared to age-matched control subjects (Mufson *et al.*, 1995; Scott *et al.*, 1995), although NGF production in the cortex is not reduced in the AD brain. Thus, defects in NGF retrograde transport appear to exist in AD and could result in cholinergic neuronal degeneration. Delivery of NGF to the cholinergic neuronal cell body using *ex vivo* gene therapy could bypass a transport defect and sustain the survival and function of cholinergic neurons. Even if a primary deficiency of NGF itself does not cause cholinergic neuronal loss in AD, NGF gene delivery could potentially benefit AD by augmenting cholinergic transmission.

2. Would amelioration of the cholinergic component of cell degeneration in AD be sufficient to impact cognitive decline? Multiple neuronal systems decline in AD in addition to cholinergic neurons. However, of the neuronal systems that decline in AD, loss of cholinergic neurons correlates best with synapse loss (Terry *et al.*, 1991) and dementia severity (Perry *et al.*, 1978). Further, the cholinergic system plays a critical role in modulating cortical plasticity in the intact brain (Conner, 2003). Indeed, several clinical trials have shown a modest effect of cholinergic system augmentation by cholinesterase inhibitors, even though these oral agents only slightly augment central levels of acetylcholine (Davis *et al.*, 1992, Mayeux and Sano, 1999). Thus, preventing cholinergic degeneration in AD, especially early in the disease, could impact disease progression and quality of life.

A Phase I safety trial of *ex vivo* NGF gene delivery in humans with early stage AD began at the University of California, San Diego (UCSD) in 2001. Early stage patients were selected for two reasons. First, patients must understand the nature and potential risks of the procedure they are undergoing, as this is the first human clinical trial of gene therapy in AD. Most patients with early AD are capable of such insight. The consent of the primary caregiver is also obtained. Second, the hypothetical benefit of NGF in preventing cholinergic degeneration will be of most benefit earliest in the disease, when the greatest population of cholinergic neurons remains to be rescued.

Using the same methods described previously for our preclinical primate studies, primary autologous patient fibroblasts were genetically modified to produce and secrete human NGF *in vitro*. After verifying adequate production of human NGF, cells were placed into the human nucleus basalis of Meynert in a dose-escalation trial design. The lowest dose of cells delivered in this clinical trial was effective in rescuing cholinergic neurons in non-human primate studies, and the highest dose used in this trial was well below maximum tolerated cell volumes in monkeys. Eight subjects have undergone the gene delivery procedure, and endpoints that will be examined in the trial include general safety measures as well as cognitive function and glucose metabolism on positron emission tomography (PET) scans.

Potential risks of the study include those associated with inadvertent, widespread distribution of NGF protein, including pain, weight loss, and Schwann cell migration into the subpial space. Other risks include hemorrhage from needle passage into the brain, tumor formation from the grafted cells, and migration of the grafted cells. These adverse effects were not encountered in more than 200 NGF-secreting cell implants in preclinical primate studies, representing an extensive base of empirical safety data that mitigates the potential risk of a clinical trial. Of potentially greatest concern in this clinical trial would be the development of adverse effects resulting from unintended widespread expression of NGF. Although pain could be treated with analgesics or weight loss controlled with behavioral measures, optimally a means would be available to turn expression of the therapeutic gene "off." Such means are not currently available, and are under development (see later). Given the impressive efficacy of NGF in preventing loss of cholinergic systems, the extensive empirical safety data in five primate studies (Emerich *et al.*, 1994; Kordower *et al.*, 1994; Tuszynski *et al.*, 1994, 1996; Smith, 1999; Conner, 2001), the lack of other broadly effective therapies for AD, and the disease's relentlessly progressive course, the clinical trial has proceeded.

In vivo Gene Therapy for Alzheimer's Disease

The current *ex vivo* trial of NGF gene therapy in AD will be followed by a trial of *in vivo* gene therapy for AD. As noted previously, the available vectors for *ex vivo* gene therapy were superior to the vectors for *in vivo* gene therapy until recently; hence our clinical program advanced on an *ex vivo* design. The emergence of AAV and lentiviral vectors for *in vivo* gene delivery to the nervous system will likely replace *ex vivo* trials in nervous system disease, unless *in vivo* vectors have unexpected toxicities that have not been revealed to date by primate studies, a possibility that seems unlikely. For applications to Alzheimer's disease, *in vivo* NGF gene delivery has been shown to prevent cholinergic neuronal death in rat models using AAV vectors (Mandel, 1999; Blesch, 2002) and, preliminarily, to reverse age-related cholinergic neuronal atrophy (Klein *et al.*, 2000). Studies are ongoing with both AAV

and lentivirus systems in primates, providing a potential basis for human clinical trials.

Other Genes for AD

More than 50 nervous system growth factors have been identified to date (Tuszynski, 1999), raising the possibility that growth factors might also be used to target other degenerating neuronal populations in AD. Two other growth factors are of particular interest in this regard: brain-derived neurotrophic factor (BDNF), and basic fibroblast growth factor-2 (FGF-2). BDNF prevents the death of cortical neurons (Giehl and Tetzlaff, 1996; Lu *et al.*, 2001), and levels of BDNF are diminished in the hippocampus in AD (Phillips *et al.*, 1991). FGF-2 prevents injury-induced degeneration of entorhinal cortical neurons (Gomez-Pinilla *et al.*, 1992; Peterson *et al.*, 1996). Whereas the entire cortex would not be a practical target for growth factor gene delivery in a disorder such as AD, critical areas such as the entorhinal cortex or hippocampus might be candidates for growth factor gene delivery to prevent cell loss. This possibility is the subject of ongoing preclinical studies. In addition, cells that have been genetically modified to secrete the neurotransmitter acetylcholine have been shown to significantly ameliorate age-related cognitive decline in rats (Winkler *et al.*, 1995). A practical limitation in the application of a transmitter replacement approach for gene therapy of AD, however, is the large size of the human cortex that would need to be targeted. In contrast, the size of the basal forebrain cholinergic system that is the target of our ongoing trial of *ex vivo* NGF gene delivery in AD is only 1 cm in rostral-caudal extent.

Gene Therapy for Parkinson's Disease

Growth Factor Gene Delivery for Parkinson's Disease

Three growth factors are potential therapeutic candidates for gene delivery in the second-most common neurodegenerative disorder, Parkinson's disease (PD): glial-derived neurotrophic factor (GDNF), neurturin (NTN), and BDNF. Progressive loss of dopaminergic neurons occurs in the substantia nigra in PD, and several *in vitro* studies demonstrate that GDNF, NTN, and BDNF support the survival of developing nigral neurons. More important, *in vivo* studies in both rodent and primate models of PD show that these growth factors prevent nigral cell loss and promote functional recovery (Spina *et al.*, 1992; Frim *et al.*, 1993; Lin *et al.*, 1993; Hyman *et al.*, 1994; Gash *et al.*, 1996; Kotzbauer *et al.*, 1996; Winkler *et al.*, 1996; Connor *et al.*, 1999; Bjorklund *et al.*, 2000; Kordower *et al.*, 2000).

Based on these types of findings, a clinical trial of ICV GDNF protein infusion was performed in PD (Nutt, 2003). However, the nontargeted nature of ICV GDNF infusions induced adverse effects including nausea, weight loss, and Lhermitte signs. Not only did this broad, nontargeted delivery route cause toxicity, but GDNF failed to reach nigral neurons in significant concentrations (J. Kordower, Rush-Presbyterian Medical Center, personal communication). Once again, this outcome reinforces the concept that the adequate clinical assessment of the potential of growth factors to treat human disease will require that effective concentrations of the agents be delivered to neural structures in adequate doses and in a manner that is highly targeted and regionally-restricted, to avoid adverse effects of these potent agents.

Evidence from the preclinical literature suggested that the optimal site for delivery of growth factors to degenerating nigrostriatal circuitry might in fact be the region of the caudate/putamen, where a growth factor could induce sprouting of host dopaminergic terminals into their natural striatal targets (Kirik, 2000). GDNF, NTN, or BDNF delivered to the striatum in this manner could promote the survival of nigral neurons because the growth factors are transported in a retrograde fashion from the striatum to neurons of the substantia nigra (Mufson et al., 1996). Rodent studies using in vivo gene delivery of GDNF in fact showed significant neuroprotection and functional recovery in rats with experimentally induced PD (Connor et al., 1999; Bjorklund et al., 2000). Subsequently, Kordower and colleagues showed in a primate model of PD, MPTP-induced parkinsonism, that in vivo lentiviral GDNF gene delivery induced nearly complete anatomical recovery and functional recovery (Kordower et al., 2000). In addition, lentiviral gene delivery to aged monkeys also reversed age-related biochemical and anatomical degeneration of nigrostriatal circuitry (Kordower et al., 2000). Hence, a clear rationale has been established for pursuing gene delivery in PD, with the potential to both prevent neuronal degeneration and augment the function of remaining circuitry. In the absence of vectors that can be clinically regulated and that would allow shut-off of the gene in vivo, however, additional primate studies should be done to demonstrate convincingly the safety of this approach.

A Phase I trial of intraparenchymal infusion of GDNF protein into the striatum in PD has been reported, providing additional rationale and proof-of-principle for delivery of growth factors to treat PD (Gill et al., 2003). Five patients received continuous infusions of GDNF into a single site in the putamen via an implanted Medtronics pump, and showed signs of both clinical improvement and increased regional fluorodopa uptake on PET scan. However, hardware problems occurred in two patients, and all subjects showed an increase in MRI signal at the infusion site. Hypothetically, gene delivery could be a safer and more effective means of long-term delivery of growth factors to the nervous system compared to chronic infusions and implanted hardware. However, such conclusions cannot be made in the absence of high-quality clinical trial data demonstrating both the safety and efficacy of in vivo gene delivery.

Neurotransmitter Gene Therapy for PD

Another potential means of using gene therapy in the nervous system in PD is the replacement of neurotransmitters. Neurotransmitter replacement could provide more physiological levels of transmitter replacement in diseases such as PD to smooth out debilitating motor fluctuations that occur in later stages of oral dopamine agonist therapy. Nearly 15 years ago, Wolff and colleagues used an ex vivo gene delivery approach to provide the dopamine synthesizing enzyme tyrosine hydroxylase (TH) to rats with experimentally induced PD (Wolff et al., 1989). They showed elevated striatal levels of dopamine and functional recovery. Subsequent studies using TH replacement, or replacement of another enzyme in the dopamine synthesis pathway, dopamine decarboxylase, have continued to support the feasibility of transmitter-synthesizing gene delivery to the striatum (Kang et al., 1993; Bankiewicz et al., 1998; Azzouz, 2002). Clinical trials using this approach are actively being considered. The primary shortcoming of this approach is an uncertainty that the efficacy and safety of nonregulated gene delivery for transmitter synthesis enzymes will exceed that of simple oral agents that are currently available. And unlike growth factor gene delivery, neurotransmitter replacement would not be neuroprotective.

Surgical lesions have recently gained (or regained) substantial popularity in PD treatment, together with a more recent shift to the implantation of stimulating electrodes (deep brain stimulation) to block aberrant excitability in neural systems that contribute to PD symptoms. Taking the approach of a "chemical" lesion of the subthalamic nucleus by gene transfer of glutamic acid decarboxylase (GAD, the synthetic enzyme for the inhibitory neurotransmitter GABA), a clinical trial of in vivo AAV GAD gene delivery is ongoing for PD (Luo et al., 2002). In rodent studies, GAD gene delivery was reported to be effective in reducing the activity of subthalamic circuitry and improving PD-like symptoms. Lacking in this program is a clear rationale for proceeding with the clinical trial based on compelling data from primate studies. Thus it is difficult to objectively assess the feasibility of this clinical program. Whereas some risk is justifiable in using a growth factor gene therapy program that offers the potential of fundamentally altering the primary pathogenesis of a neurodegenerative disease such as PD by preventing cell loss, the rationale for transmitter replacement is less clear; therefore the preclinical data should be at least equally strong to that of a growth factor approach. At present, it is not.

Growth Factor Gene Therapy in ALS and Huntington's Disease

Growth factors have been identified that are neuroprotective and functionally beneficial in models of Huntington's disease and ALS, offering again the potential to improve disease treatment by reducing cell death. *Ex vivo* gene delivery of ciliary neurotrophic factor in rodent and primate models of Huntington's disease ameliorates cell loss and promotes functional recovery (Emerich *et al.*, 1996, 1997). More recently, AAV-*in vivo* gene delivery of GDNF was reported to exhibit beneficial functional effects in a rat model of Huntington's disease (McBride *et al.*, 2003). Given the lack of alternative therapies for this neurodegenerative disorder, clinical trials of growth factor gene delivery are being contemplated.

ALS is a relentlessly progressive and untreatable neurodegenerative disorder. BDNF, GDNF, ciliary neurotrophic factor, and insulin-like growth factor I have separately been reported to improve cell survival and functional outcomes in animal models of the disease. Clinical trials with protein infusions of these growth factors have been conducted and failed (Group, 1996, 1999). Yet again, it is likely the case that the routes of growth factor administration in these trials (peripheral or ICV) failed to reach motor neurons in sufficient concentrations to elicit biological stimulation of cell bodies. However, targeted delivery to motor neurons distributed along the neuraxis in ALS will be practically difficult to achieve. Two gene therapy studies, the first using adenovirus to deliver GDNF (Mohajeri *et al.*, 1999) and the second using AAV to deliver insulin-like growth factor I (Kaspar, *et al.*, 2003), reported that localized delivery of growth factors to functionally critical motor neuronal groups, such as diaphragmatic motor neurons, can prolong survival in animal models of ALS. Adenovirus and AAV are transported in a retrograde fashion to motor neurons of the spinal cord when injected into muscle. Thus, central neuronal rescue in this disease might be achievable following peripheral gene delivery. These possibilities are also the subject of ongoing deliberation with regard to initiation of clinical trials.

Gene Therapy for Inborn Errors of Metabolism

A significant potential for the field of gene delivery is to replace absent or mutant enzymes in inborn errors of metabolism. Beneficial effects of *in vivo* gene transfer in several animal models of inborn errors of metabolism affecting the nervous system have been reported, including lysosomal storage disease (mucopolysaccharidosis) (Li and Davidson, 1995; Daly *et al.*, 1999; Stein *et al.*, 1999), metachromatic leukodystrophy (Consiglio *et al.*, 2001), α-sarcoglycan-deficient mice (Allamand *et al.*, 2000), arylsulfatase A deficiency (Learish, *et al.*, 1996), β-hexosaminidase deficiency (the gene affected in Tay-Sachs disease) (Lacorazza *et al.*, 1996), galactosialidosis (Hahn *et al.*, 1998), and Fabry disease (Takenaka *et al.*, 2000). Gene replacement to the bone marrow early in life, or to the neuraxis either early or at slightly later time points, could provide biochemically and symptomatically beneficial therapies for many of these diseases that are otherwise untreatable. Some active work is ongoing in an attempt to bring these potential therapies to clinical trials.

The first human gene therapy trial in the nervous system in fact attempted to deliver the enzyme aspartoacylase, which is deficient throughout the neuraxis, to children with Canavan's disease (Leone *et al.*, 2000). AAV vectors coding for aspartoacylase were injected intraventricularly in two children. This route of ICV injection rendered remote the probability that the desired gene product would be widely accessible to the brain parenchyma. Another clinical trial in Canavan's disease is underway in the United States, using multiple intraparenchymal injections of *in vivo* AAV expressing aspartoacylase.

Gene Therapy for Nervous System Trauma

Another potential application of gene therapy is to locally deliver growth-promoting substances, such as growth factors, to focal sites of CNS injury to promote cell survival and axon growth. For example, growth factor gene delivery in models of spinal cord injury has stimulated the growth of injured axons and generated modest functional recovery in some cases (Grill *et al.*, 1997; Liu *et al.*, 1999; Tuszynski, 2002). Growth factor effects are sustained in the larger spinal cords of primates (Tuszynski, 2002). Nonetheless, restoring neural circuitry after injury remains a great challenge. An important challenge for the practical implementation of growth factor gene delivery in spinal cord injury is the need for regulated gene delivery. Continuous activation of growth factor production in sites of spinal cord injury leads to failure of axons to regenerate beyond the lesion site because they become "trapped" within the growth factor source. There is a need to turn growth factor gene expression sequentially "on" and then "off" at an injury site to promote axonal growth into, through, and beyond the lesion site. Such studies are ongoing. The prospect of gene delivery to prevent neuronal loss after brain trauma is also a possibility (Lu *et al.*, 2001). Most applications for growth factor delivery for trauma will require brief bursts of gene expression, or, perhaps more likely, short-term growth factor infusion.

Gene Therapy for Nervous System Cancer

Gene therapy has been investigated in models of intracranial glioma for some time (Short *et al.*, 1990). In one set of studies, death-inducing genes such as thymidine kinase can be introduced into rapidly dividing tumor cells *in vivo*, or passively transferred by injecting cells that express thymidine kinase into the tumor. On addition of the drug gancyclovir, cells that have incorporated thymidine kinase are killed, together with neighboring cells (Barba *et al.*, 1994; Alavi and Eck, 2001). Early clinical trials are underway. In another approach, genetically modified herpes viruses that express genes to augment presentation of tumor antigens to the host immune system are injected into tumors (Ram *et al.*, 1997; Shand *et al.*, 1999; Cohen *et al.*, 2000; Rainov, 2000; Alavi and Eck, 2001; Markert *et al.*, 2001; Nanda *et al.*, 2001a; Todo *et al.*, 2001). Clinical trials with herpes virus-based approaches to gliomas are in Phase III clinical trials.

Optimizing Gene Delivery for the Future

Regulatable Vectors

An important safety feature that will optimize the future of gene therapy will be the ability to control gene expression *in vivo*, turning gene therapy expression systems "on" and "off" at will. Such regulatable expression systems have been under development for some time (Gossen *et al.*, 1994). Ultimately, it is hoped that regulatable vectors will be able not only to stop gene delivery should an adverse event occur, but allow the physician to adjust the therapeutic dose of a delivered gene product *in vivo* much in the way that pharmacological drug doses can be adjusted.

The most thoroughly studied regulatable expression system is probably the tetracycline-sensitive system, developed by Bujard and colleagues (Gossen *et al.*, 1994). In this controlled expression system, an operon is normally bound to a promoter and allows gene expres-

sion to occur. When tetracycline is present, however, it binds to the operon, causing it to dissociate from the promoter, turning gene expression off ("tet-off" system). The inverse of this regulatable scheme has also been constructed, wherein the operon binds to the promoter only in the presence of tetracycline, turning gene expression on ("tet-on" system). Using the tet-off system, we have shown that the ability of NGF gene delivery to rescue cholinergic neurons can be completely eliminated by administering an oral tetracycline analog, doxycycline (Blesch *et al.*, 2001). When tested in primates, however, the tetracycline-regulated system lost efficacy over time because of the formation of host antibodies to its regulatory components (Favre *et al.*, 2002). Hence, this system requires further development before becoming practical for clinical use. Other systems for regulating gene expression *in vivo* have been developed, including systems that are regulated in the presence of the steroid ecdysone (Suhr *et al.*, 1998), and "suicide" systems that can kill genetically modified cells if an "escape" from gene therapy is needed (Nanda *et al.*, 2001b). Regulatable systems are the subject of ongoing studies.

Convection-Enhanced Gene Delivery

Although localized delivery of a therapeutic gene is a clear strength of gene therapy in the nervous system, the practical distance over which genes of interest can be delivered to the brain can sometimes be limiting. For example, in Parkinson's disease, the entire striatum may need to be targeted to effectively deliver growth factor or neurotransmitter genes. "Convection-enhanced" delivery methods have been developed for the delivery of cytotoxic agents to brain tumors, and the same techniques have been adopted for use in delivering genes to broader areas of the brain such as the striatum (Bankiewicz *et al.*, 1998). Similar approaches could be useful in inborn errors of metabolism, where broad distribution of a therapeutic gene might be needed to achieve phenotypic correction.

CONCLUSIONS AND PROSPECTS

The development of molecularly based therapies in medicine moves inevitably forward. As knowledge regarding the human genome, mechanisms of gene and cell function, and the development of practical gene delivery systems move rapidly forward, so too does the potential for developing more effective therapies for neurological disease. Together with this remarkable potential for altering the landscape of medical therapy comes a need for responsibility on the part of investigators to implement their clinical trials in a rationale and evidence-based manner. Upon this strong foundation of science and responsibility, it is not unlikely that the fate of people facing untreatable diseases will be markedly improved within our lifetimes.

Gene therapy is part of the new era of molecular medicine. Merging the potential of growth factors to rescue neurons with gene therapy technology that can deliver therapeutic substances across the blood-brain barrier into specific and restricted brain regions, the possibility exists that neuronal degeneration could be significantly reduced in neurological disorders. Gene therapy's potential is broader still, with the possibility of replacing deficient or mutant genes in inborn errors of metabolism or replacing deficient neurotransmitters in the brain. The newest generation of *in vivo* gene delivery vectors significantly improved the safety profile of gene therapy, although the need for regulatable expression vectors remains.

ACKNOWLEDGMENTS

Many thanks to Armin Blesch for helpful suggestions. Supported by the National Institutes for Health, the Veterans Administration, and the Alzheimer's Association. The author is a co-founder of Ceregene, Inc.

References

Alavi JB, Eck SL. (2001). Gene therapy for high grade gliomas, *Expert Opin Biol Ther* 1:239-252.

Allamand V *et al.* (2000). Early adenovirus-mediated gene transfer effectively prevents muscular dystrophy in alpha-sarcoglycan-deficient mice, *Gene Ther* 7:1385-1391.

Anderson WF, Fletcher JC. (1980). Sounding boards. Gene therapy in human beings: when is it ethical to begin? *N Engl J Med* 303:1293-1297.

Angeletti RH, Hermodson MA, Bradshaw RA. (1973). Amino acid sequences of mouse 2.5S nerve growth factor. II. Isolation and characterization of the thermolytic and peptic peptides and the complete covalent structure, *Biochemistry* 12:100-115.

Azzouz M *et al.* (2002). Multicistronic lentiviral vector-mediated striatal gene transfer of aromatic L-amino acid decarboxylase, tyrosine hydroxylase, and GTP cyclohydrolase I induced sustained transgene expression, dopamine production, and functional improvement in a rat model of Parkinson's disease, *J Neurosci* 22:10302-10312.

Bankiewicz KS *et al.* (1998). Application of gene therapy for Parkinson's disease: nonhuman primate experience, *Adv Pharmacol* 42:801-806.

Barba D, Hardin J, Sadelain M, Gage FH. (1994). Development of antitumor immunity following thymidine kinase-mediated killing of experimental brain tumors, *Proc Natl Acad Sci U S A* 91:4348-4352.

Bartus R, Dean RL, Beer C, Lippa AS. (1982). The cholinergic hypothesis of geriatric memory dysfunction, *Science* 217:408-417.

Bjorklund A *et al.* (2000). Towards a neuroprotective gene therapy for Parkinson's disease: use of adenovirus, AAV and lentivirus vectors for gene transfer of GDNF to the nigrostriatal system in the rat Parkinson model, *Brain Res* 886:82-98.

Blesch A, Conner JM, Tuszynski MH. (2001). Modulation of neuronal survival and axonal growth in vivo by tetracycline-regulated neurotrophin expression, *Gene Ther* 8:954-960.

Blesch A *et al.* (2002). Lentiviral NGF gene transfer rescues medial septal cholinergic neurons from lesion-induced degeneration, *Abstr Soc Neurosci* 237:9.

Candy JM *et al.* (1983). Pathological changes in the nucleus basalis of Meynert in Alzheimer's and Parkinson's diseases, *J Neurosci* 54:277-289.

Cartmell T *et al.* (1999). Interleukin-1 mediates a rapid inflammatory response after injection of adenoviral vectors into the brain. *J Neurosci* 19:1511-1523.

Chen KS, Gage FH. (1995) Somatic gene transfer of NGF to the aged brain: behavioral and morphological amelioration, *J Neurosci* 15:2819-2825.

Cohen JL *et al.* (2000). Preservation of graft-versus-infection effects after suicide gene therapy for prevention of graft-versus-host disease, *Hum Gene Ther* 11:2473-2481.

Conner JM *et al.* (2003). Lesions of the basal forebrain cholinergic system impair task acquisition and abolish cortical plasticity associated with motor skill learning, *Neuron* 38:819-829.

Conner JM, Darracq MA, Roberts J, Tuszynski MH. (2001). Nontropic actions of neurotrophins: subcortical NGF gene delivery reverses age-related degeneration of primate cortical cholinergic innervation, *Proc Natl Acad Sci* 98:1941-1946.

Connor B *et al.* (1999). Differential effects of glial cell line-derived neurotrophic factor (GDNF) in the striatum and substantia nigra of the aged Parkinsonian rat, *Gene Ther* 6:1936-1951.

Consiglio A *et al.* (2001). In vivo gene therapy of metachromatic leukodystrophy by lentiviral vectors: correction of neuropathology and protection against learning impairments in affected mice, *Nat Med* 7:310-316.

Culver KW, Anderson WF, Blaese RM. (1991). Lymphocyte gene therapy, *Hum Gene Ther* 2:107-109.

Daly TM *et al.* (1999). Neonatal gene transfer leads to widespread correction of pathology in a murine model of lysosomal storage disease, *Proc Natl Acad Sci U S A* 96:2296-2300.

Davis KL *et al.* (1992). A double-blind, placebo-controlled multicenter study of tacrine for Alzheimer's disease. The Tacrine Collaborative Study Group, *N Engl J Med* 327:1253-1259.

Dekker AJ, Langdon DJ, Gage FH, Thal LJ. (1991). NGF increases cortical acetylcholine release in rats with lesions of the nucleus basalis, *Neuroreport* 2:577-580.

Emerich DF *et al.* (1996). Implants of encapsulated human CNTF-producing fibroblasts prevent behavioral deficits and striatal degeneration in a rodent model of Huntington's disease, *J Neurosci* 16:5168-5181.

Emerich DF *et al.* (1997). Protective effect of encapsulated cells producing neurotrophic factor CNTF in a monkey model of Huntington's disease, *Nature* 386:395-399.

Emerich DW *et al.* (1994). Implants of polymer-encapsulated human NGF-secreting cells in the nonhuman primate: rescue and sprouting of degenerating cholinergic basal forebrain neurons, *J Comp Neurol* 349:148-164.

Emmett CJ *et al.* (1996). Distribution of radioiodinated recombinant human nerve growth factor in primate brain following intracerebroventricular infusion, *Exp Neurol* 140:151-160.

Eriksdotter Jonhagen M *et al.* (1998). Intracerebroventricular infusion of nerve growth factor in three patients with Alzheimer's disease, *Dement Geriatr Cogn Disord* 9:246-257.

Favre D *et al.* (2002). Lack of an immune response against the tetracycline-dependent transactivator correlates with long-term doxycycline-regulated transgene expression in nonhuman primates after intramuscular injection of recombinant adeno-associated virus, *J Virol* 76:11605-11611.

Fischer W *et al.* (1987). Amelioration of cholinergic neuron atrophy and spatial memory impairment in aged rats by nerve growth factor, *Nature* 329:65-68.

Friedmann T, Roblin R. (1972). Gene therapy for human genetic disease?, *Science* 175:949-955.

Frim DM *et al.* (1993). Effects of biologically delivered NGF, BDNF and bFGF on striatal excitotoxic lesions, *Neuroreport* 4:367-370.

Gash DM *et al.* (1996). Functional recovery in parkinsonian monkeys treated with GDNF, *Nature* 380:252-255.

Giehl KM, Tetzlaff W. (1996). BDNF and NT-3, but not NGF, prevent axotomy-induced death of rat corticospinal neurons in vivo, *Eur J Neurosci* 8:1167-1175.

Gill SS *et al.* (2003). Direct brain infusion of glial cell line-derived neurotrophic factor in Parkinson disease, *Nat Med* 9:589-595.

Glorioso JC, DeLuca NA, Fink DJ. (1995). Development and application of herpes simplex virus vectors for human gene therapy, *Annu Rev Microbiol* 49:675-710.

Gomez-Pinilla F, Lee JW, Cotman CW. (1992). Basic fibroblast growth factor in adult rat brain: cell distribution and response to entorhinal lesion and fimbria-fornix transection, *J Neurosci* 12:345-355.

Gossen M, Bonin AL, Freundlieb S, Bujard H. (1994). Inducible gene expression systems for higher eukaryotic cells, *Curr Opin Biotechnol* 5:516-520.

Grill R *et al.* (1997). Cellular delivery of neurotrophin-3 promotes corticospinal axonal growth and partial functional recovery after spinal cord injury, *J Neurosci* 17:5560-5572.

ALS-CNTF Treatment study Group. (1996). A double-blind placebo-controlled trial of subcutaneous recombinant human ciliary neurotrophic factor (rhCNTF) in amyotrophic lateral sclerosis, *Neurology* 46:1244-1249.

The BDNF Study Group. (1999). A controlled trial of recombinant methionyl human BDNF in ALS: The BDNF Study Group (Phase III), *Neurology* 52:1427-1433.

Hahn CN *et al.* (1998). Correction of murine galactosialidosis by bone marrow-derived macrophages overexpressing human protective protein/cathepsin A under control of the colony-stimulating factor-1 receptor promoter, *Proc Natl Acad Sci U S A* 95:14880-14885.

Hefti F. (1986). Nerve growth factor (NGF) promotes survival of septal cholinergic neurons after fimbrial transection, *J Neurosci* 6:2155-2162.

Hefti F, Weiner WJ. (1986). Nerve growth factor and Alzheimer's disease, *Ann Neurol* 20:275-281.

Heister T, Heid I, Ackermann M, Fraefel C. (2002). Herpes simplex virus type 1/adeno-associated virus hybrid vectors mediate site-specific integration at the adeno-associated virus preintegration site, AAVS1, on human chromosome 19, *J Virol* 76:7163-7173.

Holtzman DM *et al.* (1993). Nerve growth factor reverses neuronal atrophy in a Down Syndrome model of age-related neurodegeneration. *Neurology* 43:2668-2673.

Hyman C *et al.* (1994). Brain derived neurotrophic factor is a neurotrophic factor for dopaminergic neurons of the substantia nigra, *Nature* 350:230-232.

Isaacson LG, Saffran BN, Crutcher KA. (1990) Intracerebral NGF infusion induces hyperinnervation of cerebral blood vessels, *Neurobiol Aging* 11:51-55.

Jacobs A, Breakefield XO, Fraefel C. (1999). HSV-1-based vectors for gene therapy of neurological diseases and brain tumors: part I. HSV-1 structure, replication and pathogenesis, *Neoplasia* 1:387-401.

Kang UJ *et al.* (1993). Regulation of dopamine production by genetically modified primary fibroblasts, *J Neurosci* 13:5203-5211.

Kaspar BK *et al.* (2003). Retrograde viral delivery of IGF-1 prolongs survival in a mouse ALS model, *Science* 301:839-842.

Kirik D, Rosenblad C, Bjorklund A, Mandel R. (2000). Long-term rAAV-mediated gene transfer of GDNF in the rat Parkinson's model: Intrastriatal but not intranigral transduction promotes functional regeneration in the lesioned nigrostriatal system, *J Neurosci* 20:4686-4700.

Klein RL *et al.* (2000). NGF gene transfer to intrinsic basal forebrain neurons increases cholinergic cell size and protects from age-related, spatial memory deficits in middle-aged rats, *Brain Res* 875:144-151.

Koliatsos VE *et al.* (1990). Mouse nerve growth factor prevents degeneration of axotomized basal forebrain cholinergic neurons in the monkey, *J Neurosci* 10:3801-3813.

Kordower JH *et al.* (1994). The aged monkey basal forebrain: Rescue and sprouting of axotomized basal forebrain neurons after grafts of encapsulated cells secreting human nerve growth factor, *Proc Natl Acad Sci* 91:10898-10902.

Kordower JH *et al.* (2000). Neurodegeneration prevented by lentiviral vector delivery of GDNF in primate models of Parkinson's disease, *Science* 290:767-773.

Kotzbauer PT *et al.* (1996). Neurturin, a relative of glial-cell-line derived neurotrophic factor, *Nature* 384:467-470.

Kromer LF. (1987). Nerve growth factor treatment after brain injury prevents neuronal death, *Science* 235:214-216.

Lacorazza HD, Flax JD, Snyder EY, Jendoubi M. (1996) Expression of human beta-hexosaminidase alpha-subunit gene (the gene defect of Tay-Sachs disease) in mouse brains upon engraftment of transduced progenitor cells, *Nat Med* 2:424-429.

Learish R *et al.* (1996). Retroviral gene transfer and sustained expression of human arylsulfatase A, *Gene Ther* 3:343-349.

Leone P *et al.* (2000). Aspartoacylase gene transfer to the mammalian central nervous system with therapeutic implications for Canavan disease, *Ann Neurol* 48:27-38.

Levi-Montalcini R. (1987). The nerve growth factor 35 years later, *Science* 237:1154-1162.

Levi-Montalcini R, Hamburger V. (1951). Selective growth stimulating effects of mouse sarcoma on the sensory and sympathetic nervous system of the chick embryo, *J Exp Zool* 116:321-362.

Li T, Davidson BL. (1995). Phenotype correction in retinal pigment epithelium in murine mucopolysaccharidosis VII by adenovirus-mediated gene transfer, *Proc Natl Acad Sci U S A* 92:7700-7704.

Lin LF *et al.* (1993). GDNF: A glial cell-line derived neurotrophic factor for midbrain dopaminergic neurons, *Science* 260:1130-1132.

Liu Y *et al.* (1999). Transplants of fibroblasts genetically modified to express BDNF promote regeneration of adult rat rubrospinal axons and recovery of forelimb function, *J Neurosci* 19:4370-4387.

Lu P, Blesch A, Tuszynski MH. (2001). Neurotrophism without neurotropism: BDNF promotes survival but not growth of lesioned corticospinal neurons, *J Comp Neurol* 436:456-470.

Luo J *et al.* (2002). Subthalamic GAD gene therapy in a Parkinson's disease rat model, *Science* 298:425-429.

Maguir-Zeis KA, Bowers WJ, Federoff HJ. (2001). HSV vector-mediated gene delivery to the central nervous system, *Curr Opin Mol Ther* 3:482-490.

Mandel RJ *et al.* (1999). Nerve growth factor expressed in the medial septum following in vivo gene delivery using a recombinant adeno-associated viral vector protects cholinergic neurons from fimbria-fornix lesion-induced degeneration, *Exp Neurol* 155:59-64.

Markert JM, Parker JN, Gillespie GY, Whitley RJ. (2001). Genetically engineered human herpes simplex virus in the treatment of brain tumours, *Herpes* 8:17-22.

Mayeux R, Sano M. (1999). Treatment of Alzheimer's disease, *N Engl J Med* 341:1670-1679.

McBride JL *et al.* (2003). Structural and functional neuroprotection in a rat model of Huntington's disease by viral gene transfer of GDNF, *Exp Neurol* 181:213-223.

McCormack MP, Rabbitts TH. (2004). Activation of the T-cell oncogene LMOZ after gene therapy for X-linked severe combined immunodeficiency. *N Engl J Med* 350:913–922.

McKay R. (1992). Reconstituting animals from immortal precursors, *Curr Opin Neurobiol* 2:582-585.

Miller DA. (1990). Retrovirus packaging cells, *Hum Gene Ther* 1:5-14.

Mohajeri MH, Figlewicz DA, Bohn MC. (1999). Intramuscular grafts of myoblasts genetically modified to secrete glial cell line-derived neurotrophic factor prevent motoneuron loss and disease progression in a mouse model of familial amyotrophic lateral sclerosis, *Hum Gene Ther* 10:1853-1866.

Mufson EJ, Conner JM, Kordower JH. (1995). Nerve growth factor in Alzheimer's disease: defective retrograde transport to nucleus basalis, *Neuroreport* 6:1063-1066.

Mufson EJ *et al.* (1996). Intrastriatal and intraventricular infusion of brain-derived neurotrophic factor in the cynomologous monkey: distribution, retrograde transport and co-localization with substantia nigra dopamine-containing neurons, *Neuroscience* 71:179-191.

Muzyczka N. (1992). Use of anedo-associated virus as a general transduction vector for mammalian cells, *Curr Top Microbiol Immunol* 158:97-129.

Naldini L *et al.* (1996a). Efficient transfer, integration, and sustained long-term expression of the transgene in adult rat brains injected with a lentiviral vector, *Proc Natl Acad Sci U S A* 93:11382-11388.

Naldini L *et al.* (1996b). In vivo gene delivery and stable transduction of nondividing cells by a lentiviral vector,. *Science* 272:263-267.

Nanda D, Driesse MJ, Sillevis Smitt PA. (2001a). Clinical trials of adenoviral-mediated suicide gene therapy of malignant gliomas, *Prog Brain Res* 132:699-710.

Nanda D *et al.* (2001b). Treatment of malignant gliomas with a replicating adenoviral vector expressing herpes simplex virus-thymidine kinase, *Cancer Res* 61:8743-8750.

Nutt JG *et al.* (2003). Randomized, double-blind trial of glial cell line-derived neurotrophic factor (GDNF) in PD, *Neurology* 60:69-73.

Osterman JV, Waddell A, Aposhian HV. (1971). Gene therapy systems: the need, experimental approach, and implications, *Ann N Y Acad Sci* 179:514-519.

Perry EK, Perry RH, Blessed G, Tomlinson BE. (1977). Necropsy evidence of central cholinergic deficits in senile dementia, *Lancet* 2:143.

Perry EK *et al.* (1978). Correlation of cholinergic abnormalities with senile plaques and mental test scores in senile dementia, *Br Med J* 2:1457-1459.

Peterson DA *et al.* (1996). Fibroblast growth factor-2 protects entorhinal layer II glutamatergic neurons from axotomy-induced death, *J Neurosci* 16:886-898.

Phelps CH *et al.* (1989). Potential use of nerve growth factor to treat Alzheimer's disease, *Neurobiol Aging* 10:205-207.

Phillips HS *et al.* (1991). BDNF is decreased in the hipocampus of individuals with Alzheimer's disease, *Neuron* 7:695-702.

Rabinowitz JE, Samulski J. (1998). Adeno-associated virus expression systems for gene transfer, *Curr Opin Biotechnol* 9:470-475.

Rabinowitz JE, Samulski RJ. (2000). Building a better vector: the manipulation of AAV virions, *Virology* 278:301-308.

Rainov NG. (2000). A phase III clinical evaluation of herpes simplex virus type 1 thymidine kinase and ganciclovir gene therapy as an adjuvant to surgical resection and radiation in adults with previously untreated glioblastoma multiforme, *Hum Gene Ther* 11:2389-2401.

Ram Z *et al.* (1997). Therapy of malignant brain tumors by intratumoral implantation of retroviral vector-producing cells, *Nat Med* 3:1354-1361.

Rosenberg MB *et al.* (1988). Grafting genetically modified cells to the damaged brain: restorative effects of NGF expression, *Science* 242:1575-1578.

Scott SA *et al.* (1995). Nerve growth factor in Alzheimer's disease: increased levels throughout the brain coupled with declines in nucleus basalis, *J Neurosci* 15:6213-6221.

Shand N *et al.* (1999). A phase 1-2 clinical trial of gene therapy for recurrent glioblastoma multiforme by tumor transduction with the herpes simplex thymidine kinase gene followed by ganciclovir. GLI328 European-Canadian Study Group, *Hum Gene Ther* 10:2325-2335.

Short MP *et al.* (1990). Gene delivery to glioma cells in rat brain by grafting of a retrovirus packaging cell line, *J Neurosci Res* 27:427-439.

Smith DE, Roberts J, Gage FH, Tuszynski MH. (1999). Age-associated neuronal atrophy occurs in the primate brain and is reversible by growth factor gene therapy, *Proc Natl Acad Sci* 96:10893-10898.

Sofroniew MV *et al.* (1990). Survival of adult basal forebrain cholinergic neurons after loss of target neurons. *Science* 247:338–342.

Spina MB, Hyman C, Squinto S, Lindsay RM. (1992) Brain-derived neurotrophic factor protects dopaminergic cells from 6-hydroxydopamine toxicity, *Ann N Y Acad Sci* 648:348-350.

Stein CS, Ghodsi A, Derksen T, Davidson BL. (1999). Systemic and central nervous system correction of lysosomal storage in mucopolysaccharidosis type VII mice, *J Virol* 73:3424-3429.

Suhr ST, Gil EB, Senut MC, Gage FH. (1998). High level transactivation by a modified Bombyx ecdysone receptor in mammalian cells without exogenous retinoid X receptor, *Proc Natl Acad Sci U S A* 95:7999-8004.

Takenaka T *et al.* (2000). Long-term enzyme correction and lipid reduction in multiple organs of primary and secondary transplanted Fabry mice receiving transduced bone marrow cells, *Proc Natl Acad Sci U S A* 97:7515-7520.

Terry RD *et al.* (1991). Physical basis of cognitive alterations in Alzheimer's disease: synapse loss is the major correlate of cognitive impairment, *Ann Neurol* 30:572-580.

Todo T, Martuza RL, Rabkin SD, Johnson PA. (2001). Oncolytic herpes simplex virus vector with enhanced MHC class I presentation and tumor cell killing, *Proc Natl Acad Sci U S A* 98:6396-6401.

Tuszynski MH. (1999). Neurotrophic factors. In: *CNS regeneration: basic science and clinical advances*, Edited by Tuszynski MH and Kordower JH. San Diego: Academic Press.

Tuszynski MH. (2002). Gene therapy for neurodegenerative disorders, *Lancet Neurol* 2002:51-57.

Tuszynski MH, H.-S. U, Gage FH. (1991). Recombinant human nerve growth factor infusions prevent cholinergic neuronal degeneration in the adult primate brain, *Ann Neurol* 30:625-636.

Tuszynski MH, H. -S. U, Amaral DG, Gage FH. (1990). Nerve growth factor infusion in primate brain reduces lesion-induced cholinergic neuronal degeneration, *J Neurosci* 10:3604-3614.

Tuszynski MH, Senut MC, Ray J, H-S. U, Gage FH. (1994). Somatic gene transfer to the adult primate CNS: In vitro and in vivo characterization of cells genetically modified to secrete nerve growth factor, *Neurobiol Dis* 1:67-78.

Tuszynski MH *et al.* (1996). Gene therapy in the adult primate brain: intraparenchymal grafts of cells genetically modified to produce nerve growth factor prevent cholinergic neuronal degeneration, *Gene Ther* 3:305-314.

Tuszynski MH *et al.* (2002). Spontaneous and augmented growth of axons in the primate spinal cord: Effects of local injury and nerve growth factor-secreting cells, *J Comp Neurol* 448:88-101.

Ullrich A, Gray A, Berman C, Dull TJ. (1983). Human beta-nerve growth factor gene sequence highly homologous to that of mouse, *Nature* 303:821-825.

Wang Y *et al.* (2002). Herpes simplex virus type 1/adeno-associated virus rep(+) hybrid amplicon vector improves the stability of transgene expression in human cells by site-specific integration, *J Virol* 76:7150-7162.

Whitehouse PJ *et al.* (1981). Alzheimer's disease: evidence for selective loss of cholinergic neurons in the nucleus basalis, *Ann Neurol* 10:122-126.

Whitehouse PJ *et al.* (1982). Alzheimer's disease and senile dementia: loss of neurons in the basal forebrain, *Science* 215:1237-1239.

Whittemore SR, Eaton MJ, Onifer SM. (1997). Gene therapy and the use of stem cells for central nervous system regeneration, *Adv Neurol* 72:113-119.

Williams LR. (1991). Hypophagia is induced by intracerebroventricular administration of nerve growth factor, *Exp Neurol* 113:31-37.

Williams LR *et al.* (1986). Continuous infusion of nerve growth factor prevents basal forebrain neuronal death after fimbria-fornix transection, *Proc Natl Acad Sci U S A* 83:9231-9235.

Winkler C, Sauer H, Lee CS, Bjorklund A. (1996). Short-term GDNF treatment provides long-term rescue of lesioned nigral dopaminergic neurons in a rat model of Parkinson's disease, *J Neurosci* 16:7206-7215.

Winkler J *et al.* (1995). Essential role of neocortical acetylcholine in spatial memory, *Nature* 375:484-487.

Winkler J *et al.* (1997). Reversible Schwann cell hyperplasia and sprouting of sensory and sympathetic neurites after intraventricular administration of nerve growth factor, *Ann Neurol* 41:82-93.

Wolff JA *et al.* (1989). Grafting fibroblasts genetically modified to produce L-dopa in a rat model of Parkinson disease, *Proc Natl Acad Sci* 86:9011-9014.

Yant SR *et al.* (2002). Transposition from a gutless adeno-transposon vector stabilizes transgene expression in vivo, *Nat Biotechnol* 20:999-1005.

16

Therapeutics Development for Hereditary Neurological Diseases

Jill Heemskerk, PhD
Kenneth H. Fischbeck, MD

This chapter discusses the identification of neurological disease genes by positional cloning and the characterization of disease mechanisms through cell culture and animal models (Figure 16.1). The use of these insights in the development of pharmacological therapy is covered in detail. Other chapters by Tuszynski, Goldman, Gage, and Isacson discuss the development of other treatment modalities: gene therapy and stem cell therapy, respectively.

POSITIONAL CLONING

Much of biomedical research is based on the premise that finding the cause of a disease is the best way to develop effective treatment. Identifying the primary defect responsible for a disease also brings new insight

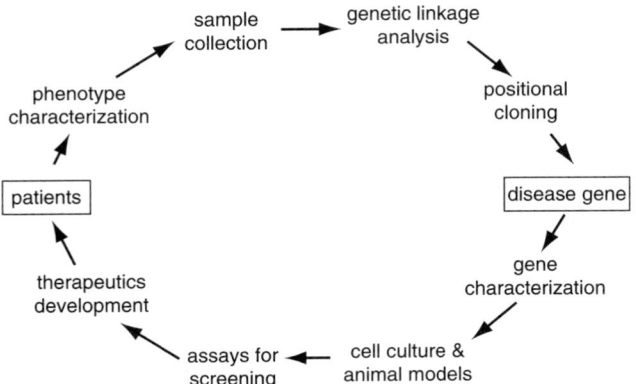

FIGURE 16.1 Events from the identification of a disease gene, beginning with studies in patients, to the development of a therapy based on the identified gene.

into normal biological structure and function. Since the mid-1980s, the strategy of positional cloning has been enormously useful in finding the causes of hereditary diseases (Figure 16.1). The path first followed with disorders such as Duchenne muscular dystrophy and cystic fibrosis—gathering DNA samples from clinically characterized patients and family members, studying these samples with genetic markers of known chromosomal location to determine which markers segregate with the disease, and analyzing candidate genes in the chromosomal interval defined by genetic linkage to find the gene that is specifically mutated in the disease—has now been followed with hundreds of other hereditary diseases. Disease gene identification allows accurate diagnostic testing in patients and prenatal, presymptomatic, and carrier testing in unaffected family members; it allows the development of cell culture and animal models that can be used to delineate the disease mechanism; and it leads to the identification of molecular targets for therapeutic intervention.

The fruits of the Human Genome Project have greatly facilitated positional cloning in recent years (see Chapter 13). Millions of genetic markers—simple sequence repeat and single nucleotide polymorphisms—have been identified and mapped, and comprehensive sequence analysis has led to detailed maps of genes of known and unknown function throughout the human genome. Advances in genotyping and sequencing technology have sped up the process of gene mapping and identification considerably. Nonetheless, there are many more disease genes yet to be found; of more than 14,000 defined hereditary diseases and genetic loci in the Online Mendelian Inheritance in Man (OMIM) database, only about 1500 have been identified to date (http://www.ncbi.nlm.nih.gov/Omim/). Many of the diseases with genes still to be found are neurological, and each of these offers the possibility of new insights into nervous system structure and function and new opportunities for refined diagnosis and therapeutic development. For the neurological disease genes that have already been identified, more needs to be done to connect genotype to phenotype, to work out the details of the disease mechanism, and to develop effective treatment.

POLYGLUTAMINE DISEASE

In 1991, the primary defect responsible for X-linked spinal and bulbar muscular atrophy (SBMA) was identified as an expanded trinucleotide (CAG) repeat in the androgen receptor gene (La Spada *et al.*, 1991). SBMA is an adult-onset, slowly progressive disease with progressive weakness of the bulbar and extremity muscles resulting from degeneration of motor neurons in the brainstem and spinal cord. The repeat expansion mutation leads to an expanded polyglutamine tract in the amino terminal domain of the androgen receptor protein. The androgen receptor is a nuclear receptor, a member of the steroid and thyroid hormone receptor family, and the amino terminal domain is likely involved in interaction with other nuclear proteins and is separate from the hormone and DNA binding domains. Although the mutation causes some loss of normal androgen receptor function, as evidenced by functional studies (Mhatre *et al.*, 1993; Brooks *et al.*, 1997) and by the signs and symptoms of androgen insensitivity that often occur in patients with SBMA, the primary effect of the mutation is to cause a toxic gain of function in the receptor protein (i.e., it alters the protein structure such that the receptor becomes toxic to motor neurons). This conclusion is based on the finding that animals and humans with other mutations that cause a partial or complete loss of androgen receptor function have a different phenotype, with feminization but not the progressive weakness and motor neuron loss of SBMA. Further support for a toxic gain of function mechanism comes from the finding that the mutant protein is toxic and reproduces features of the disease phenotype when introduced into cultured cells and transgenic animals (Merry *et al.*, 1998; Abel *et al.*, 2001).

When the gene responsible for Huntington's disease was identified (Huntington's Disease Collaborative Research Group, 1993), it was found to have the same kind of CAG/polyglutamine expansion mutation as had been found in SBMA. Although the clinical manifestations of Huntington's disease (chorea, psychological problems, and cognitive decline) differ from the progressive weakness of SBMA, and the cell type most suscepti-

ble to degeneration is the medium spiny neuron of the caudate nucleus rather than the primary motor neuron in the brainstem and spinal cord, the diseases are similar in that they are both adult-onset, neurodegenerative disorders. In subsequent years, seven more neurodegenerative diseases were found to be caused by CAG/polyglutamine expansion, including six spinocerebellar ataxias and dentatorubral-pallidoluysian atrophy (Zoghbi and Orr, 2000). Although caused by mutations in different genes and involving different neuronal cell populations, these disorders likely share the same basic mechanism of length-dependent polyglutamine toxicity. The protein context likely determines the neuronal specificity, but the toxicity is inherent to the polyglutamine tract. These diseases may all thus be seen as variants of the same disorder: polyglutamine expansion neurodegenerative disease.

Various mechanisms have been proposed for the neuronal toxicity of expanded polyglutamine (Ross, 2002). The expanded proteins are prone to aggregation, with a similar repeat length threshold for aggregation and disease. There is evidence of mitochondrial dysfunction and excitotoxicity. Mutant polyglutamine may impair proteasomal function, with consequent effects on protein turnover. Loss of normal protein function may contribute to the pathology in Huntington's disease as well as SBMA. Nuclear inclusions of mutant protein are seen in autopsied patients and in animal models of polyglutamine disease; and in model systems, toxicity can be enhanced by nuclear localization, indicating that the cell nucleus may be an important site of toxic interaction. There is evidence to support early transcriptional dysregulation and sequestration/depletion of important nuclear proteins in the disease mechanism. These mechanisms are not mutually exclusive. The challenge now in polyglutamine disease research is to identify the critical pathways in the pathogenesis and to develop treatment aimed at targets along these paths. Well-established cell culture and animal models are available for this purpose.

SPINAL MUSCULAR ATROPHY

Spinal muscular atrophy (SMA) is an autosomal recessive form of motor neuron disease, with onset in infancy or childhood. SMA patients have marked proximal muscle weakness and in severe cases early death owing to respiratory failure. SMA is caused by deletion and other mutations of the survival motor neuron (SMN) gene on chromosome 5 (Lefebvre *et al.*, 1995). Disease severity correlates with the decrease in SMN protein levels (Lefebvre *et al.*, 1997). SMN is a ubiquitously expressed protein that plays a role in the assembly and stability of spliceosomes, and SMN deficiency leads to defects in mRNA splicing (Pellizzoni *et al.*, 1998).

The SMN gene is normally present in two copies on chromosome 5: *SMN1*, which is mutated in SMA, and *SMN2*. *SMN2* differs from *SMN1* at a single nucleotide in exon 7 that leads to exon skipping in a majority of transcripts from this gene and a failure to produce full length protein. Cell culture and mouse models have been developed for SMA (Wang and Dreyfuss, 2001; Monani et al., 2000). An attractive approach to SMA treatment is to increase levels of expression or promote correct splicing of transcripts from *SMN2*. Drug screens have led to several candidates that up-regulate *SMN2*, including sodium butyrate, a histone deacetylase inhibitor (Chang et al., 2001). The main questions in SMA research now are: (1) How does SMN deficiency cause selective degeneration of motor neurons, and (2) What can be done to mitigate the manifestations of the disease?

HEREDITARY NEUROPATHY AND MOTOR NEURON DISEASE

The characteristic clinical phenotype of hereditary motor and sensory neuropathy, or Charcot-Marie-Tooth disease (CMT), includes progressive weakness, muscle atrophy, and sensory and reflex loss most prominent in the distal extremities (see Chapter 24). The same clinical features can occur with either demyelination or axonal degeneration in the peripheral nerves. There are more than 30 genetic loci for hereditary neuropathy, and at least 18 genes have now been identified (http://molgen-www.uia.ac.be/CMTMutations/). The more common demyelinating neuropathy genes were found in the early 1990s, but only recently have the rarer axonal neuropathy genes begun to be identified. Axonal neuropathy genes found recently encode proteins important in axon structure and transport, including kinesin family member 1B (CMT2A), lamin A (CMT2B1), and neurofilament light chain (CMT2E). The recently discovered CMT2D gene suggests a different mechanism (Antonellis et al., 2003). More work is needed to identify additional axonal neuropathy genes and in determining how mutations in these genes cause axonal degeneration. Each new gene that is found is another piece of the jigsaw puzzle that represents axonal biology and pathophysiology.

Hereditary motor neuron disease includes SMA and SBMA (discussed previously) and familial amyotrophic lateral sclerosis (ALS). About 20% of familial ALS patients have mutations in the gene for SOD1, with inexorably progressive weakness and spasticity due to degeneration of both upper and lower motor neurons, similar to what is seen in sporadic ALS patients (Rosen et al., 1993). Other patients with hereditary motor neuron disease have a more benign course (e.g., ALS2, caused by mutations in the gene for alsin) (Yang et al., 2001;

Hadano et al., 2001); dominantly inherited distal SMA, which resembles CMT2D and results from mutations in the same gene (Antonellis et al., 2003); ALS4, which is caused by mutations in the senataxin gene (Chen et al., 2004); and distal SBMA with vocal fold paralysis, which is caused by mutation in the gene for a subunit of dynactin (Puls et al., 2003). As with hereditary neuropathy, more motor neuron disease genes remain to be found, and for those that have been found, more needs to be learned about the pathophysiology. The new insights that have been gained into the causes and mechanisms of hereditary motor neuron disease need to be used to develop effective treatment for patients with these and other sporadic forms of motor neuron disease.

FRIEDREICH'S ATAXIA

Friedreich's ataxia (FRDA) is an autosomal recessive disease, the most common inherited form of ataxia. Clinical manifestations usually start in childhood or adolescence and include distal weakness, sensory loss, and cardiomyopathy, as well as loss of coordination and gait difficulty. Pathologically, there is degeneration of peripheral nerve and spinocerebellar pathways; the primary sensory neurons in the dorsal root ganglia are particularly vulnerable. The genetic cause is mutation, most commonly an intronic repeat expansion, in the gene for frataxin, a nuclear encoded mitochondrial protein (Campuzano et al., 1996; Koutnikova et al., 1997). Loss of frataxin function leads to iron accumulation in mitochondria and oxidative damage (Puccio & Koenig, 2000). Small clinical trials have indicated that patients with cardiomyopathy benefit from treatment with the antioxidant idebenone (Rustin et al., 2002; Mariotti et al., 2003). It remains to be determined whether treatment with higher doses over a longer period of time has a beneficial effect on the neurological manifestations that are the major problem in FRDA.

MUSCULAR DYSTROPHY

The muscular dystrophies are a set of over twenty hereditary diseases that involve progressive degeneration of muscle. Duchenne muscular dystrophy, the most common and severe of these diseases, is caused by mutations in the gene for the muscle structural protein dystrophin. Becker's dystrophy is a less severe disease caused by other mutations in the same gene. Four limb girdle dystrophies (which resemble Duchenne and Becker's dystrophies except that they have autosomal rather than X-linked recessive inheritance) are caused by mutations in genes for the sarcoglycans, which are structurally linked to dystrophin and associated with the muscle plasma membrane. Therapeutic strategies for Duchenne and Becker's dystrophies have included up-regulation of the dystrophin-related protein utrophin (similar to the up-regulation of *SMN2* to substitute for *SMN1*, described previously), replacement of muscle with transplanted myoblasts or stem cells, gene replacement with naked DNA or viral vectors, and transcript correction with oligonucleotides that promote exon skipping. The gene for dystrophin is too large for currently available viral delivery systems (although truncated versions with preserved function may fit); the sarcoglycan genes are smaller and suitable for packaging into adenovirus and adeno-associated virus. Gene replacement and the other strategies have shown some promise in animal experiments, but safety and efficacy have not yet been demonstrated in patients.

PHARMACOLOGICAL DRUG DISCOVERY FOR NEUROLOGICAL DISORDERS

Although a critical step toward finding a therapy, the identification of disease-causing genes is a very early stage in the drug discovery process. After a target for drug discovery has been identified, it may be years before a compound is identified that has the desired effect at that target. Even once an active compound is identified, the industry-wide estimate for time to the start of clinical testing is more than 4 years, with clinical testing and regulatory approval requiring an additional 7.5 years on average (DiMasi *et al.*, 2003). As we discuss later, the genes linked to neurological disorders are not necessarily drug targets in and of themselves, and additional time is likely to be necessary to boot strap from the causative gene to a suitable target molecule. The following sections outline the standard industry approach to target-based drug development, describe the particular challenges for genetic neurological disorders, and present novel approaches for rising to meet these challenges.

COMMERCIAL DRUG DISCOVERY

A now standard approach to drug discovery in the pharmaceutical and biotechnology industries is the use of high throughput drug screening (HTS, Fox *et al.*, 2002). HTS involves testing hundreds of thousands, even millions, of chemical compounds to find the rare ones that have activity against a molecular target of interest. This is made possible by the use of multiwell plates (e.g., 384-well plates) and automated, assembly line handling and analysis. The assays used in automated screening are necessarily simple, such as enzyme activity

assays or cell viability assays; thus a simple molecular or cellular model must be developed for HTS to be practical in any disease area. The compounds that are screened in HTS approaches are chosen largely for their chemical diversity. The aim is to represent as many chemical structures as possible to maximize the chances of finding a novel activity. The pharmaceutical properties of the compounds screened are defined only very generally if at all, with the goal of empirical discovery of novel chemical activities.

Depending on the assay, HTS may yield an initial success rate of 0.1% to 1% of the compounds screened; thus screening of 100,000 compounds might yield 100 or 1000 positive compounds, or hits. However, 90% or more of these initial hits may be false-positive results: while the large scale, automated screening approach allows testing of many compounds simultaneously, the necessary simplicity of the experimental design gives an inherently high error rate. These hits must therefore undergo a rigorous confirmation process that includes dose response testing and verifying the activities in other related assays. Prioritization of hit compounds also involves assessment of the potential suitability for pharmaceutical development of the compounds, based on the pharmacological properties of active compounds and the structure-activity relationship with other active compounds. At the end of this process, there will perhaps be only 1 in 100 hit compounds warranting further consideration.

It is important to realize that, although compounds that reach this stage are a refined set compared to the starting compounds, even these confirmed hits are not yet viable leads. In general, most of the hit compounds will prove to be impractical for drug development. All chemical compounds have some level of inherent toxicity, and the potency of activity of a compound must be high enough that the dose required for clinical benefit would be well below that at which intolerable toxicity or side effects would occur. Iterative rounds of medicinal chemistry are required to modify the compounds in an attempt to increase the potency of a compound and decrease its toxicity before it can safely enter clinical testing.

The final stages of drug development before the start of clinical testing are detailed pharmacological and toxicological studies (FDA Guidance, 1997). Pharmacological studies are done in rodents and other animals to determine whether the compound can get to its intended target at sufficient levels and be broken down and eliminated without toxicity. The rigorous toxicology testing required to begin human testing involves short- and long-term evaluation in multiple organ systems of both rodents and larger animals. *In vitro* genotoxicity studies are also generally required.

Given the effort and time required to transform a chemical compound into a drug, it is not surprising that industry tends to focus attention on disease areas with large patient populations and well-defined pathways to useful therapies. This is even less surprising considering that only one compound in five that enter clinical trials ever becomes a marketed drug (PhRMA Profile, 2003). The U.S. Food and Drug Administration (FDA) has instituted orphan drug incentives that have encouraged some biotechnology companies to take on development of drugs for small disease areas (Milne, 2002), but most of the drug development burden for genetic neurological diseases has thus far been taken on by private disease foundations and government funded academic researchers. As described next, this has led to the development of new approaches that may have broad impact on drug discovery in general (Verkman, 2004).

CHALLENGES OF THE GENETIC NEUROLOGICAL DISORDERS

One of the most important considerations in how readily the identification of a gene will lead to a therapeutic strategy is the nature of the gene product. The range of drug targets used in industry is relatively narrow, and many pharmaceutical and biotechnology companies favor target classes that have formed the basis of prior drug development efforts (Knowles and Gromo, 2003). These include G-protein-coupled receptors, nuclear hormone receptors, serine/threonine and tyrosine kinases, serine proteases, phosphodiesterases, and ion channels (Hopkins and Groom, 2002). As discussed earlier in this chapter, the genes linked to most genetic neurological disorders do not fall into the industry definition of "druggable" targets.

One strategy for making these diseases more amenable to drug discovery on the industry model is to identify targets involved in the disease mechanism that may be more amenable to therapeutics development, such as kinases or receptors. However, much basic mechanistic work will be required to understand the causative pathways in these diverse diseases. Many of the genes identified by positional cloning, such as frataxin, are novel, and their function is as yet incompletely understood (Puccio and Koenig, 2000). Others, such as mutant SOD1, carry mutations that confer unknown toxic functions on an otherwise well-characterized protein (Cleveland and Rothstein, 2001). Although indicators of oxidative stress and other cellular dysfunction may correlate with the presence of the disease-causing mutations, it is not yet known where in the hierarchy of cellular events these mechanisms occur. Further, because the mutations are present throughout development and life, it is necessary to determine the period required for the gene function to cause the disease phenotype in order to know when an intervention might produce clinical improvement. Thus an important challenge that remains after the identification of a disease-causing gene is the need to understand its mechanism of action.

NOVEL PROTEIN TARGETS

With the current lack of standard drug targets, academic investigators are developing novel strategies for drug discovery. Screening efforts with atypical protein targets, such as expanded polyglutamine tracts and α-synuclein are underway with encouraging results (Conway et al., 2001, Heemskerk et al., 2002). These in vitro screens are based on the propensity of purified mutant proteins to form aggregates in solution, and chemical compounds are screened for their ability to disrupt or prevent this aggregation. This approach identifies compounds that exert their effect through direct interaction with the disease-causing proteins. From a therapeutic standpoint, the underlying hypothesis that preventing aggregation will be clinically beneficial is untested, and as such it may not be appealing for commercial drug discovery efforts. However, active compounds can be used to test this idea in the academic setting, and the development of tools for hypothesis testing is one of the goals of such screens. Active compounds can be evaluated in mouse models to determine whether the in vitro effects of a compound can be reproduced in vivo and, if so, whether the effects are beneficial. Thus chemical compound screening can provide research tools to determine the disease mechanisms and to test whether an approach holds promise for drug discovery.

WHOLE CELL TARGET-BASED ASSAYS

Cellular assays have also been developed based on neurogenetic targets. In these assays, the behavior of a disease-linked protein is measured in the context of the whole cell. Examples are aggregation of proteins with expanded polyglutamine tracts and mutant SOD1 (Pollitt et al., 2003, Heemskerk et al., 2002), and activation of caspase 3 in response to mutant proteins (Piccioni et al., 2004). Although effects on a specific protein or metabolic pathway are measured in these assays, an active compound could in principle exert its effect through interaction with any protein in the cell. This strategy allows the identification of compounds based on targets that have not yet been precisely identified. As discussed later, active compounds can then be used to identify interacting proteins, thus providing new targets for further drug discovery.

PHENOTYPIC ASSAYS

Assays have also been developed that make no assumptions about the mechanism of action of a disease-causing protein. These assays measure the phenotypic effects of expression of disease proteins, such as cellular toxicity caused by expression of expanded polyglutamine and mutant SOD1 (Igarashi *et al.*, 2003; Heemskerk *et al.*, 2002) or by reduced function of SMN (Wang and Dreyfuss, 2001). As with other cell-based assays, compounds active in these assays may act at any point in a disease cascade and thus provide an opportunity to identify new components of the disease pathways in an approach known as chemical genetics (Stockwell, 2002; Lokey, 2003). Because the assay does not rely on specific knowledge of the mechanism of action of the proteins that cause disease, these assays have the further advantage that active compounds can be used to determine the underlying causes of toxicity.

Such assays tend not to be favored in industry drug discovery efforts because the medicinal chemistry required to optimize leads is hampered by the lack of a known target. The design of chemical modifications to maximize the potency of an interaction is most efficient when guided by knowledge of both interacting partners (i.e., target and ligand). This limitation can be overcome by (1) screening with collections of clinically tested compounds that may have known targets and are already optimized for human use or (2) using the active compounds from phenotypic screens to identify their cellular targets. These strategies are discussed further next.

SCREENS WITH CLINICALLY TESTED COMPOUNDS

Although medicinal chemistry is possible without a known target, the process is expedited when the starting compound has favorable pharmaceutical features such as low toxicity and availability *in vivo*. Bioactive compounds, including clinically tested or approved drugs, natural products, and narcotics, are compounds that have been optimized, either through medicinal chemistry or natural selection, for activity in biological systems. In the event that hits with such compounds are not sufficiently potent in the assay of interest, chemical analogs with similar pharmacological properties are often available that can be tested for increased potency. A second advantage of screening with clinically approved drugs and known bioactive compounds is that these compounds have been observed to give a higher hit rate than compounds chosen for chemical diversity (Stockwell, 2003); therefore, confirmed hits can be identified from starting collections as small as several hundred compounds (Korth *et al.*, 2001;

Conway *et al.*, 2001; Piccioni *et al.*, 2004). Because of the high success rate with a relatively small collection of compounds, the bioactive screening approach is applicable for more complex models such as *C. elegans* (Heemskerk *et al.*, 2002) and *Drosophila* (Muqit and Feany, 2002; Heemskerk *et al.*, 2002). Screening of bioactive compound collections is ideally suited to academia where the desire to make new biological discoveries outweighs the desire to claim novel intellectual property for commercial purposes. Screens with clinically approved drugs can also be commercially viable as a drug discovery approach, because the FDA orphan drug act provides commercial incentives for finding new rare disease indications for previously marketed drugs (Milne, 2002).

TARGET IDENTIFICATION

Various methods have been developed to identify the molecular targets of compounds identified in cell-based and phenotypic screens (Lokey, 2003). Once identified, these targets can be used in further screens to identify additional active compounds. This stepwise approach to target selection has the advantage that two key features of the target are already known: first, that the target can be modulated to ameliorate the disease phenotype, at least in a model system; and second, that the activity of the target can be modified by interaction with small molecules. Methods for identifying the protein partners of active compounds include biochemical approaches using labeled or immobilized compounds and probing cDNA expression cloning systems with labeled compounds (reviewed in Lokey, 2003). Given the somewhat promiscuous nature of bioactive compounds (Stockwell, 2003), targets identified by *in vitro* binding must be further validated. The possibility of identifying the cellular targets of small molecules in models of genetic neurological diseases provides hope that tractable drug development strategies for neurological disorders will emerge and will broaden the focus of commercial target-based drug discovery.

From an industry perspective, the lack of reliable drug targets and predictive models for neurological disorders results in an extremely high level of economic risk (Choi, 2002). Despite the large potential markets for treatments of conditions such as stroke and Alzheimer's disease, pharmaceutical and biotech companies have limited their efforts in neurotherapeutics development because of the high rate of failure in clinical testing of drug candidates for these disorders (Lee *et al.*, 1999; Fillit *et al.*, 2002). Thus, defining novel therapeutic targets and strategies for the genetic neurological disorders presents a valuable opportunity as well as a challenge. In addition to finding treatments for these diseases, the identification of new drug discovery approaches is likely to expand the industry definition of druggable targets, thus increasing the variety of approaches to finding therapies in all disease areas.

References

Abel A *et al.* (2001). Expression of expanded repeat androgen receptor produces neurologic disease in transgenic mice, *Hum Mol Genet* 10:107-116.

Antonellis A *et al.* (2003). Glycyl tRNA synthetase mutations are responsible for Charcot-Marie-Tooth disease type 2D and distal spinal muscular atrophy type V, *Am J Hum Genet* 72:1293-1299.

Brooks BP *et al.* (1997). Characterization of an expanded glutamine repeat androgen receptor in a neuronal cell culture system, *Neurobiol Dis* 4:313-323

Campuzano V *et al.* (1996). Friedreich's ataxia: autosomal recessive disease caused by an intronic GAA triplet repeat expansion, *Science* 271:1423-1427.

Chen YZ *et al.* (2004). DNA/RNA helicase gene mutations in a form of juvenile amyotrophic lateral sclerosis (ALS4) *Am J Hum Genet* 74:1128-1135.

Chang JG *et al.* (2001). Treatment of spinal muscular atrophy by sodium butyrate, *Proc Natl Acad Sci U S A* 98:9808-9813.

Choi DW. (2002). Exploratory clinical testing of neuroscience drugs, *Nat Neurosci Suppl* 5:1023-1025.

Cleveland DW, Rothstein JD. (2001). From Charcot to Lou Gehrig: deciphering selective motor neuron death in ALS, *Natl Rev Neurosci* 2: 806-819.

Conway KA, Rochet JC, Bieganski RM, Lansbury PT Jr. (2001). Kinetic stabilization of the alpha-synuclein protofibril by a dopamine-alpha-synuclein adduct, *Science* 294:1346-1349.

DiMasi JA, Hansen RW, Grabowski HG. (2003). The price of innovation: new estimates of drug development costs, *J Health Econ* 22:151-185.

FDA Guidance for Industry. (1997). M3 nonclinical safety studies for the conduct of human clinical trials for pharmaceuticals, *Fed Reg* 62:62922-62925.

Fillit HM *et al.* (2002). Barriers to drug discovery and development for Alzheimer disease, *Alzheimer Dis Assoc Disord* 16 Suppl 1:S1-S8.

Fox S, Wang H, Sopchak L, Farr-Jones S. (2002). High throughput screening 2002: moving toward increased success rates, *J Biomol Screen* 7:313-316.

Hadano S *et al.* (2001). A gene encoding a putative GTPase regulator is mutated in familial amyotrophic lateral sclerosis 2, *Nat Genet* 29:166-173.

Heemskerk J, Tobin AJ, Bain LJ. (2002). Teaching old drugs new tricks, *Trends Neurosci* 25:494-496.

Hopkins AL, Groom CR. (2002). The druggable genome, *Nat Rev Drug Disc* 1:727-730.

Huntington's Disease Collaborative Research Group. (1993). A novel gene containing a trinucleotide repeat that is expanded and unstable on Huntington's disease chromosomes, *Cell* 72:971-983.

Igarashi S *et al.* (2003). Inducible PC12 cell model of Huntington's disease shows toxicity and decreased histone acetylation, *Neuroreport* 14:565-568.

Knowles J, Gromo G. (2003). Target selection in drug discovery, *Nat Rev Drug Disc* 2:63-69.

Korth C, May BC, Cohen FE, Prusiner SB. (2001). Acridine and phenothiazine derivatives as pharmacotherapeutics for prion disease, *Proc Natl Acad Sci U S A* 98:9836-9841.

Koutnikova H *et al.* (1997). Studies of human, mouse and yeast homologues indicate a mitochondrial function for frataxin, *Nat Genet* 16:345-351.

La Spada A *et al.* (1991). Androgen receptor gene mutations in X-linked spinal and bulbar muscular atrophy, *Nature* 352:77-79

Lee JM, Zipfel GJ, Choi DW. (1999). The changing landscape of ischaemic brain injury mechanisms, *Nature* 399(6738 Suppl):A7-A14.

Lefebvre S *et al.* (1995). Identification and characterization of a spinal muscular atrophy-determining gene, *Cell* 80:155-165.

Lefebvre S *et al.* (1997). Correlation between severity and SMN protein level in spinal muscular atrophy, *Nat Genet* 16:265-269.

Lokey RS. (2003). Forward chemical genetics: progress and obstacles on the path to a new pharmacopoeia, *Curr Opin Chem Biol* 7:91-96.

Mariotti C *et al.* (2003). Idebenone treatment in Friedreich patients: one-year-long randomized placebo-controlled trial, *Neurology* 60:1676-1679.

Merry DE *et al.* (1998). Cleavage, aggregation and toxicity of the expanded androgen receptor in spinal and bulbar muscular atrophy, *Hum Mol Genet* 7:693-701.

Mhatre AN *et al.* (1993). Reduced transcriptional regulatory competence of the androgen receptor in X-linked spinal and bulbar muscular atrophy, *Nat Genet* 5:184-188.

Milne C-P. (2002). Orphan products: pain relief for clinical development headaches, *Nat Biotechnol* 20:780-784.

Monani UR *et al.* (2000). The human centromeric survival motor neuron gene (SMN2) rescues embryonic lethality in Smn(-/-) mice and results in a mouse with spinal muscular atrophy, *Hum Mol Genet* 9:333-339.

Muqit MM, Feany MB. (2002). Modelling neurodegenerative diseases in Drosophila: a fruitful approach? *Nat Rev Neurosci* 3:237-243.

Pellizzoni L, Kataoka N, Charroux B, Dreyfuss G. (1998). A novel function for SMN, the spinal muscular atrophy disease gene product, in pre-mRNA splicing, *Cell* 95:615-624.

Pharmaceutical Research and Manufacturers of America (PhRMA). (2003). Washington DC: Pharmaceutical Industry Profile.

Piccioni F, Roman BR, Fischbeck KH, Taylor JP. (2004). A screen for drugs that protect against the cytotoxicity of polyglutamine-expanded androgen receptor, *Hum Mol Genet* 13:437-446.

Pollitt SK *et al.* (2003). A rapid cellular FRET assay of polyglutamine aggregation identifies a novel inhibitor, *Neuron* 40:685-694.

Puccio H, Koenig M. (2000). Recent advances in the molecular pathogenesis of Friedreich ataxia, *Hum Mol Genet* 9:887-892.

Puls I *et al.* (2003). Mutant dynactin in motor neuron disease, *Nat Genet* 33:455-456.

Rosen DR *et al.* (1993). Mutations in Cu/Zn superoxide dismutase gene are associated with familial amyotrophic lateral sclerosis, *Nature* 362:59-62.

Ross CA. (2002). Polyglutamine pathogenesis: emergence of unifying mechanisms for Huntington's disease and related disorders, *Neuron* 35:819-822.

Rustin P, Rotig A, Munnich A, Sidi D. (2002). Heart hypertrophy and function are improved by idebenone in Friedreich's ataxia, *Free Radic Res* 36:467-469.

Stockwell BR. (2002). Chemical genetic screening approaches to neurobiology, *Neuron* 36:559-562.

Stockwell BR. (2003). Biological mechanism profiling using an annotated compound library, *Chem Biol* 10: 881-992.

Verkman AS. (2004). Drug discovery in academia, *Am J Physiol Cell Physiol* 286:C465-C474.

Wang J, Dreyfuss G. (2001). A cell system with targeted disruption of the SMN gene: functional conservation of the SMN protein and dependence of Gemin2 on SMN, *J Biol Chem* 276:9599-9605.

Yang Y *et al.* (2001). The gene encoding alsin, a protein with three guanine-nucleotide exchange factor domains, is mutated in a form of recessive amyotrophic lateral sclerosis, *Nat Genet* 29:160-165.

Zoghbi HY, Orr HT. (2000). Glutamine repeats and neurodegeneration, *Annu Rev Neurosci* 23:217-247.

Inherited Channelopathies of the CNS: Lessons for Clinical Neurology

Dimitri M. Kullmann, MD, PhD

A major development in neurology in the 1990s was the realization that several inherited central nervous system (CNS) diseases, mainly characterized by paroxysmal movement disorders or epilepsy, are caused by mutations of ion channels (Kullmann, 2002). The number of ion channels linked to neurological diseases continues to increase at a rate of one or two a year, and there are hints that the mutations associated with neurological diseases inherited in a mendelian fashion are only the tip of the iceberg: Subtle derangements of ion channel function may be at the root of idiopathic generalized epilepsy.

This chapter summarizes the clinical features and genetic mechanisms underlying the known CNS channelopathies, assesses the state of understanding of the cellular mechanisms that explain the manifestations, and tentatively points the way to further progress in understanding the full impact of this fast-moving area for neurological therapeutics. The chapter concentrates on the genetic channelopathies and does not discuss autoimmune diseases affecting ion channels. Nor does it consider the possibility that acquired defects in ion channel transcription contribute to disease mechanisms in a vast range of other areas of neurology. Moreover, muscle channelopathies are considered separately in this book, and are not discussed here, except where they shed light on CNS diseases caused by mutations of closely related ion channels.

THE CNS CHANNELOPATHIES: A BRIEF SUMMARY OF THE CHANNELS KNOWN TO BE AFFECTED

The CNS channelopathies span a broad range of "classical" ion channels. These include *voltage-gated channels* known to underlie action potential initiation

and repolarization, channels involved in membrane potential stabilization, and channels that mediate Ca^{++} ion flux across membranes. The classical ion channels also include *ligand-gated ion channels* (ionotropic receptors) that mediate communication among neurons by opening an ion-selective pore in the plasma membrane in response to binding of neurotransmitter molecules.

The classification of the neurological channelopathies is problematic because distinct mutations of the same ion channel can cause different clinical disorders and because, conversely, mutations of several different channels can converge onto a common clinical presentation. Moreover, some phenotypes that have been identified in association with ion channel mutations are indistinguishable from disorders that may have arisen from other mechanisms.

Table 17.1 shows a genetic classification of those channelopathies known to be transmitted in a mendelian fashion and where the causative mutations have been identified. This implies that the mutation is sufficient to cause the phenotype and generally indicates dominant inheritance when it is present together with the wild-type allele, or recessive inheritance when it is homozygous. It must be borne in mind, however, that some dominantly inherited mutations (e.g., of nicotinic receptor subunit genes) show incomplete penetrance. Moreover, some recessive mutations in glycine receptor subunits can cause disease through compound heterozygosity. That is, an affected individual can inherit two different mutant alleles from clinically unaffected parents in whom each mutation behaves recessively.

K+ Channelopathies

Voltage-gated K+ channels are encoded by a very large family of genes that show considerable homology,

TABLE 17.1   A Classification of Monogenic CNS Channelopathies

	Type	Gene	Channel	Disease
Voltage-gated	Na$^+$	SCN1A	α subunit of Na$_V$1.1	Generalized epilepsy with febrile seizures plus Severe myoclonic epilepsy of infancy
		SCN2A	α subunit of Na$_V$1.2	Generalized epilepsy with febrile seizures plus
		SCN1B	β$_1$ subunit	
	K$^+$	KCNA1	α subunit of Kv1.1	Episodic ataxia type 1
		KCNQ2 KCNQ3	Kv7.2 and Kv7.3 subunits (M current)	Benign familial neonatal convulsions
	Ca^{2+}	CACNA1A	α$_1$ subunit of Ca$_V$2.1 (P/Q channel)	Familial hemiplegic migraine Episodic ataxia type 2 Spinocerebellar ataxia type 6
	Cl	CLCN2	ClC-2	Idiopathic generalized epilepsy
Ligand-Gated	Nicotinic ACh	CHRNA4 CHRNB2	α$_4$ subunit β$_2$ subunit	Autosomal dominant nocturnal frontal lobe epilepsy
	Glycine	GLRA1	α$_1$ subunit	Familial hyperekplexia
	GABA$_A$	GABRG2	γ$_2$ subunit	Generalized epilepsy with febrile seizures plus
		GABRA1	α$_1$ subunit	Juvenile myoclonic epilepsy

although they only account for a subset of all known K$^+$ channels. They generally open slowly upon membrane depolarization and mediate an outward K$^+$ flux that contributes to repolarizing neurons and/or stabilizing their membrane potential. These channels are composed of four pore-forming α subunits, together with four cytoplasmic accessory β subunits. They share a common motif, with intracellular N and C terminals, six transmembrane segments (of which the fourth has several positively charged residues that are critical for voltage-dependent activation), and a pore-lining region between the fifth and sixth segments. Mutations have been identified in three different trans-membrane subunits: Kv1.1 (encoded by KCNA1), Kv7.2 (KCNQ2), and Kv7.3 (KCNQ3). Kv1.1 is the archetypal delayed rectifier channel. That is, its properties are well described by Hodgkin-Huxley kinetics, and it was the first K$^+$ channel to be cloned from a mutant *Drosophila* strain that showed temperature-dependent shaking, hence its alternative name "Shaker channel." It is thought to play an important role in repolarizing axons. As for Kv7.2 and Kv7.3, they have slow kinetics and open at relatively negative membrane potentials. Their physiological role has only recently been resolved, to a great extent as a result of progress in understanding the channelopathies (see later).

Missense mutations of KCNA1 occur in association with the autosomal dominant disorder episodic ataxia type 1 (EA1) (Browne *et al.*, 1994). This disorder is characterized by brief paroxysms of cerebellar incoordination, frequently triggered by stress or exertion. In between attacks patients are generally unaffected, although many have myokymia (spontaneous discharges of motor units) implying hyperexcitability of motor axon terminals. More recently, the phenotypic range of this disorder has been broadened by the finding of patients with isolated episodic ataxia, or myokymia and cramp without ataxia (Eunson *et al.*, 2000). Moreover, some families appear to have a relatively severe form of EA1 that does not respond to the usual treatments (carbamazepine or acetazolamide). Finally, epilepsy is found at a greater than expected rate in EA1 families.

The other K$^+$ channels, Kv7.2 and Kv7.3, are both associated with benign familial neonatal convulsions (BFNC) (Biervert *et al.*, 1998; Charlier *et al.*, 1998; Singh *et al.*, 1998). This syndrome is also autosomal dominant and consists of frequent partial and generalized seizures beginning in the first week of life and generally resolving spontaneously by 6 weeks. A variety of electroencephalography (EEG) disturbances have been reported, but the interictal EEG is usually normal. The discovery of these channel subunits genes was prompted by linkage studies, which identified two loci for BFNC. Positional cloning led to the discovery of two closely related genes, KCNQ2 and KCNQ3. Upon heterologous co-expression, the two subunits assemble to form a heteromultimeric K$^+$ channel that has a much higher current density than when either Kv7.2 or Kv7.3 is expressed alone (Wang *et al.*, 1998). The channel composed of both subunits has the kinetic and pharmacological properties of the so-called M current, an important conductance modulated by muscarinic receptors. In common with EA1, the phenotypic spectrum of KCNQ2 mutations has recently been widened, with the discovery of a family with the combination of neonatal convulsions and myokymia (Dedek *et al.*, 2001).

Na$^+$ Channelopathies

Voltage-gated Na$^+$ channels underlie the upstroke of fast-action potentials throughout the CNS. This family of channels is considerably smaller than the voltage-gated K$^+$ channels. Na$^+$ channels are made up of a single pore-lining α subunit, which resembles four voltage-gated K$^+$ channels in tandem (from which the Na$^+$ channel family probably evolved through twofold gene duplication). This subunit mediates the ion flux and contains the voltage-sensing and principal gating mechanisms. Na$_V$1.1 and Na$_V$1.2 are distributed in complementary patterns in many CNS neurons, with Na$_V$1.1 principally located to dendrites, where it probably underlies dendritic action potential, and Na$_V$1.2 in the axon, where it underlies action potential initiation and propagation. Both channels have fast kinetics. They open very rapidly on depolarization, and then rapidly inactivate in the presence of sustained depolarization. Two accessory subunits (β$_1$ and β$_2$), containing single trans-membrane segments, are associated with each α subunit. The β$_1$ subunit further accelerates fast inactivation, when the function of coexpressed α subunits is compared to the α subunit expressed alone.

The first mutation to be found in a CNS Na$^+$ channel subunit gene was found in SCN1B, which encodes β$_1$ in association with a highly pleomorphic dominantly inherited syndrome, generalized epilepsy with febrile seizures plus (GEFS+) (Wallace *et al.*, 1998). This disorder is unusual in that it cannot be diagnosed in a single patient, but only when it is present in several members of a family. This occurs because individual members of affected families can have very different patterns of seizures, including febrile and nonfebrile convulsions persisting beyond childhood, generalized tonic-clonic seizures, but also myoclonic, absence and atonic seizures. Some members are also affected by myoclonic-astatic epilepsy, a malignant pediatric syndrome. The boundaries of this syndrome continue to expand, while the minimal diagnostic criteria remain uncertain because neither febrile seizures nor epilepsy per se occurs universally.

Mutations associated with GEFS+ or similar familial syndromes (generalized febrile and nonfebrile seizures)

have since been discovered in SCN1A (Na$_V$1.1) (Escayg *et al.*, 2000b) and SCN2A (Na$_V$1.2) (Sugawara *et al.*, 2001; Heron *et al.*, 2002). Further mutations have been discovered in association with a ligand-gated channel, the α$_1$ subunit of ionotropic γ-aminobutyric (GABA$_A$) receptors (see later).

Ca^{++} Channelopathies

Voltage-gated Ca^{++} channels activate upon membrane depolarization and inactivate relatively slowly in comparison with Na$^+$ channels. They mediate an inward flux of Ca^{++} ions, which, in some neurons (for instance in the thalamic reticular nucleus or in Purkinje cells of the cerebellum), can be sufficient to sustain a regenerative depolarization (a Ca^{++} action potential). Another ubiquitous role for Ca^{++} channels is to trigger a number of intracellular signaling cascades that depend on Ca^{++}, in particular neurotransmitter release at synapses. The pore-forming α$_1$ subunits of Ca^{++} channels have a similar trans-membrane topology as Na$^+$ channels, with four repeated domains each containing six trans-membrane segments; however, the mechanisms of inactivation are different, as are the accessory subunits. At least two subunits are associated with CNS Ca^{++} channels: the α$_2$δ subunit, most of which is extracellular, but which also contains a single trans-membrane segment, and the β subunit, which is cytoplasmic. These subunits exist in several isoforms encoded by different genes.

The pharmacological and biophysical profile of Ca^{++} channels is principally determined by the identity of the α$_1$ subunit, which itself undergoes alternative splicing. Thus, the CACNA1A gene encodes the α$_1$A subunit of Ca$_V$2.1 channels, which, depending on the splice variant and together with accessory subunits, mediate a high-threshold, slowly inactivating current that is sensitive to certain invertebrate toxins. The older name P/Q-type current, which developed before the molecular classification of Ca^{++} channels, reflects the variable inactivation rates and toxin sensitivities of the individual splice variants of this gene product. Different mutations of CACNA1A give rise to dominantly inherited CNS disorders with a wide phenotypic spectrum. This includes familial hemiplegic migraine (FHM), a severe variant of migraine with aura, inherited in an autosomal dominant manner, where the aura can consist of focal weakness that mimics a stroke (Ophoff *et al.*, 1996). It also includes episodic ataxia type 2 (EA2), which differs from EA1 in that paroxysms are more prolonged and often include migrainous features, and patients can develop a slowly progressive cerebellar degeneration. A pure progressive cerebellar ataxia without obvious fluctuations in severity is also recognized (spinocerebellar ataxia type 6, SCA6) (Zhuchenko *et al.*, 1997). Some preliminary evidence suggests that mutations of the β_4 auxiliary subunit may also be associated with CNS disease (Escayg et al., 2000a).

Cl⁻ Channelopathies

In contrast to muscle fibers, Cl⁻ channels play only a minor role in setting the resting membrane potential of neurons. These channels have a quite different structure from the cation-selective channels listed previously. Their function and subunit composition are relatively poorly understood, but at least some of them exist as homodimers, with two Cl⁻ pores, which operate relatively independently. They are tonically active at rest, but allow Cl⁻ ions to enter relatively more effectively at depolarized membrane potentials. Three different mutations of the ClC-2 channel (encoded by CLCN2) have recently been reported to cosegregate with autosomal dominant epilepsy (Haug et al., 2003). Although the range of epileptic syndromes is narrower than in GEFS+, it includes all forms of idiopathic generalized epilepsy (IGE): childhood and juvenile absence epilepsy, juvenile myoclonic epilepsy, and epilepsy with grand mal seizures on awakening.

Nicotinic Receptor Channelopathies

These receptors are heteropentameric ligand-gated cation channels that are activated by acetylcholine (the endogenous agonist) and nicotine (a widely used exogenous agonist). CNS nicotinic receptors are biophysically similar to the well-characterized receptors that mediate excitation at the neuromuscular junction; however, the subunit composition of CNS nicotinic receptors is different. An abundant subtype occurring in the thalamus and cortex is composed of α_4 and β_2 subunits. Its role is unclear. Although nicotinic receptors of this subtype are at least partially located to presynaptic structures and appear to enhance the release of other neurotransmitters, it is not clear under which circumstances this phenomenon occurs physiologically. Part of the difficulty of studying these receptors is that both acetylcholine and nicotine not only activate them but also powerfully desensitize them. Whether a similar form of desensitization occurs in response to synaptically released acetylcholine is not known.

Mutations in both α_4 and β_2 (encoded by CHRNA4 and CHRNB2) have been found in families affected by autosomal dominant nocturnal frontal lobe epilepsy (ADNFLE) (Steinlein et al., 1995; De Fusco et al., 2000; Phillips et al., 2001). This disorder is characterized by clusters of convulsions arising from sleep, often with preserved consciousness. The attacks are sometimes misdiagnosed as parasomnias, partly because EEG abnormalities can be missed. The discovery of disease-associated CHRNA4 mutations, followed by CNRNB2 mutations, followed a similar sequence as that of KCNQ2 and KCNQ3 mutations in association with BFNC. As mentioned previously, although within an individual family the mutation cosegregates with the disease, there is evidence for incomplete penetrance, implying that other factors, genetic or environmental, affect the seizure threshold.

GABA_A Receptor Channelopathies

By comparison with nicotinic receptors, far more is known about the role of $GABA_A$ receptors in mediating fast GABAergic transmission in the CNS. They are abundantly expressed throughout the brain and spinal cord, and are frequently, although not exclusively, located in the postsynaptic membrane opposite GABA release sites of inhibitory interneurons. Although they have a similar heteropentameric composition as nicotinic receptors (and share a common trans-membrane topology and gating mechanisms), they are permeable to anions (Cl⁻ and HCO_3^-) instead of cations. Depending on the electrochemical trans-membrane gradient for these ions, their activation leads either to hyperpolarization or to depolarization of the neuronal membrane. These effects are accompanied by a shunting effect on membrane resistance, which means that even when they depolarize neurons, $GABA_A$ receptors can attenuate the excitatory effects of other synaptic inputs. $GABA_A$ receptors are thus principally inhibitory, at least in adult tissue. $GABA_A$ receptors are made up of a wide variety of subunits, but two common constituents, which probably coassemble in many receptors, are α_1 and γ_2. Both subunits contribute to determining the kinetics, affinity, desensitization rate, and pharmacological profile of the receptors (i.e., the sensitivity to benzodiazepine, alcohol, and barbiturates among clinically relevant agents). The γ_2 subunit also contributes to synaptic targeting of $GABA_A$ receptors.

Mutations of the γ_2 subunit, encoded by GABRG2, have been identified in several families with GEFS+ or closely associated syndromes (Baulac et al., 2001; Wallace et al., 2001). And one family with juvenile myoclonic epilepsy (JME) has been reported to carry a mutation in the gene for α_1 GABRA1 (Cossette et al., 2002).

Glycine Receptor Channelopathies

All of the channelopathies listed previously are dominantly transmitted. The exception is that familial hyperekplexia (FH) can arise from either dominant or recessive mutations of the GLRA1 gene, which encodes the α_1 subunit of glycine receptors (Shiang *et al.*, 1993; Brune *et al.*, 1996). This receptor also belongs to the nicotinic family, together with nicotinic acetylcholine and GABA$_A$ receptors. Although glycine receptors are abundant throughout the CNS, α_1 containing receptors are thought to mediate fast inhibitory transmission only in the spinal cord and brainstem. The receptors are heteropentameric, composed of α_1 and β subunits.

Patients with FH can present in infancy or later, with stiffness and exaggerated startle responses triggered by sounds or other unexpected stimuli.

THE NATURE OF THE MOLECULAR LESION: CAN IT EXPLAIN THE PHENOTYPE?

Mutations cause disease through several different genetic mechanisms, many of which occur among the channelopathies. Most of the mutations that have been identified are missense. However premature stop codons and splice site mutations that are predicted to truncate the peptide sequence also occur. SCA6 often arises from an expansion in a CAG repeat in the CACNA1A gene (Zhuchenko *et al.*, 1997), but this is the only known example of a disease caused by a polyglutamine tract in an ion channel gene.

Ion channels are almost unique in that the consequences of mutations can be characterized in great detail with biophysical methods. Foremost among these are electrophysiological recordings from cells expressing mutant ion channels, in particular oocytes of *Xenopus laevis* and transfected mammalian cultures. This approach goes far beyond the level of analysis possible with conventional enzyme activity assays. In particular, voltage clamp recordings potentially allow an insight into the details of the voltage-dependence and ligand binding kinetics, opening probability, single channel conductance, and mean current density (a reflection of the population behavior of channels expressed in a whole cell). In addition, interactions with wild-type subunits, or accessory subunits can be assayed. Additional information comes from examining the subcellular distribution of mutated ion channels, which can be achieved either by immunocytochemistry or by tagging subunits with fluorescent molecules.

Bearing in mind that altered channel properties may well give rise to compensatory alterations in the expression of these and other proteins and also possibly to developmental alterations, the results of heterologous expression studies have begun to yield an insight into disease mechanisms.

An alternative top-down approach to understand the disease mechanisms is to ask whether the channel is expected to enhance or decrease the excitability of neu-

ronal circuits known to be involved in particular clinical disorders. Thus a simple view of epilepsy and paroxysmal and/or hyperkinetic movement disorders (hyperekplexia, episodic ataxia, hemiplegic migraine) is that they may arise from a failure of the mechanisms that normally repolarize and/or stabilize neuronal membranes, or an increase in mechanisms that depolarize them. This gives rise to the prediction that such disorders can result from deficient voltage K$^+$ channel function, failure of GABAergic or glycinergic inhibition, or gain of function of Na$^+$ and possibly Ca^{++} channels.

In some cases such top-down predictions are in general agreement with the results of heterologous expression studies. Thus, missense mutations associated with KCNQ2 or KCNQ3 that have been expressed have been found to cause partial loss of function; the current density is reduced compared to wild-type (Biervert *et al.*, 1998; Schroeder *et al.*, 1998; Lerche *et al.*, 1999). Kinetics, however, are apparently unchanged. Channels in an individual carrying a mutation in KCNQ2 potentially consist of three types of subunits: mutant and wild-type Kv7.2 and wild type Kv7.3. When this situation was reproduced *in vitro* by coexpressing all three DNA species, the current density reduction was decreased by only about 25% (Schroeder *et al.*, 1998). This implies that relatively modest degrees of loss of function are sufficient to give rise to the disease. It also explains the dominant inheritance of BFNC. Why most patients cease to have seizures after about 6 weeks of age is unclear, but this may reflect compensatory up regulation of other subunits, or a general decrease in the excitability of the neuronal circuitry where seizures initially arose as a result of impaired M current function.

Missense KCNA1 mutations associating with EA1 also cause loss of function (Adelman *et al.*, 1995). This loss ranges from very mild changes in the voltage dependence or rate of activation and inactivation through to completely nonfunctional channels. Some of the mutations are not only unable to assemble to form functional channels, but also exert a dominant negative effect on the coexpressed wild-type K$^+$ channel subunits, possibly by coassembling with them and sequestering them within intracellular organelles (Rea *et al.*, 2002a). Mutations associated with a more severe clinical syndrome (drug resistance and/or association with epilepsy) also cause more profound K$^+$ current reductions and/or dominant negative effects (Eunson *et al.*, 2000). Why patients with these mutations have paroxysms of ataxia is unclear. The strong expression of Kv1.1 in the cerebellum (where they occur in axons of GABAergic inhibitory neurons) hints at a failure of normal regulation of Purkinje cell output, a possibility that is supported by studies of mice with targeted deletion of the KCNA1 gene (Zhang *et al.*, 1999). A knock-in mouse model has lent further support to this view, in that the frequency of spontaneous GABAergic synaptic signals was increased in Purkinje cells, possibly because of hyperexcitability of the axons of interneurons (Herson *et al.*, 2003). Similar processes may underlie the occurrence of spontaneous action potentials in motor axons, providing a possible explanation for myokymia. Little is known of the events underlying the initiation and termination of a paroxysm of ataxia, although it has been suggested that the cerebellar cortex may be susceptible to a phenomenon analogous to cortical spreading depression (Ebner and Chen, 2003).

Another example in which predictions from known functions of channels agree with the results of heterologous expression is the association of GABA$_A$ receptor mutations with epilepsy. Several missense mutations of the γ_2 subunit cause partial loss of function, either by impairing the maximal opening probability of channels containing this subunit or by affecting its membrane expression (Baulac *et al.*, 2001; Wallace *et al.*, 2001; Bianchi *et al.*, 2002). The relatively modest predicted decrease in GABAergic transmission may be sufficient to explain the dominant inheritance. In common with other genetic causes of GEFS+, however, it is not known why different members of the same family exhibit highly disparate types of seizures.

Finally, GLRA1 mutations in FH also impair the normal function of glycine receptors. Dominantly inherited mutations tend to encode glycine receptors that are functional, albeit with reduced apparent affinity for the ligand and/or decreased maximal current (Langosch *et al.*, 1994; Lewis *et al.*, 1998). In contrast, recessively inherited mutations tend to be nonfunctional. These include both missense and nonsense mutation (Brune *et al.*, 1996; Rea *et al.*, 2002b). This implies that the product of one wild-type allele is sufficient to mediate normal glycinergic inhibition. Haploinsufficiency does not appear to play a role in this disorder, in striking contrast to BFNC.

An alternative mechanism that can potentially give rise to paroxysmal excessive neuronal discharges is gain of function of channels that underlie action potentials, namely Na$^+$ channels. Although some progress has been made in relating Na$^+$ channel mutations to GEFS+, here too, there are some major difficulties. The original SCN1B mutation discovered in a large kindred with GEFS+ apparently causes complete loss of function of the β_1 subunit (Wallace *et al.*, 1998). Because this subunit normally accelerates the rate of fast inactivation, this is equivalent to a gain of function for Na$^+$ channels: they may remain open for longer than when the α subunit coassembles with the wild-type β_1. This effect is similar to that seen for some mutations of the muscle Na$^+$ channel SCN4A associated with hyperkalemic periodic paralysis. Some, but not all, heterologous expression studies that

have examined missense mutations of SCN1A in GEFS+ have also reported that they impair fast inactivation of Na$^+$ channels (Lossin *et al.*, 2002). Other reports, however, have drawn attention to changes in voltage-dependent kinetics (Alekov *et al.*, 2001).

An unexpected challenge to the view that enhanced Na$^+$ currents underlie seizures is the observation that nonsense mutations are a common cause of the malignant pediatric syndrome Severe Myoclonic Epilepsy of infancy (Claes *et al.*, 2001). Many of these mutations are predicted to result in truncation of large and essential parts of the α subunit, and therefore should result in loss of function. A possible resolution of this paradox that remains to be explored is that the mutation results in compensatory overexpression of the wild-type allele, or of other Na$^+$ channels.

Some channelopathies remain difficult to explain by asking whether a simple gain or loss of function can be expected to increase or decrease the excitability of the brain. The different disorders associated with Cav2.1 mutations are especially problematic. Different mutations associated with FHM have been reported to either increase or decrease Ca^{++} current density on heterologous expression (Kraus *et al.*, 1998; Hans *et al.*, 1999). Moreover, the mutations have different effects on the voltage dependence of activation and inactivation and single channel conductance, without a consistent pattern. Given the persistent uncertainty as to the site (let alone mechanism) of initiation of a migraine attack, it is difficult to know whether gain or loss of function of Ca^{++} channels is expected from a top-down approach to the problem. An interesting recent suggestion is that, although the current density, which reflects population behavior of the channels, is affected in different ways, at least one aspect of the behavior of single channels is consistently changed in the same direction by the different FHM mutations. The Ca^{++} flux through an individual channel (which is the product of single channel conductance and open channel probability) is consistently increased when relatively modest depolarizing steps are applied (Tottene *et al.*, 2002). This could mean that the disease arises from gain of Ca^{++} channel function, assuming that other confounding phenomena, such as trafficking of the channels to the membrane, are not relevant to the role of the channels in sites critically involved in migraine initiation. This hypothesis is consistent with the "microdomain" hypothesis of Ca^{++}-triggered transmitter release, which holds that only one or a few channels may be sufficient to mediate the Ca^{2+} influx necessary to trigger vesicle exocytosis.

Other mutations of Cav2.1 also present some difficulties in interpretation. In particular, conflicting reports exist about the effect of the SCA6-associated polyglutamine expansion on current density (Matsuyama *et al.*, 1999; Restituito *et al.*, 2000; Piedras-Renteria *et al.*, 2001). Given that some other polyglutamine disorders are thought to result from abnormal handling of the expanded region and intranuclear protein deposition, it is not clear whether any effect on channel function is relevant at all. Indeed, some neuropathological data suggest that surviving Purkinje cells of the cerebellum accumulate cytoplasmic inclusions, which are reminiscent of intranuclear inclusions seen in other polyglutamine expansion diseases (Ishikawa et al., 2001). On the other hand, arguing against a cytotoxic effect of the expanded sequence is the finding that an apparently identical purely progressive ataxia can arise as a result of missense mutations of the same gene (Friend et al., 1999).

As for EA2, all the mutations that have been studied result in partial or complete loss of function (Guida et al., 2001; Wappl et al., 2002). It remains to be established why this disorder is paroxysmal and whether the impaired Ca^{++} channel function causes ataxia via a direct effect on cerebellar neurons, or by altering transmitter release. A limitation of the heterologous expression studies reported hitherto is that they have not explored systematically the importance of alternative splicing of the affected channel proteins.

Finally, CLCN2 mutations also present a paradox. Of the three reported mutations associated with IGE, two are loss of function while the third appears to increase Cl^- flux by shifting the activation threshold for the channel (Haug et al., 2003). Given that the functions of this channel in setting the Cl^- gradient and/or the resting membrane potential of neurons are somewhat mysterious, there is clearly much work to be done to understand this channelopathy.

MECHANISMS OF DISEASE: WHAT REMAINS TO BE UNDERSTOOD?

The preceding discussion has considered the general principle that seizures and hyperkinetic movement disorders result from gain of function mutations of Na^+ channels and loss of function mutations of K^+ or glycine or $GABA_A$ receptors. It ignores some important principles of brain organization, which greatly complicate the relationship between ion channel function and the function of the brain in situ. These include the trivial fact that opposite consequences may be expected from an alteration in ion channel function depending on whether it is principally expressed in an inhibitory neuron or an excitatory cell. Moreover, the "safety factor" by which a population of neurons tolerates a change in ion channel function may differ. For instance, relatively subtle decreases in Kv7.2/Kv7.3 channel function seem to be sufficient to give rise to neonatal convulsions, and yet functional loss of a GLRA1 allele is without effect. A

possible explanation why BFNC is a benign syndrome may simply be that the safety factor increases with age. Further, as hinted previously, compensatory changes in the expression of some ion channel genes may occur. These secondary developmental effects may either mask the effect of the molecular lesion in some brain circuits, or reveal unexpected effects. An overcompensatory increase in homologous currents could explain why nonsense mutations of SCN1A cause SMEI, or reconcile the hypothesis that neurotoxic effects of Ca^{++} influx explain cerebellar degeneration in EA2 with the evidence that CACNA1A mutations result in loss of function.

Ultimately, these complications call for physiological studies of neuronal circuits in situ. Because human tissue is generally inaccessible, several animal models may provide some answers. The best is the knock-in mouse, where the native channel gene has been substituted with a gene (human or murine ortholog) harboring the disease-associated mutation. As mentioned previously, heterozygous mice expressing an EA1-associated mutation of Kv1.1 reproduce some but not all features of the human disorder: impaired motor coordination, but only after isoprenaline injection, and the animals do not obviously have neuromyotonia (Herson et al., 2003). Nevertheless, these mice do allow examination of candidate neuronal circuits that are involved in ataxia. The frequency of spontaneous GABAergic synaptic currents in Purkinje cells was found to be higher than in wild-type littermates, implying that a defective "brake" on GABAergic inhibition in the cerebellar cortex may be involved in the ataxia. However, much work remains to be done to find an explanation for the initiation of paroxysms in patients affected by this disorder.

Other potentially useful models are transgenic mice overexpressing mutant channels or mice in which the native channel has been disrupted (to mimic a nonsense mutation). Finally, a variety of spontaneous mouse mutants have been identified where the ortholog is affected, generally by mutations different from those occurring in humans. Some of these mutants reproduce the human disorder quite well, but some unexpected results call for caution in relying on such models. Thus, for instance, in contrast to homozygous human nonsense mutations of GLRA1 (which only result in hyperekplexia but probably a normal life span), a nonsense mutation of the same gene in the mouse leads to death by 3 weeks (Buckwalter et al., 1994).

Another example in which deleting the murine gene appears to have different effects from the human nonsense mutation is CACNA1A, which encodes $Ca_V2.1$. Although heterozygous patients have episodic ataxia, heterozygous mice are unaffected, although homozygous knock-out mice have severe ataxia and behavioral arrest suggestive of absence seizures (Jun et al., 1999; Fletcher et al., 2001). Several spontaneous homozygous murine strains with spontaneous mutations of the same gene also have the combination of ataxia and behavioral arrest (and associated EEG changes similar to those seen in human absence epilepsy) (Fletcher *et al.*, 1996; Lorenzon *et al.*, 1998). This prompts the question whether human absence epilepsy may also be linked to the same gene. Supporting such an association, one *de novo* nonsense mutation has been identified in a patient affected by a combination of absence epilepsy and EA2 (Jouvenceau *et al.*, 2001).

CHANNELOPATHIES AND THERAPEUTIC ADVANCES

The full impact of ion channel mutations for epilepsy and other paroxysmal disorders remains to be determined. Most patients do not show a clear mendelian pattern of inheritance, although concordance is elevated among family members for most paroxysmal neurological disorders. Screening of sporadic hemiplegic migraine cases for CACNA1A mutations has revealed coding mutations in a minority of cases (Terwindt *et al.*, 2002). However, screening patients with SMEI for SCN1A mutations has a relatively high yield (Claes *et al.*, 2001; Sugawara *et al.*, 2002). The genetics of idiopathic epilepsies and migraine are consistent with polygenic mechanisms. Subtle effects of sequence variants in promoters or in introns on gene expression or splicing patterns may be insufficient to cause an overt phenotype on their own, but may act in combination with similar phenomena affecting other genes. Clearly, given that ion channel genes play such an important role in monogenic paroxysmal neurological disorders (and indeed account for 10 of 12 monogenic epilepsies), they are also the best candidate genes for explaining complex inheritance. This is likely to be an active and fruitful field of research in the near future.

It is reasonable to speculate that the neuronal circuit alterations that occur in the channelopathies are relevant to common paroxysmal neurological diseases with indistinguishable or similar phenotypes. If so, what clues do they provide for new therapeutic avenues? Perhaps disappointingly, several of these avenues are already well trodden as a result of serendipitous or screening-led drug discovery programs. Thus, Na$^+$ channels and GABA$_A$ receptors are already the targets of several of the most widely used antiepileptic drugs. An interesting coincidence is that Kv7.2/Kv7.3 channel function is relatively selectively enhanced by the experimental antiepileptic drug retigabine (Main *et al.*, 2000; Wickenden *et al.*, 2000). Although development of this drug appears to have been suspended, it provides a unique opportunity to attempt a pharmacological correction of the decreased M current density thought to underlie BFNC. However,

because most affected individuals cease to have seizures after a few weeks of life, treatment is usually not required. A few individuals do continue to have persistent seizures, and it would be interesting to know whether the drug is especially effective in these patients.

One *in vitro* pharmacological study has also suggested that mutations of nicotinic receptor subunits may enhance their sensitivity to carbamazepine (Picard *et al.*, 1999). However, it is difficult to design an experiment to determine whether this phenomenon explains the effectiveness of this drug in treating ADNFLE, let alone other epilepsies. The role of the channel *in vivo* is still far from understood, and different disease-associated mutations have a variety of effects on nicotinic receptor current density, affinity, and Ca^{++} permeability.

A poorly understood feature of several neurological channelopathies is that they are responsive to the carbonic anhydrase inhibitors acetazolamide and dichlorphenamide. Although this is especially noticeable for the periodic paralyses (which are not addressed in this chapter), it is also reported for EA2, and possibly also for FHM and EA1 (Kullmann *et al.*, 2001). Among possible explanations is that a metabolic acidosis reduces the excitability of neuronal circuits. Although carbonic anhydrase inhibitors were at one time commonly prescribed antiepileptic drugs, they have been displaced by drugs thought to target other systems, including Na$^+$ channels and GABA$_A$ receptors. Nevertheless, inhibition of carbonic anhydrase has been reported to be an important mechanism of action of the new antiepileptic drug topiramate (Dodgson *et al.*, 2000). The beneficial effect of acetazolamide in some channelopathies has prompted open-label trials in sporadic migraine with and without aura, with some anecdotal reports of success (Haan *et al.*, 2000).

CONCLUSION

Although the monogenic channelopathies are likely to remain relatively uncommon, further therapeutic advances in the treatment of common paroxysmal neurological diseases may still derive directly or indirectly from this area of research. This progress will depend heavily on establishing whether subtle functional variants of ion channels are risk factors for sporadic epilepsy and related disorders. Moreover, some of the neuronal circuits implicated in the monogenic channelopathies may also be involved in common disorders with similar clinical manifestations. In addition, of course, it will be necessary to develop more selective drugs. It may ultimately be important to identify agents that discriminate between closely related channels (or even splice variants) that perform different functions at different sites, such as Ca^{++} channels that trigger the release of excitatory or inhibitory transmitters and also mediate postsynaptic Ca^{++} signaling.

References

Adelman JP, Bond CT, Pessia M, Maylie J. (1995). Episodic ataxia results from voltage-dependent potassium channels with altered functions, *Neuron* 15:1449-1454.

Alekov AK *et al.* (2001). Enhanced inactivation and acceleration of activation of the sodium channel associated with epilepsy in man, *Eur J Neurosci* 13:2171-2176.

Baulac S *et al.* (2001). First genetic evidence of GABA(A) receptor dysfunction in epilepsy: a mutation in the gamma2-subunit gene, *Nat Genet* 28:46-48.

Bianchi MT, Song L, Zhang H, Macdonald RL. (2002). Two different mechanisms of disinhibition produced by GABAA receptor mutations linked to epilepsy in humans, *J Neurosci* 22:5321-5327.

Biervert C *et al.* (1998). A potassium channel mutation in neonatal human epilepsy, *Science* 279:403-406.

Browne DL *et al.* (1994). Episodic ataxia/myokymia syndrome is associated with point mutations in the human potassium channel gene, KCNA1 [see comments], *Nat Genet* 8:136-140.

Brune W *et al.* (1996). A GLRA1 null mutation in recessive hyperekplexia challenges the functional role of glycine receptors, *Am J Hum Genet* 58:989-997.

Buckwalter MS *et al.* (1994). A frameshift mutation in the mouse alpha 1 glycine receptor gene (Glra1) results in progressive neurological symptoms and juvenile death, *Hum Mol Genet* 3:2025-2030.

Charlier C *et al.* (1998). A pore mutation in a novel KQT-like potassium channel gene in an idiopathic epilepsy family [see comments], *Nat Genet* 18:53-55.

Claes L *et al.* (2001). De novo mutations in the sodium-channel gene scn1a cause severe myoclonic epilepsy of infancy, *Am J Hum Genet* 68:1327-1332.

Cossette P *et al.* (2002). Mutation of GABRA1 in an autosomal dominant form of juvenile myoclonic epilepsy, *Nat Genet* 31:184-189.

De Fusco M *et al.* (2000). The nicotinic receptor beta 2 subunit is mutant in nocturnal frontal lobe epilepsy, *Nat Genet* 26:275-276.

Dedek K *et al.* (2001). Myokymia and neonatal epilepsy caused by a mutation in the voltage sensor of the KCNQ2 K+ channel, *Proc Natl Acad Sci U S A* 98:12272-12277.

Dodgson SJ, Shank RP, Maryanoff BE. (2000). Topiramate as an inhibitor of carbonic anhydrase isoenzymes, *Epilepsia* 41:S35-39.

Ebner TJ, Chen G. (2003). Spreading acidification and depression in the cerebellar cortex, *Neuroscientist* 9:37-45.

Escayg A *et al.* (2000a). Coding and noncoding variation of the human calcium-channel beta4-subunit gene CACNB4 in patients with idiopathic generalized epilepsy and episodic ataxia, *Am J Hum Genet* 66:1531-1539.

Escayg A *et al.* (2000b). Mutations of SCN1A, encoding a neuronal sodium channel, in two families with GEFS+2 [In Process Citation], *Nat Genet* 24:343-345.

Eunson LH *et al.* (2000). Clinical, genetic, and expression studies of mutations in the potassium channel gene KCNA1 reveal new phenotypic variability [In Process Citation], *Ann Neurol* 48:647-656.

Fletcher CF *et al.* (1996). Absence epilepsy in tottering mutant mice is associated with calcium channel defects, *Cell* 87:607-617.

Fletcher CF *et al.* (2001). Dystonia and cerebellar atrophy in Cacna1a null mice lacking P/Q calcium channel activity, *FASEB J* 15:1288-1290.

Friend KL *et al.* (1999). Detection of a novel missense mutation and second recurrent mutation in the CACNA1A gene in individuals with EA-2 and FHM, *Hum Genet* 105:261-265.

Guida S *et al.* (2001). Complete loss of P/Q calcium channel activity caused by a CACNA1A missense mutation carried by patients with episodic ataxia type 2, *Am J Hum Genet* 68:759-764.

Haan J, Sluis P, Sluis LH, Ferrari MD. (2000). Acetazolamide treatment for migraine aura status, *Neurology* 55:1588-1589.

Hans M *et al.* (1999). Functional consequences of mutations in the human alpha1A calcium channel subunit linked to familial hemiplegic migraine, *J Neurosci* 19:1610-1619.

Haug K *et al.* (2003). Mutations in CLCN2 encoding a voltage-gated chloride channel are associated with idiopathic generalized epilepsies, *Nat Genet* 33:527-532.

Heron SE *et al.* (2002). Sodium-channel defects in benign familial neonatal-infantile seizures, *Lancet* 360:851-852.

Herson PS *et al.* (2003). A mouse model of episodic ataxia type-1, *Nat Neurosci* 6:378-383.

Ishikawa K *et al.* (2001). Cytoplasmic and nuclear polyglutamine aggregates in SCA6 Purkinje cells, *Neurology* 56:1753-1756.

Jouvenceau A *et al.* (2001). Human epilepsy associated with dysfunction of the brain P/Q-type calcium channel, *Lancet* 358:801-807.

Jun K *et al.* (1999). Ablation of P/Q-type Ca(2+) channel currents, altered synaptic transmission, and progressive ataxia in mice lacking the alpha(1A)-subunit, *Proc Natl Acad Sci U S A* 96:15245-15250.

Kraus RL *et al.* (1998). Familial hemiplegic migraine mutations change alpha1A Ca2+ channel kinetics, *J Biol Chem* 273:5586-5590.

Kullmann DM. (2002). The neuronal channelopathies, *Brain* 125:1177-1195.

Kullmann DM, Rea R, Spauschus A, Jouvenceau A. (2001). The inherited episodic ataxias: how well do we understand the disease mechanisms?, *Neuroscientist* 7:80-88.

Langosch D *et al.* (1994). Decreased agonist affinity and chloride conductance of mutant glycine receptors associated with human hereditary hyperekplexia, *EMBO J* 13:4223-4228.

Lerche H *et al.* (1999). A reduced K+ current due to a novel mutation in KCNQ2 causes neonatal convulsions, *Ann Neurol* 46:305-312.

Lewis TM *et al.* (1998). Properties of human glycine receptors containing the hyperekplexia mutation alpha1(K276E), expressed in Xenopus oocytes, *J Physiol* 507:25-40.

Lorenzon NM, Lutz CM, Frankel WN, Beam KG. (1998). Altered calcium channel currents in Purkinje cells of the neurological mutant mouse leaner, *J Neurosci* 18:4482-4489.

Lossin C *et al.* (2002). Molecular basis of an inherited epilepsy, *Neuron* 34:877-884.

Main MJ *et al.* (2000). Modulation of KCNQ2/3 potassium channels by the novel anticonvulsant retigabine, *Mol Pharmacol* 58:253-262.

Matsuyama Z *et al.* (1999). Direct alteration of the P/Q-type Ca2+ channel property by polyglutamine expansion in spinocerebellar ataxia 6, *J Neurosci* 19:RC14.

Ophoff RA *et al.* (1996). Familial hemiplegic migraine and episodic ataxia type-2 are caused by mutations in the Ca2+ channel gene CACNL1A4, *Cell* 87:543-552.

Phillips HA *et al.* (2001). CHRNB2 is the second acetylcholine receptor subunit associated with autosomal dominant nocturnal frontal lobe epilepsy, *Am J Hum Genet* 68:225-231.

Picard F, Bertrand S, Steinlein OK, Bertrand D. (1999). Mutated nicotinic receptors responsible for autosomal dominant nocturnal frontal lobe epilepsy are more sensitive to carbamazepine, *Epilepsia* 40:1198-1209.

Piedras-Renteria ES *et al.* (2001). Increased expression of alpha 1A Ca2+ channel currents arising from expanded trinucleotide repeats in spinocerebellar ataxia type 6, *J Neurosci* 21:9185-9193.

Rea R *et al.* (2002a). Variable K+ channel subunit dysfunction in inherited mutations of KCNA1, *J Physiol* 538:5-23.

Rea R *et al.* (2002b). Functional characterisation of compound heterozygosity for GlyR 1 mutations in the startle disease hyperekplexia, *Eur J Neurosci* 16:186-196.

Restituito S *et al.* (2000). The polyglutamine expansion in spinocerebellar ataxia type 6 causes a beta subunit-specific enhanced activation of P/Q-type calcium channels in *Xenopus* oocytes, *J Neurosci* 20:6394-6403.

Schroeder BC, Kubisch C, Stein V, Jentsch TJ. (1998). Moderate loss of function of cyclic-AMP-modulated KCNQ2/KCNQ3 K+ channels causes epilepsy, *Nature* 396:687-690.

Shiang R *et al.* (1993). Mutations in the alpha 1 subunit of the inhibitory glycine receptor cause the dominant neurologic disorder, hyperekplexia, *Nat Genet* 5:351-358.

Singh NA *et al.* (1998). A novel potassium channel gene, KCNQ2, is mutated in an inherited epilepsy of newborns [see comments], *Nat Genet* 18:25-29.

Steinlein OK *et al.* (1995). A missense mutation in the neuronal nicotinic acetylcholine receptor alpha 4 subunit is associated with autosomal dominant nocturnal frontal lobe epilepsy, *Nat Genet* 11:201-203.

Sugawara T *et al.* (2002). Frequent mutations of SCN1A in severe myoclonic epilepsy in infancy, *Neurology* 58:1122-1124.

Sugawara T *et al.* (2001). A missense mutation of the Na+ channel alpha II subunit gene Na(v)1.2 in a patient with febrile and afebrile seizures causes channel dysfunction, *Proc Natl Acad Sci U S A* 98:6384-6389.

Terwindt G *et al.* (2002). Mutation analysis of the CACNA1A calcium channel subunit gene in 27 patients with sporadic hemiplegic migraine, *Arch Neurol* 59:1016-1018.

Tottene A *et al.* (2002). Familial hemiplegic migraine mutations increase Ca(2+) influx through single human CaV2.1 channels and decrease maximal CaV2.1 current density in neurons, *Proc Natl Acad Sci U S A* 99:13284-13289.

Wallace RH *et al.* (2001). Mutant GABA(A) receptor gamma2-subunit in childhood absence epilepsy and febrile seizures, *Nat Genet* 28:49-52.

Wallace RH *et al.* (1998). Febrile seizures and generalized epilepsy associated with a mutation in the Na+-channel beta1 subunit gene SCN1B, *Nat Genet* 19:366-370.

Wang HS *et al.* (1998). KCNQ2 and KCNQ3 potassium channel subunits: molecular correlates of the M-channel, *Science* 282:1890-1893.

Wappl E *et al.* (2002). Functional consequences of P/Q-type Ca2+ channel Cav2.1 missense mutations associated with episodic ataxia type 2 and progressive ataxia, *J Biol Chem* 277:6960-6966.

Wickenden AD *et al.* (2000). Retigabine, a novel anti-convulsant, enhances activation of KCNQ2/Q3 potassium channels, *Mol Pharmacol* 58:591-600.

Zhang CL, Messing A, Chiu SY. (1999). Specific alteration of spontaneous GABAergic inhibition in cerebellar Purkinje cells in mice lacking the potassium channel Kv1.1, *J Neurosci* 19:2852-2864.

Zhuchenko O *et al.* (1997). Autosomal dominant cerebellar ataxia (SCA6) associated with small polyglutamine expansions in the alpha 1A-voltage-dependent calcium channel, *Nat Genet* 15:62-69.

Inherited Channelopathies of Muscle: Implications for Therapy

Theodore R. Cummins, PhD
Robert L. Ruff, MD, PhD

The electrical activity of muscle and nerves is controlled by ion channels, membrane proteins with gated pores whose opening and closing is controlled by changes in the voltage gradient across the membrane or by ligand binding. Ion channel mutations have been associated with many different skeletal muscle disorders. Periodic paralysis has been linked to mutations in sodium, calcium, and potassium channels. Myotonia has been linked to mutations in sodium and chloride channels. Mutations in voltage-gated and ligand-gated calcium channels can cause malignant hyperthermia and central core disease. Finally, congenital myasthenia syndromes can be caused by mutations in the acetylcholine receptor, as well as the voltage-gated sodium channel. These and other disorders that are caused by ion channel defects are often referred to as channelopathies. This chapter examines how disease mutations alter the function of different muscle ion channels and considers how these changes in ion channel properties contribute to the pathophysiological mechanisms underlying muscle disorders. Although channelopathies are typically rare, the insights gained from studying these genetic mutations should aid in the development of better treatments for excitability disorders.

ACETYLCHOLINE RECEPTOR MUTATIONS CAN UNDERLIE CONGENITAL MYASTHENIC SYNDROMES

The first step in the excitation of muscle fibers is when acetylcholine (ACh) released from the presynaptic nerve terminals diffuses across the synaptic cleft and binds to acetylcholine receptors (AChRs) in the muscle sarcolemma, opening an intrinsic ion channel. The AChR ion channel allows sodium along with some calcium ions to flow into the muscle, thereby depolarizing the membrane potential, which triggers an action potential in the muscle. Genetic mutations in the AChR have been identified as one of the underlying causes of congenital myasthenic syndromes (CMS). This group of disorders typically presents at birth or early childhood and is characterized by muscle weakness, often involving cranial muscles, and a fatiguing or decremental response to low-frequency stimulation of the motor nerves. CMS can arise from defects in presynaptic, synaptic, or postsynaptic proteins and different syndromes show autosomal or recessive inheritance patterns. The most common cause for postsynaptic CMS is a defect in the AChR. The nicotinic AChR of skeletal muscle, an integral membrane protein, is perhaps one of the best-characterized ligand-gated ion channels (Corringer *et al.*, 2000). The AChR has a pentameric structure, with all of the subunits contributing to the pore structure. In immature and denervated muscle the receptors are formed by two α, one β, one γ and one δ subunits. In mature innervated muscle, the γ is replaced by an ε subunit. This developmental difference seems to have important implications in the molecular pathophysiology of CMS. All of the different AChR subunits show a significant degree of homology and are probably derived from a common ancestor. Each AChR has two ACh binding sites, one formed at the interface between an α and δ subunit and the other by the second α subunit and either the γ or ε subunit. Binding of ACh is believed to produce a conformational change, altering the positions of the M2 domains of each subunit that contributes to the pore, which increases the diameter of the pore and allows the cationic flux to occur (Unwin, 1995).

Several relatively distinct syndromes are caused by mutations in the muscle AChR (Engel *et al.*, 2003). Slow channel syndrome (SCS) is a rare disorder that is typically associated with progressive muscle atrophy and characteristic repetitive compound muscle action potentials that are elicited in response to single stimuli (Engel *et al.*, 1982). As with most other congenital muscle disorders, however, significant clinical variability is observed. At least 16 different missense SCS AChR mutations (involving single-amino acid substitutions) have been identified (Croxen *et al.*, 1997, 2002; Engel *et al.*, 1996; Gomez *et al.*, 2002; Milone *et al.*, 1997; Ohno *et al.*, 1995; Sine *et al.*, 1995; Wang *et al.*, 1997). SCS usually shows a classic autosomal dominant inheritance pattern; however, some mutations can exhibit variable penetrance, and one SCS mutation showed a recessive inheritance pattern in a family studied by Croxen *et al.* (2002). SCS AChR mutations have been identified in all of the subunits of the mature AChR, with eight in the α subunit, five in the ε subunit, three in the β subunit, and one in the δ subunit. Half the mutations occur in the M2 transmembrane domain that contributes to the pore of the channel. At least two occur at residues that are important for ACh binding (αG153S and αV156M). All of the mutations associated with SCS are gain-of-function mutations, and in general the SCS mutations result in endplate potentials with slower decay kinetics. Several of the SCS AChR mutations enhance the affinity of the AChR for ACh (Sine *et al.*, 1995; Wang *et al.*, 1997). This enhanced affinity results in slower dissociation of ACh and prolonged channel activation as a result of longer episodes of channels reopening (Sine *et al.*, 1995). Many of the other SCS mutations, especially those in the M2

domain, enhance the gating efficiency of the channel (Engel *et al.*, 1996; Milone *et al.*, 1997; Ohno *et al.*, 1995). Channels with these mutations typically exhibit prolonged open times and a higher incidence of spontaneous channel openings (Figure 18.1), indicating that the closed-state is destabilized and the open-state is stabilized by the mutations. The changes in gating efficiency and/or ACh affinity seem to play a crucial role in the progressive nature of the disease. The prolonged channel activation associated with these mutations can allow increased calcium entry into the muscle, and this is believed to contribute to the progressive degeneration of the junctional folds and the increased muscle weakness that occurs in patients with SCS (Croxen *et al.*, 1997; Engel *et al.*, 1996). Zhou *et al.* (1999) found that many of the SCS mutant channels could be abnormally activated by choline, suggesting that serum choline could result in continuous channel activity and thus exacerbate the muscle damage. Acetylcholinesterase inhibitors, which are useful in other types of CMS, are countertherapeutic with SCS, as they can increase the decay of the already prolonged endplate potential. However, the functional characterization of the SCS mutations indicated that AChR blockers might be therapeutically effective (Fukudome *et al.*, 1998). Some patients treated with quinidine sulfate, a long-lived AChR blocker, exhibited significant clinical improvement (Harper and Engel, 1998). Fluoxetine, another long-lived AChR blocker, also has been used successfully in SCS patients (Harper *et al.*, 2003). Another approach that might eventually be successful for the treatment of SCS is the use of RNA interference to silence the expression of mutant AChR subunits (Abdelgany *et al.*, 2003).

Two other types of AChR mutations are also associated with CMS, but these mutations have different functional

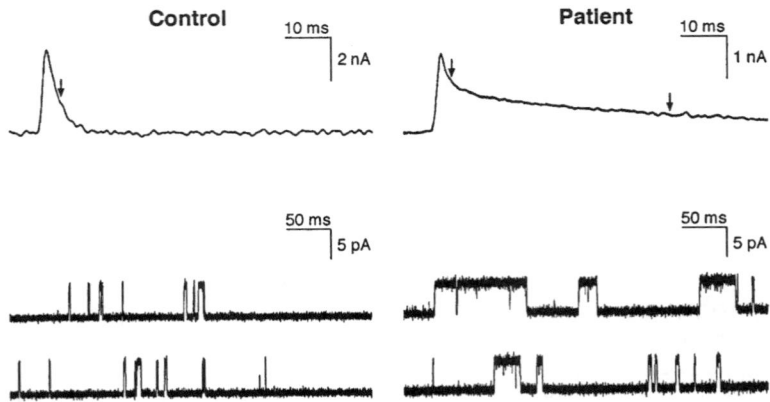

FIGURE 18.1   Endplate recordings from control muscle and muscle with AChRs containing the CMS αV249F mutation. The top traces show miniature endplate currents and the bottom traces show channel events. The endplate currents and channel events are markedly prolonged in the recordings from the muscle with the mutant AChRs (*right traces*) compared to those recorded from the control muscle (*left traces*). Reprinted with permission from Milone *et al.* (1997) with permission from the Society for Neuroscience (copyright 1997).

consequences than the SCS mutations and the mechanism by which they cause CMS are clearly distinct from that of the SCS mutations. One group of mutations, referred to as fast-channel mutations, alter the functional properties of the AChR. The other group is made up of null or low-expressor mutations that significantly reduce or eliminate expression of the mutant subunit. Both of these groups of mutations are associated with recessive CMS disorders with overlapping phenotypes. Some mutations both reduce subunit expression and alter the functional properties of the AChR (Shen *et al.*, 2002). The fast-channel mutations contribute to fast-channel syndrome (FCS), which is characterized by mild to severe muscular weakness. FCS has some clinical overlap with SCS and autoimmune myasthenia gravis. However, biopsied muscle from patients with FCS typically shows rapidly decaying, low-amplitude endplate currents (Ohno *et al.*, 1996). In contrast to SCS, FCS is not usually associated with muscle degeneration, but AChR endplate density is usually reduced in FCS (Engel *et al.*, 1993). FCS shows a recessive inheritance pattern and typically the fast-channel mutation on one allele is accompanied by a null mutation on the second allele. However, individuals with homozygous fast-channel mutations have been identified (Wang *et al.*, 2000). Approximately 10 different FCS mutations have been identified (Engel *et al.*, 2003; Shen *et al.*, 2002; Sine *et al.*, 2002). Mutations have been identified in the α, δ, and ε subunits, and most of these are missense mutations. The main functional consequences of the FCS mutations are reduced steady-state affinity for ACh, shorter open times, and reduced frequency and duration of channel bursting. The reduced channel activity and smaller endplate currents can directly impair neuromuscular transmission (Ohno *et al.*, 1996). Acetylcholinesterase inhibitors and agents that increase the size or number of the quanta released by nerve impulses have been effective in treating FCS (Shen *et al.*, 2002, 2003).

More than 60 low expressor or null mutations have been identified in AChR subunits (Engel *et al.*, 2003). These mutations involve missense, frameshift, and splice-site mutations, as well as promoter mutations (Croxen *et al.*, 1998, 2001; Nichols *et al.*, 1999; Ohno *et al.*, 1997). The vast majority of these mutations are in the ε subunit, and it is thought that this is because low levels of expression of the fetal type γ subunit can partially compensate for the loss of ε subunit expression (Croxen *et al.*, 2001). The primary result of AChR low expressor or null mutations is a reduced AChR density, and the clinical phenotype varies from very mild to very severe (Ealing *et al.*, 2002; Ohno *et al.*, 1997; Sieb *et al.*, 2000). Deficiencies in the AChR endplate density can also result from defects in other proteins that regulate AChR density such as Rapsyn (Ohno *et al.*, 2002). As with the related FCS, patients with AChR deficiency typically respond to acetylcholinesterase inhibitors and agents that increase the size or number of the quanta released by nerve impulses such as 3,4-diaminopyridine.

Understanding the molecular pathophysiologies of CMS can enhance our ability to treat these disorders. Even though SCS and FCS are both CMS disorders that result from missense mutations in the same ion channel, the FCS and SCS mutations have different effects on the activity of AchRs; therefore these closely related syndromes require distinct therapeutic strategies.

LIGAND-GATED CALCIUM CHANNEL MUTATIONS

Genetic defects in a different skeletal muscle ligand-gated ion channel, the ryanodine receptor (RyR1), have been shown to be a common cause of malignant hyperthermia (MH). MH is a potentially fatal disease that predisposes patients to muscle rigidity and sudden rises in body temperature after exposure to volatile anesthetics, such as halothane, or depolarizing muscle relaxants, such as succinylcholine (Jurkat-Rott *et al.*, 2000a). These agents trigger an increase in intracellular calcium in skeletal muscle of patients with MH, which can directly cause the muscle rigidity. Because individuals with MH do not exhibit clinical symptoms unless exposed to triggering agents, it has been difficult to identify individuals that are susceptible to MH. Assessment of the response of biopsied muscle to halothane or caffeine is an effective diagnostic tool; hypersensitivity to these agents *in vitro* correlates well with the patient's susceptibility to MH (Brandt *et al.*, 1999), but a noninvasive test for routine diagnosis is not available. It is hoped that a better understanding of the molecular basis of MH may lead to better diagnostic tools (Robinson *et al.*, 2003).

MH associated with RyR1 mutations show autosomal dominant inheritance patterns with significant clinical variability, including incomplete penetrance. RyR1, perhaps one of the largest functional proteins, acts as a calcium release channel in the sarcoplasmic reticulum of skeletal muscle and is formed as a tetramer from four identical 564 kDa subunits. Each human RyR1 subunit is 5032 amino acids long (Zorzato *et al.*, 1990), with much of the protein comprising a large N-terminal cytoplasmic domain, referred to as the myoplasmic foot. More than 40 different missense mutations and at least one single-amino acid deletion of RyR1 have been identified in patients with MH (Manning *et al.*, 1998; McCarthy *et al.*, 2000; McWilliams *et al.*, 2002; Quane *et al.*, 1994; Tammaro *et al.*, 2003; Yang *et al.*, 2003), and many of the disease mutations are clustered in two regions of the myoplasmic foot (Figure 18.2). The myoplasmic foot

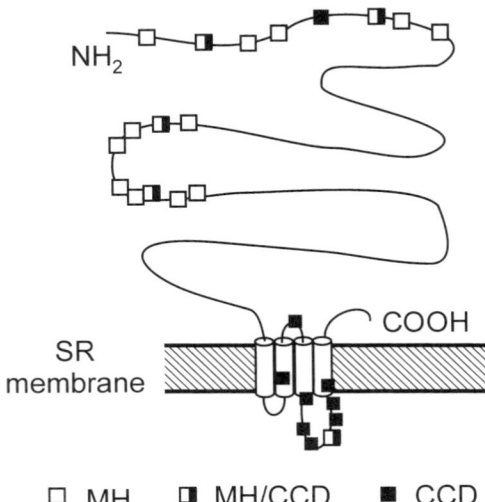

□ MH   ◨ MH/CCD   ■ CCD

FIGURE 18.2 Diagrammatic representation of the RyR1 structure showing the locations of mutations associated with MH and CCD. MH mutations tend to be located in two regions in the large myoplasmic foot, and CCD mutations are predominantly found in the C-terminal region associated with the transmembrane segments and channel pore.

seems to play a crucial role in coupling voltage-dependent gating of L-type calcium channels in the sarcolemma to RyR1 (Proenza et al., 2002), although this interaction may not be direct and may be mediated by other proteins that bind to RyR1. In the myoplasmic foot 10 mutations have been identified between amino acids 35 and 614, and another 23 mutations have been identified between amino acids 2129 and 2458. It is not entirely clear how mutations in these regions are involved in the abnormal rise in cytoplasmic calcium after exposure to triggering agents, but the myoplasmic foot contains the majority of the ligand binding sites. Dantrolene, a skeletal muscle relaxant that depresses excitation-contraction coupling by suppressing the release of calcium from intracellular stores in muscle, binds specifically to residues 590-609 of RyR1 (Paul-Pletzer et al., 2002). Dantrolene has been successfully used in treating MH. An additional nine mutations have been identified between amino acids 4643 and 4906 in the C-terminal portion of the protein that forms the transmembrane segments and the ion channel pore. Functional characterization of the different RyR1 mutations has been limited by the intracellular location of the RyR1 channels and the difficulty in obtaining current recordings from intracellular membranes. However, studies of pigs with MH caused by a mutation that corresponds to the human MH mutation R614C indicated that this mutation increases RyR1 open times and decreases closed times (Fujii et al., 1991). This mutation is sufficient to allow clinically relevant levels of halothane to induce a rapid increase in cytoplasmic cal-

cium (Otsu et al., 1994). A study of six different RyR1 human mutations (R163C, G341R, R614C, V2168M, R2458H, and T4826I) caused similar changes in the functional properties of RyR1 channels expressed in dyspedic myotubes (Yang et al., 2003). All of these mutations increased the sensitivity of the myotubes to potassium depolarization-induced calcium release by a similar amount when compared to wild-type RyR1 channels, suggesting that MH RyR1 mutations might alter the interaction between RyR1 and voltage-gated L-type calcium channels. In addition, despite being located in different regions of the channel sequence, all of these mutations increased the sensitivity to caffeine and reduced the inhibition of RyR1 channels by calcium and magnesium, suggesting that a reduction in negative feedback control of RyR1 might contribute to the clinical phenotype of MH. Thus the MH RyR1 mutations seem to have similar effects on the functional properties of the RyR1 channels. As with other genetic muscle disorders, however, there is significant heterogeneity in the MH clinical phenotype. Yang et al. (2003) reported significant differences in the magnitude of the effects that the different mutations had on RyR1 function, and these difference could contribute to the interfamilial heterogeneity. However, intrafamilial heterogeneity suggests that environmental and/or genetic modifiers also contribute to the clinical phenotype of MH (Robinson et al., 2000).

Mutations in the RyR1 channel have also been identified in patients with central core disease (CCD). CCD is a rare myopathy that typically shows an autosomal-dominant inheritance pattern (Shy and Magee, 1956). Affected individuals present with proximal muscle weakness and hypertonia in infancy. Although the clinical severity of CCD is highly variable, CCD usually is not progressive. As many as 20 different mutations in RyR1 have been linked to CCD (Davis et al., 2003; McCarthy et al., 2000). Some of the MH RyR1 mutations also cause CCD in some, but not all, individuals (Robinson et al., 2002). Although the CCD-associated RyR1 mutations have been found in all three of the hotspot regions of the RyR1 sequence described for the MH mutations, mutations in the third region, which comprises the transmembrane and pore segments of the channel, are more likely to be associated with CCD (Davis et al., 2003; Monnier et al., 2001; Tilgen et al., 2001).

Two distinct mechanisms have been proposed to explain how the RyR1 mutations lead to CCD (Dirksen and Avila, 2002). Some mutations, including those that also cause MH, seem to result in leaky RyR1 channels that allow the sarcoplasmic reticulum to lose calcium. This depletion of the intracellular calcium stores can impair excitation-contraction coupling and thus contribute to muscle weakness. Although many of the RyR1 mutations involve single amino acid substitutions, some of the patients with CCD have deletions in region 3 of RyR1 that cause the channels to become leaky (Avila and Dirksen, 2001; Zorzato et al., 2003). More recently a second mechanism for CCD has been proposed. A set of CCD mutations in region 3 of RyR1 have been identified that do not reduce the calcium content of the sarcoplasmic reticulum, but instead seem to disrupt voltage-gated calcium release, thus effectively uncoupling excitation-contraction coupling (Avila et al., 2001, 2003). In some families, CCD inheritance shows an autosomal-recessive pattern (Romero et al., 2003). In two families this involved compound heterozygous mutations. In one family, the R614C mutation, which is classically associated with a strict MH phenotype, was identified on one allele, whereas a novel G215E mutation was identified on the other allele in two affected family members. It is not clear how these two distinct RyR1 mutations might interact to cause CCD.

VOLTAGE-GATED CALCIUM CHANNEL MUTATIONS

A mutation in the voltage-gated L-type calcium channel of skeletal muscle has also been linked to MH. The L-type calcium channel, also known as the dihydropyridine receptor, is a heteroligomeric complex, but the main functional subunit is the α_1 subunit, which forms the channel pore and contains the voltage sensors (Catterall, 2000a). All of the different voltage-gated calcium channel α_1 subunits have the same basic overall structure, with 24 putative transmembrane segments arranged in four pseudosubunits or domains. Each domain consists of six transmembrane segments (S1-S6) and the SS1-SS2 regions that form the channel pore. The S4 segment of each domain is thought to be involved in voltage sensing and activation gating. The α_1 subunit of the skeletal muscle voltage-gated L-type calcium channel is encoded by the CACNL1A3 gene (Hogan et al., 1994). An arginine to histidine substitution (R1086H) in the cytoplasmic loop-linking domains III and IV of CACNL1A3 has been associated with MH (Monnier et al., 1997). This cytoplasmic linker may play a role in binding to RyR1 (Leong and MacLennan, 1998); therefore this mutation may directly impact excitation-contraction coupling and intracellular calcium release. It is predicted that this mutation would lead to an increase in calcium influx through the L-type calcium channel or alter the coupling of L-type calcium channel and the ryanodine receptor only in the presence of MH triggering agents; however, this has not been established in experimental systems.

Three other missense mutations in CACNL1A3 (Jurkat-Rott et al., 1994; Lehmann-Horn et al., 1995; Ptacek et al., 1994) have been identified as causing hypokalemic periodic paralysis (HypoPP). Individuals with HypoPP experience episodes of muscle weakness that are often associated with a low serum potassium level. Lowered extracellular potassium has the paradoxical effect of depolarizing the membrane potential of muscle from HypoPP individuals. The muscle weakness, which can last many hours and often occurs in the morning, typically affects the limbs, but facial and respiratory muscle weakness can also occur. Although the CACNL1A3 mutations show an autosomal-dominant mode of inheritance, significant interfamilial and intrafamilial variability is observed (Elbaz et al., 1995). For example, the R528H mutation is associated with a mild phenotype, with reduced penetrance in both males and females in some families (Sillen et al., 1997), whereas in other families, complete penetrance is observed with a severe clinical prognosis (Caciotti et al., 2003). This suggests the presence of important genetic or environmental modifiers that affect the severity of the disease phenotype.

The three CACNL1A3 missense HypoPP mutations all involve altering a positively charged residue in the S4 subunits of domain II (R528H) and domain IV (R1239H/G). Although these mutations are located in the voltage sensing segments, the predominant effect appears to be a reduction of current density. Changes in cytoplasmic calcium levels play a crucial role in muscle contraction, but it is not clear how a reduction of calcium current could underlie the sporadic episodes of muscle paralysis or the paradoxical depolarizing effect of lowered extracellular potassium on muscle membrane potential that is characteristic of HypoPP. Some studies have reported small shifts in the voltage-dependence of activation and/or steady-state inactivation with the CACNL1A3 HypoPP mutations (Jurkat-Rott et al., 1998; Morrill and Cannon, 1999), although the R528H mutation shifts activation in the negative direction and the R1239H and R1239G mutations shift activation in the positive direction.

In addition to functioning as a calcium channel, CACNL1A3 also plays a critical role as a calcium-independent voltage-sensor in excitation-contraction coupling to the ryanodine receptor. Thus it is possible that these mutations affect functional properties that have not yet been identified. It is interesting to note that in patients with the R528H mutation, the reduced excitability of the muscle fibers was associated with a significant reduction in sodium current amplitude (Ruff and Al-Mudallal, 2000). Reductions in inward rectifier potassium currents (Ruff, 1999) and adenosine triphosphate potassium currents (Tricarico et al., 1999) have also been associated with HypoPP associated with calcium channel mutations. These studies indicate that the calcium channel mutations might indirectly cause HypoPP by altering the properties of other ionic channel and suggest that HypoPP resulting from CACNL1A3 mutations might represent an indirect channelopathy (Ruff, 2000). Although MH and HypoPP are thought to be distinct disorders, several patients with HypoPP, including a patient with the R528H mutation, have been reported to also exhibit MH (Caciotti *et al.*, 2003; Lambert *et al.*, 1994; Rajabally and El Lahawi, 2002).

PERIODIC PARALYSIS DISORDERS ASSOCIATED WITH POTASSIUM CHANNEL MUTATIONS

Several different potassium channel mutations have been identified as causing periodic paralysis disorders, including Andersen's syndrome (AS). Individuals with AS exhibit periodic paralysis, cardiac arrhythmias, and dysmorphic features (Tawil *et al.*, 1994). Thus, unlike most of the other skeletal muscle channelopathies, which primarily alter muscle activity, this disorder affects skeletal muscle activity, cardiac muscle activity, and developmental signaling. As with other channelopathies, however, marked clinical variability can be observed, and some family members may exhibit only one or two of the three major clinical characteristics. AS shows an autosomal-dominant inheritance pattern, but can also occur sporadically. Periodic paralysis associated with AS can be hyperkalemic, hypokalemic, or normokalemic (Sansone *et al.*, 1997) and is present in about two thirds of individuals with AS. Rest after exercise is the most common trigger of attacks and, as with most other forms of periodic paralysis, acetazolamide can be effective in reducing the frequency and severity of attacks. Because a prolonged QT interval is the most common cardiac abnormality observed in patients with AS, evaluating the consequences of changing serum potassium levels on muscle weakness is not generally recommended with AS. Six missense mutations and two deletion mutations were identified in families or individuals with AS in the initial study, showing that mutations in the inward rectifying potassium channel Kir2.1 can cause AS (Plaster *et al.* 2001). A ninth mutation (R67W) was identified by Andelfinger *et al.* (2002). In the kindred that this particular mutation was identified, female subjects typically exhibited ventricular arrhythmia (but not prolonged QT intervals) and male subjects exhibited periodic paralysis. None of the family members in this kindred carrying the R67W mutation displayed all three of the classic AS characteristics: 32% displayed two and 20% did not exhibit any. In two unrelated individuals with the R67W mutation, however, the complete clinical triad of skeletal muscle, cardiac, and developmental abnormalities was observed.

At least 21 different Kir2.1 mutations have now been associated with AS. Mutations have been identified in the C (13/21) and N (4/21) termini, as well as the pore loop (3/21), but only one mutation (Δ95-98) has been identified in a transmembrane domain. Roughly half of the mutations affect residues that are important to the binding of phosphatidylinosital-4,5-bisphosphate (PIP$_2$) by Kir2.1 (Ai *et al.*, 2002; Donaldson *et al.*, 2003; Lopes *et al.*, 2002). PIP$_2$ binding by Kir2.1 is critical for the expression of functional currents (Soom *et al.*, 2001). All of the Kir2.1 AS mutations characterized to date seem to completely eliminate current expression when the mutant channels are expressed alone in heterologous expression systems. However, inward rectifier potassium channels are tetramers formed with four subunits. When expressed with wild-type Kir2.1 channels, different AS mutants resulted in different magnitudes of current reduction (Figure 18.3). With some mutants it appeared that one mutant subunit was sufficient to eliminate current expression, indicating a dominant negative effect of the mutation, whereas with others the reduction showed varying degrees of severity (Plaster *et al.*, 2001; Andelfinger *et al.*, 2002; Ai *et al.*, 2002; Tristani-Firouzi *et al.*, 2002; Lange *et al.*, 2003). However, Tristani-Firouzi *et al.* (2002) were unable to find a correlation between the degree of dominant-negative suppression that a particular mutation exerted and the clinical severity of AS for subjects expressing Kir2.1 mutations.

Kir2.1 channels are predominantly expressed in heart, muscle, and brain tissue, where they contribute to setting and stabilizing the resting membrane potential and regulating cell excitability (Jan and Jan, 1997). Blocking inward rectifier potassium currents in skeletal muscle can significantly depolarize the muscle fibers, thereby leading to inactivation of skeletal muscle sodium channels and contributing to muscle inexcitability and paralysis. The possibility that mutant Kir2.1 subunits form heteromultimers with Kir2.2 and Kir2.3 subunits (Preisig-Muller *et al.*, 2002) makes it difficult to fully predict the impact of the AS mutants on excitability. Kir2.1 mutations appear to account for about two thirds of the individuals with AS. It is not clear what causes the remaining occurrences.

Mutations in a subunit of a different type of potassium channel have also been associated with periodic paralysis. A mutation (R83H) in the gene (KCNE1) coding for the potassium channel accessory subunit MiRP2 was identified in several individuals with periodic paralysis (Abbott *et al.*, 2001). Paralysis was associated with rest after exercise and a low serum potassium level in one affected individual, but with normal potassium levels and sleep for an unrelated individual with the same MiRP2 mutation. Myotonia has not been observed in individuals with this mutation. Dias Da Silva *et al.*

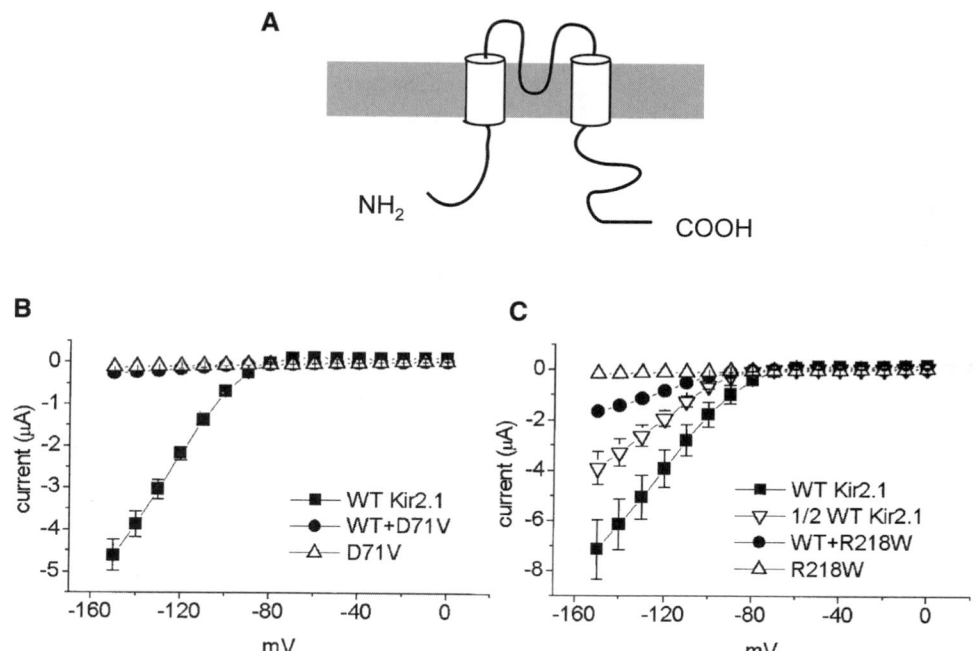

FIGURE 18.3 Mutations in Kir2.1 associated with Andersen's syndrome alter potassium current properties. **A:** Diagrammatic representation of the membrane topology of a subunit of the Kir2.1 inward rectifying potassium channel. **B:** Instantaneous current-voltage relationships indicate that the D71V mutation is sufficient to eliminate functional currents even when coexpressed with wild-type (WT) subunits. **C:** the R218W mutation only reduces the magnitude of currents when expressed with wild-type subunits. Reprinted from Plaster *et al.* (2001) with permission from Elsevier (copyright 2001).

(2002) subsequently identified the R83H MiRP2 mutation in a single patient who had been diagnosed with thyrotoxic hypokalemic periodic paralysis (THypoPP). Abbott *et al.* found that wild-type MiRP2 associated with and altered the properties of Kv3.4 potassium channels, which generate a slowly inactivating voltage gated potassium current in skeletal muscle. The R83H mutation reduced the current density produced by MiRP2-Kv3.4 channel complexes and the R83H-MiRP2 mutant depolarized C2C12 muscle cell resting membrane potential by more than 10 mV. Thus, as with the AS Kir2.1 mutations, it appears that the R83H-MiRP2 mutation contributes to paralysis by depolarizing the muscle membrane potential, leading to sodium channel inactivation and muscle inexcitability.

MYOTONIA CONGENITA IS CAUSED BY MUTATIONS IN THE VOLTAGE-GATED CHLORIDE CHANNEL

Voltage gated chloride channels also play an important role in regulating the membrane potential of muscle. More than 65 different mutations in the ClC-1 chloride channel, which is predominantly expressed in skeletal mus-

cle, have been identified in patients with congenital myotonia. Two forms of general myotonia congenita (GMC) are associated with chloride channel mutations, an autosomal dominant form (Thomsen's disease) and a recessive form (Becker's disease) (George *et al.*, 1993, 1994; Heine *et al.*, 1994; Koch *et al.*, 1992). Both forms exhibit a similar phenotype, in which myotonia (muscle stiffness) can often be triggered by exercise after a period of rest. Muscle fibers from affected individuals are hyperexcitable, may exhibit higher input resistances, depolarized resting membrane potentials, and reduced chloride conductances. At least 41 different missense mutations, 16 nonsense mutations, and 9 splice mutations have been identified (see Pusch, 2002); and the majority of these myotonia-causing ClC-1 mutations show an autosomal-recessive inheritance pattern. Only about half of the ClC-1 mutations have been functionally characterized in heterologous expression systems, and many of the mutant channels that have been studied do not produce functional currents.

It is important to understand the stoichiometry and structure-function relationships of ClC-1 channels in order to understand how different mutations might contribute to the clinical phenotype and the inheritance patterns observed with GMC. Nine different ClC chloride channels have been identified in mammals, and these voltage-gated chloride channels have a different structure from other voltage-gated ion channels. ClC channels are homodimers, but each ClC subunit in the homodimer forms a distinct ion conducting pathway, giving the channels a double-barreled structure (Ludewig et al., 1996). The two separate ClC-1 ion pores seem to have separate fast gates that control their activity independently (Figure 18.4). These fast gates increase the probability of channel openings with depolarization on a millisecond or sub-millisecond time scale. In ClC-0 channels (cloned from the electric organ of Torpedo mamorata), it was found that in addition to the fast gate there was also a slow gate that affected the activity of both conduction pathways in a coordinated manner. The slow gate of ClC-0 channels decreases the probability of channel opening with depolarization on the time course of seconds to minutes. ClC-1 channels also seem to have a common slow gate that affects the activity of both conductance pathways in parallel; however, the slow gate of ClC-1 operates on a millisecond to tens of milliseconds time scale, and the slow gate of ClC-1 closes at hyperpolarized potentials (Accardi and Pusch, 2000). Saviane et al. (1999) proposed that mutations that altered the common slow gate might be more likely to be associated with the dominant form of GMC, whereas those that affect the fast gate or the conductance of individual subunits might be more likely to be associated with the recessive form. Although the activity of ClC channels is voltage-dependent, the mechanism by which voltage gates ClC channels is rather different, in that there is no obvious voltage-sensing structure in ClC channels that is analogous to the S4 segments of the voltage-gated sodium, potassium, and calcium channels. Chloride, the permeating anion in ClC channels, itself is thought to play a crucial role in the voltage-dependent gating of ClC channels. Thus mutations that affect residues that line the pore of ClC-1 can affect permeation, gating, or both (Fahlke et al., 1997; Zhang et al. 2000b). ClC-1 GMC mutations have been reported to induce a depolarizing shift in the voltage-dependence of opening, increase the cation permeability, and even invert the voltage-dependence of ClC-1 gating so that channels act as rectifiers (Fahlke et al., 1995, 1997; Pusch et al., 1995; Wollnik et al., 1997; Zhang et al., 2000a). Most of the mutant channels either eliminate or drastically decrease the chloride conductance at voltages near typical muscle resting membrane potential when expressed alone. When expressed with wild-type ClC-1 channels, however, some mutants exert a dominant negative effect, whereas with others, normal currents are recorded (Steinmeyer et al., 1994; Kubisch et al., 1998). For example, Kubisch et al. (1998) found that several mutations identified in patients with dominant GMC imparted a depolarizing voltage-

shift on the gating of mutant-wild-type heteromeric channels while two mutations identified in patients with recessive GMC had little or no effect on the currents recorded from cells coexpressing wild-type channels (Figure 18.4). Although this might explain why some mutations are associated with a dominant inheritance pattern and others with a recessive inheritance pattern, it does not explain why several GMC ClC-1 mutations show mixed dominant/recessive patterns (Koty et al., 1996; Plassart-Schiess et al., 1998; Zhang et al., 1996). However, several studies have indicated that a greater than 50% reduction in the chloride conductance of skeletal muscle is required to produce myotonia, and thus genetic or environmental factors that modulate the expression of ClC-1 channels could determine whether a particular mutation can have a dominant effect (Meyer-Kleine et al., 1995).

Because the myotonia associated with GMC ClC-1 mutations is clearly associated with a reduced chloride conductance, which leads to membrane depolarization, therapeutic strategies targeted at increasing the chloride conductance in muscle cells might be the most direct approach. Currently there are no specific activators of ClC-1 channels; however, the recent crystal structure of the bacterial ClC homolog (Dutzler et al., 2002) may provide clues regarding how to design specific activators (Estevez et al., 2003). Functional RNA repair using a trans-splicing ribozyme have also shown some success in vitro (Rogers et al., 2002). GMC is usually not a major problem for many patients, and symptoms can be controlled using drugs, such as mexiletine, that reduce sodium channel activity. However, some studies have indicated that in myotonic dystrophy type 1 (DM1), which is caused by a CTG expansion in the gene for dystrophia myotonica protein kinase, loss of chloride current as a result of aberrant splicing of the ClC-1 pre-mRNA increases membrane hyperexcitability; thus reduced chloride channel expression may play a crucial role in the pathophysiology of DM1 (Charlet et al., 2002; Mankodi et al., 2002). Treatments that increase the activity and expression of ClC-1 channels could benefit other more common muscular disorders.

VOLTAGE-GATED SODIUM CHANNEL MUTATIONS

Voltage-gated sodium channels underlie the rapid upstroke of the action potential in skeletal muscle and are crucial to muscle excitability. The first skeletal muscle ion channel mutations to be associated with a human disease were hyperkalemic periodic paralysis (HyperPP) mutations in the voltage-gated sodium channel (Lehmann-Horn et al., 1987; Ptacek et al., 1991; Rojas et al., 1991).

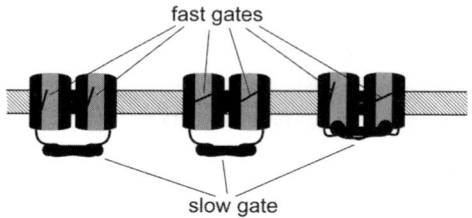

fast gates

slow gate

FIGURE 18.4 Schematic representation of the double-barreled structure of ClC-1 chloride channels. Fast gates act independently to control the activity of two distinct ionic pathways. A common slow gate controls the activity of the two ionic pathways simultaneously. GMC mutations can alter either fast or slow gating.

Since then more than 30 different mutations in the skeletal muscle sodium channel (Nav1.4) have been linked to autosomal-dominant inherited disorders in humans. The primary functional unit of the fast-activating, fast-inactivating sodium channel is a 220-260kD polypeptide α subunit, which has the same basic structure as the voltage-gated calcium channel α_1 subunit (Catterall, 2000b). The SS1-SS2 regions of each of the four domains form the channel pore, the S4 region of each domain is thought to be involved in activation gating, and the cytoplasmic linker between domains III and IV is believed to be important in fast inactivation. Although mutations have been identified in all four of the domains, almost half of the mutations identified are in domain 4 and only two mutations have been identified in domain 1 (Figure 18.5A). The majority of mutations are in transmembrane segments, with the most of the remainder occurring in cytoplasmic loops. Only one mutation has been identified in an extracellular loop, and none have been identified in the pore segments or the N and C termini.

Nav1.4 mutations have been identified in patients with HyperPP, HypoPP, paramyotonia congenital (PMC), and potassium-aggravated myotonia (PAM). HyperPP is associated with attacks of generalized weakness that can be provoked by potassium loading or resting after exercise. The attacks of weakness are usually mild but can last several hours. Many patients with HyperPP also exhibit myotonia. With PMC paradoxical myotonia that is worsened rather than reduced with warm-up is the characteristic symptom. In patients with PMC cold can exacerbate or trigger episodes of myotonia and sometimes attacks of weakness. Individuals with PAM exhibit myotonia that can be induced by potassium loading but not by cold and do not experience attacks of weakness. However it is often difficult to classify a patient with a distinct disorder because of clinical variability and overlapping symptoms. For example, patients with the T704M mutation, one of the most common Nav1.4 mutations, have been diagnosed as having classic HyperPP as well as a mixed HyperPP/PMC phenotype (Brancati *et al.*, 2003; Feero *et al.*,

1993; Ptacek *et al.*, 1991). As with other disease-associated channel mutations, significant intrafamilial and interfamilial variability has been reported for patients with sodium channel mutations (Plassart *et al.*, 1994), indicating that other genetic and/or environmental factors can affect the clinical phenotype.

Almost all of the Nav1.4 mutations associated with human disease have been studied using recombinant channels in heterologous expression systems. Initial studies were done using Nav1.4 cloned from rat (Trimmer *et al.*, 1989), but many of the more recent studies have been done with the human Nav1.4 clone (George *et al.*, 1992). Channels have been expressed and characterized in *Xenopus* oocytes and mammalian cell lines by many different groups. Studies have examined a range of different aspects of channel gating. However, it can be difficult to compare the effects of different mutations on channel properties because of the use of different protocols, as well as different expression systems. Initial studies with rat Nav1.4 channels suggested that HyperPP mutations were associated with an increase in persistent noninactivating sodium currents (Cannon and Strittmatter, 1993). Typically Nav1.4 currents inactivate completely within several milliseconds, but Cannon and Strittmatter reported that rat channels containing mutations corresponding to the human T704M and M1592V mutations continued to allow sodium current to flow even after tens of milliseconds. Computer simulations and pharmacological studies showed that enhanced persistent sodium currents could indeed lead to the development of hyperexcitability and inexcitability (Cannon and Corey, 1993; Cannon *et al.*, 1993) depending on the conditions. However, although the development of persistent sodium currents resulting from impaired fast inactivation may underlie myotonia and paralysis, other mechanisms have also been proposed. Another early study using the rat Nav1.4 channel to study the effects of the T704M mutation reported that this mutation shifted the voltage dependence of activation in the negative direction (Cummins *et al.*, 1993). It was proposed that this shift in activation could lead to a lowered threshold for action potential generation, which could support myotonic activity. The shift in activation could also lead to enhanced window currents, which are persistent currents that occur in a limited voltage range and arise from overlaps in the voltage dependence of activation and fast inactivation. Subsequent studies of the T704M and M1592V mutations in the human Nav1.4 channels showed that both of these HyperPP mutations shift activation in the negative direction by 5 to 12 mV (Bendahhou *et al.*, 1999b; Rojas *et al.*, 1999; Yang *et al.*, 1994), but that neither impaired fast inactivation. Although studies on muscle biopsied from individuals with HyperPP have clearly shown that when persistent sodium currents were increased in fibers that expressed HyperPP mutant sodium channels (Lehmann-Horn *et al.*, 1987), it was not possible to discern the mechanism that generates the persistent currents in native muscle. Furthermore, while enhanced persistent currents, either as a result of impaired fast inactivation or enhanced activation and the development of window currents, are probably sufficient to induce muscle depolarization and paralysis, it is believed that they are not sufficient to account for the prolonged paralysis that can last several hours in patients with HyperPP. Ruff (1994) proposed that HyperPP sodium channel mutations must also impair sodium channel slow inactivation, a distinct inactivation process that occurs on the order of seconds to minutes, or muscle function would recover relatively quickly. Subsequent studies of slow inactivation clearly demonstrated that slow inactivation was indeed severely impaired by several of the mutations associated with HyperPP (Cummins and Sigworth, 1996; Hayward *et al.*, 1997). The L689I, I693T, T704M, and M1592V mutations, all of which have been associated with HyperPP or a mixed HyperPP/PMC phenotype in which the predominant symptom is episodic weakness and myotonia is absent or rare, impair slow inactivation and shift the voltage dependence of activation in the negative direction (Figure 18.5B) (Bendahhou *et al.*, 1999b, 2002; Hayward *et al.*, 1999). However, other HyperPP mutations do not follow this pattern. The A1156T mutation, which is associated with a mixed HyperPP/PMC phenotype with frequent myotonic discharges, did not alter either slow inactivation or activation, but slowed the rate of fast inactivation (Hayward *et al.*, 1999; Yang *et al.*, 1994). The I1495F mutation, which is associated with a HyperPP phenotype without evidence of myotonic discharge, shifted activation in the negative direction, but enhanced both fast and slow inactivation (Bendahhou *et al.*, 1999b). Clearly not all HyperPP mutations have the same functional consequences on sodium channel gating, despite contributing to similar clinical phenotypes.

Many, but not all, of the PMC Nav1.4 mutations have been characterized. Six of these mutations (T1313M, L1433R, R1448C/H/P/S) have similar effects on Nav1.4 properties to the A1156T mutation (Bendahhou *et al.*, 1999a; Chahine *et al.*, 1994; Lerche *et al.*, 1996; Yang *et al.*, 1994). These mutations all decrease the rate of fast inactivation and increase the rate of recovery from inactivation. However, they have mixed effects on the voltage-dependence of fast inactivation, with some causing positive shifts and others negative shifts. The direction of the shift may influence the clinical phenotype, as mutations that cause negative shifts in the voltage dependence of inactivation are more likely to be identified in PMC

patients that experience episodes of weakness (Lerche *et al.*, 1996). None of the PMC mutations are known to affect slow inactivation, and none of them significantly alter activation, but several have been shown to significantly slow deactivation (the rate at which opened channels return to the closed state during repolarizations) (Richmond *et al.*, 1997a, 1997b). The combination of slowed inactivation and slowed deactivation is thought to be crucial to the development of myotonic activity and may play a role in the cold sensitivity of PMC (Featherstone *et al.*, 1998). The increased rate of recovery from inactivation, which can allow more rapid repetitive firing of action potentials, is also likely to contribute to muscle hyperexcitability. The M1360V mutation, which is associated with a mixed HyperPP/PMC phenotype, also decreased the rate of fast inactivation and increased the rate of recovery from inactivation but did not alter slow inactivation (Hayward *et al.*, 1997; Wagner *et al.*, 1997). Thus many PMC mutations exhibit similar functional defects in heterologous expression systems. The exception is the V1293I mutation, which shifts the voltage dependence of activation in the negative direction but does not significantly alter fast or slow inactivation (Green *et al.*, 1998). Only one report indicates that a PMC mutation enhances persistent sodium currents (T1313M, Hayward *et al.*, 1996), and other studies of this mutation have not confirmed this finding (Yang *et al.*, 1994).

Eight PAM mutations have been described (L266V, V445M, S804F, I1160V, G1306A/V/E, and V1589M). Many, but not all, of these mutations slow the rate of fast inactivation and many, but not all, enhance the rate of recovery from inactivation (Green *et al.*, 1998; Mitrovic *et al.*, 1994, 1995; Wang *et al.*, 1999; Wu *et al.*, 2001). With the exception of G1306E, these mutations do not significantly alter the voltage dependence of activation, but four of five mutations tested slowed the rate of deactivation. Five of the eight mutations shifted the voltage dependence of fast inactivation in the positive direction, but none of the mutations have been reported to impair slow inactivation (although two may enhance slow inactivation) (Hayward *et al.*, 1996; Takahashi and Cannon, 1999). Thus PAM mutations have similar functional consequences to PMC mutations. However, at least four of the PAM mutations seem to increase the size of persistent currents by roughly threefold. None of the Nav1.4 mutations seem to endow intrinsic potassium sensitivity to the sodium channels, and it is now believed that the effect of potassium on muscle activity is indirect.

Hypokalemic periodic paralysis has also been linked to Nav1.4 mutations in some patients (Bulman *et al.*, 1999; Sternberg *et al.*, 2001). Four different mutations at two nearby residues have been identified. Although these all involve the neutralization of a charge in one of the

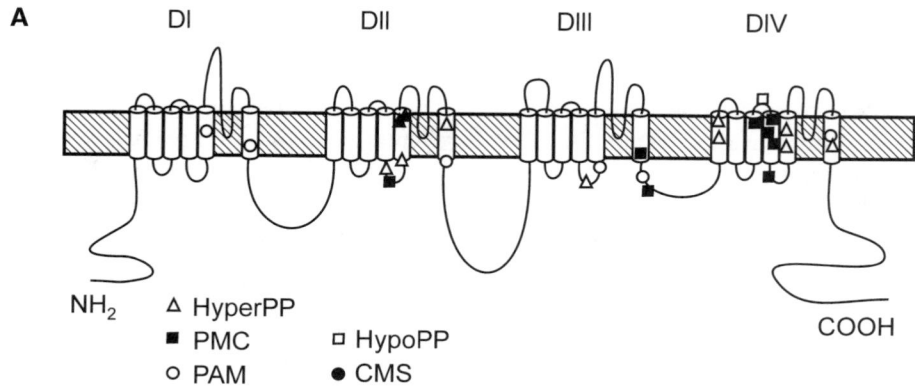

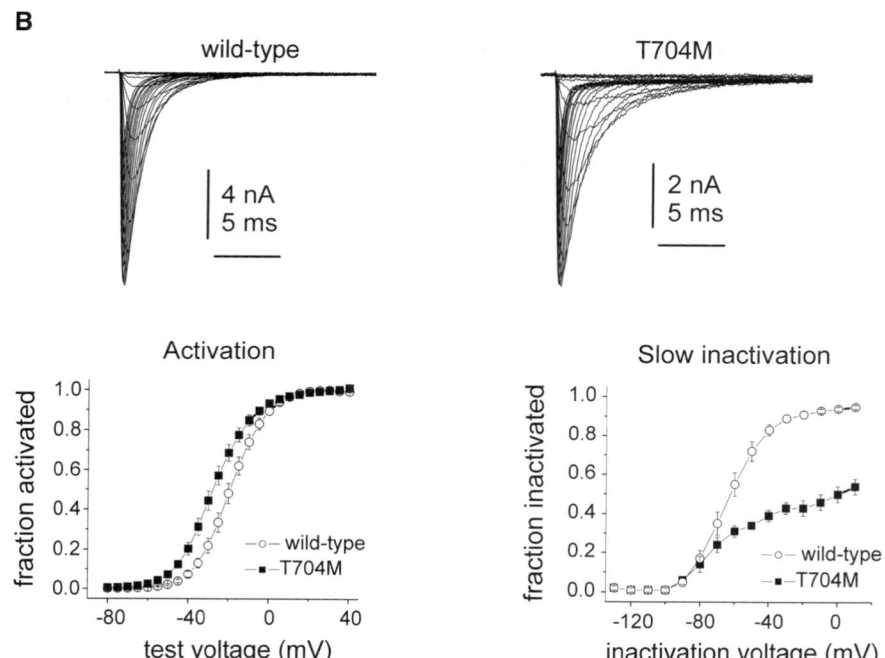

FIGURE 18.5 Mutations in $Na_v1.4$ are associated with several muscle disorders. **A:** Schematic representation of the voltage-gated sodium channel α-subunit. Locations of mutations associated with HyperPP, PMC, PAM, HypoPP, and CMS are shown. **B:** The HyperPP T704M mutation alters the voltage dependence of activation (*bottom left*) and impairs slow inactivation (*bottom right*) of $Na_v1.4$ sodium currents.

putative voltage-sensors of the sodium channel, little effect is seen on activation. These mutations seem to either enhance fast inactivation (Jurkat-Rott *et al.*, 2000b), enhance slow inactivation (Struyk *et al.*, 2000), or enhance both (Bendahhou *et al.*, 2001). While the majority of Nav1.4 disease-causing mutations involve a gain of function (enhanced activation or impaired inactivation), the HypoPP mutations reduce the availability of sodium channels at depolarized potentials. Thus in individuals with HypoPP membrane depolarization causes paralysis by triggering sodium channel inactivation. The

selective enhancement of inactivation explains why membrane hyperexcitability and myotonia are not associated with HypoPP. Thus a reduced sodium current density is a common feature of HypoPP. However, as with the calcium channel HypoPP mutations, it is not clear why hypokalemia paradoxically depolarizes the muscle fibers in patients with these mutations.

Finally, a skeletal muscle sodium channel mutation has been implicated in CMS (Tsujino *et al.*, 2003). The V1442E mutation, the only Nav1.4 mutation identified in an extracellular loop, markedly enhanced the rate of fast inactivation and shifted the voltage dependence of inactivation by more than 30 mV in the negative direction without altering the voltage dependence of activation. By contrast to the other Nav1.4 mutations, which do not seem to affect the respiratory muscles despite that fact that Nav1.4 is the predominant sodium channel in respiratory muscles, the patient with the V1442E mutation did exhibit significant episodes of respiratory paralysis in addition to the paralysis of other muscles. The V1442E mutation had a significantly greater effect on reducing sodium channel availability than the HypoPP mutations; therefore action potential generation could fail in muscle from the affected patient, even at normal membrane resting potential. However, as with many of the other Nav1.4 mutations, acetazolamide has been beneficial in preventing attacks of muscle paralysis in the patient with the V1442E.

All of the Nav1.4 sodium channel mutations reported to date in humans involve single amino acid substitutions rather than truncation or deletion of parts of the channel. None of the mutations that have been tested have eliminated the expression of functional currents, although a myriad of effects on channel properties have been reported. It has been difficult to study the effects of these mutants in native muscle cells; therefore it is not always clear if the profiles generated in heterologous expression systems are accurate representations of the consequences of these mutations in native tissue. However, these studies suggest that treatments that enhance the rate of sodium channel inactivation might be effective for treating PAM and PMC, and treatments that enhance slow inactivation or shift the voltage dependence of activation might be effective in treating some patients with HyperPP.

CONCLUSIONS

More than 250 distinct mutations that contribute to congenital muscle disorders have been identified in seven different muscle ion channels. The muscle disorders caused by ion channel mutations typically show significant variability in the phenotypic expression of the associated disease, often showing intrafamilial and interfamilial variability. A specific ion channel mutation can contribute to what appear to be two distinct disorders (such as CCD and MH) or exhibit different patterns of inheritance (dominant or recessive) in different families. Thus it is clear that genetic and environmental modifiers play important roles in determining the clinical phenotype. Many of the ion channel mutations are associated with a gain of function but others cause a loss of function. Some, such as the calcium channel mutations, which may cause HypoPP by reducing sodium and potassium current densities, may represent indirect chan-

nelopathies. Some of the inherited muscle disorders, such as the myotonic sodium channel disorders, are treated by agents that largely target symptoms but not the cause of the disease, but others such as MH caused by RyR1 mutations can be treated by agents that specifically target the defective molecule. Understanding the molecular pathophysiology that underlies different skeletal muscle channelopathies is aiding the development of rational therapies for these disorders and could help develop strategies for the treatment of related disorders of excitability.

References

Abbott GW *et al.* (2001). MiRP2 forms potassium channels in skeletal muscle with Kv3.4 and is associated with periodic paralysis, *Cell* 104:217-231.

Abdelgany A, Wood M, Beeson D. (2003). Allele-specific silencing of a pathogenic mutant acetylcholine receptor subunit by RNA interference, *Hum Mol Genet* 12:2637-2644.

Accardi A, Pusch M. (2000). Fast and slow gating relaxations in the muscle chloride channel CLC-1, *J Gen Physiol* 116:433-444.

Ai T *et al.* (2002). Novel KCNJ2 mutation in familial periodic paralysis with ventricular dysrhythmia, *Circulation* 105:2592-2594.

Andelfinger G *et al.* (2002). KCNJ2 mutation results in Andersen syndrome with sex-specific cardiac and skeletal muscle phenotypes, *Am J Hum Genet* 71:663-668.

Avila G, Dirksen RT. (2001). Functional effects of central core disease mutations in the cytoplasmic region of the skeletal muscle ryanodine receptor, *J Gen Physiol* 118:277-290.

Avila G, O'Brien JJ, Dirksen RT. (2001). Excitation: contraction uncoupling by a human central core disease mutation in the ryanodine receptor, *Proc Natl Acad Sci U S A* 98:4215-4220.

Avila G, O'Connell KM, Dirksen RT. (2003). The pore region of the skeletal muscle ryanodine receptor is a primary locus for excitation-contraction uncoupling in central core disease, *J Gen Physiol* 121:277-286.

Bendahhou S *et al.* (1999a). Characterization of a new sodium channel mutation at arginine 1448 associated with moderate Paramyotonia congenita in humans, *J Physiol* 518(Pt 2):337-344.

Bendahhou S *et al.* (1999b). Activation and inactivation of the voltage-gated sodium channel: role of segment S5 revealed by a novel hyperkalaemic periodic paralysis mutation, *J Neurosci* 19:4762-4771.

Bendahhou S *et al.* (2001). Sodium channel inactivation defects are associated with acetazolamide-exacerbated hypokalemic periodic paralysis, *Ann Neurol* 50:417-420.

Bendahhou S *et al.* (2002). Impairment of slow inactivation as a common mechanism for periodic paralysis in DIIS4-S5, *Neurology* 58:1266-1272.

Brancati F *et al.* (2003). Severe infantile hyperkalaemic periodic paralysis and paramyotonia congenita: broadening the clinical spectrum associated with the T704M mutation in SCN4A, *J Neurol Neurosurg Psychiatry* 74:1339-1341.

Brandt A *et al.* (1999). Screening of the ryanodine receptor gene in 105 malignant hyperthermia families: novel mutations and concordance with the in vitro contracture test, *Hum Mol Genet* 8:2055-2062.

Bulman DE *et al.* (1999). A novel sodium channel mutation in a family with hypokalemic periodic paralysis, *Neurology* 53:1932-1936.

Caciotti A *et al.* (2003). Severe prognosis in a large family with hypokalemic periodic paralysis, *Muscle Nerve* 27:165-169.

Cannon SC, Corey, DP. (1993). Loss of Na+ channel inactivation by anemone toxin (ATX II) mimics the myotonic state in hyperkalaemic periodic paralysis, *J Physiol* 466:501-520.

Cannon SC, Strittmatter SM. (1993). Functional expression of sodium channel mutations identified in families with periodic paralysis, *Neuron* 10:317-326.

Cannon SC, Brown RH Jr, Corey DP. (1993). Theoretical reconstruction of myotonia and paralysis caused by incomplete inactivation of sodium channels, *Biophys J* 65:270-288.

Catterall WA. (2000a). Structure and regulation of voltage-gated Ca2+ channels, *Annu Rev Cell Dev Biol* 16:521-555.

Catterall WA. (2000b). From ionic currents to molecular mechanisms: the structure and function of voltage-gated sodium channels, *Neuron* 26:13-25.

Chahine M *et al.* (1994). Sodium channel mutations in paramyotonia congenita uncouple inactivation from activation, *Neuron* 12:281-294.

Charlet BN *et al.* (2002). Loss of the muscle-specific chloride channel in type 1 myotonic dystrophy due to misregulated alternative splicing, *Mol Cell* 10:45-53.

Corringer PJ, Le Novere N, Changeux JP. (2000). Nicotinic receptors at the amino acid level, *Annu Rev Pharmacol Toxicol* 40:431-458.

Croxen R *et al.* (1997). Mutations in different functional domains of the human muscle acetylcholine receptor alpha subunit in patients with the slow-channel congenital myasthenic syndrome, *Hum Mol Genet* 6:767-774.

Croxen R *et al.* (1998). A single nucleotide deletion in the epsilon subunit of the acetylcholinereceptor (AChR) in five congenital myasthenic syndrome patients with AChR deficiency, *Ann N Y Acad Sci* 841:195-198.

Croxen R *et al.* (2001). End-plate gamma- and epsilon-subunit mRNA levels in AChR deficiency syndrome due to epsilon-subunit null mutations, *Brain* 124:1362-1372.

Croxen R *et al.* (2002). Recessive inheritance and variable penetrance of slow-channel congenital myasthenic syndromes, *Neurology* 59:162-168.

Cummins TR, Sigworth FJ. (1996). Impaired slow inactivation in mutant sodium channels, *Biophys J* 71:227-236.

Cummins TR *et al.* (1993). Functional consequences of a Na+ channel mutation causing hyperkalemic periodic paralysis, *Neuron* 10:667-678.

Davis MR *et al.* (2003). Principal mutation hotspot for central core disease and related myopathies in the C-terminal transmembrane region of the RYR1 gene, *Neuromuscul Disord* 13:151-157.

Dias Da Silva MR, Cerutti JM, Arnaldi LA, Maciel RM. (2002). A mutation in the KCNE3 potassium channel gene is associated with susceptibility to thyrotoxic hypokalemic periodic paralysis, *J Clin Endocrinol Metab* 87:4881-4884.

Dirksen RT, Avila G. (2002). Altered ryanodine receptor function in central core disease: leaky or uncoupled Ca(2+) release channels?, *Trends Cardiovasc Med* 12:189-197.

Donaldson MR *et al.* (2003). PIP(2) binding residues of Kir2.1 are common targets of mutations causing Andersen syndrome, *Neurology* 60:1811-1816.

Dutzler R *et al.* (2002). X-ray structure of a ClC chloride channel at 3.0 A reveals the molecular basis of anion selectivity, *Nature* 415:287-294.

Ealing J *et al.* (2002). Mutations in congenital myasthenic syndromes reveal an epsilon subunit C-terminal cysteine, C470, crucial for maturation and surface expression of adult AchR, *Hum Mol Genet* 11:3087-3096.

Elbaz A *et al.* (1995). Hypokalemic periodic paralysis and the dihydropyridine receptor (CACNL1A3): genotype/phenotype correlations for two predominant mutations and evidence for the absence of a founder effect in 16 Caucasian families, *Am J Hum Genet* 56:374-380.

Engel AG *et al.* (1982). A newly recognized congenital myasthenic syndrome attributed to a prolonged open time of the acetylcholine-induced ion channel, *Ann Neurol* 11:553-569.

Engel AG *et al.* (1993). Congenital myasthenic syndromes. I. Deficiency and short open-time of the acetylcholine receptor, *Muscle Nerve* 16:1284-1292.

Engel AG *et al.* (1996). New mutations in acetylcholine receptor subunit genes reveal heterogeneity in the slow-channel congenital myasthenic syndrome, *Hum Mol Genet* 5:1217-1227.

Engel AG, Ohno K, Sine SM. (2003). Congenital myasthenic syndromes: progress over the past decade, *Muscle Nerve* 27:4-25.

Estevez R *et al.* (2003). Conservation of chloride channel structure revealed by an inhibitor binding site in ClC-1, *Neuron* 38:47-59.

Fahlke C *et al.* (1995). An aspartic acid residue important for voltage-dependent gating of human muscle chloride channels, *Neuron* 15:463-472.

Fahlke C, Beck CL, George AL Jr. (1997). A mutation in autosomal dominant myotonia congenita affects pore properties of the muscle chloride channel, *Proc Natl Acad Sci U S A* 94:2729-2734.

Featherstone DE, Fujimoto E, Ruben PC. (1998). A defect in skeletal muscle sodium channel deactivation exacerbates hyperexcitability in human paramyotonia congenita, *J Physiol* 506:627-638.

Feero WG *et al.* (1993). Hyperkalemic periodic paralysis: rapid molecular diagnosis and relationship of genotype to phenotype in 12 families, *Neurology* 43:668-673.

Fujii J *et al.* (1991). Identification of a mutation in porcine ryanodine receptor associated with malignant hyperthermia, *Science* 253:448-451.

Fukudome T, Ohno K, Brengman JM, Engel AG. (1998). AChR channel blockade by quinidine sulfate reduces channel open duration in the slow-channel congenital myasthenic syndrome, *Ann N Y Acad Sci* 841:199-202.

George AL Jr *et al.* (1993). Molecular basis of Thomsen's disease (autosomal dominant myotonia congenita), *Nat Genet* 3:305-310.

George AL Jr *et al.* (1994). Nonsense and missense mutations of the muscle chloride channel gene in patients with myotonia congenita, *Hum Mol Genet* 3: 2071-2072.

George AL Jr, Komisarof J, Kallen RG, Barchi RL. (1992). Primary structure of the adult human skeletal muscle voltage-dependent sodium channel, *Ann Neurol* 31:131-137.

Gomez CM *et al.* (2002). Novel delta subunit mutation in slow-channel syndrome causes severe weakness by novel mechanisms, *Ann Neurol* 51:102-112.

Green DS, George AL Jr, Cannon SC. (1998). Human sodium channel gating defects caused by missense mutations in S6 segments associated with myotonia: S804F and V1293I, *J Physiol* 510:685-694.

Harper CM, Engel AG. (1998). Quinidine sulfate therapy for the slow-channel congenital myasthenic syndrome, *Ann Neurol* 43:480-484.

Harper CM, Fukodome T, Engel AG. (2003). Treatment of slow-channel congenital myasthenic syndrome with fluoxetine, *Neurology* 60:1710-1713.

Hayward LJ, Brown RH Jr, Cannon SC. (1996). Inactivation defects caused by myotonia-associated mutations in the sodium channel III-IV linker, *J Gen Physiol* 107:559-576.

Hayward LJ, Brown RH Jr, Cannon SC. (1997). Slow inactivation differs among mutant Na channels associated with myotonia and periodic paralysis, *Biophys J* 72:1204-1219.

Hayward LJ, Sandoval GM, Cannon SC. (1999). Defective slow inactivation of sodium channels contributes to familial periodic paralysism, *Neurology* 52:1447-1453.

Heine R *et al.* (1994). Proof of a non-functional muscle chloride channel in recessive myotonia congenita (Becker) by detection of a 4 base pair deletion, *Hum Mol Genet* 3:1123-1128.

Hogan K, Powers PA, Gregg RG. (1994). Cloning of the human skeletal muscle alpha 1 subunit of the dihydropyridine-sensitive L-type calcium channel (CACNL1A3), *Genomics* 24:608-609.

Jan LY, Jan YN. (1997). Voltage-gated and inwardly rectifying potassium channels, *J Physiol* 505:267-282.

Jurkat-Rott K *et al.* (1994). A calcium channel mutation causing hypokalemic periodic paralysis, *Hum Mol Genet* 3:1415-1419.

Jurkat-Rott K *et al.* (1998). Calcium currents and transients of native and heterologously expressed mutant skeletal muscle DHP receptor alpha1 subunits (R528H), *FEBS Lett* 423:198-204.

Jurkat-Rott K, McCarthy T, Lehmann-Horn F. (2000a). Genetics and pathogenesis of malignant hyperthermia, *Muscle Nerve* 23:4-17.

Jurkat-Rott K *et al.* (2000b). Voltage-sensor sodium channel mutations cause hypokalemic periodic paralysis type 2 by enhanced inactivation and reduced current, *Proc Natl Acad Sci U S A* 97:9549-9554.

Koch MC *et al.* (1992). The skeletal muscle chloride channel in dominant and recessive human myotonia, *Science* 257:797-800.

Koty PP *et al.* (1996). Myotonia and the muscle chloride channel: dominant mutations show variable penetrance and founder effect, *Neurology* 47:963-968.

Kubisch C *et al.* (1998). ClC-1 chloride channel mutations in myotonia congenita: variable penetrance of mutations shifting the voltage dependence, *Hum Mol Genet* 7:1753-1760.

Lambert C *et al.* (1994). Malignant hyperthermia in a patient with hypokalemic periodic paralysis, *Anesth Analg* 79:1012-1014.

Lange PS, Er F, Gassanov N, Hoppe UC. (2003). Andersen mutations of KCNJ2 suppress the native inward rectifier current IK1 in a dominant-negative fashion, *Cardiovasc Res* 59:321-327.

Lehmann-Horn F *et al.* (1987). Adynamia episodica hereditaria with myotonia: a non-inactivating sodium current and the effect of extracellular pH, *Muscle Nerve* 10:363-374.

Lehmann-Horn F *et al.* (1995). Altered calcium currents in human hypokalemic periodic paralysis myotubes expressing mutant L-type calcium channels, *Soc Gen Physiol Ser* 50:101-113.

Leong P, MacLennan DH. (1998). The cytoplasmic loops between domains II and III and domains III and IV in the skeletal muscle dihydropyridine receptor bind to a contiguous site in the skeletal muscle ryanodine receptor, *J Biol Chem* 273:29958-29964.

Lerche H, Mitrovic N, Dubowitz V, Lehmann-Horn F. (1996). Paramyotonia congenita: the R1448P Na+ channel mutation in adult human skeletal muscle, *Ann Neurol* 39:599-608.

Lopes CM *et al.* (2002). Alterations in conserved Kir channel-PIP2 interactions underlie channelopathies, *Neuron* 34:933-944.

Ludewig U, Pusch M, Jentsch TJ. (1996). Two physically distinct pores in the dimeric ClC-0 chloride channel, *Nature* 383:340-343.

Mankodi A *et al.* (2002). Expanded CUG repeats trigger aberrant splicing of ClC-1 chloride channel pre-mRNA and hyperexcitability of skeletal muscle in myotonic dystrophy, *Mol Cell* 10:35-44.

Manning BM *et al.* (1998). Identification of novel mutations in the ryanodine-receptor gene (RYR1) in malignant hyperthermia: genotype-phenotype correlation, *Am J Hum Genet* 62:599-609.

McCarthy TV, Quane KA, Lynch PJ. (2000). Ryanodine receptor mutations in malignant hyperthermia and central core disease, *Hum Mutat* 15:410-417.

McWilliams S *et al.* (2002). Novel skeletal muscle ryanodine receptor mutation in a large Brazilian family with malignant hyperthermia, *Clin Genet* 62:80-83.

Meyer-Kleine C *et al.* (1995). Spectrum of mutations in the major human skeletal muscle chloride channel gene (CLCN1) leading to myotonia, *Am J Hum Genet* 57:1325-1334.

Milone M *et al.* (1997). Slow-channel myasthenic syndrome caused by enhanced activation, desensitization, and agonist binding affinity attributable to mutation in the M2 domain of the acetylcholine receptor alpha subunit, *J Neurosci* 17:5651-5665.

Mitrovic N *et al.* (1994). K(+)-aggravated myotonia: destabilization of the inactivated state of the human muscle Na+ channel by the V1589M mutation, *J Physiol* 478:395-402.

Mitrovic N *et al.* (1995). Different effects on gating of three myotonia-causing mutations in the inactivation gate of the human muscle sodium channel, *J Physiol* 487:107-114.

Monnier N, Procaccio V, Stieglitz P, Lunardi J. (1997). Malignant-hyperthermia susceptibility is associated with a mutation of the alpha 1-subunit of the human dihydropyridine-sensitive L-type voltage-dependent calcium-channel receptor in skeletal muscle, *Am J Hum Genet* 60:1316-1325.

Monnier N *et al.* (2001). Familial and sporadic forms of central core disease are associated with mutations in the C-terminal domain of the skeletal muscle ryanodine receptor, *Hum Mol Genet* 10:2581-2592.

Morrill JA, Cannon SC. (1999). Effects of mutations causing hypokalaemic periodic paralysis on the skeletal muscle L-type Ca2+ channel expressed in *Xenopus laevis* oocytes, *J Physiol* 520:321-336.

Nichols P *et al.* (1999). Mutation of the acetylcholine receptor epsilon-subunit promoter in congenital myasthenic syndrome, *Ann Neurol* 45:439-443.

Ohno K *et al.* (1995). Congenital myasthenic syndrome caused by prolonged acetylcholine receptor channel openings due to a mutation in the M2 domain of the epsilon subunit, *Proc Natl Acad Sci U S A* 92:758-762.

Ohno K *et al.* (1996). Congenital myasthenic syndrome caused by decreased agonist binding affinity due to a mutation in the acetylcholine receptor epsilon subunit, *Neuron* 17:157-170.

Ohno K *et al.* (1997). Congenital myasthenic syndromes due to heteroallelic nonsense/missense mutations in the acetylcholine receptor epsilon subunit gene: identification and functional characterization of six new mutations, *Hum Mol Genet* 6:753-766.

Ohno K *et al.* (2002). Rapsyn mutations in humans cause endplate acetylcholine-receptor deficiency and myasthenic syndrome, *Am J Hum Genet* 70:875-885.

Otsu K *et al.* (1994). The point mutation Arg615→Cys in the Ca2+ release channel of skeletal sarcoplasmic reticulum is responsible for hypersensitivity to caffeine and halothane in malignant hyperthermia, *J Biol Chem* 269:9413-9415.

Paul-Pletzer K *et al.* (2002). Identification of a dantrolene-binding sequence on the skeletal muscle ryanodine receptor, *J Biol Chem* 277:34918-34923.

Plassart E *et al.* (1994). Mutations in the muscle sodium channel gene (SCN4A) in 13 French families with hyperkalemic periodic paralysis and paramyotonia congenita: phenotype to genotype correlations and demonstration of the predominance of two mutations, *Eur J Hum Genet* 2:110-124.

Plassart-Schiess E *et al.* (1998). Novel muscle chloride channel (CLCN1) mutations in myotonia congenita with various modes of inheritance including incomplete dominance and penetrance, *Neurology* 50:1176-1179.

Plaster NM *et al.* (2001). Mutations in Kir2.1 cause the developmental and episodic electrical phenotypes of Andersen's syndrome, *Cell* 105:511-519.

Preisig-Muller R *et al.* (2002). Heteromerization of Kir2.x potassium channels contributes to the phenotype of Andersen's syndrome, *Proc Natl Acad Sci U S A* 99:7774-7779.

Proenza C *et al.* (2002). Identification of a region of RyR1 that participates in allosteric coupling with the alpha(1S) (Ca(V)1.1) II-III loop, *J Biol Chem* 277:6530-6535.

Ptacek LJ *et al.* (1994). Dihydropyridine receptor mutations cause hypokalemic periodic paralysis, *Cell* 77, 863-868.

Ptacek LJ *et al.* (1991). Identification of a mutation in the gene causing hyperkalemic periodic paralysis, *Cell* 67:1021-1027.

Pusch M. (2002). Myotonia caused by mutations in the muscle chloride channel gene CLCN1, *Hum Mutat* 19:423-434.

Pusch M, Steinmeyer K, Koch MC, Jentsch TJ. (1995). Mutations in dominant human myotonia congenita drastically alter the voltage dependence of the ClC-1 chloride channel, *Neuron* 15:1455-1463.

Quane KA *et al.* (1994). Mutation screening of the RYR1 gene in malignant hyperthermia: detection of a novel Tyr to Ser mutation in a pedigree with associated central cores, *Genomics* 23:236-239.

Rajabally YA, El Lahawi M. (2002). Hypokalemic periodic paralysis associated with malignant hyperthermia, *Muscle Nerve* 25:453-455.

Richmond JE, Featherstone DE, Ruben PC. (1997a). Human Na+ channel fast and slow inactivation in paramyotonia congenita mutants expressed in *Xenopus laevis* oocytes, *J Physiol* 499:589-600.

Richmond JE et al. (1997b). Defective fast inactivation recovery and deactivation account for sodium channel myotonia in the I1160V mutant, *Biophys J* 73:1896-1903.

Robinson RL et al. (2000). Multiple interacting gene products may influence susceptibility to malignant hyperthermia, *Ann Hum Genet* 64:307-320.

Robinson RL et al. (2002). RYR1 mutations causing central core disease are associated with more severe malignant hyperthermia in vitro contracture test phenotypes, *Hum Mutat* 20:88-97.

Robinson RL et al. (2003). Recent advances in the diagnosis of malignant hyperthermia susceptibility: How confident can we be of genetic testing?, *Eur J Hum Genet* 11:342-348.

Rogers CS, Vanoye CG, Sullenger BA, George AL Jr. (2002). Functional repair of a mutant chloride channel using a trans-splicing ribozyme. *J Clin Invest* 110:1783-1789.

Rojas CV et al. (1999). Hyperkalemic periodic paralysis M1592V mutation modifies activation in human skeletal muscle Na+ channel, *Am J Physiol* 276:C259-266.

Rojas CV et al. (1991). A Met-to-Val mutation in the skeletal muscle Na+ channel alpha-subunit in hyperkalaemic periodic paralysis, *Nature* 354:387-389.

Romero NB et al. (2003). Dominant and recessive central core disease associated with RYR1 mutations and fetal akinesia, *Brain* 126:2341-2349.

Ruff RL. (1994). Slow Na+ channel inactivation must be disrupted to evoke prolonged depolarization-induced paralysis, *Biophys J* 66:542-545.

Ruff RL. (1999). Insulin acts in hypokalemic periodic paralysis by reducing inward rectifier K+ current, *Neurology* 53:1556-1563.

Ruff RL. (2000). Skeletal muscle sodium current is reduced in hypokalemic periodic paralysis, *Proc Natl Acad Sci U S A* 97:9832-9833.

Ruff RL, Al-Mudallal A. (2000). Reduced skeletal muscle membrane excitability in hypokalemic periodic paralysis (HypoPP) is due to reduced expression of Na channels, *Neurology* 54: A270.

Sansone V et al. (1997). Andersen's syndrome: a distinct periodic paralysis, *Ann Neurol* 42:305-312.

Saviane C, Conti F, Pusch M. (1999). The muscle chloride channel ClC-1 has a double-barreled appearance that is differentially affected in dominant and recessive myotonia, *J Gen Physiol* 113:457-468.

Shen XM et al. (2002). Congenital myasthenic syndrome caused by low-expressor fast-channel AChR delta subunit mutation, *Neurology* 59:1881-1888.

Shen XM et al. (2003). Mutation causing severe myasthenia reveals functional asymmetry of AChR signature cystine loops in agonist binding and gating, *J Clin Invest* 111:497-505.

Shy GM, Magee KR. (1956). A new congenital non-progressive myopathy, *Brain* 79:610-621.

Sieb JP et al. (2000). Severe congenital myasthenic syndrome due to homozygosity of the 1293insG epsilon-acetylcholine receptor subunit mutation, *Ann Neurol* 48:379-383.

Sillen A et al. (1997). Identification of mutations in the CACNL1A3 gene in 13 families of Scandinavian origin having hypokalemic periodic paralysis and evidence of a founder effect in Danish families, *Am J Med Genet* 69:102-106.

Sine SM et al. (1995). Mutation of the acetylcholine receptor alpha subunit causes a slow-channel myasthenic syndrome by enhancing agonist binding affinity, *Neuron* 15:229-239.

Sine SM et al. (2002). Naturally occurring mutations at the acetylcholine receptor binding site independently alter ACh binding and channel gating, *J Gen Physiol* 120:483-496.

Soom M et al. (2001). Multiple PIP2 binding sites in Kir2.1 inwardly rectifying potassium channels, *FEBS Lett* 490:49-53.

Steinmeyer K et al. (1994). Multimeric structure of ClC-1 chloride channel revealed by mutations in dominant myotonia congenita (Thomsen), *EMBO J* 13:737-743.

Sternberg D et al. (2001). Hypokalaemic periodic paralysis type 2 caused by mutations at codon 672 in the muscle sodium channel gene SCN4A, *Brain* 124:1091-1099.

Struyk AF, Scoggan KA, Bulman DE, Cannon SC. (2000). The human skeletal muscle Na channel mutation R669H associated with hypokalemic periodic paralysis enhances slow inactivation, *J Neurosci* 20:8610-8617.

Takahashi MP, Cannon SC. (1999). Enhanced slow inactivation by V445M: a sodium channel mutation associated with myotonia, *Biophys J* 76:861-868.

Tammaro A et al. (2003). Scanning for mutations of the ryanodine receptor (RYR1) gene by denaturing HPLC: detection of three novel malignant hyperthermia alleles, *Clin Chem* 49:761-768.

Tawil R et al. (1994). Andersen's syndrome: potassium-sensitive periodic paralysis, ventricular ectopy, and dysmorphic features, *Ann Neurol* 35:326-330.

Tilgen N et al. (2001). Identification of four novel mutations in the C-terminal membrane spanning domain of the ryanodine receptor 1: association with central core disease and alteration of calcium homeostasis, *Hum Mol Genet* 10:2879-2887.

Tricarico D et al. (1999). Impairment of skeletal muscle adenosine triphosphate-sensitive K+ channels in patients with hypokalemic periodic paralysis, *J Clin Invest* 103:675-682.

Trimmer JS et al. (1989). Primary structure and functional expression of a mammalian skeletal muscle sodium channel, *Neuron* 3:33-49.

Tristani-Firouzi M et al. (2002). Functional and clinical characterization of KCNJ2 mutations associated with LQT7 (Andersen syndrome), *J Clin Invest* 110:381-388.

Tsujino A et al. (2003). Myasthenic syndrome caused by mutation of the SCN4A sodium channel, *Proc Natl Acad Sci U S A* 100:7377-7382.

Unwin N. (1995). Acetylcholine receptor channel imaged in the open state, *Nature* 373:37-43.

Wagner S et al. (1997). A novel sodium channel mutation causing a hyperkalemic paralytic and paramyotonic syndrome with variable clinical expressivity, *Neurology* 49:1018-1025.

Wang DW et al. (1999). Functional consequences of a domain 1/S6 segment sodium channel mutation associated with painful congenital myotonia, *FEBS Lett* 448:231-234.

Wang HL et al. (1997). Mutation in the M1 domain of the acetylcholine receptor alpha subunit decreases the rate of agonist dissociation, *J Gen Physiol* 109:757-766.

Wang HL et al. (2000). Fundamental gating mechanism of nicotinic receptor channel revealed by mutation causing a congenital myasthenic syndrome, *J Gen Physiol* 116:449-462.

Wollnik B, Kubisch C, Steinmeyer K, Pusch M. (1997). Identification of functionally important regions of the muscular chloride channel ClC-1 by analysis of recessive and dominant myotonic mutations, *Hum Mol Genet* 6:805-811.

Wu FF et al. (2001). A new mutation in a family with cold-aggravated myotonia disrupts Na(+) channel inactivation, *Neurology* 56:878-884.

Yang N et al. (1994). Sodium channel mutations in paramyotonia congenita exhibit similar biophysical phenotypes in vitro, *Proc Natl Acad Sci U S A* 91:12785-12789.

Yang T, Ta TA, Pessah IN, Allen PD. (2003). Functional defects in six ryanodine receptor isoform-1 (RyR1) mutations associated with malignant hyperthermia and their impact on skeletal excitation-contraction coupling, *J Biol Chem* 278:25722-25730.

Zhang J *et al.* (1996). Mutations in the human skeletal muscle chloride channel gene (CLCN1) associated with dominant and recessive myotonia congenita, *Neurology* 47:993-998.

Zhang J, Bendahhou S, Sanguinetti MC, Ptacek LJ. (2000a). Functional consequences of chloride channel gene (CLCN1) mutations causing myotonia congenita, *Neurology* 54:937-942.

Zhang J, Sanguinetti MC, Kwiecinski H, Ptacek LJ. (2000b). Mechanism of inverted activation of ClC-1 channels caused by a novel myotonia congenita mutation, *J Biol Chem* 275:2999-3005.

Zhou M, Engel AG, Auerbach A. (1999). Serum choline activates mutant acetylcholine receptors that cause slow channel congenital myasthenic syndromes, *Proc Natl Acad Sci U S A* 96:10466-10471.

Zorzato F *et al.* (1990). Molecular cloning of cDNA encoding human and rabbit forms of the Ca2+ release channel (ryanodine receptor) of skeletal muscle sarcoplasmic reticulum, *J Biol Chem* 265:2244-2256.

Zorzato F *et al.* (2003). Clinical and functional effects of a deletion in a COOH-terminal lumenal loop of the skeletal muscle ryanodine receptor, *Hum Mol Genet* 12:379-388.

Transcriptional Channelopathies of the Nervous System: New Targets for Molecular Medicine

Stephen G. Waxman, MD, PhD

The rational design of new therapeutics often begins with the identification of disease mechanisms and of the molecules that are involved. Targeting these molecules with appropriate chemical entities or biological therapies can alter disease-producing pathways in ways that are clinically helpful. Because ion channels play important roles in the nervous system, it is not surprising that they have emerged as important molecular targets and that the channelopathies, that is, disorders in which abnormal function of ion channels results in clinical signs and symptoms (Figure 19.1), are of major interest to neurotherapeutics investigators.

Before the past several years, attention focused on several broad groups of channelopathies. In the *genetic channelopathies*, channel protein structure is altered so that channels fail to function or function abnormally as a result of *mutations of ion channel genes* (Rose and Griggs, 2001). Examples are provided by GEFS +2 (Generalized Epilepsy with Febrile Seizures Plus, Type 2), which is caused by a mutation of the $Na_v1.1$ sodium channel (Escayg *et al.,* 2000), and by mutations of the $Na_v1.2$ sodium channel, which can produce epilepsy in animal models (Kearney *et al.,* 2001) and in humans (Sugawara *et al.,* 2001). Another example is provided by familial hemiplegic migraine, which is associated with mutations of the CACNA1A calcium channel gene (Ophoff *et al.,* 1996). In contrast to the genetic channelopathies, which are inherited, some channelopathies are acquired. The acquired channelopathies include the *autoimmune and toxic channelopathies* in which channel function is perturbed by antibodies or toxins. One particularly well-studied autoimmune channelopathy is the Lambert-Eaton myasthenic syndrome in which antibodies against presynaptic calcium channels interfere with transmitter release at the neuromuscular junction (Newsom-Davis, 1997).

Another example of an autoimmune channelopathy is provided by Rasmussen's encephalitis in which autoantibodies bind to glutamate receptors and affect their function in neurotransmission (Gahring and Rogers, 1998). An example of a toxic channelopathy is provided by poisoning with ciguatoxin, which activates sodium channels, causing spontaneous repetitive firing, synaptic fatigue, and swelling of axons and nerve terminals (Benoit *et al.,* 1996).

Recent studies have provided evidence for the existence of yet another group of channelopathies, which have been termed the *transcriptional channelopathies* (Waxman, 2001). In these disorders, there is *dysregulated production of an aberrant repertoire of channels whose protein structure is not abnormal*, due to changes in the transcription of (non-mutated) channel genes; and the altered ensemble of channels (which are not mutated or affected by antibodies or toxins) perturbs neuronal function. Recognition of the transcriptional channelopathies has focused attention on aberrantly expressed channels as molecular targets in neurological disease and has identified new therapeutic opportunities. In this chapter we will discuss several examples of transcriptional channelopathies involving the nervous system (peripheral nerve injury and peripheral neuropathies, spinal cord injury, multiple sclerosis), as well as changes in channel transcription within neurons in diabetes, the functional implications of which (adaptive vs maladaptive) are still being studied.

SODIUM CHANNEL GENE TRANSCRIPTION: A DYNAMIC PROCESS

Recognition that dysregulated expression of ion channel genes can produce neurological disorders has been propelled by identification and characterization of the

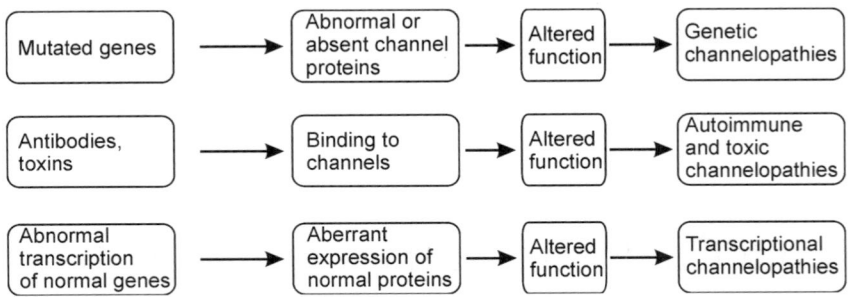

FIGURE 19.1 The channelopathies occur as a result of several types of molecular pathology. *Genetic channelopathies* are the result of mutations in the genes encoding channel proteins. In the *autoimmune* and *toxic channelopathies*, the binding of autoantibodies or toxins to channels alters their function. *Transcriptional channelopathies* are the result of dysregulated expression of non-mutated genes, which results in the production of an abnormal repertoire of channels whose protein structure is not abnormal.

channel genes and gene products, and by the discovery that the transcription of ion channel genes is a complex and *dynamic* process. Voltage-gated sodium channels are among the most completely studied channels in this respect and thus will be used as examples in this chapter.

Neurophysiological doctrine has traditionally referred to *the* sodium channel as if it were a singular entity. An additional layer of complexity has been added, however, by analysis at the molecular level, which has shown that at least nine different genes encode distinct voltage-gated sodium channels with a common overall structural motif but with different amino acid sequences (Catterall, 1992; Plummer and Meisler, 1999; Goldin *et al.*, 2000). The functional properties of most of these channels have now been studied, and it has become apparent that different sodium channels can exhibit different voltage-dependencies, activation and inactivation kinetics, and recovery properties. Not surprisingly, the selective expression of different ensembles of sodium channels contributes to the different functional characteristics of different types of neurons. Properties as fundamental for neuronal function as threshold, refractory period, and the temporal patterning of action potentials are all affected by the type(s) of sodium channels that are expressed within a given neuron (Waxman, 2000). It is thus not unexpected that changes in sodium channel transcription can have significant effects on neuronal function.

The transcription of sodium channels within neurons is not a static process. On the contrary, it is dynamic. Levels of transcription for some sodium channels (e.g., $Na_v1.6$) increase during the course of development of some types of neurons, while transcription of others (e.g., $Na_v1.3$) decreases (Beckh *et al.*, 1989; Brysch *et al.*, 1991; Felts *et al.*, 1997). Multiple factors appear to participate in the regulation of channel transcription, translation, and deployment. Neurotrophic factors exert a complex set of effects on channel expression. For exam-

ple, both nerve growth factor (Black *et al.*, 1997; Dib-Hajj *et al.*, 1998; Fjell *et al.*, 1999a; Cummins *et al.*, 2000) and glial-derived neurotrophic factor (GDNF) (Cummins *et al.*, 2000; Fjell *et al.*, 1999b; Boucher *et al.*, 2000) up-regulate transcription of the $Na_v1.8$ and $Na_v1.9$ sodium channel genes, while down-regulating transcription of the $Na_v1.3$ gene in spinal sensory neurons. Electrical activity may also modulate the transcription of sodium channels within myocytes (Offord and Catterall, 1989) and neurons (Sashihara *et al.*, 1996; Sashihara *et al.*, 1997; Klein *et al.*, 2003). There are also strong regulatory controls on post-transcriptional aspects of sodium channel expression. For example, an oligodendrocyte-derived soluble factor contributes to the regulation of clustering of $Na_v1.2$ channels along axons (Kaplan *et al.*, 2001), while myelination is necessary for $Na_v1.6$ clustering (Boiko *et al.*, 2001; Kaplan *et al.*, 2001).

Plasticity in the transcription of ion channel genes within the uninjured nervous system is not limited to development. The dynamic nature of ion channel gene expression in the normal adult brain is illustrated by the magnocellular neurons within the hypothalamic supraoptic nucleus. These neurosecretory neurons function as part of an osmoregulatory system that controls thirst, diuresis, and other aspects of water balance. They are relatively quiescent in their basal state, firing irregularly at low frequencies (<3 impulses per second). But when there are increases in the osmolarity of extracellular milieu, these neurons convert to a bursting mode in which they generate high-frequency bursts of action potentials that trigger the release of vasopressin from their terminals within the pituitary gland (Andrew and Dudek, 1983; Inenaga *et al.*, 1993; Li and Hatton, 1996).

The two modes of operation of supraoptic magnocellular neurons (quiescent and bursting), and the ability to drive them from one state to another by noninvasive maneuvers such as oral salt-loading, which results in a change in brain osmolarity, made them a useful model system in which to ask: when a neuron converts from a quiescent to a bursting state, does it merely use its (pre-existing) ion channels differently, or is there a change in ion channel expression? To answer this question, Tanaka *et al.* (1999) used *in situ* hybridization, immunocytochemistry, and patch clamp methods to examine the expression of various sodium channel subtypes in these neurons in control rats and following salt loading. Their *in situ* hybridization results clearly showed that, in association with the transition of the magnocellular neurons to the bursting state, transcription of mRNA for the $Na_v1.2$ and $Na_v1.6$ sodium channels (but not for $Na_v1.1$ or $Na_v1.3$) is up-regulated. Immunocytochemical studies showed that the increase in mRNA was accompanied by an increase in channel protein. Moreover, patch clamp recordings showed that the newly produced channels had been inserted into the neuronal cell membrane and were functional. The amplitudes of both fast, transient currents (which are produced by both $Na_v1.2$ and $Na_v1.6$ and which underlie the depolarizing upstroke of the action potential) and persistent currents (which are produced by $Na_v1.6$ and can be activated by small depolarizations close to resting potential, thus contributing to the lowering of threshold for action potential generation) are increased (Tanaka *et al.,* 1999). Thus these neurons rebuild their cell membranes, increasing their electrogenicity, in association with the transition to the bursting state. The molecular remodeling of these neurons within the normal nervous system provides an example of *adaptive* plasticity via the production and deployment of a new repertoire of ion channels, which retunes these cells and converts them to a different functional state.

Considering the dynamic nature of channel expression in the normal nervous system, it is not surprising that pathological events can trigger changes in ion channel gene expression within neurons. Changes in channel expression within the injured nervous system can be adaptive (see Craner *et al.,* 2003a), but this is not always the case. We now discuss three examples of changes in the transcription of sodium channel genes within the injured nervous system which appear to be *maladaptive*, that is, the channelopathies associated with peripheral nerve injury, spinal cord injury, and multiple sclerosis (MS). We also discuss changes in sodium channel expression whose adaptive/maladaptive consequences are not yet fully understood, within the magnocellular neurons of the hypothalamus in untreated diabetes mellitus.

PERIPHERAL NERVE INJURY

Growing evidence from animal models and humans indicates that changes in sodium channel transcription can occur as a result of peripheral nerve injury, and suggests that altered sodium channel expression can contribute to clinically significant phenomena including paraesthesia and neuropathic pain. Paresthesia and neuropathic pain arise, in part, from abnormal firing of injured axons and the spinal sensory neurons (dorsal root ganglion or DRG neurons) from which they arise (Devor and Seltzer, 1999; Zhang *et al.,* 1997; Wood and Perl, 1999). Even before the use of patch clamp recording for the study of neuropathic pain, electrophysiological studies using sharp microelectrodes suggested that abnormal activity of sodium channels can contribute to the hyperexcitability of injured afferents (Kocsis and Waxman, 1983; Kocsis *et al.,* 1983). The intra-axonal recording in Figure 19.2A, from a previously transected, long-term regenerated axon in rat sciatic nerve, illustrates the tendency for injured axons to produce aberrant repetitive action potential activity which is not seen in uninjured axons. The bursting activity is abnormal in itself since healthy axons do not respond in this way. In addition, it

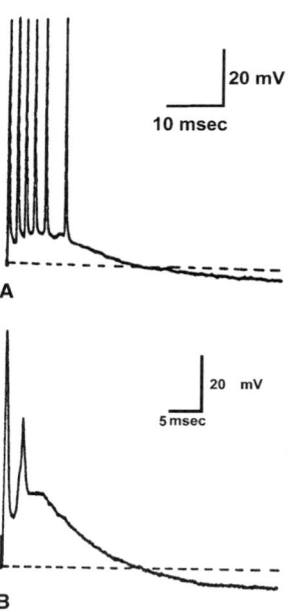

FIGURE 19.2 Primary spinal sensory (dorsal root ganglion; DRG) neurons and their axons can become hyperexcitable after nerve injury. **A:** Intra-axonal recording showing repetitive action potential activity in a previously transected, regenerating axon from rat sciatic nerve (1 year post-crush), following block of potassium channels with 4-aminopyridine. The abnormal burst activity arises from a prolonged depolarization that follows the first action potential. This bursting, and the slow depolarization, are not seen in uninjured axons. Modified from Kocsis and Waxman (1983). **B:** Intra-axonal recording from an axon within a human sural nerve obtained via biopsy from a patient with a painful peripheral neuropathy. Repetitive activity, which is not present in normal axons, arises from an abnormal delayed depolarization. Modified from Kocsis and Waxman (1987).

arises from a prolonged or "slow" depolarization, which lasts for more than 10 milliseconds. The burst and underlying depolarization have an envelope similar to the paroxysmal depolarizing shift seen in epileptic neurons, but are generated along an injured axon trunk where synaptic activity is absent. The configuration of the burst and depolarization were interpreted in early studies as suggesting a contribution of kinetically slow or persistent sodium channels, not seen in normal axons, to firing in injured axons (Kocsis et al., 1983). However, observation of this abnormal mode of electrogenesis in injured axons did not distinguish whether it was the result of a different mode of function of pre-existing ion channels or of the activity of newly expressed channels. Moreover, the early intracellular recordings could not provide information about the molecular identity of the channels that were involved.

Recent studies at the molecular level have provided answers to these questions. These studies have used patch clamp and molecular methods to demonstrate that nerve injury does, in fact, trigger dysregulation of sodium channel gene transcription and have provided insights into the molecular identity of the channels. The changes triggered by nerve injury include the turning-off of expression of some previously active sodium channel genes, and the turning-on (or turning-up) of expression of another sodium channel gene that is normally relatively quiescent in spinal sensory neurons. Figure 19.3 shows that gene expression for the tetrodotoxin (TTX)-resistant sodium channels $Na_v1.8$ (SNS) and $Na_v1.9$ (NaN) is down-regulated within primary spinal sensory neurons following transection of their peripheral axons by nerve injury (Dib-Hajj et al., 1996). In parallel with this transcriptional change, as shown in Figure 19.4A and B, the currents produced by $Na_v1.8$ and $Na_v1.9$ channels within these neurons are attenuated (Sleeper et al., 2000). Changes in expression of $Na_v1.9$ may be especially important because $Na_v1.9$ produces a persistent current (Cummins et al., 1999), which contributes to the setting of resting potential (Herzog et al., 2001) and threshold (Baker et al., 2003), and amplifies small depolarizing inputs such as generator potentials (Herzog et al., 2001; Baker et al., 2003).

In contrast to $Na_v1.8$ and $Na_v1.9$, which are down-regulated, transcription of the $Na_v1.3$ (type III) sodium channel gene is up-regulated in axotomized DRG neurons (Dib-Hajj et al., 1996; Sleeper et al., 2000; Waxman et al., 1994). The newly formed $Na_v1.3$ mRNA is translated so that $Na_v1.3$ sodium channel protein, which is not detectable in uninjured spinal sensory neurons or their axons, is produced and is deployed to distal parts of transected axons (Figure 19.5), which are sites of abnormal impulse generation within neuromas (Black et al., 1999a). Concomitant with the up-regulation of $Na_v1.3$

expression, patch-clamp studies show the emergence of a new sodium current, characterized by rapid repriming (i.e., rapid recovery from inactivation) within the axotomized neurons (Cummins and Waxman, 1997) (Figure 19.4C–E). Confirmation that $Na_v1.3$ channels produce at least a significant part of this current has been provided by patch clamp recordings on heterologously expressed $Na_v1.3$ channels, and on $Na_v1.3$ channels experimentally expressed in DRG neurons (Cummins et al., 2001). The accelerated repriming of the inappropriately expressed $Na_v1.3$ channels produces a decrease in refractory period which leads to high frequency action potential activity, thus contributing to hyperexcitability of DRG neurons following nerve injury (Cummins and Waxman, 1997).

Although the factor(s) that trigger changes in sodium channel expression after nerve injury are not entirely understood, several lines of evidence point to loss of access to peripheral pools of neurotrophic factors including NGF and GDNF as contributing factors: i) DRG neurons respond differently to axonal transection within peripheral nerves (which project to target tissues that produce neurotrophins), compared to transection within dorsal roots (which project to the spinal cord). Up-regulated transcription of $Na_v1.3$ (Black et al., 1999a) and down-regulated transcription of $Na_v1.8$ and $Na_v1.9$ (Sleeper et al., 2000) are seen following transection of the peripherally directed branch of DRG neurons within the sciatic nerve, but not after central axotomy within the dorsal root. ii) Experimental delivery of NGF (Dib-Hajj et al., 1998) and GDNF (Cummins et al., 2000) to peripherally axotomized DRG neurons up-regulates $Na_v1.8$ and $Na_v1.9$ mRNA and protein, as well as the sodium currents that are produced by these channels. These neurotrophic factors also down-regulate $Na_v1.3$ expression in spinal sensory neurons (Black et al., 1997; Boucher et al., 2000; Cummins et al., 2001; Leffler et al., 2002). iii) NGF deprivation reduces $Na_v1.8$ transcription in DRG neurons of adult rats in vivo (Fjell et al., 1999a), a finding that explains the critical role of NGF in the phenotypic maintenance (i.e., maintenance of appropriate physiological properties) of these neurons (Lewin et al., 1992; Ritter and Mendell, 1992). Taken together, these observations indicate that changes in sodium channel transcription following nerve injury are at least partly due to a loss of access to peripheral sources of neurotrophic factors.

Experimental evidence for behavioral consequences of these changes has been provided by observations, in experimental nerve injury models in rats, that allodynia and hyperalgesia are associated with similar changes in transcription of the $Na_v1.8$ (down-regulated), $Na_v1.9$ (down-regulated), and $Na_v1.3$ (up-regulated) sodium channel genes and in the currents produced by these

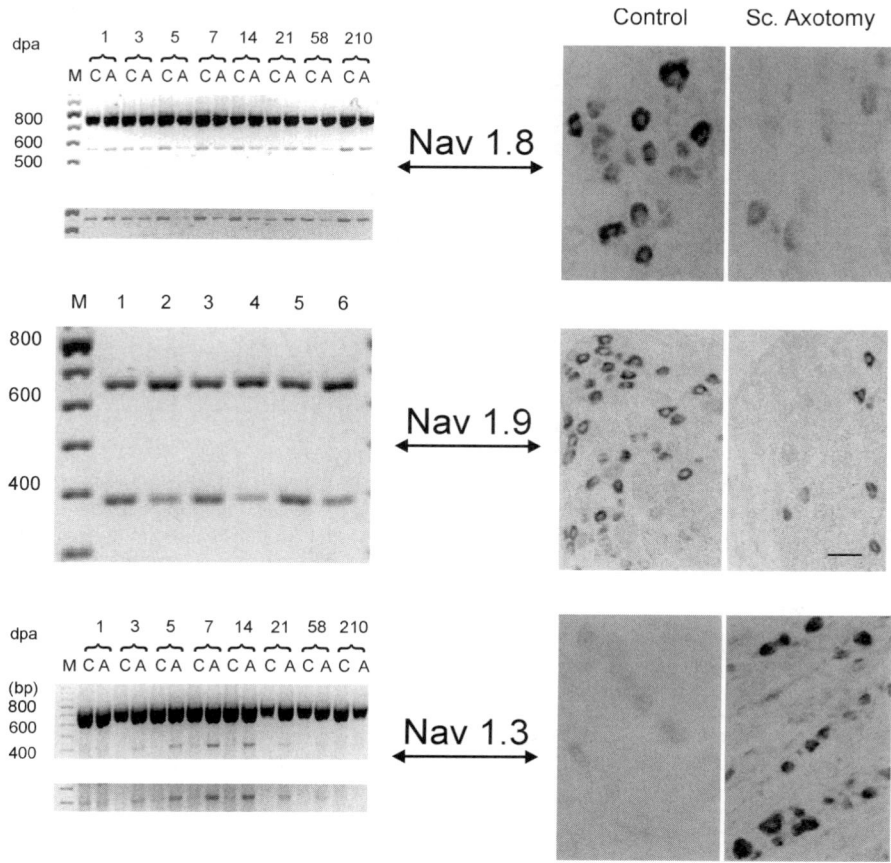

FIGURE 19.3 Sodium channel gene expression is altered in DRG neurons following transection of their axons within the sciatic nerve. $Na_v1.8$ **(top row)** and $Na_v1.9$ **(middle row)** sodium channel gene transcription is down-regulated, while $Na_v1.3$ sodium channel gene transcription is up-regulated **(lower row)** following axonal transection. Micrographs (right column) show *in situ* hybridizations in control DRG and 5 to 7 days following axotomy. Gels **(left)** show RT-PCR products from controls (C) and after axotomy (A) following co-amplification of $Na_v1.8$ **(top left)** and $Na_v1.3$ mRNA **(bottom left)** together with β-actin transcripts (days post-axotomy indicated above gels), with computer-enhanced images of amplification products shown below gels. Co-amplification of $Na_v1.9$ and glyceraldehyde-3-phosphate dehydrogenase (GAPDH) (middle row, left) shows decreased expression of $Na_v1.9$ mRNA at 7 days post-axotomy (lanes 2, 4, 6) compared to uninjured controls (lanes 1, 3, 5). Upper and lower panels modified from Dib-Hajj S *et al.* (1996). Down-regulation of transcripts for Na channel a-SNS in spinal sensory neurons following axotomy. *Proc Natl Acad Sci* 93:14950-14954. Middle panels from Dib-Hajj SD *et al.* (1998). Rescue of alpha-SNS/PN3 sodium channel expression in small dorsal root ganglion neurons after axotomy by nerve growth factor in vivo. *J Neurophysiol* 79:2668-2676.

channels (Dib Hajj *et al.,* 1999). Intrathecal delivery of GDNF (which, as described previously, plays a role in the regulation of $Na_v1.3$, $Na_v1.8$, and $Na_v1.9$ expression in spinal sensory neurons) partially reverses the down-regulation of $Na_v1.8$ and $Na_v1.9$ mRNA, and up-regulation of $Na_v1.3$ mRNA in a rodent model of chronic neuropathic pain (Boucher *et al.,* 2000). Accompanying its down-regulating effect on levels of $Na_v1.3$ mRNA, intrathecally administered GDNF also reverses the acceleration of repriming of sodium currents in axotomized DRG neurons (Leffler *et al.,* 2002)

and has been reported to prevent and reverse sensory abnormalities in rodent neuropathic pain models (Boucher *et al.,* 2000). These results demonstrate, in experimental models, that nerve injury—in part by interrupting access to peripheral pools of trophic factors—can produce changes in the pattern of sodium channel gene transcription in DRG neurons that alter the excitability of these cells, and indicate that these changes in channel gene expression and the resultant changes in neuronal activity participate in the generation of neuropathic pain.

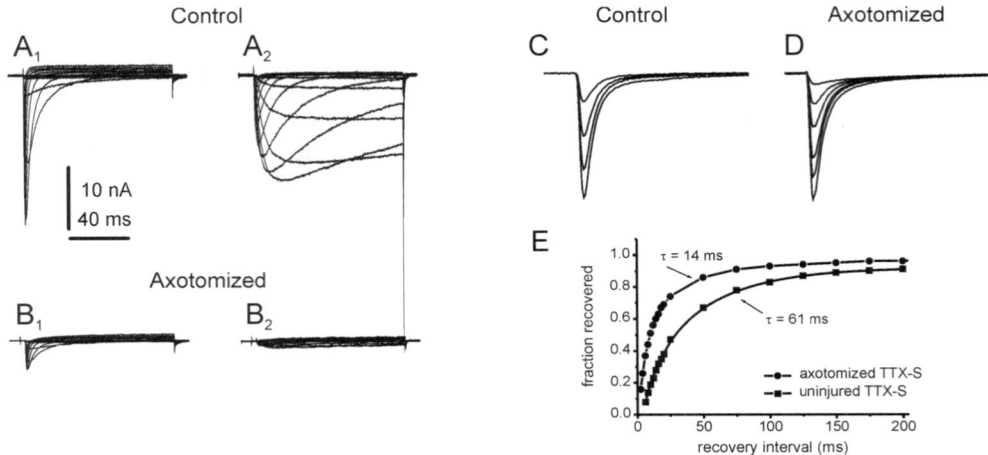

FIGURE 19.4 Changes in the number of functional channels within the neuronal cell membrane are reflected by changes in amplitudes of the currents produced by Na$_v$1.8, Na$_v$1.9, and Na$_v$1.3 channels following transection of the axons of DRG neurons within the sciatic nerve. **A, B:** Panels A1 and A2 show patch clamp records of typical Na$_v$1.8 and Na$_v$1.9 currents in control DRG neurons. Panels B1 and B2 show reduced currents from similar neurons 7 days after sciatic nerve axotomy. From Cummins *et al.* (2000). Glial-derived neurotrophic factor upregulates expression of functional SNS and NaN sodium channels and their currents in axotomized dorsal root ganglion neurons. *J Neurosci* 20:8754-8761. **C–E:** A rapidly repriming tetrodotoxin-sensitive current is produced by Na$_v$1.3 in DRG neurons following peripheral axotomy. **C:** Family of TTX-sensitive sodium current traces (with recovery times indicated) showing time course of recovery from inactivation at −80 mV, from a control DRG neuron. **D:** Similar family of traces showing accelerated recovery 7 days after peripheral axotomy. **E:** Single exponential fits showing accelerated recovery from inactivation following sciatic nerve transection. Modified from Black *et al.* (1999a).

PERIPHERAL NEUROPATHIES

There is also experimental evidence for changes in sodium channel expression that are associated with pain in experimental diabetic neuropathy (Craner *et al.*, 2002). Allodynia and mechanical hyperalgesia develop within weeks following the onset of hyperglycemia in rats with experimental (streptozotocin [STZ]-induced) diabetes

FIGURE 19.5 Na$_v$1.3 sodium channel immunostaining is upregulated, as a result of the deployment of Na$_v$1.3 channel protein within the tips of injured axons following transection of the sciatic nerve (arrows indicate site of transection and ligature, which prevents regeneration). Control nerves do not display immunoreactivity for Na$_v$1.3. Scale bar, 100 μm. Modified from Black *et al.* (1999a).

mellitus. *In situ* hybridization and immunocytochemistry demonstrate that, in association with painful STZ-induced diabetic neuropathy, there is a significant up-regulation of mRNA and protein for the Na$_v$1.3, Na$_v$1.6, and Na$_v$1.9 sodium channels and a down-regulation of Na$_v$1.8 mRNA and protein (Craner *et al.*, 2002). The changes in expression of Na$_v$1.9 are especially interesting because they include injury-induced expression in non-nociceptive cells that normally do not express this channel. In the uninjured nervous system Na$_v$1.9 is expressed preferentially within pain-signaling DRG neurons (Black and Waxman, 2002; Fang *et al.*, 2002). In rats with experimental painful diabetic neuropathy, there is up-regulation of Na$_v$1.9 mRNA and protein not only in C-type DRG neurons, but also in larger DRG neurons (Craner *et al.*, 2002). There is evidence for sprouting after nerve injury of non-nociceptive myelinated (Aβ afferent) fibers onto dorsal horn neurons that normally would receive exclusively Aδ and C-nociceptive projections (Woolf *et al.*, 1992; Kohama *et al.*, 2000) and for a role of increased responsiveness of Aδ fibers (Field *et al.*, 1999; Ahlgren *et al.*, 1992) and Aβ fibers (Campbell *et al.*, 1988) in the generation of dysesthesia and pain after nerve injury. Craner *et al.* (2002) have suggested that up-regulation of Na$_v$1.3, Na$_v$1.6, and Na$_v$1.9 within C-type DRG neurons may contribute to hyperexcitability of these cells, and have speculated that the appearance of $Na_v1.9$ protein may lower threshold or confer a molecular signature normally associated with nociceptive function on some non-nociceptive Aβ DRG neurons in painful diabetic neuropathy.

Although less well studied, there is evidence for a transcriptional channelopathy in human nerve injury. Figure 19.2B shows an intra-axonal recording from an axon within a sural nerve biopsy from a patient who was referred for evaluation because of a painful peripheral neuropathy (Kocsis and Waxman, 1987). Axons in this nerve displayed abnormal spontaneous activity as well as inappropriate bursting in response to brief stimuli, with repetitive firing arising from a slow depolarization that is not present in normal human nerve fibers. By analogy to the studies on experimentally injured nerves described above, it would be predicted that this is due to dysregulation of sodium channel expression. Recent immunocytochemical studies have in fact provided evidence for abnormal expression patterns of $Na_v1.8$ and $Na_v1.9$ sodium channel proteins, consistent with the altered transcription observed in animal models, in DRG neurons from patients with nerve injuries and neuropathic pain (Coward et al., 2000). Taken together, the results from human tissue as well as animal models provide evidence that indicates that axonal injury can trigger dysregulated expression of sodium channel genes. Abnormally expressed channels including the $Na_v1.3$ channel, which is sparsely expressed in the normal central nervous system (CNS), and $Na_v1.9$ and $Na_v1.8$, which are preferentially expressed in nociceptors (Akopian et al., 1996; Dib-Hajj et al., 1998; Fang et al., 2002), may represent especially tractable targets for pain therapies.

SPINAL CORD INJURY

Clinically significant pain is experienced by more than 50% of patients with spinal cord injury and is often refractory, or only partially responsive, to standard clinical interventions (Balzy, Park, 1992; Turner et al., 2001). There is thus a need for information about the molecular mechanisms underlying pain after spinal cord injury. Long-lasting central neuropathic pain occurs following experimental contusion spinal cord injury in rodents (Hulsebosch et al., 2000; Lindsey et al., 2000; Hains et al., 2001; Mills et al., 2001) and is associated with hyperexcitability of dorsal horn neurons (Hao et al., 1991; Yezeiersky and Park, 1993; Hains et al., 2003a,b). Because the rapidly repriming $Na_v1.3$ sodium channel has been indicted as a potential contributor to peripheral neuropathic pain, we recently hypothesized that dysregulated expression of $Na_v1.3$ might contribute to pain following

spinal cord injury (Hains et al., 2003c). We tested this hypothesis in rats, which developed mechanical allodynia and thermal hyperalgesia by 28 days following contusion spinal cord injury at lower thoracic levels. In situ hybridization demonstrated up-regulated expression of $Na_v1.3$ mRNA within lumbar spinal cord dorsal horn neurons after spinal cord injury (Figure 19.6A and B), and immunocytochemistry with an isoform-specific antibody revealed up-regulation of $Na_v1.3$ protein (Figure 19.6C and D). $Na_v1.3$ co-localized with the NK1 receptor, which is localized within second-order nociceptive neurons within the dorsal horn. While spontaneous background activity rates of dorsal horn neurons were unchanged following spinal cord injury, the majority of multireceptive neurons within the dorsal horn after spinal cord injury were hyperexcitable and fired at substantially increased rates following stimulation compared to uninjured controls. (Figure 19.6E and F) (Hains et al., 2003c).

To further examine the role of $Na_v1.3$ in dorsal horn sensory neuron hyperexcitability and pain following spinal cord injury, we next studied spinal cord–injured rats following intrathecal administration of antisense oligonucleotides that targeted $Na_v1.3$ (Hains et al., 2003c). As shown in Figure 19.6G and H, the expression of $Na_v1.3$ mRNA in dorsal horn neurons following spinal cord injury was substantially reduced after 4 days of antisense treatment. Similarly, there was a decrease in the level of $Na_v1.3$ protein when dorsal horn neurons were treated, after spinal cord injury, with $Na_v1.3$ antisense. (Figure 19.6I and J). Accompanying the knockdown of $Na_v1.3$ expression, evoked discharge rates in dorsal horn neurons were significantly reduced in spinal cord–injured animals that had received antisense, returning to levels close to controls (Figure 19.6K and L). Administration of $Na_v1.3$ antisense also had a rapid and significant effect on pain-related behavior, reversing the decreases in mechanical and thermal thresholds (Figure 19.7). Moreover, following cessation of antisense administration, $Na_v1.3$ protein levels and dorsal horn neuron excitability returned to pre-antisense levels, and mechanical and thermal thresholds returned to pre-antisense levels in spinal cord-injured rats (Figure 19.7), adding additional evidence for a role of $Na_v1.3$ in pain following spinal cord injury (Hains et al., 2003c).

In addition to demonstrating that there are changes in sodium channel expression within neurons following spinal cord injury, these studies establish a link between dysregulated expression of $Na_v1.3$ within dorsal horn neurons, hyperexcitability of these nociceptive cells, and pain following spinal cord injury. Near-complete attenuation of the increased evoked activity of dorsal horn neurons associated with spinal cord injury and reduction in pain following selective knock-down of $Na_v1.3$, and

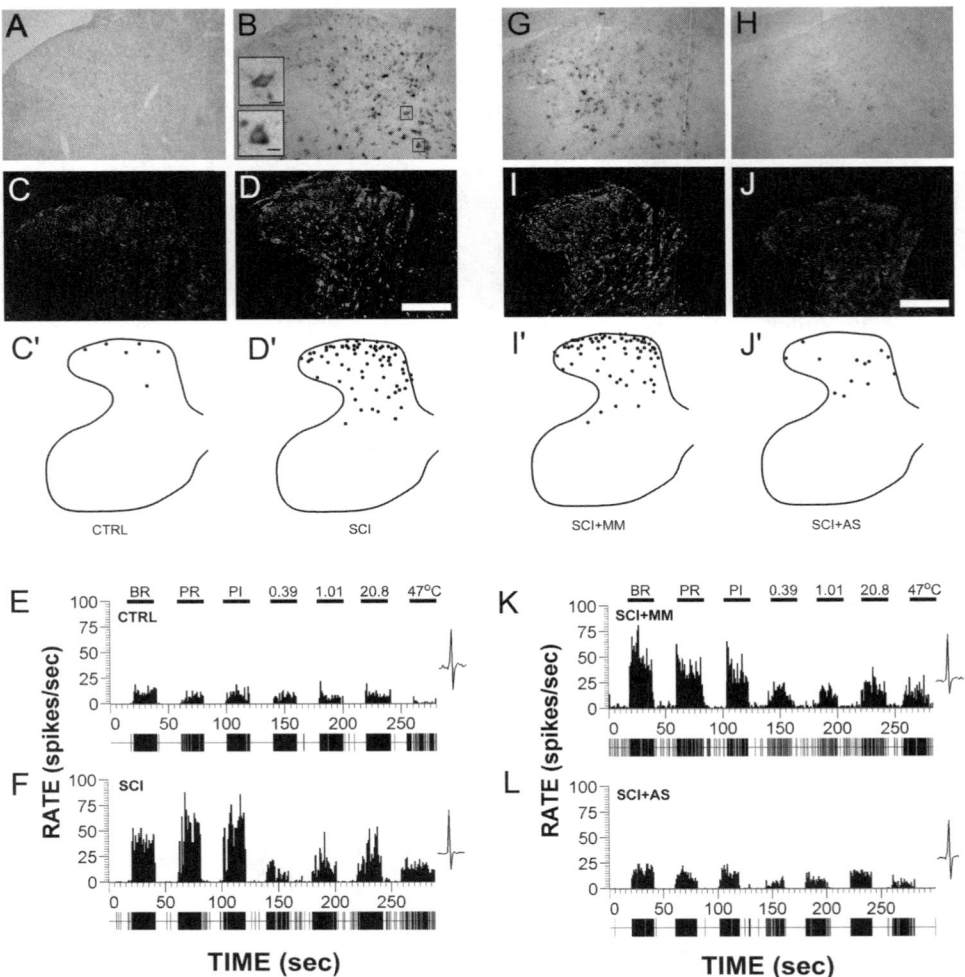

FIGURE 19.6 Sodium channel Na$_v$1.3 is up-regulated within dorsal horn neurons, and contributes to hyperexcitability and pain following spinal cord injury. **A, B:** *In situ* hybridization showing up-regulation of Na$_v$1.3 mRNA in dorsal horn neurons within laminae I-V of lumbar spinal cord, 28 days following contusion injury at T-9 (**B**), compared to sham-operated controls (**A**). Higher magnification images (insets, **B**) show Na$_v$1.3 signal within individual cells. Scale bar = 300 μm (**A, B**), 10 μm (insets in **B**). **C, D:** immunocytochemical localization of Na$_v$1.3 protein showing up-regulation within lumbar dorsal horn neurons in superficial and deep laminae 28 days following contusion injury (**D**), compared to sham-operated controls (**C**). Localization of immunopositive profiles from representative sections (**C', D'**) shows up-regulation of Na$_v$1.3 within all laminae following spinal cord injury. Scale bar = 300 μm. **E, F:** representative peristimulus time histograms (spikes/1 sec. bin) of multireceptive neurons, recorded extracellularly from L3 to L5 in response to innocuous and noxious peripheral stimuli: brush (BR), press (PR), pinch (PI), increasing intensity von Frey filaments (0.39, 1.01, and 20.8g) and thermal (47° C). Single unit waveforms are shown to the right. Injured animals demonstrate evoked hyperexcitability (**F**) compared to controls (**E**). **G–L:** Na$_v$1.3 antisense oligodeoxynucleotides, intrathecally administered for 4 days beginning on day 28 after spinal cord injury, result in decreased expression of Na$_v$1.3 mRNA and protein, reduced hyperexcitability of multireceptive dorsal horn neurons, and attenuated mechanical allodynia and thermal hyperalgesia after SCI. **G:** *In situ* hybridization for Na$_v$1.3 following spinal cord injury shows widely distributed Na$_v$1.3 signal in animals treated with mismatch oligodeoxynucleotide (similar to untreated spinal cord injury, panel A). **H:** In contrast, in antisense treated animals, Na$_v$1.3 hybridization signal is markedly decreased. **I:** Immunocytochemical localization reveals widely distributed Na$_v$1.3 immunopositive cell profiles throughout the dorsal horn, similar to untreated spinal cord injury, in animals that received mismatch oligodeoxynucleotide injections. **J:** In contrast, in animals that received Na$_v$1.3 antisense injections, the number of Na$_v$1.3 immunopositive cell profiles was markedly decreased. **K, L:** Representative peristimulus time histograms, similar to those shown in panels E and F, demonstrate evoked hyperexcitability in animals that received mismatch oligonucleotides (**K**), whereas evoked activity was decreased in animals that received antisense injections following spinal cord injury (**L**). Modified from Hains *et al.* (2003c).

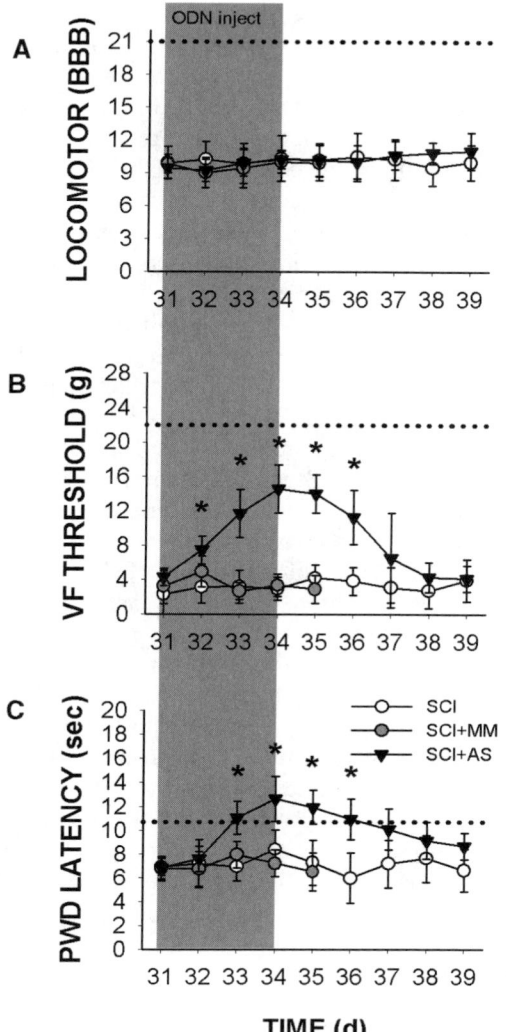

FIGURE 19.7 Behavioral changes following T-9 spinal cord injury and intrathecal administration (days 31–34, shaded) of antisense (AS; filled triangles) or mismatch (MM; filled circles) sequences generated against Na$_v$1.3. The dotted line in each panel indicates sham-operated control thresholds. **A:** BBB open-field locomotor scores, which were impaired after spinal cord injury, did not change during the course of oligodeoxynucleotide administration. **B:** von Frey filament threshold scores to hind limb paw withdrawal show that, by 28 days after SCI, withdrawal thresholds are significantly decreased compared to sham-operated controls, indicating the development of mechanical allodynia; these mechanical thresholds increase in antisense-treated animals, but not in mismatch oligodeoxynucleotide animals. Cessation of antisense administration on day 34 resulted in re-emergence of mechanical allodynia; **C:** paw withdrawal latency to radiant thermal stimuli for hind limbs following spinal cord injury demonstrates development of thermal hyperalgesia by 28 days after injury; by day 2 of antisense administration, paw withdrawal latency increased significantly, and was not significantly different from sham-operated control latencies. Cessation of antisense administration resulted in restoration of thermal hyperalgesia. * = $P < .05$. Modified from Hains *et al.* (2003c).

return of hyperexcitability and pain after cessation of Na$_v$1.3 knock-down, firmly establish a relationship between Na$_v$1.3 expression and pain.

It is not known, at this time, whether changes in Na$_v$1.3 expression after spinal cord injury are solely responsible for dorsal horn neuron hyperexcitability, because these neurons express multiple types of sodium channels, and it is possible that changes in the expression of these or other channels contribute to changes in threshold, refractory period, or the ability to generate and conduct high-frequency impulse trains. Irrespective of this, these new results implicate Na$_v$1.3 as a significant contributor to dorsal horn neuron hyperexcitability and pain following spinal cord injury.

MULTIPLE SCLEROSIS

Recent evidence indicates that, in addition to demyelination and axonal degeneration, a transcriptional channelopathy may also contribute to neuronal dysfunction in MS (Waxman, 2002). Demyelination of an axon might be expected to impose an increased drive on channel expression in the neuronal cell body from which it arises. In healthy myelinated axons, sodium channels are clustered within the axon membrane at the node of Ranvier, and only a low density of channels (too low to support secure impulse conduction) is present in the internodal axon membrane beneath the myelin (Waxman, 1977; Ritchie and Rogart, 1977). Conduction block in demyelinated axons occurs, in part, because injury to the myelin exposes sodium channel–poor membrane, which imposes a capacitative load, thus draining current from the axon without generating an action potential. Recovery of action potential conduction nonetheless occurs in some chronically demyelinated axons (Bostock and Sears, 1978), and this has been shown to be due to the acquisition of a higher-than-normal density of sodium channels within the bared axon membrane (Foster *et al.*, 1980; England *et al.*, 1991; Felts *et al.*, 1997). Recent studies by Craner *et al.* (2003a) demonstrate that this adaptive change includes a reversion from expression of Na$_v$1.6, the major sodium channel at mature nodes of Ranvier (Caldwell *et al.*, 2000) to expression of Na$_v$1.2, which is the major channel along premyelinated axons (Kaplan *et al.*, 2001; Boiko *et al.*, 2001). The deployment of Na$_v$1.2 along extensive regions of axons that have been demyelinated is accompanied by up-regulated transcription of Na$_v$1.2 mRNA within the neuronal cell bodies, which give rise to these axons (Craner *et al.*, 2003a). This transcriptional change appears to serve an adaptive role, providing a substrate for the continuous conduction of impulses along demyelinated axons.

There is also evidence which suggests that there are maladaptive changes in ion channel transcription in animal models of MS and in humans with MS. To determine whether there are changes in sodium channel transcription in inflammatory demyelinating diseases, Black *et al.* (1999) initially studied the mutant *Taiep* rat, a model in which myelin initially ensheaths CNS axons in a normal manner but subsequently degenerates due to an abnormality of oligodendrocytes (Duncan *et al.*, 1992). The early studies focused on expression of the $Na_v1.8$ sodium channel (originally termed SNS, Sensory Neuron Specific), which is normally detectable only in spinal sensory neurons and trigeminal neurons (Akopian *et al.*, 1996; Sangameswaren *et al.*, 1996) because, as described previously, the transcription of $Na_v1.8$ within these neurons changes markedly after axon transection. *In situ* hybridization showed that, following loss of myelin within the cerebellar white matter of *Taiep* rats, there was markedly enhanced expression of $Na_v1.8$ mRNA within Purkinje cells. Immunocytochemical studies of *Taiep* rats showed that the up-regulation of $Na_v1.8$ mRNA is accompanied by the production of $Na_v1.8$ protein in Purkinje cells.

In a more recent study, Black *et al.* (2000) examined the expression of $Na_v1.8$ in the brains of mice with chronic-relapsing experimental allergic encephalomyelitis (CR-EAE), an inflammatory model of MS, and in postmortem brain tissue from humans with MS. The Biozzi mouse CR-EAE model (Baker *et al.*, 1990) was used because it reproducibly provides lesions within the cerebellum, a characteristic that is not commonly observed in monophasic EAE (Baker *et al.*, 2000). As shown in Figure 19.8A–D, *in situ* hybridization demonstrated significantly increased expression of $Na_v1.8$ mRNA within Purkinje cells in CR-EAE. Expression of $Na_v1.8$ protein within Purkinje cells was also up-regulated as shown in Figure 19.8E–G (Black *et al.*, 2000). Up-regulation transcription of $Na_v1.8$ was not part of a global increase in expression of sodium channels, because expression of $Na_v1.9$, another TTX-resistant sodium channel that is normally expressed in a preferential pattern, like $Na_v1.8$, in DRG and trigeminal ganglion neurons, was not up-regulated.

The increased expression of $Na_v1.8$ is not limited to animal models of MS. Study of postmortem brain tissue, obtained from patients with disabling progressive MS with cerebellar deficits on neurological examination, has clearly demonstrated up-regulation of $Na_v1.8$ mRNA and protein within human Purkinje cells (Figure 19.9) (Black *et al.*, 2000).

Taken together, these observations in two animal models of MS, and in humans with MS, show that expression of $Na_v1.8$ is up-regulated in Purkinje neurons, producing $Na_v1.8$ sodium channel protein, which is not normally present in these cells. Studies extended the evidence for a channelopathy in MS by demonstrating (Figure 19.10) that annexin II/p11 [a protein that binds to the N-terminus of $Na_v1.8$ and facilitates the insertion of

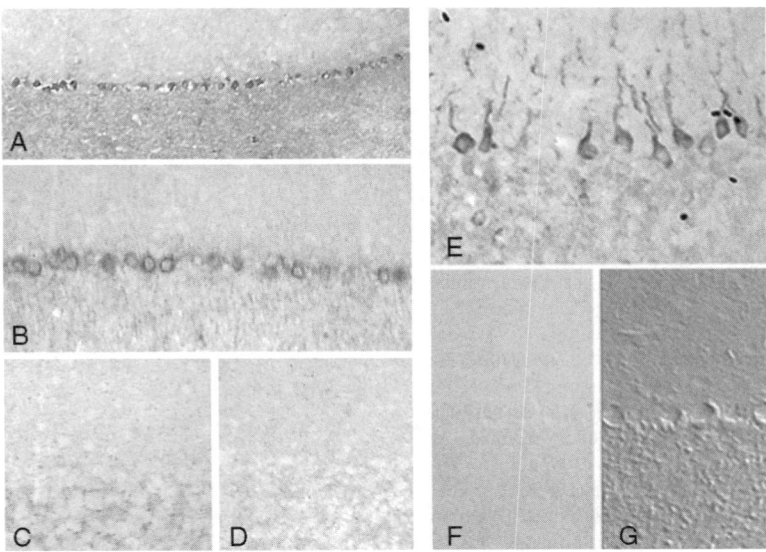

FIGURE 19.8 Sodium channel $Na_v1.8$ (originally termed SNS, Sensory Neuron Specific) is abnormally expressed within cerebellar Purkinje neurons in mice with EAE. **A–D:** *In situ* hybridization showing $Na_v1.8$ mRNA within Purkinje cells in EAE (**A, B**) but not in healthy control mice (**C**) or after hybridization with sense riboprobes (**D**). **E–G:** Immunostaining with $Na_v1.8$-specific antibodies, showing up-regulated expression of $Na_v1.8$ protein in EAE (**E**) compared to controls (**F**, bright field; **G**, Nomarski; image). **A**, x 120; **B–D**, x 200; **E–G**, x 220. From Black *et al.* (2000).

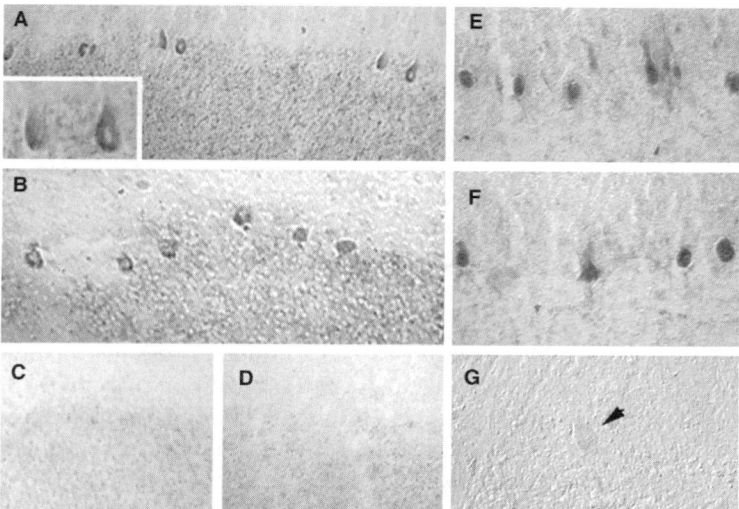

FIGURE 19.9 Expression of the Sensory Neuron Specific (SNS) sodium channel Na$_v$1.8 is up-regulated in cerebellar Purkinje neurons in patients with MS. *In situ* hybridization with Na$_v$1.8 specific antisense riboprobes demonstrates increased Na$_v$1.8 mRNA in Purkinje cells from MS patients obtained at postmortem (**A, B**), compared to controls without neurological disease (**C**). No signal is present following hybridization with sense riboprobe (**D**). Immunocytochemistry with SNS-specific antibodies demonstrates up-regulation of Na$_v$1.8 channel protein in Purkinje cells from MS patients (**E, F**) compared to controls (**G**; arrowhead indicates Purkinje cell). **A:** x 120, inset x 280; **B–D:** x 165; **E–G:** x 175. From Black JA *et al.* (2000).

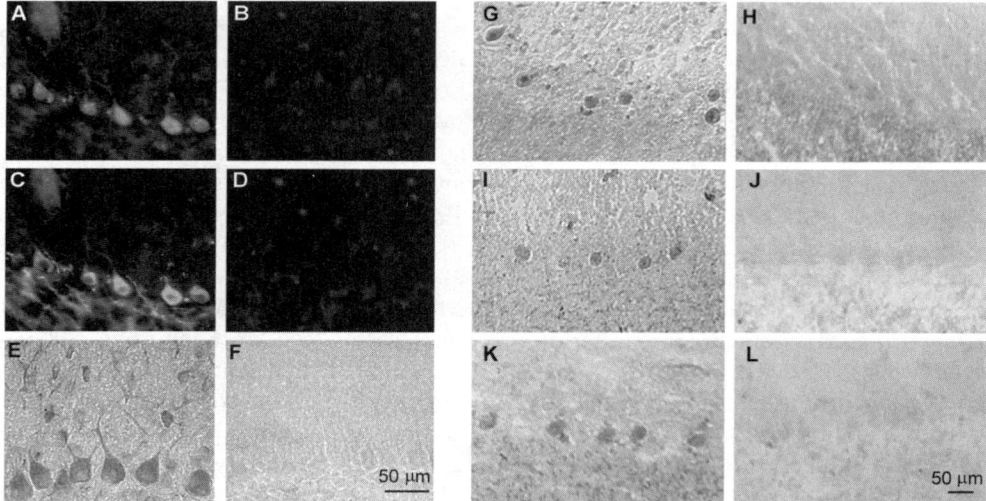

FIGURE 19.10 Annexin, which facilitates insertion of functional Na$_v$1.8 channels into the cell membrane, is up-regulated and co-expressed with Na$_v$1.8 in Purkinje cells in EAE and MS. **Left Panel:** Na$_v$1.8 (**A**) and annexinII/p11 (**B**) are up-regulated in Purkinje neurons in EAE vs control (**B** and **D**, respectively). AnnexinII/p11 immunostaining extends along the proximal portion of the dendritic tree in EAE (**E**) but not in control (**F**). **Right Panels:** Up-regulation of Na$_v$1.8 and annexinII/p11 in human MS tissue. Na$_v$1.8 is up-regulated in human MS Purkinje neurons (**G**) vs control (**H**). Up-regulated expression of annexinII/p11 is shown for two MS cases (**I** and **K**) in comparison to controls (**J** and **L**). Modified from Craner MJ *et al.* (2003b).

functional channels in the neuronal cell membrane (Okuse *et al.,* 2002)] is up-regulated within Purkinje cells, and is co-localized with $Na_v1.8$ in EAE and MS (Craner *et al.,* 2003b).

Expression of $Na_v1.8$/SNS and annexin11/p11 is also up-regulated in retinal ganglion cells (which give rise to axons that travel in the optic nerve, a site that is frequently affected by MS) in EAE (Craner, Black, and Waxman, unpublished). Nav1.8 protein is not present, however, along the axons of Purkinje cells within the cerebellar white matter (Black and Waxman, unpublished) or the axons of retinal ganglion cells within the optic nerve (Craner *et al.,* 2003a).

The functional consequences of up-regulated $Na_v1.8$ transcription and translation have not yet been studied *in vivo* in patients with MS. One possibility that deserves consideration is that $Na_v1.8$ up-regulation is an adaptive change, providing a substrate for the restoration of action potential conduction along demyelinated Purkinje cell axons. Renganathan *et al.* (2001) found that $Na_v1.8$ channels produce a substantial fraction of the inward current that flows during the upstroke of the action potential in the DRG neurons in which these channels are normally present. Thus, if $Na_v1.8$ channels were inserted into the demyelinated axon membrane, they might contribute to restoration of action potential conduction in demyelinated axons. However, $Na_v1.8$ protein has not been detected along demyelinated Purkinje cell or optic nerve axons, although $Na_v1.2$ and $Na_v1.6$ protein can be clearly detected (Craner *et al.,* 2003a). This argues against an adaptive role of $Na_v1.8$ in restoration of conduction along demyelinated axons.

It is more likely that up-regulated expression of $Na_v1.8$ in EAE and MS is a maladaptive change; that is, represents a transcriptional channelopathy. Sodium channels play a crucial role in electrogenesis within Purkinje cells (Llinas and Sugimori, 1980; Stuart and Hausser, 1994; Raman and Bean, 1997). Mutations of sodium channels that are expressed in Purkinje cells have been shown to produce changes in patterns of impulse generation in these cells, which can result in clinical signs of cerebellar dysfunction such as ataxia (Raman *et al.,* 1997; Kohrman *et al.,* 1996). The physiological signature of $Na_v1.8$ is unique in that it exhibits depolarized voltage-dependence of inactivation, slow development of inactivation (Akopian *et al.,* 1996; Sangameswaren *et al.,* 2000), and rapid recovery from inactivation (Elliott and Elliott, 1993; Dib-Hajj *et al.,* 1997). As a result of these properties, $Na_v1.8$ channels are available over a wider range of dynamic activity and membrane potential than other sodium channels (Schild and Kunze, 1997), and cells expressing $Na_v1.8$ should be more slowly adapting than cells lacking $Na_v1.8$ (Elliott and Elliott, 1993).

Several observations support the prediction that abnormal $Na_v1.8$ expression in Purkinje cells perturbs the temporal pattern of electrical activity in the cerebellum. Renganathan *et al.,* (2001) used voltage-clamp and current clamp recording to examine the pattern of electrogenesis in DRG neurons from transgenic $Na_v1.8$ –/– DRG neurons in which functional $Na_v1.8$ channels are not present (Akopian *et al.,* 1999), and to compare it with electrogenesis in $Na_v1.8$ +/+ neurons. Renganathan *et al.* (2001) showed that the presence of $Na_v1.8$ channels within DRG neurons markedly influences both the configuration of the action potentials [a change that can cause activation of N-type channels and thus transmitter release (Scroggs and Fox, 1992) if $Na_v1.8$ is deployed to axon terminals] and the temporal pattern of firing in response to depolarizing stimuli. Consistent with the suggestion that cells expressing $Na_v1.8$ should be slowly adapting (see previous section), this study demonstrated that $Na_v1.8$ +/+ C-type DRG neurons produce sustained pacemaker-like trains of action potentials in response to depolarizing stimuli, which are not present within $Na_v1.8$ –/– C-type DRG neurons.

Renganathan *et al.* (2003) used biolistic techniques to transfect Purkinje cells *in vitro* with $Na_v1.8$ and studied these cells by patch clamp. Voltage-clamp recording from these cells demonstrated that $Na_v1.8$ channels can be functionally expressed at physiological levels (with current amplitudes similar to those observed within DRG neurons where $Na_v1.8$ is normally expressed) within Purkinje cells. Current clamp recordings demonstrated that the expression of $Na_v1.8$ within Purkinje cells alters the activity of these neurons in three ways: first, by increasing the amplitude and duration of action potentials (compare Figure 19.11B and 19.11A); second, by decreasing the proportion of action potentials that are conglomerate and the number of spikes per conglomerate action potential (again, compare Figure 19.11B and 19.11A); and third, by supporting sustained, pacemaker-like impulse trains in response to depolarization, which are not seen in the absence of $Na_v1.8$ (Figure 19.12). More recently, we studied Purkinje cell firing patterns in mice with CR-EAE (Saab *et al.,* 2004), and observed a reduction in the number of secondary spikes per conglomerate action potential, and breakdown of the conglomerate action potentials with irregularity of timing of the secondary spikes. We also observed bursts of non-adapting, high-frequency action potentials in Purkinje cells from mice with CR-EAE.

These results indicate that the aberrant expression of $Na_v1.8$ within Purkinje cells distorts their pattern of electrical activity, both *in vitro* and *in vivo*. This perturbation in Purkinje cell activity should interfere with normal cerebellar functioning. Whether the abnormal expression of $Na_v1.8$ leads to degeneration of Purkinje cells, in

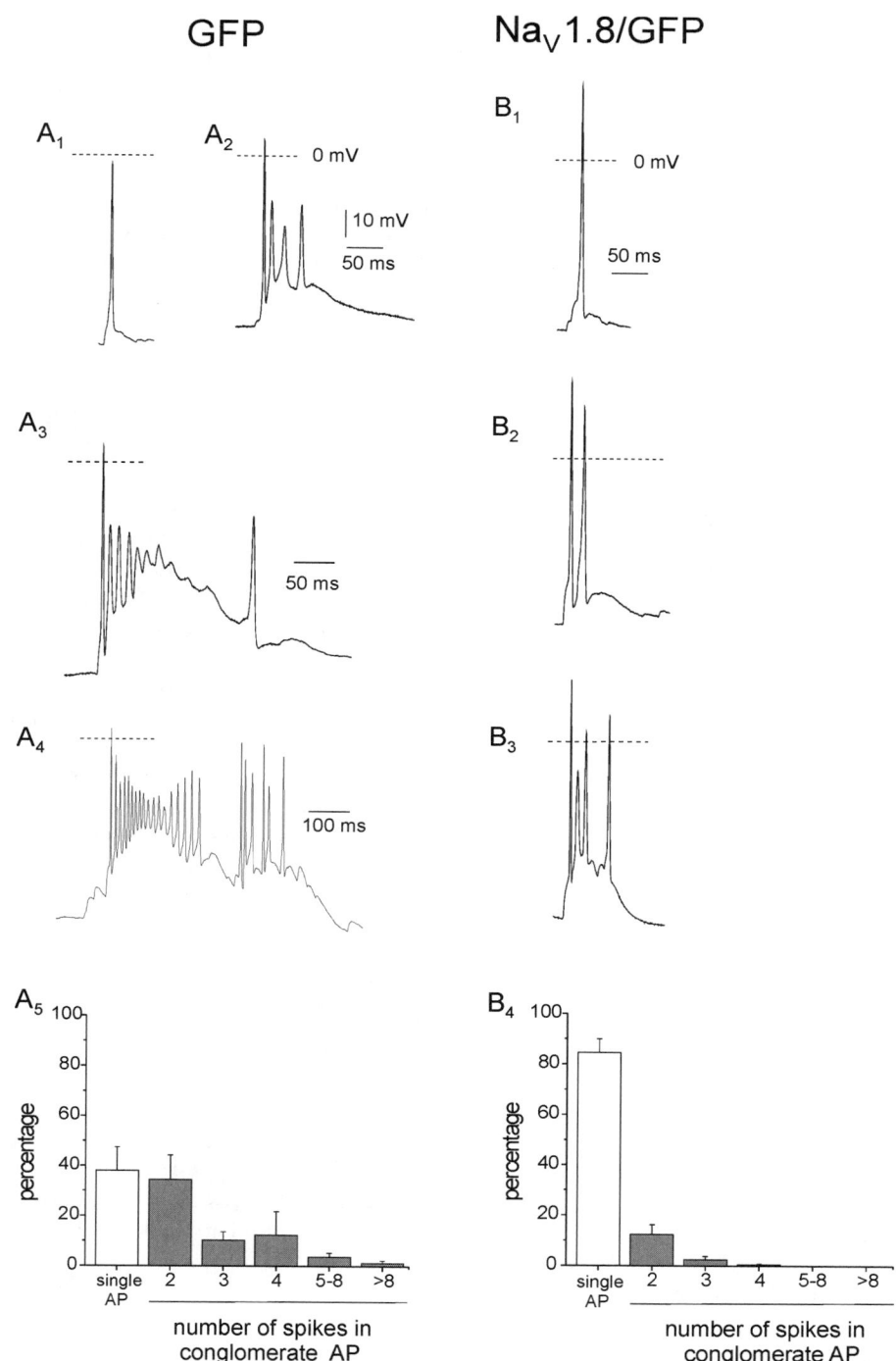

FIGURE 19.11 Expression of $Na_v1.8$ alters action potential electrogenesis within Purkinje cells. These current clamp recordings show spontaneous action potentials recorded in Purkinje neurons two days after biolistic expression of GFP (with or without $Na_v1.8$) which provided a marker of transfection. A_1–A_4: Action potentials in control neurons lacking $Na_v1.8$ show little if any overshoot (dotted lines indicate 0 mV) and tend to be conglomerate (62%; A_2–A_4). B_1–B_3: Action potentials in neurons expressing $Na_v1.8$ display larger overshoot. Conglomerate action potentials are less common after expression of $Na_v1.8$ (15%) and, when present, tend to consist of doublets (B_2), only rarely consisting of >2 spikes (B_3). Time calibration in A_2 applies to A_1; time calibration in B_1 applies to B_1 – B_3. The mV calibration in A_2 applies to all panels. A_5, B_4: Percentage of action potentials that were single, or conglomerate with 2, 3, 4, 5–8 or >8 spikes, in Purkinje cells lacking $Na_v1.8$ ($n = 12$) and with $Na_v1.8$ ($n = 14$), respectively. There is a lower percentage of conglomerate action potentials and smaller number of spikes per conglomerate in Purkinje neurons expressing $Na_v1.8$. Modified from Renganathan *et al.* (2003).

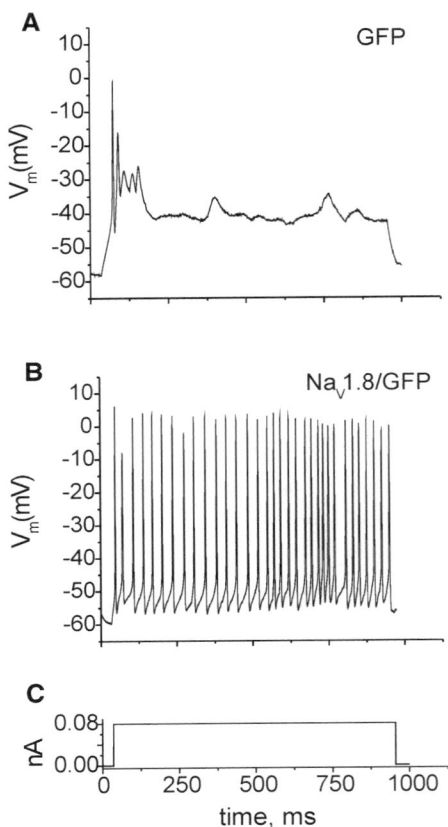

FIGURE 19.12 Purkinje neurons transfected with $Na_v1.8$ show sustained repetitive firing, not present in the absence of $Na_v1.8$, on injection of depolarizing current. **A:** Control Purkinje neuron lacking $Na_v1.8$ produces a conglomerate action potential consisting of five spikes, but no sustained firing, in response to a sustained depolarizing stimulus (80 pA, 1 second). **B:** Purkinje neuron expressing $Na_v1.8$ produces larger-amplitude action potentials and shows sustained pacemaker-like activity in response to identical stimulus. The current pulse protocol is shown in **C.** Modified from Renganathan *et al.* (2003).

addition to changes in their pattern of firing, is not known at this time.

Because biopsy of cerebellar tissue is usually not performed in MS, it will not be easy to directly establish whether these physiological changes occur in Purkinje cells in humans with MS. Some support for this suggestion is, however, provided by clinical observations. Most clinical observers regard cerebellar signs in MS as persistent (non-remitting) even if developing early in the disease course (see, e.g., Matthews *et al.*, 1991), and this is not readily explained by inflammation or demyelination per se. Consistent with a role of $Na_v1.8$ in producing cerebellar dysfunction, Craner *et al.* (2003c) observed, in a relapsing-remitting model of EAE, that the level of $Na_v1.8$ expression within Purkinje cells increases progressively and is correlated with the severity of non-

remitting (including cerebellar) deficits. In addition, the clinical observation of occasional patients of MS with cerebellar deficits on examination, without apparent cerebellar lesions in neuroimaging, is consistent with the hypothesis that molecular changes, too subtle to be detected by currently available imaging techniques, can contribute to cerebellar dysfunction. Finally, paroxysmal ataxia has been well described in MS and, in many cases, responds to treatment with carbamazepine (Andermann *et al.*, 1959; Espir *et al.*, 1966). The temporal profile of the sudden and brief attacks, which are similar to the paroxysmal episodes that are associated with episodic ataxias, are not easily explained by demyelination or axonal degeneration; and the therapeutic response to carbamazepine suggests that sodium channels participate in their pathogenesis.

Up-regulated expression of $Na_v1.8$ has, as noted previously, been observed within retinal ganglion neurons in EAE (Craner, Black, and Waxman, unpublished) and, if similar changes occur in MS, may also have clinical consequences. As in Purkinje cells, voltage-gated sodium channels contribute significantly to electrogenesis in retinal ganglion cells (Lipton and Tauck, 1986; Skaliora *et al.*, 2002). Consistent with an abnormality of retinal neurons, there is evidence for retinal dysfunction in MS, even in patients without a history of optic neuritis (Plant *et al.*, 1986). Study of human retinal tissue will be required to determine whether $Na_v1.8$ is expressed in retinal ganglion neurons in MS.

The hypothesis that $Na_v1.8$ channels, aberrantly expressed within Purkinje neurons and possibly within retinal ganglion neurons, contribute to symptomatology in MS would be strengthened if it could be shown that $Na_v1.8$ blocking drugs ameliorate ataxia, other cerebellar abnormalities, and/or visual loss in MS. $Na_v1.8$-specific channel blocking drugs are not yet available, but this is likely to change since the deployment of $Na_v1.8$ within nociceptive neurons has made it an attractive molecular target. When $Na_v1.8$-specific blocking drugs are developed, a next step will be to examine the effect of these drugs on animal models of MS such as EAE and, if indicated, in humans with MS.

A CENTRAL NERVOUS SYSTEM CHANNELOPATHY IN DIABETES?

The role of the CNS in the pathophysiology of diabetes remains incompletely understood. There is some evidence suggesting that prolonged elevated levels of vasopressin, which is produced within the magnocellular neurons of the hypothalamic supraoptic nucleus, are a risk factor for the development of diabetic nephropathy, which can lead to end-stage renal disease (Bardoux *et al.*,

1999; Ahloulay *et al.*, 1999). As noted previously, vasopressin is released as a result of action potential firing by supraoptic magnocellular neurons. Because changes in the transcription of sodium channel genes can be evoked within supraoptic magnocellular neurons by elevations in osmolality within the normal brain (Tanaka *et al.*, 1999), as described at the beginning of this chapter, Klein *et al.* (2002) recently hypothesized that the hyperosmolality associated with diabetic hyperglycemia can trigger a change in sodium channel expression in these cells. To test this hypothesis, they used *in situ* hybridization, immunocytochemistry, and patch clamp recording to study sodium channel expression within magnocellular neurons from the hypothalamus of rats with STZ-induced diabetes. As shown in Figure 19.13, *in situ* hybridization and immunocytochemical analysis demonstrated significant up-regulation of Na$_v$1.2 and Na$_v$1.6 sodium channel mRNA and protein in the magnocellular neurons within weeks of the development of hyperglycemia, but there was no change in sodium channel expression within other nearby nuclei or the surrounding neuropil. To determine whether the up-regulated mRNA and protein resulted in the insertion of functional channels within the membranes of magnocellular neu-

rons, acutely dissociated magnocellular neurons were studied using patch clamp. Although the voltage-dependence of activation and steady-state inactivation (Figure 19.14) and the time constants for inactivation were similar for diabetic and control neurons, the peak transient current amplitude was 130% greater ($P < .001$) in diabetic magnocellular neurons; peak current density in diabetic neurons was 65% greater ($P < .05$) than that in controls. Persistent sodium currents, which are known to contribute to neuronal bursting and are increased in magnocellular neurons exposed to hyperosmolar conditions (Tanaka *et al.*, 1999), were also compared. Tetrodotoxin-sensitive persistent currents that were activated at potentials close to threshold (−65 to −55mV) were significantly larger in diabetic neurons, with a current density that was increased by approximately 50% compared to that in control neurons ($P < .05$; Klein *et al.*, 2002).

This study demonstrates that, in experimental diabetes, increased transcription results in an increased density of functional sodium channels within magnocellular neurosecretory neurons of the hypothalamus, a change that will increase excitability of these cells and evoke vasopressin release from them. Because prolonged

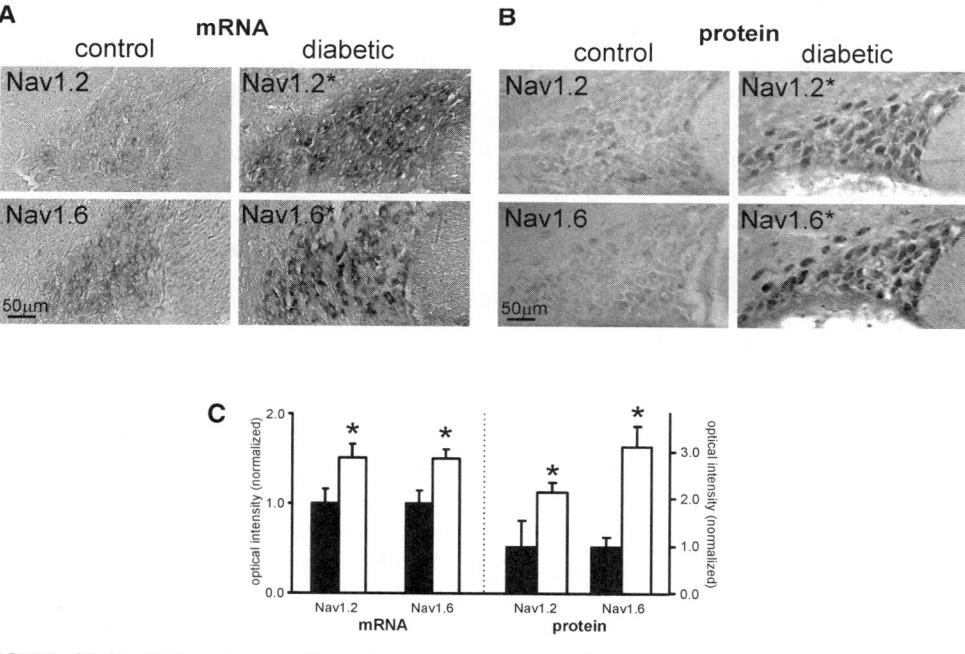

FIGURE 19.13 Voltage-gated sodium channel gene expression in hypothalamic magnocelluar neurons is up-regulated in diabetes. **A:** *In situ* hybridization shows up-regulated sodium channel Na$_v$1.2 and Na$_v$1.6 mRNA levels in the SON after 6 weeks of STZ-induced diabetes. **B:** Immunocytochemistry using subtype-specific antibodies shows up-regulated Na$_v$1.2 and Na$_v$1.6 protein after 6 weeks of STZ-induced diabetes, demonstrating that the newly produced mRNA is translated into sodium channel protein. **C:** Quantification of optical intensities of individual SON neurons shows increases in Na$_v$1.2 and Na$_v$1.6 mRNA and protein. mRNA and protein data shown in **A** and **B** are represented in the histogram in **C**; black bars = control, white bars = diabetic, data plotted as mean ± SE, * = $P < .05$ by non-paired t-test. Modified from Klein *et al.* (2002).

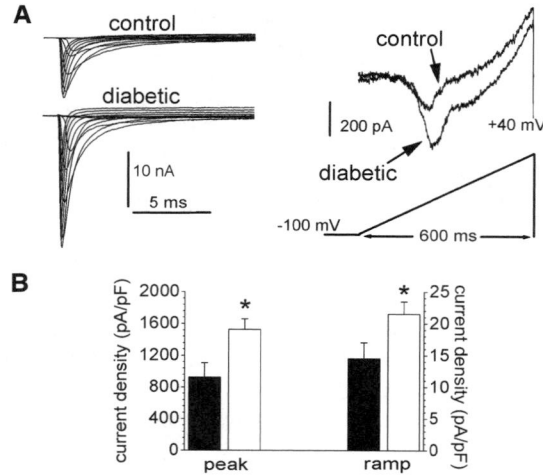

FIGURE 19.14 Patch clamp recordings of acutely dissociated MNCs from diabetic rats show increased peak and ramp currents (**A**, left and right, respectively) and current densities (**B**), indicating that increased numbers of functional sodium channels have been inserted within the membranes of diabetic MNCs; black bars = control, white bars = diabetic, data plotted as mean ± SE, * = $P < .05$ by non-paired t-test. Reproduced from Klein *et al.* (2002).

elevation of serum vasopressin levels, which increases blood pressure and increases filtration demand to the kidneys, is a factor that leads to the development of diabetic nephropathy (Bardoux *et al.,* 1999; Ahloulay *et al.,* 1999), Klein *et al.,* (2002) hypothesized that the changes in sodium channel expression within the supraoptic nucleus may contribute in a long-term manner to the pathogenesis of diabetic nephropathy. If this speculation is correct, altered sodium channel expression within supraoptic neurons associated with diabetes may be viewed as a channelopathy and may be amenable to therapeutic approaches that might delay the development of end-organ complications such as nephropathy. On the other hand, it is possible that the changes are adaptive, contributing to homeostatic mechanisms in untreated or undertreated diabetes. Further study is clearly needed. Irrespective of whether the changes are maladaptive or adaptive, the changes in the diabetic hypothalamus provide another example of the dynamic nature of ion channel expression within the CNS.

OTHER CHANNELS? OTHER NEURONS? OTHER TRIGGERS?

Transcriptional channelopathies of the nervous system are a growing group of disorders that includes other classes of channels and other types of neurons. Increased $K_v1.1$ and $K_v1.2$ potassium channel (Wang *et al.,* 1999)

expression, for example, have been reported along dysmyelinated axons within the brain of the *Shiverer* mouse, in which myelin fails to compact due to a defect in myelin basic protein. Recent studies provide evidence for the ectopic expression of the α_{1B} pore-forming subunit of N-type calcium channels within dystrophic (presumably degenerating) axons in EAE and MS (Kornek *et al.,* 2000) and for the presence of mGluR1α glutamate receptor immunoreactivity along axons in MS (Geurts *et al.,* 2003), although it is not clear whether the receptors are inserted into the axon membrane. Immunocytochemical (Ishikawa *et al.,* 1999) and electrophysiological (Everill and Kocsis, 1999) studies have provided evidence for down-regulation of potassium channels and up-regulation of some calcium channels (Baccei and Kocsis, 2000) in axotomized dorsal root ganglion cells. There also is evidence for altered $GABA_A$ receptor expression in axotomized DRG neurons (Oyelese and Kocsis, 1996; Oyelese *et al.,* 1997).

Although less well studied, there is also evidence suggesting that channel transcription may be perturbed in other disorders of the nervous system. It is possible, for example, that transcriptional channelopathies contribute to some types of epilepsy. Increased expression of $Na_v1.2$ and $Na_v1.3$ sodium channel mRNA has been observed in rat hippocampus following kainate-induced seizures (Gastaldi *et al.,* 1997); it is not known whether these transcriptional changes, or a change in properties of pre-existing channels, for example, by phosphorylation, produce the increased sodium current density and shifted voltage-dependence that are seen in hippocampal neurons after kindling epileptogenesis (Vruegdenhil *et al.,* 1998). At this time it is also not known whether diffuse axonal injury associated with head trauma or corticospinal axon transection due to spinal cord injury can trigger changes in ion channel expression, for example, in cortical motor neurons, that might contribute to posttraumatic epilepsy or to loss of these neurons.

The factor(s) that provoke dysregulated ion channel in injured neurons need further study. A silencer element (RE1) has been identified within the 5′ promoter region of the $Na_v1.2$ sodium channel gene (Kraner *et al.,* 1992; Mori *et al.,* 1992), and it may play a role in some transcriptional channelopathies. Whether similar regulatory elements occur within other channel genes is not yet known. There is a need for more information about the effects of changes in electrical activity (Offord and Catterall, 1989; Sashihara *et al.,* 1996; Sashihara *et al.,* 1997; Klein *et al.,* 2003), and in levels of neurotrophic factors (including NGF and GDNF) (Dib-Hajj *et al.,* 1998; Cummins *et al.,* 2000; Toledo-Aral *et al.,* 1995) on channel expression, both in the intact nervous system and after various insults. Inflammation within the terminal field of axons is known to result in a change in Na$_v$1.8 expression in DRG neurons (Tanaka *et al.,* 1998), raising the question of whether it can trigger a channelopathy in disorders such as MS.

POSSIBLE THERAPEUTIC OPPORTUNITIES

With further progress from molecular and physiological studies *in vitro* and *in vivo,* we will undoubtedly learn more about the transcriptional channelopathies. A more complete understanding of these channelopathies may point to new therapeutic strategies, because the unique amino acid sequences of different channels may facilitate specific targeting. The preferential expression of some channels within specific groups of neurons may, moreover, facilitate their manipulation with minimal side effects. The transcriptional channelopathies, and the molecular targets they present, may offer new opportunities for treatment of a range of disorders affecting the nervous system.

ACKNOWLEDGMENTS

Research in the author's laboratory has been supported, in part, by grants from the National Multiple Sclerosis Society, the Medical Research Service and Rehabilitation Research Service, Department of Veterans Affairs, the Paralyzed Veterans of America, the United Spinal Association, and the Nancy Davis Foundation.

References

Ahlgren SC, White DH, Levine JD. (1992). Increased responsiveness of sensory neurons in the saphenous nerve of the streptozotocin-diabetic rat. *J Neurophysiol* 68:2077-2085.

Ahloulay M, Schmitt F, Dechaux M, Bankir L. (1999). Vasopressin and urinary concentrating activity in diabetes mellitus. *Diabetes Metab* 25:213-222.

Akopian AN, Sivilotti L, Wood JN. (1996). A tetrodotoxin-resistant voltage-gated sodium channel expressed by sensory neurons. *Nature* 379:257-262.

Akopian AN et al. (1999). The tetrodotoxin-resistant sodium channel SNS has a specialized function in pain pathways. *Nat Neurosci* 2:541-548.

Andermann F, Cosgrove JBR, Lloyd-Smith D, Walters AM. (1959). Paroxysmal dysarthria and ataxia in multiple sclerosis. *Neurology* 9:21-216.

Andrew RD, Dudek FE. (1983). Burst discharge in mammalian neuroendocrine cells involves an intrinsic regenerative mechanism. *Science* 221:1050-1052.

Baccei ML, Kocsis JD. (2000). Voltage-gated calcium currents in axotomized adult rat cutaneous afferent neurons. *J Neurophysiol* 83:2227-2238.

Baker D et al. (1990). Induction of chronic relapsing experimental allergic encephalomyelitis in biozzi mice. *J Neuroimmunology* 28:261-270.

Baker D et al. (2000). Cannabinoids control spasticity and tremor in a multiple sclerosis model. *Nature* 404:84-87.

Baker MD et al. (2003). GTP-induced tetrodotoxin-resistant Na$^+$ current regulates excitability in mouse and rat small diameter sensory neurons. *J Physiol (Lond)*, 548.2:373-383.

Balazy TE. (1992). Clinical management of chronic pain in spinal cord injury. *Clin J Pain* 8:102-110.

Bardoux P et al. (1999). Vasopressin contributes to hyperfiltration, albuminuria, and renal hypertrophy in diabetes mellitus: study in vasopressin-deficient Brattleboro rats. *Proc Natl Acad Sci USA* 96:10397-10402.

Beckh S, Noda M, Lubbert H, Numa, S. (1989). Differential regulation of three sodium channel messenger RNAs in the rat central nervous system during development. *EMBO J* 8:3611-3616.

Benoit E, Juzans P, Legrand AM, Molgo J. (1996). Nodal swelling produced by ciguatoxin-induced selective activation of sodium channels in myelinated nerve fibers. *Neuroscience* 71:1121-1131.

Black JA et al. (1997). NGF has opposing effects on Na$^+$ channel III and SNS gene expression in spinal sensory neurons. *NeuroReport* 8:2331-2335.

Black JA et al. (1999a). Upregulation of a silent sodium channel after peripheral, but not central, nerve injury in DRG neurons. *J Neurophysiol* 82:2776-2785.

Black JA et al. (1999b). Abnormal expression of SNS/PN3 sodium channel in cerebellar Purkinje cells following loss of myelin in the taiep rat. *NeuroReport* 10:913-918.

Black JA et al. (2000). Sensory neuron specific sodium channel SNS is abnormally expressed in the brains of mice with experimental allergic encephalomyelitis and humans with multiple sclerosis. *Proc Natl Acad Sci* 97:11598-11602.

Black JA, Waxman SG. (2002). Molecular identities of two tetrodotoxin-resistant sodium channels in corneal axons. *Exp Eye Res* 75:193-199.

Boiko T et al. (2001). Compact myelin dictates the differential targeting of two sodium channel isoforms in the same axon. *Neuron* 30:91-104.

Bostock H, Sears TA. (1978). The internodal axon membrane: electrical excitability and continuous conduction in segmental demyelination. *J Physiol (Lond)* 280:273-301.

Boucher TJ et al. (2000). Potent analgesic effects of GDNF in neuropathic pain states. *Science* 290:124-127.

Brysch W et al. (1991). Regional and temporal expression of sodium channel messenger RNAs in the rat brain during development. *Exp Brain Res* 86:562-567.

Campbell JN, Raja SN, Meyer RA, Mackinnon SE. (1988). Myelinated afferents signal the hyperalgesia associated with nerve injury. *Pain* 32:89-94.

Catterall WA. (1992). Cellular and molecular biology of voltage-gated sodium channels. *Physiol Rev* 72:515-548.

Coward K et al. (2000). Immunolocalization of SNS/PN3 and NaN/SNS2 sodium channels in human pain states. *Pain* 85:41-50.

Craner MJ, Lo AC, Black JA, Waxman SG. (2003a). Abnormal sodium channel distribution in optic nerve axons in a model of inflammatory demyelination. *Brain*, 126:1552-1561.

Craner MJ et al. (2003b). Annexin II/p11 is up-regulated in Purkinje cells in EAE and MS. *NeuroReport* 14:555-558.

Craner MJ et al. (2003c). Temporal course of upregulation of Nav1.8 in Purkine neurons parallels the progression of clinical deficit in EAE. *J Neuropathol Exp Neurol*, 62:968-976.

Cummins TR, Black JA, Dib-Hajj SD, Waxman SG. (2000). Glial-derived neurotrophic factor upregulates expression of functional SNS and NaN sodium channels and their currents in axotomized dorsal root ganglion neurons. *J Neurosci* 20:8754-8761.

Cummins TR et al. (2001). Na$_v$1.3 sodium channels: rapid repriming and slow closed-state inactivation display quantitative differences following expression in a mammalian cell line and in spinal sensory neurons. *J Neurosci* 21:5952-5961.

Cummins TR, Waxman SG. (1997). Down-regulation of tetrodotoxin-resistant sodium currents and up-regulation of a rapidly reprimng tetrodotoxin-sensitive sodium current in small spinal sensory neurons following nerve injury. *J Neurosci* 17:3503-3514.

Devor M, Seltzer Z. (1999). The pathophysiology of damaged peripheral nerves in relation to chronic pain. In Wall PD, Melzack R, editors. *Textbook of pain,* 4th ed. Edinburgh: Churchill Livingstone, pp 129-164.

Dib-Hajj SD *et al.* (1998). Rescue of alpha-SNS/PN3 sodium channel expression in small dorsal root ganglion neurons after axotomy by nerve growth factor in vivo. *J Neurophysiol* 79:2668-2676.

Dib-Hajj S, Black JA, Felts P, Waxman SG. (1996). Down-regulation of transcripts for Na channel α-SNS in spinal sensory neurons following axotomy. *Proc Natl Acad Sci* 93:14950-14954.

Dib-Hajj SD *et al.* (1999). Plasticity of sodium channel expression in DRG neurons in the chronic constriction injury model of neuropathic pain. *Pain* 83:591-600.

Dib-Hajj SD, Ishikawa I, Cummins TR, Waxman SG. (1997). Insertion of an SNS-specific tetrapeptide in the S3-S4 linker of D4 accelerates recovery from inactivation of skeletal muscle voltage-gated Na channel f1 in HEK293 cells. *FEBS Lett* 416:11-14.

Dib-Hajj SD, Tyrrell L, Black JA, Waxman SG. (1998). NaN, a novel voltage-gated Na channel preferentially expressed in peripheral sensory neurons and down-regulated following axotomy. *Proc Natl Acad Sci* 95:8963-8968.

Duncan ID *et al.* (1992). The taiep rat: a myelin mutant with an associated oligodendrocyte microtubular defect. *J Neurocytol* 21:870-884.

Elliott AA, Elliott JR. (1993). Characterization of TTX-sensitive and TTX-resistant sodium currents in small cells from adult rat dorsal root ganglia. *J Physiol (Lond)* 463:39-56.

England JD, Gamboni F, Levinson SR. (1991). Increased numbers of sodium channels form along demyelinated axons. *Brain Res* 548:334-337.

Escayg A *et al.* (2000). Mutations of SCN1A, encoding a neuronal sodium channel, in two families with GEFS+2. *Nature Genetics* 24:343-345.

Espir MLE, Watkins SM, Smith HV. (1966). Paroxysmal dysarthria and other transient neurological disturbances in MS. *J Neurol Neurosurg Psychiatry* 29:323-330.

Everill B, Kocsis JD. (1999). Reduction in potassium currents in identified cutaneous afferent dorsal root ganglion neurons after axotomy. *J Neurophysiol* 82:700-708.

Fang X *et al.* (2002). The presence and role of the TTX resistant sodium channel Na$_v$1.9 (NaN) in nociceptive primary afferent neurons. *J Neurosci* 22:7425-7433.

Felts PA *et al.* (1997). Sodium channel α-subunit mRNAs I, II, III, NaG, Na6 and hNE: Different expression patterns in developing rat nervous system. *Mol Brain Res* 45:71-83.

Felts PA, Baker TA, Smith KJ. (1997). Conduction in segmentally demyelinated mammalian central axons. *J Neurosci* 17:7267-7277.

Field MJ, Bramwell S, Hughes J, Singh L. (1999). Detection of static and dynamic components of mechanical allodynia in rat models of neuropathic pain: are they signaled by distinct primary sensory neurons?, *Pain* 83:303-311.

Fjell J *et al.* (1999a). In vivo NGF deprivation reduces SNS/PN3 expression and TTX-R sodium currents in IB4-negative DRG neurons. *J Neurophysiol* 81:803-911.

Fjell J *et al.* (1999b). Differential role of GDNF and NGF in the maintenance of two TTX-resistant sodium channels in adult DRG neurons. *Mol Brain Res* 67:267-282.

Foster RE, Whalen CC, Waxman SG. (1980). Reorganization of the axonal membrane of demyelinated nerve fibers: morphological evidence. *Science* 210:661-663.

Gahring LC, Rogers SW. (1998). Autoimmunity to glutamate receptors in Rasmussen's encephalitis: a rare finding or the tip of an iceberg?, *The Neuroscientist* 4:373-379.

Gastaldi M *et al.* (1997). Increase in mRNAs encoding neonatal II and III sodium channel alpha-isoforms during kainate-induced seizures in adult rat hippocampus. *Mol Brain Res* 44:179-190.

Geurts JJ *et al.* (2003). Altered expression patterns of group I and II metabotropic glutamate receptors in multiple sclerosis. *Brain* 126:1755-1766.

Goldin AL *et al.* (2000). Nomenclature of voltage-gated sodium channels. *Neuron* 2:365-368.

Hains BC, Yucra JA, Hulseboch CE. (2001). Selective COX-2 inhibition with NS-398 preserves spinal parenchyma and attenuates behavioral deficits following spinal contusion injury. *J Neurotrauma* 18:409-423.

Hains BC, Eaton MJ, Willis WD, Hulseboch CE. (2003a). Engraftment of serotonergic precursors amends hyperexcitability of dorsal horn neurons after spinal hemisection-induced central sensitization. *Neuroscience* 116:1097-1110.

Hains BC, Willis WD, Hulseboch CE. (2003b). Temporal plasticity of dorsal horn somatosensory neurons after acute and chronic spinal cord hemisection in rat. *Brain Res* 970:238-241.

Hains BC *et al.* (2003c). Upregulation of sodium channel Na$_v$1.3 and functional involvement in neuronal hyperexcitability associated with central neuropathic pain after spinal cord injury. *J Neurosci* 23:8881-8892.

Hao JX *et al.* (1991). Allodynia-like effects in rat after ischaemic spinal cord injury photochemically induced by laser irradiation. *Pain* 45:175-185.

Herzog RI, Cummins TR, Waxman SG. (2001). Persistent TTX-resistant Na$^+$ current affects resting potential and response to depolarization in simulated spinal sensory neurons. *J Neurophysiol* 86:1351-1364.

Inenaga K, Nagamoto T, Kannan H, Yamashita H. (1993). Inward sodium current involvement in regenerative bursting activity of rat magnocellular supraoptic neurons in vitro. *J Physiol Lond* 465:289-301.

Ishikawa K, Tanaka M, Black JA, Waxman SG. (1999). Changes in expression of voltage-gated potassium channels in dorsal root ganglion neurons following axotomy. *Muscle Nerve* 22:502-507.

Kearney JA *et al.* (2001). A gain-of-function in the sodium channel gene Scn2a results in seizures and behavioral abnormalities. *Neuroscience* 102:307-317.

Klein JP *et al.* (2002). Sodium channel expression in hypothalamic osmosensitive neurons in experimental diabetes. *NeuroReport* 13:1481-1485.

Klein JP *et al.* (2003). Patterned electrical activity modulates sodium channel expression in sensory neurons. *J Neurosci Res,* 74:192-199.

Kocsis JD, Waxman SG. (1983). Long-term regenerated nerve fibres retain sensitivity to potassium channel blocking agents. *Nature* 304:640-642.

Kocsis JD, Ruiz JA, Waxman SG. (1983). Maturation of mammalian myelinated fibers: changes in action potential characteristics following 4-aminopyridine application. *J Neurophysiol* 50:449-463.

Kocsis JD, Waxman SG. (1987). Ionic channel organization of normal and regenerating mammalian axons. *Progr Brain Res* 71:89-102.

Kohama I, Ishikawa K, Kocsis JD. (2000). Synaptic reorganization in the substantia gelatinosa after peripheral nerve neuroma formation: aberrant innervation of lamina II neurons by Aβ afferents. *J Neurosci* 20:1538-1549.

Kohrman DC *et al.* (1996). A missense mutation in the sodium channel Scn8a is responsible for cerebellar ataxia in the mouse mutant jolting. *J Neurosci* 16:5993-5999.

Kornek B *et al.* (2001). Distribution of a calcium channel subunit in dystrophic axons in multiple sclerosis and experimental autoimmune encephalomyelitis. *Brain* 124:1114-1124.

Kraner SD, Chong JA, Tsay H-J, Mandel G. (1992). Silencing the type II sodium channel gene: a model for neural-specific gene regulation. *Neuron* 9:37-44.

Leffler A *et al.* (2002). Glial-derived neurotrophic factor and nerve growth factor reverse changes in repriming of TTX-sensitive Na$^+$ currents following axotomy of dorsal root ganglion neurons. *J Neurophysiol* 88:650-660.

Lewin GR, Ritter AM, Mendell LM. (1992). On the role of nerve growth factor in the development of myelinated nociceptors. *J Neurosci* 12:1896-1905.

Li Z, Hatton GI. (1996). Oscillatory bursting of physically firing rat supraoptic neurones in low-Ca^{2+} medium: Na+ influx, cytosolic Ca^{2+} and gap junctions. *J Physiol Lond* 496:397-394.

Lindsey AE *et al.* (2000). An analysis of changes in sensory thresholds to mild tactile and cold stimuli after experimental spinal cord injury in the rat. *Neurorehabil Neural Repair* 14:287-300.

Lipton S, Tauck D. (1996). Voltage-dependent conductances of ganglion cells from rat retina. *J Physiol* 385:361-391.

Llinas R, Sugimori M. (1980). Electrophysiological properties of in vitro Purkinje cell somata in mammalian cerebellar slices. *J Physiol* 305:171-195.

Matthews WB, Compston A, Allen IV, Martyn CN. (1991). *McAlpine's Multiple Sclerosis*. New York: Churchill Livingstone.

Mills CD, Hains BC, Johnson KM, Hulseboch CE. (2001). Strain and model dependence of locomotor deficits and development of chronic central pain after spinal cord injury. *J Neurotrauma* 18:743-756.

Mori N, Schoenherr C, Vandenbergh DJ, Anderson DJ. (1992). A common silencer element in the SCG10 and type II Na$^+$ channel genes binds a factor present in nonneuronal cells but not in neuronal cells. *Neuron* 9:45-54.

Newsom-Davis J. (1997). Autoantibody-mediated channelopathies at the neuromuscular junction. *The Neuroscientist* 3:337-346.

Offord J, Catterall WA. (1989). Electrical activity, cAMP, and cytosolic calcium regulate mRNA encoding sodium channel α subunits in rat muscle cells. *Neuron* 2:1447-1452.

Okuse K *et al.* (2002). Annexin II light chain regulates sensory neuron-specific sodium channel expression. *Nature* 47:653-656.

Ophoff RA *et al.* (1996). Familial hemiplegic migraine and episodic ataxia type-2 are caused by mutations in the Ca2+ channel gene CACNLA4. *Cell* 87:543-552.

Oyelese AA, Kocsis JD. (1996). GABA$_A$-receptor-mediated conductance and action potential waveform in cutaneous and muscle afferent neurons of the adult rat: differential expression and response to nerve injury. *J Neurophysiol* 76:2383-2392.

Oyelese AA, Rizzo MA, Waxman SG, Kocsis JD. (1997). Differential effects of NGF and BDNF on axotomy-induced changes in GABA$_A$-receptor-mediated conductance and sodium currents in cutaneous afferent neurons. *J Neurophysiol* 78:31-42.

Plant GT, Hess RF, Thomas SJ. (1986). The pattern evoked electroretinogram in optic neuritis. A combined psychophysical and electrophysiological study. *Brain* 109:469-490.

Plummer W, Meisler MH. (1999). Evolution and diversity of sodium channel genes. *Genomics* 57:323-331.

Raman IM, Bean BP. (1997). Resurgent sodium current and action potential formation in dissociated cerebellar Purkinje neurons. *J Neurosci* 17:4517-4526.

Raman IM, Sprunger LK, Meisler MH, Bean BP. (1997). Altered subthreshold sodium currents and disrupted firing patterns in Purkinje neurons of Scn81 mutant mice. *Neuron* 19:881-891.

Renganathan M, Cummins TR, Waxman SG. (2001). The contribution of Na$_v$1.8 sodium channels to action potential electrogenesis in DRG neurons. *J Neurophysiol* 86:629-640.

Renganathan M, Gelderblom M, Black JA, Waxman SG. (2003). Expression of Na$_v$1.8 sodium channels perturbs the firing patterns of cerebellar Purkinje cells. *Brain Res* 959:235-243.

Ritchie JM, Rogart RB. (1977). The density of sodium channels in mammalian myelinated nerve fibers and the nature of the axonal membrane under the myelin sheath. *Proc Natl Acad Sci USA* 74:211-215.

Ritter AM, Mendell LM. (1992). Soma membrane properties of physiologically identified sensory neurons in the rat: effects of nerve growth factor. *J Neurophysiol* 68:2033-2041.

Rose M, Griggs RE, editors. (2001). *Channelopathies of the nervous system*. Oxford: Butterworth-Heinemann.

Saab CY, Craner MJ, Kataoka Y, Waxman SG. (2004). Abnormal Purkinje cell activity *in vivo* in experimental allergic encephalomyelitis. *Exper Brain Res*, in press.

Sangameswaren L *et al.* (1996). Structure and function of a novel voltage-gated tetrodotoxin-resistant sodium channel specific to sensory neurons. *J Biol Chem* 271:5953-5956.

Sashihara S, Greer CA, Oh Y, Waxman SG. (1996). Cell specific differential expression of Na+ channel a1 subunit mRNA in the olfactory system during postnatal development and following denervation. *J Neurosci* 16:702-714.

Sashihara S, Waxman SG, Greer CA. (1997). Down-regulation of Na+ channel mRNA following sensory deprivation of tufted cells in the neonatal rat olfactory bulb. *Neuro Report* 8:1289-1293.

Schild JH, Kunze DL. (1997). Experimental and modeling study of Na$^+$ current heterogeneity in rat nodose neurons and its impact on neuronal discharge. *J Neurophysiol* 78:3198-3209.

Scroggs RS, Fox AP. (1992). Multiple Ca^{2+} currents elicited by action potential waveforms in acutely isolated adult rat dorsal root ganglion neurons. *J Neurosci* 12:1789-1801.

Skaliora I, Scobey R, Chalupa L. (2002). Prenatal development of excitability in retinal ganglion cells: action potentials and sodium currents. *J Neurosci* 13:313-323.

Sleeper AA *et al.* (2000). Changes in expression of two tetrodotoxin-resistant sodium channels and their currents in dorsal root ganglion neurons after sciatic nerve injury but not rhizotomy. *J Neurosci* 20:7279-7289.

Stuart G, Hausser M. (1994). Initiation and spread of sodium action potentials in cerebellar Purkinje cells. *Neuron* 13:703-712.

Stys PK, Waxman SG. (2004). Ischemic white matter injury. In Lazzarini R, editor: *Myelin and its diseases*. New York: Academic Press.

Sugawara T *et al.* (2001). A missense mutation of the Na$^+$ channel α$_{II}$ subunit gene Na$_v$1.2 in a patient with febrile and afebrile seizures causes channel dysfunction. *Proc Natl Acad Sci* 98:6384-6389.

Tanaka M *et al.* (1999). Molecular and functional remodeling of electrogenic membrane of hypothalamic neurons in response to changes in their input. *Proc Natl Acad Sci USA* 96:1088-1093.

Tanaka M *et al.* (1998). SNS Na$^+$ channel expression increases in dorsal root ganglion neurons in the carrageenan inflammatory pain model. *Neuro Report* 9:967-972.

Tanaka M *et al.* (1999). Molecular and functional remodeling of electrogenic membrane of hypothalamic neurons in response to changes in their input. *Proc Natl Acad Sci* 96:1088-1093.

Toledo-Aral JJ, Brehm P, Halegoua S, Mendell G. (1995). A single pulse of nerve growth factor triggers long-term neuronal excitability through sodium channel gene induction. *Neuron* 14:607-611.

Turner JA, Cardenas DD, Warms CA, McClellan CB. (2001). Chronic pain associated with spinal cord injuries: a community survey. *Arch Phys Med Rehabil* 82:501-509.

Vreugdenhil M, Faas GC, Wadman WJ. (1998). Sodium currents in isolated rat CA1 neurons after kindling epileptogenesis. *Neuroscience* 86:99-107.

Wang H *et al.* (1999). Hypomyelination alters K$^+$ channel expression in mouse mutants *Shiverer* and *Trembler*. *Neuron* 15:1337-1347.

Waxman SG. (2000). The neuron as a dynamic electrogenic machine: modulation of sodium channel expression as a basis for functional plasticity in neurons. *Phil Trans Roy Soc Lond B* 355:199-213.

Waxman SG, Kocsis JD, Black JA. (1994). Type III sodium channel mRNA is expressed in embryonic but not adult spinal sensory neurons, and is re-expressed following axotomy. *J Neurophysiol* 72:466-471.

Waxman SG. (1977). Conduction in myelinated, unmyelinated, and demyelinated fibers. *Arch Neurol* 34:585-590.

Waxman SG. (2001). Transcriptional channelopathies: an emerging class of disorders. *Nat Rev Neurosci* 2:652-659.

Waxman SG. (2002). Ion channels and neuronal dysfunction in multiple sclerosis. *Arch Neurol* 59:1377-1380.

Wood JN, Perl ER. (1999). Pain. *Curr Opin Genet Dev* 9:328-332.

Woolf CJ, Shortland P, Coggeshall RE. (1992). Peripheral nerve injury triggers central sprouting on myelinated afferents. *Nature* 355:75-78.

Yezierski RP, Park SH. (1993). The mechanosensitivity of spinal sensory neurons following intraspinal injections of quisqualic acid in the rat. *Neurosci Lett* 157:115-119.

Zhang J-M, Donnelly DF, Song X-J, LaMotte RH. (1997). Axotomy increases the excitability of dorsal root ganglion cells with unmyelinated axons. *J Neurophysiol* 78:2790-2794.

20

New Molecular Targets for the Treatment of Neuropathic Pain

John N. Wood, PhD, ScD
Stephen G. Waxman, MD, PhD

Pain resulting from damage to peripheral nerves is a major clinical problem. The broad range of first-line treatments often focusing on antidepressants and opioids, and the fact that even the most useful drugs only alleviate pain in a minority of recipients highlight our lack of understanding of these pain syndromes. Fortunately, new insights into the pathophysiological mechanisms that underlie neuropathic pain are changing this situation. Over the past decade, a number of useful animal models of neuropathic pain have been developed. The ability to manipulate gene expression in transgenic animals and to examine the consequence of gene ablation or overexpression in such models has provided a wealth of new potentially interesting drug targets for neuropathic pain. In addition, analysis of the efficacy of drugs in the treatment of neuropathic pain in both animal models and the clinic has suggested new directions for effective drug development. In this chapter we review recent insights into the molecular mechanisms of neuropathic pain, focusing on alterations in the excitability of damaged peripheral nerves, and the receptors and regulatory molecules that underlie this process.

THE PROBLEM

Neuropathic pain may arise as a consequence of diseases that lead to peripheral nervous system degeneration such as diabetes, herpes zoster, or acquired immunodeficiency syndrome (AIDS). Between 2 and 3 million patients in the United States suffer such pain, with many referrals to pain clinics arising from the million people a year in whom herpes zoster, also known as shingles, develops. Post-herpetic neuralgia as a result of permanently damaged nerves develops in approximately 20% of shingles sufferers. At least 2% of the Western world is affected by diabetes and 50%, according to some estimates, eventually develop diabetic neuropathy. Of the many millions of AIDS patients, at least 10% develop a neuropathic pain syndrome as a result of viral or drug-induced damage to peripheral nerves. Apart from these clinical conditions, peripheral nerve injury as a result of trauma or surgical complications may also lead to neuropathic pain states. Up to 10% of amputees may develop phantom limb pain. Damage to the spinal cord and brain may also lead to chronic central pain states. Chronic pain is, in fact, reported in a majority of patients after spinal cord injury.

The substantial figures of neuropathic pain incidence may be an underestimate. For example, US surveys suggest that the lifetime incidence of lower back pain (LBP) ranges from 60% to 90% with a 5% annual incidence. No consensus exists concerning the mechanism or most appropriate treatment and management of mechanical LBP. To date there are no comprehensive figures outlining the incidence of various chronic pain syndromes. Approximately 38% of the general population in Denmark suffer from chronic pain (Andersen and Worm-Pedersoen, 1989). Similarly a survey of chronic pain in Sweden found that 45% of all adults have experienced recurrent or persistent pain, while 8% endure severe persistent pain (Korff *et al.*, 1990). A telephone survey found that 7% of a large random sample of adults in the United Kingdom were in substantial pain (Bowsher *et al.*, 1991). A more detailed study of chronic pain in the United Kingdom found that 50% of those surveyed reported chronic pain or discomfort, including 16% with back pain and 16% with arthritis. In 16% of those surveyed, chronic pain was severe (Elliott *et al.*, 1999). Because the mechanisms underlying these chronic pain syndromes are mainly unknown, it is safe to assume that neuropathic pain incidence figures are likely to be underestimates of a vast and troublesome clinical problem. Such pain can have a substantial impact on the quality of life not only of the chronic pain sufferers, but also their families. The consequences of long-term chronic pain are not only physical but may also include social isolation, marital breakdown, loss of employment with consequent financial difficulties and, unsurprisingly, depression.

PRESENT TREATMENTS AND ANIMAL MODELS

In a retrospective analysis of placebo-controlled trials, Sindrup and Jensen (1999) extracted the data on the efficacy of various drugs in treating neuropathic pain. A useful measure is to estimate the number of patients that need to be treated (NNT) for one person to obtain 50% pain relief. Thus the smaller the number, the more effective the drug treatment. An important caveat in this historical survey is that newer drugs such as gabapentin may not have been tested in as much depth as antidepressants that have been known to be useful for some time. In addition, N-methyl-D-aspartate (NMDA) antagonists and a variety of other potentially useful drugs are not included in the study. In diabetic neuropathy, NNT was 2.4 for tricyclic antidepressants. The NNT was 6.7 for selective serotonin reuptake inhibitors; 3.3 for the anti-epileptic carbamazepine; 10.0 for a sodium channel blocker, mexiletine; 3.7 for gabapentin, which probably targets nociceptor Cav2.2 calcium channels; 1.9 for dextromethorphan and 3.4 for tramadol, both opioid-like drugs; and 5.9 for capsaicin, the exogenous activator of the TrpV1 noxious heat receptor.

In post-herpetic neuralgia, opioids were inactive, but the NNT was similar for other drugs, for example, 2.3 for tricyclic antidepressants, 3.2 for gabapentin, 2.5 for oxycodone (a weak opioid), and 5.3 for capsaicin.

In peripheral nerve injury, NNT was again 2.5 for tricyclics and lower (3.5) for capsaicin. In central pain, NNT was 2.5 for tricyclic antidepressants, whereas selective serotonin reuptake inhibitors, mexiletine, and dextromethorphan were inactive. These figures are interesting from a variety of perspectives. First, the most effective drugs do not target the site or underlying mechanism of neuropathic pain development, but rather target the integrative assessment of the pain, and its effect through the use of antidepressants. The distinct patterns of drug efficacy also suggest that different mechanisms may underlie various subtypes of neuropathic pain.

Recapitulating the human peripheral nerve injuries that lead to neuropathic pain in animal models has proved extremely useful for mechanistic studies. Usually such models are named after the principal author on the first papers in which these models were described. Most models focus on partial nerve injury to sciatic or sural nerves, which allow altered hind-limb pain sensitivity to thermal and mechanical insults to be measured and compared with the uninjured contralateral paw (Figure 20.1).

For technical reasons, most groups have focused on neuropathic pain models in inbred strains of rat, where the surgery is easier and behavioral read-outs are fairly robust. Yoon et al. (1999) investigated the differences between different strains of rat in terms of neuropathic pain behavior. Despite very similar normal pain thresholds, neuropathic pain behavior showed major differences between rat strains. After injury produced by tightly ligating the left L5 and L6 spinal nerves, behavioral signs representing mechanical allodynia, cold allodynia, and spontaneous

Animal Models of Neuropathic Pain

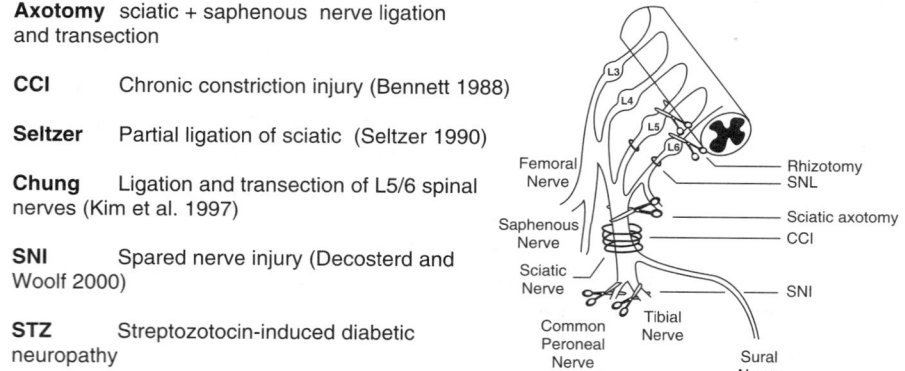

Axotomy sciatic + saphenous nerve ligation and transection

CCI Chronic constriction injury (Bennett 1988)

Seltzer Partial ligation of sciatic (Seltzer 1990)

Chung Ligation and transection of L5/6 spinal nerves (Kim et al. 1997)

SNI Spared nerve injury (Decosterd and Woolf 2000)

STZ Streptozotocin-induced diabetic neuropathy

FIGURE 20.1 Ligation and transection of peripheral nerves innervating the hind paw are two of the most commonly used approaches to modeling neuropathic pain in rodents. STZ (streptozotocin)-induced general nerve damage is used as a model of diabetic neuropathy.

pain were investigated over a period of weeks. Surprisingly, there were differences even between supposedly identical rat strains from different suppliers, with a relative resistance to neuropathic pain behavior in some rat strains.

As transgenic technology has been increasingly applied to studies of pain mechanisms, the emphasis on animal models has switched to the study of mice. Almost all transgenic mice are made using embryonic stem cell lines from 129/Sv mice that are introduced into C57Bl6 mothers. The advantage of this approach is that coat color can indicate whether the engineered stem cells have integrated into the host to create a chimera. The progeny are thus not inbred. This genetic heterogeneity has had important consequences for analyzing pain phenotypes because, as with the rat, there are dramatic differences in pain behavior between different strains of mice. Mogil et al. (1999) have embarked on a comprehensive analysis of the pain behavior of different mouse strains. Such studies can provide important information about common mechanisms between different types of pain behavior, as well as allowing the mapping of candidate genes that may play important roles in pain pathways. These studies also highlight the potential variability in responses between different human populations to pain therapeutics. Mogil et al. tested 11 inbred mouse strains (129/J, A/J, AKR/J, BALB/cJ, C3H/HeJ, C57BL/6J, C58/J, CBA/J, DBA/2J, RIIIS/J, and SM/J) using 12 measures of pain behavior, including thermal nociception (hot plate, Hargreaves' test, tail withdrawal), mechanical nociception (von Frey filaments), chemical nociception (abdominal constriction, carrageenan, formalin), and neuropathic pain (autotomy, Chung model peripheral nerve injury). Strain differences in the various assays could be as much as 50-fold, emphasizing the importance of appropriate littermate controls for the analysis of transgenic mouse pain behavior.

The first commonly used model of rodent neuropathic pain (Bennett model of chronic constriction injury, CCI) relied on the tight ligation of the sciatic nerve using thread soaked in chrome alum, which in itself imparted a low pH insult to the local tissue (Bennett et al., 1992). Kim et al. (1997) ligated tightly both L5 and L6 (or L5 alone) spinal nerves (Chung model), and compared the evoked behavior with that found in the Bennett model. Both thermal and mechanical thresholds were affected, with a long-lasting hyperalgesia to noxious heat (at least 5 weeks) and mechanical allodynia (at least 10 weeks) of the affected foot. In addition, there were behavioral signs of the presence of spontaneous pain. The partial nerve injury caused in this model was an important development in allowing functional changes in both damaged and adjacent neurons to be investigated. In a similar fashion, Seltzer (1990) ligated about half of the sciatic nerve close to the spinal cord. Within a few hours the rats

developed guarding and licking behavior of the ipsilateral hind paw, suggesting the presence of spontaneous pain. This behavior continued for many months. There was a decrease in the withdrawal thresholds in response to repetitive von Frey hair stimulation at the plantar side. Both allodynia and mechanical hyperalgesia were apparent in this model, as was thermal hyperalgesia

Given the variation in procedures between different laboratories, a useful single laboratory comparison of the characteristics of chronic constriction injury by loose ligation of the sciatic nerve, tight ligation of the partial sciatic nerve (PSL), and tight ligation of spinal nerves (SNL) was made by Kim et al. (1997). All three methods of peripheral nerve injury produced behavioral signs of both ongoing and evoked pain with similar time courses. However, mechanical allodynia was greatest in the SNL injury and smallest in the CCI model. On the other hand, ongoing pain was much more prominent in the CCI model than in the other two.

More recently, a further variation on the theme of partial nerve injury has been developed with the spared nerve injury model (SNI) (Decosterd and Woolf, 2000). This involves a lesion of two of the three terminal branches of the sciatic nerve (tibial and common peroneal nerves) leaving the remaining sural nerve intact. The spared nerve injury model is unique in restricting contact between intact and degenerating axons, and this allows behavioral testing of the non-injured skin territories next to the denervated areas. Erichsen and Blackburn-Munro (2002) investigated the pharmacological sensitivity of the SNI model to various potential analgesic drugs measuring reflex nociceptive responses to mechanical and cold stimulation after systemic administration of opioids, sodium channel blockers, and other drugs. They found that morphine attenuated mechanical hypersensitivity in response to von Frey hair and pinprick stimulation and cold hypersensitivity in response to ethyl chloride. The sodium channel blocker, mexiletine, relieved both cold allodynia and mechanical hyperalgesia, but the most distinct and prolonged effect was observed on mechanical allodynia. Gabapentin significantly alleviated mechanical allodynia, but perhaps surprisingly had no effect on mechanical hyperalgesia. In contrast, the NMDA receptor antagonist MK-801 and the AMPA receptor antagonist NS1209 did not provide any relief.

The Chung model has also been characterized in terms of responses to different type of analgesic drugs that have some utility in the clinic (Abdi et al., 1998). Amitriptyline (an antidepressant), gabapentin, and lidocaine (a local anesthetic) were effective in increasing the threshold for mechanical allodynia Amitriptyline and lidocaine reduced the rate of continuing discharges of injured afferent fibers, although gabapentin did not influence these discharges, presumably acting centrally. The differences in the pharmacological characteristics of the various animal models of neuropathic pain are likely to reflect different mechanisms that are relevant to human pathophysiology. It is heartening that effective drugs found serendipitously in the clinic (e.g., gabapentin) are effective in animal models, thus validating studies of animal models as relevant to analgesic drug development.

CANDIDATE GENES FOR CLINICAL INTERVENTION

Classical genetic studies involving mapping of inbred strains of animals are time-consuming. Now that the complete sequence of the mouse and human genomes are available, it is possible to scan essentially all expressed genes in normal and pathophysiological states to try to find candidate transcripts that may be involved in the underlying pathology. Microarray technology now generally uses oligonucleotide arrays of 50 to 60 bases (rather than cDNA copies) immobilized on glass slides, that are hybridized with tagged cDNA copies of RNA extracted from control or experimental samples. Analysis of triplicate samples is usually considered necessary to obtain meaningful figures on alterations in transcript levels.

A number of limitations to this powerful and expensive technology should be borne in mind. First, changes in gene expression of less than twofold are hard to detect. Second, splice variant differences, particularly involving small exon substitutions, may not be detectable with present arrays. In addition, dramatic alterations in mRNA expression in a small subset of neurons may be swamped by background levels of the transcript in other tissues. Perhaps most importantly, transcriptional regulation may not be the principal site of gene dysfunction. For example, the redistribution of voltage-gated channels within the membrane of a damaged nerve may produce major excitability changes that are not a consequence of altered transcriptional regulation.

There are a number of analyses of altered gene expression patterns characteristic of axotomy in the literature. Wang *et al.* (2003) found 178 genes to be dysregulated 3 days after axotomy. Using Northern blot analysis, quantitative slot blots, and *in situ* hybridization, they confirmed these data for 83% of the genes. Xiao *et al.* (2002) carried out a similar analysis 2 weeks after axotomy, and cataloged altered genes into a number of different categories. Costigan *et al.* (2002) found effects similar to those reported by Wang *et al.* (2003) with approximately 240 genes dysregulated 3 days after axotomy. This wealth of information brings its own problems. How can one test the significance of these altered levels of expression

functionally? The relative merits of antisense, siRNA, and gene deletion technologies for the analysis of these candidate genes are discussed subsequently.

RECEPTORS AND CHANNELS IMPLICATED IN THE PATHOGENESIS OF NEUROPATHIC PAIN

A combination of gene regulation studies, knockouts, and pharmacological insights have provided us with a range of new targets for neuropathic pain that are cataloged subsequently. Although a survey of the literature provides an apparently bewildering range of different mechanisms in the pathogenesis of neuropathic pain, there are common themes and interrelationships between many of the different deficits cataloged in animal models. First, damage to sensory neurons or their axons can lead to alterations in responses to trophic factors and cytokines (e.g., nerve growth factor [NGF], glial-derived neurotrophic factor (GDNF), tumor necrosis factor [TNF]), leading to altered patterns of gene expression (e.g., the re-expression of genes normally expressed in early development such as the sodium channel $Na_v1.3$). Second, damage to axons leads to altered subcellular patterns of channel expression (voltage-gated sodium and potassium channels) that lead to hyperexcitability or spontaneous electrical activity. Thus large diameter sensory neurons express high levels of the HCN pacemaker channel and become hyperexcitable. This in turn leads to altered patterns of input into the dorsal horn involving voltage-gated calcium channel subunits (e.g., $\alpha 2\delta$-1) and changes in second order sensory neuron synaptic efficacy involving NMDA receptors. Presynaptic or postsynaptic P2X receptors within the dorsal horn also appear to be necessary for the establishment of neuropathic pain. The alterations in synaptic organization within the dorsal horn are accompanied by local increases in microglia that appear to further potentiate neuropathic pain behavior by the release of soluble factors. Molecular targets that are reasonably well established as contributing to neuropathic pain are shown in Table 20.1.

Sodium Channels

Voltage-gated sodium channels comprise a family of ten structurally related genes that are expressed in spatially and temporally distinct patterns in the mammalian nervous system. It has long been known that sodium channel blockers are powerful analgesics when delivered at low concentrations (Strichartz, 2002). Evidence of a role for these channels in neuropathic pain has come from studies of neuronal excitability, analysis of patterns of expression of channel isoforms in animal models of neuropathic pain, and antisense and knockout studies.

TABLE 20.1 Neuropathic Pain Targets

Channels and Receptors	Drug	Established Role?
Sodium channels		
$Na_v1.3$	Tetrodotoxin/ lidocaine	correlative and pharmacological,
$Na_v1.8$		antisense, knockout
$Na_v1.9$		correlative
β-3 subunits		correlative
Calcium Channels		
$Ca_v2.2$	N-type Calcium channel blockers	knockout, pharmacological
α2δ	Gabapentin	pharmacological
Potassium channels		
Kv1.4		correlative
KCNQ	Retigabine	correlative and pharmacological
Pacemakers		
HCN	ZD7288	correlative and pharmacological
Ligand-gated		
TrpV1	Capsazepine	pharmacological
P2X3	A317491	antisense and pharmacological
P2X4		antisense and pharmacological
NMDA-NR2b	CP-101,606	pharmacological
mGluR1		
mGluR2/3	LY379268	pharmacological - agonists
Galanin		transgenics
CB1	WIN55212-2	pharmacological
CB2		microglia?
Neurotrophins/Cytokines		
NGF	receptor-bodies	transgenics
BDNF	receptor-bodies	
GDNF	receptor-bodies	pharmacological
IL-6		
TNF	thalidomide	pharmacological

Two sodium channels, $Na_v1.8$ and $Na_v1.9$, are selectively expressed within the peripheral nervous system, predominantly in nociceptive sensory neurons, and these particular isoforms have attracted attention as analgesic drug targets. In addition, an embryonic channel $Na_v1.3$ and a beta subunit, β-3, are up-regulated in DRG neurons in neuropathic pain states.

Craner et al. (2002) showed that in experimental diabetes, there is a significant up-regulation of mRNA and protein for the $Na_v1.3$, $Na_v1.6$, and $Na_v1.9$ sodium channels and a down-regulation of $Na_v1.8$ mRNA that accompanies allodynia. In most neuropathic pain models exemplified by axotomy, however, $Na_v1.8$ and 1.9 are down-regulated, while $Na_v1.3$ is induced (Waxman et al., 1994; Dib-Hajj et al., 1996, 1998; Okuse et al., 1997) (Figure 20.2, Table 20.2).

$Na_v1.3$ is widely expressed in the adult central nervous system (CNS) but is normally absent, or present at low levels, in the adult peripheral nervous system. Axotomy or other forms of nerve damage lead to the re-expression of $Na_v1.3$ and the associated beta-3 subunit in sensory neurons, but not in primary motor neurons (Waxman et al., 1994; Dib-Hajj et al., 1996; Hains et al., 2002). This event can be reversed in vitro and in vivo by treatment with high levels of exogenous GDNF. $Na_v1.3$ is known to recover (reprime) rapidly from inactivation (Cummins et al., 2001). Axotomy induces the expression of rapidly repriming tetrodotoxin (TTX)-sensitive sodium channels in damaged neurons, and this event can also be reversed by the combined actions of GDNF and NGF (Leffler et al., 2002). Concomitant with the reversal of $Na_v1.3$ expression by GFNF, ectopic action potential generation is diminished and thermal and mechanical pain-related behavior in a rat CCI model is reversed (Boucher et al., 2000). Moreover, $Na_v1.3$ is up-regulated in multireceptive nociceptive dorsal horn neurons following experimental spinal cord injury; this up-regulation is associated with hyperexcitability of these nociceptive neurons and pain; antisense knockdown of $Na_v1.3$ attenuates the dorsal horn neuron hyperexcitability and the pain behaviors in animals with spinal-cord injury (Hains et al., 2003). It therefore seems likely that $Na_v1.3$ re-expression plays a significant role in increasing neuronal excitability, thus contributing to neuropathic pain after nerve and spinal cord injury.

$Na_v1.8$ is mainly expressed in nociceptive neurons (Akopian et al., 1996; Djouhri et al., 2003). This channel contributes a majority of the sodium current underlying the depolarizing phase of the action potential in cells in which it is present (Renganathan et al., 2001). Functional expression of the channel is regulated by inflammatory mediators, including NGF (Dib-Hajj et al., 1998b), and both antisense and knockout studies support a role for the channel in contributing to inflammatory pain (Khasar et al., 1998; Akopian et al., 1999). Antisense studies have also suggested a role for this protein in the development of neuropathic pain (Lai et al., 2002), and a deficit in ectopic action propagation has been described in the $Na_v1.8$ null mutant mouse (Roza et al., 2003). However, neuropathic pain behavior at early time points seems to be normal in the $Na_v1.8$ null mutant mouse (Kerr et al., 2000).

Identification of annexin II/p11, which binds to $Na_v1.8$ and facilitates the insertion of functional channels in the cell membrane (Okuse et al., 2002), may

Structure of voltage-gated sodium channels

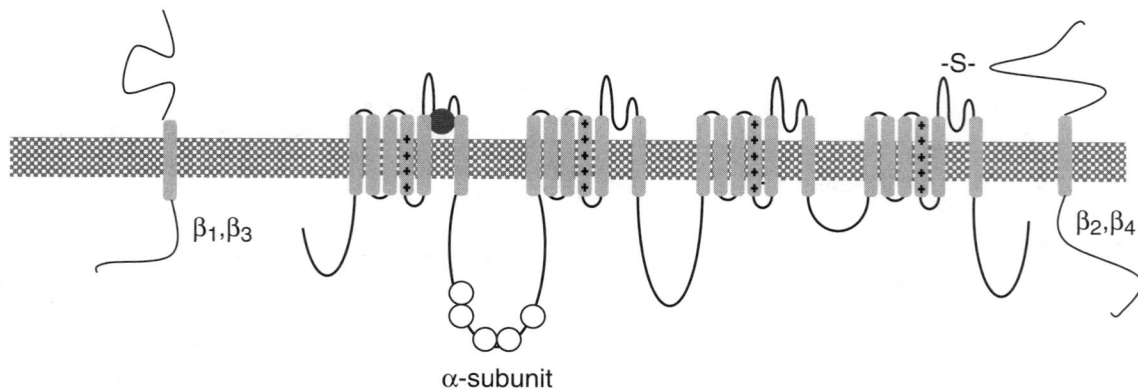

FIGURE 20.2 Voltage-gated sodium channels comprise a single functional α-subunit that is tethered by associated accessory subunits to subcellular locations determined by both extracellular and intracellular proteins. Phosphorylation sites are shown on intracellular loop 2, and the serine residue in domain 1 that confers tetrodotoxin sensitivity is highlighted.

provide a target that can be used to modulate the expression of $Na_v1.8$ and hence the level of $Na_v1.8$ current in nociceptive neurons.

$Na_v1.9$ is also expressed selectively in nociceptive neurons (Dib-Hajj *et al.*, 1998a; Fang *et al.*, 2003) and underlies a persistent sodium current with substantial overlap between activation and steady-state inactivation (Cummins *et al.*, 1999) that has a probable role in setting thresholds of activation (Dib-Hajj *et al.*, 2002; Baker *et al.*, 2003), suggesting that blockade of $Na_v1.9$ might be useful for the treatment of pain. Conversely, it has been suggested that $Na_v1.9$ activators might alleviate pain

because $Na_v1.9$ is down-regulated after axotomy (Dib-Hajj *et al.*, 1998a; Cummins *et al.*, 2000) and the resultant loss of the $Na_v1.9$ persistent current and its depolarizing influence on resting potential (Cummins *et al.*, 1999) might remove resting inactivation from other sodium channels (Cummins and Waxman, 1997). In the absence of $Na_v1.9$ null mutants or selective blockers, there is incomplete information on the role of this channel in neuropathic pain, although it is known that normal level of expression seems to be dependent on the supply of NGF or GDNF (Cummins *et al.*, 2000).

Present evidence thus makes sodium channels highly attractive analgesic drug targets, but specific antagonists for $Na_v1.3$, 1.8, and 1.9 have yet to be tested in the clinic.

Calcium Channels

The evidence of an important role for voltage-gated calcium channels in the pathogenesis of neuropathic pain is strong. A variety of drugs targeted at calcium channel subtypes are effective analgesics, and mouse null mutants of N-type $Ca_v2.2$ calcium channels show dramatic diminution in neuropathic pain behavior in response to both mechanical and thermal stimuli. In addition, two highly effective analgesic drugs used in neuropathic pain conditions selectively target calcium channel subtypes. The conotoxin ziconotide blocks $Ca_v2.2$ alpha subunits, and the widely prescribed drug gabapentin binds with high affinity to $\alpha2\delta$ subunits of calcium channels (Figure 20.3).

Voltage-gated calcium channels comprise a single alpha subunit and show structural homology with sodium channels, but the accessory subunits associated with these channels are more complex. The functional

TABLE 20.2 Voltage-gated Sodium Channel Isoforms as Pain Targets

	Channel	TTXs	DRG	Pain Target	
$Na_v1.1$	Type I	+	+	No	broadly present in CNS
$Na_v1.2$	Type II	+	+	No	broadly present in CNS
$Na_v1.3$	Type III	+	−/+	Yes	in damaged adult neurons
$Na_v1.4$	SkM1	+	−	No	muscle only
$Na_v1.5$	SkM II	−	−	No	heart and CNS
$Na_v1.6$	NaCh6	+	−	?	broad distribution
$Na_v1.7$	PN1	+	+	Yes	only peripheral neurons
$Na_v1.8$	SNS/PN3	−	−	Yes	KO mouse analgesic
$Na_v1.9$	NaN/SNS2	−	−	Yes	mainly in nociceptors
Na_x	NaG	+	−	No	salt homeostasis

Structure of voltage-gated Ca++ Channels

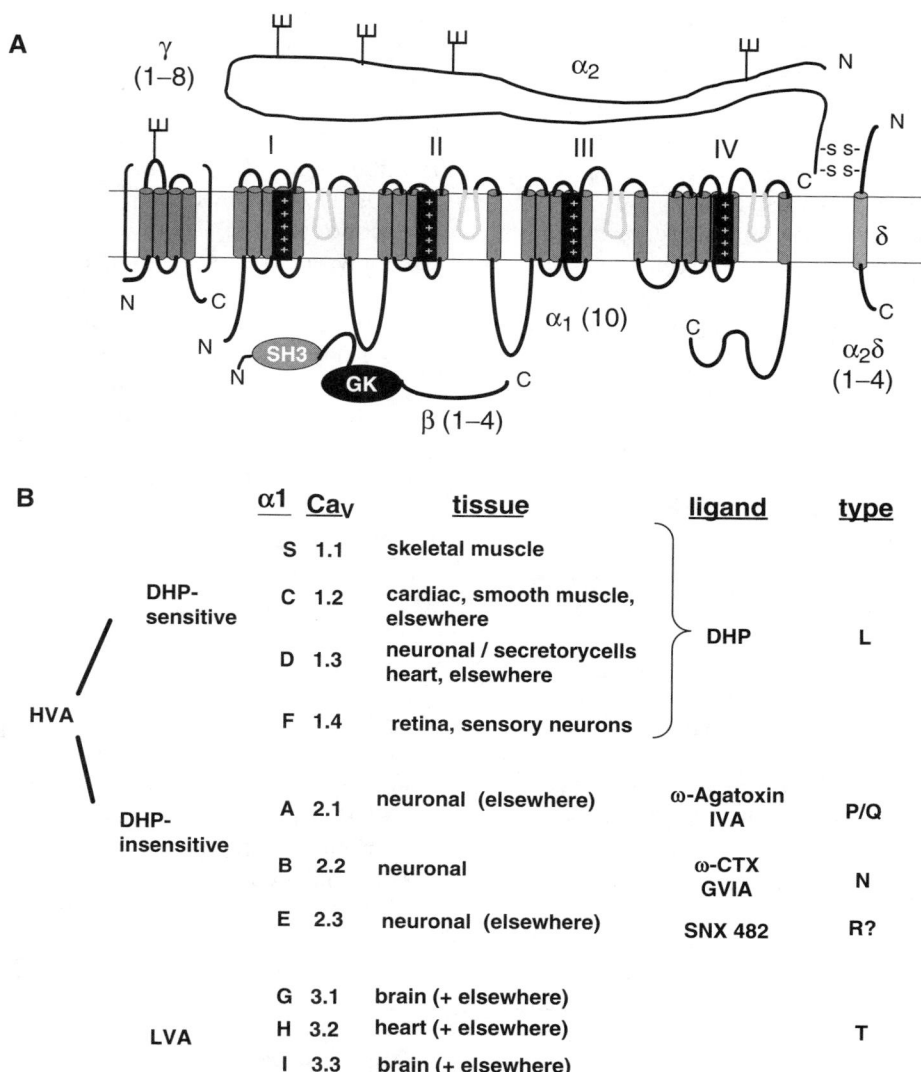

FIGURE 20.3 Voltage-gated calcium channels, like sodium channels, comprise a single functional α-subunit, but the accessory subunits (α, β, γ, δ, α2δ) associated with these channels are more complex. Channels are divided into HVA (high voltage activated) or LVA (low voltage activated) and DHP (dihydropyridine) sensitive or insensitive. The α-subunits are now classified using a numerical designation.

calcium channel complexes contain four proteins: α1 (170 kDa), α2 (150 kDa), β (52 kDa), δ (17–25 kDa), and γ (32 kDa). Four α2δ subunit genes have now been cloned. Message and protein for the α2δ-1 subunit is highly expressed in sensory neurons but is also found fairly ubiquitously in other tissues (Gong *et al.,* 2001). The gene structure is similar for all α2δ subunits, with 39 exons and a number of splice variants. All α2δ subunits have a predicted N terminal signal sequence, indicating that the N terminus is extracellular, with an intracellular C terminus and potential transmembrane region. There

are up to 14 conserved cysteines throughout the α2δ-1, 2, and 3 sequences, six of which are within δ, providing additional evidence that α2 and δ are disulfide-bonded. Following the identification of α2δ subunits as components of skeletal muscle calcium channels, they have also been shown to be associated with neuronal N and P/Q type channels. The α2δ-1 subunit has been shown to bind to extracellular regions including Domain III on the $Ca_v1.2$ subunit (Felix *et al.,* 1997).

Gabapentin binds to high affinity sites in the brain, and the target binding site has been identified as the α2δ-1 subunit. Transient transfection of cells with α2δ-1 increased the number of gabapentin binding sites (Gee et al., 1996). Subsequently gabapentin has been found to bind to two isoforms of α2δ subunits (the α2δ-1 and α2δ-2 isoforms, but not α2δ-3 or α2δ-4) (Gee et al., 1996; Gong et al., 2001). This binding may involve the Cache and other domains (Anantharaman and Aravind, 2000), but it is unknown how gabapentin exerts its action. The effects of gabapentin on native calcium currents are controversial, with some but not all authors reporting small inhibitions of calcium currents in different cell types. Gabapentin could interfere with α2-δ binding to the α1 subunit, thus destabilizing the heteromeric complex. However, the skeletal muscle calcium channel complex was not affected by gabapentin. Interestingly, α2δ-1 up-regulation in neuropathic pain correlates well with gabapentin sensitivity (Luo et al., 2002), suggesting that the α2δ-1 isoform is the most likely site of action of gabapentin. The up-regulation of α2-δ subunits does not occur in all animal models of neuropathic pain that result in allodynia. Luo et al. (2002) compared DRG and spinal cord α2δ-1 subunit levels and gabapentin sensitivity in allodynic rats with mechanical nerve injuries (sciatic nerve chronic constriction injury, spinal nerve transection, or ligation), a metabolic disorder (diabetes), or chemical neuropathy (vincristine neurotoxicity). Allodynia occurred in all types of nerve injury investigated, but DRG and/or spinal cord α2δ-1 subunit up-regulation and gabapentin sensitivity coexisted only in mechanical and diabetic neuropathies. This may partially explain why gabapentin is ineffective in some patients with neuropathic pain.

Further support for calcium channels as useful drug targets in neuropathic pain comes from an analysis of the characteristics of the $Ca_v2.2$ null mutant mouse generated by Saegusa et al. (2001). The same authors (Saegusa et al., 2002) have compared the $Ca_v2.2$ and 2.3 null mutants in a variety of pain models. Despite the widespread expression of $Ca_v2.2$, it has proved possible to demonstrate major deficits in inflammatory and in particular neuropathic pain in this transgenic mouse using the Seltzer model. Thermal and mechanical thresholds were dramatically stabilized in this mutant mouse. The relationship between α2δ subunits and $Ca_v2.2$ has not been investigated in detail, so it is possible that gabapentin has sites of action on calcium channels other than $Ca_v2.2$.

A role for $Ca_v2.2$ in chronic pain is consistent with a known analgesic role for N-type calcium channel blockers. Ziconotide, a toxin derived from marine snails, blocks $Ca_v2.2$ channels with high affinity and has been found to have analgesic actions in animal models and humans. In a study of the antinociceptive properties of ziconotide, morphine, and clonidine in a rat model of postoperative pain, heat hyperalgesia and mechanical allodynia were induced in the hind paw (Wang et al., 2000). Intrathecal ziconotide blocked established heat hyperalgesia in a dose-dependent manner and caused a reversible blockade of established mechanical allodynia. Intrathecal ziconotide was found to be more potent, longer acting, and more specific in its actions than intrathecal morphine in this model of postsurgical pain.

Brose (1997) provided dramatic evidence of the utility of ziconotide (SNX-111) in the clinic in a case of a patient with 23 years of intractable severe deafferentation pain. Intrathecal ziconotide provided complete pain relief with elimination of hyperesthesia and allodynia in a dose-dependent manner, although side effects were experienced. This type of study has led to clinical use of ziconotide in the treatment of intractable pain in patients with late stage cancer. Recent evidence that omega cono-toxins also target P2X3 receptors (Lalo et al., 2001) suggests that ziconotide may act on a broader range of targets than first suspected.

Potassium Channels

Neuronal excitability is enhanced by lowered levels of functional expression of voltage-gated potassium channels. Evidence has been obtained that potassium channel transcripts are differentially regulated at the transcriptional level in animal models of neuropathic pain. Using RT-PCR, Ishikawa et al. (2002) found that, in a chronic constriction injury model of neuropathic pain, $K_v1.2$, 1.4, 2.2, 4.2, and 4.3 mRNA levels in the ipsilateral DRG were reduced to 63% to 73% of the contralateral sides of the same animal at 3 days and to 34% to 63% at 7 days following CCI. In addition, $K_v1.1$ mRNA levels declined to approximately 72% of the contralateral level at 7 days. No significant changes in $K_v1.5$, 1.6, 2.1, 3.1, 3.2, 3.5, and 4.1 mRNA levels were detectable in the ipsilateral DRG at either time. Interestingly, of the K_v channels present in DRG, $K_v1.4$ seems to be the only channel expressed in small diameter sensory neurons, and the expression levels of this channel are much reduced in a Chung model of neuropathic pain (Rasband et al., 1999). Passmore et al. (2003) provided evidence that KCNQ potassium currents (responsible for the M-current) may also play a role in setting pain thresholds. Retigabine potentiates M-currents, and leads to a diminution of nociceptive input into the dorsal horn of the spinal cord in both neuropathic and inflammatory pain models in the rat.

Nicotinic and GABA Receptors

Rashid and Ueda (2002) found that intrathecal administration of (−)nicotine and (+)epibatidine, at doses without undesirable effects, had no antinociceptive action in sham-operated mice but completely reversed thermal and mechanical hyperalgesia in mice with partial sciatic nerve injury. The GABA(A) agonist muscimol (administered intrathecally) also produced neuropathy-specific analgesic action, giving analgesia only in nerve-injured mice. The GABA(A) antagonists bicuculline and picrotoxin blocked the analgesic effect of muscimol as well as that of nicotine and epibatidine. Thus, neuropathy-specific analgesic actions of nicotinic agonists may be due to stimulation of the GABAergic system whose inhibitory tone had been reduced because of injury.

Pacemaker Channels

Cyclic nucleotide–regulated hyperpolarization-activated cation channels (HCN channels) play an important role in cardiac function and are expressed within sensory neurons. Chaplan and collaborators (2003) have described a novel role for HCN channels in touch-related pain and spontaneous neuronal discharge originating in the damaged dorsal root ganglia. Nerve injury markedly increased pacemaker currents in large-diameter dorsal root ganglion neurons and resulted in pacemaker-driven spontaneous action potentials in the ligated nerve. Pharmacological blockade of HCN activity using the specific inhibitor ZD7288 reversed hypersensitivity to light touch and decreased the firing frequency of ectopic discharges originating in A-β and A-δ fibers by 90% and 40%. Targeting these channels appears problematic, however, given their other roles in the periphery.

NMDA and Glutamate Receptor Systems

Glutamate is the principal excitatory signaling molecule within the nervous system, and nociceptive input into the spinal cord is down-regulated by molecules such as ketamine that block NMDA receptors. The main problem with targeting the glutamate signaling system is the presence of side effects, because of the global expression of glutamate receptors throughout the nervous system. However, progress has been made in minimizing side effects by targeting NMDA receptor subtypes to induce analgesia. Conditional *knockout* of NR1 within dorsal horn reduces injury-induced pain (South *et al.,* 2003). Parsons (2001) focused on the functional inhibition of NMDA receptors that can be achieved through actions on the glycine binding site, the NR2B polyamine site, or the phencyclidine site located inside the NMDA channel. Moderate-affinity channel blockers such as glycine(B) and NR2B selective antagonists show a much better side effect profile in animal models than high-affinity channel blockers or competitive NMDA receptor antagonists. Boyce *et al.* (1999) found that the NR2B subunit had a restricted distribution in laminas I and II of the dorsal horn, suggesting a presynaptic location on primary afferent fibers. The selective NR2B antagonist CP-101,606 caused no motor impairment or stimulation in rats at doses far in excess of doses inhibiting allodynia in neuropathic rats. NR2B antagonists may therefore have clinical utility for the treatment of neuropathic pain conditions in humans, although the best efforts of the pharmaceutical community have yet to achieve this objective (Figure 20.4).

Glutamate-gated AMPA receptors

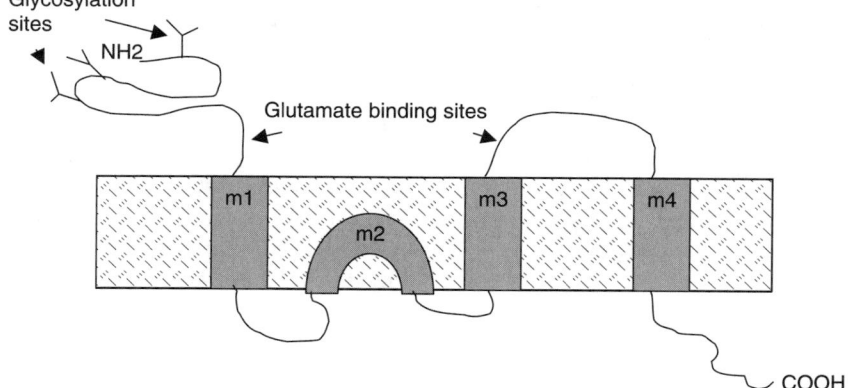

FIGURE 20.4 Glutamate receptors are involved in fast synaptic transmission throughout the central nervous system. Prolonged depolarization and glutamate release can lead to activation of NMDA receptors, a specialized coincidence detector for glutamate implicated in alterations in synaptic plasticity. Thus prolonged noxious input into the dorsal horn can lead to central changes in synaptic efficacy that have been implicated in chronic pain.

Malmberg *et al.* (2003) also exploited the channel blocking properties of conotoxins to compare the analgesic effects of conotoxins targeted against the NR2B (Conantokin B) or NR2A and NR2B (Conantokin G) subtypes of the NMDA receptor with the effects of ziconotide in a rat sciatic nerve ligation model. Con-G reversed the allodynia produced by nerve injury, with greater potency on thermal sensitivity than on mechanical allodynia. The therapeutic window between analgesic effects and motor impairment was greater for NMDA receptor blockers than ziconotide in these studies.

Other components of glutamate signaling including metabotropic receptors and transporters have been implicated in the development of neuropathic pain. Simmons *et al.* (2002) found that agonists directed at group II mGluR receptors were effective in attenuating neuropathic pain. In the L5/L6 spinal nerve ligation model of neuropathic pain in rats, LY379268 significantly reversed mechanical allodynia behavior in a dose-related manner but had no significant effects on acute pain tests such as tail flick or paw withdrawal. These results suggest a useful action of selective group II mGluR agonists for the treatment of persistent pain. By contrast, antisense studies suggest that mGluR1 antagonists may be useful. Fundytus (2001) used antisense oligonucleotides to knock down mGluR1 receptors and found reduced cold hyperalgesia, heat hyperalgesia, and mechanical allodynia in the injured hind paw of neuropathic rats.

Glutamate transporters play an important role in glutamate homeostasis and the suppression of excitotoxicity. A differential expression of glutamate transporters in the spinal cord has been described (Sung *et al.*, 2003) in a rat CCI model. Here an initial functional transporter up-regulation up to day 5 within the ipsilateral spinal cord dorsal horn was followed by a down-regulation at days 7 and 14 following nerve injury. The mitogen-activated protein kinase inhibitor PD98059 almost abolished transporter up-regulation with concomitant exacerbated thermal hyperalgesia and mechanical allodynia, although some of these effects are likely to be due to other actions of the map kinase inhibitor.

Taken together, plasticity in the glutamate signaling system seems to be central to the development of neuropathic pain, but the problems of targeting chronic pain without side effects have yet to be fully resolved.

Purinergic Receptors

The ATP-gated cation channel P2X3 is relatively specifically expressed by nociceptive neurons, and has been assessed as an analgesic target for antisense studies, the generation of null mutant mice, and the development of specific pharmacological antagonists (North 2003).

There appears to be a strong case that this receptor plays a role in the pathogenesis of neuropathic pain. Barclay *et al.* (2002) used antisense oligonucleotides administered intrathecally to functionally down-regulate P2X3 receptors. After 7 days of treatment, P2X3 protein levels were reduced in the primary afferent terminals in the dorsal horn. After partial sciatic ligation, both inhibition of the development of mechanical hyperalgesia and significant reversal of established hyperalgesia were observed within 2 days of antisense treatment. The time course of the reversal of hyperalgesia was consistent with down-regulation of P2X3 receptor protein and function. Despite these observations there is no evidence that P2X3 receptors are up-regulated in neuropathic pain. There does in fact seem to be down-regulation of P2X3 following L5/L6 spinal nerve ligation in rats (Kage *et al.*, 2002). A significantly reduced number of small diameter neurons exhibited a response to α, β–methyleneATP (a P2X3 selective agonist), but large diameter neurons and some small neurons retained their expression of functional P2X3 receptors (Figure 20.5).

TNP-ATP is a potent antagonist of P2X3 receptors but is metabolically unstable and also acts on P2x1-4 subtypes. Nevertheless, TNP-ATP is capable of completely reversing tactile allodynia, albeit in a transient fashion over an hour (Tsuda *et al.*, 1999). More recently a potent stable antagonist of P2X3 and P2X2/3 heteromultimers has been developed. This compound, A317491 (Jarvis *et al.*, 2002), reverses mechanical allodynia and thermal sensitivity in a rat neuropathic pain model. Interestingly, ziconotide also blocks P2X3 receptors, so some of its analgesic actions may also be mediated through this receptor (Lalo *et al.*, 2001). P2Y receptors may also play a regulatory role in neuropathic pain. Ogada (2003) showed that intrathecal administration of P2Y receptor agonists UTP and UDP produced significant antiallodynic effects in a rat sciatic nerve ligation model.

Vanilloid Receptors

The polymodal nature of responses to noxious stimuli demonstrated by the capsaicin receptor TrpV1 and the usefulness of capsaicin in treating some chronic pain syndromes suggest that members of this receptor class may play a role in chronic pain. There is indeed evidence that TrpV1 is up-regulated in the DRG neurons adjacent to those that have been damaged by spinal nerve ligation (Fukuoka *et al.*, 2002) while being down-regulated in damaged neurons. The up-regulated TrpV1 expression seems to be associated with A-fiber sensory neurons (Hudson *et al.*, 2001). These new receptors have been suggested to be important for the analgesic effects of capsaicin cream (Rashid *et al.*, 2003). There are also claims that TrpV1 antagonists may be useful in treating neuropathic

P2X receptors – cation channels gated by ATP

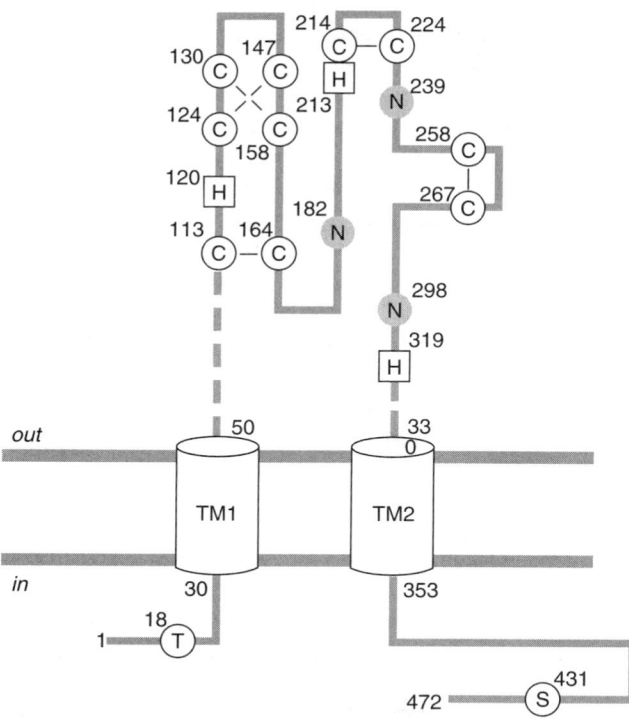

FIGURE 20.5 P2X receptors comprise two-pass transmembrane monomers that can heteromultimerise to form cation-selective ion channels. Glycosylation, phosphorylation, and possible disulfide bonding of P2X$_2$ receptors are shown. Circles (N) indicate sites that are glycosylated in the native P2X$_2$ receptor. Open circles (T, S) indicate the positions of Thr-18 phosphorylated by protein kinase C and Ser-431 phosphorylated by protein kinase A. Open circles (C) indicate the 10 conserved cysteines found in all P2X receptors. Squares (H) indicate histidine residues involved in zinc and proton binding. Redrawn from North (2003). The P2X3 subunit: a molecular target in pain therapeutics. *Curr Opin Invest Drugs* 4(7):833-840.

pain. Capsazepine, a non-competitive antagonist of TrpV1, markedly attenuated mechanical hyperalgesia in a guinea pig sciatic nerve injury model (Walker *et al.*, 2003), but this effect was absent in rodents. Thus, a case for an important role of Trp channels in neuropathic pain pathogenesis is missing. However, evidence of a role for TrpV4 in noxious mechano-sensation (Suzuki *et al.*, 2003) may stimulate further investigation of this class of receptors as pain targets.

Cannabinoids

Cannabinoid-1 (CB1) receptors both on sensory neurons and within the CNS are known to be useful targets for agonists with analgesic activity in neuropathic pain (Fox *et al.*, 1999). In the partial sciatic ligation model of neuropathic pain, CB-selective agonists WIN55,212-2,

CP-55,940, and HU-210 produced complete reversal of mechanical hyperalgesia within 3 hours of subcutaneous administration. Zhang *et al.* (2003) showed that chronic pain models associated with peripheral nerve injury, but not peripheral inflammation, induce CB2 receptor expression in a highly restricted and specific manner within the lumbar spinal cord. The appearance of CB2 expression coincided with the appearance of activated microglia, and it is likely that CB2 agonists may be acting through these immune system cells.

Opioids and Other Neuropeptides

Conventional opioid drugs have been shown unequivocally to be useful in treating certain neuropathic pain conditions such as diabetic neuropathy (Rowbotham *et al.*, 2003). More controversial is the role of the nociceptin/orphanin FQ system in regulating neuropathic pain. Initial reports suggested that nociceptin had analgesic effects in neuropathic pain models (Halo *et al.*, 1998). In contrast, Mabuchi *et al.* (2001) used a nociceptin/orphanin FQ antagonist, JTC-801, to demonstrate attenuation of thermal hyperalgesia in neuropathic pain models. Levels of endomorphins, novel opioid agonists that act on mu receptors, increase in the periaqueductal gray matter of rats with sciatic nerve ligation (Sun *et al.*, 2001). The role of these novel opioid ligands in neuropathic pain is thus still unclear.

Holmes *et al.* (2003) provided evidence that overexpression of galanin attenuated mechanical allodynia compared with controls after nerve damage using a spared nerve-injury model. These results support an inhibitory role for galanin in the modulation of nociception both in intact animals and in neuropathic pain states. By contrast, NK1 antagonists rather than agonists have some analgesic actions, suggesting that substance P does play a role in neuropathic pain induction.

Neurotrophins and Cytokines

The breakdown of axonal transport of neurotrophic factors has been associated with some of the alterations in DRG gene expression that are associated with neuropathic pain. For example, the sodium channels Na$_v$1.8 and Na$_v$1.9 require basal NGF levels for normal patterns of expression (see, for example, Cummins *et al.*, 2000). Counterintuitively, NGF has algogenic activity in neuropathic pain. Overexpression of NGF in glia, driven by a GFAP-promoter, caused enhanced ipsilateral responses to thermal and mechanical stimulation following CCI compared to wild-type mice. Following CCI, sympathetic neurons sprouted into ipsilateral and to a lesser extent contralateral DRG with the sprouting much greater in mice than in controls (Ramer *et al.*, 1999). Consistent with this overexpression model, Zhou *et al.* (2000) found that injecting neutralizing anti-NGF antibodies into injured DRG could lead to significant analgesic effects in terms of mechanical thresholds measured with von Frey hairs. Svensson *et al.* (2003) extended these observations to the clinic, injecting the human masseter muscle with NGF and evoking a long-lasting mechanical allodynia and hyperalgesia in the jaw-closing muscles as measured by pressure pain thresholds and pressure tolerance thresholds.

Some of the actions of NGF seem to be affected by downstream release of BDNF. At the molecular level, acting through its high affinity receptor TrkB, BDNF seems to be able to regulate the activity of some ion channels, thus modulating synaptic efficacy. A role for BDNF in regulating pain pathways has long been proposed (Thompson *et al.*, 1999), and pharmacological evidence that BDNF can act to regulate NMDA receptor activity important in spinal cord pain pathways has been presented. The evidence of a role for BDNF as a neuromodulator in the spinal cord includes its presence in vesicles in nociceptive fiber terminals, increases in dorsal horn levels of BDNF in inflammatory pain, analgesic effects of Trk-B receptor bodies that are considered to act as scavengers of BDNF released from primary sensory neurons, and excitatory effects on dorsal horn neurons. BDNF is up-regulated in uninjured DRG neurons in sciatic nerve ligation models (Fukuoka *et al.*, 2001), but the role of this neurotrophin in neuropathic pathogenesis is uncertain.

By contrast, the withdrawal of trophic support by GDNF seems to play an important role in changing pain thresholds. Takahashi *et al.* (2000) found that behavioral changes in rat neuropathic pain models were accompanied by decreased expression of GDNF in the DRG and the sciatic nerve on the injured side on the fourteenth day after the surgery. Boucher *et al.* (2000) normalized neuropathic pain behavior in CCI rats by the intrathecal administration of GDNF. Some of these effects may be due to reversing changes in sodium channel expression. Trophic factors of the GDNF family acting through the c-ret family of receptor kinases are thus potentially interesting therapeutic agents. Interleukin-6 is another intriguing target. Murphy *et al.* (1999) showed that in constriction nerve injury, interleukin-6 (IL-6) mRNA was induced in a subset of rat primary sensory neurons. Mice with null mutations of the IL-6 gene lacked hypersensitivity to cutaneous heat and pressure after CCI injury.

Tumor Necrosis Factor

TNF is known to alter mechanical but not thermal pain thresholds when applied exogenously to the nerve trunk (Sorkin and Doom, 2000). Transgenic animals overexpressing TNF under the control of a GFAP promoter (that would express TNF in Schwann cells and astrocytes) also show greater mechanical allodynia in an L5 transection nerve injury model than control mice (de Leo *et al.*, 2000). In neuropathic pain models such as the sciatic nerve ligation model, the expression of TNF receptors type 1 and 2 is up-regulated in both damaged and uninjured DRG neurons as measured immunocytochemically (Schafers *et al.*, 2003). In addition to increased receptor expression, immunoreactive TNF itself is induced. Okamoto (2001) found that gene expression in injured rat sciatic nerve was significantly increased at day 7 for IL-1beta and IL-6 and at day 14 for TNF. A role for TNF is strongly supported by the therapeutic actions of compounds that block TNF action. Thalidomide blocks the production of TNF by activated macrophages. Sommer *et al.* (1998) found that in rats pretreated with thalidomide, CCI resulted in diminished mechanical allodynia and thermal hyperalgesia, although treatment with thalidomide at a time point when hyperalgesia was already present did not alter the course of the pain-related behavior. Similarly, a metalloprotease inhibitor, TAPI, that blocks maturation of TNF (Sommer *et al.*, 1997) leads to a reduction of thermal hyperalgesia and mechanical allodynia of up to 50%. Thus, TNF seems to be involved in the early development of neuropathic pain, but postinjury block of TNF action does not appear therapeutically useful.

PHYSICAL CHANGES IN NEUROPATHIC PAIN STATES

Peripheral nerve damage is associated with the invasion of damaged tissues and spinal cord by immune system cells, potential rewiring of central synapses, and altered behavior in supporting glia. Hanani *et al.* (2002) found that after axotomy, Schwann cells became extensively coupled to other Schwann cells that enveloped other neurons, apparently by gap junctions. Such connections were absent in control ganglia. The number of gap junctions increased 6.5-fold after axotomy.

There is renewed debate about the possibility of central sprouting after nerve injury within the dorsal horn. Early studies using cholera toxin B as a marker are now suspect because of phenotypic changes in central terminals that occur after nerve injury. Bao *et al.* (2002) showed that after nerve injury, a small number of A-fibers do indeed sprout into inner lamina II, a region normally innervated by C-fibers, but not into outer lamina II or lamina I. Neuropeptide Y (NPY) was found in these sprouts in inner lamina II, an area very rich in NPY1 receptor-positive processes. Thus, large NPY immunoreactive DRG neurons sprout into the NPY1 receptor-rich inner lamina II after peripheral nerve damage.

We have already seen that CB2 receptor-bearing microglia invade the dorsal horn adjacent to injured peripheral neurons (Zhang *et al.*, 2003). Abbadie *et al.* (2003) found that CCR2 chemokine receptor null mutant mice had dramatic deficits in neuropathic pain. In a model of neuropathic pain, the development of mechanical allodynia was totally abrogated in CCR2 knockout mice. In response to nerve ligation, persistent and marked up-regulation of CCR2 mRNA was normally evident in the nerve and dorsal root ganglion. Chronic pain also resulted in the appearance of activated CCR2-positive microglia in the spinal cord. Thus macrophages and microglia bearing the chemokine receptor CCR2 (and P2X4 receptors) appear to play an important role in neuropathic pain pathogenesis and represent an exciting new prospect for drug development.

TARGET VALIDATION METHODS IN THE STUDY OF NEUROPATHIC PAIN

Dissecting the role of potential targets in regulating pain thresholds has often relied upon the use of supposedly selective pharmacological agents (e.g., MK801—an antagonist for NMDA receptors), but for many signaling systems there are no specific antagonists to analyze their physiological function. In such situations, three genetic approaches can prove informative: the use of antisense oligonucleotides, the specific down-regulation of mRNA

using small interfering RNA (siRNA), and the generation of mice with targeted mutations. Each of these approaches has advantages and disadvantages. Antisense technology is cheap, but specificity is a problem because high concentrations of oligonucleotide may have some cellular toxicity, and may also target structurally related transcripts. SiRNA technology is still being developed, but has already revolutionized *C.elegans* genetics, where the specificity of siRNA action and the catalytic nature of RNA degradation mean that very low concentrations of dsRNA can be used. SiRNA is effective *in vitro* and *in vivo* in primary sensory neurons, where a 21bp complementary double-stranded RNA can be used to specifically degrade cognate RNA sequences through the formation of a complex with ribonucleases (Figure 20.6).

siRNA acts transiently and catalytically, and may not lead to long-lived RNA degradation (part of its attraction), but for animal models where neuropathic pain is modelled over a period of weeks, this is a major difficulty. Null mutants do not share this problem, but the problems of developmental compensatory mechanisms and death during development have often provided obstacles to interpretation of phenotype. It is also desirable to generate mice where tissue-specific deletions can be carried out, and ideally postnatal activation of Cre recombinase should be possible. Thanks to the work of Sauer and collaborators (Le and Sauer, 1999), who have exploited the recombinase activity of a bacteriophage enzyme Cre, to delete DNA sequences that are flanked by lox-P sites recognized by this enzyme, it was possible to

RNA interference as a route to blocking gene expression

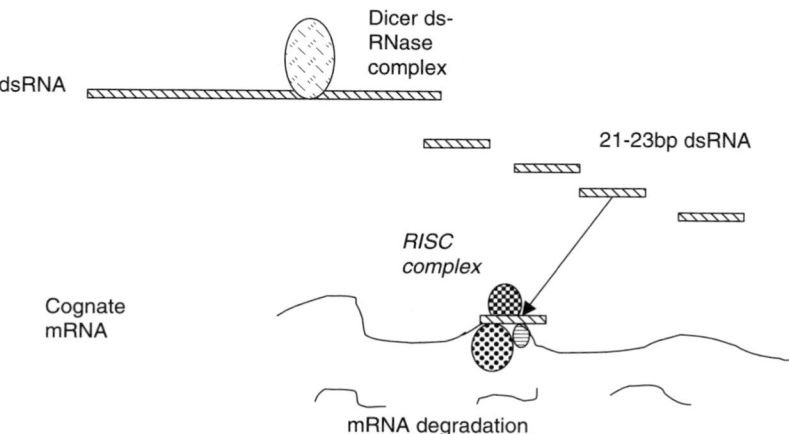

FIGURE 20.6 In the first step, input double-stranded (ds) RNA is processed by the enzyme DICER into 21–23-nucleotide "guide sequences." In mammals, these dsRNAs may also be introduced into the cell after chemical synthesis by standard transfection techniques. The dsRNAs are incorporated into a nuclease complex, called the RNA-induced silencing complex (RISC), which acts in the second effector step to destroy mRNAs that are recognized by the guide RNAs through base-pairing interactions.

generate tissue-specific null mutants. An analogous system exploits the Flp recombinase that recognizes frt sites.

To ablate genes in sensory ganglia, it is necessary to produce mice in which functional Cre recombinase is driven by sensory neuron-specific promoters. To analyze the effectiveness of expressed Cre in excising lox-P-flanked genes, a reporter mouse using the β-galactosidase expressing gene with a floxed (lox-P flanked) stop signal can be used. Where Cre removes the stop signa, β-galactosidase activity can be analyzed histochemically. It will also be useful to generate transgenic mice expressing Cre in subsets of sensory neurons and drug-activatable Cre isoforms. Recently a tamoxifen-activatable form of Cre recombinase has been developed. This form of Cre recombinase comprises a fusion protein between Cre and a human mutated estrogen receptor. The addition of tamoxifen but not endogenous steroids allows the Cre recombinase to assume an active conformation. This method allows the excision of genes at defined periods in adulthood.

This powerful technology is likely to be applied increasingly over the next few years, and together with siRNA promises to speed up target validation strategies in animal models of neuropathic pain. DRG-specific CRE-recombinase mice have been made by a number of mice, and an increasing number of floxed target genes (e.g., BDNF) are also now available for this type of analysis (Figure 20.7).

TRANSLATING TARGETS INTO TREATMENTS

Because a complex interplay between many types of molecules within damaged neurons, glia, and cells of the immune system underlies the establishment of chronic neuropathic pain, there is now a wealth of potential molecular targets which may be useful in the treatment of pain. Huge strides in understanding the molecular basis for pain have recently been made. A key to the development of useful analgesic drugs is to avoid unnecessary side effects. Importantly, as yet there is not a single target that has been shown to be uniquely associated with the establishment of neuropathic pain. Thus we may be faced with the challenge of developing magic shrapnel rather than a magic bullet. Nonetheless, the concept of a magic bullet for the treatment for pain remains an important one, since some molecules may play predominant (although not solitary) roles in the production of pain, and as noted previously, some molecules such as specific types of sodium channels are selectively expressed in nociceptive neurons. As yet, many of the most attractive targets (e.g., sodium channels) do not have specific pharmacological blockers or activators. However, the intense focus of academia and of the pharmaceutical industry on the problem of neuropathic pain suggests that progress will be made in short order, with important consequences for clinical treatment.

ACKNOWLEDGMENTS

We regret the omission of many important papers for reasons of space. We thank our colleagues for helpful insights and suggestions. Research in the authors' laboratories was supported in part by the MRC and in part by the Wellcome Trust (JNW) and by the Dept. of Veterans Affairs, the National Multiple Sclerosis Society, the Paralyzed Veterans of America, and the United Spinal Association (SGW).

Tissue-specific gene deletion

Cre LoxP system *in vivo*

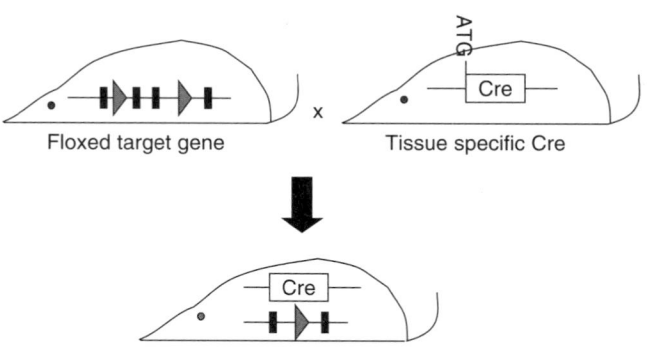

FIGURE 20.7 Mice expressing Cre-recombinase are crossed with reporter mice where β-galactosidase activity is expressed after Cre action (see Le and Sauer, 2000). This allows the distribution of active Cre to be assessed prior to tissue-specific deletion of other floxed target genes.

References

Abdi S, Lee DH, Chung JM. (1998). The anti-allodynic effects of amitriptyline, gabapentin, and lidocaine in a rat model of neuropathic pain. *Anesth Analg* 87(6):1360-1366.

Abbadie C *et al.* (2003). Impaired neuropathic pain responses in mice lacking the chemokine receptor CCR2. *Proc Natl Acad Sci USA* 100(13):7947-7952.

Abe M *et al.* (2002). Changes in expression of voltage-dependent ion channel subunits in dorsal root ganglia of rats with radicular injury and pain. *Spine* 27(14):1517-1524.

Akopian AN *et al.* (1999). The TTX-R sodium channel SNS has a specialized function in pain pathways, *Nature Neurosci* 2:5481-5489.

Akopian AN, Sivilotti L, Wood JN. (1996). A tetrodotoxin-resistant voltage-gated sodium channel expressed by sensory neurons. *Nature* 379:257-262.

Anantharaman V, Aravind L. (2001). The CHASE domain: a predicted ligand-binding module in plant cytokinin receptors and other eukaryotic and bacterial receptors. *Trends Biochem Sci* (10):579-582.

Andersen S, Worm-Pedersen J. (1987). The prevalence of persistent pain in a Danish population. *Pain* S4:S332.

Baker MD *et al.* (2003). GTP-induced tetrodotoxin-resistant Na+ current regulates excitability in mouse and rat small diameter sensory neurones. *J Physiol* 548:373-382.

Bao L *et al.* (2002). Peripheral axotomy induces only very limited sprouting of coarse myelinated afferents into inner lamina II of rat spinal cord. *Eur J Neurosci* 16(2):175-185.

Barclay J *et al.* (2002). Functional downregulation of P2X3 receptor subunit in rat sensory neurons reveals a significant role in chronic neuropathic and inflammatory pain. *J Neurosci* 22(18):8139-8147.

Bennett DL, French J, Priestley JV, McMahon SB. (1996). NGF but not NT-3 or BDNF prevents the A fiber sprouting into lamina II of the spinal cord that occurs following axotomy. *Mol Cell Neurosci* 8(4):211-220.

Bennett GJ, Xie YK. (1988). A peripheral mononeuropathy in rat that produces disorders of pain sensation like those seen in man. *Pain* 33:87-107.

Black JA *et al.* (2003). Tetrodotoxin-resistant sodium channels Na(v)1.8/SNS and Na(v)1.9/NaN in afferent neurons innervating urinary bladder in control and spinal cord injured rats. *Brain Res* 963:132-138.

Boucher TJ *et al.* (2000). Potent analgesic effects of GDNF in neuropathic pain states. *Science* 290(5489):124-127.

Bowsher D, Rigge M, Sopp L. (1991). Prevalence of chronic pain in the British population: A telephone survey of 1037 households. *Pain Clin* 4:223-230.

Boyce S *et al.* (1999). Selective NMDA NR2B antagonists induce antinociception without motor dysfunction: correlation with restricted localisation of NR2B subunit in dorsal horn. *Neuropharmacology* 38(5):611-623.

Brose WG *et al.* (1997). Use of intrathecal SNX-111, a novel, N-type, voltage-sensitive, calcium channel blocker, in the management of intractable brachial plexus avulsion pain. *Clin J Pain* 13(3):256-259.

Chaplan SR *et al.* (2003). Neuronal hyperpolarization-activated pacemaker channels drive neuropathic pain. *J Neurosci* 23(4):1169-1178.

Costigan M *et al.* (2002). Replicate high-density rat genome oligonucleotide microarrays reveal hundreds of regulated genes in the dorsal root ganglion after peripheral nerve injury. *BMC Neurosci* 3(1):16.

Craner MJ, Klein JP, Black JA, Waxman SG. (2002). Preferential expression of IGF-I in small DRG neurons and down-regulation following injury. *Neuroreport* 13(13):1649-1652.

Craner MJ *et al.* (2002). Changes of sodium channel expression in experimental painful diabetic neuropathy. *Ann Neurol* 52(6):786-792.

Cummins TR, Waxman SG. (1997). Down-regulation of tetrodotoxin-resistant sodium currents and up-regulation of a rapidly repriming tetrodotoxin-sensitive sodium current in small spinal sensory neurons following nerve injury. *J Neurosci* 17:3503-3514.

Cummins TR *et al.* (1999). A novel persistent tetrodotoxin-resistant sodium current in SNS-null and wild-type small primary sensory neurons. *J Neurosci* 19:RC 43 (1-6).

Cummins TR *et al.* (2001). Na_v1.3 sodium channels: rapid repriming and slow closed-state inactivation display quantitative differences following expression in a mammalian cell line and in spinal sensory neurons. *J Neurosci* 21:5952-5961.

Cummins TR, Black JA, Dib-Hajj SD, Waxman SG. (2000). GDNF up-regulates expression of functional SNS and NaN sodium channels and their currents in axotomized DRG neurons. *J Neurosci* 20:8754-8761.

Decosterd I, Woolf CJ. (2000). Spared nerve injury: an animal model of persistent peripheral neuropathic pain. *Pain* 87(2):149-158.

DeLeo JA, Rutkowski MD, Stalder AK, Campbell IL. (2000). Transgenic expression of TNF by astrocytes increases mechanical allodynia in a mouse neuropathy model. *Neuroreport* 11(3):599-602.

Dib-Hajj S, Black JA, Cummins TR, Waxman SG. (2002). NaN/Nav1.9: a sodium channel with unique properties. *Trends Neurosci* 25(5):253-259.

Dib-Hajj S, Black JA, Felts P, Waxman SG. (1996). Down-regulation of transcripts for Na channel SNS in spinal sensory neurons following axotomy. *Proc Natl Acad Sci* 93:14950-14954.

Dib-Hajj SD, Tyrrell L, Black JA, Waxman SG. (1998a). NaN, a novel voltage-gated Na channel preferentially expressed in peripheral sensory neurons and down-regulated following axotomy. *Proc Natl Acad Sci* 95:8963-8968.

Dib-Hajj SD *et al.* (1998b). Rescue of Nav1.8 sodium channel expression in small dorsal root ganglion neurons after axotomy by nerve growth factor in vitro. *J Neurophysiol* 79:2668-2677.

Djouhri L *et al.* (2003). The TTX-resistant sodium channel Nav1.8 (SNS/PN3): expression and correlation with membrane properties in rat nociceptive primary afferent neurons. *J Physiol* 550(Pt 3):739.

Elliott AM *et al.* (1999). The epidemiology of chronic pain in the community. *Lancet* 354:1248-1252.

Fang X *et al.* (2002). The presence and role of the TTX resistant sodium channel Na_v1.9 (NaN) in nociceptive primary afferent neurons. *J Neurosci* 22:7425-7433.

Felix R, Gurnett CA, De Waard M, Campbell KP. (1997). Dissection of functional domains of the voltage-dependent Ca2+ channel alpha2delta subunit. *J Neurosci* 17(18):6884-6891.

Fox A *et al.* (2001). The role of central and peripheral Cannabinoid1 receptors in the antihyperalgesic activity of cannabinoids in a model of neuropathic pain. *Pain* 92:91-100.

Fukuoka T *et al.* (2001). Brain-derived neurotrophic factor increases in the uninjured dorsal root ganglion neurons in selective spinal nerve ligation model. *J Neurosci* 21(13):4891-4900.

Fukuoka T *et al.* (2002). VR1, but not P2X(3), increases in the spared L4 DRG in rats with L5 spinal nerve ligation. *Pain* 99:111-120.

Fundytus ME *et al.* (2001). Knockdown of spinal metabotropic glutamate receptor 1 (mGluR(1)) alleviates pain and restores opioid efficacy after nerve injury in rats. *Br J Pharmacol* 132(1):354-367.

Gee NS *et al.* (1996). The novel anticonvulsant drug, gabapentin (Neurontin), binds to the alpha2delta subunit of a calcium channel. *J Biol Chem* 271:5768-5776.

Gong HC *et al.* (2001). Tissue-specific expression and gabapentin-binding properties of calcium channel alpha2delta subunit subtypes. *J Membr Biol* 184(1):35-43.

Gonzalez MI, Field MJ, Hughes J, Singh L. (2000). Evaluation of selective NK(1) receptor antagonist CI-1021 in animal models of inflammatory and neuropathic pain. *J Pharmacol Exp Ther* 294(2):444-450.

Hains BC *et al.* (2003). Up-regulation of sodium channel Na$_v$1.3 and functional involvement in neuronal hyperexcitability associated with central neuropathic pain after spinal cord injury. *J Neurosci* 26:8881-8892.

Hanani M *et al.* (2002). Glial cell plasticity in sensory ganglia induced by nerve damage. *Neuroscience* 114(2):279-283.

Hao JX, Xu IS, Wiesenfeld-Hallin Z, Xu XJ. (1998). Anti-hyperalgesic and anti-allodynic effects of intrathecal nociceptin/orphanin FQ in rats after spinal cord injury, peripheral nerve injury and inflammation. *Pain* 76(3):385-393.

Holmes FE *et al.* (2003). Transgenic overexpression of galanin in the dorsal root ganglia modulates pain-related behavior. *Proc Natl Acad Sci USA* 100(10):6180-6185.

Hudson LJ *et al.* (2001). VR1 protein expression increases in undamaged DRG neurons after partial nerve injury. *Eur J Neurosci* 13(11):2105-2114.

Ishikawa K, Tanaka M, Black JA, Waxman SG. (1999). Changes in expression of voltage-gated potassium channels in dorsal root ganglion neurons following axotomy. *Muscle Nerve* 22:502-507.

Jarvis MF *et al.* (2002). A-317491, a novel potent and selective non-nucleotide antagonist of P2X3 and P2X2/3 receptors, reduces chronic inflammatory and neuropathic pain in the rat. *Proc Natl Acad Sci USA* 99(26):17179-17184.

Jorum E, Warncke T, Stubhaug A. (2003). Cold allodynia and hyperalgesia in neuropathic pain: the effect of N-methyl-D-aspartate (NMDA) receptor antagonist ketamine—a double-blind, cross-over comparison with alfentanil and placebo. *Pain* 101(3):229-235.

Kage K *et al.* (2002). Alteration of dorsal root ganglion P2X3 receptor expression and function following spinal nerve ligation in the rat. *Exp Brain Res* 147(4):511-519.

Kerr BJ, Souslova V, McMahon SB, Wood JN. (2001). A role for the TTX-resistant sodium channel Nav 1.8 in NGF-induced hyperalgesia, but not neuropathic pain. *Neuroreport* 12(14):3077-3080.

Khasar SG, Gold MS, Levine JD. (1998). A tetrodotoxin-resistant sodium current mediates inflammatory pain in the rat. *Neurosci Lett* 256(1):17-20.

Kim CH, Oh Y, Chung JM, Chung K. (2001). The changes in expression of three subtypes of TTX sensitive sodium channels in sensory neurons after spinal nerve ligation. *Mol Brain Res* 95(1-2):153-161.

Kim DS, Choi JO, Rim HD, Cho HJ. (2002). Downregulation of voltage-gated potassium channel alpha gene expression in dorsal root ganglia following chronic constriction injury of the rat sciatic nerve. *Mol Brain Res* 105(1-2):146-152.

Kim KJ, Yoon YW, Chung JM. (1997). Comparison of three rodent neuropathic pain models. *Exp Brain Res* 113(2):200.

Kingery WS *et al.* (2000). The alpha(2A) adrenoceptor and the sympathetic postganglionic neuron contribute to the development of neuropathic heat hyperalgesia in mice. *Pain* 85(3):345-358.

Korff M, Dworkin SF, Le Resche L. (1990). Graded chronic pain status: an epidemiologic evaluation. *Pain* 40:279-229.

Lai J *et al.* (2002). Inhibition of neuropathic pain by decreased expression of the tetrodotoxin-resistant sodium channel, Nav1.8. *Pain* 95(1-2):143-152.

Lalo UV, Pankratov YV, Arndts D, Krishtal OA. (2001). Omega-conotoxin GVIA potently inhibits the currents mediated by P2X receptors in rat DRG neurons. *Brain Res Bull* 54(5):507-512.

Le Y, Sauer B. (2001). Conditional gene knockout using Cre recombinase. *Mol Biotechnol* 17(3):269-275.

Leffler A *et al.* (2002). GDNF and NGF reverse changes in repriming of TTX-sensitive Na(+) currents following axotomy of dorsal root ganglion neurons. *J Neurophysiol* 88(2):650-658.

Luo ZD *et al.* (2002).Injury type-specific calcium channel alpha 2 delta-1 subunit up-regulation in rat neuropathic pain models correlates with antiallodynic effects of gabapentin. *J Pharmacol Exp Ther* 303(3):1199-1205.

Mabuchi T *et al.* (2003). Attenuation of neuropathic pain by the nociceptin/orphanin FQ antagonist JTC-801 is mediated by inhibition of nitric oxide production. *Eur J Neurosci* 17(7):1384-1392.

Madiai F *et al.* Upregulation of FGF-2 in reactive spinal cord astrocytes following unilateral lumbar spinal nerve ligation. *Exp Brain Res* 148(3):366-376.

Malcangio M, Tomlinson DR. (1998). A pharmacologic analysis of mechanical hyperalgesia in streptozotocin/diabetic rats. *Pain* 76(1-2):151.

Malmberg AB, Gilbert H, McCabe RT, Basbaum AI. (2003). Powerful antinociceptive effects of the cone snail venom-derived subtype-selective NMDA receptor antagonists conantokins G and T. *Pain* 101(1-2):109-116.

Malmberg *et al.* (1997). Diminished inflammation and nociceptive pain in mice with preservation of neuropathic pain in mice with a targeted mutation of the type I regulatory subunit of cAMP-dependent protein kinase. *J Neurosci* 17(19):7462-7470.

Miki K *et al.* (2000). Differential effect of brain-derived neurotrophic factor on high-threshold mechanosensitivity in a rat neuropathic pain model. *Neurosci Lett* 278(1-2):85-88.

Mogil JS *et al.* (1999). Heritability of nociception I: responses of 11 inbred mouse strains on 12 measures of nociception. *Pain* 80(1-2): 67-82.

Murphy PG *et al.* (1999). Endogenous interleukin-6 contributes to hypersensitivity to cutaneous stimuli and changes in neuropeptides associated with chronic nerve constriction in mice. *Eur J Neurosci* 11(7):2243-2253.

Nicholson B. (2000). Gabapentin use in neuropathic pain syndromes. *Acta Neurol Scand* 101(6):359-371.

North RA. (2003). The P2X3 subunit: a molecular target in pain therapeutics. *Curr Opin Invest Drugs* 4(7):833-840.

Okada M, Nakagawa T, Minami M, Satoh M. (2002). Analgesic effects of intrathecal administration of P2Y nucleotide receptor agonists UTP and UDP in normal and neuropathic pain model rats. *J Pharmacol Exp Ther* 303(1):66-73.

Okamoto K *et al.* (2001). Pro- and anti-inflammatory cytokine gene expression in rat sciatic nerve chronic constriction injury model of neuropathic pain. *Exp Neurol* 169(2):386-391.

Okuse K *et al.* (1997). Regulation of expression of the sensory neuron-specific sodium channel SNS in inflammatory and neuropathic pain. *Mol Cell Neurosci* 10(3/4):196-207.

Okuse K *et al.* (2002). Annexin II light chain regulates sensory neuron-specific sodium channel expression. *Nature* 47:653-656.

Park SK, Chung K, Chung JM. (2000). Effects of purinergic and adrenergic antagonists in a rat model of painful peripheral neuropathy. *Pain* 87:171-179.

Parsons (2001). NMDA receptors as targets for drug action in neuropathic pain. *Eur J Pharmacol* 429(1-3):71-78.

Passmore GM *et al.* (2003). KCNQ/M currents in sensory neurons: significance for pain therapy, *J Neurosci* 23(18):7227-7236.

Rasband MN *et al.* (2001). Distinct potassium channels on pain-sensing neurons. *Proc Natl Acad Sci USA* 98(23):13373-13378.

Rashid MH *et al.* (2003). Novel expression of vanilloid receptor 1 on capsaicin-insensitive fibers accounts for the analgesic effect of capsaicin cream in neuropathic pain. *J Pharmacol Exp Ther* 304(3): 940-948.

Rashid MH, Ueda H. (2002). Neuropathy-specific analgesic action of intrathecal nicotinic agonists and its spinal GABA-mediated mechanism. *Brain Res* 953(1-2):53-62.

Renganathan M, Cummins TR, Waxman SG. (2001). Contribution of Na$_v$1.8 sodium channels to action potential electrogenesis in DRG neurons. *J Neurophysiol* 86:629-640.

Ro LS, Chen ST, Tang LM, Jacobs JM. (1999). Effect of NGF and anti-NGF on neuropathic pain in rats following chronic constriction injury of the sciatic nerve. *Pain* 79(2-3):265-274.

Rowbotham MC *et al.* (2003). Oral opioid therapy for chronic peripheral and central neuropathic pain. *N Engl J Med* 348:1223-1232.

Roza C *et al.* (2003). The tetrodotoxin-resistant Na+ channel Nav1.8 is essential for the expression of spontaneous activity in damaged sensory axons of mice. *J Physiol* 550(Pt 3):921-926.

Saegusa H *et al.* (2001). Suppression of inflammatory and neuropathic pain symptoms in mice lacking the N-type Ca2+ channel. *EMBO J* 20(10):2349-2356.

Saegusa H, Matsuda Y, Tanabe T. (2002). Effects of ablation of N- and R-type Ca(2+) channels on pain transmission. *Neurosci Res* 43(1):1-7.

Schafers M, Sorkin LS, Geis C, Shubayev VI. (2003). Spinal nerve ligation induces transient upregulation of tumor necrosis factor receptors 1 and 2 in injured and adjacent uninjured dorsal root ganglia in the rat. *Neurosci Lett* 347(3):179-182.

Seltzer Z, Dubner R, Shir Y. (1990). A novel behavioral model of neuropathic pain disorders produced in rats by partial sciatic nerve injury. *Pain* 43(2):205-218.

Shamash S, Reichert F, Rotshenker S. (2002). The cytokine network of Wallerian degeneration: tumor necrosis factor-alpha, interleukin-1alpha, and interleukin-1beta. *J Neurosci* 22(8):3052-3060.

Simmons RM, Webster AA, Kalra AB, Iyengar S. (2002). Group II mGluR receptor agonists are effective in persistent and neuropathic pain models in rats. *Pharmacol Biochem Behav* 73(2):419-427.

Sindrup SH, Jensen TS. (1999). Efficacy of pharmacological treatments of neuropathic pain: an update and effect related to mechanism of drug action. *Pain* 83(3):389-400.

Sleeper AA *et al.* (2000). Changes in expression of two tetrodotoxin-resistant sodium channels and their currents in dorsal root ganglion neurons after sciatic nerve injury but not rhizotomy. *J Neurosci* 20(19):7279-7289.

Sokal DM, Chapman V. (2003). Inhibitory effects of spinal baclofen on spinal dorsal horn neurones in inflamed and neuropathic rats in vivo. *Brain Res* 987(1):67-75.

Sommer C, Schmidt C, George A, Toyka KV. (1997). A metalloprotease-inhibitor reduces pain associated behavior in mice with experimental neuropathy. *Neurosci Lett* 237(1):45-48.

Sommer C, Marziniak M, Myers RR. (1998). The effect of thalidomide treatment on vascular pathology and hyperalgesia caused by chronic constriction injury of rat nerve. *Pain* 74(1):83-91.

Sorkin LS, Doom CM. (2000). Epineurial application of TNF elicits an acute mechanical hyperalgesia in the awake rat. *J Peripher Nerv Syst* 5(2):96-100.

South SM *et al.* (2003). A conditional deletion of the NR1 subunit of NMDA receptor in adult spinal cord dorsal horn reduces NMDA currents and injury-induced pain. *J Neurosci* 23:5031-5040.

Strichartz GR, Zhou Z, Sinnott C, Khodorova A. (2002). Therapeutic concentrations of local anaesthetics unveil the potential role of sodium channels in neuropathic pain. *Novartis Foundation Symposium* 241:189-201.

Sun RQ *et al.* (2001). Changes in brain content of nociceptin/orphanin FQ and endomorphin 2 in a rat model of neuropathic pain. *Neurosci Lett* 311(1):13-16.

Sung B, Lim G, Mao J. (2003). Altered expression and uptake activity of spinal glutamate transporters after nerve injury contribute to the pathogenesis of neuropathic pain in rats. *J Neurosci* 23(7): 2899-2910.

Suzuki M, Mizuno A, Kodaira K, Imai M. (2003). Impaired pressure sensation in mice lacking TRPV4. *J Biol Chem* 278(25): 22664-22668.

Svensson P, Cairns BE, Wang K, Arendt-Nielsen L. (2003). Injection of nerve growth factor into human masseter muscle evokes long-lasting mechanical allodynia and hyperalgesia. *Pain* 104(1-2):241-247.

Takahashi N, Nagano M, Suzuki H, Umino M. (2003). Expression changes of glial cell line-derived neurotrophic factor in a rat model of neuropathic pain. *J Med Dent Sci* 50(1):87-92.

Thompson SW, Bennett DL, Kerr BJ, Bradbury EJ, McMahon SB. (1999). Brain-derived neurotrophic factor is an endogenous modulator of nociceptive responses in the spinal cord. *Proc Natl Acad Sci USA* 96(14):7714-7718.

Tsuda M *et al.* (2003). P2X4 receptors induced in spinal microglia gate tactile allodynia after nerve injury. *Nature* 424(6950):778-783.

Turner JA, Cardenas DD, Warms CA, McClellan CB (2001). Chronic pain associated with spinal cord injuries: a community survey. *Arch Phys Med Rehabil* 82:501-509.

Wallace VC, Cottrell DF, Brophy PJ, Fleetwood-Walker SM. (2003). Focal lysolecithin-induced demyelination of peripheral afferents results in neuropathic pain behavior that is attenuated by cannabinoids. *J Neurosci* 23(8):3221-3233.

Wang H *et al.* (2003). Chronic neuropathic pain is accompanied by global changes in gene expression and shares pathobiology with neurodegenerative diseases. *J Peripher Nerv Syst* (2):128-133.

Wang YX *et al.* (2000). Effects of intrathecal administration of ziconotide, a selective neuronal N-type calcium channel blocker, on mechanical allodynia and heat hyperalgesia in a rat model of post-operative pain. *Pain* 84(2-3):151-158.

Waxman SG, Cummins TR, Dib-Hajj, Fjell J, Black JA. (1999). Sodium channels, excitability of primary sensory neurons, and the molecular basis of pain. *Muscle Nerve* 22:1177-1187.

Waxman SG, Kocsis JK, Black JA. (1994). Type III sodium channel mRNA is expressed in embryonic but not adult spinal sensory neurons, and is re-expressed following axotomy. *J Neurophysiol* 72:466-471.

Xiao HS *et al.* (2002). Identification of gene expression profile of dorsal root ganglion in the rat peripheral axotomy model of neuropathic pain. *Proc Natl Acad Sci USA* 99(12):8360-8365.

Yoon YW *et al.* (1999). Different strains and substrains of rats show different levels of neuropathic pain behaviors. *Exp Brain Res* 129(2):167-171.

Zhang J *et al.* (2003). Induction of CB2 receptor expression in the rat spinal cord of neuropathic but not inflammatory chronic pain models. *Eur J Neurosci* (12):2750-2754.

Zhou XF, Deng YS, Xian CJ, Zhong JH. (2000). Neurotrophins from dorsal root ganglia trigger allodynia after spinal nerve injury in rats. *Eur J Neurosci* 12(1):100-105.

Therapeutic Transfer and Regeneration of Neural Progenitors or Neurons in Gray Matter of the Adult Brain

Ole Isacson, MD

TRANSFER OF APPROPRIATE NEURONS AND GLIA FOR CELL REPAIR AND REPLACEMENT IN NEUROLOGICAL DISEASES

Degenerative neurological diseases have multiple roots and interactive pathogeneses, and many are characterized by the loss of relatively vulnerable groups of neurons. If possible, it is reasonable to initiate protective measures before onset and to prevent further degeneration, once the disease appears. In reality, most patients are free of symptoms until most of the relatively vulnerable cells are dysfunctional or dead. For this reason, and since any retardation of degeneration is unlikely to be absolute, it is rational to develop cell replacement for the lost neurons and glia. Such "live cell" replacement therapies are conceptually different from classical pharmacology. Novel cell-based therapies therefore have generated many new questions in the neurological community and skepticism. These novel cell repair and regeneration therapies are experimental, with limited data accumulated on the best therapeutic cell compositions and application to the potentially responsive patient groups. In Parkinson's disease (PD), there are patients who have experienced significant and long-lasting benefits after fetal cell transfer to gray matter. These benefits are physiologically verified (fluorodopa PET scans and functional MRI), long-lasting (beyond 14 years), and clinically meaningful (50% to 60% reduction in clinical scores off DA drug therapy) (Piccini et al., 1999; Piccini et al., 2000). Transfer of fetal dopamine (DA) neuron–containing suspension (about 5% DA cell content) into putamen (Mendez et al., 2002; Piccini et al., 1999; Piccini et al., 2000) has so far yielded the best results in patients with PD (Figure 21.1; see also the Color Plate section). There appears to be less benefit to patients, and substantially higher risk for side effects, after implantation of solid tissue-piece fetal DA cells (also approximate 5% cell DA content) (Freed et al., 2001; McNaught et al., 2002). To understand the hurdles of developing a neurological cell therapy, it is instructive to look at the history of other live cell therapies. Surgical cell replacement (cell transplantation) instead of insulin injection for patients with type 1 diabetes mellitus was initially heralded as a potential breakthrough, but a number of biological and technical hurdles have slowed progress (Ryan et al., 2002; Weir et al., 1990).

As for diabetes mellitus, neuronal replacement therapies for PD have so far not successfully addressed all the complexities to make the cell transfer therapies work in patients with PD. Although fetal DA cell transfer has been validated and effective in some patients with PD, recent surgical applications of the techniques have not established the most effective parameters for cell preparation, transplantation, immunological treatments, or patient selection. Modern double-blind pharmacological trial designs (which are indeed very good tools as *final* tests to establish the safety and clinical efficacy of well-designed drugs) were perhaps applied too early to evaluate these prototype treatments. Nonetheless, much can still be learned from these early trials, particularly about side effects (off-medication dyskinesia in PD) and the need for appropriate patient selection based on preoperative evaluations of drug responses, as well as disease severity and stage. As an example of differences between drug trials and cell-based studies, a double-blind cell transplantation trial used a 1-year endpoint as the conclusion of the blind study, and final analysis of data

Pre-operative   Post-operative

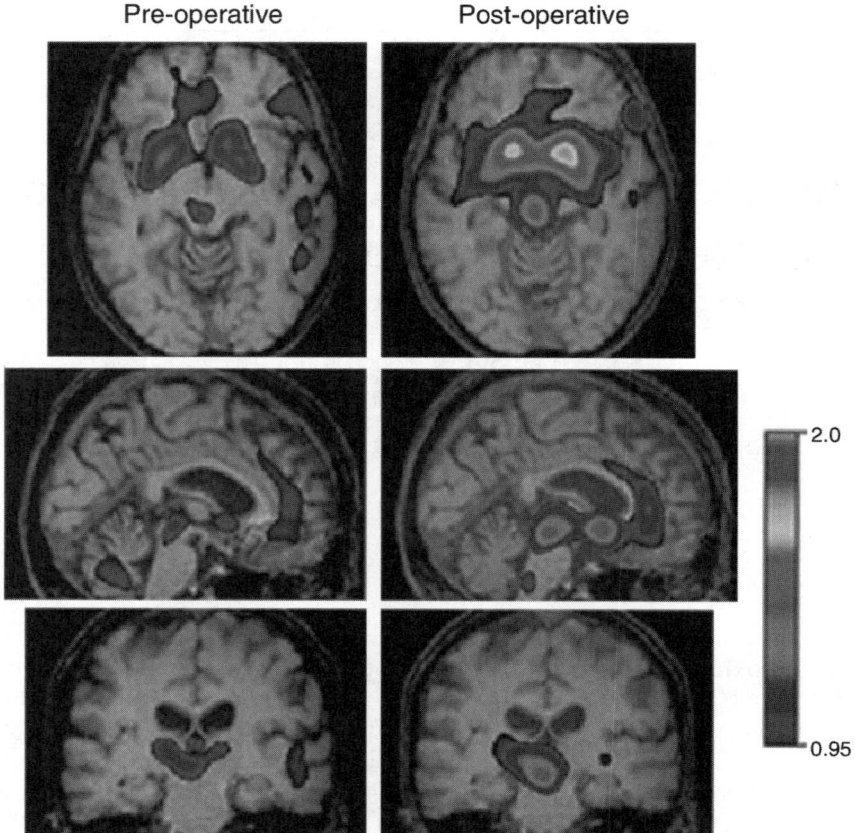

FIGURE 21.1 Preoperative and postoperative fluorodopa PET scans obtained in a patient with Parkinson's disease who received simultaneous intrastriatal and intranigral fetal VM grafts. These images consist of parametric maps of fluorodopa K_i transformed into stereotactic space and overlaid on the patient's MR image (also in stereotactic space). The parametric maps were smoothed using an 8-mm gaussian filter. The axial, coronal, and sagittal sections demonstrate an increase in K_i in the midbrain and putamen bilaterally, likely resulting from surviving grafted dopamine neurons. From Mendez *et al.* (2002), used with permission. (See also the Color Plate section.)

(Freed *et al.*, 2001). The 1-year time point using live immature DA cells is insufficient for human fetal DA cells to fully grow and establish functional effects in animal models or patients (Isacson *et al.*, 2001). Consequently, after the study was published, many patients continued to improve during year 2 and 3 after surgery, as presumably the fetal cells continued to mature, integrate, adapt, and help improve the functional status of the patients with PD (Piccini *et al.*, 1999).

DERIVING OR REGENERATING OPTIMAL PROGENITOR DOPAMINE NEURONS FOR PARKINSON'S DISEASE

Midbrain lateral substantia nigra DA loss of function creates parkinsonism. This specific DA loss produces akinesia, rigidity, and tremor. During brain development, newborn DA cells migrate into fetal midbrain and send axons to targets in the emerging striatum and cortex. The DA neuron is generated by cell signals emanating from factors released around the ventral midbrain from neural ectoderm or ES cells (Hynes and Rosenthal, 2000) (Figure 21.2; see also the Color Plate section). Remarkably, of the three original germ layers, the neural ectoderm and the brain develop partly through a so-called default pathway (Munoz-Sanjuan and Brivanlou, 2002). Similarly, after embryonic stem (ES) cells or embryoid body cells have become neural precursors, many will spontaneously acquire a neuronal midbrain-hindbrain identity, including DA cell specificity (Deacon *et al.*, 1998; Lee *et al.*, 2000b; Tropepe *et al.*, 2001) (see Figure 21.2). When there is an absence or block of mesodermal and endodermal signals, the ES cells will become primitive and eventually differentiated neuronal cell types. This process was described initially in frogs and in knockout gene cell systems, where bone morphogenetic protein

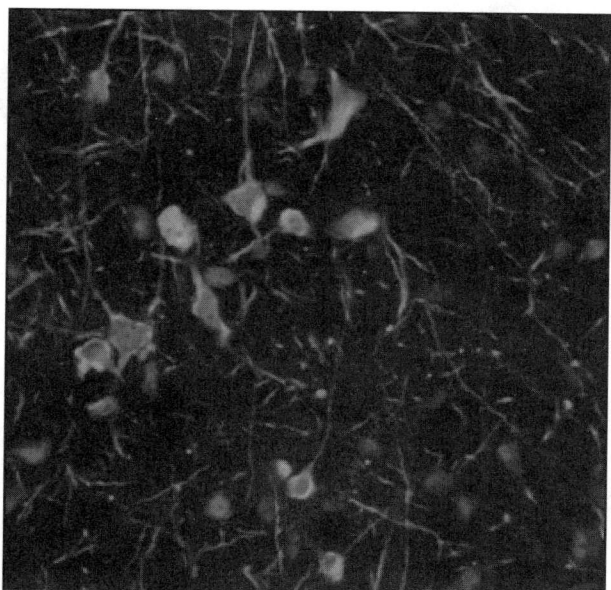

FIGURE 21.2 Dopamine neurons derived *in vivo* from mouse embryonic stem (ES) cells. This confocal image shows numerous dopamine cells in the striatum of a 6-OHDA–lesioned rat, 11 weeks' post-transplantation of undifferentiated ES cells (~2000 cells). Dopamine neurons co-expressed the synthetic enzymes tyrosine hydroxylase (TH, green) and aromatic amino acid decarboxylase (AADC, blue) and the neuronal nuclei marker (NeuN, red). Animals with ES cells with derived dopamine neurons in the grafts showed recovery of motor asymmetry. (See also the Color Plate section.)

(BMP) and activin receptors (including vertebrate TFG-β receptor-related signaling pathways) were implicated (Munoz-Sanjuan and Brivanlou, 2002). One of the cell types that is derived from such spontaneous neural differentiation is the DA cell type (Deacon *et al.*, 1998; Lee *et al.*, 2000b). Our work in 1998 (Deacon *et al.*, 1998) showed that this default pathway also operated in mammalian cells after transplantation, or in a cell culture dish (Lee *et al.*, 2000b; Tropepe *et al.*, 2001). It was demonstrated that when cells are dissociated and transplanted at very low concentrations (simulating a low access cell-mediated contact and presumably an absence of BMP signaling) into tissues, a neuronal identity was the default cell choice—which in the mouse included DA and serotonergic neurons (Bjorklund *et al.*, 2002) (Figures 21.2 and 21.3; see also the Color Plate section). The actual genetic subprograms controlling the midbrain DA cell type development appear to include sequential and parallel action of several transcription factors. Following proliferation in the ventricular zone, neuroblasts will migrate down toward the central and lateral ventral mesencephalon. The FGF8-rich zone and territories reached by SHH protein (Hynes and Rosenthal, 2000) induce in the progenitor cells a sequence and group of transcription factors and enzymes that establish most of the cell character of the midbrain neurons that share the DA trans-

mitter type (see Figure 21.3). Engrail genes, *Nurr1* and *PitX3*, are some of the most typical transcription factors of this set. *Nurr1* certainly activates most of the transmitter-related genes and some of the trophic signaling pathways (Kim *et al.*, 2003), but *PitX3* is also critical to the construction and survival of DA neurons that can reach motor control regions of the caudate-putamen (Burbach *et al.*, 2003). In fact, the functional absence of *PitX3* during development will cause the A9 DA neuron to fail to establish in the midbrain, whereas the A10 VTA neuron can still survive, grow, and function in the limbic circuitry. In our experiments, the ES cell–derived DA neurons reinnervated the brain and restored DA transmission (Figure 21.4; see also the Color Plate section). This was shown by DA release, behavioral correction of a motor syndrome, and functional integration (blood flow and restored activity in cerebral cortex) (Bjorklund *et al.*, 2002) (see Figure 21.4).

More recent works confirmed these findings and also enhanced the design of ES cells to more readily and reliably generate necessary fetal DA cells as replacement donor cells in PD (Chung *et al.*, 2002b; Kim *et al.*, 2002). It is possible to enhance the percent of ES cells that become DA neurons by stably inserting and expressing the gene transcription factor *Nurr1*, which drives genes associated with the DA cell type—for instance, tyrosine hydroxylase (the rate-limiting enzyme in the synthesis of dopamine) and the dopamine transporter, responsible for transmitter reuptake (see Figure 21.3). Dopa-decarboxylase present in dopaminergic and serotonergic neurons is also induced by *Nurr1*, along with several other marker and trophic support genes typical of the midbrain, including the *c-ret* components (Grf-α subunit) of the GDNF receptor (Wallen *et al.*, 2001). It has also been demonstrated that the ES cells can be differentiated into DA neurons in cell culture, and this process is enhanced by overexpression of *Nurr1* (Chung *et al.*, 2002a; Kim *et al.*, 2002) (see Figure 21.3). Transplanting such cells at their fetal DA equivalent stage into the rat brain produces reinnervation of the degenerated host striatum, and creates a behavioral recovery from deficits associated with the dopamine deficiency (Kim *et al.*, 2002). In conclusion, genetic engineering of stem cells, cell sorting, and selection technologies are viable methods to generate the potentially therapeutic DA cells.

CLINICAL APPLICATION OF THE EMERGING CELL-BASED TREATMENT APPROACHES FOR PARKINSON'S DISEASE

New therapeutic non-pharmacological methods involve cell and synaptic renewal or replacement in the living brain to restore function of neuronal systems, including the DA system in PD (Isacson *et al.*, 2003). While recent laboratory work has focused on using stem

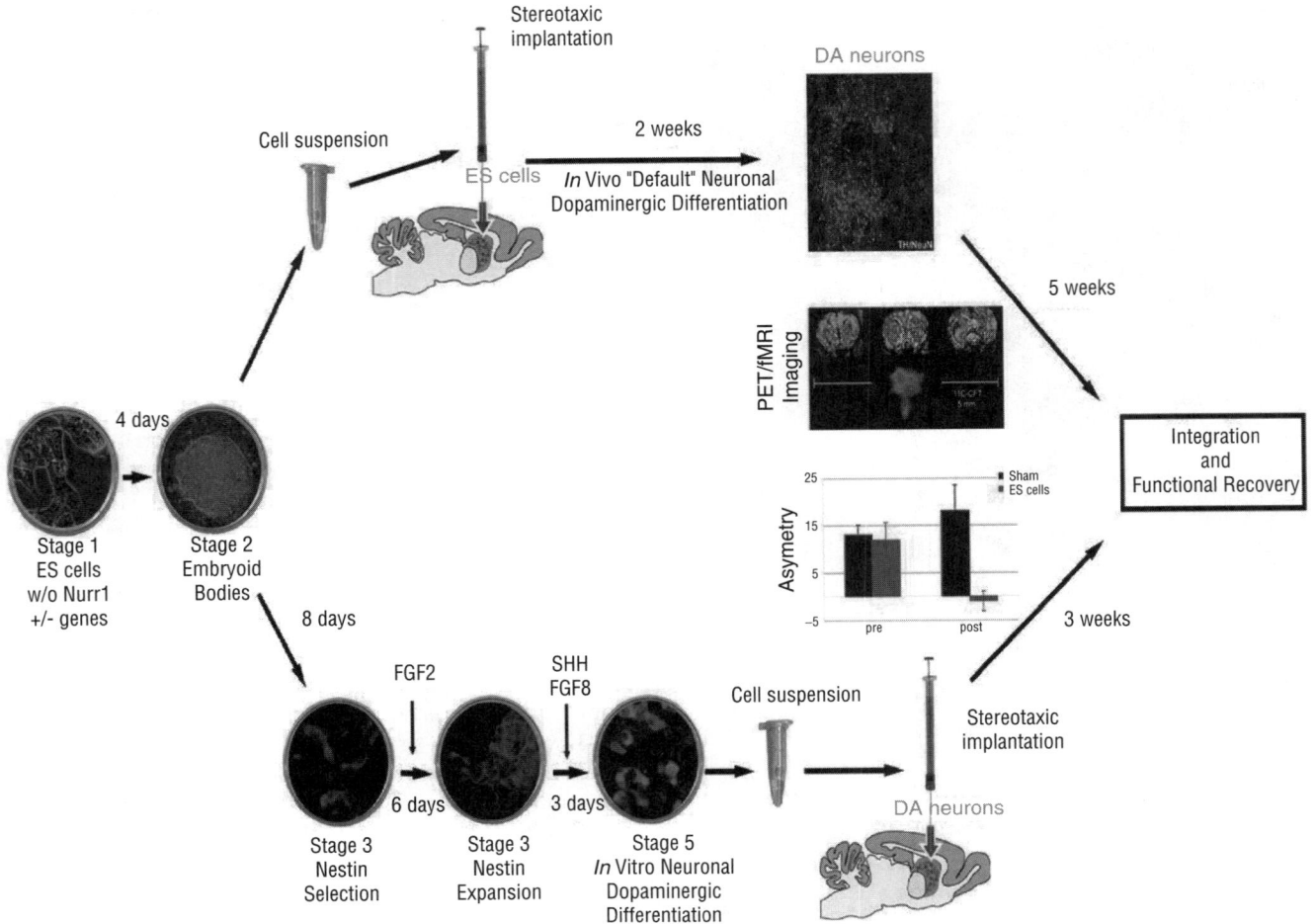

FIGURE 21.3 Schematic drawing of the process of obtaining dopamine or cholinergic neurons from embryonic pluripotent stem cells. In the case of mouse embryonic stem cells, these cells can grow on feeder layers in leukemia inhibitor factors (LIF) in stage 1. At this stage, various genes can be added or deleted. In the case of Nurr1, it is a positive drive for transcription of dopamine and neuron-related genes. At stage 2, embryoid bodies are formed containing similarly pluripotent cells. In the left panel, as shown by Kim *et al.* (2002), the embryonic stem cells are brought in five stages to their fetal stage, in which they are then made into a cell suspension and transplanted directly into the Parkinson rat model. Similarly to the default pathway, these embryonic stem cell derived fetal neurons will create a functional recovery on amphetamine rotation, as well as a number of spontaneous tasks including reaching tasks and other evidence of corrected dopamine deficiency. In the upper panel (Bjorklund *et al.*, 2002), the cells are dissociated at specific cell concentrations and can be placed directly in the brain on low concentrations and will then, after a few weeks' time, develop over the species-specific program towards dopaminergic and serotonergic neurons. In previous experiments, such neurons would, over time, produce sufficient dopamine to correct the behavioral syndrome of Parkinson's disease asymmetry, such that the correction can be normalized if sufficient cells are reinnervating the brain compared to sham cells (Bjorklund *et al.*, 2002). From Isacson (2003); used with permission. (See also the Color Plate section.)

cells as a starting point for exogenous or endogenous derivation of the optimal DA cells for repair, DA cell therapy using fetal DA neurons has been explored in patients with PD (Freed *et al.*, 2001; Isacson *et al.*, 2001; Mendez *et al.*, 2002; Piccini *et al.*, 1999; Piccini *et al.*, 2000) (see Figure 21.1). It has been demonstrated that functional motor deficits associated with PD can be reduced after application of this new technology. Evidence shows that

the underlying disease process does not destroy the transplanted fetal DA cells, although the patient's original DA system degeneration progresses (Piccini *et al.*, 1999; Piccini *et al.*, 2000). The optimal DA cell regeneration system would reconstitute a normal network capable of restoring feedback-controlled release of DA in the nigrostriatal system (Bjorklund and Isacson, 2002). The success of cell therapy for neurological diseases is limited

A   11C-CFT   TH-IR

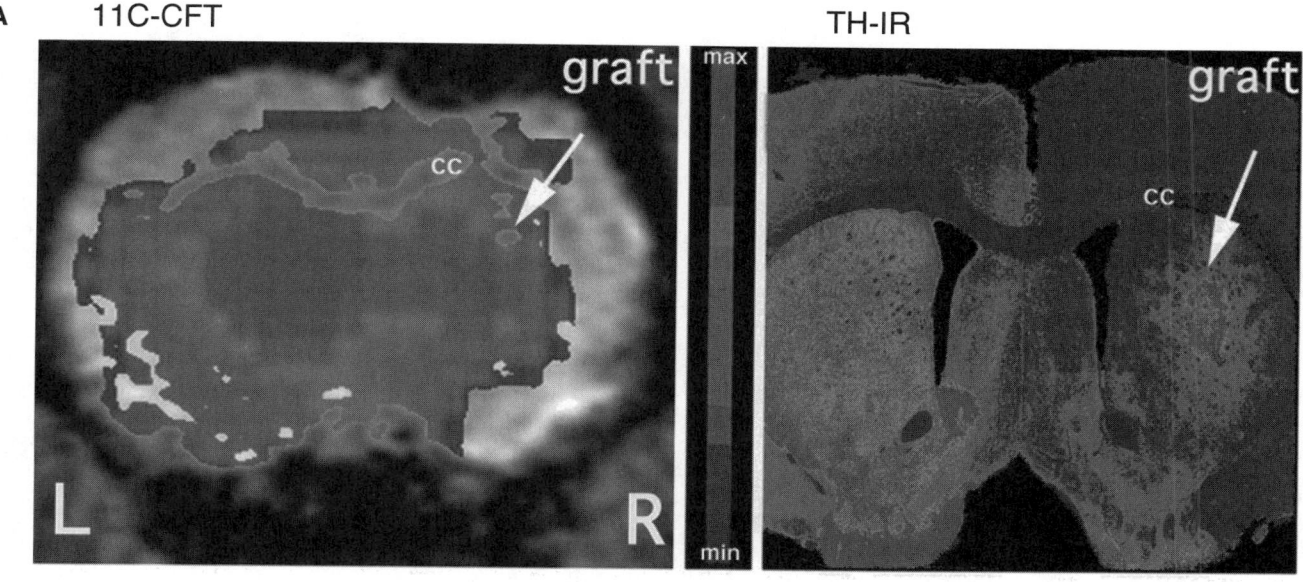

Amphetamine-induced change in CBV

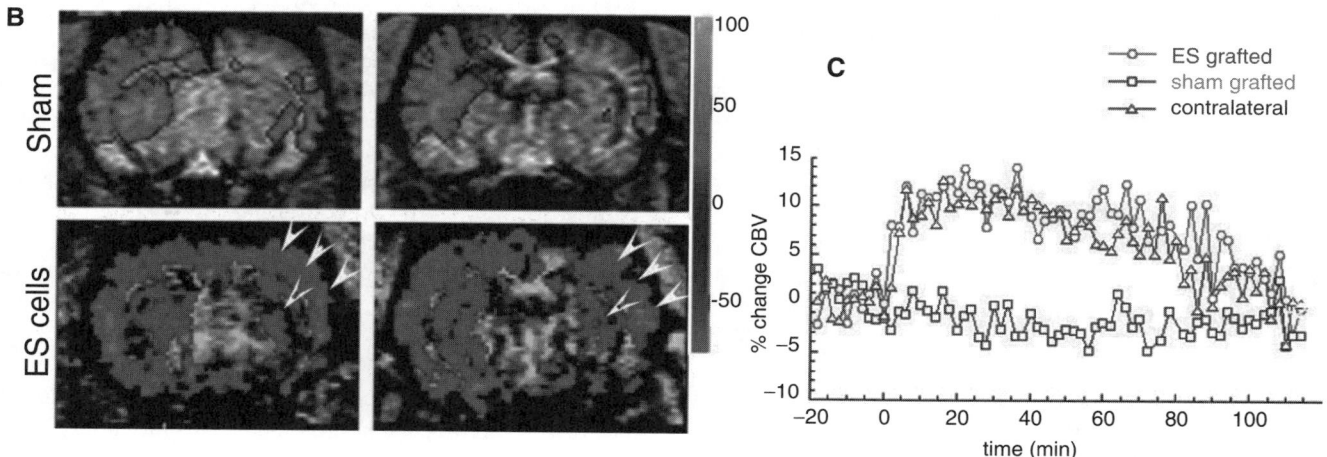

FIGURE 21.4   **A:** Using PET and the specific dopamine transporter (DAT) ligand carbon-11-labeled 2β-car-bomethoxy-3β-(4-fluorophenyl) tropane (^{11}C-CFT), we identified specific binding in the right grafted striatum, as shown in this brain slice (**A, left panel**) acquired 26 min after injection of the ligand into the tail vein (acquisition time was 15 seconds). Color-coded (activity) PET images were overlaid with MRI images for anatomical localiza-tion. The increase in ^{11}C-CFT binding in the right striatum was correlated with the postmortem presence of TH-immunoreactive (IR) neurons in the graft (**A, right panel**). **B:** Neuronal activation mediated by DA release in response to amphetamine (2 mg/kg) was restored in animals receiving ES grafts. Color-coded maps of percent change in relative cerebral blood volume (rCBV) are shown at two striatal levels for control (upper panel) and an ES cell derived DA graft (lower panel). 6-OHDA lesion results in a complete absence of CBV response to amphet-amine on striatum and cortex ipsilateral to the lesion (upper panel). Recovery of signal change in motor and somato-sensory cortex (arrows), and to a minor extent in the striatum was observed only in ES grafted animals. **C:** Graphic representation of signal changes over time in the same animal shown in **B**. The response on the grafted (red line) and normal (blue line) striata was similar in magnitude and time course while no changes were observed in sham grafted animals (green line). Baseline was collected for 10 minutes before and 10 minutes after MION injection and amphetamine was injected at time 0. cc, corpus callosum (Bjorklund *et al.*, 2002). (See also the Color Plate section.)

by access to preparation and development of highly specialized dopaminergic neurons found in the A9 and A10 regions of the substantia nigra (SN) in the ventral mesencephalon, as well as technical and surgical steps associated with transplantation.

In clinical trials, despite technical shortcomings, human fetal dopamine specified phenotypic ventral mesencephalic (VM) neurons have shown functional capacity in patients with PD (Freed *et al.*, 2001; Lindvall and Hagell, 2000; Mendez *et al.*, 2002; Piccini *et al.*, 1999; Piccini *et al.*, 2000; Ramachandran *et al.*, 2002). Unregulated production of DA from grafted fetal neurons is the cause of unwanted dyskinesias seen in a subset of patients with PD who have undergone transplantation (Freed *et al.*, 2001; Olanow, 2002). This is quite possible in a scenario of a "primed" dyskinetic circuitry produced by previous L-dopa treatment in a patient. Nonetheless, preclinical and clinical transplantation work demonstrates that striatal DA terminal release from appropriately grafted fetal DA neurons is controlled by both cell intrinsic and extrinsic synaptic and autoreceptor mechanisms and therefore, at least theoretically, the risk for dyskinesias should be less than that with drug treatments (Bjorklund and Isacson, 2002; Piccini *et al.*, 1999). In animal models, fetal DA-grafted neurons can reduce L-dopa–induced peak dyskinesias (Lee *et al.*, 2000a). In grafted DA neurons, presynaptic DA autoreceptors regulate excess DA release, and *in vivo* infusion of the full DA agonist apomorphine can block spontaneous DA release in the striatum (Bjorklund and Isacson, 2002; Strecker *et al.*, 1987; Zetterstrom and Ungerstedt, 1984).

IS IT NECESSARY TO OBTAIN STRUCTURAL RECONSTITUTION OF TERMINAL SYNAPTIC FUNCTION AND CONTROLLED DOPAMINE RELEASE BY NEW DOPAMINE NEURONS?

The scientific and clinical field of neural and cell repair has experienced major conceptual problems, perhaps typical of all new technologies. The initial success of L-dopa for PD, and growth of pharmacology as a medical discipline, in some ways prevented a paradigm shift to new molecular, cell biological, and regenerative strategies. Pharmacology-like reasoning suggested that a "biological pump" of dopamine would be sufficient for functional recovery in PD. This misconception led to several unsuccessful clinical trials, for example those involving autologous transplantation of catecholamine-containing adrenal medulla cells (Backlund, 1985; Madrazo, 1988). The lack of recovery, the low adrenal medulla graft survival, and the reported morbidity of patients with PD strengthened, however, the rationale

for using dopamine-releasing fetal neural mesencephalic donor cells (instead of adrenal medulla that produces minute nonsynaptic DA even after nerve growth factor stimulation). Again, however, the cell therapy using fetal DA neurons is in a "prototype" technology phase, since the quality of cell implantation technology and cell preparations using fetal DA cells is variable, and highly experimental in PD (Isacson, 2003). Thus, cell transfer and brain repair using immature neurons and glia is innovative both from the technical and biological standpoints, and will require much work to optimize. The scaling up of this method from rodents to primates and humans has proved very demanding, particularly in obtaining an acceptable, abundant, and reliable donor cell source and preparation and consistent surgical results. Early studies indicated some motor improvement associated with increased fluorodopa uptake (Widner *et al.*, 1992). Longitudinal and validated data from Lindvall and colleagues indicate stable DA cell survival and function in patients for over a decade after surgery and substantially reduced need for pharmacological substitution by dopaminergic drugs (Piccini *et al.*, 1999; Piccini *et al.*, 2000). The transplantation of non-dissociated (solid) human VM tissue pieces has also provided benefits to some patients (Freed *et al.*, 1992; Freeman *et al.*, 1995). In a series of pilot transplantation studies carried out by Olanow and colleagues in the United States, autopsy from two bilaterally transplanted (6.5- to 9-week human fetal VM) patients who died 18 to 19 months after surgery showed over 200,000 surviving DA neurons, which *partially* reinnervated the right putamen (about 50%) and the left putamen (only about 25%) (Kordower *et al.*, 1996). Electron microscopy revealed axo-dendritic and occasional axo-axonic synapses between graft and host was possible, and analysis of tyrosine hydroxylase (TH) mRNA revealed higher expression within the fetal neurons than within the residual host nigral cells (Kordower *et al.*, 1996). Autopsy of another patient in this surgical group showed over 130,000 surviving DA neurons, reinnervating almost 80% of the putamen (Kordower *et al.*, 1998). Notably, both patients had shown major improvements in motor function and increases in fluorodopa uptake in the putamen on positron emission tomography (PET) scanning. Nevertheless, these and other detailed observations indicate that many specific regions of the human putamen may not be innervated by such nonspecific VM grafts, unless they contain the appropriate number and type of the A9 dopamine neuronal phenotype (Haque *et al.*, 1997; Mendez *et al.*, 2003). More critically, the clinical trials using fetal "noodle" and solid mesencephalic tissue also created significant side effects in the form of dyskinesias typical of DA excess and striatal lesions

(Freed *et al.*, 2002; Hantraye *et al.*, 1992; Olanow, 2002).

The most important factor in obtaining optimal functional effects (and minimal side effects) in PD by brain repair is probably the presence of new terminals and DA transmission that adequately adapt to the local milieu, and provide physiologically appropriate DA release in the host caudate-putamen and substantia nigra (Bjorklund and Isacson, 2002; Isacson, 2003; Piccini *et al.*, 1999; Piccini *et al.*, 2000) and synaptic feedback control, such as DAT and DA D2 receptors. Fetal DA neurons typically grow to establish functional connections with mature host striatal neurons. Synaptic contacts between transplanted fetal DA cells and host cells, as well as afferent contact by host neurons to transplanted cells, have been observed ultrastructurally (Doucet *et al.*, 1989; Mahalik *et al.*, 1985). The critical insight is that pharmacological delivery of DA into the striatum may in the end not be as effective in ameliorating the motor symptom of PD, as regulated, synaptic release provided with transplanted DA neurons (Isacson *et al.*, 2003). First, when DA is directly administered into the ventricles of patients with PD, serious psychosis and motor abnormalities develop (Venna *et al.*, 1984). Second, after high DA levels *in vivo*, experiments show an abnormal up-regulation of a large number of genes within the striatum (Gerfen *et al.*, 1998). Complications associated with unregulated DA levels are also obvious when observing effects of long-term L-dopa administration in patients. As PD progresses, and the midbrain DA neuron and its synapses continue to degenerate, the non-physiological levels of DA within the striatum and abnormal downstream activity in the basal ganglia produce severe motor abnormalities, such as dyskinesias. In neurophysiological recordings, it is clear that dopamine provides a modulatory role for the glutamate-mediated transmission, so that it appears to have a form of a gating function at that important striatal synapse (Freeman *et al.*, 1985; Grace, 1991; Johnson *et al.*, 1992; Ljungberg *et al.*, 1992; Nutt *et al.*, 2000; Onn *et al.*, 2000; Strecker and Jacobs, 1985). Physiologically appropriate DA functions can be achieved by normal DA synapses or, alternatively, cells that express the complete set of feedback elements required to regulate release and uptake of DA (Isacson and Deacon, 1997; Strecker *et al.*, 1987; Zetterstrom *et al.*, 1986). DA-mediated involvement in the striatal neuronal network is such that unless there is tonic release of DA that is finely regulated at the synaptic cleft by afferents, glutamatergic synapses will be less effective in control of the striatal GABAergic output neurons (Nutt *et al.*, 2000; Onn *et al.*, 2000). Discontinuous stimulation of striatal dopamine receptors following loss of DA terminals

or excessive L-dopa treatment is likely a major cause of dyskinesia induction. Normally, basal changes in firing by mesencephalic dopamine neurons is limited, at least as has been tested in animal models (Ljungberg *et al.*, 1992; Strecker and Jacobs, 1985). More importantly, such frequency stimulation does not raise the extracellular concentration of putaminal dopamine because the synaptic network and terminals work to reduce fluctuations in DA concentration by reuptake mechanisms (DAT) and possibly other autoreceptor-mediated functions (D2) (Freeman *et al.*, 1985; Grace, 1991; Johnson *et al.*, 1992; Ljungberg *et al.*, 1992; Nutt *et al.*, 2000; Onn *et al.*, 2000; Strecker and Jacobs, 1985). Indeed, fetal DA transplants have been shown to reduce the incidence of L-dopa–induced peak dyskinesias in animal models (Lee *et al.*, 2000a). Several clinical studies have also shown normalized metabolic and brain functional activity throughout the basal ganglia after DA neural transplantation (Piccini *et al.*, 1999; Piccini *et al.*, 2000).

PET and carbon-11–labeled 2B-carbomethoxy-3B-(4-fluorophenyl)tropane (11C-CFT) can be used to visualize and quantify striatal presynaptic DA transporters in PD patients and degeneration models. In one such study of a unilateral lesion of the SN DA system in rodents, the binding ratio was initially reduced to 15–35% of an intact side. After fetal DA neuronal transplantation, behavioral recovery occurred gradually when the 11C-CFT binding ratio had increased to 75–85% of the intact side, revealing a threshold for functional recovery in the lesioned nigrostriatal system after neural transplantation that fits our understanding of the normal DA motor system requirements (Brownell *et al.*, 1998). Importantly, autoregulation of DA release and metabolism by intrastriatal DA cell containing grafts has been shown by *in vivo* microdialysis in the striatum. Infusion of a nonselective DA agonist (apomorphine) almost abolishes endogenous DA release in the DA neuron-grafted striatum (Galpern *et al.*, 1996; Strecker *et al.*, 1987), showing a near-normal autoregulation of DA levels by the implanted DA neurons. The formation of effective DA terminals and synapses with adequate DA release and control have been determined in transplanted rodents after dyskinesia inducing L-dopa injections (Gaudin *et al.*, 1990; Lee *et al.*, 2000a). L-dopa–induced peak-dose dyskinesias in non-human primates are also much reduced after fetal DA cell transplantation (H. Widner, personal communication). These data indicate that DA levels within the transplanted striatum can be regulated in a functional manner by correctly transplanted DA neurons if they act as the normal functional cellular regulators of DA neurotransmission in their normal target areas (Bjorklund and Isacson, 2002; Isacson and Deacon, 1997).

TECHNICAL DEVELOPMENTS OF A NEW TECHNOLOGY FOR REGENERATION OF NEURAL FUNCTION AND PATHWAYS IN PATIENTS

There are a number of critical variables that determine the outcome of cell therapy for PD. Most of these variables have not been systematically evaluated in primates or patients. This is not because of passivity on anybody's part, but rather a reflection of the technical and cellular challenges that the new cell therapies present. Small exploratory (and lately a couple of placebo-controlled) trials have shown both encouraging results in some patients (validated by analytic imaging and postmortem studies) and the potential problems with getting the right cells to the right brain regions and those patients who may benefit the most. The specific preparations of donor cells and associated procedures are very important in cell-based therapies and there are major differences in cell preparations for transplantation in PD in clinical trials to date. The most successful method involves freshly dissected fetal tissue pieces (minute cubic millimeter pieces) that can be treated with proteolytic enzymes, then dissociated into a cell suspension (Brundin et al., 1985; Lindvall and Hagell, 2000; Mendez et al., 2002; Piccini et al., 1999; Piccini et al., 2000; Ramachandran et al., 2002). Other less effective procedures have included untreated tissue pieces, minced from the ventral mesencephalon of aborted fetuses (Freeman et al., 1995; Henderson et al., 1991; Kordower et al., 1995; Olanow, 2002). A third type, the so-called noodle technique (Freed et al., 2001) includes a cell culture step. Such culture steps may alter the cells and select for cell types that are different than the populations obtained by fresh preparations. These three different techniques produce different grafts. Only the cell suspension grafts grow to slender deposits that do not disrupt or displace the host target tissue (Brundin et al., 1985; Mendez et al., 2002; Piccini et al., 1999; Piccini et al., 2000; Ramachandran et al., 2002). In any event, all the current PD transplantation studies have employed rather crude cell preparations, because the starting point for all work are the pieces dissected from fetal tissue, which contain only approximately 10% newborn DA neurons; the rest are cells of types not generally relevant to PD degeneration (Brundin et al., 1985), except the progenitors of glial support cells. In addition, while there is selective degeneration of substantia nigra pars compacta (SNc; A9) neurons, with a relative sparing of ventral tegmental area (VTA; A10) neurons in PD (Gibb, 1992; Iacopino and Christakos, 1990; Ito et al., 1992; Yamada et al., 1990), both of these DA cell groups are currently transplanted as a mixture (Freed et al., 2001; Hauser et al., 1999; Kordower et al., 1996; Lindvall et al., 1988). Notably, these two subpopulations of DA neurons within the SN have very different functions and project to different brain areas (importantly also within the SN, through dendritic release). The medial- to midline-positioned DA neurons (known as area A10) (Dahlstrom and Fuxe, 1964) selectively send axons to limbic and cortical regions (Gerfen et al., 1987) while the adjacent lateral A9 DA neurons (the major dysfunction in PD) grow selectively to putamen areas controlling motor function (Damier et al., 1999). Thus, the differences between A10 and A9 DA neurons (Costantini et al., 1997; Haque et al., 1997) are potentially very important for understanding how to make cell replacement strategies work in patients with PD. In summary, while current work demonstrates a clear capacity of fetal dopamine cell transplants to repair PD brains (Mendez et al., 2000b; Piccini et al., 1999; Piccini et al., 2000), there are technical limitations with available prototypes of this new technique. Both animal model experiments and clinical trials are limited by today's mixed cell preparations of low DA neuronal yield, inappropriate surgical placements in the brain, and potentially the loss of DA cell subpopulations of therapeutic interest during tissue incubation, cell culture, and preparation steps. Practically and ethically it is not possible to use fetal tissue as a source for transplantation for PD in more than rare experimental situations. For example, to replace a sufficient number of DA neurons, one needs six to eight fetal tissue pieces per patient, primarily because of low post-transplantational survival of the grafted fetal DA neurons. In addition, current surgical techniques are inconsistent with respect to placement, volume, and type of cells grafted. However, it is encouraging that recent research in stem cell biology may provide a solution to this problem of low cell access and yield (Bjorklund et al., 2002; Kim et al., 2002).

CAN WE PRODUCE BETTER THERAPIES WITH FEWER SIDE EFFECTS BY ACCOMPLISHING A MORE SPECIFIC CELLULAR AND SYNAPTIC DOPAMINE REPLACEMENT?

Recently it has been postulated that excess and unregulated production of DA from grafted fetal neurons could be responsible for some unwanted side effects seen in patients with PD who have undergone transplantation (Freed et al., 2001). However, it has been convincingly demonstrated that striatal DA release normally is tightly regulated by both intrinsic and extrinsic mechanisms. Presynaptic dopamine auto-receptors regulate excess DA release and administration of the DA agonist apomorphine can inhibit spontaneous DA release up to 100% (Zetterstrom and Ungerstedt, 1984). It is believed that approximately 50% of this inhibition can be contributed to a direct effect on DA autoreceptors and 50% involves postsynaptic neurons engaged in short- and long-distance feedback circuits (Zetterstrom and Ungerstedt, 1984). Subsequently it has been shown that grafted fetal DA neurons also display similar auto-inhibition mechanisms as demonstrated by a 40% reduction in spontaneous DA release when grafted animals are treated with apomorphine (Bjorklund and Isacson, 2002; Strecker *et al.*, 1987). These basic studies are important, because they provide evidence contrary to current theories that implanted DA neurons may release excess amounts of DA in an unregulated manner and thereby contribute to side effects such as dyskinesias (Freed *et al.*, 2001). Instead, such side effects may have several alternative explanations (Isacson *et al.*, 2001). DA released from nerve terminals into the synaptic cleft is normally rapidly taken up and transported back into the terminals by the dopamine transporter (DAT). This mechanism is important for appropriate temporal regulation of DA concentration in the synaptic cleft. When the DAT is blocked by agents such as cocaine, DA will remain in the synaptic cleft in higher concentrations and for longer time periods than normal thus allowing increased binding and activation of postsynaptic receptors. When fetal ventral midbrain is dissected prior to transplantation, most published protocols do not make any distinction between the DA neurons residing in the VTA (A10) and those in the SNc (A9) (Freed *et al.*, 2001; Hauser *et al.*, 1999; Kordower *et al.*, 1996; Lindvall *et al.*, 1988). These two midbrain subpopulations of DA neurons express different levels of the DAT (Blanchard *et al.*, 1994), project to different areas (Graybiel *et al.*, 1990), and show different response to growth factors (Johansson *et al.*, 1995; Meyer *et al.*, 1999). In addition, patients with PD show a relative sparing of DA neurons in the VTA compared to the SNc indicating that the VTA DA neurons are less vulnerable compared to their SNc counterparts (Abe, 1992; Gibb, 1992; Iacopino and Christakos, 1990; Yamada *et al.*, 1990). One possible explanation for the reported dyskinesias in five of 37 patients in the study by Freed *et al.* (2001) and 13 of 23 patients in the study by Olanow *et al.* (2002) could be that unselective preparations of fetal tissue pieces containing fetal dopamine neurons before implantation may result in selective survival of the less vulnerable, but inappropriate VTA DA neurons (Isacson *et al.* 2001). When such neurons are implanted into the putamen of patients with PD, they may form inappropriate connections with, or avoid the projection neurons of the putamen, because these are not their normal targets (Isacson *et al.*, 2001; Isacson and Deacon, 1997). In addition, differences in DAT expression between subpopulations of DA neurons (Blanchard *et al.*, 1994; Ciliax *et al.*, 1999; Sanghera *et al.*, 1997) may also result in abnormal dendritic DA release in the putamen (Falkenburger *et al.*, 2001) and uptake patterns that could cause suboptimal DA transmission. Another likely

reason for uncontrolled motor response after transplantation is that location and size of tissue pieces implanted may in some cases create small lesions in the putamen with subsequent dysregulation of the GABAergic output neurons (as seen in Huntington's chorea). This theory is supported by the fact that small lesions in the striatum of primates (modeling Huntington's disease) make these animals severely dyskinetic in response to DA agonist treatment (Burns *et al.*, 1995; Hantraye *et al.*, 1990; Hantraye *et al.*, 1992), that would also occur in a situation of diffuse release from inappropriate DA neurons placed as tissue pieces in the lateral putamen (Freed *et al.*, 2001; Olanow, 2002). Supporting such a view is the relative absence of side effects from transplanted fetal DA cells to humans, when donor cells are prepared and placed as liquid cell suspension into PD putamen (Lindvall and Hagell, 2000; Mendez *et al.*, 2000a; Ramachandran *et al.*, 2002; Schumacher *et al.*, 2000) (see Figure 21.1). Such grafts typically reach less destructive size, but are better integrated in the host with more extensive axonal and synaptic outgrowth into appropriate target zones (Mendez *et al.* 2004, submitted).

Finally, of significance, the importance of appropriate cellular and biochemical characteristics of transplanted DA cells has also been shown by behavioral experiments. In a rodent model of parkinsonism, recovery from movement asymmetry is correlated with the rate of cellular maturation of the *donor* species (Bjorklund and Isacson, 2002; Isacson and Deacon, 1997). ES cells generating DA neurons also abide by such biological principles (Bjorklund and Isacson, 2002). Multiple anatomical analyses have demonstrated that specific axon guidance and cell differentiation factors remain in the adult and degenerating brain, providing growth and axonal guidance cues for fetal or ES cells (Haque *et al.*, 1997; Isacson and Deacon, 1996; Isacson *et al.*, 1995).

SUBPOPULATIONS OF MIDBRAIN DOPAMINERGIC NEURONS PERFORM DIFFERENT FUNCTIONS AND REACH DIFFERENT TARGETS: ITS POTENTIAL RELEVANCE TO REPAIR OF PD BRAINS

An important question in neural transplantation is the capacity for specific neuronal cell types to selectively reinnervate denervated host target regions (Isacson, 2003; Isacson and Deacon, 1996; Isacson *et al.*, 1995; Nilsson, 1988; Schultzberg *et al.*, 1984). Neurons developed from fetal or ES cell stages, when transplanted into host targets display a relative specificity in axonal outgrowth into regions typical of their mature phenotype (Isacson and Deacon, 1996; Isacson *et al.*, 1995; Nilsson, 1988; Schultzberg *et al.*, 1984). Midbrain DA neurons can be clearly categorized into subpopulations based on differences in the expression of certain proteins such as the dopamine transporter, TH, calbindin, and cholecystokinin (CCK) and molecules performing target neuron selection (Blanchard *et al.*, 1994; Haber, 1995; Schultzberg *et al.*, 1984). The biochemical markers used reflect specific projection areas reached by the different categories of DA SN neurons. In rodents, Schultzberg *et al.* (1984) first demonstrated by grafting embryonic VM tissue into the dorsal deafferented striatum, that subsets of SN DA neurons have specific final patterns of axonal terminal networks. CCK-negative/TH (A9) cells extend their axons into the motor striatum, while CCK-positive/TH neurons (A10) do not. Instead, A10 DA neurons project their axons in cortical areas and limbic striatal regions. Another interesting protein is the retinoic acid-generating enzyme, aldehyde dehydrogenase 2 (AHD2), which is expressed only in a subset of DA neurons in the substantia nigra (and VTA in some species) (McCaffery and Drager, 1994). Most AHD2 DA neurons project to the dorsal and rostral regions of the striatum and have a reduced DA terminal density gradient ventrally. Transplanted DA neurons at first appear to extend fibers in a nonspecific heliocentric fashion around the graft; however we have shown that the subset of transplanted DA fibers expressing AHD2 preferentially reinnervate the dorsolateral part of the rat striatum (Haque *et al.*, 1997). Histological observations of postmortem patient tissue (Isacson, unpublished) also indicate that transplanted human DA neurons will only seek their normal targets; and actively avoid innervation of other caudate-putamen regions according to their specific neuronal phenotype. Interestingly, the AHD2 enzyme is very sensitive to oxidation which may correlate with increased vulnerability of A9 neurons seen in PD (McCaffery and Drager, 1994). Recently, we have found with our collaborators doing clinical trials, that cell surface markers such as G-protein linked inward rectifying potassium channels type 2 (GIRK2) are selectively expressed in ventral and midbrain A9 DA neurons (not expressing calbindin). These new data allow a more in-depth analysis of the cell types and their growth characteristics while repairing the DA system in PD. Our clinical work (Mendez *et al.*, 2003) now indicates that a substantial proportion of type A9 GIRK2-positive DA neurons can be obtained by fetal cell suspension grafts. Because only 2% to 10% of neurons normally in the dissected VM are of a phenotype DA, hypothetically after transplantation they may compete for trophic support with the majority of the transplanted non-DA neurons. It is well known that midbrain DA neurons are positively influenced by growth factors such as bFGF (Mayer *et al.*, 1993) and GDNF (Granholm *et al.*, 1997; Meyer *et al.*, 1999). Such trophic dependency may vary between midbrain DA cell subpopulations, and

it is likely that a limited supply of trophic molecules influences DA cell survival and growth. In addition, 20% to 80% of the SN cell population in rodent laboratory studies of programmed cell death between P2 and P14 (Janec and Burke, 1993). In summary, transplanted fetal VM-derived DA neurons develop extensive axonal terminal networks, in the host striatum and nucleus accumbens in a normal, dense, homogeneous fashion (Bjorklund *et al.*, 1980; Schultzberg *et al.*, 1984). By contrast, identical phenotypic DA neurons (from fetal or stem cell derivation) only extend a few axons into the host brain when transplanted into non-target areas such as cortex, thalamus and hypothalamus (Abrous, 1988). Ultrastructural data show that grafted DA neurons are able to form appropriate and abundant synaptic connections with medium-sized spiny striatal neurons, which are the primary target of the mesencephalic DA afferents (Clarke, 1988). The molecular mechanisms defining anatomical specificity of the DA projections have not been fully elucidated but they are likely to involve target-derived trophic signals and recognition of target-specific cell surface molecules by growth-cones (Isacson and Deacon, 1997).

Is this cellular specificity of relevance to all cell therapy paradigms? We have demonstrated (Haque *et al.*, 1997) that a nonlinear regression function exists between DA neuron survival and extent of functional motor recovery. For example, approximately 590 rat-derived DA cells in the graft were necessary to obtain a 50% reduction in motor asymmetry using solid piece VM transplants. A plateau in motor recovery was reached at about 1200 TH-positive neurons. Interestingly, using pieces (non-dissociated grafts), the number of cells required for a 50% reduction in rotation is higher than comparable values in the literature for cell suspension grafts (Haque *et al.*, 1997). There may be several reasons for differences in efficacy between single cell suspension grafts and tissue pieces. Single cell suspension grafts become more vascularized than multiple-fragment transplants (Leigh, 1994). Further, Nikkah *et al.* (1994) showed that smaller suspension grafts show relatively better axonal coverage of the striatum compared to larger grafts. Large grafts, whether transplanted as cell suspensions or as tissue pieces, may actually limit the direct access to host tissue cues causing grafted cells to have less interaction with the host and integrate more within the actual transplanted tissue. These data and reasons indicate that single fetal cells in smaller deposits or possibly endogenous progenitor cells integrate better into the host brain and elicit behavioral recovery more efficiently (Brundin *et al.*, 1985; Hernit-Grant, 1996; Sotelo, 1987). Co-grafting of embryonic striatal tissue with fetal VM show that such grafts result in enhanced DA neuronal survival and function through trophic support (Costantini and

Snyder-Keller, 1997). Thus, graft-host interaction and graft-graft interaction are both factors regulating fetal VM transplant development. Interestingly, we established that the percentage reduction in rotations at 10 weeks' post-transplantation is mostly associated with the proportion of AHD2-positive A9 DA neurons in the grafts (Haque *et al.*, 1997). An inverse tendency was also apparent with the presence of A10 neurons. This observation suggests that the greater the proportion of A9 neurons in cell transfer or repair, the greater the degree of symptomatic recovery of motor signs or abnormalities. The importance of proper development of appropriate molecular and phenotypic cell characteristics can be observed in the functional effects of transplanted xenogeneic DA neurons. In xenografting experiments in a rat model of parkinsonism, recovery from amphetamine-induced motor asymmetry is tightly linked to the normal developmental rate of the species from where the grafted DA neurons are derived (Bjorklund and Isacson, 2002; Isacson and Deacon, 1997). Galpern *et al.* (1996) reported that between 80 and 120 porcine TH-positive neurons were required to see the same statistical effect using 120 to 140 rodent DA neurons, which is consistent with larger species-determined axonal growth areas in animals with larger brains.

In summary, the pathology of PD reveals a relatively selective loss of DA neurons in the substantia nigra pars compacta (SNc; A9) with a relative sparing of ventral tegmental area (VTA; A10) neurons. Our preliminary studies and reasoning suggest that selective repair by A9 DA cells in the putamen is more likely to contribute to symptomatic relief and prevent motor side-effects compared to implantation of the A10 DA cell group (Brownell *et al.*, 1998; Haque *et al.*, 1997; Mendez *et al.*, 2003). For example, transplanted A9 (AHD2+, GIRK2+ and dopamine D2 receptor+) neurons selectively reinnervate the appropriate motor regions of the affected putamen and correlate better with improved behavioral function in rodent models (Haque *et al.*, 1997) than non-A9 SN DA neurons. Because the A9 DA neurons have a better capacity to down-regulate dopamine release than A10 DA neurons by synaptic mechanisms, the A9 neurons are also a priori less likely to have unregulated DA release, that may cause too high or low and unstable DA levels provoking motor side effects, such as dyskinesia in patients as typically seen after long-term L-dopa or DA agonist treatment (Isacson *et al.*, 2001; Olanow and Obeso, 2000; Olanow and Tatton, 2000).

GENERATION OF A STEM CELL–DERIVED THERAPY FOR PARKINSON'S DISEASE

Previous work has shown that DA neurons can be expanded in a cell culture dish from growth factor expanded embryonic day 12 (E12) VM precursor cells (Studer *et al.*, 1998) or ES cells (Kawasaki *et al.*, 2000; Lee *et al.*, 2000b), as well as from *Nurr1* transfected mouse progenitor clones cultured with midbrain type 1 astrocytes (Wagner *et al.*, 1999). Remarkably, by injecting the undifferentiated ES cell directly into living tissue (brain or kidney capsule) ES cells or embryoid body cells can spontaneously differentiate into neurons (Deacon *et al.*, 1998) (see Figure 21.4). We have recently demonstrated that naïve ES cells implanted in low numbers into a DA-depleted striatum can develop and function as replicas of the DA neurons lost in PD. These new cells restore amphetamine induced motor symmetry and cortical activation (Bjorklund and Isacson, 2002) lost both in PD and PD animal models (see Figure 21.4). Such findings suggest that ES cells are a reasonable cell source for PD transplantation and could overcome the problem associated with using fetal primary neuron (pre-differentiated) or the fetal VM expanded precursor cells (Studer *et al.*, 1998), which both show low *in vivo* survival rates after transplantation (3% to 5% of the grafted DA neurons) (Brundin and Bjorklund, 1998; Studer *et al.*, 1998). In contrast, the direct ES cell transfer to brain has a high *in vivo* yield (up to fourfold) expansion and differentiation into DA neurons from a small number of implanted cells (1000 ES cells) (Bjorklund and Isacson, 2002). Vigorous basic research that defines the developmental sequences and repair mechanisms involved in fetal and stem cell–derived DA neuron cell therapy will likely provide a future modality for the treatment of PD. Nonetheless, the growth potential of the cell source (ES) also needs to be tempered and controlled (Bjorklund and Isacson, 2002). The risk for growth of non-neural tissues from the endodermal and mesodermal germ-layers needs to be eliminated (Bjorklund and Isacson, 2002). This can be accomplished by blocking activation or influence of meso- and endodermal cell fate inducers (Bjorklund and Isacson, 2002).

HOW CAN STEM CELL BIOLOGY RESEARCH HELP PARKINSON PATIENTS?

Most living systems undergo continuous growth. In humans, bone marrow stem cells are capable of dividing into most of the cells necessary for blood and immune systems. Even entire tissues or organs, such as the liver, can be regrown. Cells in the lining of the gut are shed and replaced on a daily basis, and in the skin, the basal cell layers of the dermis provide a continuous supply of growth. In adult mature mammalian cell systems, it is however typical that only organ-specific and specialized progenitor cells divide to maintain growth of organ systems in the body, while pathological processes can limit such replenishment. A fertilized cell is capable of cell divisions that grow logarithmically and in the early cluster of cells (in the range of 250 cells) each cell is capable of (or appears to be) forming any germ layer and part of the body plan (Hemmati-Brivanlou and Melton, 1997). This type of cell is therefore denoted stem cell, or in this case, ES cells. Recently, such divisible (yet non-malignant and non-carcinogenic) cells have gained increased attention. In a more limited scientific context, the concept and methodology of stem cell derived dopamine cells intrigues both neurobiologists and clinically oriented scientists, insofar as it could both explain the biology of the developing dopamine cells and what controls its function, as well as generating a procedure for obtaining such cells in abundance for clinical applications.

Starting at the genetic level, a number of genes related to the development or control of dopaminergic precursor proliferation, identity, and specialization (e.g., sonic hedgehog protein, LMX1b, *PitX3*, *Nurr1*, Engrail genes) act in concert with other transcription factors to activate specific transmitter enzymes (for instance TH, DAT, and dopa-decarboxylase) in dopaminergic neurons (Mandel *et al.*, 1998; Marsden, 1982; Montgomery *et al.*, 1996) (see Figure 21.3). An alternate or different route to cell repair is the possibility of manipulating inherent neurogenesis in the adult brain (Rakic, 2002). For example, in the adult brain exist neural precursor cells embedded in the subventricular zone (SVZ) that are capable of migration and differentiation into different neural cell types (Alvarez-Buylla and Garcia-Verdugo, 2002). These neural progenitor cells can be expanded and studied, possibly even as a neuronal repair tool (Flax *et al.*, 1998). The expansion of SVZ neuroprecursor cells is stimulated by delivery of basic fibroblast growth factor (bFGF), brain-derived neurotrophic factor (BDNF) (Benraiss *et al.*, 2000), noggin (Lim *et al.*, 2000), ciliary neurotrophic factor (CNTF) (Shimazaki *et al.*, 2001), or epidermal growth factor (EGF) (Craig *et al.*, 1996) to the CSF. Migration into the parenchyma can be stimulated by infusion of transforming growth factor-alpha (TGF-α) into a target region (Fallon *et al.*, 2000). Neuronal differentiation from precursors is enhanced by GDNF, BDNF (Zurn *et al.*, 2001) and natural BMP receptor antagonists (Lim *et al.*, 2000). Expression of a dopaminergic phenotype can be driven by transcription factors such as *PitX3* (Lebel *et al.*, 2001) and *Nurr1* which drive genes of the full DA neuronal phenotype such as TH, DAT and components of trophic factor receptors (Saucedo-Cardenas *et al.*, 1998; Tornqvist *et al.*, 2002; Wallen *et al.*, 2001). Further, release of GDNF or BDNF by genetically modified cells in the caudate-putamen can increase survival of precursor cells and of dopaminergic neurons in the SN following lesions (Akerud *et al.*, 2001), and the programmed cell death process can also be temporarily suspended by anti-apoptotic factors such as XIAP (Eberhardt *et al.*, 2002).

Interestingly for novel therapies, genetically modified cells can also serve as biologic pumps to produce growth factors that sustain neurons on site in the striatum or projecting dopaminergic neurons from the SN (Akerud *et al.*, 2001; Brownell, 1994; Schumacher *et al.*, 1991).

In summary, stem cell biology is valuable in a discovery process of several new and potential treatments for Parkinson's disease. For example, research on ES cells may demonstrate the genetic transcription factors that control the specific genes participating in DA orchestrated neuronal function and cell development. Genetic programs have dynamic components; for example, concerted actions of growth factors or neurotrophic factors that act as molecular switches required for initiating and maintaining the function of a DA neuron. The knowledge of how DA neurons can be formed would allow a reasonable process to be established for industrially producing a large number of such cells for transplantation or their regeneration inside the living brain (Isacson, 2003). The stem cell biology related to brain development and repair can be approached by methods using either adult or embryonic stem cells potentially capable of generating new neurons, after selective engineering, expansion in cell culture systems or in the living brain. The investigation of ES cells also provides a system in which transcription factors responsible for directing the typical dopaminergic cell fate in the nigrostriatal systems can be determined. This allows a sophisticated understanding of the factors controlling the specialization and health of such cells. This research may yield a stable and renewable cell source for therapeutic use in neurodegenerative diseases.

In conclusion, recent discoveries elucidating the cell biology of dopaminergic neurons allow both sequential and parallel strategies for protection of remaining cells and treating with new cells to restore function in Parkinson's patients (Check, 2002). The rapidly developing understanding of pathological mechanisms in Parkinson's disease and the life cycle of the dopamine neuron from stem cells, via progenitor cells, to adult and later aging dopamine neurons provides reasonable opportunities for new interventions to reverse the effects of this disease. Nonetheless, detailed knowledge is necessary about (1) the appropriate neurons, (2) correct brain locations for repair, and (3) responsive PD patients for these regeneration therapies to become successful.

THE CHALLENGES OF DEVELOPING NOVEL THERAPEUTICS USING LIVING CELLS AS AGENTS FOR PARKINSON'S DISEASE

Neural disorders caused by acute and widespread damage, such as spinal cord trauma or stroke, pose formidable conceptual and therapeutic challenges because of the many cells and cell types required to reconstitute such tissues (Rossi and Cattaneo, 2002). Nevertheless, recent use of neural progenitors in animal models has produced some evidence for direct or induced regeneration associated with functional recovery. Given the major efforts placed on understanding biological and therapeutic factors involved, there is also hope for new cell based procedures in such cases. However, even for the best candidate neurological diseases for cell therapy (such as PD), where neuronal repair is conceptually and practically proven to work in some patients, there is an apparent underestimation of the biological, technical, and cellular sophistication required to make the treatment reliable and safe. Although neural repair for PD may be a more challenging therapeutic development than cell-based therapy for type 1 diabetes mellitus, that experience may provide insight and perspective. About 25 years ago, the first reports were made of cell-based delivery of insulin as effective in rats with induced diabetes (Warnock et al., 1983). The rapid application of this completely new technology led to many disappointing clinical trials—where islet cells did not survive or function well. In the late 1990s, researchers revisited the conceptually elegant and useful idea of providing the patient with normal islet cells, with biofeedback to respond and release controlled levels of insulin (Weir et al., 1990). By the recent work of clinical and basic research groups, between 150 and 200 patients with diabetes mellitus now have functioning islet cell transplants that in most cases have obviated the need for insulin injections, and probably the associated side effects of insulin injection (Ryan et al., 2002). Likewise, understanding and accomplishing procedures that generate, select, and transfer appropriate cells in brains of responsive patients will be a requirement for the success of future cell and cell-based therapies. The development of brain cell transplantation with embryonic neurons and glia is innovative both from a technical and biological standpoint and thus will require much work to optimize the transplant protocol. The scaling up of this method from rodents to primates has proved very challenging, particularly in obtaining an acceptable, abundant, and reliable donor cell source and tissue preparation.

In a realistic future perspective, the most important factor to consider in trying to obtain optimal functional effects (and minimal side effects) is probably the establishment of new synapses and DA transmission that adequately adapts to the local milieu and provides physiologically appropriate DA release in the nigrostriatal system. Notably, transplanted fetal DA neurons have been shown to grow synaptic connections with mature host striatal neurons (Doucet et al., 1989; Mahalik et al., 1985). Current evidence indicates that DA-mediated regulation of striatal neuronal and network interactions is of critical importance for normal motor control, and that if DA release is not regulated, glutamatergic synapses are less able to provide normal control of striatal GABAergic output neurons (Nutt et al., 2000; Onn et al., 2000). Physiological DA function is achieved by normal DA release and uptake. Implanted cells that express the complete set of feedback elements required to regulate the release and uptake of DA may function well (Isacson and Deacon, 1997; Strecker et al., 1987; Zetterstrom et al., 1986). Fetal DA transplants have been shown to reduce the incidence of L-dopa–induced dyskinesias in rodents with 6-OHDA lesions (Lee et al., 2000a). Similarly, L-dopa–induced dyskinesias in non-human primates also are reduced after fetal DA cell transplantation (Bjorklund and Isacson, 2002). In several cases after transplantation of fetal mesencephalic cell mixtures, patients with PD have been able to eliminate their dopamine medication altogether (Lindvall and Hagell, 2000; Mendez et al., 2002). However, such cases are rare, and while transplants can survive for at least 15 years, there is limited understanding of optimal cell transfer parameters and patient selection. For the patients who have received cell transplants in Sweden, Canada, and the United States, it is not clear why some transplants work and others do not (Freed et al., 2001; Lindvall and Hagell, 2000; Mendez et al., 2002; Olanow, 2002). The current clinical studies have noted several mild-to-severe side effects caused by the implantation of dopaminergic cells. The primary problem has been the development of "off medication" dyskinesias in some patients with otherwise successful dopamine graft survival (not *successful* function) (Freed et al., 2001; Olanow, 2002). The mechanism(s) that underlie these "off medication" dyskinesias are unknown. It has been shown that despite good graft viability, there is suboptimal reinnervation of the host target areas and suboptimal dopamine cell composition of the transplanted and surviving grafts. In addition, some transplanted neurons may in effect produce too much and/or continuous dopamine for the denervated striatum, thereby producing dyskinesia during off periods, analogous to the well-known dopamine-induced dyskinesia during patients' ON periods. The dopamine neuron is a highly specialized functional unit that has evolved to release dopamine in a very specific way in target structures. In this way, it is very different than a transmitter pump or any other cell that could release and supply dopamine. More optimal cell therapy for PD may therefore include only cells that function like A9 DA neurons, which are indeed specialized for function in putamen (excluding A10 cell types that work in mesolimbic and cortical settings). There are risks for other side effects and dyskinesia that depend on the final size and location of transplants; because posterior putamen lesions when exposed to dopamine will generate dyskinesias (perhaps by space-occupying lesions such as a grafts) (Hantraye et al., 1992). These latest exploratory clinical transplantation studies have found some evidence that the stage of the disease and the responsiveness of the patient to L-dopa will influence the outcome of the transplantation procedures and function of the new dopamine cells. Some Parkinson's plus syndromes are not L-dopa responsive. It has become apparent in recent controlled clinical trials in PD demonstrating that the severity of PD influences transplant effects, or lack thereof. Below the level of 50 point UPDRS, patients had significantly better responses than those with above 50 score on UPDRS, before surgery (Olanow, 2002). In summary, identifying responsive patients and the generation and production of appropriate neurons (specificity principle) are necessary criteria towards successful future cell therapies against neurological diseases. Additional criteria to provide patients with cell technology require information about surgical and procedural applications, including cell implantation locations and cell dosage, cell preparations and trophic factors, and immunological and connectivity variables to allow functional reconstitution neurocircuitry. Combined methods of stem cell design, cell genesis, and selection to repair spinal cord and cortical motor system in amyotrophic lateral sclerosis, the cortico-striatal system in Huntington's disease, or the motor circuitries involving the DA system in Parkinson's Disease, may then lead to new regenerative and restorative therapies.

Therefore, the current state-of-the-art cell therapy for PD appears to require transfer of appropriate and selectively placed DA neurons and glia in patients who are responsive to DA substitution therapy. Possibly, the emergence of stem cell generated DA neurons in concert with improved surgical and technical approaches may provide a more optimal intervention.

ACKNOWLEDGMENTS

The author's research is supported by a Udall Parkinson's Disease Research Center of Excellence grant from NINDS, R01 grants for research on Parkinson's, Huntington's and Alzheimer's disease from the NIH, DOD USAMRMC, and a grant from the Harvard Center for Neurodegeneration and Repair. The ALS Association is also sponsoring this work on new initiatives in regeneration therapies.

References

Abe T et al. (1992). Molecular characterization of a novel metabotropic glutamate receptor, mGluR5, coupled to inositol phosphate/Ca2+ signal transduction. J Biol Chem 267:13361-13368.

Abrous N et al. (1988). Development of intracerebral dopaminergic grafts: A combined immunohistochemical and autoradiographic study of its time course and environmental influences. J Comp Neurol 273:26-41.

Akerud P et al. (2001). Neuroprotection through delivery of glial cell line-derived neurotrophic factor by neural stem cells in a mouse model of Parkinson's disease. J Neurosci 21:8108-8118.

Alvarez-Buylla A, Garcia-Verdugo JM. (2002). Neurogenesis in adult subventricular zone. J Neurosci 22:629-234.

Backlund E, Granberg P, Hamberger B. (1985). Transplantation of adrenal medullary tissue to striatum in parkinsonism. J Neurosurg 62:169-173.

Benraiss A et al. (2000). Adenoviral transduction of the ventricular wall with a BDNF expression vector induces neuronal recruitment from endogenous progenitor cells in the adult forebrain. Mol Ther 1:S35-S36.

Bjorklund A et al. (1980). Reinnervation of the denervated striatum by substantia nigra transplants: functional consequences as revealed by pharmacological and sensorimotor testing. Brain Res 199:307-333.

Bjorklund LM, Isacson O. (2002). Regulation of dopamine cell type and transmitter function in fetal and stem cell transplantation for Parkinson's disease. Prog Brain Res 138:411-420.

Bjorklund LM, et al. (2002). Embryonic stem cells develop into functional dopaminergic neurons after transplantation in a Parkinson rat model. Proc Natl Acad Sci USA 99:2344-2349.

Blanchard V et al. (1994). Differential expression of tyrosine hydroxylase and membrane dopamine transporter genes in subpopulations of dopaminergic neurons of the rat mesencephalon. Brain Res Mol Brain Res 22:29-38.

Brownell AL et al. (1994). PET- and MRI-based assessment of glucose utilization, dopamine receptor binding, and hemodynamic changes after lesions to the caudate-putamen in primates. Exp Neurol 125:41-51.

Brownell AL et al. (1998). In vivo PET imaging in rat of dopamine terminals reveals functional neural transplants. Ann Neurol 43:387-390.

Brundin P, Bjorklund A. (1998). Survival of expanded dopaminergic precursors is critical for clinical trials. Nat Neurosci 1:537.

Brundin P et al. (1985). Monitoring of cell viability in suspensions of embryonic CNS tissue and its use as a criterion for intracerebral graft survival. Brain Res 331:251-259.

Burbach JP et al. (2003). Transcription factors in the development of midbrain dopamine neurons. Ann NY Acad Sci 991:61-68.

Burns L H et al. (1995). Selective putaminal excitotoxic lesions in non-human primates model the movement disorder of Huntington disease. Neuroscience 64:1007-1017.

Check E. (2002). Parkinson's patients show positive response to implants. Nature 416:666.

Chung S et al. (2002a). Analysis of different promoter systems for efficient transgene expression in mouse embryonic stem cell lines. Stem Cells 20:139-145.

Chung S et al. (2002b). Genetic engineering of mouse embryonic stem cells by Nurr1 enhances differentiation and maturation into dopaminergic neurons. Eur J Neurosci 16:1829-1838.

Ciliax BJ et al. (1999). Immunocytochemical localization of the dopamine transporter in human brain. J Comp Neurol 409:38-56.

Clarke DJ et al. (1988). Human fetal dopamine neurons grafted in a rat model of Parkinson's disease: ultrastructural evidence for synapse formation using tyrosine hydroxylase immunocytochemistry. Exp Brain Res 73:115-126.

Costantini LC et al. (1997). Medial fetal ventral mesencephalon: a preferred source for dopamine neuron grafts. Neuroreport 8:2253-2257.

Costantini LC, Snyder-Keller A. (1997). Co-transplantation of fetal lateral ganglionic eminence and ventral mesencephalon can augment function and development of intrastriatal transplants. Exp Neurol 145:214-227.

Craig CG et al. (1996). In vivo growth factor expansion of endogenous subependymal neural precursor cell population in the adult mouse brain. J Neurosci 16:2649-2658.

Dahlstrom A, Fuxe K. (1964). Localization of monoamines in the lower brain stem. *Experientia* 20:398-399.

Damier P *et al.* (1999). The substantia nigra of the human brain. I. Nigrosomes and the nigral matrix, a compartmental organization based on calbindin D(28K) immunohistochemistry. *Brain* 122:1421-1436.

Deacon T *et al.* (1998). Blastula-stage stem cells can differentiate into dopaminergic and serotonergic neurons after transplantation. *Exp Neurol* 149:28-41.

Doucet G *et al.* (1989). Host afferents into intrastriatal transplants of fetal ventral mesencephalon. *Exp Neurol* 106:1-19.

Eberhardt O *et al.* (2002). Protection by synergistic effects of adenovirus-mediated X-linked chromosome-linked inhibitor of apoptosis and glial-derived neurotrophic factor gene transfer in the 1-methyl-4-phenyl-1,2,3,6-tetrahydropyridine model of Parkinson's disease. *J Neurosci* 20:9126-9134.

Falkenburger BH *et al.* (2001). Dendrodendritic inhibition through reversal of dopamine transport. *Science* 293:2465-2470.

Fallon J *et al.* (2000). In vivo induction of massive proliferation, directed migration, and differentiation of neural cells in the adult mammalian brain. *Proc Natl Acad Sci USA* 97:14686-14691.

Flax JD *et al.* (1998). Engraftable human neural stem cells respond to developmental cues, replace neurons, and express foreign genes. *Nat Biotechnol* 16:1033-1039.

Freed CR *et al.* (2002). Transplants of embryonic dopamine cells show progressive histologic maturation for at least eight years and improve signs of Parkinson's up to the maximum benefit of L-dopa preoperatively. *Neurology, Suppl. 3, Am Acad Neurol Mtg (abstract S31.006)* 58:A242.

Freed CR *et al.* (1992). Survival of implanted fetal dopamine cells and neurologic improvement 12 and 46 months after transplantation for Parkinson's disease. *N Engl J Med* 327:1549-1555.

Freed CR *et al.* (2001). Transplantation of embryonic dopamine neurons for severe Parkinson's disease. *N Engl J Med* 344:710-719.

Freeman AS *et al.* (1985). Firing properties of substantia nigra dopaminergic neurons in freely moving rats. *Life Sci* 36:1983-1994.

Freeman TB *et al.* (1995). Bilateral fetal nigral transplantation into the postcommissural putamen in Parkinson's disease. *Ann Neurol* 38:379-388.

Galpern WR *et al.* (1996). Xenotransplantation of porcine fetal ventral mesencephalon in a rat model of Parkinson's disease: functional recovery and graft morphology. *Exp Neurol* 140:1-13.

Gaudin D *et al.* (1990). Fetal dopamine neuron transplants prevent behavioral supersensitivity induced by repeated administration of L-dopa in the rat. *Brain Res* 506:166-168.

Gerfen C *et al.* (1998). Dopamine-mediated gene regulation in the striatum. *Adv Pharmacol* 42:670-673.

Gerfen CR *et al.* (1987). The neostriatal mosaic: II. Patch- and matrix-directed mesostriatal dopaminergic and non-dopaminergic systems. *J Neurosci* 7:3915-3934.

Gibb WR. (1992). Melanin, tyrosine hydroxylase, calbindin and substance P in the human midbrain and substantia nigra in relation to nigrostriatal projections and differential neuronal susceptibility in Parkinson's disease. *Brain Res* 581:283-291.

Grace AA. (1991). Phasic versus tonic dopamine release and the modulation of dopamine system responsivity: A hypothesis for the etiology of schizophrenia. *Neuroscience* 41:1-24.

Granholm A *et al.* (1997). Glial cell line-derived neurotrophic factor improves survival of ventral mesencephalic grafts to the 6-hydroxydopamine lesioned striatum. *Exp Brain Res* 116:29-38.

Graybiel AM *et al.* (1990). Patterns of cell and fiber vulnerability in the mesostriatal system of the mutant mouse weaver. I. Gradients and compartments. *J Neurosci* 10:720-733.

Haber SN, Ryoo H, Cox C, Lu W. (1995). Subsets of midbrain dopaminergic neurons in monkeys are distinguished by different levels of mRNA for the dopamine transporter: comparison with the mRNA for the D2 receptor, tyrosine hydroxylase and calbindin immunoreactivity. *J Comp Neurol* 362:400-410.

Hantraye P *et al.* (1990). A primate model of Huntington's disease: behavioral and anatomical studies of unilateral excitotoxic lesions of the caudate-putamen in the baboon. *Exp Neurol* 108:91-104.

Hantraye P *et al.* (1992). Intrastriatal transplantation of cross-species fetal striatal cells reduces abnormal movements in a primate model of Huntington disease. *Proc Natl Acad Sci USA* 89:4187-4191.

Haque N *et al.* (1997). Differential dissection of the rat E16 ventral mesencephalon and survival and reinnervation of the 6-OHDA-lesioned striatum by a subset of aldehyde dehydrogenase-positive TH neurons. *Cell Transplantation* 6:239-248.

Hauser RA, *et al.* (1999). Long-term evaluation of bilateral fetal nigral transplantation in Parkinson disease. *Arch Neurol* 56:179-187.

Hemmati-Brivanlou A, Melton D. (1997). Vertebrate embryonic cells will become nerve cells unless told otherwise. *Cell* 88:13-17.

Henderson BT *et al.* (1991). Implantation of human fetal ventral mesencephalon to the right caudate nucleus in advanced Parkinson's disease. *Arch Neurol* 48:822-827.

Hernit-Grant CS, Macklis JD. (1996). Embryonic neurons transplanted to regions of targeted photolytic cell death in adult mouse somatosensory cortex re-form specific callosal projections. *Exp Neurol* 139:131-142.

Hynes M, Rosenthal A. (2000). Embryonic stem cells go dopaminergic. *Neuron* 28:11-14.

Iacopino AM, Christakos S. (1990). Specific reduction of calcium-binding protein (28-kilodalton calbindin-D) gene expression in aging and neurodegenerative diseases. *Proc Natl Acad Sci USA* 87:4078-4082.

Isacson O. (2003). The production and use of cells as therapeutic agents in neurodegenerative diseases. *Lancet Neurol* 2:417-424.

Isacson O *et al.* (2001). Parkinson's disease: interpretations of transplantation study are erroneous. *Nat Neurosci* 4:553.

Isacson O *et al.* (2003). Towards full restoration of synaptic and terminal function of the dopaminergic system in Parkinson's disease from regeneration and neuronal replacement by stem cells. *Ann Neurol* 53:135-148.

Isacson O, Deacon TW. (1996). Specific axon guidance factors persist in the mature rat brain: evidence from fetal neuronal xenografts. *Neuroscience* 75:827-837.

Isacson O, Deacon TW. (1997). Neural transplantation studies reveal the brain's capacity for continuous reconstruction. *Trends Neurosci* 20:477-482.

Isacson O *et al.* (1995). Transplanted xenogeneic neural cells in neurodegenerative disease models exhibit remarkable axonal target specificity and distinct growth patterns of glial and axonal fibres. *Nat Med* 1:1189-1194.

Ito H *et al.* (1992). Calbindin-D28k in the basal ganglia of patients with parkinsonism. *Ann Neurol* 32:543-550.

Janec E, Burke RE. (1993). Naturally occurring cell death during postnatal development of the substantia nigra pars compacta of rat. *Mol Cell Neurosci* 4:30-35.

Johansson M *et al.* (1995). Effects of glial cell line-derived neurotrophic factor on developing and mature ventral mesencephalic grafts in oculo. *Exp Neurol* 134:25-34.

Johnson SW *et al.* (1992). Burst firing in dopamine neurons induced by N-methyl-D-aspartate: role of electrogenic sodium pump. *Science* 258:665-667.

Kawasaki H, *et al.* (2000). Induction of midbrain dopaminergic neurons from ES cells by stromal cell-derived inducing activity. *Neuron* 28:31-40.

Kim J-H *et al.* (2002). Dopamine neurons derived from embryonic stem cells function in an animal model of Parkinson's disease. *Nature* 418:50-56.

Kim KS *et al.* (2003). The orphan nuclear receptor Nurr1 directly transactivates the promotor of the tyrosine hydroxylase gene, but not the dopamine b-hydroxylase gene, in a cell-specific manner. *J Neurochem* 85:622-634.

Kordower J *et al.* (1998). Fetal nigral grafts survive and mediate clinical benefit in a patient with Parkinson's disease. *Mov Disord* 13:383-393.

Kordower J *et al.* (1996). Functional fetal nigral grafts in a patient with Parkinson's disease: chemoanatomic, ultrastructural, and metabolic studies. *J Comp Neurol* 370:203-230.

Kordower JH *et al.* (1995). Neuropathological evidence of graft survival and striatal reinnervation after the transplantation of fetal mesencephalic tissue in a patient with Parkinson's disease. *N Engl J Med* 332:1118-1124.

Lebel M *et al.* (2001). PitX3 activates mouse tyrosine hydroxylase promoter via a high-affinity binding site. *J Neurochem* 77:558-567.

Lee CS *et al.* (2000a). Embryonic ventral mesencephalic grafts improve levodopa-induced dyskinesia in a rat model of Parkinson's disease. *Brain* 123:1365-1379.

Lee SH *et al.* (2000b). Efficient generation of midbrain and hindbrain neurons from mouse embryonic stem cells. *Nat Biotechnol* 18:675-679.

Leigh K, Elisevich K, Rogers KA. (1994). Vascularisation and microvascular permeability in solid versus cell-suspension embryonic neural grafts. *J Neurosurg* 81:272-283.

Lim DA *et al.* (2000). Noggin antagonizes BMP signaling to create a niche for adult neurogenesis. *Neuron* 28:713-726.

Lindvall O, Hagell P. (2000). Clinical observations after neural transplantation in Parkinson's disease. *Prog Brain Res* 127:299-320.

Lindvall O *et al.* (1988). Fetal dopamine-rich mesencephalic grafts in Parkinson's disease. *Lancet* 2:1483-1484.

Ljungberg T *et al.* (1992). Responses of monkey dopamine neurons during learning of behavioral reactions. *J Neurophysiol* 67:145-163.

Madrazo I, Leon V, Torres C. (1988). Transplantation of fetal substantia nigra and adrenal medulla to the caudate putamen in two patients with Parkinson's disease. *N Engl J Med* 318:51.

Mahalik T *et al.* (1985). Substantia nigra transplants into denervated striatum of the rat: ultrastructure of graft and host interconnections. *J Comp Neurol* 240:60-70.

Mandel R *et al.* (1998). Characterization of intrastriatal recombinant adeno-associated virus-mediated gene transfer of human tyrosine hydroxylase and human GTP-cyclohydrolase I in a rat model of Parkinson's disease. *J Neurosci* 18:4271-4284.

Marsden CD. (1982). Basal ganglia disease. *Lancet* 2:1141-1147.

Mayer E *et al.* (1993). Mitogenic effect of basic fibroblast growth factor on embryonic ventral mesencephalic dopaminergic neurone precursors. *Dev Brain Res* 72:253-258.

McCaffery P, Drager U. (1994). High levels of a retinoic acid-generating dehydrogenase in the meso-telencephalic dopamine system. *Proc Natl Acad Sci USA* 91:7772-7776.

McNaught KS *et al.* (2002). Selective loss of 20S proteasome alpha-subunits in the substantia nigra pars compacta in Parkinson's disease. *Neurosci Lett* 326:155-158.

Mendez I *et al.* (2000a). Simultaneous intraputaminal and intranigral fetal dopaminergic grafts in Parkinson's disease: first clinical trials. *Exp Neurol* 164:464.

Mendez I *et al.* (2002). Simultaneous intrastriatal and intranigral fetal dopaminergic grafts in patients with Parkinson disease: a pilot study. Report of three cases. *J Neurosurg* 96:589-596.

Mendez I *et al.* (2000b). Enhancement of survival of stored dopaminergic cells and promotion of graft survival by exposure of human fetal nigral tissue to glial cell line–derived neurotrophic factor in patients with Parkinson's disease. Report of two cases and technical considerations. *J Neurosurg* 92:863-869.

Mendez I *et al.* (2004). Function and cell-specific analysis of fetal dopamine cell suspension transplants placed in the striatum and substantia nigra of patients with Parkinson's disease. *Nature Med*, submitted.

Meyer M *et al.* (1999). GDNF increases the density of cells containing calbindin but not of cells containing calretinin in cultured rat and human fetal nigral tissue. *Cell Transplant* 8:25-36.

Montgomery R *et al.* (1996). Herpes simplex virus-1 entry into cells mediated by a novel member of the TNF/NGF receptor family. *Cell* 87:427-436.

Munoz-Sanjuan I, Brivanlou AH. (2002). Neural induction, the default model and embryonic stem cells. *Nat Rev Neurosci* 3:271-280.

Nikkah G *et al.* (1994). Improved graft survival and striatal reinnervation by microtransplantation of fetal nigral cell suspensions in the rat Parkinson model. *Brain Res* 633:133-143.

Nilsson OG, Clarke DJ, Brundin P, Bjorklund A. (1988). Comparison of growth and reinnervation properties of cholinergic neurons from different brain regions grafted to the hippocampus. *J Comp Neurol* 268:204-222.

Nutt JG *et al.* (2000). Continuous dopamine-receptor stimulation in advanced Parkinson's disease. *Trends Neurosci* 23:S109-S115.

Olanow CW. (2002). Transplantation for Parkinson's disease: Pros, cons, and where do we go from here?, *Mov Disord* 17:S15.

Olanow CW, Obeso JA. (2000). Preventing levodopa-induced dyskinesias. *Ann Neurol* 47:167-178.

Olanow CW, Tatton WG. (2000). Etiology and pathogenesis of Parkinson's disease. *Annu Rev Neurosci* 22:123-144.

Onn S-P *et al.* (2000). Dopamine-mediated regulation of striatal neuronal and network interactions. *Trends Neurosci* 23:S45-56.

Piccini P *et al.* (1999). Dopamine release from nigral transplants visualized in vivo in a Parkinson's patient. *Nat Neurosci* 2:1137-1140.

Piccini P *et al.* (2000). Delayed recovery of movement-related cortical function in Parkinson's disease after striatal dopaminergic grafts. *Ann Neurol* 48:689-695.

Rakic P. (2002). Adult neurogenesis in mammals: An identity crisis. *J Neurosci* 22:614-618.

Ramachandran AC *et al.* (2002). A multiple target neural transplantation strategy for Parkinson's disease. *Rev Neurosci* 13:243-256.

Rossi F, Cattaneo E. (2002). Opinion: neural stem cell therapy for neurological diseases: dreams and reality. *Nat Rev Neurosci* 3:401-409.

Ryan EA *et al.* (2002). Successful islet transplantation: continued insulin reserve provides long-term glycemic control. *Diabetes* 51:2148-2157.

Sanghera MK *et al.* (1997). Dopamine transporter mRNA levels are high in midbrain neurons vulnerable to MPTP. *Neuroreport* 8:3327-3331.

Saucedo-Cardenas O *et al.* (1998). Nurr1 is essential for the induction of the dopaminergic phenotype and the survival of ventral mesencephalic late dopaminergic precursor neurons. *Proc Natl Acad Sci USA* 95:4013-4018.

Schultzberg M *et al.* (1984). Dopamine and cholecystokinin immunoreactive neurons in mesencephalic grafts reinnervating the neostriatum: evidence for selective growth regulation. *Neuroscience* 12:17-32.

Schumacher J *et al.* (2000). Transplantation of embryonic porcine mesencephalic tissue in patients with Parkinson's disease. *Neurology* 54:1042-1050.

Schumacher JM *et al.* (1991). Intracerebral implantation of nerve growth factor-producing fibroblasts protects striatum against neurotoxic levels of excitatory amino acids. *Neuroscience* 45:561-570.

Shimazaki T *et al.* (2001). The ciliary neurotrophic factor/leukemia inhibitory factor/gc130 receptor complex operates in the maintenance of mammalian forebrain neural stem cells. *J Neurosci* 21:7642-7653.

Sotelo C, Alvarado-Mallart RM. (1987). Embryonic and adult neurons interact to allow Purkinje cell replacement in mutant cerebellum. *Nature* 327:421-423.

Strecker R *et al.* (1987). Autoregulation of dopamine release and metabolism by intrastriatal nigral grafts as revealed by intracerebral dialysis. *Neuroscience* 22:169-178.

Strecker RE, Jacobs BL. (1985). Substantia nigra dopaminergic unit activity in behaving cats: Effects of arousal on spontaneous discharge and sensory evoked activity. *Brain Res* 361:339-350.

Studer L *et al.* (1998). Transplantation of expanded mesencephalic precursors leads to recovery in parkinsonian rats. *Nat Neurosci* 1:290-295.

Tornqvist N *et al.* (2002). Generation of tyrosine hydroxylase-immunoreactive neurons in ventral mesencephalic tissue of Nurr1 deficient mice. *Brain Res Dev Brain Res* 133:37-47.

Tropepe V *et al.* (2001). Direct neural fate specification from embryonic stem cells: a primitive mammalian neural stem cell stage acquired through a default mechanism. *Neuron* 30:65-78.

Venna N *et al.* (1984). Treatment of severe Parkinson's disease by intraventricular injection of dopamine. *Appl Neurophysiol* 47:62-64.

Wagner J *et al.* (1999). Induction of a midbrain dopaminergic phenotype in Nurr1-overexpressing neural stem cells by type 1 astrocytes. *Nat Biotechnol* 17:653-659.

Wallen AA *et al.* (2001). Orphan nuclear receptor Nurr1 is essential for Ret expression in midbrain dopamine neurons and in the brain stem. *Mol Cell Neurosci* 18:649-663.

Warnock GL *et al.* (1983). Normoglycemia after reflux of islet-containing pancreatic fragments into the splenic vascular bed in dogs. *Diabetes* 32:452-459.

Weir GC *et al.* (1990). Islet mass and function in diabetes and transplantation. *Diabetes* 39:401-405.

Widner H *et al.* (1992). Bilateral fetal mesencephalic grafting in two patients with severe parkinsonism induced by MPTP. *N Engl J Med* 327:1556-1563.

Yamada T *et al.* (1990). Relative sparing in Parkinson's disease of substantia nigra dopamine neurons containing calbindin-D28K. *Brain Res* 526:303-307.

Zetterstrom T *et al.* (1986). In vivo measurement of spontaneous release and metabolism of dopamine from intrastriatal nigral grafts using intracerebral dialysis. *Brain Res* 362:344-349.

Zetterstrom T, Ungerstedt U. (1984). Effects of apomorphine on the in vivo release of dopamine and its metabolites, studied by brain dialysis. *Eur J Pharmacol* 97:29-36.

Zurn AD *et al.* (2001). Sustained delivery of GDNF: towards a treatment for Parkinson's disease. *Brain Res* 36:222-229.

22

Cellular Plasticity of the Adult Human Brain

Steven A. Goldman, MD, PhD

Neural stem cells and more restricted neuronal and glial progenitor cells are dispersed widely throughout the adult vertebrate brain. Within the adult ventricular zone, self-renewing and multipotential neural stem cells are widely dispersed throughout the adult ventricular wall (Kirschenbaum and Goldman, 1995; Morshead et al., 1994; Weiss et al., 1996), and are especially abundant in the forebrain, in which they have been identified in species ranging from mice to man (Gage, 2000; Goldman, 1998). In addition, phenotypically committed neuronal progenitor cells remain widely distributed throughout the ventricular wall (Kirschenbaum and Goldman, 1995), especially within its rostral extensions to the olfactory bulb, and caudally through the subgranular zone to the dentate gyrus of the hippocampus (Doetsch and Alvarez-Buylla, 1996; Doetsch et al., 1997; Palmer et al., 1997). Besides these neurogenic progenitor cell populations, glial progenitors also reside within both the ventricular zone and tissue parenchyma (Roy et al., 1999). These include a subpopulation of multipotential parenchymal progenitors, which though nominally glial may generate neurons as well as astrocytes and oligodendrocytes (Kondo and Raff, 2000; Nunes et al., 2003; Palmer et al., 1999). In humans, these cells are dispersed throughout the subcortical white matter, just as in other infraprimate mammals (Nunes et al., 2003). Together, these different classes of progenitors constitute the major known categories of neural precursor cells in the adult human central nervous system (CNS) (Figure 22.1; see also the Color Plate section) (reviewed in Alvarez-Buylla and Garcia-Verdugo, 2002; Goldman, 2001, 2003; Goldman et al., 2002).

ADULT PROGENITORS MAY BE CATEGORIZED AS DISTINCT POOLS OF TRANSIT-AMPLIFYING CELLS

The neuronal and glial progenitor cells of the adult human brain may be considered akin to "transit amplifying cells," which have now been described as such in a variety of solid tissues. As initially defined in the skin and GI mucosae, transit amplifying cells comprise the phenotypically biased, still-mitotic progeny of uncommitted stem cells (Loeffler and Potten, 1997; Niemann and Watt, 2002; Potten and Loeffler, 1990; Watt, 2001). As stem cell progeny depart these localized regions of stem cell expansion, their daughters may commit to more restricted lineages, phenotypically delimited but still mitotic, which comprise the transit-amplifying pools. Although these cells proliferate to expand discrete lineages, they do not exhibit unlimited multilineage expansion, as distinct from their parental stem cells.

By this definition, the neuronal and glial progenitor cells of the adult brain may be considered distinct transit-amplifying derivatives of a common ventricular zone stem cell (Doetsch et al., 2002; Goldman, 2003). The neuronal progenitor cell of the forebrain subependyma was first proposed as a transit amplifying cell type, on the basis of its neuronal bias during mitotic expansion, and its ability to replenish the stem cell pool under appropriate mitotic stimulation. (Garcia-Verdugo et al., 1998). However, because the neuronally committed progenitor cells of the rostral migratory stream, and those of the subgranular zone of the dentate gyrus, both continue to divide while migrating (Menezes et al., 1995); they too comprise transit amplifying phenotypes (Doetsch et al., 2002). Indeed, even the glial progenitor

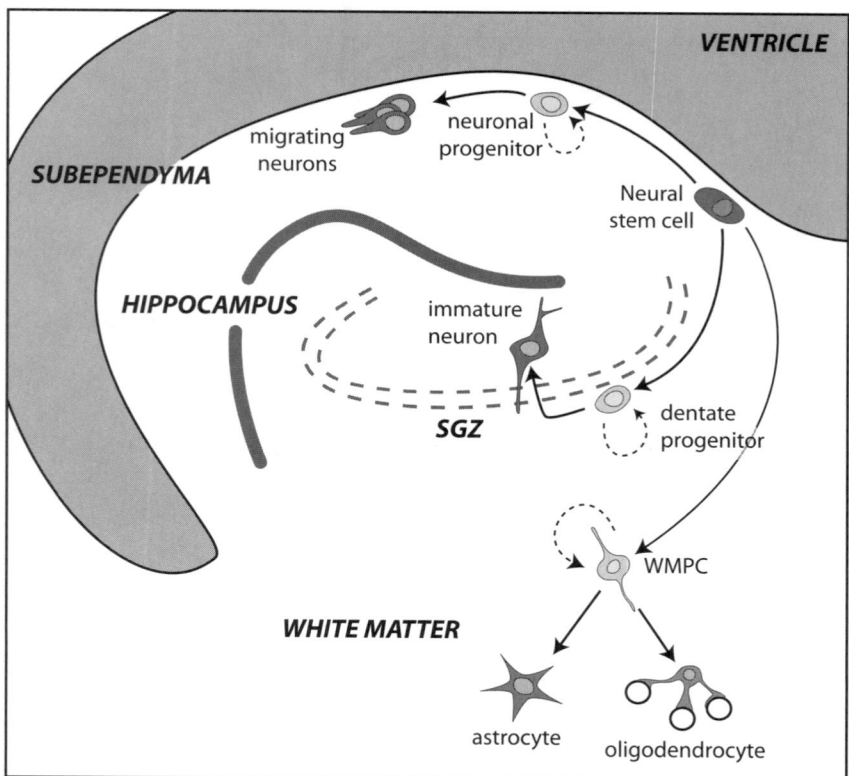

FIGURE 22.1 Progenitors of the adult human brain. This schematic illustrates the basic categories of progenitor cells in the adult brain, and their known interrelationships. The human temporal lobe is schematized by way of example. It includes both ventricular zone neural stem cells (*red*), which in turn generate at least three populations of potentially neurogenic transit amplifying progenitors of both neuronal and glial lineages (*yellow*). These include the neuronal progenitor cells of the ventricular subependyma, those of the subgranular zone of the dentate gyrus, and the white matter progenitor cells (WMPC) of the subcortical parenchyma, which although nominally glial, may remain potentially neurogenic. The latter may be represented in the gray matter parenchyma as well, although the relationships of the parenchymal gray and white matter progenitors have not yet been elucidated. Each transit amplifying pool may then give rise to differentiated progeny appropriate to their locations, including neurons (*purple*), oligodendrocytes (*green*) and parenchymal astrocytes (*blue*). Adapted from Goldman (2003). (See also the Color Plate section.)

of the adult white matter may now be considered a type of transit-amplifying cell, able to divide and yield variably restricted daughters, still mitotic but possessed of neither unbiased multipotentiality nor self-renewal capacity (Nunes *et al.*, 2003).

The nominally glial progenitor cell of the brain parenchyma now appears to comprise by far the most abundant progenitor cell phenotype of the adult brain. In adult humans, in whom the subependymal zone is but a discontinuous monolayer, periventricular zone neural stem cells appear to comprise a relatively scarce pool (Pincus *et al.*, 1998), although the subependymal astrocytes from which they may derive may be more widely distributed (Sanai *et al.*, 2004). Similarly, the neuronal progenitor pool of the olfactory bulb appears to be similarly vestigial in humans, although the neuronal progenitors of the adult human hippocampus may be more abundant (Eriksson *et al.*, 1998; Roy *et al.*, 2000). As a result, in adult humans the major transit-amplifying pool would appear to reside within the parenchyma itself, in which large numbers of widely dispersed glial progenitors provide an abundant reservoir of cycling, multipotential progenitors that, although restricted to glial phenotype *in vivo*, are multipotential and neurogenic. These cells may comprise as many as 3% of all cells in the adult white matter, yielding remarkably high estimates for their absolute incidence (Nunes *et al.*, 2003; Roy *et al.*, 1999; Scolding *et al.*, 1999). Although the incidence of analogous parenchymal progenitor cells in the adult human gray matter has not yet been rigorously evaluated, these cells have already been described as abundant in the cortical gray matter of adult rodents (Palmer *et al.*, 1999), and there is no reason to think that they are any less abundant in humans.

PROGENITOR CELLS ARE PARCELED INTO PERMISSIVE NICHES FOR NEURONAL AND GLIAL PRODUCTION

The largely parallel studies of adult neurogenesis and adult neural progenitor cells conducted over the past decade led to a fundamental paradox: Neural stem cells appear widely dispersed throughout the adult subependyma, and neurogenic progenitors can be harvested throughout the adult lateral ventricular wall. Nonetheless, neurogenesis occurs at only a few discrete and invariant sites, rostrally in the striatal wall and olfactory subependyma, and caudally in the dentate gyrus. In all other regions of the neuraxis, subependymal progenitors, despite their neurogenic capacity *in vitro*, appear to generate glia *in vivo*. This process is mirrored in lower species, such as the songbird, in which widely dispersed progenitor pools are associated with neurogenesis in only a few discrete brain regions (Goldman, 1998). Across phylogeny then, persistent progenitor cells appear to be tonically inhibited from neuronal production, save for discrete regions of permissive neurogenesis (Gage, 2000).

This paradox of a regional restriction of neurogenesis, despite more widely dispersed if not ubiquitous ventricular zone progenitor pools, has proven highly instructive. Specifically, it led to the concept of permissive niches for neurogenesis, borrowing from the hematopoietic system the concept of permissive niches for both stem cell expansion and phenotypic restriction. At the most basic level, neurogenic (and gliogenic) niches are both established and defined by reciprocal paracrine interactions between astrocytes, endothelial cells, and ependymal cells. It is the largely humoral products of these interactions that appear to determine whether resident precursor cells divide at all, and if they do, whether they then give rise to neurons or glia.

In most areas of the adult ventricular neuroepithelium, astrocyte-derived bone morphogenetic proteins (BMPs) typically suppress local neurogenesis, by directing astrocytic differentiation from competent neural progenitors (Gross *et al.*, 1996). In addition, nitric oxide synthase appears to tonically inhibit progenitor turnover in these regions, and its inhibition can mobilize resident progenitors to divide (Packer *et al.*, 2003). Yet in the specifically neurogenic regions of the adult olfactory stream and hippocampal dentate gyrus, both ependymal cells and progenitors express noggin, a competitive antagonist of the BMPs (Chmielnicki *et al.*, 2003; Lim *et al.*, 2000). The local production and parenchymal sequestration of noggin may then permit subependymal neurogenesis, by limiting the access of astrocyte-derived BMPs to neural stem and progenitor cells (Lim *et al.*, 2000; Paine-Saunders *et al.*, 2001).

In these specifically neurogenic regions, astrocytes also appear to support neurogenesis (Lim and Alvarez-Buylla, 1999; Song *et al.*, 2002), doing so through both IGF1 (Aberg *et al.*, 2000; Jiang *et al.*, 1998) and FGF (Palmer *et al.*, 1995) receptor-dependent activation of cycling progenitors. In addition, astrocyte-released adenosine triphosphate (ATP) and adenosine diphosphate (ADP), acting through P2Y G-protein–coupled purine receptors, may mediate glial regulation of mitotic neurogenesis in the adult ventricular zone and hippocampus (Braun *et al.*, 2003). Thus, astrocytes appear to be elaborating both positive and negative influences on neurogenesis, in a spatial pattern that may be both organized by ependymal noggin, and sustained in part by noggin's matrix sequestration (Paine-Saunders *et al.*, 2002). This mechanism may permit the establishment of neurogenic niches not only in the striatal subependyma, but also in the dentate gyrus and olfactory bulb, each of which expresses high levels of noggin protein (Chmielnicki *et al.*, 2004; Chmielnicki and Goldman, 2002). As such, the distribution and degree of noggin production may be appropriate targets for modulating both the geography and extent of ventricular zone neurogenesis in adult animals (Chmielnickl *et al.*, 2004).

NEUROGENIC NICHES MAY BE DEFINED BY FOCI OF LOCAL ANGIOGENESIS

Permissive niches for neurogenesis may be supported by the paracrine interactions of neural precursor cells with the cerebral microvasculature. In the adult hippocampus, neuronal progenitor cells are spatially associated with mitotic endothelial cells, in discrete foci of concurrent angiogenesis and neurogenesis (Palmer, 2002; Palmer *et al.*, 2000). Both endothelial cells and neural progenitors express the VEGFR2/KDR receptor for vascular endothelial growth factor (VEGF) (Jin *et al.*, 2002), suggesting that the two phenotypes may be co-activated by stimuli inducing local VEGF secretion or release. Mammalian endothelial cells can express brain-derived neurotrophic factor (BDNF), which acts as a potent differentiation and survival factor for neurons generated from adult subependymal progenitor cells (Ahmed *et al.*, 1995; Kirschenbaum and Goldman, 1995; Pincus *et al.*, 1998; Shetty and Turner, 1998). Indeed, human brain endothelial cells can support the migration and survival of neurons arising from cultures of the adult striatal ventricular zone, and their support of ventricular zone neurogenesis can be suppressed by trkB-Fc, a scavenger of secreted BDNF (Leventhal *et al.*, 1999). It would seem likely that hippocampal angiogenesis might also support local neurogenesis in a BDNF-dependent fashion, although this has yet to be directly demonstrated.

This concept of endothelial support of adult neurogenesis finds additional support in studies of persistent neurogenesis in the adult songbird forebrain, in which angiogenesis per se appears to be a potent trigger for neurogenesis and neuronal recruitment. In the neostriatal vocal control nucleus of these animals, testosterone stimulates the production and release of VEGF. This local increase in VEGF elicits a burst of mitotic angiogenesis, which is followed by the production of BDNF by the stimulated microvascular cells (Louissaint et al., 2002). The local release of BDNF then acts as both a migratory and survival cue for neurons departing the VZ, permitting their parenchymal integration as new neurons. In this system as well as in the mammalian hippocampus and subependyma (Palmer, 2002), multiple non-neuronal cell types thus collaborate to provide a locally instructive environment for neuronal differentiation.

PROGENITOR CELLS PERSIST WITHIN THE BRAIN PARENCHYMA AS WELL AS ITS GERMINAL ZONES

Besides the subependymal pools of neurogenic progenitors, cycling progenitor cells of apparent glial lineage and fate persist throughout the brain parenchyma. These too may constitute a population of transit amplifying cells, from which astrocytes and oligodendrocytes arise, both tonically and reactively. Remarkably though, at least some of these cycling cells appear to persist as uncommitted progenitors, and are able to generate neurons as well as glia once freed from the local environment of the adult white matter (Belachew et al., 2003; Nunes et al., 2003; Palmer et al., 1995). As such, the nominally oligodendrocyte progenitor cells of the adult human brain may now best be viewed as a pool of uncommitted neural progenitor cells, that are biased toward glial production by their environment, although still fundamentally multipotent. Indeed, the very absence of neurogenesis in most regions of the adult brain, despite the ubiquity of neurogenic progenitor cells in the brain parenchyma, argues that the environment of the adult brain actively suppresses neuronal mitogenesis. Instead, most regions of adult brain parenchyma appear to sustain an environment specifically instructive for central gliogenesis, so that gliogenesis is essentially the default fate of adult parenchymal progenitor cells.

Transplant studies have supported this point. Spinal progenitors, which although multipotential in vitro are typically gliogenic when implanted back into spinal cord, can generate neurons when transplanted to the pro-neurogenic environment of the hippocampus (Horner et al., 2000). Conversely, highly neurogenic hippocampal and olfactory progenitors typically quickly cease neurogene-

sis once transplanted to non-neurogenic regions of brain. Similarly, otherwise neurogenic progenitor cells implanted into the sublethally irradiated hippocampus generate glia (Monje et al., 2002; Monje and Palmer, 2003). Thus, the irradiated hippocampus ceases neurogenesis not because of the loss of resident progenitor cells, but rather because radiation appears to render the hippocampal microenvironment nonpermissive for neurogenesis. As such, progenitor activation in the irradiated hippocampus yields gliosis rather than neurogenesis, because newly generated cells follow astrocytic rather than neuronal differentiation paths. Such is the power of the microenvironment in dictating the differentiated phenotypes of its progeny.

GLIA AS NEURAL PROGENITOR CELLS

Alvarez-Buylla and colleagues first reported that adult neural stem cells comprised a subpopulation of subependymal glial cells, in which the glial fibrillary acidic protein promoter was transcriptionally active (Doetsch et al., 1999). On this basis, these investigators proposed that neural stem cells were of astrocytic lineage (Alvarez-Buylla and Garcia-Verdugo, 2002; Alvarez-Buylla et al., 2001). However, it is important to note that neither GFAP promoter activation nor GFAP protein expression are sufficient to define a cell as an astrocyte (Barres, 2003; Goldman, 2003). Rather, astrocytes may best be defined functionally—by their participation in the blood-brain barrier and gliovascular unit, their high-capacity uptake of glutamate, and their strongly hyperpolarized resting membrane potential (Nedergaard et al., 2003). The antigenic markers that correspond to these functional traits include the water transport molecules aquaporin 4 and 9 (Rash et al., 1998), the calcium binding protein S100β, and the glutamate transporters GLAST, EAAT2 and 4 (Anderson and Swanson, 2000). These are all markers of astrocytic phenotype, and yet none has been shown to have a necessary correspondence to GFAP expression patterns, either within the subependyma or otherwise. Thus, if neural stem cells are indeed astrocytes, then one might expect both the aquaporins and glutamate transporters to be expressed by neural stem cells. Similarly, if neural stem cells were indeed astrocytes, one would expect to identify their anatomic relationship with the vasculature, and to see both vascular endfeet and junctional contacts with endothelial cells (Simard et al., 2003). Although this remains a topic of active investigation, to date none of these conditions has yet been met in the investigation of neural stem cells of either the fetal or adult brain. Thus, neural stem and progenitor cells may include stages during which they express GFAP, without necessarily being astrocytes as rigorously defined. This point remains controversial, as several investigators have reported lineage data that would at the very least argue for the lineal derivation of persistent neural stem cells from astrocytic founders (Zhuo et al., 2001), as well as for the ability of astrocytes to generate a neural stem cell phenotype in vitro (Kondo and Raff, 2000; Laywell et al., 2000).

From the standpoint of clinical therapeutics, the real importance of identifying the phenotype of resident neural stem cells is in defining the precise populations amenable to inductive regeneration, and in identifying the signals that will allow stem cell mobilization and differentiation without affecting parenchymal glia. Indeed, one might argue that the more similar both subependymal and parenchymal progenitor cells are to mature astrocytes, the more narrow will be the therapeutic window for their induction, and the harder it will be to design strategies that will allow their specific mobilization and directed differentiation without untoward effects on tissue astrocytes—which, after all, are by far the most abundant cell population of the adult brain.

PARENCHYMAL GLIAL PROGENITORS INCLUDE MULTIPOTENTIAL PROGENITORS

The phenotypic identification of neural stem cells was made more difficult by recent observations that besides the subependymal cell population of the ventricular zone, glial progenitor cells within the brain parenchyma might also have multilineage competence. Such multipotential precursor cells were first identified in the parenchyma of the adult rodent brain, as well as in its ventricular lining (Palmer et al., 1999; Richards et al., 1992). In the human, analogous pools were found within the cortex (Arsenijevic et al., 2001) and subcortical white matter (Nunes et al., 2003). The latter cells in particular, defined as white matter progenitor cells (WMPCs), were initially identified as oligodendrocyte progenitors on the basis of their expression of PDGFαR and the A2B5 epitope (Scolding et al., 1998; Scolding et al., 1999), their transcriptional activation of the CNP promoter (Roy et al., 1999), and their predominant differentiation as oligodendrocytes, both in vitro and upon transplantation (Windrem et al., 2002). Yet upon removal to low-density, serum-free culture, in which the cells are effectively removed from both autocrine and paracrine influences, they generate neurons as well as astrocytes and oligodendrocytes, and remain propagable for several months in vitro (Nunes et al., 2003) (Figure 22.2; see also the Color Plate section). In vivo, these cells appear as small, highly ramified cells with thin processes that lack endothelial endfeet or even contact. Most express nestin and the NG2 chondroitin sulfate proteoglycan, markers of immature neural progenitors, and S100β, a glial marker expressed by immature cells of both the astrocytic and oligodendrocytic lineages; most express neither GFAP nor aquaporin, nor do they transcriptionally activate the GFAP promoter, and hence cannot be readily characterized as astroglial (Goldman, 2003). Together, these data suggest that the parenchymal glial progenitor of the adult human white matter, nominally an oligodendrocyte progenitor cell, is in fact a multipotential neural progenitor cell, restricted to generate glia by virtue of the adult parenchymal environment, and not because of any autonomous lineage commitment.

GLIAL PROGENITORS OF THE SUBCORTICAL WHITE MATTER ARE TRANSIT-AMPLIFYING CELLS

Importantly, although the parenchymal progenitor cell seems to be fundamentally multipotential, it is subject to replicative senescence and does not express measurable telomerase (Nunes et al., 2003). As a result, it typically ceases expansion after 3–4 months in vitro, spanning no more than 18 population doublings. In light of its lack of telomerase, its self-limited expansion capacity, and its glial bias despite multilineage competence, the parenchymal progenitor cell cannot be considered a parenchymal stem cell. Rather, this phenotype may best be considered a transit-amplifying progenitor of both astrocytes and oligodendrocytes. Interestingly then, just as the transit-amplifying neuronal precursor of the adult rat ventricular zone may revert to a multipotential state in the presence of EGF (Doetsch et al., 2002), the dividing glial progenitor of the human white matter is similarly able to revert to a multilineal neurogenic precursor. Although this is especially manifest when the cells are expanded in vitro, adult glial progenitor cells may generate neurons as soon as they are removed from the local tissue environment, whether extracted on the basis of CNP promoter activation, A2B5 immunoselection (Nunes et al., 2003), or NG2 expression (Belachew et al., 2003). Accordingly, when freshly sorted parenchymal progenitors were introduced via transuterine xenograft into the fetal rat brain, all neural phenotypes were found to arise in a context-dependent manner (Nunes et al., 2003). In general terms then, removal of glial progenitor cells from all environmental influences may prove both necessary and sufficient to ensure the appearance of multipotential progenitors in their resultant isolates. As such, one may consider the parenchymal progenitor pool of the adult brain as including both multipotential and glial-restricted members, the relative proportions and lineage competencies of which are dynamically regulated by the local tissue environment. These observations may suggest a hitherto unappreciated degree of cellular plasticity within the adult brain. A salient implication of this work is that with a greater

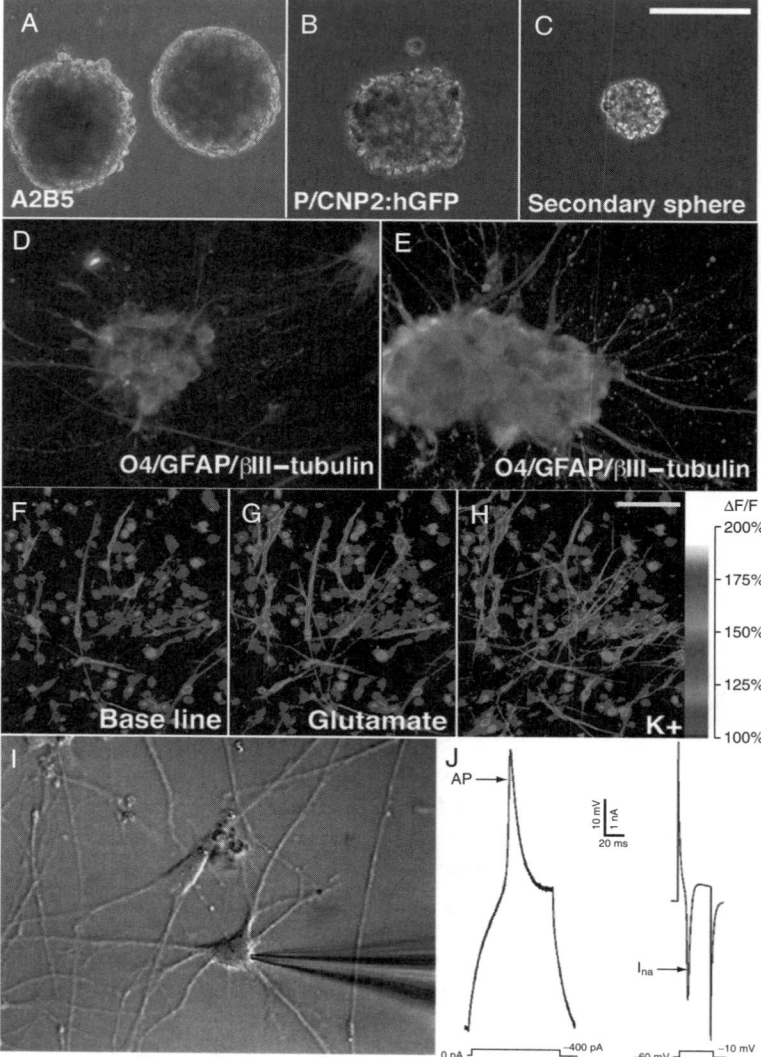

FIGURE 22.2   Resident progenitor cells of the adult human white matter can generate neurons as well as oligo-dendrocytes and astrocytes. **A:** First passage spheres generated from A2B5-sorted cells 2 weeks' post-sort. **B:** First passage spheres arising from P/CNP2:hGFP sorted cells, 2 weeks. **C:** Second passage sphere derived from an A2B5-sorted sample, at 3 weeks. **D:** Once plated onto substrate, the primary spheres differentiated as βIII-tubulin+ neurons (*red*), GFAP+ astrocytes (*blue*), and O4+ oligodendrocytes (*green*). **E:** Neurons (*red*), astrocytes (*blue*), and oligodendrocytes (*green*) similarly arose from spheres derived from P/CNP2:GFP-sorted WMPCs. **F–H:** *WMPC-derived neurons developed neuronal Ca^{2+} responses to depolarization.* **D** shows an image of WMPC-derived cells loaded with the calcium indicator dye fluo-3, 10 days after plating of first passage spheres derived from A2B5-sorted white matter (35 DIV total); many fiber-bearing cells of both neuronal and glial morphologies are apparent. **E,** The same field after exposure to 100 μM glutamate, and **F,** after exposure to a depolarizing stimulus of 60 mM KCl. The neurons displayed rapid, reversible, greater than 100% elevations in cytosolic calcium in response to K+, consistent with the activity of neuronal voltage-gated calcium channels. **I–J,** *Whole cell patch-clamp revealed voltage-gated sodium currents and action potentials in WMPC-derived neurons.* **G:** A representative cell 14 days after plating of first passage sphere derived from A2B5-sorted white matter. The cell was patch-clamped in a voltage-clamped configuration, and its responses to current injection recorded. **H:** The fast negative deflections noted after current injection are typical of the voltage-gated sodium currents of mature neurons. Action potentials were noted only at I$_{Na}$ greater than 800 pA. Scale: **A–E,** 100 μm; **F–H,** 80 μm. Adapted from Nunes *et al.* (2003). (See also the Color Plate section.)

understanding of the necessary and sufficient conditions for establishing neurogenic niches, we might expect to be able to modulate the local environment to encourage the production of new neurons from resident glial progenitor cells. Achieving and then directing this capability to therapeutic endpoints may prove a strategy of great import.

NON-NEURAL PROGENITORS MAY RARELY CONTRIBUTE TO BRAIN ARCHITECTURE

Another challenge ripe with promise is that of using non-neural sources of cells for therapeutic engraftment to the CNS. A number of studies had earlier posited that non-neural progenitor cells, including both bone marrow- and umbilical cord–derived stromal cells, might be competent to give rise to neurons and glial cells in the adult CNS (Krause, 2001; Mezey *et al.*, 2000). From these studies arose the concept of trans-differentiation of non-neural progenitors to neural stem cells, and then to neurons and both oligodendrocytes and astrocytes. This issue has engendered considerable controversy, in part due to the methodological limitations of these studies, in which the expression of defined neural markers by donor stromal cells was often viewed as synonymous with functional phenotypic trans-differentiation. Yet despite the lack of clear data supporting the differentiation of implanted stromal cells into functionally competent neurons and myelinogenic oligodendrocytes, several especially compelling instances of apparent neuronal differentiation from donor stromal cells had begged explanation. These included several reports of cerebellar Purkinje neurons of unmistakable morphology and distribution that expressed genetic tags indicating their derivation from stromal donor cells (Priller, 2001; Weimann *et al.*, 2003). Against this background, studies of hematopoietic and stromal stem cell treatment of hepatic insufficiency then led to the discovery that the fusion of donor stromal cells with resident hepatocytes could lead to the formation of donor-recipient heterokaryons (Vassilopoulos *et al.*, 2003; Wang, 2003). This might readily have led to the false impression of donor cell differentiation to nonmesenchymal phenotypes.

In a follow-up to these studies, Alvarez-Buylla and colleagues then demonstrated that cerebellar Purkinje neurons, along with both hepatocytes and cardiomyocytes, may be particularly susceptible to such fusion events (Alvarez-Dolado *et al.*, 2003). As a result, the initial reports of neuronal trans-differentiation from bone marrow stromal cells, including those of Purkinje neuronal differentiation, now appear to have been artifactual. In retrospect, these reports appear to have been based on fusion events, in which host neurons and glia incorporated the genetic tags imparted by their stromal

fusion partners (Ying *et al.*, 2002). Nonetheless, these studies succeeded in raising the possibility that endogenous bone marrow–derived mesenchymal progenitors, once mobilized to the bloodstream, might be able to fuse with autologous CNS partners, thereby conferring new information to resident cells. Whether this occurs in the natural history of either normal adult humans, or in response to disease or injury, is presently unknown. Yet whatever the natural history of fusion events within the CNS, the administration of stromal donor cells for the purpose of fusion-mediated gene delivery to the CNS may prove a useful and broadly applicable therapeutic strategy.

THERAPEUTIC TARGETS FOR IMPLANTED NEURAL PROGENITOR CELLS

The potential therapeutic targets of adult progenitor cells are manifold and are the subject of other reviews, both in this volume and elsewhere (Chmielnicki and Goldman, 2002; Goldman *et al.*, 2002; Goldman *et al.*, 2001). Briefly, these cells may be viewed as vectors for cell replacement via transplantation, or as targets for the endogenous induction of resident progenitor cells. Whereas the former represents the broad category of cell-based therapy of the nervous system, the latter comprises gene therapeutic approaches to progenitor induction and repair. For cell replacement, neuronal progenitors of relatively homogeneous phenotype are just beginning to become available, from both tissue-derived and embryonic stem cell–derived neural progenitor cells. As both the directed induction (Studer *et al.*, 2000; Wichterle *et al.*, 2002) and promoter-based isolation (Keyoung *et al.*, 2001; Li *et al.*, 1998; Roy *et al.*, 2000a; Roy *et al.*, 1999; Roy *et al.*, 2000b; Wang *et al.*, 1998) of therapeutically desired phenotypes become more refined, we may envisage the acquisition of progenitors for all major phenotypes of the human CNS. Indeed, our ability to do so may be expected to advance in parallel with our growing understanding of the developmental ontogeny of these cells. Such positionally restricted and transmitter-defined neuronal progenitors already appear amenable to non-oncogenic immortalization using telomerase (Roy *et al.*, 2004), which should anticipate their being generated in clinically useful numbers and purities. As a result, neurodegenerative diseases as diverse as Parkinson's, Huntington's, and Alzheimer's may prove feasible targets for cell-based therapy, as progenitors restricted to midbrain dopaminergic, striatal GABAergic, and forebrain cholinergic fate, respectively, become available.

As a therapeutic modality, the transplantation of neuronal progenitors would seem of greatest potential therapeutic efficacy precisely for neurodegenerative diseases such as Parkinson's and Huntington's, as well as for both the ischemic and metabolic oligodendrocytic disorders, since each of these is largely attributable to the dysfunction or loss of a single neural phenotype. As a corollary to this point, it may be fair to predict that diseases of polyphenotypic loss, such as stroke, trauma, and spinal cord injury, may be far less amenable to cell-based implant strategies, precisely because of the many different types of cells lost in these processes and our lack of understanding of how to reconstruct vectorially specified neural networks from their constituent cellular elements. Although several studies have reported some degree of benefit from the implantation of undifferentiated neural precursor cells into stroke and traumatic cord injury, none has convincingly demonstrated any restoration of normal microanatomy or structural integrity to these more profound disruptions of neural architecture.

Glial pathologies may lend themselves to cell-based therapy even more readily than neuronal disorders, given the relative homogeneity and accessibility of the major glial phenotypes, oligodendrocytes and astrocytes, and the abundance of their parental progenitors. As noted, oligodendrocyte progenitors may be isolated to purity and used for therapeutic remyelination. This may be done in both adult targets of acquired demyelination (Windrem et al., 2002), and perinatally in disorders of myelin formation or maintenance, such as the congenital leukodystrophies (Windrem et al., 2004) (Figure 22.3; see also the Color Plate section). The latter include disorders of glycosaminoglycan metabolism such as the mucopolysaccharidoses, as well as the gangliosidoses and other lysosomal lipid storage disorders (Kaye, 2001). As a group, these disorders may be amenable to astrocyte-mediated cell therapy as well as to oligodendrocyte progenitor-based strategies, because astroglia may be sufficiently migratory in the CNS to ensure the widespread delivery of wild-type enzymes and therapeutic transgenes to recipient brain tissue. From the standpoint of structural repair, as opposed to enzyme replacement and metabolic restitution, disease targets as diverse as the vascular leukoencephalopathies in adults, and cerebral palsy in children, may prove amenable to glial cell–based therapies; each of these disease categories is predominantly characterized by glial loss. Indeed, one may expect rapid clinical translation of glial cell–based therapy, as more is learned of the disease stages and attendant therapeutic windows at which human glial progenitor cell transplants may be most effectively used.

THE THERAPEUTIC INDUCTION OF ENDOGENOUS PROGENITOR CELLS

Our emphasis here on cell therapy–based repair reflects the reality that most studies in restorative neurol-

ogy of the past decade have focused on the use of transplanted neurons and glia, and more recently of stem and progenitor cells, as therapeutic agents. Relatively less effort has been devoted to using or mobilizing endogenous progenitor populations. Yet as we have discussed, several major populations of accessible progenitor populations persist in the adult brain, including that of the human. These pools are individually accessible, and may be mobilized through a variety of both pharmacological and gene therapeutic strategies that result in the mitotic expansion of resident stem and progenitor cells. Epidermal and fibroblast growth factors (Craig et al., 1996; Kuhn et al., 1996; Kuhn et al., 1997), tumor growth factor–α (Fallon et al., 2000), nitric oxide synthase inhibitors (Packer et al., 2003), insulin-like growth factor (Aberg et al., 2000), BDNF (Benraiss et al., 2001; Pencea et al., 2001), and noggin (Chmielnicki et al., 2004; Chmielnicki and Goldman, 2002), among others, have each been shown to mobilize endogenous subependymal progenitor cell populations. Parenchymal glial progenitors have also been the target of mobilization strategies, a prominent goal of which has been the induction of oligoneogenesis for induced remyelination of demyelinated foci (Levine et al., 2001). Parallel studies have sought to generate therapeutically useful neuronal populations from resident parenchymal progenitors; for example, dopaminergic nigrostriatal neurons from parenchymal progenitors of the adult substantia nigra (Zhao et al., 2003).

Such studies have been predicated on the logic that progenitors may be destined, by virtue of early homeodomain encoding, to give rise to neurons appropriate to the area of CNS from which those progenitors are derived. For instance, neural stem cells derived from the ventral mesencephalic ventricular zone appear biased toward dopaminergic neurogenesis (Sawamoto et al., 2001). In addition, discrete subpopulations of progenitors competent to give rise to specific neuronal phenotypes, such as the striatal or dentate neuronal progenitor populations, may be stimulated through the use of cognate growth factors, whether delivered as purified proteins or gene therapeutics. For instance, viral overexpression of BDNF may be used to stimulate generation of new medium spiny neurons (Benraiss et al., 2001; Chmielnicki et al., 2004; Chmielnicki and Goldman, 2002), a possible therapeutic strategy in Huntington's disease. Such strategies of inducing resident progenitor cells to regenerate those neuronal and glial cell populations lost to specific diseases have only begun to bear fruit experimentally. Realistically, a great number of both conceptual and operational hurdles must be overcome before induced neurogenesis can be considered a practical or imminently feasible therapeutic option. Nonetheless, the stimulation of endogenous progenitor cells to replace both neurons and glia lost to injury or

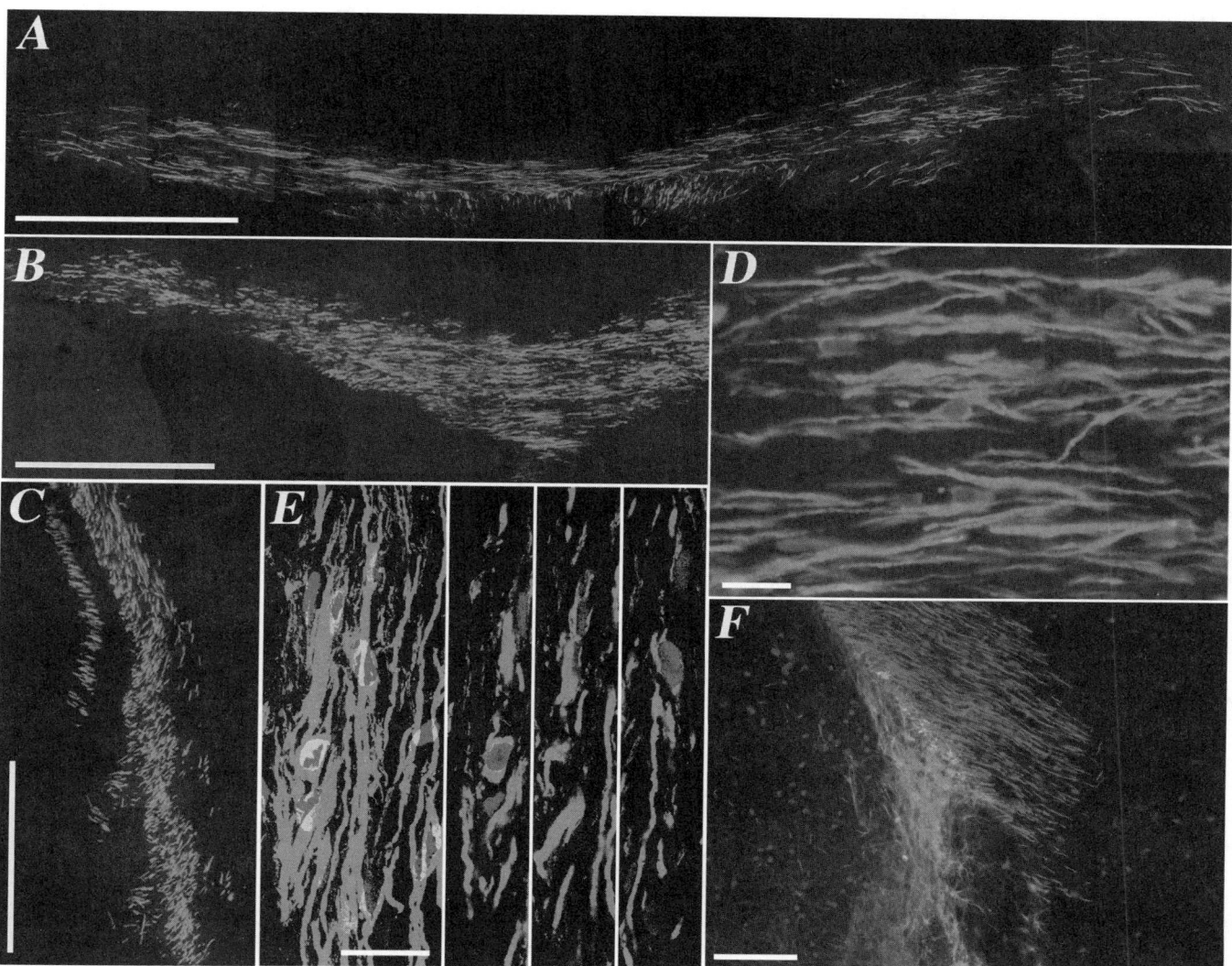

FIGURE 22.3 Engrafted human OPCs myelinate an extensive region of the forebrain. **A–B:** Extensive myelin basic protein expression by sorted human fetal OPCs, implanted into homozygote shiverer mice as neonates, indicates that large regions of the corpus callosum (**A** and **B**, 2 different mice) have myelinated by 12 weeks (MBP, *green*). **C:** Human OPCs also migrated to and myelinated fibers throughout the dorsoventral extents of the internal capsules, manifesting widespread remyelination of the forebrain after a single perinatal injection. **D:** MBP expression (*green*), in an engrafted shiverer callosum 3 months after perinatal xenograft, is associated with donor cells, identified by human nuclear antigen (hNA, in *red*). Both the engrafted human cells and their associated myelin were invariably found to lay parallel to callosal axonal tracts. **E:** Optical sections of implanted shiverer corpus callosum, with hNA+ human cells surrounded by MBP. Human cells (*arrows*) found within meshwork of MBP+ fibers (**E**, merged image of **E1-3**). **F:** OPCs were recruited as oligodendrocytes or astrocytes in a context-dependent manner. This photo shows the striatocallosal border of a shiverer brain, 3 months after perinatal engraftment with human fetal OPCs (hNA in *blue*). Donor-derived MBP+ oligodendrocytes and myelin (*red*) is evident in the corpus callosum, while donor-derived GFAP+ (*green*) astrocytes predominate on the striatal side.

Continued

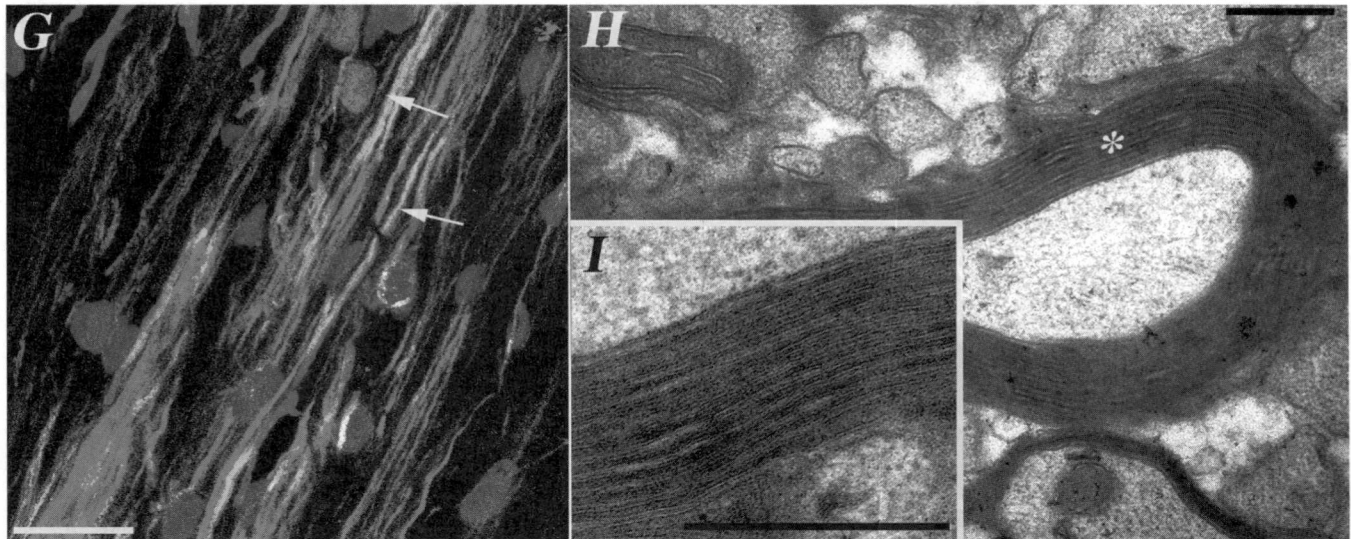

FIGURE 22.3, cont'd   **G:** A confocal optical section exhibiting axonal ensheathment and myelin compaction by engrafted human progenitor cells. This section of engrafted corpus callosum has been triple-immunostained for MBP (*red*), human ANA (*blue*), and neurofilament protein (*green*). All MBP immunostaining is derived from the human OPCs, whereas the NF[+] axons are those of the mouse host. *Arrows* identify segments of murine axons surrounded by human oligodendrocytic MBP. **H–I:** An electron micrograph of a 16-week-old shiverer homozygote implanted with human OPCs shortly after birth, showing host axons with densely compacted, donor-derived myelin sheaths. The *asterisk* indicates the field enlarged in the inset, which shows the major dense lines of mature myelin. Scale: **A–C,** 1 μm; **D,** 100 μm; **E,** 20 μm; **F,** 200 fm. Adapted from Windrem M. *et al.* (2004). (See also the Color Plate section.)

disease, and the direction of this process to achieve precise and functionally competent restoration of disrupted neural networks, is very much a principal goal of the new neurology that we now approach.

References

Aberg M *et al.* (2000). Peripheral infusion of IGF-1 selectively induces neurogenesis in the adult rat hippocampus, *J Neurosci* 20:2896-2903.

Ahmed S, Reynolds BA, Weiss S. (1995). BDNF enhances the differentiation but not the survival of CNS stem cell-derived neuronal precursors, *J Neurosci* 15:5765-5778.

Alvarez-Buylla A, Garcia-Verdugo JM. (2002). Neurogenesis in adult subventricular zone, *J Neurosci* 22:629-634.

Alvarez-Buylla A, Garcia-Verdugo JM, Tramontin A. (2001). A unified hypothesis on the lineage of neural stem cells, *Nat Rev Neurosci* 2:287-293.

Alvarez-Dolado M *et al.* (2003). Fusion of bone marrow-derived cells with Purkinje neurons, cardiomyocytes and hepatocytes, *Nature* 425:968-974.

Anderson C, Swanson R. (2000). Astrocyte glutamate transport: review of properties, regulation, and physiological functions, *Glia* 32:1-14.

Arsenijevic Y *et al.* (2001). Isolation of multipotent neural precursors residing in the cortex of the adult human brain, *Exp Neurol* 170:48-62.

Barres B. (2003). What is a glial cell, *Glia* 43:4-5.

Belachew S *et al.* (2003). Postnatal NG2 proteoglycan-expressing progenitor cells are intrinsically multipotent and generate functional neurons. *J Cell Biol* 161:169-186.

Benraiss A *et al.* (2001). Adenoviral brain-derived neurotrophic factor induces both neostriatal and olfactory neuronal recruitment from endogenous progenitor cells in the adult forebrain, *J Neurosci* 21:6718-6731.

Braun N *et al.* (2003). Expression of the ecto-ATPase NTPDase2 in the germinal zones of the developing and adult rat brain, *Eur J Neurosci* 17:1355-1364.

Chmielnicki E, Benraiss A, Economides A, Goldman SA. (2004). Adenovirally-expressed noggin and BDNF cooperate to induce new medium spiny neurons from resident progenitor cells in the adult striatal ventricular zone, *J Neurosci* 24:2133-2142.

Chmielnicki E, Goldman SA. (2002). Induced neurogenesis by endogenous progenitor cells in the adult mammalian brain, *Progr Brain Res* 138:451-464.

Craig CG *et al.* (1996). In vivo growth factor expansion of endogenous subependymal neural precursor cell populations in the adult mouse brain, *J Neurosci* 16:2649-2658.

Doetsch F, Alvarez-Buylla A. (1996). Network of tangential pathways for neuronal migration in adult mammalian brain, *Proc Natl Acad Sci USA* 93:14895-14900.

Doetsch F *et al.* (1999). Subventricular zone astrocytes are neural stem cells in the adult mammalian brain, *Cell* 97:703-716.

Doetsch F, Garcia-Verdugo JM, Alvarez-Buylla A. (1997). Cellular composition and three-dimensional organization of the subventricular germinal zone in the adult mammalian brain, *J Neurosci* 17:5046-5061.

Doetsch F *et al.* (2002). EGF converts transit-amplifying neurogenic precursors in the adult brain into multipotent stem cells, *Neuron* 36:1021-1034.

Eriksson PS *et al.* (1998). Neurogenesis in the adult human hippocampus, *Nat Med* 4:1313-1317.

Fallon J *et al.* (2000). In vivo induction of massive proliferation, directed migration, and differentiation of neural cells in the adult mammalian brain, *Proc Natl Acad Sci USA* 97:14686-14691.

Gage F. (2000). Mammalian neural stem cells, *Science* 287:1433-1438.

Garcia-Verdugo J, Doetsch F, Wichterle H, Alvarez-Buylla A. (1998). Architecture and cell types of the adult subventricular zone: in search of the stem cells, *J Neurobiol* 36:234-248.

Goldman S. (1998). Adult neurogenesis: From canaries to the clinic, *J Neurobiol* 36:267-286.

Goldman S. (2003). Glia as neural progenitor cells, *Trends Neurosci* 26:590-596.

Goldman S *et al.* (2002). Isolation and induction of adult neural progenitor cells, *Clin Neurosci Res* 2:70-79.

Goldman SA. (2001). Neural progenitor cells of the adult human forebrain. In Rao M, editor: *Stem cells and CNS development*, New York: Humana, pp. 177-206.

Gross RE *et al.* (1996). Bone morphogenetic proteins promote astroglial lineage commitment by mammalian subventricular zone progenitor cells, *Neuron* 17:595-606.

Horner PJ *et al.* (2000). Proliferation and differentiation of progenitor cells throughout the intact adult rat spinal cord, *J Neurosci* 20:2218-2228.

Jiang J, McMurtry J, Niedzwiecki D, Goldman SA. (1998). Insulin-like growth factor-1 is a radial cell-associated neurotrophin that promotes neuronal recruitment from the adult songbird ependyma/subependyma, *J Neurobiol* 36:1-15.

Jin, K *et al.* (2002). VEGF stimulates neurogenesis in vitro and in vivo. *Proc Natl Acad Sci* 99:11946-11950.

Kaye E. (2001). Update on genetic disorders affecting white matter, *Pediatr Neurol* 24:11-24.

Keyoung HM *et al.* (2001). Specific identification, selection and extraction of neural stem cells from the fetal human brain, *Nat Biotechnol* 19:843-850.

Kirschenbaum B, Goldman SA. (1995). Brain-derived neurotrophic factor promotes the survival of neurons arising from the adult rat forebrain subependymal zone, *Proc Natl Acad Sci USA* 92:210-214.

Kondo T, Raff M. (2000). Oligodendrocyte precursor cells reprogrammed to become multipotential CNS stem cells, *Science* 289:1754-1757.

Krause D. (2001). Multi-organ, multi-lineage engraftment by a single bone marrow-derived stem cell, *Cell* 105:369-377.

Kuhn G, Dickinson-Anson H, Gage F. (1996). Neurogenesis in the dentate gyrus of the adult rat: age related decrease of neuronal progenitor proliferation, *J Neurosci* 16:2027-2033.

Kuhn HG *et al.* (1997). Epidermal growth factor and fibroblast growth factor-2 have different effects on neural progenitors in the adult rat brain, *J Neurosci* 17:5820-5829.

Laywell E *et al.* (2000). Identification of a multipotent astrocytic stem cell in the immature and adult mouse brain, *Proc Natl Acad Sci USA* 97.

Leventhal C *et al.* (1999). Endothelial trophic support of neuronal production and recruitment from the adult mammalian subependyma, *Mol Cell Neurosci* 13:450-464.

Levine JM, Reynolds R, Fawcett JW. (2001). The oligodendrocyte precursor cell in health and disease, *Trends Neurosci* 24:39-47.

Li M, Pevny L, Lovell-Badge R, Smith A. (1998). Generation of purified neural precursors from embryonic stem cells by lineage selection, *Curr Biol* 8:971-974.

Lim D *et al.* (2000). Noggin antagonizes BMP signaling to create a niche for adult neurogenesis, *Neuron* 28:713-726.

Lim DA, Alvarez-Buylla A. (1999). Interaction between astrocytes and adult subventricular zone precursors stimulates neurogenesis, *Proc Natl Acad Sci USA* 96:7526-7531.

Loeffler M, Potten CS. (1997). Stem cells and cellular pedigrees. In Pottten CS, editor: *Stem cells*, San Diego: Academic Press, pp. 1-28.

Louissaint A, Rao S, Leventhal C, Goldman SA. (2002). Coordinated interaction of angiogenesis and neurogenesis in the adult songbird brain, *Neuron* 34:945-960.

Menezes JR, Smith CM, Nelson KC, Luskin, MB (1995). The division of neuronal progenitor cells during migration in the neonatal mammalian forebrain, *Mol Cell Neurosci* 6:496-508.

Mezey E *et al.* (2000). Turning blood into brain: cells bearing neuronal antigens generated in vivo from bone marrow, *Science* 290:1779-1782.

Monje M, Mizumatsu S, Fike J, Palmer T. (2002). Irradiation induces neural precursor cell dysfunction, *Nat Med* 8:955-962.

Monje M, Palmer T. (2003). Radiation injury and neurogenesis, *Curr Opin Neurol* 16:129-134.

Morshead CM *et al.* (1994). Neural stem cells in the adult mammalian forebrain: a relatively quiescent subpopulation of subependymal cells, *Neuron* 13:1071-1082.

Nedergaard, M, Ransom, B, Goldman, S.A. (2003). A new role for astrocytes: redefining the functional architecture of the brain. *Trends Neurosci* 26:523-529.

Niemann C, Watt FM. (2002). Designer skin: lineage commitment in postnatal epidermis, *Trends Cell Biol* 12:185-192.

Nunes MC *et al.* (2003). Identification and isolation of multipotential neural progenitor cells from the subcortical white matter of the adult human brain, *Nat Med* 9:439-447.

Packer M *et al.* (2003). Nitric oxide negatively regulates mammalian adult neurogenesis, *Proc Natl Acad Sci* 100:9566-9571.

Paine-Saunders S, *et al.* (2002). Heparan sulfate proteoglycans retain noggin at the cell surface: A potential mechanism for shaping BMP gradients, *J Biol Chem* 277:2089-2096

Palmer T. (2002). Adult neurogenesis and the vascular Nietzsche, *Neuron* 34:856-858.

Palmer T, Willhoite A, Gage F. (2000). Vascular niche for adult hippocampal neurogenesis, *J Comp Neurol* 425:479-494.

Palmer TD *et al.* (1999). Fibroblast growth factor-2 activates a latent neurogenic program in neural stem cells from diverse regions of the adult CNS, *J Neurosci* 19:8487-8497.

Palmer TD, Ray J, Gage FH. (1995). FGF-2-responsive neuronal progenitors reside in proliferative and quiescent regions of the adult rodent brain, *Mol Cell Neurosci* 6:474-486.

Palmer TD, Takahashi J, Gage FH. (1997). The adult rat hippocampus contains primordial neural stem cells, *Mol Cell Neurosci* 8:389-404.

Pencea V, Bingaman KD, Wiegand SJ, Luskin MB. (2001). Infusion of brain-derived neurotrophic factor into the lateral ventricle of the adult rat leads to new neurons in the parenchyma of the striatum, septum, thalamus, and hypothalamus, *J Neurosci* 21:6706-6717.

Pincus DW *et al.* (1998). Fibroblast growth factor-2/brain-derived neurotrophic factor-associated maturation of new neurons generated from adult human subependymal cells, *Ann Neurol* 43:576-585.

Potten CS, Loeffler M. (1990). Stem cells: attributes, cycles, spirals, pitfalls and uncertainties. Lessons for and from the crypt, *Development* 110:1001-1020.

Priller J. (2001). Neogenesis of cerebellar Purkinje neurons from gene-marked bone marrow cells in vivo, *J Cell Biol* 155:733-738.

Rash J, Yasumura T, Hudson C, Agre P, Nielsen S. (1998). Direct immunogold labeling of aquaporin-4 in square arrays of astrocyte and ependymocyte plasma membranes in rat brain and spinal cord, *Proc Natl Acad Sci USA* 95:11981-11986.

Richards LJ, Kilpatrick TJ, Bartlett PF. (1992). De novo generation of neuronal cells from the adult mouse brain, *Proc Natl Acad Sci USA* 89:8591-8595.

Roy N et al. (2004). Telomerase-immortalization of neuronal progenitor cells derived from the human fetal spinal cord, *Nat Biotechnol* 22:289-305.

Roy NS et al. (2000a). Promoter-targeted selection and isolation of neural progenitor cells from the adult human ventricular zone, *J Neurosci Res* 59:321-331.

Roy NS et al. (1999). Identification, isolation, and promoter-defined separation of mitotic oligodendrocyte progenitor cells from the adult human subcortical white matter, *J Neurosci* 19:9986-9995.

Roy NS et al. (2000b). In vitro neurogenesis by progenitor cells isolated from the adult human hippocampus, *Nat Med* 6:271-277.

Sanai, N et al. (2004). Unique astrocyte ribbon in adult human brain contains neural stem cells but lacks chain migration. *Nature* 427:740-743.

Sawamoto K et al. (2001). Generation of dopaminergic neurons in the adult brain from mesencephalic precursor cells labeled with a nestin-GFP transgene, *J Neurosci* 21:3895-3903.

Scolding N et al. (1998). Oligodendrocyte progenitors are present in the normal adult human CNS and in the lesions of multiple sclerosis, *Brain* 121:2221-2228.

Scolding NJ, Rayner PJ, Compston DA. (1999). Identification of A2B5-positive putative oligodendrocyte progenitor cells and A2B5-positive astrocytes in adult human white matter, *Neuroscience* 89:1-4.

Shetty AK, Turner DA. (1998). In vitro survival and differentiation of neurons derived from epidermal growth factor-responsive postnatal hippocampal stem cells: inducing effects of brain-derived neurotrophic factor, *J Neurobiol* 35:395-425.

Simard M et al. (2003). Signaling at the gliovascular interface, *J Neurosci* 23:9254-62.

Song H, Stevens CF, Gage FH. (2002). Astroglia induce neurogenesis from adult neural stem cells, *Nature* 417:39-44.

Studer L et al. (2000). Enhanced proliferation, survival and dopaminergic differentiation of CNS precursors in lowered oxygen, *J Neurosci* 20:7377-7738.

Vassilopoulos G, Wang P, Russell D. (2003). Transplanted bone marrow regenerates liver by cell fusion, *Nature* 422:901-904.

Wang S et al. (1998). Isolation of neuronal precursors by sorting embryonic forebrain transfected with GFP regulated by the T alpha 1 tubulin promoter [published erratum appears in *Nat Biotechnol* 1998 May;16(5):478], *Nat Biotechnol* 16:196-201.

Wang X. (2003). Cell fusion is the principal source of bone marrow-derived hepatocytes, *Nature* 422:897-901.

Watt FM. (2001). Stem cell fate and patterning in mammalian epidermis, *Curr Opin Genet Dev* 11:410-417.

Weimann J et al. (2003). Contribution of transplanted bone marrow cells to Purkinje neurons in human adult brains, *Proc Natl Acad Sci USA* 100:2088-2093.

Weiss S et al. (1996). Multipotent CNS stem cells are present in the adult mammalian spinal cord and ventricular neuroaxis, *J Neurosci* 16:7599-7609.

Wichterle H, Lieberam I, Porter J, Jessell T. (2002). Directed differentiation of embryonic stem cells into motor neurons, *Cell* 110:385-397.

Windrem M et al. (2004). Fetal and adult human oligodendrocyte progenitor cell isolates myelinate the congenitally dysmyelinated brain, *Nat Med* 10:93-97.

Windrem M et al. (2002). Progenitor cells derived from the adult human subcortical white matter disperse and differentiate as oligodendrocytes within demyelinated regions of the rat brain, *J Neurosci Res* 69:966-975.

Ying QL, Nichols J, Evans EP, Smith AG. (2002). Changing potency by spontaneous fusion, *Nature* 416:545-548.

Zhao M et al. (2003). Evidence for neurogenesis in the adult mammalian substantia nigra, *Proc Natl Acad Sci USA* 100:7925-7930.

Zhuo L et al. (2001). hGFAP-cre transgenic mice for manipulation of glial and neuronal function in vivo, *Genesis* 31:85-94.

Harnessing Endogenous Stem Cells for Central Nervous System Repair

D. Chichung Lie, MD
Sophia A. Colamarino, PhD
Hongjun Song, PhD
Fred H. Gage, PhD

The adult central nervous system (CNS) displays only limited regenerative capacity. It was long believed that the predominant repair mechanisms did not extend beyond postmitotic events, such as sprouting of axon terminals, changes in neurotransmitter-receptor expression and synaptic reorganization. The generation of new neurons in the adult brain was reported for the first time in the mid-1960s. Using tritiated thymidine autoradiography, Altman and Das found that dividing cells survived in the adult brain and generated cells with the morphological characteristics of neurons (Altman and Das, 1965). Over subsequent years, Altman and his colleagues continued to provide evidence for this phenomenon. However, due to technical restrictions, such as the lack of definitive phenotypic markers, and conceptual constraints posed by the existing dogma that denied the possibility of regeneration in the adult CNS, the idea of adult mammalian neurogenesis was met with major skepticism and enthusiasm remained low. Several years later, Kaplan and colleagues revisited the idea and provided strong ultrastructural evidence for the neuronal identity of newly generated cells in the adult hippocampal dentate gyrus, one of the regions where neurogenesis persists into adulthood (Kaplan and Hinds, 1977). In the mid-1980s, the developing field of adult neurogenesis was further stimulated by the work of Nottebohm and co-workers, who studied seasonal neuronal cell death and generation of new neurons in the adult canary brain and postulated that this process was involved in the acquisition of new memory (Goldman and Nottebohm, 1983). At the beginning of the 1990s, the focus of the field shifted back from adult avian neurogenesis to mammalian systems. Technical advances,

such as the characterization and visualization of newborn neurons by newly developed markers and the ability to isolate, propagate, and manipulate cells that are considered to be the origin of the newly generated neurons, have fueled the progress in this field (reviewed in Gage, 2000). And although the basic mechanisms of adult mammalian neurogenesis are far from understood, recent suggestions of a potential to generate new neurons in lesioned areas of the adult brain have raised our hopes that, in the future, we might be able to use this process to repair the diseased CNS.

NEURAL STEM CELLS ARE POTENTIALLY PRESENT THROUGHOUT THE ADULT MAMMALIAN CENTRAL NERVOUS SYSTEM

During development of the mammalian CNS, neurons and glia arise from immature proliferating cells in a stereotypical sequence: first neurons are generated, primarily during the embryonic period, followed by glia, the majority of which differentiate after most neurons are born. Within the developing CNS the most immature cell in this lineage is the neural stem cell. Neural stem cells are defined as cells that have the ability to self-renew and to give rise to the three major cell types of the CNS: neurons, astrocytes, and oligodendrocytes. Progenitor cells are defined as cells that are lineage-restricted and/or show limited proliferative capacity. Through a sequence of symmetric and asymmetric divisions, in which neural stem cells give rise to lineage-restricted cells which, in turn, differentiate into mature CNS cells, the major CNS phenotypes are generated while a pool of undifferentiated stem cells is maintained (reviewed in Temple, 2001) (Figure 23.1; see also the Color Plate section).

In contrast to the hematopoietic system, in which stem and lineage-restricted progenitor cells can be identified and prospectively isolated based on their expression of different cell surface antigens, there are no markers that will unequivocally identify stem or progenitor cells in the adult CNS (Weissman et al., 2001). Thus these cell types are currently identified in retrospect, based on their behavior after isolation, expansion, and manipulation in vitro. Cells that can be propagated in the presence of mitogens, such as fibroblast growth factor 2 and/or epidermal growth factor, have been isolated from multiple adult CNS regions, including the cortex, spinal cord, substantia nigra, and hippocampus of many different mammalian species, including humans (Reynolds and Weiss, 1992; Gage et al., 1995; Palmer et al., 1995; Weiss et al., 1996; Palmer et al., 1997; Johansson et al., 1999b; Johansson et al., 1999a; Palmer et al., 1999; Arsenijevic et al., 2001b; Palmer et al., 2001; Yamamoto et al., 2001;

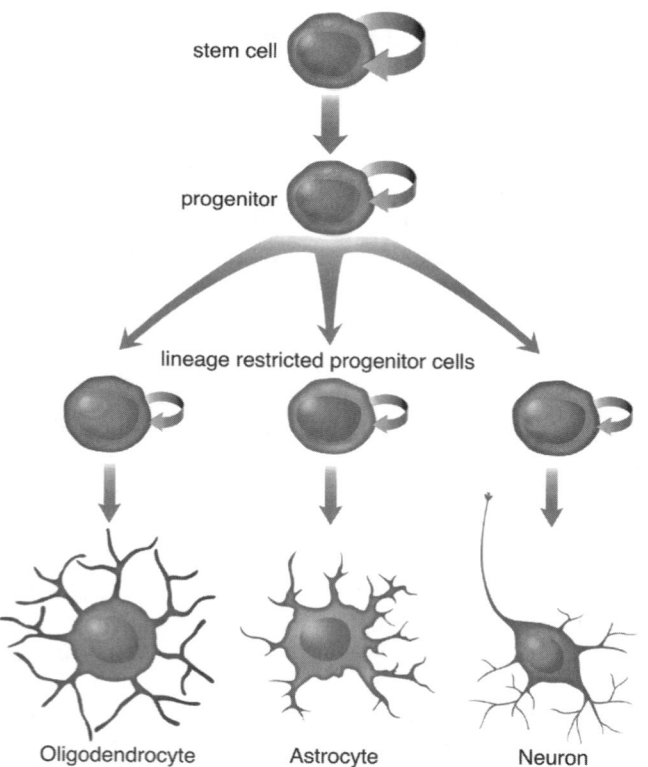

FIGURE 23.1 Model for the generation of differentiated neural cells from stem cells. Multipotent, self-renewing neural stem cells generate progeny with restricted proliferation and differentiation potential (i.e., progenitor cells). Progenitor cells differentiate into postmitotic neurons, oligodendrocytes, and astrocytes. (See also the Color Plate section.)

Lie et al., 2002; Nunes et al., 2003). Analysis of the differentiation potential of cells cultured under these conditions has shown that these cells are able to give rise to neurons and glia. Moreover, analysis of the fate of single cells and their progeny (i.e, clonal analysis) has revealed that some cells in these cultures are multipotent, with the potential to differentiate into all three CNS lineages. These findings suggest that adult neural stem cells are present throughout the entire neuraxis. One caveat to this interpretation, however, is that proliferating cells derived from the adult CNS have been primarily analyzed after expansion in the presence of high concentrations of mitogens that can potentially lead to changes in the characteristics of the isolated cells (Palmer et al., 1997; Palmer et al., 1999; Kondo and Raff, 2000). Thus, the observed differentiation pattern might not reflect the potential that the cells have in vivo. Therefore, future studies need to focus on the development of strategies to identify and prospectively isolate pure populations of potential candidates for neural stem cells, making it possible to investigate the differentiation potential of these populations without prior exposure to mitogens. Recently, protocols have been developed that allow the enrichment of adult CNS-derived proliferating cells in culture, thereby allowing in vitro characterization soon after isolation (Palmer et al., 1999; Rietze et al., 2001). In vitro studies using these methods have confirmed that in vivo proliferating cells from gliogenic regions have the ability to give rise to neurons in culture without exposure to growth factors/mitogens (Lie et al., 2002), thereby providing additional support for the idea of a broad presence of neural stem cells in the adult mammalian CNS.

NEUROGENESIS IS RESTRICTED TO SPECIALIZED AREAS OF THE ADULT BRAIN

Dividing cells and their progeny can be labeled by injection of 3H-thymidine or of the thymidine analog bromodeoxyuridine (BrdU). Both markers will be incorporated into the DNA of dividing cells during S-Phase and will be passed on to the progeny of these cells. 3H-thymidine can then be visualized by autoradiography, and BrdU can be detected by immunostaining. With use of these methods, dividing cells have been found along the entire neuraxis (Kaplan and Hinds, 1980; Kuhn et al., 1997; Horner et al., 2000; Lie et al., 2002). Consistent with the possibility of isolating proliferating cells with stem cell characteristics from many regions of the adult CNS, the majority of the dividing cells express proteins that, during development or in culture, are associated with neural stem cells, for example, nestin, or lineage-restricted progenitor cells, for example, NG2, suggesting that neural stem and progenitor cells reside throughout the entire neuraxis and continue to divide.

What is the fate of the dividing cells in the adult CNS? Do they generate differentiated neural progeny in the adult CNS? The use of BrdU in combination with markers for differentiated neural cells allows us to phenotype the progeny of dividing cells in the CNS. In most areas of the adult CNS, dividing cells give rise to new astrocytes and oligodendrocytes (Kuhn *et al.*, 1997; Horner *et al.*, 2000; Lie *et al.*, 2002). However, in two restricted areas of the adult CNS, the subgranular zone (SGZ) of the hippocampal dentate gyrus and the subventricular zone (SVZ) of the lateral ventricle, new neurons arise from neural stem cells through the generation of rapidly dividing progenitor cells and through differentiation of these progenitor cells into immature neurons (reviewed in Kempermann and Gage, 2000; Alvarez-Buylla and Garcia-Verdugo, 2002) (Figure 23.2; see also the Color Plate section).

Many of the immature neurons born in the SGZ die within the first 2 weeks after birth. The surviving neurons migrate into the molecular layer of the dentate gyrus and differentiate into granule neurons (Kempermann *et al.*, 2003). Within 1 month, they send axonal projections into the CA3 region and project dendrites into the outer molecular layer in a way that is similar to the pre-existing dentate granule neurons (Hastings and Gould, 1999; Markakis and Gage, 1999; Seri *et al.*, 2001; van Praag *et al.*, 2002).

Immature neurons born in the SVZ migrate long distances through the rostral migratory pathway (RMP) towards the olfactory bulb. In the RMP, new neurons form chains and migrate through tubular structures formed by specialized astrocytes (Lois *et al.*, 1996; Wichterle *et al.*, 1997). Around 2 weeks after birth, the first neurons have reached the olfactory bulb and start to migrate radially to their final positions in the glomerular and periglomerular layers, where they differentiate into interneurons (Winner *et al.*, 2002; Carleton *et al.*, 2003).

Due to the development of new techniques that allow the labeling of new neurons with live-cell markers, the electrophysiological properties of new neurons in the hippocampus and the olfactory bulb can now be examined. Recent studies have shown that new neurons in both areas display the electrophysiological properties of mature neurons, such as the ability to generate action potentials and receive synaptic input (van Praag *et al.*, 2002; Carleton *et al.*, 2003). These important studies indicate that the new neurons are functional and suggest that they are integrated into the existing circuitry.

The function of adult neurogenesis is currently not known. It has been proposed that, because neurons in the dentate gyrus and the olfactory bulb undergo programmed cell death, the new neurons compensate for this continued cell loss (Biebl *et al.*, 2000; Petreanu and Alvarez-Buylla, 2002). Correlative evidence suggests that new hippocampal neurons are involved in spatial learning and memory (reviewed in Kempermann, 2002), whereas new olfactory interneurons might contribute to olfactory discrimination (Gheusi *et al.*, 2000; Petreanu and Alvarez-Buylla, 2002).

It has been suggested that new neurons are also born in other areas of the adult brain, such as the cortex (Gould *et al.*, 1999a) and the substantia nigra (Zhao *et al.*, 2003). However, these findings are controversial and currently lack confirmation (Kay and Blum, 2000; Kornack and Rakic, 2001; Lie *et al.*, 2002).

REGULATION OF ADULT NEUROGENESIS

Environmental Control of Adult Neural Stem Cell Fate

How is neurogenesis restricted to the SGZ and SVZ, if neural stem cells are ubiquitously present in the adult CNS? Transplantation of neural stem/progenitor cells derived from neurogenic (hippocampus) or non-neurogenic (spinal cord, substantia nigra) regions of the adult brain have provided evidence for the role of environmental factors in the neuronal differentiation of adult neural stem cells. When transplanted into neurogenic regions, neural stem cells from adult neurogenic and gliogenic CNS regions give rise to neurons *in vivo*. In contrast, when transplanted into gliogenic regions, adult neural stem cells, regardless of their origin, will differentiate into glial cells (Suhonen *et al.*, 1996; Shihabuddin *et al.*, 2000; Lie *et al.*, 2002). These findings indicate that adult neural stem cells from different regions are not fate-restricted by intrinsic programs but that extrinsic cues derived from the local environment control their differentiation.

Cellular Elements Regulating Neurogenesis

What are the cellular elements that control the proliferation and fate choice of neural stem cells? Functional studies in songbirds and anatomical studies in rodents have implied that vasculature plays an important role in providing a neurogenic environment. Stem/progenitor cells in the hippocampus, but not in non-neurogenic regions, proliferate in close proximity to blood vessels and dividing endothelial cells, suggesting that common signals, probably derived from the vasculature, regulate neurogenesis and vasculogenesis (Palmer *et al.*, 2000). Indeed factors that promote endothelial cell proliferation also increase neurogenesis, further indicating an important relationship between these two processes (Jin *et al.*, 2002). In the adult songbird brain, a recent study showed

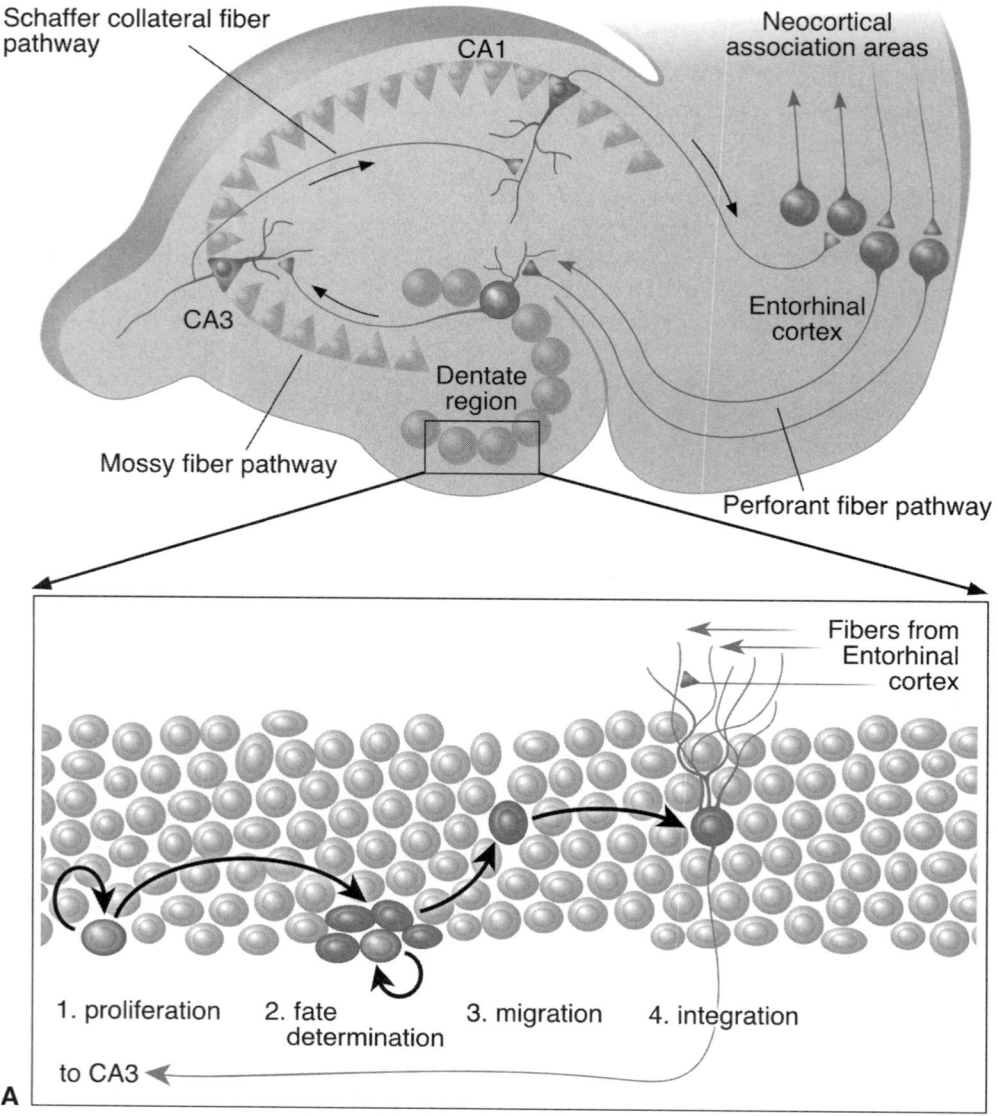

FIGURE 23.2 Neurogenic zones of the adult mammalian CNS. **A:** Hippocampal dentate gyrus. 1. Proliferation and fate determination: Stem cells (beige) in the subgranular zone of the dentate gyrus give rise to transit amplifying cells that differentiate into immature neurons. 2. Migration: Immature neurons migrate into the granule cell layer of the dentate gyrus. 3. Integration: Immature neurons mature into new granule neurons, receive input from the entorhinal cortex, and extend projections into CA3. (See also the Color Plate section.)

Continued

that increased vasculogenesis stimulates neurogenesis by an increase in endothelial cell–derived growth factors (Louissaint *et al.*, 2002).

In the SVZ, neural stem and progenitor cells proliferate and differentiate into immature neurons in proximity to ependymal cells. Ependymal cells have been shown to contribute to the neurogenic environment in the SVZ by secreting factors that inhibit glial differentiation of neural stem cells (Lim *et al.*, 2000) (see subsequent discussion).

Astrocytes are also in intimate contact with proliferating stem and progenitor cells in the SVZ and the SGZ. Astrocytes in the hippocampus and the SVZ provide signals that not only stimulate the proliferation of neural stem cells but also instruct these cells to adopt a neuronal fate (Lim and Alvarez-Buylla, 1999; Song *et al.*, 2002). Interestingly, astrocytes from non-neurogenic regions such as the adult spinal cord do not stimulate the neuronal differentiation of neural stem cells. These findings

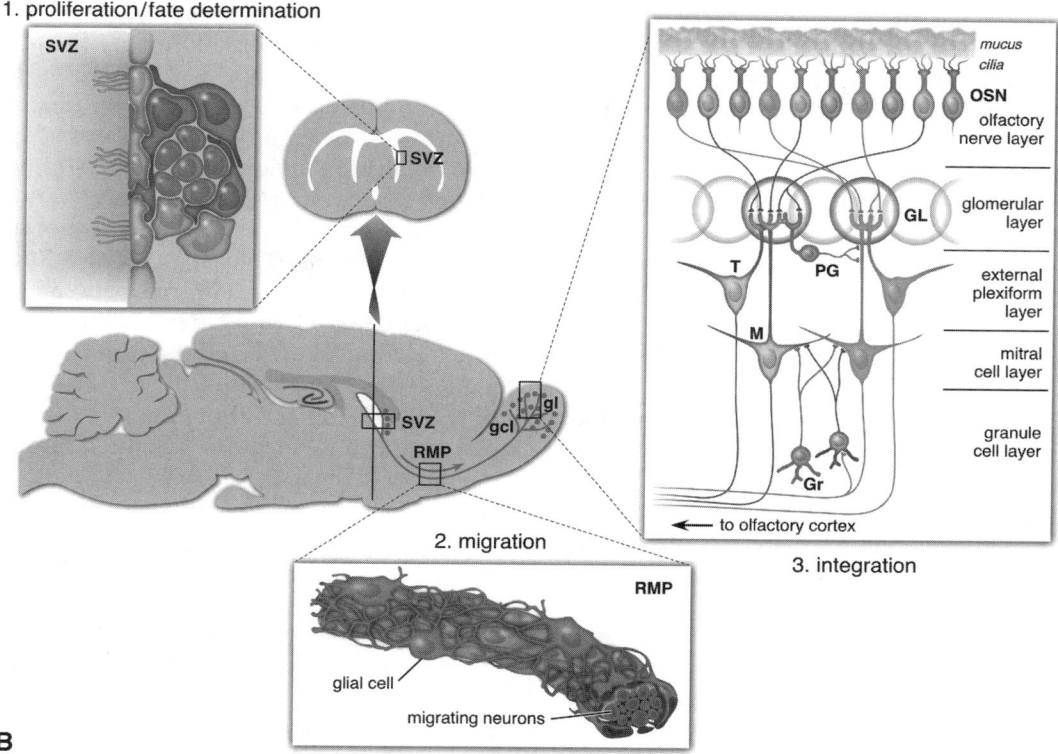

1. proliferation/fate determination

2. migration

3. integration

FIGURE 23.2, cont'd **B:** Subventricular zone (SVZ)/olfactory bulb system. 1. Proliferation and fate determination: Stem cells in the SVZ of the lateral ventricle (*blue*) give rise to transit amplifying cells (*green*) that differentiate into immature neurons (*red*). Adjacent ependymal cells (*light brown*) of the lateral ventricle are essential for the neuronal fate determination by providing inhibitors of glial differentiation. 2. Migration: Immature neurons (*red*) migrate along each other in chains through the rostral migratory pathway (RMP). The migrating neurons are ensheathed by astrocytes (*blue*). 3. Integration: Immature neurons differentiate local interneurons (*red*) in the granule cell layer and the periglomerular layer. Olfactory sensory neurons (OSN); tufted neurons (T); mitral neurons (M); granule neurons (Gr); periglomerular neurons (PG). Figure reproduced from Lie *et al.* (2004). Copyright by Annual Reviews. (See also the Color Plate section.)

indicate that, in the areas of adult neurogenesis, regionally specialized astrocytes contribute to the creation of a neurogenic environment.

Molecular Regulation of Neurogenesis

The molecular mechanisms that control adult neurogenesis are poorly understood. Multiple growth factors, hormones, and neurotransmitters have been implicated in the regulation of neurogenesis, based on their ability to influence proliferation, differentiation, and survival of neural stem cells *in vitro* and *in vivo*. The relationship between these different factors is currently not understood and it is not clear whether all of them play a physiological role in the regulation of neurogenesis. However, given their potential to affect the behavior of neural stem cells *in vivo*, these factors and their receptors might represent candidate molecules and targets, respectively, for the recruitment of endogenous stem cells for CNS repair.

Factors Affecting Proliferation

Epidermal growth factor (EGF) and fibroblast growth factor–2 (FGF-2) are commonly used as mitogens for the maintenance of neural stem cells *in vitro* (reviewed in Gage, 2000). Intracerebroventricular infusion of either factor increases the proliferation of cells in the SVZ but not in the hippocampus (Kuhn *et al.*, 1997; Wagner *et al.*, 1999). Other growth factors, such as vascular endothelial growth factor (VEGF) (Jin *et al.*, 2002), Sonic hedgehog (SHH) (Lai *et al.*, 2003), brain-derived neurotrophic factor (BDNF) (Benraiss *et al.*, 2001; Pencea *et al.*, 2001), insulin-like growth factor–1 (IGF-1) (Aberg *et al.*, 2000; Arsenijevic *et al.*, 2001a), FGF-8 (Lie *et al.*, 2002), and Amphiregulin (Falk and Frisen, 2002), have also been shown to be sufficient to propagate neural stem cells *in vitro* or to enhance proliferation *in vivo*. Currently, there is no direct evidence that any of these growth factors is an endogenous regulator of neural stem cell proliferation.

Moreover, the potential interaction of these growth factors with steroid hormones (estrogens, testosterone, glucocorticoids) (Cameron and Gould, 1994; Cameron *et al.*, 1998a; Duman *et al.*, 2001b) and neurotransmitters (glutamate, serotonin) (Cameron *et al.*, 1995; Cameron *et al.*, 1998b; Brezun and Daszuta, 1999; Banasr *et al.*, 2001; Duman *et al.*, 2001a), which have been shown to influence the proliferation of neural stem cells, remains to be determined.

Factors Controlling Fate Specification of Adult Neural Stem Cells

To date, only a few factors have been identified that control the fate of adult neural stem cells. Activated Notch 1 and Notch 3 have been found to restrict adult neural stem cells toward an astroglial lineage *in vitro* (Tanigaki *et al.*, 2001). In addition, members of the bone morphogenetic protein (BMP) family instruct adult neural stem cells to adopt a glial cell fate (Lim *et al.*, 2000). Interestingly BMPs are expressed by adult neural stem cells themselves. This autocrine gliogenic signal is inhibited by noggin, a secreted protein that blocks BMP signaling by binding to BMPs. In the neurogenic SVZ, noggin is secreted by ependymal cells in the lateral ventricle, thereby blocking the gliogenic effects of BMPs and permitting neuronal differentiation of adult neural stem cells. However, blocking of BMP by itself is not sufficient to induce the neuronal differentiation of adult neural stem cells; that requires additional factors, potentially derived from local astrocytes.

There are currently no physiological inducers of neuronal cell fate known. *In vitro*, differentiation of adult neural stem cells into all three neural lineages is induced by withdrawal of mitogens. The neuronal differentiation can be enhanced by retinoic acid and increased cyclic adenosine monophosphate (cAMP) levels (Palmer *et al.*, 1997; Takahashi *et al.*, 1999; Nakagawa *et al.*, 2002). In addition, neurotrophins have limited influence on the neurotransmitter phenotype expressed *in vitro* (Takahashi *et al.*, 1999). *In vivo*, increased generation of neurons has been observed following infusion/overexpression of BDNF (Benraiss *et al.*, 2001; Pencea *et al.*, 2001) and IGF-1 (Aberg *et al.*, 2000). However, it is not clear whether this increase is caused by enhanced neuronal differentiation of adult neural stem cells or by enhanced survival of new neurons.

Neuronal Migration, Nerve Guidance, Synapse Formation, and Survival

Factors controlling later steps in neurogenesis, such as functional maturation, synapse formation and integration into the neuronal circuit and survival, are currently unknown. Some mechanisms and molecules that control the remarkable long-distance migration of newly generated neurons through RMP have been described. Migrating neurons interact with their environment through expression of the polysialated glycoprotein neural cell adhesion molecule (PSA-NCAM), which is essential for proper migration, as a null mutation for NCAM or the deletion of the polysialic acid moiety results in migratory defects (Ono *et al.*, 1994; Rousselot *et al.*, 1995). Interactions via integrins between migrating neurons themselves (Jacques *et al.*, 1998), as well as environment-derived short- and long-range chemorepulsive factors such as members of the ephrin-B family (Conover *et al.*, 2000) and Slit (Wu *et al.*, 1999), have also been demonstrated to direct the migration through the rostral migratory pathway.

Influence of Genetics and Behavior on Neurogenesis

Interestingly, genetic background and behavioral modifications influence the rate of neurogenesis. Mouse strains differ significantly in their baseline rate of neurogenesis, due to differences in proliferation, differentiation and survival of newly generated cells (Kempermann *et al.*, 1997a). In addition, the strain differences influence how neurogenesis is affected by environmental factors. Thus, exposure to an enriched environment consisting of toys, larger housing, and opportunity for learning, social interaction, and physical activity was shown to significantly increase hippocampal neurogenesis by enhancing the survival of new neurons (Kempermann *et al.*, 1997b). In a different mouse strain, the same environment stimulated hippocampal neurogenesis by increasing proliferation (Kempermann *et al.*, 1997b).

With respect to the stimulation of hippocampal neurogenesis, increased physical activity appears to be the most important component of the enriched environment (van Praag *et al.*, 1999). The suggestion that neurogenesis is stimulated by spatial learning, a hippocampus-dependent task, remains controversial. Although one laboratory reported an increased rate of neurogenesis in animals exposed to a spatial learning task due to enhanced survival of newborn neurons (Gould *et al.*, 1999b), our group has not found any evidence for enhanced neurogenesis in the context of this task (van Praag *et al.*, 1999).

Neurogenesis in the SVZ/olfactory bulb is not affected by running and enriched environment (Brown *et al.*, 2003). However, odor enrichment, that is, exposure to an environment with daily changing sources of odor (Rochefort *et al.*, 2002), stimulates neurogenesis in this system, indicating that specific environmental and behavioral stimuli differentially influence neural stem cell behavior in the two neurogenic areas. Whether this differential effect is due to locally specific expression of common molecular regulators for both regions or to region-specific responsiveness of stem cells to broadly expressed factors remains to be determined.

Although the immediate molecular consequences of running, enriched environment, and odor enrichment are only poorly characterized, the finding that certain behavioral modifications can specifically increase the rate of neurogenesis in defined CNS areas may be important for future CNS repair strategies, potentially in combination with defined pharmacological therapies.

ADULT NEURAL STEM CELLS AND CENTRAL NERVOUS SYSTEM REPAIR

Due to their ability to proliferate and to give rise to differentiated progeny, stem cells are considered promising candidates for use in the treatment of currently intractable diseases. Cell replacement therapy has been pursued in Parkinson's disease over the past three decades, involving transplantation of primarily fetus-derived midbrain dopaminergic precursors (Lindvall and Hagell, 2001). More recently, methods have been developed that allow the directed differentiation of embryonic stem cells into different neural subtypes that can be used for exogenous cell replacement strategies in CNS diseases (Kim et al., 2002). However, the use and manipulation of fetal tissue and embryonic stem cells have raised a number of ethical concerns and political restrictions. These restrictions do not apply to the use of endogenous adult neural stem cells for cell replacement therapy. In addition, these cells offer a unique advantage over other, exogenous cell sources: immunological reactions are avoided. However, the mobilization of endogenous neural stem cells for cell replacement poses multiple problems that need to be overcome.

ENDOGENOUS NEURAL STEM CELLS IN DISEASE AND LESION MODELS

The recruitment of endogenous neural stem cells for cellular replacement was initially observed in neurogenic regions. A broad variety of lesions lead to transient increases in cell proliferation in the SGZ and in the rate of hippocampal neurogenesis. Similarly, precursor cell proliferation in the SVZ and the number of migrating neurons in the RMP are temporarily augmented in defined lesion paradigms.

Limited *de novo* neurogenesis in a region that under physiological circumstances is non-neurogenic was observed for the first time in a lesion model with no

direct clinical correlate (Magavi et al., 2000). This particular model is based on the photoactivation of a chromophore, which results in the generation of reactive oxygen species that in turn lead to apoptotic cell death. Following induced apoptosis of neocortical neurons by this method, approximately 2% of the newborn cells in the lesioned neocortex expressed neuronal markers. These new neurons not only survived for at least 6 months but also appeared to have formed appropriate, long-distance projections. These data suggested for the first time that new neurons could potentially contribute to repair in non-neurogenic regions.

Following this report, there has been a growing number of reports of neurogenesis in multiple CNS regions in models for human diseases.

DISEASE MODELS IN WHICH POTENTIAL NEUROGENESIS HAS BEEN DESCRIBED

Seizures

Seizures induce neuronal cell death in various CNS regions. Increased precursor proliferation and neuronal differentiation have been observed following chemoconvulsant-induced seizure in the hippocampal dentate gyrus and the SVZ/olfactory bulb system (reviewed in Parent and Lowenstein, 2002), indicating that increased neurogenesis might contribute to repair by compensation of seizure-induced neuronal cell death in these circuits. However, some new neurons displayed abnormal migratory behavior and were found in ectopic locations (Parent et al., 1997; Parent et al., 2002a). Moreover, they contributed to aberrant network organization and in some cases showed altered physiological properties (Parent et al., 1997; Scharfman et al., 2000).

Ischemic Brain Insults

Global forebrain ischemia, a model for cardiac arrest or coronary artery occlusion, leads to selective death of defined vulnerable neuronal populations. In contrast to other hippocampal structures, such as the CA1 region, the dentate gyrus is relatively resistant to this insult. Interestingly, several groups have found that global forebrain ischemia promotes progenitor cell proliferation and neurogenesis in the adult dentate gyrus (Liu et al., 1998; Kee et al., 2001; Yagita et al., 2001), suggesting that the resistance of the dentate gyrus formation to global ischemia is related to its ability to generate new dentate granule neurons to replace degenerating ones.

Occlusion of the middle cerebral artery (MCAO) is a commonly used stroke model that causes infarction in the striatum and the parietal cortex. Under physiological circumstances, no neurogenesis is observed in these two regions. Following MCAO, however, several laboratories have observed newborn, immature neurons in the infarcted striatum as early as 2 weeks after ischemia (Arvidsson et al., 2002; Parent et al., 2002b). Interestingly, at this early time point, the majority of these neurons expressed markers of developing striatal neurons, suggesting that they were differentiating into the neuronal phenotype that was destroyed and that they might be able to contribute to cell replacement and repair following striatal infarction.

The observations as to whether de novo cortical neurogenesis occurs following stroke are inconsistent. It has been reported that new cells are generated in the cortex following photothrombotic lesion or MCAO (Gu et al., 2000; Jiang et al., 2001; Zhang et al., 2001). One group reported that, preferentially in the peri-infarct area, some new cells expressed markers consistent with a neuronal phenotype (Gu et al., 2000). However, studies from other laboratories failed to observe this phenomenon (Zhang et al., 2001; Arvidsson et al., 2002; Parent et al., 2002b).

Parkinson's Disease Model

6-Hydroxydopamine and 1-methyl-4-phenyl-1,2,3,6, tetrahydropyridine are neurotoxins that can induce the rapid death of dopaminergic neurons in the substantia nigra pars compacta. Due to their relative specificity for the dopaminergic system, these toxins are commonly used to create animal models for Parkinson's disease. Both lesions enhance the appearance of newborn cells in the substantia nigra (Kay and Blum, 2000; Lie et al., 2002; Zhao et al., 2003). One study reported increased dopaminergic neurogenesis following MPTP lesion and also suggested a basal turnover of dopaminergic neurons with constant dopaminergic cell death and dopaminergic neurogenesis under physiological conditions (Zhao et al., 2003). However, other laboratories have not found any evidence for local neurogenesis following degeneration of neurons in the substantia nigra (Kay and Blum, 2000; Lie et al., 2002).

Glial Cell Replacement in Lesion Models

In addition to degenerating neurons, glial cells are potentially regenerated following lesion. In human demyelinating diseases such as multiple sclerosis, spontaneous partial remyelination of lesions has been observed (Prineas et al., 1993). Demyelinating lesions can be induced in animals by lysolecithin injection. In these lesions, new myelinating oligodendrocytes are generated from immature, locally dividing cells, suggesting that remyelination in multiple sclerosis is potentially achieved

by recruitment of endogenous neural stem/progenitor cells (Gensert and Goldman, 1997). Increased proliferation of local stem/progenitor cells and enhanced generation of new astrocytes have been observed following traumatic lesions in the spinal cord (Johansson et al., 1999b; Yamamoto et al., 2001). Whether these cells are contributing to repair or are actually inhibitory for regenerative processes such as the growth of axons is currently not known.

RESTRICTIONS ON REGENERATION FROM ENDOGENOUS NEURAL STEM/PROGENITOR CELLS

The observation that new neurons and glial cells are generated in some CNS regions following injury suggests a potential for self-repair in the adult CNS by cell replacement from endogenous stem cells. Thus far, however, the extent of de novo neurogenesis that has been described is very limited, and in some lesion models, for example, spinal cord injury, no de novo neurogenesis has been found.

The restrictions on repair by neural stem cells in the adult CNS are highlighted in the study by Arvidsson and colleagues (Arvidsson et al., 2002), who studied neuronal cell replacement following ischemia by MCAO. In this study, neuronal replacement was observed in the ischemic striatum but not in the ischemic cortex, suggesting that potential differences exist in the regenerative capacity of different CNS regions. The percentage of replaced striatal neurons was estimated at 0.2%, a rate probably too low to have a significant impact on functional recovery.

Several factors contribute to this restricted neuronal replacement. First, the initial number of new neurons observed within the first 2 weeks after ischemia was already very low compared to the number of neurons lost, suggesting that there is insufficient proliferation or neuronal fate commitment of neural stem cells. Second, a large proportion of newborn neurons die, and only 20% of the initially generated new neurons in the post-ischemia striatum survive longer than 14 days. Third, only half of these new neurons appear to differentiate into a mature striatal phenotype. Fourth, the majority of new neurons appear to be derived from stem cells in the SVZ that have migrated into the ischemic striatum, not from resident striatal stem/progenitor cells, suggesting that resident stem cells were also compromised by the lesion. These findings illustrate that the process of lesion-induced neurogenesis is inhibited at multiple levels, including proliferation of stem cells, neuronal fate commitment, survival, and maturation of newborn neurons. In addition, the important questions of whether the new neurons are functional and integrate into neuronal circuits remain unanswered.

STIMULATORY AND INHIBITORY FACTORS FOR LESION-INDUCED NEUROGENESIS

What signals stimulate *de novo* neurogenesis in lesioned areas? And what are the factors that limit the extent of lesion-induced neurogenesis? While the exact identity of the molecular signals underlying stimulation and inhibition of lesion-induced neurogenesis are not known, studies in related fields have provided some insight into the potential origin and nature of these factors.

A number of signals regulating proliferation, neuronal migration, differentiation, survival, and connectivity during embryogenesis are reactivated in the injured adult environment. Interneurons surrounding dying neurons in the adult neocortex show increased expression of neurotrophins (Wang *et al.*, 1998) and glial cells in diseased areas can acquire properties of radial glia (Leavitt *et al.*, 1999), which act as a substrate for neuronal migration during development. Moreover, an increased expression of mitogens for neural precursor cells is found in some lesions (Yoshimura *et al.*, 2001; Jin *et al.*, 2002).

Transplantation of fetal immature neurons and neonatal and adult neural stem cells has been used to probe the lesioned environment of the adult CNS for its ability to recruit neural precursor cells for neuronal cell replacement. These experiments have shown that, in some specialized lesion models, the environment is providing signals that are sufficient to stimulate the migration of transplanted fetal- and neonatal-derived precursor cells toward the lesioned area, their differentiation into neurons, and the establishment of synaptic contacts (Hernit-Grant and Macklis, 1996; Snyder *et al.*, 1997; Shin *et al.*, 2000; Fricker-Gates *et al.*, 2002). In contrast, glial differentiation, but not neuronal differentiation, of transplanted adult neural stem cells has been observed in the lesion models tested so far (Vroemen *et al.*, 2003) (Dziewczapolski *et al.*, 2003), suggesting that environment-derived signals that can recruit adult neural stem cells are limited or that inhibitory factors for adult neural stem cells are predominant.

In acute injuries such as stroke and spinal cord injury, factors that promote reactive gliosis and scar formation are up-regulated, and it is possible that they prevent neuronal differentiation of endogenous stem cells by promoting their glial fate choice (for review see Horner and Gage, 2000). The inhibitory effects of the glial scar (Stichel and Muller, 1998; Zuo *et al.*, 1998; Pasterkamp *et al.*, 1999), of extracellular matrix molecules (Haas *et al.*, 1999; Jaworski *et al.*, 1999; Lemons *et al.*, 1999; Jones and Tuszynski, 2002; Jones *et al.*, 2002) and of myelin-derived factors such as Nogo, MAG, and OMgp (Domeniconi *et al.*, 2002; Liu *et al.*, 2002; Wang *et al.*, 2002a; Wang *et al.*, 2002b) on axonal outgrowth and regeneration in spinal cord injury are well documented. It is possible that the same factors interfere with the formation of functional synaptic connections of newborn neurons and their integration into existing circuits, thereby depriving them of target-derived trophic support and decreasing their survival rate. Given that axonal outgrowth and cell migration share similar pathways (Park *et al.*, 2002), these inhibitory factors might also interfere with the migration of stem cells/immature neurons into the lesion.

ENHANCING THE RECRUITMENT OF ENDOGENOUS STEM CELLS FOR REPAIR

How can we overcome these hurdles and augment the repair from endogenous stem cells? Given our incomplete knowledge of how adult cell genesis is regulated under physiological conditions, and of what factors are promoting or inhibiting lesion-induced cell genesis, it is difficult currently to devise logical strategies to enhance this process. Nevertheless, recent studies have suggested that some growth factors might be able to either promote neurogenesis in otherwise gliogenic regions or enhance the repair from neural stem cells.

BDNF delivery to the adult forebrain by either continuous infusion or adenoviral overexpression in the ependymal layer has been shown to result in enhanced proliferation in the SVZ and increased addition of neurons in the olfactory bulb, demonstrating that BDNF stimulates neurogenesis in the SVZ/olfactory bulb system *in vivo* (Benraiss *et al.*, 2001; Pencea *et al.*, 2001). Interestingly, in these studies the appearance of potentially newborn neurons in the striatum, septum, and hypothalamus was described in BDNF-treated animals but not in controls. Some of the newborn cells in the striatum also appeared to have differentiated into mature striatal spiny neurons, suggesting that BDNF might not only have the potential to recruit stem cells for neurogenesis in otherwise gliogenic regions but might also support their differentiation into region-specific phenotypes.

The effects of growth factors on regeneration from endogenous stem cells have also been tested in lesion models. In 6-OHDA-lesioned animals, transforming growth factor α (TGFα) infusion into the striatal parenchyma concomitant with or 2 weeks after the lesion leads to behavioral recovery (Fallon *et al.*, 2000). Moreover, increased proliferation of precursor cells, migration from the SVZ toward the striatum, and differentiation into new dopaminergic neurons in the striatum were described in the TGFα-treated animals but not in the control animals. Based on these results, the authors concluded that TGFα, in conjunction with signals arising from injury, recruited endogenous stem cells to generate new dopaminergic neurons in the striatum, which restored dopaminergic neurotransmission and caused behavioral recovery.

A second, more detailed study investigated the effects of EGF and FGF-2 infusions following transient global ischemia (Nakatomi *et al.*, 2002). In rodents, this lesion produces selective degeneration of hippocampal CA1 pyramidal neurons. Growth factor–treated animals, however, showed approximately 40% recovery of the total number of CA1 pyramidal neurons lost by ischemia (Figure 23.3; see also the Color Plate section). BrdU labeling, retroviral lineage tracing, and inhibition of proliferation by administration of an antimitotic drug shortly after the lesion provided evidence that a large number of the recovered pyramidal neurons were generated *de novo* in response to the combination of growth factor treatment and lesion (Figure 23.4; see also the Color Plate section). Moreover, some data suggested that these neurons were derived from progenitor cells in the periventricular region that had migrated into the lesioned CA1 region. Importantly, in this study evidence was also found that functional synapses were formed onto the regenerated neurons and that growth factor–treated animals performed significantly better in spatial learning tasks, suggesting that the recovered pyramidal neurons were participating in the hippocampal circuitry and contributed to behavioral recovery.

Recovery of CNS function through recruitment of endogenous stem cells has also been suggested in aged animals. Hippocampal neurogenesis decreases in aging rodents (Kuhn *et al.*, 1996) and is potentially linked to

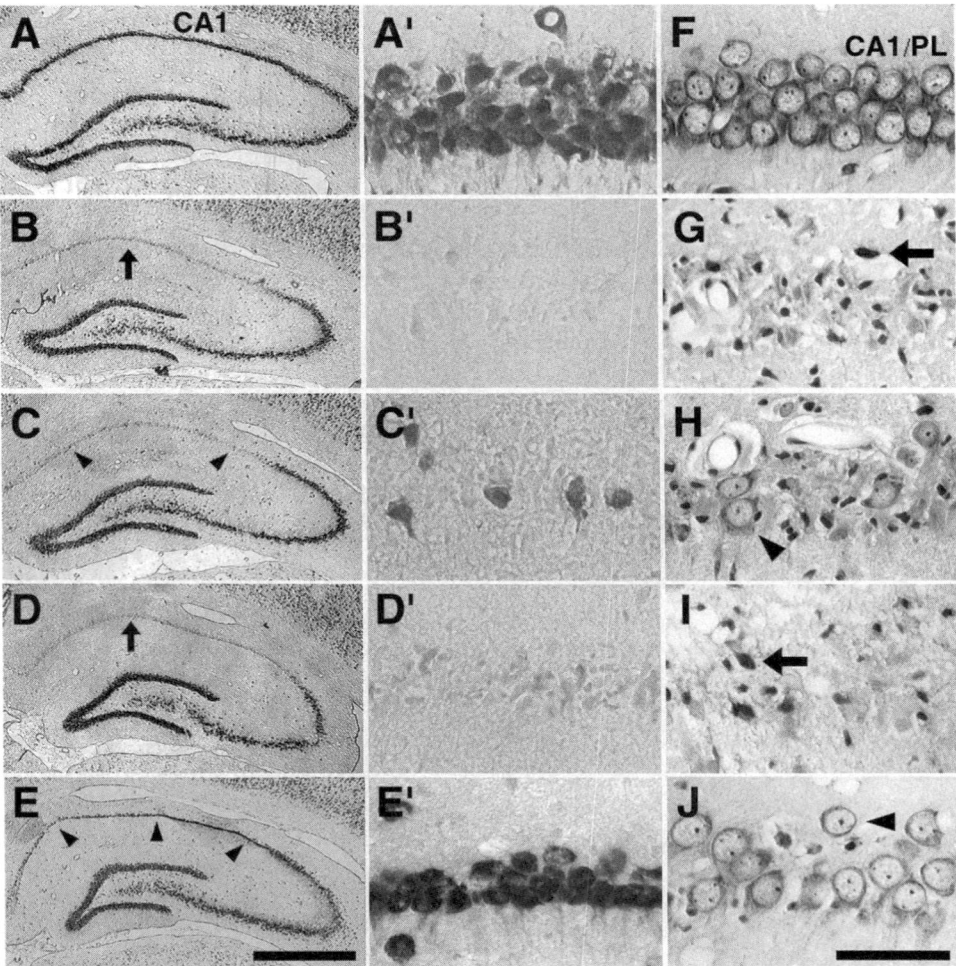

FIGURE 23.3 **A:** Animals treated with EGF and FGF2 following ischemia show significant regeneration of CA1 pyramidal neurons. **A–E:** Neurons are stained with the neuronal marker NeuN. **F–J:** Cresyl violet staining of the hippocampal CA1 pyramidal layer. (A, A′, and F) Intact animals, (B, B′, and G) untreated animals at day 7 after ischemia, (C, C′, and H) untreated animals at day 28 after ischemia, (D, D′, and I) EGF/FGF2 treated animals at day 7 after ischemia, (E, E′, and J) EGF/FGF2 treated animals at day 28 after ischemia. Arrows and arrowheads in **G–J** indicate cells with pyknotic and intact morphology, respectively. Figure reproduced from Nakatomi *et al.* (2002). Regeneration of hippocampal pyramidal neurons after ischemic brain injury by recruitment of endogenous neural progenitors, *Cell* 110:429-441. Copyright by Cell Press. (See also the Color Plate section.)

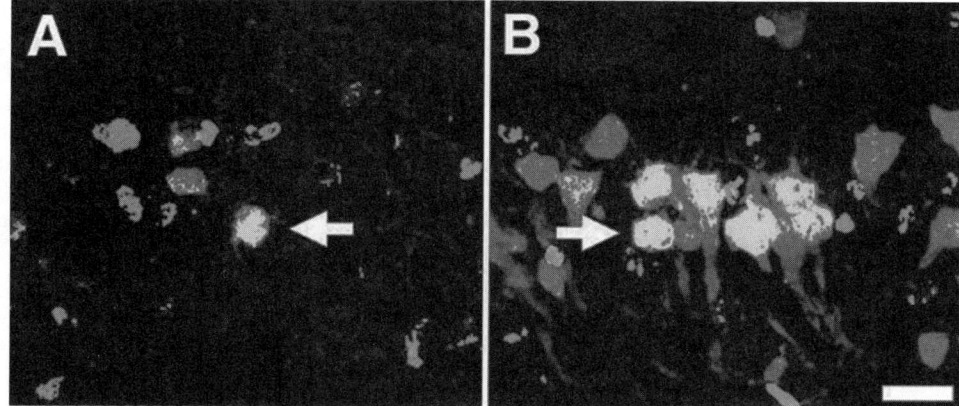

FIGURE 23.4 Regenerated CA1 pyramidal neurons are derived from proliferating cells. The thymidine-analogue BrdU was administered to untreated **(A)** and EGF/FGF2 treated **(B)** ischemic animals. A large number of pyramidal neurons, identified by the expression of the neuronal marker NeuN (*red*), are positive for BrdU (*green*), indicating that they are the progeny of previously dividing cells. Figure reproduced from Nakatomi *et al.* (2002). Regeneration of hippocampal pyramidal neurons after ischemic brain injury by recruitment of endogenous neural progenitors, *Cell* 110:429-441. Copyright by Cell Press. (See also the Color Plate section.)

the age-related decline in spatial learning. Similar to their effects in young animals, FGF-2 and EGF infusion can significantly increase cell proliferation in the SGZ and SVZ in aged rodents (Jin *et al.*, 2003). In addition, it has been demonstrated that exposure to an enriched environment increases the rate of hippocampal neurogenesis by approximately fivefold and is accompanied by improved performance in spatial learning tasks (Kempermann *et al.*, 1998; Kempermann *et al.*, 2002). While the molecular consequences of environmental enrichment are far from understood, these findings present us with the intriguing possibility that behavioral and environmental influences in conjunction with growth factor delivery could be beneficial for the recruitment of adult neural stem cells for CNS repair.

CAVEATS: THE NEED TO LINK NEW NEURONS TO RECOVERY

The observation that endogenous stem cells can be recruited to generate new neurons following lesion is highly exciting and reinforces our efforts to develop strategies that aim at harnessing endogenous stem cells for repair of the adult CNS. However, the observation that new neurons generated in response to epileptic injury in the hippocampus can be found in heterotopic locations (Parent *et al.*, 1997; Parent *et al.*, 2002a), can contribute to aberrant networks (Parent *et al.*, 1997), and can show altered physiological properties (Scharfman *et al.*, 2000) reminds us that the uncontrolled generation of new neurons can also cause pathology (Parent and Lowenstein, 2002). Moreover, the contribution of new

neurons to repair and functional improvement has not been unequivocally demonstrated but has been assumed, primarily based on correlations of morphological and behavioral data. The study by Nakatomi and colleagues regarding the effects of FGF-2 and EGF in global ischemia provides important evidence for a direct link between *de novo* neurogenesis and behavioral improvement (Nakatomi *et al.*, 2002). However, the contribution of other factors, such as trophic support of injured neurons, axonal sprouting, and synaptic modulation by the growth factors, remains to be determined.

Finally, we need to improve and develop alternative methods for the detection of newborn cells. The current standard method is labeling of proliferating cells and their progeny by the thymidine analogue BrdU. However, the concerns regarding BrdU incorporation in DNA repair and into dying cells and its potential accumulation into other cellular compartments have not been completely removed (Gilmore *et al.*, 2000; Nowakowski and Hayes, 2000; Cooper-Kuhn and Kuhn, 2002); alternative labeling methods could therefore greatly increase our confidence in the existence and significance of neuronal replacement in CNS repair.

NEURAL CELL REPLACEMENT FROM ENDOGENOUS STEM CELLS: A PROCESS RELEVANT IN HUMANS?

Neuronal cell replacement in the mammalian CNS following lesion has been demonstrated in rodents. But what is the evidence that this process is also occurring in humans? And what is the evidence that neural stem cells exist in the adult human CNS? Cells that proliferate in response to growth factors and have the ability to give rise to neurons and/or glia *in vitro* have been isolated from surgical specimens and postmortem tissue of donors of a broad age range. Initially, such progenitor cells were isolated from the subependymal zone of the lateral ventricle and the hippocampus, that is, the CNS regions that had previously been shown to be neurogenic in other mammalian species (Pincus *et al.*, 1998; Kukekov *et al.*, 1999; Palmer *et al.*, 2001). More recently, Nunes and colleagues succeeded in isolating progenitor cells from human subcortical white matter, based on the activity of a promoter or the expression of a surface marker that had previously been associated with glial precursors (Nunes *et al.*, 2003). Interestingly, cells purified by these methods not only differentiate into glial cells but also into functionally mature neurons *in vitro*, suggesting that these cells might be multipotent rather than glia-restricted progenitor cells. These progenitor cells were also transplanted into the developing rat CNS. One month after transplantation, neurons and glia derived from human progenitor cells were found in the mature rat brain, indicating that the neurogenic potential of these cells was also present *in vivo*. Importantly, the ability to differentiate into neurons *in vitro* and *in vivo* was not dependent on prolonged culturing in the presence of growth factors, suggesting that white matter–derived progenitor cells possess an intrinsic neurogenic potential. Taken together, these data suggest that neural stem/progenitor cells may be widely present in the adult human CNS. Are these progenitor cells quiescent or do they contribute to the integrity of the adult human brain? Spontaneous remyelination occurs in multiple sclerosis lesions. Platelet-derived growth factor–α (PDGF-α) receptor has been widely used in rodent studies to identify oligodendrocyte precursors. Interestingly one study found that PDGF-α receptor positive cells were present in acute and chronic lesions of patients with multiple sclerosis, suggesting that these cells might potentially be involved in myelin repair (Scolding *et al.*, 1998). However, the definite proof that these PDGF-α receptor expressing cells indeed correspond to human oligodendrocyte progenitor cells and that they differentiate into myelinating mature oligodendrocytes is currently lacking.

One study that examined postmortem brain tissue from cancer patients who received BrdU injections for diagnostic purposes has provided evidence that hippocampal neurogenesis is also occurring in humans (Eriksson *et al.*, 1998). The tissue samples varied in the postinfusion intervals (16–781 days) as well as in the age of the donor (57–72 years). BrdU-positive cells were found in the SGZ and the granular cell layer of the hippocampus of all specimens, indicating that in these regions new cells are continuously generated. The number of BrdU-positive cells declined with increasing postinfusion interval, suggesting that new cells in the hippocampus might have limited lifetimes. Interestingly, up to one fourth of the BrdU-positive cells co-labeled for the neuronal markers NeuN, calbindin, and neuron-specific enolase (Figure 23.5; see also the Color Plate section). Approximately the same percentage of newly generated cells co-labeled for the glial marker GFAP. These findings indicate that new neurons and astrocytes are continuously generated in the human dentate gyrus. BrdU-positive cells were also found in the SVZ of the lateral ventricle. The morphology of these BrdU-labeled nuclei resembled that of progenitor cells in the rodent SVZ, suggesting that these proliferating cells might correspond to the neurogenic progenitor cells in the rodent SVZ that migrate and differentiate into mature interneurons in the olfactory bulb.

Taken together, the current data suggest that, similar to other mammalian species, humans retain the capacity to generate new neural cells throughout adulthood. Ethical concerns and lack of adequate techniques restrict the study of cell genesis in the adult human brain; thus there is currently no evidence that neurons or glia are regenerated in the diseased human CNS. However, the possibility of isolating cells with neural stem cell–like characteristics from different areas of the adult human brain fuels our hope that we will be able to stimulate endogenous cells to repair the diseased human CNS.

CHALLENGES AND FUTURE DIRECTIONS

At present, our knowledge about adult neural stem cell biology is very limited. It may be premature to evaluate the potential of endogenous stem cells to contribute to functional repair of the human CNS. There are several outstanding challenges that need to be overcome to bring us closer to defining the potential of endogenous stem cells for repair and to proposing experimental strategies for the recruitment of these cells in the diseased CNS:

1. *Prospective identification and isolation of adult neural stem cells.* The prospective isolation of stem cells constitutes a key step in the study of hematopoiesis (reviewed in Weissman *et al.*, 2001). Identification and separation of hematopoietic stem cells from other cell types based on their expression of surface markers allowed the detailed characterization of this cell population. Thus far, our insight into the biology of adult neural stem cells has been primarily based on the analysis of cultured cells with stem cell–like characteristics and retrospective analysis *in vivo*. At this point, the identity and location of neural stem cells *in vivo* remains elusive and there exists no firm link between *in vivo* neural stem cells and cultured stem-like cells,

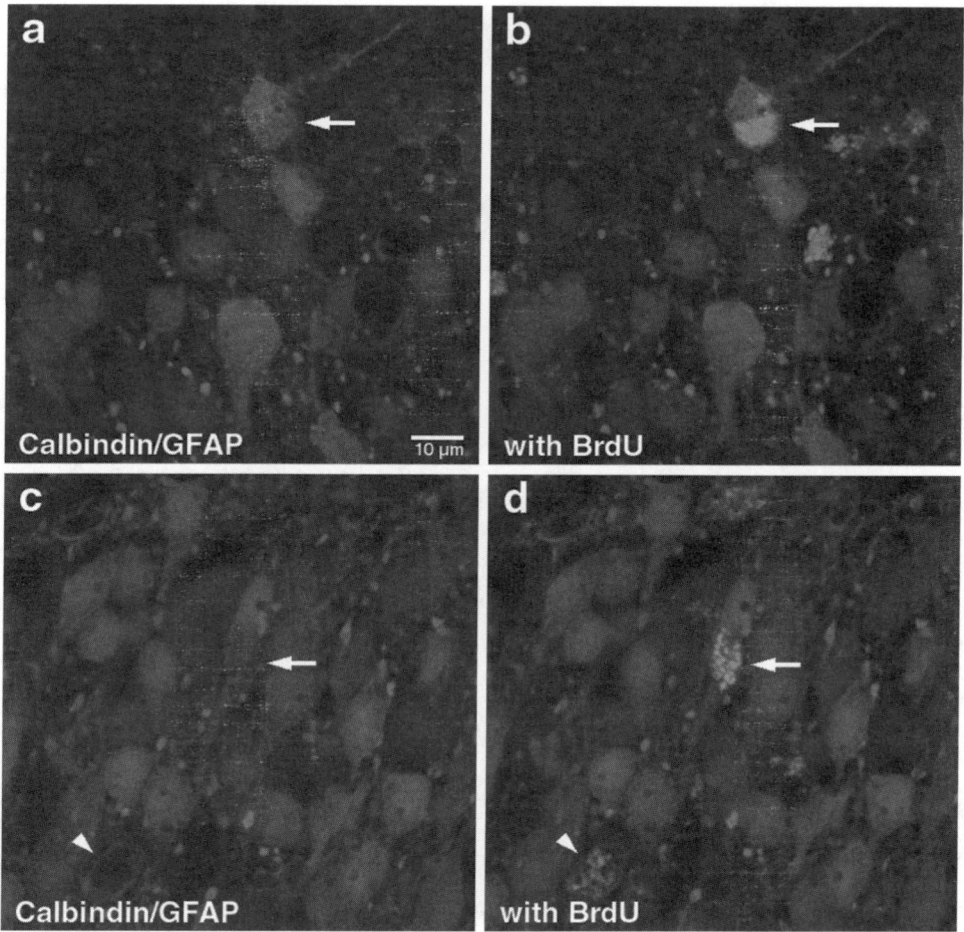

FIGURE 23.5 Neurogenesis in the adult human hippocampal dentate gyrus. Human hippocampal tissue was obtained at autopsy from cancer patients that had received BrdU for assessment of proliferative activity of tumor cells. Staining for BrdU (*green*), the neuronal marker calbindin (*red*) and the glial marker GFAP (*blue*) demonstrated that new neurons (arrows) and astrocytes (arrowheads) are generated in the human dentate gyrus. Figure reproduced from Eriksson *et al.* (1998). Neurogenesis in the adult human hippocampus, *Nat Med* 4:1313-1317. Copyright by Nature Medicine. (See also the Color Plate section.)

which raises concerns about the *in vivo* relevance of the *in vitro* findings. The development of techniques to prospectively identify and isolate adult neural stem cells will provide a powerful tool to reevaluate our hypotheses and to characterize adult neural stem cells in detail.

2. *Differentiation potential of adult neural stem cells.* Cells with neural stem–like characteristics have been isolated from several regions of the adult CNS. Do adult neural stem cells have different characteristics depending on their region of origin? Do adult neural stem cells have broad potential and differentiate into different neuronal subtypes or are they restricted by intrinsic determinants in the types of neurons they can generate? Until now it has only been demonstrated that adult neural stem cells can form functional interneurons in the olfactory bulb and local projection neurons

in the hippocampal dentate gyrus (van Praag *et al.*, 2002; Carleton *et al.*, 2003). However, one important requirement for therapeutic strategies using endogenous stem cells is the ability to specifically generate the class of nerve cell that is affected by the disease.

3. *Define the regulatory mechanisms underlying cell genesis.* Our current knowledge on the regulatory mechanisms of cell genesis in the adult CNS is very incomplete and primarily descriptive. With the help of new techniques such as large-scale expression analysis it should be possible to identify molecular players that govern the proliferation, fate choice, migration, integration and survival of new cells. The identification of regulators may provide potential targets for the recruitment of endogenous stem cells for repair.

4. *Characterization of the disease environment.* How does disease alter the environment surrounding

endogenous stem cells? Which CNS cell types are affected by the disease process? It is important to recognize that not only defined subpopulations of neurons but also glial cells are killed by, for example, injury and stroke. Until recently, neurodegenerative diseases such as Parkinson's disease, have been thought of as diseases affecting specialized classes of neurons. However, it is becoming evident that neurodegeneration also elicits a glial reaction that potentially perpetuates neuronal cell death and it is possible that the neurodegenerative process also affects glial cells. The behavior of neural stem cells is not only controlled by cell-autonomous signals but is greatly influenced by environment-derived factors. In addition, it is evident that the function of mature nerve cells is highly dependent on an intact environment that is created by many different cell types. It is therefore important to focus not only on the replacement of dying neurons but also on the creation of a cellular environment that supports the development and function of the new neurons.

Finally, advances in related fields that deepen our understanding of the pathophysiology of neurological diseases and enhance our ability to safely manipulate the diseased CNS, will have a significant impact on endogenous stem cell–based therapies by allowing us to identify candidate diseases and by guiding us toward strategies for this experimental therapy.

ACKNOWLEDGMENTS

The authors thank Dr. Jenny Hsieh, Mary-Lynne Gage, and Heather Lansford for helpful suggestions and critical reading of the manuscript; Linda Kitabayashi for artwork in Figure 23.5, and Jamie Simon for artwork in Figures 23.1 and 23.2.

References

Aberg MA, *et al.*, (2000). Peripheral infusion of IGF-I selectively induces neurogenesis in the adult rat hippocampus, *J Neurosci* 20:2896-2903.

Altman J, Das GD. (1965). Autoradiographic and histological evidence of postnatal hippocampal neurogenesis in rats, *J Comp Neurol* 124:319-335.

Alvarez-Buylla A, Garcia-Verdugo JM. (2002). Neurogenesis in adult subventricular zone, *J Neurosci* 22:629-634.

Arsenijevic Y, Weiss S, Schneider B, Aebischer P. (2001a). Insulin-like growth factor-I is necessary for neural stem cell proliferation and demonstrates distinct actions of epidermal growth factor and fibroblast growth factor-2, *J Neurosci* 21:7194-7202.

Arsenijevic Y *et al.* (2001b). Isolation of multipotent neural precursors residing in the cortex of the adult human brain, *Exp Neurol* 170: 48-62.

Arvidsson A *et al.* (2002). Neuronal replacement from endogenous precursors in the adult brain after stroke, *Nat Med* 8:963-970.

Banasr M, Hery M, Brezun JM, Daszuta A. (2001). Serotonin mediates oestrogen stimulation of cell proliferation in the adult dentate gyrus, *Eur J Neurosci* 14:1417-1424.

Benraiss A *et al.* (2001). Adenoviral brain-derived neurotrophic factor induces both neostriatal and olfactory neuronal recruitment from endogenous progenitor cells in the adult forebrain, *J Neurosci* 21:6718-6731.

Biebl M, Cooper CM, Winkler J, Kuhn HG. (2000). Analysis of neurogenesis and programmed cell death reveals a self-renewing capacity in the adult rat brain, *Neurosci Lett* 291:17-20.

Brezun JM, Daszuta A. (1999). Depletion in serotonin decreases neurogenesis in the dentate gyrus and the subventricular zone of adult rats, *Neuroscience* 89:999-1002.

Brown J *et al.* (2003). Enriched environment and physical activity stimulate hippocampal but not olfactory bulb neurogenesis, *Eur J Neurosci* 17:2042-2046.

Cameron HA, Gould E. (1994). Adult neurogenesis is regulated by adrenal steroids in the dentate gyrus, *Neuroscience* 61:203-209.

Cameron HA, McEwen BS, Gould E. (1995). Regulation of adult neurogenesis by excitatory input and NMDA receptor activation in the dentate gyrus, *J Neurosci* 15:4687-4692.

Cameron HA, Tanapat P, Gould E. (1998a). Adrenal steroids and *N*-methyl-*D*-aspartate receptor activation regulate neurogenesis in the dentate gyrus of adult rats through a common pathway, *Neuroscience* 82:349-354.

Cameron HA, Hazel TG, McKay RD. (1998b). Regulation of neurogenesis by growth factors and neurotransmitters, *J Neurobiol* 36: 287-306.

Carleton A *et al.* (2003). Becoming a new neuron in the adult olfactory bulb, *Nat Neurosci* 6:507-518.

Conover JC *et al.* (2000). Disruption of Eph/ephrin signaling affects migration and proliferation in the adult subventricular zone, *Nat Neurosci* 3:1091-1097.

Cooper-Kuhn CM, Kuhn HG. (2002). Is it all DNA repair? Methodological considerations for detecting neurogenesis in the adult brain, *Brain Res Dev Brain Res* 134:13-21.

Domeniconi M *et al.* (2002). Myelin-associated glycoprotein interacts with the Nogo66 receptor to inhibit neurite outgrowth, *Neuron* 35:283-290.

Duman RS, Nakagawa S, Malberg J. (2001a). Regulation of adult neurogenesis by antidepressant treatment, *Neuropsychopharmacology* 25:836-844.

Duman RS, Malberg J, Nakagawa S. (2001b). Regulation of adult neurogenesis by psychotropic drugs and stress, *J Pharmacol Exp Ther* 299:401-407.

Dziewczapolski G *et al.* (2003). Survival and differentiation of adult rat-derived neural progenitor cells transplanted to the striatum of hemiparkinsonian rats, *Exp Neurol* 183:653-664.

Eriksson PS *et al.* (1998). Neurogenesis in the adult human hippocampus, *Nat Med* 4:1313-1317.

Falk A, Frisen J. (2002). Amphiregulin is a mitogen for adult neural stem cells, *J Neurosci Res* 69:757-762.

Fallon J *et al.* (2000). In vivo induction of massive proliferation, directed migration, and differentiation of neural cells in the adult mammalian brain, *Proc Natl Acad Sci USA* 97:14686-14691.

Fricker-Gates RA *et al.* (2002). Late-stage immature neocortical neurons reconstruct interhemispheric connections and form synaptic contacts with increased efficiency in adult mouse cortex undergoing targeted neurodegeneration, *J Neurosci* 22:4045-4056.

Gage FH. (2000). Mammalian neural stem cells, *Science* 287: 1433-1438.

Gage FH, Ray J, Fisher LJ. (1995). Isolation, characterization, and use of stem cells from the CNS, *Annu Rev Neurosci* 18:159-192.

Gensert JM, Goldman JE. (1997). Endogenous progenitors remyelinate demyelinated axons in the adult CNS, *Neuron* 19:197-203.

Gheusi G *et al.* (2000). Importance of newly generated neurons in the adult olfactory bulb for odor discrimination, *Proc Natl Acad Sci USA* 97:1823-1828.

Gilmore EC, Nowakowski RS, Caviness VS, Jr, Herrup K. (2000). Cell birth, cell death, cell diversity and DNA breaks: how do they all fit together? *Trends Neurosci* 23:100-105.

Goldman SA, Nottebohm F. (1983). Neuronal production, migration, and differentiation in a vocal control nucleus of the adult female canary brain, *Proc Natl Acad Sci USA* 80:2390-2394.

Gould E, Reeves AJ, Graziano MS, Gross CG. (1999a). Neurogenesis in the neocortex of adult primates, *Science* 286:548-552.

Gould E *et al.* (1999b). Learning enhances adult neurogenesis in the hippocampal formation, *Nat Neurosci* 2:260-265.

Gu W, Brannstrom T, Wester P. (2000). Cortical neurogenesis in adult rats after reversible photothrombotic stroke, *J Cereb Blood Flow Metab* 20:1166-1173.

Haas CA *et al.* (1999). Entorhinal cortex lesion in adult rats induces the expression of the neuronal chondroitin sulfate proteoglycan neurocan in reactive astrocytes, *J Neurosci* 19:9953-9963.

Hastings NB, Gould E. (1999). Rapid extension of axons into the CA3 region by adult-generated granule cells, *J Comp Neurol* 413: 146-154.

Hernit-Grant CS, Macklis JD. (1996). Embryonic neurons transplanted to regions of targeted photolytic cell death in adult mouse somatosensory cortex re-form specific callosal projections, *Exp Neurol* 139:131-142.

Horner PJ, Gage FH. (2000). Regenerating the damaged central nervous system, *Nature* 407:963-970.

Horner PJ *et al.* (2000). Proliferation and differentiation of progenitor cells throughout the intact adult rat spinal cord, *J Neurosci* 20: 2218-2228.

Jacques T. S *et al.* (1998). Neural precursor cell chain migration and division are regulated through different beta1 integrins, *Development* 125:3167-3177.

Jaworski DM, Kelly GM, Hockfield S. (1999). Intracranial injury acutely induces the expression of the secreted isoform of the CNS-specific hyaluronan-binding protein BEHAB/brevican, *Exp Neurol* 157:327-337.

Jiang W *et al.* (2001). Cortical neurogenesis in adult rats after transient middle cerebral artery occlusion, *Stroke* 32:1201-1207.

Jin K *et al.* (2002). Vascular endothelial growth factor (VEGF) stimulates neurogenesis in vitro and in vivo, *Proc Natl Acad Sci USA* 99:11946-11950.

Jin K *et al.* (2003). Neurogenesis and aging: FGF-2 and HB-EGF restore neurogenesis in hippocampus and subventricular zone of aged mice, *Aging Cell* 2:175-183.

Johansson CB *et al.* (1999a). Neural stem cells in the adult human brain, *Exp Cell Res* 253:733-736.

Johansson CB *et al.* (1999b). Identification of a neural stem cell in the adult mammalian central nervous system, *Cell* 96:25-34.

Jones LL, Tuszynski MH. (2002). Spinal cord injury elicits expression of keratan sulfate proteoglycans by macrophages, reactive microglia, and oligodendrocyte progenitors, *J Neurosci* 22: 4611-4624.

Jones LL Yamaguchi Y, Stallcup WB, Tuszynski MH. (2002). NG2 is a major chondroitin sulfate proteoglycan produced after spinal cord injury and is expressed by macrophages and oligodendrocyte progenitors, *J Neurosci* 22:2792-2803.

Kaplan MS, Hinds JW. (1977). Neurogenesis in the adult rat: electron microscopic analysis of light radioautographs, *Science* 197: 1092-1094.

Kaplan MS, Hinds JW. (1980). Gliogenesis of astrocytes and oligodendrocytes in the neocortical grey and white matter of the adult rat: electron microscopic analysis of light radioautographs, *J Comp Neurol* 193:711-727.

Kay JN, Blum M. (2000). Differential response of ventral midbrain and striatal progenitor cells to lesions of the nigrostriatal dopaminergic projection, *Dev Neurosci* 22:56-67.

Kee NJ, Preston E, Wojtowicz JM. (2001). Enhanced neurogenesis after transient global ischemia in the dentate gyrus of the rat, *Exp Brain Res* 136:313-320.

Kempermann G. (2002). Why new neurons? Possible functions for adult hippocampal neurogenesis, *J Neurosci* 22:635-638.

Kempermann G, Gage FH. (2000). Neurogenesis in the adult hippocampus, *Novartis Found Symp* 231:220-235; discussion 235-241, 302-226.

Kempermann G, Kuhn HG, Gage FH. (1997a). Genetic influence on neurogenesis in the dentate gyrus of adult mice, *Proc Natl Acad Sci USA* 94:10409-10414.

Kempermann G, Kuhn HG, Gage FH. (1997b). More hippocampal neurons in adult mice living in an enriched environment, *Nature* 386:493-495.

Kempermann G, Kuhn HG, Gage FH. (1998). Experience-induced neurogenesis in the senescent dentate gyrus, *J Neurosci* 18:3206-3212.

Kempermann G, Gast D, Gage FH. (2002). Neuroplasticity in old age: sustained fivefold induction of hippocampal neurogenesis by long-term environmental enrichment, *Ann Neurol* 52:135-143.

Kempermann G *et al.* (2003). Early determination and long-term persistence of adult-generated new neurons in the hippocampus of mice, *Development* 130:391-399.

Kim JH *et al.* (2002). Dopamine neurons derived from embryonic stem cells function in an animal model of Parkinson's disease, *Nature* 418:50-56.

Kondo T, Raff M. (2000). Oligodendrocyte precursor cells reprogrammed to become multipotential CNS stem cells, *Science* 289:1754-1757.

Kornack DR, Rakic P. (2001). Cell proliferation without neurogenesis in adult primate neocortex, *Science* 294:2127-2130.

Kuhn HG, Dickinson-Anson H, Gage FH. (1996). Neurogenesis in the dentate gyrus of the adult rat: age-related decrease of neuronal progenitor proliferation, *J Neurosci* 16:2027-2033.

Kuhn HG, Winkler J, Kempermann G, Thal LJ, Gage FH. (1997). Epidermal growth factor and fibroblast growth factor-2 have different effects on neural progenitors in the adult rat brain, *J Neurosci* 17:5820-5829.

Kukekov VG *et al.* (1999). Multipotent stem/progenitor cells with similar properties arise from two neurogenic regions of adult human brain, *Exp Neurol* 156:333-344.

Lai K, Kaspar BK, Gage FH, Schaffer DV. (2003). Sonic hedgehog regulates adult neural progenitor proliferation in vitro and in vivo, *Nat Neurosci* 6:21-27.

Leavitt BR, Hernit-Grant CS, Macklis JD. (1999). Mature astrocytes transform into transitional radial glia within adult mouse neocortex that supports directed migration of transplanted immature neurons, *Exp Neurol* 157:43-57.

Lemons ML, Howland DR, Anderson DK. (1999). Chondroitin sulfate proteoglycan immunoreactivity increases following spinal cord injury and transplantation, *Exp Neurol* 160:51-65.

Lie DC *et al.* (2002). The adult substantia nigra contains progenitor cells with neurogenic potential, *J Neurosci* 22:6639-6649.

Lie *et al.* (2004). Neurogenesis in the adult brain: New strategies for central nervous system diseases, *Ann Rev Pharmacol Toxicol* 44:399-421.

Lim DA, Alvarez-Buylla A. (1999). Interaction between astrocytes and adult subventricular zone precursors stimulates neurogenesis, *Proc Natl Acad Sci USA* 96:7526-7531.

Lim DA *et al.* (2000). Noggin antagonizes BMP signaling to create a niche for adult neurogenesis, *Neuron* 28:713-726.

Lindvall O, Hagell P. (2001). Cell therapy and transplantation in Parkinson's disease, *Clin Chem Lab Med* 39:356-361.

Liu BP, Fournier A, GrandPre T, Strittmatter SM. (2002). Myelin-associated glycoprotein as a functional ligand for the Nogo-66 receptor, *Science* 297:1190-1193.

Liu J, Solway K, Messing RO, Sharp FR. (1998). Increased neurogenesis in the dentate gyrus after transient global ischemia in gerbils, *J Neurosci* 18:7768-7778.

Lois C, Garcia-Verdugo JM, Alvarez-Buylla A. (1996). Chain migration of neuronal precursors, *Science* 271:978-981.

Louissaint A, Jr, Rao S, Leventhal C, Goldman SA. (2002). Coordinated interaction of neurogenesis and angiogenesis in the adult songbird brain, *Neuron* 34:945-960.

Magavi SS, Leavitt BR, Macklis JD. (2000). Induction of neurogenesis in the neocortex of adult mice, *Nature* 405:951-955.

Markakis EA, Gage FH. (1999). Adult-generated neurons in the dentate gyrus send axonal projections to field CA3 and are surrounded by synaptic vesicles, *J Comp Neurol* 406:449-460.

Nakagawa S *et al.* (2002). Regulation of neurogenesis in adult mouse hippocampus by cAMP and the cAMP response element-binding protein, *J Neurosci* 22:3673-3682.

Nakatomi H *et al.* (2002). Regeneration of hippocampal pyramidal neurons after ischemic brain injury by recruitment of endogenous neural progenitors, *Cell* 110:429-441.

Nowakowski RS, Hayes NL. (2000). New neurons: extraordinary evidence or extraordinary conclusion? *Science* 288:771.

Nunes MC *et al.* (2003). Identification and isolation of multipotential neural progenitor cells from the subcortical white matter of the adult human brain, *Nat Med* 9:439-447.

Ono K, Tomasiewicz H, Magnuson T, Rutishauser U. (1994). N-CAM mutation inhibits tangential neuronal migration and is phenocopied by enzymatic removal of polysialic acid, *Neuron* 13:595-609.

Palmer TD, Ray J, Gage FH. (1995). FGF-2-responsive neuronal progenitors reside in proliferative and quiescent regions of the adult rodent brain, *Mol Cell Neurosci* 6:474-486.

Palmer TD, Takahashi J, Gage FH. (1997). The adult rat hippocampus contains primordial neural stem cells, *Mol Cell Neurosci* 8:389-404.

Palmer TD, Willhoite AR, Gage FH. (2000). Vascular niche for adult hippocampal neurogenesis, *J Comp Neurol* 425:479-494.

Palmer TD *et al.* (1999). Fibroblast growth factor-2 activates a latent neurogenic program in neural stem cells from diverse regions of the adult CNS, *J Neurosci* 19:8487-8497.

Palmer TD *et al.* (2001). Cell culture. Progenitor cells from human brain after death, *Nature* 411:42-43.

Parent JM, Lowenstein DH. (2002). Seizure-induced neurogenesis: are more new neurons good for an adult brain? *Prog Brain Res* 135:121-131.

Parent JM, Valentin VV, Lowenstein DH. (2002a). Prolonged seizures increase proliferating neuroblasts in the adult rat subventricular zone-olfactory bulb pathway, *J Neurosci* 22:3174-3188.

Parent JM *et al.* (2002b). Rat forebrain neurogenesis and striatal neuron replacement after focal stroke, *Ann Neurol* 52:802-813.

Parent JM *et al.* (1997). Dentate granule cell neurogenesis is increased by seizures and contributes to aberrant network reorganization in the adult rat hippocampus, *J Neurosci* 17:3727-3738.

Park HT, Wu J, Rao Y. (2002). Molecular control of neuronal migration, *Bioessays* 24:821-827.

Pasterkamp RJ *et al.* (1999). Expression of the gene encoding the chemorepellent semaphorin III is induced in the fibroblast component of neural scar tissue formed following injuries of adult but not neonatal CNS, *Mol Cell Neurosci* 13:143-166.

Pencea V, Bingaman KD, Wiegand SJ, Luskin MB. (2001). Infusion of brain-derived neurotrophic factor into the lateral ventricle of the adult rat leads to new neurons in the parenchyma of the striatum, septum, thalamus, and hypothalamus, *J Neurosci* 21:6706-6717.

Petreanu L, Alvarez-Buylla A. (2002). Maturation and death of adult-born olfactory bulb granule neurons: role of olfaction, *J Neurosci* 22:6106-6113.

Pincus DW *et al.* (1998). Fibroblast growth factor-2/brain-derived neurotrophic factor-associated maturation of new neurons generated from adult human subependymal cells, *Ann Neurol* 43:576-585.

Prineas JW *et al.* (1993). Multiple sclerosis: remyelination of nascent lesions, *Ann Neurol* 33:137-151.

Reynolds BA, Weiss S. (1992). Generation of neurons and astrocytes from isolated cells of the adult mammalian central nervous system, *Science* 255:1707-1710.

Rietze RL *et al.* (2001). Purification of a pluripotent neural stem cell from the adult mouse brain. *Nature* 412:736-739.

Rochefort C, Gheusi G, Vincent JD, Lledo PM. (2002). Enriched odor exposure increases the number of newborn neurons in the adult olfactory bulb and improves odor memory, *J Neurosci* 22: 2679-2689.

Rousselot P, Lois C, Alvarez-Buylla A. (1995). Embryonic (PSA) N-CAM reveals chains of migrating neuroblasts between the lateral ventricle and the olfactory bulb of adult mice, *J Comp Neurol* 351:51-61.

Scharfman HE, Goodman JH, Sollas AL. (2000). Granule-like neurons at the hilar/CA3 border after status epilepticus and their synchrony with area CA3 pyramidal cells: functional implications of seizure-induced neurogenesis, *J Neurosci* 20:6144-6158.

Scolding N *et al.* (1998). Oligodendrocyte progenitors are present in the normal adult human CNS and in the lesions of multiple sclerosis, *Brain* 121 (Pt 12):2221-2228.

Seri B, Garcia-Verdugo JM, McEwen BS, Alvarez-Buylla A. (2001). Astrocytes give rise to new neurons in the adult mammalian hippocampus, *J Neurosci* 21:7153-7160.

Shihabuddin LS, Horner PJ, Ray J, Gage FH. (2000). Adult spinal cord stem cells generate neurons after transplantation in the adult dentate gyrus, *J Neurosci* 20:8727-8735.

Shin JJ *et al.* (2000). Transplanted neuroblasts differentiate appropriately into projection neurons with correct neurotransmitter and receptor phenotype in neocortex undergoing targeted projection neuron degeneration, *J Neurosci* 20:7404-7416.

Snyder EY, Yoon C, Flax JD, Macklis JD. (1997). Multipotent neural precursors can differentiate toward replacement of neurons undergoing targeted apoptotic degeneration in adult mouse neocortex, *Proc Natl Acad Sci USA* 94:11663-11668.

Song H, Stevens CF, Gage FH. (2002). Astroglia induce neurogenesis from adult neural stem cells, *Nature* 417:39-44.

Stichel CC, Muller HW. (1998). The CNS lesion scar: new vistas on an old regeneration barrier, *Cell Tissue Res* 294:1-9.

Suhonen JO, Peterson DA, Ray J, Gage FH. (1996). Differentiation of adult hippocampus-derived progenitors into olfactory neurons in vivo, *Nature* 383:624-627.

Takahashi J, Palmer TD, Gage FH. (1999). Retinoic acid and neurotrophins collaborate to regulate neurogenesis in adult-derived neural stem cell cultures, *J Neurobiol* 38:65-81.

Tanigaki K *et al.* (2001). Notch1 and Notch3 instructively restrict bFGF-responsive multipotent neural progenitor cells to an astroglial fate, *Neuron* 29:45-55.

Temple S. (2001). The development of neural stem cells, *Nature* 414:112-117.

van Praag H, Kempermann G, Gage FH. (1999). Running increases cell proliferation and neurogenesis in the adult mouse dentate gyrus, *Nat Neurosci* 2:266-270.

van Praag H *et al.* (2002). Functional neurogenesis in the adult hippocampus, *Nature* 415:1030-1034.

Vroemen M, Aigner L, Winkler J, Weidner N. (2003). Adult neural progenitor cell grafts survive after acute spinal cord injury and integrate along axonal pathways, *Eur J Neurosci* 18:743-751.

Wagner JP, Black IB, DiCicco-Bloom E. (1999). Stimulation of neonatal and adult brain neurogenesis by subcutaneous injection of basic fibroblast growth factor, *J Neurosci* 19:6006-6016.

Wang KC et al. (2002a). P75 interacts with the Nogo receptor as a co-receptor for Nogo, MAG and OMgp, *Nature* 420:74-78.

Wang KC et al. (2002b). Oligodendrocyte-myelin glycoprotein is a Nogo receptor ligand that inhibits neurite outgrowth, *Nature* 417:941-944.

Wang Y, Sheen VL, Macklis JD. (1998). Cortical interneurons upregulate neurotrophins in vivo in response to targeted apoptotic degeneration of neighboring pyramidal neurons, *Exp Neurol* 154:389-402.

Weiss S et al. (1996). Multipotent CNS stem cells are present in the adult mammalian spinal cord and ventricular neuroaxis, *J Neurosci* 16:7599-7609.

Weissman IL, Anderson DJ, Gage F. (2001). Stem and progenitor cells: origins, phenotypes, lineage commitments, and transdifferentiations, *Annu Rev Cell Dev Biol* 17:387-403.

Wichterle H, Garcia-Verdugo JM, Alvarez-Buylla A. (1997). Direct evidence for homotypic, glia-independent neuronal migration, *Neuron* 18:779-791.

Winner B et al. (2002). Long-term survival and cell death of newly generated neurons in the adult rat olfactory bulb, *Eur J Neurosci* 16:1681-1689.

Wu W et al. (1999). Directional guidance of neuronal migration in the olfactory system by the protein Slit, *Nature* 400:331-336.

Yagita Y et al. (2001). Neurogenesis by progenitor cells in the ischemic adult rat hippocampus, *Stroke* 32:1890-1896.

Yamamoto S et al. (2001). Proliferation of parenchymal neural progenitors in response to injury in the adult rat spinal cord, *Exp Neurol* 172:115-127.

Yoshimura S et al. (2001). FGF-2 regulation of neurogenesis in adult hippocampus after brain injury, *Proc Natl Acad Sci USA* 98:5874-5879.

Zhang RL, Zhang ZG, Zhang L, Chopp M. (2001). Proliferation and differentiation of progenitor cells in the cortex and the subventricular zone in the adult rat after focal cerebral ischemia, *Neuroscience* 105:33-41.

Zhao M et al. (2003). Evidence for neurogenesis in the adult mammalian substantia nigra, *Proc Natl Acad Sci USA* 100: 7925-7930.

Zuo J et al. (1998). Degradation of chondroitin sulfate proteoglycan enhances the neurite-promoting potential of spinal cord tissue, *Exp Neurol* 154:654-662.

24

Neuropathies—Translating Causes into Treatments

Christiane Massicotte, DMV, PhD
Steven S. Scherer, MD, PhD

INHERITED NEUROPATHIES

Peripheral neuropathy, or simply neuropathy, is the term used for any disease of peripheral nerves. As depicted in Table 24.1, the causes can be separated according to whether they are inherited or acquired, part of a syndrome, and "axonal" or "demyelinating" (whether the primary abnormality appears to affect axons/neurons or myelinating Schwann cells). Non-syndromic inherited neuropathies (Table 24.2) are called Charcot-Marie-Tooth disease (CMT) or hereditary motor and sensory neuropathy (HMSN). Different kinds are recognized clinically, aided by electrophysiological testing of peripheral nerves (Lupski and Garcia, 2001; Kleopas and Scherer, 2002). If the forearm motor nerve conduction velocities (NCVs) are greater or less than 38 m/sec, the neuropathy is traditionally considered to be "axonal" (CMT2/HMSN II) or "demyelinating" (CMT1/HMSN I), respectively. CMT1 is more common, has an earlier age of disease-onset (first or second decade of life), and nerve biopsies show segmental demyelination and remyelination as well as axonal loss. CMT2 typically has a later onset and is associated with loss of myelinated axons, without much evidence of demyelination. Besides CMT, there are other inherited neuropathies that have been traditionally given different names. These may be milder (hereditary neuropathy with liability to pressure palsies; HNPP), or more severe (Dejerine-Sottas syndrome [DSS] and congenital hypomyelinating neuropathy [CHN]). If the neuropathy is clinically recognized at birth, it is often called CHN; if it is recognized later in infancy, it is called DSS. Patients with CHN or DSS typically have NCVs less than 10 m/sec, and dysmyelinated axons characterized by improperly formed myelin sheaths.

Beginning in the early 1990s, the molecular causes of many inherited neuropathies were discovered, as summarized in Table 24.2. The molecular diagnoses are often difficult to reconcile with the traditional classification schemes, and the relationships between the genotypes of phenotypes of CMT are complex and incompletely understood. As these issues have been thoroughly reviewed (Lupski and Garcia, 2001; Kleopas and Scherer, 2002), a few salient points will be emphasized here:

- CMT1, CMT2, and DSS/CHN are caused by mutations in different genes; for DSS/CHN, both dominant and recessive causes are known. Conversely, different mutations in the same gene cause different phenotypes.
- Some dominant mutations cause haplotype insufficiency. In this case, the phenotype results from the loss of the gene product of one allele. For example, the deletion of *peripheral myelin protein 22kDa* (*PMP22*) causes HNPP, a loss-of-function *KIF1B* mutation causes CMT2A, and loss-of-function *myelin protein zero* (*MPZ*) mutations cause CMT1B.
- Most dominant mutations have a "toxic-gain-of-function"—this term includes dominant-negative effects on the wild-type allele, as well as other effects such as the consequences of a misfolded mutant protein. Dominant mutations that cause a gain-of-function are not necessarily informative about the normal function of the wild-type protein.
- For most CMT genes, there is compelling evidence that the more severe phenotypes are caused by gain-of-function alleles, and that (heterozygous) loss-of-function alleles are associated with milder phenotypes.

TABLE 24.1   Causes of Neuropathies[a]

	Syndromic	Non-syndromic	
Inherited	Waardenburg IV	CMT1	**demyelinating**
	MLD (recessive)	CMT4	
	FAP 1-4	CMT2	**axonal**
	GAN (recessive)	CMT2B	
Acquired	Osteosclerotic myeloma	AIDP, CIDP	**demyelinating**
	Immunoglobulin amyloidosis	AMAN; most toxins	**axonal**

[a]Neuropathies are classified by whether they are inherited or acquired, part of a syndrome, and by their primary pathological cause (axonal or demyelinating). Examples of each are provided.
MLD: metachromatic leukodystrophy; FAP: familial amyloid polyneuropathy; GAN: giant axonal neuropathy; AIDP/CIDP: acute/chronic inflammatory demyelinating polyneuropathy; AMAN: acute motor axonal neuropathy.

Mutations of *PMP22*, *MPZ*, and neurofilament light (*NEFL*) are examples.

- Demyelinating neuropathies are caused by mutations in genes expressed by myelinating Schwann cells, and are first manifested in the myelinating Schwann cells themselves. Conversely, axonal neuropathies are caused by mutations in genes expressed by neurons, and are first manifested in axons. There are possible exceptions to this scheme, such as *NEFL* mutations that appear to cause a demyelinating neuropathy, and *MPZ* mutations that appear to cause an axonal neuropathy.

- Animal models provide the proper cellular context for examining the effects of mutations, including the key means of analyzing the development of neuropathy (Martini, 1997).

Neurons and Axonal Neuropathies

Peripheral nervous system (PNS) neurons have axons outside of the central nervous system (CNS). These include motor, sensory, autonomic, and enteric neurons, as well as the special senses except for vision. Even though these neurons serve diverse functions, their shared molecular characteristics join them in certain genetic diseases. For example, the relevant sensory and autonomic neurons are absent in individuals with congenital insensitivity to pain and anhidrosis (CIPA) because the gene encoding the TrkA receptor is mutated. TrkA binds nerve growth factor (NGF), a trophic factor required for the survival of these neurons during their early development (Bibel and Barde, 1999).

Similarly, a shared vulnerability for the same molecular defect may explain why hearing loss, retinitis pigmentosa, and/or other CNS manifestations are associated with certain syndromic neuropathies (for examples, see website: http://www.neuro.wustl.edu/neuromuscular/).

In most neuropathies, the clinical features tend to have a distal predilection, both in first appearance and in ultimate severity. This suggests that axonal length itself is the primary determinant of neuropathy. Axons have a prominent cytoskeleton composed of neurofilaments and microtubules. Neurofilaments are composed of three subunits, termed heavy, medium, and light. Dominant mutations in *NEFL*, the gene encoding the light subunit, cause an axonal neuropathy, CMT2E (see Table 24.2). Because heterozygous *Nefl*-null mice do not develop a CMT2-like neuropathy (Zhu et al., 1997), dominant *NEFL* mutations are probably not the result of haplotype insufficiency caused by loss-of-function alleles. Introduction of an L394P mutation disrupts neurofilament assembly and causes a severe, dominantly inherited neuropathy/ neuronopathy in transgenic mice (Lee et al., 1994). Similarly, a dominant *Nefl* mutation in quail causes neuropathy with profoundly reduced axonal calibers (Ohara et al., 1993). Thus, missense mutations may disorganize neurofilament assembly, causing neuropathy and even altered axonal caliber. In agreement, gene transfer experiments in cultured neurons suggest that CMT2E-associated mutations disrupt the assembly and axonal transport of neurofilaments (Brownlees et al., 2002; Perez-Olle et al., 2002).

A distal predilection, however, does not mean that the defect necessarily lies in the axons themselves; it could just as well represent a primary neuron cell body abnormality. Neurons with the longest axons could be the most vulnerable to a given insult. For instance, large doses of pyridoxine promptly kill large primary sensory neurons, whereas smaller doses cause only subtle shrinkage of these neurons, and mild distal axonal degeneration (Xu et al., 1989). Thus, a modest neuronal abnormality may result in distal axonopathy, but a more severe insult of the same type may cause the neuron itself to degenerate as the primary event. The neuropathological appearance of some inherited axonal neuropathies consists of primary nerve cell body degeneration associated with anterograde degeneration of the axons. Some inherited axonal neuropathies, such as the hereditary motor neuropathies and hereditary sensory (and autonomic) neuropathies (see Table 24.2), could be considered as primary neuronopathies rather than primary axonopathies, but it is difficult to make this distinction.

PNS axons are the longest cells in the body, and contain much more cytoplasm than their cell bodies themselves. Axonal proteins and organelles, including the cytoskeletal components, synaptic vesicle precursors,

TABLE 24.2 Non-syndromic Inherited Neuropathies with a Genetically Identified Cause[a]

Disease (MIM)	Genetic Mutations	Clinical Features
Autosomal or X-linked dominant demyelinating neuropathies		
HNPP (162500)	Usually deletion of one *PMP22* allele	Episodic mononeuropathies at typical sites of nerve compression; underlying mild demyelinating neuropathy.
CMT1A (118220)	Usually duplication of one *PMP22* allele	Onset first to second decade. Weakness, atrophy, and sensory loss, beginning in the feet and progressing proximally.
CMT1B (118200)	*MPZ* mutations	Similar to CMT1A, but severity varies according to mutation, from "severe" to "mild" CMT1.
CMT1C (601098)	*LITAF/SIMPLE* mutations	Similar to CMT1A. Motor NCVs about 20 m/s.
CMT1(D)	*EGR2* mutations	Similar to CMT1A, but severity varies according to mutation, from "severe" to "mild" CMT1.
CMT1X (302800)	Mainly loss of function *GJB1* mutations	Similar to CMT1A, but distal atrophy more pronounced At every age, men are more affected than are women.
Autosomal dominant axonal neuropathies		
CMT2A (118210)	*KIF1Bβ* mutations	Onset of neuropathy by 10 y, progression to weakness, atrophy in distal legs; mild sensory disturbance.
CMT2B (600882)	*RAB7* mutations	Onset second to third decade; severe sensory loss with distal ulcerations; also length-dependent weakness.
CMT2D (601472)	*GARS* mutations	Arm more than leg weakness; onset of weakness second to third decade; sensory axons involved.
CMT2E (162280)	*NEFL* mutations	Variable onset and severity, ranging from DSS-like to CMT2 phenotype.
CMT2-P$_0$ (118200)	*MPZ* mutations	Late onset (30y or older), but progressive neuropathy with pain, hearing loss, and abnormally reactive pupils.
Severe demyelinating neuropathies (autosomal dominant or recessive; "CMT3 or HMSN III")		
DSS (Dejerine-Sottas Syndrome; 145900)	Dominant (*PMP22, MPZ, GJB1, EGR2, NEFL*) and recessive (*MTMR2, PRX*) mutations	Delayed motor development before 3 y; severe weakness and atrophy in distal more than proximal muscles; severe sensory loss particularly of modalities subserved by large myelinated axons.
CHN (Congenital Hypomyelinating Neuropathy; 605253)	Dominant (*EGR2, PMP22, MPZ*) and recessive *(EGR2)* mutations	Clinical picture often similar to that of Dejerine-Sottas syndrome, but hypotonic at birth.
Autosomal recessive demyelinating neuropathies ("CMT4")		
CMT4A (214400)	*GDAP1* mutations	Early childhood onset, progressing to wheelchair-dependency; mixed demyelinating and axonal features.
CMT4B1 (601382)	*MTMR2* mutations	Early childhood onset, may progress to wheelchair-dependency; focally folded myelin sheaths.
CMT4B2 (604563)	*MTMR13* mutations	Childhood onset, progression to assistive devices for walking; focally folded myelin sheaths; glaucoma.
CMT4D (601455)	*NDRG1* mutations	Childhood onset; progression to severe disability by 50y; hearing loss and dysmorphic features.
CMT4F (605260)	*PRX* mutations	Childhood onset; usually progression to severe disability; prominent sensory symptoms and findings.
CMT4	*EGR2* mutations	Infantile onset; progressing to wheelchair-dependency.
Autosomal recessive axonal neuropathies ("AR-CMT2" or "CMT2B")		
AR-CMT2A (605588)	Recessive *LMNA* mutations	Onset of neuropathy in second decade, but progresses to severe weakness and atrophy in distal muscles.
Hereditary Motor Neuropathies (HMN or "distal SMA")		
SMARD1 (604320)	Recessive *IGHMBP2* mutations	Distal infantile spinal muscular atrophy with diaphragm paralysis.
HMN 5 (600794)	Dominant *GARS* mutations	Arm more than leg weakness; onset of weakness second to third decade; no sensory involvement.

Continued

TABLE 24.2   Non-syndromic Inherited Neuropathies with a Genetically Identified Cause—(continued)

Hereditary Sensory (and Autonomic) Neuropathies/Neuronopathies (HSN or HSAN)

HSN-1 (162400)	Dominant *SPTLC1* mutations	Onset second to third decade; severe sensory loss with distal ulcerations; also length-dependent weakness.
HSN 3 (Riley-Day syndrome; 223900)	Recessive *IKBKAP* mutations	Dysautonomia and absence of fungiform papilla.
HSN 4 (CIPA; 256800)	Recessive *TRKA* mutations	Dysautonomia and loss of pain sensation caused by congenital absence of sensory and sympathetic neurons.
Other		
Cold-induced sweating syndrome (272430)	Recessive *CRLF1* mutations	Poor sucking in infancy, cold-induced sweating, scoliosis.

[a]The neuropathies are classified by MIM (http://www.ncbi.nlm.nih.gov/Omim/); the references for the individual mutations are compiled in the CMT mutation database (http://molgen-www.uia.ac.be/CMTMutations/DataSource/MutByGene.cfm). For syndromic neuropathies, see http://www.neuro.wustl.edu/neuromuscular/.

and mitochondria, are transported down the axon. Molecular motors, kinesins, mediate orthograde axonal transport, using microtubules as tracks. Kinesins are a large gene family, and each kinesin transports specific organelles. A dominant mutation in the gene encoding kinesin KIF1Bβ causes an inherited axonal neuropathy, probably by haplotype insufficiency. Mutations affecting another kinesin, KIF5A, cause a dominantly inherited form of hereditary spastic paraparesis, possibly by disrupting microtubule-dependent slow axonal transport, including the transport of neurofilaments (Xia *et al.,* 2003). Retrograde axonal transport is mediated by the dynein-dynactin complex. If this complex is disrupted in transgenic mice, a progressive motor neuron disease is observed that is caused by deficient retrograde transport and neurofilament accumulation (LaMonte *et al.,* 2002). In accord with these findings, a dominant mutation in the p150 subunit of dynactin has been found in a family with a dominantly inherited motor neuron disease. Finally, recessive mutations in gigaxonin cause giant axonal neuropathy, a syndromic condition in which CNS and PNS neurons are affected, possibly both in a length-dependent pattern. Gigaxonin binds to microtubule-associated protein 1B (MAP1B), and enhances microtubule stability (Ding *et al.,* 2002). One anticipates that disrupted axonal transport may be a consequence of other genetic mutations that cause neuropathy and/or neuronopathy. Deficient axonal transport has long been considered to be a most attractive hypothesis to account for length-related axonal degeneration (Griffin and Watson, 1988).

Myelinating Schwann Cells and Demyelinating Neuropathies

Myelin is a multilamellar spiral of specialized cell membrane that ensheathes axons larger than 1 μm in diam-
eter (Scherer and Arroyo, 2002). By reducing the capacitance and/or increasing the resistance, myelin reduces current flow across the internodal axonal membrane, thereby facilitating saltatory conduction at nodes. The myelin sheath itself can be divided into two domains—compact and noncompact myelin—each containing a nonoverlapping set of proteins (Figure 24.1; see also Color Plate section). Compact myelin forms the bulk of the myelin sheath. It is largely composed of lipids (including cholesterol, sphingolipids, galactocerebroside, and sulfatide) as well as the proteins—P0, PMP22, and myelin basic protein (MBP). Noncompact myelin is found in paranodes and incisures and contains a mutually exclusive set of molecules, including the gap junction protein connexin 32 (Cx32), myelin-associated glycoprotein (MAG), and E-cadherin. The outermost/abaxonal cell membrane, apposing the basal lamina, contains dystroglycan and α6β4 integrin, which are receptors of laminin-2. Dystroglycan is linked to a complex that contains periaxin, which may, in turn, be linked to compact myelin.

PMP22 Mutations Cause HNPP, CMT1A, DSS, and CHN

PMP22 is a small intrinsic membrane protein (see Figure 24.1). PMP22 is involved in early steps of myelination and plays a structural role in compact myelin. How deletion and duplication of *PMP22* cause HNPP and CMT1A, respectively, is not completely settled, but the deleterious effects are related to gene dosage, which in turn, is correlated with the relative amounts of PMP22 protein found in compact myelin (Lupski and Garcia, 2001; Suter and Scherer, 2003). PMP22 can form dimers and even multimers and may also interact with P0 (Tobler *et al.,* 2002), so that an altered PMP22/P0 ratio might destabilize the myelin sheath.

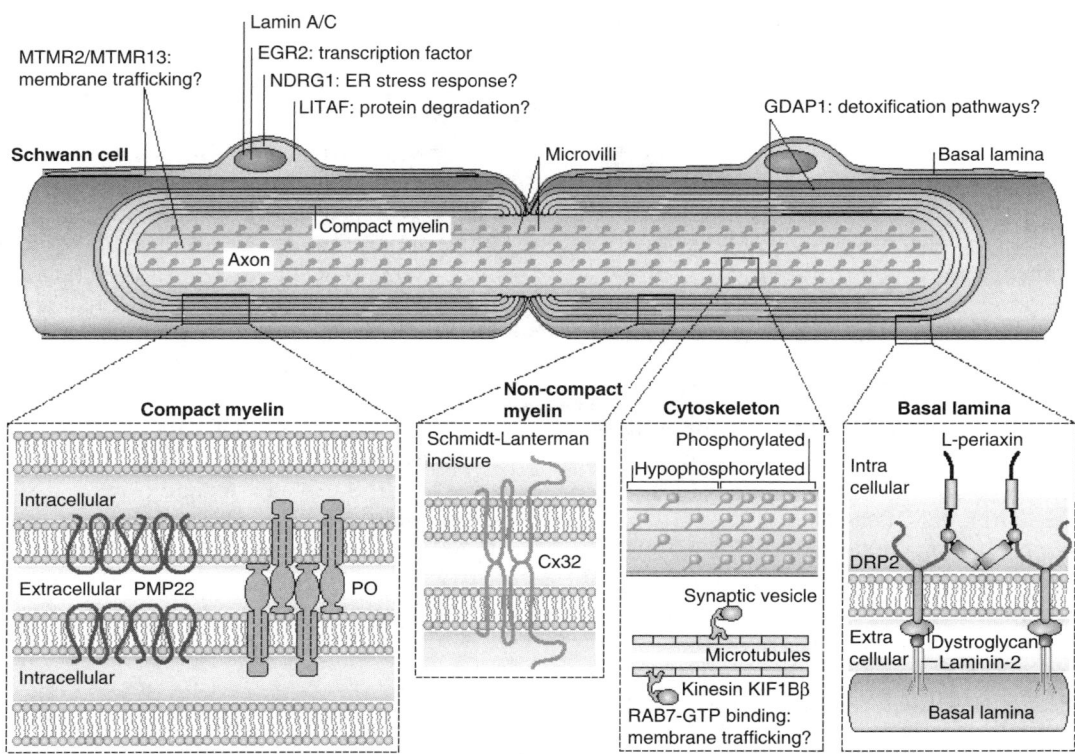

FIGURE 24.1 Schematic overview of the molecular organization of myelinated axons highlighting the proteins affected in CMT. The middle panel depicts a myelinated axon; the upper panel is an enlargement of the molecules associated with the dystroglycan receptor; the lower panel is an enlargement of the molecules associated with the myelin sheath or the axonal cytoskeleton. From Suter and Scherer (2003). Disease mechanisms in inherited neuropathies, *Nat Neurosci Rev* 4:714-726, with permission of *Nature*. (See also the Color Plate section.)

Besides gene duplications and deletions, more than 40 different *PMP22* mutations cause amino acid substitutions (missense mutations), premature stops (nonsense mutations), or frameshifts. Except for the patients with the HNPP phenotype, most patients with *PMP22* point mutations, including those with a CMT1 phenotype, are more severely affected than those carrying the duplication or deletion (Lupski and Garcia, 2001; Kleopas and Scherer, 2002). The inference that these mutations cause a gain-of-abnormal function is strongly supported by the finding that the neuropathy in heterozygous *Trembler* (*Tr*; G150D) and *Trembler-J* (*Tr-J*; L16P) mice is more severe than that in *Pmp22* heterozygous null (*Pmp22^{+/-}*) mice (Adlkofer *et al.*, 1997). If the mutant PMP22 could block a proportion of wild-type PMP22 from transport to the cell membrane by a dominant-negative effect, this should result in a phenotype between a heterozygous and a homozygous *Pmp22*-null (*Pmp22^{-/-}*) mouse. The severe phenotype of *Tr* and *Tr-J* mice, however, suggests that the mutant protein has another toxic gain-of-function (Adlkofer *et al.*, 1997).

PMP22 dominant mutant proteins could have several effects. In cultured cells, they are retained intracellularly in the ER and/or the intermediate compartment, and may aggregate abnormally (Tobler *et al.*, 2002). Epitope-tagged versions of *Tr* and *Tr-J* in developing rat nerves via adenoviral gene transfer confirms the inability of these mutant proteins to reach the myelin sheath (Colby *et al.*, 2000). Accumulation of mutant PMP22 could trigger the unfolded protein response (UPR) in the ER as described for proteolipid protein mutants (Southwood *et al.*, 2002). However, markers of the UPR are not grossly altered in *Tr-J* nerves (Dickson *et al.*, 2002). Instead, wild-type PMP22 and L16P (Tr-J) mutant proteins form a complex with the chaperon calnexin in the ER. L16P has an increased association time with calnexin, suggesting that the sequestration of calnexin might contribute to the disease mechanism by affecting the general protein folding quality control pathway (Dickson *et al.*, 2002).

MPZ Mutations Cause CMT1B, DSS, CHN, and a CMT2-like Phenotype

P0 is the most abundant protein in compact PNS myelin. It has a cytoplasmic C-terminus, a transmembrane domain, and an extracellular domain similar to immuno-globulin variable chain, including a disulfide bond. The crystal structure of P0 reveals that it forms tetramers that bind other tetramers both in *cis* (in the plane of the membrane) and in *trans* (on the apposed membrane), thereby holding the extracellular surfaces of the myelin sheath together. The positively charged cytoplasmic tail of P0 likely interacts with the negatively charged phospholipids of the membrane, thereby augmenting the role of MBP in holding the apposed cytoplasmic surfaces of the membranes together. These functions have been confirmed in transfection experiments, and in mice lacking P0 (*Mpz$^{-/-}$* mice), which have uncompacted myelin and severe demyelination. Like PMP22, the "dose" of P0 in myelin appears to be critical: reduced P0 causes myelin instability (Martini *et al.*, 1995), and overexpression of P0 causes severe dysmyelination that correlates with the level of expression (Wrabetz *et al.*, 2000; Yin *et al.*, 2000).

More than 80 different *MPZ* mutations have been identified, including missense, nonsense, and frameshift mutations. Most cause a CMT phenotype that is typically more severe than CMT1A; others cause a severe dysmyelinating/demyelinating neuropathy (DSS/CHN) or even a late onset, CMT2-like. Milder phenotypes (CMT1B) may be caused by haplotype insufficiency; severe phenotypes are caused by a toxic-gain-of-function. How do different *MPZ* mutations result in such different phenotypes? Diminished adhesion is one possibility; this has been observed for mutations affecting functional P0 domains, including a PKC phosphorylation site (Xu *et al.*, 2001). More severe phenotypes could be caused by dominant-negative effects of the mutant protein. Whether mutants have other toxic effects, such as inducing an UPR, remains to be determined.

GJB1 Mutations Cause CMT1X

Cx32 belongs to a family of about 20 highly homologous proteins, the connexins. Six connexins form a hemi-channel, and two apposed hemi-channels form a functional channel. Gap junctions mediate the diffusion of ions and small molecules between adjacent cells, but in the myelin sheath, they allow diffusion between the layers of the same cell (Balice-Gordon *et al.*, 1998), as depicted in Figure 24.1. Such a radial pathway—directly across the layers of the myelin sheath—would be advantageous as it provides a much shorter pathway (up to 1000-fold) than a circumferential route. Disruption of this radial pathway may be the reason that *GJB1/Gjb1* mutations cause demyelination in humans and in mice. However, this pathway did not appear to be different between *Gjb1*-null and wild-type mice, implying that another connexin also forms functional gap junctions

in PNS myelin sheaths. This connexin may be Cx29/Cx31.3 (Altevogt *et al.*, 2002).

CMT1X is caused by mutations in *GJB1*, the gene that encodes Cx32. More than 240 mutations have been identified, affecting the promoter, 3′ untranslated region, and every domain of the protein. In transfected mammalian cells, most mutants accumulate in the ER or Golgi apparatus and hence are nonfunctional (Yum *et al.*, 2002). The subset of mutants that reach the cell membrane may form functional channels, often with abnormal characteristics (Abrams *et al.*, 2000). How the mutants that form functional channels in cultured cells cause diseases *in vivo* remains to be determined. Other manifestations of CMT1X have been reported, including relapsing stroke-like episodes that resolve. These "CNS mutants" probably have an addition toxic effect, possibly by interacting with Cx47, which is also expressed by oligodendrocytes, as mice lacking both Cx32 and Cx47 have a severe CNS demyelination (Menichella *et al.*, 2003; Odermatt *et al.*, 2003).

Other Inherited Demyelinating Neuropathies

The aforementioned kinds of CMT are the most common and are caused by known mutations in "structural proteins" of the myelin sheath. The molecular defects of one third of CMT1 families remain to be discovered (Boerkoel *et al.*, 2002). Even these less common causes should be sought vigorously, as they will potentially provide key insights into the function of myelinated axons (see Figure 24.1).

Mutations in *LITAF* are associated with dominantly inherited CMT1C. *LITAF* (lipopolysaccharide-induced TNF-alpha factor) encodes a widely expressed transcript. *LITAF* mutations appear to cluster (G112S, T115N, W116G), potentially defining a putative domain of the LITAF protein. The function of LITAF, its role in myelinating Schwann cells, and the mechanism of dominant mutations all remain to be determined.

Mutations in two transcription factors, *ERG2* and *SOX10*, cause demyelinating neuropathy. Both are expressed in myelinating Schwann cells. Dominant EGR2 mutants likely affect the expression of myelin-related genes (Nagarajan *et al.*, 2001), possibly by directly affecting activation of the *MPZ* and *GJB1* promoters (Kleopas and Scherer, 2002). Mutations affecting the zinc finger domain of ERG2 impair DNA binding, resulting in residual binding that correlates with disease severity (Warner *et al.*, 1999). A recessive R1 domain mutation prevents interaction of EGR2 with the NAB corepressors and thereby increases transcriptional activity. Heterozygous loss-of-function *SOX10* mutations cause Waardenburg-Shah/Waardenburg type IV syndrome, while mutations associated with dysmyelination seem to act through a dominant-negative mechanism (Bondurand *et al.*, 2001).

Rare, autosomal recessive forms of demyelinating CMT have been collectively designated CMT4 (see Table 24.2). Recessive mutations in *ganglioside-induced differentiation-associated protein-1* (*GDAP1*) cause CMT4A, initially characterized by a severe demyelinating neuropathy. However, recessive *GDAP1* mutations have also been described in severe axonal neuropathy (AR-CMT2). Except for the relative preservation of NCV, some affected individuals could be called DSS. This variability raises the possibility that *GDAP1* mutations affect both neurons and Schwann cells. Moreover, it remains to be determined whether these effects are cell type–autonomous or caused by altered axon-Schwann cell interactions. The GDAP1 protein is predicted to have two transmembrane domains and a glutathione S transferase domain, suggesting a role in antioxidant pathways or detoxification. If such a function can be experimentally confirmed, damage due to reactive oxygen species would be an attractive hypothesis for the underlying disease mechanism.

Recessive mutations in *myotubularin-related protein-2* (*MTMR2*) cause CMT4B1. Nerve biopsy specimens have a distinctive feature, irregular folding and redundant loops of myelin, suggesting the primary defect occurs in myelinating Schwann cells. Neuronal expression may also contribute to the phenotype. MTMR2 is one of about a dozen members of the highly conserved family of myotubularin-related dual specific phosphatases, whose principal physiological substrates are the phospholipids phosphatidylinositol-3-phosphate (PI(3)P) and phosphatidylinositol-3,5-phosphate (PI(3,5)P2). These phospholipids are dephosphorylated by MTM1, MTMR2, MTMR3, and MTMR6 at the D3 position (Berger *et al.*, 2002). *MTMR2* mutations affect different parts of the protein, although reduced or absent phosphatase activity appears to be a common denominator (Berger *et al.*, 2002). Because phosphoinositides regulate intracellular membrane trafficking, CMT4B1 might be the result of abnormal membrane recycling, and/or endocytic or exocytotic processes, and/or disturbed membrane-mediated transport pathways. In addition, the neurofilament light subunit may be a substrate for MTMR2.

Mutations in *MTMR13/SBF2*, which is related to *MTMR2*, cause CMT4B2, which has similar pathological features to CMT4B1. Some mutations cause neuropathy alone, whereas other mutations also cause glaucoma. *MTMR13* encodes a protein lacking a functional phosphatase domain, but it may interact directly with MTMR2, just as MTM1 and MTMR2 can interact with the pseudophosphatases MTMR12 and MTMR5, respectively (Kim *et al.*, 2003). By this analogy, MTMR13 may regulate the phosphatase activity and/or alter the

subcellular localization of MTMR2. MTMR13 mRNA is broadly expressed, including in the brain, spinal cord and sciatic nerve (see Figure 24.1). Given their possible interactions, it will be imperative to determine the cellular/subcellular localizations and activities of MTMR2, MTMR5, and MTMR13, and correlate these findings with the various potential disease mechanisms outlined previously.

Recessive mutations in *N-myc downstream-regulated gene-1* (*NDRG1*) cause CMT4D. CMT4D is a syndromic neuropathy, as affected individuals also have hearing loss and dysmorphic features. It was initially described in Gypsies from Lom, Bulgaria, but has been subsequently found in Gypsies in other countries. The disease has a clinical onset in childhood and progresses to severe disability by the fifth decade. Biopsy specimens show demyelination, onion bulbs, and cytoplasmic inclusions in Schwann cells. Many cell types, including Schwann cells, express NDGR1 as an intracellular protein (see Figure 24.1 and the Color Plate section), but its function is unknown.

Recessive mutations in the *periaxin* (*PRX*) cause CMT4F. These mutations are expected to cause loss-of-function. Affected members usually have a "DSS phenotype" with delayed motor milestones, progressing to severe distal weakness. Milder, "CMT-like," cases are less common. Sensory loss, to the point of sensory ataxia, is more prominent than in other inherited demyelinating neuropathies. These findings are consistent with the phenotype of *prx*-null mice, in which myelin sheaths are initially formed, then fold abnormally, then break down (Gillespie *et al.*, 2000). Periaxin is a membrane-associated protein with a PDZ domain, exclusively expressed in myelinating Schwann cells (see Figure 24.1). Abaxonal periaxin interacts with the dystroglycan complex through dystrophin-related protein-2 (DRP-2), thereby linking laminin-2 in the basal lamina to the actin cytoskeleton and possibly other components in the cytoplasm (Sherman *et al.*, 2001). It is surprising that disrupting this complex results in a severe demyelinating neuropathy, as Schwann cells express also two other dystrophin complexes—one containing utrophin and another an isoform of dystrophin. The indispensable nature of the dystroglycan-DRP-2-periaxin complex may be related to the peripheral neuropathy of leprosy, as the binding of *Mycobacterium leprae* to α-dystroglycan can cause demyelination (Rambukkana *et al.*, 2002).

THERAPEUTIC CONSIDERATIONS—GENERAL

In the case of toxic neuropathies, prevention is the goal. Agents without a clinical use, such as acrylamide, arsenic, or thallium, should be avoided. Useful medications that rarely cause neuropathy, such as statins, may be discontinued if neuropathy develops. For clinically important drugs that cause neuropathy in a dose-related manner, such as colchicine, thalidomide, vincristine, paclitaxel/taxol, or cisplatin, vigilant observation is required to minimize the amount of damage. For this last group of drugs, it is highly desirable to develop countermeasures, so that neuropathy—often their dose-limiting toxic effect—could be ameliorated or avoided. Concomitant treatment with trophic factors (see below) is such an approach.

Prevention is also a feasible strategy for neuropathies that are the result of certain diseases. In these cases, treating the underlying disease should limit the amount of neuropathy. Antibiotics for leprosy, dialysis for uremia, and rigorous control of serum glucose levels are all examples. Similarly, correcting certain vitamin deficiencies (B_1/thiamine, B_6/pyridoxine, and B_{12}/cobalamine) should prevent the worsening of the corresponding deficiency-related neuropathy. Autoimmune neuropathies are another important, treatable group of neuropathies. These include acute (e.g., Guillain-Barré syndrome) and chronic (chronic inflammatory demyelinating polyneuropathy) diseases in which myelinated axons appear to be the target of an immune response, as well as various kinds of vasculitis, in which peripheral nerves suffer ischemic damage. Effective immune-modulating drugs exist for these diseases (see Chapter 10 by Asbury in this volume). Moreover, to the degree that the immune system mediates the severity of inherited neuropathies (Maurer et al., 2002), immune-modulating drugs might also be considered for these diseases.

Therapeutic Considerations—Inherited Neuropathies

Understanding the molecular pathogenesis of inherited neuropathies is essential for the development of rational therapies. While much remains to be learned, it is clear that most are caused by the expression of a mutant allele(s) in myelinating Schwann cells or neurons. For recessive neuropathies, one could, in principle, "replace" the defective gene by introducing a normal version. This has been done for Cx32—a transgenic mouse expressing the normal human *GJB1/Cx32* gene prevents the development of demyelinating in *Gjb1/cx32*-null mice (Scherer, unpublished observations). For dominant neuropathies, the situation is more complex, as these are likely to be caused by a toxic gain-of-function that is not necessarily related to the normal function of the gene product. Nevertheless, for dominant demyelinating neuropathies caused by altered gene dosage, reestablishing the normal level of gene expression might be of benefit—a transgenic mouse provides

a proof-of-concept that demyelination can be reversed by normalizing expression of PMP22. In the case of CMT1A, where overexpression of PMP22 in Schwann cells leads to segmental demyelination and ultimately axonal loss, the role of gene therapy will be to reduce the expression of this gene. This could be achieved by delivery of a recombinant adenovirus expressing PMP22 antisense messenger RNA or oligonucleotide antisense therapy. Transfection of PMP22 has already been demonstrated in myelinating Schwann cells using a replication-defective adenovirus in sciatic nerves of rats (Colby et al., 2000). For PMP22 mutant proteins that form an abnormal complex with calnexin resulting in its sequestration in the endoplasmic reticulum, increasing calnexin levels might be therapeutic.

Knowing the molecular basis of inherited diseases prompts immediate consideration of gene therapy. Because germ line transmission is not an option for treating human diseases, viral vectors have been widely used. Schwann cells and neurons can be infected, and numerous vectors and ways of delivering them have been developed (see Chapter 15 by Tuszynski in this volume). However, there are no proven ways of targeting vectors to Schwann cells or PNS neurons selectively, and their wide distribution and isolation by blood-brain, blood-nerve, and perineurial barriers are additional logistical problems. Retroviral vectors (Feltri et al., 1992), adenoviral vectors (Shy et al., 1995; Sorensen et al., 1998; Guénard et al., 1999; Colby et al., 2000) and adeno-associated viral (AAV) vectors (Ralph et al., 2000) can infect Schwann cells, and one presumes that other vectors can, too. After injecting replication-deficient adenovirus into peripheral nerves, PMP22 and P0 cDNAs can be expressed from viral promoters, and the corresponding proteins can be detected (Guénard et al., 1999; Colby et al., 2000). To date, however, not a single genetically affected Schwann cell or neuron has been cured in this manner. Progress in this area demands more sophisticated delivery systems and more knowledge of the molecular pathogenesis of neuropathies. Theoretically, one could use a promiscuous vector and restrict gene expression to myelinating Schwann cells or PNS neurons.

The neuron-restrictive silencer element (NRSE) provides an illuminating example. Originally identified within the promoter of the type II sodium channel gene, NRSE is a negative-acting regulatory element that prevents the expression of genes in non-neuronal cell types. The addition of an NRSE restricts the expression of a reporter gene (luciferase) to neuronal cells following intramuscular or intracerebral injections of adenoviral vectors (Millecamps et al., 2000). Inducible promoter systems are available, such as a tetracycline-inducible transgene, and these can be incorporated into viral vectors. When combined with cell type–specific promoters, such as the synapsin-1 or GFAP promoter, gene expression can be specifically targeted to neuronal or glial populations (Ralph *et al.*, 2000; Chtarto *et al.*, 2003). Regulating the level of gene expression by inducible promoters in specific cell types is an attractive idea for disease caused by altered gene dosage.

It is possible that targeting other cell types could be beneficial in the treatment of certain neuropathies. For example, percutaneous intramuscular gene transfer of vascular endothelial gene factor protects neurons from ischemic peripheral neuropathy by promoting angiogenesis (Schratzberger *et al.*, 2001). Cotransfecting hepatocyte growth factor and prostacyclin synthase is more effective than either one alone in stimulating angiogenesis and improving experimental diabetic neuropathy (Koike *et al.*, 2003). Using vectors as a means of increasing the systemic levels of trophic factors appears to work. The subcutaneous administration of a herpes simplex virus–mediated expression of neurotrophin-3 (NT-3) ameliorates the development of pyridoxine/B_6-toxic neuropathy (Chattopadhyay *et al.*, 2002). Electroporation-mediated gene transfer following intramuscular injection of plasmid containing murine NT-3 cDNA reduces cisplatin-induced neuropathy (Pradat *et al.*, 2001).

Preventing Axonal Loss

How do molecular defects that originate in Schwann cells and result in demyelination affect axons? Regardless of the cause of demyelination, the myelin sheath has pronounced effects on axonal caliber, axonal transport, the phosphorylation and packing of neurofilaments, and the organization of ion channels in the axonal membrane (Scherer and Arroyo, 2002). The degree of demyelination can be correlated with axonal loss (Martini, 2001), but pronounced axonal pathology has been observed even in genetic models in which axons are associated with normal appearing myelin sheaths. How do myelinating glia communicate with axons (and vice versa), and how does demyelination alter axon-Schwann cell interactions? Possibilities include increased energetic costs of propagating action potentials, decreased trophic support from Schwann cells, abnormal signaling emanating from the altered myelin sheath itself, and the loss of signals from the adaxonal glial membrane and/or cytoplasm, especially in the paranodal region. Other contributing factors to demyelination and/or axonal loss include inflammatory changes initiated by demyelination (Maurer *et al.*, 2002).

Determining the cause of axonal loss in inherited demyelinating neuropathies is of great importance because it is a prominent feature of many diseases, and likely causes clinical disabilities. These diseases include neuropathies and some neuronopathies, as well as more pervasive neurodegenerative disorders (neuroaxonal dystrophy, giant axonal neuropathy, and hereditary spastic paraparesis) and inflammatory demyelinating diseases, including multiple sclerosis. Existing animal models should be key in this regard (Sancho *et al.*, 1999). One anticipates that such studies will elucidate whether the mechanisms are the same in different demyelinating as well as axonal neuropathies and reveal why axonal loss is length-dependent. The most attractive hypothesis to account for why axonal degeneration is maximal at the distal ends is altered axonal transport (Griffin and Watson, 1988).

Wallerian degeneration may be related to length-dependent axonal loss. Wallerian degeneration refers to changes that occur distal to an interrupted axon, including the degeneration of axons, the proliferation and de-differentiation of Schwann cells, the invasion of the nerve by macrophages, and degradation of myelin (Scherer and Salzer, 2001). Although usually studied following mechanical injury in rodents, the identical features are found in the nerves of patients with nontraumatic neuropathies. The available evidence supports the idea that the same pathophysiological mechanisms occur in all neuropathies (Wang *et al.*, 2000; Coleman and Perry, 2002). For example, elevated intracellular Ca^{2+} is required for the degeneration of PNS and CNS axons in a variety of experimental models, including axotomy, blunt trauma, and hypoxia/ischemia. Increased Ca^{2+} activates calpains, proteases that degrade the axonal cytoskeleton, whereas lowering intracellular Ca^{2+} or inhibiting calpains reduces axonal degeneration in these models (Glass *et al.*, 2002). Similarly, Wld^S mice have slowed Wallerian degeneration and are relatively resistant to toxic neuropathy and even the axonal loss *Mpz*-null mice (Wang *et al.*, 2002; Samsam *et al.*, 2003). A chromosomal alteration in Wld^S mice generates a novel chimeric protein, and expression of this protein alone accounts for the phenotype (Mack *et al.*, 2001; Wang *et al.*, 2001).

Trophic Factors

Trophic factors are a most promising approach to treat established neuropathies or prevent toxin-related neuropathies. The trophic dependence of developing neurons and glia is well established (Bibel and Barde, 1999; Jessen and Mirsky, 1999), and the same trophic factors typically affect the corresponding cells in adult animals. The main conceptual problem with this approach is the diversity of both trophic factors and PNS neurons, especially because different kinds of neurons respond to different trophic factors (Table 24.3). Three families of trophic factors are particularly important for PNS neurons—the neurotrophin family, the glial-derived

TABLE 24.3 Trophic Factor Responsiveness of PNS Neurons

Neuronal type	Trophic Factor	Receptor
motoneurons	BNDF, NT-4	TrkB
	NT-3	trkC
	GDNF	GFRα1+ret
	LIF	LIFβR+gp130
	CNTF	CNTFRα+gp130
pre-ganglionic autonomic	NT-4	TrkB
	GDNF	GFRα1+ret
	TGF-betas	Type I & II
sympathetic	NGF	trkA
parasympathetic	NTN	GFRα2+ret
muscle spindle afferents	NT-3	trkC
trkA nociceptive	NGF	trkA
IB4 nociceptive	GNDF	GFRα1+ ret
mechanoreceptors	NT-3	trkC
enteric neurons	GDNF	GFRα1+ ret
	neurturin	GFRα2+ ret

neurotrophic factor (GDNF) family, and the ciliary neurotrophic factor (CNTF) family of cytokines (see Figure 24.2). Neuregulin-1 is singularly important for Schwann cells.

NGF, brain-derived neurotrophic factor (BDNF), neurotrophin-3 (NT-3), and NT-4 constitute the neurotrophin family (Bibel and Barde, 1999). They are secreted as homodimers and cleaved by a protease to their mature form. All bind to the p75 receptor, but the uncleaved form of NGF binds with highest affinity. Each of the neurotrophins binds to protein tyrosine kinase (Trk) receptors, causing dimerization, autophosphorylation, and downstream signaling (see Figure 24.2). All Trks interact with p75, and p75 can also signal independently of the Trks.

GDNF, neurturin, artemin, and persephin constitute the GDNF family, which belongs to the transforming growth factor beta superfamily. Unlike the other members of this superfamily, members of the GDNF family bind at least one glycosyl phosphatidylinositol (GPI)-linked protein known as GFRα1-4 (see Figure 24.2), all of which bind to a common signal transducing subunit, the protein tyrosine kinase receptor, c-ret (Rosenthal, 1999). Binding causes dimerization, autophosphorylation, and downstream signaling.

CNTF, leukemia inhibitory factor (LIF), interleukin (IL)-6, IL-11, cardiotrophin-1 (CT-1), cardiotrophin-like cytokine (CLC), and oncostatin-M are a family of structurally related cytokines that share a common set of receptors (Bravo and Heath, 2000; Plun-Favreau et al., 2001). An intrinsic membrane protein, gp130, is the signal transducing subunit common to all of these cytokines. The LIF receptor is composed of gp130 and the LIFβ receptor (LIFβR) subunit, also an integral membrane protein. IL-6Rα is a soluble receptor for IL-6. CNTFRα has a GPI anchor, but may also exist in a soluble form; it is a receptor for CNTF and CLC. The IL-6/IL-6R complex interacts with gp130, and the CNTF/CNTFRα and CLC/CLF/CNTFRα complexes interact with LIFβR and gp130. Cells that express gp130 and LIFβR respond similarly to CNTF and LIF, and signal through the JAK-STAT pathway, in which the cytoplasmic STAT proteins translocate into the nucleus where they act as transcription factors.

Neuregulin-1 is the founding member of the neuregulin family, which belongs to epidermal growth factor (EGF) superfamily (Lemke, 1996). Alternative promoters and alternative splicing generates soluble and membrane-bound proteins of different potency. Like all members of the EGF superfamily, neuregulin-1 binds to members of the ErbB receptor tyrosine kinase family—ErbB1, ErbB2, ErbB3, and ErbB4. Ligand binding results in receptor dimerization, tyrosine phosphorylation, and subsequent activation of intracellular signaling pathways, notably the PI3-K and ras/MAP kinase pathways. ErbB2 and ErbB3 are the neuregulin receptors of Schwann cells, and both are required for signal transduction (see Figure 24.2).

In addition to their effects on developing neurons, trophic factors have effects on mature PNS neurons (see Table 24.3). Neurite outgrowth in vitro and on axonal regeneration in vivo are the best documented (Terenghi, 1999), but many of these growth factors also modulate other aspects of the neuronal response to axotomy (Scherer and Salzer, 2001). Neuronal receptors probably mediate these effects—only neurons with the proper receptors typically respond to a given growth factor. Receptor-mediated, retrograde axonal transport delivers the trophic factors to neuronal cell bodies, where they likely exert their effects. NGF, BDNF, NT-3, NT-4, an engineered "pan-neurotrophin," CNTF, IL-6, and insulin-like growth factor (IGF I and II) have all been reported to enhance axonal regeneration. Schwann cells themselves express most of these trophic factors, but overexpressing a given trophic factor in neurons, Schwann cells, or other cells using viral vectors has been reported to enhance regeneration (Gravel et al., 1997).

Trophic factors can also prevent or ameliorate neuropathy. In animal models, NT-3 reduces neuropathy caused by B6/pyridoxin or cisplatin; this makes sense because these toxins principally affect large sensory neurons that express the NT-3 receptor, TrkC (Pradat

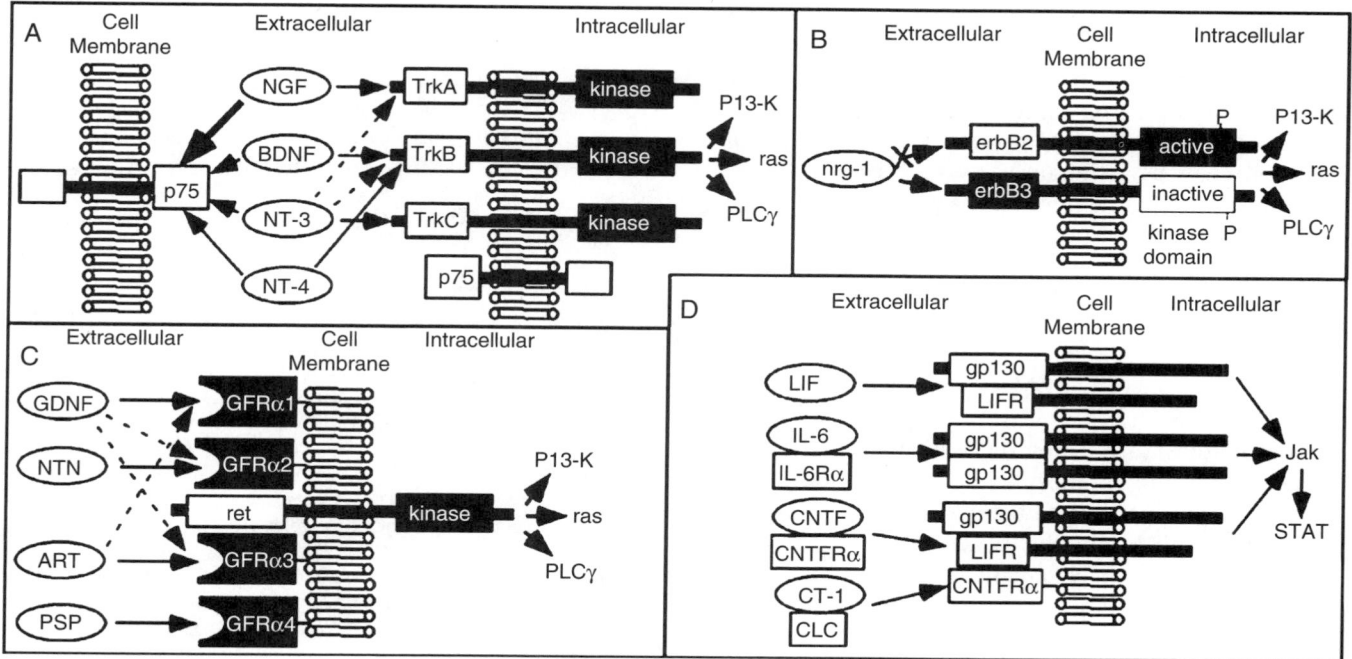

FIGURE 24.2 Ligand-receptor interactions in the PNS. Panel **A** depicts the binding of neuregulin-1 (nrg-1) to erbB3, the phosphorylation of the erbB2/erbB3 heterodimer, and the downstream pathways—*ras*/MAP kinase (ras), phosphatidylinositol 3-kinase (PI3-K), and phospholipase Cγ (PLCγ). Panel **B** shows the receptors for LIF, IL-6, CNTF and Cardiotrophin-1(CT-1), and the downstream JAK/STAT pathway. Panel **C** shows that TGF-betas bind to a heterodimer composed of high-affinity type I (R-I) and II (R-II) receptors as well as a low-affinity type III receptor, betaglycan. Panel **D** depicts the relative affinities of GDNF, neurturin (NTN), persephin (PSP), and artemin (ART) for their GPI-linked receptors GFRα1-4. Panel **E** depicts the relative affinities of the neurotrophins for the trks, as well as their mutual affinity for p75.

et al., 2001). Interestingly, therapeutic effects of NT-3 for cisplatin-treated mice or human diabetic patients were not dose-dependent and the beneficial effects could be seen at the lowest doses. NGF, however, also ameliorates cisplatin-induced neuropathy—likely an indirect effect, because the "rescued" cells do not express TrkA. This discrepancy remains to be clarified. GDNF specifically prevents the loss of central terminals of IB4-positive sensory neurons in the spinal cord of diabetic mice; it does not affect those of TrkA-positive neurons, which are NGF-responsive (Akkina 2001). Just as these IB-4 central terminals are lost after axotomy, experimental diabetes causes a functional, if not structural, axotomy. IGF-I and NT-3 ameliorate diabetic neuropathy, and IGF-1 prevents vincristine neuropathy; these are probably direct effects on PNS neurons. These trophic factors have been administered exogenously and/or by infecting cells with an appropriate viral construct. In one large clinical trial, however, exogenous NGF did not alter the progression of human diabetic neuropathy. A combination of trophic factors, or engineered "pan-neurotrophic factors" (Funakoshi *et al.,* 1998), might be more beneficial than a single factor.

The rationale for treating neuropathy with trophic factors pertains to the treatment of neuronopathies. Motoneuron diseases have been particularly well studied in this regard. Local injection of CNTF- and/or BDNF-expressing adenoviruses into the facial nucleus improves the survival of motoneurons for at least 5 weeks post-axotomy, even though less than 10% of the motoneurons are infected (Gravel *et al.,* 1997). Treating *pmn* mice, a model of inherited motoneuron disease, by intramuscular or intravenous injection of CNTF-expressing adenovirus increased life span by 25% and reduced the degeneration of phrenic myelinated axons (Haase *et al.,* 1999). Finally, in a genetically authentic animal model of spinal muscular atrophy, intramuscular injections of CT-1–expressing adenovirus improves median survival, delays motor defects, and reduces the loss of myelinated axons (Lesbordes *et al.,* 2003). Despite impressive results both in cultured motoneurons and in animal models (Sendtner *et al.,* 2000), large human trials using BDNF, CNTF, or IGF-I for amyotrophic lateral sclerosis have not provided proof of efficacy.

Schwann cells have also adapted to enhance axonal regeneration and remyelination (Scherer and Salzer,

2001), and their expression can also be manipulated. Axonal regeneration in chronically denervated distal nerve stumps can be enhanced by treating Schwann cells with TGF-β (Sulaiman and Gordon, 2002). Further, neuregulin-1 enhances axonal regeneration (Chen *et al.*, 1998), probably by acting on Schwann cells as PNS neurons do not express neuregulin receptors. Desert hedgehog, a morphogen expressed by Schwann cells, helps prevent the progression of experimental diabetic neuropathy (Calcutt *et al.*, 2003).

SUMMARY

Neuropathies are common, but their causes are diverse. There are promising prospects for developing safe and effective therapies for peripheral neuropathies, particularly with trophic factors. Although trophic factors can be produced by gene therapy, this is unlikely to be clinically applicable, because purified factors can be given. The potential roles of gene therapy include replacing genes in diseases caused by loss-of-function mutations, and modifying the levels of expression in dominantly inherited neuropathies caused by altered gene dosage. Many obstacles, however, will need to be overcome to make gene therapy a feasible strategy, including an efficient means of targeting genes to myelinating Schwann cells and/or neurons without significant cellular toxicity. Understanding the molecular basis of neuropathies will bring new therapeutic strategies for a number of neuropathic and neurodegenerative disorders.

ACKNOWLEDGMENTS

We apologize to our colleagues whose work we could not cite owing to space restrictions. The work in our laboratory is supported by an advanced postdoctoral fellowship award from the Muscular Dystrophy Association (CM) and by the NIH (SSS).

References

Abrams CK, Oh S, Ri Y, and Bargiello TA. (2000). Mutations in Connexin 32: the molecular and biophysical bases for the X-linked form of Charcot-Marie-Tooth disease, *Brain Res Rev* 32:203-214.

Adlkofer K, Naef R, Suter U. (1997). Analysis of compound heterozygous mice reveals that the Trembler mutation can behave as a gain-of-function allele, *J Neurosci Res* 49:671-680.

Altevogt BM *et al.* (2002). Cx29 is uniquely distributed within myelinating glial cells of the central and peripheral nervous systems, *J Neurosci* 22:6458-6470.

Balice-Gordon RJ, Bone LJ, Scherer SS. (1998). Functional gap junctions in the Schwann cell myelin sheath, *J Cell Biol* 142:1095-1104.

Berger P *et al.* (2002). Loss of phosphatase activity in myotubularin-related protein 2 is associated with Charcot-Marie-Tooth disease type 4B1, *Hum Mol Genet* 11:1569-1579.

Bibel M, Barde Y.-A. (1999). Neurotrophins: key regulators of cell fate and cell shape in the vertebrate nervous system, *Genes Dev* 14:2919-2937.

Boerkoel CF *et al.* (2002). Charcot-Marie-Tooth disease and related neuropathies: mutation distribution and genotype-phenotype correlation, *Ann Neurol* 51:190-201.

Bondurand N *et al.* (2001). Human connexin 32, a gap junction protein altered in the X-linked form of Charcot-Marie-Tooth disease, is directly regulated by the transcription factor SOX10, *Hum Mol Genet* 10:2783-2795.

Bravo J, Heath JK. (2000). Receptor recognition by gp130 cytokines, *EMBO J* 19:2399-2411.

Brownlees J *et al.* (2002). Charcot-Marie-Tooth disease neurofilament mutations disrupt neurofilament assembly and axonal transport, *Hum Mol Genet* 11:2837-2844.

Calcutt NA *et al.* (2003). Therapeutic efficacy of sonic hedgehog protein in experimental diabetic neuropathy, *J Clin Invest* 111:507-514.

Chattopadhyay M *et al.* (2002). In vivo gene therapy for pyridoxine-induced neuropathy by herpes simplex virus-mediated gene transfer of neurotrophin-3, *Ann Neurol* 51:19-27.

Chen LE *et al.* (1998). Recombinant human glial growth factor 2 (rhGGF 2) improves functional recovery of crushed peripheral nerve (a double-blind study), *Neurochem Int* 33:341-351.

Chtarto A *et al.* (2003). Tetracycline-inducible transgene expression mediated by a single AAV vector, *Gene Ther* 10:84-94.

Colby J *et al.* (2000). PMP22 carrying the Trembler or Trembler-J mutation is intracellularly retained in myelinating Schwann cells, *Neurobiol Dis* 7:561-573.

Coleman MP, Perry VH. (2002). Axon pathology in neurological disease: a neglected therapeutic target, *Trends Neurosci* 25:532-537.

Dickson KM *et al.* (2002). Association of calnexin with mutant peripheral myelin protein-22 ex vivo: a basis for "gain-of-function" ER diseases, *Proc Natl Acad Sci USA* 99:9852-9857.

Ding JQ *et al.* (2002). Microtubule-associated protein 1B: a neuronal binding partner for gigaxonin, *J Cell Biol* 158:427-433.

Feltri ML *et al.* (1992). Mitogen-expanded Schwann cells retain the capacity to myelinate regenerating axons after transplantation into rat sciatic nerve, *Proc Natl Acad Sci USA* 89:8827-8831.

Funakoshi H *et al.* (1998). Targeted expression of a multifunctional chimeric neurotrophin in the lesioned sciatic nerve accelerates regeneration of sensory and motor axons, *Proc Natl Acad Sci USA* 95:5269-5274.

Gillespie CS *et al.* (2000). Peripheral demyelination and neuropathic pain behavior in periaxin-deficient mice, *Neuron* 26:523-531.

Glass JD, Culver DG, Levey AI, Nash NR. (2002). Very early activation of m-calpain in peripheral nerve during Wallerian degeneration, *J Neurol Sci* 196:9-20.

Gravel C, Gotz R, Lorrain A, Sendtner M. (1997). Adenoviral gene transfer of ciliary neurotrophic factor and brain derived neurotrophic factor leads to long-term survivla of axotomized motor neurons, *Nat Med* 3:765-770.

Griffin JW, Watson DF. (1988). Axonal transport in neurologic disease, *Ann Neurol* 23:3-13.

Guénard V *et al.* (1999). Effective gene transfer of lacZ and P0 into Schwann cells of P0-deficient mice, *Glia* 25:165-178.

Haase G *et al.* (1999). Therapeutic benefit of ciliary neurotrophic factor in progressive motor neuronopathy depends on the route of delivery, *Ann Neurol* 45:296-304.

Jessen KR, Mirsky R. (1999). Schwann cells and their precursors emerge as major regulators of nerve development, *Trends Neurosci* 22:402-410.

Kim SA *et al.* (2003). Regulation of myotubularin-related (MTMR) 2 phosphatidylinositol phosphatase by MTMR5, a catalytically inactive phosphatase, *Proc Natl Acad Sci USA* 100:4492-4497.

Kleopas KA, Scherer SS. (2002). Inherited neuropathies, *Neurol Clin North Am* 20:679-709.

Koike H *et al.* (2003). Enhanced angiogenesis and improvement of neuropathy by cotransfection of human hepatocyte growth factor and prostacyclin synthase gene, *FASEB J* 17:779-781.

LaMonte BH *et al.* (2002). Disruption of dynein/dynactin inhibits axonal transport in motor neurons causing late-onset progressive degeneration, *Neuron* 34:715-727.

Lee MK, Marzalek JR, Cleveland DW. (1994). A mutant neurofilament subunit causes massive, selective motor neuron death: implications for the pathogenesis of human motor neuron disease, *Neuron* 13:975-988.

Lemke G. (1996). Neuregulins in development, *Mol Cell Neurosci* 7:247-262.

Lesbordes J-C *et al.* (2003). Therapeutic benefits of cardiotrophin-1 gene transfer in a mouse model of spinal muscular atrophy, *Hum Mol Genet* 12:1233-1239.

Lupski JR, Garcia CA. (2001). Charcot-Marie-Tooth peripheral neuropathies and related disorders. In Scriver CR *et al.*, editors: *The metabolic & molecular basis of inherited disease.* pp. 5759-5788.New York: McGraw-Hill, pp. 5759-5788.

Mack TGA *et al.* (2001). Wallerian degeneration of injured axons and synapses is delayed by a Ube4b/Nmnat chimeric gene, *Nat Neurosci* 4:1199-1206.

Martini R. (1997). Animal models for inherited peripheral neuropathies, *J Anat* 191:321-336.

Martini R. (2001). The effect of myelinating Schwann cells on axons, *Muscle Nerve* 24:456-466.

Martini R *et al.* (1995). Protein zero (P0)-deficient mice show myelin degeneration in peripheral nerves characteristic of inherited human neuropathies, *Nat Genet* 11:281-285.

Maurer M *et al.* (2002). Role of immune cells in animal models for inherited neuropathies: facts and visions, *J Anat* 200:405-414.

Menichella DM *et al.* (2003). Connexins are critical for normal myelination in the central nervous system, *J Neurosci* 23:5963-5973.

Millecamps S *et al.* (2000). Neuron-restrictive silencer elements mediate neuron specificity of adenoviral gene expression, *Nat Biotechnol* 17:865-869.

Nagarajan R *et al.* (2001). EGR2 mutations in inherited neuropathies dominant-negatively inhibit myelin gene expression, *Neuron* 30:355-368.

Odermatt B *et al.* (2003). Connexin 47 (Cx47)-deficient mice with enhanced green fluorescent protein reporter gene reveal predominant oligodendrocytic expression of Cx47 and display vacuolized myelin in the CNS, *J Neurosci* 23:4549-4559.

Ohara O *et al.* (1993). Neurofilament deficiency in quail caused by nonsense mutation in neurofilament-L gene, *J Cell Biol* 121:387-395.

Perez-Olle R, Leung CL, Liem RKH. (2002). Effects of Charcot-Marie-Tooth-linked mutations of the neurofilament light subunit on intermediate filament formation, *J Cell Sci* 115:4937-4946.

Plun-Favreau H *et al.* (2001). The ciliary neurotrophic factor receptor alpha component induces the secretion of and is required for functional responses to cardiotrophin-like cytokine, *EMBO J* 20:1692-1703.

Pradat PF *et al.* (2001). Partial prevention of cisplatin-induced neuropathy by electroporation-mediated nonviral gene transfer, *Hum Gene Ther* 12:367-375.

Ralph GS *et al.* (2000). Targeting of tetracycline-regulatable transgene expression specifically to neuronal and glial cell populations using adenoviral vectors, *Mol Neurosci* 11:2051-2055.

Rambukkana A, Zanazzi G, Tapinos N, Salzer J L. (2002). Contact-dependent demyelination by *Mycobacterium leprae* in the absence of immune cells, *Science* 296:927-931.

Rosenthal A. (1999). The GDNF protein family: gene ablation studies reveal what they really do and how, *Neuron* 22:201-217.

Samsam M *et al.* (2003). The *Wld^S* mutation delays robust loss of motor and sensory axons in a genetic model for myelin-related axonopathy, *J Neurosci* 23:2833-2839.

Sancho S, Magyar JP, Aguzzi A, Suter U. (1999). Distal axonopathy in peripheral nerves of PMP22 mutant mice, *Brain* 122:1563-1577.

Scherer SS, Arroyo EJ. (2002). Recent progress on the molecular organization of myelinated axons, *J Peripher Nerv Syst* 7:1-12.

Scherer SS, Salzer J. (2001). Axon-Schwann cell interactions in peripheral nerve degeneration and regeneration. In Jessen KR, Richardson WD, editors: *Glial cell development.* Oxford: Oxford University Press, pp. 299-330.

Schratzberger P *et al.* (2001). Reversal of experimental diabetic neuropathy by VEGF gene transfer, *J Clin Invest* 107:1083-1092.

Sendtner M *et al.* (2000). Developmental motoneuron cell death and neurotrophic factors, *Cell Tissue Res* 301:71-84.

Sherman DL, Fabrizi C, Gillespie CS, Brophy PJ. (2001). Specific disruption of a Schwann cell dystrophin-related protein complex in a demyelinating neuropathy, *Neuron* 30:677-687.

Shy ME *et al.* (1995). Towards the gene therapy of Charcot-Marie-Tooth disease type 1: an adenoviral vector can transfer lacZ into Schwann cells in culture and in sciatic nerve, *Ann Neurol* 38:429-436.

Sorensen J *et al.* (1998). Gene transfer to Schwann cells after peripheral nerve injury: a delivery system for therapeutic agents, *Ann Neurol* 43:205-211.

Southwood CM, Garbern J, Jiang W, Gow A. (2002). The unfolded protein response modulates disease severity in Pelizaeus-Merzbacher disease, *Neuron* 36:585-596.

Sulaiman OAR, Gordon T. (2002). Transforming growth factor-beta and forskolin attenuate the adverse effects of long-term Schwann cell denervation on peripheral nerve regeneration in vivo, *Glia* 37:206-218.

Suter U, Scherer SS. (2003). Disease mechanisms in inherited neuropathies, *Nat Neurosci Rev* 4:714-726.

Terenghi G. (1999). Peripheral nerve regeneration and neurotrophic factors, *J Anat* 194:1-14.

Tobler AR, Liu N, Mueller L, Shooter EM. (2002). Differential aggregation of the Trembler and TremblerJ mutants of peripheral myelin protein 22, *Proc Natl Acad Sci USA* 99:483-488.

Wang MS, Davis AA, Culver DG, Glass JD. (2002). Wld(S) mice are resistant to paclitaxel (Taxol) neuropathy, *Ann Neurol* 52:442-447.

Wang MS *et al.* (2001). The Wld(S) protein protects against axonal degeneration: a model of gene therapy for peripheral neuropathy, *Ann Neurol* 50:773-779.

Wang MS, Wu Y, Culver DG, Glass JD. (2000). Pathogenesis of axonal degeneration: Parallels between Wallerian degeneration and vincristine neuropathy, *J Neuropathol Exp Neurol* 59:599-606.

Warner LE, Svaren J, Milbrandt J, Lupski JR. (1999). Functional consequences of mutations in the early growth response 2 gene (EGR2) correlate with severity of human myelinopathies, *Hum Mol Genet* 8:1245-1251.

Wrabetz L *et al.* (2000). P0 glycoprotein overexpression causes congenital hypomyelination of peripheral nerves, *J Cell Biol* 148:1021-1033.

Xia CH *et al.* (2003). Abnormal neurofilament transport caused by targeted disruption of neuronal kinesin heavy chain KIF5A, *J Cell Biol* 161:55-66.

Xu WB *et al.* (2001). Mutations in the cytoplasmic domain of P0 reveal a role for PKC-mediated phosphorylation in adhesion and myelination, *J Cell Biol* 155:439-445.

Xu Y, Sladky JT, Brown MJ. (1989). Dose-dependent expression of neuronopathy after experimental pyridoxine intoxication, *Neurology* 39:1077-1083.

Yin X *et al.* (2000). Schwann cell myelination requires timely and precise targeting of P0 protein, *J Cell Biol* 148:1009-1020.

Yum SW, Kleopa KA, Shumas S, Scherer SS. (2002). Diverse trafficking abnormalities for connexin32 mutants causing CMTX, *Neurobiol Dis* 11:43-52.

Zhu Q, Couillard-Despres S, Julien JP. (1997). Delayed maturation of regenerating myelinated axons in mice lacking neurofilaments, *Exp Neurol* 148:299-316.

25

Therapeutic Potential of Neurotrophic Factors

Stephen McMahon, PhD
Beth Murinson MD, PhD

The word *trophic* literally means *nourish* and a neurotrophic factor is therefore one that nourishes or supports the growth of neurons. There is a very well-described role for neurotrophic factors in the development of the nervous system and in this context the term is used mostly to refer to the ability of a factor to promote survival and growth of discrete subpopulations of neurons. (The restriction to discrete subpopulations excludes general nutrients such as glucose from the definition.) In the mature nervous system, the stabilization and maintenance of neuronal populations over an extended period are also recognized as important neurotrophic effects. That is, neurotrophic factors may be neuroprotective and neurorestorative against injury, disease, and damage. It is this possibility, in particular, that has generated interest in the potential use of such factors as therapeutic agents when neuronal function is compromised.

Endogenous neurotrophic factors are mostly proteins but there are now a number of synthetic small molecules that mimic the actions of these endogenous factors. Neurotrophic factors, by definition, affect neurons. Normally this occurs because of a direct effect of the factor on specific neurotrophic factor receptors expressed on the neurons. But, because there are neurotrophic factor receptors on some non-neuronal cells, indirect actions of factors on neurons are possible.

The developmental role of neurotrophic factors has been studied experimentally. Details of these effects go beyond the scope of this article, and the interested reader is referred to the literature (Bennet *et al.*, 2002) for a recent review. However, one important principle to emerge from these developmental studies is the neurotrophic hypothesis. This states that the survival of

populations of neurons depends absolutely at some stages of development on the availability of particular factors from the targets of those neurons. This is an efficient mechanism for both regulating the number of neurons in a particular circuit and eliminating spurious or misguided projections (Bennet *et al.*, 2002). Particularly clear examples of the neurotrophic hypothesis are apparent for neurons of all the branches of the peripheral nervous system—sensory, motor, and autonomic. The prototypical neurotrophic factor, nerve growth factor (NGF) was identified because of its ability to promote growth of sympathetic axons and nearly all post-ganglionic sympathetic neurons are lost embryonically in mice with a null mutation in the NGF gene, or indeed in the gene for its receptor, trkA. Motoneurons also die in the absence of trophic support normally provided by the muscles that they innervate, although there continues to be uncertainty about the factor(s) concerned.

In the mature nervous system, the critical dependence for survival of peripheral neurons on target-derived neurotrophic factors is largely lost. Nonetheless, as we will review in this chapter, neurotrophic factors may still protect neurons against cytotoxic insults, and, moreover, neurotrophic factors may offer important therapeutic opportunities that go beyond promoting survival. We will also consider some of the neurobiological roles of neurotrophic factors in mature animals and discuss how, in some contexts—notably some persistent pain states—a dysregulation of neurotrophic factor expression may be an important contributor to the pathophysiological state. In these circumstances, a therapeutic potential exists for blocking (rather than promoting) particular neurotrophic effects.

NEUROTROPHIC FACTORS AND THEIR RECEPTORS

There are a number of families of neurotrophic factors, but two in particular have been studied for their neuroprotective effects. These are neurotrophins and the glial cell line–derived neurotrophic factor (GDNF) family, which we will consider in turn.

Neurotrophin Family

This family consists of NGF, the prototypical neurotrophic factor that was first purified from mouse submandibular gland by Levi-Montalcini, Booker, and Cohen (see Levi-Montalcini, 1987); brain-derived neurotrophic factor (BDNF); neurotrophin-3 (NT3); neurotrophin-4/5 (NT4/5); and neurotrophin-6 (NT6), which is only present in teleost fish (Lindsay, 1996; Gotz *et al.*, 1994). The neurotrophins are expressed as large precursors which, when processed, yield the highly basic mature proteins of approximately 14 kDa (approximately 120 amino acids). The different family members share around 50% amino acid homology and at physiological concentrations the neurotrophins exist as homodimers.

Two different classes of neurotrophin receptor have now been characterized. The first to be identified was the so-called p75 or low-affinity nerve growth factor receptor (LNGFR), which binds all the neurotrophins equally with relatively low affinity (Chao *et al.*, 1986). Additionally there is a family of high-affinity receptors, trks, which are tyrosine kinase receptors (Kaplan *et al.*, 1991). There are three known members of the trk family of receptors, termed trkA, trkB, and trkC, which show different specificities for the neurotrophins, as illustrated in Figure 25.1. NGF is the preferred ligand for trkA,

BDNF and NT4/5 the preferred ligands for trkB, and NT3 the preferred ligand for trkC. NT3 appears to be somewhat less specific than the other neurotrophins and at least *in vitro* can activate all of the trk receptors, albeit with different potencies.

The p75 receptor contains a single transmembrane segment flanked by extracellular and intracellular domains. The exact role of the p75 receptor is still a matter of contention. The p75 receptor appears to interact with trk receptors and promote the specificity and sensitivity of their interaction with the neurotrophins (Horton *et al.*, 1997). More recently it has been found that the precursors of mature NGF (preproNFG) specifically bind to p75 receptors and are retrogradely transported in neurons bearing p75 (Yiangou *et al.*, 2002), which may indicate novel neurotrophic roles for both p75 and higher molecular weight forms of the neurotrophins.

There is now considerable information regarding the nature of signalling via neurotrophin receptors. Most of this derives from studies of PC12 cells, where neurotrophin binding to appropriate trk receptors leads to receptor dimerization—the critical step for receptor activation. Autophosphorylation of tyrosine residues of dimerized trk receptors leads to activation of a wide number of intracellular pathways (see Kaplan and Miller, 2000 for review), which mediate neurotrophin responses both by altering gene expression and by post-translational modification of proteins in the target neuron. NGF binding to p75 can also lead to the activation of a number of transcription factors.

Glial Cell Line–derived Neurotrophic Factor Family

A second well-studied family of neurotrophic factors forms a branch of the transforming growth factor–β (TGF-β) superfamily. The founding member was GDNF (Lin *et al.*, 1993). Other related members are neurturin (NTN) (Kotzbauer *et al.*, 1996), artemin (ART; also called neublastin) (Baloh *et al.*, 1998), and persephin (PSP) (Milbrandt *et al.*, 1998). Like the neurotrophin family, all these proteins share significant structural homology. The different factors promote survival of different subpopulations of neurons, although the functional role of persephin remains poorly defined.

Members of the GDNF family signal via a receptor complex. This first consists of RET, a transmembrane tyrosine kinase receptor that acts as a signal transducing domain and, second, a member of the GFRα family of GPI-linked receptors (GFRα-1–4), which act as ligand binding domains (Figure 25.2). Either GFRα-1 or GFRα-2 in conjunction with RET can mediate GDNF or NTN signalling (although GDNF is thought to bind preferentially to GFRα-1 and NTN to GFRα-2). ART binds to GFRα-3. It is less clear if RET is always

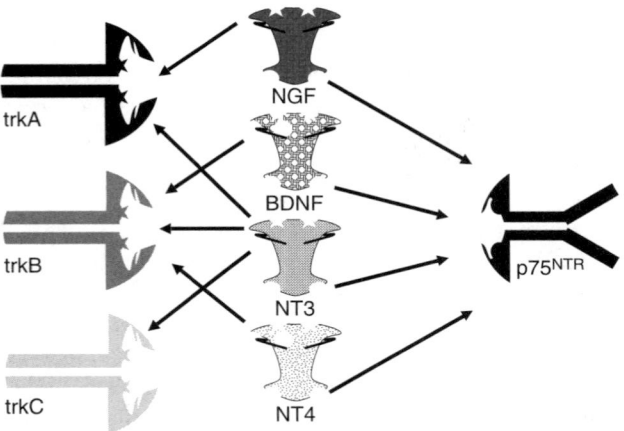

FIGURE 25.1 The different members of the neurotrophin family, and their affinities for high-affinity trk receptors.

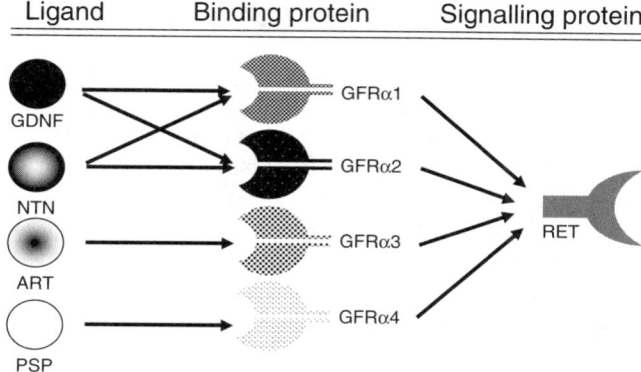

Ligand	Binding protein	Signalling protein

FIGURE 25.2 The GDNF family of neurotrophic factors and their receptors.

NEUROTROPHIC FACTORS AND REGENERATION

Peripheral Nervous System

Although often described as successful when compared with the central nervous system (CNS), regeneration in the peripheral nervous system (PNS) is in reality slow and limited. Proximal nerve injuries in the human arm or leg are rarely followed by reinnervation of distal structures such as the hand or foot. This is, in part, because the normal rate of axonal regeneration (approximately 1 mm/day) is so slow that irreversible muscle atrophy takes place before regenerating axons arrive. Given the profound disability resulting from this type of injury, the therapeutic opportunities in this area are substantial. Research to date has identified three major challenges to successful peripheral nerve repair: 1) promoting selective nerve reconnection to target tissues, 2) bridging gaps that occur as a consequence of nerve injury, and 3) increasing the number of axons that regenerate over long distances.

Most large peripheral nerves are mixed nerves. This means that they contain motor, sensory, and autonomic component axons. These axons typically travel together for some distance and then separate to reach different target tissues. When mixed peripheral nerves are disrupted, the routing of the various component axons must be restored for functional recovery to occur. Re-establishing appropriate target innervation is an area of ongoing research discussed subsequently.

Traumatic nerve injuries are classically described in three forms: neuropraxis, a temporary stunning of the nerve that is not associated with physical discontinuity; axonotmesis, discontinuity of the axon with preservation of the surrounding structures; and neurotmesis, complete transection of the nerve. This classification is useful in guiding treatment and predicting recovery in that neuropraxis typically recovers most quickly while axonotmesis requires axonal regrowth but not axonal retargeting and can recover well. Neurotmesis injuries require surgical repair and are often associated with discontinuities between the proximal nerve stump and the distal nerve end. Neurotmesis injuries have received the most attention experimentally but crush injury models (axonotmesis) are frequently used as well.

A second potentially important application of neurotrophic factors (NTFs) in the peripheral nervous system is in the treatment of peripheral neuropathies, as discussed subsequently.

Neurotrophic Factors in Repair of Traumatic Nerve Injury

One of the most promising strategies for promoting peripheral nerve repair is the implantation of NTF-enriched conduits. Conduits are especially attractive for

necessary for signal transduction after ART binding (Bennett *et al.*, 2000). PSP selectively binds to GFRα-4. Recently, developing striatum has been observed to express PSP, and midbrain dopaminergic neurons GFRα-4 (Akerud *et al.*, 2002). All of the GFRαs show some promiscuity in their ability to bind different GDNF family members and PSP appears to be the only ligand with very selective binding preference to just one receptor.

Other Factors

A variety of neurotrophic factors have been studied to a more modest degree in the context of neuroprotection. The family of neuropoietic cytokines are defined by their binding to the common receptor, gp130. Members include ciliary-derived neurotrophic factor (CNTF), leukemia inhibitory factor (LIF), interleukin-6 (IL-6), interleukin-11 (IL-11), oncostatin M, and cardiotrophin-1 (CT-1). Some of these factors use GPI-linked binding proteins (like the GDNF family) and all use gp130 and/or the LIF-receptor β for signalling. The fibroblast growth factors (FGFs) also promote growth and survival of distinct populations of developing neurons. Interestingly, recent data suggest that some of the growth-promoting effects of FGFs are mediated by release of endogenous cannabinoids and activation of neuronal CB1 receptors (Williams *et al.*, 2003). There are also interesting parallels between cell adhesion molecules and neurotrophic factors. The FGF receptor has been proposed to act as a receptor for growth promoted by some adhesion molecules (Williams *et al.*, 2003) and, conversely, NCAM has been suggested to form an alternative receptor for GDNF (Paratcha *et al.*, 2003). One of the potential limitations of the therapeutic uses of some of these neurotrophic factors, such as FGF, is their known mitogenic effects on a number of cell types.

nerve repair applications because they can guide regenerating fibers into a physical structure resembling the original nerve, they can be attached to both proximal and distal nerve stumps restoring continuity with the target tissues, and they may be constructed of materials conducive to successful regeneration. This strategy has been used with several NTFs (Terenghi, 1999). Treatment of the sciatic nerve with NGF has been reported to enhance regeneration (Fine *et al.*, 2002; Pu *et al.*, 1999; Santos *et al.*, 1999), but there is little support for positive effects on functional recovery (Young *et al.*, 2001). Indeed, one of the complications of using NGF is that there is a large body of evidence that it causes pain and hyperalgesia, as discussed subsequently. GDNF-enriched nerve conduit repairs produced remarkable regeneration of facial nerve (Barras *et al.*, 2002). Six weeks after repair, the number of axons in GDNF-treated nerves was 75% that of naïve control nerves while unenriched conduits showed no axons in this model. Myelination approximated the normal pattern with GDNF treatment. The ratio of axonal diameter to fiber outside diameter (G ratio) was high and resembled a juvenile pattern (Midroni and Bilbao, 1995). In contrast, animals treated with NT3-enriched conduits showed less than 10% of the number of axons present in naïve nerve, the pattern of myelination was less uniform, and the total myelin thickness was less than that in GDNF-treated animals. Retrograde labelling of motoneurons that was absent in un-enriched conduit repairs was present in NT3-enriched repairs and markedly increased in GDNF-enriched repairs. These results are very encouraging but further information about the effects of various NTFs on restoration of function is still needed.

Moreover, successful recovery depends on re-establishing connections with specific target tissues as well as vigorous axonal extension. A pure motor nerve, such as the facial nerve, may require a simpler repertoire of extension-promoting factors but nonetheless represents a complex targeting challenge; that is, correctly re-innervating a number of target muscles. This highlights the potential importance of a further observation: the finding that neutralizing antibodies to some NTFs reduced axonal branching after nerve injury. Decreasing the number of axonal branches in regenerating axons may improve functional recovery because excessive branching may be associated with reduced efficiency of target finding. Although preclinical data reveal the potential of NTFs to promote peripheral regeneration, it is less clear whether this can be translated to effective clinical therapies. One large clinical trial, to assess the potential of NGF to promote sensory regeneration in diabetic neuropathy, failed, perhaps because of the need to reduce doses to minimize pain-related side effects (Quasthoff and Hartung, 2001).

Other Neurotrophic Factor Strategies for Peripheral Nerve Repair

The transfer of Schwann cells, particularly pre-denervated Schwann cells, is a strategy that may promote regeneration. Experimental evidence indicates that there is an optimal period after nerve injury (between 3 and 10 days after axotomy) when Schwann cells and factors in the distal stump are especially conducive to nerve regeneration. Cells and factors harvested in the acute period are not as effective; materials obtained at late time periods (weeks to months after injury) are also not as supportive of regrowth (Marcol *et al.*, 2003). The transplantation of stem cells is a strategy that may be especially useful when nerve continuity is preserved. The idea of introducing stem cells at the regenerating nerve tip is based on the observation that stem cells or juvenile cells produce factors, including NTFs, that will promote nerve regeneration. In addition to this, stem cells can be genetically modified to produce increased levels of specific NTFs.

Clinical and Experimental Models of Peripheral Neuropathy

The implementation of the skin biopsy for assessment of epidermal nerve fiber density has revolutionized clinical assessment of peripheral neuropathy. A 3-mm full-thickness skin punch biopsy is obtained, stained for the pan-axonal marker PGP9.5 and systematically evaluated. The incision site heals spontaneously with few complications. The introduction of a method for rapid and repeatable assessment of small myelinated and unmyelinated fibers in humans has opened a new diagnostic window for peripheral nerve disease. Small fiber neuropathies, most commonly diabetic and pre-diabetic neuropathies, have been highly resistant to traditional methods of evaluating nerve disease: Nerve conduction studies assess large fiber function and fail to detect small-fiber predominant neuropathies until they are far advanced. Other methods are even less ideal; nerve biopsies are invasive and cumbersome; microneurography is potentially damaging and restricted to expert research studies; and quantitative sensory testing, while useful, is not consistently predictive of neuropathology.

In humans, evaluation of epidermal nerve fiber density by skin biopsy (Figure 25.3) has shown that there is a normal gradient of innervation density. Proximal structures such as back and proximal thigh have a higher number of epidermal nerve fibers than distal ankle; additionally there is a gradual decrease in fiber density with advancing age (McCarthy *et al.*, 1995). In patients with small fiber neuropathy, epidermal nerve fibers are reduced or absent, more so distally. Two main approaches have been taken to study regeneration using

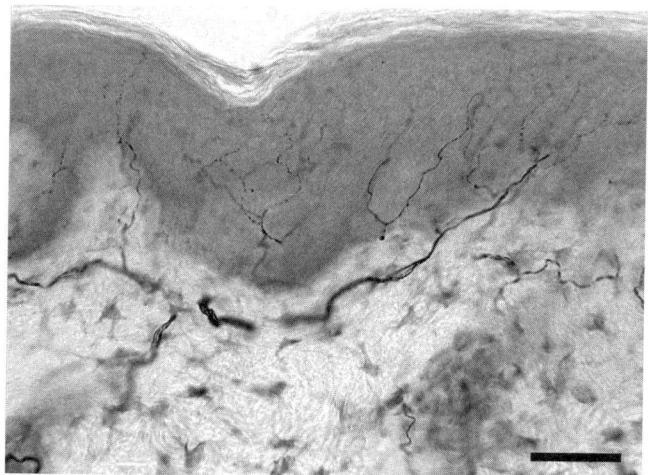

FIGURE 25.3 Epidermal nerve fibers in a human skin biopsy specimen. Photomicrograph courtesy of Mr. Peter Haner and Dr. Justin McArthur. Bar equals 75 microns.

skin biopsy: incision-reincision and capsaicin denervation. In the incision-reincision model, a 3-mm skin punch is followed weeks later by a 5-mm skin punch that includes the first biopsy site. Studies of these tissues show that epidermal fibers extend into the injury site along the dermal-epidermal junction. Repair from deeper dermal structures is limited by dense collagen scarring. In the capsaicin denervation model, an occlusive dressing of capsaicin cream remains in place for 24 hours. This is well tolerated in clinical trials and produces a patch of anesthetic skin with no epidermal nerve fibers. There is some depletion of dermal nerve fibers as well. Using this technique, recent studies have shown that diabetics regenerate nerve less quickly after a denervating injury. Several animal studies have shown that NGF is able to promote an apparently complete reinnervation of skin after capsaicin treatment (Schicho *et al.*, 1999, and references therein). However, clinical trials that used skin biopsy as an endpoint for NGF treatment of peripheral neuropathy failed to demonstrate any improvement in epidermal nerve fiber density, again perhaps because of the necessity to limit NGF dose (Quasthoff and Hartung, 2001).

Neuropathic pain frequently arises because of lesions to the peripheral nervous system. There are many precipitating causes, some metabolic (e.g., diabetes mellitus), some infectious (e.g., human immunodeficiency virus), some iatrogenic (e.g., secondary to cancer chemotherapy), and some traumatic (for instance, following accident or surgery). There has been extensive experimental study of painful neuropathic states, particularly those created by traumatic nerve injuries that result in a partial denervation of a major target such as the hindlimb (see McMahon, 2002, for review). We have shown that GDNF treatment can prevent the development

of neuropathic pain and reverse an established painful state in several rodent models (Boucher *et al.*, 2000). Those experiments suggested one key mechanism of action resulted from the ability of GDNF to reverse the upregulation of a sodium channel transcript, $Na_v1.3$, in damaged nerve, and there are several other reports also showing the ability of neurotophic factors to regulate sodium channel expression in primary sensory neurons (Cummins *et al.*, 2000; Leffler *et al.*, 2002). Abnormal sodium channel expression itself is likely to contribute to the emergence of ectopic activity in damaged neurons that is known to be essential for the normal expression of neuropathic behavior. Several other groups have subsequently reported similar anti-neuropathic actions of GDNF (Hao *et al.*, 2003; Wang *et al.*, 2003). Another member of the GDNF family, artemin, has also been reported to have anti-allodynic effects in preclinical models of neuropathic pain (Gardell *et al.*, 2003).

Central Nervous System

The growth-promoting effects of neurotrophic factors on neurons of the peripheral nervous system naturally raise the question of whether these same factors might be of use in promoting regeneration in the central nervous system. Whereas the peripheral nervous system will spontaneously regenerate (albeit frequently with only limited functional recovery, as described above) the central nervous system does not, at least in mature mammals. And while there is increased understanding of the reasons for this failure of CNS regeneration, the absence of existing clinical therapies for these disorders acts a strong stimulus for the development of novel strategies. Here we will review the preclinical data relating to the potential therapeutic benefits of neurotrophic factor treatment in two contexts of CNS regeneration: regeneration of sensory neurons at the dorsal root entry zone and regeneration within the damaged spinal cord. The potential use of these factors in neurodegenerative disease is briefly reviewed later in this chapter. The reader is also referred to Chapter 15 by Tuszynski in this volume for a fuller discussion of emerging therapies for these conditions.

Regeneration at the Dorsal Root Entry Zone

The dorsal root entry zone (DREZ), also known as the transitional zone (TZ), is the specialized interface where the dorsal roots join the spinal cord (Fraher, 2000; Ramer *et al.*, 2001b; Ramer *et al.*, 2001b). At the DREZ, peripheral sensory axons penetrate the glia limitans—the astrocyte layer that covers the surface of the CNS—to enter the spinal cord. The DREZ delimits the characteristic myelin-forming cells of the CNS (oligodendrocytes) from those of the PNS (Schwann cells). During development, growing or regenerating sensory axons are able to cross the DREZ, but they fail to do so in the adult (Carlstedt *et al.*, 1988). The reason for failure is believed to be the postnatal expression of inhibitory molecules within the DREZ (Pindzola *et al.*, 1993; Zhang *et al.*, 1995) as well as developmental changes in the response of sensory axons to such molecules (Golding *et al.*, 1997). Traumatic injuries, frequently associated with motorcycle accidents, can cause dorsal root avulsions in humans. In these cases, most commonly involving the brachial plexus, sensory input is lost and this additionally limits motor performance. Many avulsion injuries also lead to quite intractable chronic pain. Significant numbers of babies suffer from root avulsion injuries during childbirth—more than suffer from spina bifida, for instance—and these palsies usually leave persistent deficits. There is clearly an unmet clinical need associated with these conditions.

Several attempts to bypass the inhibitory DREZ have been explored. These include the use of peripheral nerve grafts (Oudega *et al.*, 1994; Saiz-Sapena *et al.*, 1997), collagen guidance channels (Liu *et al.*, 1997; Liu *et al.*, 1998c), and grafts of fetal spinal cord (Itoh *et al.*, 1992; Itoh *et al.*, 1993) as bridges between severed dorsal roots and the spinal gray matter. These attempts are reported to result in some limited regeneration of adult sensory axons across the DREZ, but the degree of recovery is probably insufficient to be of any clinical benefit. Transplants of olfactory ensheathing cells (a specialized type of CNS glia which, uniquely, support the spontaneous turnover of CNS olfactory neurons) into the dorsal horn are also reported to promote the ingrowth of sensory axons and restitution of spinal reflexes (Navarro *et al.*, 1999; Ramon-Cueto and Nieto-Sampedro, 1994). More recently, others have not been able to show any positive effects of this treatment (Riddell *et al.*, 2002). Conditioning lesions to a peripheral nerve, which switch on the neuron's regenerative machinery, accelerate regeneration within the dorsal root (Richardson and Verge, 1987) and result in the penetration of the DREZ by a few axons (Chong *et al.*, 1999). These experiments indicate that neutralization of inhibitory CNS molecules is not necessarily required for penetration of the DREZ by sensory axons.

We and others have therefore investigated whether neurotrophic factors can promote the regeneration of damaged dorsal root axons across the DREZ. In our experiments, multiple cervical dorsal root crushes (Ramer *et al.*, 2000) or cuts (Ramer *et al.*, 2001b) were carried out and neurotrophic factors were delivered intrathecally by osmotic minipump. We observed that neurotrophic factors do indeed promote a robust regeneration of damaged dorsal root axons across the DREZ, and that the populations of axons that regenerate correspond to the known pattern of neurotrophic factor

receptor expression (Ramer *et al.*, 2000). That is, NGF selectively promoted the in-growth of small diameter CGRP-expressing axons, NT3 promoted the in-growth of large diameter NF200-expressing axons, and GDNF promoted the regeneration of both small diameter P2X3-expressing axons and a subset of large diameter NF200-positive fibers. An example is shown in Figure 25.4.

We have recently confirmed this regeneration into CNS tissue in a stereological electron microscopy (EM) study (Ramer *et al.*, 2002), which indicated that NT3 treatment produced a growth of approximately 40% of axons across the DREZ. It also showed that some of the centrally regenerated fibers became re-myelinated. Functional connectivity of the regenerated primary afferent fibers with postsynaptic dorsal horn neurons was also demonstrated in electrophysiological experiments. NGF and GDNF treatment resulted in the reappearance of long-latency postsynaptic responses, reflective of the regeneration of small caliber afferent fibers, while NT3 and GDNF restored short-latency potentials, consistent with regeneration and re-connection of large caliber A-fibers with dorsal horn neurons. Most importantly, the regeneration associated with trophic factor treatment improved the sensory behavior of the animals: rats treated with NGF and GDNF recovered behavioral sensitivity to noxious thermal and mechanical stimulation. Rats treated with NT3 and GDNF showed improvements in some sensory-motor tasks, such as locomotion across narrow beams.

The B fragment of cholera toxin (CTB) binds to the GM1 ganglioside expressed specifically by myelinated axons and CTB can be used to trace the anatomical projections of these large caliber axons. Myelinated axons normally terminate in lamina I and III-X of the dorsal horn, and are excluded from lamina II. NT3 promotes the regeneration of CTB-labelled axons through the entry zone and into the superficial part of the dorsal horn 1 week following rhizotomy. Many axons can also be found growing along the pia mater, possibly as a result of a tropic effect of NT3 combined with the growth-permissive meningeal surface of the spinal cord. With EM, CTB-labelled axons and vesicle-filled terminals appear within the superficial laminae of the gray matter as early as 1 week following rhizotomy and NT3 treatment. Normally, large diameter afferent fibers terminate not only in the dorsal horn, but they also send ascending branches toward the brainstem in the posterior columns, to terminate in the dorsal column nuclei. Following NT3 treatment, rather than growing along their usual course, regenerating axons grow rostro-caudally in the superficial gray matter. This is probably due to the preference of growing axons for gray matter rather than white matter, as has been shown in tissue culture experiments (Savio and Schwab, 1989), and/or the presence of degenerating

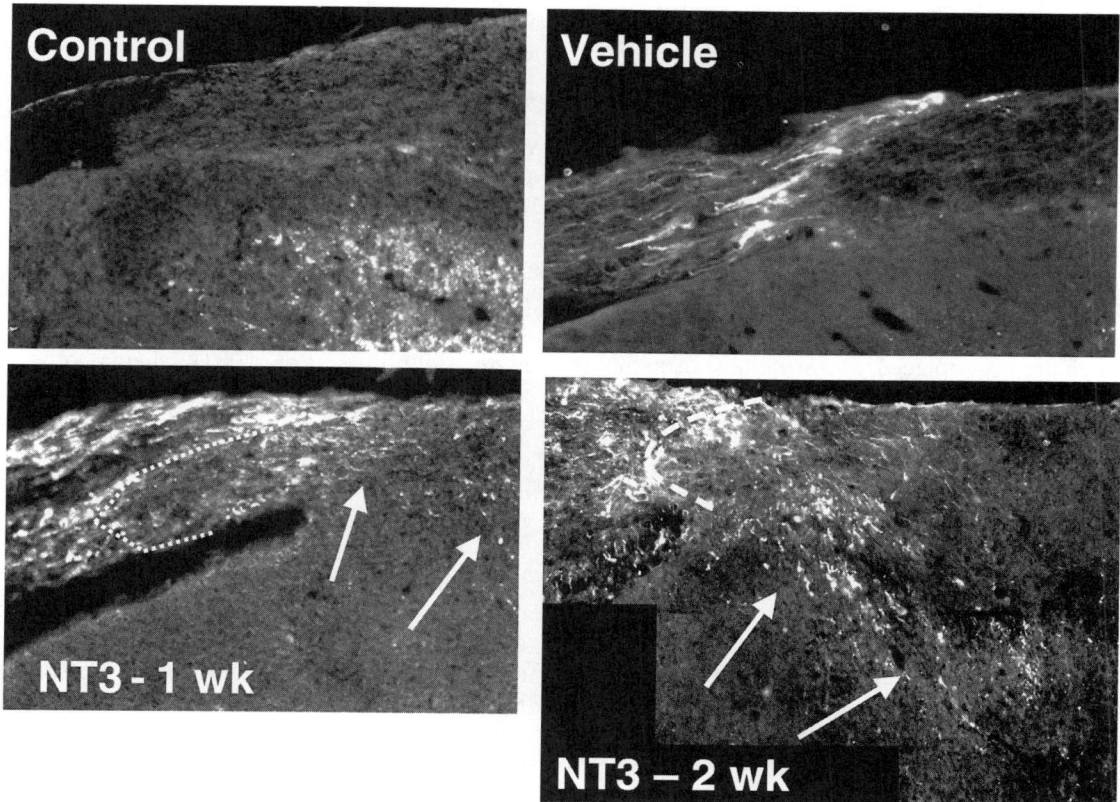

FIGURE 25.4 Regeneration of sensory neurons under the influence of NT3. Sensory axons were labelled with an anterograde tracer, shown in white. In intact animals (top left), the label shows the normal central termination pattern of myelinated fibers in the deeper laminae of the dorsal horn. After lesions of the dorsal root (top right) axons regenerate to the DREZ but do not cross it. When rhizotomy is combined with intrathecal NT3 treatment (bottom), many myelinated fibers are induced to grow across the DREZ (yellow dotted lines) and grow into the spinal cord. (Data from Ramer *et al*, 2001b)

myelin within the dorsal columns. Other laboratories have also seen sensory regeneration across the adult DREZ under the influence of neurotrophic factors. Romero *et al.* (Romero *et al.*, 2000; Romero *et al.*, 2001) also reported that trophic factors induced the ingrowth of sensory neurons to the spinal cord following rhizotomy. In these experiments, adenoviral vectors were used to transduce spinal cord cells (principally astrocytes) with a transgene making one of several factors. The ingrowth of sensory axons seen with this approach was extensive, and some regenerated fibers were found at ectopic locations.

Together, these data strongly suggest the possible use of such factors in the repair of dorsal root avulsion injuries in man. However, there are several caveats. Neurotrophic factors may induce sprouting of intact sensory neurons, especially when delivered at high doses, and by this or other mechanisms (McMahon and Bennett, 1999) might result in abnormal pain sensitivity if nociceptive afferent fibers are targeted. This approach might be more useful for restoring tactile and proprioceptive

information. We found that if NT3 treatment is delayed after a dorsal rhizotomy, axons are able to cross the DREZ, but their progress through the degenerating white matter portion of the root is more restricted (Ramer *et al.*, 2001a) and many regenerating axons form large dystrophic end-bulbs, and they show an attenuated course. Analysis of the time course of glial events following rhizotomy indicated that the emergence of this abortive growth most closely matched the appearance within the white matter of mature ED1-expressing phagocytic cells, rather than astrogliosis or microglial proliferation, both of which are well under way by the second day following a dorsal root lesion (Liu *et al.*, 1998b). Because the peripheral part of the root is heavily laden with macrophages, the abortive regeneration is not likely to be a direct result of the mere presence of phagocytic cells. In the periphery, macrophages rapidly clear myelin, and thus aid in the regenerative process. In the CNS, myelin clearance is a much more protracted process (George and Griffin, 1994), probably a result of failure of complement system activation (Liu *et al.*, 1998a). Once the process of myelin clearance commences in the cord, it appears to create a second inhibitory boundary for regenerating sensory neurons, and one which is not so amenable to neurotrophic factor treatment. Thus, these factors alone may be most effective if delivered immediately after injury.

Regeneration in the Spinal Cord

Damage to the spinal cord may result in the death of neurons, which are not replaced, and severance of axons, which do not regenerate. A delayed phase of cell death is seen after many injuries that may be secondary to inflammatory products produced at the injury site, or secondary to the failure of regeneration of damaged axons and the associated long-term loss of trophic support normally provided by the targets of these axonal projections. One reason for the failure of regeneration is that the CNS, as opposed to the PNS, is rich in several myelin associated inhibitory molecules, such as NOGO and MAG (McGee and Strittmatter, 2003). A second important reason is that direct damage to the spinal cord is associated with the formation of a glial scar, in which reactive astrocytes, in particular, create a dense and highly inhibitory barrier at the site of injury. Nonetheless, given the very positive effects of trophic factors described previously, it is reasonable to ask if these factors might be neuroprotective or neurorestorative in the context of spinal cord injuries.

This question has been addressed experimentally quite intensively over the past 5 years or so. There are consistent reports that several neurotrophic factors, delivered in different ways and following different types of spinal cord injury can prevent the atrophy and cell death of a variety of neuronal subpopulations (Baumgartner and Shine, 1998; Baumgartner and Shine, 1997; Blesch and Tuszynski, 2001; Giehl and Tetzlaff, 1996; Kobayashi *et al.*, 1997a; Lu *et al.*, 2001; Liu *et al.*, 2002; Novikova *et al.*, 2002; Bradbury *et al.*, 1998; Koda *et al.*, 2002). The most successfully used trophic factors are NT3, BDNF, and GDNF. These factors have been delivered to the sites of spinal injury or to the cell bodies of damaged neurons, by various techniques including single depot provision, continuous infusion, genetically altered cells, or following transduction with viral vectors. Both ascending and descending spinal tracts, including the corticospinal and rubrospinal systems, are protected. The protection can last for many weeks or months.

The ability of the same factors to promote regeneration of damaged neurons (rather than protect against cell death) has also been extensively studied, but with mixed results. The most positive studies report a local proliferation of axonal sprouts at the injury sites, or an increased

regeneration of CNS axons into a more favorable environment—typically a peripheral nerve segment grafted into the CNS lesion site (Grill *et al.*, 1997; Hiebert *et al.*, 2002; Kobayashi *et al.*, 1997b; Blits *et al.*, 2000; Hiebert *et al.*, 2002; Weidner *et al.*, 1999; Tuszynski *et al.*, 2002a; Bradbury *et al.*, 1999; Oudega and Hagg, 1999; Jin *et al.*, 2002; Novikova *et al.*, 2002). The ability of neurotrophic factors to induce long-range regeneration of damaged spinal tracts across a lesion site and into and through intact tissue, appears very much more limited. Some studies have failed to see any such growth (Bradbury *et al.*, 1999; Blits *et al.*, 2003), whereas others have seen for the most part growth of very small numbers of fibers (Grill *et al.*, 1997; Oudega and Hagg, 1999; Tuszynski *et al.*, 2003; Jin *et al.*, 2002; Cheng *et al.*, 1996; Schnell *et al.*, 1994). The failure to induce robust long range regeneration does not appear to be a reflection of limited tests of different factors, models, neuronal subpopulations, etc., because equivalent experiments have shown neuroprotection or local (non-regenerative) sprouting of damaged axons. Notwithstanding this relatively poor anatomical response, neurotrophic factors have been reported to improve behavioral performance in many studies. One possible explanation is that small numbers of regenerated axons support significant behavioral response. Another is that the trophic factors are inducing functional change in undamaged neuronal systems. Recently, there has been some enthusiasm for notion that these factors (and indeed other treatments) induce sprouting of intact neuronal systems that might contribute to behavioral improvements (e.g., Jeffery and Fitzgerald, 2001; Zhou *et al.*, 2003; Zhou and Shine, 2003). The full extent of this sprouting is only now being explored, and indeed, whether such sprouts make maladaptive as well as adaptive connections is yet to be tested. Another possible effect of trophic factor treatment is an indirect action via non-neuronal cells. This is suggested in some of the cases cited above where factors appear to promote sprouting or growth of neuronal populations that do not have the appropriate neurotrophic factor receptor, and also for instance by the anti-apoptotic effects on oligodendrocytes (Koda *et al.*, 2002).

The highly consistent, if limited, positive effects of neurotrophic factors, on spinal cord injuries, suggest they may find some use in the clinic. However, the existing preclinical data would not suggest that alone these factors will promote dramatic recovery. It seems much more likely that effective spinal cord injury treatments will require multiple interventions, and these are only now being explored at a preclinical level. There are other potential problems particularly associated with trophic factors. The overwhelming use of rodents as preclinical models (but see Tuszynski *et al.*, 2002b, for an example of a primate study) may not translate well to humans where the absolute distances needed for restitution of damaged tracts are so much greater. There is also the possibility that high concentrations of trophic factors may act as a sink for growing axons, promoting growth but trapping the regrowing axons in the vicinity of treatment. Finally, it may not prove easy to use the natural proteins as therapeutics because of problems associated with delivery.

NEUROTROPHIC FACTORS AND CHRONIC PAIN

Nerve Growth Factor as a Peripheral Inflammatory Mediator

The administration of small doses of NGF to adult animals and humans can produce pain and hyperalgesia. In rodents, a thermal hyperalgesia is present within 30 minutes of systemic NGF administration and both a thermal and a mechanical hyperalgesia after a couple of hours (Lewin *et al.*, 1993). In humans, intravenous injections of very low doses of NGF produce widespread aching pains in deep tissues and hyperalgesia at the injection site (Petty *et al.*, 1994). The rapid onset of some of these effects, and their localization to the injection site, strongly suggests that they arise at least in part from a local effect on the peripheral terminals of nociceptors. This has been substantiated by the observation that acute administration of NGF can sensitize nociceptive afferent fibers to thermal and chemical stimuli (Rueff and Mendell, 1996). Much of this sensitization appears to be a direct action on primary afferent nociceptors, many of which express trkA. The intracellular pathways concerned have been much studied but with conflicting results (see Bonnington and McNaughton, 2003). However, there is evidence that some of the sensitizing effects of NGF are indirect, via mast cells, sympathetic efferent neurons, and neutrophils (see McMahon and Bennett, 1999 for review).

NGF is a potent regulator of gene expression in sensory neurons. Some sensory neuropeptides, which are released with activity from central nociceptor terminals, are strongly upregulated by NGF. These include CGRP and substance P. NGF has also been shown to produce a dramatic upregulation of BDNF in trkA-expressing DRG cells, and there is now growing evidence that BDNF may serve as a central regulator of excitability, as discussed subsequently. NGF regulates the expression of some of the receptors expressed by nociceptors. Capsaicin sensitivity is increased *in vivo* by NGF, which results from the regulation of the heat transducer TRPV1 (formerly VR1). Finally, several ion channels, such as $Na_v1.8$ are strongly regulated by NGF availability.

The dramatic effects of exogenously administered NGF on pain signaling systems does not reveal the role of endogenous NGF. To ascertain if any of the pharmacological effects of NGF truly reflect those of endogenous NGF, one must perform experiments in which the biological actions of endogenous NGF are somehow blocked. This has been achieved by two major experimental approaches: targeted recombination in embryonic stem cells to selectively knock out either NGF or its receptor trkA, and the administration of proteins that inhibit the bioactivity of NGF. Each of these techniques has advantages and disadvantages, but studies using them largely confirm an important role for endogenous NGF in regulating pain sensitivity. The knock-out approach has provided information particularly regarding the developmental role of NGF and has confirmed that essentially all spinal nociceptive afferent fibers require this factor for survival in the perinatal period.

Because mice with NGF or trkA deletions rarely survive past the first postnatal week, most of what we know about endogenous NGF function in the adult has been determined by the use of blocking agents. There have been a number of studies using a synthetic chimeric protein consisting of the extracellular domain of trkA fused to the Fc tail of human immunoglobulin (trkA-IgG). Local infusion of trkA-IgG into the rat hindpaw leads to thermal hypoalgesia and a decrease in CGRP content in those DRG neurons projecting to the infused area (McMahon *et al.*, 1995). These changes take several days to develop. In addition, there is a decrease in the thermal and chemical sensitivity of nociceptors projecting to the area and a decrease in the epidermal innervation density. These results provide strong evidence that NGF continues to play an important role in regulating the function of the small, peptidergic sensory neurons in the adult.

Finally, there is now compelling evidence that the biology of NGF described previously is particularly important in many forms of inflammation. NGF is found in many cell types in tissues subject to inflammatory insult, and much evidence now supports the hypothesis that up-regulation of NGF levels is a common component of the inflammatory response that relates to hyperalgesia. Elevated NGF levels have been found in a variety of inflammatory states in both humans and animal models (see McMahon and Bennett, 1999). There is now widespread agreement that blocking NGF bioactivity (either systemically or locally) largely blocks the effects of inflammation on sensory nerve function. For instance, intraplantar injection of carrageenan produces an acute inflammatory reaction, which has previously been widely used in the study of the analgesic effects of nonsteroidal anti-inflammatory drugs. When trkA-IgG was coadministered with carrageenan, it could largely prevent the development of the thermal hyperalgesia that normally develops (McMahon *et al.*, 1995), and the sensitization of primary afferent nociceptors. There are now many other similar examples. IL-1β and TNFα have been shown to mediate changes in NGF expression during inflammation *in vivo* and the hyperalgesia produced by these cytokines can be prevented by NGF antagonism.

Brain-derived Neurotrophic Factor as a Central Pain Neuromodulator

Neurotrophic factors are best known for their roles as secreted factors, both during development (as target-derived survival factors) and in the adult. But there is now a growing body of evidence that at least one of the neurotrophins, brain-derived neurotrophic factor (BDNF), may act as a neuromodulator. That is, in some neurons, BDNF does not enter the secretory pathway but instead is packaged in synaptic vesicles and is released with neuronal activity to modulate postsynaptic neurons. Such a role has been suggested in hippocampus, cortex, cerebellum, and spinal cord (McAllister *et al.*, 1999; Malcangio and Lessmann, 2003).

Here we will consider the role of BDNF in nociceptive processing. This protein is constitutively expressed in a significant minority of small and medium-sized sensory neurons of the dorsal root ganglion (DRG) (Zhou and Rush, 1996; Barakat-Walter, 1996) where it is contained in dense-core vesicles. Moreover, it is dramatically up-regulated in many small diameter nociceptive sensory neurons in models of inflammatory pain and in larger neurons in some animal models of neuropathic pain (Michael *et al.*, 1999; Michael *et al.*, 1997; Zhou *et al.*, 1999; Ha *et al.*, 2001). One important trigger for the increased synthesis of BDNF in nociceptive neurons is the peripheral increase in NGF availability that is a common feature of inflammation (see previous discussion). BDNF is known to be released from the central terminals of sensory neurons with activity, and in a frequency-dependent manner (Balkowiec and Katz, 2000). In normal tissue, release is associated specifically with activity in nociceptors, either induced by chemical algogens such as capsaicin, or by electrical stimulation of, specifically, small diameter sensory neurons. Interestingly, the patterns of electrical activity necessary for BDNF are different from those producing release of other sensory neuron transmitters such as glutamate and substance P (Lever *et al.*, 2001).

High-affinity receptors for BDNF (the tyrosine kinase receptor trkB) are widely expressed on spinal neurons including spinal projection neurons. We have recently shown that nociceptor activation in a variety of experimental preparations leads to activation (phosphorylation) of dorsal horn trkB receptors, indicating that BDNF can be released from nociceptors and activate postsynaptic neurons (Pezet *et al.*, 2002a). We have also shown that intrathecally injected BDNF can also induce ERK phosphorylation in laminae I–II neurons of the spinal cord (Pezet *et al.*, 2002a), and this kinase is known to be an important second messenger in mediating some changes in spinal nociceptive processing associated with persistent pain states (Ji *et al.*, 1999). In other experiments, we finally showed that endogenous BDNF release from nociceptors accounts for approximately one third of the activation of ERK (Pezet *et al.*, 2002b). Thus, there is a body of biochemical evidence implicating BDNF as a pain-related central modulator.

There are also functional studies demonstrating the neuromodulatory role of BDNF in the spinal cord. Exogenous BDNF selectively enhances sensory neuron–evoked spinal reflex activity and NMDA-induced depolarization of *in vitro* preparation of rat spinal cord (Kerr *et al.*, 1999; Thompson *et al.*, 1999). Recently, BDNF null mutant mice have been shown to display a selective deficit in the ventral root potentials evoked by nociceptive primary afferent fibers (Heppenstall and Lewin, 2001). In adult rats, intrathecal injection of anti-BDNF antibodies or sequestering fusion molecule TrkB-IgG prevented the development of thermal hyperalgesia associated with acute peripheral inflammation (Kerr *et al.*, 1999) or neuropathic pain (Fukuoka *et al.*, 2001), respectively. It is not clear what the intracellular mechanism is of BDNF-induced modulation. One straightforward possibility is that synaptically released BDNF activates the map kinase ERK in second-order cells and this second messenger produces a post-translational change in NMDA receptor properties. Whatever the mechanism, there is a clearly documented pathway linking nociceptor activity with BDNF-mediated central modulation of sensory transmission. It is not clear at present when this pathway is active. The data so far suggest that it is one important mediator of central sensitization and might therefore contribute to a variety of hyperalgesic states. BDNF therefore joins a list of promising new targets for analgesic drug development.

OTHER POTENTIAL APPLICATIONS OF NEUROTROPHIC FACTORS

The role of NTFs in the treatment of CNS neurodegenerative disorders is varied. In some disorders, such as Parkinson's disease, GDNF holds clear promise. In Alzheimer's disease, the potential benefit of NTFs is unclear; in Huntington's disease, BDNF has been implicated in the disease process; and in amyotrophic lateral sclerosis, there is hope that selected NTFs may forestall or reverse disease progression. All of these hopes spring from the basic biology of neurotrophic factors—that they are neuroprotective and neurorestorative for selected populations of neurons. Despite the potent effects that have been demonstrated in culture or preclinical models, it remains a sobering fact that there are no current clinically proven applications for neurotrophic factors.

References

Akerud P *et al.* (2002). Persephin-overexpressing neural stem cells regulate the function of nigral dopaminergic neurons and prevent their degeneration in a model of Parkinson's disease, *Mol Cell Neurosci* 21:205-222.

Balkowiec A, and Katz DM. (2000). Activity-dependent release of endogenous brain-derived neurotrophic factor from primary sensory neurons detected by ELISA in situ, *J Neurosci* 20:7417-7423.

Barakat-Walter I. (1996). Brain-derived neurotrophic factor-like immunoreactivity is localized mainly in small sensory neurons of rat dorsal root ganglia, *J Neurosci Methods* 68:281-288.

Barras FM, Pasche P, Bouche N, Aebischer P, Zurn AD. (2002). Glial cell line-derived neurotrophic factor released by synthetic guidance channels promotes facial nerve regeneration in the rat. *J Neurosci Res* 70:746-755.

Baumgartner BJ, and Shine HD. (1997). Targeted transduction of CNS neurons with adenoviral vectors carrying neurotrophic factor genes confers neuroprotection that exceeds the transduced population, *J Neurosci* 17:6504-6511.

Baumgartner BJ, and Shine HD. (1998). Permanent rescue of lesioned neonatal motoneurons and enhanced axonal regeneration by adenovirus-mediated expression of glial cell-line-derived neurotrophic factor, *J Neurosci Res* 54:766-777.

Bennet MR, Gibson WG, and Lemon G. (2002). Neuronal cell death, nerve growth factor and neurotrophic models: 50 years on, *Auton Neurosci* 95:1-23.

Bennett DL *et al.* (2000). The glial cell line-derived neurotrophic factor family receptor components are differentially regulated within sensory neurons after nerve injury, *J Neurosci* 20:427-437.

Blesch A, and Tuszynski MH. (2001). GDNF gene delivery to injured adult CNS motor neurons promotes axonal growth, expression of the trophic neuropeptide CGRP, and cellular protection, *J Comp Neurol* 436:399-410.

Blits B *et al.* (2000). The use of adenoviral vectors and ex vivo transduced neurotransplants: towards promotion of neuroregeneration, *Cell Transplant* 9:169-178.

Blits B *et al.* (2003). Adeno-associated viral vector-mediated neurotrophin gene transfer in the injured adult rat spinal cord improves hind-limb function, *Neuroscience* 118:271-281.

Bonnington JK, and McNaughton PA. (2003). Signalling pathways involved in the sensitisation of mouse nociceptive neurones by nerve growth factor, *J Physiol* 551:433-446.

Boucher TJ *et al.* (2000). Potent analgesic effects of GDNF in neuropathic pain states, *Science* 290:124-127.

Bradbury EJ *et al.* (1999). NT-3 promotes growth of lesioned adult rat sensory axons ascending in the dorsal columns of the spinal cord, *Eur J Neurosci* 11:3873-3883.

Bradbury EJ *et al.* (1998). NT-3, but not BDNF, prevents atrophy and death of axotomized spinal cord projection neurons, *Eur J Neurosci* 10:3058-3068.

Carlstedt T, Cullheim S, Risling M, and Ulfhake B. (1988). Mammalian root-spinal cord regeneration, *Prog Brain Res* 78:225-229.

Chao MV *et al.* (1986). Gene transfer and molecular cloning of the human NGF receptor, *Science* 232:518-521.

Cheng H, Cao Y, and Olson L. (1996). Spinal cord repair in adult paraplegic rats: partial restoration of hind limb function, *Science* 273:510-513.

Chong MS, Woolf CJ, Haque NS, and Anderson PN. (1999). Axonal regeneration from injured dorsal roots into the spinal cord of adult rats, *J Comp Neurol* 410:42-54.

Cummins TR, Black JA, Dib-Hajj SD, and Waxman SG. (2000). Glial-derived neurotrophic factor upregulates expression of functional SNS and NaN sodium channels and their currents in axotomized dorsal root ganglion neurons, *J Neurosci* 20:8754-8761.

Fine EG *et al.* (2002). GDNF and NGF released by synthetic guidance channels support sciatic nerve regeneration across a long gap, *Eur J Neurosci* 15:589-601.

Fraher JP. (2000). The transitional zone and CNS regeneration, *J Anat* 196(Pt 1):137-158.

Fukuoka T *et al.* (2001). Brain-derived neurotrophic factor increases in the uninjured dorsal root ganglion neurons in selective spinal nerve ligation model, *J Neurosci* 21:4891-4900.

Gardell LR *et al.* (2003). Multiple actions of systemic artemin in experimental neuropathy, *Nat Med* 9:1383-1389.

George R, and Griffin JW. (1994). Delayed macrophage responses and myelin clearance during Wallerian degeneration in the central nervous system: the dorsal radiculotomy model, *Exp Neurol* 129:225-236.

Giehl KM, and Tetzlaff W. (1996). BDNF and NT-3, but not NGF, prevent axotomy-induced death of rat corticospinal neurons in vivo, *Eur J Neurosci* 8:1167-1175.

Golding J, Shewan D, and Cohen J. (1997). Maturation of the mammalian dorsal root entry zone—from entry to no entry, *Trends Neurosci* 20:303-308.

Gotz R *et al.* (1994). Neurotrophin-6 is a new member of the nerve growth factor family, *Nature* 372:266-269.

Grill R *et al.* (1997). Cellular delivery of neurotrophin-3 promotes corticospinal axonal growth and partial functional recovery after spinal cord injury, *J Neurosci* 17:5560-5572.

Ha SO *et al.* (2001). Expression of brain-derived neurotrophic factor in rat dorsal root ganglia, spinal cord and gracile nuclei in experimental models of neuropathic pain, *Neuroscience* 107:301-309.

Hao S *et al.* (2003). HSV-mediated gene transfer of the glial cell-derived neurotrophic factor provides an antiallodynic effect on neuropathic pain, *Mol Ther* 8:367-375.

Heppenstall PA, and Lewin GR. (2001). BDNF but not NT-4 is required for normal flexion reflex plasticity and function, *Proc Natl Acad Sci USA* 98:8107-8112.

Hiebert GW *et al.* (2002). Brain-derived neurotrophic factor applied to the motor cortex promotes sprouting of corticospinal fibers but not regeneration into a peripheral nerve transplant, *J Neurosci Res* 69:160-168.

Horton A *et al.* (1997). NGF binding to p75 enhances the sensitivity of sensory and sympathetic neurons to NGF at different stages of development, *Mol Cell Neurosci* 10:162-172.

Itoh Y, Sugawara T, Kowada M, and Tessler A. (1992). Time course of dorsal root axon regeneration into transplants of fetal spinal cord: I. A light microscopic study, *J Comp Neurol* 323:198-208.

Itoh Y, Sugawara T, Kowada M, and Tessler A. (1993). Time course of dorsal root axon regeneration into transplants of fetal spinal cord: an electron microscopic study, *Exp Neurol* 123:133-146.

Jeffery ND, and Fitzgerald M. (2001). Effects of red nucleus ablation and exogenous neurotrophin-3 on corticospinal axon terminal distribution in the adult rat, *Neuroscience* 104:513-521.

Ji RR, Baba H, Brenner GJ, and Woolf CJ. (1999). Nociceptive-specific activation of ERK in spinal neurons contributes to pain hypersensitivity, *Nat Neurosci* 2:1114-1119.

Jin Y, Fischer I, Tessler A, and Houle JD. (2002). Transplants of fibroblasts genetically modified to express BDNF promote axonal regeneration from supraspinal neurons following chronic spinal cord injury, *Exp Neurol* 177:265-275.

Kaplan DR *et al.* (1991). The trk proto-oncogene product: a signal transducing receptor for nerve growth factor, *Science* 252:554-558.

Kaplan DR, and Miller FD. (2000). Neurotrophin signal transduction in the nervous system, *Curr Opin Neurobiol* 10:381-391.

Kerr BJ *et al.* (1999). Brain-derived neurotrophic factor modulates nociceptive sensory inputs and NMDA-evoked responses in the rat spinal cord. *J Neurosci.* 19, 5138-5148.

Kobayashi NR, Fan DP, Giehl KM, Bedard AM, Wiegand SJ, and Tetzlaff W. (1997a). BDNF and NT-4/5 prevent atrophy of rat rubrospinal neurons after cervical axotomy, stimulate GAP-43 and Talpha1-tubulin mRNA expression, and promote axonal regeneration, *J Neurosci* 17:9583-9595.

Kobayashi NR *et al.* (1997b). BDNF and NT-4/5 prevent atrophy of rat rubrospinal neurons after cervical axotomy, stimulate GAP-43 and Talpha1-tubulin mRNA expression, and promote axonal regeneration, *J Neurosci* 17:9583-9595.

Koda M *et al.* (2002). Brain-derived neurotrophic factor suppresses delayed apoptosis of oligodendrocytes after spinal cord injury in rats, *J Neurotrauma* 19:777-785.

Kotzbauer PT *et al.* (1996). Neurturin, a relative of glial-cell-line-derived neurotrophic factor, *Nature* 384:467-470.

Leffler A *et al.* (2002). GDNF and NGF reverse changes in repriming of TTX-sensitive Na(+) currents following axotomy of dorsal root ganglion neurons, *J Neurophysiol* 88:650-658.

Lever IJ *et al.* (2001). Brain-derived neurotrophic factor is released in the dorsal horn by distinctive patterns of afferent fiber stimulation, *J Neurosci* 21:4469-4477.

Levi-Montalcini R. (1987). The nerve growth factor 35 years later, *Science* 237:1154-1162.

Lewin GR, Ritter AM, Mendell LM. (1993). Nerve growth factor-induced hyperalgesia in the neonatal and adult rat. *J Neurosci* 13:2136-2148.

Lin LF *et al.* (1993). GDNF: a glial cell line-derived neurotrophic factor for midbrain dopaminergic neurons, *Science* 260:1130-1132.

Lindsay RM. (1996). Role of neurotrophins and trk receptors in the development and maintenance of sensory neurons: an overview, *Philos Trans R Soc Lond B Biol Sci* 351:365-373.

Liu L, Persson JK, Svensson M, andAldskogius H. (1998a). Glial cell responses, complement, and clusterin in the central nervous system following dorsal root transection, *Glia* 23:221-238.

Liu L, Persson JK, Svensson M, andAldskogius H. (1998b). Glial cell responses, complement, and clusterin in the central nervous system following dorsal root transection, *Glia* 23:221-238.

Liu S *et al.* (1998c). Axonal regrowth through a collagen guidance channel bridging spinal cord to the avulsed C6 roots: functional recovery in primates with brachial plexus injury, *J Neurosci Res* 51:723-734.

Liu S *et al.* (1997). Axonal regrowth through collagen tubes bridging the spinal cord to nerve roots, *J Neurosci Res* 49:425-432.

Liu Y *et al.* (2002). Grafts of BDNF-producing fibroblasts rescue axotomized rubrospinal neurons and prevent their atrophy, *Exp Neurol* 178:150-164.

Lu P, Blesch A, and Tuszynski MH. (2001). Neurotrophism without neurotropism: BDNF promotes survival but not growth of lesioned corticospinal neurons, *J Comp Neurol* 436:456-470.

Malcangio M, and Lessmann V. (2003). A common thread for pain and memory synapses? Brain-derived neurotrophic factor and trkB receptors, *Trends Pharmacol Sci* 24:116-121.

Marcol W *et al.* (2003). Regeneration of sciatic nerves of adult rats induced by extracts from distal stumps of pre-degenerated peripheral nerves, *J Neurosci Res* 72:417-424.

McAllister AK, Katz LC, and Lo DC. (1999). Neurotrophins and synaptic plasticity, *Annu Rev Neurosci* 22:295-318.

McCarthy BG *et al.* (1995). Cutaneous innervation in sensory neuropathies: evaluation by skin biopsy, *Neurology* 45:1848-1855.

McGee AW, and Strittmatter SM. (2003). The Nogo-66 receptor: focusing myelin inhibition of axon regeneration, *Trends Neurosci* 26:193-198.

McMahon SB. (2002). Neuropathic pain mechanisms. In Giamberardino MA, editor: *Proceedings of the 10th World Congress on Pain.* San Diego: IASP Press, pp. 155-163.

McMahon SB, and Bennett DLH. (1999). Trophic factors and pain. In *Textbook of Pain.* New York: Churchill Livingstone, pp. 105-128.

McMahon SB, Bennett DL, Priestley JV, and Shelton DL. (1995). The biological effects of endogenous nerve growth factor on adult sensory neurons revealed by a trkA-IgG fusion molecule, *Nat Med* 1:774-780.

Michael GJ *et al.* (1997). Nerve growth factor treatment increases brain-derived neurotrophic factor selectively in TrkA-expressing dorsal root ganglion cells and in their central terminations within the spinal cord, *J Neurosci* 17:8476-8490.

Michael GJ, Averill S, Shortland PJ, Yan Q, and Priestley JV. (1999). Axotomy results in major changes in BDNF expression by dorsal root ganglion cells: BDNF expression in large trkB and trkC cells, in pericellular baskets, and in projections to deep dorsal horn and dorsal column nuclei, *Eur J Neurosci* 11:3539-3551.

Midroni G, and Bilbao JM. (1995). Quantitative techniques. In Midroni G, Bilbao JM, editors: *Biopsy diagnosis of peripheral neuropathy.* andToronto: Butterworth-Heinemann, pp. 35-44.

Milbrandt J *et al.* (1998). Persephin, a novel neurotrophic factor related to GDNF and neurturin, *Neuron* 20:245-253.

Navarro X *et al.* (1999). Ensheathing glia transplants promote dorsal root regeneration and spinal reflex restitution after multiple lumbar rhizotomy, *Ann Neurol* 45:207-215.

Novikova LN, Novikov LN, and Kellerth JO. (2002). Differential effects of neurotrophins on neuronal survival and axonal regeneration after spinal cord injury in adult rats, *J Comp Neurol* 452:255-263.

Oudega M, and Hagg T. (1999). Neurotrophins promote regeneration of sensory axons in the adult rat spinal cord, *Brain Res* 818:431-438.

Oudega M, Varon S, and Hagg T. (1994). Regeneration of adult rat sensory axons into intraspinal nerve grafts: promoting effects of conditioning lesion and graft predegeneration, *Exp Neurol* 129:194-206.

Paratcha G, Ledda F, and Ibanez CF. (2003). The neural cell adhesion molecule NCAM is an alternative signaling receptor for GDNF family ligands, *Cell* 113:867-879.

Petty BG *et al.* (1994). The effect of systemically administered recombinant human nerve growth factor in healthy human subjects, *Ann Neurol* 36:244-246.

Pezet S *et al.* (2002a). Noxious stimulation induces Trk receptor and downstream ERK phosphorylation in spinal dorsal horn, *Mol Cell Neurosci* 21:684-695.

Pezet S, Malcangio M, and McMahon SB. (2002b). BDNF: a neuromodulator in nociceptive pathways? *Brain Res Brain Res Rev* 40:240-249.

Pindzola RR, Doller C, and Silver J. (1993). Putative inhibitory extracellular matrix molecules at the dorsal root entry zone of the spinal cord during development and after root and sciatic nerve lesions, *Dev Biol* 156:34-48.

Pu LL *et al.* (1999). Effects of nerve growth factor on nerve regeneration through a vein graft across a gap, *Plast Reconstr Surg* 104:1379-1385.

Quasthoff S, and Hartung HP. (2001). [Nerve growth factor (NGF) in treatment of diabetic polyneuropathy. One hope less?], *Nervenarzt* 72:456-459.

Ramer MS *et al.* (2002). Neurotrophin-3-mediated regeneration and recovery of proprioception following dorsal rhizotomy, *Mol Cell Neurosci* 19:239-249.

Ramer MS, Duraisingam I, Priestley JV, and McMahon SB. (2001a). Two-tiered inhibition of axon regeneration at the dorsal root entry zone, *J Neurosci* 21:2651-2660.

Ramer MS, McMahon SB, and Priestley JV. (2001b). Axon regeneration across the dorsal root entry zone, *Prog Brain Res* 132:621-639.

Ramer MS, Priestley JV, and McMahon SB. (2000). Functional regeneration of sensory axons into the adult spinal cord, *Nature* 403:312-316.

Ramon-Cueto A, and Nieto-Sampedro M. (1994). Regeneration into the spinal cord of transected dorsal root axons is promoted by ensheathing glia transplants, *Exp Neurol* 127:232-244.

Richardson PM, and Verge VM. (1987). Axonal regeneration in dorsal spinal roots is accelerated by peripheral axonal transection, *Brain Res* 411:406-408.

Riddell JS, Enriquez-Denton M, Toft A, and Barnett SC. (2002). Do pure olfactory ensheathing cell grafts promote functional regeneration of afferent fibres following dorsal root lesions? Abstract Viewer/Itinerary Planner. Washington, DC: Society for Neuroscience , 635.2.

Romero MI, Rangappa N, Garry MG, and Smith GM. (2001). Functional regeneration of chronically injured sensory afferents into adult spinal cord after neurotrophin gene therapy, *J Neurosci* 21:8408-8416.

Romero MI *et al.* (2000). Extensive sprouting of sensory afferents and hyperalgesia induced by conditional expression of nerve growth factor in the adult spinal cord, *J Neurosci* 20:4435-4445.

Rueff A, and Mendell LM. (1996). Nerve growth factor NT-5 induce increased thermal sensitivity of cutaneous nociceptors in vitro, *J Neurophysiol* 76:3593-3596.

Saiz-Sapena N, Vanaclocha V, Insausti R, and Idoate M. (1997). Dorsal root repair by means of an autologous nerve graft: experimental study in the rat, *Acta Neurochir (Wien)* 139:780-786.

Santos X, Rodrigo J, Hontanilla B, and Bilbao G. (1999). Regeneration of the motor component of the rat sciatic nerve with local administration of neurotrophic growth factor in silicone chambers, *J Reconstr Microsurg* 15:207-213.

Savio T, and Schwab ME. (1989). Rat CNS white matter, but not gray matter, is nonpermissive for neuronal cell adhesion and fiber outgrowth, *J Neurosci* 9:1126-1133.

Schicho R, Skofitsch G, and Donnerer J. (1999). Regenerative effect of human recombinant NGF on capsaicin-lesioned sensory neurons in the adult rat, *Brain Res* 815:60-69.

Schnell L *et al.* (1994). Neurotrophin-3 enhances sprouting of corticospinal tract during development and after adult spinal cord lesion, *Nature* 367:170-173.

Terenghi G. (1999). Peripheral nerve regeneration and neurotrophic factors, *J Anat* 194 (Pt 1):1-14.

Thompson SW *et al.* (1999). Brain-derived neurotrophic factor is an endogenous modulator of nociceptive responses in the spinal cord, *Proc Natl Acad Sci USA* 96:7714-7718.

Tuszynski MH *et al.* (2003). NT-3 gene delivery elicits growth of chronically injured corticospinal axons and modestly improves functional deficits after chronic scar resection, *Exp Neurol* 181:47-56.

Tuszynski MH *et al.* (2002b). Spontaneous and augmented growth of axons in the primate spinal cord: effects of local injury and nerve growth factor-secreting cell grafts, *J Comp Neurol* 449:88-101.

Tuszynski MH *et al.* (2002a). Spontaneous and augmented growth of axons in the primate spinal cord: effects of local injury and nerve growth factor-secreting cell grafts, *J Comp Neurol* 449:88-101.

Wang R *et al.* (2003). Glial cell line-derived neurotrophic factor normalizes neurochemical changes in injured dorsal root ganglion neurons and prevents the expression of experimental neuropathic pain, *Neuroscience* 121:815-824.

Weidner N, Blesch A, Grill RJ, and Tuszynski MH. (1999). Nerve growth factor-hypersecreting Schwann cell grafts augment and guide spinal cord axonal growth and remyelinate central nervous system axons in a phenotypically appropriate manner that correlates with expression of L1, *J Comp Neurol* 413:495-506.

Williams EJ, Walsh FS, and Doherty P. (2003). The FGF receptor uses the endocannabinoid signaling system to couple to an axonal growth response, *J Cell Biol* 160:481-486.

Yiangou Y *et al.* (2002). Molecular forms of NGF in human and rat neuropathic tissues: decreased NGF precursor-like immunoreactivity in human diabetic skin, *J Peripher Nerv Syst* 7:190-197.

Young C, Miller E, Nicklous DM, and Hoffman JR. (2001). Nerve growth factor and neurotrophin-3 affect functional recovery following peripheral nerve injury differently, *Restor Neurol Neurosci* 18:167-175.

Zhang Y *et al.* (1995). Tenascin-C expression by neurons and glial cells in the rat spinal cord: changes during postnatal development and after dorsal root or sciatic nerve injury, *J Neurocytol* 24:585-601.

Zhou L *et al.* (2003). Neurotrophin-3 expressed in situ induces axonal plasticity in the adult injured spinal cord, *J Neurosci* 23:1424-1431.

Zhou L, and Shine HD. (2003). Neurotrophic factors expressed in both cortex and spinal cord induce axonal plasticity after spinal cord injury, *J Neurosci Res* 74:221-226.

Zhou XF *et al.* (1999). Injured primary sensory neurons switch phenotype for brain-derived neurotrophic factor in the rat, *Neuroscience* 92:841-853.

Zhou XF, and Rush RA. (1996). Endogenous brain-derived neurotrophic factor is anterogradely transported in primary sensory neurons, *Neuroscience* 74:945-953.

26

Promoting the Regeneration of Axons within the Central Nervous System

James H. Park, AB
Stephen M. Strittmatter, MD, PhD

During development, central nervous system (CNS) neurons extend exuberantly. Their growth cones sense attractive and repulsive cues from the extracellular milieu and guide growing axons to appropriate partners. Once the critical period for the formation of specific connections ends, CNS plasticity within higher vertebrates is greatly diminished. A clinical consequence of limited adult CNS axonal extension is that patients with spinal cord injury (SCI) generally experience little functional recovery and suffer from permanent functional deficits. In SCI, axonal interruption at one spinal level is the primary cause of dysfunction. Loss of function in lower spinal segments occurs despite the survival of rostral and caudal neurons controlling these segments. Disconnection is the proximate cause of clinical deficits in SCI.

Similarly, limited CNS axonal growth and plasticity are likely to restrict recovery from brain trauma, stroke, and chronic progressive multiple sclerosis. For head trauma, a range of mechanisms explains persistent functional deficits. Focal parenchymal damage and hemorrhage cause direct neuronal cell loss. In addition, diffuse axonal injury produces neuronal disconnection. For stroke, recovery of function over months following ischemia is largely dependent on the success or failure of plasticity in the brain. For chronic progressive multiple sclerosis, axonal disconnection rather than demyelination is now recognized as the best clinical correlate of disability. This review focuses on SCI.

In contrast to the CNS, peripheral nervous system (PNS) axons maintain their plasticity beyond the developmental phase and remain capable of axonal regeneration after injury. *A priori*, the difference between PNS

and CNS regenerative capacities may be attributed to the innate differences between CNS and PNS neurons, to the environmental differences presented by CNS glia and PNS Schwann cells or a combination of both (see Figure 26.1). To address the relative importance of these factors, Aguayo and colleagues grafted sciatic nerve segments as bridges between the medulla and spinal cord and reported that CNS axons can grow through a PNS bridge (David and Aguayo, 1981). This landmark experiment proves that the CNS environment suppresses axonal regeneration, but in a permissive PNS environment, many CNS axons can regenerate. Recently, progress has been made in identifying the molecular determinants promoting and inhibiting CNS axonal regeneration. As we better understand how these determinants function, pharmacological agents can be screened and medical treatments can be devised for the new therapeutic modality of axon regeneration in neuro-recovery.

CENTRAL NERVOUS SYSTEM REGENERATION SIGNALS

CNS regeneration can result from the promotion or antagonism of four interdependent signal classes—extrinsic (glial) stimulatory, extrinsic (glial) inhibitory, intrinsic (neuronal) stimulatory, and intrinsic (neuronal) inhibitory. In the design of a therapeutic strategy for SCI, the modulation of these four signal types should be considered with respect to their relative contributions, accessibility to modulating agents, and side effects. Transplantation studies, pharmacological agents disrupting a specific signaling pathway, knockout and treatment animal models have

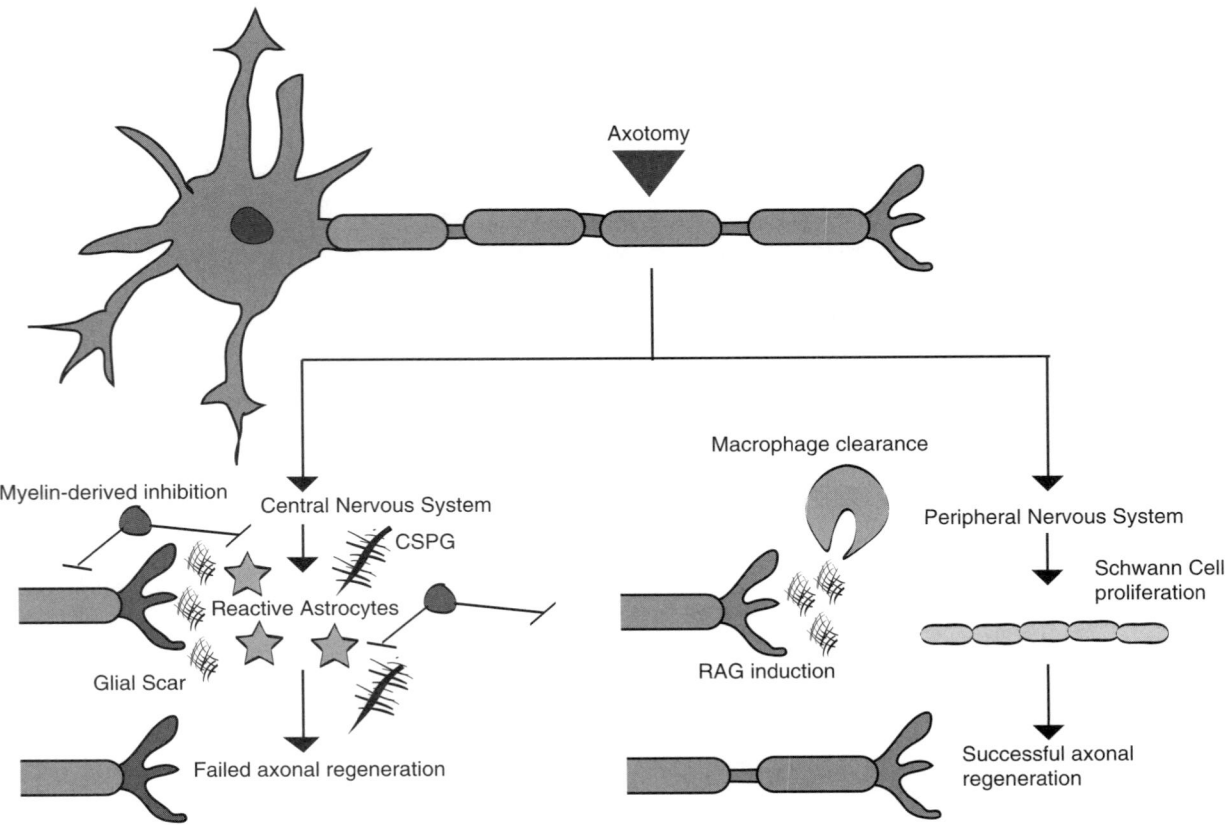

FIGURE 26.1   Cellular basis of the CNS and PNS response to axotomy. Following injury, central nervous system and peripheral nervous system axons have markedly different axon regenerative responses. CNS axons do not regenerate due to inhibition by myelin and glial scars. In addition, CNS axons lack the RAG induction, macrophage clearance, and Schwann cell proliferation found in PNS regeneration.

shed insight into their relative weights and which combination may be the most promising to pursue for clinical development (Figure 26.2).

PROMOTING EXTRINSIC STIMULATORY FACTORS

Diffusible Neurotrophic Factors

Because the CNS and PNS differ in their expression of neurotrophic factors after axotomy, it has been hypothesized that if PNS expression of neurotrophic factors (such as brain-derived neurotrophic factor [BDNF], nerve growth factor [NGF], and NT-3) is mimicked in the CNS, then regeneration of injured CNS axons will be promoted. Of the several neurotrophic factors such as BDNF and NGF characterized in *in vitro* studies, NT-3 has shown beneficial *in vivo* effects in SCI animal injury models (Schnell *et al.*, 1994). In adult rats with lesioned spinal cord tracts, local injection of NT-3

promotes corticospinal (CST) sprouting and in some cases long-distance regeneration. One proposed mechanism by which NT-3 promotes regeneration is the up regulation of regeneration-associated gene (RAG) expression, which will be discussed later, within the cell bodies of injured axons.

Extracellular Matrix–Associated Factors

The majority of extracellular matrix (ECM)/regeneration studies have focused on laminin-1 (LM1), the classic growth-promoting matrix molecule, mediating cell attachment and stimulating motility (Condic and Lemons, 2002). Laminin is a trimeric extracellular matrix molecule consisting of α, β, and γ subunits. Numerous genes encoding laminin have been characterized, with greater than 15 unique laminin trimers expressed in tissue. However, some laminin isoforms, such as s-laminins, do not promote neurite extension and act as inhibitory signals. It is now recognized that different laminin

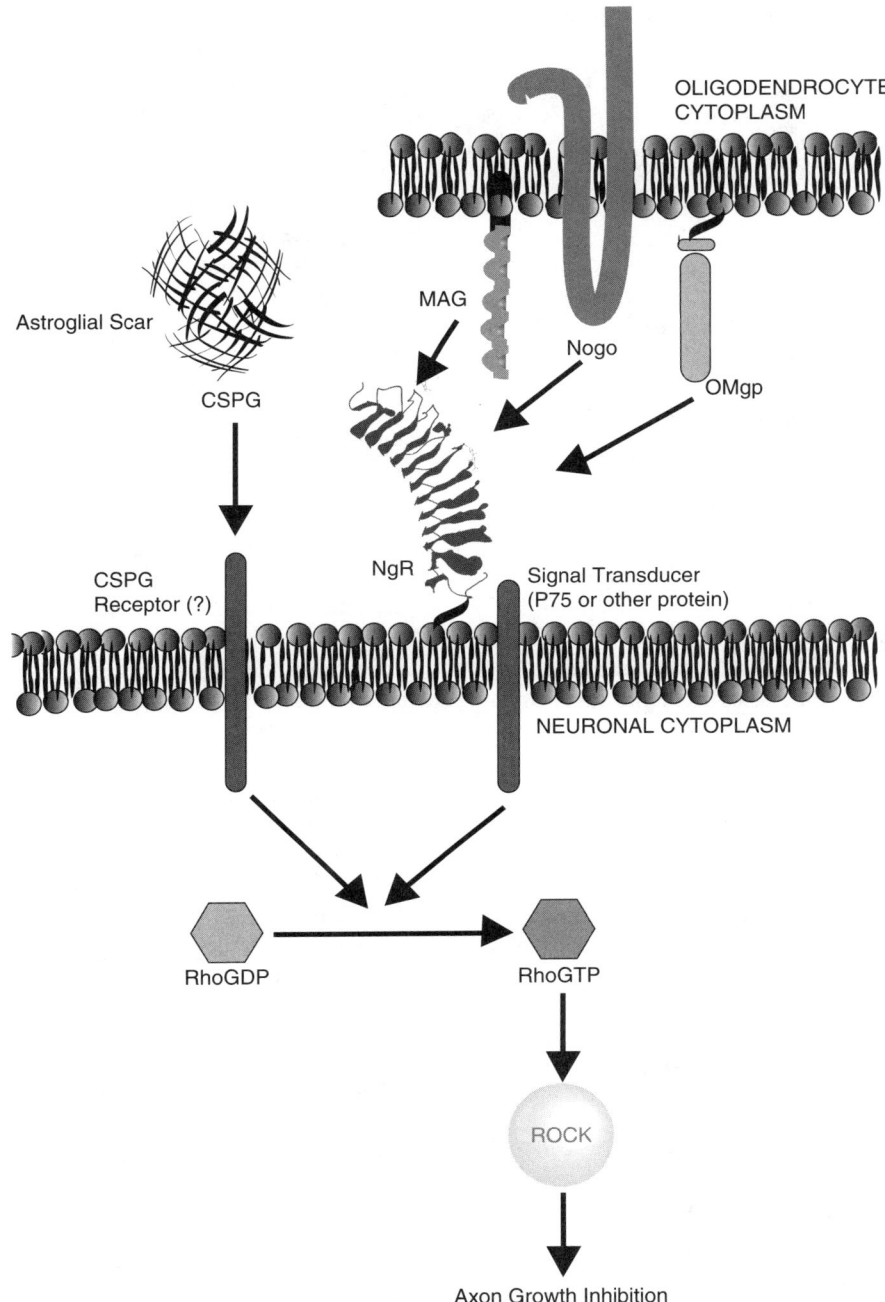

FIGURE 26.2 Molecular inhibition of axonal regeneration in the CNS. Many of the molecular determinants responsible for limited CNS plasticity have been characterized. Nogo, MAG, and OMgp converge at NgR to activate Rho/ROCK, inhibiting axon growth. CSPGs from the glial scar also activate Rho/ROCK signaling. Therapeutic disruption of these interactions is being tested.

isoforms signal combinatorially to promote and inhibit neurite outgrowth (Condic and Lemons, 2002). Data indicate that laminin's growth-promoting characteristics can partially counter the effects of CNS inhibitors from oligodendrocytes or astrocytes.

Clinical Applications

For the previously mentioned extrinsic factors, adequate delivery and penetration play a critical role in their efficacy. Recent projects have used injections and pumps to deliver neurotrophic factors to the lesion site. Other successful strategies include implanting fibroblasts engineered to secrete neurotrophic factors such as NT-3 directly into the lesion site (Grill *et al.*, 1997; Bradbury *et al.*, 1999) or using gene therapy with viral constructs containing neurotrophic factor genes (Ruitenberg *et al.*, 2003). Novel materials such as hydrogels primed with neurotrophic growth factors also stimulate recovery in animal injury models.

These findings highlight the potential of identifying an optimal combination of neurotrophic factors and delivery methods for enhancing growth of injured CNS axons. However, the challenge in all these treatment modalities is managing extrinsic inhibition from glial scar tissue formation and myelin. Optimal axon growth through an implant requires growth into normal CNS tissue to restore function.

OVERCOMING EXTRINSIC INHIBITORY FACTORS

The classic nerve graft mentioned at the outset emphasizes the critical role of extrinsic inhibition of CNS regeneration. Two principal extrinsic factors recognized as inhibitors of CNS axonal regeneration are the glial scar (Davies *et al.*, 1997) and CNS myelin (Niederost *et al.*, 1999) itself.

Glial Scar Formation

Following most if not all CNS injuries, proliferating astrocytes create a glial scar over several weeks that mechanically and biochemically blocks axonal regeneration. Tightly interdigitating astrocytic processes are thought to create a mechanical barrier for axonal regeneration (Fawcett and Asher, 1999). Chondroitin sulfate proteoglycans (CSPG) act as the principal known biochemical barrier.

Chondroitin Sulfate Proteoglycans

CSPGs are either membrane-bound or extracellular matrix glycoproteins that inhibit neurite growth *in vitro* and are up-regulated after CNS injury (Snow *et al.*, 1990). Specific CSPGs that increase after injury include NG2, neurocan, versican, aggrecan, and brevican (Lemons *et al.*, 1999). NG2 appears to be the major CSPG constituent at SCI sites (Fidler *et al.*, 1999; Dou and Levine, 1994; Fidler *et al.*, 1999). One study attributes CSPG inhibitory activity to the core protein. However, assigning the functional region of CSPG may not be as straightforward because similar sugar structures and chondroitinase digestion experiments emphasize the need for glycosaminoglycan (GAG) side chains for axonal inhibition (Jones *et al.*, 2003).

Interestingly, growing axons after injury *in vivo* can be associated with inhibitory CSPG substrates. Schwann cells and endothelial cells not only produce inhibitory CSPGs but also permissive cell adhesion molecules such as laminin that promote axonal regeneration. This suggests that axonal regeneration is dependent on a balance between positive and negative cues at localized sites. Presently, the molecular basis of CSPG action and a neuronal receptor for CSPG, if it exists, have not yet been described.

Myelin-Associated Inhibitors

Age Matters

During development, embryonic neurons are sensitive to the guidance class molecules such as semaphorins, slits, netrins, and ephrins (Wang and Anderson, 1997) but are not sensitive for CNS myelin inhibition. Postnatally, neurons become more sensitive to CNS myelin–derived inhibitors, of which three, Nogo, MAG, and OMgp, have been well characterized. The age-dependent myelin sensitivity has been explained both by postnatal decrease in cyclic adenosine monophosphate (cAMP) and an increase in CNS myelin inhibitor receptors (NgR, see subsequent discussion). Although the relative contribution of each of these myelin-associated inhibitors has been only partially assessed, multiple inhibitors may have developed during evolution to ensure CNS axonal pattern stabilization. Therefore blockade of any single inhibitory system may be expected to result in partial relief from myelin-based inhibition of axon growth.

Three Myelin-associated Inhibitors and Their Receptor

Nogo

In 1988, Schwab and colleagues raised a monoclonal antibody (IN-1) against an SDS fraction of myelin that exerted strong inhibitory activity in both 3T3 fibroblast spreading and neurite outgrowth (Caroni and Schwab, 1988a). IN-1 recognizes two inhibitory bands from rat brain myelin with molecular weights of 250 kD (NI (Neurite outgrowth inhibitor)-250) and 35 kD (NI-35) (Caroni and Schwab, 1988b).

Twelve years later, three groups independently cloned the gene for NI-250 and termed it *nogo* (Huber and Schwab, 2000; GrandPre *et al.*, 2000; Chen *et al.*, 2000). Three Nogo protein isoforms, which are all expressed in neurons, have been identified—Nogo A (1192 residues), B (373 residues), and C (199 residues). Of the three isoforms, Nogo-A is exclusively expressed in oligodendrocytes, whereas Nogo B and Nogo C can also be found in kidney/lung and skeletal muscle, respectively. Because of

Nogo-A's concentration in oligodendrocytes, it has been the focus for investigating myelin-associated inhibition. Histological analysis reveals Nogo-A to be mainly localized in the endoplasmic reticulum but is also expressed at the surface of oligodendrocytes. It has been hypothesized that CNS injury increases the exposure of Nogo-A to the extracellular environment (Oertle *et al.*, 2003).

Primary sequence analysis indicates that Nogo-A lacks a signal sequence and contains two internal hydrophobic domains. Functional studies demonstrate the presence of at least two inhibitory domains: a Nogo-A specific N-terminal domain and a pan-isoform 66 residue loop domain separating the two hydrophobic domains, termed Nogo-66 (Fournier *et al.*, 2001). The Nogo-A specific N-terminal domain serves as growth cone collapsing agent and an inhibitory substrate for both neuronal (primary neurons and PC-12 cells) and

non-neuronal (fibroblasts) cells (Chen *et al.*, 2000; Prinjha *et al.*, 2000).

Although sequence and immunohistochemical analyses predict the topology of the Nogo-A to have the 66-residue loop at the extracellular or luminal side and the rest of the protein in the cytoplasm, alternative topologies have been proposed. Experimental results show that Nogo A can have at least two different membrane topologies, pointing to the possibility of multiple functions at the membrane and within the cytoplasm.

Using differing gene deletion methods, three groups generated Nogo-A –/– mice and reported varying levels of regeneration and recovery (Zheng *et al.*, 2003; Kim *et al.*, 2003; Simonen *et al.*, 2003). Strittmatter and colleagues (Kim *et al.*, 2003) observed substantial regeneration and functional recovery in about half of young mice lacking Nogo A/B. In contrast, Tessier-Lavigne's group

A Nogo-A/B knockout mice

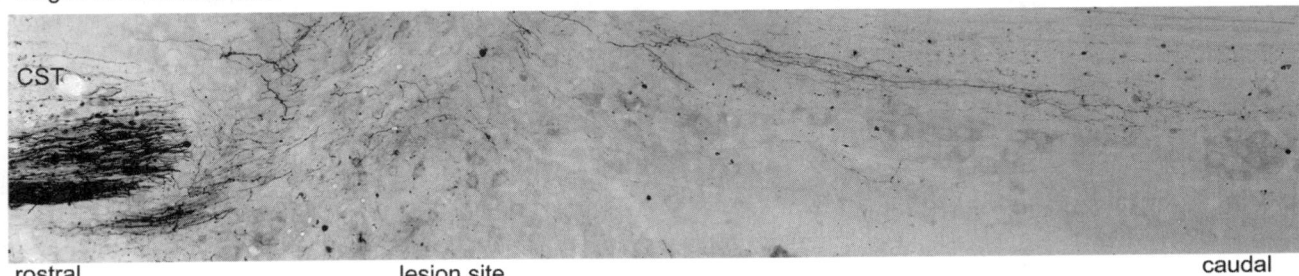

rostral   lesion site   caudal

B NEP1-40 treated rat caudal spinal cord

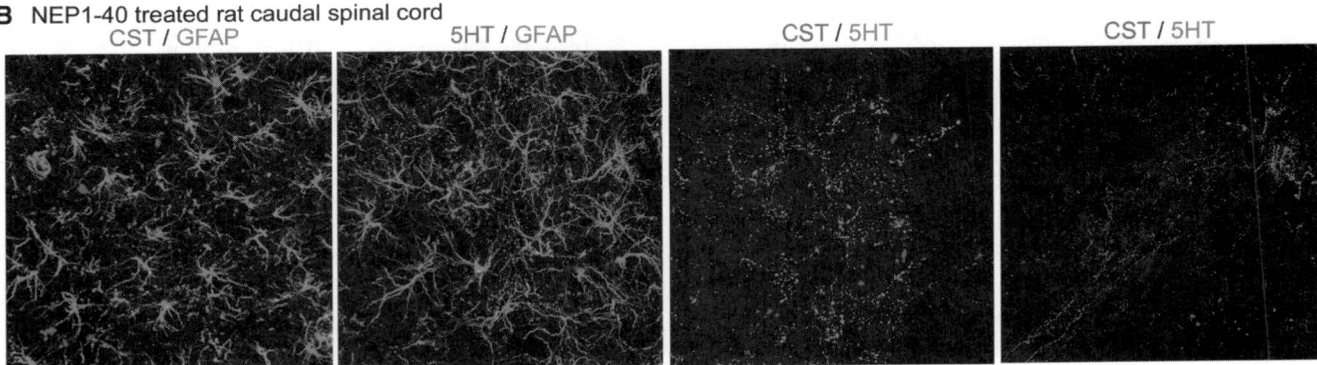

CST / GFAP   5HT / GFAP   CST / 5HT   CST / 5HT

FIGURE 26.3 Axonal regeneration after SCI in animals lacking Nogo-A/B or treated with NgR antagonist peptide. **A:** Mice lacking a functional Nogo-A/B gene were subjected to mid-thoracic spinal hemisection injury and the corticospinal tract (CST) was traced by cortical injection of biotin dextran amine (Kim *et al.*, 2003). Note fibers growing across the injury site into the caudal spinal cord. Wild-type mice have no CST fibers crossing such lesions. **B:** Rats were treated with the NEP1-40 antagonist peptide intrathecally after mid-thoracic spinal cord hemisection and the CST was traced (GrandPre *et al.*, 2002). Descending raphespinal fibers were detected by anti-serotonin immunohistology (5HT) and reactive astrocytes by anti-GFAP immunohistology (GFAP), as indicated. Sections from the spinal cord caudal to the lesion site demonstrate regenerating fibers despite normal level of reactive astrocytosis. Untreated animals exhibit similar astrocytosis but no CST or 5HT regeneration. (See also Color Plate section.)

studied two mice, a pan-Nogo –/– and Nogo A/B –/–, and detected no evidence of recovery in either knockout (Zheng *et al.*, 2003). Finally, Schwab's group noted axonal regeneration but no functional recovery in mice lacking Nogo-A selectively (Simonen *et al.*, 2003). Both the Tessier-Lavigne's and Strittmatter's groups note CNS myelin is less inhibitory without Nogo-A. These studies confirm some contribution of Nogo-A to limiting axon regeneration. However, they also demonstrate that other factors such as including genetic background, age, and mutation penetrance may play a role (Figure 26.3; see also the Color Plate section).

Myelin-associated Glycoprotein

Another myelin component that has potent *in vitro* inhibitory activity is myelin-associated glycoprotein (MAG) (McKerracher *et al.*, 1994; Mukhopadhyay *et al.*, 1994). Identified by chromatographic separation of myelin proteins, MAG contains five IgG domains and an intracellular domain. MAG acts as a bifunctional molecule (Mukhopadhyay *et al.*, 1994) during development. Before postnatal day 4, MAG promotes outgrowth during development, but the protein inhibits axonal growth in older neurons. Unlike Nogo, immunohistochemistry reveals that MAG is in both the CNS (in the periaxonal membrane and regions of uncompacted myelin) and the PNS (on outermost myelin loops).

MAG –/– mice do not exhibit improved CNS regeneration after spinal cord injury, suggesting that MAG plays a more minor role in deterring CNS regeneration post-injury (Bartsch *et al.*, 1995). However, MAG –/– mice show delayed myelination, myelin splitting, myelin redundancy, and decreased thickness of the periaxonal cytoplasmic collar of oligodendrocytes, indicating that MAG's physiological function may be more relevant in maintaining myelin integrity (Johnson *et al.*, 1989).

Oligodendrocyte-myelin Glycoprotein

He and colleagues identified oligodendrocyte-myelin glycoprotein (OMgp) when they screened phosphoinositol-specific phospholipase C (PIPLC) released proteins for the ability to inhibit neurite outgrowth (Wang *et al.*, 2002b). OMgp was also determined to be responsible for the inhibitory activity of a previously described myelin fraction termed arretin. Structural analysis reveals that OMgp is heavily glycosylated protein, composed of a cysteine-rich amino terminus, five leucine-rich repeat (LRR) domains, and a carboxy-terminal domain with serine/threonine repeats followed by a glycophosphatidylinositol (GPI)-anchor. Like Nogo, OMgp is expressed in both oligodendrocytes and neurons.

Heretofore, an OMgp null phenotype has been described.

Nogo Receptor

The Nogo receptor (NgR) was identified in an expression cloning strategy using AP-Nogo-66 binding (Fournier *et al.*, 2001). *In vitro*, NgR is required for Nogo-66 inhibition of axonal growth. NgR was later found to bind MAG and OMgp with high affinity and to mediate their inhibitory functions (Liu *et al.*, 2002; Domeniconi *et al.*, 2002).

Immunostaining reveals NgR to be expressed in nearly all neurons and in high levels in CNS white matter in profiles consistent with axons. Structurally NgR contains a signal sequence followed by eight leucine-rich repeat (LRR) domains, a carboxy-terminus cysteine-rich domain, and a GPI anchor, localizing NgR to lipid rafts. Because GPI-linked proteins do not span the membrane, a co-receptor is required for NgR signal transduction, in a manner reminiscent of glial cell line–derived neurotrophic factor (GDNF) receptor (Tansey *et al.*, 2000). One candidate co-receptor is p75-NTR, the low-affinity neurotrophin receptor (Wang *et al.*, 2002a). Despite *in vitro* data supporting this role, *in situ* studies reveal that p75-NTR expression patterns in the adult do not match well to patterns of neuronal responsiveness to myelin inhibition. p75-NTR may or may not be the only NgR co-receptor.

As part of an effort to understand how NgR interacts with three structurally unrelated ligands, two groups solved NgR's crescent shaped crystal structure (He *et al.*, 2003; Barton *et al.*, 2003). Understanding the molecular basis behind NgR binding to its *cis*- and *trans*-partners would allow for selective and potentially therapeutic disruption of this ligand-receptor complex.

Signaling downstream of the NgR is discussed in the section that characterizes intrinsic inhibitory factors. The phenotype of the NgR –/– mouse is under study in our laboratory.

Clinical Applications

Chondroitinase ABC

Fawcett, McMahon, and colleagues intrathecally administered chondroitinase ABC (ChABC) to digest CSPGs (Bradbury *et al.*, 2002). CSPG digestion allows crushed adult rat dorsal columns to sprout and grow both ascending sensory projections and descending CST axons after injury. Increased sprouting promotes functional recovery of locomotor and proprioceptive behaviors. Because chondroitin sulfate is the major biochemical inhibitor of the glial scar, a principal regeneration impediment, novel SCI therapies may incorporate CSPG neutralization strategies (Moon *et al.*, 2001).

Anti-Nogo-A Antibody

Of the three Nogo isoforms, Nogo-A has been the principal target in *in vivo* studies. Schwab and colleagues implanted hybridoma cells that produce a monoclonal antibody against Nogo-A, IN-1, into the cortex of young rats and reported that the treatment improves long-distance sprouting and regeneration after corticospinal and optic nerve injuries (Merkler *et al.*, 2001). Functionally, chronic exposure to IN-1 improves both sensorimotor reflex and locomotor function (Bregman *et al.*, 1995). Reassuringly, neurogenic pain caused by inappropriate fiber growth was not detected in a tail flick assay (Merkler *et al.*, 2001). These results indicate that rewiring of motor systems' functional recovery without neurogenic pain is possible after Nogo-A neutralization.

NEP 1-40

Because Nogo-66 induces inhibition via NgR, a Nogo-66 antagonist should reverse its action. Analysis of the minimal Nogo domain for binding to NgR reveals a short segment consisting of 40 residues that binds but does not stimulate the receptor (GrandPre *et al.*, 2002). NEP 1-40 peptide (NgR antagonist) systemically administered to spinal-injured rats promotes CST regeneration and ambulation (Li and Strittmatter, 2003). The ability of systemic NEP1-40 to induce regeneration supports the notion that the blood-brain barrier is permissive after trauma. It is also noted that antagonist treatment could be delayed for up to 1 week without limiting the benefit to motor function recovery. This may prove to be useful in the clinical setting where SCI therapy cannot begin until the patient is transported from the trauma scene and is medically stable. Because NEP1-40 blocks Nogo-66 but not MAG or OMgp, a more general NgR antagonist may be more effective in promoting recovery.

Ecto-NgR

Because NEP 1-40 is a selective Nogo-66 antagonist, Strittmatter and colleagues considered the soluble ectodomain of NgR (Ecto-NgR) as an universal blocker (Fournier *et al.*, 2002). This reagent stimulates a greater degree of axon growth over myelin *in vitro* than does NEP 1-40. *In vivo* studies suggest that this reagent increases axon regeneration to twice the extent of NEP 1-40. However, Ecto-NgR is unable to overcome the inhibitory activity of chondroitin sulfate proteoglycan, suggesting that Ecto-NgR disruption is specific for myelin inhibition. This reagent is undergoing *in vivo* SCI testing in rodents.

Myelin Vaccination

Because other uncharacterized myelin inhibitory activities may exist, McKerracher and colleagues have used a myelin vaccination approach to promote injured CST regeneration (Huang *et al.*, 1999). Mice immunized with CNS myelin show extensive CST fiber regeneration and hindlimb motor recovery, without immune response side effects such as experimental autoimmune encephalomyelitis (EAE). It is likely that a substantial axonal regeneration will require the simultaneous alleviation of multiple inhibitors. For clinical use, this approach of stimulating the immune system to block CNS inhibition will require refinement to ensure that no autoimmune disease results. As the authors note, identifying the blocking antibodies could lead to the development of passive immunization approaches to axon regeneration. In a related approach, Schwartz and colleagues obtain similar results by administration of myelin-primed immune cells.

Surgical Transplantation: Olfactory Ensheathing Cells

Because several factors (blocking myelin-associated inhibitors, controlling fibrosis, blocking inhibitory proteoglycans, preventing tissue necrosis and cavitation) need to be addressed to overcome extrinsic inhibition, one approach that obviates these issues is transplantation of exogenous cells to create a new regenerative environment (Davies *et al.*, 1997). The olfactory ensheathing cells (OECs) have been tested in this regard because they normally ensheathe axons that capable of regenerative growth in the CNS and can form myelin (Resnick *et al.*, 2003). Schwann cells or embryonic stem cells have also been used in various experiments. Using OECs, Avila and colleagues report an impressive motor function improvement 3 to 7 months after a complete spinal cord transection in adult rats (Ramon-Cueto *et al.*, 2000). Several other independent groups also report improved function with transplantation of adult-derived cultured OECs. For further OEC discussion, refer to the chapter by J. Kocsis, Cell Transplantation and Repair of Injured White Matter.

OVERCOMING INTRINSIC INHIBITORY FACTORS

Extrinsic inhibitory signaling through NgR leads to Rho-A activation and consequent ROCK activation. Therefore, Rho-A and ROCK represent potential therapeutic loci for intervention.

Rho-A Activation Is Responsible for NgR Signaling

Data from several laboratories confirm that Rho family of GTPases, specifically Rho-A is the downstream effector of myelin and NgR inhibitory signaling (Jin and Strittmatter, 1997; Lehmann *et al.*, 1999; Dergham *et al.*, 2002; Niederost *et al.*, 2002; Fournier *et al.*, 2003). As a member of the small GTPase Rho family, Rho-A is activated by guanine nucleotide exchange factors (GEFs) that enhance GTP binding or inhibited by GTPase activating proteins (GAPs) that promote GTP hydrolysis (Hall, 1998). Most cases of receptor-mediated activation of monomeric G-proteins occur via GEF proteins. However, one hypothesis suggests that inhibitory extracellular cues such as myelin and Nogo promote Rho-A activation by inactivating Rho guanine dissociation inhibitor (GDI), a molecule that sequesters inactive Rho-GDP from GEFs (Yamashita *et al.*, 1999).

Rock Inhibition Promotes Growth on Inhibitory Substrates

Several studies suggest that Rho-A's downstream effector in axonal growth inhibition by myelin is the serine/threonine Rho-associated kinase (ROCK) (Fournier *et al.*, 2003). As a Rho-A effector, ROCK causes rapid growth cone collapse, neurite retraction, and neurite growth inhibition. In a stripe assay, incubation of retinal explants with Y27632, a specific pyridine derived inhibitor of ROCK, also reduces the CSPG's inhibition (Monnier *et al.*, 2003). Because Rho and ROCK are convergent signaling points, the added benefit of blocking ROCK activity is that both CSPG and myelin-associated inhibition can be reduced.

Clinical Applications

C3

An enzyme from *Clostridium botulinum*, C3 transferase, blocks Rho A, B, and C function by ADP ribosylation on asparagine 41, without affecting Rac or Cdc42. C3 treatment of SCI rodents has produced mixed results (Dergham *et al.*, 2002; Fournier *et al.*, 2003). In one study, a single local dose of C3 produces long-distance regeneration of anterogradely labeled CST axons and improvement in locomotion by open-field testing. In contrast, a second study reports that C3 treatment only increases scarring without functional improvement. The discrepancy may be explained by variable C3 accessibility to the neuronal cytoplasm.

Y-27632

Administration of the ROCK inhibitor, Y-27632, has produced regenerative CST fiber sprouting after spinal cord injury in two studies. Treated animals recover hindlimb use for locomotion to significant extent over control rodents.

In comparing the effects of C3 and Y27632, McKerracher and colleagues conclude that C3 is more effective than Y27632. They speculate that unidentified Rho effectors uninhibited by Y27632, such as protein kinase N and rhotekin, may explain the difference (Winton *et al.*, 2002).

PROMOTING INTRINSIC STIMULATORY FACTORS

Because there are intrinsic differences between the PNS and CNS response to axonal injury, a number of studies have investigated changes in gene expression to identify intrinsic stimulatory factors. Two intrinsic stimulatory factors are RAGs and cAMP.

Regeneration-Associated Genes

A landmark study by Richardson, Aguayo, and colleagues demonstrates that dorsal root ganglion (DRG) cells are 100 times more likely to regenerate into peripheral nerve grafts if their peripheral axons were also cut (Richardson and Issa, 1984). In a follow-up study, Richardson and colleagues show that DRG cells could regenerate their central axons into PNS grafts only if the axon had been previously cut. Recently, Woolf and colleagues report that a conditioning peripheral lesion allows ascending dorsal column axons to extensively sprout into the spinal injury site (Neumann and Woolf, 1999). This set of observations suggests that a coordinated nerve regeneration program at the cell body exists and needs to be turned on with conditioning.

At the molecular level, there is mounting evidence for a coordinated regeneration program. Like myelin-associated inhibitors, in recent years several program components have been identified, characterized, and dubbed as RAGs [GAP-43 (Caroni, 1997; Bomze *et al.*, 2001), CAP-23 (Frey *et al.*, 2000), SPRR1A (Bonilla *et al.*, 2002), and Fn14 (Tanabe *et al.*, 2003)]. Two characterized RAGs that will be discussed here are GAP-43 and SPRR1A.

GAP-43

Regarded as the classic RAG, GAP-43 was identified as a rapidly transported axonal protein that is highly up-regulated after sciatic nerve injury. GAP-43 is localized to growth cones associated with neuropil areas. RAGs, such as GAP-43, are highly expressed during nervous system developmental and regenerative axon growth (Kalil and Skene, 1986).

Constitutive GAP-43 overexpression in lesioned adult mouse neurons leads to enhanced sprouting and reinnervation (Aigner *et al.*, 1995), whereas the GAP-43 –/– mouse exhibits high postnatal mortality and retinal ganglion axonal pathfinding defects, without changes in nerve growth rate (Strittmatter *et al.*, 1995), suggesting that the presence of GAP-43 promotes outgrowth, but the absence does not exclude outgrowth. Transgenic overexpression of both CAP-23 and GAP-43 allows a limited degree of CNS sensory fiber regeneration (Bomze *et al.*, 2001). Methods for pharmacologically mimicking GAP-43 action may have potential as axon regenerative therapeutics (Strittmatter *et al.*, 1994).

SPRR1A

Identified by microarray analysis for up-regulated genes after mouse sciatic nerve injury, SPRR1A is a member of the small proline-rich family (Bonilla *et al.*, 2002). SPRR1A expression after axotomy is unexpected because SPRR1A, found in the cross-linked cornified epithelium, is a highly specific marker for the differentiation of keratinocytes and squamous epithelial cells. Structurally, SPRR1A is composed of a repeating proline-rich octapeptide sequence. *In vitro*, SPRR1A expressing neurons exhibit outgrowth comparable to conditioning axotomy supporting the idea that acute SPRR1A induction is a part of a peripheral nerve regeneration program. Unlike GAP-43, SPRR1A is not detected in developing neurons, suggesting that there may be differences between gene expression during developmental axonal extension and regenerative axonal growth. As for GAP-43, a SPRR1A-mimetic may be a potential therapeutic class to regenerate axons. Further understanding will come from how RAG signaling converges to promote outgrowth.

cAMP

Filbin and colleagues discovered that endogenous cAMP levels influence the responsiveness of dorsal root ganglion to myelin inhibition (Cai *et al.*, 2001; Cai *et al.*, 1999). In addition, inhibiting a downstream cAMP effector, protein kinase A (PKA), prevents myelin and MAG inhibition of neurite outgrowth. Therefore, cAMP modulation represents another control point for promote axonal extension (Song *et al.*, 1998).

Protein Kinase A

Two lines of evidence supporting cAMP's role in modulating inhibition are that endogenous DRG cAMP levels decrease postnatally and dorsal root ganglion cAMP levels increase (Cai *et al.*, 2001). The first line of evidence is consistent with the notion that high levels of endogenous cAMP in young DRG neurons account for the neurite outgrowth promotion (Qiu *et al.*, 2002). The second line suggests that reconstitution of some regeneration program may be mediated by cAMP. Investigation into downstream cAMP effects have led to two potential effectors, PKA and Arginase I (Cai *et al.*, 2002). Increased cAMP levels have the potential to directly alter the actin cytoskeleton by coupling momomeric G proteins (Bos, 2003). Second, transient cAMP increases with protein kinase A (PKA) activation and CREB have been implicated in gene regulation. A gene that has been shown to be up-regulated after cAMP elevation is Arginase I (Arg I).

Arginase I and Polyamines

cAMP elevation by direct injection into DRG overcomes MAG and myelin inhibition and results in a moderate degree of dorsal column axon regeneration for mice lesioned 1 week later. This process involves PKA, but eventually becomes PKA-independent, suggesting possible transcriptional changes. One potential explanation for this change is that cAMP up-regulates Arginase I, a key enzyme involved in polyamine synthesis. Both Arginase I overexpression and exogenous polyamines addition can overcome MAG and myelin inhibition, *in vitro*, whereas blocking Arginase I or ornithine decarboxylase (Odc) prevents elevated cAMP from overcoming inhibition (Cai *et al.*, 2002; Chu *et al.*, 1995).

Clinical Application

Until gene therapy becomes a safer reality, therapies that directly modulate RAG expression such as GAP-43 need to overcome many technical and regulatory obstacles. Therefore, murine models that overexpress intrinsic stimulatory factors with poor CNS accessibility are good starting points to identify relevant targets, but for clinical applications, agents that modulate accessible targets such as db-cAMP are more relevant.

db-cAMP Modulation

Several compounds that increase cAMP levels *in vivo* by blocking specific cAMP phosphodiesterase have been developed. Cell-permeable dibutyryl cAMP mimics cAMP itself, and promotes the regeneration of ascending sensory fibers across a spinal injury site when injected near the DRG cell body (Neumann *et al.*, 2002; Qiu *et al.*, 2002). Rolipram is a specific inhibitor of the brain-enriched type 4 cAMP phosphodiesterase. It was originally developed as an antidepressant, but it may provide a means to elevate cAMP in axotomized CNS neurons. Ongoing tests will explain their efficacy and specificity for promoting axon regeneration *in vivo*.

POTENTIAL SIDE EFFECTS OF AXONAL REGENERATION

If one or more of the above approaches leads to clinical axon regeneration, then there must be concern for potential side effects. A primary concern has to relate to the aberrant and deleterious connections. Such connections could potentially lead to neuropathic pain and/or epilepsy. To date, such side effects have not been detected in any of the rodent spinal injury studies despite the increase in axon regrowth. For various novel methods, continued vigilance for such side effects is warranted.

Other potential side effects relate to alterations in biology other than axon growth. *A priori*, upstream disruptors such as IN-1, NEP 1-40, Ecto-NgR, and ChABC appear least likely to cause unintended side effects. Downstream inactivation of the Rho signaling pathway, alterations in RAGs or modulation of cAMP seem more prone to alter the function of other cellular events and other cell types. A full evaluation of their side effects is necessary.

CONCLUSION

More than two decades have passed since Aguayo and colleagues performed their classic transplantation studies, and since then the spinal cord regeneration field has produced promising targets for drug discovery. Early successes in promoting axon regeneration by Nogo/NgR blockade, CSPG digestion, ROCK inhibition, cAMP regulation, and cell transplantation are encouraging. It is quite likely that optimal future approaches will involve combinatorial treatments that remove endogenous regeneration roadblocks and encourage the regenerative capacity of injured and non-injured axons. It is reasonable to hope that, in the not too distant future, neurologists and neurosurgeons will be able to offer their patients with SCI new restorative treatments.

ACKNOWLEDGMENTS

This work was supported by grants to S.M.S. from the NIH and the McKnight Foundation for Neuroscience. S.M.S. is an Investigator of the Patrick and Catherine Weldon Donaghue Medical Research Foundation.

References

Aigner L *et al.* (1995). Overexpression of the neural growth-associated protein GAP-43 induces nerve sprouting in the adult nervous system of transgenic mice, *Cell* 83:269-278.

Barton WA *et al.* (2003). Structure and axon outgrowth inhibitor binding of the Nogo-66 receptor and related proteins, *EMBO J* 22:3291-3302.

Bartsch U *et al.* (1995). Lack of evidence that myelin-associated glycoprotein is a major inhibitor of axonal regeneration in the CNS, *Neuron* 15:1375-1381.

Bomze HM *et al.* (2001). Spinal axon regeneration evoked by replacing two growth cone proteins in adult neurons, *Nat Neurosci* 4:38-43.

Bonilla IE, Tanabe K, Strittmatter SM. (2002). Small proline-rich repeat protein 1A is expressed by axotomized neurons and promotes axonal outgrowth, *J Neurosci* 22:1303-1315.

Bos JL. (2003). Epac: a new cAMP target and new avenues in cAMP research, *Nat Rev Mol Cell Biol* 4:733-738.

Bradbury EJ *et al.* (1999). NT-3 promotes growth of lesioned adult rat sensory axons ascending in the dorsal columns of the spinal cord, *Eur J Neurosci* 11:3873-3883.

Bradbury EJ *et al.* (2002). Chondroitinase ABC promotes functional recovery after spinal cord injury, *Nature* 416:636-640.

Bregman BS *et al.* (1995). Recovery from spinal cord injury mediated by antibodies to neurite growth inhibitors, *Nature* 378:498-501.

Cai D *et al.* (2002). Arginase I and polyamines act downstream from cyclic AMP in overcoming inhibition of axonal growth MAG and myelin in vitro, *Neuron* 35:711-719.

Cai D *et al.* (2001). Neuronal cyclic AMP controls the developmental loss in ability of axons to regenerate, *J Neurosci* 21:4731-4739.

Cai D *et al.* (1999). Prior exposure to neurotrophins blocks inhibition of axonal regeneration by MAG and myelin via a cAMP-dependent mechanism, *Neuron* 22:89-101.

Caroni P. (1997). Overexpression of growth-associated proteins in the neurons of adult transgenic mice, *J Neurosci Methods* 71:3-9.

Caroni P, Schwab ME. (1988a). Antibody against myelin-associated inhibitor of neurite growth neutralizes nonpermissive substrate properties of CNS white matter, *Neuron* 1:85-96.

Caroni P, Schwab ME. (1988b). Two membrane protein fractions from rat central myelin with inhibitory properties for neurite growth and fibroblast spreading, *J Cell Biol* 106:1281-1288.

Chen MS *et al.* (2000). Nogo-A is a myelin-associated neurite outgrowth inhibitor and an antigen for monoclonal antibody IN-1, *Nature* 403:434-439.

Chu PJ, Saito H, Abe K. (1995). Polyamines promote regeneration of injured axons of cultured rat hippocampal neurons, *Brain Res* 673:233-241.

Condic ML, Lemons ML. (2002). Extracellular matrix in spinal cord regeneration: getting beyond attraction and inhibition, *Neuroreport* 13:A37-48.

David S, Aguayo AJ. (1981). Axonal elongation into peripheral nervous system "bridges" after central nervous system injury in adult rats, *Science* 214:931-933.

Davies SJ *et al.* (1997). Regeneration of adult axons in white matter tracts of the central nervous system, *Nature* 390:680-683.

Dergham P *et al.* (2002). Rho signaling pathway targeted to promote spinal cord repair, *J Neurosci* 22:6570-6577.

Domeniconi M *et al.* (2002). Myelin-associated glycoprotein interacts with the Nogo66 receptor to inhibit neurite outgrowth, *Neuron* 35:283-290.

Dou CL, Levine JM. (1994). Inhibition of neurite growth by the NG2 chondroitin sulfate proteoglycan, *J Neurosci* 14:7616-7628.

Fawcett JW, Asher RA. (1999). The glial scar and central nervous system repair, *Brain Res Bull* 49:377-391.

Fidler PS *et al.* (1999). Comparing astrocytic cell lines that are inhibitory or permissive for axon growth: the major axon-inhibitory proteoglycan is NG2, *J Neurosci* 19:8778-8788.

Fournier AE, Gould GC, Liu BP, Strittmatter SM. (2002). Truncated soluble Nogo receptor binds Nogo-66 and blocks inhibition of axon growth by myelin, *J Neurosci* 22:8876-8883.

Fournier AE, GrandPre T, Strittmatter SM. (2001). Identification of a receptor mediating Nogo-66 inhibition of axonal regeneration, *Nature* 409:341-346.

Fournier AE, Takizawa BT, Strittmatter SM. (2003). Rho kinase inhibition enhances axonal regeneration in the injured CNS, *J Neurosci* 23:1416-1423.

Frey D *et al.* (2000). Shared and unique roles of CAP23 and GAP43 in actin regulation, neurite outgrowth, and anatomical plasticity, *J Cell Biol* 149:1443-1454.

GrandPre T, Li S, Strittmatter SM. (2002). Nogo-66 receptor antagonist peptide promotes axonal regeneration, *Nature* 417:547-551.

GrandPre T, Nakamura F, Vartanian T, Strittmatter, S. M. (2000). Identification of the Nogo inhibitor of axon regeneration as a Reticulon protein, *Nature* 403:439-444.

Grill R *et al.* (1997). Cellular delivery of neurotrophin-3 promotes corticospinal axonal growth and partial functional recovery after spinal cord injury, *J Neurosci* 17:5560-5572.

Hall A. (1998). G proteins and small GTPases: distant relatives keep in touch, *Science* 280:2074-2075.

He XL *et al.* (2003). Structure of the nogo receptor ectodomain: a recognition module implicated in myelin inhibition, *Neuron* 38:177-185.

Huang DW, McKerracher L, Braun PE, David S. (1999). A therapeutic vaccine approach to stimulate axon regeneration in the adult mammalian spinal cord, *Neuron* 24:639-647.

Huber AB, Schwab ME. (2000). Nogo-A, a potent inhibitor of neurite outgrowth and regeneration, *Biol Chem* 381:407-419.

Jin Z, Strittmatter SM. (1997). Rac1 mediates collapsin-1-induced growth cone collapse, *J Neurosci* 17:6256-6263.

Johnson PW *et al.* (1989). Recombinant myelin-associated glycoprotein confers neural adhesion and neurite outgrowth function, *Neuron* 3:377-385.

Jones LL, Sajed D, Tuszynski MH. (2003). Axonal regeneration through regions of chondroitin sulfate proteoglycan deposition after spinal cord injury: a balance of permissiveness and inhibition, *J Neurosci* 23:9276-9288.

Kalil K, Skene JH. (1986). Elevated synthesis of an axonally transported protein correlates with axon outgrowth in normal and injured pyramidal tracts, *J Neurosci* 6:2563-2570.

Kim JE *et al.* (2003). Axon regeneration in young adult mice lacking Nogo-A/B, *Neuron* 38:187-199.

Lehmann M *et al.* (1999). Inactivation of Rho signaling pathway promotes CNS axon regeneration, *J Neurosci* 19:7537-7547.

Lemons ML, Howland DR, Anderson DK. (1999). Chondroitin sulfate proteoglycan immunoreactivity increases following spinal cord injury and transplantation, *Exp Neurol* 160:51-65.

Li S, Strittmatter SM. (2003). Delayed systemic Nogo-66 receptor antagonist promotes recovery from spinal cord injury, *J Neurosci* 23:4219-4227.

Liu BP, Fournier A, GrandPre T, Strittmatter SM. (2002). Myelin-associated glycoprotein as a functional ligand for the Nogo-66 receptor, *Science* 297:1190-1193.

McKerracher L *et al.* (1994). Identification of myelin-associated glycoprotein as a major myelin-derived inhibitor of neurite growth, *Neuron* 13:805-811.

Merkler D *et al.* (2001). Locomotor recovery in spinal cord-injured rats treated with an antibody neutralizing the myelin-associated neurite growth inhibitor Nogo-A, *J Neurosci* 21:3665-3673.

Monnier PP *et al.* (2003). The Rho/ROCK pathway mediates neurite growth-inhibitory activity associated with the chondroitin sulfate proteoglycans of the CNS glial scar, *Mol Cell Neurosci* 22:319-330.

Moon LD, Asher RA, Rhodes KE, Fawcett JW. (2001). Regeneration of CNS axons back to their target following treatment of adult rat brain with chondroitinase ABC, *Nat Neurosci* 4:465-466.

Mukhopadhyay G *et al.* (1994). A novel role for myelin-associated glycoprotein as an inhibitor of axonal regeneration, *Neuron* 13:757-767.

Neumann S, Bradke F, Tessier-Lavigne M, Basbaum AI. (2002). Regeneration of sensory axons within the injured spinal cord induced by intraganglionic cAMP elevation, *Neuron* 34:885-893.

Neumann S, Woolf CJ. (1999). Regeneration of dorsal column fibers into and beyond the lesion site following adult spinal cord injury, *Neuron* 23:83-91.

Niederost B *et al.* (2002). Nogo-A and myelin-associated glycoprotein mediate neurite growth inhibition by antagonistic regulation of RhoA and Rac1, *J Neurosci* 22:10368-10376.

Niederost BP, Zimmermann DR, Schwab ME., Bandtlow CE. (1999). Bovine CNS myelin contains neurite growth-inhibitory activity associated with chondroitin sulfate proteoglycans, *J Neurosci* 19:8979-8989.

Oertle T *et al.* (2003). Nogo-A inhibits neurite outgrowth and cell spreading with three discrete regions, *J Neurosci* 23:5393-5406.

Prinjha R *et al.* (2000). Inhibitor of neurite outgrowth in humans, *Nature* 403:383-384.

Qiu J *et al.* (2002). Spinal axon regeneration induced by elevation of cyclic AMP, *Neuron* 34:895-903.

Ramon-Cueto A, Cordero MI, Santos-Benito FF, Avila J. (2000). Functional recovery of paraplegic rats and motor axon regeneration in their spinal cords by olfactory ensheathing glia, *Neuron* 25:425-435.

Resnick DK *et al.* (2003). Adult olfactory ensheathing cell transplantation for acute spinal cord injury, *J Neurotrauma* 20:279-285.

Richardson PM, Issa VM. (1984). Peripheral injury enhances central regeneration of primary sensory neurones, *Nature* 309:791-793.

Ruitenberg MJ *et al.* (2003). Ex vivo adenoviral vector-mediated neurotrophin gene transfer to olfactory ensheathing glia: effects on rubrospinal tract regeneration, lesion size, and functional recovery after implantation in the injured rat spinal cord, *J Neurosci* 23:7045-7058.

Schnell L *et al.* (1994). Neurotrophin-3 enhances sprouting of corticospinal tract during development and after adult spinal cord lesion, *Nature* 367:170-173.

Simonen M *et al.* (2003). Systemic deletion of the myelin-associated outgrowth inhibitor nogo-a improves regenerative and plastic responses after spinal cord injury, *Neuron* 38:201–211.

Snow DM *et al.* (1990). Sulfated proteoglycans in astroglial barriers inhibit neurite outgrowth in vitro, *Exp Neurol* 109:111-130.

Song H *et al.* (1998). Conversion of neuronal growth cone responses from repulsion to attraction by cyclic nucleotides, *Science* 281:1515-1518.

Strittmatter SM *et al.* (1995). Neuronal pathfinding is abnormal in mice lacking the neuronal growth cone protein GAP-43, *Cell* 80:445-452.

Strittmatter SM, Igarashi M, Fishman MC. (1994). GAP-43 amino terminal peptides modulate growth cone morphology and neurite outgrowth, *J Neurosci* 14:5503-5513.

Tanabe K, Bonilla I, Winkles JA, Strittmatter SM. (2003). Fibroblast growth factor-inducible-14 is induced in axotomized neurons and promotes neurite outgrowth, *J Neurosci* 23:9675-9686.

Tansey MG, Baloh RH, Milbrandt J, Johnson EM Jr. (2000). GFRalpha-mediated localization of RET to lipid rafts is required for effective downstream signaling, differentiation, and neuronal survival, *Neuron* 25:611-623.

Wang HU, Anderson DJ. (1997). Eph family transmembrane ligands can mediate repulsive guidance of trunk neural crest migration and motor axon outgrowth, *Neuron* 18:383-396.

Wang KC *et al.* (2002a). P75 interacts with the Nogo receptor as a co-receptor for Nogo, MAG and OMgp, *Nature* 420:74-78.

Wang KC *et al.* (2002b). Oligodendrocyte-myelin glycoprotein is a Nogo receptor ligand that inhibits neurite outgrowth, *Nature* 417:941-944.

Winton MJ *et al.* (2002). Characterization of new cell permeable C3-like proteins that inactivate Rho and stimulate neurite outgrowth on inhibitory substrates, *J Biol Chem* 277:32820-32829.

Yamashita T, Tucker KL, Barde YA. (1999). Neurotrophin binding to the p75 receptor modulates Rho activity and axonal outgrowth, *Neuron* 24:585-593.

Zheng B *et al.* (2003). Lack of enhanced spinal regeneration in nogo-deficient mice, *Neuron* 38:213–224.

27

Alzheimer's Disease: Clinical Features, Neuropathological and Biochemical Abnormalities, Genetics, Models, and Experimental Therapeutics

Donald L. Price, MD
David R. Borchelt, PhD
Philip C. Wong, PhD

INTRODUCTION

This review focuses on Alzheimer's disease (AD), the most common age-related dementia, and on the value of experimental models for understanding disease mechanisms and for insights into experimental therapeutics. AD affects more than 4 million people in the United States. Because of increased life expectancy and the postwar baby boom, the elderly are the fastest growing segment of our society, and during the next 25 years, the number of people with AD in the United States will triple, as will the cost (Mayeux 2003; Petersen 2000; Brookmeyer et al., 1998). Because of its high prevalence, cost, lack of mechanism-based treatments, and impact on individuals and caregivers, AD is one of the most challenging diseases in medicine and development of new effective therapies will have a significant impact on the health and care of the elderly (Brookmeyer et al., 1998; Mayeux, 2003; Selkoe, 2001; Price et al., 1998b; Wong, et al., 2002).

The classical clinical phenotype of AD, a late life progressive disorder of memory and cognition, is quite distinct and results from pathology associated with dysfunction and death of specific populations of neurons, particularly those systems involved in the functions impaired in AD (Price et al., 1998b). Characteristic intracellular and extracellular protein aggregates (tau and Aβ-related abnormalities, respectively), implicated in pathogenic processes, are critical elements of this pathology (Wong et al., 2002; Hardy et al., 2002; Lee et al., 2001; Selkoe 2001). Genetic evidence indicates that inheritance of mutations in several genes causes autosomal dominant familial AD (FAD), while the presence of certain alleles of other genes, particularly ApoE, are significant risk factors for putative sporadic disease (Mayeux 2003; Tanzi et al., 2001; Price et al., 1998b). This information has been used to create models of disease (i.e., mice expressing mutant transgenes). In parallel, targeting of genes encoding proteins thought to be implicated in disease pathways has provided new understanding of the roles of specific gene products in AD and the potential of these proteins as therapeutic targets. The value of these targets for new treatment strategies is being tested in model systems and, once safety and efficacy are assured, in human trials.

To illustrate some of these concepts, we first describe the clinical, pathological, biochemical, and genetic features of the human illness. Subsequently, we discuss selected aspects: the biology of proteins implicated in pathogenesis of disease; the value of genetically engineered models; the identification of new therapeutic targets; and experimental treatments in models.

CLINICAL FEATURES

AD is the most common cause of dementia occurring in the elderly (Mayeux 2003; Selkoe 2001; Price et al., 1998b; Wong et al., 2002), a term that refers to a syndrome associated with memory loss and cognitive impairments of sufficient severity to interfere with social, occupational, and personal functions (Petersen 2000; Petersen et al., 2001a; Morris et al., 2001b; Petersen 2003). AD often initially manifests as a syndrome termed mild cognitive impairment (MCI), which is usually characterized by: a memory complaint (corroborated by informant); memory impairments on formal testing; intact general cognition; preserved activities of daily events; and absence of overt dementia (Petersen, 2003). MCI, particularly of the amnesic variety, is regarded as a transitional stage between normal aging and early AD (Petersen 2003; Petersen *et al.*, 1999; Petersen *et al.*, 2001a; Morris *et al.*, 2001b; Petersen *et al.*, 2001b; Petersen 2003). The clinical manifestations of AD include difficulties in memory and in domains of other cognitive functions (executive functions, language, attention, judgment, etc.) (Petersen *et al.*, 2000; Albert *et al.*, 2001; Petersen *et al.*, 2001a; Daly *et al.*, 2000; Price *et al.*, 1999; Morris *et al.*, 2001b; Morris *et al.*, 2001a; Killiany *et al.*, 2000; Killiany *et al.*, 2002; Petersen *et al.*, 1999; Petersen *et al.*, 2001b; Petersen, 2000). Psychotic symptoms develop in some patients, such as depression/hallucinations and delusions (Cummings, 2003; Lyketsos *et al.*, 2002). Over time, mental functions and activities of daily living are increasingly impaired, and in the late stages, these individuals become profoundly demented and bedridden and usually die of intercurrent illness.

To make a diagnosis of AD, clinicians rely on histories from patients and informants, physical, neurological, and psychiatric examinations, neuropsychological testing (Daly *et al.*, 2000; Albert *et al.*, 2001), laboratory studies (Sunderland *et al.*, 2003), and a variety of other approaches, including neuroimaging studies (see subsequent discussion). In the cerebrospinal fluid cases of AD, the levels of Aβ peptide are often low, and levels of tau may be higher than those in controls (Sunderland *et al.*, 2003). Values vary between individuals and single measures may not be great diagnostic value. However, future research should disclose whether serial measures of CSF markers may have diagnostic utility (Petersen 2000). Over several examinations, the clinical profile, in concert with these laboratory assessments, allows the clinician to make a diagnosis of possible or probable AD (McKhann *et al.*, 1984).

Imaging approaches promise to be increasingly valuable for aid in diagnosis and for assessing outcomes of new therapies. In cases of AD, magnetic resonance imaging (MRI) often discloses regional brain atrophy, particularly involving hippocampus (Jack *et al.*, 2000) and entorhinal cortex) (Killiany *et al.*, 2002; Petersen *et al.*, 2000). Rates of hippocampal atrophy appear to correlate with changes in clinical status (Jack *et al.*, 2000) and may have predictive value for diagnosis of AD (Killiany *et al.*, 2000). In AD, positron emission tomography (PET) and single photon emission computerized tomography (SPECT) commonly demonstrate decreased regional blood flow in the parietal and temporal lobes with involvement of other cortical areas at later stages (Klunk *et al.*, 2001; Jack, *et al.*, 2000). Recently, a Pittsburgh-Swedish consortium has used Pittsburgh Compound B (PIB), a brain penetrant ^{18}F-labeled uncharged thioflavin

derivative, which binds to Aβ with high affinity, to visualize, via PET, the Aβ burden in the brains of affected individuals (Klunk *et al.*, 2001; Klunk *et al.*, 2003; Klunk *et al.*, 2004). In comparison with controls, subjects with AD show marked retention of PIB in areas of brain known to accumulate amyloid (Klunk *et al.*, 2004). This approach (and others under development) should eventually prove useful for enhancing accuracy of diagnosis of early AD and for assessing the efficacies of anti-amyloid therapeutics in both animal models and human patients. Although AD is a disorder for which only symptomatic treatments are available (Petersen, 2003), we are on the threshold of new mechanism-based therapies, and, in the future, our ability to make early and accurate diagnoses of AD will become increasingly important for identifying the clinical populations most likely to respond to specific treatments.

NEUROPATHOLOGY AND BIOCHEMISTRY

Brain Regions/Neural Systems

The clinical signs of AD reflect that pathology associated with the selective dysfunction and death of populations of neurons in brain regions/systems critical for memory, learning, and cognitive performance. These circuits include the basal forebrain cholinergic system, hippocampus, entorhinal cortex, limbic cortex, and neocortex (Whitehouse *et al.*, 1981; Beach *et al.*, 2000; Morris *et al.*, 2001b; Price *et al.*, 1999; Hyman *et al.*, 1984; Braak *et al.*, 1996; Gomez-Isla *et al.*, 1996; Braak *et al.*, 1991). Degeneration of neurons in these regions/circuits are reflected by the presence of cytoskeletal abnormalities in these cells as manifested by neurofibrillary pathology and dystrophic neurites (Lee *et al.*, 2001); the presence of neuritic Aβ-containing plaques (sites of synaptic disconnection) in brain regions receiving inputs from these nerve cells (Selkoe 2001; Sisodia *et al.*, 2002; Price *et al.*, 1999); reductions in both generic and transmitter specific synaptic markers of these transmitter systems in the target fields of these neurons (Beach *et al.*, 2000; Sze *et al.*, 1997; Price *et al.*, 1998a; Masliah *et al.*, 1994; Terry *et al.*, 1991); loss of neurons in these regions (Mattson 2001; Whitehouse *et al.*, 1981; Roth 2001); and local glial/inflammatory reactions (particularly associated with plaques) (Akiyama *et al.*, 2000). Disruption of synaptic communication in these circuits, associated eventually with degeneration of neurons, has profound clinical consequences. Abnormalities that damage the circuits involving the entorhinal cortex, hippocampus, and medial temporal cortex are presumed to make critical contributions to memory impairments. Pathology in the neocortex is reflected by deficits in higher cognitive functions, such as disturbances in language, calculation, problem solving, and judgment. Alterations in the basal forebrain cholinergic system may contribute to memory difficulties and attention deficits, while the behavioral and emotional disturbances may reflect involvement of the limbic cortex, amygdala, thalamus, and monoaminergic systems.

Amyloid and Plaques

Aβ, a 4-kD peptide, is derived by cleavages of the amyloid precursor protein (APP) (Selkoe 2001; Hardy et al., 2002; Sisodia et al., 2002; Price et al., 1998b) to generate Aβ1-40, 42 and 11-40, 42 amyloid peptides (Cai et al., 2001). These extracellular peptides accumulate in the neuropil of the neocortex and hippocampus in cases of AD (Price et al., 1998b; Wong et al., 2002; Hardy et al., 2002; Sisodia et al., 2002; Selkoe 2001). In neurons, APP is rapidly transported in axons (Buxbaum et al., 1998; Koo et al., 1990; Sisodia et al., 1993), and Aβ appears to be principally produced at terminals by several cleavage activities (see subsequent section) (Wong et al., 2001; Wolfe 2002; Sisodia et al., 2002). Aβ is secreted as monomers, but oligomers and multimers, particularly involving Aβ42 (Iwatsubo et al., 1994) appear to be the toxic moiety (Klein et al., 2001; Selkoe 2002; Walsh et al., 2002). These peptides adopt β-pleated sheet conformations and assemble into protofilaments and fibrils (Caughey et al., 2003); peptide aggregates are birefringent when stained with Congo Red (or thioflavine) and viewed in polarized light (or fluorescence). Neurites are swollen axons/terminals containing filamentous and membranous organelles; they occur in proximity to amyloid deposits (Price et al., 1998b; Sisodia, et al., 2002). Synaptic terminals are thought to be a major source of APP that gives rise to Aβ (Wong et al., 2001); Thus, synapses release Aβ and are, in turn, damaged by elevated levels of Aβ peptides; presynaptic terminals become disconnected from targets and form the neurites in plaques.

APP, a single pass transmembrane protein, is encoded by a gene localized to the mid-portion of the long arm of human chromosome 21 (Price et al., 1999; Price et al., 1998b; Sisodia et al., 2002). Enriched in neurons, APP is transported anterograde in axons to terminals (Buxbaum et al., 1998; Koo et al., 1990; Sisodia et al., 1993). It is hypothesized that the proteins responsible for β- and γ-secretase activities are also present at terminals in CNS circuits. This concept is supported by immunocytochemical studies interpreted to show colocalization of APP with several secretase proteins in neurites (Sheng et al., 2003). At these sites, β-secretase (see below) cleaves APP to form amyloidogenic C-terminal derivatives (Buxbaum et al., 1998), which are then further cleaved by γ-secretase to generate Aβ 40, 42, 43 peptides. Released at terminals,

Aβ may serve as a synaptic modulator depressing activity at excitatory, glutaminergic synapses via the NMDA receptor (Kamenetz et al., 2003). It has been suggested that accumulation of Aβ oligomers and multimers (Walsh et al., 2002; Selkoe 2002; Klein et al., 2001; Gong et al., 2003), released in proximity to synapses, act as toxic disrupting synaptic functions, including long-term potentiation (LTP) (Walsh et al., 2002; Hsia et al., 1999). Eventually, these terminals degenerate. As described previously, the disconnected terminals form neurites around Aβ fibrillar deposits. Thus, neuritic amyloid plaques, a classical feature of AD, are complex structures, representing sites of synaptic disconnection. Surrounding plaques are astrocytes and microglia which produce cytokines, chemokines, and other factors (including complement components) involved in inflammatory processes (Akiyama et al., 2000).

Neurofibrillary Pathology

Neurofibrillary tangles (NFT) are fibrillar intracytoplasmic inclusions (in cell bodies/proximal dendrites of affected neurons), while neuropil threads and neurites are swollen filament-containing dendrites and distal axons/terminals, respectively. The principal intracellular component of these lesions are paired helical filaments (PHF) comprised of poorly soluble hyperphosphorylated isoforms of tau, a low-molecular-weight microtubule-associated protein (Lee et al., 2001; Goedert et al., 1992). In mature human brains, the six isoforms of tau are derived by alternative splicing from a single gene on chromosome 17 (Lee et al., 2001; Edbauer et al., 2002; Goedert et al., 1992). The six tau isoforms consist of three isoforms of three repeat tau (3-R tau), and three isoforms of four repeat (4-R) tau, the latter derived by inclusion of exon 10 in the transcript (Lee et al., 2001). Normally, tau is synthesized in neuronal cell bodies and transported anterograde in axons (Mercken et al., 1995), where it acts, via repeat regions that interact with tubulin, to stabilize tubulin polymers critical for microtubule assembly and stability (Lee et al., 2001; Edbauer et al., 2002). The post-translationally modified tau underlying the cellular pathology in some neurodegenerative disorders differs somewhat in the different tauopathies: in cases of AD, the PHF are comprised of six isoforms of tau; the inclusions occurring in cases of progressive supranuclear palsy and cortical basal degeneration (PSP and CBD) are characterized by 4-R tau; and the inclusions seen in individuals with Pick disease are enriched in 3-R tau (Lee et al., 2001).

All of the AD-linked fibrillar inclusions (NFT, neurites and threads) appear to result from common mechanisms (Lee et al., 2001). However, in spite of considerable clinical, pathological research and experimental work, the links between tau pathology and Aβ are not fully defined (Busciglio *et al.*, 1995; Mattson, 2001; Lee *et al.*, 2001; Delacourte *et al.*, 1998; Giasson *et al.*, 2003; Hardy *et al.*, 2002). Because the cytoskeleton is essential for maintaining cell geometry and for the intracellular trafficking and transport of proteins and organelles, it is likely that disturbances of the cytoskeleton are associated with alterations in axonal transport, which in turn could compromise the functions and viability of neurons. Eventually, affected nerve cells die (possibly by apoptosis) (Gastard *et al.*, 2003; Whitehouse *et al.*, 1981; Mattson, 2001; Troncoso *et al.*, 1996; Roth, 2001), and extracellular neurofibrillary tangles are left behind as "tombstones" of the nerve cells destroyed by disease.

GENETICS

Genetic risk factors for AD (Tanzi *et al.*, 2001; Mayeux, 2003; Sisodia *et al.*, 2002; Price *et al.*, 1998b) include mutations in the *presenilin 1* (*PS1*) (chromosome 14); mutations in the *presenilin 2* (*PS2*) (chromosome 1); and mutations in the *APP* (chromosome 21); and different alleles of ApoE, which is positioned on the proximal long arm of chromosome 19. The presence of autosomal dominant mutations in *APP*, *PS1*, or *PS2* tends to cause disease earlier than occurs in sporadic cases. Individuals with trisomy 21 (Down's syndrome) have an extra copy of *APP* and AD pathology develops early in life. ApoE E4 predisposes to later onset AD and some cases of late-onset familial AD (Strittmatter *et al.*, 1993; Corder *et al.*, 1994). Recent research has identified other loci that confer risk (Tanzi *et al.*, 2001; Mayeux, 2003; Ertekin-Taner *et al.*, 2000; Myers *et al.*, 2000; Bertram *et al.*, 2000).

APP, a type I transmembrane protein existing as several isoforms, is abundant in the nervous system, but specific functions of APP remain to be defined. Expressed in many different cell types, APP is particularly abundant in neurons. The protein is cleaved by activities of BACE1 (*β-site APP cleaving enzyme 1*) and the γ-secretase complex (see subsequent section), which generate the N- and C-termini of Aβ peptides, respectively (Vassar *et al.*, 2000; Sisodia *et al.*, 2002; Vassar *et al.*, 1999; Cai *et al.*, 2001; Vassar *et al.*, 2000; Sisodia *et al.*, 2002). The majority of mutations in *APP*, *PS1*, and *PS* influence these cleavages to increase the levels of all Aβ species or the relative amounts of toxic Aβ42 (Tanzi *et al.*, 2001; Sisodia *et al.*, 2002; Sisodia *et al.*, 2002). For example, the *APPswe* mutation, a double mutation involving codons 670 and 671, enhances many fold BACE1 cleavage at the N-terminus of Aβ; the result is substantial elevation in levels of all Aβ 1-40, 42 peptides. In several families, the normally occurring valine residue at position 717 (of APP-770) is replaced with Ile, Gly, or Phe; these

mutations promote γ-secretase activities and lead to increased secretion of Aa1- and 11-42,43. In a hereditary disease associated with Aβ deposition around blood vessels and cerebral hemorrhages, an *APP* mutation at position 693 leads to a Glu-Gln substitution (corresponding to amino acid 22 of Aβ), which, in turn, is associated with formation of Aβ peptide species that are more prone to aggregate into fibrils. Thus, some of the mutations linked to FAD can change the processing of APP and influence the biology of Aβ by increasing the production of Aβ peptides or the amounts of the longer, more toxic Aβ42, while other mutations may promote fibril formation.

PS1 and *PS2* encode two highly homologous, and conserved 43- to 50-kD multipass transmembrane proteins (Doan *et al.*, 1996; Sherrington *et al.*, 1995; Sisodia *et al.*, 2002; Price *et al.*, 1998b) that are involved in Notch 1 signaling pathways critical for cell fate decisions (Kopan *et al.*, 2002; Sisodia *et al.*, 2002). PS1 and PS2 are endoproteolytically cleaved to an N-terminal ~28-kDa fragment and a C-terminal ~18-kDa fragment (Thinakaran *et al.*, 1997; Doan *et al.*, 1996), both of which are critical components of the γ-secretase complex (Takasugi *et al.*, 2003; Sisodia *et al.*, 2002, Wolfe, 2002). Other components of the γ-secretase complex include: Nicastrin (Nct), a type I transmembrane glycoprotein (Sisodia *et al.*, 2002; Kopan *et al.*, 2002; De Strooper, 2003; Edbauer *et al.*, 2002; Li *et al.*, 2003; Kopan *et al.*, 2002; Yu *et al.*, 2000); and Aph1 and Pen2, two multipass transmembrane proteins (Francis *et al.*, 2002; Steiner *et al.*, 2002; Goutte *et al.*, 2002).

Significantly, γ-secretase cleaves both Notch-1 and APP (Σ site), generating intracellular peptides termed *Notch1 intracellular domain* (NICD) and *APP Intracellular domain* (AICD) (Kopan *et al.*, 2002; Cao *et al.*, 2001). These intracellular peptides traffic to the nucleus where they are involved in activation of transcription (Cao *et al.*, 2001). For example, AICD interacts with FE65, a cytosolic adapter, and moves into the nucleus where it influences transcription (Cao *et al.*, 2001). Studies of targeting of *PS1* and *Nct* in mice (Wong *et al.*, 1997; Shen *et al.*, 1997; Liet al., 2003; Kopan *et al.*, 2002) are consistent with the concept that PS1 and Nct are, along with Aph1 and Pen2 (Takasugi *et al.*, 2003; De Strooper, 2003; Capell *et al.*, 2000), critical components of the γ-secretase complex. As discussed subsequently, the phenotypes of targeted *PS1* and *Nct* are the result of failed Notch1 signaling.

Nearly 50% of cases of early-onset familial AD are linked to the *PS1* gene (chromosome 14), which has been reported to harbor >85 different FAD mutations (Tanzi *et al.*, 2001; Sisodia *et al.*, 2002; Sherrington *et al.*, 1995; Sisodia *et al.*, 2002; Tanzi *et al.*, 2001; Price *et al.*, 1998b). A small number of mutations in the *PS2* gene cause autosomal dominant AD in several FAD pedigrees (Price *et al.,* 1998b; Tanzi *et al.,* 2001). The majority of abnormalities in *PS* genes are missense mutations that result in single amino acid substitutions, which enhance γ-secretase activities and increase the levels of the Aβ42 peptides.

ApoE, a glycoprotein that carries cholesterol and other lipids in the blood, has been implicated as a risk factor for the disease (Mayeux, 2003; Corder *et al.,* 1994; Strittmatter *et al.,* 1993; Tanzi *et al.,* 2001). At the single ApoE locus there are three alleles: ApoE2, apoE3, and apoE4 (Holtzman, 2001; Corder *et al.,* 1994). The apoE3 allele is most common in the general population (frequency of 0.78), whereas the allelic frequency of apoE4 is 0.14. However, in clinic-based studies, patients with late-onset disease (>65 years of age) have an apoE4 allelic frequency of 0.50; thus, the risk for AD is increased by the presence of apoE4 (Corder *et al.,* 1994; Strittmatter *et al.,* 1993). The mechanisms whereby the apoE allele type elevates the risk for late-onset disease are not known. ApoE has no effect on Aβ synthesis, and it appears that ApoE and other proteins may influence extracellular Aβ metabolism in the CNS (DeMattos *et al.,* 2002b; Holtzman, 2001). Significant differences exist in the abilities of ApoE isoforms to bind Aβ and these features of the individual protein are hypothesized to differentially influence aggregation, deposition and/or clearance of Aβ by ApoE.

Although the presence of the ApoE-4 allele confers risk in late-onset disease (Tanzi *et al.,* 2001), ApoE genotyping, a useful research tool, does not appear helpful for routine diagnostic purposes.

APP PROCESSING BY SECRETASES

As indicated previously, APP is processed by secretase enzymes resulting in release of the ectodomain of APP, the production of AICD, and the generation of several Aβ peptide fragments. These enzyme activities and the secretase protein responsible for the cleavages are discussed subsequently.

β-Secretase Activity

BACE1 and BACE2, encoded by genes on chromosomes 11 and 21, respectively, are transmembrane aspartyl proteases that are directly involved in the cleavages of APP (Vassar, 2001; Vassar *et al.,* 1999; Cai *et al.,* 2001; Luo *et al.,* 2001; Sinha *et al.,* 1999; Yan *et al.,* 1999; Bennett *et al.,* 2000; Farzan *et al.,* 2000; Haniu, 2000). BACE1 mRNA levels are quite high in many regions of brain, and BACE1 protein is present in the CNS. In contrast, levels of BACE1 in PNS appear to be low.

Moreover, although BACE1 mRNA is present in a variety of tissues, levels of this protein are low in most non-neural tissues. Significantly, in the pancreas, BACE1mRNA is high but the transcript is alternatively spliced to produce a smaller protein incapable of cleaving APP (Bodendorf *et al.,* 2001). BACE2 mRNA, present in a variety of systemic organs, is very low in neural tissues, except for scattered nuclei in the hypothalamus and brainstem (Bennett *et al.,* 2000). Moreover, BACE2 activity appears to be virtually undetected in brain regions involved in AD.

BACE1, an aspartyl protease (Hong *et al.,* 2000; Vassar *et al.,* 1999; Vassar, 2001) preferentially cleaves APP at the +11 > +1 sites of Aβ in APP (Cai *et al.,* 2001), and this enzyme is essential for the generation of Aβ (Cai *et al.,* 2001). Significantly, APPswe is cleaved perhaps 100-fold more efficiently at the +1 site than is wildtype APP. Thus, the presence of this mutation greatly increases BACE1 cleavage and accounts for the elevation of Aβ species in the presence of this mutation. Analyses of cells and brains from *BACE1*-deficient mice, which show no overt developmental phenotype (Cai *et al.,* 2001; Luo *et al.,* 2001; Luo *et al.,* 2003), disclosed that Aβ1-40/42 and Aβ11-40/42 is not secreted in these samples (Cai *et al.,* 2001; Luo *et al.,* 2001). Thus, BACE1 is the principal neuronal β-secretase and is responsible for the critical penultimate proamyloidogenic cleavages. In contrast, BACE2 is responsible, along with α-secretase (Sisodia *et al.,* 1990), for anti-amyloidogenic cleavages at +19/+20 of Aβ (Farzan *et al.,* 2000; Haniu, 2000). Thus, BACE2 acts like α-secretase (Sisodia *et al.,* 2002; Sisodia *et al.,* 1990). In the CNS, BACE1, along with APP and, presumably, components of γ-secretase complex, is present at some, if not all, synapses. These observations are consistent with the idea that Aβ, derived from neuronal APP and released in synaptic terminals, may influence synaptic activity (Kamenetz *et al.,* 2003).

γ-Secretase

These enzyme activities, essential for the regulated intramembraneous proteolysis of a variety of transmembrane proteins, depend on a multiprotein catalytic complex, which includes PS and several other transmembrane proteins (Sisodia *et al.,* 2002; Takasugi *et al.,* 2003; Selkoe *et al.,* 2003; Li *et al.,* 2003). PS1 may act as an aspartyl protease itself; function as a co-factor critical for the activity of γ-secretase; or play a role in trafficking of APP or proteins critical for enzyme activity to the proper compartment for γ-secretase cleavage (Wolfe *et al.,* 1999; Naruse *et al.,* 1998; Esler *et al.,* 2000; De Strooper *et al.,* 1998; Wolfe, 2002). These concepts are supported by several observations: PS1 is isolated with γ-secretase under specific detergent-soluble conditions (Li *et al.,* 2000a);

PS1 is selectively cross-linked or photoaffinity-labeled by transition state inhibitors (Li *et al.*, 2000b; Esler *et al.*, 2000); substitutions of aspartate residues at D257 in TM6 and at D385 in TM7 have been reported to reduce secretion of Aβ and cleavage of Notch1 *in vitro* (Wolfe *et al.*, 1999); cells in which *PS1* has been targeted show decreased levels of secretion of Aβ (De Strooper, 2003; Takasugi *et al.*, 2003; Li *et al.*, 2003; Naruse *et al.*, 1998). Several other proteins, including Nct, Aph1, and Pen2, also play roles in the multiprotein catalytic complex (Yu *et al.*, 2000; Francis *et al.*, 2002; Li *et al.*, 2003; Takasugi *et al.*, 2003; Edbauer *et al.*, 2002; Li *et al.*, 2003). As described below, studies of *Nct* null mice and *NCT*[−/−] fibroblasts have shown that NCT is an integral member of the γ-secretase complex (Li *et al.*, 2003), and partial decreases in the level of NCT significantly reduce the secretion of Aβ (Li *et al.*, 2003). Aph-1 and Pen-2 (De Strooper, 2003; Takasugi *et al.*, 2003; Francis *et al.*, 2002; Goutte *et al.*, 2002) are novel transmembrane proteins: Aph1 has seven predicted transmembrane domains; while Pen2 has two predicted transmembrane regions (De Strooper, 2003; Takasugi *et al.*, 2003). These proteins interact with sel-12/PS and aph2/nicastrin and their inactivation decreases γ-secretase cleavages of APP and Notch1 and reduces levels of processed PS. For these reasons, Aph1 and Pen2 are now recognized to be critical components of γ-secretase complex.

MODELS OF AB AMYLOIDOSIS IN THE CENTRAL NERVOUS SYSTEM

For a number of autosomal dominant neurological diseases, transgenic approaches have been used to model features of neurodegenerative diseases in mice and other species (Zoghbi *et al.*, 2002; Shulman *et al.*, 2003; Bonini *et al.*, 2003; Duff *et al.*, 2001; Wong *et al.*, 2002). In mice, expression of *APPswe* or *APP717* minigenes (with or without mutant *PS1*) leads to an Aβ amyloidosis in the CNS (Mucke *et al.*, 2000; Borchelt *et al.*, 2002; Whittemore, 1999; Calhoun *et al.*, 1998; Duff *et al.*, 2001; Holtzman *et al.*, 2000; Hsia *et al.*, 1999; Bacskai *et al.*, 2001; Bacskai *et al.*, 2003; Borchelt *et al.*, 1997; Borchelt *et al.*, 1996). The nature and levels of the expressed transgene and the specific mutation influence the severity of the pathology. Mice expressing both mutant *PS1* and mutant *APP* develop accelerated disease. In these animals, levels of Aβ in brain are elevated, and diffuse Aβ deposits and neuritic plaques appear in the hippocampus and cortex. For example, in transgenic mice generated at Hopkins (Borchelt *et al.*, 1996; Borchelt *et al.*, 1997; Lesuisse *et al.*, 2001), the pathology evolves in stages: levels of Aβ in brain increase with age; synaptic markers are altered as diffuse Aβ deposits appear (although there is

some disagreement as to the sequence); neurites become apparent in proximity to deposits; and neuritic plaques are associated glial responses. In our mice, aggregated tau and neurofibrillary tangles are not present (Xu *et al.*, 2002). In some lines, the density of synaptic terminals is reduced and these abnormalities appear linked to a deficiency in synaptic transmission (Hsia *et al.*, 1999; Chapman *et al.*, 1999). Some lines of mice show degeneration of neurons (Calhoun *et al.*, 1998) and reductions in transmitter markers (Boncristiano *et al.*, 2002).

Learning deficits, problems in object recognition memory, and difficulties performing tasks assessing spatial reference and working memory have been identified in some of the lines of mutant mice with high levels of mutant transgene expression (Savonenko *et al.*, 2003a; Savonenko *et al.*, 2003b; Chapman *et al.*, 1999; Chen *et al.*, 2000). Behavioral deficits show some correlation with physiological properties identified in hippocampal circuits (Chapman *et al.*, 1999). Although the mutant mice do not fully recapitulate the complete phenotype of AD, these animals are very useful subjects for research designed to examine disease mechanisms and to test novel therapies designed to ameliorate the amyloid in related pathology and behavioral abnormalities.

The paucity of tau abnormalities in these various lines of mutant transgenic mice may be explained by differences between species. Mice express three copies of the murine variant of 4-R tau while humans express three copies of both 3-R tau and 4-R tau. Early efforts to express mutant *tau* transgenes in mice did not lead to striking clinical phenotypes or pathology (Duff *et al.*, 2000). Subsequent studies produced *tau*-overexpressing mice with clinical signs. For example, overexpression of *tau* in some lines of mice led to weakness thought to reflect on degeneration of motor axons and evidence of accumulation of hypophosphorylated tau in neurons of the spinal cord, brainstem, and cortex (Lee *et al.*, 2001). When the prion or Thy1 promoters are used to drive P3O1L/*tau* (a mutation linked to autosomal dominant fronto-temporal dementia with parkinsonism), neurofibrillary tangles developed in neurons of the brain and spinal cord (Lewis *et al.*, 2000; Gotz *et al.*, 2001). Mice expressing *APPswe/tau* P301L show enhanced NFT pathology in limbic system and olfactory cortex (Lewis *et al.*, 2001). Moreover, when Aβ42 fibrils are injected into specific brain regions of *tau P301L* mice, the number of tangles was increased in those neurons projecting to sites of injection (Götz *et al.*, 2001). More recently, a triple transgenic mouse (3xTg-AD) was created by microinjecting *APPswe* and *P3o1L/tau* into single cells derived from monozygous PS1 M146V knock in mice (Oddo *et al.*, 2003). These mice develop age-related plaques and tangles. Deficits in LTP appear to antedate overt pathology (Oddo *et al.*, 2003) However, mice bearing both mutant *tau* and *APP* (or *APP/PS1*) or mutant *tau* mice injected with Aβ are not complete models of FAD because the presence of the tau mutation alone is associated with the development of tangles.

Although mice expressing wt human *tau* transgene do not develop pathology, introduction of human *tau* into tau⁻/⁻ mice leads to accumulations of tau immunoreactivity in neurons, including the presence of certain phosphorylated and conformational epitopes occurring in human disease (Dr. Peter Davies, personal communication). These observations have been interpreted to suggest that, in mice, endogenous mouse tau may interfere with the formation of tangles related to human tau. Thus, a more appropriate model of AD might be achieved by co-expression of mutant *FAD*-linked genes and all six isoforms of wild-type human *tau* on a mouse *tau* null background.

GENE TARGETING

To begin to understand the functions of some of the proteins thought to play roles in AD, investigators have targeted a variety of genes encoding the proteins, including *APP*, amyloid precursor–like proteins (*APLPs*), *BACE1*, *PS1*, *Nct*, and *Aph1*.

APP and APLPs

Homozygous *APP⁻/⁻* mice are viable and fertile, but appear to have subtle decreases in locomotor activity and forelimb grip strength (Zheng *et al.*, 1995). The absence of substantial phenotypes in *APP⁻/⁻* mice is thought to be related to functional redundancy of two amyloid precursor–like proteins, APLP1 and APLP2, homologous to APP (Wasco *et al.*, 1993; Slunt *et al.*, 1994). *APLP2⁻/⁻* mice appear relatively normal (von Koch *et al.*, 1997), while APLP1 mice exhibit a postnatal growth deficit (Heber *et al.*, 2000). APLP⁻/⁻ mice are viable but, *APP/APLP2* null mice and *APLP1* and *APLP2* null mice do not survive the perinatal period (Heber *et al.*, 2000). These observations support the concept that redundancy exists between APLP2 and other members of this interesting family of proteins (Heber *et al.*, 2000).

BACE1

BACE1 null mice are viable and healthy, have no obvious phenotype or pathology, and can mate successfully (Vassar *et al.*, 2000; Cai *et al.*, 2001; Luo *et al.*, 2001, 2003; Cai *et al.*, 2001). Importantly, cortical neurons cultured from *BACE1⁻/⁻* embryos, do not show cleavages at the +1 and +11 sites of Aβ, and the secretion of Aβ

peptides is abolished even in the presence of elevated levels of exogenous wt or mutant *APP* (Cai *et al.*, 2001). The brains of *BACE1* null mice appear morphologically normal and Aβ peptides are not produced (Luo *et al.*, 2001; Cai *et al.*, 2001). These results establish that BACE1 is the neuronal β-secretase required to cleave APP to generate the N-termini of Aβ (Cai *et al.*, 2001). At present, a consensus has not been reached with regards to other substrates cleaved by BACE1. Behavioral studies of the null mice, *in vitro* and *in vivo* challenges of *BACE1* neural cells, and two dimensional gels of organs from BACE1⁺/⁺, BACE⁺/⁻, and BACE⁻/⁻ mice will be critical in determining the consequences of the absence of BACE1 on behavior, susceptibility to challenge/disease, and influence on other substrates that may be cleaved by this enzyme. At present, BACE1 appears to be an outstanding target for development of an anti-amyloidogenic therapy.

PS1 and -2

In contrast to *BACE1* null mice, *PS1⁻/⁻* mice do not survive beyond the early postnatal period and show severe developmental abnormalities of the axial skeleton, ribs, and spinal ganglia, features resembling a partial Notch 1⁻/⁻ phenotype (Wong *et al.*, 1997; Shen *et al.*, 1997). These abnormalities occur because PS1 (along with Nct, Aph1, and pen2) are components of the γ-secretase complex that carries out the S3 intramembranous cleavage of Notch1, a receptor protein involved in critical cell-fate decisions during development (De Strooper, 2003; Takasugi *et al.*, 2003; Esler *et al.*, 2002; Edbauer *et al.*, 2002; Huppert *et al.*, 2000; Struhl *et al.*, 2001; De Strooper *et al.*, 1999; Li *et al.*, 2003; Francis *et al.*, 2002). Without this cleavage, the Notch1 intracellular domain (NICD) is not released from the plasma membrane and does not reach the nucleus to initiate transcriptional processes essential for cell fate decisions. In cell culture, absence of *PS1* or substitution of particular aspartate residues leads to reduced levels of γ-secretase cleavage products and levels of Aβ (De Strooper *et al.*, 1998; Wolfe *et al.*, 1999). *PS2* null mice are viable and fertile, although they develop age-associated mild pulmonary fibrosis and hemorrhage (Donoviel *et al.*, 1999). Mice lacking *PS1* and heterozygous for *PS2* die midway through gestation with full *Notch1* null-like phenotype (Donoviel *et al.*, 1999). To study the role of PS1 *in vivo* in adult mice, investigators generated conditional *PS1*-targeted mice lacking PS1 expression in the forebrain after embryonic development (Feng *et al.*, 2001). As expected, the absence of PS1 resulted in decreased generation of Aβ, providing support for the concept that PS1 is critical for γ-secretase activity in the brain.

Nct, Aph1, and Pen2

Another critical member of the γ-secretase complex is Nct, a type I transmembrane glycoprotein, which forms high-molecular-weight complexes with presenilins (Yu *et al.*, 2000) and binds to the membrane-tethered form of Notch1 (Chen *et al.*, 2001). Recent studies have indicated that Nct is required for Notch1 signaling and APP processing (Yu *et al.*, 2000; Chung *et al.*, 2001; Edbauer *et al.*, 2002; Hu *et al.*, 2002; Lopez-Schier *et al.*, 2002; Li *et al.*, 2003; De Strooper, 2003; Takasugi *et al.*, 2003). Nct may function to stabilize PS and appears to be critical for the trafficking of PS to the cell surface (Li *et al.*, 2003; Edbauer *et al.*, 2002). Likewise, PS is required for the maturation and cell surface accumulation of Nct (Edbauer *et al.*, 2002; Li *et al.*, 2003). These results have suggested that Nct and PS form functional components of a multimeric complex required for the intramembraneous proteolysis of both Notch1 and APP.

To determine whether Nct is required for proteolytic processing of Notch1 and APP in mammals and the role of Nct in the assembly of the γ-secretase complex, nicastrin-deficient ($NCT^{-/-}$) mice were generated and fibroblasts were derived from $NCT^{-/-}$ embryos (Li *et al.*, 2003). *Nct*-null embryos die by embryonic day 10.5 and exhibit several developmental patterning defects, including abnormal somite segmentation; the phenotype closely resembles that seen in embryos lacking *Notch1* or both *PS*. Importantly, secretion of Aβ peptides is abolished in $NCT^{-/-}$ fibroblasts, whereas it is reduced by approximately 50% in $NCT^{+/-}$ cells (Li *et al.*, 2003). The failure to generate Aβ peptides in $NCT^{-/-}$ cells is accompanied by destabilization of the presenilin/γ-secretase complex and accumulation of APP C-terminal fragments. Moreover, analysis of APP trafficking in $NCT^{-/-}$ fibroblasts revealed a significant delay in the rate of APP reinternalization compared with that of control cells. Thus, Nct, along with Aph1 and Pen2, is a critical component of the γ-secretase complex. These proteins are critical for stabilizing PS and for generation of Aβ (Francis *et al.*, 2002; Li *et al.*, 2003; De Strooper *et al.*, 1999; De Strooper, 2003; Takasugi *et al.*, 2003; Edbauer *et al.*, 2002). These results establish that several proteins are essential components of the multimeric γ-secretase complex in mammals required for both γ-secretase activity and APP trafficking and suggest that these proteins may be a valuable therapeutic target for anti-amyloidogenic treatments (De Strooper, 2003; Li *et al.*, 2003; Takasugi *et al.*, 2003).

WHY IS ALZHEIMER'S DISEASE A NEURONAL DISEASE?

Among the most challenging mysteries of AD is the identification of factors that render the brain particularly susceptible to the extracellular deposition Aβ peptides.

Recent research has begun to provide new clues concerning reasons why Aβ accumulates in the brain. Aβ is generated from cleavages of APP by BACE1 and by an γ-secretase. *In vivo*, BACE1 is the principal β-secretase necessary to cleave APP to generate Aβ. In contrast, APP can also be cleaved within the Aβ sequence by α-secretases (*Tumor Necrosis Factor Converting Enzyme* or TACE) or by BACE2 to release the ectodomain of APP. These BACE2 and TACE cleavages within the Aβ domain of APP preclude the formation of Aβ. As subsequently outlined, the presence of APP and pro-amyloidogenic secretase in the CNS may contribute substantially to the selective vulnerability of the Aβ amyloidosis.

BACE1 and BACE2 are expressed ubiquitously, but levels of BACE1 mRNA are particularly high in brain and pancreas, whereas the levels of BACE2 mRNA are relatively low in all tissues, except in the brain, where it is nearly undetectable. Because BACE1 is the principal β-secretase in neurons and BACE2 (and TACE) limits the secretion of Aβ peptides, we propose that BACE1 is a pro-amyloidogenic enzyme, whereas BACE2 and TACE are anti-amyloidogenic enzymes. We hypothesize that the relative levels of BACE1 and the anti-amyloidogenic secretases, in concert with the abundance of APP in neurons, are major determinants of Aβ formation (Wong *et al.*, 2001). In this model, the secretion of Aβ peptides would be expected to be highest from neurons in brain, as compared with other cell types or organs because CNS neurons (not peripheral nervous system nerve cells), express relatively high levels of BACE1 coupled with low levels of BACE2 and TACE. If the ratio of the level of BACE1 to BACE2 (or TACE) is one of the critical factors that selectively predisposes the brain to the formation of Aβ, it is not surprising that AD would involve the brain selectively. Seemingly inconsistent with this hypothesis is the observation that there are high levels of BACE1 mRNA in the pancreas (Bodendorf *et al.*, 2001). However, It appears that BACE1 mRNA in pancreatic cells is alternatively spliced to generate a BACE1 isoform that is incapable of cleaving APP (Bodendorf *et al.*, 2001); thus, Aβ is not deposited in the pancreas.

Whether neurons are the source of APP giving rise to Aβ and has been debated (Wong *et al.*, 2001). Significantly, key participants in Aβ amyloidosis, particularly APP and BACE1, are synthesized by neurons, and appear to be colocalized within neurites in immediate proximity to sites of Aβ deposition in brain (Sheng *et al.*, 2003). This observation provides circumstantial evidence consistent with a neuronal origin for Aβ in CNS. However, it is established that APP is cleaved by BACE1 in distal axon terminals to generate C-terminal amyloidogenic peptides (Buxbaum *et al.*, 1998); after local γ-secretase cleavage, Aβ appears in terminal folds. Consistent with this concept is the demonstration that lesions of entorhinal cortex or perforant pathway reduce the levels of Aβ and the numbers of plaques in terminal fields of these neurons (Sheng *et al.*, 2002; Lazarov *et al.*, 2002).

EXPERIMENTAL THERAPEUTICS FOR ALZHEIMER'S DISEASE

Experimental therapeutic efforts have focused on influencing Aβ production, aggregation, clearance, and downstream neurotoxicity. Although mutant transgenic mice do not recapitulate the full phenotype of AD, they represent excellent models of Aβ amyloidosis and are highly suitable for identification of therapeutic targets and for testing new treatments *in vitro*. In this short review, it is not possible to discuss all therapeutic efforts in mice. Therefore, we focus on a few examples.

Secretase Activities

Although both β- and γ-secretase activities represent therapeutic targets for the development of novel protease inhibitors, at present BACE1 appears to be the most attractive target. It is the principal β-secretase in neurons (Luo *et al.*, 2001; Cai *et al.*, 2001; Wong *et al.*, 2001; Wolfe, 2002; Vassar *et al.*, 2000; Vassar *et al.*, 1999). In contrast to $PS1^{-/-}$ and $Nct^{-/-}$ mice (Wong *et al.*, 1997; Shen *et al.*, 1997; De Strooper *et al.*, 1998; Li *et al.*, 2003), the $BACE1^{-/-}$ mice are viable and live a normal life span without developing an obvious phenotype (Cai *et al.*, 2001). Significantly, *in vitro*, BACE1-deficient neurons fail to secrete Aβ even when co-expressing the *APPswe* and mutant *PS1* genes (Cai *et al.*, 2001; Luo *et al.*, 2001). Moreover, mutant *APP/PS1* mice lacking BACE1 do not produce Aβ and do not develop Aβ plaques in the brain (Cai *et al.*, submitted; Luo *et al.*, 2003). These findings provide excellent rationale for focusing on the design of novel therapeutics to inhibit BACE1 activity in brain.

PS, Nct, Aph1, and Pen2 are required for γ-secretase activity (Yu *et al.*, 2000; Li *et al.*, 2003; De Strooper, 2003; Francis *et al.*, 2002; Takasugi *et al.*, 2003; Wolfe, 2002). Recent work suggests that PS1 is essential for activity, but may not be the protease performing γ cleavage (at least for Notch1) (Taniguchi *et al.*, 2002). Because of the role of the γ-secretase complex in Notch processing, attempts are being made to try to design therapeutics that inhibit selectively γ-secretase activity with regard to APP while minimizing inhibition of the activity involved in Notch1 processing. This approach is being pursued in part, because several populations of cells, particularly hematopoietic stem cells, use Notch 1 signaling for cell-fate decisions even in adults.

Many pharmaceutical and biotechnology companies and some academic laboratories are using high-throughput screen and molecular modeling strategies to discover compounds that inhibit these enzyme activities (Wolfe, 2002; Vassar *et al.*, 2000; Vassar, 2001). Once lead compounds are identified, medicinal chemists will modify the agents to enhance efficacy, allow passage through the blood-brain barrier, and reduce any potential toxicities.

Immunotherapy

In both prevention and treatment trials in mutant mice, both Aβ immunization (with Freund's adjuvant) and passive transfer of Aβ antibodies reduce levels of Aβ and plaque burden in mutant APP transgenic mice (Soderberg *et al.*, 2003; Monsonego *et al.*, 2003; Solomon, 2001; Wilcock *et al.*, 2001, 2003; Bard *et al.*, 2000; DeMattos *et al.*, 2001, 2002a; Kotilinek *et al.*, 2002; Morgan *et al.*, 2000; Dodart *et al.*, 2002; Schenk *et al.*, 1999). Efficacy seems to be related to antibody titer. The mechanisms of enhanced clearance are not certain, but two not mutually exclusive hypotheses have been suggested: (1) a small amount of Aβ antibody reaches the brain, binds to Aβ peptides, promotes the disassembly of fibrils, and, via the Fc antibody domain, activated microglia enter lesion and remove Aβ (Schenk *et al.*, 1999); or (2) serum antibodies serve as a sink to draw the amyloid peptides from the brain into the circulation, thus changing the equilibrium of Aβ in different compartments and promoting removal from the brain (DeMattos *et al.*, 2001). These are not mutually exclusive mechanisms. Significantly, immunotherapy in transgenic mice is successful in partially clearing Aβ and attenuates learning and behavioral deficits in at least two cohorts of mutant *APP* mice (Kotilinek *et al.*, 2002; Morgan *et al.*, 2000; Dodart *et al.*, 2002).

In humans, although phase 1 trials with Aβ peptide and adjuvant were not associated with any adverse events, phase 2 trials were suspended because of severe adverse reactions (meningoencephalitis) in a subset of patients (Monsonego *et al.*, 2003; Nicoll *et al.*, 2003). These events illustrate the challenges of extrapolating outcomes in mice to trials with humans. The pathology in this single case, consistent with T-cell meningitis (Nicoll *et al.*, 2003), was interpreted to show some clearance of Aβ deposits (Nicoll *et al.*, 2003). Specifically, some areas of cortex showed very few plaques, yet these regions contained a relatively high density of tangles, neuropil threads, and vascular amyloid deposits. There was a paucity of plaque-associated dystrophic neurites and astrocytic clusters. In some regions with low plaque densities, Aβ immunoreactivity was associated with microglia. T-cells were conspicuous in subarachnoid space and around some vessels (Nicoll *et al.*, 2003).

Interestingly, assessment of cognitive functions in 30 patients who received vaccination and booster immunizations disclosed that patients who generated antibodies had a slower decline in several functional measures over time (Hock *et al.*, 2003). Investigators have continued to pursue the passive immunization approach and passive immunization and attempting to make antigens that do not stimulate T-cell–mediated immunologic attacks (Monsonego *et al.*, 2003).

Clearance of Aβ Peptides

It is established that amyloid burden can be significantly reduced by a variety of approaches, including immunotherapy, reducing BACE1 activity, etc. Reviewing the neuronal source of APP and secretases allows clearance to proceed with a new supply of protein critical for formation of Aβ. For example, lesions of the entorhinal cortex or the perforant pathway reduce the levels of Aβ and the number of plaques in the terminal fields (molecular dentate gyrus) of the ablated inputs (Sheng *et al.*, 2002; Lazarov *et al.*, 2002). This work clearly shows that APP is major source of the Aβ peptide in the brain and that Aβ is in a dynamic state capable of dramatic clearance (at least in mice).

In another approach to reduce the amyloid burden, a Lentiviral vector system was used to express human neprilysin (NEP), a putative Aβ-degrading enzyme, in the brains of transgenic mice with amyloid deposits (Marr *et al.*, 2003). Unilateral intercerebral injections of lentivirus expressing NEP reduce Aβ deposits and may attenuate neurodegenerative processes (Marr *et al.*, 2003).

Another approach has been to target copper and zinc, which are enriched in Aβ deposits in the brains of individuals with AD (Cherny *et al.*, 2001). It has been suggested that these metals play a role in aggregation of Aβ and local H_2O_2 production and thus, their removal might attenuate the deposition of Aβ and its consequences. In one study, clioquinol, a brain-penetrant antibiotic that binds zinc and copper, appeared to reduce the levels of these metals in the brain (Cherny *et al.*, 2001). Nine-week therapy was associated with a modest increase in soluble Aβ and a reduction in Aβ deposition in the brain. A human trial is in progress. A variety of other illustrations of therapeutic interaction to reduce Aβ burden are described in the literature.

CONCLUSIONS

The identification of genes mutated or deleted in the inherited forms of many neurodegenerative diseases has allowed investigators to create *in vivo* and *in vitro* model systems relevant to a wide variety of human neurological disorders (Wong *et al.*, 2002; Duff, 1998; Bonini *et al.*, 2003; Shulman *et al.*, 2003; Zoghbi *et al.*, 2002; Lee *et al.*, 2001; Price *et al.*, 1998a, 1998b). In this review, we emphasized the value of transgenic and gene targeted models and the lessons they provided for understanding mechanisms of AD, for identifying therapeutic targets, and for testing new treatments in models. For illustrative purposes, we focused on studies of the progeny of $BACE1^{-/-}$ mice and mutant *APP/PS1* transgenic mice, which provides a dramatic example of the impact of reducing activity of secretases. Over the past several years, research has provided extraordinary new information about the mechanisms of Aβ amyloidogenesis, the reasons for vulnerability of brain to amyloidosis, and the proteins that are potential therapeutic targets. New models using conditional systems will allow investigators to examine the molecular mechanisms by which mutant proteins cause selective dysfunction and death of neurons and to delineate pathogenic pathways that can be tested by breeding strategies on RNAi injection methods that reduce levels of specific proteins. The results of these approaches will provide us with a better understanding of the mechanisms that lead to diseases and help us to design new treatments.

Moreover, studies of mice and other mammalian species will be enormously helped by investigations of invertebrate organisms, such as *Caenorhabditis elegans* or *Drosophila melanogaster*, which provide excellent genetically manipulatable model systems for examining many aspects of neurodegenerative diseases (Shulman *et al.*, 2003; Bonini *et al.*, 2003; Zoghbi *et al.*, 2002). Invertebrates harboring mutant transgenes can be generated rapidly. The nervous systems of the animals are relatively simple and behaviors can be scored readily. Because these invertebrate models reproduce aspects of pathologies occurring in human neurodegenerative diseases, studies of these systems should prove very useful in delineating disease pathways, identifying suppressors and enhancers of these processes, and in testing the influences of drugs on phenotypes in flies and worms.

In summary, investigations of genetically engineered (transgenic or gene targeted) mice have disclosed participants in pathogenic pathways, have reproduced some of the features of human neurodegenerative disorders, have provided important new information about the disease mechanisms, and allowed identification of new therapeutic targets. This information has provided the basis for new therapeutic strategies that can be tested in models. These lines of research have made spectacular progress over the past few years, and we anticipate that future discoveries will lead to design of promising mechanism-based therapies that can be tested in models of this devastating disease.

ACKNOWLEDGMENTS

The authors wish to thank the many colleagues who have worked at JHMI as well as those at other institutions for their contributions to some of the original work cited in this review and for their helpful discussions. Aspects of this work were supported by grants from the U.S. Public Health Service (AG05146) as well as the Metropolitan Life Foundation, Adler Foundation, CART Foundation, and Bristol-Myers Squibb Foundation.

References

Akiyama H *et al.* (2000). Inflammation and Alzheimer's disease, *Neurobiol Aging* 21:383-421.

Albert MS, Moss MB, Tanzi R, Jones K. (2001). Preclinical prediction of AD using neuropsychological tests, *J Int Neuropsychol Soc* 7:631-639.

Bacskai BJ *et al.* (2003). Four-dimensional multiphoton imaging of brain entry, amyloid binding, and clearance of an amyloid-beta ligand in transgenic mice, *Proc Natl Acad Sci USA* 100:12462-12467.

Bacskai BJ *et al.* (2001). Imaging of amyloid-beta deposits in brains of living mice permits direct observation of clearance of plaques with immunotherapy, *Nat Med* 7:369-372.

Bard F *et al.* (2000). Peripherally administered antibodies against amyloid beta-peptide enter the central nervous system and reduce pathology in a mouse model of Alzheimer disease, *Nat Med* 6:916-919.

Beach TG *et al.* (2000). The cholinergic deficit coincides with A-beta deposition at the earliest histopathologic stages of Alzheimer disease, *J Neuropathol Exp Neurol* 59:308-313.

Bennett BD *et al.* (2000). Expression analysis of BACE2 in brain and peripheral tissues, *J Biol Chem* 275:20647-20651.

Bertram L *et al.* (2000). Evidence for genetic linkage of Alzheimer's disease to chromosome 10q, *Science* 290:2302-2303.

Bodendorf U *et al.* (2001). A splice variant of beta-secretase deficient in the amyloidogenic processing of the amyloid precursor protein, *J Biol Chem* 276:12019-12023.

Boncristiano S *et al.* (2002). Cholinergic changes in the APP23 transgenic mouse model of cerebral amyloidosis, *J Neurosci* 22:3234-3243.

Bonini NM, Fortini ME. (2003). Human neurodegenerative disease modeling using *Drosophila*, *Annu Rev Neurosci* 26:627-656.

Borchelt DR *et al.* (2002). Accumulation of proteolytic fragments of mutant presenilin 1 and accelerated amyloid deposition are co-regulated in transgenic mice, *Neurobiol Aging* 23:171-177.

Borchelt DR *et al.* (1997). Accelerated amyloid deposition in the brains of transgenic mice coexpressing mutant presenilin 1 and amyloid precursor proteins, *Neuron* 19:939-945.

Borchelt DR *et al.* (1996). Familial Alzheimer's disease-linked presenilin 1 variants elevate Abeta1-42/1-40 ratio in vitro and in vivo, *Neuron* 17:1005-1013.

Braak H, Braak E. (1991). Neuropathological staging of Alzheimer-related changes, *Acta Neuropathol* 82:239-259.

Braak H *et al.* J. (1996). Pattern of brain destruction in Parkinson's and Alzheimer's diseases, *J Neural Transm* 103:455-490.

Brookmeyer R, Gray S, Kawas C. (1998). Projections of Alzheimer's disease in the United States and the public health impact of delaying disease onset, *Am J Public Health* 88:1337-1342.

Busciglio J, Lorenzo A, Yeh J, Yankner BA. (1995). β-amyloid fibrils induce tau phosphorylation and loss of microtubule binding, *Neuron* 14:879-888.

Buxbaum JD *et al.* (1998). Alzheimer amyloid protein precursor in the rat hippocampus: transport and processing through the perforant path, *J Neurosci* 18:9629-9637.

Cai H *et al.* (2001). BACE1 is the major beta-secretase for generation of Abeta peptides by neurons, *Nat Neurosci* 4:233-234.

Calhoun ME *et al.* (1998). Neuron loss in APP transgenic mice, *Nature* 395:755-756.

Cao X, Sudhof TC. (2001). A transcriptively active complex of APP with Fe65 and histone acetyltransferase Tip60, *Science* 293:115-120.

Capell A *et al.* (2000). Presenilin-1 differentially facilitates endoproteolysis of the beta-amyloid precursor protein and Notch, *Nat Cell Biol* 2:205-211.

Caughey B, Lansbury PT. (2003). Protofibrils, pores, fibrils, and neurodegeneration: separating the responsible protein aggregates from the innocent bystanders, *Annu Rev Neurosci* 26:267-298.

Chapman PF *et al.* (1999). Impaired synaptic plasticity and learning in aged amyloid precursor protein transgenic mice, *Nat Neurosci* 2:271-276.

Chen F *et al.* (2001). Nicastrin binds to membrane-tethered Notch, *Nat Cell Biol* 3:751-754.

Chen G *et al.* (2000). A learning deficit related to age and β-amyloid plaques in a mouse model of Alzheimer's disease, *Nature* 408:975-979.

Cherny RA *et al.* (2001). Treatment with a copper-zinc chelator markedly and rapidly inhibits β-amyloid accumulation in Alzheimer's disease transgenic mice, *Neuron* 30:665-676.

Chung HM, Struhl G. (2001). Nicastrin is required for Presenilin-mediated transmembrane cleavage in *Drosophila*, *Nat Cell Biol* 3:1129-1132.

Corder EH *et al.* (1994). Protective effect of apolipoprotein E type 2 allele for late onset Alzheimer disease, *Nat Genet* 7:180-184.

Cummings JL. (2003). Toward a molecular neuropsychiatry of neurodegenerative diseases, *Ann Neurol* 54:147-154.

Daly E *et al.* (2000). Predicting conversion to Alzheimer disease using standardized clinical information, *Arch Neurol* 57:675-680.

De Strooper B. (2003). Aph-1, Pen-2, and Nicastrin with Presenilin generate an active gamma-secretase complex, *Neuron* 38:9-12.

De Strooper B *et al.* (1999). A presenilin-1-dependent gamma-secretase-like protease mediates release of Notch intracellular domain, *Nature* 398:518-522.

De Strooper B *et al.* (1998). Deficiency of presenilin-1 inhibits the normal cleavage of amyloid precursor protein, *Nature* 391:387-390.

Delacourte A *et al.* (1998). The biochemical pathway of neurofibrillary degeneration in aging and Alzheimer's disease, *Neurology* 52:1158-1165.

DeMattos RB *et al.* (2001). Peripheral anti-Ab antibody alters CNS and plasma Ab clearance and decreases brain Ab burden in a mouse model of Alzheimer's disease, *Proc Natl Acad Sci USA* 98:8850-8855.

DeMattos RB *et al.* (2002a). Brain to plasma amyloid-beta efflux: a measure of brain amyloid burden in a mouse model of Alzheimer's disease, *Science* 295:2264-2267.

DeMattos RB *et al.* (2002b). Clusterin promotes amyloid plaque formation and is critical for neuritic toxicity in a mouse model of Alzheimer's disease, *Proc Natl Acad Sci USA* 99:10843-10848.

Doan A *et al.* (1996). Protein topology of presenilin 1, *Neuron* 17:1023-1030.

Dodart JC *et al.* (2002). Immunization reverses memory deficits without reducing brain Abeta burden in Alzheimer's disease model, *Nat Neurosci*

Donoviel DB *et al.* (1999). Mice lacking both presenilin genes exhibit early embryonic patterning defects, *Genes Dev* 13:2801-2810.

Duff K. (1998). Recent work Alzheimer's disease transgenics, *Curr Opin Biotechnol* 9:561-564.

Duff K *et al.* (2000). Characterization of pathology in transgenic mice over-expressing human genomic and cDNA tau transgenes, *Neurobiol Dis* 7:87-98.

Duff K, Rao MV. (2001). Progress in the modeling of neurodegenerative diseases in transgenic mice, *Curr Opin Neurol* 14:441-447.

Edbauer D, Winkler E, Haass C, Steiner H. (2002). Presenilin and nicastrin regulate each other and determine amyloid beta-peptide production via complex formation, *Proc Natl Acad Sci USA* 99:8666-8671.

Ertekin-Taner N *et al.* (2000). Linkage of plasma Ab42 to a quantitative locus on chromosome 10 in late-onset Alzheimer's disease pedigrees, *Science* 290:2303-2304.

Esler WP *et al.* (2000). Transition-state analogue inhibitors of gamma-secretase bind directly to presenilin-1, *Nat Cell Biol* 2:428-434.

Esler WP *et al.* (2002). Activity-dependent isolation of the presenilin-γ-secretase complex reveals nicastrin and a γ substrate, *Proc Natl Acad Sci USA* 14:1-6.

Farzan M *et al.* (2000). BACE2, a β-secretase homolog, cleaves at the β site and within the amyloid-β region of the amyloid-β precursor protein, *Proc Natl Acad Sci USA* 97:9712-9717.

Feng R *et al.* (2001). Deficient neurogenesis in forebrain-specific Presenilin-1 knockout mice is associated with reduced clearance of Hippocampal memory traces, *Neuron* 32:911-926.

Francis R *et al.* (2002). Aph-1 and pen-2 are required for notch pathway signaling, gamma-secretase cleavage of betaAPP, and presenilin protein accumulation, *Dev Cell* 3:85-97.

Gastard MC, Troncoso JC, Koliatsos VE. (2003). Caspase activation in the limbic cortex of subjects with early Alzheimer's disease, *Ann Neurol* 54:393-398.

Giasson BI *et al.* (2003). Initiation and synergistic fibrillization of tau and alpha-synuclein, *Science* 300:636-640.

Goedert M, Spillantini MG, Cairns NJ, Crowther RA. (1992). Tau proteins of Alzheimer paired helical filaments: abnormal phosphorylation of all six brain isoforms, *Neuron* 8:159-168.

Gomez-Isla T *et al.* (1996). Profound loss of layer II entorhinal cortex neurons occurs in very mild Alzheimer's disease, *J Neurosci* 16:4491-4500.

Gong Y *et al.* (2003). Alzheimer's disease-affected brain: presence of oligomeric A beta ligands (ADDLs). suggests a molecular basis for reversible memory loss, *Proc Natl Acad Sci USA* 100:10417-10422.

Götz J, Chen F, Barmettler R, Nitsch RM. (2001). Tau filament formation in transgenic mice expressing P301L tau, *J Biol Chem* 276:529-534.

Gotz J, Chen F, Van Dorpe J, Nitsch RM. (2001). Formation of neurofibrillary tangles in P301l tau transgenic mice induced by Abeta fibrils, *Science* 293:1491-1495.

Goutte C, Tsunozaki M, Hale VA, Priess JR. (2002). APH-1 is a multipass membrane protein essential for the Notch signaling pathway in *Caenorhabditis elegans* embryos, *Proc Natl Acad Sci USA* 99:775-779.

Haniu M. (2000). Characterization of alzheimer's beta-secretase protein BACE. a pepsin family member with unusual properties, *J Biol Chem* 275:21099-21106.

Hardy J, Selkoe DJ. (2002). The amyloid hypothesis of Alzheimer's disease: progress and problems on the road to therapeutics, *Science* 297:353-356.

Heber S *et al.* (2000). Mice with combined gene knock-outs reveal essential and partially redundant functions of amyloid precursor protein family members, *J Neurosci* 20:7951-7963.

Hock C *et al.* (2003). Antibodies against beta-amyloid slow cognitive decline in Alzheimer's disease, *Neuron* 38:547-554.

Holtzman DM. (2001). Role of apoE/A beta interactions in the pathogenesis of Alzheimer's disease and cerebral amyloid angiopathy, *J Mol Neurosci* 17:147-155.

Holtzman DM *et al.* (2000). Apolipoprotein E isoform-dependent amyloid deposition and neuritic degeneration in a mouse model of Alzheimer's disease, *Proc Natl Acad Sci USA* 97:2892-2897.

Hong L *et al.* J. (2000). Structure of the protease domain of memapsin 2 (beta-secretase). complexed with inhibitor, *Science* 290:150-153.

Hsia AY *et al.* (1999). Plaque-independent disruption of neural circuits in Alzheimer's disease mouse models, *Proc Natl Acad Sci USA* 96:3228-3233.

Hu Y, Ye Y, Fortini ME. (2002). Nicastrin is required for gamma-secretase cleavage of the *Drosophila* Notch receptor, *Dev Cell* 2:69-78.

Huppert SS *et al.* (2000). Embryonic lethality in mice homozygous for a processing-deficient allele of Notch1, *Nature* 405:966-970.

Hyman BT, Van Hoesen GW, Damasio AR, Barnes CL. (1984). Alzheimer's disease: cell-specific pathology isolates the hippocampal formation, *Science* 225:1168-1170.

Iwatsubo T *et al.* (1994). Visualization of A beta 42(43) and A beta 40 in senile plaques with end-specific A beta monoclonals: evidence that an initially deposited species is A beta 42(43), *Neuron* 13:45-53.

Jack CR Jr *et al.* (2000). Rates of hippocampal atrophy correlate with change in clinical status in aging and AD, *Neurology* 55:484-489.

Kamenetz F *et al.* (2003). APP processing and synaptic function, *Neuron* 37:925-937.

Killiany RJ *et al.* (2000). Use of structural magnetic resonance imaging to predict who will get Alzheimer's disease, *Ann Neurol* 47:430-439.

Killiany RJ *et al.* (2002). MRI measures of entorhinal cortex vs hippocampus in preclinical AD, *Neurology* 58:1188-1196.

Klein WL, Krafft GA, Finch CE. (2001). Targeting small Ab oligomers: the solution to an Alzheimer's disease conundrum? *Trends Neurosci* 24:219-223.

Klunk WE *et al.* (2004). Imaging brain amyloid in Alzheimer's disease using the novel positron emission tomography tracer, Pittsburgh compund-B, *Ann Neurol* 55:1-14.

Klunk WE *et al.* (2001). Uncharged thioflavin-T derivatives bind to amyloid-beta protein with high affinity and readily enter the brain, *Life Sci* 69:1471-1484.

Klunk WE *et al.* (2003). The binding of 2-(4'-methylaminophenyl) benzothiazole to postmortem brain homogenates is dominated by the amyloid component, *J Neurosci* 23:2086-2092.

Koo EH *et al.* (1990). Precursor of amyloid protein in Alzheimer disease undergoes fast anterograde axonal transport, *Proc Natl Acad Sci USA* 87:1561-1565.

Kopan R, Goate A. (2002). Aph-2/Nicastrin: an essential component of gamma-secretase and regulator of notch signaling and presenilin localization, *Neuron* 33:321-324.

Kotilinek LA *et al.* (2002). Reversible memory loss in a mouse transgenic model of Alzheimer's disease, *J Neurosci* 22:6331-6335.

Lazarov O, Lee M, Peterson DA, Sisodia SS. (2002). Evidence that synaptically released beta-amyloid accumulates as extracellular deposits in the hippocampus of transgenic mice, *J Neurosci* 22:9785-9793.

Lee VM, Goedert M, Trojanowski JQ. (2001). Neurodegenerative tauopathies, *Annu Rev Neurosci* 24:1121-1159.

Lesuisse C *et al.* (2001). Hyper-expression of human apolipoprotein E4 in astroglia and neurons does not enhance amyloid deposition in transgenic mice, *Hum Mol Gen* 10:2525-2537.

Lewis J *et al.* (2001). Enhanced neurofibrillary degeneration in transgenic mice expressing mutant tau and APP, *Science* 293:1487-1491.

Lewis J *et al.* (2000). Neurofibrillary tangles, amyotrophy and progressive disturbance in mice expressing mutant (P301L) tau protein, *Nat Genet* 25:402-405.

Li T *et al.* (2003). Nicastrin is required for assembly of presenilin/gamma-secretase complexes to mediate notch signaling and for processing and trafficking of beta-amyloid precursor protein in mammals, *J Neurosci* 23:3272-3277.

Li YM *et al.* (2000a). Presenilin 1 is linked with gamma-secretase activity in the detergent solubilized state, *Proc Natl Acad Sci USA* 97:6138-6143.

Li YM *et al.* (2000b). Photoactivated gamma-secretase inhibitors directed to the active site covalently label presenilin 1, *Nature* 405:689-694.

Lopez-Schier H, St Johnston D. (2002). *Drosophila* nicastrin is essential for the intramembranous cleavage of notch, *Dev Cell* 2:79-89.

Luo Y *et al.* (2003). BACE1 (beta-secretase) knockout mice do not acquire compensatory gene expression changes or develop neural lesions over time, *Neurobiol Dis* 14:81-88.

Luo Y *et al.* (2001). Mice deficient in BACE1, the Alzheimer's β-secretase, have normal phenotype and abolished β-amyloid generation, *Nature* 4:231-232.

Lyketsos CG *et al.* (2002). Prevalence of neuropsychiatric symptoms in dementia and mild cognitive impairment: results from the cardiovascular health study, *JAMA* 288:1475-1483.

Marr RA *et al.* (2003). Neprilysin gene transfer reduces human amyloid pathology in transgenic mice, *J Neurosci* 23:1992-1996.

Masliah E *et al.* (1994). Synaptic and neuritic alterations during the progression of Alzheimer's disease, *Neurosci Lett* 174:67-72.

Mattson M. (2001). Apoptosis in neurodegenerative disorders, *Nature* 1:120-129.

Mayeux R. (2003). Epidemiology of neurodegeneration, *Annu Rev Neurosci* 26:81-104.

McKhann G *et al.* (1984). Clinical diagnosis of Alzheimer's disease: report of the NINCDS-ADRDA Work Group under the auspices of the Department of Health and Human Services Task Force on Alzheimer's Disease, *Neurology* 34:939-944.

Mercken M, Fischer I, Kosik KS, Nixon RA. (1995). Three distinct axonal transport rates for tau, tubulin, and other microtubule-associated proteins: evidence for dynamic interactions of tau with microtubules *in vivo*, *J Neurosci* 15:8259-8267.

Monsonego A, Weiner HL. (2003). Immunotherapeutic approaches to Alzheimer's disease, *Science* 302:834-838.

Morgan D *et al.* (2000). Ab peptide vaccination prevents memory loss in an animal model of Alzheimer's disease, *Nature* 408:982-985.

Morris JC, Price JL. (2001a). Pathologic correlates of nondemented aging, mild cognitive impairment, and early-stage Alzheimer's disease, *J Mol Neurosci* 17:101-118.

Morris JC *et al.* (2001b). Mild cognitive impairment represents early-stage Alzheimer disease, *Arch Neurol* 58:397-405.

Mucke L *et al.* (2000). High-level neuronal expression of Ab$_{1-42}$ in wild-type human amyloid protein precursor transgenic mice: synaptotoxicity without plaque formation, *J Neurosci* 20:4050-4058.

Myers A *et al.* (2000). Susceptibility locus for Alzheimer's disease on chromosome 10, *Science* 290:2304-2305.

Naruse S *et al.* (1998). Effects of PS1 deficiency on membrane protein trafficking in neurons, *Neuron* 21:1213-1221.

Nicoll JA *et al.* (2003). Neuropathology of human Alzheimer disease after immunization with amyloid-beta peptide: a case report, *Nat Med* 9448-9452.

Oddo S *et al.* (2003). Triple-transgenic model of Alzheimer's disease with plaques and tangles: intracellular Abeta and synaptic dysfunction, *Neuron* 39:409-421.

Petersen RC. (2000). Aging, mild cognitive impairment, and Alzheimer's disease, *Neurol Clin* 18:789-806.

Petersen RC. (2003). Mild cognitive impairment clinical trials, *Nat Rev Drug Discov* 2:646-653.

Petersen RC *et al.* (2001a). Current concepts in mild cognitive impairment, *Arch Neurol* 58:1985-1992.

Petersen RC *et al.* (2000). Memory and MRI-based hippocampal volumes in aging and AD, *Neurology* 54:581-587.

Petersen RC *et al.* (1999). Mild cognitive impairment: clinical characterization and outcome, *Arch Neurol* 56:303-308.

Petersen RC *et al.* (2001b). Practice parameter: early detection of dementia: mild cognitive impairment (an evidence-based review). Report of the Quality Standards Subcommittee of the American Academy of Neurology, *Neurology* 56:1133-1142.

Price DL, Sisodia SS. (1998a). Mutant genes in familial Alzheimer's disease and transgenic models, *Annu Rev Neurosci* 21:479-505.

Price DL, Tanzi RE, Borchelt DR, Sisodia SS. (1998b). Alzheimer's disease: Genetic studies and transgenic models, *Ann Rev Genet* 32:461-493.

Price JL, Morris JC. (1999). Tangles and plaques in nondemented aging and "preclinical" Alzheimer's disease, *Ann Neurol* 45:358-368.

Roth KA. (2001). Caspases, apoptosis, and Alzheimer disease: causation, correlation, and confusion, *J Neuropathol Exp Neurol* 60:829-838.

Savonenko A *et al.* On the nature of cognitive deficits in a transgenic mouse model of Alzheimer's disease: working versus reference memory and the role of β-amyloid deposition. Submitted.

Savonenko AV *et al.* (2003b). Normal cognitive behavior in two distinct congenic lines of transgenic mice hyperexpressing mutant APPswe, *Neurobiol Dis* 12:194-211.

Schenk D *et al.* (1999). Immunization with amyloid-beta attenuates Alzheimer disease-like pathology in the PDAPP mouse, *Nature* 400:173-177.

Selkoe D, Kopan R. (2003). Notch and Presenilin: regulated intramembrane proteolysis links development and degeneration, *Annu Rev Neurosci* 26:565-597.

Selkoe DJ. (2001). Alzheimer's disease: genes, proteins, and therapy, *Physiol Rev* 81:741-766.

Selkoe DJ. (2002). Alzheimer's disease is a synaptic failure, *Science* 298:789-791.

Shen J *et al.* (1997). Skeletal and CNS defects in Presenilin-1-deficient mice, *Cell* 89:629-639.

Sheng JG, Price DL, Koliatsos VE. (2002). Disruption of corticocortical connections ameliorates amyloid burden in terminal fields in a transgenic model of Abeta amyloidosis, *J Neurosci* 22:9794-9799.

Sheng JG, Price DL, Koliatsos VE. (2003). The beta-amyloid-related proteins presenilin 1 and BACE1 are axonally transported to nerve terminals in the brain, *Exp Neurol* 184:1053-1057.

Sherrington R *et al.* (1995). Cloning of a gene bearing missense mutations in early-onset familial Alzheimer's disease, *Nature* 375:754-760.

Shulman JM, Shulman LM, Weiner WJ, Feany MB. (2003). From fruit fly to bedside: translating lessons from *Drosophila* models of neurodegenerative disease, *Curr Opin Neurol* 16:443-449.

Sinha S *et al.* (1999). Purification and cloning of amyloid precursor protein beta-secretase from human brain, *Nature* 402:537-540.

Sisodia SS, George-Hyslop PH. (2002). gamma-Secretase, Notch, Abeta and Alzheimer's disease: where do the presenilins fit in? *Nat Rev Neurosci* 3:281-290.

Sisodia SS *et al.* (1990). Evidence that β-amyloid protein in Alzheimer's disease is not derived by normal processing, *Science* 248:492-495.

Sisodia SS *et al.* (1993). Identification and transport of full-length amyloid precursor proteins in rat peripheral nervous system, J. *Neurosci* 13:3136-3142.

Slunt HH *et al.* (1994). Expression of a ubiquitous, cross-reactive homologue of the mouse β-amyloid precursor protein (APP), *J Biol Chem* 269:2637-2644.

Soderberg L *et al.* J. (2003). Molecular identification of AMY, an Alzheimer disease amyloid-associated protein, *J Neuropathol Exp Neurol* 62:1108-1117.

Solomon B. (2001). Immunotherapeutic strategies for prevention and treatment of Alzheimer's disease, *DNA Cell Biol* 20:697-703.

Steiner H *et al.* (2002). PEN-2 is an integral component of the gamma-secretase complex required for coordinated expression of presenilin and nicastrin, *J Biol Chem* 277:39062-39065.

Strittmatter WJ *et al.* (1993). Apolipoprotein E: high-avidity binding to β-amyloid and increased frequency of type 4 allele in late-onset familial Alzheimer disease, *Proc Natl Acad Sci USA* 90:1977-1981.

Struhl G, Greenwald I. (2001). Presenilin-mediated transmembrane cleavage is required for Notch signal transduction in *Drosophila*, *Proc Natl Acad Sci USA* 98:229-234.

Sturchler-Pierrat C *et al.* (1997). Two amyloid precursor protein transgenic mouse models with Alzheimer disease-like pathology, *Proc Natl Acad Sci USA* 94:13287-13292.

Sunderland T *et al.* (2003). Decreased beta-amyloid1-42 and increased tau levels in cerebrospinal fluid of patients with Alzheimer disease, *JAMA* 289:2094-2103.

Sze C-I *et al.* (1997). Loss of the presynaptic vesicle protein synaptophysin in hippocampus correlates with early cognitive decline in aged humans, *J Neuropathol Exp Neurol* 56:933-944.

Takasugi N *et al.* (2003). The role of presenilin cofactors in the gamma-secretase complex, *Nature* 422:438-441.

Taniguchi Y *et al.* (2002). Notch receptor cleavage depends on but is not directly executed by presenilins, *Proc Natl Acad Sci USA* 99:4014-4019.

Tanzi RE, Bertram L. (2001). New frontiers in Alzheimer's disease genetics, *Neuron* 32:181-184.

Terry RD *et al.* (1991). Physical basis of cognitive alterations in Alzheimer's disease: synapse loss is the major correlate of cognitive impairment, *Ann Neurol* 30:572-580.

Thinakaran G *et al.* (1997). Evidence that levels of presenilins (PS1 and PS2) are coordinately regulated by competition for limiting cellular factors, *J Biol Chem* 272:28415-28422.

Troncoso JC, Sukhov RR, Kawas CH, Koliatsos VE. (1996). In situ labeling of dying cortical neurons in normal aging and in Alzheimer's disease: correlations with senile plaques and disease progression, *J Neuropathol Exp Neurol* 55:1134-1142.

Vassar R. (2001). The beta-secretase, BACE: a prime drug target for Alzheimer's disease, *J Mol Neurosci* 17:157-170.

Vassar R *et al.* (1999). β-secretase cleavage of Alzheimer's amyloid precusor protein by the transmembrane aspartic protease BACE, *Science* 286:735-741.

Vassar R, Citron M. (2000). Ab-generating enzymes: recent advances in β- and γ-secretase research, *Neuron* 27:419-422.

von Koch CS *et al.* Generation of APLP2 KO mice and early postnatal lethality in APLP2/APP double KO mice, *Neurobiol Aging* 1997.

Walsh DM *et al.* (2002). Naturally secreted oligomers of amyloid β-protein potently inhibit hippocampal LTP *in vivo*, *Nature*

Wasco W *et al.* (1993). Isolation and characterization of *APLP2* encoding a homologue of the Alzheimer's associated amyloid β protein precursor, *Nat Genet* 5:95-99.

Whitehouse PJ *et al.* (1981). Alzheimer disease: evidence for selective loss of cholinergic neurons in the nucleus basalis, *Ann Neurol* 10:122-126.

Whittemore SR. (1999). Neuronal replacement strategies for spinal cord injury, *J Neurotrauma* 16:667-673.

Wilcock DM *et al.* (2003). Intracranially administered anti-Abeta antibodies reduce beta-amyloid deposition by mechanisms both independent of and associated with microglial activation, *J Neurosci* 23:3745-3751.

Wilcock DM *et al.* (2001). Number of Abeta inoculations in APP+PS1 transgenic mice influences antibody titers, microglial activation, and congophilic plaque levels, *DNA Cell Biol* 20:731-736.

Wolfe MS. (2002). Therapeutic strategies for Alzheimer's disease, *Nat Rev Drug Discov* 1:859-866.

Wolfe MS *et al.* (1999). Two transmembrane aspartates in presenilin-1 required for presenilin endoproteolysis and gamma-secretase activity, *Nature* 398:513-517.

Wong PC, Cai H, Borchelt DR, Price DL. (2002). Genetically engineered mouse models of neurodegenerative diseases, *Nat Neurosci* 5:633-639.

Wong PC, Price DL, Cai H. (2001). The brain's susceptibility to amyloid plaques, *Science* 293:1434-1435.

Wong PC *et al.* (1997). Presenilin 1 is required for Notch1 and DII1 expression in the paraxial mesoderm, *Nature* 387:288-292.

Xu G, Gonzales V, Borchelt DR. (2002). Abeta deposition does not cause the aggregation of endogenous tau in transgenic mice, *Alzheimer Dis Assoc Disord* 16:196-201.

Yan R *et al.* (1999). Membrane-anchored aspartyl protease with Alzheimer's disease β-secretase activity, *Nature* 402:533-537.

Yu G *et al.* (2000). Nicastrin modulates presenilin-mediated notch/glp-1 signal transduction and betaAPP processing, *Nature* 407:48-54.

Zheng H *et al.* (1995). β-amyloid precursor protein-deficient mice show reactive gliosis and decreased locomotor activity, *Cell* 81:525-531.

Zoghbi HY, Botas J. (2002). Mouse and fly models of neurodegeneration, *Trends Genet* 18:463-471.

28

Behavioral Intervention and Recovery from CNS Damage

Bernhard Voller, MD
Mark Hallett, MD

PRINCIPLES OF RECOVERY

Theories of Recovery

In the first days following injury to the brain, recovery might be related to resolution of the damaged tissue and the surrounding edema. At remote cortical sites, other suppressing events occur, a process known as *remote functional depression* or diaschisis (Feeney and Baron, 1986; Seitz *et al.*, 1999). At least part of the recovery is thought to result from the gradual reversal of diaschisis. By studying diaschisis it became evident that it may continue even after significant recovery has occurred. In addition to hypometabolism and inhibition, persistent remote effects include hyperexcitability and disinhibition. Another part of recovery is a result of behavioral compensation. The use of compensatory strategies is particularly seen in moderately to severely affected patients (Cirstea and Levin, 2000). One theory of recovery, known as vicariation of function, in which adjacent or remote structures from the damaged area are thought to take over the function of the damaged area, has been revived over the past decade due to several findings of functional plasticity after cortical injury (Xerri *et al.*, 1998). Plasticity is now known as the main underlying mechanism for recovery and the known modalities affecting plasticity as well as rehabilitative interventions modulating plasticity will be discussed in detail. Interventions of repair will be mentioned briefly.

Plasticity

Brain plasticity is the concept that the brain is able to change (Hallett, 1999, 2000, 2001a, 2001b). Over the past decade, newly developed noninvasive mapping studies such as neuroimaging, electroencephalography (EEG), magnetoencephalography, or transcranial magnetic stimulation (TMS) in humans, corresponding to findings from neurophysiological and neuroanatomical studies in animals, provided substantial evidence that the adult brain is capable of significant functional plasticity. Neurons and neural networks can change their function in several ways. Knowledge about the cellular mechanisms that are likely to be responsible for plasticity in humans is important to set new goals for rehabilitation purposes. Cellular physiology has been studied in different model systems, and we still do not know which mechanism applies to which phenomenon. First, a change in the balance of excitation and inhibition can happen very quickly. This process depends on the fact that neurons or neural pathways have a much larger region of anatomical connectivity than their usual territory of functional influence. A remarkable feature of the projecting pyramidal neurons is that they all have extensive local collateral branches in addition to the afferent axon (Gosh and Porter, 1988). The branches synapse in the vicinity of the neuron and typically have substantial horizontal connection systems (Hess and Donoghue, 1994). Together with the large amount of intracortical interneurons they build and connect to different networks or zones. Some zones may be kept in check by constant inhibition. If the inhibition is removed, the region of influence can be quickly augmented or enlarged in a process often called unmasking (Jacobs and Donoghue, 1991). A second process that can also be relatively fast is the activity dependent change of the efficacy of synapses by strengthening or weakening. These processes, such as long-term potentiation

(LTP) or long-term depression (LTD), were particularly investigated in the horizontal connections of the rat motor cortex (Hess et al., 1996, Hess and Donoghue, 1996). This modification of synaptic strength in fact would also be necessary for a continually adjustable connectional network that would depend on ongoing levels and patterns of activity. In other words, the ability to flexibly change the connections may be one substrate for reorganization. A third process is a change in neuronal membrane excitability by intracellular modification of sodium channels throughout the soma and axon (Halter et al., 1995). A fourth process is anatomical changes, which need a longer time, between days and weeks. Specific anatomical changes include sprouting of new axon terminals and formation of new synapses (Toni et al., 1999). Results from animal studies indicate the presence of dendritic growth, dendritic branching, synapse formation and changes in the specific structure of synaptic connections in the cortical tissue surrounding an injury and in the unlesioned hemisphere (Nudo et al., 2001).

All these processes operate in different periods and are not mutually exclusive. Certainly, one can be followed by another. Furthermore, it is not only the case that these processes can occur, they are constantly occurring. In general, brain plasticity is often involved following injury but not less important in normal motor control.

Repair

Growth factors, such as nerve growth factor (NGF), basic fibroblast growth factor (bFGF), and osteogenic protein-1 (OP-1), enhance recovery of sensorimotor function in rats (Kawamata et al., 1997, Kawamata et al., 1998, Kolb et al., 1997). The natural metabolite inosine, a purine nucleoside, produces actual rewiring in the rat brain, probably by expressing genes associated with axon growth (Chen et al., 2002). Experimental treatments for spinal cord injury (SCI) include antibodies to block axonal growth inhibitors (e.g., the anti Nogo A antibody IN-1), gangliosides to augment neuritic growth (e.g., chondroitinase ABC), implanting genetically pretreated fibroblasts that secret trophic factors such as NT-3, administration of 4-aminopyridine to enhance axonal conduction through demyelinated nerve fibers, implantation of fetal tissue for bridging completed lesions, and implantation of olfactory glia cells that migrate into the host tissue and carry regenerating axons with them (Fawcett, 2002, Kirshblum et al., 2002).

Transplanted cells may preserve existing host cells and connections by secreting trophic factors, establishing local connections that enhance synaptic activity, providing a bridge for host axonal regeneration, or actually replacing cellular elements (Wechsler and Kondziolka,

2003). There is some evidence in rat models that implantation of new cells might lead to restitution of lost tissue and rewiring (Veizovic et al., 2001). Although a primate model is still missing, there are already human trials in stroke using cultured human neuronal cells derived from a teratocarcinoma cell line. The safety and feasibility of transplantation was studied in 12 patients with basal ganglia stroke and fixed motor deficits (Kondziolka et al., 2000). Serial evaluations over an 18-month period showed no adverse effects. Although there was a benefit in some of the 12 patients in the European Stroke Scale score and an improvement of fluorodeoxyglucose uptake in the positron emission tomography (PET) scans at the implant site, this study is still preliminary and controversial (Meltzer et al., 2001). It is unclear whether and how the implanted cells contributed to the outcome.

REHABILITATIVE INTERVENTIONS MODULATING PLASTICITY

Enhancement of Function by Use of a Body Part

Physical Exercise Therapy

Cortical representation areas increase or decrease depending on use. For instance, Braille readers use their fingers many hours daily in the skilled task of interpreting the Braille characters. Detailed mapping of the motor cortical areas was performed with TMS targeting the first dorsal interosseus (FDI) and the adductor digiti minimi (ADM) muscles bilaterally in Braille readers and blind controls (Pascual-Leone et al., 1995). The FDI is used for reading, while the ADM is largely passive. In the controls, motor representations of right and left FDI and ADM were not significantly different. However, in the proficient Braille readers, the representation of the FDI in the reading hand was significantly larger than that in the nonreading hand or in either hand of the controls. In addition, the representation of the ADM in the reading hand was significantly smaller than that in the nonreading hand or in either hand of the controls, suggesting that the cortical representation for the reading finger in proficient Braille readers is enlarged at the expense of the representation of other fingers. In another study looking at cortical plasticity in patients who had a unilateral immobilization of the ankle joint with no peripheral nerve lesion, the motor cortex area of the inactivated tibialis anterior muscle diminished compared to the unaffected leg (Liepert et al., 1995).

It is assumed that different interventions such as physical exercise therapy, occupational therapy, and speech therapy facilitate such changes. Specific evidence demonstrating the values of specific interventions is sparse. Comparisons between different methods currently in use have failed so far to show that one strategy is highly superior to the other (Woldag and Hummelsheim, 2002). However, the concept of intensive physical exercise therapy after stroke has been best demonstrated by the constraint-induced movement therapy (CI) (Miltner *et al.*, 1999, Taub *et al.*, 1993, Taub *et al.*, 1999). This method forces use of the hemiplegic limb by constraining the good limb. In a number of clinical trials, there has been behavioral improvement even in patients with chronic and apparently stable deficits. It was shown that TMS maps of the weakened muscles increase in size in this circumstance showing that the expected cortical changes are occurring (Liepert *et al.*, 2000, Liepert *et al.*, 1998, Liepert *et al.*, 2000). In experiments at the National Institutes of Health, we studied 16 patients a year or more following stroke randomized to either constraint-induced movement therapy or control therapy (Wittenberg *et al.*, 2003, Wittenberg *et al.*, 2000). Clinical assessments performed before and after the intervention and at 6 months after the intervention included the Wolf Motor Function Test (WMFT), the Motor Activity Log (MAL), and the Assessment of Motor and Process Skills (AMPS). MAL was significantly increased in the CI group, while the other measures did not differ significantly. Furthermore, cerebral activation assessed with PET during a motor task decreased significantly and TMS motor map size increased in the affected motor cortex in CI patients only (Figure 28.1).

Other techniques that likely use the same principle of physical exercise therapy are robot-enhanced training, virtual reality training, mental practice, music therapy, bilateral arm training, and electrical stimulation.

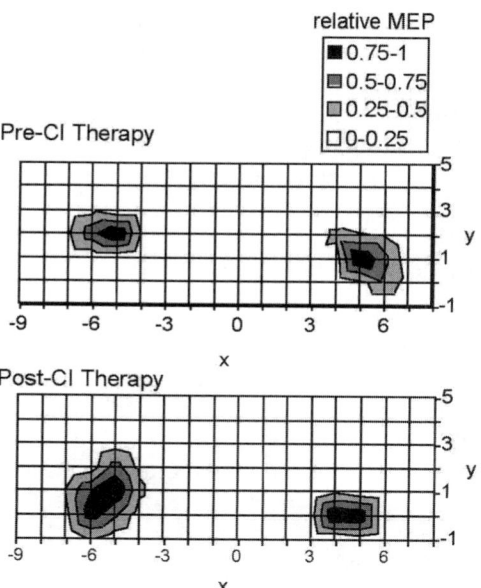

FIGURE 28.1 TMS maps of a stroke patient before and after CI therapy. The relative abductor pollicis brevis (APB) MEP size of both sides is displayed in a contour plot over the scalp surface as viewed from above, with x presenting the medial-lateral axis and y the anterior-posterior axis. Positive x is to the right, y anterior, and the origin (0,0) is the scalp vertex (Cz in the 10–20 EEG system). The size of the cortical representation of the APB muscle mapped by TMS is increased in the affected left side after CI therapy. Adapted from Wittenberg *et al.* (2000). Task-related and resting regional cerebral blood flow changes after constraint-induced movement therapy, *Neurology* 54(Suppl 3):A8.

Robotic Devices and Virtual Reality Systems

Robotic devices may be used to help with physical exercise training. In robot-aided sensorimotor training, the subject is given a goal-directed task; the robotic device measures the movement and, if necessary, guides the limb to its target. In one study, 56 patients with stroke and hemiparesis or hemiplegia received standard post-stroke multidisciplinary rehabilitation, and were randomly assigned either to receive robotic training (at least 25 hours) or exposure to the robotic device without training (Volpe *et al.*, 2000). By the end of treatment, the robot-trained group demonstrated improved motor outcome for the trained shoulder and elbow that did not generalize to the untrained wrist and hand. The robot-treated group also demonstrated significantly improved functional outcome. In another study, four modes of robot-assisted movements patterned after exercises currently used in therapy were promoted: passive mode,

active-assisted mode, active-constraint mode, and bimanual mode (Lum *et al.*, 2002). Compared with the control group which received conventional therapy, the robot group had larger improvements in the proximal movement portion of the Fugl-Meyer test, larger gains in strength and larger increases in reach extent for up to two months. Although it is unclear whether robot-assisted movement has unique therapeutic effects that cannot be provided by a human therapist some advantages may exist, such as a minimized need for supervision, the possibility of higher intensity of individualized therapy, and better applicability to more severely impaired subjects.

Virtual reality systems are also being applied to rehabilitation (Holden *et al.*, 2001). For instance, a personal computer-based desktop virtual reality system can be used to rehabilitate hand function in stroke patients (Jack *et al.*, 2001). The system uses two input devices, a CyberGlove and a Rutgers Master II-ND (RMII) force feedback glove, allowing user interaction with a virtual environment. Specific virtual reality exercises were created in which the intensity of practice and the feedback were enhanced to create the most efficient and individualized motor learning approach. Pilot clinical trials were conducted with 6 chronic stroke patients in two case series (Merians *et al.*, 2002). Objective measurements showed that each patient improved on most of the hand parameters, lasting in one study for over three weeks.

Mental Practice and Music Therapy

Because mental practice activates much of the same circuitry in the brain as does real action, it is not surprising that even mental practice has been demonstrated to be useful in motor rehabilitation (Deiber *et al.*, 1998; Page *et al.*, 2001; Pascual-Leone *et al.*, 1995). The interaction of rhythm, an essential structure of music, and the human motor system, is obvious. However, until recently this had never been substantiated by neurophysiological research (Thaut *et al.*, 1999). The beneficial effect of rhythmic facilitation on gait parameters (e.g., velocity, stride length) has now been demonstrated in stroke patients (Thaut *et al.*, 1997).

Methods are all presumably effective not only due to the increased use of the affected body part, but also because of attempted learning of new skills, attention to the task, and reward.

Ipsilateral Pathways

General Findings

Although rather weak in humans, there are ipsilateral pathways that could theoretically contribute to motor recovery. In classical functional anatomy studies, it was suggested that these pathways are more prominent in operating the proximal part of the limbs, but they can be documented even in normal humans even in distal muscles (Brinkman and Kuypers, 1973; Gazzaniga, 1970). From TMS studies, two candidates in the CNS were proposed to play a role in ipsilateral conduction: first, the ipsilateral uncrossed direct pyramidal tract; second, the bilateral cortico-reticulospinal pathway (Lawrence and Kuypers, 1968, 1968).

Ipsilateral pathways are certainly involved in recovery in patients with hemispherectomy, often used to treat intractable epilepsy or neoplasms. It is unclear by which neural mechanisms these patients regain reasonable motor function in muscle groups contralateral to the cerebral damage. TMS of the remaining hemisphere induces bilateral activation of proximal arm muscles, demonstrating the existence of ipsilateral pathways in this condition (Benecke *et al.*, 1991; Cohen *et al.*, 1991). It is possible, although controversial, that these pathways are also relevant in stroke recovery (Hallett, 2001a, 2001b). The clinical observation that unilateral hemi-

spheric lesions in patients with stroke can induce bilateral motor deficits also points to a possible bilateral cerebral control of the motor function of one side of the body (Colebatch and Gandevia, 1989; Jones *et al.*, 1989; Kim *et al.*, 2003).

After stroke, the intact hemisphere plays a crucial role in recovery from dysphagia (Hamdy *et al.*, 1998). In another TMS study, functional recovery of movements of the tongue after stroke was strongly related to the function of the intact hemisphere (Muellbacher *et al.*, 1999).

Recruitment of the ipsilateral unaffected sensorimotor cortex after brain injury has also been identified using single-photon emission tomography (Sabatini *et al.*, 1994), PET (Chollet *et al.*, 1991; Weiller *et al.*, 1992; Weiller *et al.*, 1993), and fMRI (Cao *et al.*, 1998; Cramer *et al.*, 1997). In these studies, profound ipsilateral hemispheric activation during contraction of the affected (recovered) hand in stroke patients was documented. In contrast to normal patterns of activation after movements of the affected hand, increased activation was mostly found in the areas of ipsilateral sensorimotor cortex, premotor cortex, supplementary motor area (SMA), insula, inferior parietal cortex, and contralateral cerebellum. In one study, in addition to the activation of an ipsilateral motor network in the unaffected hemisphere, two other possible processes of motor recovery were suggested: increased degree of SMA activation and activation of foci along the rim of a cortical infarct (Cramer *et al.*, 1997). In a recent review, it was suggested from results of MRI and PET studies that following stroke, the main mechanism underlying motor recovery involves enhanced activity of preexisting networks, with the concept that the greater the involvement of the ipsilateral motor network, the better the recovery (Calautti and Baron, 2003).

Another possible influence from the undamaged hemisphere could come from transcallosal connections. Some of these transcallosal connections are inhibitory and these may be inhibited by training, in particular by bilateral training. One fMRI study shows increased blood flow in the damaged hemisphere when movements are made bilaterally which is consistent with the idea that activity of the ipsilateral hemisphere may be supportive of the damaged hemisphere (Figure 28.2; see also Color Plate section) (Staines *et al.*, 2001).

Bilateral Arm Training

Although not fully understood, it does seem that attempts to perform bilateral symmetrical arm movement are helpful in motor rehabilitation via the mechanism of interhemispheric disinhibition. In a report on 12

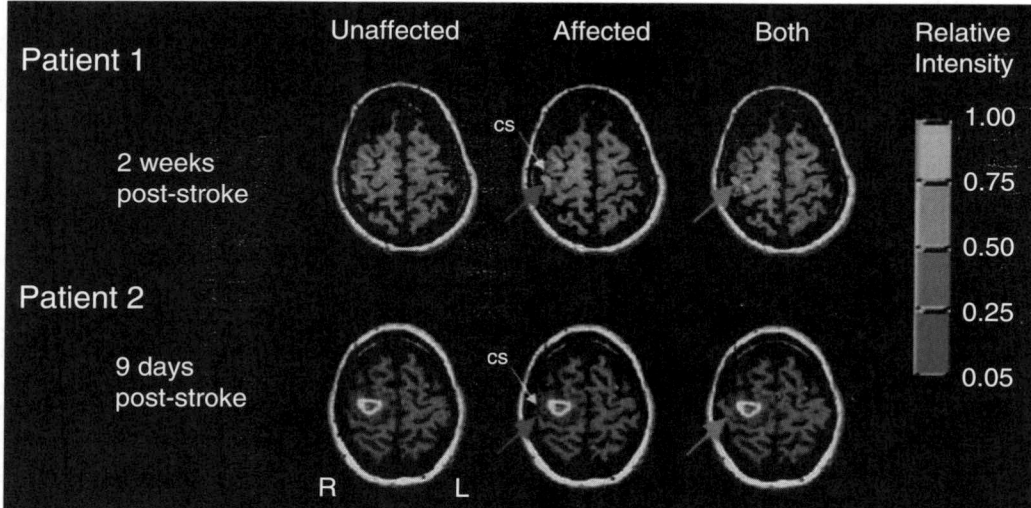

FIGURE 28.2 Functional MRI during repetitive unilateral or bilateral hand grips of two patients with right hemispheric stroke. The first column illustrates activation associated with repetitive gripping with the right (unaffected) hand. The second column illustrates activation associated with repetitive gripping with the left (affected) hand. The third column illustrates activation associated with repetitive gripping with both the affected and the anaffected hands simultaneously. Any significant activation of the right motor cortex when gripping with the unrecovered left hand (green arrow) was enhanced during bilateral movement (blue arrow). CS = central sulcus; R/L = right/left hemisphere. Adapted from Staines *et al.* (2001). Bilateral movement enhances ipsilesional cortical activity in acute stroke: A pilot functional mri study, *Neurology* 56:401-404. (See also Color Plate section.)

single-case experiments, improvement was shown following bilateral training, at least in the trained task (Mudie and Matyas, 2000). In the unimpaired brain during unilateral action, the ipsilateral hemisphere is inhibited, as unilateral actions are executed only by contralateral pathways. In unilateral stroke, there is inhibition by diaschisis in the damaged hemisphere during movement of the affected limb as well as inhibition coming from the undamaged side through the corpus callosum. The latter is thought to be more active in stroke patients than in healthy persons, because the undamaged side is hyperactive after stroke. This also occurs because the undamaged hemisphere controls the undamaged one less by inhibition.

It is theorized that the strong interhemispheric inhibition originating in the undamaged hemisphere may be overcome by bimanual movements. Three different pathways that could be activated during bimanual training were proposed. First, it is known that motor neurons of the ipsilateral pathway are disinhibited during bilateral action. In the damaged hemisphere, the spared cells of the ipsilateral pathway could be encouraged to directly interact with the adjacent spared cells of the contralateral pathway. Second, indirect corticospinal pathways could be recruited, such as cortico-reticulospinal pathways. Third, ipsilateral pathways from the undamaged hemisphere serving the affected limb could be utilized because they are potentially freer to be involved in the concurrent activity with contralateral pathways of the damaged hemisphere.

Regardless of how bilateral training works, another version that seems promising is repetitive bilateral arm training with rhythmic auditory cueing (BATRAC). In a study of 14 patients with chronic hemiparetic stroke, 6 weeks of BATRAC were performed (Whitall *et al.*, 2000). Four 5-minute periods per session (three times per week) of BATRAC were administered using a custom-designed arm training machine. The patients showed significant and potentially long-lasting increases in the Fugl-Meyer Upper Extremity Motor Performance Test of impairment, the Wolf Motor Function Test, and the University of Maryland Arm Questionnaire for Stroke. Isometric strength and active and passive ranges of motion improved in several joints of the paretic arm. These benefits were largely sustained 8 weeks after training concluded.

Mirror Therapy

A version of sensory feedback that is also related to bilateral arm movement is mirror therapy. Mirror therapy was reported as effective in a patient with poor functional use of an upper extremity, due mainly to somatosensory deficits (Sathian *et al.*, 2000). The subject underwent rehabilitation of the weak arm by being asked to move both arms in a symmetrical fashion while receiving feedback in the weak arm from a mirror reflection of the intact arm. Thus, there was illusory visual feedback of movement. This therapy was reported to be useful in functional recovery.

Sensory Stimulation

Somatosensory feedback is required for precise motor performance. In stroke patients, the finding that somatosensory deficits are usually associated with a slower recovery of motor function provides further evidence for the important role of somatosensory input (Reding and Potes, 1988). Techniques able to induce brain plasticity via sensory input may also help to improve function.

Electric Stimulation

A prolonged period of peripheral nerve stimulation increases excitability of related muscle representations in the motor cortex using TMS (Ridding et al., 2000). In some circumstances, even the motor mapping changes (Ridding et al., 2001). Sensory stimulation may be given in different ways, from passive movement to cutaneous and transcutaneous electrical nerve stimulation and even acupuncture. While there is some evidence for the efficacy of transcutaneous electrical nerve stimulation (Tekeoglu et al., 1998) and acupuncture (Wong et al., 1999), there is also contrary evidence (Hummelsheim et al., 1997; Johansson et al., 2001). In isolated cases, electrical stimulation may improve function in part by improving inattention (Mackenzie-Knapp, 1999). Stimulation of the pharynx may improve swallowing function (Hamdy et al., 2001).

Recently, a new technique combining peripheral and central stimulation was performed (Stefan et al., 2000). In healthy volunteers, high-frequency peripheral stimulation induced prolonged facilitation for up to 40 minutes (Pitcher et al., 2003). Another technique is transcranial direct current stimulation. It was shown that weak anodal stimulation of M1 resulted in increased implicit learning in healthy subjects, whereas stimulation of other motor areas or stimulation with opposite current flow did not improve performance (Nitsche et al., 2003).

Neuromuscular Electrical Stimulation

In a controlled study, neuromuscular electrical stimulation (NMES) was electromyographically triggered on the wrist and finger extension muscles in 11 chronic stroke patients (Cauraugh et al., 2000). The voluntary initiation of the electrical activity in the wrist extensor muscle served as a stimulus for the onset of electrical

stimulation. These movements produced externally enhanced proprioceptive feedback, theoretically completing the sensorimotor cycle and rendering the treatment effect. Subjects completed 12 30-minute treatment sessions. After the rehabilitation treatment, the experimental group moved significantly more blocks in the Box and Block test and displayed a higher isometric force impulse in a force generation task. In another study, it was shown that NMES may be useful even if not electromyographically triggered (Powell et al., 1999). In a study similar to the previous NMES study (Cauraugh et al., 2000), a second, simultaneous intervention— bilateral arm movement triggered by NMES—was added on 25 stroke patients (Cauraugh and Kim, 2002). In the coupled protocol group, results for the numbers of blocks moved in the functional task, reaction times to initiate movements and sustained muscle contraction capabilities were superior to those in the unilateral NMES or the control group. This study promotes a new approach in rehabilitation research, because two different mechanisms for activating the sensorimotor cortex to improve recovery are combined.

Reduction of Inhibition

Deafferentation as a Modulator of Cortical Plasticity

Amputation and transient deafferentation (e.g., by nerve blocks) are associated with plastic changes in the human motor system (Brasil-Neto et al., 1993; Cohen et al., 1991; Fuhr et al., 1992; Kew et al., 1994; Ridding and Rothwell, 1995). Muscles located close to the stump or to the deafferented limb are activated by TMS more easily and from more scalp positions than the same muscles in the control side in amputees or in normal experimentally pre-deafferented controls (Chen et al., 2002). A PET study showed increased regional cerebral blood flow (rCBF) in the contralateral sensorimotor cortex with movement of the amputated side compared to the normal side (Kew et al., 1994). These results are compatible with the development of more (sprouting) or stronger (unmasking) corticomotoneuronal connections targeting muscles close to the stump or area of experimental deafferentation, and indicate that deafferentation is an adequate trigger of cortical plasticity. Additionally, it is known that the deafferented motor cortex, during a forearm ischemic nerve block, becomes modifiable by inputs that are normally subthreshold for inducing changes in excitability. Plastic changes in the deafferented cortex can be up-regulated by direct transcranial stimulation of the plastic cortex, and down-regulated by TMS stimulation of the opposite cortex; that is, it can be modulated. This noninvasive modulation of deafferentation-induced plasticity might be used to facilitate plasticity when it is primarily beneficial, or to suppress it when it is predominantly maladaptive (Ziemann *et al.*, 1998).

A combination of motor training and experimental deafferentation can lead to enhanced cortical plasticity. Motor practice of the upper arm in the presence of experimental distal deafferentation substantially increases the electrophysiological measures of cortical plasticity in normal volunteers (Ziemann *et al.*, 2001). These results indicate that use-dependent plasticity is substantially potentiated in the human motor cortex after experimental deafferentation.

The effect of sensory deafferentation of the proximal arm combined with practice on recovery of hand function after chronic stroke was investigated in six patients with monohemispheric stroke. Brachial plexus blockade through the interscalene approach has been proven to be a safe and effective method that selectively blocks the upper portion of the brachial plexus branches without influencing hand function. Although patients showed significant enhancement of pinch power and acceleration together with motor-evoked potential (MEP) amplitudes after repetitive sessions of practices alone, these practice-related changes were further enhanced under local anesthesia of the proximal arm (Muellbacher *et al.*, 2002).

Repetitive Transcranial Magnetic Stimulation

TMS can produce a brief excitation or inhibition of the brain. When delivered repetitively, there may be longer effects, some of which appear to outlast the period of stimulation. Repetitive TMS (rTMS) has been tested for several years as a nonpharmacological, noninvasive method of directly influencing patients' cortical functions. A growing number of studies have utilized the modulatory effects of rTMS on cortical excitability (Wassermann and Lisanby, 2001). In general, high-frequency rTMS enhances cortical excitability and slow-frequency rTMS reduces excitability. High-frequency rTMS seems to improve motor function when applied to the primary motor areas in Parkinson patients (Pascual-Leone *et al.*, 1994; Siebner *et al.*, 2000; Sommer *et al.*, 2002). Strategies to decrease the inhibitory effects in the brain after stroke would focus on enhancing the process of plasticity after neural injury by changing the efficacy of already existing pathways and synapses, i.e., by the mechanisms of unmasking or LTP. It seems reasonable, for example, that the combination of motor training and decrease of inhibition by rTMS could lead to enhanced cortical plasticity.

Pharmacological Interventions

Another method to improve rehabilitation is combining use of drug therapy with motor training. Studies in general indicate that the combination is important. Although the mechanisms of neurotransmitter-modulated recovery are unknown, the best documented influence is with amphetamine and related noradrenergic agents (Feeney, 1997; Feeney *et al.*, 1993; Gladstone and Black, 2000). Amphetamine analogs (Table 28.1) or α_2-adrenergic receptor agonists accelerate motor recovery after experimental brain injury (Goldstein, 2000). Amphetamines may produce increased blood pressure, behavioral arousal, and hypermotility. In addition to the changes in rCBF, the actions of amphetamine may be mediated through noradrenergic, dopaminergic, or serotoninergic neurons. While it is possible that amphetamine has its effect by relieving diaschisis, it also seems to enhance plastic changes in motor learning situations in animals and humans (Lee and Ma, 1995; Soetens *et al.*, 1995). There is also evidence for enhanced neural sprouting and synaptogenesis (Stroemer *et al.*, 1998). There are several clinical trials demonstrating that amphetamine together with physical therapy is more effective than physical therapy alone (Crisostomo *et al.*, 1988; Walker-Batson *et al.*, 1995). However, other studies have shown no effect (Sonde *et al.*, 2001).

Some antidepressants and dopamine agonists have also been shown to be useful (see Table 28.1). The attention of clinicians involved in stroke rehabilitation has been particularly drawn to three prospective, randomized, double-blind, placebo-controlled studies. In the first study, the effect of methylphenidate was studied in 21 patients (Grade *et al.*, 1998). Treatment was given for 3 weeks in conjunction with physical therapy. Methylphenidate appeared to be both safe and effective. Second, levodopa was studied in 53 patients (Scheidtmann *et al.*, 2001). For the first 3 weeks, patients received single doses of levodopa (100 mg) or placebo daily in combination with physiotherapy, and for the second 3 weeks patients had only physiotherapy. Motor recovery was significantly improved after 3 weeks of drug intervention in the levodopa group and was maintained at 6 weeks. Third, the effect of fluoxetine and maprotiline was studied in 52 patients (Dam *et al.*, 1996). Patients randomly either received one of the two drugs or placebo over 3 months with physical therapy. Functional indices improved most in the fluoxetine group. All three studies seemed to confirm some results of animal studies, although they remain controversial.

On the other hand, some drugs have been demonstrated to impair recovery in animals and humans (see Table 28.1). These are antihypertensives, such as clonidine and prazosin; major tranquilizers, such as haloperidol, fluanisone, and droperidol; antidepressants, such as trazodone; anxiolytics, such as diazepam or phenobarbital; and anticonvulsants, such as phenytoin (Goldstein, 1995, 1997, 2000).

TABLE 28.1 Possible Mechanisms of Potential Positive and Negative Interference of Drugs on Motor Recovery after Brain Injury

Potential Good Drug?	Class	Possible Mechanism
(Dextro)amphetamine[a]	Psychostimulant, sympathomimetic	Noradrenergic, dopaminergic, serotoninergic, increases norepinephrine release?
Phentermine	Psychostimulant, sympathomimetic	Noradrenergic, dopaminergic, serotoninergic, increases norepinephrine release?
Phenylpropanolamine	Psychostimulant, sympathomimetic	Noradrenergic, dopaminergic, serotoninergic, increases norepinephrine release?
Methylphenidate[a]	Psychostimulant, sympathomimetic	Noradrenergic, dopaminergic, serotoninergic, increases norepinephrine release?
Yohimbine	α_2-adrenergic receptor agonists	Increases norepinephrine release
Idaxozan	α_2-adrenergic receptor agonists	Increases norepinephrine release
Fluoxetine[a]	Antidepressant	Serotonine reuptake inhibition
Trazodone[a]	Tranquilizer serotoninergic	Noradrenergic
Desipramine	Antidepressant	Noradrenergic
Bromocriptine	Antiparkinsonian	Dopamine agonist
Levodopa	Antiparkinsonian	Dopamine agonist

Potantial Bad Drug?	Class	Possible mechanism
Clonidine[a]	Antihypertensive, α_2-adrenergic receptor agonists	Reduces norepinephrine release
Prazosin[a]	Antihypertensive, α_2-adrenergic receptor antagonists	Reduces norepinephrine release
Phenoxybenzamine	Antihypertensive, α_2-adrenergic receptor antagonists	Reduces norepinephrine release
Haloperidol[a]	Neuroleptic	Noradrenergic receptor antagonist?, dopamine antagonist
Droperidol	Neuroleptic	Noradrenergic receptor antagonist?, dopamine antagonist
Fluanisone	Neuroleptic	Noradrenergic receptor antagonist?, dopamine antagonist
Diazepam[a]	Tranquilizer, anticonvulsant, muscle relaxant	GABA agonist
Phenytoin[a]	Antiepileptic	GABA agonist
Phenobarbital	Tranquilizer, hypnotic, sedative	GABA agonist
Trazodone	Tranquilizer serotoninergic	Noradrenergic

Probably no Interference?

Popanolol	β-adrenergic receptor antagonists
Amitryptiline	Antidepressant
Maprotiline[a]	Antidepressant
Nortriptylene[a]	Antidepressant
Fluoxetine	Antidepressant
Trazodone	Tranquilizer serotoninergic
Lamotrigine	Antiepileptic
Carbamazepine	Antiepileptic
Bromocriptine[a]	Antiparkinsonian
Clozapine	Atypical neuroleptic

[a] Including human studies.

In summary, clinical data do not yet conclusively justify widespread use of any of the above-mentioned potentially beneficial drugs. On the other, it seems logical to avoid drugs suspected of impairing recovery when alternatives are available in the clinical situation (e.g., hypertension and mood disturbances).

Restorative Devices

Whereas robotic devices are used for physical exercise therapy, other prosthetic devices are used for restoring function only. Prostheses can be entirely robotic or control a patient's limb by providing functional electrical stimulation to remaining muscle groups. Weak attempts at contraction may be detected and enhanced purely peripherally either with surface stimulation electrodes (Handmaster NMS-1, Bionic Glove, ETHZ-ParaCare neuroprosthesis and others) (Popovic *et al.*, 2001; Prochazka *et al.*, 1997; Snoek *et al.*, 2000) or by electrodes implanted in muscles (Freehand system, NEC FESMate system) (Smith *et al.*, 1987). The available neuroprostheses for grasping enable the restoration of the two most frequently used grasping styles, the palmar and the lateral grasp (Popovic *et al.*, 2002).

For patients with severe wide-ranging neuromuscular disorders and significant preserved cognitive function, EEG signals can provide a nonmuscular channel for sending commands to an external world, namely by a brain-computer interface (BCI). Successful operations require that the user encode commands in these signals and that the BCI derives the commands from the signals. Over time the user and the BCI necessarily have to adapt to each other. It has been shown for years that EEG can be used to control cursor movements with several degrees of freedom (Wolpaw *et al.*, 2000; Wolpaw, McFarland, and Vaughan, 2000). Such a control system has been demonstrated to control a device such as hand grasp prosthesis (Lauer *et al.*, 1999). The development of a BCI for multiple and complex movements would require, however, an implantable multi-microelectrode array. This would need to be coupled with a real-time conversion of simultaneously recorded neural activity to an online command for movement. Some progress is being made in this area (Isaacs *et al.*, 2000; Lauer *et al.*, 2000). Direct control of the BCI has been used in animal studies by intracortically implanted electrodes. New developments have also been made in cortical input devices implied for sensory loss (Donoghue, 2002).

Spasticity Management (Botulinum toxin)

While plasticity often produces functional improvement after brain damage, it may also produce unwanted effects, such as the development of spasticity.

Management of spasticity may therefore also influence plasticity in a beneficial way. In spasticity, pharmacological agents such as the benzodiazepines, baclofen, tizanidine, or clonidine influence spinal cord function. Baclofen and clonidine can be given intrathecally to increase their effective dose. Muscles can be weakened with dantrolene. A special note should be made of the utility of botulinum toxin. With botulinum toxin a focal reduction in muscle tone, painful spasms, and improved functionality can be obtained with fewer side effects (Esquenazi and Mayer, 2001). Also in general, spasticity must be approached carefully, because it may sometimes be useful for function.

CONCLUSION, PERSPECTIVES, AND HORIZONS

This is an exciting time for neurorehabilitation, because there is a wide variety of treatment options available. The occasional preoccupation with hyperacute and acute intervention in brain damage, especially in stroke, while critically important, should not lead one to neglect testing interventions that might enhance recovery in more chronic stages. The challenge is to identify which of the many changes observed are important and beneficial in mediating recovery and which are accessible by various treatments and to what extent. Of course, prevention and acute treatment are to be preferred, but if cerebral damage occurs, there is now considerable hope for patients.

ACKNOWLEDGMENT

Part of this chapter is modified, revised, and updated from Hallett, M. (2001b). Plasticity of the human motor cortex and recovery from stroke, *Brain Res Rev* 36:169-174.

References

AHA. (1991). Heart and stroke facts. Report, Dallas, Texas.

Benecke R, Meyer BU, Freund H.-J. (1991). Reorganisation of descending motor pathways in patients after hemispherectomy and severe hemispheric lesions demonstrated by magnetic brain stimulation, *Exp Brain Res* 83:419-426.

Brasil-Neto JP *et al.* (1993). Rapid modulation of human cortical motor outputs following ischaemic nerve block, *Brain* 116:511-25.

Brinkman J, Kuypers HG. (1973). Cerebral control of contralateral and ipsilateral arm, hand and finger movements in the split-brain rhesus monkey, *Brain* 96:653-674.

Calautti C, Baron J-C. (2003). Functional neuroimaging studies of motor recovery after stroke in adults: A review, *Stroke* 34:1553-1566.

Cao Y *et al.* (1998). Pilot study of functional mri to assess cerebral activation of motor function after poststroke hemiparesis, *Stroke* 29:112-122.

Cauraugh J *et al.* (2000). Chronic motor dysfunction after stroke: Recovering wrist and finger extension by electromyography-triggered neuromuscular stimulation, *Stroke* 31:1360-1364.

Cauraugh JH, Kim S. (2002). Two coupled motor recovery protocols are better than one: Electromyogram-triggered neuromuscular stimulation and bilateral movements, *Stroke* 33:1589-1594.

Chen P *et al.* (2002). Inosine induces axonal rewiring and improves behavioral outcome after stroke, *PNAS* 99:9031-9036.

Chen R, Cohen LG, Hallett M. (2002). Nervous system reorganization following injury, *Neuroscience* 111:761-773.

Chollet F *et al.* (1991). The functional anatomy of motor recovery after stroke in humans: A study with positron emission tomography, *Ann Neurol* 29:63-71.

Cirstea MC, Levin MF. (2000). Compensatory strategies for reaching in stroke, *Brain* 123:940-953.

Cohen LG, Bandinelli S, Findley TW, Hallett M. (1991). Motor reorganization after upper limb amputation in man. A study with focal magnetic stimulation, *Brain* 114:615-627.

Cohen LG *et al.* (1991). Magnetic stimulation of the human cerebral cortex, an indicator of reorganization in motor pathways in certain pathological conditions, *J Clin Neurophysiol* 8:56-65.

Colebatch JG, Gandevia SC. (1989). The distribution of muscular weakness in upper motor neuron lesions affecting the arm, *Brain* 112:749-763.

Cramer SC *et al.* (1997). A functional MRI study of subjects recovered from hemiparetic stroke, *Stroke* 28:2518-2527.

Crisostomo EA *et al.* (1988). Evidence that amphetamine with physical therapy promotes recovery of motor function in stroke patients, *Ann Neurol* 23:94-97.

Dam M *et al.* (1996). Effects of fluoxetine and maprotiline on functional recovery in poststroke hemiplegic patients undergoing rehabilitation therapy, *Stroke* 27:1211-1214.

Deiber M-P *et al.* (1998). Cerebral processes related to visuomotor imagery and generation of simple finger movements studied with positron emission tomography, *NeuroImage* 7:73-85.

Donoghue JP. (2002). Connecting cortex to machines: Recent advances in brain interfaces, *Nat Neurosci* 5:1085-1088.

Esquenazi A, Mayer N. (2001). Botulinum toxin for the management of muscle overactivity and spasticity after stroke, *Curr Atheroscler Rep* 3:295-298.

Fawcett J. (2002). Repair of spinal cord injuries: Where are we, where are we going? *Spinal Cord* 40:615-623.

Feeney DM. (1997). From laboratory to clinic: Noradrenergic enhancement of physical therapy for stroke or trauma patients, *Adv Neurol* 73:383-394.

Feeney DM, Baron JC. (1986). Diaschisis, *Stroke* 17:817-830.

Feeney DM, Weisend MP. (1993). Noradrenergic pharmacotherapy, intracerebral infusion and adrenal transplantation promote functional recovery after cortical damage, *J Neural Transplant Plast* 4:199-213.

Fuhr P *et al.* (1992). Physiological analysis of motor reorganization following lower limb amputation, *Electroencephalogr Clin Neurophysiol* 85:53-60.

Gazzaniga MS. (1970). *The bisected brain.* New York: Appleton-Century-Crofts.

Gladstone DJ, Black SE. (2000). Enhancing recovery after stroke with noradrenergic pharmacotherapy: A new frontier? *Can J Neurol Sci* 27:97-105.

Goldstein LB. (1995). Common drugs may influence motor recovery after stroke. The Sygen in Acute Stroke Study Investigators. *Neurology* 45:865-871.

Goldstein LB. (1997). Influence of common drugs and related factors on stroke outcome, *Curr Opin Neurol* 10:52-57.

Goldstein LB. (2000). Effects of amphetamines and small related molecules on recovery after stroke in animals and man, *Neuropharmacology* 39:852-859.

Goldstein LB. (2000). Rehabilitation and recovery after stroke, *Curr Treat Options Neurol* 2000:4.

Gosh S, Porter R. (1988). Morphology of pyramidal neurones in monkey motor cortex and the synaptic actions of their intracortical axon collaterals, *J Physiol* 400:593-615.

Grade C *et al.* (1998). Methylphenidate in early poststroke recovery: A double-blind, placebo-controlled study, *Arch Phys Med Rehab* 79:1047-1050.

Hallett M. (1999). Plasticity in the human motor system, *Neuroscientist* 5:324-332.

Hallett M. (2000). Plasticity (chapter 23). In Mazziotta JC, Toga Aw, Frackowiak RS, editors: *Brain mapping; the disorders*, pp. 569-586. San Diego: Academic Press.

Hallett M. (2001a). Functional reorganization after lesions of the human brain: Studies with transcranial magnetic stimulation, *Rev Neurol (Paris)* 157:822-826.

Hallett M. (2001b). Plasticity of the human motor cortex and recovery from stroke, *Brain Res Rev* 36:169-174.

Halter JA, Carp JS, Wolpaw JR. (1995). Operantly conditioned motoneuron plasticity: Possible role of sodium channels, *J Neurophysiol* 73:867-871.

Hamdy S *et al.* (1998). Recovery of swallowing after dysphagic stroke relates to functional reorganization in the intact motor cortex, *Gastroenterology* 115:1104-1112.

Hamdy S, Aziz Q, Thompson DG, Rothwell JD. (2001). Physiology and pathophysiology of the swallowing area of human motor cortex, *Neural Plast* 8:91-97.

Hess G, Aizenman CD, Donoghue JP. (1996). Conditions for the induction of long-term potentiation in layer ii/iii horizontal connections of the rat motor cortex, *J Neurophysiol* 75:1765-1778.

Hess G, Donoghue JP. (1994). Long-term potentiation of horizontal connections provides a mechanism to reorganize cortical motor maps, *J Neurophysiol* 71:2543-2547.

Hess G, Donoghue JP. (1996). Long-term depression of horizontal connections in rat motor cortex, *Eur J Neurosci* 8:658-665.

Holden MK *et al.* (2001). Retraining movement in patients with acquired brain injury using a virtual environment, *Stud Health Technol Inform* 81:191-198.

Hummelsheim H, Maier-Loth ML Eickhof C. (1997). The functional value of electrical muscle stimulation for the rehabilitation of the hand in stroke patients, *Scand J Rehabil Med* 29:3-10.

Isaacs RE, Weber DJ, Schwartz AB. (2000). Work toward real-time control of a cortical neural prosthesis, *IEEE Trans Rehabil Eng* 8:196-198.

Jack D *et al.* (2001). Virtual reality-enhanced stroke rehabilitation, *IEEE Trans Neural Syst Rehabil Eng* 9:308-318.

Jacobs KM, Donoghue JP. (1991). Reshaping the cortical motor map by unmasking latent intracortical connections, *Science* 251:944-947.

Johansson BB *et al.* (2001). Acupuncture and transcutaneous nerve stimulation in stroke rehabilitation: A randomized, controlled trial, *Stroke* 32:707-713.

Jones RD, Donaldson IM, Parkin PJ. (1989). Impairment and recovery of ipsilateral sensory-motor function following unilateral cerebral infarction, *Brain* 112:113-132.

Kawamata T *et al.* (1997). Intracisternal basic fibroblast growth factor enhances functional recovery and up-regulates the expression of a molecular marker of neuronal sprouting following focal cerebral infarction, *PNAS* 94:8179-8184.

Kawamata T *et al.* (1998). Intracisternal osteogenic protein-1 enhances functional recovery following focal stroke. *Neuroreport* 9:1441-1445.

Kew JJ *et al.* (1994). Reorganization of cortical blood flow and transcranial magnetic stimulation maps in human subjects after upper limb amputation, *J Neurophysiol* 72:2517-2524.

Kim SH *et al.* (2003). Ipsilateral deficits of targeted movements after stroke, *Arch Phys Med Rehabil* 84:719-724.

Kirshblum SC *et al.* (2002). Spinal cord injury medicine. 1. Etiology, classification, and acute medical management, *Arch Phys Med Rehabil* 83:S50-57, S90-98.

Kolb B, Cote S, Ribeiro-da-Silva A, Cuello AC. (1997). Nerve growth factor treatment prevents dendritic atrophy and promotes recovery of function after cortical injury, *Neuroscience* 76:1139-1151.

Kondziolka D *et al.* (2000). Transplantation of cultured human neuronal cells for patients with stroke, *Neurology* 55:565-569.

Lauer RT, Peckham PH, Kilgore KL. (1999). Eeg-based control of a hand grasp neuroprosthesis, *Neuroreport* 10:1767-1771.

Lauer RT, Peckham PH, Kilgore KL, Heetderks WJ. (2000). Applications of cortical signals to neuroprosthetic control: A critical review, *IEEE Trans Rehabil Eng* 8:205-208.

Lawrence DG, Kuypers HG. (1968). The functional organization of the motor system in the monkey. I. The effects of bilateral pyramidal lesions, *Brain* 91:1-14.

Lawrence DG, Kuypers HG. (1968). The functional organization of the motor system in the monkey. II. The effects of lesions of the descending brain-stem pathways, *Brain* 91:15-36.

Lee EH, Ma YL. (1995). Amphetamine enhances memory retention and facilitates norepinephrine release from the hippocampus in rats, *Brain Res Bull* 37:411-416.

Liepert J *et al.* (2000). Treatment-induced cortical reorganization after stroke in humans, *Stroke* 31:1210-1216.

Liepert J *et al.* (1998). Motor cortex plasticity during constraint-induced movement therapy in stroke patients, *Neurosci Lett* 250:5-8.

Liepert J, Storch P, Fritsch A, Weiller C. (2000). Motor cortex disinhibition in acute stroke, *Clin Neurophysiol* 111:671-676.

Liepert J, Teghenthoff M, Malin JP. (1995). Changes of cortical motor area size during immobilization, *Electroencephalogr Clin Neurophysiol* 97:382-386.

Lum PS *et al.* (2002). Robot-assisted movement training compared with conventional therapy techniques for the rehabilitation of upper-limb motor function after stroke, *Arch Phys Med Rehabil* 83:952-959.

Mackenzie-Knapp M. (1999). Electrical stimulation in early stroke rehabilitation of the upper limb with inattention, *Aust J Physiother* 45:223-227.

Meltzer CC *et al.* (2001). Serial [18f] fluorodeoxyglucose positron emission tomography after human neuronal implantation for stroke, *Neurosurgery* 49:586-591: discussion, 591-592.

Merians AS *et al.* (2002). Virtual reality-augmented rehabilitation for patients following stroke, *Phys Ther* 82:898-915.

Miltner WHR *et al.* (1999). Effects of constraint-induced movement therapy on patients with chronic motor deficits after stroke: A replication, *Stroke* 30:586-592.

Mudie MH, Matyas TA. (2000). Can simultaneous bilateral movement involve the undamaged hemisphere in reconstruction of neural networks damaged by stroke? *Disabil Rehabil* 22:23-37.

Muellbacher W, Artner C, Mamoli B. (1999). The role of the intact hemisphere in recovery of midline muscles after recent monohemispheric stroke, *J Neurol* 246:250-256.

Muellbacher W *et al.* (2002). Improving hand function in chronic stroke, *Arch Neurol* 59:1278-1282.

Nitsche MA *et al.* (2003). Facilitation of implicit motor learning by weak transcranial direct current stimulation of the primary motor cortex in the human, *J Cogn Neurosci* 15:619-626.

Nudo RJ, Plautz EJ, Frost SB. (2001). Role of adaptive plasticity in recovery of function after damage to motor cortex, *Muscle Nerve* 24:1000-1019.

Page SJ, Levine P, Sisto SA, Johnston MV. (2001). Mental practice combined with physical practice for upper-limb motor deficit in subacute stroke, *Phys Ther* 81:1455-1462.

Pascual-Leone A *et al.* (1995). Modulation of muscle responses evoked by transcranial magnetic stimulation during the acquisition of new fine motor skills, *J Neurophysiol* 74:1037-1045.

Pascual-Leone A *et al.* (1994). Akinesia in parkinson's disease. Ii. Effects of subthreshold repetitive transcranial motor cortex stimulation, *Neurology* 44:892-898.

Pitcher JB, Ridding MC, Miles TS. (2003). Frequency-dependent, bi-directional plasticity in motor cortex of human adults, *Clin Neurophysiol* 114:1265-1271.

Popovic MR *et al.* (2001). Surface-stimulation technology for grasping and walking neuroprosthesis, *IEEE Eng Med Biol Mag* 20:82-93.

Popovic MR, Popovic DB, Keller T. (2002). Neuroprostheses for grasping, *Neurol Res* 24:443-452.

Powell J *et al.* (1999). Electrical stimulation of wrist extensors in poststroke hemiplegia, *Stroke* 30:1384-1389.

Prochazka A, Gauthier M, Wieler M, Kenwell Z. (1997). The bionic glove: An electrical stimulator garment that provides controlled grasp and hand opening in quadriplegia, *Arch Phys Med Rehabil* 78:608-614.

Reding MJ, Potes E. (1988). Rehabilitation outcome following initial unilateral hemispheric stroke. Life table analysis approach, *Stroke* 19:1354-1358.

Ridding MC *et al.* (2000). Changes in muscle responses to stimulation of the motor cortex induced by peripheral nerve stimulation in human subjects, *Exp Brain Res* 131:135-143.

Ridding MC, McKay DR, Thompson PD, Miles TS. (2001). Changes in corticomotor representations induced by prolonged peripheral nerve stimulation in humans, *Clin Neurophysiol* 112:1461-1469.

Ridding MC, Rothwell JC. (1995). Reorganisation in human motor cortex, *Can J Physiol Pharmacol* 73:218-222.

Sabatini U *et al.* (1994). Motor recovery after early brain damage. A case of brain plasticity, *Stroke* 25:514-517.

Sathian K, Greenspan AI, Wolf SL. (2000). Doing it with mirrors: A case study of a novel approach to neurorehabilitation, *Neurorehabil Neural Repair* 14:73-76.

Scheidtmann K, Fries W, Muller F, Koenig E. (2001). Effect of levodopa in combination with physiotherapy on functional motor recovery after stroke: A prospective, randomised, double-blind study, *Lancet* 358:787-790.

Seitz RJ *et al.* (1999). The role of diaschisis in stroke recovery, *Stroke* 30:1844-1850.

Siebner HR *et al.* (2000). Short-term motor improvement after sub-threshold 5-hz repetitive transcranial magnetic stimulation of the primary motor hand area in parkinson's disease, *J Neurol Sci* 178:91-94.

Smith P, Peckham PH, Keith MW, Roscoe DD. (1987). An externally powered, multichannel, implantable stimulator for versatile control of paralyzed muscle, *IEEE Trans Biomed Eng* 34:499-508.

Snoek GJ *et al.* (2000). Use of the ness handmaster to restore hand function in tetraplegia: Clinical experiences in ten patients, *Spinal Cord* 38:244-249.

Soetens E, Casaer S, D'Hooge R, Hueting JE. (1995). Effect of amphetamine on long-term retention of verbal material, *Psychopharmacology (Berl)* 119:155-162.

Sommer M *et al.* (2002). Repetitive paired-pulse transcranial magnetic stimulation affects corticospinal excitability and finger tapping in parkinson's disease, *Clin Neurophysiol* 113:944-950.

Sonde L *et al.* (2001). A double-blind placebo-controlled study of the effects of amphetamine and physiotherapy after stroke, *Cerebrovasc Dis* 12:253-257.

Staines WR, McIlroy WE, Graham SJ, Black SE. (2001). Bilateral movement enhances ipsilesional cortical activity in acute stroke: A pilot functional mri study, *Neurology* 56:401-404.

Stefan K *et al.* (2000). Induction of plasticity in the human motor cortex by paired associative stimulation, *Brain* 123: 572-584.

Stroemer RP, Kent TA, Hulsebosch CE, Feeney DM. (1998). Enhanced neocortical neural sprouting, synaptogenesis, and behavioral recovery with d-amphetamine therapy after neocortical infarction in rats (editorial comment), *Stroke* 29:2381-2395.

Taub E *et al.* (1993). Technique to improve chronic motor deficit after stroke, *Arch Phys Med Rehabil* 74:347-354.

Taub E, Uswatte G, Pidikiti R. (1999). Constraint-induced movement therapy: A new family of techniques with broad application to physical rehabilitation—a clinical review, *J Rehabil Res Dev* 36:237-251.

Tekeoglu Y, Adak B, Goksoy T. (1998). Effect of transcutaneous electrical nerve stimulation (tens) on barthel activities of daily living (adl) index score following stroke, *Clin Rehabil* 12:277-280.

Thaut MH, Kenyon GP, Schauer ML, McIntosh G.C. (1999). The connection between rhythmicity and brain function, *IEEE Eng Med Biol Mag* 18:101-108.

Thaut MH, McIntosh GC, Rice RR. (1997). Rhythmic facilitation of gait training in hemiparetic stroke rehabilitation, *J Neurol Sci* 151:207-212.

Toni N *et al.* (1999). Ltp promotes formation of multiple spine synapses between a single axon terminal and a dendrite, *Nature* 402:421-425.

Veizovic T *et al.* (2001). Resolution of stroke deficits following contralateral grafts of conditionally immortal neuroepithelial stem cells, *Stroke* 32:1012-1019.

Volpe BT *et al.* (2000). A novel approach to stroke rehabilitation: Robot-aided sensorimotor stimulation, *Neurology* 54:1938-1944.

Walker-Batson D *et al.* (1995). Amphetamine paired with physical therapy accelerates motor recovery after stroke: Further evidence, *Stroke* 26:2254-2259.

Wassermann EM, Lisanby SH. (2001). Therapeutic application of repetitive transcranial magnetic stimulation: A review, *Clin Neurophysiol* 112:1367-1377.

Wechsler LR, Kondziolka D. (2003). Cell therapy: Replacement, *Stroke* 34:2081-2082.

Weiller C *et al.* (1992). Functional reorganization of the brain in recovery from striatocapsular infarction in man, *Ann Neurol* 31:463-472.

Weiller C *et al.* (1993). Individual patterns of functional reorganization in the human cerebral cortex after capsular infarction, *Ann Neurol* 33:181-189.

Whitall J, Waller SM, Silver KHC, Macko RF. (2000). Repetitive bilateral arm training with rhythmic auditory cueing improves motor function in chronic hemiparetic stroke, *Stroke* 31:2390-2395.

Wittenberg GF *et al.* (2003). Constraint-induced therapy in stroke: Magnetic-stimulation motor maps and cerebral activation, *Neurorehabil Neural Repair* 17:48-57.

Wittenberg GF *et al.* (2000). Task-related and resting regional cerebral blood flow changes after constraint-induced movement therapy, *Neurology* 54(Suppl 3):A8.

Woldag H, Hummelsheim H. (2002). Evidence-based physiotherapeutic concepts for improving arm and hand function in stroke patients: A review, *J Neurol* 249:518-528.

Wolpaw JR *et al.* (2000). Brain-computer interface technology: A review of the first international meeting, *IEEE Trans Rehabil Eng* 2000:2.

Wolpaw JR, McFarland DJ, Vaughan TM. (2000). Brain-computer interface research at the wadsworth center, *IEEE Trans Rehabil Eng* 8:222-226.

Wong AM *et al.* (1999). Clinical trial of electrical acupuncture on hemiplegic stroke patients, *Am J Phys Med Rehabil* 78: 117-122.

Xerri C, Merzenich MM, Peterson BE, Jenkins W. (1998). Plasticity of primary somatosensory cortex paralleling sensorimotor skill recovery from stroke in adult monkeys, *J Neurophysiol* 79:2119-2148.

Ziemann U, Corwell B, Cohen LG. (1998). Modulation of plasticity in human motor cortex after forearm ischemic nerve block, *J Neurosci* 18:1115-1123.

Ziemann U, Muellbacher W, Hallett M, Cohen LG. (2001). Modulation of practice-dependent plasticity in human motor cortex, *Brain* 124:1171-1181.

What Functional Brain Imaging Will Mean for Neurology

Richard S.J. Frackowiak, MD, DSc, FRCP, FMedSci
John C. Mazziotta, MD, PhD

Functional brain imaging was introduced in the late 1970s by an adaptation of principles established by Hounsfield (1973) and Cormack (1963) for noninvasive tomographic scanning of the human brain. The notion that the distribution of radio labeled tracers would provide functional information regionally throughout the brain was quickly applied using positron-emitting radiotracers (Phelps *et al.*, 1975). The detection of positron emitting isotopes with positron emission tomography (PET) cameras was quantitative and the unit of function could be calibrated in terms of radioactivity deposited in a volume of tissue over time. These techniques were rapidly used to investigate pathophysiology, especially in relation to energy metabolism with the use of oxygen-15–labeled water, oxygen, and fluorine-18–labeled glucose (Frackowiak, 1989). In the middle 1980s, multi-slice tomographs with rapid response characteristics enabled dynamic measurement of local cerebral blood flow (Fox *et al.*, 1988). This ability was exploited to make repetitive measurements of local brain function using flow as an indirect index of synaptic activity. The era of making dynamic measurements of specific human brain functions in a reasonable time period was upon us.

In the early 1990s, functional magnetic resonance imaging (fMRI) using the blood oxygen level–dependent (BOLD) technique appeared (Kwong *et al.*, 1992). The major advantage was that it eschewed the use of radioisotopes, thus making it safer and repeatable on innumerable occasions in normal volunteers and patients. A major disadvantage was the inability to calibrate the BOLD signal and an incomplete understanding of its origins. It is certainly largely dependent on the balance between oxyhemoglobin and de-oxyhemoglobin,

which have different magnetic properties, at sites where local brain activity changes, but it is also associated with local increases of perfusion and blood volume. In fact, the BOLD signal is an indirect index of local oxygen extraction—when the BOLD signal is high, oxygen extraction is low. Recently, empirical evidence indicates that the BOLD signal accurately reflects local field potentials in the brain (Logothetis *et al.*, 2001). This observation confirms the idea that activation reflects neural activity in projecting axons arriving in the dendritic arbors at a site in the brain. The activation also includes local neural processing, but the dominant influence is likely to be exerted by excitatory afferents to a brain region. These advances have lead to an explosion of studies of the functional anatomy of normal human brain (Frackowiak *et al.*, 1997). It is on the basis of results from normal brains in the past 10 years and the increasing input from neuropsychologists and other neuroscientists using scanning to define systems level cerebral physiology that future applications to clinical neurology will be based.

In the first 5 years of the 1990s, novel statistical methods were introduced and systematically validated. These are sometimes collectively called Statistical Parametric Mapping. They were developed for the analysis of functional scans (Friston *et al.*, 1991; Friston *et al.*, 1995, 1995a; http://www.fil.ion.ucl.ac.uk/spm) but recently they have also been applied to the analysis of structural MRI scans providing accurate definition of local brain anatomy—so-called voxel-based morphometry (VBM) (Ashburner and Friston, 2001; Good *et al.*, 2002). Indeed, a number of sophisticated morphometric methods based on structural MRI scanning of the brain are now available (Mazziotta *et al.*, 2001a). The armamentarium of the clinical neuroscientist is now complete. Broca brought anatomy and behavior together in his pioneering observations on the patient "Tan" in 1862. The modern clinical neuroscientist interested in systems-level analysis of the brain is now able to bring together sophisticated, accurate, and bias-free measurements of local brain function, and anatomy, to add to clinical observations and behavioral manipulations. The quantitative nature of the data, the elimination of observer bias through the use of automated analytic computer algorithms, and an increasing appreciation of the principles underlying functional brain organization, both in terms of local specialization and integration between functionally specialized areas, makes this a particularly exciting time. The application of modern neuroimaging to diagnostics, prognostics, and therapeutics is upon us. It remains to be seen what the eventual impact will be.

ENABLING METHODOLOGICAL ADVANCES

Systematic Image Analysis Procedures Eliminating Observer Bias

Functional brain maps and indeed, structural MRI images, are essentially matrixes of data elements known as voxels. There are as many as 100,000 per scanned brain volume. In imaging experiments based on BOLD, many sequential brain scans may be recorded in a single session. Each scan will be collected under particular conditions or in a particular physiological or behavioral context. Scans may be collected in health and in disease. The subsequent analysis of these scans aims to detect significant areas of change of function or anatomy that may be localized or distributed among a number of regions or in a particular tissue compartment; for example, gray matter. The analytical procedure therefore requires comparisons in a statistical sense of many individual data points. To draw valid inferences from these comparisons, a clinical scientist has to protect against false-positive and false-negative results. Classically, protection against false-positive results is done with the Bonferroni correction, although typically, that modification applies to multiple independent comparisons. The voxels in a brain image are not independent. They are linked by virtue of the resolution of the camera so that the values in adjacent voxels are correlated. They are also correlated with adjacent and distant voxels by common physiological responses. The problem is therefore one of correcting for multiple *non-independent* comparisons. The general linear model and theory of Gaussian Random Fields is

a favored method of estimating the actual number of independent comparisons to make the appropriate connections. This major advance in statistics allows valid influences to be drawn from imaging data (Worsley *et al.*, 1996).

Analysis of structural scans is classically done by an expert using the eye and clinical experience. Although extremely efficient at detecting focal changes in brain structure that method is not good at detecting or interpreting diffuse changes in gray matter volume, white matter integrity, or cerebrospinal fluid (CSF) volumes. Contemporary morphometric methods of choice use observer-dependent manual delineation of brain structures and pixel counting to detect atrophy in regions of interest (e.g., the hippocampus) (Watkins *et al.*, 2002). Such methods have been useful but are subject to observer bias that frequently needs to be controlled by test-retest and receiver-operator characteristic analysis. The more modern voxel-based computerized techniques are fully automated and therefore observer bias is eliminated, which is a significant advantage.

Another major advance in structural and functional brain mapping that renders them more clinically relevant is the introduction of novel imaging test paradigms. The classical experiment of comparing brain function in one state with that in another and attributing changes in activity to differences in the two states has been supplemented by more complex and informative paradigms. Categorical comparisons are good for delineating functionally specialized brain areas such as those found in early perceptual pathways. However, brain areas may respond to changes in stimuli in a nonlinear fashion. One way of examining such response characteristics is to use a parametric design where brain function is correlated with a systematic change in the relevant stimulus. The responses of brain areas may also differ in various contexts, for example, in different pathological states. Scanning can be done so that categorical or parametric comparisons are repeated in both states and then contrasted. Such factorial designs elucidate nonlinear relationships between activities in different areas under different conditions. This type of scanning paradigm is particularly informative for studying effects of therapy. Functional scans can be collected with and without a stimulus of interest and the patterns of activation recorded in the presence of active treatment and placebo. The difference between activations in the two therapeutic conditions will show the site at which therapy is having its physiological effect (Pariente *et al.*, 2001).

Most recently, induced changes in activity between different brain regions, in functional brain systems, have been quantified using advanced techniques such as Effective Connectivity Analysis and Structural Equation Modeling (Rowe *et al.*, 2002). The ability to detect systematic and significant changes of correlated activity in a complex brain system, comprising a number of functionally specialized areas, is particularly promising for those interested in the treatment of complex behaviors.

fMRI has a relatively poor temporal resolution compared to its exquisite spatial resolution. Nevertheless, techniques are now available that permit one to distinguish activity in two different states with a resolution of approximately 1 second (Josephs and Henson, 1999). These event-related scans are to be compared with scans that are blocked. In these latter scans, the stimulus is given over a finite period, often 20–30 seconds, to generate a single activity map per stimulus type. With event related scanning, the response to each individual stimulus is recorded. There are a number of advantages that accrue from event related scanning. Scans can be classified, post hoc, on the basis of some behavioral or physiological response, task difficulty, or other parameter, and then compared. The method acknowledges that brain responses often habituate with constant stimuli and provides a way of identifying transient activity in such situations by avoiding habituating situations. A relative disadvantage of the event-related method is that it is statistically less powerful than the blocked method.

Small Study Populations with Sufficient Statistical Power

Functional brain imaging provides quantitative data on local brain activity and has sufficient statistical power to allow detection of small changes in brain function with relatively small subject groups. This capability is particularly useful for screening treatments for effectiveness, as well as defining appropriate dosage schedules before phase three trials of clinical efficacy. With the blocked method, as a rule no more than six to ten subjects are required to generate a significant result. With the Event Related Method, when the aim is to provide an answer relevant to the study population (fixed effects), a similar number of subjects or patients is required. If generalization to the population of which the study group is representative is wanted, then 16 to 24 subjects are usually necessary to account for random error. Standard power calculations of scanning studies are difficult because of the interaction and interdependence of voxels. However, a decade of experience has provided the empirical rules described above. The signal to noise characteristics of MR scans are very favorable, so that 1% changes in signal can be accurately determined using standard scanning techniques and magnets and the numbers of subjects described previously (Friston et al., 1999).

Creation of Structural and Functional Atlases of Normal Populations and Patients

Atlases and databases of the normal brain represent the logical use of digital and clinical data about brain structure and function in health and disease in the modern era. The complexity of the human brain mitigates against the continued use of traditional methods where narrative reports are published in journals which must then be laboriously collected and reviewed. An emerging approach, that is now well characterized in the literature (see Mazziotta, 2001, 2001a for review), is to develop probabilistic, digital atlases and databases of the human brain where such information can be stored for both normal individuals and patients with neurological disorders. Using four dimensions as the framework for such atlases (three in space and one in time [age]), information from *in vivo* imaging studies, post mortem data sets, as well as clinical and genetic information, can all be referenced to a particular brain location at a given point in the human lifespan. Ultimately, all clinical data could be organized in this fashion. This formidable task is far from being realized at the present time, but through research studies in normal subjects and more limited examples in patients, such approaches are taking hold and provide the opportunity to produce previously unavailable insights into structure-function relationships in the normal brain and in the course of neurological disorders in specific patient populations.

Combining information about structure and function will be a mainstay of brain mapping strategies in the years to come. Population studies from large numbers of subjects provide information about the anatomical locations and their variance for macroscopic cerebral structures including gyri, sulci, the size, shape, and position of deep nuclei as well as white matter tracts (Mazziotta *et al.*, 1995, 2001, 2001a; Steinmetz *et al.*, 1991; Paus *et al.*, 1996, 1996a; Penhume *et al.*, 1996; Geschwind and Levitsky, 1968; Ono *et al.*, 1990; Talairach and Tournoux, 1988; Thompson *et al.*, 1998). Similar population data sets have been (Dumoulin *et al.*, 2000) and will be obtained using functional techniques to provide estimates of functional anatomy in large segments of the population.

Strategies for the automated identification of microscopic cyto- and chemoarchitectural data from significant numbers of post mortem human specimens will provide probability estimates for these variables at a microscopic level (Clark and Miklossy, 1990; Amunts *et al.*, 1998, 2000; Rademacher *et al.*, 1992, 1993; Geyer *et al.*, 2000; Hayes and Lewis, 1995), an issue that has been discussed for the past century (Brodmann, 1909; Bailey and von Bonin, 1951). By combining structural and functional data sets at microscopic and macroscopic levels (Bodegård, 2000), it will be possible to understand where information about variance in these parameters is quantified, and what the important structure-function relationships in the brain actually are for the first time in large populations. For example, is there concordance or discordance between cytoarchitectural zones and major gyri and sulci in the human brain? Do chemoarchitectural zones, populated by transmitter-specific neurons, correspond to cytoarchitectural regions? Do functional activation sites in the brain correspond more closely to macroscopic anatomy such as gyri and sulci, or are they better aligned with cytoarchitectural or chemoarchitectural regions? If it turns out that functional brain responses are best correlated with cytoarchitectural or chemoarchitectural regions, standardized functional tasks may be used as surrogate markers of these more difficult to obtain and costly microscopic assays (Mazziotta et al., 2001, 2001a).

This information will also have practical importance with regard to clinical questions. For example, if one can develop a probabilistic distribution for functional activations of key sites in the brain, the targeting of therapeutic lesions or for locating the placement site of stimulation electrodes (e.g., subthalamic electrodes in patients with Parkinson's disease) will be greatly facilitated and potentially more therapeutically efficacious. The same type of information would be useful in presurgical planning for patients who are under consideration for surgical therapy of resectable cerebral lesions such as tumors or epileptic foci.

Understanding the gross anatomical distribution of white matter tracts will be similarly important. The advent of diffusion tensor imaging (DTI) (Basser et al., 1996; Conturo et al., 1999; Pierpaoli et al., 1996; Stieltjes et al., 2001) has already provided the opportunity to see the position and direction of large bundles of nerve fibers and their myelin sheaths in human subjects (Figure 29.1; see also the Color Plate section), a topic that has languished for most of the last century (Flechsig, 1920). By collecting such data in large populations, probabilistic estimates of the location and variance of fiber tracts can be developed. By evaluating changes in these tracts during maturation, development (Paus et al., 2001), and aging, it may be possible to get a better sense of how the brain responds to these natural events and the implications associated with them. This will further enhance the interpretation of activation studies performed to monitor changes in the responses of gray matter regions throughout the human age range.

While less attention has been paid to white matter anatomy and function by the brain mapping community, a resurgence of interest in this area will certainly emerge as the new techniques discussed previously mature and because questions will arise relevant to brain maturation

and aging that rely heavily on information of this sort. Already, MRI has provided an indirect measure of myelinization (Paus et al., 2001) that can provide important new insights into the relationship between gray and white matter maturation during the human life span.

Finally, a detailed understanding of white matter anatomy will be critical in preoperative and intraoperative surgical planning (Kettenbach et al., 1999) where disruption of a critical white matter connection is undoubtedly as detrimental to the patient as the resection of the gray matter sites of origin or targets for specific functional systems.

Multimodality Imaging in Disease

Individual imaging modalities provide a unique perspective on brain structure and function. Taken alone, each perspective provides a unique and important insight into cerebral organization and its alteration in disease states. When methods are combined, disadvantages of one technique can be supplemented by advantages of another. Complimentary aspects of multiple techniques provide an integrative view of brain structure and function that is greater than the sum of the parts. For example, PET and MRI can provide images of brain function with relatively high spatial resolution (i.e., 10-75 mm^3) but have limitations with regard to temporal resolution (i.e., seconds to minutes). Combining information about temporal events from techniques with high temporal resolution (e.g., electroencephalography [EEG], magnetoencephalography [MEG]) but which have relatively low spatial resolution can result in the complementary integration of such methods with an improved overall perspective on the spatial and temporal function of the brain in a given state.

By directly stimulating the brain with a magnetic field (transcranial magnetic stimulation [TMS]) and combining the effects measured using this technique with passive "observation" of intrinsic brain activity during behavioral tasks, maps of structure, function, and connectivity can be combined to evaluate physiology in the normal brain and pathophysiology in patients with neurological disorders.

Certain prerequisites are required for such studies to stand up to rigorous scientific scrutiny. The alignment and registration of data sets from multiple tomographic modalities is now a reality (Ashburner and Friston, 1997; Woods et al., 1992, 1993, 1998). Such approaches make it possible to integrate information in the same subject from data sets acquired with different tomographic techniques (e.g., CT, PET, MRI, fMRI). They can also be used for averaging data in the same subject from the same modality (e.g., serial MRI studies in patients with degenerative diseases or brain tumors to measure rates of change in atrophy or tumor growth).

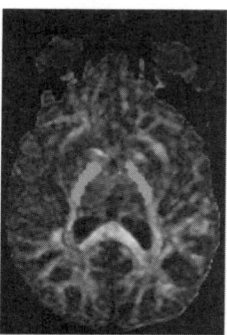

FIGURE 29.1   Human white matter anatomy. It is possible to identify white matter bundles *in vivo* in the human brain using DTI MRI. Such studies take advantage of the fact that diffusion of water molecules is compartmentalized within the CNS. As such, this diffusion is not isotropic. The anisotropic movement of water is most obvious in fiber bundles before the fibers branch and reach their targets. As such, major tracks in the brain, such as the cortico-spinal track shown in this image, can be quite clearly identified in individuals. When populations of normal subjects are imaged in this fashion, probabilistic estimates for the location of specific fiber bundles can be determined for the population in general and may be of value in understanding the variance of these structures in the human brain. In a given individual, such studies may be helpful in presurgical planning to avoid disconnection syndromes associated with the inadvertent resections of part or all of these tracks. Courtesy of A. Alexander, M. Lazar and colleagues, Waisman Center, University of Wisconsin. (See also the Color Plate section.)

One example of between modality data integration has been with the use of fMRI, EEG, and MEG data sets (Ahlfors *et al.*, 1999; Bonmassar *et al.*, 2001). When applied in the same subject using the same behavioral paradigm, fMRI provides useful information on functional localization with good spatial resolution (~3-4 mm in plane) but relatively low temporal resolution. MEG provides the opposite; that is, high temporal resolution with relative low spatial resolution. The integration of the two techniques produces a combined data set that has both high spatial and temporal resolution. Nevertheless, this type of strategy is not without its pitfalls. The source of the fMRI signal results from relative hemodynamic changes in perfusion associated with increased neuronal activity resulting from the net increase (or decrease) in cell firing in large numbers of cells over an integrated time frame. The MEG data results from synchronous firing of cells, of sufficient number to be detected by surface SQUID devices in a reliable fashion. Thus, the two signal sources are not identical and using one (i.e., MEG) to provide temporal information for the other (i.e., fMRI) where spatial resolution is excellent, may lead to erroneous interpretations. Various strategies are now underway to determine the weighting factors that each should use for selecting the sites and timing of events; for example, the specific relationship between cellular electrophysiology and fMRI responses (Logothetis *et al.*, 2001). If successful, the integrated use of both techniques may result in composite data sets that have improved quality over either of the two individually.

fMRI has also been integrated with EEG data for two purposes. In the first, the two methods have been used in an analogous fashion to the fMRI-MEG data sets discussed above, namely, to provide composite high spatial and temporal resolution data sets where timing of events is obtained from the EEG data and spatial localization from the fMRI data.

Clinically, fMRI-EEG integration has been used to identify pathologic electrophysiological events and their site of origin. In patients with epilepsy, focal "spikes" occur in the EEG record as subclinical manifestations of the propensity to have clinical seizures. Spike localization is an important aspect of epilepsy diagnostics and the selection of appropriate therapy. The poor spatial resolution of scalp EEG, however, can often make such localization quite difficult. In the mid-1990s, it was shown that it was possible, at 1.5T (Ives et al., 1993; Detre et al., 1995), to monitor the timing and location of epileptic spikes in fMRI-EEG data sets. Nevertheless, artifacts produced by the radio frequency signals from the fMRI data acquisition contaminated the EEG and movement of electrode wires in the MRI's magnetic field, caused by voluntary subject movement or the ballistocardiogram, resulted in fMRI artifacts. Since then, many of these problems have been solved or minimized. In fact, it is now possible to perform "spike mapping" using combined fMRI and EEG in MRI units that function at higher fields (e.g., 3T) (Goldman et al., 2001; Lemieux et al., 2001). This type of integrated strategy should be very useful in diagnostic applications relevant to patients with epilepsy or in studies of normal subjects where fMRI–event-related potentials (ERPs) are of interest (Bonmassar et al., 2001).

One may also use a combination of imaging techniques to examine neuronal connectivity (Fox et al., 1997; Paus et al., 1997). Three techniques are typically combined. Consider the following example: a healthy subject performs a motor task during an fMRI study to identify the hand area of the motor cortex. Three-dimensional images are reconstructed demonstrating the location of that functional area in that subject's brain. The subject is then positioned in a PET device and a TMS stimulator is targeted, using frameless stereotaxy, to the coordinates of the subject's hand area of the motor cortex. At the time of injection of the PET blood flow tracer, the TMS device is fired, artificially activating the hand area of the motor cortex, thereby causing it to appear on the functional image. Because this activation will also be propagated orthodromically and antidromically, both inputs and outputs from that subject's motor cortex will also increase their relative perfusion and facilitate depiction on the PET image. The strategy allows for the identification of regions in the brain that are functionally connected through the combination of these three brain-mapping methods. By performing such studies in large numbers of subjects and in many brain locations, considerable data can be added to probabilistic atlases, providing the important attribute of connectivity to maps of brain structure and function. Such methods can be used in conjunction with anatomical connection data derived from diffusion MRI techniques (Conturo et al., 1999; Stieltjes et al., 2001). Furthermore, these methods can be compared with functional connectivity maps (Friston et al., 1993; Hyde et al., 1995) to determine the relationship between anatomical and functional connectivity within the human brain.

Connectivity may prove to be one of the most valuable, interesting and, currently, least addressed issue of functional neuroanatomy. Taken together, information derived from effective, anatomical and other forms of connectivity (e.g., microscopic techniques in post mortem specimens) will be valuable, if not critical, components in developing better theoretical frameworks for understanding cerebral organization and primary or high level mental processes.

Preoperative and intraoperative mapping for surgical procedures is another area ripe for integration from multiple modalities. Consider a patient with a brain tumor near language areas in the dominant hemisphere. Structural imaging with MRI would provide information about basic brain anatomy as well as the size, position and distortions induced by the tumor. Functional imaging provides information about critical cortical and subcortical graygray matter areas that should be avoided to preserve language function adjacent to the tumor itself. DTI with MRI can provide an estimate of deep white matter structures, which should also be avoided in the course of the surgical resection. High density EEG and/or MEG could provide "spike maps" to indicate whether cortical regions near the tumor were epileptogenic and should also be considered for resection. Repetitive TMS stimulation in critical areas adjacent to the tumor could theoretically be used to produce a "virtual lesion" demonstrating the types and degree of deficits that the patient might experience if normal cortical regions were resected inadvertently. TMS could also be used to map functional connections of language areas that pass through white matter structures near or adjacent to the tumor itself. These combined data sets could be brought to the operating room, after their pre-operative acquisition, and displayed with a surgical navigation system that would register with probes on the surface of, or deep within the brain to aid in the execution of the procedure. Intraoperative MRI or optical imaging (Toga et al., 1995) could then be used to update functional maps as the brain shape changed in the course of resection as a result of therapeutic maneuvers (e.g., osmotic dehydration) or local pathophysiological effects (e.g., local edema). Such a suite of tools would provide the neurosurgeon with a vastly increased wealth of information about an individual patient and a greater margin of safety for performing surgical resections in or near critical brain structures.

PROSPECTS FOR DIAGNOSIS

Experience in the past 15 years has suggested that functional imaging is more likely to provide information about pathogenesis, or of importance to monitoring treatment than specifically for diagnostic purposes. The recent computerized morphometric techniques, however, appear to be particularly sensitive to the detection of diffuse cerebral change as found in dementias and degenerative diseases. The accuracy with which the component tissues of the brain can be segmented also provides the possibility that specific measurements of damage to the white matter could improve assessment of disease load in demyelinating diseases. Patterns of gray matter atrophy differ between different types of primary dementia. Combinations of gray and white matter atrophy may have even more specific diagnostic characteristics. The ability to record the EEG simultaneously with MRI imaging has already been exploited to detect cortical ectopia and relate spiking activity to sites of cortical dysgenesis. The opportunities for improved diagnosis with multi-modality imaging appear to be great.

Functional Disease

The distinction between functional and psychiatric disease is a difficult one. There have been recent studies, mostly case studies to date, that have tackled syndromes that primarily represent somatization disorders. Examples of these include hysteria and malingering. For the purposes of this chapter, paroxysmal headaches, particularly migraine and its variants, are also grouped under functional disorders because, until the functional and structural imaging studies of the late 1990s, little advance had been made in the understanding of the origin of these disorders, despite considerable advances in therapeutics.

Functional imaging will be informative in somatization disorders such as hysteria when plausible and realistic cognitive models of these syndromes can be formulated. At that stage, it will be possible to decompose the cognitive components and attempt to measure their integrity in the brains of patients. Attempts have been made to image patients with hysterical paralysis (Halligan et al., 2000). Obvious difficulties arise in trying to compare cortical activity that might be attributed to intended movement, attention to movement, or mechanisms of movement inhibition, in situations where a limb cannot actually be moved. A number of ingenious ways of overcoming this sort of problem have been described, including examining patients and normal volunteers in states of hypnosis. In hypnotized subjects, motor states can be induced that correspond to intent to deceive (malingering) or paralysis. Such studies are generating results that implicate interactions between pre-frontal, pre-motor, and motor executive regions, and provide a unique way forward in the diagnosis of these disorders.

In the headache syndromes, the way forward has been clearer. Capturing patients during an aura or in the phase of a paroxysmal headache has been difficult, but case studies of immense importance have been described (May et al., 2000; Woods et al., 1994). Alternative strategies include scanning during the induction of pain by agents such as subcutaneous capsaicin or with resolution of pain following effective treatments (for example, the imidazoles). The role of the brainstem in generating migrainous paroxysmal headaches has been indicated by a number of studies (Weiller et al., 1995), leading to the suggestion that it is a primarily neurological disease with secondary effects on the vascular system (May et al., 1999). More intriguingly, one as yet unconfirmed study suggests that there are structural abnormalities that colocalize with functional abnormalities in unilateral cluster headaches (May et al., 1999a). Independent therapeutic attempts using deep brain stimulation of the hypothalamic area, implicated by structural and functional results in chronic sufferers from cluster headache that are resistant to all forms of noninvasive therapy, have proved anecdotally successful (Leone et al., 2001). The indication of therapeutic site or target by functional and structural studies such as these is promising for those interested in the treatment of functional disorders.

Another example relates to people suffering from chronic pain. Studies in this area have concentrated on trying to determine whether pain processing is normal in such patients or is in some way pathological. This hypothesis is supported by some evidence, with indications that the anterior cingulate centered medial pain system shows abnormal processing of acute painful stimuli in patients. The concept of pathological processing of normal stimuli, without corresponding structural change, is an important one. This type of result suggests the possibility of identifying targets for pharmacological or stereotactic surgery in people suffering from sometimes intensely disabling functional syndromes.

Degenerative Disease

Imaging studies have become a mainstay for diagnosing a wide range of central nervous system disorders. The list is long and well-known to practicing clinicians. Most studies focus on structural imaging using MRI, but the prospects are ripe for functional diagnoses using PET, fMRI, and SPECT. An important aspect of the use of functional imaging to diagnose degenerative disorders is that since these disorders have an underlying biochemical basis, functional imaging demonstrates abnormalities earlier, and with more sensitivity, than structural imaging. The latter ultimately shows the hallmark of almost all degenerative diseases, namely cerebral atrophy, whether focal or generalized. Such functional changes can also precede clinical signs and symptoms of these disorders. For example, in Huntington's disease, functional changes in glucose metabolism of the caudate nucleus occur years before the onset of signs and symptoms in patient populations (Mazziotta *et al.*, 1987; Grafton *et al.*, 1990). This is followed by structural changes that can be measured with MRI as atrophy of the caudate nuclei. As such, presymptomatic individuals (i.e., those that have the pathologic genotype) could be monitored for some period of time with functional imaging, a window opportune for exploring the effects of experimental therapies. Because such presymptomatic individuals are, by definition, normal, there are no clinical measures by which to determine the efficacy of a given experimental treatment for them. Functional imaging can provide noninvasive, objective, and quantifiable information relevant to the progression of the disease and the effectiveness of the experimental treatment. Interestingly, recent VBM studies are now also showing anatomical changes in patients with presymptomatic Huntington's disease (Thieben *et al.*, 2002).

The same strategy has been developed and used for Alzheimer's disease. It was noted, early in the application of PET technology, that metabolism was reduced in the parietal and superior temporal neocortical regions in patients with signs and symptoms of Alzheimer's disease (Frackowiak *et al.*, 1981). Subsequently, using the same strategies that were discussed above for Huntington's disease, Small *et al.* (1995) demonstrated that such early changes were also demonstrable in individuals at increased risk for Alzheimer's disease, by virtue of the fact that they had the genotype ApoE 4. Subsequent studies demonstrated that there was a significant decline in glucose metabolism in parietal, temporal, and cingulate cortices in these high risk, presymptomatic individuals when compared to control subjects, again demonstrating an avenue for evaluating experimental therapies in this patient population (Small *et al.*, 2000). With fMRI, Bookheimer *et al.* (2000) used a similar experimental design in high-risk individuals (ApoE 4 genotype) who were presymptomatic for Alzheimer's disease. Reiman *et al.* (2001) obtained similar results. Thus, an easily available technique, fMRI, could be used to screen individuals for functional abnormalities as well as to monitor experimental therapies for such conditions. Differential patterns of abnormal metabolism and atrophy (Rosen *et al.*, 2002) are already providing highly sensitive ways of distinguishing and monitoring, in time, the different forms of dementia that are being recognized and described (Alzheimer's, frontotemporal, semantic, etc.)

The impact of such approaches cannot be underestimated given the large number of individuals at risk for degenerative disorders of the brain and the limited armamentarium of medical therapeutics available to treat them. As new therapeutic agents are brought into the experimental clinical domain from basic neuroscientific explorations, a vital need for sensitive, presymptomatic monitoring of the therapeutic effects of such agents would be immediately realized and required. In the area of diagnosis, similar pressures and requirements will continue to increase as the magnitude of these global neurological health problems outpaces our capacity to treat such patients.

Paroxysmal Disease

Both structural and functional imaging can be used to diagnose paroxysmal disorders of the brain as well as to monitor ongoing cerebrovascular pathophysiology in patients with ischemic brain injury. The concept of "spike mapping" has previously been discussed but structural brain imaging may also provide useful insights into the site of epileptogenic lesions in patients with seizure disorders.

Advanced analysis strategies for morphometric evaluation of brain structure have resulted in the probabilistic atlases noted above, that quantify variance within the human population (Mazziotta *et al.*, 1995, 2001, 2001a). Subpopulations drawn from these larger data sets can provide comparison populations for smaller groups or individual subjects (Thompson *et al.*, 1998, 2000). Using such an approach, combined with advanced image segmentation analysis strategies, it is possible to find subtle structural abnormalities in the brain. This has also been applied in the clinical environment. For example, small and qualitatively subtle heterotopias that produce focal epilepsy can be missed when conventional MRI studies of brain structure are qualitatively analyzed by neuroradiologists. This is because the variance in cortical thickness is large among individuals in the population. Nevertheless, when that variance can be quantified through probabilistic strategies and compared to single individuals, such heterotopias have been identified (Bastos *et al.*, 1999) because they fall outside of the confidence limits for the normal variance of the brain in a given cortical region. In some cases, patients have had as many as three or four qualitative MRI studies demonstrating no abnormality even when the EEG focus of seizures had been identified and could be used by the radiologist to bias their qualitative examination of structural MRI studies for abnormalities. Quantitative techniques can identify these "missed" heterotopias (Bastos *et al.*, 1999), in some cases leading to surgical resection of the cortical zones and improvement or resolution of a patient's seizure disorder.

In the area of ischemic brain disease, there is a critical need to develop sensitive indications of perfusion, metabolism, and prognosis for tissue at jeopardy from ischemic injury. This has led to the quest to find "tissue clocks" for ischemic brain injury (Koroshetz *et al.*, 1997). MRI and PET studies have already provided new insights into the hemodynamic and metabolic events that result in cerebral ischemia and infarction (Derdeyn *et al.*, 1999; Firlik *et al.*, 1998; Lutsep *et al.*, 1997; Marchal *et al.*, 1993; Moseley *et al.*, 1990; Gilman, 1998; Schlaug *et al.*, 1999). However, patients who arrive at hospitals with new ischemic events must be screened for the duration of time between the onset of an event and potential therapy (e.g., tPA). Currently, a 3-hour window is allowable and predictive of efficacious intervention using this approach (Hacke *et al.*, 1995; Katzan *et al.*, 2000; tTP Stroke Study Group, 1995). While this seems straightforward, the determination of this time interval is often difficult and fraught with errors. For example, a patient may awaken with a new deficit and be unable to determine when it actually occurred. The patient may be unable to report the information personally (e.g., secondary to aphasia) and family members may not have a clear or accurate estimate of when a given patient's symptoms began. Finally, the therapeutic window for the use of agents such as tPA has been determined through clinical trials where an arbitrary time frame was selected and efficacy determined in a binary fashion (i.e., it worked or it didn't). It may be that there are selected patients for which a time window determined by these strategies is inappropriate, at either end of the temporal spectrum. For all these reasons, it is highly desirable that a tissue clock (Baird and Warach, 1998; Kidwell *et al.*, 2000; Koroshetz *et al.*, 1997; Staroselskaya *et al.*, 2001) be developed using objective and noninvasive tools. It is conceivable that some combination of perfusion, diffusion, and other MRI-based techniques may, in aggregate, provide such an estimate of tissue viability and a window of therapeutic intervention. It is predicted that rigorous, multipulse sequence evaluations of such patients with MRI methods will result in just such a tissue clock, thereby enhancing the application of such therapies to a wider range of patients and providing a better understanding of why and when such methods are efficacious. In addition, functional imaging of the cortex following infarction may provide improved estimates of cortical injury that account for plastic reorganization and a better understanding of the recovery process (Crafton *et al.*, 2003; Rossini, *et al.*, 2003).

Psychiatric Disease

There are many who still think that functional imaging of the brain will be of greatest use to the analysis of psychiatric disorders. It is true that in the early days of FDG-based PET scanning, many studies showing patterns of abnormality in resting brains of people with major syndromes such as schizophrenia, obsessive-compulsive disorder, and depression appeared (see Mayberg, 2000, for review). What has been disappointing is that the resting pattern has not provided information that is more sensitive for diagnostic purposes than expert clinical examination and analysis. Nevertheless, it was of some interest that patients with depression tended to show generalized hypofunction that would reverse on anti-depressive treatment, patients with obsessive compulsive disorders showed orbital frontal abnormalities that again reversed, while patients with schizophrenia in general showed abnormalities of dorsolateral prefrontal function, although at times this was described as a hypofunction and others reported hyperactivity.

More recently, the approach taken has been to analyze the symptoms that characterize these disorders. For example, relative poverty of spontaneous action is a characteristic of retarded depression, schizophrenia with negative features, as well as some neurodegenerative disorders such as Parkinson's disease. The approach has been to examine resting state abnormalities across diagnostic categories. For example, patients with relative hypokinesia, from whatever cause, are compared to those with normal spontaneous activity. The delineation of a dorsolateral prefrontal and anterior cingulate system that is hypofunctional in hypoactive patients has been clearly demonstrated. This observation generated a hypothesis that these parts of the frontal lobes are in some way responsible for the intention to act in a given circumstance, in a manner that is different to stimulus-response activity. This hypothesis was tested in normal volunteers and found to be correct. In addition, it was determined that the frontal areas interacted with post-central motor and sensory areas relevant to the activity undertaken. Thus, spontaneous, as opposed to reactive, movement of the hand normally results in activation of the dorsolateral prefrontal cortex and the anterior cingulate cortex with hypoactivity of the hand area of the motor cortex. Reflex activity results in activation of the hand area of the motor cortex. On the other hand, speech and articulation result in hypoactivation of the face area and the auditory cortex during self-initiated speech, in addition to the hyperactivity of dorsolateral prefrontal and anterior cingulate cortex.

The description of a normal cortical system, which includes two components, a frontal supra-modal system and a post-central modality specific area, gives a basis for understanding what happens in hypokinetic patients. Schizophrenic patients with hypokinesia have hypoactivity of the frontal and cingulate cortices with paradoxical hyper-activation of the post central regions.

Modification of this phenomenology by dopamine modulating agents results in a normalization of frontal and post central activity (Frith *et al.*, 1995). This example introduces the concept of dysfunctional cerebral systems and how treatment of such a dysfunctional system can be followed by correlation between behavioral modification and normalization of activity patterns in the relevant brain systems.

Morphometric analyses of the major psychiatric syndromes are still in there infancy. The first use of voxel-based morphometry suggested abnormalities in the medial temporal aspects of the brain in the schizophrenia (Wright *et al.*, 1995). Generalized atrophy and focal changes have been demonstrated. It appears that schizophrenia may be too coarse a diagnosis. There is a realistic prospect that phenomenological or genetic stratification of patients combined with differential imaging patterns of atrophy or dysfunction may be a better way to classify psychotic patients in the future; especially for the purposes of early treatment.

PHARMACOLOGY AND POST-GENOMICS

PET with positron labeled tracers of drugs continues to be a viable method for assessing pharmacokinetics, drug distribution, and dose ranges in human brains. The procedure for making such labeled tracers for human investigative use is burdensome. It is still not clear whether this approach will be more efficient and less costly than conventional ways of carrying out studies to determine these characteristics of potential therapeutic compounds.

Since the mapping of the human genome and the beginning of the postgenomic era, a major problem in biology will be the determination of the functions of the proteins coded by the genome. The link between the molecular composition of cells and the system physiology of which they are a part is a difficult one to define. There are pilot studies that suggest functional and structural imaging may contribute to this endeavor in a significant way. Some of these studies are described below.

A further use for PET based ligands is in studies of the viability of cells implanted into the brain for therapeutic purposes. The prime example is the demonstration of the viability of fetal mesencephalic cells for the treatment of Parkinson's disease (Lindvall *et al.*, 1994). It is not yet clear whether such approaches will be easier or more effective than the alternative of challenging the therapeutic implant physiologically using BOLD fMRI to determine brain responses. The effectiveness of deep brain stimulation of the sub-thalamic nucleus in Parkinson's disease appears to be due to a distant influence that the stimulation has on the function of the supplementary motor area and prefrontal cortex.

Pharmacodistribution, Pharmacokinetic, and Dose-Ranging Studies

Techniques have been developed for measuring the regional cerebral uptake of labeled tracers of receptors, vesicles, and neurotransmitter reuptake sites. Tracers such as PK11195 target specific receptor molecules that are expressed only in certain pathological conditions—in this case, peripheral benzodiazepine binding sites on the surface of activated microglia in inflamed brain tissue—and can be excellent markers of specific pathogenic processes in diseases of interest (Banati, 2002). The uptake of neurotransmitter-related tracers has not proved of great importance in the differential diagnosis of degenerative basal ganglion disease, even though some differential group patterns have been demonstrated.

Tracers of specific enzymes have been used in dose ranging studies. For example, monoamine oxidase B activity can be assessed by the kinetics of uptake of carbon-11–labeled deprenyl. The previous introduction of a novel antagonist at different doses permits an analysis from the images of the differential subsequent capture of the radio-labeled deprenyl and hence calculation of the effectiveness of the new agent as a reversible inhibitor of the enzyme (Bench *et al.*, 1991; Lammertsma *et al.*, 1991). Though the principle of this type of dose-ranging study was established some time ago, it has not entered routine practice, largely because it depends on developing an appropriate tracer to test putative drugs. Similar techniques have been developed for measuring the time courses of drug binding to receptors and the kinetics of reversible receptor blockade (Bench *et al.*, 1993). However, the complexity of such studies and the need for teams of experienced chemists and pharmacologists has meant that these techniques have not been used systematically yet in the development of new drugs.

Post-Genomic Studies with Positron Emission Tomography Using Reporter Genes

A key area for the use of tracer techniques in human brain mapping experiments will be in the area of monitoring gene expression in human subjects. The ability to image gene expression *in vivo* in humans would provide an important link between the disciplines of molecular biology and phenotype assessment. These two rapidly expanding fields would then be joined with a common purpose, namely, the visualization of molecular events that cause disease and interventions, at a molecular level, that can prevent, delay, arrest, or reverse disease progression. All current approaches require tracer techniques. These use radioisotope tracers and imaging methods such as PET, single-photon emission computed tomography, or non-radioactive contrast agents visible with MRI.

The more general the solution to the problem of tracing such events, the more generalizable and useful the result. The approach is similar to a standard molecular biology experiment, in which a reporter gene and a reporter probe produce a signal that can be visualized in the cell, typically with some type of optical (e.g., fluorescence) change. Two promising approaches have been described.

To illustrate the strategy that can be used with PET, consider a situation in which the protein product of an expressed PET reporter gene is an enzyme, such as herpes simplex virus thymidine kinase, and the PET reporter probe is a fluorine-18–labeled analog of ganciclovir, ^{18}F-fluorogancyclovir (Phelps, 2000). The strategy works as follows. The PET reporter gene is incorporated into the genome of an adenovirus that is administered to the test subject or animal. Subsequently, ^{18}F-fluorogancyclovir is injected intravenously. It diffuses into cells and, if there has been no reporter gene expression, it will diffuse back out of cells and will be cleared from circulation by the kidneys. If gene expression has occurred, the ^{18}F-fluorogancyclovir will be phosphorylated by the reporter gene enzyme and trapped in the cell. One molecule of the enzyme can phosphorylate many molecules of the reporter probe, so that there is an amplification effect in this enzyme-mediated approach (Phelps, 2000).

Two MRI strategies are under consideration for demonstrating gene expression. In the first, a molecule is created with a gadolinium ion at its center, positioned in such a way that the ion has limited access to water protons and, as such, perturbs the MRI signal minimally. Site-specific placement of a portion of the molecule that can be cleaved by the product of a reporter gene then exposes the gadolinium ion, making it more accessible to water protons and changing the MRI signal at local sites of gene expression. For example, the site-specific placement of a galactopyranocyl ring in the macromolecule can be cleaved by the commonly used reporter gene beta-galactosidase (Louie et al., 2000). An alternative approach is to create particles composed of DNA and a polylysine molecule modified with a paramagnetic contrast agent such as gadolinium (Louie et al., 2000). The co-transport of the DNA and MRI contrast agent demonstrates areas of gene expression that can target specific cells in vivo as a marker of this process (Kayyem, 1995). These tools will be extremely useful in both research and clinical settings as gene therapy becomes a reality in the years to come.

Linking Structural and Functional Images to Clinical Phenotypes and Genotypes

The simultaneous collection of phenotypic information from brain mapping experiments in the human nervous system with genotype information derived from the analysis of subjects' DNA will allow for true phenotype-genotype experimentation to flourish in the future. If such studies are performed in large enough populations and include twins (both monozygotic and dizygotic), estimates of heritability for certain traits or disorders should emerge (Mazziotta et al., 2001, 2001a). There is evidence for this type of evaluation already present in the literature (Bartley et al., 1997; Bookheimer et al., 2000; Egan et al., 2001; Thompson et al., 2001). Studies that seek to determine the relationship between phenotype and genotype can use cohorts of twins to understand the contribution of genetics to human behavior while controlling for environmental factors with twins that are reared together or apart. Thus, the full gamut of genetic and environmental factors can be put in a framework that is manageable and potentially scientifically rigorous. While the uses of such information are obvious in studies of patients with specific cerebral disorders, particularly those that are inherited, a wide range of previously unanswered, or even considered, questions will emerge in the realm of basic neuroscience and for the normal human brain.

An excellent example of how behavior can be related to imaging phenotypes in genetics is represented by the study of serotonin function and anxiety disorders by Hariri et al. (2002). This work demonstrated the value of having an imaging phenotype with less variance than behavioral phenotypes provide, to identify correlations between genotype and (imaging) phenotype. The basis for this work stems from the fact that basic neuroscience studies have demonstrated that serotonin transporter promoter polymorphisms alter serotonin transporter gene (SLC 6A4) transcription and function. One or two copies of the short ("s") allele (versus the long "l" allele) result in decreased 5-HTT mRNA production and a 50% reduction in 5-HT uptake, but only slight differences in the incidence of anxiety and affective disorders in human subjects. In this experiment, 28 right-handed subjects were divided into genotype-specific ("s" or "l") groups matched for age, gender, and mean IQ. BOLD fMRI was used to measure relative cerebral blood flow with visual stimuli comparing emotional expression in faces with simple spatial comparisons (Figure 29.2; see also the Color Plate section). A working memory task was also employed as a second control state. Both genotype groups ("s" and "l") had bilateral increases in relative cerebral blood flow in the amygdalae, fusiform and inferior parietal cortices as well as the frontal eye fields. The responses were larger for the emotional comparison task for the "s" versus the "l" group on the right side. There were no group differences noted for working memory tasks or on behavioral measures (e.g., reaction time, etc.). There are also no effects related to gender. These investigators concluded that the "s" genotype may act to increase available serotonin at excitatory synapses in the amygdala because of relatively decreased transporter expression and function leading to increased neuronal activity, as measured by enhanced relative cerebral blood flow.

This is an example of how genotype-phenotype correlations can be performed in situations where behavioral aspects of the comparison have a high variance and individual inconsistencies. Nevertheless, if there is an anatomical substrate for the behavior in question, in this case the amygdala, imaging studies may prove to be a good way to quantify these relationships by providing an imaging phenotype to relate to genotype. It is most likely

that similar comparisons will be performed in a wide range of neurological diseases as well as to assess normal brain function as it relates to inherited factors that can be correlated with genotype.

Another example is the deletion mapping strategy used to define areas of the X chromosome in which there are genes that determine the development of specific regions of the brain by examining patients with Turner's and partial Turner's syndromes (Good *et al.*, 2003). Patients with Turner's syndrome have subtle behavioral and cognitive abnormalities. Patients with partial Turner's syndrome have partial deletions of the short

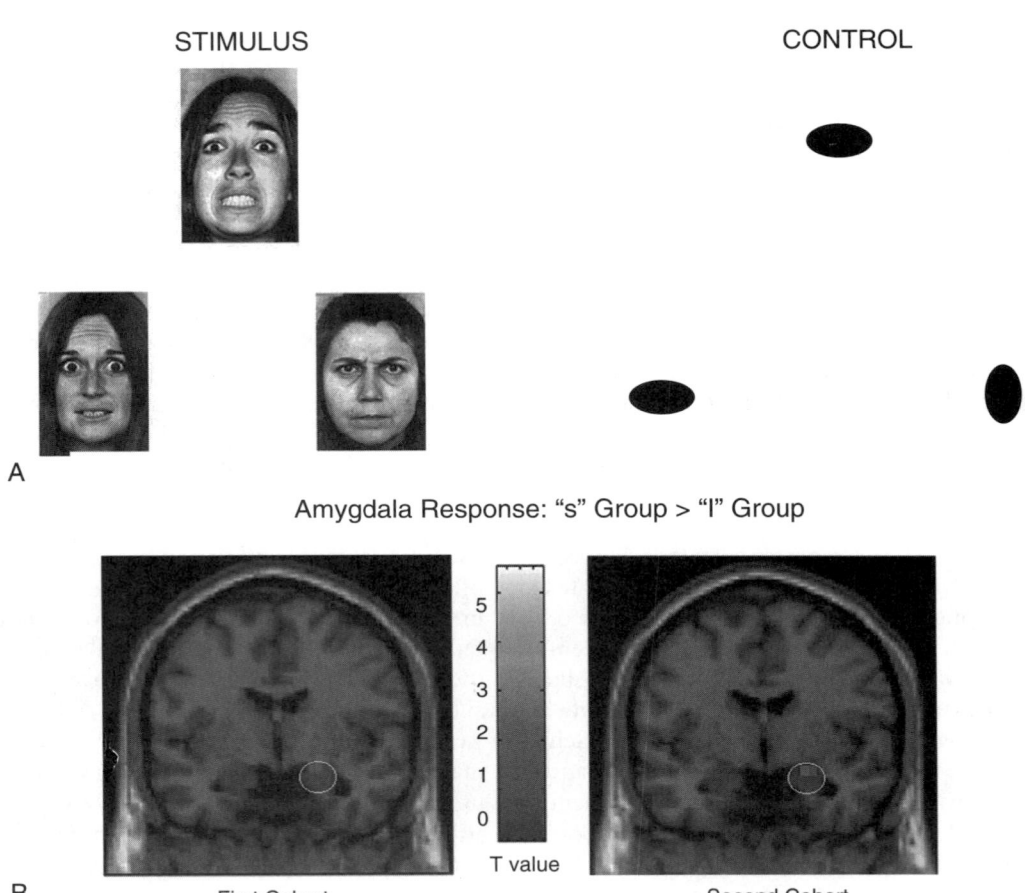

FIGURE 29.2 Imaging phenotype and genotype related to behavior. Human behavior is quite variable and may fluctuate from day to day. This is particularly true of emotional states. Thus, when a specific genotype was hypothesized to be associated with different degrees of serotonin reuptake it was thought that there would be a good correlation with anxiety and affective disorders. This was only found to a very limited degree and its significance was questioned. It had been known from basic neuroscience research that the serotonin (5-HT) reuptake transport (5-HTT) promoter could affect serotonin reuptake and that this promoter had two alleles. The short "s" allele would result in less 5-HT reuptake while the long "l" allele would produce normal 5-HT reuptake (see text for details). **A:** Emotional and spatial stimuli shown to subjects segregated by genotype ("s" or "l" alleles) but matched for age, gender and mean IQ. Subjects were asked to match the emotional expression of the face (left) in the top row with one of the two faces in the bottom row. In the control stimulus (right), the orientation of an ellipse (top row) was to be matched with the orientation of one of the two ellipses in the bottom row. **B:** It was demonstrated that the magnitude and spatial extent of the relative cerebral blood flow response of the right amygdala was greater in the group having the "s" allele (first cohort) relative to the "l" allele (second cohort). (See also the Color Plate section.)

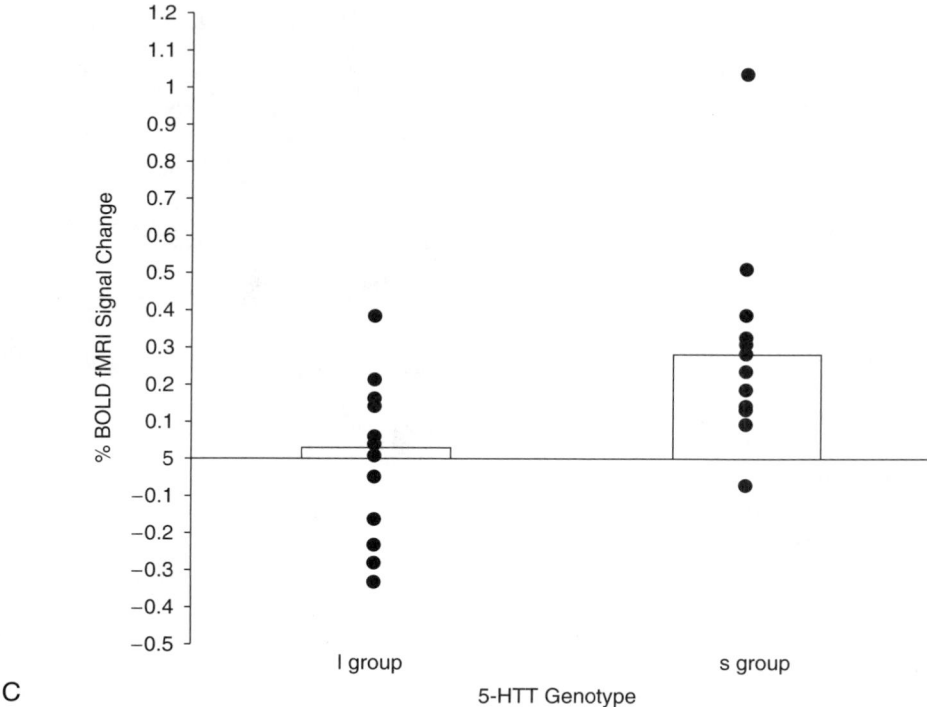

FIGURE 29.2 cont'd C: While there was overlap between the groups, there was clear separation of the mean of the fMRI signal response in the amygdala of these two groups as demonstrated in this graph. Such studies demonstrate that phenotype-genotype comparisons can often be better quantified and made more specific when a behavioral phenotype is substituted with a functional imaging parameter, in this case, response to an emotional matching task in the visual modality. Courtesy of A. Hariri and colleagues, National Institutes of Mental Health, Bethesda, Maryland. (See also the Color Plate section.)

arm of the X chromosome, rather than complete absence of one X chromosome. This suggested an imaging strategy using voxel-based morphometry that involved examining structural differences in patients and normal volunteers and correlating these with chromosome X status. Focal abnormalities with excess gray matter volume in the amygdala and orbitofrontal cortices were found. Analysis of the patients with partial Turner's syndrome indicated that a critical region of the short arm of the X chromosome contained a gene or genes that influenced the development of these two areas. Patients who retained this critical region showed no abnormalities of brain structure. Patients who lost the critical region showed the abnormalities of the full Turner syndrome. The amygdala is known from other studies to be involved in recognition of fear in the faces of others. The morphometric result suggested that the patients with partial Turner's syndrome might have abnormalities detecting fear in the faces of others that correlated with their genetic characterization. The prediction is completely confirmed with behavioral studies. This example of mapping the function of chromosomal regions using opportunities provided by genetic diseases provides tantalizing opportunities for the future.

There are studies of individuals from families with Huntington's disease (see previous discussion) who are normal but nevertheless demonstrate atrophy of caudate nucleus on structural MRI that correlates with their genetic status in terms of numbers of CAG codon repeats. This finding raises the possibility of defining endophenotypes that permit much more accurate stratification of patients and prediction of their eventual clinical outcome from analysis of brain structure or function.

SURROGATE MARKERS OF EFFECTIVENESS OF TREATMENT

Classically, the introduction of new treatments for neurological diseases, especially those based on neurodegeneration require large cohorts of patients studied in a double-blind placebo-controlled trial. These studies which are large and expensive often follow equally expensive phase 2 proof of principle studies and the whole process can take many years before a drug comes to clinical practice. There is an as yet unproved hope that structural and functional imaging studies will provide surrogate markers of therapeutic efficacy. Quite apart from potentially improving the rational choice of dose and dosing schedule, activation studies provide an opportunity to examine the direct effects of a putative treatment on dysfunctional brain systems. The rationale has been described above in relation to the treatment of hypokinesia. A potential drug and its placebo can be regarded as two contexts within which a particular system operates. A comparison of the activity within this brain system under the two contexts will reveal how, and to what extent, an agent acts. It is important to be clear that one is not examining the specific site of action of the putative drug but its overall effectiveness. Thus a drug may act on a particular set of receptors and the neurons expressing those receptors may then have downstream effects on other neurons which themselves alter interactions between other sets of neurons. This approach is fundamentally different to using a specific tracer of the receptor to determine its ability to bind and block that receptor specifically.

Structural and Functional Images as Surrogate Markers of Disease Progression

Both structural and functional imaging studies have the potential to produce a wide range of surrogate markers that allow for the objective, quantifiable and noninvasive monitoring of disease progression in both symptomatic and presymptomatic individuals. Given the magnitude of neurological disease and its economic impact, such surrogate markers will be vital in the future design of experimental clinical trials, in which both pharmacological and behavioral interventions are explored. The range of possible applications spans the gamut from drug trials in degenerative disease and stroke to strategies for effective optimization of neural plasticity in patients with acute brain injury following stroke or trauma. Already, examples of such studies are becoming more commonplace in the neuroscientific, clinical, and neuroimaging literature. Both pharmaceutical corporations as well as the biotechnology industry are interested in using such strategies to make such explorations more effective in terms of cost and time.

Experimental drug trials are both important and extremely expensive. Any approach that can be used to make such trials more economical and shorter in duration will certainly be adopted. Through the combined use of probabilistic normal and disease-specific atlases, it may be possible to use data from brain mapping methods to serve as surrogate markers in clinical trials designed to evaluate a pharmacological or behavioral therapy (Figure 29.3; see also the Color Plate section).

Consider the following example: a new experimental drug has been developed for the treatment of relapsing

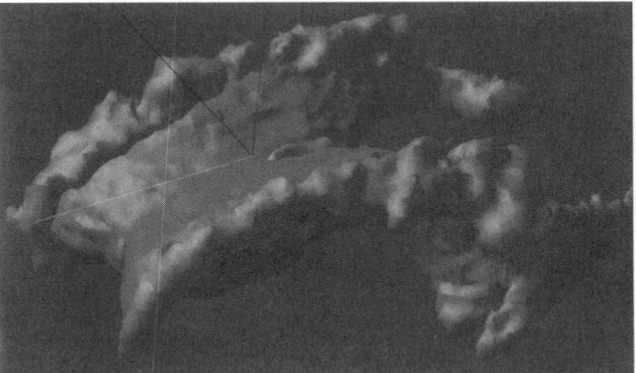

FIGURE 29.3 Probabilistic atlas of multiple sclerosis. This image is produced from 460 patients with multiple sclerosis derived from 5800 individual pulse sequences. The green areas demonstrate the probabilistic location of the actual MS plaques in the population referenced to the human ventricular system (red). This strategy gives a composite view of overall disease burden across a large population. Consider its use in a clinical trial where these patients were randomly assigned to treatment groups that included placebo, conventional therapy and experimental therapy. By observing the group for some period and obtaining serial images, an automated, quantifiable, and objective measure of the relative effect of experimental therapy versus conventional therapy and the natural history of the disease could be obtained using MRI lesions as a surrogate marker of disease burden. Such strategies should lead to more efficient and cost-effective clinical trials. Courtesy of Zijdenbos, Evans and colleagues, Montreal Neurologic Institute, Montreal, Canada. (See also the Color Plate section.)

and remitting multiple sclerosis. It is clinically important to compare the efficacy of this drug relative to conventional treatment and to the natural course of the disease. Three groups of subjects are selected, one receiving the experimental drug, one on conventional therapy, and one receiving a placebo. Subjects are serially studied using gadolinium-enhancing lesions as a surrogate marker of disease burden (Weiner et al., 2000). A duration is determined for the trial, with equally spaced MRI evaluations both before and after the onset of the proposed interventions. Probabilistic atlases are developed from the longitudinal studies in each subject for all three groups. The placebo group provides an estimate of the natural history of disease burden during the interval. The conventional therapy atlas can be used to show the impact of traditional approaches on the natural course of the disease. Lastly, the experimental therapeutic group is compared with the other two to determine, in an objective, noninvasive, and quantifiable manner, the relative efficacy of the proposed new therapy relative to the natural course of the disease and conventional treatments.

This approach should be extremely appealing because it has the potential to make clinical trials less expensive and more objective. Furthermore, because the data can be continually reanalyzed and used in comparisons with other experimental drug trials, the results of such work are additive rather than performed in isolation. This further enhances the value of the funds spent to perform such studies. Already examples of this type of work have been obtained in patients with multiple sclerosis (Evans et al., 1997; Kamber et al., 1992, 1995; Zijdenbos et al., 1996) (see Figure 29.3) and have been purposed for patients with Alzheimer's disease (Bookheimer et al., 2000). A critical factor in the utilization of this strategy is that there is a biologically significant and important marker of disease burden that can be estimated using the brain mapping method at hand. Without a strong correlation between the surrogate marker and the biological significance of the disease process, it would be unwise and ineffective to base the efficacy of a given therapeutic strategy on the surrogate marker derived from brain-mapping methods (Weiner et al., 2000). It is quite likely that such surrogate markers will be validated and can serve as an important means by which to improve the efficacy and economics of clinical trials on experimental therapeutics.

We have already discussed the possibility of using brain mapping methods, particularly MRI strategies, to develop a tissue clock for the evaluation of patients with acute ischemic injury of the brain. The same approach may be useful in other acute settings such as trauma. In patients with acute head trauma, particularly if it is severe, it is difficult or impossible to determine the ultimate outcome and quality of the patient during the first few days following injury. Extraordinary costs and emotional anguish are expended by the family for patients who might otherwise be judged as nonviable. Concomitantly, it is conceivable that treatment is withheld, delayed, or terminated in some individuals who might ultimately have a better-than-expected clinical outcome. If it were possible to reliably determine the prognosis for such patients in the acute setting, using one or more brain mapping methods, the net result will be a reduced economic burden to families and healthcare plans, as well as the important and substantial reduction in family anguish over the ambiguity of this type of clinical situation. Surrogate markers of prognosis will be an important part of the evaluations of such patients in the future.

Serial Measurement of Atrophy in Degenerative Disease

The physical world is organized in four dimensions; three in space and one in time. Tomographic brain mapping techniques produce true three-dimensional data while surface techniques such as EEG and OIS produce two-dimensional surface maps. The ability to combine temporal information with two- or three-dimensional spatial information can result in rate-of-change maps. Most commonly used with three-dimensional structural MRI data, four-dimensional rate-of-change maps have demonstrated cross-sectional and longitudinal changes of the corpus callosum during natural development (Thompson et al., 2000a) (Figure 29.4A; see also Color Plate section). Similarly, these same strategies have been applied in pathologic conditions to observe the four-dimensional growth of tumors, either as a measure of their natural course or to evaluate selective treatment strategies (Haney et al., 2001, 2001a). The same approach has been used to monitor tissue loss (i.e., atrophy) in neurodegenerative disorders such as Alzheimer's disease (Thompson et al., 1998, 2000) (Figure 29.4B; see also Color Plate section). It is anticipated that four-dimensional rate-of-change maps will be applied to functional information measured during the processes of learning, functional loss in degenerative disease, and recovery of function following acute injury such as trauma and stroke, with and without the applications of neurorehabilitation strategies.

Probabilistic atlases based on diseased populations (Thompson et al., 2000) show enormous promise in advancing our understanding of disease. As imaging studies expand into ever-larger patient populations, population-based brain atlases offer a powerful framework to synthesize the results of disparate imaging studies. Disease-specific atlases, for example, are a type of probabilistic atlas specialized to represent a particular clinical group (see Thompson et al., 2000, for a review). A disease-specific atlas of brain in Alzheimer's disease has recently been generated to reflect the unique anatomy and physiology of this subpopulation (Thompson et al., 1997, 1998, 1999, 2000; Mega et al., 1997, 1998, 1999). Based on well-characterized patient groups, this atlas contains thousands of structure models, as well as composite maps, average templates, and visualizations of structural variability, asymmetry, and group-specific differences. It also correlates the structural, metabolic, molecular, and histological hallmarks of the disease (Mega et al., 1997, 1999, 2000). Additional algorithms use information stored in the atlas to recognize anomalies and label structures in new patients. Because they retain information on group anatomical variability, the resulting atlases can identify patterns of altered structure or function, and can guide algorithms for knowledge-based image analysis, automated image labeling (Collins et al., 1994; Pitiot et al., 2002), tissue classification (Zijdenbos and Dawant, 1994), and functional image analysis (Dinov et al., 2000). At the core of the atlas is an average MRI data set based on a population of subjects with early dementia (Figure 29.5; see also Color Plate section). Using specialized mathematical approaches for averaging cortical anatomy, the resulting average MRI

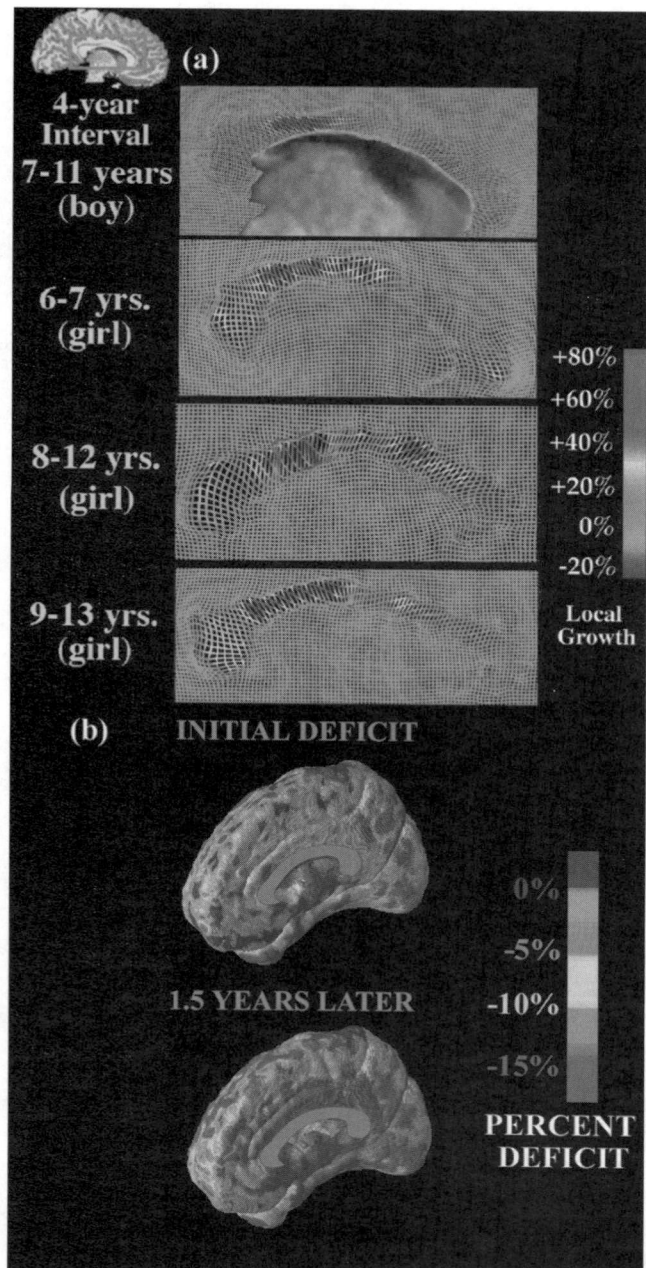

FIGURE 29.4 Dynamic maps of changing brain structure. If follow-up (longitudinal) images are available from the same subject, the dynamics of brain change can be measured with *tensor mapping* approaches. These compute an elastic deformation field that reshapes the baseline image to its anatomic configuration at follow-up. From these deformations, maps are generated to visualize local volumetric change, showing local rates of tissue growth or loss (red colors denote fastest growth). **A:** In children, scanned serially with MRI around the age of puberty, fastest growth is detected in the isthmus of the *corpus callosum*, which connects the language regions of the two cerebral hemispheres. **B:** *Progression of Alzheimer's Disease.* These maps show the average profile of gray matter loss in a group of 17 patients with mild-to-moderate Alzheimer's disease. Average percent reductions in the local amount of gray matter are plotted, relative to the average values in a group of 14 healthy age- and gender-matched elderly controls. Over 1.5 years of progressive cognitive decline from mild to moderate dementia, gray matter deficits in these patients spread through the limbic system to encompass the majority of the cortex. Courtesy of Paul Thompson and Arthur Toga, UCLA, Los Angeles, California. (See also the Color Plate section.)

individual patient against a normative reference population, for early diagnosis or for clinical trials. Since individual variations in cortical patterning complicate the comparison of gray matter profiles across subjects, an elastic matching technique can be used (driven by 84 structures per brain) that elastically deforms each brain into the group mean geometric configuration (Figure 29.5C). By averaging a measure of gray matter across corresponding regions of cortex, these shape differences are factored out. The net reduction in gray matter, in a large patient population relative to controls ($N = 46$), can then be plotted as a statistical map in the atlas (Figure 29.5B; Thompson *et al.*, 2000). This type of analysis uncovers important systematic trends, with an early profile of severe gray matter loss detected in temporoparietal cortex, consistent with the early distribution of neuronal loss, metabolic change and perfusion deficits at this stage of Alzheimer's disease. Finally, this local encoding of information on cortical variation can also be exploited to map abnormal atrophy in an individual patient (Thompson *et al.*, 1997). Figure 29.5D illustrates the use of a probabilistic atlas to identify a region of abnormal atrophy in the frontal cortex of a dementia patient. Severe abnormality is detected with a color code used to indicate the significance of the abnormality (*red colors*). As expected, corresponding regions in a matched elderly control subject are signaled as normal (Figure 29.5D, *right panel*).

Hypertrophy as a Measure of Compensation or Plasticity

Recent excitement in the neuroscience community about stem cells and the potential for neuronal regeneration in

template has a well-resolved cortical pattern (Figure 29.5A), with the mean geometry of the patient group (Thompson *et al.*, 2000). Surfaces for the cortex can include the external hull, the gray-white matter interface, or an average of the full cortical thickness. Figure 29.5A represents the external hull.

Resolving the average profile of early gray matter loss in an Alzheimer's disease population is one such example of application for this type of atlas. It would be ideal, for example, to calibrate the profile of gray matter loss in an

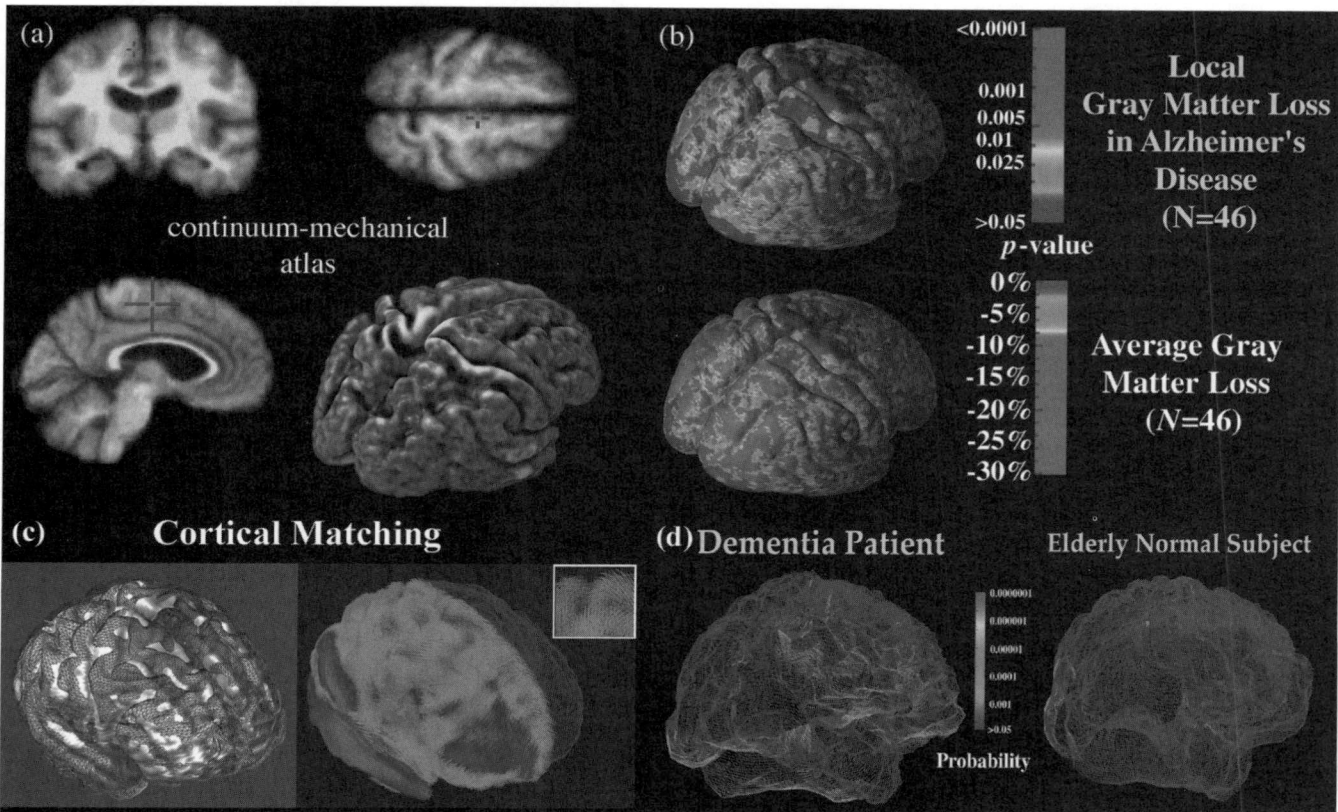

FIGURE 29.5 Disease-specific brain atlases. Disease-specific brain atlases reflect the unique anatomy and physiology of a clinical population, in this case an Alzheimer's disease population. Using mathematical strategies to average cortical anatomy across subjects (Thompson *et al.*, 2000), an average MRI template can be generated for a specific patient group; in this case, nine patients with mild-to-moderate Alzheimer's disease **(A)**. The cortical pattern indicates clear sulcal widening and atrophic change, especially in temporo-parietal cortices. By averaging a measure of gray matter across corresponding cortical regions, the average profiles of gray matter loss can also be mapped **(B)**. The net reduction in gray matter, in a large patient population relative to controls (Thompson *et al.*, 2000), can then be plotted as a statistical map in the atlas **((B)**, *upper panel*). This type of analysis uncovers profiles of early anatomical change in disease. Panel **(C)** shows an individual's cortex (*brown mesh*) overlaid on an average cortical model for a group. Differences in cortical patterns are encoded by computing a three-dimensional elastic deformation **((C)**; pink colors: large deformation) that reconfigures the average cortex into the shape of the individual, matching elements of the gyral pattern exactly. These deformation fields **(C)** provide detailed information on individual deviations. These statistical data can be used to detect patterns of abnormal anatomy in new subjects. By encoding variations in gyral patterns and gray matter distribution, algorithms can detect a region of abnormal atrophy in the frontal cortex of a dementia patient **(D)**. Severe abnormality is detected (red colors) while corresponding regions in a matched elderly control subject are signaled as normal **((D)**, *right panel*). Courtesy of Paul Thompson and Arthur Toga, UCLA, Los Angeles, California. (See also the Color Plate section.)

adult life has led to a number of interesting, if possibly inconclusive, studies. These have sought to identify changes in brain structure associated with skills, practice, and repetitive behavior. The first demonstration was in London taxi drivers (Maguire *et al.*, 2000). They were examined because of their intense use of visuospatial memory. Functional studies in normal humans had demonstrated the importance of the posterior hippocampus of the non-dominant hemisphere in tasks requiring visuospatial skills. Activity in this part of the brain correlated with the accuracy with which subjects

were able to perform visuospatial tasks. The investigators therefore sought to find evidence of structural change associated with this skill and thus used the taxi drivers whose occupation required that they practice it intensively in their daily work. The astonishingly encouraging result was that there is a relative increase in the size of the posterior part of the non-dominant hippocampus in the taxi drivers compared to normal controls, co-locating with the functionally active part. Furthermore, the relative change in size appeared to correlate with number of years spent in the profession, or intensity of use. There have been further small studies of this nature that appear to suggest that this is not a unique finding. The cause of the structural changes is entirely unknown. It could be due to local changes in neuronal numbers, but equally to changes in glia, synapses, dendritic branching, and the like. Pathological studies will be required to elucidate the cause. In any event, if due to remodeling of the neuronal architecture, such investigations may have considerable implications for those studying the recovery of functions following brain injury.

TREATMENT: TRANSCRANIAL MAGNETIC STIMULATION TO DISTURB FUNCTION AND RESTORE IT

We tend to think of brain-mapping methods as passive and observational. They provide information about the structure and function of the brain with varying degrees of invasiveness. Nevertheless, some brain mapping techniques can also serve a therapeutic role. A good example is repeated stimulation with TMS (rTMS) and its effects on brain function (Epstein et al., 1996; Pascual-Leone et al., 1994; Paus et al., 1997). Single pulse TMS enhances neuronal excitability in local circuits. As such, a single TMS pulse over the motor cortex at sufficient threshold will induce a motor contraction in the appropriate contralateral musculature of the body. Subclinical thresholds for stimulation will change the peripheral motor-evoked potential. Repeated stimulation with TMS causes a temporary deficit of function and can be used experimentally to remove cortical nodes from complicated networks subserving specific behaviors. In essence, such stimulation produces virtual patients in which normal subjects can have transient focal obliteration of function that can be observed during functional imaging or behaviorally.

Daily sessions with rTMS have also been used as a therapeutic intervention. Focal and repeated daily TMS over time may, in fact, have specific therapeutic benefits analogous to the way electroconvulsive shock therapy can be beneficial in a global and less refined strategy for the treatment of intractable depression. rTMS can provide such regional stimulation, optimized to a given subject's functional neuroanatomy, based on previous functional imaging.

Already, this technique has been used as a potential treatment for depression that has proven to be intractable to pharmacological and behavioral therapy (George et al., 1995, 1999; Levkovitz, 2001; Pascual-Leone et al., 1996; Post et al., 1999). It is likely that such strategies of focal, repetitive transcranial magnetic stimulation will prove successful in the treatments of certain disorders, including depression and some types of focal epilepsies and dystonias. In addition, the use of other energy sources, such as focused ultrasound, may find a place in the treatment of lesions of the brain such as tumors or arteriovenous malformations.

CONCLUSION

The methods of functional and structural imaging with MRI using computerized methods of image analysis provide new and fascinating tools for clinical neuroscientists. They considerably expand the ability to investigate human brain function in diseased states and under different therapeutic regimes. The method of PET, although reliant on radioactivity, has been largely supplanted in this field of inquiry. However, it continues to provide a unique way of measuring the distribution, kinetics, and effective dose of putative therapeutic agents. However, there are now new approaches using candidate drugs in pharmacological rather than trace doses to examine their effects on brain function.

Functional and structural patterns of abnormality may be of diagnostic or prognostic importance. In diseases where there is a significant genetic component, such patterns may turn out to be accurate endophenotypes permitting more precise genetic characterization, especially if patterns of atrophy or dysfunction can be shown to be predictive of particular behavioral abnormalities. This is a program of research for the future.

There is no doubt that much fundamental information has been gained in the past 20 years about human cerebral disease with imaging techniques. This information has been additional to the proven importance of diagnostic brain scanning. It remains to be seen whether the generalization of specialized image acquisition and analysis techniques will bring advantages in turn to the treatment and prognosis of common cerebral disorders. The catalogue of knowledge obtained to date seems very promising. It may well be that combining the techniques of MRI with methods for measuring rapid changes of brain activity such as EEG and MEG or transient irreversible impairment of that activity with transient transcranial stimulation will provide additional ways of accessing the brain and revealing the mysteries of its organization, health, and disease with even greater precision.

ACKNOWLEDGMENTS

RSJF was funded by the Wellcome Trust. JCM received partial support for this work was provided by a grant from the Human Brain Project (P20-MHDA52176), the National Institute of Mental Health, National Institute for Drug Abuse, National Cancer Institute and the

National Institute for Neurologic Disease and Stroke. For generous support the authors also wish to thank the Brain Mapping Medical Research Organization, Brain Mapping Support Foundation, Pierson-Lovelace Foundation, The Ahmanson Foundation, Tamkin Foundation, Jennifer Jones-Simon Foundation, Capital Group Companies Charitable Foundation, Robson Family, Northstar Fund, and the National Center for Research Resources grants RR12169, RR13642 and RR08655. The authors would also like to thank Kami Yaden for preparation of the manuscript and Andrew Lee for preparation of the graphic materials.

References

Ahlfors SP *et al.* (1999). Spatiotemporal activity of a cortical network for processing visual motion revealed by MEG and fMRI, *J Neurophysiol* 82:2545-2555.

Amunts K *et al.* (1998). Cytoarchitectonic definition of Broca's region and its role in functions different from speech, *NeuroImage* 7:8.

Amunts K *et al.* (2000). Brodmann's areas 17 and 18 brought into stereotaxic space – where and how variable? *NeuroImage* 11:66-84.

Ashburner J, Friston KJ. (2000). Voxel-based morphometry – the methods, *NeuroImage* 11:805-821.

Ashburner J Friston KJ. (1997). Multimodal image coregistration and partitioning: A unified framework, *NeuroImage* 6:209-217.

Bailey P, von Bonin G. (1951). The Isocortex of Man, In "*Urbana*" University Press.

Baird AE, Warach S. (1998). Magnetic resonance imaging of acute stroke, *J Cereb Blood Flow Metab* 18(6):583-609.

Banati RB. (2002). Visualising microglial activation in vivo, *Glia* 40:206-217.

Bartley AJ, Jones DW, Weinberger DR. (1997). Genetic variability of human brain size and cortical gyral patterns, *Brain* 120:257-269.

Basser PJ, Pierpaoli C. (1996). Microstructural and physiological features of tissues elucidated by quantitative-diffusion-tensor MRI, *J Magn Reson* 3:209-219.

Bastos AC *et al.* (1999). Diagnosis of subtle focal dysplastic lesions: Curvilinear reformatting from three-dimensional magnetic resonance imaging, *Ann Neurol* 46:88-94.

Bench CJ *et al.* (1991). Measurement of human cerebral monoamine oxidase type B (MAO-B) activity with positron emission tomography: A dose ranging study with the reversible inhibitor Ro 19-6327, *Eur J Clin Pharmacol* 40:169-173.

Bench CJ *et al.* (1993). Dose dependent occupancy of central dopamine D2 receptors by the novel neuroleptic CP-88059-01: A study using positron emission tomography and 11C-raclopride, *Psychopharmacology* 112:308-314.

Bodegård A *et al.* (2000). Somatosensory areas in man activated by moving stimuli: Cytoarchitectonic mapping and PET, *Neuroreport* 11:187-191.

Bonmassar G *et al.* (2001). 7 Tesla interleaved EEG and fMRI recordings: BOLD measurements, *NeuroImage* 13(6):S4.

Bookheimer SY *et al.* (2000). Patterns of brain activation in people at risk for Alzheimer's disease, *N Engl J Med* 343(7):450-456.

Brodmann K. (1909). *Vergleichende Lokalisationslehre der Groahirnrinde in ihren Prinzipien dargestellt auf Grund des Zellenbaues*. Leipzig: Barth JA.

Clark S, Miklossy J. (1990). Occipital cortex in man: Organization of callosal connections, related myelo- and cytoarchitecture, and putative boundaries of functional visual areas, *J Comp Neurol* 298:188-214.

Collins DL, Neelin P, Peters TM, Evans AC. (1994). Automatic 3D registration of MR volumetric data in standardized Talairach space, *J Comp Assist Tomogr* 18(2):192-205.

Conturo T *et al.* (1999). Tracking neuronal fiber pathways in the living human brain, *PNAS* 96:10422-10427.

Cormack AMJ. (1964). *Appl Physics* 35:2908.

Crafton KR, Mark AN, Cramer SC. (2003). Improved understanding of cortical injury by incorporating measures of functional anatomy, *Brain* 126:1650-1659.

Derdeyn CP, Grubb RL Jr, Powers WJ. (1999). Cerebral hemodynamic impairment: Methods of measurement and association with stroke risk, *Neurology* 53:251-259.

Detre JA *et al.* (1995). Localization of subclinical ictal activity by functional magnetic resonance imaging: Correlation with invasive monitoring, *Ann Neurol* 38:618-624.

Dinov ID *et al.* (2000). Analyzing functional brain images in a probabilistic atlas: A validation of subvolume thresholding, *J Comp Assist Tomogr* 24(1):128-138.

Dumoulin SO *et al.* (2000). A new neuroanatomical landmark for the reliable identification of human area V5/MT: A quantitative analysis of sulcal patterning, *Cerebral Cortex* 10:454-463.

Egan MF *et al.* (2001). Effect of COMT Val108/158 Met genotype on frontal lobe function and risk for schizophrenia, *Proc Natl Acad Sci* 98(12):6917-6922.

Epstein CM *et al.* (1996). Optimum stimulus parameters for lateralized suppression of speech with magnetic brain stimulation, *Neurology* 47:1590-1593.

Evans AC, Frank JA, Antel J, Miller DH. (1997). The role of MRI in clinical trials of multiple sclerosis: Comparison of image processing techniques, *Ann Neurol* 41:125-132.

Firlik AD, Rubin G, Yonas H, Wechsler LR. (1998). Relation between cerebral blood flow and neurologic deficit resolution in acute ischemic stroke, *Neurology* 51:177-182.

Flechsig P. (1920). *Anatomie des menschlichen Gehirns und Ruckenmarks*. Leipzig: Thieme.

Fox PT, Mintun MA, Reiman EM, Raichle ME. (1988). Enhanced detection of focal brain responses using intersubject averaging and change-distribution analysis of subtracted PET images, *J Cereb Blood Flow Metab* 8:642-653.

Fox PT *et al.* (1997). Imaging human intra-cerebral connectivity by PET during TMS, *NeuroReport* 8(12):2787-2791.

Frackowiak RS *et al.* (1981). Regional cerebral oxygen supply and utilization in dementia, *Brain* 104:753-778.

Frackowiak RS. (1989). A short introduction to positron emission tomography, *Semin Neurol* 9(4):275-280.

Frackowiak RSJ *et al.* (1997). *Human Brain Function*, San Diego: Academic Press, pp 1-528.

Friston KJ, Frith CD, Liddle PF, Frackowiak RSJ. (1991). Comparing functional images: The assessment of significant change, *J Cereb Blood Flow Metabol* 11:690-699.

Friston KJ, Frith CD, Frackowiak RS J. (1993). Time-dependent changes in effective connectivity measured with PET, *Hum Brain Mapp* 1(1):69-79.

Friston KJ, Frith CD, Frackowiak RSJ, Turner R. (1995). Characterizing dynamic brain responses with fMRI: A multivariate approach, *NeuroImage* 2:166-172.

Friston KJ *et al.* (1995a). Statistical parametric maps in functional imaging: A general linear approach, *Hum Brain Mapping* 2:189-210.

Friston KJ, Holmes AP, Worsley KJ. (1999). How many subjects constitute a study? *NeuroImage* 10:1-5.

Frith CD *et al.* (1995). Regional brain activity in chronic schizophrenic patients during the performance of a verbal fluency task, *Br J Psychiatry* 167:343-349.

George MS *et al.* (1995). Daily repetitive transcranial magnetic stimulation (rTMS) improves mood in depression, *Neuroreport* 6:1853-1856.

George MS, Lisanby SH, Sackeim HA. (1999). Transcranial magnetic stimulation: Applications in neuropsychiatry, *Arch Gen Psychiatry* 56(4):300-311.

Geschwind N, Levitsky W. (1968). Human brain: Left-right asymmetries in temporal speech region, *Science* 161:186-187.

Geyer S, Schormann T, Mohlberg H, Zilles K. (2000). Areas 3a, 3b, and 1 of human primary somatosensory cortex: 2. Spatial normalization to standard anatomical space, *NeuroImage* 11:684-696.

Gilman S. (1998). Medical progress: Imaging the brain: Second of two parts, *N Engl J Med* 338:889-898.

Goldman R, Stern J, Engel J, Cohen M. (2001). Tomographic mapping of alpha rhythm using simultaneous EEG/fMRI, *NeuroImage* 13(6):S1291.

Good CD et al. (2002). Automatic differentiation of anatomical patterns in the human brain: validation with studies of degenerative dementias, *NeuroImage* 17:29-46.

Good CD et al. (2003). Dosage sensitive X-linked locus influences amygdala development and fear recognition in humans, *Brain* 126:2341-2446.

Grafton ST et al. (1990). A comparison of neurologic, metabolic, structural and genetic evaluations in persons at risk for Huntington's disease, *Ann Neurol* 28:614-621.

Hacke W et al. (1995). Intravenous thrombolysis with recombinant tissue plasminogen activator for acute hemispheric stroke, *JAMA* 274:1017-1025.

Halligan PW, Athwal BS, Oakley DA, Frackowiak RSJ. (2000). Imaging hypnotic paralysis: Implications for conversion hysteria, *Lancet* 355:986-987.

Haney S et al. (2001). Tracking tumor growth rates in patients with malignant gliomas: A test of two algorithms, *Am J Neuroradiol* 22:73-82.

Haney S et al. (2001a). Mapping therapeutic responses in a patient with malignant glioma, *J Comp Assist Tomogr* 25(4):529-536.

Hariri AR et al. (2002). Serotonin transporter genetic variation and the responses of the human amygdala, *Science* 297(5580):400-403.

Hayes TL, Lewis DA. (1995). Anatomical specialization of the anterior motor speech area: Hemispheric differences in magnopyramidal neurons, *Brain Language* 49:289-308.

Hounsfield GN. (1973). Computerised transverse axial scanning (tomography) I. Description of system, *Br J Radiol* 46:1016-1022.

Hyde JS, Biswal B, Yetkin FZ, Haughton VM. (1995). Functional connectivity determined from analysis of physiological fluctuations in a series of echo-planar images, *Hum Brain Mapping* S1:287.

Ives J et al. (1993). Monitoring the patient's EEG during echo planar MRI, *Electroencephalogr Clin Neurophysiol* 87:417-420.

Josephs O, Henson RNA. (1999). Event-related functional magnetic resonance imaging; modelling, inference and optimisation, *Phil Trans R Soc Lond B* 354:1215-1218.

Kamber M et al. (1992). Model-based 3D segmentation of multiple sclerosis lesions in dual-echo MRI data, *Proc SPIE Vis Biomed Comp* 1808:590-600.

Kamber M et al. (1995). Model-based 3D segmentation of multiple sclerosis lesions in magnetic resonance brain images, *IEEE Trans Med Imag* 14(3):442-453.

Katzan IL et al. (2000). Use of tissue-type plasminogen activator for acute ischemic stroke, *JAMA* 283:1151-1158.

Kayyem JF, Kumar RM, Fraser SE, Meade TJ. (1995). Receptor-targeted co-transport of DNA and magnetic resonance contrast agents, *Chem Biol* 2:615-620.

Kettenbach J et al. (1999). Computer-based imaging and interventional MRI: Applications for neurosurgery, *Comput Med Imag Graphics* 23(5):245-258.

Kidwell CS et al. (2000). Thrombolytic reversal of acute human cerebral ischemic injury shown by diffusion/perfusion magnetic resonance imaging, *Ann Neurol* 47(4):462-469.

Koroshetz WJ, Gonzalez G. (1997). Diffusion-weighted MRI: An ECG for "brain attack?" *Ann Neurol* 41:565-566.

Kwong KK et al. (1992). Dynamic images of human brain activity during primary sensory stimulation, *Proc Natl Acad Sci USA* 89:5675-5679.

Lammertsma AA et al. (1991). Measurement of cerebral monoamine oxidase B activity using L-(11C)deprenyl and dynamic positron emission tomography, *J Cereb Blood Flow Metab* 11:545-556.

Lemieux L et al. (2001). Event-related fMRI with simultaneous and continuous EEG: Description of the method and initial case report, *NeuroImage* 14:780-787.

Leone M, Franzini A, Bussone G. (2001). Stereotactic stimulation of posterior hypothalamic gray matter in a patient with intractable cluster headache, *N Engl J Med* 345:1428-1429.

Levkovitz Y. (2001). Transcranial magnetic stimulation and antidepressive drugs share similar cellular effects in rat hippocampus, *Neuropsychopharmacology* 24(6):608-616.

Lindvall O et al. (1994) Evidence for long-term survival and function of dopaminergic grafts in progressive Parkinson's disease, *Ann Neurol* 35:172-180.

Logothetis NK et al. (2001). Neurophysiological investigation of the basis of the fMRI signal, *Nature* 412:150-157.

Louie A et al. (2000). In vivo visualization of gene expression using magnetic resonance imaging, *Nat Biotech* 18(3):321-325.

Lutsep HL et al. (1997). Clinical utility of diffusion-weighted magnetic resonance imaging in the assessment of ischemic stroke, *Ann Neurol* 41:574-580.

Maguire EA et al. (2000). Navigation-related structural change in the hippocampi of taxi drivers, *Proc Natl Acad Sci USA* 97:4398-4403.

Marchal G et al. (1993). PET imaging of cerebral perfusion and oxygen consumption in acute ischemic stroke: Relation to outcome, *Lancet* 341:925-927.

May A et al. (1999) Intracranial vessels in trigeminal transmitted pain: A PET study, *NeuroImage* 9:453-460.

May A et al. (1999a). Correlation between structural and functional changes in brain in an idiopathic headache syndrome, *Nat Med* 5:836-838.

May A et al. (2000). PET and MRA findings in cluster headache and MRA in experimental pain, *Neurology* 55:1328-1335.

Mayberg H. (2000). Psychiatric disorders. In Mazziotta J, Toga A, Frackowiak R, editor: *brain mapping: the disorders*, vol. 1, pp. 485-508. San Diego: Academic Press.

Mazziotta JC et al. (1987). Cerebral glucose utilization reductions in clinically asymptomatic subjects at risk for Huntington's disease, *N Engl J Med* 316:357-362.

Mazziotta JC et al. (1995). A probabilistic atlas of the human brain: Theory and rationale for its development, *NeuroImage* 2:89-101.

Mazziotta JC et al. (2001). A four-dimensional probabilistic atlas of the human brain, *J Am Med Inform Assn* 8(5):401-430.

Mazziotta JC et al. (2001a). A probabilistic atlas and reference system for the human brain, *Philos T Roy Soc B* 356(1412):1293-1322.

Mega MS et al. (1997). Mapping pathology to metabolism: coregistration of stained whole brain sections to PET in Alzheimer's disease, *NeuroImage* 5:147-153.

Mega MS et al. (1998). Sulcal variability in the Alzheimer's brain: Correlations with cognition, *Neurology* 50(1):145-151.

Mega MS et al. (1999). Mapping biochemistry to metabolism: FDG-PET and amyloid burden in Alzheimer's disease, *Neuroreport* 10(14):2911-2917.

Mega MS et al. (2000). Cerebral correlates of psychotic symptoms in Alzheimer's disease, *J Neurol Neurosurg Psychiatry* 69(2):167-171.

Moseley M et al. (1990). Diffusion-weighted MR imaging of acute stroke: Correlation of T2 weighted and magnetic susceptibility-enhanced MR imaging in cats, *Am J Neuroradiol* 11:423-429.

Ono M, Kubik S, Abernathy C. (1990). *Atlas of the cerebral sulci.* Stuttgart, Germany: Thieme Medical Publishers..

Pariente J *et al.* (2001). Fluoxetine modulates motor performance and cerebral activation of patients recovering from stroke, *Ann Neurol* 50:718-729.

Pascual-Leone A *et al.* (1994). Transcranial magnetic simulation: A new tool for the study of higher cognitive functions in humans. In Boller F, Graffman J, editors: *Handbook of neuropsychology,* New York: Elsevier.

Pascual-Leone A, Rubio B, Pallardo F, Catala MD. (1996). Rapid-rate transcranial magnetic stimulation of left dorsolateral prefrontal cortex in drug-resistant depression, *Lancet* 358:234-237.

Paus T *et al.* (1996). In-vivo morphometry of the intrasulcal gray-matter in the human cingulate, paracingulate and superior-rostral sulci: Hemispheric asymmetries, gender differences, and probability maps, *J Comp Neurol* 376:664-673.

Paus T *et al.* (1996a). Human cingulate and paracingulate sulci: Pattern, variability, asymmetry, and probabilistic map, *Cerebral Cortex* 6:207-214.

Paus T *et al.* (1997). Transcranial magnetic stimulation during positron emission tomography: A new method for studying connectivity of the human cerebral cortex, *J Neurosci* 17:3178-3184.

Paus T *et al.* (2001). Maturation of white matter in the human brain: A review of magnetic resonance studies, *Brain Res Bull* 54(3):255-266.

Penhune VB, Zatorre RJ, MacDonald JD, Evans AC. (1996). Interhemispheric anatomical differences in human primary auditory cortex: Probabilistic mapping and volume measurement from MR scans, *Cerebral Cortex* 6(5):617-672.

Phelps ME. (2000). PET: The merging of biology and imaging into molecular imaging, *J Nucl Med* 41(4):661-681.

Phelps ME *et al.* (1975). Application of annihilation coincidence detection to transaxial reconstruction tomography, *J Nucl Med* 16:210-224.

Pierpaoli C *et al.* (1996). Diffusion tensor MR imaging of the human brain, *Radiology* 3:637-648.

Pitiot A, Toga AW, Thompson PM. (2002). Spatially and temporally adaptive elastic template matching, *IEEE Trans Med Imaging* 21(8):910-923.

Post RM *et al.* (1999). Repetitive transcranial magnetic stimulation as a neuropsychiatric tool: present status and future potential, *JECT* 15(1):39-59.

Rademacher J *et al.* (1992). Human cerebral cortex: Localization, parcellation, and morphometry with magnetic resonance imaging, *J Cogn Neurosci* 4:352-374.

Rademacher J, Caviness VS, Steinmetz H, Galaburda AM. (1993). Topographical variation of the human primary cortices: Implications for neuroimaging, brain mapping and neurobiology, *Cereb Cortex* 3(4):313-329.

Reiman EM *et al.* (2001). Declining brain activity in cognitively normal apolipoprotein E epsilon 4 heterozygotes: A foundation for using positron emission tomography to efficiently test treatments to prevent Alzheimer's disease, *Proc Natl Acad Sci USA* 98:3334-3339.

Rosen HJ *et al.* (2002). Patterns of brain atrophy in frontotemporal dementia and semantic dementia, *Neurology* 58:198-208.

Rossini PM, Calautti C, Pauri F, Baron J-C. (2003). Post-stroke plastic reorganization in the adult brain, *Lancet Neurol* 2:493-502.

Rowe J *et al.* (2002). Attention to action in Parkinson's disease: Impaired effective connectivity among frontal cortical regions, *Brain* 125:276-289.

Schlaug G *et al.* (1999). The ischemic penumbra: Operationally defined by diffusion and perfusion MRI, *Neurology* 53(7):1528-1537.

Small GW *et al.* (1995). Apolipoprotein E type 4 allele and cerebral glucose metabolism in relatives at risk for familial Alzheimer's disease, *JAMA* 273:942-947.

Small GW *et al.* (2000). Cerebral metabolic and cognitive decline in persons at genetic risk for Alzheimer's disease, *PNAS* 97:6037-6042.

Staroselskaya I *et al.* (2001). Relationship between magnetic resonance arterial patency and perfusion-diffusion mismatch in acute ischemic stroke and its potential clinical use, *Arch Neurol* 58:1069-1074.

Steinmetz H, Volkman J, Jancke L, Freund H. (1991). Anatomical left-right asymmetry of language-related temporal cortex is different in left and right handers, *Ann Neurol* 29:315-319.

Stieltjes B *et al.* (2001). Diffusion tensor imaging and axonal tracking in the human brainstem, *NeuroImage* 14:723-735.

Talairach J, Tournoux P. (1988). Principe et technique des études anatomiques. In *co-planar stereotaxic atlas of the human brain -3-dimensional proportional system: an approach to cerebral imaging,* New York: Thieme Medical Publishers.

Thieben MJ *et al.* (2002). The distribution of structural neuropathology in pre-clinical Huntington's disease, *Brain* 125:1815-1828.

Thompson PM *et al.* (1997). Detection and mapping of abnormal brain structure with a probabilistic atlas of cortical surfaces, *J Comp Assist Tomogr* 21(4):567-581.

Thompson PM *et al.* (1998). Cortical variability and asymmetry in normal aging and Alzheimer's disease, *Cereb Cortex* 8(6):492-509.

Thompson P, Mega MS, Toga AW. (2000). Disease-specific brain atlases. In Mazziotta J, Toga A, Frackowiak R, editors: *Brain mapping: the disorders,* vol. 1, pp. 131-177. San Diego: Academic Press.

Thompson PM *et al.* (2000a). Growth patterns in the developing brain detected by using continuum mechanical tensor maps, *Nature* 404(6774):190-193.

Thompson PM, Woods RP, Mega MS, Toga AW. (2000b). Mathematical/ computational challenges in creating deformable and probabilistic atlases of the human brain, *Hum Brain Mapping* 9(2):81-92.

Thompson P *et al.* (2001). Genetic influences on brain structure, *Nature Neurosci* 4(12):1253-1258.

Tissue plasminogen activator for acute ischemic stroke. (1995). The National Institute of Neurological Disorders and Stroke rt-PA Stroke Study Group. *N Engl J Med* 333(24):1581-1587.

Toga A, Cannestra A, Black K. (1995). The temporal/spatial evolution of optical signals in human cortex, *Cereb Cortex* 5(6):561-565.

Watkins KE *et al.* (2002). MRI analysis of an inherited speech and language disorder: structural brain abnormalities, *Brain* 125:465-478.

Weiller C *et al.* (1995). Brain stem activation in spontaneous human migraine attacks, *Nat Med* 7:658-660.

Weiner HL *et al.* (2000). Serial magnetic resonance imaging in multiple sclerosis: Correlation with attacks, disability, and disease stage, *J Neuroimmunol* 104(2):164-173.

Woods RP, Iacoboni M, Mazziotta JC. (1994). Brief report: Bilateral spreading cerebral hypoperfusion during spontaneous migraine headache, *N Engl J Med* 331:1689-1692.

Worsley KJ *et al.* (1996). A unified statistical approach for determining significant signals in images of cerebral activation. *Hum Brain Mapping* 4:58-73.

Wright IC *et al.* (1995). A voxel-based method for the statistical analysis of gray and white matter density applied to schizophrenia, *NeuroImage* 2:244-252.

Zijdenbos AP, Dawant BM. (1994). Brain segmentation and white matter lesion detection in MR images, *Crit Rev Biomed Engin* 22(5-6):401-465.

Zijdenbos AP *et al.* (1996). Automatic quantification of multiple sclerosis lesion volume using stereotaxic space. In Proceedings of the 4th International Conference of Visualization Biomedical Computed VBC '96, pp. 439-448, Hamburg, Germany.

Index

Note: *Page numbers followed by the letter f denote figures, t denote tables.*

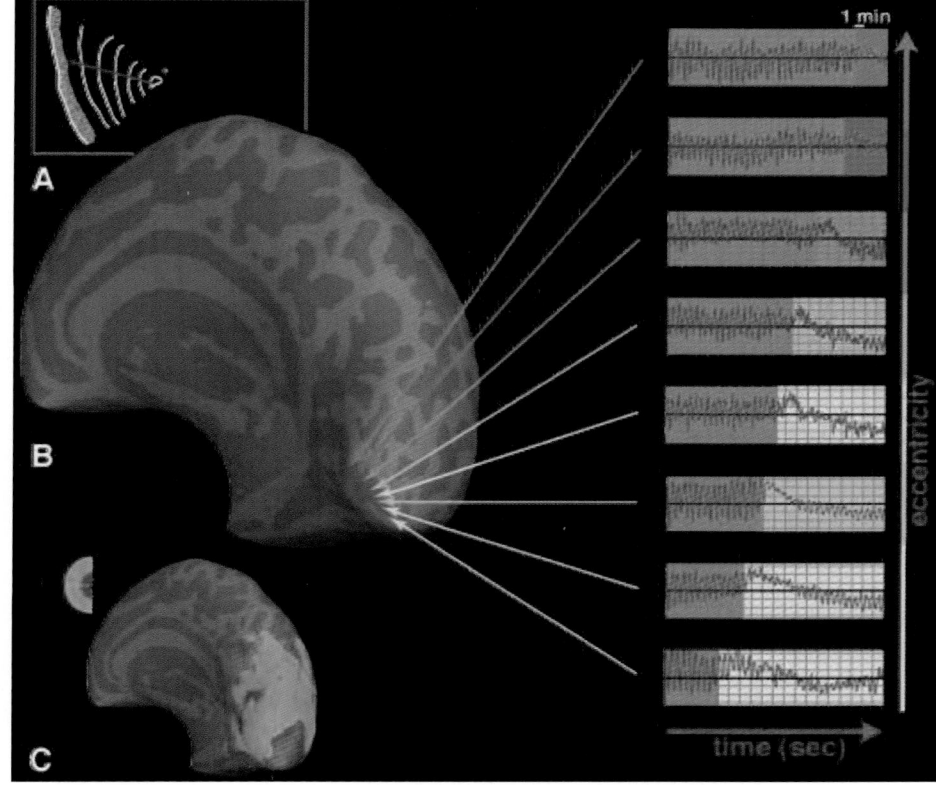

FIGURE 6.1 Spreading suppression of cortical activation during migraine visual aura. **A:** Drawing (*top*) showing the propagation (*red arrow*) of scintillations and scotoma in the left hemifield (*red arrow*) over 20 minutes during visual aura; cross indicates the fixation point. **B:** An inflated reconstruction of the migraineur's brain showing the cortical sulci (*dark*) and gyri (*lighter gray*). fMRI signal changes are shown on the right sampled from primary visual cortex from occipital pole anteriorly during the visual aura. Initial focal increase in BOLD signal followed by diminished BOLD response to light stimulation was observed in all extrastriate areas starting from occipital pole (central visual fields represented) toward more anterior regions representing central and peripheral eccentricities, respectively, and differing only in time of onset. The retinotopic progression of the BOLD signal was congruent with spread of visual aura (**A** and **C**). **D:** The maps of retinotopic eccentricities from the same subject acquired interictally. Activation within fovea are coded in red; parafoveal eccentricities are shown in blue and more peripheral eccentricities are shown in green (Hadjikhani *et al.*, 2001).

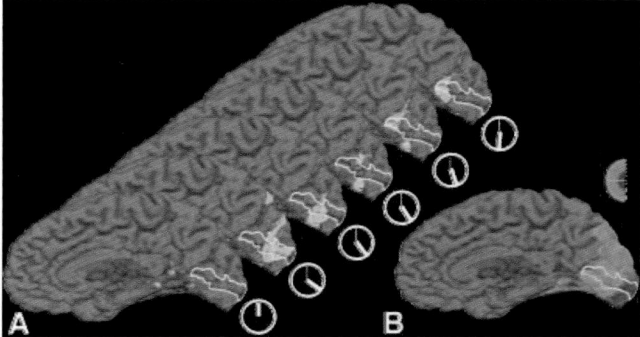

FIGURE 6.2 Progression of scintillations in the dark without visual stimuli during visual aura. **A:** A series of MR images over time (clock) showing posterior to anterior propagation of enhanced BOLD signal. The primary visual cortex (V1) is outlined with a white line. Initially no activation is seen in V1. However, with the onset of scintillations (after 20 minutes, see clock), activation appears that progresses from the parafoveal representation to more peripheral representations, paralleling the progression of the scintillations described by the subject. **B:** A medial view of the subject's brain with the MR maps of retinotopic eccentricity acquired during interictal scans (see Figure 6.1) (Hadjikhani *et al.*, 2001).

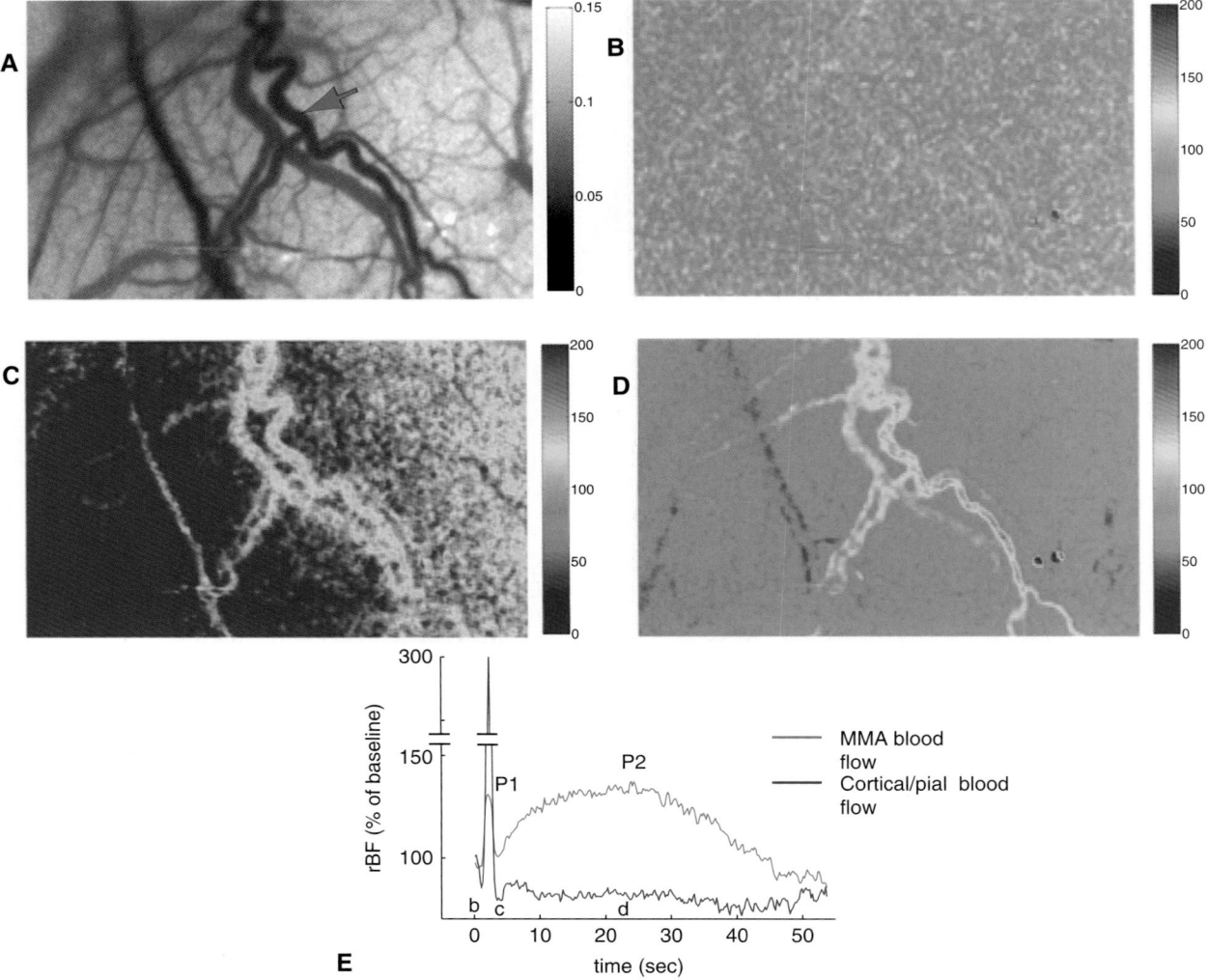

FIGURE 6.6 Blood flow images of MMA following CSD. Laser speckle images of blood flow in middle meningeal artery (MMA) and underlying parietal cortex and pial vessels following a single CSD. **A,** Speckle contrast image demonstrating the spatial flow heterogeneities across the imaged area at a single time illustrating MMA (*arrow*; proximal vessel is at top) and associated dural vein plus pial vessels and cerebral cortex. The darker values correspond to higher blood flow. **B-D:** Relative blood flow maps expressed as percentage of baseline, during the temporal evolution of CSD and represented by pseudo-colored images at the time points shown in **E**. **B:** Imaged blood flow after CSD induction, but before propagation of cortical hyperemia into the imaged area showing no immediate changes. **C:** Cortical hyperemia (*dark red*) spreading from left to right during CSD. Note that the marked flow increase in cortex and overlying pial vessels was comparable during CSD. A smaller transient rise in MMA blood flow (125% of baseline) was detected at this time. **D:** Blood flow map taken 20 minutes after CSD induction showing selective elevation of flow in MMA (140% of baseline), as well as in a dural vein. The relative blood flow in pial vessels and cortex is below baseline at this time (80% of baseline). **E:** The time course of the change in relative blood flow in MMA (*red line*) and cerebral cortex (*blue line*) demonstrates that CSD induces biphasic blood flow elevations within MMA with an early transient peak (P1) coincident with cortical hyperemia. A slower and more sustained increase in MMA blood flow (P2) is observed for approximately 45 minutes. These images clearly demonstrate that CSD-induced trigeminal activation causes long-lasting vasodilation selectively in dura mater; oligemia was prominent in cerebral cortex (Bolay *et al.*, 2002).

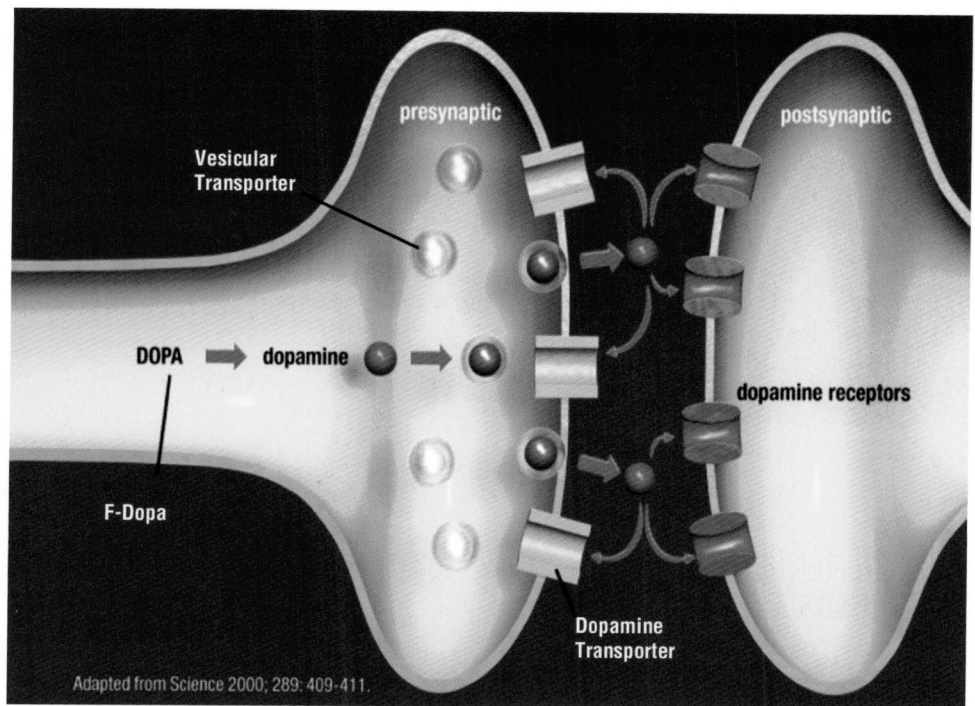

FIGURE 8.2 Dopamine synapse showing the presynaptic dopamine imaging targets (Marek and Seibyl, 2000).

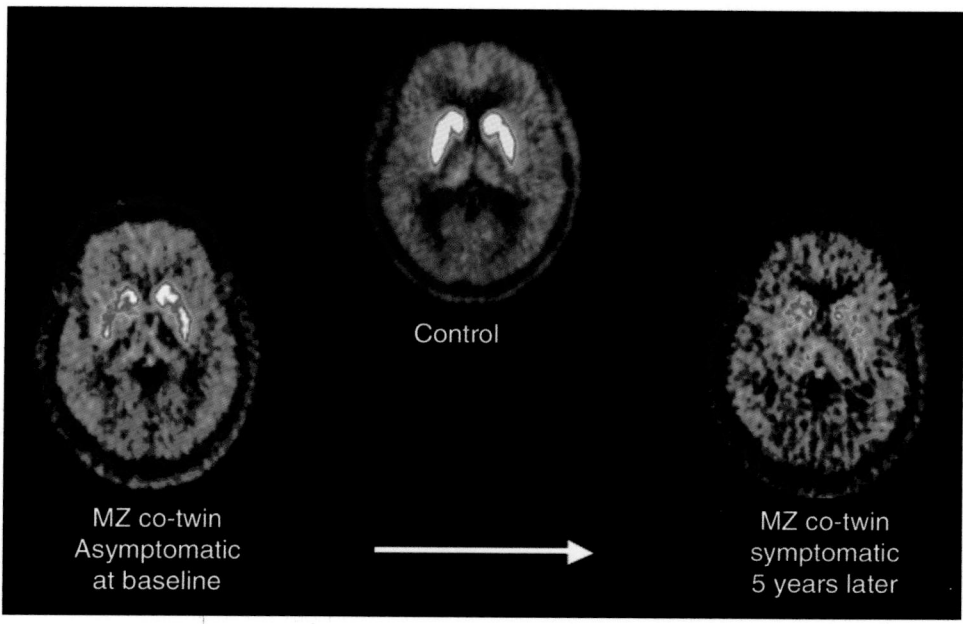

FIGURE 8.4 [18F]Dopa PET of an initially asymptomatic co-twin of a PD patient at baseline and after 5 years. Note that baseline scan shows loss of [18F]Dopa uptake compared to control and that scans show progressive loss of [18F]Dopa uptake activity consistent with onset of symptoms of PD at time of follow-up scan (Figure courtesy of Prof. David Brooks, Hammersmith Hospital).

A. SPECT IMAGING

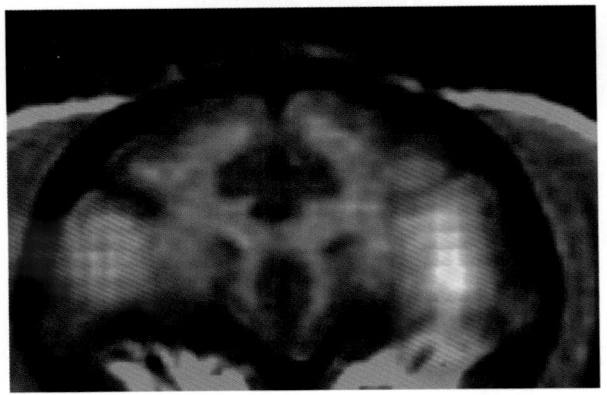

FIGURE 11.4 A: A single photon emission computed tomography (SPECT) imaging study of rhesus monkeys performing a spatial working memory task found that guanfacine significantly increased regional cerebral blood flow in the principal sulcal prefrontal cortex. The difference between the guanfacine and saline SPECT scans is overlaid on a coronal section of the structural MRI for a representative rhesus monkey. (Adapted from Avery *et al.*, 2000). **B:** Many prefrontal cortical neurons exhibit increased firing during the delay period while monkeys perform a delayed response task. These cells are often tuned (e.g., firing only if the cue had been on the right but not on the left side). Bao Ming Li has shown that either systemic or iontophoretic application of clonidine significantly increases delay-related firing of prefrontal cortical neurons without altering spontaneous firing rate (Li *et al.*, 1999). This finding is schematically depicted in Figure 11.4B.

B. PHYSIOLOGY

SALINE

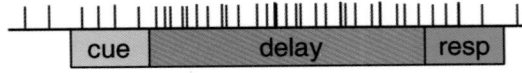

α2 AGONIST

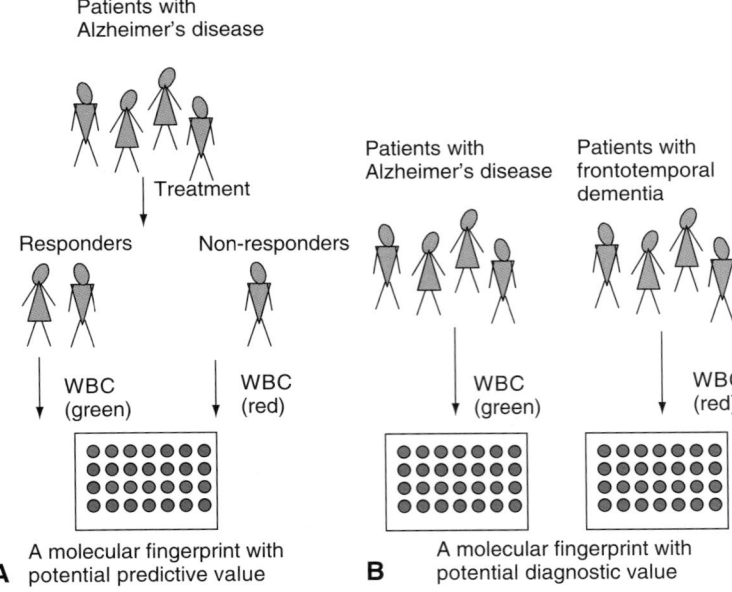

A A molecular fingerprint with potential predictive value

B A molecular fingerprint with potential diagnostic value

FIGURE 13.2 The use of microarrays for disease classification. **A:** Microarrays can be used to detect single nucleotide differences (polymorphisms) or identify gene expression patterns that predict treatment outcome or adverse affects in patients. **B:** Oligonucleotide arrays that allow polymorphism detection in a number of candidate genes simultaneously, such as cytochrome p450, that may affect drug metabolism, hence predisposing to treatment failure or adverse affects, will likely soon be widely adopted. Microarrays may also be used to augment diagnostic protocols. In the speculative example shown, mRNA from peripheral blood leukocytes is hybridized onto a cDNA array and the pattern indicates whether the patient is more likely to have Alzheimer's disease or frontotemporal dementia. Perhaps, more readily adopted into clinical practice will be the use of arrays to classify tumor samples from patients to both aid in diagnosis and predict optimal treatment. Figure and legend reprinted with permission from *Lancet Neurology*, 2003, 2:275-282, Elsevier Press.

Pre-operative   **Post-operative**

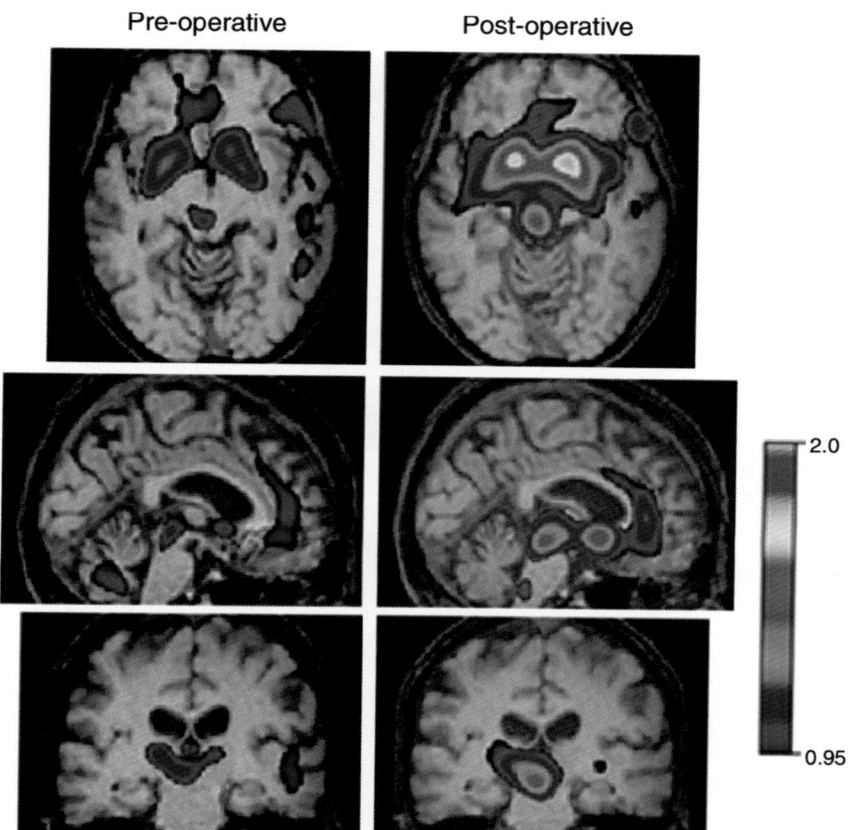

FIGURE 21.1 Preoperative and postoperative fluorodopa PET scans obtained in a patient with Parkinson's disease who received simultaneous intrastriatal and intranigral fetal VM grafts. These images consist of parametric maps of fluorodopa K_i transformed into stereotactic space and overlaid on the patient's MR image (also in stereotactic space). The parametric maps were smoothed using an 8-mm gaussian filter. The axial, coronal, and sagittal sections demonstrate an increase in K_i in the midbrain and putamen bilaterally, likely resulting from surviving grafted dopamine neurons. From Mendez *et al.* (2002). Used with permission.

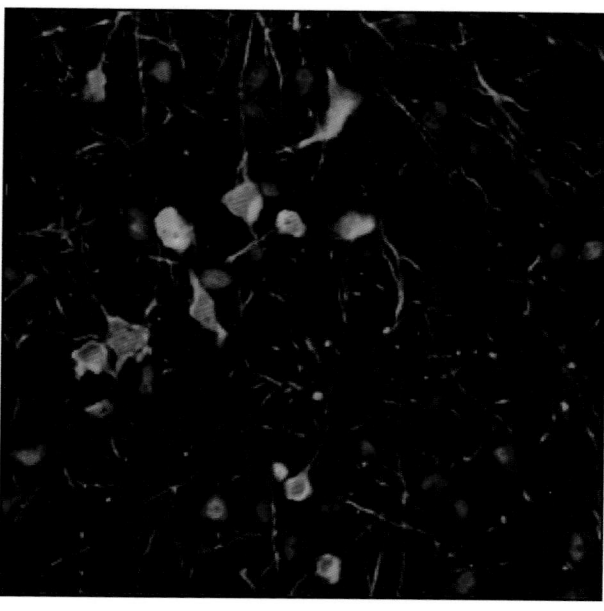

FIGURE 21.2 Dopamine neurons derived *in vivo* from mouse embryonic stem (ES) cells. This confocal image shows numerous dopamine cells in the striatum of a 6-OHDA–lesioned rat, 11 weeks post-transplantation of undifferentiated ES cells (~2000 cells). Dopamine neurons co-expressed the synthetic enzymes tyrosine hydroxylase (TH, green) and aromatic amino acid decarboxylase (AADC, blue) and the neuronal nuclei marker (NeuN, red). Animals with ES cells with derived dopamine neurons in the grafts showed recovery of motor asymmetry.

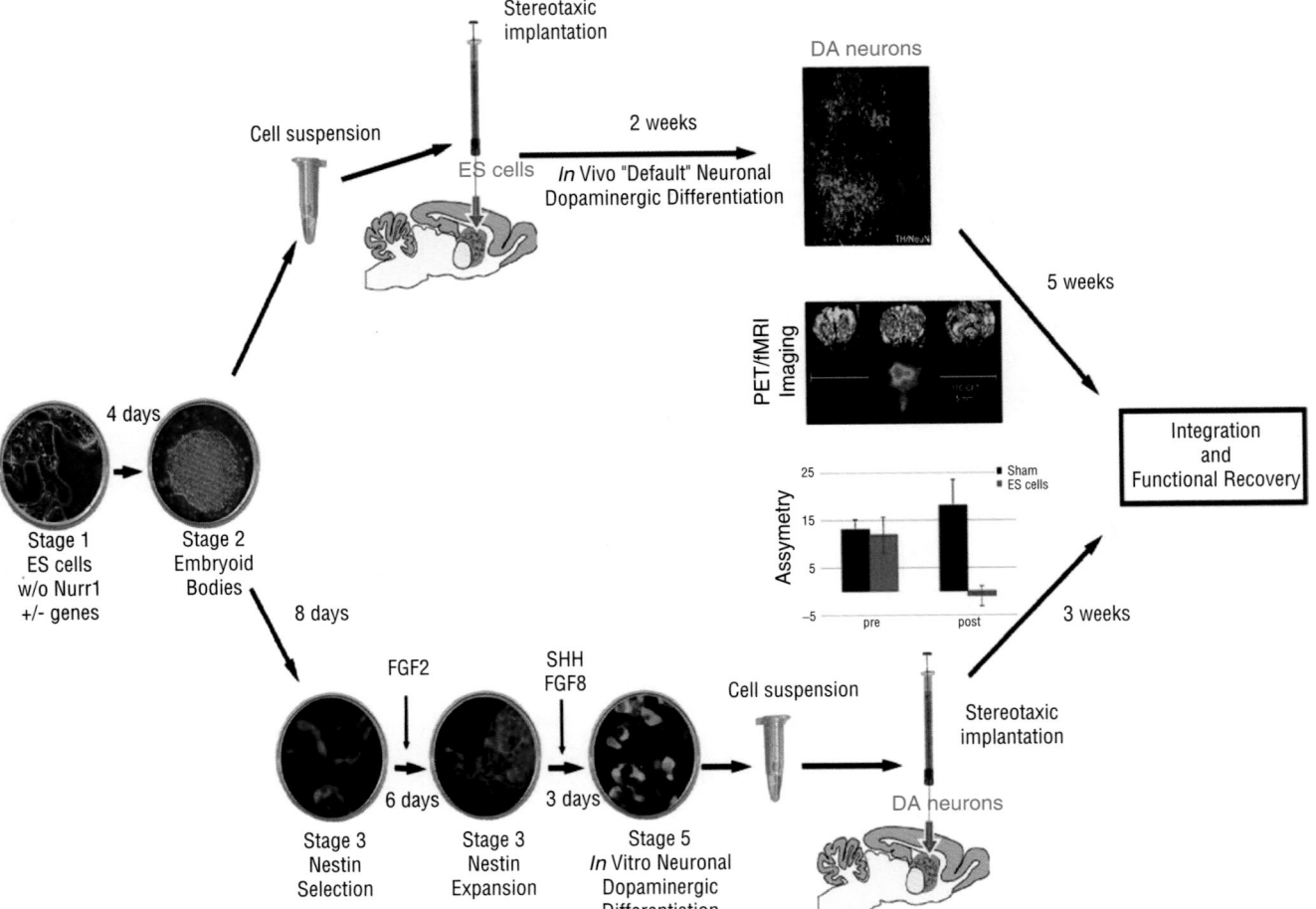

FIGURE 21.3 Schematic drawing of the process of obtaining dopamine or cholinergic neurons from embryonic pluripotent stem cells. In the case of mouse embryonic stem cells, these cells can grow on feeder layers in leukemia inhibitor factors (LIF) in stage 1. At this stage, various genes can be added or deleted. In the case of Nurr1, it is a positive drive for transcription of dopamine and neuron-related genes. At stage 2, embryoid bodies are formed containing similarly pluripotent cells. In the left panel, as shown by Kim *et al.* (2002), the embryonic stem cells are brought in five stages to their fetal stage, in which they are then made into a cell suspension and transplanted directly into the Parkinson rat model. Similarly to the default pathway, these embryonic stem cell derived fetal neurons will create a functional recovery on amphetamine rotation, as well as a number of spontaneous tasks including reaching tasks and other evidence of corrected dopamine deficiency. In the upper panel (Bjorklund *et al.*, 2002), the cells are dissociated at specific cell concentrations and can be placed directly in the brain on low concentrations and will then, after a few weeks' time, develop over the species-specific program towards dopaminergic and serotonergic neurons. In previous experiments, such neurons would, over time, produce sufficient dopamine to correct the behavioral syndrome of Parkinson's disease asymmetry, such that the correction can be normalized if sufficient cells are reinnervating the brain compared to sham cells (Bjorklund *et al.*, 2002). From Isacson (2003); used with permission.

A

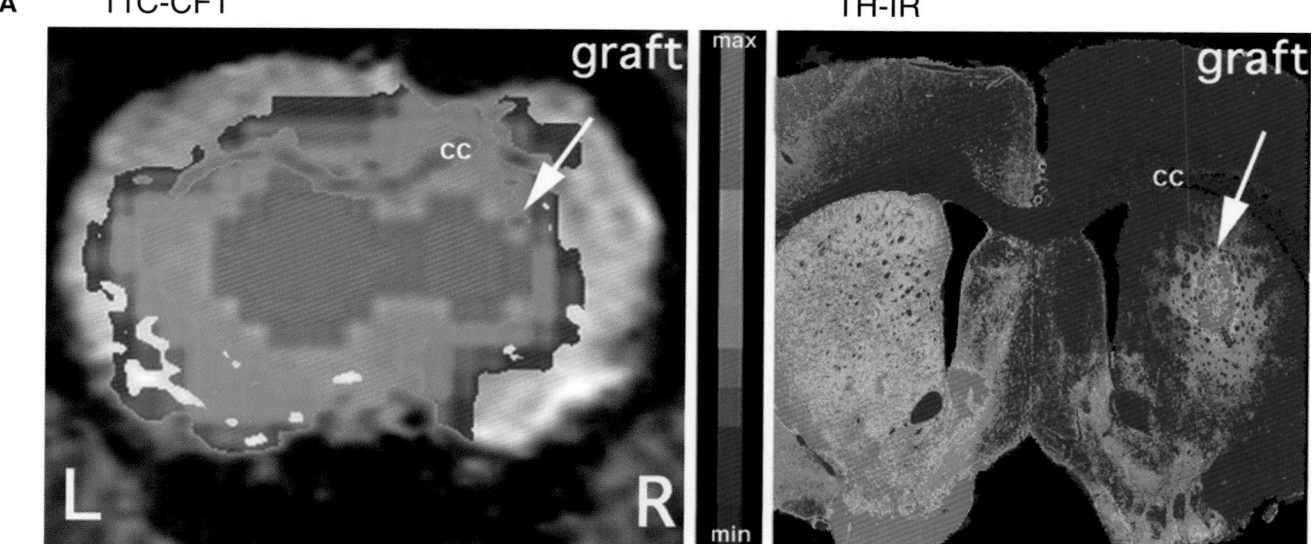

11C-CFT　　　　　　　　　　　　　　　TH-IR

Amphetamine-induced change in CBV

B

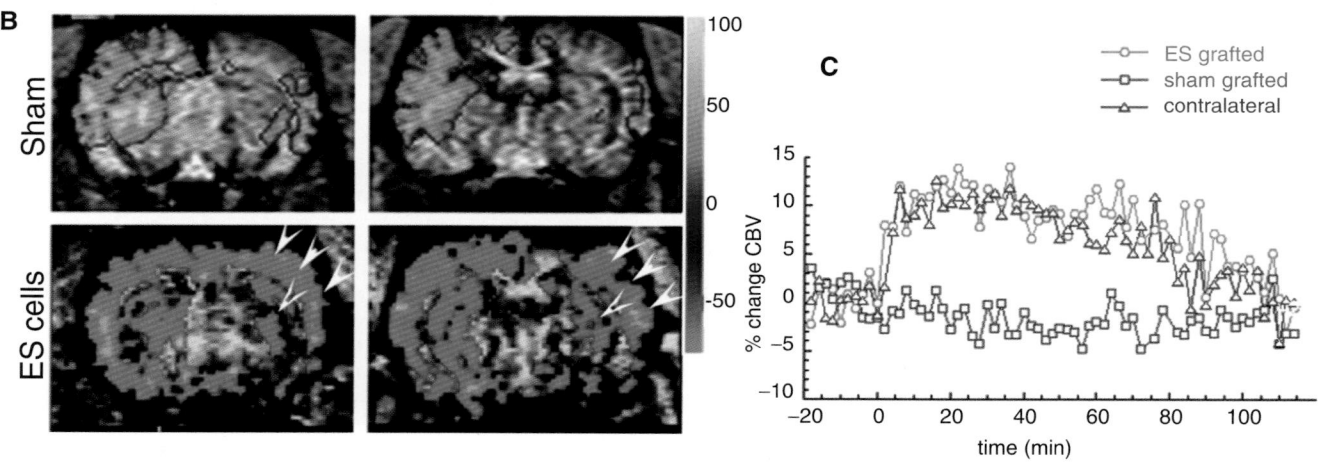

FIGURE 21.4 **A:** Using PET and the specific dopamine transporter (DAT) ligand carbon-11-labeled 2β-car-bomethoxy-3β-(4-fluorophenyl) tropane (^{11}C-CFT), we identified specific binding in the right grafted striatum, as shown in this brain slice (**A, left panel**) acquired 26 min after injection of the ligand into the tail vein (acquisition time was 15 seconds). Color-coded (activity) PET images were overlaid with MRI images for anatomical localiza-tion. The increase in ^{11}C-CFT binding in the right striatum was correlated with the postmortem presence of TH-immunoreactive (IR) neurons in the graft (**A, right panel**). **B:** Neuronal activation mediated by DA release in response to amphetamine (2 mg/kg) was restored in animals receiving ES grafts. Color-coded maps of percent change in relative cerebral blood volume (rCBV) are shown at two striatal levels for control (upper panel) and an ES cell derived DA graft (lower panel). 6-OHDA lesion results in a complete absence of CBV response to amphet-amine on striatum and cortex ipsilateral to the lesion (upper panel). Recovery of signal change in motor and somato-sensory cortex (arrows), and to a minor extent in the striatum was observed only in ES grafted animals. **C:** Graphic representation of signal changes over time in the same animal shown in **B**. The response on the grafted (red line) and normal (blue line) striata was similar in magnitude and time course while no changes were observed in sham grafted animals (green line). Baseline was collected for 10 minutes before and 10 minutes after MION injection and amphetamine was injected at time 0. cc, corpus callosum (Bjorklund et al., 2002).

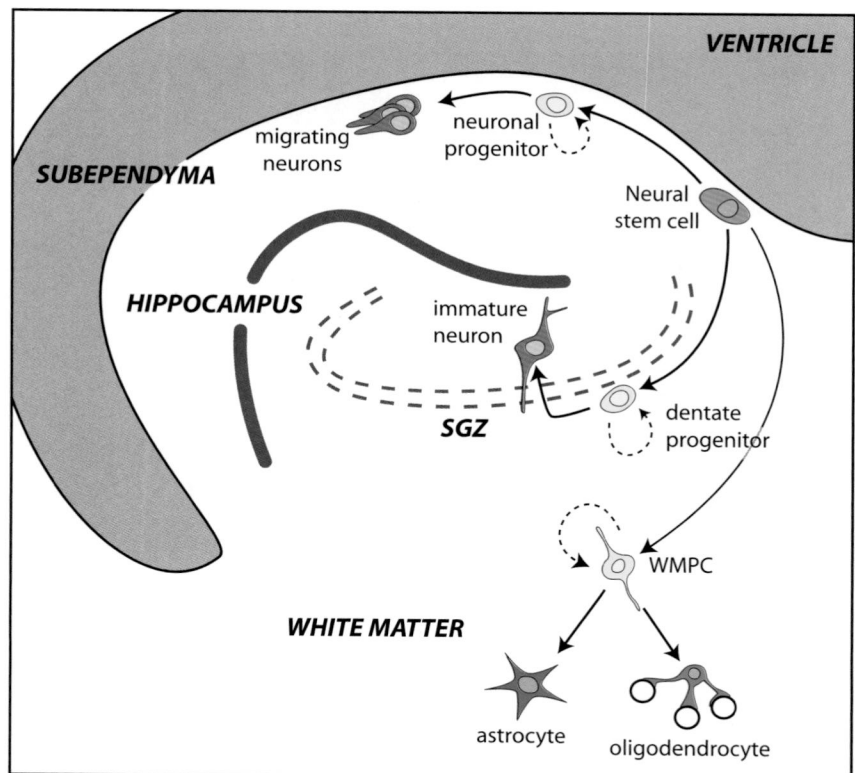

FIGURE 22.1 Progenitors of the adult human brain. This schematic illustrates the basic categories of progenitor cells in the adult brain, and their known interrelationships. The human temporal lobe is schematized by way of example. It includes both ventricular zone neural stem cells (*red*), which in turn generate at least three populations of potentially neurogenic transit amplifying progenitors of both neuronal and glial lineages (*yellow*). These include the neuronal progenitor cells of the ventricular subependyma, those of the subgranular zone of the dentate gyrus, and the white matter progenitor cells (WMPC) of the subcortical parenchyma, which although nominally glial, may remain potentially neurogenic. The latter may be represented in the gray matter parenchyma as well, although the relationships of the parenchymal gray and white matter progenitors have not yet been elucidated. Each transit amplifying pool may then give rise to differentiated progeny appropriate to their locations, including neurons (*purple*), oligodendrocytes (*green*) and parenchymal astrocytes (*blue*). Adapted from Goldman (2003).

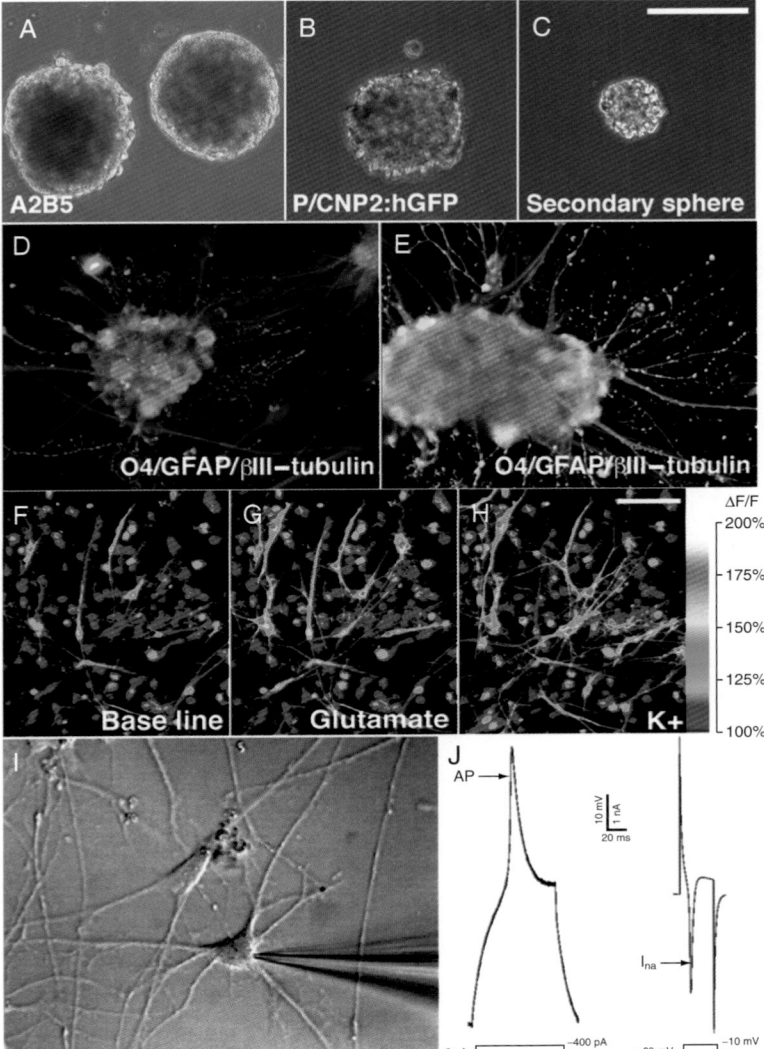

FIGURE 22.2 Resident progenitor cells of the adult human white matter can generate neurons as well as oligo-dendrocytes and astrocytes. **A:** First passage spheres generated from A2B5-sorted cells 2 weeks post-sort. **B:** First passage spheres arising from P/CNP2:hGFP sorted cells, 2 weeks. **C:** Second passage sphere derived from an A2B5-sorted sample, at 3 weeks. **D:** Once plated onto substrate, the primary spheres differentiated as βIII-tubulin⁺ neurons (*red*), GFAP⁺ astrocytes (*blue*), and O4⁺ oligodendrocytes (*green*). **E:** Neurons (*red*), astrocytes (*blue*), and oligodendrocytes (*green*) similarly arose from spheres derived from P/CNP2:GFP-sorted WMPCs. **F–H:** *WMPC-derived neurons developed neuronal Ca²⁺ responses to depolarization.* **D** shows an image of WMPC-derived cells loaded with the calcium indicator dye fluo-3, 10 days after plating of first passage spheres derived from A2B5-sorted white matter (35 DIV total); many fiber-bearing cells of both neuronal and glial morphologies are apparent. **E,** The same field after exposure to 100 μM glutamate, and **F,** after exposure to a depolarizing stimulus of 60 mM KCl. The neurons displayed rapid, reversible, greater than 100% elevations in cytosolic calcium in response to K⁺, consistent with the activity of neuronal voltage-gated calcium channels. **I–J,** *Whole cell patch-clamp revealed voltage-gated sodium currents and action potentials in WMPC-derived neurons.* **G:** A representative cell 14 days after plating of first passage sphere derived from A2B5-sorted white matter. The cell was patch-clamped in a voltage-clamped configuration, and its responses to current injection recorded. **H:** The fast negative deflections noted after current injection are typical of the voltage-gated sodium currents of mature neurons. Action potentials were noted only at I_Na greater than 800 pA. Scale: **A–E,** 100 μm; **F–H,** 80 μm. Adapted from Nunes *et al.* (2003).

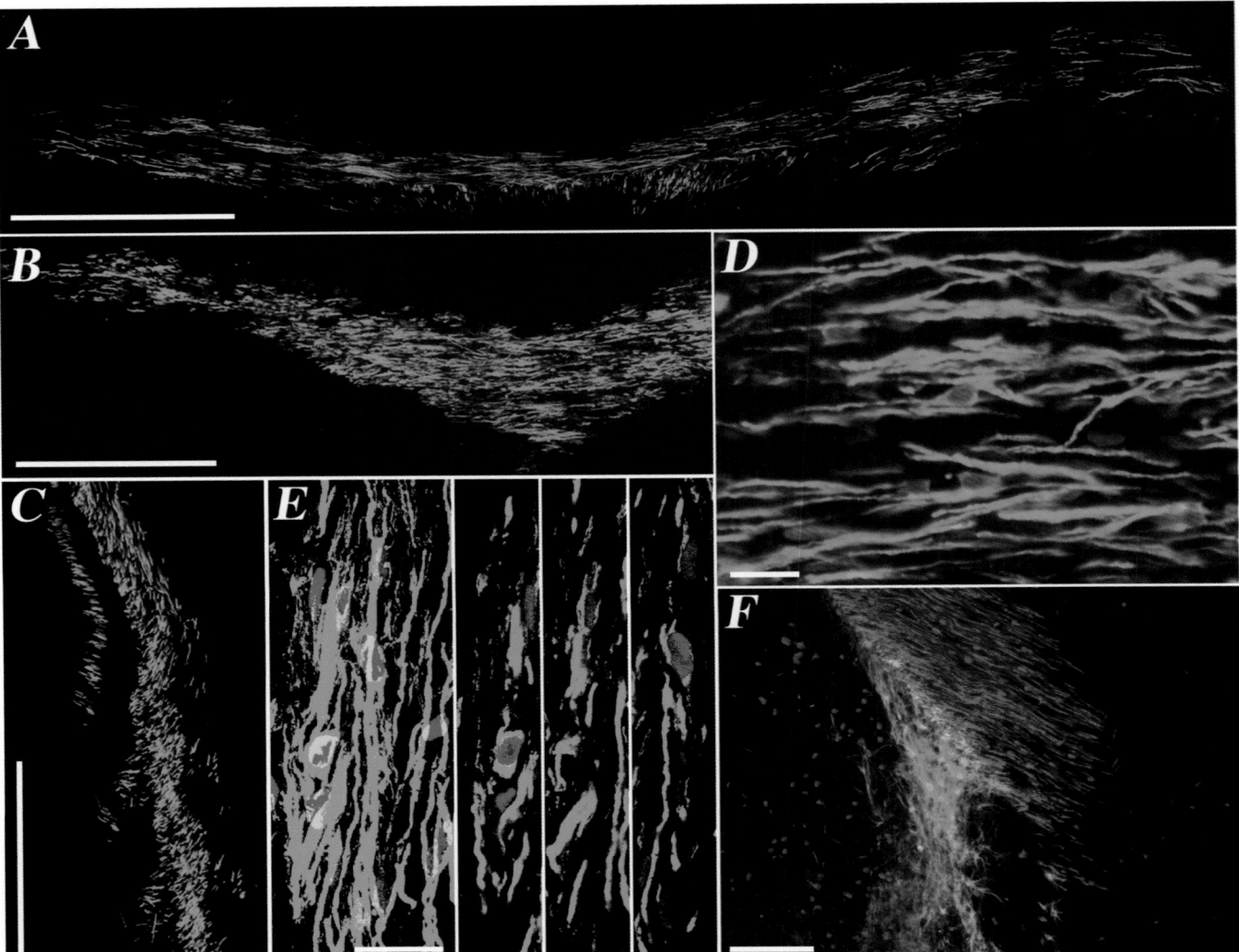

FIGURE 22.3 Engrafted human OPCs myelinate an extensive region of the forebrain. **A–B:** Extensive myelin basic protein expression by sorted human fetal OPCs, implanted into homozygote shiverer mice as neonates, indicates that large regions of the corpus callosum (**A** and **B**, 2 different mice) have myelinated by 12 weeks (MBP, *green*). **C:** Human OPCs also migrated to and myelinated fibers throughout the dorsoventral extents of the internal capsules, manifesting widespread remyelination of the forebrain after a single perinatal injection. **D:** MBP expression (*green*), in an engrafted shiverer callosum 3 months after perinatal xenograft, is associated with donor cells, identified by human nuclear antigen (hNA, in *red*). Both the engrafted human cells and their associated myelin were invariably found to lay parallel to callosal axonal tracts. **E:** Optical sections of implanted shiverer corpus callosum, with hNA+ human cells surrounded by MBP. Human cells (*arrows*) found within meshwork of MBP+ fibers (**E**, merged image of **E1-3**). **F:** OPCs were recruited as oligodendrocytes or astrocytes in a context-dependent manner. This photo shows the striatocallosal border of a shiverer brain, 3 months after perinatal engraftment with human fetal OPCs (hNA in *blue*). Donor-derived MBP+ oligodendrocytes and myelin (*red*) is evident in the corpus callosum, while donor-derived GFAP+ (*green*) astrocytes predominate on the striatal side.

Continued

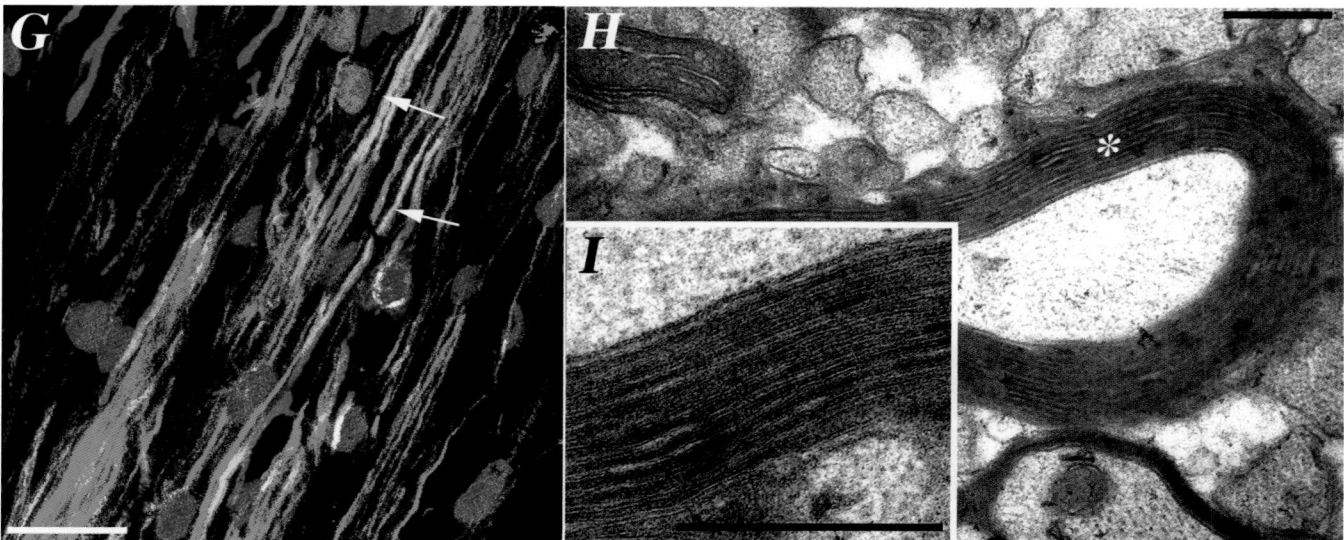

FIGURE 22.3, cont'd   G: A confocal optical section exhibiting axonal ensheathment and myelin compaction by engrafted human progenitor cells. This section of engrafted corpus callosum has been triple-immunostained for MBP (*red*), human ANA (*blue*), and neurofilament protein (*green*). All MBP immunostaining is derived from the human OPCs, whereas the NF+ axons are those of the mouse host. *Arrows* identify segments of murine axons surrounded by human oligodendrocytic MBP. **H–I:** An electron micrograph of a 16-week-old shiverer homozygote implanted with human OPCs shortly after birth, showing host axons with densely compacted, donor-derived myelin sheaths. The *asterisk* indicates the field enlarged in the inset, which shows the major dense lines of mature myelin. Scale: A–C, 1 μm; D, 100 μm; E, 20 μm; F, 200 fm. Adapted from Windrem M. et al. (2004). Fetal and adult human oligodendrocyte progenitor cell isolates myelinate the congenitally dysmyelinated brain. *Nat Med*, in press.

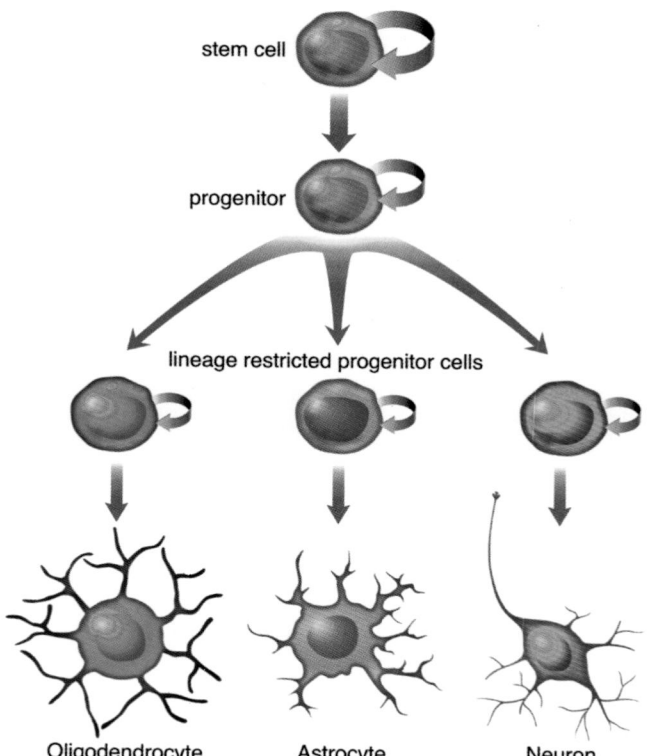

stem cell

progenitor

lineage restricted progenitor cells

Oligodendrocyte   Astrocyte   Neuron

FIGURE 23.1   Model for the generation of differentiated neural cells
from stem cells. Multipotent, self-renewing neural stem cells generate
progeny with restricted proliferation and differentiation potential
(i.e., progenitor cells). Progenitor cells differentiate into postmitotic
neurons, oligodendrocytes, and astrocytes.

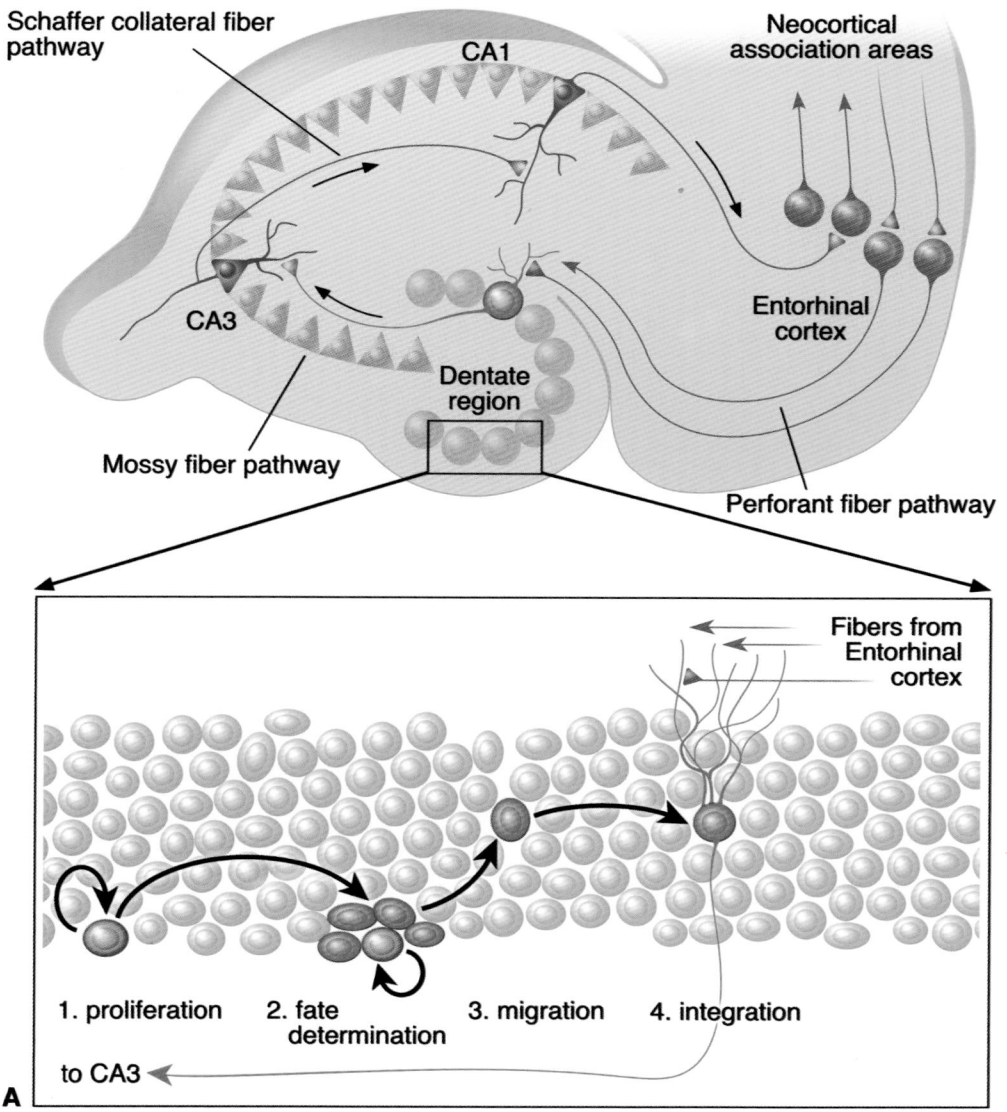

FIGURE 23.2 Neurogenic zones of the adult mammalian CNS. **A:** Hippocampal dentate gyrus. 1. Proliferation and fate determination: Stem cells (beige) in the subgranular zone of the dentate gyrus give rise to transit amplifying cells that differentiate into immature neurons. 2. Migration: Immature neurons migrate into the granule cell layer of the dentate gyrus. 3. Integration: Immature neurons mature into new granule neurons, receive input from the entorhinal cortex and extend projections into CA3.

Continued

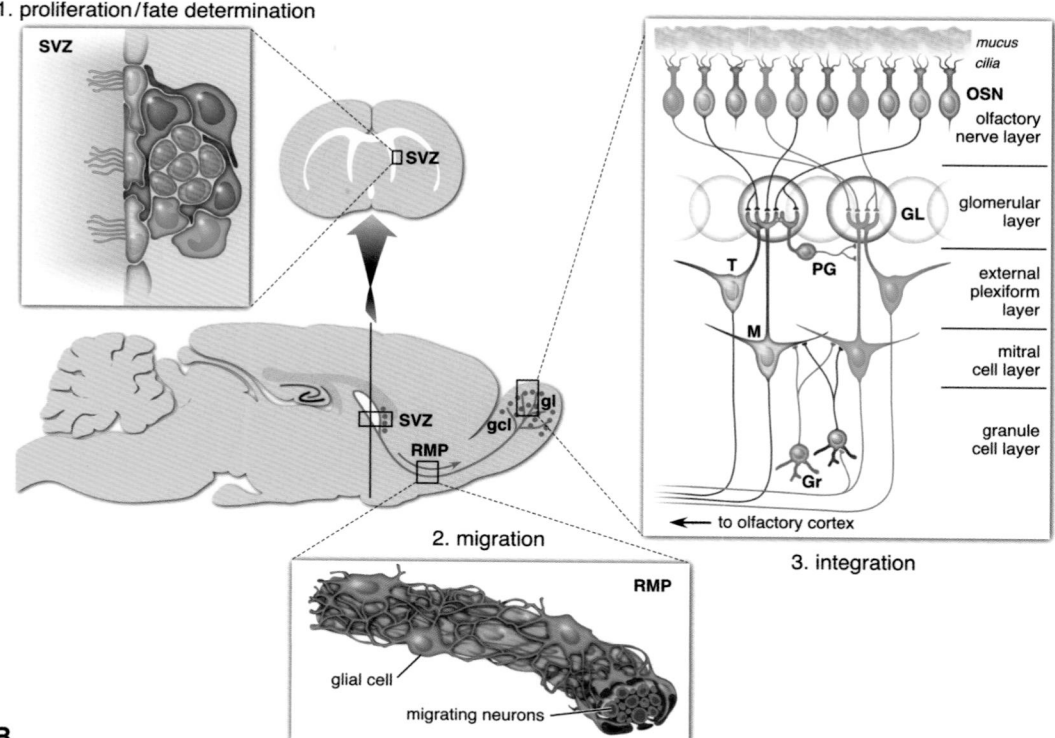

B

FIGURE 23.2, cont'd   **B:** Subventricular zone (SVZ)/olfactory bulb system. 1. Proliferation and fate determination: Stem cells in the SVZ of the lateral ventricle (*blue*) give rise to transit amplifying cells (*green*) that differentiate into immature neurons (*red*). Adjacent ependymal cells (*light brown*) of the lateral ventricle are essential for the neuronal fate determination by providing inhibitors of glial differentiation. 2. Migration: Immature neurons (*red*) migrate along each other in chains through the rostral migratory pathway (RMP). The migrating neurons are ensheathed by astrocytes (*blue*). 3. Integration: Immature neurons differentiate local interneurons (*red*) in the granule cell layer and the periglomerular layer. Olfactory sensory neurons (OSN); tufted neurons (T); mitral neurons (M); granule neurons (Gr); periglomerular neurons (PG). Figure reproduced from Lie *et al.,* 2004. Copyright by Annual Reviews.

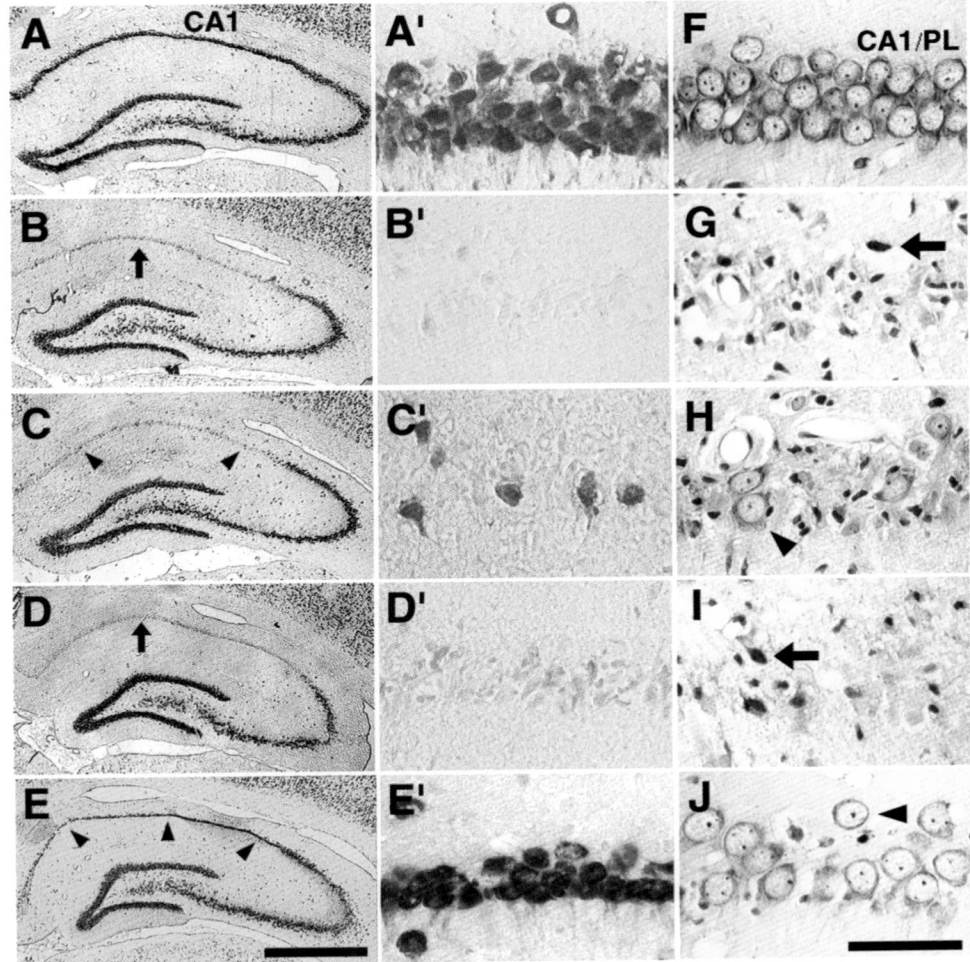

FIGURE 23.3 **A:** Animals treated with EGF and FGF2 following ischemia show significant regeneration of CA1 pyramidal neurons. **A–E:** Neurons are stained with the neuronal marker NeuN. **F–J:** Cresyl violet staining of the hippocampal CA1 pyramidal layer. (A, A′, and F) Intact animals, (B, B′, and G) untreated animals at day 7 after ischemia, (C, C′, and H) untreated animals at day 28 after ischemia, (D, D′, and I) EGF/FGF2 treated animals at day 7 after ischemia, (E, E′, and J) EGF/FGF2 treated animals at day 28 after ischemia. Arrows and arrowheads in **G–J** indicate cells with pyknotic and intact morphology, respectively. Figure reproduced from Nakatomi *et al.* (2002). Regeneration of hippocampal pyramidal neurons after ischemic brain injury by recruitment of endogenous neural progenitors, *Cell* 110:429–441. Copyright by Cell Press.

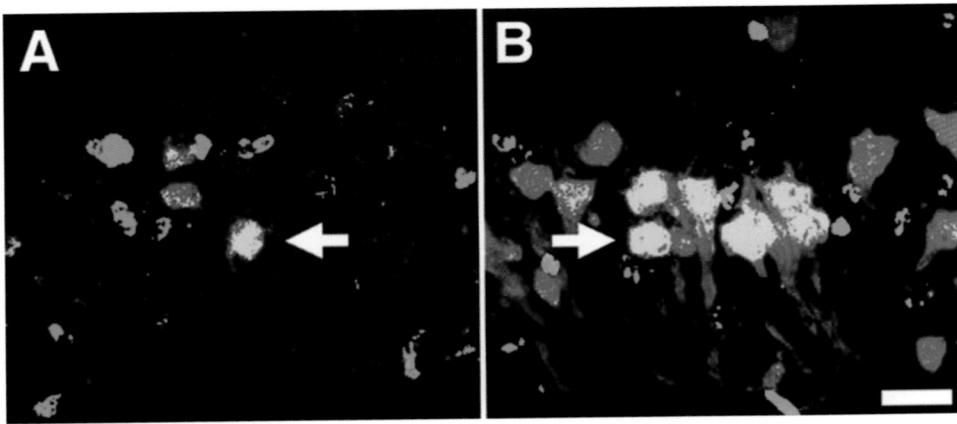

FIGURE 23.4 Regenerated CA1 pyramidal neurons are derived from proliferating cells. The thymidine-analogue BrdU was administered to untreated **(A)** and EGF/FGF2 treated **(B)** ischemic animals. A large number of pyramidal neurons, identified by the expression of the neuronal marker NeuN (*red*), are positive for BrdU (*green*), indicating that they are the progeny of previously dividing cells. Figure reproduced from Nakatomi *et al.* (2002). Regeneration of hippocampal pyramidal neurons after ischemic brain injury by recruitment of endogenous neural progenitors, *Cell* 110:429–441. Copyright by Cell Press.

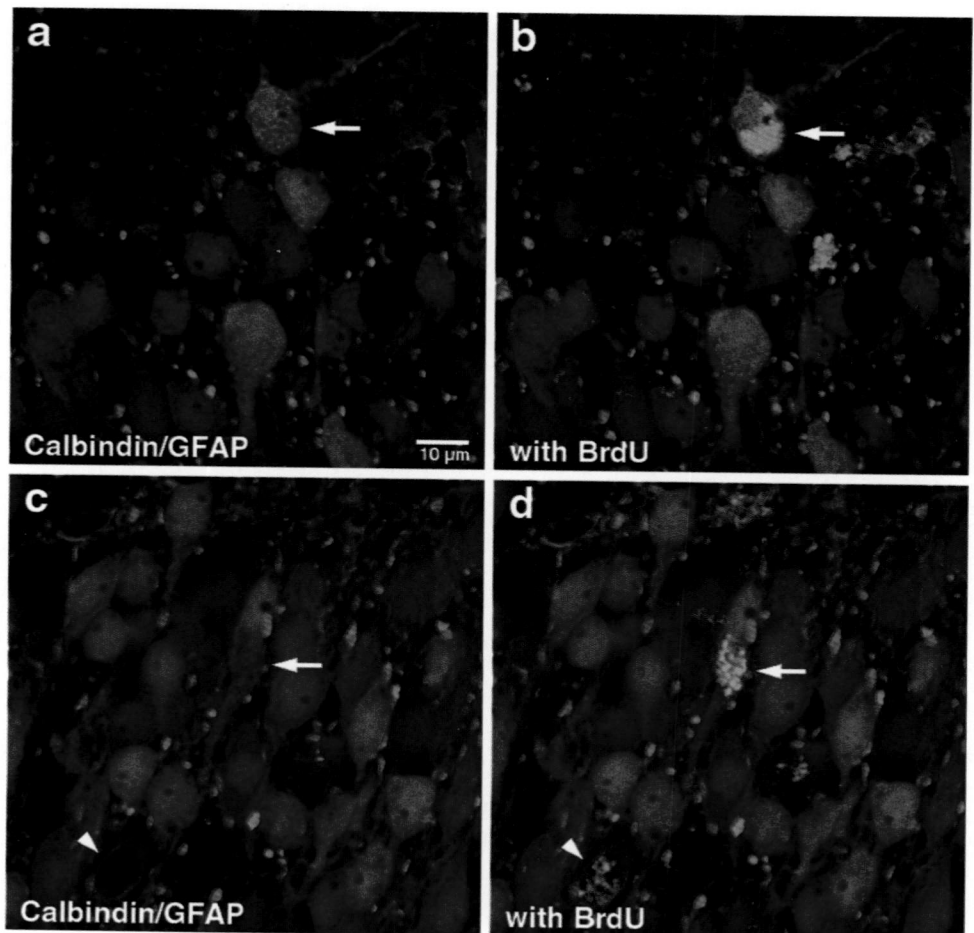

FIGURE 23.5 Neurogenesis in the adult human hippocampal dentate gyrus. Human hippocampal tissue was obtained at autopsy from cancer patients that had received BrdU for assessment of proliferative activity of tumor cells. Staining for BrdU (*green*), the neuronal marker calbindin (*red*) and the glial marker GFAP (*blue*) demonstrated that new neurons (arrows) and astrocytes (arrowheads) are generated in the human dentate gyrus. Figure reproduced from Eriksson *et al.* (1998). Neurogenesis in the adult human hippocampus, *Nat Med* 4:1313-1317. Copyright by Nature Medicine.

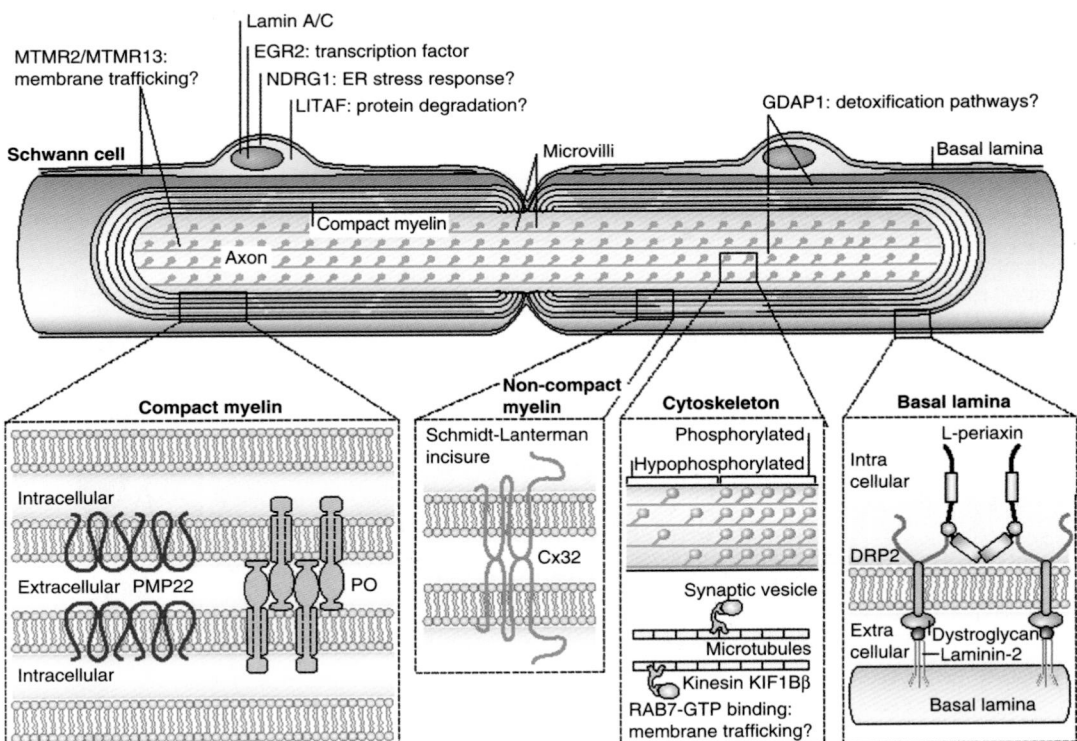

FIGURE 24.1 Schematic overview of the molecular organization of myelinated axons highlighting the proteins affected in CMT. The middle panel depicts a myelinated axon; the upper panel is an enlargement of the molecules associated with the dystroglycan receptor; the lower panel is an enlargement of the molecules associated with the myelin sheath or the axonal cytoskeleton. From Suter and Scherer (2003). Disease mechanisms in inherited neuropathies, *Nat Neurosci Rev* 4:714-726, with permission of *Nature*.

A Nogo-A/B knockout mice

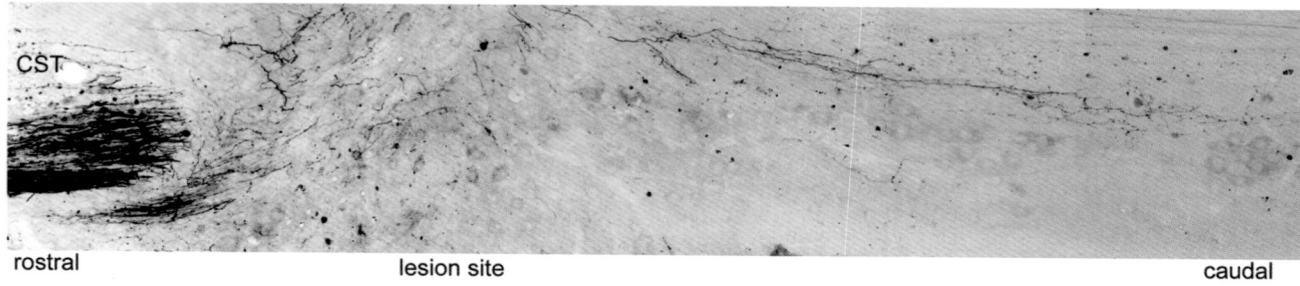

CST

rostral   lesion site   caudal

B NEP1-40 treated rat caudal spinal cord

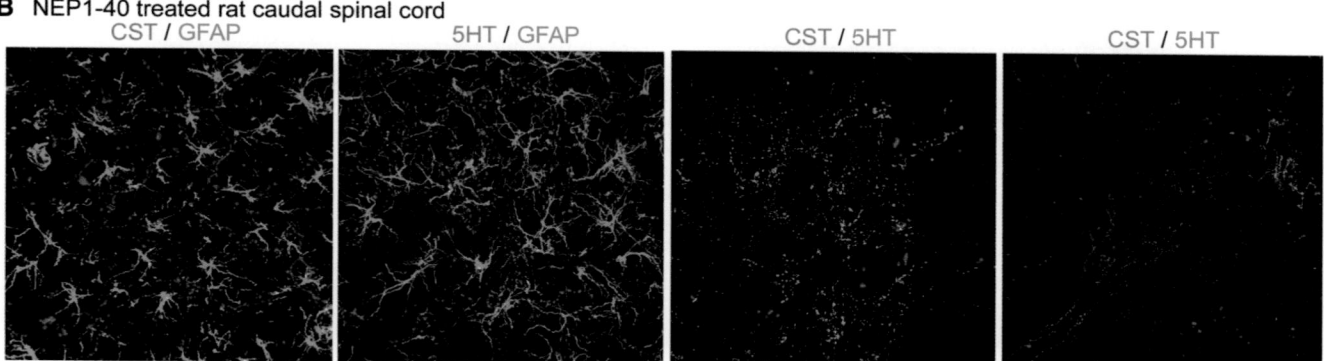

CST / GFAP   5HT / GFAP   CST / 5HT   CST / 5HT

FIGURE 26.3 Axonal regeneration after SCI in animals lacking Nogo-A/B or treated with NgR antagonist peptide. **A:** Mice lacking a functional Nogo-A/B gene were subjected to mid-thoracic spinal hemisection injury and the corticospinal tract (CST) was traced by cortical injection of biotin dextran amine (Kim *et al.*, 2003). Note fibers growing across the injury site into the caudal spinal cord. Wild-type mice have no CST fibers crossing such lesions. **B:** Rats were treated with the NEP1-40 antagonist peptide intrathecally after mid-thoracic spinal cord hemisection and the CST was traced (GrandPre *et al.*, 2002). Descending raphespinal fibers were detected by anti-serotonin immunohistology (5HT) and reactive astrocytes by anti-GFAP immunohistology (GFAP), as indicated. Sections from the spinal cord caudal to the lesion site demonstrate regenerating fibers despite normal level of reactive astrocytosis. Untreated animals exhibit similar astrocytosis but no CST or 5HT regeneration.

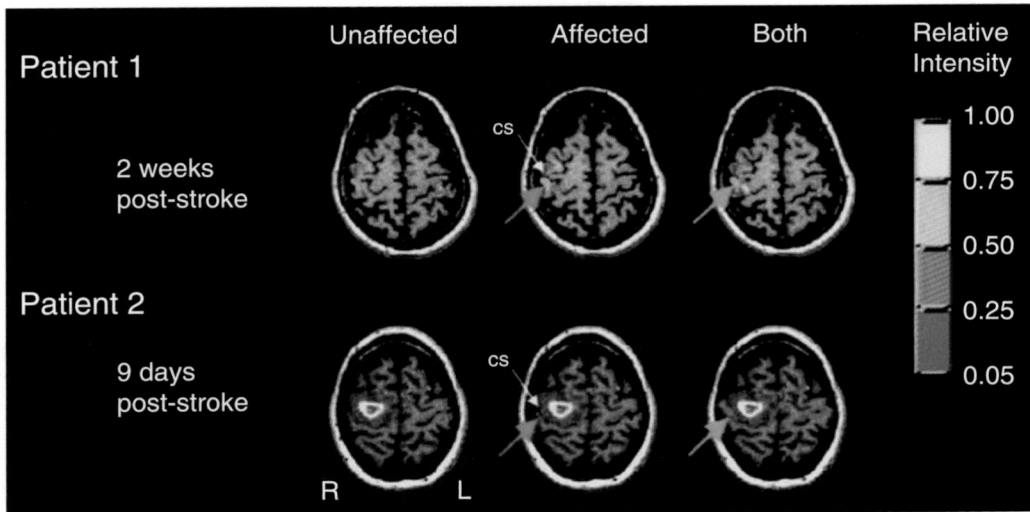

FIGURE 28.2 Functional MRI during repetitive unilateral or bilateral hand grips of two patients with right hemispheric stroke. The first column illustrates activation associated with repetitive gripping with the right (unaffected) hand. The second column illustrates activation associated with repetitive gripping with the left (affected) hand. The third column illustrates activation associated with repetitive gripping with both the affected and the anaffected hands simultaneously. Any significant activation of the right motor cortex when gripping with the unrecovered left hand (green arrow) was enhanced during bilateral movement (blue arrow). CS = central sulcus; R/L = right/left hemisphere. Adapted from Staines *et al.* (2001). Bilateral movement enhances ipsilesional cortical activity in acute stroke: A pilot functional mri study, *Neurology* 56:401-404.

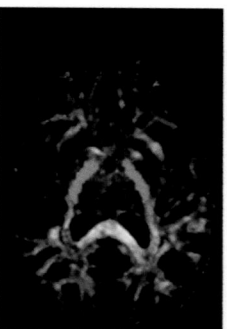

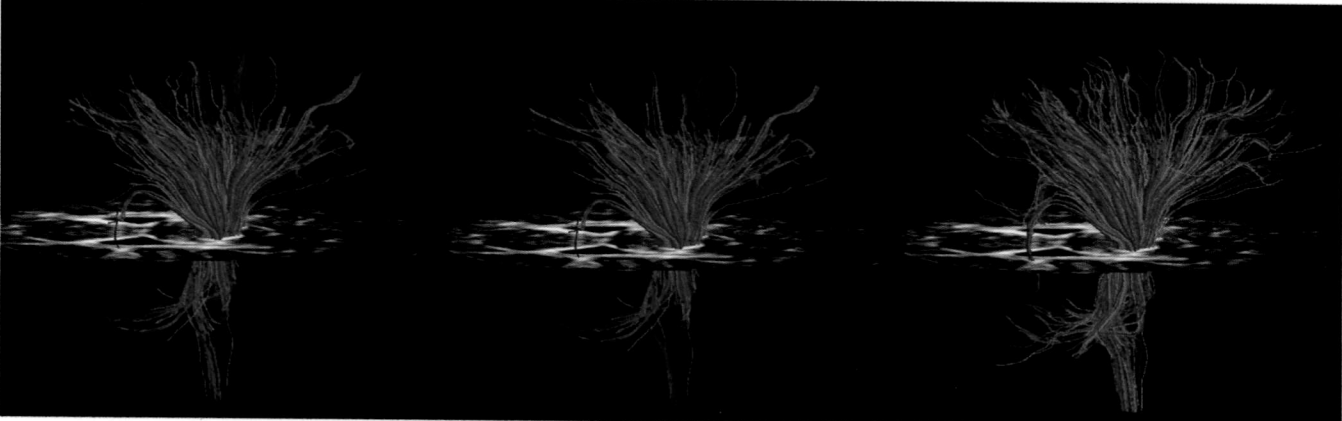

FIGURE 29.1 Human white matter anatomy. It is possible to identify white matter bundles *in vivo* in the human brain using DTI MRI. Such studies take advantage of the fact that diffusion of water molecules is compartmentalized within the CNS. As such, this diffusion is not isotropic. The anisotropic movement of water is most obvious in fiber bundles before the fibers branch and reach their targets. As such, major tracks in the brain, such as the cortico-spinal track shown in this image, can be quite clearly identified in individuals. When populations of normal subjects are imaged in this fashion, probabilistic estimates for the location of specific fiber bundles can be determined for the population in general and may be of value in understanding the variance of these structures in the human brain. In a given individual, such studies may be helpful in presurgical planning to avoid disconnection syndromes associated with the inadvertent resections of part or all of these tracks. Courtesy of A. Alexander, M. Lazar and colleagues, Waisman Center, University of Wisconsin.

STIMULUS   CONTROL

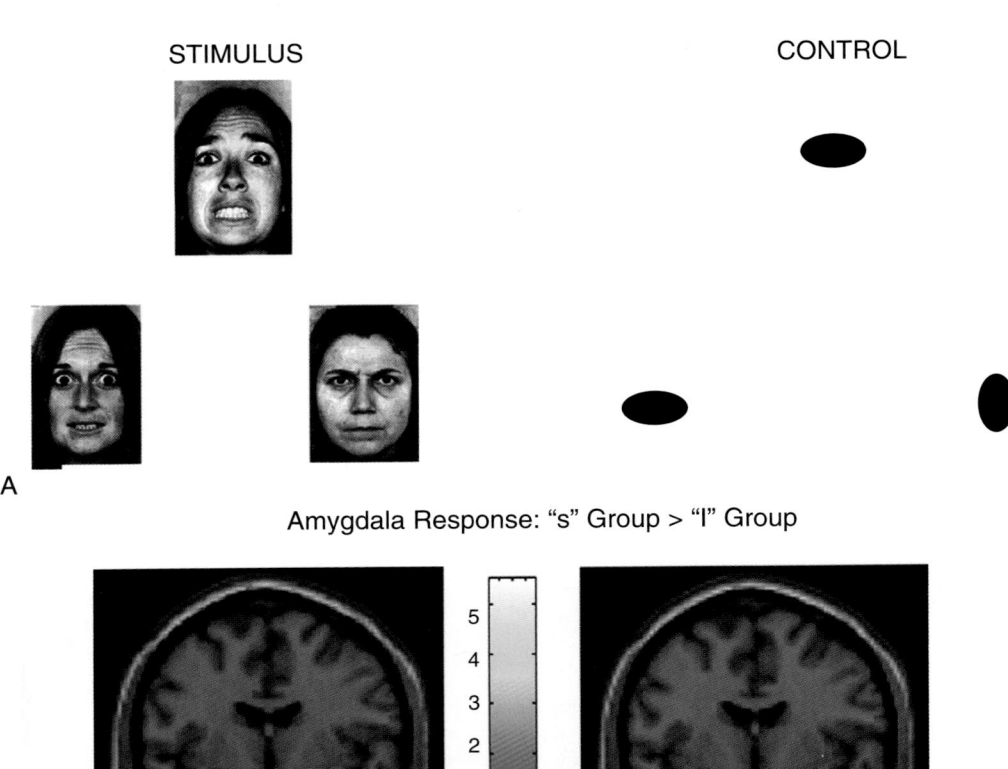

A

Amygdala Response: "s" Group > "l" Group

B   First Cohort   Second Cohort

FIGURE 29.2 Imaging phenotype and genotype related to behavior. Human behavior is quite variable and may fluctuate from day to day. This is particularly true of emotional states. Thus, when a specific genotype was hypothesized to be associated with different degrees of serotonin reuptake it was thought that there would be a good correlation with anxiety and affective disorders. This was only found to a very limited degree and its significance was questioned. It had been known from basic neuroscience research that the serotonin (5-HT) reuptake transport (5-HTT) promoter could affect serotonin reuptake and that this promoter had two alleles. The short "s" allele would result in less 5-HT reuptake while the long "l" allele would produce normal 5-HT reuptake (see text for details). **A:** Emotional and spatial stimuli shown to subjects segregated by genotype ("s" or "l" alleles) but matched for age, gender and mean IQ. Subjects were asked to match the emotional expression of the face (left) in the top row with one of the two faces in the bottom row. In the control stimulus (right), the orientation of an ellipse (top row) was to be matched with the orientation of one of the two ellipses in the bottom row. **B:** It was demonstrated that the magnitude and spatial extent of the relative cerebral blood flow response of the right amygdala was greater in the group having the "s" allele (first cohort) relative to the "l" allele (second cohort).

Continued

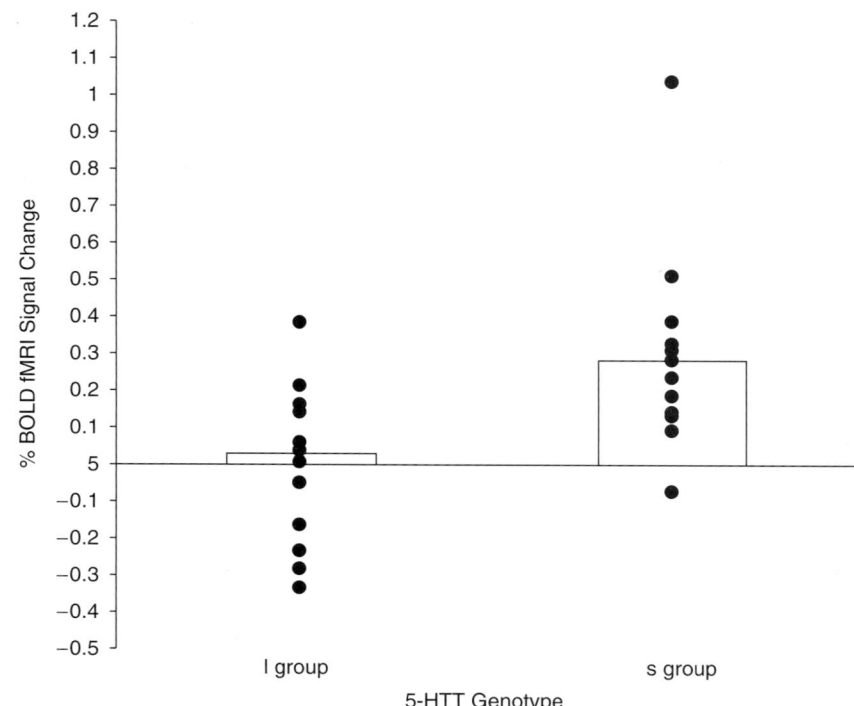

C

FIGURE 29.2 cont'd **C:** While there was overlap between the groups, there was clear separation of the mean of the fMRI signal response in the amygdala of these two groups as demonstrated in this graph. Such studies demonstrate that phenotype-genotype comparisons can often be better quantified and made more specific when a behavioral phenotype is substituted with a functional imaging parameter, in this case, response to an emotional matching task in the visual modality. Courtesy of A. Hariri and colleagues, National Institutes of Mental Health, Bethesda, Maryland.

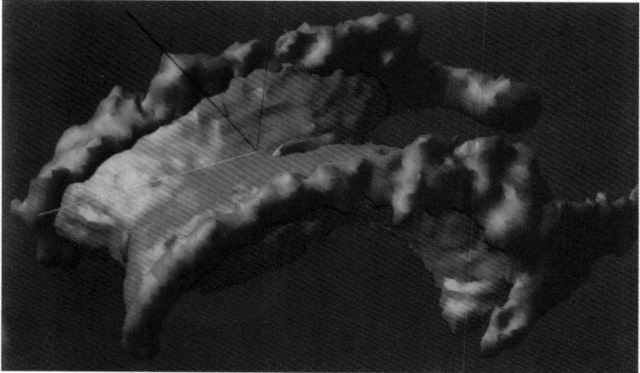

FIGURE 29.3 Probabilistic atlas of multiple sclerosis. This image is produced from 460 patients with multiple sclerosis derived from 5800 individual pulse sequences. The green areas demonstrate the probabilistic location of the actual MS plaques in the population referenced to the human ventricular system (red). This strategy gives a composite view of overall disease burden across a large population. Consider its use in a clinical trial where these patients were randomly assigned to treatment groups that included placebo, conventional therapy and experimental therapy. By observing the group for some period and obtaining serial images, an automated, quantifiable, and objective measure of the relative effect of experimental therapy versus conventional therapy and the natural history of the disease could be obtained using MRI lesions as a surrogate marker of disease burden. Such strategies should lead to more efficient and cost-effective clinical trials. Courtesy of Zijdenbos, Evans and colleagues, Montreal Neurologic Institute, Montreal, Canada.

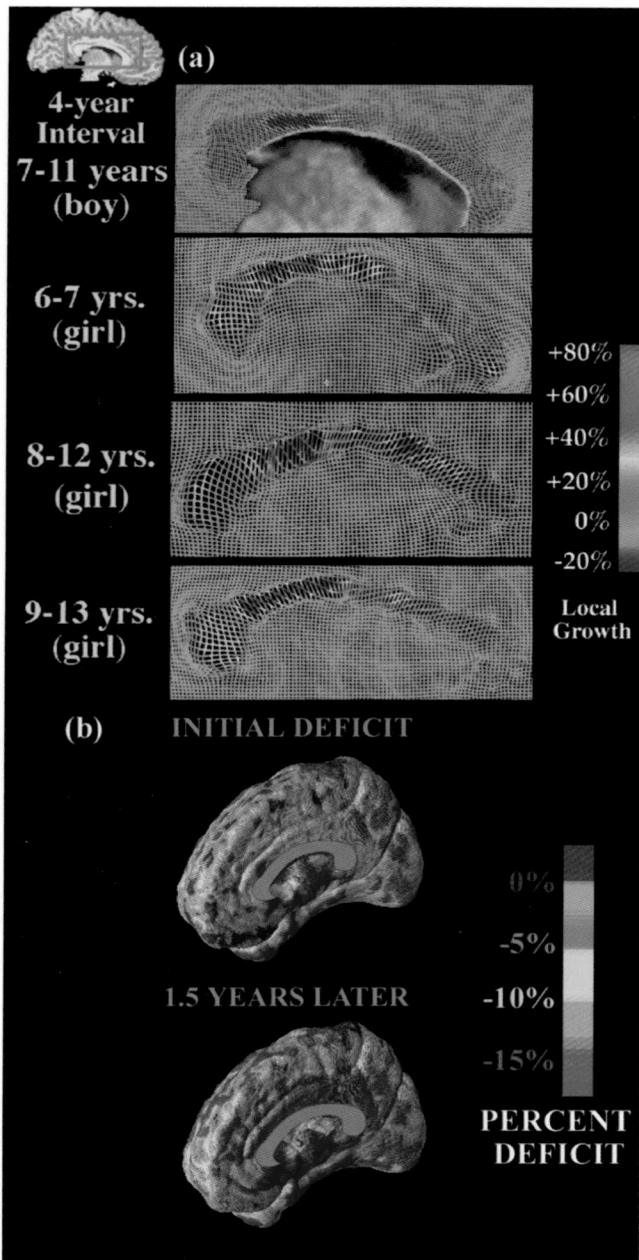

(a)

4-year
Interval
7-11 years
(boy)

6-7 yrs.
(girl)

8-12 yrs.
(girl)

9-13 yrs.
(girl)

+80%
+60%
+40%
+20%
0%
-20%

Local
Growth

(b) INITIAL DEFICIT

1.5 YEARS LATER

0%
-5%
-10%
-15%

PERCENT
DEFICIT

FIGURE 29.4 Dynamic maps of changing brain structure. If follow-up (longitudinal) images are available from the same subject, the dynamics of brain change can be measured with *tensor mapping* approaches. These compute an elastic deformation field that reshapes the baseline image to its anatomic configuration at follow-up. From these deformations, maps are generated to visualize local volumetric change, showing local rates of tissue growth or loss (red colors denote fastest growth). **A:** In children, scanned serially with MRI around the age of puberty, fastest growth is detected in the isthmus of the *corpus callosum*, which connects the language regions of the two cerebral hemispheres. **B:** *Progression of Alzheimer's Disease.* These maps show the average profile of gray matter loss in a group of 17 patients with mild-to-moderate Alzheimer's disease. Average percent reductions in the local amount of gray matter are plotted, relative to the average values in a group of 14 healthy age- and gender-matched elderly controls. Over 1.5 years of progressive cognitive decline from mild to moderate dementia, gray matter deficits in these patients spread through the limbic system to encompass the majority of the cortex. Courtesy of Paul Thompson and Arthur Toga, UCLA, Los Angeles, California.

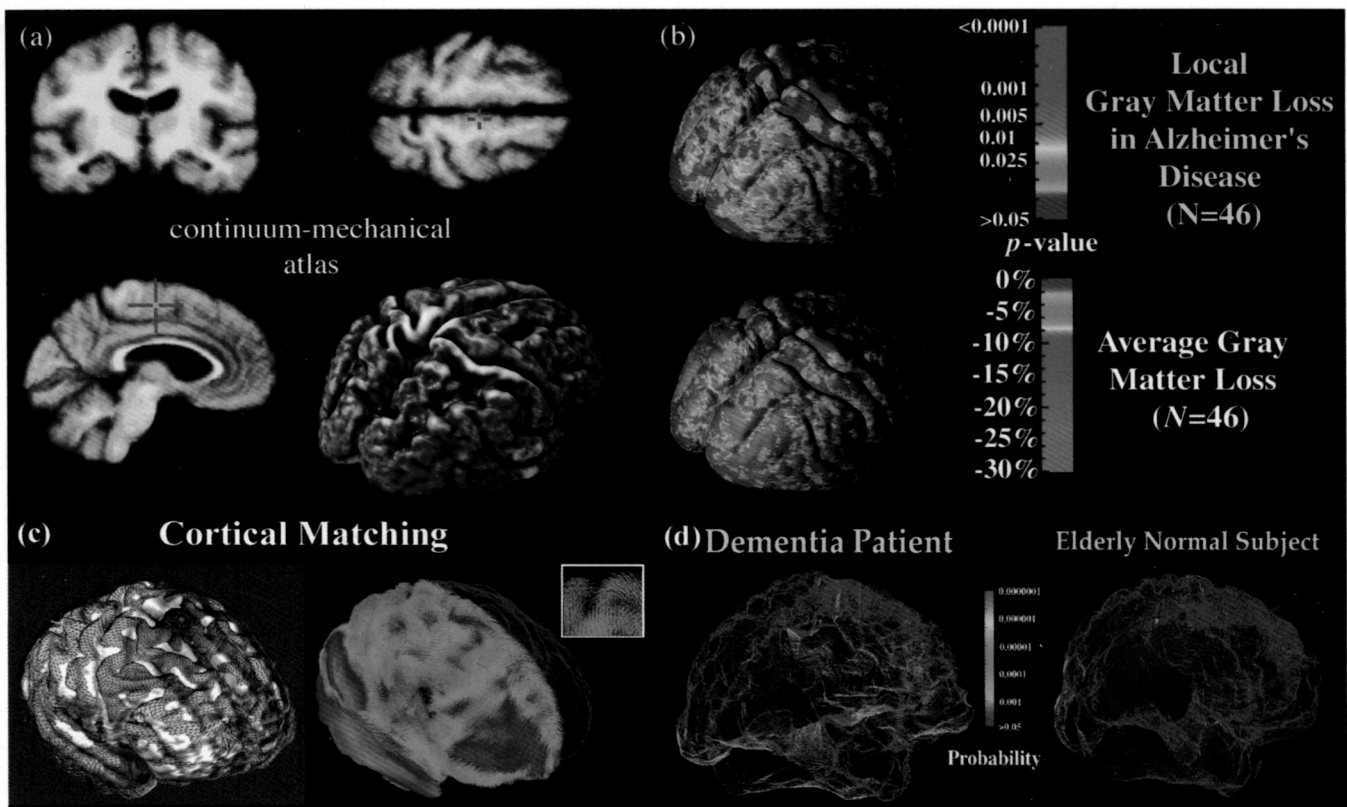

FIGURE 29.5 Disease-specific brain atlases. Disease-specific brain atlases reflect the unique anatomy and physiology of a clinical population, in this case an Alzheimer's disease population. Using mathematical strategies to average cortical anatomy across subjects (Thompson *et al.*, 2000), an average MRI template can be generated for a specific patient group; in this case, nine patients with mild-to-moderate Alzheimer's disease **(A)**. The cortical pattern indicates clear sulcal widening and atrophic change, especially in temporo-parietal cortices. By averaging a measure of gray matter across corresponding cortical regions, the average profiles of gray matter loss can also be mapped **(B)**. The net reduction in gray matter, in a large patient population relative to controls (Thompson *et al.*, 2000), can then be plotted as a statistical map in the atlas (**(B)**, *upper panel*). This type of analysis uncovers profiles of early anatomical change in disease. Panel **(C)** shows an individual's cortex (*brown mesh*) overlaid on an average cortical model for a group. Differences in cortical patterns are encoded by computing a three-dimensional elastic deformation (**(C)**; pink colors: large deformation) that reconfigures the average cortex into the shape of the individual, matching elements of the gyral pattern exactly. These deformation fields **(C)** provide detailed information on individual deviations. These statistical data can be used to detect patterns of abnormal anatomy in new subjects. By encoding variations in gyral patterns and gray matter distribution, algorithms can detect a region of abnormal atrophy in the frontal cortex of a dementia patient **(D)**. Severe abnormality is detected (red colors) while corresponding regions in a matched elderly control subject are signaled as normal (**(D)**, *right panel*). Courtesy of Paul Thompson and Arthur Toga, UCLA, Los Angeles, California.